CONCEPTS AND CONTROVERSIES IN TIDAL MARSH ECOLOGY

Concepts and Controversies in Tidal Marsh Ecology

Edited by

Michael P. Weinstein
New Jersey Marine Sciences Consortium,
Fort Hancock, NJ, USA

and

Daniel A. Kreeger
Academy of Natural Sciences,
Philadelphia, PA, USA

SPRINGER-SCIENCE+BUSINESS MEDIA, B.V.

Library of Congress Cataloging-in-Publication Data

ISBN 978-94-017-4080-7 ISBN 978-0-306-47534-4 (eBook)
DOI 10.1007/978-0-306-47534-4

Printed on acid-free paper

Cover photograph courtesy of
Dr. Ronald T. Kneib, University of Georgia, USA

DEDICATION

This volume is dedicated to Dr. Eugene P. Odum and Dr. John M. Teal:

For Pioneering Work in Salt Marsh Research
and Inspiring a New Generation of Salt Marsh Ecologists

FOREWORD

In 1968 when I forsook horticulture and plant physiology to try, with the help of Sea Grant funds, wetland ecology, it didn't take long to discover a slim volume published in 1959 by the University of Georgia and edited by R.A. Ragotzkie, L.R. Pomeroy, J.M. Teal, and D.C. Scott, entitled ''Proceedings of the Salt Marsh Conference'' held in 1958 at the Marine Institute, Sapelo Island, Ga. Now forty years later, the Sapelo Island conference has been the major intellectual impetus, and another Sea Grant Program the major backer, of another symposium, the ''International Symposium: Concepts and Controversies in Tidal Marsh Ecology''. This one re-examines the ideas of that first conference, ideas that stimulated four decades of research and led to major legislation in the United States to conserve coastal wetlands. It is dedicated, appropriately, to two then young scientists – Eugene P. Odum and John M. Teal – whose inspiration has been the starting place for a generation of coastal wetland and estuarine research.

I do not mean to suggest that wetland research started at Sapelo Island. In 1899 H.C. Cowles described successional processes in Lake Michigan freshwater marsh ponds. There is a large and valuable early literature about northern bogs, most of it from Europe and the former USSR, although Eville Gorham and R. L. Lindeman made significant contributions to the American literature before 1960. V.J. Chapman published ''Salt Marshes and Salt Deserts of the World'' in 1960 after two decades of writings on the subject. J.H. Davis published a description of the ecology and geology of mangroves in 1940. However, an important distinction of the Sapelo Island conference was the focus on process – productivity, trophic structure, energy flow – in fact, on the way whole ecosystems function. E.P. Odum, H.T. Odum, J.M. Teal, L. R. Pomeroy, A.C. Redfield, and others pioneered this approach to salt marshes, which differed sharply from the predominantly descriptive earlier studies.

From these fairly small (in terms of numbers, not ideas) beginnings in 1958, coastal ecology has grown to a large, diverse and healthy enterprise at the end of the 21st century. This volume gives some indication of strength of the discipline. Whereas most of the attendees at the 1958 Sapelo Island conference worked along the east coast of the U.S., the authors of this volume come from all over the States and from Canada and Europe. The lead authors were selected by the symposium conveners for their research contributions, and they read like a who's who of salt marsh ecology. Of 97 authors, 26 are from the northeast, 39 from the southeast, 12 from the Gulf coast, and 7 from the West Coast. Eight are from three European countries, one from Canada. In terms of educational institutions and laboratories represented, 8 are from the northeast, 14 the southeast, 4 the Gulf States, and 5 the west coast. Five institutions can claim six or more contributors – Rutgers University, Woods Hole Oceanographic Institute, University of Maryland, University of Georgia, and Louisiana State University. Several of these institutions did not have marine programs or marine laboratories in 1958.

The National Sea Grant Program developed on a parallel course with the growth of coastal research, and was in many cases instrumental in the development of strong programs. Certainly that was the case at Louisiana State University, where I have personal experience. Sea Grant has grown from a small nucleus of programs when it was authorized and funded as a federal program in 1969, to a multimillion dollar enterprise that includes Sea Grant Colleges in 29 states. Its role in encouraging the symposium that has resulted in this volume is only one example of its many productive activities.

The chapters in this volume are all major syntheses of the current understanding of salt marsh ecology. They are not merely literature reviews, they are syntheses that meld thousands of individual research efforts into coherent summaries of the state of salt marsh ecology today. They are rich in ideas and hypotheses. As is fitting, many of the chapters address, directly or indirectly, the two major paradigms that rose from the first salt marsh conference; first, the Detrital paradigm which states that the base of the food web is marsh macrophyte production that is microbially decomposed before it becomes available as food to invertebrate and vertebrate organisms; and second , the Outwelling paradigm, that salt marshes export surplus production to coastal waters, thus supplementing the coastal phytoplankton food source.

The detrital paradigm comes under considerable attack in some chapters, but the overall conclusion seems to be one of modification,, not discrediting of the original hypothesis. Edaphic algae and phytoplankton are much more important in the food web than initially thought, not so much for the quantity of their production as for their food quality. Secondly, enormous strides have been made in documenting the decomposition of dead macrophyte tissue, including the importance of epibenthic fungi, the complexity of the benthic microbial process, and specific links to invertebrate meiobenthic and macrobenthic organisms.

The Outwelling paradigm is similarly modified, from Teal's (1962) hypothesis that as much as one half of marsh macrophyte production is exported as detritus, to the documentation of much smaller fluxes of dissolved and particulate organic material (if indeed efflux from the marsh occurs at all), and the realization of the importance of geomorphology and hydrology in these fluxes. Recent research has focused on not only organic fluxes but also nutrient exchanges, has clarified the confusion that arises when the source and sink are not clearly defined, and has documented the importance of benthic and pelagic fish and shellfish in the transfer of energy from marsh to coastal waters.

The scientific understanding of coastal, and more broadly environmental processes in general, and their links to human values, has led in the past 40 years to major environmental legislation. Most relevant for wetlands has been passage of the National Environmental Policy Act of 1969, the Clean Water Act in 1972, and the Coastal Zone Management Act in 1972 (see Chapter 17). In particular, although wetlands are not mentioned in the Clean Water Act, the rules implementing it require a permitting system for development activities in wetlands. In 1988 the National Wetlands Policy

Forum set a goal of no net wetlands loss for the United States. Although this goal has no legislative mandate, it has been embraced by the federal and state agencies that manage wetlands. Wetland conversion is still permitted when it can be economically justified, but usually the permitted action requires restoration or creation of equivalent (or more) wetlands than that destroyed. Thus wetland engineering – active management, restoration, or creation – has skyrocketed. The final sections of this book show the high level of concern among the participants as to whether engineered wetlands achieve functional equivalence with natural systems. This question is leading to a great deal of discussion:

1) What characteristics of wetlands are important in measuring functional equivalence?
2) How can these characteristics be measured economically?
3) Do existing projects create functional wetlands? and
4) What should design criteria for wetland restoration and creation include? This volume discusses these issues, usually in the context of the entire coastal ecosystem.

This International Symposium volume is a fitting tribute to the success of the first salt marsh conference forty years ago, and a worthy successor.

– *James G. Gosselink, Ph.D.*

CONTENTS

Retrospective on the Salt Marsh Paradigm

Sources and Patterns of Production

Fate of Production Within Marsh Food Webs

Habitat Value: Food and/or Refuge

Biogeochemical Processes

Modeling Nutrient and Energy Flux

Tidal Marsh Restoration: Fact or Fiction?

Ecological Engineering of Restored Marshes

Measuring Function of Restored Tidal Marshes

Success Criteria for Tidal Marsh Restoration

PREFACE

While more than 50% of the nation's coastal wetlands have been claimed for agriculture, drained to reduce pestilence, or developed for human occupancy, more than 3.5 million ha remain in varying degrees of 'health'. Worldwide recognition that these habitats provide important environmental services has helped to reverse the trend in loss and degradation, but there is clearly a long way to go. Tidal marshes are no exception, and the public now generally perceives that marshes are important habitats for animals and plants of substantial commercial value. Moreover, tidal marshes are increasingly recognized for their role as crucial buffers between the land and sea.

For more than a century, estuaries and their fringing marshes have been classified as essential habitat for finfish and shellfish. Up to 80% of marine recreational and commercial species are believed to have estuarine dependent life stages, the majority of which use tidal salt marshes as primary nurseries for feeding and refuge. Although this view may be based more on perception than fact, it is so ingrained in the psyche of the public, managers and regulatory bodies that substantial legislation and policy have evolved to conserve and protect these habitats. Much of this legislation remains in effect today. The relatively young science of ecological engineering has also emerged, and there are now attempts to reverse centuries old losses by encouraging sound wetland restoration practices. Today, tens of thousands of hectares of degraded and/or isolated coastal wetlands are being restored worldwide. Whether restored wetlands reach functional equivalency to "natural" systems is the subject of heated debate.

Equally debatable is the paradigm that depicts tidal salt marshes as the 'engine' that drives much of the secondary production in coastal waters. This view was questioned in the early 1980s by investigators who noted that total carbon export was much lower than originally thought, on the order of 100 to 200 gC $m^{-2}y^{-1}$. These scientists also recognized that some marshes were either net importers of carbon, or showed no net exchange. Thus, the notion of 'outwelling' has become but a single element in an evolving view of marsh function and the link between primary and secondary production. The 'revisionist' movement was launched in 1979 when stable isotopic ratios of macrophytes and animal tissues were found to be 'mismatched.' Some 20 years later, the view of marsh function is still undergoing modification, and we are slowly unraveling the complexities of biogeochemical cycles, 'secret gardens,' nutrient exchange, and the links between primary producers and the marsh/estuary fauna. Although much important research has been published since Teal's 1962 paper, scientists still have much to do to understand how marshes 'work.' If anything, the story is far more complicated than originally thought. After more than four decades of intense research, it is still not certain *how* salt marshes function as essential habitat, or their relative contribution to secondary production, both *in situ* and in the open waters of the estuary.

Despite heightened interest in tidal marsh ecology and a wealth of new research during the past 35 years, there have been few attempts to synthesize what has been learned and to clearly articulate questions that remain unanswered. Conferring with many other marsh ecologists, we found unequivocal support for development of a long overdue reference text that would summarize the state of ecological research in tidal marshes. Early in the discussions, it became clear that to be of greatest value, the book would have to be globally relevant, expanded beyond the traditional focus on salt marshes to include tidal freshwater systems, and to include the relatively new topics of ecological engineering and wetland restoration. To develop such a multidisciplinary reference work, we recognized that we would need to hold a specially focused scientific conference. This meeting, ''Concepts and Controversies in Tidal Marsh Ecology,'' was held in Vineland, New Jersey, USA, in April 1998. More than 40 invited presentations were given at the meeting, and it was attended by more than 400 participants from around the world.

The chapters in this book are organized into the following topics:

- Retrospective on the Salt Marsh Paradigm (Chapters 1 and 2)
- Sources and Patterns of Production (Chapters 3 to 8)
- Fate of Production Within Marsh Food Webs (Chapters 9 to 11)
- Habitat Value: Food and/or Refuge (Chapters 12 to 16)
- Biogeochemical Processes (Chapters 17 to 21)
- Modeling Nutrient and Energy Flux (Chapters 22 to 24)
- Tidal Marsh Restoration: Fact or Fiction? (Chapters 25 to 28)
- Ecological Engineering of Restored Marshes (Chapters 29-30)
- Measuring Function of Restored Tidal Marshes (Chapters 31 to 33)
- Success Criteria for Tidal Marsh Restoration (Chapters 34 to 37)

Our objective was to prepare a comprehensive synthesis of tidal marsh ecology research. To create a book with greatest value to a wide audience of scientists, students, regulatory personnel, and natural resource managers, we sought to maximize coverage of the major issues of widespread interest while minimizing overlap among contributions. Hence, each chapter covers a specific area of inquiry and is meant to briefly review past and current research as well as identify future research needs. Authors were granted considerable freedom to use examples from their own research, but the context was meant to be broader than the typical scientific paper.

This book would not have been possible without the dedicated editorial expertise and assistance of Lisa S. Young, who spent countless hours preparing the camera-ready manuscripts. For assistance in organizing, and coordinating the symposium, we are deeply indebted to Barbara Kieffer, Heidi Hertler, Kim Kosko, Steve Litvin, Dan Snyder, and Roger Thomas. John Tiedemann helped coordinate the manuscript reviews. Finally, we thank all the participants who contributed to a stimulating and enlightening symposium, and more than 100 external reviewers who selflessly gave of their time to improve the quality of this book. Of course, any errors or omissions are solely the responsibility of the Editors.

MICHAEL P. WEINSTEIN
DANIEL A. KREEGER

This International Symposium and publication of *Concepts and Controversies in Tidal Marsh Ecology* would not have been possible without the generous support of the following institutions:

Connecticut Sea Grant College Program
Cumberland County College
Delaware River Bay Authority
Delaware Sea Grant College Program
Georgia Sea Grant College Program
Louisiana Sea Grant College Program
Maryland Sea Grant College Program
WHOI Sea Grant College Program
National Marine Fisheries Service
National Sea Grant College Program Office
New Jersey Marine Sciences Consortium
New Jersey Sea Grant College Program
North Carolina Sea Grant College Program
Rhode Island Sea Grant College Program
Port Authority of NY&NJ
Public Service Electric and Gas Company
Texas Sea Grant College Program
The Academy of Natural Sciences of Philadelphia
United States Environmental Protection Agency

RETROSPECTIVE ON THE SALT MARSH PARADIGM

TIDAL MARSHES AS OUTWELLING/PULSING SYSTEMS

EUGENE P. ODUM
Institute of Ecology
University of Georgia
Athens, GA 30602 USA

Abstract
Now that we are beginning to understand that the balance of nature is a pulsing one and not a steady-state as is the case at the organism level (i.e., homeorhesis rather than homeostasis) estuaries become important sites for research because they pulse so strongly. The external tidal pulses interface in a complex manner with internal biological and life history pulses. I review the concept that productivity of near shore ocean waters can be enhanced not only by upwelling of nutrients from deeper waters but also by outwelling of nutrients and organic matter from fertile estuaries. I conclude 1) that the extent of the latter depends on the level of production within the estuary, the tidal amplitude and the geomorphology of the estuarine landscape and 2) the outwelling from tidal marshes where it occurs is often intermittent and largest during rain storms and storm tides.

1. Introduction

In this paper I review the history of the concept that productivity of nearshore waters can be enhanced not only by "upwelling" of nutrients from deeper waters but also by "outwelling" of nutrients, organic matter and organisms from fertile estuaries. As a background for assessing this possibility, we need to consider two holistic or ecosystem-level concepts: the pulsing paradigm and the source-sink paradigm.

2. The Pulsing Paradigm

As first pointed out by Patten and Odum (1981) cybernetics at levels of organization above that of the organism is different from that operating at the cellular and organism levels in that there are no set-point controls such as chemostats, genes, hormones, etc. that maintain tight control over growth, development and metabolic functions at the organism levels (and also in human engineered servomechanisms). But there are positive and negative feedbacks at the ecosystem and above levels that maintain a looser control. Accordingly, the so called balance of nature is a pulsing one rather than an equilibrium or steady-state one. Odum et al. (1995) have reviewed the pulsing paradigm with special reference to tidal systems. In that paper, it was suggested that maximum power is achieved when internal biological pulses such as predator-prey or life cycles are coordinated with external pulses such as tides. Recently, I have suggested that we restrict the use of the word ''homeostasis'' to cybernetics at the cell and organism level and use the term ''homeorhesis'' (i.e., maintaining the flow) for the pulsing cybernetics of ecosystems (Odum 1997).

3. Source-Sink Energetics

Energetic ''hot spots,'' areas where activity is much more intense than in the surrounding cooler matrix, are characteristic of most natural as well as human dominated landscapes. Cities, of course, are very intensive ''hot spots'' in the much less energetic countryside. In natural environments ''hot spots'' are frequent. For example, as much as 90% of the activity of soil organisms may occur in small aggregates and root zones that constitute less than 10% of the total soil volume (Coleman 1995). The source-sink concept refers to situations where excess production by one ecosystem or patch (the source) is exported to another less productive ecosystem or patch (the sink). At the species level it is not uncommon for a population in one area to produce more offspring than are needed to maintain it, with the surplus moving to an adjacent population that otherwise would not be self-sustaining (Pulliam 1988).

With these basic ecological concepts in mind let us now consider the questions of whether, when or where salt marshes are pulsing ''sources'' or ''hot spots'' that ''outwell'' to adjacent waters.

4. Chronological History of the Outwelling Notion

1962. John Teal's mass balance energy budget for Georgia salt marshes at Sapelo Island indicated that primary production was greater than community respiration ($P/R > 1$), and he assumed that the excess was exported along with shrimp and other organisms that use the marsh as nursery grounds.

1966. J.P. (Jim) Thomas made C- 14 measurements along the Georgia coast finding that high off shore productivity was associated with nearby marshes rather than with large river plumes.

1968. In a commentary published in the Proceedings of a Sea Grant Conference, I suggested that most fertile zones in coastal areas capable of supporting expanded fisheries result either from ''upwelling'' of nutrients from deep water or from ''outwelling'' of nutrients and organic detritus from shallow water hot spots such as reefs, banks, seaweed or seagrass beds and salt marshes.

1976 and 1977. Based on stable carbon isotope ratios in the biota, soils and tidal water, Evelyn B. Haines cast doubt on the hypothesis that estuarine food chains were mostly supported by marsh grass detritus from bordering salt marshes; algae were suggested as often more important.

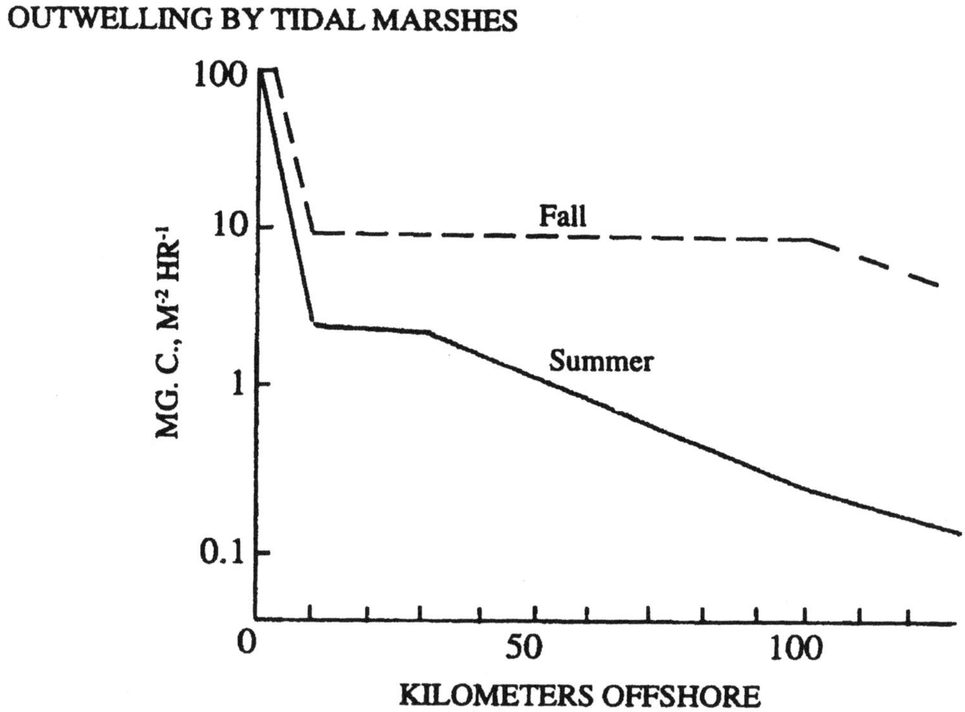

Figure 1. Pattern of productivity off shore of southeastern marsh estuaries. (After Turner et al. 1979)

1979. R. Eugene Turner, W. Woo and H.R. Jitts in an article in *Science* reported that off shore productivity measurements along both South Atlantic and Gulf coasts supports outwelling. As shown in Fig. 1, they found that primary productivity was often an order of magnitude higher within 10 km of estuaries as compared to further offshore.

1979. William E. Odum, J.S. Fisher and J. C. Pickrel suggested that coastal geomorphology could be a major factor in controlling the flux of particulate organic carbon from estuarine wetlands. As shown in Fig. 2, outwelling would be more likely where marshes are more open to the sea as in B and C.

1980. Scott Nixon failed to find any evidence for outwelling from New England marshes. In fact, many of these marshes appeared to be importing rather than exporting carbon. It is important to point out that compared with the South Atlantic coast, New England salt marshes are much less extensive, tidal amplitude is less and connections with the sea generally more restricted as shown in Fig. 2A.

1985. Charles S. Hopkinson and coworkers conducted extensive metabolic measurements just offshore of the Georgia barrier islands finding that respiration in the entire water column (benthic and pelagic) exceeds *in-situ* production (i.e., the zone is heterotrophic) indicating that organic matter is being imported from the marsh estuaries.

1985. Alice G. Chalmers, R.G. Wiegert and P.L. Wolf found that the excess organic matter produced in the Georgia marshes was in constant flux in and out of the marsh due to deposition and resuspension with each tidal cycle on the marsh and tidal creek surfaces.

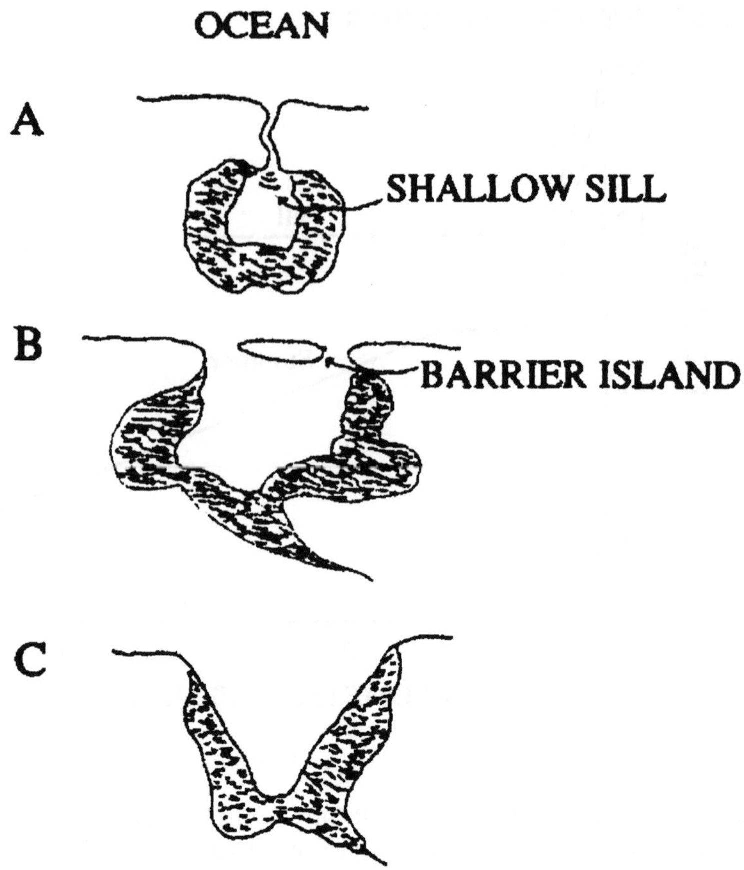

Figure 2. Three types of estuarine basins ringed by intertidal marshes that may affect export-import dynamics. (After W.E. Odum et al. 1979)

Large exports to the sea were found to occur mainly during rain storms at low tide, or during high spring tides when these surfaces were eroded by the strong water flows.

5. Conclusions

Based on these reviews there is no doubt that outwelling occurs in the South Atlantic bight where salt marshes are extensive and extremely productive, and tidal amplitudes large, and also in Louisiana where coastal marshes are extensive and open to the sea. In these areas tidal marshes are definitely exporting ''hot spots''. Such enrichment of offshore waters may be less important or may not occur in other areas of the Atlantic and Gulf coasts.

Export pulses of organic matter and nutrients from marshes to the sea do not necessarily occur with every tidal cycle but may be intermittent associated with rain storms and high spring tides. However, since a wave of small fishes comes in with each tide to ''graze'' on detritus, microbes, microfauna and algae on the marsh and estuary surfaces it may be that marsh productivity is often ''outwelled'' as organisms rather than as organic matter and nutrients.

Overall, we conclude that the extent of outwelling is related to the level of productivity and extent of marsh cover within the estuary, the tidal amplitude and the geomorphology of the estuarine landscape. While some export may occur with each tidal cycle, large output pulses tend to occur during rain storms and high spring tides.

6. Literature Cited

Chalmers, A. G., R. G. Wiegert and P. L. Wolf. 1985. Carbon balance is a salt marsh: interaction of diffusive export, tidal deposition and rainfall caused erosion, Estuarine, Coastal and Shelf Science **21**: 757-371.

Coleman, David C. 1995. Energetics of detrivory and microbivory in soil in theory and practice. Pages 39-50 *in* G. A. Polis and K. O. Winemiller, editors. Food webs, integration of patterns and dynamics, Chapman and Hall, New York, New York, USA.

Haines, E. B. 1976. Stable carbon isotope ratios in biota, soils and tidal water of a Georgia salt marsh, Estuarine and Coastal Marine Science **4**: 609-616.

—— 1977. The origin of detritus in Georgia salt marsh estuary. Oikos **29**: 254-260.

Hopkinson, C. S. 1985. Shallow water benthic and pelagic metabolism: evidence for heterotrophy in the nearshore, Marine Biology **87**: 19-32.

Nixon, S. W. 1980. Between coastal marshes and coastal water — a review of twenty years of speculation and research in the role of salt marshes in estuarine productivity and water chemistry. Pages 437-525 *in* P. Hamilton and K.B. MacDonald, editors. Wetland processes with emphasis on modelling. Plenum Press, New York, New York, USA.

Odum, E. P. 1968. A research challenge: evaluating the productivity of coastal and estuarine water. Pages 6364 *in* Proceedings of the second Sea Grant congress, University of Rhode Island, Graduate School of Oceanography, Kingston, Rhode Island, USA.

—— 1997. Ecology. A Bridge Between Science and Society, Sinauer Associates, Sunderland, Massachusetts, USA.

Odum, W. E. J. S. Fisher and J. C. Pickrel. 1979. Factors controlling the flux of particulate organic carbon from estuarine wetlands. Pages 69-79 *in* R. J. Livingston, editor. Ecological processes in coastal and marine systems, Plenum Press, New York, New York, USA.

Odum, W. E., E. P. Odum and H. T. Odum. 1995. Nature's pulsing paradigm. Estuaries **18**: 547-555.

Patton, B. C. and E. P. Odum. 1981. The cybernetic nature of ecosystems. American Naturalist **118**: 886-895.

Pulliam, H. R. 1988. Sources, sinks and population regulation. American Naturalist **132**: 652-661.

Teal, J. M. 1962. Energy flow in the salt marsh ecosystem of Georgia. Ecology 43: 614-624.

Thomas, J. P. 1966. The influence of the Altamaha River on primary production beyond the mouth of the river. Thesis, University of Georgia, Athens, Georgia, USA.

Turner, R. E. S., W. Woo and H. R. Jitts. 1979. Estuarine influences on a continental shelf plankton community. Science 206: 218-220.

SALT MARSH VALUES: RETROSPECTION FROM THE END OF THE CENTURY

JOHN M. TEAL
Woods Hole Oceanographic Institution
Woods Hole, MA 02543, USA, and
Teal Ltd., Rochester, MA 02770 USA

BRIAN L. HOWES
Center for Marine Science and Technology
University of Massachusetts
New Bedford, MA 02744 USA

Abstract
Two of the greatest problems in coastal waters are eutrophicaton and rapid decline in populations of important fish species. Salt marshes are important in combating both these problems. A paradigm for salt marsh function: marshes import inorganic nutrients and export organic nutrients and, as a result, grow fish. As ground and tidal water flow through salt-water wetlands, plants, bacteria and algae produce or transform the organic matter of the food chain that supports fish and shellfish populations. While salt marshes modify the principal plant nutrients, N and P, some of the pathways result in removal of nutrients from biologically active systems. Nitrogen is removed primarily either by being trapped in refractory organic matter that contributes to marsh maintenance through accretion or through loss to the atmosphere (as N_2) by denitrification. Salt marshes along the Atlantic coast of the United States have changed during the past century; the number of hectares has declined and the nutrient loading per hectare has increased. We examine data on the correlation between fish catch and various marsh features from Long Island, New York in 1880. We review research on the ways salt marshes reduce both the level and rate of eutrophication of coastal waters by intercepting nitrate in discharging groundwater. Finally, we consider how these functions have changed with the decrease in area of salt marshes along the Atlantic coast from Georgia to Maine.

1. Introduction

Thirty-six years ago, in 1962, John Teal published a paper in *Ecology* (Teal 1962), based upon research done on Sapelo Island, Georgia. This early work helped set the stage for thinking about salt marsh functions *vis a vis* organic production and the role of marshes in the estuarine complex. The *Ecology* paper built on the 1959 proceedings of the first salt marsh conference promoted by Dr. Alfred Redfield and held at Sapelo Island. When Dr. Michael Weinstein conceived the idea for a salt marsh meeting in April 1998, of which this paper is a part, he suggested we revisit the ideas in that earlier work and add other salt marsh ecology principles that have evolved over the intervening four decades.

The main point of the 1962 paper was that salt marshes produce more organic matter than they consume. Teal reached that conclusion by adding up the production processes on the marsh and subtracting all the consumptive processes. The latter accounted for only a little more than half of the former, leading to the conclusion that there was excess marsh production that was exported from the marsh plain and therefore was available to support secondary production in the surrounding estuary.

From what we know today, there are obvious shortcomings in the early Sapelo Island study. It was conducted on a mature marsh system that was no longer expanding, accumulating sediment only to maintain surface elevations in balance with sea level rise. Most importantly, it did not take into account the role of the fish in the marsh. The piscivorous (and carnivorous) fishes that come into the marshes at high tides could be considered as belonging to estuarine rather than marsh faunas. This cannot be said of fishes such as *Fundulus heteroclitus*, mummichogs, which spend almost their entire lives within the salt marsh (Kneib 1994). The basic conclusion of "excess" marsh production in Teal's 1962 paper has been supported by subsequent research. However, the principal export mechanism Teal hypothesized for organic matter and energy export from the marsh plain was detrital movement from the marsh surface. More recent evidence of overlapping food chains as described by Kneib and Wagner (1994) have supplemented and partially replaced detrital movement as the principal marsh export mechanism.

Why was this of interest? The paper suggested that salt marshes contributed to estuarine food chains beyond their borders and had a greater ecological (and economic) value beyond just being there as open space. The suggestion that salt marshes had other values became one of the driving forces for salt marsh protection laws passed by most coastal states over the last 30 years. Furthermore, it led to identification and quantification of these values, providing a major underpinning to salt marsh research during this period.

Salt marsh protection legislation does not derive exclusively from the idea that there is excess marsh production. Other functions are also recognized. In the words of the Massachusetts Wetland Protection Act (Mass. Gen. Laws 1987), wetlands are:

> significant to public or private water supply, to the ground water supply, to flood control, to storm damage prevention, to prevention of pollution, to protection of land containing shellfish, to the protection of fisheries and to the protection of wildlife habitat.

All of these wetland functions are performed by salt marshes, except that the water supply functions are limited.

Salt marsh ecology has been a subject of academic research throughout this century. But it has only been in the latter half of the century that most of our knowledge of marsh functions has been gathered. Some of these functions are obvious to anyone who spends much time on or about salt marshes, although in some cases quantification has been difficult. In contrast, other functions are not obvious to casual observation. Establishing their very existence requires research. To make matters more complex, some of the functions of salt marshes may not be constant within a single marsh but may change through time as the landscape around them is altered. The group of obvious functions includes observations that marsh creeks contain a lot of fishes, that marsh grasses grow abundantly, and that ducks, geese, rails and herons feed in marshes. If animals congregate in marshes, then there must be a reason they are there—they must be deriving some benefit either from the marsh as a food source, a nursery area, or just a resting-place. For fish like the mummichog, which spend their entire lives in the marsh, they must be satisfying all their requirements there.

Among the non-obvious functions are the cycling of plant nutrients, nitrogen and phosphorus, and interactions of marsh sediments, plants and animals with pollutants. Odum's observation that periodic events such as upland flooding drive marsh processes is another non-obvious feature of salt marshes (Odum, this volume). The seasonal shifts in the marsh's nitrogen balance, including the autumnal export of nitrogen as the marsh plants senesce, is a function that requires sustained rather than simple observation (Valiela and Teal 1979).

We have chosen to look further into two aspects of marsh function from the list above: 1) prevention of degradation of estuarine habitat from nutrient pollution, and 2) protection of fisheries. In the first case, we consider the role of salt marshes in nutrient cycling with special attention to nitrogen and denitrification. For the second, we examine the contribution of salt marshes to fishery yields by using data from around Long Island, New York from 1880.

2. Plant Nutrients

Phosphorus has not been intensively studied in salt marshes though it may limit production in some circumstances (Pomeroy 1959, Smart and Barko 1980, Valiela et al. 1982). In oxic water, phosphorus is generally found as insoluble salts and so it is transported to marshes attached to particles. Most of the transport occurs during storms that stir up marsh and estuarine sediments or when high river flows deposit sediments on the marsh surfaces. Inorganic phosphates are not transported to any appreciable extent by most oxic ground waters due to its reactivity with aquifer solids to form insoluble complexes. It is exported from salt marshes to the extent that plant tissues and sediments containing it are exported.

Nitrogen cycling is a much more dynamic story. Coastal marshes, like most other coastal marine ecosystems, tend to be nitrogen limited. With increasing nitrogen supply, marshes show greater primary productivity by both grasses and algae. Unlike some coastal systems, salt marshes can withstand very large additions of nitrogen without severe damage. Plant production increases and plant species composition may also change, but the marsh ecosystem survives, frequently with enhanced secondary production. In addition to increased plant mass, added nitrogen increases the plants' food value for herbivores (Buchsbaum et al. 1981). Food value is enhanced both through increased tissue nitrogen and decreased concentrations of anti-herbivory compounds. As a result, populations of geese or insect herbivores may increase. Secondary effects may also result from alterations of the physical structure of the marsh plain habitat as plant density decreases with increasing plant height (Vince et al.

1976). This structural change can facilitate access to marsh invertebrate prey species by aquatic predators such as blue crabs and fish during high tide. There is also an indication that eutrophication, which is typically driven by enhanced external nitrogen loading, plays a role in determining the plant species composition, for instance by reducing the viability of *Phragmites* by delaying translocation of reserves to the rhizomes in autumn (Kühl and Kohl 1993).

Nitrogen additions to the vegetated marsh plain increase both primary and secondary production. However, long term nitrogen enrichment of marsh plots in Great Sippewissett Marsh, Massachusetts has shown two things: 1) diminishing returns, and 2) the fate of the added nitrogen changes as the nitrogen load increases. It appears that, like the macrobiological community, the microbial community is also stimulated by nitrogen additions. At low levels of nitrogen addition, plants take up most of the nitrogen and primary production is enhanced. At higher levels, the microbial denitrification pathway is able to out-compete the plants for additional nitrogen uptake to the extent that at the highest levels of nitrogen addition, 7.5-g m^{-2} wk^{-1}, almost two thirds of the added nitrogen left the marsh as N_2 through denitrification (Howes et al. 1996). The shift from plant uptake to microbial denitrification is consistent with exceedingly high nitrogen loading tolerance noted for the vegetated marsh plain by various researchers. The importance of this process to estuarine nitrogen cycling is relatively limited except in those regions where tidal waters have been highly polluted with wastewater effluent or agricultural runoff. In these situations, both plant uptake and denitrification serve to remove nitrogen from tidal waters. In contrast, at the low nitrogen levels typical of most coastal waters, tidal water nitrogen is generally in balance with a relatively small seasonal uptake and release of nitrogen by the marsh plain. It is interesting that mechanisms supporting the increasing nitrogen removal through denitrification as nitrogen loading increases provide for the use of engineered marsh systems as tertiary treatment systems (Peterson and Teal 1996).

Similar to the vegetated marsh plain, tidal creek bottoms support active microbial denitrification. Unlike the vegetated marsh, however, much of the organic substrate supporting this heterotrophic respiration appears not to be produced *in situ*, but is imported from the adjacent vegetated marsh surface. Thus, denitrification of externally derived nitrogen within the marsh ecosystem may be primarily fueled by plant production. Denitrification within salt marshes is predominantly controlled by the availability of nitrate. In pristine marshes, denitrification is driven primarily by the coupled nitrification-denitrification of ammonium supplied in the decomposition of organic matter within the marsh soils. As ammonium is released in the anoxic portion of the sediments, that portion which is not taken up by plants, is generally oxidized to nitrate in the surficial soil layer or around plant roots (Reddy et al. 1989). This nitrate is then available to support the heterotrophic respiration of denitrifying bacteria naturally occurring throughout marsh systems. Under these circumstances, only a small portion of the nitrogen annually cycling through marshes is denitrified, the majority is recycled into new plant biomass (White and Howes 1994). But, with increasing development in coastal zones worldwide, the supply of externally derived nitrogen to salt marshes is increasing and entering the wetland nitrogen cycle.

How does this increased nitrogen load effect salt marshes? As stated above, the increased nitrogen input can affect both productivity and denitrifying activity. The pathway of input of this increasing nitrogen load, surface/tidal waters versus groundwater, structures both the marsh response and fate of the added nitrogen. The crux of the issue is that marsh biological systems can only process nitrogen as they can access the load.

Nitrogen entering the marsh in tidal or surface waters has limited access to the vegetated marsh areas, reaching many of them only during spring tides. While surface waters may contain a large nitrogen load, the concentrations are typically low. An additional transport issue is involved in denitrification of surface water transported nitrogen in that nitrate must reach the lower anoxic portion of the sediments that support this process. The result is that, unless the surface waters become significantly enriched in nitrogen, the ability of salt marshes to lower the nitrogen burden on the adjacent estuary is limited.

In contrast, external nitrogen entering marshes through groundwater discharge typically supports proportionally higher levels of denitrification. In most coastal regions, groundwater has become enriched due to wastewater discharges and loading through a variety of sources in the watersheds contributing to estuaries. The result has been increasing nitrate concentrations in groundwater discharging to fringing salt marshes. This nitrogen source predominantly enters the nitrogen cycle at the anoxic creek bottom rather than the oxic vegetated marsh plain. Groundwater discharges into most mature salt marshes, which tend to have thick (>0.5m) organic rich soils, at the upland/marsh boundary, at the head of creeks or directly via seepage through creek bottom sediments (Howes et al. 1996). The more than two orders of magnitude difference in hydraulic conductivity of marsh sediments between the vegetated marsh and creek bottom areas explain this pattern.

The contact of creek bottom sediments with groundwater nitrate results in significant stimulation of sediment denitrification that is directly controlled by nitrate concentration (Fig. 1). The typical pattern is that most of the nitrate load enters in those creeks closest to the adjacent upland. The result is a pattern of nitrate removal by sediment denitrification highest near the creek headwaters and diminishing with distance as the nitrate levels become depleted or diluted with draining tidal waters. This removal is, therefore, primarily a low tide phenomenon, since at high tide the nitrate concentrations are diluted by the low nitrate floodwaters. In addition to denitrification, the groundwater nitrate also stimulates epibenthic algal production that can reach quite high levels and serve as a significant food source for secondary production (Sullivan and Currin, this volume).

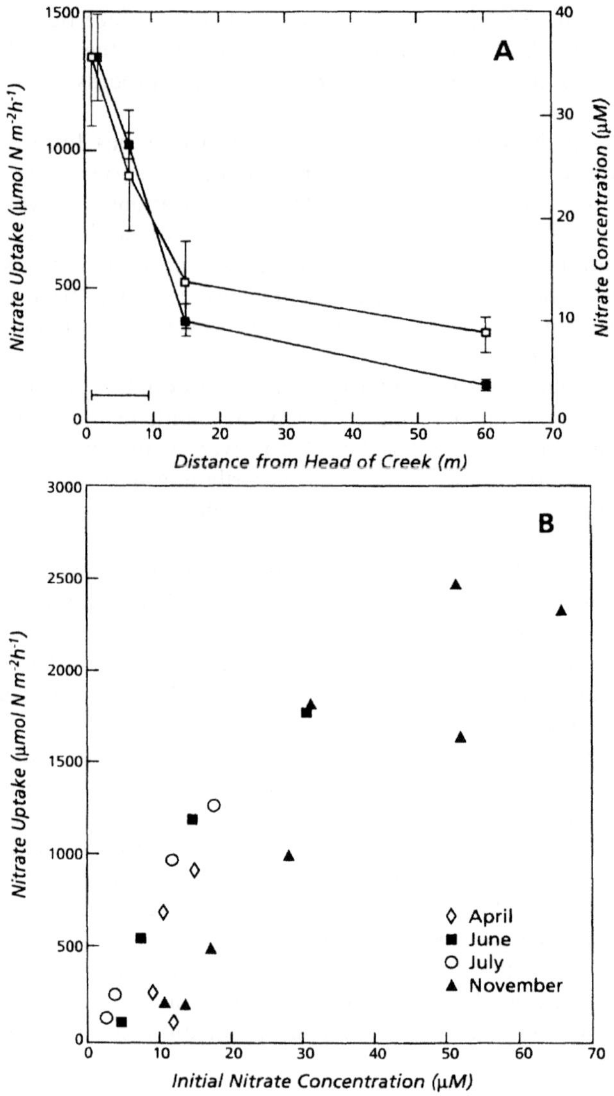

Figure 1. (A) Low tide nitrate concentrations in creek water and the rate of denitrification in the sediments of a tidal creek in Great Sippewissett Marsh, MA. The creek is perpendicular to the adjacent upland and is primarily accreting fine-grained organic sediments. (B) Seasonal pattern of the rate of denitrification in the sediments of a tidal creek (see above) in relation to the concentration of nitrate in the overlying receding groundwater. At low tide the water within the creeks is primarily fresh (2-4) and high in nitrate compared to the flooding tidal waters (salinities of 25-30) which are typically low in nitrate.

To illustrate the importance of these processes for intercepting the nitrogen loading from the upland to estuaries, data are available from an ecosystem level study of a Cape Cod salt marsh. Namskaket marsh is on the north, cold side of Cape Cod with a nine-foot tidal range (Weiskel et al. 1996). The nitrate levels in this marsh flow in groundwater from the residential areas around this site; however, higher levels may arrive in groundwater in the future. There is a septage treatment plant in Orleans that is permitted to discharge nitrate to groundwater via an infiltration system. The background nitrate concentration in groundwater entering the marsh is now about 35 μM and 44% of this has been denitrified by the time the water passes 100 m down the tidal creek from the upland edge of the marsh. When the nitrate concentration was experimentally increased to 240 to 620 μM, 30 to 34% was denitrified in the first 100 m of travel. These numbers are from the warmer 6 months of the year when microbial activity is highest. It is also during this period that the adjacent coastal waters are most sensitive to degradation from nitrogen overloading. Most of this action occurs during low tide because the hydraulic barrier of high tide prevents much groundwater discharge at high water. The data is from both field surveys and dark chamber experiments. We also have data from Great Sippewissett Salt Marsh that indicate denitrification in the tidal creeks of nitrate entering the marsh as nitrate in groundwater of 2 Mm^{-2} y^{-1}. This accounts for the entire nitrate uptake by the marsh system (Howes et al. 1996). These data are also from dark chamber experiments and from field surveys.

What is the meaning of this new information? There is an additional value to be attributed to salt marshes, one that is increasing in both magnitude and importance as nitrogen loading to the coastal oceans increases. Since marshes of more than a few hundred meters width will denitrify most of the nitrate entering from groundwater, they have the potential for intercepting a significant fraction of terrestrially derived nitrogen in groundwater-dominated watersheds. In addition, their rate of nitrogen removal appears to increase with increasing loading providing a buffer for eutrophication of coastal waters. In recent experiments, the denitrifying capacity of organic rich creek bottom sediment was not saturated by nitrate concentrations in overlying water more than 200 times the current concentrations typically observed. In many areas, organic rich salt marsh sediments are providing free tertiary treatment of nitrogen discharges from land prior to entry to open water. In the process, production is also enhanced. Historically, most of the shoreline of the East Coast of the U.S. supported salt marsh. Within this century filling or fragmentation of fringing marsh has lowered the nitrogen buffering capacity of many estuaries during the same period when the terrestrial loading rates are accelerating.

3. Do Marshes Grow Fish?

Intertidal marshes are net exporters of organic material in the form of detritus and of animals (Teal 1962, Valiela and Teal 1979, Deegan and Garritt 1997). Marshes located in restricted basins or basins newly opened to the sea that are net importers of sediments may be exporting most of their detritus only to the adjacent, sediment-accumulating basin (Nixon 1980, Odum et al. 1979). But the fish and birds feeding in the marsh ecosystem are not restricted to an adjacent basin; they represent a true export mechanism. Even in a system that exhibits export of all aspects of marsh production, much of the detrital export is probably in the form of organic compounds resistant to degradation. While this might contribute to bacterial and fungal metabolism somewhere else, it would not greatly contribute to the food webs that propel us to protect marshes.

It is obvious that fish make use of marshes. *Fundulus* and *Cyprinodon*, permanent marsh residents, are found everywhere in marshes within their ranges. During the summer they are found feeding in the marshes, marsh pools, and associated marsh waters. Young fish of non-permanent-resident species spend part of their lives in marshes. Weinstein and O'Neil (1986) and Weinstein et al. (1984) showed that spot (*Leiostomus xanthurus*) in Virginia stayed in the marsh and marsh creeks for months at a time. Werme (1981) found that a variety of non-resident species were found in a marsh in Massachusetts throughout the warm summer months. Gut contents of fish caught in marsh creeks show they are eating things from the marsh, e.g., Able et al. (this volume). Striped bass (*Morone saxatilis*) caught in marsh creeks often have guts full of mummichogs.

The initial stable isotope work of Haines and Montague (1979) at Sapelo Island seemed to contradict the inferences from these observations and to indicate that estuarine animals were not getting nutrition from the marsh. The story from more recent isotope work, for example by Deegan and Garritt (1997) and Deegan et al. (this volume), indicates that if the animals are captured closed to a *Spartina* marsh, they have a signal that combines *Spartina* and benthic algae. If they are close to a *Phragmites* marsh or far up in upland streams they have the signal characteristic of upland plants. This seems straightforward and just what one would expect. Fish eat what is available.

A much more ambitious sampling and analysis program would be necessary to determine how much of the total fish production in an area is directly attributable to the production in a marsh compared to other marshes or open water areas of an estuary. People doing the most intensive fish "sampling" are fishermen who make their living catching and selling fish. Nixon suggested to us that the survey of fish catch from Long Island published by Mather (1887) would be a good data set for examining possible correlation's between fish catch and marshes. Most fishermen in the Long Island villages in the late 19th century used small boats, beach seines, clam tongs and rakes. They did not appear to venture far from their villages. In a few cases, Mather reported that some of the catch came from more distant grounds or from the ocean sides of barrier beaches. In these cases, we did not use the data from those ports.

We used U.S. Coast and Geodetic Survey charts from 1897 to 1910 to planimeter the areas of water shallower and deeper than five feet, the marshes and the length of the water-marsh edges in the inlets and bays around Long Island. We also used marsh areas planimetered by Mr. Alfred Church Lane for Shaler (1886). Inaccuracies in data arise from the scale of the charts that make it difficult to measure the areas, possible inconsistencies in depicting all the marsh creeks of whatever size, and uncertainties in the depiction of the landward edges of marshes. The chart makers did not need to be accurate on the landward edge since they were producing navigational charts.

Most of the correlations were unimpressive, influenced by a single high value. The relationship between shellfish harvest and shallow water had an r^2 of 0.88 that was not improved by considering the marsh areas. Fish catch for each port was poorly related to total marsh area in the local bay or to area of water in the bay. There was one relationship (Fig. 2) with a significant correlation in both the statistical and biological senses. Fish catch in the ports was correlated with marsh edge in the surrounding bay with an r^2 of 0.74. This makes sense because commercially valuable species that rely on marsh productivity would have access to production at the boundary between the marsh and estuarine water.

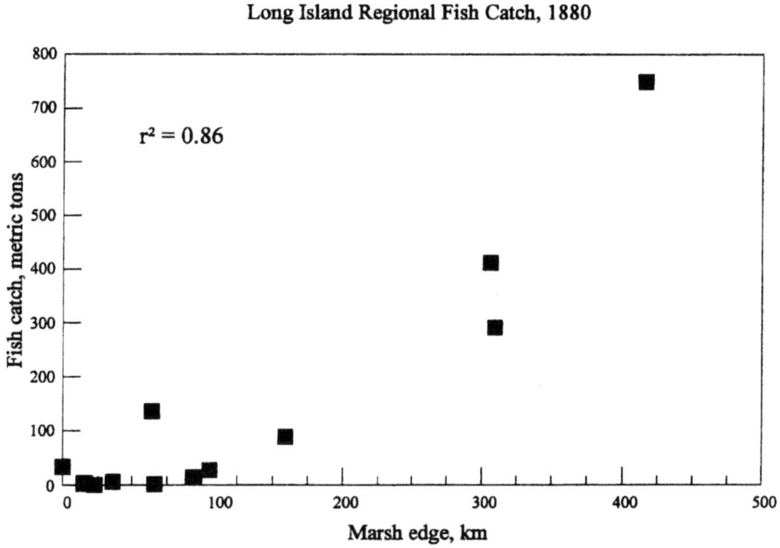

Figure 2. The relationship between fish catch in Long Island, New York ports in 1880 and the length of the edge between salt marsh and open water in the vicinity of that port.

4. Conclusions

Salt marshes along the Atlantic coast of the U.S. have changed during the past century; the number of hectares has declined by over 20% and the nutrient loading per hectare has increased. In this paper we reviewed the mechanisms by which salt marshes trap phosphorus, reduce ammonia in surface water, and reduce both the level and rate of eutrophication of coastal waters by intercepting nitrate in discharging groundwater. We note the rate of nitrogen removal appears to increase with increasing loading. We conclude that, while the East Coast has lost salt marsh area, the relationship between area lost and nitrogen removal capacity is not linear. We reviewed studies of the value of salt marshes to fish production and conclude that both marsh and estuarine species depend upon production from salt marshes for all or part of their food. We examined data on the correlation among fish catch and various marsh features from Long Island, New York in 1880. We conclude that marsh edge is an important feature for fish using the marsh production as food source. Finally, we conclude that marshes import and export nutrients and marshes grow fish.

5. Literature Cited

Buchsbaum, R., I. Valiela and J.M Teal. 1981. Grazing by Canada Geese and related aspects of the chemistry of salt marsh grasses. Colonial Waterbirds **4**: 126-131.

Deegan, L.A. and R.H. Garritt. 1997. Evidence for spatial variability in estuarine food webs. Marine Ecology Progress Series **147**: 31-47.

Haines, E.B. and C.L. Montague. 1979. Food sources of estuarine invertebrates analyzed using carbon 12/ carbon 13 ratios. Ecology **60**: 48-56.

Howes, B.L., P.K. Weiskel, D.D. Geohringer and J.M. Teal. 1996. Interception of freshwater and nitrogen transport from uplands to coastal waters: the role of saltmarshes. Pages 287-310 *in* K.F. Nordstrom and C.T. Roman, editors. Estuarine shores: evolution, environments and human alterations. John Wiley & Sons, New York, New York, USA.

Kneib, R.T. 1994. Spatial pattern, spatial scale and feeding in fishes. Pages 171-185 *in* D.J. Stouder and R.J. Feller, editors. Theory and application in fish feeding ecology. University of South Carolina Press, Columbia, South Carolina, USA.

Kneib, R.T. and S.L. Wagner. 1994. Nekton use of vegetated marsh habitats at different stages of tidal inundation. Marine Ecology Progress Series **106**: 227-238.

Kühl, H. and J. G. Kohl. 1993. Seasonal nitrogen dynamics in reed beds (*Phragmites australis*) (Cav.Trin.ex. Steudel) in relation to productivity. Hydrobiologica **251**: 1-12.

Massachusetts General Laws, Chapter 131, Section 40. 1987.

Mather, F. 1887. New York and its fisheries. Pages 341-377 *in* G.B. Goode, editor. The fisheries and fishery industries of the United States, U.S. Commission of Fish and Fisheries, United States Government Printing Office, Washington, District of Columbia, USA.

Nixon, S.W. 1980. Between coastal marshes and coastal waters—a review of twenty years of speculation and research on the role of salt marshes in estuarine productivity and water chemistry. Pages 437-525 *in* P. Hamilton and K.B. Macdonald, editors. Estuarine and wetlands processes with emphasis on modeling. Plenum Press, New York, New York, USA.

Odum, W.E., J.S. Fisher and J.C. Pickral. 1979. Pages 69-80 *in* R.C. Livingston, editor. Factors controlling the flux of particulate organic carbon from estuarine wetlands. Ecological processes in coastal and marine systems. Plenum Press, New York, New York, USA.

Peterson, S.B. and J.M. Teal. 1996. The role of plants in ecologically engineered wastewater treatment systems. Ecological Engineering **6**: 137-148.

Pomeroy, L.R. 1959. Algal productivity in salt marshes. Limnology and Oceanography **4**: 386-397.

Reddy, K.R., W.H. Patrick, Jr. and C.W. Lindau. 1989. Nitrification-denitrification at the plant root-sediment interface in wetlands. Limnology and Oceanography **34**: 1004-1013.

Shaler, N.S. 1886. Preliminary report on sea coast swamps of the Eastern United States. U. S. Geological Survey Annual Report **6**: 353-398.

Smart, R.M. and J. W. Barko. 1980. Nitrogen nutrition and salinity tolerance of *Distichlis spicata* and *Spartina alterniflora.* Ecology **61**: 630-638.

Teal, J.M. 1962. Energy flow in the salt marsh ecosystem of Georgia. Ecology **43**: 614-624.

Valiela, I., B. Howes, R. Howarth, A. Giblin, K. Foreman, J.M. Teal and J.E. Hobbie. 1982. Regulation of primary production and decomposition in a salt marsh ecosystem. Pages 151-168 *in* B. Gopal, R.E. Turner, R.G. Wetzel and D.F. Whigham, editors, Wetlands: ecology and management. National Institute of Ecology and International Scientific Publications, Jaipur, India.

Valiela, I. and J.M Teal. 1979. The nitrogen budget of a salt marsh ecosystem. Nature **280**: 652-656.

Valiela, I., J.M Teal, S. Volkmann, D. Shafer and E.J. Carpenter. 1978. Nutrient and particulate fluxes in a salt marsh ecosystem: tidal exchangers and inputs by precipitation and groundwater. Limnology and Oceanography **23**:798-812.

Vince, S., I. Valiela, N. Backus and J.M. Teal. 1976. Predation by the salt marsh killifish *Fundulus heteroclitus* (L.) in relation to prey size and habitat structure: consequences for prey distribution and abundance. Journal Experimental Marine Biology Ecology **23**: 255-266.

Weinstein, M.P. and S.P. O'Neil. 1986. Exchange of marked juvenile spots between adjacent tidal creeks in the York River Estuary, Virginia. Transactions American Fisheries Society **115**: 93-97.

Weinstein, M.P., L. Scott, S.P. O'Neil, R.C.I. Seigfried and S.T. Szedlmayer. 1984. Populations dynamics of spot, *Leiostomus xanthurus*, in polyhaline tidal creeks of the York River Estuary, Virginia. Estuaries 7: 444-450.

Weiskel, P.K., L.A. DeSimone and B.L. Howes. 1996. Transport of wastewater nitrogen through a coastal aquifer and marsh, Orleans, MA. U.S. Geological Survey, Open-File Report **96-111**, Reston, Virginia, USA.

White, D.S. and B.L. Howes. 1994. Long-term ^{15}N-nitrogen retention in the vegetated sediments of a New England salt marsh. Limnology and Oceanography **39**: 1878-1892.

SOURCES AND PATTERNS OF PRODUCTION

ROLE OF SALT MARSHES AS PART OF COASTAL LANDSCAPES

IVAN VALIELA
MARCI L. COLE
JAMES MCCLELLAND[a]
JENNIFER HAUXWELL
JUST CEBRIAN[b]
Boston University Marine Program,
Marine Biological Laboratory, Woods Hole MA 02543 USA

SAMANTHA B. JOYE
Department of Marine Sciences,
University of Georgia, Athens GA 30602 USA

Abstract
Salt marshes are located between land and coastal water environments, and nutrient and production dynamics within salt marshes interact with those of adjoining ecosystems. Salt marshes tend to export materials to deeper waters, as shown by mass balance and stable isotopic studies. Salt marshes also intercept land-derived nutrients, and thus modify the potential response of phytoplankton, macroalgae, and seagrasses in the receiving estuarine waters. In particular, the maintenance of eelgrass meadows seems to depend on the ability of fringing salt marshes to intercept land-derived nitrogen. The bulk of the interception of land-derived nitrogen is likely to be the result of relatively high rates of denitrification characteristic of salt marshes. Thus, through exports of energy-rich materials, and interception of limiting nutrients, salt marsh parcels interact in quantitatively important ways with adjoining units of landscape. These interactions are of importance in understanding the basic functions of these mosaics of different coastal systems, as well as provide information needed to manage estuaries, as for example, in conservation of valuable eelgrass meadows.

1. Introduction

Coastal zone land/seascapes are composed of a diversity of ecosystems coupled to varying degrees by material exchanges and transformations taking place within and between the ecosystem units. Salt marshes in particular are one type of ecosystem whose presence alters the nature of neighboring units within the ecological land/seascape.

Much has been written about whether or not salt marshes export materials to deeper waters (many papers summarized in Nixon 1980, Valiela 1983, Taylor and Allanson

[a] Present address: School of Biology, Georgia Inst. of Technology, Atlanta GA 30332.
[b] Present address: Dauphin Island Sea Lab, Dauphin Island, AL 36528 USA.

1995), about the degree to which salt marsh-produced organic matter is transferred to other receiving systems (Haddad and Martens 1987, Moran and Hodson 1994, Taylor and Allanson 1995), and about interceptions and transformations that occur in salt marsh ecosystems (Valiela and Teal 1979, Johnston 1991, Johnson et al. 1994). These features have largely been considered separately, and consequently there is a gap in our understanding about the links among these features. These links are manifestations of a larger issue: salt marshes interact with adjoining types of environments, in powerful couplings that link the different and adjoining coastal tesserae (mosaic pieces). The units of such linked tesserae include terrestrial land parcels with covers of diverse types, salt marshes, macroalgal beds, seagrass meadows, and phytoplankton-dominated waters.

In this paper we attempt to cover the gap in understanding by briefly reviewing the evidence for export of materials from salt marsh ecosystems, assessing information on the degree to which material produced in salt marshes enters other ecosystems, and examining the effects of interception and transformations of nitrogen in fringing salt marshes on the structure of producer assemblages in adjoining aquatic ecosystems. Our overall goal, then, is to show that salt marshes are an integral part of the coastal land/ seascape and may be involved in the natural maintenance and protection of other vulnerable estuarine habitats.

In response to widespread concern with the pervasive losses of salt marshes in the U.S. and elsewhere (Mitsch and Gosselink 1993), we also examine some features of salt marshes that justify the concern that salt marsh destruction could result in further changes in estuarine systems.

2. Export of Materials from Salt Marshes to Deeper Waters

Three major lines of evidence (biogeochemical analyses of sediments, metabolic measurements, and mass balance data) suggest that there is often significant export of materials from salt marshes to neighboring ecosystems.

Measurement of stable isotopic ratios and of concentrations of lignin derivatives in surface sediments of receiving estuaries suggests that modest amounts of salt marsh-derived organic matter exist there. Sediment derived from *Spartina*, the dominant salt marsh grass, comprised only 1- 4% of the top 1 cm in Buzzards Bay (Wilson et al. 1985) and 2% of sediment organic matter in Cape Lookout Bight (Haddad and Martens 1987). These values appear small, but need to be taken in context, for which we need to consider two points. First, the acreages of open water such as Buzzards Bay and Cape Lookout Bight are almost always much larger than those of salt marshes fringing the shore, hence salt marsh-derived organic matter is dispersed onto a larger surface area and concentrations will necessarily be lower. Second, the materials left in sediments are likely the refractory fraction remaining from much larger amounts of organic matter. For example, lignin makes up at most 5% of weight of *Spartina* organic matter, and some of the lignin is labile, so that the truly refractory amount — the fraction that would accumulate in sediments — is a rather small amount of the total weight of organic matter that actually was exported. Hence, contributions from salt marshes to the organic matter pool in coastal waters (and sediments below) may be substantially larger than indicated by the small percentages cited above.

Measurements of ecosystem metabolism in coastal waters off salt marsh-dominated coasts showed that, in fact, the coastal waters were heterotrophic (Nixon 1980, Dankers et al. 1984, Hopkinson 1985). The magnitude and the timing of the changes in respiration of coastal planktonic systems were such that Hopkinson (1985) concluded that organic matter export from the fringing salt marshes was the most reasonable source for the organic materials that apparently supported a significant fraction of metabolism taking place within the coastal water column.

Most studies of salt marsh export are based on mass balance measurements in which export and import of various materials were assessed. Nixon (1980) reviewed the mass balance measurements available (Table 1) and was impressed with the heterogeneity of results; it was evident that not all salt marshes studied showed the same features. Estimates of total N or C exports and imports are seldom available since researchers may not measure all of the major forms of N or C. This incompleteness prompted us to examine whether each component was exported or imported (Table 1). We then take the aggregate results for the several marshes as representing the overall export or import. On aggregate we can compute the proportion of salt marshes that exported, imported, or showed no net exchange (''E'', ''I'', and ''O'', respectively, in Table 1) of the different materials. A majority of salt marshes exported materials. For all materials, with the single exception of nitrate, which will be discussed below, the aggregate data suggest that an important majority of salt marshes export materials. We will return to the matter of nitrate below.

It might be possible to further explain differences in export characteristics among different marshes. As one example, we sorted the different salt marshes into ''young'' and ''mature'' categories, based on a qualitative ratio of vegetated to open water (Redfield 1972) and simplicity or complexity of the outlet to the sea (Odum et al. 1979). We qualitatively assigned specific salt marshes to the ''young'' category if they had much open water and relatively simple tidal outlets to the sea. Salt marshes whose surfaces were largely covered with vegetation and whose channels and inlets were relatively more complex were assigned to the ''mature'' category. This typology was based on the shift from relatively open water bays to vegetated and sediment-filled wetlands that describe the historical changes of many salt marshes in the eastern seacoast of the U. S. The differences in exports between these two relatively simple categories suggest that salt marshes, as they mature and fill in with vegetation and sediments, gradually export more materials than young marshes. Other ways to examine the variability in export-import properties that Nixon (1980) observed might be to see if tidal excursion and hypsometry make a difference. The difficulty with making such detailed comparisons is in finding a sufficient number of salt marshes, for which all necessary data exist, to include in these studies.

In any case, the aggregate data of Table 1 can be used to convincingly argue that, within the inevitable variability associated with most ecological comparisons, there is a remarkable consistency in the results: export of most materials is a feature of most salt marshes. Further study of the exceptions might teach us more about how exports are controlled. Recent advances in methods using Ra isotopes as tracers do confirm nutrient exports from salt marshes (Krest et al. 2000).

There seems to be, therefore, consistent comparative evidence that most salt marshes export energy-containing reduced materials. A more difficult question to address, however, is whether or not the materials exported from salt marshes represent a substantial percentage of the autochthonous production in the receiving ecosystems. For example, ammonium export, even if it is small in absolute terms, may have a large impact on primary production in coastal waters that are nitrogen limited (Howarth 1988). Carbon export to these same waters, on the other hand, could be negligible relative to local production. This issue should be kept in mind as we examine the coupling between salt marshes and coastal food webs below.

3. Salt Marsh-produced Organic Matter in Coastal Food Webs

The most compelling information on the penetration of organic matter produced in salt marshes into subtidal food webs comes from stable isotopic studies (Peterson et al. 1986, Currin et al. 1995, Deegan and Garritt 1997). Through the use of carbon, nitrogen, and sulfur isotopes, these studies showed that the isotopic signatures of consumers reflect different mixes of *Spartina* and algae in their diets depending on availability in the specific habitat in which they are found. This approach provides the single most useful approach to understanding marsh food webs.

There is, however, some ambiguity associated with interpretation of stable isotope data in salt marsh food-web studies. This ambiguity arises for two main reasons. First, stable isotope ratios of *Spartina* can change during decomposition (Benner et al. 1987, Currin et al. 1995). Second, there are often multiple combinations of food sources that can account for the isotopic signature of a consumer. Isotopic examination of the estuarine food web of Sage Lot Pond, Cape Cod, MA, highlights both of the above concerns (Fig. 1).

To interpret the data in Fig. 1, remember that points for consumers are expected to fall near points for producers that are dominant food sources (or half-way between two producers if they contribute equally to the diet of a consumer). Several combinations of producers could lead to the pattern of consumer isotope values that we see (Fig. 1). In fact, at face value, live *Spartina* seems to be the only producer that can be largely ruled out as a major food source for consumers (Fig. 1). When changes associated with decomposition of *Spartina* are accounted for, however, *Spartina* detritus becomes potentially a major food source (Fig. 1). It turns out, however, that for Sage Lot Pond the potential role of *Spartina* is small because of the small tidal range and the level topography of the marsh areas (McClelland and Valiela 1998), so it is unlikely to play a large trophic function for this system.

Fortunately, isotope data are rarely interpreted without the aid of ancillary information (primary production data and acreage of relevant vegetation, for example) about the system under study. With this added context, much of the ambiguity can be resolved. The above example, however, simply demonstrates that, although the stable isotopic approach is extremely helpful, we are still struggling to find a completely unambiguous way to quantitatively understand the salt marsh contribution to estuarine food webs.

TABLE 1. Summary of studies of exports and imports of ammonium (NH_4), nitrate (NO_3), dissolved organic nitrogen (DON), particulate nitrogen (PON), total nitrogen (TN), dissolved organic carbon (DOC), particulate organic carbon (POC), and total carbon (TC) from salt marshes to coastal waters[1].

Estuary	NH_4	NO_3	DON	PON	TN	DOC	POC	TC
				Young[3] marshes				
Canary Creek, DE	I[2]	I	E	E	E	E	E	E
Crommet Creek, NH	I	O						
Ems-Dollard Marsh, Netherlands						E	I	I
Flax Pond, NY	E	I				O	I	I
Sapelo Island, GA	O	O	E	I		E	I	
Stroodorpe Marsh, Netherlands				I			I	
				Mature[3] marshes				
Barataria Bay, LA						E	E	E
Bly Creek, SC						E	I	E
Branford Marsh, CT	E	I						
Carter Creek, VA	I	I	E	I	E	E	E	E
Coon Creek, TX						E	O	E
Dill Creek, SC							E	
Gotts' Creek, MD	E	E	E	E			E	
Great Sippewissett Marsh, MA	E	E	E	E	E		E	E
Kariega Marsh, S. Africa						E	O	E
North Inlet, SC	E	E	E	E		E	E	E
Tijuana Estuary, CA	E	E		E			E	
(two marshes)	E	E		E			E	
Ware Creek, VA	E	I	E	E	E	E	E	E
				Percent of marshes that export materials				
"young" marshes:	25	0	100	33	100	75	20	33
"mature" marshes:	89	67	100	88	100	100	75	100
All marshes:	77	46	100	72	100	91	59	82

[1]Data from many sources, summarized by Nixon (1980), Valiela (1983), Dame et al. (1991), and Taylor and Allanson (1995).
[2]E: export, I: import, O: no net transport.
[3]See text for definition of "young" and "mature".

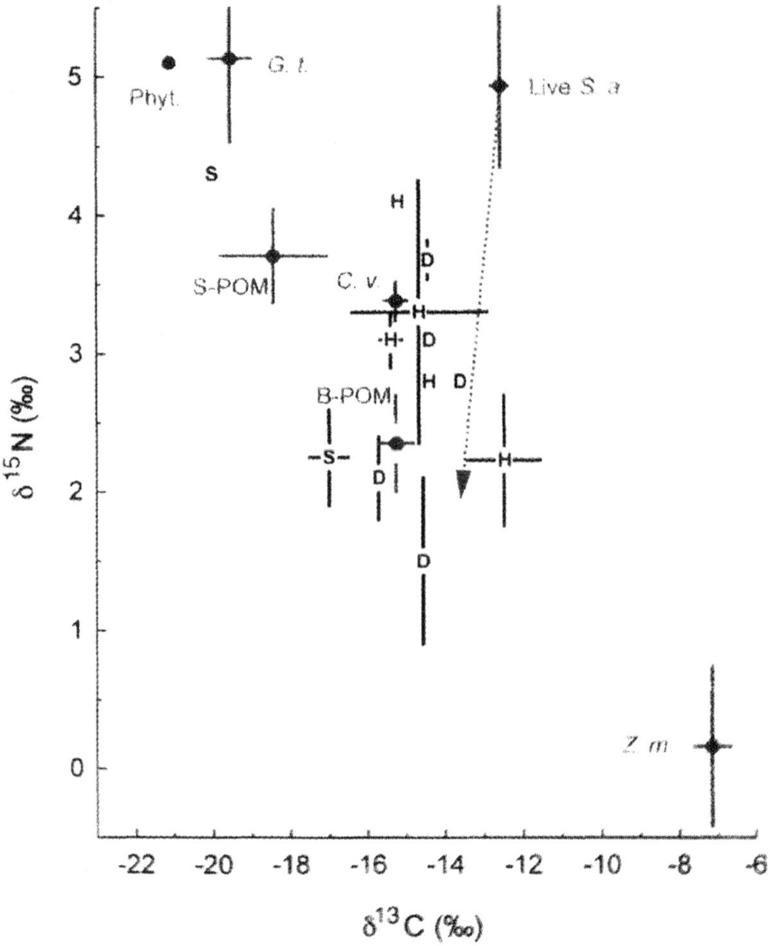

Figure 1. $\delta^{13}C$ versus $\delta^{15}N$ of consumers (adjusted for trophic fractionation, see below) and potential producer food sources in Sage Lot Pond. Consumers (black) are suspension feeders (S), herbivores (H), or deposit/detritus feeders (D). Producers are phytoplankton (Phyt.), macroalgae [*Gracilaria tikvahiae* (*G. t.*) and *Cladophora vagabunda* (*C. v.*)], eelgrass [*Zostera marina* (*Z. m.*)], salt marsh cordgrass [*Spartina alterniflora* (*S. a.*)], and particulate organic matter [suspended (S-POM) and benthic (B-POM)]. To correct for trophic fractionation, 1‰ has been subtracted from the $\delta^{13}C$ and $\delta^{15}N$ values of amphipods (herbivores) (Macko et al. 1982). For all other consumers, 1‰ has been subtracted from $\delta^{13}C$ and 3‰ has been subtracted from $\delta^{15}N$. These values reflect an average fractionation effect of a trophic step based on a broad range of consumer types (Peterson and Fry 1987, Michener and Schell 1994). The red arrow indicates the position where a point for decomposed cordgrass might be expected to fall based on Benner et al. (1987) and Currin et al. (1995).

4. Salt Marshes as Units Within Coastal Mosaics

Discussion of exports from salt marshes, and penetration of salt marsh-produced organic matter into subtidal food webs already suggest that we cannot consider these coastal tesserae as isolated entities, but rather that there are likely to be powerful links among these adjoined landscape units. These couplings occur through tidal exchanges with deeper waters, and via ground- and streamwater transport between terrestrial and estuarine units.

There are, in addition, other links mediated by biogeochemical transformations, which, although less well-defined, might be as influential as the better-known tidal and freshwater transports. Interception of land-derived nitrogen within salt marshes might be one such biogeochemical transformation that affects the structure of producer communities in subtidal ecosystems.

Before discussing the role of fringing salt marshes in the context of land/seascapes, we need to introduce the importance of land-derived nitrogen loads in structuring assemblages of coastal producers. In the Waquoit Bay estuarine system we identified a series of subwatersheds and corresponding estuaries that receive specific nitrogen loads from each subwatershed (Valiela et al. 1992, Valiela et al. 1997a). We developed the Waquoit Bay Nitrogen Loading Model (Valiela et al. 1997a) to estimate nitrogen loads provided by wastewater disposal, fertilizer use, and atmospheric deposition. We then verified predictions of the model in two different ways. First, we compared model estimates to empirically measured nitrogen loads (Valiela et al. 2000) obtained by multiplication of annual groundwater recharge rates times basin-weighted concentrations of nitrogen within groundwater at the seepage face to the estuaries. Second, since $\delta^{15}N$ in nitrate derived from wastewater differs from that of nitrate derived from atmospheric deposition and fertilizer (McClelland et al. 1997), we would expect that $\delta^{15}N$ values in bulk groundwater would become heavier as the percentage of the nitrogen load entering the groundwater derived from wastewater increases. In fact, the NLM predictions do fit the measured values well, and they also agree with the $\delta^{15}N$ data (Valiela et al. 2000). We conclude that the model reasonably captures the complexities of multiple nitrogen sources passing through watersheds with different land cover mosaics, and through soil, vadose zone, and aquifer.

The value of having identified land parcels that deliver different nitrogen loads to their receiving waters is that we can then ask whether the vegetation within the estuaries differs in estuaries subject to different nitrogen loads. Indeed, we found that as nitrogen load to estuaries increased, production of phytoplankton (Fig. 2 top row, left panel) and biomass of macroalgae (Fig. 2 top row, middle panel) increased significantly. In contrast, eelgrass biomass decreased sharply as nitrogen loads increased (Fig. 2 top row, right panel).

The reduction in eelgrass associated with even small increases in nitrogen loads appears to be a general pattern. For example, we have mapped the distribution of eelgrass meadows in the central part of Waquoit Bay across decades, and found decreases that were synchronous with the relative urbanization of the watersheds from the late 1960s to more recent years (Fig. 3). High sensitivity of seagrasses to even small increases in nitrogen loads is well-known (Sand-Jensen and Borum 1991, Duarte

1995). Increases in nitrogen loading from watersheds to estuaries is accompanied by loss of eelgrass and increases in phytoplankton, epiphytes, and macroalgal canopies (Nienhuis 1983, Cambridge and McComb 1984, Orth and Van Montfrans 1984, Borum 1985, Giesen et al. 1990, Valiela et al. 1992, Thybo-Christesen et al. 1993, Short et al. 1993, Lyons et al. 1995, Short and Burdick 1996).

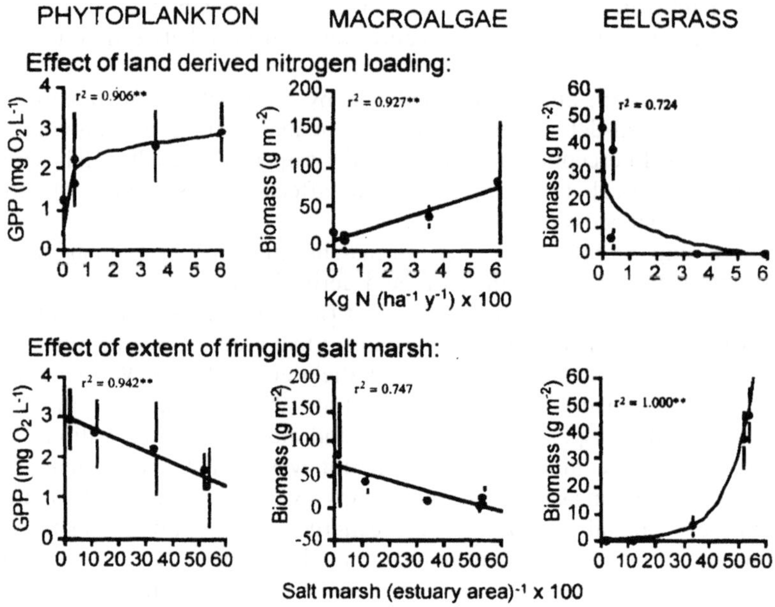

Figure 2. Effect of land-derived N loading (top panels) and of extent of fringing salt marsh (bottom panels) on phytoplankton gross production (left panels), benthic macroalgae biomass (middle panels), and eelgrass biomass (right panels) in 5 Waquoit Bay estuaries.

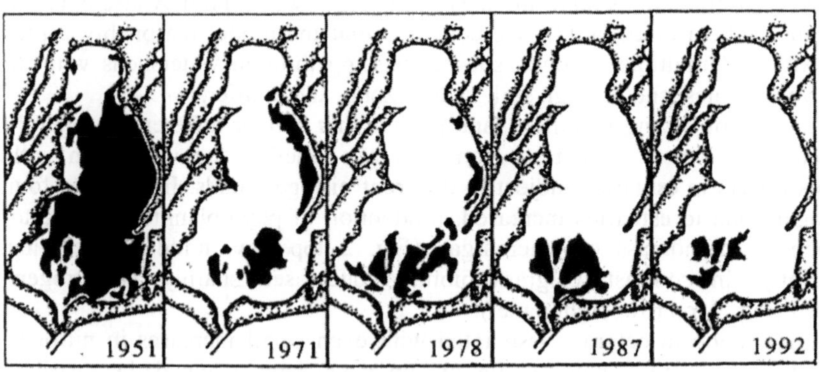

Figure 3. Black areas represent the extent of eelgrass beds within Waquoit Bay during different years, 1951-1992. Data from 1951-1987 by J. Costa, published in Valiela et al. (1992), and 1992 data from Short and Burdick (1996).

Two types of mechanisms, one indirect (shading) and one direct (nitrate toxicity), might be responsible for the decline of eelgrass under increased nitrogen loads. The indirect mechanism might be that nitrate loads increase micro- and macroalgae, which in turn shade eelgrass. Growth of eelgrass is largely light-limited (Dennison and Alberte 1982, Zimmerman et al. 1987). Shading by increased biomass of other nitrogen-limited producers that result from increased nitrogen availability (Twilley et al. 1985, Sand-Jensen and Borum 1991, Lapointe et al. 1994, Duarte 1995, Valiela et al. 1997b), could impair growth of eelgrass. Laboratory and mesocosm experiments demonstrate that likelihood of shading by 1) water column phytoplankton, 2) unattached macroalgae (Hauxwell et al. 2000), and 3) micro- or macroalgal epiphytes on eelgrass increases with nutrient enrichment (Harlin and Thorne-Miller 1981, Neckles et al. 1993, Short et al. 1993, Short et al. 1995, Short and Burdick 1996).

Increases in nitrate concentrations may also act directly on eelgrass by toxic effects at concentrations higher than 10 μM (Burkholder et al. 1992, 1994). Mesocosm experiments, in which other producers were excluded, showed that eelgrass exposed to nitrate concentrations representative of estuaries undergoing cultural eutrophication had 35% lower shoot production than controls (Burkholder et al. 1992, 1994). We are not certain whether direct or indirect mechanisms are more or less important, but we are sure that increased nitrogen loading exerts a powerful influence on eelgrass meadows.

There is compelling evidence, therefore, from Waquoit Bay and elsewhere, that differences in nitrogen loads may substantially alter the community of different producers found in shallow coastal waters. But there is more to this issue than that, and here is where fringing salt marshes may play a key function: it may be that salt marshes intercept substantial proportions of the land-derived nitrogen loads that affect eelgrass meadows. Evidence of this issue can be garnered from plots of the producer data used in the top row of panels of Fig. 2 versus the percentage of the estuary area that is represented by fringing salt marsh habitat (Fig. 2 bottom row). Data for salt marsh area were obtained from aerial photos. This depiction of the data suggests that in estuaries where there was proportionately more salt marsh habitat, there was significantly less production and biomass of phytoplankton and macroalgae, respectively (Fig. 2 bottom row, left and middle). Again in contrast, where there was more salt marsh acreage, crops of eelgrass were significantly higher (Fig. 2 bottom row, right). Incidentally, the correlation between nitrogen load and area of salt marsh was weak; the r^2 between these two variables explained only 23% of the variation. These results suggest that the more salt marsh, the better for eelgrass meadows.

The results shown in Fig. 2 (bottom) are what has been called a space-for-time substitution, in which miscellaneous differences among sites may confound differences among loading rates, the variable of interest here. To independently corroborate our space-for-time results, we sought historical data for each estuary. In each estuary we mapped the extent of eelgrass meadows during 1997. These values were then compared to data obtained from aerial photos, first-hand observer reports, and earlier publications (Curley et al. 1971, Short and Burdick 1996), and we calculated the percentage loss of eelgrass habitat that took place in each estuary from the mid 1960s to 1997.

The percentage loss of eelgrass meadows increased sharply even with small increases in nitrogen loads (Fig. 4 top). Losses of eelgrass became near total beyond the lower third of the range in nitrogen loads to which these estuaries were exposed.

There is some ambiguity in this data presentation because the loads are present-day loads, and are to some extent higher than those occurring during the 1960s and 1970s (Sham et al. 1995). Much as in the case of the space-for-time substitution, the actual loss results suggest that losses of eelgrass meadows were significantly reduced in estuaries where there were larger relative acreages of salt marsh (Fig. 4 bottom). In this case, the relationship is linear, suggesting a constant effect of the fringing salt marsh.

Both the space-for-time data and the historical loss data show that eelgrass meadows diminish markedly where exposed to increased nitrogen loads, and that in some fashion, larger areas of fringing salt marshes counter the effect of nitrogen loads and preserve eelgrass meadows. At least two mechanisms might account for the role of salt marshes in reducing the impact of land-derived nitrogen loads: denitrification, and burial in salt marsh sediments. Denitrification rates in salt marshes (Table 2) are high compared to those in most aquatic habitats (Valiela and Teal 1979, Howarth 1988, Seitzinger 1988). Salt marshes do accumulate sediments so that nitrogen is buried in the process of marsh sediment accretion. We are unsure of the actual magnitude of these processes relative to the rates of land-derived nitrogen loads, but we predict that they will be qualitatively significant relative to inputs of terrestrial nitrogen.

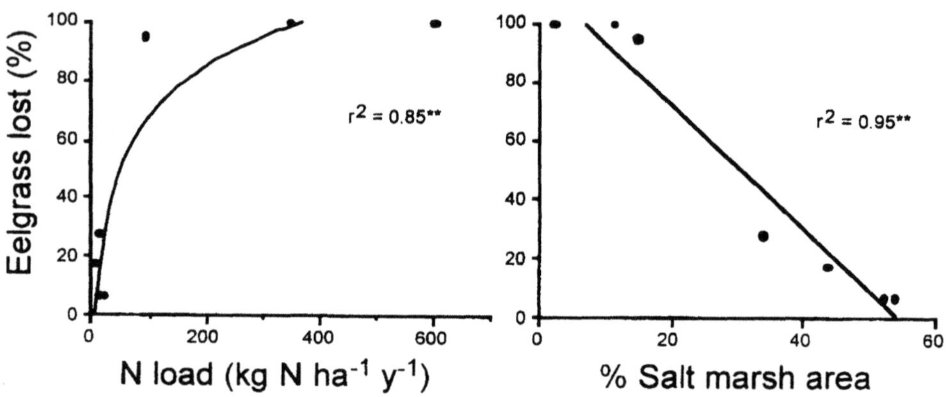

Figure 4. Percentage of eelgrass habitat area lost in the Waquoit estuaries in relation to the N load received by the estuaries (left) and to the percentage of the estuary area that is covered by salt marsh habitats (right).

Of all the substances measured in the export/import studies, nitrate was the only one that was not consistently exported from salt marshes (Table 1). We speculate that denitrification within salt marshes is large enough to intercept a significant portion of land-derived nitrate, thereby preventing nitrate from moving into deeper waters.

To get a rough idea of the magnitude of denitrification and burial relative to land-derived N loads, we can make use of published information (Table 2). Ranges of rates of denitrification and burial of nitrogen were determined by White and Howes (1994) in a ^{15}N tracer experiment in which they measured losses of the added nitrogen in Great Sippewissett marsh in Cape Cod. They found that up to 20 kg N ha^{-1} y^{-1} may be buried, and up to 50 kg N ha^{-1} y^{-1} may be denitrified. For comparison, the rates of land-derived nitrogen to the estuaries of Waquoit Bay ranged from 10 to 600 kg N ha^{-1} y^{-1}. The magnitudes of these rates suggest that, first, denitrification is likely to be more significant than burial as a mechanism for interception of land-derived nitrogen in salt marshes. Second, the ability of salt marshes to intercept land-derived nitrogen may be qualitatively significant in situations where the land-derived N loads are reasonably low (perhaps up to 100 kg N ha^{-1} y^{-1}), but as land-derived nitrogen loads continue to increase, the beneficial function of salt marshes cannot keep up with the anthropogenic loads, and estuarine eutrophication necessarily increases.

TABLE 2. Daily denitrification rates measured in salt marsh sediments reported in the literature for different locations and using different methods.

Producer taxon associated with measurement	Location	Denitrification rate ($mg\ N\ m^{-2}\ d^{-1}$)	Method	Reference
Spartina alterniflora	MA	12.0-31.0	N_2/Ar, gas partitioning	Kaplan et al. (1979)
S. alterniflora	MA	94	N_2 flux	Johnson et al. (1994)
S. alterniflora	MA	25.2	^{15}N tracer	White (1994), White and Howes (1994)
S. alterniflora	GA	33	N_2 sediment profiles, diffusion model	Haines et al. (1977)
Puccinellia/Halimione	UK	2.8	^{15}N tracer	Abd Aziz and Nedwell (1986)
Halimione	UK	1.7-40*	Acetylene	Koch et al. (1992)
Salicornia	CA	0.3-6.6	Acetylene	Joye and Paerl (1994)
S. alterniflora	NC	0.8-4.4*	Acetylene	Thompson et al. (1995)

*Shown as 2X the published value. We based the corrections on the results of Seitzinger et al. (1993), who found that use of the acetylene reduction technique underestimated rates of total denitrification by about half.

These are speculations; we need to better define all these values. In addition, we need to ascertain whether there is a load-dependent response of denitrification and burial rates. Preliminary measurements in Cape Cod marshes do not clearly show whether denitrification rates increase as external nitrogen loads increase (Kaplan 1978, Lee et al. 1997). In estuaries in general there is ambiguous evidence about the response of denitrification to external loads (Jorgensen and Sorensen 1985, Seitzinger 1994, Law et al. 1991, Valiela 1995).

In any case, the results of the space-for-time substitution and the calculation of actual losses across recent decades both suggest that as we look at adjoining parcels of the land/ seascape (Fig. 5) we need to realize that different kinds of habitats in the coastal zone—both terrestrial and aquatic—are not isolated tesserae, but rather that these units are coupled by powerful linkages and the couplings can strongly influence the vegetation (and we may presume, the entire food webs) of the adjoining habitats, both land and sea.

Recent decades have witnessed marked losses of coastal wetlands. Many arguments have been advanced for the preservation of salt marshes as useful parcels of coastal land/ seascape (Vince et al. 1981, Mitsch and Gosselink 1993): salt marshes export materials important to food webs of deeper waters, act as nurseries for many species of commercially important fisheries stocks, provide sources of harvestable shellfish and sites for aquaculture, intercept toxic contaminants, stabilize shorelines, provide waterfowl refuges and nesting areas and stopover for migratory birds, intercept land-derived nutrients, and protect water quality. The finding that fringing salt marshes appear to intercept a substantial fraction of land-derived nitrogen loads and hence protect the quality of valued eelgrass habitats provides yet another reason for the preservation of coastal salt marshes.

Figure 5. A coastal zone with a well-established salt marsh fringe interposed between terrestrial habitats and a flourishing eelgrass meadow. Taken in the Waquoit Bay estuarine system by Jennifer Hauxwell and Just Cebrian.

5. Acknowledgments

This work is part of the Waquoit Bay Land Margin Ecosystems Research project, and was supported by a grant from the National Science Foundation's Land Margin Ecosystems Research initiative, by the U.S. Environmental Protection Agency's Region 1, and by the National Oceanic and Atmospheric Administration's Sanctuaries and Reserves Division. The synthesis of much of the data was supported by the National Center for Environmental Assessment, Office of Research and Development, U.S. Environmental Protection Agency, and by the Woods Hole Oceanographic Institution's Sea Grant Program. Part of the work was carried out at the Waquoit Bay National Estuarine Research Reserve, and we appreciate the cooperation of the reserve manager, Christine Gault, and her staff.

6. Literature Cited

Abd Aziz, S. A. and D. B. Nedwell. 1986. The nitrogen cycle of an east coast, U.K. salt marsh. 2. Nitrogen fixation, nitrification, denitrification, tidal exchange. Estuarine, Coastal and Shelf Science **22**:689-704.

Benner, R., M. L. Fogel, E. K. Sprague and R. E. Hodson. 1987. Depletion of ^{13}C in lignin and its implications for stable carbon isotope studies. Nature **329**:708-710.

Borum, J. 1985. Development of epiphytic communities on eelgrass (*Zostera marina*) along a nutrient gradient in a Danish estuary. Marine Biology **87**:211-218.

Burkholder. J. M., H. B. Glasgow, Jr. and J. E. Cooke. 1994. Comparative effects of water-column nitrate enrichment on eelgrass *Zostera marina*, shoalgrass *Halodule wrightii* and widgeon grass *Ruppia maritima*. Marine Ecology Progress Series **105**:121-138.

Burkholder, J. M., K. M. Mason and H. B. Glasgow, Jr. 1992. Water-column nitrate enrichment promotes decline of eelgrass *Zostera marina* L.: evidence from seasonal mesocosm experiments. Marine Ecology Progress Series **61**:163-178.

Cambridge, M. L. and A. J. McComb. 1984. The loss of seagrasses in Cockburn Sound, Western Australia. 1. The time course and magnitude of seagrass decline in relation to industrial development. Aquatic Botany **20**:229-243.

Curley, J. R., R. P. Lawton, J. M. Hickey and J. D. Fiske. 1971. A study of the marine resources of the Waquoit Bay-Eel Pond Estuary. Monograph Series Number 9. Division of Marine Fisheries. Department of Natural Resources. The Commonwealth of Massachusetts.

Currin, C. A., S. Y. Newell and H. W. Paerl. 1995. The role of standing dead *Spartina alterniflora* and benthic microalgae in salt marsh food webs: Considerations based on multiple stable isotope analysis. Marine Ecology Progress Series **121**:99-116.

Dame, R. F., J. Spurrier, T. Williams, B. Kjerfve, R. Zingmark, T. Wolaver, T. Chrzanowski, H. N. McKellar, and F. J. Vernberg. 1991. Annual material processing by a salt marsh-estuarine basin in South Carolina, USA. Marine Ecology Progress Series **72**: 153-166.

Dankers, N., M. Binsbergen, K. Zegers, R. Laane and M. R. van der Loeff. 1984. Transportation of water, particulate and dissolved organic and inorganic matter between a salt marsh and the Ems-Dollard estuary, The Netherlands. Estuarine, Coastal and Shelf Science **19**:143-166.

Deegan, L. A. and R. H. Garritt. 1997. Evidence for spatial variability in estuarine food webs. Marine Ecology Progress Series **147**:31-47.

Dennison, W. C. and R. S. Alberte. 1982. Photosynthetic responses of *Zostera marina* L. (eelgrass) to *in situ* manipulations of light intensity. Oecologia **55**:137-144.

Duarte, C. M. 1995. Submerged aquatic vegetation in relation to different nutrient regimes. Ophelia **41**:87-112.

Giesen, W. B. J. T., M. M. van Katwijk and C. den Hartog. 1990. Eelgrass condition and turbidity in the Dutch Wadden Sea. Aquatic Botany **37**:71-85.

Haddad, R. I. and C. S. Martens. 1987. Biogeochemical cycling in an organic-rich coastal marine basin. 9. Sources and accumulation rates of vascular plant-derived organic matter. Geochimica et Cosmochimica Acta **51**:2991-3001.

Haines, E. B., A. Chalmers, R. Hanson and B. Sherr. 1977. Nitrogen pools and fluxes on a Georgia salt marsh. Pages 241-254 *in* M. Wiley, editor. Estuarine processes, Volume 2. Academic Press, New York, New York, USA.

Harlin, M. M. and B. Thorne-Miller. 1981. Nutrient enrichment of seagrass beds in a Rhode Island coastal lagoon. Marine Biology **65**:221-229.

Hauxwell, J., J. Cebrián, C. Furlong, and I. Valiela. 2000. Macroalgal canopies contribute to celgrass (*Zostera marina*) decline in temperate estuarine ecosystems. Ecology, in press.

Hopkinson, Jr., C. S. 1985. Shallow-water benthic and pelagic metabolism: evidence of heterotrophy in the nearshore Georgia Bight. Marine Biology **87**:19-32.

Howarth, R. 1988. Nutrient limitation of net primary production in marine ecosystems. Annual Review of Ecology and Systematics **19**:89-110.

Johnson, R., M. LaMontagne and I. Valiela. 1994. Rate of denitrification in submerged salt marsh sediments. Biological Bulletin **187**:289-290.

Johnston, C. A. 1991. Sediment and nutrient retention by freshwater wetlands: effects on surface water quality. CRC Critical Reviews in Environmental Control **21**:491-565.

Jorgensen, B. B. and J. Sorensen. 1985. Seasonal cycles of O_2, NO_3^- and SO_4^{2-} reduction in estuarine sediments: the significance of an NO_3^- reduction maximum in spring. Marine Ecology Progress Series **24**:65-74.

Joye, S. B. and H. W. Paerl. 1994. Nitrogen cycling in microbial mats: rates and patterns of denitrification and nitrogen fixation. Marine Biology **119**:285-295.

Kaplan, W. 1978. Denitrification in a Massachusetts salt marsh. Thesis. Boston University, Boston, Massachusetts, USA.

Kaplan, W., I. Valiela and J. M. Teal. 1979. Denitrification in a salt marsh ecosystem. Limnology and Oceanography **24**:726-734.

Koch, M. S., E. Maltby, G. A. Oliver and S. A. Bakker. 1992. Factors controlling denitrification rates of tidal mudflats and fringing salt marshes in Southwest England. Estuarine, Coastal and Shelf Science **34**:471-485.

Krest, J. M., W. S. Moore, L. R. Gardner, and J. T. Morris. 2000. Marsh nutrient export supplied by groundwater discharge: evidence from radium measurements. Global Biogeochemical Cycles **14**: 167-176.

Lapointe, B. E., D. A. Tomasko and W. R. Matzie. 1994. Eutrophication and trophic state classification of seagrass communities in the Florida Keys. Bulletin of Marine Science **54**:696-717.

Law, C. S., A. P. Rees and N. J. P. Owens. 1991. Temporal variability of denitrification in estuarine sediments. Estuarine, Coastal and Shelf Science **33**:37-56.

Lee, R. Y., S. B. Joye, B. J. Roberts and I. Valiela. 1997. Release of N_2 and N_2O from salt-marsh sediments subject to different land-derived nitrogen loads. Biological Bulletin **193**:292-293.

Lyons, J., J. Ahern, J. McClelland and I. Valiela. 1995. Macrophyte abundances in Waquoit Bay estuaries subject to different nutrient loads and the potential role of fringing salt marsh in groundwater nitrogen interception. Biological Bulletin **189**:255-256.

Macko, S. A., W. Y. Lee and P. L. Parker. 1982. Nitrogen and carbon isotope fractionation by two species of marine amphipods: laboratory and field studies. Journal of Experimental Marine Biology and Ecology **63**:145-49.

McClelland J. and I. Valiela. 1998. Linking nitrogen in estuarine producers to land-derived sources. Limnology and Oceanography **43**:577-585.

McClelland, J., I. Valiela and R. Michener. 1997. Nitrogen stable isotope signatures in estuarine food webs: a record of increasing urbanization in coastal watersheds. Limnology and Oceanography **42**:930-937.

Michener, R. H. and D. M. Schell. 1994. Stable isotope ratios as tracers in marine aquatic food webs. Pages 138-157 *in* K. Lajtha and R. H. Michener, editors. Stable isotopes in ecology and environmental science. Blackwell Scientific Publications, Oxford, England.

Mitsch, W. J. and J. G. Gosselink. 1993. Wetlands. Van Nostrand Reinhold, New York, New York, USA.

Moran, M. A. and R. E. Hodson. 1994. Dissolved humic substances of vascular plant origin in a coastal marine environment. Limnology and Oceanography **39**:762-771.

Neckles, H. A., R. L. Wetzel and R. J. Orth. 1993. Relative effects of nutrient enrichment and grazing on epiphyte-macrophyte (*Zostera marina*) dynamics. Oecologia **93**:285-295.

Nienhuis, P. H. 1983. Temporal and spatial patterns of eelgrass (*Zostera marina* L.) in a former estuary in the Netherlands, dominated by human activities. Marine Technology Society Journal **17**:69-77.

Nixon, S. W. 1980. Between coastal marshes and coastal waters — a review of twenty years of speculation and research on the role of salt marshes in estuarine productivity and water chemistry. Pages 437-526, *in* P. Hamilton and K. B. MacDonald, editors. Estuarine and wetland processes with emphasis on modeling. Plenum Press, New York, New York, USA.

Odum, W. E., J. S. Fisher and J. C. Pickral. 1979. Factors controlling the flux of particulate organic carbon from estuarine wetlands. Pages 69-82, *in* R. J. Livingston, editor. Ecological processes in coastal and marine systems. Plenum Press, London, England.

Orth, R. J. and J. Van Montfrans. 1984. Epiphyte-seagrass relationships with an emphasis on the role of micrograzing: a review. Aquatic Botany **18**:43-69.

Peterson, B. J. and B. Fry. 1987. Stable isotopes in ecosystem studies. Annual Review of Ecology and Systematics **18**:293-320.

Peterson, B. J., R. W. Howarth and R. H. Garritt. 1986. Sulfur and carbon isotopes as tracers of salt marsh organic matter flow. Ecology **67**:865-874.

Redfield, A. C. 1972. Development of a New England salt marsh. Ecological Monographs **42**:201-237.

Sand-Jensen, K. and J. Borum. 1991. Interactions among phytoplankton, periphyton and macrophytes in temperate freshwaters and estuaries. Aquatic Botany **41**:137-175.

Seitzinger, S. P. 1988. Denitrification in freshwater and coastal marine ecosystems: ecological and geochemical significance. Limnology and Oceanography **30**:1332-1339.

Seitzinger, S. P. 1994. Linkages between organic matter mineralization and denitrification in eight riparian wetlands. Biogeochemistry **25**:19-39.

Seitzinger, S. P., L. P. Nielsen, J. Caffrey and P. B. Christiansen. 1993. Denitrification measurements in aquatic sediments: a comparison of three methods. Biogeochemistry **23**:147-167.

Sham, C. H., J. W. Brawley and M. A. Moritz. 1995. Quantifying septic nitrogen loadings to receiving waters: Waquoit Bay, Massachusetts. International Journal of Geographic Information Systems **9**:463-473.

Short, F. T. and D. M. Burdick. 1996. Quantifying eelgrass habitat loss in relation to housing development and nitrogen loading in Waquoit Bay, Massachusetts. Estuaries **19**:730-739.

Short, F. T., D. M. Burdick and J. E. Kaldy III. 1995. Mesocosm experiments quantify the effects of eutrophication on eelgrass, *Zostera marina*. Limnology and Oceanography **40**:740-749.

Short, F. T., D. M. Burdick, J. S. Wolf and G. E. Jones. 1993. Eelgrass in estuarine research reserves along the East Coast, U.S.A., Part I. Declines from pollution and disease and Part II. Management of eelgrass meadows. National Oceanic and Atmospheric Administration. Coastal Ocean Program Publication, Rockville, Maryland.

Taylor, D. I. and B. R. Allanson. 1995. Organic carbon fluxes between a high marsh and estuary and the inapplicability of the Outwelling Hypothesis. Marine Ecology Progress Series **120**:263-270.

Thompson, S. P., H. W. Paerl and M. C. Go. 1995. Seasonal patterns of nitrification and denitrification in a natural and a restored salt marsh. Estuaries **18**:399-408.

Thybo-Christesen, M., M. B. Rasmussen and T. H. Blackburn. 1993. Nutrient fluxes and growth of *Cladophora sericea* in a shallow Danish bay. Marine Ecology Progress Series **100**:273-281.

Twilley, R. R., W. M. Kemp, K. W. Staver, J. C. Stevenson and W. R. Boynton. 1985. Nutrient enrichment of estuarine submersed vascular plant communities. 1. Algal growth and effects on production of plants and associated communities. Marine Ecology Progress Series **23**:179-191.

Valiela, I. 1983. Nitrogen in salt marsh ecosystems. Pages 649-678 *in* E. J., Carpenter and D. G. Capone, editors. Nitrogen in the marine environment. Academic Press, New York, New York, USA.

Valiela, I. 1995. Marine ecological processes. Second edition. Springer Verlag. New York, New York, USA.

Valiela, I., G. Collins, J. Kremer, K. Lajtha, M. Geist, B. Seely, J. Brawley and C. H. Sham 1997a. Nitrogen loading from coastal watersheds to receiving estuaries: new method and application. Ecological Applications **7**:358-380.

Valiela, I., K. Foreman, M. LaMontagne, D. Hersh, J. Costa, P. Peckol, B. DeMeo-Anderson, C. D'Avanzo, M. Babione, C. H. Sham, J. Brawley and K. Lajtha. 1992. Couplings of watersheds and coastal waters: sources and consequences of nutrient enrichment in Waquoit Bay, Massachusetts. Estuaries **15**:443-457.

Valiela, I, M. Geist, J. McClelland, and G. Tomasky. 2000. Nitrogenloading from watersheds to estuaries: Verification of the Waquoit Bay nitrogen loading model. Biogeochemistry, in press.

Valiela, I., J. McClelland, J. Hauxwell, P. J. Behr, D. Hersh and K. Foreman. 1997b. Macroalgal blooms in shallow estuaries: controls and ecophysiological and ecosystem consequences. Limnology and Oceanography **42**:1105-1118.

Valiela, I. and J. M. Teal. 1979. The nitrogen budget of a salt marsh ecosystem. Nature **280**:652-656.

Vince, S. W., I. Valiela and J. M. Teal. 1981. An experimental study of the structure of herbivorous insect communities in a salt marsh. Ecology **62**:1662-1678.

White, D. S. 1994. ^{15}N studies of nitrogen retention in the vegetated sediments of a New England salt marsh. Thesis. Boston University, Boston Massachusetts, USA.

White, D. S. and B. L. Howes. 1994. Long term ^{15}N-nitrogen retention in the vegetated sediments of a New England salt marsh. Limnology and Oceanography **39**:1878-1892.

Wilson, J. O., I. Valiela and T. Swain. 1985. Sources and concentrations of vascular plant material in sediments of Buzzards Bay, Massachusetts. Marine Biology **90**:129-137.

Zimmerman, R. C., R. D. Smith and R. S. Alberte. 1987. Is growth of eelgrass nitrogen limited? A numerical simulation of the effects of light and nitrogen on the growth dynamics of *Zostera marina*. Marine Ecology Progress Series **41**:167-176.

SPATIAL VARIATION IN PROCESS AND PATTERN IN SALT MARSH PLANT COMMUNITIES IN EASTERN NORTH AMERICA

MARK D. BERTNESS
Department of Ecology and Evolutionary Biology
Brown University
Providence, Rhode Island 02912 USA

STEVEN C. PENNINGS
University of Georgia Marine Institute
Sapelo Island, Georgia 31327 USA

Abstract

While we have learned a great deal about the structure and organization of salt marsh plant communities in the past two decades, this understanding is based on experimental studies conducted at just a handful of study sites. How general are these results and how far can we extrapolate from them to understand other marsh systems? In this paper, we argue that the zonation of eastern North American marsh plant communities may be strongly influenced by both eutrophication and climate, and that spatial variation in these factors may limit our ability to uncritically generalize between marshes. The striking zonation of marsh plant communities has been explained to be the product of competitively superior plants dominating physically mild habitats and displacing competitively subordinate plants to physically harsh habitats. At higher latitudes, this typically results in competitively dominant plants monopolizing high marsh elevations while competitively subordinate plants are limited to lower elevations. Recent studies, however, have suggested that both nutrient supply and thermal stress can influence this simple scenario. Increased nutrient availability, a typical consequence of eutrophication, may alleviate below ground competition for nutrients and lead to above ground competition for light dictating competitive dominance among marsh plants. In marsh systems that historically have been nutrient limited, this may influence plant competitive dominance hierarchies and lead to major shifts in plant zonation patterns. Similarly, climate may have important, but largely unrecognized effects on marsh plant community organization. In cool temperate marshes, low soil salinities result in salinity playing only a minor role in maintaining marsh plant distributional patterns. In contrast, at lower latitudes, hotter climates lead to salt accumulation, elevated soil salinities, and marsh zonation patterns that are strongly driven by soil salinity patterns. We suggest that our current understanding of marsh zonation patterns is oversimplified, and that the processes creating these patterns may vary in importance between marshes. Systematic experimental studies of this spatial variation will be necessary to provide a general understanding of the forces influencing marsh plant community structure.

1. Introduction

Our understanding of ecological processes has greatly advanced over the past three decades with the common use of manipulative experimental field studies. Results of experimental studies, however, are often haunted by questions of generality (Underwood and Denley 1984, Diamond 1986, Keddy 1989). When can the results of experiments done at a single study site, frequently at small spatial scales, be extrapolated across study sites and spatial scales? While this is an important question that is central to the success of ecology as a predictive science (Goldberg 1990, Reader et al. 1994), it is frequently ignored. The problem with assessing the robustness of field studies is complicated by evidence that the results of field experiments, more often than not, change as a function of abiotic conditions (Moloney 1990, Dunson and Travis 1991, Goldberg and Barton 1992, Bertness and Shumway 1993, Bertness and Hacker 1994). The dangers of over-generalizing limited experimental results into ecological paradigms are well-known (Underwood and Denley 1984). But, because field experiments are often labor-intensive and limited in number, the temptation to overstate and over-extrapolate our understanding of community processes remains, and our appreciation of variation in community processes is largely speculative.

In the few cases in which similar experimental field studies have been well replicated across time and space, literature surveys have generated considerable insight into the issue of generality (Connell 1983, Schoener 1983, Goldberg and Barton 1992). One of the main lessons of examining issues of generality through literature reviews, however, is that when similar questions are asked by different researchers in different locations and at different times, it is often difficult to compare results due to differences in experimental design. Researcher bias in the questions asked and the systems chosen to ask them with, as well as the bias in what is ultimately published, can also strongly taint our view of the processes that are important in natural communities. Even when experiments are coordinated and performed identically to explicitly address the issue of generality, as recently advocated by Reader et al. (1994) and Goldberg (1990), they may lack a priori predictions about how results should vary geographically, and thus may do little more than support an already established idea or generate post-hoc hypotheses.

Our current understanding of the organization of salt marsh plant communities is vulnerable to these criticisms. Although salt marshes have been intensively studied in terms of understanding nutrient flows (e.g., Teal 1962, Gallagher 1975, Valiela and Teal 1979, Pomeroy and Wiegert 1981), most of our understanding of the community ecology of marsh plants is based on studies from only a few, heavily-studied, sites (e.g., Rhode Island: Bertness and Ellison 1987, Bertness 1988, 1991a,b, 1992, Maryland: Furbish and Albano 1994, North Carolina: Silander and Antonovics 1982, Alaska: Snow and Vince 1984, Southern California: Pennings and Callaway 1992, 1996, Callaway 1994). Consequently, we are constrained in knowing how general the mechanisms that generate pattern in marsh plant communities are.

Elucidating the organization of salt marsh plant communities is not simply an intellectual exercise, but is directly relevant to practical issues facing ecosystem managers. Without knowledge of the physical and biotic processes that interact to generate the distribution and abundance of plants across marsh landscapes, applied ecologists will be unable to predict how marshes may change in response to anthropogenic influences such as eutrophication and climate change. Moreover, without a fairly complete understanding of marsh plant community dynamics, marsh restoration efforts are reduced to simple trial and error (Zedler 1995). Thus, understanding how general our conceptual models of marsh plant community organization are should be a high priority.

In this paper, we argue that our current understanding of marsh plant community organization is likely oversimplistic. We begin by summarizing the results of studies that have examined causes of the striking tidal height zonation of plants across marsh landscapes. We then argue that it would be naive to over-generalize from these results. In particular we present preliminary data and arguments that both nutrient supply and climate may strongly influence the zonation and organization of marsh plant communities. We close by suggesting that further experimental studies examining the robustness of models of marsh plant community organization are needed before we can confidently extrapolate from our current models.

2. Zonation in Marsh Plant Communities

The zonation of plants across tidal height in salt marshes is one of the most striking features of these habitats and has long attracted the attention of researchers (Johnson and York 1915, Chapman 1940, Ranwell 1971). Typically, plants in salt marsh habitats are restricted to particular tidal heights leading to pronounced bands of specific species of plants paralleling marsh shorelines at specific elevations. In southern New England, for example, the cordgrass *Spartina alterniflora* usually dominates low marsh habitats that are flooded daily by tides, the marsh hay, *Spartina patens* dominates intermediate elevations, the black rush, *Juncus gerardi* dominates high marsh elevations, while the terrestrial fringe of the marsh is dominated by the woody shrub, *Iva frutescens* (Fig. 1). Similarly pronounced tidal height zonation schemes are characteristic of most salt marshes throughout the world (Chapman 1974).

Intertidal salt marsh habitats occur across strong gradients in physical stress which have long been thought to be responsible for the pronounced zonation of marsh plants. Salt marshes are physically stressful habitats for vascular plants, and the plants that live in marshes are highly adapted to cope with these stresses. Salt marsh plants experience physical stresses of flooding and salinity (Chapman 1974, Adam 1990), both of which vary markedly across intertidal gradients. Regularly flooded soils at lower elevations are waterlogged and contain less oxygen than infrequently flooded soils at higher elevation (Howes et al. 1981, 1986). Thus, only plants capable of living in anoxic soils occur at low marsh elevations. Salt stress is an equally severe problem for vascular plants since high soil salt concentration osmotically draws water from plants and only the halophytic plants that can manage the osmotic problems of high soil salinities can live in salt marsh habitats.

Whereas ecologists have long speculated about the roles of physical stress and plant competition in generating the elevational zonation of marsh plant systems (Miller and Egler 1950, Adams 1963, MacDonald and Barbour 1974), it has been only in the last two decades that they have experimentally examined the proximate causes of marsh plant zonation (e.g. Silander and Antonovics 1982, Snow and Vince 1984, Bertness and Ellison 1987, Bertness 1991a, b, Bertness et al. 1992, Pennings and Callaway 1992).

These experimental studies, primarily using transplant techniques, have found that there is a trade-off in marsh plants between competitive ability and the ability to deal with physical stress. This trade-off typically leads to competitively subordinate marsh plants dominating physically stressful habitats, while their competitive dominants monopolize physically benign habitats. In northeastern North American marshes, which are the best studied with respect to this issue, this translates into competitively subordinate plants living in regularly-flooded low marsh habitats, while competitively dominant plants monopolize less frequently-flooded, high marsh habitats.

The competitively subordinate plants that dominate physically stressful habitats have repeatedly been shown to be capable of invading and thriving in physically benign habitats (if competitors are removed). Thus, competitive subordinates live in harsh habitats because they are displaced from benign habitats by dominant spatial competitors. In contrast, competitive dominants are typically unable to live in physically harsh marsh habitats with or without neighbors, so are constrained by physical stresses. The strong elevational zonation of marsh plant communities is thus thought to result from a combination of the strong gradient in physical factors across marsh landscapes and of strong competitive displacement operating across this gradient.

While these simple community organization or assembly rules explain most marsh plant zonation patterns, anything that changes the physical stress gradient across marsh habitats or that influences the competitive relationship among plants may modify and influence the zonation of marsh plant communities. Below, we focus on eutrophication and climate, arguing that both are likely to affect process and pattern in salt marshes by mediating competitive ability and/or physical stress.

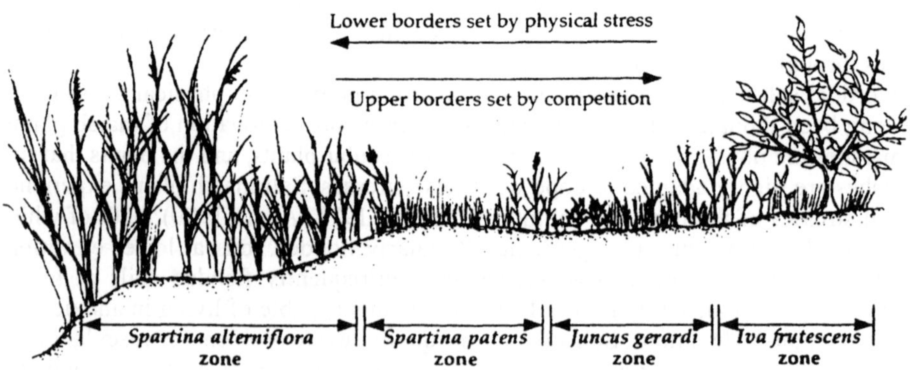

Figure 1. Zonation of vegetation in southern New England salt marshes (after Bertness 1991a, b and Bertness et al. 1992)

3. The Role of Nitrogen Supply in Salt Marsh Plant Zonation

One of the most widely accepted paradigms of salt marsh ecology is that nitrogen supply plays a critical role in marsh ecosystems. As has been found for most marine plants (Valiela 1984), the primary production of marsh plants is typically thought to be nitrogen limited. Experimental addition of nitrogen generally increases marsh plant production (Valiela and Teal 1974, Sullivan and Daiber 1974, Gallagher 1975, Mendelssohn 1979a, b). Moreover, natural variation in the production of marsh plants has been shown to be correlated with natural variation in nitrogen supply (Nixon and Oviatt 1978).

Since salt marsh plant communities are thought to be nitrogen limited, they may be particularly sensitive (responsive) to nitrogen enrichment effects resulting from human activity. Global supplies of biologically useful nitrogen have increased dramatically over the past century due to anthropogenic causes. The burning of fossil fuels and the discovery and rampant use of artificial nitrogen fertilizers has more than doubled the annual supply rate of useable nitrogen globally over natural ambient levels (Vitousek et al. 1997). This has resulted in nutrient saturation with dramatic consequences on many plant communities (Hiel and Diemont 1983, Tilman 1987, Berendse and Elberse 1990, Bobbink 1991, Berendse et al. 1993). In estuarine systems, human impacts on the global nitrogen cycle have been manifested as dramatic and increasing eutrophication of coastal waters (Neilson and Cronin 1981, Howarth 1988, Peierls et al. 1991, Turner and Rabalais 1991, Holligan and Reiners 1992).

The potential role played by increased nutrient supplies on marsh plant production (Valiela et al. 1985) and food webs (Vince et al. 1981) has been considered. The potential role that increased nutrient supplies could play in affecting the distribution and abundance of plants across marsh landscapes, however, has not been explicitly addressed until recently.

To initially address the question ''Does nitrogen supply influence the competitive relations of marsh plants?'' Levine and colleagues (1998) fertilized quarter meter square experimental plots containing natural mixtures of perennial marsh turfs in a typical southern New England salt marsh. The plots were haphazardly located on zonal boundaries and compared with nearby controls that were not fertilized. After two field seasons, the results of this simple experiment were dramatic. Without exception, fertilization led to the increased success of what previous research had determined to be the competitive subordinate and decreased success of the competitive dominant. In other words, fertilization entirely reversed the competitive relations of Southern New England marsh plants (Fig. 2). While the marsh hay *Spartina patens* is known to competitively displace the cordgrass *Spartina alterniflora* to low marsh habitats, in fertilized plots cordgrass increased in abundance while marsh hay decreased. Similarly, fertilization led to *Spartina patens* dominating *Juncus gerardi*, and *Distichlis spicata* dominating both *Spartina patens* and *Juncus gerardi*.

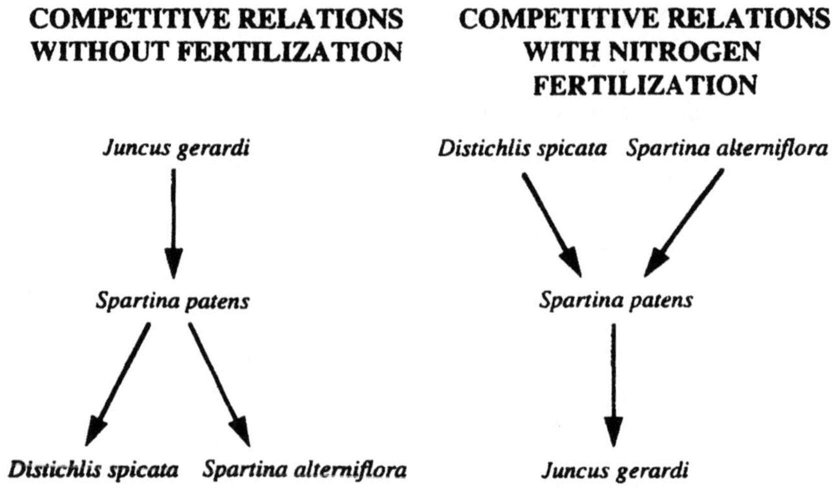

Figure 2. Competitive hierarchies of New England plants with and without nitrogen fertilization (after Levine et al. 1998)

What could cause such striking results? While it is possible that our results reflect metabolic differences in the nitrogen utilization efficiency of New England marsh plants leading to a shift in plant dominance (see Arp and Drake 1991, Arp et al. 1993), we suspect that fertilization dramatically shifted the competitive balance among these plants because they were initially nutrient-limited and competing for below ground resources and adding nutrients eliminated competition for nutrients. The competitive dominants in New England marsh plant communities (*Spartina patens* and *Juncus gerardi*) are both clonal turfs that invest heavily in below ground, nutrient-harvesting biomass at the expense of investing less in above ground, light-harvesting biomass. In contrast, the competitive subordinates (*Spartina alterniflora* and *Distichlis spicata* and the solitary forbs) allocate much less to below ground biomass and relatively more to above ground biomass (Brewer et al. 1998). We suggest that adding nutrients alleviated below ground competition for nutrients and led to increased competition for light. Thus, we hypothesize that adding nutrients switched the dominant arena of competition among marsh plants from below ground to above ground, leading to a shift in dominance from superior below ground to superior above ground competitors. Since above- and below ground competitive ability may be simply reflected in a straight forward trade-off between allocation to nutrient-gathering roots or light-gathering leaves, nutrient supply could tip the balance of competitive dominance among these plants.

We are currently testing this hypothesis with a field experiment where we are evaluating the effect of nutrients on above and below ground components of competition between pairs of marsh plant species. In our design (Fig. 3), size-standardized cores of each species are transplanted into naturally occurring monocultures of a second test species. One third of the transplants are assigned as full (above and below ground) competition treatments. A second third of the transplants are assigned as no above ground competition treatments, where the above ground biomass of neighbors is pinned back to minimize above ground competition while they continue to compete below ground. The last third of the replicates are assigned as no competition treatments and neighboring vegetation in a 10 cm radius of the target core is routinely clipped with regular maintenance to ground level. To examine the effect of nutrients on these components of competition, half of the full competition, no above ground competition and no competition replicates were randomly assigned as fertilized replicate, while the remaining replicates were left as unfertilized controls. This design is similar to that used by Wilson and Tilman (1993) to tease apart above- and below ground components of plant competition (also see Twolan-Strutt and Keddy 1996). Initial results of this experiment support the hypothesis that fertilization shifts competition from below- to above ground, but are too preliminary to present.

Figure 3. Design of transplant experiment to examine above and below ground components of competition among marsh plants. Circle represents core transplanted into a matrix of second species with the illustrated treatments replicated with and without fertilization.

Our findings that nitrogen supply can dictate the competitive relations of numerically dominant marsh plants have important potential consequences on the abundance and distribution of plants across marsh landscapes (Levine et al. 1998). As already discussed, while the lower tidal boundaries of marsh plant distributions are generally set by physical stress and therefore would not be predicted to be influenced by nutrient levels, the upper boundaries of marsh plants are typically set by competitive exclusion. Thus, if nitrogen supply reverses competitive dominance rankings among marsh plants, nitrogen supply may impact the elevation of zonal borders in marshes. In southern New England where the study described above has been done, high nitrogen supply is predicted to allow the cordgrass *Spartina alterniflora* to move to higher tidal heights displacing the marsh hay, *Spartina patens*. *Distichlis spicata* is predicted to increase in abundance across the high marsh, and marsh hay is predicted to move to higher elevations, displacing *Juncus gerardi*. This simple graphical model (Fig. 4) predicts that high nitrogen supplies, like those occurring with eutrophication, may lead to major shifts in marsh plant zonation with historically lower marsh species displacing higher marsh plants.

The generality of these results and whether these small-scale nutrient addition results can be extrapolated to larger landscape spatial scales remains to be tested. Preliminary results in Georgia (S. Pennings unpublished) and Mississippi (J. S. Brewer unpublished) marshes indicate that species borders are sensitive to nitrogen supply in southern marshes in the United States.

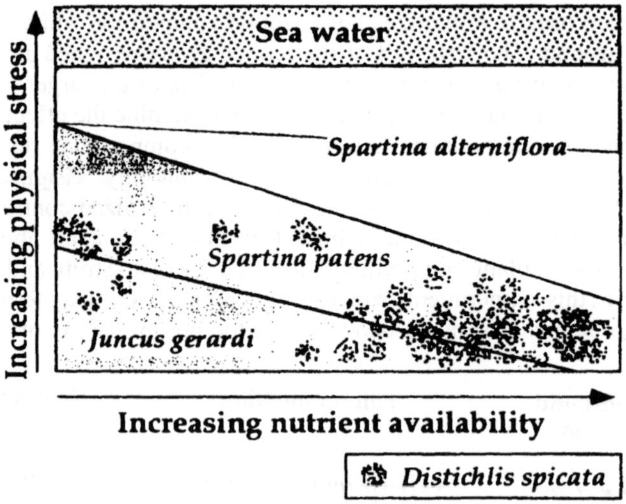

Figure 4. Conceptual model of the effects of nutrients on marsh plant zonation pattern (after Levine et al. 1998)

Moreover, on the east coast of North America the recent range expansion of the common reed, *Phragmites australis* may also be due to nutrient enrichment. *Phragmites* is a strong above ground competitor and the unprecedented invasion of *Phragmites* into marshes over the last few decades may be a reflection of increased nitrogen supply (Minchinton and Bertness unpublished data). This is particularly interesting since eutrophication has often been suggested along with other anthropogenic factors as a cause for the decline of *Phragmites australis* in Europe (Ostendorp 1989). Our results also suggest that the zonation patterns described by ecologists early in this century and still observed in many habitats may only occur under low nutrient conditions. We hypothesize that, in marsh systems like those found in New England that are nitrogen-limited and where plants compete intensely for nitrogen, eutrophication likely has important consequences for plant community organization and zonation. Whether these results can be extrapolated to other marsh systems will depend on the generality of the hypothesized links between zonation, plant height and competitive ability, and on the extent of historical nutrient limitation in these systems.

4. Climate And Marsh Plant Zonation

Like the potential role played by nitrogen supply, climate is a largely unexplored aspect of marsh plant community organization. Climate might influence marsh plant community structure in many ways. Temperature affects decomposition rates, nutrient cycling and ultimately peat accumulation, which likely has strong impacts on zonation patterns (Bertness 1988). Climate affects photosynthesis and transpiration rates (DeLucia et al.

1994, Friend et al. 1989, Holt 1990, Lajtha and Getz 1993, Salisbury and Ross 1992), and sets constraints on plant phenology, both of which could affect biomass production (Turner 1976) and mediate competitive interactions between plants. Biogeographic patterns in herbivore pressure and plant defenses (MacArthur 1972, Vermeij 1978, Coley and Aide 1991, Pennings unpublished) could lead to an increased role of consumers in community organization at lower latitudes. Although all these effects of climate are likely, in this paper we will limit our focus to the potential role of climate influencing marsh plant zonation patterns by affecting physical gradients. In particular, by controlling evaporative processes and the potential accumulation of salt in marsh soils, climate may determine the importance of soil salinity in influencing the distribution and abundance of plants across marsh habitats.

A potentially productive approach to exploring the linkage between climate and the organization of marsh plant communities is to examine latitudinal variation in marshes. Along the east coast of North America, for example, salt marsh plant communities are a common shoreline habitat from the Canadian Maritime provinces to central Florida, where marshes give way to mangrove forests, the tropical analog to salt marshes. Whereas marshes north of central Maine differ from more southerly marshes due to the heavy, chronic impact of winter ice, marshes from southern Maine to Florida are composed of a similar suite of plants and are a good model system to examine the effects of climate on marsh plant community organization.

Latitudinal variation in temperature along the east coast of North America is substantial (Fig. 5). For example, near two sites we work at in New England, mean monthly temperatures are over 10°C only from June through September, while mean maximum air temperatures of near 30°C only occur in July and August. In contrast, near two southern sites we work at in Georgia and Alabama, close to the southern latitudinal limit of salt marsh vegetation, mean monthly air temperatures are always above 10°C and mean maximum daily air temperatures of 30°C or higher occur from June through September. Heavier summer precipitation at southern than at northern sites may moderate salt build-up at southern sites, but since temperatures remain relatively high year-round at southern sites, differences in soil salinities between northern and southern marshes are substantial (Pennings and Bertness 1999).

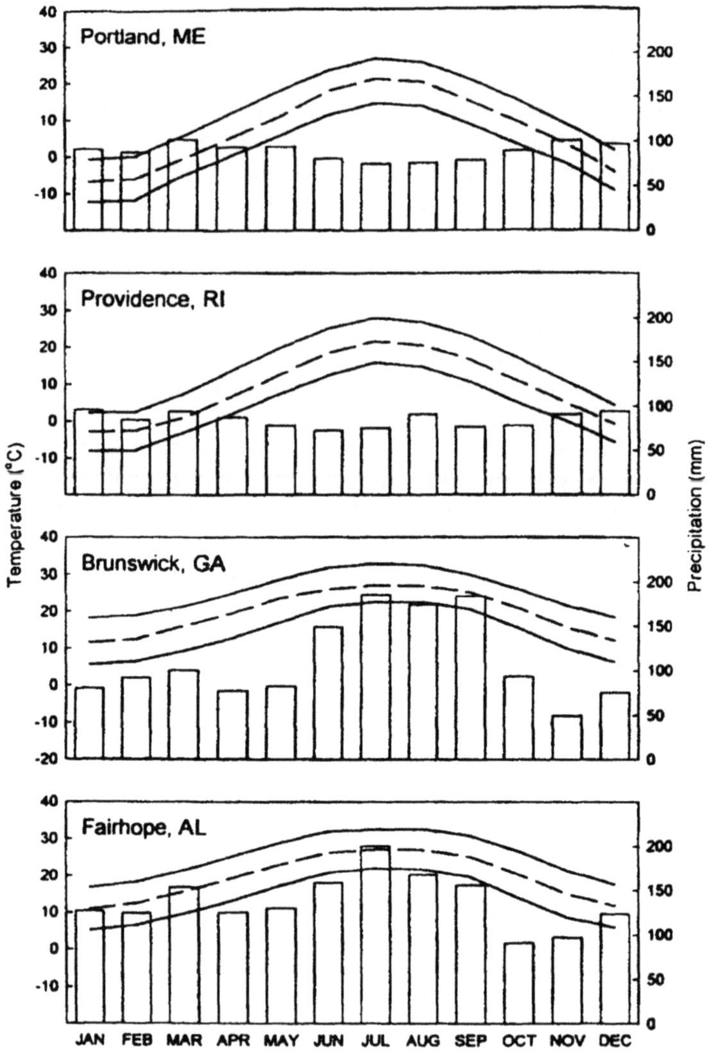

Figure 5. Mean monthly temperatures (minimum, maximum and average) and rainfall (bars) for northern and southern sites on the Atlantic coast near our study sites. Data are 90 to 120 year averages from the National Weather Service

The salinity of marsh soils is determined by a balance between the salinity of ambient tidal water, the frequency of tidal flooding and amount of precipitation which limit salt accumulation, and the intensity of solar radiation which dictates evaporation and the potential extent of salt accumulation (Fig. 6). At low marsh elevations, frequent tidal flooding prevents the evaporative build-up of salt by flushing soils regularly. In contrast, at high marsh elevations infrequent exposure to salt water, and large freshwater inputs from rain and runoff limit soil salinity build-up. At intermediate marsh elevations, however, solar radiation and soil heating can lead to the evaporation of pore water and elevated soil salinities. A variety of factors likely influence this balance. Porous marsh soils, where water rapidly moves through soils, flushing away salts, minimize salt buildup. Conversely, nonporous soils that limit percolation typically maximize soil salt accumulation. High tidal amplitudes also likely minimize the occurrence of hypersaline conditions by increasing the flushing of marsh soils. Climate, however, can strongly influence this balance by affecting the strength and duration of solar radiation, and thus the potential for evaporative water loss and soil salt accumulation.

Northern and southern marshes on the east coast of North America appear to have very different salinity profiles that are largely driven by latitudinal variation in climate (Pennings and Bertness 1999). In New England, salinity decreases from the water's edge to the terrestrial border in many marshes, but disturbance-generated bare patches in the high marsh can become hypersaline due to increased evaporation in the absence of vegetation (Bertness 1991a, Bertness et al. 1992). In contrast, marshes in the southern United States experience higher evapotranspiration rates, leading to hypersaline soil being typical at mid-marsh elevations even in undisturbed stands of vegetation (Stout 1984, Wiegert and Freeman 1990).

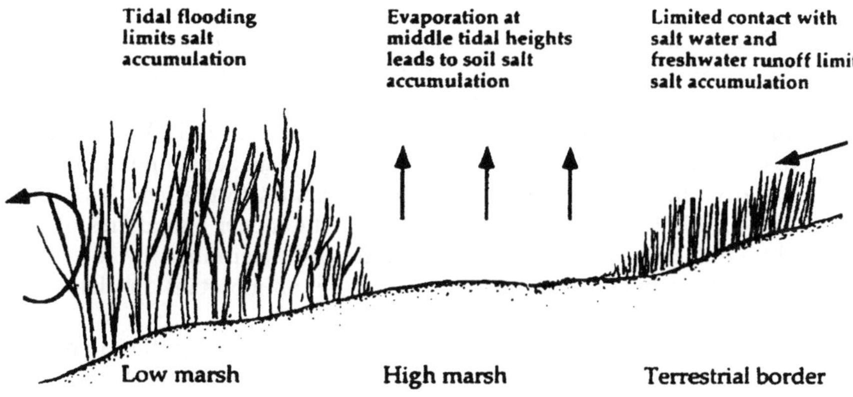

Figure 6. Factors influencing evaporative salt accumulation in tidal salt marshes

We suggest that climate plays a major role in the structure and organization of marsh plant communities by dictating the role of soil salinity in affecting the distribution and abundance of plants across marsh habitats. Below, we describe four hypotheses about the link between climate and process and pattern in marsh plant communities that we are currently testing (Fig. 7).

Hypothesis I. The mechanisms determining the zonation of salt marsh plants change as a function of climate. In northern, colder sites, lower limits are always set by the tolerance of plants to flooding. In southern, hotter sites, particular at middle elevations, zonal limits are set by the tolerance of plants to elevated salinities.

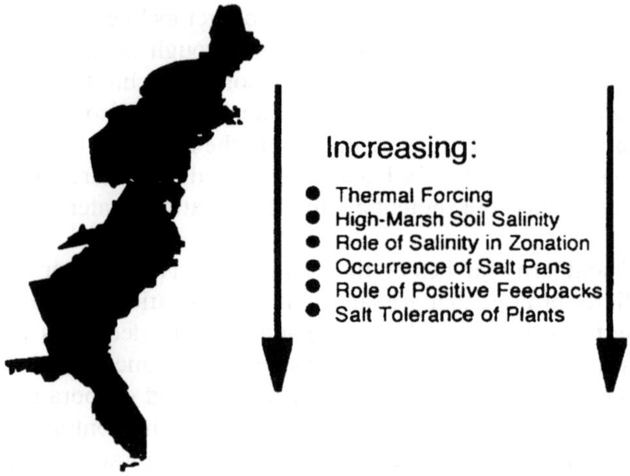

Figure 7. Hypotheses developed in this paper linking climate and process and pattern in salt marsh plant communities

The roles of physical and biological factors in determining the striking elevational zonation of salt marsh plant communities have been widely debated (Adams 1963, Cooper 1982, Snow and Vince 1984, Bertness 1991a, b, Pennings and Callaway 1992). The failure of this debate to be resolved may at least in part be due to a lack of recognition that the relative importance of the factors governing marsh plant zonation may vary with climate. Manipulative field experiments in northern marshes (Snow and Vince 1984, Bertness and Ellison 1987) strongly indicate that while the upper boundaries of marsh plants are set by competition, their lower boundaries are determined by tolerance to flooding. In northern marshes salinity appears to be unimportant in determining zonation patterns.

Similar work at lower latitudes, however, hints that soil salinities may be an important determinant of the zonation patterns of southern marshes (Pennings and Callaway 1992). We suggest that the hot climate experienced by southern marshes leads to hypersaline conditions in middle marsh zones and that these hypersaline conditions are important determinants of zonal boundaries in southern marshes. We are testing this hypothesis by watering zonal boundaries in southern and northern marshes with the prediction that alleviating salt stress will lead to shifting zonal borders in southern, but not northern marshes.

Hypothesis II. The presence of salt pans in southern, but not northern marshes is driven by latitudinal variation in climate. Increasing temperatures and solar radiation at lower latitudes leads to hypersaline soil conditions so severe that no plants can tolerate them leading to the formation of permanent bare areas.

One of the most characteristic features of southern salt marshes is bare, unvegetated areas at mid-marsh elevations with little or no plant cover. These salt pan areas are permanent features of these systems and typically have very high soil salinities in excess of 150 ‰. In northern salt marshes, permanent mid-marsh salt pans are rare and disturbance-generated bare patches in these marshes are transitional, successional features that normally close by the clonal invasion of surrounding plants within a few years (Bertness and Ellison 1987).

We hypothesize that permanent salt pans occur in southern marshes because the hot climate at lower latitudes leads to elevated mid-marsh salinities that no plants can tolerate. Thus, we suggest that there is a salinity threshold above which plants can not persist and that above this threshold, bare patches occur. Furthermore, we hypothesize that once this threshold is crossed, the system moves into a new stable state because the lack of shading by vegetation promotes extremely hypersaline soils that can kill surrounding vegetation and prevent invasions (Fig. 8). We recently began testing this climate-driven salt pan hypothesis by watering southern bare patches. We predict that watering these salt pans will lead to plant colonization and pan closure.

Hypothesis III. Strong positive feedbacks between neighboring plants due to the amelioration of hypersaline soil conditions by plant shading occur predictably within marshes as a function of physical stress and increase in strength and importance with decreasing latitude.

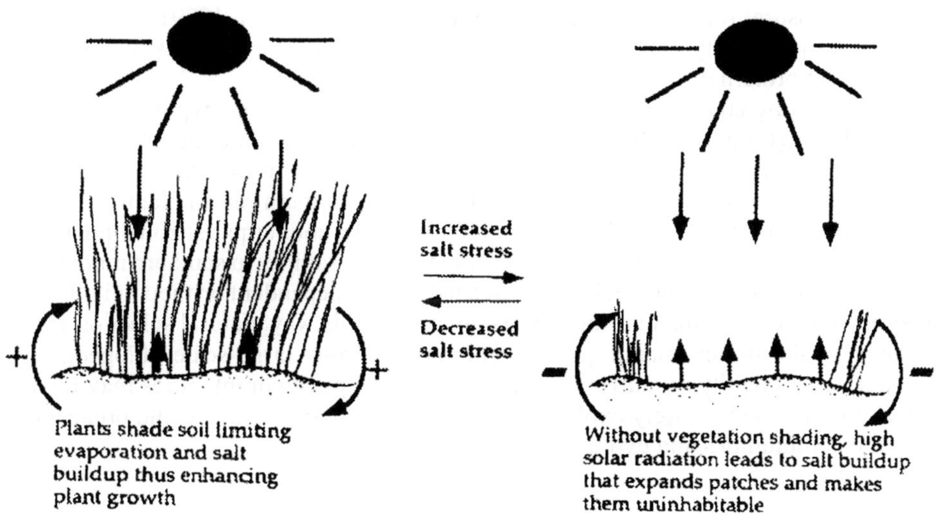

Figure 8. Feedbacks involved in the maintenance of vegetation or salt pans at intermediate marsh elevations. We suggest that thermal thresholds dictate which of these alternate stable states occur.

Elevated soil salinities are often an important physical stress in intertidal salt marshes. The severity of evaporation and the development of hypersaline soil conditions, however, is powerfully affected by plant cover since plants shade the substrate and limit evaporation and enhance the development of hypersaline soil conditions (Zedler 1982, Bertness et al. 1992). As a result of this feedback between plant cover and reduced soil salinity, positive interactions among plant neighbors have been shown to be a predictable feature of potentially hypersaline habitats, but not habitats where the potential for hypersaline soil conditions is low. In southern New England where this phenomenon has been investigated, secondary succession (Bertness and Shumway 1993), seedling establishment (Bertness and Yeh 1994), zonal boundaries (Bertness and Hacker 1994) and plant species diversity (Hacker and Bertness 1998) have all been shown to be affected by positive neighbor effects in potentially harsh habitats.

We suggest that these salt-stress-driven positive interactions between plants are even more common and stronger in southern marshes. In northern marshes, salt accumulation and neighbor buffering would be predicted to be minimal due to cooler climates. In southern marshes with hotter climates, in contrast, we predict that positive neighbor effects due to salt stress habitat amelioration become more important and are likely mandatory for much of the middle tidal height vegetation to persist. Thus, we hypothesize a latitudinal gradient in the role of positive interactions in marsh plant communities driven by climate.

Hypothesis IV. Increased salinity in southern marshes has led to strong selection for highly salt tolerant plants. As a consequence of this, both within and among species, southern plants are more salt tolerant than northern plants.

We have argued elsewhere that low latitude marshes have a suite of extremely salt tolerant plants not found in high latitude marshes that are the product of intense selection pressure for salt tolerance (Pennings and Bertness 1999). We are also currently testing the hypothesis that, within a species, southern plants are more salt tolerant than northern ones.

An interesting consequence of this evolutionary salt tolerance hypothesis is that, if correct, the first three ecological hypotheses just discussed predicting community organization shifts based on climate could be weakened or nullified by the response of community members to selection for salt tolerance. If southern marsh plants, in general and uniformly, are more salt tolerant than their northern counterparts, zonal boundaries would not be predicted to be more strongly influenced by salinity in southern marshes and southern marsh plants would not be predicted to be more dependent on their neighbors for salt stress amelioration than northern marsh plants. In this way, natural selection for salt tolerance could buffer marsh plant communities from large-scale, biogeographic variation in community organization. Thus, salt marsh plant communities may be ideal systems to examine the rarely explored interface between the structure and organization of natural communities and evolution.

5. Towards a Predictive Understanding of Process and Pattern in Salt Marsh Plant Communities

While we currently have a basic understanding of the forces that lead to the pronounced tidal height zonation in salt marsh plant communities, much remains to be learned. Without a firm grasp of the mechanisms responsible for generating pattern in these communities we will not be able to predict how they will respond to future environmental changes. We further suggest that we will only be able to appreciate the forces affecting marsh community organization by taking a manipulative field experiment approach to elucidate the mechanisms that generate pattern in these communities.

Identifying the mechanisms responsible for pattern generation in marsh plant communities is essential if we are to ever attain a predictive level of understanding of these systems. The necessity for understanding mechanism in community and ecosystem level studies is simple. If the mechanisms responsible for community pattern generation are understood, system responses to novel external disturbances can be reasonably predicted on the basis of whether or not such disturbances influence pattern-generating variables. Alternatively, if patterns are described, but not understood mechanistically, responses to novel perturbations can not be anticipated. Both potential nutrient and climate effects on marsh plant community organization are good examples of the power of understanding mechanisms in elucidating community organization issues.

Without a mechanistic understanding of competition among marsh plants and an appreciation for the above- and below ground components of plant competition, it would be impossible to predict the drastic response of New England marsh plant communities to nutrient additions. Similarly, without a knowledge of the specific physical constraints on marsh plants, predicting potential shifts in marsh organization due to variation in climate would not be realistic.

Attaining a mechanistic understanding of process and pattern in marsh plant communities will be impossible without taking a hypothesis testing, manipulative experimental approach to addressing these questions in the field. While correlative sampling programs are important descriptors of pattern in natural communities, and laboratory or greenhouse studies can help to focus accurate hypotheses (and we have used both extensively in our work) neither can replace the role of properly controlled field experiments in understanding mechanisms of process and pattern in natural communities. Even the best correlative data is still correlative data, and the most elegant laboratory experiments only test the variables isolated in the experiment. If important variables are overlooked or underestimated, the results of laboratory experiments can be entirely misleading.

A good example of the value of field experimentation in elucidating the mechanisms responsible for generating pattern in natural communities is found in the history of rocky shore ecology. Forty to fifty years ago zonation schemes of rocky shores around the world had been well described (Stephenson and Stephenson 1949, 1971) and correlated with physical stress gradients in these habitats and the stress tolerances of the resident organisms (Barnes and Barnes 1957, Doty 1946, Newell 1979, Lewis 1964). At this point, most workers thought that physical stress explained the majority of the pattern in intertidal communities; however, once hypothesis-driven field experimentation was undertaken in these habitats (Connell 1961, 1972, Paine 1966, Dayton 1971, Menge 1976), the important role of biological interactions, and particularly of the interplay between physical and biological factors became clear. Our understanding of the organization of salt marsh plant communities is in need of a similar infusion of experimental fieldwork. In particular, since we have gained a deep knowledge of a few sites, we now need studies that examine the issues of generality and variability between sites.

6. Acknowledgments

Our research on the biogeography of marsh plant communities is funded by the U. S. Department of Energy's National Institute for Global Environmental Change and the Andrew W. Mellon Foundation. Financial support does not constitute an endorsement by DOE of the views expressed in this article. We thank Kelly Benoit for the illustrations and Michael Weinstein, Tatyana Rand, Pat Ewanchuk and two anonymous reviewers for comments on the manuscript.

7. Literature Cited

Adam, P. 1990. Saltmarsh Ecology. Cambridge University Press, Cambridge, England.

Adams, D. A. 1963. Factors influencing vascular plant zonation in North Carolina salt marshes. Ecology **44**:445-456.

Arp, W. J. and B. G. Drake. 1991. Increased photosynthetic capacity of *Scirpus olneyi* after four years of exposure to elevated CO_2. Plant, Cell Environment **14**:1003-1006.

Arp, W. J., B. G. Drake, W. T. Pockman, P. S. Curtis and D. F. Whigam. 1993. Effects of four years exposure to elevated atmospheric CO_2 on competition between C_3 and C_4 salt marsh plants. Vegetatio **104**: 133-143.

Barnes, H. and M. Barnes. 1957. Resistance to desiccation in intertidal barnacles. Science **126**: 358.

Bertness, M.D. 1988. Peat accumulation and the success of marsh plants. Ecology **69**:703-713.

—— 1991a. Interspecific interactions among high marsh perennials in a New England salt marsh. Ecology **72**: 125-137.

—— 1991b. Zonation of *Spartina patens* and *Spartina alterniflora* in a New England salt marsh. Ecology **72**: 138-148.

—— 1992. The ecology of New England salt marsh plant communities. American Scientist **80**:260-268.

Bertness, M. D. and A. M. Ellison. 1987. Determinants of pattern in a New England marsh plant community. Ecological Monographs **57**: 129-147.

Bertness, M.D., T. Chatkupt and K. Wikler. 1992. Flood tolerances and the distribution of *Iva frutescens* across New England salt marshes. Oecologia **91**:171-178.

Bertness, M.D., L. Gaugh and S. W. Shumway. 1992. Salt tolerances and the distribution of plants across a New England salt marsh. Ecology **72**:1842-1851.

Bertness, M.D. and S. W. Shumway. 1993. Competition and facilitation in marsh plants. American Naturalist **142**:718-724.

Bertness, M. D. and S. D. Hacker. 1994. Physical stress and positive associations among plants. American Naturalist **144**:363-372.

Bertness, M. D. and S. M. Yeh. 1994. Cooperative and competitive interactions in the recruitment of marsh elders. Ecology **75**:2416-2429.

Bertness, M. D. and R. Callaway. 1994. Positive interactions in communities. Trends in Ecology and Evolution **9**:191-193.

Berendse, F. and W. T. Elberse. 1990. Competition and nutrient availability in heathland and grassland ecosystems. Pages 93-116 *in* J. B. Grace and D. Tilman, editors. Perspectives on plant competition. Academic Press, San Diego, California, USA.

Berendse, F., R. Aerts and R. Bobbink. 1993. Atmospheric nitrogen deposition and its impact on terrestrial ecosystems. Pages 104-121 *in* C. C. Vos and P. Opdam, Landscape ecology of a stressed environment. Chapman and Hall, London, England.

Bobbink, R. 1991. Effects of nutrient enrichment in Dutch chalk grassland. Journal of Applied Ecology **28**:28-41.

Brewer, J. S., J. M. Levine and M. D. Bertness. 1997. Effects of biomass removal and elevation on plant species richness in a New England salt marsh. Oikos **80**: 333-341.

Callaway, R. M. 1994. Facilitative and interfering effects of *Arthrocnemum subterminale* on winter annuals. Ecology **75**:681-686.

Chapman, V. J. 1940. Studies in salt-marsh ecology: comparisons with marshes on the east coast of North America. Journal of Ecology **28**: 118-152.

Chapman, V. J. 1974. Salt marshes and salt deserts of the world. Interscience Press. New York, New York, USA.

Coley, P. D. and T. M. Aide. 1991. Comparison of herbivory and plant defenses in temperate and tropical broad-leaved forests. Pages 25-49 *in* P. W. Price, T. M. Lewinsohn, G. W. Fernandes and W. W. Benson, editors. Plant-animal interactions: evolutionary ecology in tropical and temperate regions. John Wiley and Sons, New York, New York, USA.

Connell, J. H. 1961. The influence of interspecific competition and other factors on the distribution of the barnacle *Chthamalus stellatus*. Ecology **42**: 710-723.

Connell, J. H. 1972. Community interactions on marine rocky shores. Annual Review of Ecology and Systematics **3**:169-192.

Connell, J. H. 1983. On the prevalence and relative importance of interspecific competition: evidence from field experiments. American Naturalist **122**: 661-696.

Cooper, A. 1982. The effects of salinity and waterlogging on the growth and cation uptake of salt marsh plants. New Phytologist **90**: 263-275.

Dayton, P. K. 1971. Competition, disturbance and community organization: the provision and subsequent utilization of space in a rocky intertidal community. Ecological Monographs **41**: 351-389.

DeLucia, E. H., R. M. Callaway and W. H. Schlesinger. 1994. Offsetting changes in biomass allocation and photosynthisis in ponderosa pine in response to climate change. Tree Physiology **14**:669-677.

Diamond, J. 1986. Laboratory experiments, field experiments and natural experiments. Pages 3-22 *in* J. Diamond and T. Case, editors. Community ecology. Harper and Row, New York, New York, USA.

Doty, M. S. 1946. Critical tide factors that are correlated with the vertical distribution of marine algae and other organisms along the Pacific coast. Ecology **27**: 315-328.

Dunson, W. A. and J. Travis. 1991. The role of abiotic factors in community organization. American Naturalist **138**: 1067-1097.

Friend, A. D., F. I. Woodward and V. R. Switsur. 1989. Field measurements of photosynthesis, stomatal conductance, leaf nitrogen and ^{13}C along altitudinal gradients in Scotland. Functional Ecology **3**:117-122.

Furbish, C. E. and M. Albano. 1994. Selective herbivory and plant community structure in a mid-Atlantic salt marsh. Ecology **75**:1015-1022.

Gallagher, J. L. 1975. Effect of an ammonium nitrate pulse on the growth and elemental composition of stands of *Spartina alterniflora* and *Juncus roemerianus*. American Journal of Botany **62**: 644-648.

Goldberg, D. E. 1990. Components of resource competition in plant communities. Pages 27-50 *in* J. Grace and D. Tilman, editors. Perspectives on plant competition. Academic Press, New York, New York, USA.

Goldberg, D. E. and A. M. Barton. 1992. Patterns and consequences of interspecific competition in natural communities: a review of field experiments with plants. American Naturalist **139**: 771-801.

Hacker, S.D. and M.D. Bertness. 1995. Morphological and physiological consequences of a positive plant interaction. Ecology **76**:2165-2175.

—— 1998. The role of positive interactions on the plant species diversity of salt marsh plant communities. Ecology, in press.

Hiel, G. W. and W. H. Diemont. 1983. Raised nutrient levels change heathland into grassland. Vegetatio **53**:113-120.

Holligan, P. M. and W. A. Reiners. 1992. Predicting the responses of the coastal zone to global change. Advances in Ecological Research **22**:211-255.

Holt, R. D. 1990. The microevolutionary consequences of climate change. Trends in Ecology and Evolution **5**:311-315.

Howarth, R. W. 1988. Nutrient limitation of net primary production in marine ecosystems. Annual Review of Ecology **19**:89-110.

Howes, B. L., R. W. Howarth, J. M. Teal and I. Valiela. 1981. Oxidation-reduction potentials in a salt marsh: spatial patterns and interactions with primary production. Limnology and Oceanography **26**: 350-360.

Howes, B. L., J. W. H. Dacey and D. D. Goehringer. 1986. Factors controlling the growth form of *Spartina alterniflora*: feedbacks between above-ground production, sediment oxidation, nitrogen and salinity. Journal of Ecology **74**: 881-898.

Johnson, D. S. and H. H. York. 1915. The relation of plants to tidal level. Carnegie Institute, Washington, District of Columbia, USA.

Keddy, P. 1989. Competition. Chapman and Hall, London, England.
Lajtha, K. and J. Getz. 1993. Photosynthesis and water use efficiency in pinyon-juniper communities along an elevational gradient in New Mexico. Oecologia **95**:59-101.
Lewis, J. R. 1964. The ecology of rocky shores. English University Press, London, England.
Levine J., S. Brewer and M. D. Bertness. 1998. Nutrient availability and the zonation of marsh plant communities. Journal of Ecology **86**: 285-292.
MacArthur, R. H. 1972. Geographical ecology: patterns in the distribution of species. Harper and Row, New York, New York, USA.
MacDonald, K. B. and M. C. Barbour. 1974. Beach and salt marsh vegetation of the North American Pacific coast. Pages 263-294 *in* R. J. Reimold and W. H. Queen, editors. Ecology of halophytes. Wiley, New York, New York, USA.
Mendelssohn, I. A. 1979a. Nitrogen metabolism in the height forms of *Spartina alterniflora* in North Carolina. Ecology **60**: 106-112.
—— 1979b. The influence of nitrogen level, form and application method on the growth response of *Spartina alterniflora* in North Carolina. Estuaries **2**: 106-112.
Menge, B. A. 1976. Organization of the New England rocky intertidal community: role of predation, competition and environmental heterogeneity. Ecological Monographs **46**: 355-393.
Miller, W. B. and F. E. Egler. 1950. Vegetation of the Wequetequock-Pawcatuck tidal marshes, Connecticut. Ecological Monographs **20**:143-172.
Moloney, K. A. 1990. Shifting demographic control of a perennial bunchgrass along a natural habitat gradient. Ecology **71**: 1133-1143.
Neilson, B. J. and L. E. Cronin. 1981. Estuaries and nutrients. Humana Press, Clifton, New Jersey, USA.
Newell, R. C. 1979. Biology of intertidal animals. Marine ecological surveys. Haversham, Kent, England.
Nixon, S. W. and C. A. Oviatt. 1978. Analysis of local standing crop of *Spartina alterniflora*. Botanica Marina **26**:103-109.
Ostendorp, W. 1989. ''Die-back'' of reeds in Europe—a critical review of literature. Aquatic Botany **35**: 5-26.
Paine, R. T. 1966. Food web complexity and species diversity. American Naturalist **100**: 65-75.
Paine, R. T. 1974. Intertidal community structure: experimental studies on the relationship between a dominant competitor and its principal predator. Oecologia **15**: 93-120.
Peierls, B. L., N. F. Caraco, M. L. Pace, J. J. Cole. 1991. Human influence on river nitrogen. Nature **350**:386-387.
Pennings, S. C. and R. M. Callaway. 1992. Salt marsh plant zonation: the relative importance of competition and physical factors. Ecology **73**:681-690.
—— 1996. Impact of a parasitic plant on the structure and dynamics of salt marsh vegetation. Ecology **77**: 1410-1419.
Pennings, S. C. and M. D. Bertness. 1999. Using latitudinal variation to examine effects of climate on coastal salt marsh pattern and process. Wetlands Biogeochemistry **3**:100-111.
Pomeroy, L. R. and R. G. Weigert. 1981. The Ecology of a salt marsh. Springer-Verlag, New York, New York, USA.
Ranwell, D. S. 1971. Ecology of salt marshes and sand dunes. Chapman and Hall, London, England.
Reader, R. J., S. D. Wilson, J. W. Belcher, I. Wisheu, P. A. Keddy, D. Tilman, E. C. Morris, J. B. Grace, J. B. McGraw, H. Olff, R. Turkington, E. Klein, Y. Leung, B, Shipley, R. Van Hulst, M. E. Johanson, C. Nilsson, J, Gurevitch, K. Grigulis and B. E. Beisner. 1994. Plant competition in relation to neighbor biomass: an intercontinental study with *Poa pratensis*. Ecology **75**: 1753-1760.
Salisbury, F. B. and C. W. Ross. 1992. Plant physiology. Wadsworth Publishing, Belmont, California, USA.
Schoener, T. W. 1983. Field experiments with interspecific competition. American Naturalist **122**: 240-285.
Silander, J. A. and J. Antonovics. 1982. Analysis of interspecific interactions in a coastal plant community: a perturbation approach. Nature **298**: 557-560.
Snow, A. and S. Vince. 1984. Plant zonation in an Alaskan salt marsh. II. An experimental study of the role of edaphic conditions. Journal of Ecology **72**: 669-684.
Stephenson, T. A. and A. Stephenson. 1949. The universal features of zonation between tidemarks on rocky coasts. Journal of Ecology **38**: 289-305.
—— 1971. Life between tidemarks on rocky shores. W. H. Freeman and Co., San Francisco, California, USA.
Stout, J. P. 1984. The Ecology of irregularly flooded salt marshes of the northeastern Gulf of Mexico: a

community profile. U. S. Fish and Wildlife Service. Biological Reports **85**:1- 98.
Sullivan, M. and F. C. Daiber. 1974. Response in production of the cordgrass. *Spartina alterniflora* to inorganic nitrogen and phosphorus fertilization. Chesapeake Science **15**: 121-123.
Teal, J. M. 1962. Energy flow in the saltmarsh ecosystem of Georgia. Ecology **43**: 614-624.
Tilman, D. 1987. Secondary succession and the pattern of plant dominance along experimental nitrogen gradients. Ecological Monographs **57**:189-214.
Turner, R. E. 1976. Geographic variations in salt marsh macrophyte production: a review. Contributions in Marine Science **20**:47-68
Turner, R. E. and N. N. Rabalais. 1991. Changes in Mississippi River water quality in this century. BioScience **41**:140-147.
Twolan-Strutt, L. and P. K. Keddy. 1996. Above- and belowground competition intensity in two contrasting wetland plant communities. Ecology **77**:259-270.
Underwood, A. J. and E. J. Denley. 1984. Paradigms, explanations and generalizations in models for the structure of intertidal communities on rocky shores. Pages 151-180 *in* D. Strong, D. Simberloff, L. Abele and A. Thistle, editors. Ecological communities. Princeton University Press, Princeton, New Jersey, USA.
Valiela, I. and J. M. Teal. 1979. The nitrogen budget of a salt marsh ecosystem. Nature **280**: 652-656.
—— 1974. Nutrient limitation of salt marsh vegetation. Pages 547-563 *in* R. J. Reimold and W. H. Queen, editors. Ecology of halophytes. Academic Press. New York, New York, USA.
Valiela, I., J. M. Teal, C. Cogswell, J. Hartman, S. Allen, R. V. Etten and D. Goehringer. 1985. Some long-term consequences of sewage contamination in salt marsh ecosystems. Pages 301-316 *in* P. J. Godfrey, E. R. Kaynor, S. Pelezarski and J. Benforado, editors. Ecological considerations in wetlands treatment of municipal wastewaters. Van Nostrand Reinhold Co., New York, New York, USA.
Valiela, I. 1984. Marine ecological processes. Springer-Verlag, New York, New York, USA.
Vermeij, G. J. 1978. Biogeography and adaptation. Harvard University Press, Cambridge, Massachusetts, USA.
Vince, S. W., I. Valiela and J. M. Teal. 1981. An experimental study of the structure of herbivorous insect communities in a salt marsh. Ecology **62**: 1662-1678.
Vitousek, P. M., J. D. Aber, R. W. Howarth, G. E. Likens, P. A. Matson, D. W. Schindler, W. H. Schlesinger and D. G. Tilman. 1997. Human alteration of the global nitrogen cycle: sources and consequences. Ecological Applications **7**: 737-750.
Wiegert, R. G. and B. J. Freeman. 1990. Tidal salt marshes of the southeast Atlantic coast: A community profile. U. S. Fish and Wildlife Service. Biological Report **85**(7.29).
Wilson, S. D. and D. Tilman. 1993. Plant competition and resource availability in response to disturbance and fertilization. Ecology **74**: 599-611.
Zedler, J. B. 1982. The ecology of southern California coastal salt marshes. United States Fish and Wildlife Service, FWS-OBS-81-54, Washington, D. C., USA.
Zedler, J. B. 1995. Salt marsh restoration: lessons from California. Pages 75-95 *in* J. Cairns, Jr., editor. Rehabilitating damaged ecosystems, Second Edition. Lewis Publishers, Boca Raton, Florida, USA.

ECO-PHYSIOLOGICAL CONTROLS ON THE PRODUCTIVITY OF *SPARTINA ALTERNIFLORA* LOISEL

IRVING A. MENDELSSOHN
Wetland Biogeochemistry Institute and
Department of Oceanography and Coastal Sciences
Louisiana State University
Baton Rouge, LA 70803 USA

JAMES T. MORRIS
Department of Biological Sciences
University of South Carolina
Columbia, SC 29208 USA

Abstract
The intertidal salt marshes of the Atlantic and Gulf coasts of the United States are dominated by the perennial grass, *Spartina alterniflora* Loisel. The ecology of salt marshes in which this species dominates has been extensively investigated because of the documented biogeochemical functions that these ecosystems perform and the resulting societal values they provide. Since many of the salt marsh-derived values originate, either directly or indirectly, from the presence of a vegetated marsh and its primary productivity, it has long been a major goal of salt marsh ecology to elucidate the determinants of the growth of *Spartina*. This paper reviews the interaction of the abiotic environment with key eco-physiological processes controlling the growth of this important plant species. The productivity of *Spartina* can vary on both spatial and temporal scales. Spatial differences in productivity on a local scale are primarily determined by abiotic factors, particularly the interaction of soil anoxia, soluble sulfide, and salinity, with plant nitrogen uptake and assimilation. Also, *Spartina* can induce a positive feedback on productivity by enhancing substrate aeration. The growth-enhancing effects of marsh infauna, e.g., fiddler crabs, are mediated through these interacting abiotic variables. Productivity differences on regional scales are largely dependent on geographical differences in climate, tidal amplitude, and soil parent material. Temporal variation results from seasonal and annual variation in climatic and tidal controls that may influence marsh salinity and/or inundation. The concerted research of a large number of scientists has provided one of the most comprehensive and ecologically-relevant analyses of determinants of the primary productivity of any non-agricultural plant species.

1. Introduction

Intertidal salt marshes are well recognized for their important ecological functions and societal values (Mitsch and Gosselink 1993). *Spartina alterniflora* (hereafter cited as *Spartina*), a facultative halophyte in the family Gramineae, dominates intertidal salt marshes along the Atlantic and Gulf of Mexico coastal zones of the United States. The ecology of this species has been well studied, especially with respect to determinants of primary productivity, because of its importance, directly and indirectly, to the health and productivity of coastal fisheries (Odum 1961, Gosselink et al. 1973, Turner 1977). The goal of this paper is to review some of the key eco-physiological processes controlling the growth of this species and how the abiotic environment modulates these processes.

Although *Spartina* marshes are ranked among the most productive natural systems in the world (Mitsch and Gosselink 1993, Dawes 1998) the primary productivity of this species can vary greatly on local, geographic (latitudinal), and temporal scales. Geographic variation of aboveground primary productivity in salt marshes of North America is striking, ranging from values averaging less than 60 g C m^{-2} yr^{-1} in northern Canada and Alaska to values averaging as high as 812 g C m^{-2} yr^{-1} in the north central Gulf of Mexico (Fig. 1). This variation in productivity is closely associated with latitude (Turner 1976) and thus likely due to geographic differences in climate and length of the growing season. Variations in tidal amplitude between regions may also influence *Spartina* productivity (Steever et al. 1976) and modulate climatic effects.

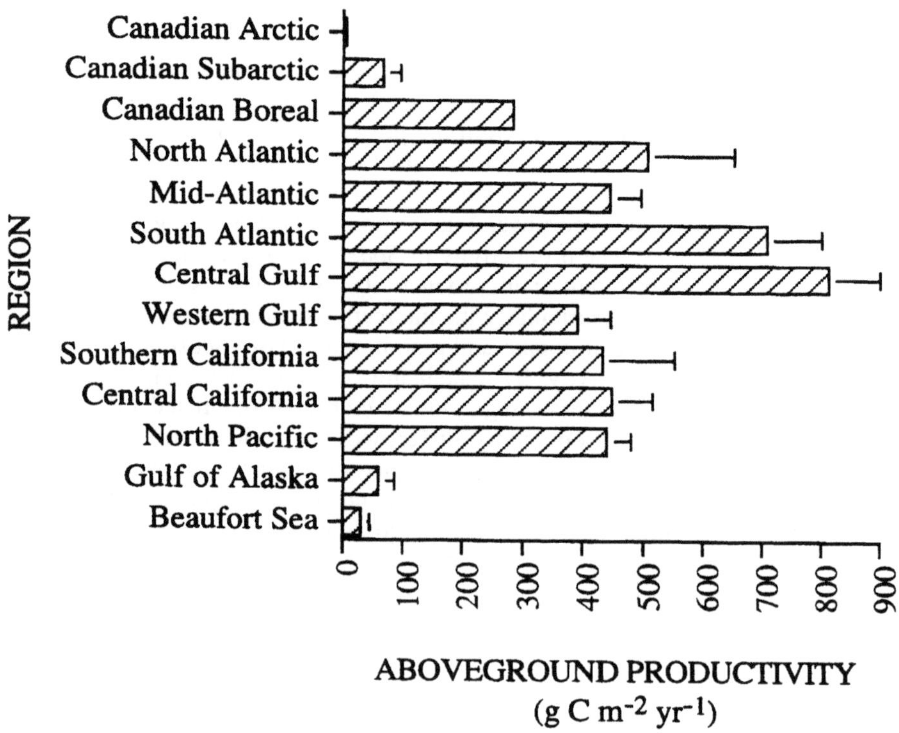

Figure 1. Salt marsh primary productivity by region in North America, north of Mexico.

Most estimates of salt marsh primary productivity have been based on incremental increases in biomass determined by the harvest technique (Smalley 1960, Linthurst and Reimold 1978, Shew et al. 1981, Kaswadji et al. 1990). Variations in this procedure can result in very different calculated rates of production for the same salt marsh. For example, depending on the particular harvest approach used, Linthurst and Reimold (1978) found as much as a 6-fold range in NAPP (net aerial primary productivity) for a *Spartina* marsh in Maine and Shew et al. (1981) demonstrated a five-fold difference in NAPP in North Carolina. Hence, it is likely that some of the differences in primary productivity among marshes cited in the literature are due, in part, to differences in methodology. Nonetheless, within any individual *Spartina*-dominated salt marsh 3-fold or greater differences in productivity have been documented, even when the same harvest methodologies have been used (Kirby and Gosselink 1976, Linthurst and Reimold 1978, Gallagher et al. 1980). Much of the salt marsh productivity literature has addressed controls on this within-marsh variation, especially relative to the so-called "height forms" of *Spartina alterniflora* (see reviews such as Anderson and Treshow 1980, Mendelssohn et al. 1982, Smart 1982). This review will emphasize eco-physiological controls of within marsh variation in primary productivity of *Spartina alterniflora*, especially within the context of flooding-induced constraints on plant nitrogen utilization and growth.

2. Within-Marsh Variation in Primary Productivity

A ubiquitous characteristic of *Spartina* marshes is the visually striking gradient in plant height evident along the transition from tidal creekbanks into the marsh interior. Along this gradient, *Spartina* productivity can vary from exceptionally high to exceptionally low levels and occur as relatively distinct height forms referred to as tall, medium and short along the Atlantic coast of the United States and streamside and inland along the Gulf of Mexico coast. Although the tall and most productive height form of *Spartina* occurs primarily along the creekbanks of the frequently flooded low marsh, the short form may occur either within the low marsh, just inland of the tall form, or in the infrequently flooded high marsh, further landward of the low marsh short *Spartina* (Fig. 2).

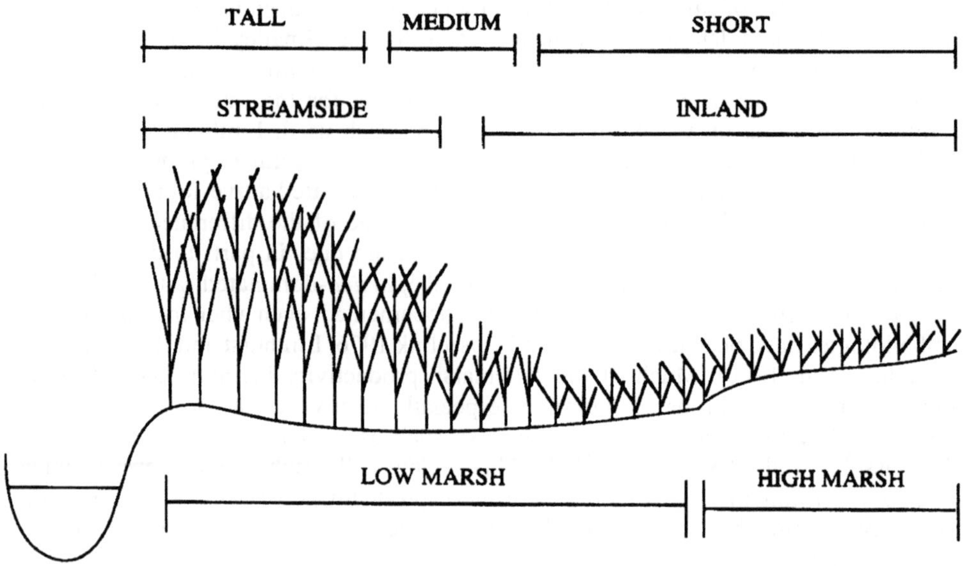

Figure 2. Schematic of within-marsh variation in the height of *Spartina alterniflora* and the terms used to describe these height forms.

3. The Soil Nitrogen Paradox and *Spartina* Growth

Considerable attention in salt marsh ecology, especially during the 1970's, was directed to understanding what nutrients limit the growth of *Spartina* (Sullivan and Daiber 1974, Valiela and Teal 1974, Broome et al. 1975, Gallagher 1975, Patrick and DeLaune 1976, Mendelssohn 1979a). Fertilization experiments have consistently demonstrated that in areas of low productivity, short *Spartina* can be stimulated by the addition of inorganic nitrogen, i.e., ammonium or nitrate (Fig. 3, Mendelssohn 1979a). Phosphorus can be the primary nutrient limiting the productivity of *Spartina* growing on sandy substrates (Broome et al. 1975) and can be secondarily limiting in some marshes. For example, in a South Carolina salt marsh, there was no growth response from *Spartina* in plots treated only with phosphorus, but the combination of phosphorus and nitrogen stimulated growth to a greater extent than nitrogen alone, which indicates that phosphorus becomes limiting when nitrogen loading exceeds a threshold (Morris 1988). Although some earlier investigations have suggested that iron may limit *Spartina* growth (Adams 1963), experimental documentation is lacking (Broome et al. 1975). Thus, by the mid to late 1970's, the scientific dogma was that nitrogen deficiencies limit the growth of *Spartina* in the same generic way that nitrogen scarcities limit the growth of phytoplankton in marine and estuarine environments (Valiela 1995). However, at this same time a paradox was emerging. Investigations quantifying plant available inorganic nitrogen concentrations along the *Spartina* productivity gradient documented that interstitial ammonium, the dominant form of inorganic nitrogen in most salt marshes, was often an order of magnitude higher in the less productive inland (short) *Spartina* zone than in the more productive streamside (tall) zone (Mendelssohn 1979b, Craft et al. 1991). Furthermore, research on the nitrogen budgets of marsh ecosystems was beginning to show that salt marshes exported more inorganic nitrogen through tidal exchange than was imported (Valiela and Teal 1979). The pertinent question then became: Why wasn't *Spartina* utilizing this available nitrogen?

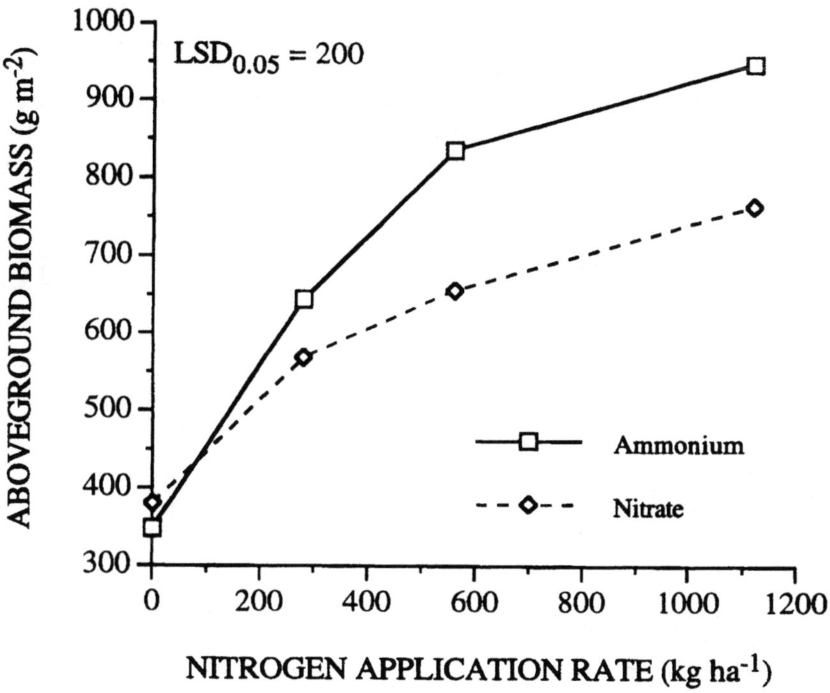

Figure 3. The response in aboveground biomass of *Spartina alterniflora* to ammonium and nitrate fertilization (adapted from Mendelssohn 1979a).

4. Soil Water Drainage — A Critical Factor for Optimal Growth

Simultaneous with the discovery of the soil nitrogen paradox was the observation that soil water movement was much more dynamic in the streamside marsh zones (tall and medium height forms) compared to that in the short *Spartina* inland zones (Mendelssohn and Seneca 1980, Howes et al. 1981). This spatial difference in soil water drainage in salt marshes was described, in part, by Chapman (1978) many years earlier, but was never identified as a potential determinant of *Spartina* differential growth. At low tide, streamside soils drain of pore water much more than the inland soils, a 20 cm difference in water table may occur between the two zones

(Mendelssohn and Seneca 1980, Howes et al. 1986). In fact, the soils of short *Spartina* within the low marsh exhibit little horizontal water movement (Osgood and Zieman 1993), and much of the soil aeration that occurs is a result of evapotranspirational water losses and subsequent air entry into the soil (Dacey and Howes 1984). This differential soil water drainage results in a dramatic difference in soil redox potential (Eh) between the creekbank soils, which are the most biochemically oxidized, and the inland soils, which are the most biochemically reduced (Fig. 4). The lower Eh conditions in the inland marsh are associated with higher soluble sulfide concentrations compared to the streamside marsh (Fig. 5) (King et al. 1982, DeLaune et al. 1983, Mendelssohn and McKee 1988). Hydrogen sulfide is a known phytotoxin (Okajima and Takagi 1953, Goodman and Williams 1961, Allam and Hollis 1972, Joshi et al. 1975) that can accumulate in waterlogged soils, especially where sulfate introduction, e.g., from seawater, is prevalent. Although reduced soils and excessive sulfide pose potential stresses to plants, *Spartina*, like other flood-tolerant plants, possess adaptations for life in an oxygen deficient and high sulfide environment.

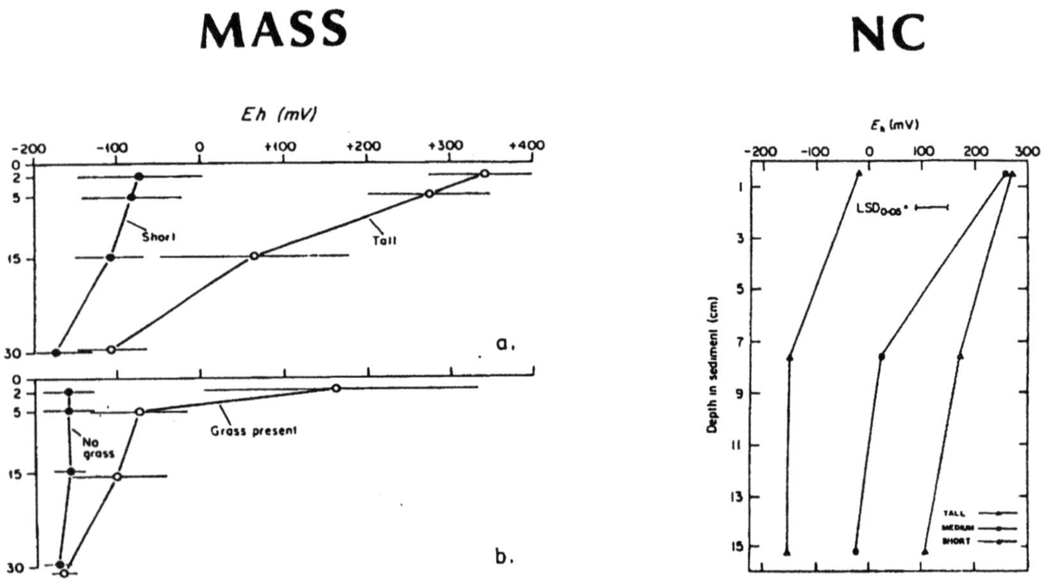

Figure 4. Redox potential (Eh) of the soil supporting different height forms of *Spartina alterniflora* in Massachusetts and North Carolina [Mass: from Howes, B.L., R.W. Howarth, J.M. Teal, I. Valiela (1981) and NC: from Mendelssohn and Seneca 1980].

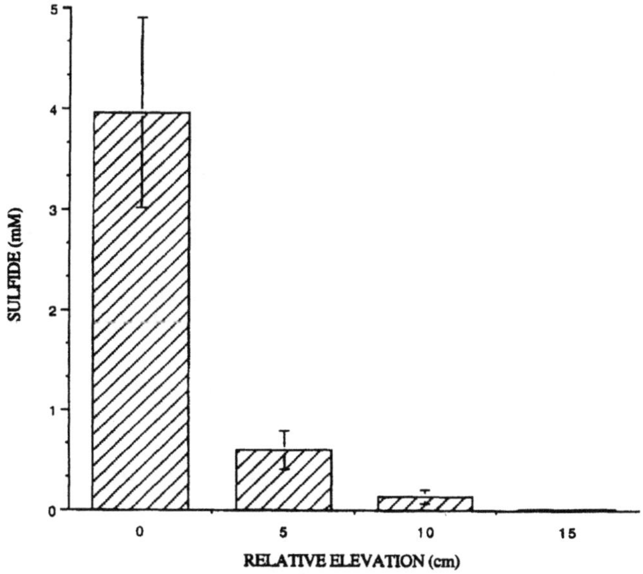

Figure 5. Sulfide concentration in a *Spartina alterniflora* salt marsh in southeast Louisiana as a function of marsh elevation (0 cm = inland marsh elevation and 15 cm = streamside marsh elevation). (Adapted from Mendelssohn and McKee 1988.)

5. Adaptations to Soil Anoxia

Nutrient absorption across root membranes, an essential process controlling plant growth, is an active metabolic function that requires the consumption of oxygen (Epstein 1972). The anatomy of *Spartina* allows for diffusion of oxygen to the root system (Teal and Kanwisher 1966, Arenovski and Howes 1992, Howes and Teal 1994). This is an adaptation that is prevalent in wetland plants (Armstrong 1979). In *Spartina*, the movement of oxygen into the root system is aided by the process of hygrometric pressurization (Hwang and Morris 1991). However, in the absence of a supply of oxygen in the soil, the rate of internal oxygen transport within the plant is apparently insufficient to promote a highly efficient nutrient uptake mechanism (Morris and Dacey 1984), and root anaerobic metabolism may occur (Mendelssohn et al. 1981, Mendelssohn and McKee 1983). Under anoxic conditions, the production of energy via aerobic root respiration is impaired, and root alcoholic fermentation, the dominant pathway of anaerobic carbon metabolism in plants (Ap Rees 1974), becomes the primary energy source in flood-adapted plants. Thus, the ability of *Spartina* to maintain high rates of alcoholic fermentation during root hypoxia/anoxia is of paramount importance to its growth and survival in flooded environments.

Although the oxygen supply to the whole root system may pose limits to nutrient uptake, there is clear evidence that *Spartina* is able to transport oxygen to the root and to maintain an oxidized rhizosphere around at least a portion of its root system, as evidenced by the formation of ferric iron plaques that are visible on the root surfaces (Mendelssohn and Postek 1982). This relatively oxidized interface between the root and its edaphic environment may protect the plant from toxins. e.g., hydrogen sulfide, in the surrounding sediment. Also, the enzymatic and non-enzymatic oxidation of sulfide by *Spartina* roots has been documented (Lee et al. 1999). The oxidized rhizosphere may also help scavenge nutrients, like phosphate and iron that will precipitate onto the root surface where they can be assimilated after solubilization by root exudates. This has been observed in other species (Masaoka et al. 1993), but experimental evidence in *Spartina* is lacking. Although flood tolerant plants are adapted to reduced environments, excessive soil reduction and the resulting accumulation of high concentrations of products of soil microbial metabolism, like sulfide, may impact plant growth. What evidence exists for such effects?

6. The Soil Drainage-Sulfide-Growth Relationship

Field experiments have demonstrated that impaired soil water drainage, resulting in low soil Eh (Mendelssohn and Seneca 1980) and elevated sulfides (King et al. 1982), inhibit *Spartina* growth. Conversely, improving soil drainage lowers interstitial sulfide concentrations and concurrently increases the biomass production of this species (Wiegert et al. 1983). The presence of infaunal organisms, such as fiddler crabs, that increase soil aeration via their borrows also promote higher rates of primary production (Bertness 1985). In addition, relationships between *Spartina* biomass with marsh elevation, soil Eh and sulfide have been identified (King et al. 1982, DeLaune et al. 1983), suggesting that sulfide may by the cause for the within-marsh productivity differences of *Spartina*. Furthermore, field reciprocal transplant experiments demonstrated that when streamside sods of marsh are transplanted into the inland zone, interstitial sulfide increases and *Spartina* biomass decreases, while the reverse is true for inland sods transplanted into the streamside zone (Mendelssohn and McKee 1988). Under more controlled greenhouse conditions, Linthurst (1979) found that poor growth of *Spartina* was associated with high concentrations of interstitial sulfide while *Spartina* production was greatest in aerated substrates where little sulfide accumulated. These field and greenhouse results further support the hypothesis that interstitial sulfide can negatively impact biomass production. However, none of these studies had demonstrated the cause and effect relationship between sulfide and *Spartina* growth.

7. The Sulfide-*Spartina* Dose Response

Negative effects of increasing sulfide concentrations on the growth response of a number of salt marsh species have been demonstrated (Ingold and Havill 1984, Havill et al. 1985, Van Diggelen et al. 1986, Pearson and Havill 1988, Bradley and Dunn 1989, Cantilli 1989). Specifically for *Spartina*, Bradley and Dunn (1989) and Koch et al. (1990)

demonstrated that at concentrations above 1 mM total soluble sulfide growth is negatively impacted. This 1 mM threshold level was further supported by Koch and Mendelssohn (1989) and Mendelssohn and McKee (unpublished, Fig. 6).

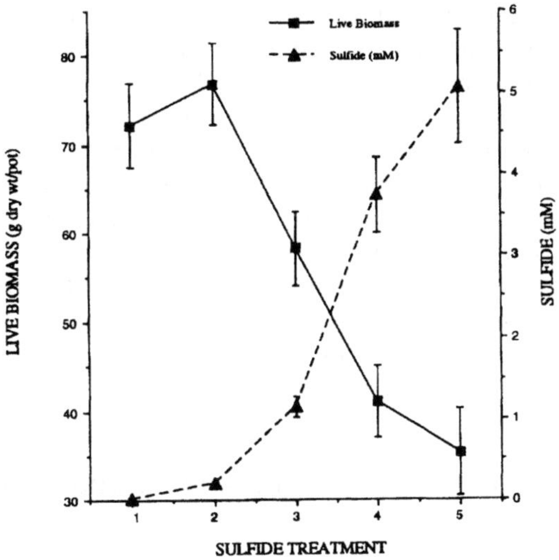

Figure 6. The dose-response relationship of Spartina alterniflora biomass to sulfide concentration (Mendelssohn and McKee, unpublished).

While high sulfide concentrations undoubtedly inhibit *Spartina* growth (see references cited previously), a low sulfide concentration appears to stimulate growth due to secondary nutrient effects or possibly an energy subsidy. In greenhouse sand cultures there was a consistent trend toward increased growth at higher sulfide concentrations up to 1 mM (Morris et al. 1996). There was a fourfold increase in relative growth rate as sulfide level increased from 0 to 1 mM (sulfate was supplied in excess). On the other hand, there is little doubt that sulfide concentrations in excess of 1 mM are toxic (Bradley and Dunn 1989, Koch and Mendelssohn 1989, Koch et al. 1990). This dose-response may depend on the culture conditions. For example, in water culture (Bradley and Dunn 1989) there is little chance for the establishment of an oxidized rhizosphere due to turbulence, whereas an oxidized zone could form around the roots and buffer the plant against external toxins in a stagnant sand culture (Morris et al. 1996). In fact, considerable indirect evidence suggests that oxygen release from roots of plants can oxidize wetland soils and lower sulfide concentrations (e.g., McKee et al. 1988). However, the isotopic composition of sulfur in the tissues of *Spartina* from the field is like that of sulfide and not like that of sulfate (Carlson and Forrest 1982), which indicates that sulfide is the *in-situ* sulfur source and that an oxidized rhizosphere probably does not completely prevent the uptake of sulfide. Thus, although oxygen release from *Spartina* roots likely moderates sulfide accumulation in salt marsh soils, sulfide uptake appears to continue to some degree.

Sulfide may in fact be the preferred sulfur source, in low concentrations, because of favorable energetics, requirements of bacterial symbionts, or secondary effects involving other nutrients. Like chemolithotrophic bacteria, perhaps *Spartina*, which is known to fix CO_2 in its root system (Hwang and Morris 1992), is able to utilize sulfide as an energy source directly or indirectly. Lee et al. (1999) also suggest that mitochondrial sulfide oxidation in *Spartina* roots may be coupled to oxidative phosphorylation. *Spartina* may also host chemolithotrophic bacterial symbionts. Moreover, the bacterial community of the rhizosphere may influence plant mineral nutrition in a variety of ways. Sulfide may have secondary effects on plant mineral nutrition because of its effect on the solubility and availability of nutrients (Leeper 1952, Engler and Patrick 1975, Howarth et al. 1983, Giblin et al. 1986, Luther et al. 1986). Van Diggelen et al. (1987) speculated that the greater growth of *S. anglica* that they observed at low sulfide concentrations could have been caused by an iron deficiency in the 0 mM sulfide treatment as plants grown without sulfide were chlorotic and had lower concentrations of Fe and P in their tissues. Although the exact mechanism for this response is unclear, the effect of sulfide on the redox potential of the rooting medium may have been involved.

8. Effects on *Spartina* Growth: Soil Anoxia and Sulfide

High sulfide concentrations can impact plant growth in two general ways: 1) indirectly through an increase in soil reducing conditions caused by greater electron availability from the presence of higher concentrations of free sulfide and 2) directly from sulfide toxicity. In the case of the former, as the availability of free sulfide increases, the greater the electron availability and the more negative the redox potential (Koch et al. 1990). The lower the soil redox potential, the greater the likelihood that root oxygen deficiencies will develop because of the high oxygen demand of a strongly reducing rooting medium. Thus, not only is there a greater potential for sulfide toxicity as free sulfide concentrations increase, but the potential for root oxygen deficiency stress also increases. Because sulfide can accumulate in soils that are flooded and therefore not exposed to atmospheric oxygen, salt marsh plants can experience root oxygen deficiency stress, sulfide stress or both. Thus, the relative importance of root oxygen deficiency stress versus sulfide stress in controlling growth is of interest.

8.1 SOIL ANOXIA AND *SPARTINA* GROWTH

Microbial and root respiration quickly deplete soil oxygen in waterlogged soils, and in the absence of any further input of atmospheric oxygen due to soil saturation and flooding, the soils remains virtually devoid of oxygen (Turner and Patrick 1968, Gambrell et al. 1991). Because most living cells require oxygen for maximum energy efficiency and growth, anoxic soils have the potential for limiting *Spartina* growth.

Koch and Mendelssohn (1989) found significantly higher total biomass (culm+rhizome+root) in flooded nonaerated sods of *Spartina* (average interstitial sulfide: 0.25 mM to 0.5 mM) compared to flooded aerated sods (ca. interstitial sulfide: 0.03 mM). Thus, biomass production was greater under more reduced soil conditions when sulfide was present, but concentrations were low (Koch and Mendelssohn 1989). Similarly, *Spartina* sods exposed to a simulated semi-diurnal tide that twice daily drained the soil porewater exhibited less growth than when the sods remained waterlogged at low tide (Mendelssohn and Seneca 1980). The greater plant growth under flooded, compared to aerated, soil conditions may relate to greater nutrient availability with greater soil reduction (Ponnamperuma 1972, Ponnamperuma 1977a, Ponnamperuma 1977b, Gambrell and Patrick 1978) or, as mentioned previously, to the use of sulfide as an energy source.

Although these results suggest that soil anoxia, per se, does not negatively affect *Spartina* growth, *Spartina* cultured hydroponically in an hypoxic rooting environment exhibited significant reductions in growth (Koch et al. 1990). Differences in response between soil and hydroponic culture may in part be due to the degree of biochemical reduction achieved by the particular treatments; i.e., strongly reduced soil conditions will have a greater potential to impact growth than less reduced systems, even if both are anoxic (DeLaune et al. 1990, Brix and Sorrell 1996).

8.2 SOIL ANOXIA: MECHANISMS OF IMPACT

Spartina was once thought to avoid root oxygen deficiencies because of its extensive aerenchyma (air space) system (Teal and Kanwisher 1966), which allows oxygen movement from the atmosphere through aboveground plant organs and into roots and rhizomes (Anderson 1974, Howes and Teal 1994). However, subsequent research has demonstrated that, depending on the degree of soil reduction, *Spartina* roots can exhibit root oxygen deficiencies as evidenced by high alcohol dehydrogenase (ADH) activity, the terminal enzyme in alcoholic fermentation and an indicator of anaerobic root metabolism. In hydroponic culture, nitrogen purging of the solution culture surrounding the roots resulted in a six-fold increase in ADH activity compared to aerated controls (Mendelssohn and McKee 1983). Flooded soil conditions can also result in elevated root ADH levels compared to drained wetland soils (Mendelssohn et al. 1981, Mendelssohn and McKee 1983). Furthermore, the degree of ADH activity is positively related to the intensity of soil reduction with low to no activity at high redox potentials (>300 mV) and increasing activity as soil reducing intensity increases [Fig. 7, Mendelssohn and McKee 1992)]. This inverse relationship between ADH activity and soil Eh is also observed along the streamside to inland gradient in natural *Spartina* marshes (Mendelssohn et al. 1981). Root ADH activity is lowest in the streamside marshes where the soil is most oxidized and increases to a maximum in the inland zone were the soil is most reduced (Mendelssohn et al. 1981). Interestingly, however, ADH activity decreases to significantly lower values in the sparsely vegetated and low productivity die-back zones, which are common to salt marshes in the north central Gulf of Mexico. Thus, it is now well recognized that *Spartina* can experience root oxygen deficiencies that may result in alcoholic fermentation, and, in turn, may cause a reduction in growth. The mechanism for impaired growth caused specifically by soil anoxia-induced root oxygen deficiency, although still not completely elucidated for *Spartina*, is likely due in part to a considerable loss in carbon from the plant in the form of ethanol as well as increased carbon consumption in the roots under anoxia (see Mendelssohn et al. 1981, Mendelssohn and McKee 1983).

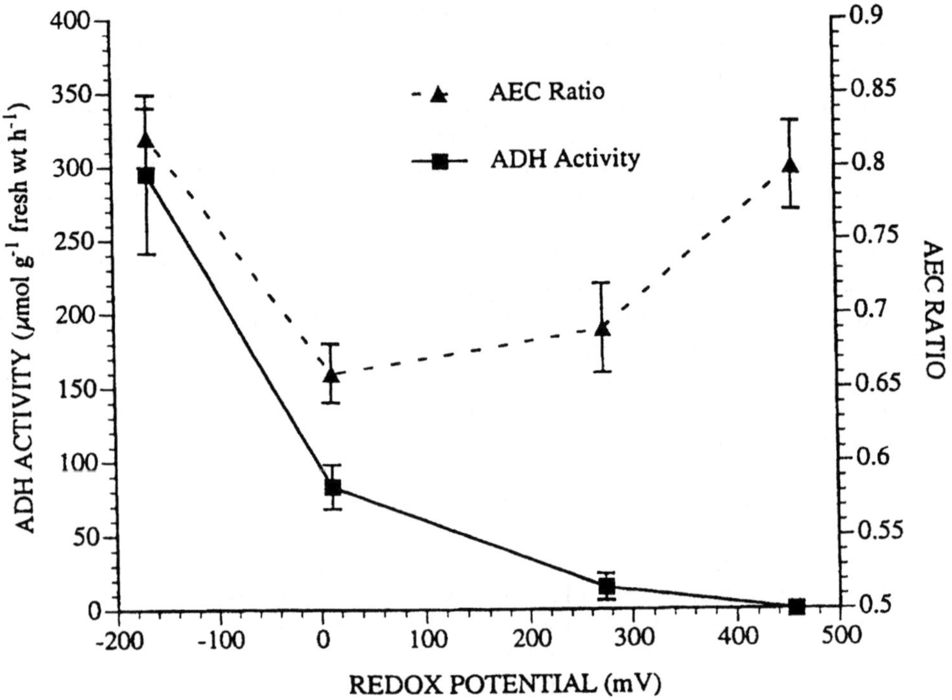

Figure 7. The relationship of between soil redox and root alcohol dehydrogenase (ADH) activity and adenylate energy charge (AEC) ration of *Spartina alterniflora* (adapted from Mendelssohn and McKee 1983).

The low productivity areas of *Spartina* marshes generally exhibit both highly reduced soils and high sulfide levels. Assuming that highly reduced soils can impart a root oxygen stress to *Spartina*, then the question arises as to whether the sulfide effect on plant growth is due to sulfide toxicity, per se, or the more reduced soil conditions created by higher concentrations of a reducing agent like sulfide. In several greenhouse and field experiments cited previously, lower soil aeration resulted in both a decrease in soil Eh and an increase in free sulfide. Thus, it was not possible in these experiments to determine which factor, soil reduction or sulfide, primarily controlled the growth response. However, Koch et al. (1990) in a short term hydroponic experiment demonstrated that sulfide concentrations as low as 0.5 mM inhibited alcohol dehydrogenase activity, the enzyme important in alcoholic fermentation and the anaerobic generation of energy (Fig. 8). With increased sulfide concentrations up to 4 mM, ADH activity continued to decrease even though the solution culture Eh became more negative (approaching -200 mV at 4 mM sulfide). If sulfide was primarily having its effect on growth through root oxygen deficiencies, we would expect to observe increasing ADH activity to some plateau with increasing sulfide. However, the opposite was true and ADH decreased to levels equivalent to the aerobic control at the highest sulfide concentrations (Fig. 8). These results explain the low ADH activity in the high sulfide-inland and die-back *Spartina* zones in the northern Gulf of Mexico

(Mendelssohn et al. 1981). Although the soils in these areas are highly reduced, ADH activity is relatively low, due probably to sulfide inhibition of ADH (Koch et al. 1990). Therefore, it appears that the primary impact of sulfide, at growth limiting concentrations, is its direct effect on plant metabolism rather than an indirect effect mediated through soil reduction intensity. Further research is required to unequivocally address these separate but related issues.

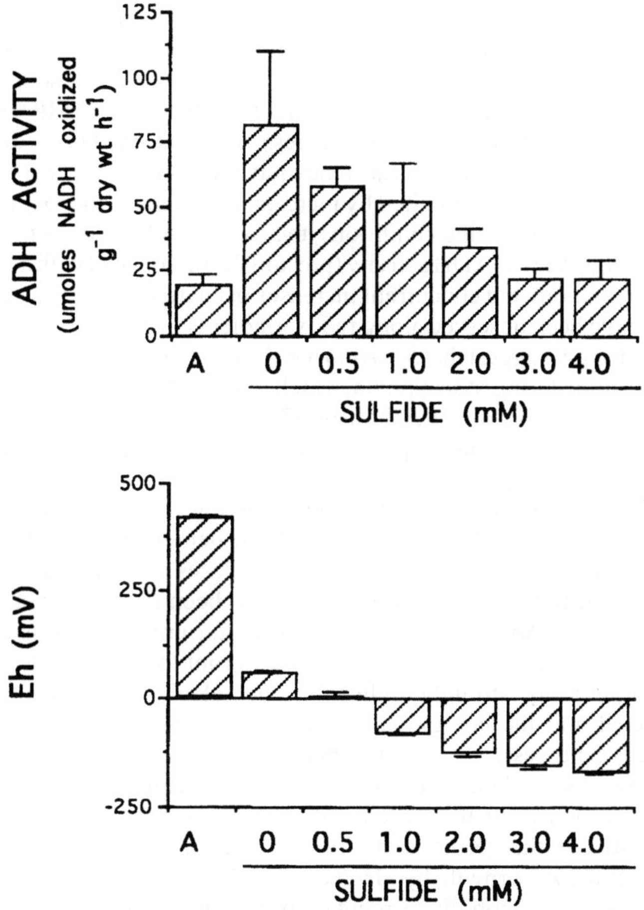

Figure 8. The effect of sulfide on root alcohol dehydrogenase (ADH) activity (top) of *Spartina alterniflora* and rooting medium redox potential (Eh) (bottom) during a six day hydroponic experiment (treatments: aerated (A), hypoxic (0.0 mM sulfide), and hypoxic with sulfide additions of 0.5, 1.0, 2.0, 3.0, and 4.0 mM). (From Koch et al. 1990.)

8.3 SULFIDE: MECHANISM OF IMPACT

As mentioned previously, it is well documented that soil waterlogging/low oxygen conditions stimulate ADH activity, indicating that the roots are switching from aerobic metabolism to anaerobic metabolism in response to oxygen deficiencies. Highly flood tolerant plants, such as rice, exhibit under root oxygen deficiencies an accelerated rate of root carbon metabolism (Pasteur effect), via alcoholic fermentation, that is concurrent with higher ADH activity (Turner 1960, Vartapetyan 1982, Mayne and Kende 1986). This higher rate of root alcoholic fermentation results in accelerated energy production per unit time and helps to compensate for the decrease in energy yield per mole of glucose in alcoholic fermentation compared to aerobic respiration. This acceleration in energy metabolism is also observed in *Spartina* (Fig. 7). At high redox potentials, the soil is oxidized and oxygen is available. Under these conditions, root metabolism is primarily aerobic, there is little ADH activity, and the energy status of the roots, indicated by the adenylate energy charge ratio (ATP + 0.5 ADP/ AMP+ADP+ATP), is relatively high (Fig. 7). With lower soil Eh, less oxidized conditions and higher soil oxygen demand, ADH activity increases somewhat and root energy charge decreases as aerobic respiration is gradually inhibited by increasing degrees of root oxygen deficiency. However, under strongly reducing conditions, the oxygen deficiency in the root is severe enough to induce high rates of ADH activity (presumably associated with accelerated rates of alcoholic fermentation) and concurrent increases in energy charge ratio (Fig. 7). These same responses have been observed along redox gradients in the field (Mendelssohn et al. 1981), and they strongly suggest that *Spartina* has the capacity of maintaining high rates of alcoholic fermentation during root hypoxia, which to some degree compensates for energy loss in the absence of aerobic root metabolism. Thus, this species' ability to generate high rates of alcoholic fermentation is paramount in maintaining active root metabolic processes such as nutrient uptake.

Sulfide impacts the growth of *Spartina* by affecting root metabolism (Bradley and Morris 1990, Koch et al. 1990). Koch et al. (1990) demonstrated that sulfide inhibits root ADH activity even though redox intensity increases with increasing sulfide. This inhibitory effect of sulfide on ADH activity has also been demonstrated in *Phragmites* (Furtig et al. 1996) and *Spartina patens* (Ewing et al., unpublished data). Concurrent with this inhibition of ADH activity is a decrease in root total adenylate concentrations and energy charge ratio, strongly indicating that alcoholic fermentation has been negatively affected and the energy status of the root impaired. Because ammonium uptake is an active process, we might expect a decrease in ammonium uptake with decreased energy status. This was demonstrated by Koch et al. (1990). Because nitrogen limits the growth of this species, it was not surprising that growth was also affected by elevated sulfide (Koch et al. 1990). Thus, the data support our thesis that sulfide impacts *Spartina* growth by decreasing the ability of the plant to generate energy anaerobically via alcoholic fermentation, thereby affecting nitrogen uptake and plant growth. These changes in root energy metabolism likely affect the nitrogen uptake kinetics of this species (Bradley and Morris 1990). The K_m for ammonium uptake is higher with greater sulfide, indicating a reduced affinity for ammonium. Thus, higher interstitial ammonium concentrations are required to attain the same ammonium uptake rate as would occur under low sulfide conditions. Furthermore, the V_{max} is reduced with increasing sulfide, i.e., the maximum uptake rate is lower at higher free sulfide concentrations. So even at relatively high interstitial ammonium concentrations, such as found in the short *Spartina* zone, the maximum rate of ammonium uptake is less than where sulfide is low.

9. Synthesis of Growth Controls

We can now integrate these findings to explain the within-marsh spatial variation in the growth of *Spartina* (Fig. 9). As mentioned previously, the streamside or tall *Spartina* marsh is a zone of relatively more oxidized soil conditions and greater subsurface water movement resulting in a low sulfide, low salinity and low ammonium environment compared to the short *Spartina* zone where sulfide and ammonium are high and the soils are strongly reducing (Bradley and Morris 1990). Salinities may be elevated if the short *Spartina* is located in the infrequently-flooded high marsh where salt may accumulate. The environmental conditions in the low, frequently inundated marsh allow for aerobic root respiration, high root energy status, and ammonium uptake kinetics characterized by a low K_m and a high V_{max}. All these factors allow for greater ammonium uptake and plant growth. The soil conditions in the inland or short *Spartina* zone lead to anaerobic root metabolism, compromised by high sulfide, leading to low energy status and resulting in uptake kinetics characterized by a high K_m and a low V_{max} and thus low ammonium uptake and reduced plant growth (Fig. 9).

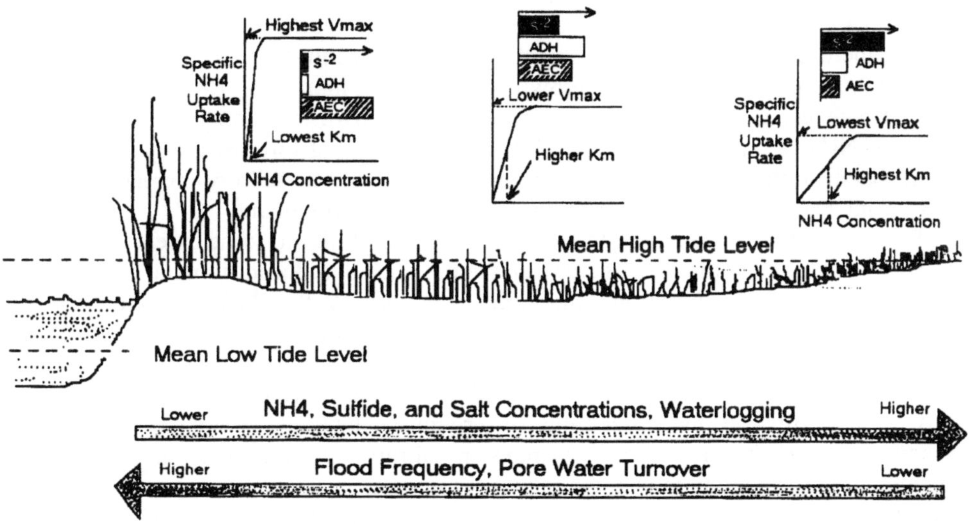

Figure 9. The tall to short *Spartina alterniflora* transition is a complex gradient characterized by increases in concentrations of interstitial ammonium, soil-reducing conditions, soluble sulfide, and, in the high marsh, salt. In contrast, inundation frequency and pore water turnover decrease. In the tall (streamside) *Spartina* zone, relatively oxidized soil conditions and low sulfide concentrations promote aerobic root respiration and limit anaerobic root metabolism (indicated by alcohol dehydrogenase activity [ADH]) and allow for high root energy status (indicated by the adenylate energy charge [AEC] ratio). Concurrently, root ammonium uptake rate is high (V_{max}) and efficient (K_m). Here, biomass production is greatest. At intermediate sulfide levels, anaerobic root metabolism is stimulated due to a more reducing rooting medium, and root energy status is only moderately affected due to high rates of alcoholic fermentation. Under these conditions, ammonium uptake kinetics are only marginally effected. However, in the short (inland) *Spartina* zone, where sulfide concentrations are extreme, anaerobic root metabolism is inhibited and root energy status low. In this part of the marsh, the ammonium uptake rates and the efficiency of uptake are severely reduced, resulting in low biomass production. Where salinities are elevated, additional declines in nitrogen uptake may occur.

The effect of sulfide and salt on nitrogen uptake kinetics explains the paradox of the nitrogen limited growth of short *Spartina* where ammonium availability is high. Sulfide and high salinity (Morris 1984) reduce the efficiency of the ammonium uptake process, thereby preventing the plant from utilizing the relatively high concentrations of ammonium present in the short zone. The higher sulfide concentrations in the interior *Spartina* zone also reduce the maximum uptake velocity. In contrast, the more oxidized soil in the tall zone prevents sulfide accumulation. This factor and a lower salinity allow for a low K_m for ammonium uptake and the ability of tall *Spartina* to utilize relatively low concentrations of ammonium with high uptake rates. A growth stimulation occurs when nitrogen fertilizer is applied to sediments supporting the short form of *Spartina* because fertilization elevates the concentration of soil ammonium to a level that compensates for the higher K_m in a high sulfide, high salt environment. Also, most nitrogen additions are to the marsh surface where the soil is more oxidized and nitrogen uptake is less likely to be impaired.

The physiological effects of salinity on *Spartina* and interactions with nitrogen nutrition are interesting and involve complex feedbacks. Salt tolerance in *Spartina* is accomplished in part by the production in the plant cells of the nitrogen-based osmoregulatory compounds proline and glycinebetaine (Cavalieri and Huang 1981, Naidoo et al. 1992). Production of these osmoregulatory compounds may explain why, with an increase in salinity, the critical nitrogen concentration (the minimum tissue concentration required to sustain growth) in *Spartina* increases (Bradley and Morris 1992). However, as discussed above, sea salts are known to competitively inhibit the uptake of NH_4+ (Bradley and Morris 1991), the prevailing form of inorganic nitrogen in salt marsh sediments. Thus, nitrogen uptake decreases as salinity increases (Morris 1984), which compromises the ability of *Spartina* to osmoregulate.

Spartina is able to compensate for an increase in salinity to some extent by exerting control over water loss through its stomata, but the price of reduced stomatal conductance is a decrease in CO_2 assimilation (Giurgevich and Dunn 1979). The response of stomatal conductance and CO_2 assimilation to salinity change is rapid, and it is persistent. That is, plants acclimated to a range of salinities also display a range of stomatal conductances and CO_2 assimilation rates that vary inversely with salinity (Hwang and Morris 1994). Thus, osmoregulation by production of organic solutes does not render the gas exchange system unresponsive to changes in salinity. In addition to the response in stomatal conductance, there is a rise in leaf respiration as salinity increases, which may result from the energy costs associated with increased salt gland activity (Levering and Thomson 1971). Consequently, salinity directly affects the growth of *Spartina* with growth greatest at salinities of 20 ‰ or less (Phleger 1971, Haines and Dunn 1976, Parrondo et al. 1978). Measurements in the laboratory (Phleger 1971, Linthurst 1980) and field (Webb 1983) indicate that the upper limit for salt tolerance is near 60 ‰, however, see Hester et. al. (1998). Drake and Gallagher (1984) found that *Spartina* biomass in the field was negatively correlated with salinity and that *Spartina* was absent at salinities greater than 75 ‰.

10. Interannual Variations in Primary Productivity

Considering that soil salinity is such an important proximate determinant of production in salt marshes, it is not surprising that the physical variables that govern salinity, like variation in rainfall amount, will also ultimately affect marsh productivity. Also, a high degree of interannual variation in *Spartina* production, as observed in a South Carolina salt marsh, is positively correlated with sea level anomalies (Morris and Haskin 1990).

Changes in mean sea level that occur without compensation in the elevation of the marsh surface affect the frequency of flooding and the solute balance of sediments (Morris 1995). The elevation of the marsh platform occurs where net sedimentation rate approaches zero, and this tends to happen at an elevation near mean high tide (Krone 1985). At this elevation, the salt marsh may not flood for several days, especially during the neap part of the tidal cycle. At these times evapotranspiration can dry the sediments and increase the salinity to levels that inhibit growth or that may be lethal. The anomalies in sea level that are critical to this process arise from variations in the solar annual cycle of sea level. This solar annual cycle of mean sea level is accounted for largely by steric (specific-volume) changes in the ocean associated with temperature fluctuations in the upper 100 m (Pattullo et al. 1955). Along the South Carolina coast, this annual cycle has a range of about 25 cm, but it is quite variable from year-to-year in both the timing and magnitude of the oscillations (Morris et al. 1990). During summers of unusually low mean sea level, soil salinities rise and primary production declines. It is during such events when rainfall will have the greatest impact on salinity and production. Thus, tidal, meteorological and climatic events have significant effects on the physical and chemical properties of salt marsh sediments, and these properties directly affect the physiology and productivity of marsh plants.

11. Summary

Investigations of the determinants of *Spartina alterniflora* productivity have revealed a complex relationship between the capacity of the plant to utilize nutrients, primarily nitrogen, and to tolerate specific abiotic stressors, most importantly sulfide and salt, that control the capacity for nitrogen acquisition. Although available nitrogen in the form of ammonium is relatively plentiful in most salt marsh environments, the capacity of the plant to uptake and utilize this nitrogen is very much controlled by the physico-chemical environment. Areas of the marsh that do not adequately drain of water at low tide frequently accumulate elevated levels of free sulfide that prevent the sufficient generation of energy needed by the plant roots to uptake the ammonium that is readily available in the soil. Where salt may accumulate, e.g., in the infrequently flooded high marsh, competitive inhibition of ammonium uptake, among other processes, also impairs ammonium nutrition. Furthermore, the anoxic soil environment that inhibits the aerobic production of energy accelerates root alcoholic fermentation that, although generating energy anaerobically, results in a considerable loss of carbon, which otherwise could be used for growth and nitrogen uptake. These eco-physiological responses result in a plant nitrogen deficiency and lower rates of growth and primary production for poorly drained, inland *Spartina* marshes.

12. Acknowledgments

We thank our research collaborators and colleagues, especially Drs. Paul Bradley, Yuan-Hsun Hwang, Marguerite Koch, and Karen McKee, for their important contributions to our research that we have herein presented. The work of the authors described in this paper was largely sponsored research funded, in part, by the National Science Foundation, Sea Grant Program, U. S. Environmental Protection Agency, U. S. Fish and Wildlife Service, and U. S. Geological Survey. We also extent our appreciation to the anonymous reviewers who have improved the manuscript and to the organizers of the Salt Marsh Ecology Conference for their support and consideration throughout the review process.

13. Literature Cited

Adams, D. A. 1963. Factors influencing vascular plant zonation in North Carolina salt marshes. Ecology **44**: 445-456.

Allam, A. I. and J. P. Hollis. 1972. Sulfide inhibition of oxidases in rice roots. Phytopathology. **62.**:634-639.

Anderson, C. and M. Treshow. 1980. A review of evironmental and genetic factors that affect height in *Spartina alterniflora* Loisel. Estuaries **3**:168-176.

Anderson, C. E. 1974. A review of structure in several North Carolina salt marsh plants. Pages 307-344 *in* R. J. Reimold and W. H. Queen, editors. Ecology of halophytes. New York Academic Press, Inc., New York, New York, USA.

Ap Rees, T. 1974. Pathways of carbohydrate breakdown in higher plants. Pages 89-127 *in* D. H. Northcote, editors. Plant Biochemistry. D. H. Northcote. Baltimore University Park Press, Baltimore, Maryland, USA.

Arenovski, A. L. and B. L. Howes. 1992. Lacunal allocation and gas transport capacity in the salt marsh grass *Spartina alterniflora*. Oecologia **90**:316-322.

Armstrong, W. 1979. Aeration in higher plants. Advances in botanical research **7**:225-332.

Bertness, M. D. 1985. Fiddler crab regulation of *Spartina alterniflora* production on a New England salt marsh. Ecology **66**:1042-1055.

Bradley, P. M. and E. L. Dunn. 1989. Effects of sulfide on the growth of three salt marsh halophytes of the southeastern United States. American Journal of Botany **76**:1707-1713.

Bradley, P. M. and J. T. Morris. 1990. Influence of oxygen and sulfide concentration on nitrogen uptake kinetics in *Spartina alterniflora*. Ecology **71**:282-287.

Bradley, P. M. and J. T. Morris. 1991. The influence of salinity on the kinetics of NH4+ uptake in *Spartina alterniflora*. Oecologia **85**:375-380.

Bradley, P. M. and J. T. Morris. 1992. Effect of salinity on the critical nitrogen concentration of *Spartina alterniflora* Loisel. Aquat. Bot. **43**:149-161.

Brix, H. and B. K. Sorrell. 1996. Oxygen stress in wetland plants: comparison of de-oxygenated and reducing root environments. Functional Ecology **10**:521-526.

Broome, S. W., W. W. Woodhouse and E. D. Seneca. 1975. The relationship of mineral nutrients to growth of *Spartina alternaiflora* in North Carolina: II. The effects of N, P, and Fe fertilizers. Soil Science Society of America Journal **39**:301-307.

Cantilli, J. F. 1989. Sulfide phytotoxicity in tidal salt marshes. Thesis, San Diego State University, San Diego, California, USA.
Carlson, P. R. and J. Forrest. 1982. Uptake of dissolved sulfide by *Spartina alterniflora*: evidence from natural sulfur isotope abundance ratios. Science. **216**:633-635.
Cavalieri, A. J. and A. H. C. Huang. 1981. Accumulation of proline and glycinebetaine in *Spartina alterniflora* Loisel. in response to NaCl and Nitrogen in the marsh. Oecologia **49**:224-228.
Chapman, V. J. 1978. The salt marsh environment. Pages 129149 *in* W. R. Waldrop, editor. Coastal Vegetation. Pergamon Press, Oxford, England.
Craft, C. B., E. D. Seneca and S. W. Broome. 1991. Porewater chemistry of natural and created marsh soils. Journal of Experimental Marine Biology and Ecology **152**:187-200.
Dacey, J. W. H. and B. L. Howes. 1984. Water uptake by roots controls water table movement and sediment oxidation in short *Spartina* marsh. Science **224**:487-489.
Dawes, C. J. 1998. Marine Botany. New York, John Wiley and Sons, Inc., New York, New York, USA.
DeLaune, R. D., S. R. Pezeshki and J. H. Pardue. 1990. An oxidation-reduction buffer for evaluating the physiological response of plants to root oxygen stress. Environmental and Experimental Botany **30**:243-247.
DeLaune, R. D., C. J. Smith and W. H. Patrick, Jr. 1983. Relationship of marsh elevation, redox potential and sulfide to *Spartina alterniflora* productivity. Soil Science Society of America Journal **47**:930-935.
Drake, B. G. and J. L. Gallagher. 1984. Osmotic potential and turgor maintenance in *Spartina alterniflora* Loisel. Oecologia **62**:368-375.
Engler, R. M. and W. H. Patrick, Jr. 1975. Stability of sulfides of manganese, iron, zinc, copper, and mercury in flooded and nonflooded soil. Soil Science. **119**:217-221.
Epstein, E. 1972. Mineral Nutrition of Plants. New York, John Wiley and Sons, Inc., New York, New York, USA.
Furtig, K., A. Ruegsegger, C. Brunold and R. Brandle. 1996. Sulphide utilization and injuries in hypoxic roots and rhizomes of common reed (*Phragmites australis*). Folia Geobotica Phytotaxonomica **31**:143-151.
Gallagher, J. L. 1975. Effect of an ammomium nitrate pulse on the growth and elemental composition of natural stands of *Spartina alterniflora* and *Juncus roemerianus*. American Journal of Botany **62**:644-648.
Gallagher, J. L., R. J. Reimold, R. A. Linthurst and W. J. Pfeiffer. 1980. Aerial production, mortality, and mineral accumulation-export dynamics in *Spartina alterniflora* and *Juncus roemerianus* plant stands in a Georgia salt marsh. Ecology **61**:303-312.
Gambrell, R. P., R. D. DeLaune and W. H. Patrick, Jr. 1991. Redox processes in soils following oxygen depletion. Pages 101-117 *in* Jackson, M. B., editor. Plant life under oxygen deprivation. M. B. Jackson. SPB Academic Publishing, Hauge, The Netherlands.
Gambrell, R. P. and W. H. Patrick, Jr. 1978. Chemical and microbiological properties of anaerobic soils and sediments. Pages 375425 *in* D. D. Hook and R. M. M. Crawford, editors. Plant life in anaerobic Environments. Ann Arbor Science, Ann Arbor, Michigan, USA.
Giblin, A. E., G. W. Luther, III, and I. Valiela. 1986. Trace metal solubility in salt marsh sediments contaminated with sewage sludge. Estuarine, Coastal and Shelf Science **23**:477-498.
Giurgevich, J. R. and E. L. Dunn. 1979. Seasonal patterms of CO_2 and water vapor exchange of the tall and short height forms of *Spartina alterniflora* Loisel in a Georgia salt marsh. Oecologia **43**:139-156.
Goodman, P. J. and W. T. Williams. 1961. Investigations into "die-back" in *spartina* townsendii agg. III. physiological correlates of "die-back". Journal of Ecology **49**:391-398.
Gosselink, J. G., E. P. Odum and R. M. Pope. 1973. The value of the tidal marsh. URDC University of Florida. Gainesville, Florida, USA.
Haines, B. L. and E. L. Dunn. 1976. Growth and resource allocation responses of *Spartina alternaiflora* Loisel to three levels of NH_4-N, Fe, and NaCl in solution culture. Botanical Gazettte **137**:224-230.
Havill, D. C., A. Ingold and J. Pearson. 1985. Sulfide tolerence in coastal halophytes. Vegetatio **62**:279-285.
Hester, M. W., I. A. Mendelssohn and K. L. McKee. 1998. Intraspecific variation in salt tolerance and morphology in *Panicum hemitomon* and *Spartina alterniflora* (Poaceae). International Journal of Plant Science **159**:127-138.
Howarth, R. W., A. Giblin, J. Gale, B. J. Peterson and G. W. Luther, III. 1983. Reduced sulfur compounds in the pore waters of a New England salt marsh. Ecological Bulletin **35**:135-152.

Howes, B. L., J. W. H. Dacey and D. D. Goehringer. 1986. Factors controlling the growth form of *Spartina alterniflora*: feedbacks between above-ground production, sediment oxidation, nitrogen and salinity. Journal of Ecology **74**:881-898.

Howes, B. L., R. W. Howarth, T. J. M. and I. Valiela. 1981. Oxidation-reduction potentials in a salt marsh: Spatial patterns and interactions with primary production. Limnology and Oceanography **26**:350-360.

Howes, B. L. and J. M. Teal. 1994. Oxygen loss from *Spartina alterniflora* and its relationship to salt marsh oxygen balance. Oecologia **97**:431-438.

Hwang, Y.H. and J. T. Morris. 1991. Evidence for hygrometric pressurization in the internal gas space of *Spartina alterniflora*. Plant Physiology **96**:166-171.

Hwang, Y.-H. and J. T. Morris. 1994. Whole-plant gas exchange responses of Spartina alterniflora (Poaceae) to a range of constant and transient salinities. American journal of botany **81**:659-665.

—— 1992. Fixation of inorganic carbon from different sources and its translocation in *Spartina alterniflora* Loisel. Aquatic Botany **43**:137-147.

Ingold, A. and D. C. Havill. 1984. The influence of sulphide on the distribution of higher plants in salt marshes. Journal of Ecology **72**:1043-1054.

Joshi, M. M., I. K. A. Ibrahim and J. P. Hollis. 1975. Hydrogen sulfide: effects on the physiology of rice plants and relation to straighthead disease. Phytopathology. **65**:1165-1170.

Kaswadji, R. F., J. G. Gosselink and R. E. Turner. 1990. Estimation of pirmary production using five different methods in a *Spartina alterniflora* salt marsh. Wetlands Ecology and Management **1**:57-64.

King, G. M., M. J. Klug, R. G. Wiegert and A. G. Chalmers. 1982. Relation of soil water movement and sulfide concentration to *Spartina alterniflora* production in a Georgia salt marsh. Science **218**:61-63.

Kirby, C. J. and J. Gosselink. 1976. Primary production in a Louisiana Gulf Coast *Spartina alterniflora* marsh. Ecology **57**:1052-1059.

Koch, M. S. and I. A. Mendelssohn. 1989. Sulphide as a soil phytotoxin: differential responses in two marsh species. Journal of Ecology **77**:565-578.

Koch, M. S., I. A. Mendelssohn and K. L. McKee. 1990. Mechanism for the hydrogen sulfide-induced growth limitation in wetland macrophytes. Limnology and Oceanography **35**:399-408.

Krone, R. B. 1985. Simulation of marsh growth under rising sea levels. Pages *in* W. R. Waldrop, editor. Hydraulics and hydrology in the small computer age. W. R. Waldrop Hydraulics Div. ASCE.

Lee, R.W., D.W. Kraus, and J.E. Doeller. 1999. Oxidation of sulfide by *Spartina alterniflora* roots. Limnology and Oceanography **44**:1155-1159.

Leeper, G. W. 1952. Factors affecting availibility of inorganic nutrients in soils with special reference to micro-nutrient metals. Annual review of Plant Physiology. **3**:1-16.

Levering, C. A. and W. W. Thomson. 1971. The ultrastructure of the salt gland of *Spartina foliosa*. Planta **97**: 183-196.

Linthurst, R. A. 1979. The effect of aeration on the growth of *Spartina alterniflora* Loisel. American Journal of Botany **66**:685-691.

Linthurst, R. A. 1980. An evaluation of aeration, nitrogen, pH and salinity as factors affecting *Spartina alterniflora* growth: a summary. Pages 235247 *in* V. S. Kennedy, editor. Estuarine perspectives. Academic Press, New York, New York, USA.

Linthurst, R. A. and R. J. Reimold. 1978. An evaluation of methods for estimating the net aerial primary productivity of estuarine angiosperms. Journal of Applied Ecology **15**:919-931.

Luther, G. W., III, T. M. Church, J. R. Scudlark and M. Cosman. 1986. Inorganic and organic sulphur cycling in salt-marsh pore waters. Science. **232**:746-749.

Masaoka, Y., M. Kojima, S. Sugihara, T. Yoshihara, M. Koshino and A. Ichihara. 1993. Dissolution of ferric phosphate by alfalfa (*Medicago sative* L.) root exudates. Plant and Soil Science **155/156**:75-78.

Mayne, R. G. and H. Kende. 1986. Glucose metabolism in anaerobic rice seedling. Plant Science **45**:31-36.

McKee, K. L., I. A. Mendelssohn and M. W. Hester. 1988. Reexamination of pore water sulfide concentrations and redox potentials near the aerial roots of *Rhizophora mangle* and *Avicennia germinans*. American Journal of Botany **75**:1352-1359.

Mendelssohn, I. A. 1979a. The influence of nitrogen level, form, and application method on the growth response of *Spartina alterniflora* in North Carolina. Estuaries **2**:106-112.

——. 1979b. Nitrogen metabolism in the height forms of *Spartina alterniflora* in North Carolina. Ecology **60**:574-584.

Mendelssohn, I. A. and K. L. McKee. 1983. Root metabolic response of *Spartina alterniflora* to hypoxia. Pages 239253 *in* Crawford, R. M. M., editor. Plant life in aquatic and amphibious habitats. British Ecological Society, Special Publication No. 5.

——. 1988. *Spartina alterniflora* die-back in Louisiana: time course investigation of soil water-logging effects. Journal of Ecology **76**:509-521.

Mendelssohn, I. A. and K. L. McKee. 1992. Indicators of environmental stress in wetland plants. Pages 603624 *in* D. H. McKenzie, D. E. Hyatt and V. J. McDonald, editors. Ecological indicators. Elsevier Applied Science, New York, New York, USA.

Mendelssohn, I. A., K. L. McKee and W. H. Patrick, Jr. 1981. Oxygen deficiency in *Spartina alterniflora* roots: metabolic adaptation to anoxia. Science **214**: 439-441.

Mendelssohn, I. A., K. L. McKee and M. T. Postek, editors. 1982. Sublethal steresses controlling *Spartina alterniflora* productivity. Wetlands: ecology and management. International Scientific Publications, Jaipur, India.

Mendelssohn, I. A. and M. T. Postek. 1982. Elemental analysis of deposits on the roots of *Spartina alterniflor*a Loisel. American Journal of Botany **69**:904-912.

Mendelssohn, I. A. and E. D. Seneca. 1980. The influence of soil drainage on the growth of salt marsh cordgrass *Spartina alterniflora* in North Carolina. Estuarine, Coastal and Marine Science **2**:27-40.

Mitsch, W. J. and J. G. Gosselink. 1993. Wetlands. Van Nostrand Reinhold, New York, New York, USA.

Morris, J. T. 1984. Effects of oxygen and salinity on ammonium uptake by *Spartina alterniflora* Loisel and *Spartina patens* (Aiton) Muhl. Journal of Experimental Marine Biology and Ecology **78**:87-98.

——. 1988. Pathways and controls of the carbon cycle in salt marshes. Pages 497-510 *in* W. H. M. D. D. Hook, Jr., H. K. Smith, J. Gregory, V. G., J. Burrell, M. R. Voe, R. E. Sojka, S. Gilbert, R. Banks, L. H. and C. B. Stolzy, T. D. Matthews, and T. H. Shear, editors. The Ecology and management of wetlands, Volume 1: Ecology of wetlands. Croom Helm Ltd., Breckenham, England.

——. 1995. The mass balance of salt and water in intertidal sediments: results from North Inlet, South Carolina. Estuaries **18**:556-567.

Morris, J. T. and J. W. H. Dacey. 1984. Effects of O_2 on ammonium uptake and root respiration by *Spartina alterniflora*. American Journal of Botany **71**: 979-985.

Morris, J. T., C. Haley and R. Krest. 1996. Effects of sulfide concentrations on growth and dimethylsulphoniopropionate (DMSP) concentration in *Spartina alterniflora.* Pages 87-95 *in* R. Kiene, P. Visscher, M.Keller and G. Kirst, editors. Biological and environmental chemistry of DMSP and related sulfonium compounds. Plenum, Press New York, New York, USA.

Morris, J. T. and B. Haskin. 1990. A 5-yr record of aerial primary production and stand characteristics of *Spartina alterniflora*. Ecology 7:2209-2217.

Morris, J. T., B. Kjerfve and J. M. Dean. 1990. Dependence of estuarine productivity on anomalies in mean sea level. Limnology and Oceanography **35**:926-930.

Naidoo, G., K. L. McKee and I. A. Mendelssohn. 1992. Anatomical and metabolic responses to waterlogging and salinity in *Spartina alterniflora* and *S. patens* (Poaceae). American Journal of Botany **79**:765-770.

Odum, E. P. 1961. The role of tidal marshes in estuarine production. The New York State Conservationist **29**:60-64.

Okajima, H. and S. Takagi. 1953. Physiological behavior of hydrogen sulfide in the rice plant. Part I: Effect of hydrogen sulfide on the absorption of nutrients. Science Reports of the Research Institutes, Tohoku University **5**:21-31.

Osgood, D. T. and J. C. Zieman. 1993. Spatial and temporal patterns of substrate physicochemical parameters in different-aged barrier island marshes. Estuarine, Coastal and Shelf Science **37**:421-436.

Parrondo, R. T., J. G. Gosselink and C. S. Hopkins. 1978. Effects of salinity and drainage on the growth of three salt marsh grasses. Botanical Gazette **139**:102-107.

Patrick, W. H., Jr. and R. D. DeLaune. 1976. Nitrogen and phosphorus utilization by *Spartina alterniflora* in a salt marsh in Barataria Bay, Louisiana. Estuarine, Coastal and Marine Science **4**:59-64.

Pattullo, J., W. Munk, R. Revelle and E. Strong. 1955. The seasonal oscillation in sea level. Journal of Marine Research **14**:88-156.

Pearson, J. and D. C. Havill. 1988. The effect of hypoxia and sulphide on culture-grown wetland and non-wetland plants. Journal of Experimental Botany **39**:363-374.

Phleger, C. F. 1971. Effect of salinity on growth of a salt marsh grass. Ecology **52**:908-911.

Ponnamperuma, F. N. 1972. The chemistry of submerged soils. Advances in Agronomy **24**:29-96.

——. 1977a. Behavior of minor elements in paddy soils. The International Rice Research Institute. Manila, Philippines. IRRI Research Paper Series No. 8.

——. 1977b. Physicochemical properties of submerged soils in relation to fertility. The International Rice Research Institute. Manila, Philippines. IRRI Research Paper Series 5.

Shew, D. M., R. A. Linthurst, and E. D. Seneca. 1981. Comparison of production computation methods in a southeastern North Carolina *Spartina alterniflora* salt marsh. Estuaries **4**:97-109.

Smalley, A. E. 1960. Energy flow of a salt marsh grasshopper population. Ecology. **41**:785-790.

Smart, R. M. 1982. Distribution and environmental control of productivity and growth form noital of *Spartina alterniflora* (Loisel). Tasks for Vegetation Science **2**:127-142.

Steever, E. Z., R. S. Warren and W. A. Niering. 1976. Tidal energy subsidy and standing crop production of *Spartina alterniflora*. Estuarine, Coastal and Marine Science **4**:473-478.

Sullivan, M. J. and F. C. Daiber. 1974. Response in production of cord grass, *Spartina alterniflora*, to inorganic nitrogen and phosphorus fertilizer. Chesapeake Science **15**:121-123.

Teal, J. M. and J. W. Kanwisher. 1966. Gas transport in the marsh grass, *Spartina alterniflora*. Journal of Experimental Botany **12**:355-361.

Turner, F. T. and W. H. Patrick, Jr. 1968. Chemical changes in waterlogged soils as a result of oxygen depletion. Transactions of the 9th International Congress of Soil Science, International Society of Soil Science and Angus and Robertson, Ltd., **4**:53-56, Sidney, Australia.

Turner, J. S. 1960. Fermentation in higher plants; its relation to respiration; the Pasteur effect. Springer-Verlag, Berlin, Germany.

Turner, R. E. 1976. Geographic variations in salt marsh macrophyte production: a review. Contributions in Marine Science **20**:47-68.

Turner, R. E. 1977. Intertidal vegetation and commercial yields of penaeid shrimp. Transactions of the American Fisheries Society **106**:411-416.

Valiela, I. 1995. Marine Ecological Processes. Springer-Verlag, New York, New York, USA.

Valiela, I. and J. M. Teal. 1974. Nutrient limitation in salt marsh vegetation. Pages 547563 *in* R. J. Reimold and W. H. Queen, editor. Ecology of halophytes. New York Academic Press, New York, New York, USA.

Valiela, I. and J. M. Teal. 1979. The nitrogen budget of a salt marsh ecosystem. Nature **280**:652-656.

Van Diggelen, J., J. Rozema and R. Broukman. 1987. Pages 260-268 *in* A. H. L. Huiskes, C. W. P. M. Blom and J. Rozema , editors. Vegetation between land and sea. W. Junk Publishers, Hauge, The Netherlands.

Van Diggelen, J., J. Rozema, D. M. J. Dickson and R. Broekman. 1986. Beta-3-dimethylsulphoniopropionate, proline and quaternary ammonium compounds in *Spartina anglica* in relation to sodium chloride, nitrogen and sulphur. New Phytologist **103**:573-586.

Vartapetyan, B. B. 1982. Anaerobiosis and the theory of physiological adaptation of plants to flooding. Soviet Plant Physiology **29**:764-771.

Webb, J. W. 1983. Soil water salinity variations and their effects on *Spartina alterniflora*. Contributions in Marine Science **26**:1-13.

Wiegert, R. G., A. G. Chalmers, and P. F. Randerson. 1983. Productivity gradients in salt marshes: the response of *Spartina alterniflora* to experimentally manipulated soil water movement. Oikos **41**:1-6.

COMMUNITY STRUCTURE AND FUNCTIONAL DYNAMICS OF BENTHIC MICROALGAE IN SALT MARSHES

MICHAEL J. SULLIVAN
Mississippi State University
Biology Department, P.O. Box GY
Mississippi State, MS 39762 USA

CAROLYN A. CURRIN
NOAA National Ocean Service
Center for Coastal Fisheries and Habitat Research
101 Pivers Island Rd.
Beaufort, NC 28516 USA

Abstract

Benthic microalgae are a ubiquitous feature in sediments directly exposed to full sunlight or shaded by a vascular plant canopy in coastal salt marshes. Diatoms, cyanobacteria, and green algae are the dominant groups. Of these, diatoms are universally present and abundant, exhibit migratory rhythms driven mainly by light, and are by far the taxonomically most diverse group. Dense mats of cyanobacteria and secondarily green algae frequently develop where light levels are high. The more abundant species of all three algal groups are widely distributed within and among salt marshes of the United States and Europe. Standing crops of benthic microalgae beneath various vascular plant canopies exhibit mean annual values of 60 to 160 mg chl *a* m^{-2}. Annual benthic microalgal production (BMP) has been shown to range from 28 g C m^{-2} y^{-1} beneath *Juncus roemerianus* to 314 g C m^{-2} y^{-1} beneath *Jaumea carnosa*. In general, BMP increases in a southerly direction in Atlantic coast marshes but is lowest in Gulf Coast marshes. In Atlantic and southern California marshes a significant portion of benthic microalgal production occurs when the overstory vascular plants are dormant. Experimental manipulations have shown that BMP and biomass beneath *Spartina alterniflora* are limited by nitrogen supplies and grazing activities. Manipulation of light appears to primarily affect the relative dominance of diatoms and cyanobacteria in the benthic microalgal assemblage. The ratio of annual BMP to net aerial production of the overstory vascular plant canopy is 10 to 60% in Atlantic and Gulf Coast marshes and 75 to 140% in a southern California marsh. The benthic microalgal portion of this two component productivity system has been shown by multiple stable isotope studies to be a major component of salt marsh food webs. Diatoms, in particular, are the preferred food item of a diverse array of invertebrate and fish species.

1. Introduction

The first image that comes to mind for most researchers when discussing coastal salt marshes is probably an extensive sward of *Spartina alterniflora* gently waving in the breeze. While such an image is not a floristically incorrect one, it is certainly a most incomplete one. Sedimentary environments on the marsh proper and those forming the sloping banks of numerous tidal creeks are typically referred to as ''unvegetated'' when vascular plants are absent. The use of this adjective is not only misleading but also inaccurate because a diverse assemblage of cyanobacteria and eukaryotic algae thrive within such sediments, as well as within ''vegetated'' sediments where the marsh proper may be shaded by a dense spermatophyte canopy. This assemblage, often called the microphytobenthos, has been referred to as a ''secret garden'' by Miller et al. (1996) and MacIntyre et al. (1996).

This chapter will review the ''secret garden'' of salt marshes but use the moniker benthic microalgae rather than microphytobenthos to emphasize the algal nature of the ''crop''. By the term ''microalgae'' is meant those organisms which contain chlorophyll *a*, evolve O_2 during photosynthesis, and whose usual dimensions are given in micrometers (μm). This last attribute cannot be precisely defined in a quantitative sense but is meant to designate those algae whose structure can only be discerned with a compound light microscope; therefore, the transition from ''micro-'' to ''macroalgae'' is largely an operational definition. By the term ''benthic'' is meant unless otherwise specified those microalgal species associated with salt marsh sediments with the following adjectives being most commonly employed in the literature: *edaphic* (general term describing microalgae within or associated with intertidal sediments), *epipelic* (microalgae that migrate up and down within intertidal sediments in response to irradiance and tidal rhythms), and *epipsammic* (microalgae attached more or less firmly to sand grains with motility absent or highly reduced). However, despite the fact that de Jonge (1985) defined 55 μm as the operational dividing line between mud (silt and clay) and sand particles, his observations that diatoms resided preferentially on the mud coatings of sand grains rather than on their bare parts led him to conclude that the descriptive terms ''epipelic'' and ''epipsammic'' were unsuitable for classifying the major components of the estuarine diatom flora. Therefore, the term ''edaphic'' will be used in this chapter when the sedimentary habitat of the benthic microalgae requires special emphasis.

2. The Benthic Microalgal Flora

The most ubiquitous component of the benthic microalgae in salt marsh sediments are the motile pennate diatoms, particularly those species belonging to the genera *Navicula, Nitzschia,* and *Amphora* (Williams 1962, Drum and Webber 1966, Sullivan 1975, 1977a, 1978). The identification of resident species is a daunting task even for the expert because of the great number of diatom taxa that have been described, the considerable morphological variability of individual taxa, and the complete absence of any monographic works for marine benthic diatoms. Although the rewards and challenges of monographing these forms would be great, funding for such an endeavor is virtually impossible to secure.

The majority of salt marsh diatom species are small forms, typically less than 30 μm in length. Virtually all the actively photosynthesizing cells are found in the upper 1 to 3 mm of sediments (Admiraal 1984, Paterson 1986, Pinckney et al. 1994b). Williams (1962) recorded edaphic diatom standing crops as high as 6.34×10^3 cells mm^{-2} beneath tall *Spartina alterniflora* in a Georgia salt marsh. Unfavorable periods (e.g. low temperatures, desiccation of sediments) are survived as vegetative cells since resting spores have never been observed in salt marsh pennate diatoms (Williams 1962). Division rates are very high; Williams (1964) recorded maximum rates of 0.6 to 3.2 divisions day^{-1} for 14 species. All species he tested proved to be highly euryhaline (i.e., most grew well over a salinity range of 1 to 68 psu with a maximum at 20 psu); this would seem to be a prerequisite for survival and growth in salt marsh sediments which constantly dry out and are rehydrated by flooding tides.

The aforementioned diatom genera and others such as *Diploneis, Gyrosigma, Pleurosigma, Caloneis,* and *Bacillaria* are all biraphid taxa with variously placed slits in their siliceous cell walls called raphes. The protoplast secretes mucilaginous strands into the raphe slits for both attachment and motility purposes. Once a diatom is attached to a sediment particle or some other substrate via mucilaginous strands, a transmembrane transport of these strands mediated by actin microfilaments is thought to occur (Edgar and Pickett-Heaps 1984). Whatever the true mechanism of diatom motility eventually proves to be, the end result is spectacular in that sunrise sees the transformation of a dull-colored sediment into one that is golden-brown (due to the dominant photosynthetic pigment in diatoms, fucoxanthin). This migration of biraphid pennate diatoms is driven mainly by light rather than tidal rhythms in salt marshes (Pomeroy 1959, Leach 1970, McIntire and Moore 1977). For details of diatom migration see Paterson (1986), who used low temperature scanning electron microscopy to produce three-dimensional images of intact benthic diatom assemblages within and on the surface of intertidal marine sediments. Different diatom species appeared at different times after sunrise, suggesting that light acts as a migration trigger and the various species have varying threshold irradiances. The copious secretion of mucilage by adhering and moving diatoms stabilizes salt marsh sediments and may pave the way for vascular plant colonization of newly created marsh habitat (see review by Sullivan 1999).

An unusual means of motility was observed by Williams (1965). Tube-dwelling diatoms, such as *Nitzschia obtusa* and *Frustulia asymmetrica*, were capable of retreating rapidly into salt marsh sediments in response to a mechanical stimulus or intense light. The movement far exceeded the rate measured for nontube-dwelling diatoms and was thought by Williams to be a possible adaption to reduce grazing pressure by fiddler crabs. Jönsson et al. (1994) recorded a similar type of motility for the large diatom *Gyrosigma balticum* whereby the individual cells formed a forest-like canopy above the sediments, being held in an upright position by a short mucilaginous tube at the base of the cell. The peak of the oxygen profile recorded by microelectrodes occurred at 250 to 300 μm above the sediment surface within the thicket of *Gyrosigma* cells. Such a habit would have clear advantages when population densities are high enough to induce self-shading within the benthic microalgal assemblage or to interfere with the diffusion of CO_2 into and O_2 out of the microalgal mat (see Admiraal et al. 1982).

Monoraphid (e.g., *Achnanthes, Cocconeis*) and araphid (e.g., *Opephora, Tabularia*) pennate diatom genera may be well represented in salt marsh sediments (Sullivan 1975, 1977a, 1978), but rarely are such forms more abundant than the biraphid taxa. Despite limited motility, these diatom genera can assure their permanence in the benthic microaglal assemblage through attachment via mucilage to large sediment particles and decaying fragments of vascular plants at the sediment surface.

Diatoms may be found on the lower 30 cm of living *Spartina alterniflora* leaves (Stowe 1982). Stowe (1980) found densities as high as 6.9×10^5 cells cm^{-2} in a Louisiana salt marsh. The red macroalgae *Bostrychia radicans* and *Caloglossa leprieurii* are frequently found attached to dead *S. alterniflora* stems in lower elevations of the marsh (Chapman 1971). A diverse epiphytic diatom flora may be found on both red algal species (Sullivan 1982a). In general, the diatom species epiphytic on *S. alterniflora* leaves and secondarily on epiphytic red algae are the same ones as those resident within the sediments below. Extensive growths of diatoms may occur on submerged aquatic plants in pools on the marsh proper such as the widgeon grass *Ruppia maritima*; the major source of colonizing diatoms also appears to be the marsh sediments (Sullivan 1977b).

Whereas large populations of edaphic diatoms are invariably present in salt marshes, cyanobacteria (formerly called blue-green algae, Cyanophyceae or Myxophyceae) are more variable in their degree of development. Extensive cyanobacterial populations develop in the salt marshes of New England (Blum 1968, Webber 1967, 1968) and Europe (Carter 1932, 1933a,b, Polderman 1975, 1978, Polderman and Polderman-Hall 1980) and the hypersaline marshes of southern California (Zedler 1980, 1982); such development typically occurs during summer although cyanobacteria are present year-round. The life forms are many but include coccoid (*Chroococcus, Anacystis*), colonial (*Merismopedia*), and particularly filamentous genera with (*Anabaena, Nostoc, Calothrix*) and without (*Oscillatoria, Lyngbya, Microcoleus, Phormidium, Schizothrix*) heterocysts. Several forms fix atmospheric nitrogen and thus may be very important to the nitrogen cycle in salt marshes where they are abundant (Currin and Paerl 1998a,b). The filamentous species may form thick mats several mm thick on the marsh sediments and grow up living vascular plants stems for distances of 5 to 10 cm (Blum 1968, Zedler 1980), although Currin and Paerl (1998a) reported cyanobacterial growths on the lower 30 cm of standing dead *Spartina alterniflora* stems. A major problem is the lack of a modern "flora" for the marine cyanobacteria and the paucity of taxonomists familiar with this group. Another major problem is a dichotomy between those that follow the Drouet system (extensive lumping of species) and those that follow the Geitler system (splitters where simple differences in filament width may justify the erection of new species). Hopefully, the much anticipated publication of the cyanobacterial volumes in *Süsswasserflora von Mitteleuropa* will do much to find a realistic median between these two positions.

Three groups of eukaroytic algae in which chlorophyll *a* completely masks the carotenoid pigments within the grass-green chloroplasts may be sporadically abundant in salt marshes. The xanthophyte (i.e., a yellow-green alga but a misnomer in this case) genus *Vaucheria* is commonly found on salt marsh sediments beneath spermatophyte canopies and on creekbanks devoid of vascular plants (Carter 1932, 1933a,b, Blum 1968, Webber 1968, Polderman 1975, 1978, Polderman and Polderman-Hall 1980). This coenocytic, variably branched alga forms thick felts anchored in the marsh sediments by colorless rhizoids. Several species are involved but their taxonomy is poorly understood. Euglenoids may be locally abundant in salt marshes but such blooms typically involve a single species such as *Euglena limosa* (Carter 1932, 1933a,b, Williams 1962). Finally, the green algae, which belong to Division Chlorophyta, may form extensive growths on salt marsh sediments and up the culms of the spermatophytes. By far the most common genera are *Rhizoclonium* and *Enteromorpha* (Carter 1932, 1933a,b, Blum 1968, Webber 1968, Polderman 1975, 1978, Polderman and Polderman-Hall 1980, Zedler 1980), although the latter is probably better viewed as a macro- rather than a microalga. The maximum development of filamentous green algae in salt marshes is during the winter months. *Rhizoclonium* species have been revised for the New England coast (Blair 1983) but systematic treatments for both green algal genera are lacking for broad geographical areas. Fallon et al. (1985) have reported that the prostrate, pseudofilamentous green alga *Pseudendoclonium submarinum* is the dominant alga found on standing dead *Spartina alterniflora* leaves in Sapelo Island salt marshes. They termed this association "phylloplane algae" and provided density data showing that such algae are far outnumbered by their edaphic counterparts on the marsh surface.

3. Community Structure of the Benthic Microalgae

The first bona fide study of benthic microalgal community structure in salt marshes was published by Carter in three separate papers (1932, 1933a,b). Edaphic algae were sampled from various zones of an English and a Welsh marsh, which were defined by the presence or absence of various spermatophyte species. Striking differences were noted between the algal and spermatophyte floras of the two marshes, and more diatom taxa (80+) were collected than all other algal species combined (63). The zonation of spermatophytes was found to be reflected to a slight degree in the benthic microalgal flora and periodicity was only striking in the zones of lowest elevation where spermatophytes were absent. Diatom distribution and that of other resident algal groups was described by a plant sociological approach where 12 communities (e.g., marginal diatom community, autumn Cyanophyceae community) were defined somewhat arbitrarily on the basis of dominant algal taxa and related to tidal levels. However, no physical factors were quantified and hence correlations between biological and physical data could only be surmised. Carter concluded that physical, rather than biological factors, largely controlled benthic microalgal distribution on the marshes under investigation.

The most informative study of European salt marsh diatoms was conducted by Round (1960) in an English marsh on the north shore of the River Dee where it empties into the Irish Sea. Twelve stations were defined and sediment samples were collected from high, middle, and low marsh in different vegetational zones. Round found that the edaphic diatom flora was dominated by euryhaline taxa typically associated with saline waters and that marine or freshwater planktonic diatoms were extremely scarce. A change in the flora was noted along a transect from low to high marsh as well as a distinct seasonal growth of the diatom assemblage as a whole at each sampling station. Distinct diatom associations in the different areas of the marsh were seen but less obvious was the restriction of certain species to definite seasons of the year. Although some interesting distributional patterns were observed, the absence of quantitative physical data only allows the reader to speculate as far as explanation of the patterns is concerned.

Williams (1962) conducted the first comprehensive study of salt marsh diatoms in the United States; however, most of this classic work in Sapelo Island marshes unfortunately was never formally published. He was unable to study community structure because he only identified the larger taxa, and according to Williams these represented less than 20% of the total number of species present. He estimated that smaller cells composed 80 to 96% of the population by number, which is consistent with personal observations of the first author made in a diversity of salt marshes. Williams made the very important observation that since diatoms of all species are constantly distributed over the marsh surface by tidal action, their absence from any region indicates an inability to survive there. This is far from a trivial point and one that all who study benthic microalgal distribution in any habitat should heed well.

The benthic microalgae of salt marshes were largely ignored following Williams' work until the mid-1970s, when a series of publications on edaphic diatoms was begun by the first author. The earliest papers by Sullivan (1975, 1977a, 1978) are descriptive accounts of edaphic diatom community structure in salt marshes of Delaware, New Jersey, and Mississippi, respectively. Sediment cores were taken beneath the canopies of the dominant spermatophytes as well as from creekbanks and salt pannes over an annual cycle. Diversity (as measured by Shannon's H' and the number of taxa in a sample S) was lowest in the open areas of the marsh: creekbanks with virtually all diatoms, and salt pannes with thick mats of cyanobacteria and the tube-forming diatom *Nitzschia obtusa* var. *scalpelliformis*. The highest H' and S values were found for the diatom assemblage beneath the canopy of *Spartina patens* in New Jersey, with maxima of 5.206 bits individual^{-1} (logs taken to the base 2) and 85 taxa in a sample, respectively. With few exceptions the more abundant taxa were widely distributed over the entire marsh surface; however, based on species/numbers relationships, associations of dominant taxa, and the restriction of some taxa to certain habitat(s) of the marsh, it was possible to recognize distinct diatom assemblages associated with each spermatophyte zone, much as Carter (1932, 1933a,b) and Round (1960) had previously done. Multiple regression showed that structural differences among the assemblages beneath different marsh spermatophytes (*Spartina alterniflora, S. patens, Distichlis spicata, Juncus roemerianus, Scirpus olneyi*), as quantified by a similarity index, were related to differences in elevation, soil temperature, moisture content of surface sediments, ammonium concentrations in the interstitial water, canopy height, and interactions between diatoms and cyanobacteria in warmer months and green algae in cooler months.

Based on the large number of species in common among Delaware, New Jersey, and Mississippi marshes, Sullivan (1978) hypothesized that "further work in salt marshes may reveal the existence, within as yet undefined limits, of a single basic edaphic diatom community indigenous to Atlantic and Gulf coast salt marshes." This may be the case for cyanobacteria in temperate North Atlantic salt marshes as originally suggested by Ralph (1977), and subsequently supported by studies of Sage and Sullivan (1978) and Maples (1982) in Mississippi and Louisiana, respectively. When coherent mats are produced, especially during summer, the dominant mat-formers are usually *Schizothrix calcicola, S. arenaria,* and *Microcoleus lyngbyaceus* (Ralph 1977, Sage and Sullivan 1978). These same three species were the major mat formers during summer in a California salt marsh studied by Zedler (1980, 1982), whereas the nitrogen-fixing species *Nostoc spumigena* largely replaced *M. lyngbyaceus* in Maple's (1982) study although the latter taxon was still abundant. Warm temperatures and high irradiance levels appear to be most conducive to cyanobacterial mat formation. These mats are capable of surviving considerable desiccation stress; Sullivan (1975) measured salinities as high as 185 psu within a "healthy" cyanobacterial mat covering a salt panne.

In returning to Sullivan's (1978) hypothesis that a single, basic diatom community exists in Atlantic and Gulf Coast salt marshes, two studies are relevant. Cook and Whipple (1982) identified 112 edaphic diatom taxa along gradients of salinity, tidal flushing, and sedimentary organic matter content in a Louisiana salt marsh. The more abundant taxa had a continuous distribution along this gradient but unfortunately no statistical correlations between the species and environmental data were made. However, species/numbers relationships (*H'* and *S*) and the species composition of the diatom flora were very similar to those reported by Sullivan (1975, 1977a, 1978), adding support to the hypothesis. Otte and Bellis (1985) compared the edaphic diatom flora of a North Carolina marsh with those studied by Sullivan (1975, 1977a, 1978) and Cook and Whipple (1982), and concluded their flora was distinct from the other four. It should be pointed out, however, that they only sampled from August to October and the nearest ocean inlet was 73 km from their brackish marsh. What can one then conclude from these five studies? Since the number of diatom taxa far exceeds that of cyanobacteria and dominance by a single taxon is much less, one should expect modifications of the "basic" diatom assemblage proposed by Sullivan (1978) to exist in different marshes. These modifications would be related to the types of spermatophyte species present and their influence on the sedimentary microenvironments. As pointed out by the data of Otte and Bellis (1985), the proximity of a marsh to high salinity coastal waters is also an important modifying factor.

Whereas diatoms are typically the dominant component of the benthic microalgae in Atlantic and Gulf Coast salt marshes, cyanobacteria are dominant in the hypersaline marshes of southern California (Zedler 1980). The most extensive development of cyanobacterial mats occurs in the lower elevations beneath the tall, dense canopy of *Spartina foliosa* and the short, open canopy of the succulent *Jaumea carnosa*. At higher elevations cyanobacteria thrive beneath the open canopy of the succulent *Batis maritima* and the dense canopy of the grass *Monanthochloe littoralis*. Based on frequency of occurrence (rather than cell counts) over an annual cycle, Zedler (1982) recorded a total of 83 species of benthic microalgae: 7 cyanobacteria, 2 greens, and 74 diatoms although only the names of 32 diatom taxa are listed. Cyanobacteria were most frequent in summer and the greens and diatoms during the cooler seasons, which agrees with observations made by Sullivan (1975) in a Delaware marsh. A chi square analysis of the frequency data for the 38 most common taxa (4 cyanobacteria, 2 greens, 32 diatoms) revealed that 37 and 36 taxa exhibited significant spatial and temporal patterns, respectively. Sørensen's similarity index was used to make all possible comparisons between the 4 spermatophyte zones and the 6 values generated ranged from 55 to 70%, indicating considerable overlap in the species composition of the benthic microalgal assemblages along the elevational gradient and beneath the different spermatophyte zones. Additional studies in salt marshes along the Pacific coast of North America are needed before any meaningful structural comparisons can made between these assemblages and their counterparts along the Atlantic and Gulf coasts.

All of the preceding studies, if statistical analysis of the data was a feature of the investigation, employed univariate statistics in an attempt to find patterns in what is a multivariate database. A given diatom assemblage is composed of several to many taxa and usually several environmental factors are measured simultaneously. Multivariate statistical methods have the potential for reducing such complex data sets into several dimensions with a minimal loss of information. These dimensions are orthogonal (uncorrelated) and biological interpretations can be made in the context of what is known about the ecosystem and the sampling strategy. McIntire (1973), working in the Yaquina Estuary, Oregon, was the first to employ multivariate statistical analyses to study the spatial and temporal distribution of marine benthic diatoms. Sullivan (1982b) first applied multivariate analysis to study the distribution of edaphic salt marsh diatoms. In this study, the database of Sullivan (1978) was subjected to a canonical correlation analysis because of the high degree of linearity characterizing the data. The database (i.e., the 26 most abundant taxa and 10 environmental variables) was collapsed into two interpretable dimensions and five distinct diatom communities were identified. Diatom distribution on Graveline Bay Marsh was primarily regulated by elevation and height of the spermatophyte canopy. Redundancy values for the first and second canonical variables were very similar to those obtained by McIntire (1978) and Amspoker and McIntire (1978) for epilithic and sedimentary diatom assemblages, respectively, in Yaquina Estuary. Striking horizontal and vertical gradients exist in Yaquina Estuary. However, such gradients are not present in salt marshes that may be described as a mosaic type of environment where the different spermatophyte zones exist as islands whose contiguous shorelines represent potential barriers to the distribution of diatom species. Canonical correlation allowed the insight that these barriers are somehow related to elevation and canopy height, and that the relative abundances of the various diatom species were affected differently by these factors in Graveline Bay Marsh.

A decade later, Sullivan and Moncreiff (1988*a*) sampled the same spermatophyte zones as did Sullivan (1978, 1982b) with the exception of the *Spartina patens* zone. As in previous studies, virtually all the more abundant diatom taxa were widely distributed over the marsh surface. Canonical correlation analysis collapsed the complex data base into two interpretable, orthogonal dimensions and identified benthic microalgal primary production and chlorophyll *a* and soil moisture as potentially related to the distribution of edaphic diatoms in Graveline Bay Marsh. Canopy height showed only weak correlations with CV1 and CV2 in contrast to the earlier data set (Sullivan 1982b) and elevation was not considered in this later study. Discriminant analysis showed the diatom assemblage beneath the canopy of *Scirpus olneyi* to be structurally most distinct and the derived functions employing the relative abundances of the 33 (out of a total of 155) most abundant diatom taxa as discriminating variables correctly classified 70% of all cases.

A comparison of the 1976-77 and 1985-86 diatom assemblages revealed that although species/numbers relationships had not changed, the taxonomic character of the flora and spatial distribution of the more abundant taxa exhibited marked changes. However, two major hurricanes directly impacted Graveline Bay Marsh during the latter study and the modifying effect of these events could not be separated out.

There has been only a handful of experimental studies in salt marshes where the subject has been the benthic microalgae and all have involved manipulation of irradiance and nutrients. Irradiance was varied by either clipping the spermatophyte canopy down to the mud-water interface or suspending shade cloths above the intact canopy, while nutrients were broadcast by hand on to the marsh surface in the form of various nitrogen- and phosphorus-containing fertilizers. Results due to clipping have been more consistent than those due to nutrient enrichment. Sullivan clipped a monotypic stand of dwarf *Spartina alterniflora* in a Delaware marsh (1976) and *Distichlis spicata* in a Mississippi marsh (1981). Species diversity (*H'*) and the number of diatom taxa in a sample (*S*) were greatly decreased. Clipping also caused pre-existing diatom taxa to be eliminated and introduced new diatom taxa into the assemblage. The removal of the protective spermatophyte canopy caused the surface sediments to become hypersaline (> 45 psu) and salinities as high as 150 psu were recorded. In the Delaware study macroscopic mats of filamentous cyanobacteria (and to a lesser extent greens) formed in the clipped plots but were conspicuously absent from clipped plots in the Mississippi study. Therefore, the filamentous algae did not take up the "slack" in productivity caused by removal of the highly productive grass canopy as is typical for Atlantic coastal marshes (Estrada et al. 1974, Sullivan and Daiber 1975). In addition, interactions (i.e., competition) between diatoms and filamentous cyanobacteria and green algae appear to be unimportant in at least the Gulf Coast marsh investigated. In the Delaware marsh, species/numbers relationships (*H'* and *S*) decreased moderately but significantly in treatments where the suspended shade cloths reduced irradiance reaching the top of the dwarf *S. alterniflora* canopy by 30%, but were unaffected by a 60% reduction in irradiance. Similarity comparisons of the experimental assemblages with those studied earlier by Sullivan (1975) in the same marsh showed that shading of dwarf *S. alterniflora* did not produce any shift in benthic microalgal assemblage structure towards that found beneath the denser canopies of tall *S. alterniflora* or *D. spicata*; therefore, light is not a factor accounting for structural differences among these three spermatophyte zones on this marsh. However, clipping the dwarf *S. alterniflora* canopy produced a pronounced shift in benthic microalgal assemblage structure towards that characteristic of a salt panne algal mat.

In a Massachusetts salt marsh, Van Raalte et al. (1976a) applied urea and sewage sludge containing 10% nitrogen (N) on a long-term basis to sediments populated by *Spartina alterniflora*. Both organic N enrichments significantly decreased species numbers/relationships (*H'* and *S*) of the edaphic diatom assemblages. The relative abundance of the biraphid diatom *Navicula salinarum* was 5 to 9% in control plots but increased to 20 to 25% in enriched plots. Sullivan (1976) studied this same microalgal assemblage beneath the same spermatophyte in a Delaware salt marsh but added inorganic nitrogen (NH_4NO_3) or phosphorus ($CaH_4(PO_4)_2$) instead of urea or sewage sludge. Phosphorus (P) enrichment significantly decreased both *H'* and *S* whereas N enrichment significantly decreased only *S* (however the decrease in *H'* was nearly statistically significant). Either N or P enrichment had a stimulatory effect on the relative abundance of *Navicula salinarum*; this agrees with work by Admiraal (1977) and Van Raalte et al. (1976a) for N enrichment. Experimental work by Underwood et al. (1998) has shown the potentially powerful role of ammonium-N in structuring the edaphic diatom assemblages of an English salt marsh. Moving south to a Mississippi salt marsh, Sullivan (1981) fertilized the marsh sediments beneath *Distichlis spicata* with NH_4Cl. Responses of the edaphic diatom assemblage were quite different in that N enrichment had virtually no effect; only the relative abundance of *Nitzschia perversa* was significantly affected (i.e., increased) by N enrichment. Comparison of control and enriched assemblages by a similarity index revealed that they shared between 73 and 95% of the maximum similarity possible over an annual cycle. In contrast to their Atlantic counterparts, the edaphic diatom assemblages of this Mississippi marsh may thus be largely ''resistant'' to (see Christian et al. 1978) or unaffected by additions of inorganic N. It should be noted that in all three studies cited above, the overstory grasses significantly increased their aboveground standing crops in response to N enrichment, and this may have been a confounding factor.

4. Biomass of the Benthic Microalgae

The biomass or standing crop of benthic microalgae is typically estimated by determining chlorophyll (chl) *a* concentrations, since this is the main photosynthetic pigment in all cyanobacteria and eukaryotic algae. Table 1 lists mean annual chl *a* concentrations in the sediments beneath overstory vascular plant canopies for two Atlantic and one Gulf Coast salt marsh. All studies used standard spectrophotometric methods except Sullivan and Moncreiff (1988*b*), who employed the hexane/acetone partitioning technique of Whitney and Darley (1979). Pinckney et al. (1994a) have shown that standard spectrophotometric methods overestimate chl *a* concentrations by 16% when compared to an HPLC extraction; however, the relationship between the two methods was constant. Annual mean values in Table 1 range from a low of 57 mg chl *a* m^{-2} beneath the *Distichlis spicata* canopy to a high of 160 mg chl *a* m^{-2} beneath the relatively open canopy of *Scirpus olneyi* in Mississippi. Values for the different height forms of *Spartina alterniflora* are remarkably similar over a broad geographical range and range from 73 to 127 mg chl *a* m^{-2}. Only one study has determined chl *a* concentrations for the benthic microalgae beneath *Juncus roemerianus* where the annual mean was at the low end of the scale (59 mg chl *a* m^{-2}). Unfortunately, we are aware of no published comparative data for Pacific coast salt marshes.

In a short-term (June to August) study, Piehler et al. (1998) measured chl *a* concentrations in transplanted and natural *Spartina alterniflora* marshes in North Carolina. Monthly means ranged from 32 to 58 and 10 to 24 mg chl *a* m^{-2} in the one year-old transplanted and natural marsh, respectively. These low values were probably a result of the passage of Hurricane Bertha over the study area on 12 July 1996. However, more extensive sampling of the transplanted marsh on 26 July produced chl *a* concentrations which ranged from 8 to 169 mg m^{-2}. A significant correlation was reported between chl *a* and organic N content in the upper 0.5 cm of marsh sediments.

In areas of the marsh devoid of vascular plants, chl *a* concentrations tend to be lower or higher than those found beneath the shading canopies. Gallagher (1971) measured an annual mean of 31 mg chl *a* m^{-2} for a creekbank adjacent to a stand of tall *Spartina alterniflora* in Delaware, while Pinckney and Zingmark (1993a) recorded a mean value of 77 mg chl *a* m^{-2} for a mudflat in a South Carolina salt marsh. The mean annual chl *a* concentration was 416 mg m^{-2} for a salt panne in Gallagher's (1971) study, where thick mats of cyanobacteria and to a lesser extent green algae were resident year-round. Monthly means ranged from 247 to nearly 800 mg chl *a* m^{-2} in this habitat.

In general, chl *a* concentrations beneath vascular plant canopies tend to peak in late winter/early spring (Gallagher 1971, Sullivan and Moncreiff 1988b, Pinckney and Zingmark 1993a). The situation is somewhat similar in "open" habitats as the bare bank and salt panne studied by Gallagher (1971) both had peak chl *a* concentrations in November.

Experimental manipulations carried out in the field have provided insight into the regulation of benthic microalgal biomasss in salt marshes by environmental and biological factors. Estrada et al. (1974) made use of the same plots fertilized by Van Raalte et al. (1976a,b) in Massachusetts. However, it is important to note that Estrada et al. sampled only during the summer months (July-September). Neither urea nor sewage increased chl *a* concentrations in the sediments beneath *Spartina alterniflora* (low marsh) or *S. patens*/*Distichlis spicata* (high marsh) because the added nitrogen (N) but not phosphorus (P) stimulated growth of the grasses, reducing light at the sediment surface to 1 to 12% of that incident at the canopy top. Sullivan and Daiber (1975) measured chl *a* concentrations beneath dwarf *S. alterniflora* utilizing the same plots and manipulations (i.e., different light levels crossed with inorganic N and P enrichment) as did Sullivan (1976) (see above). Nitrogen enrichment increased the standing crop of benthic microalgae only in spring whereas P enrichment produced increases in fall/winter and spring. This was basically consistent with known cycles of N and P availability in this Delaware marsh, except for N during summer. However, the experimental treatment consisting of a 30% reduction in light over the canopy and N enrichment inhibited the growth of the cord grass but increased the edaphic chl *a* concentration in the sediments below. Hence, this response and results of Estrada et al. (1974) provide good evidence that the benthic microalgae in these two northeastern marshes are severely limited by N supplies during summer; however, the dense canopy of grass produced by nutrient enrichment reduces light to levels where the exogenously added N cannot be fully utilized by the benthic microalgae. In the Delaware marsh, clipping the dwarf *S. alterniflora* canopy greatly increased chl *a* whereas shading the grass canopy (30% or 60% reduction in light) had no effect. A gradient in algal composition was produced: as light levels decreased from full sunlight through natural levels to those produced by 30% and 60% shade the diatom component of the benthic microalgal assemblage became more important. In all plots with 60% shade, virtually all living cells in the assemblage were diatoms. Darley et al. (1981) carried out short-term experiments in the dwarf and tall *S. alterniflora* zones of a Georgia salt marsh where they enriched sediment cores daily with NH_4Cl and excluded macrograzers (mainly fiddler crabs). During summer and winter, N enrichment caused chl *a* levels to increase significantly only beneath dwarf *S. alterniflora*. This is consistent with N-limitation of the benthic microalgae in the dwarf but not the tall *S. alterniflora* zone,

TABLE 1. Comparison of mean annual chlorophyll (chl) *a* concentrations (mg m^{-2}) in sediments beneath vascular plant canopies in different salt marshes of the U.S.

State	Vascular Plant	Chl *a* conc.	Reference
Delaware	dwarf *Spartina alterniflora*	127	Gallagher (1971)
	tall *Spartina alterniflora*	111	
	Distichlis spicata	92	
	dwarf *Spartina alterniflora*	95	Sullivan and Daiber (1975)
South Carolina	dwarf *Spartina alterniflora*	73	Pinckney and Zingmark (1993*a*)
	tall *Spartina alterniflora*	102	
Mississippi	*Spartina alterniflora*	95	Sullivan and Moncreiff (1988*b*)
	Distichlis spicata	57	
	Juncus roemarianus	59	
	Scirpus olneyi	160	

where more frequent tidal inundation provides adequate N supplies. Not surprisingly, exclusion of grazers produced large increases in the biomass of the benthic microalgae in both zones.

Sullivan and Moncreiff (1988b) carried out multiple regressions for each of the four vascular plant zones of a Mississippi marsh using chl *a* concentrations as the dependent variable. The standing crop of benthic microalgae was negatively correlated with pheophytin *a* (degradation product of chl *a* where the central Mg atom has been lost) and salinity and positively correlated with soil moisture and canopy height. R^2 values for the four vascular plant zones ranged from 0.43 to 0.68. Irradiance reaching the marsh surface appeared in only two of the equations, was the last variable to enter, and produced only minor increases in R^2. This is consistent with results from the shading experiments of Sullivan and Daiber (1975).

5. Primary Production of the Benthic Microalgae

Before proceeding to a lengthy discussion of benthic microalgae production within salt marsh sediments, their counterparts above the marsh surface will first be considered. Jones (1980) measured ^{14}C uptake rates of microalgae epiphytic on dead *Spartina alterniflora* stems in a tidal creek of a Georgia salt marsh in April. Diatoms were the dominant algal group. In the laboratory, epiphytic microalgal production was highest at the higher light level employed in the incubations (1,300 vs 250 μE m^{-2} h^{-1}) and on the lower stem section (0-7 vs 17-24 cm). Although these differences were significant, production rates at the lower irradiance level and more than 17 cm above the mud surface were still considerable. Average epiphytic production was 25 mg C (m^2 of stem)$^{-1}$ h^{-1}, which was stated to be at the lower end of production values for the benthic microalgae of salt marsh sediments. Jones argued that epiphyte production could considerably augment production of the latter following sloughing off of dead *S. alterniflora* leaves and before appreciable development of the canopy occurs in summer. In contrast to Jones' assertion, Stowe and Gosselink (1985) concluded that epiphytic algal production on living *S. alterniflora* culms in Barataria Bay was "one of quality rather than quantity." Mean net production (light/dark bottle technique) over an annual cycle at a shoreline station was 26 mg C (m^2 culm)$^{-1}$ h^{-1} but only 1.5 m inland decreased to 3 mg C (m^2 culm)$^{-1}$ h^{-1}. The red algae *Bostrychia* and *Polysiphonia* and diatoms were significantly more abundant on the lower 10 cm of *S. alterniflora* culms at the shoreline station. It would have been of interest if the authors had reported gross primary production and turnover rates and had quantified the impact of grazing on the epiphytic algae.

Measuring the primary production rates of the benthic microalgae in sediments is problematical. The simple act of taking a core introduces a host of unknown artifacts into the measurement technique. The most common techniques employed have been oxygen changes in bell jars over the sediments or cores in light/dark bottles, ^{14}C uptake in water or air, and, more recently, oxygen microelectrodes. While it is beyond the scope of this paper to discuss the technical merits of these methods, Pinckney and Zingmark (1993a) point out the advantages of using microelectrodes and the importance of considering the vertical migration periodicities of the benthic microalgae in making primary production estimates. They state that O_2 and ^{14}C techniques may underestimate benthic microalgal production by as much as 75%. Furthermore, in an earlier paper, Pinckney and Zingmark (1991) presented a simulation model to predict benthic microalgal production which incorporated tidal angles (a measure of tidal stage) and sun angles (a measure of time of day) to account for hourly variability in production. This approach is clearly the way of the future for more accurately and reliably measuring benthic microalgal production in salt marshes and other intertidal and shallow subtidal habitats. For now, however, we have no choice but to compare rates measured by different investigators using a diverse array of methodologies.

TABLE 2. Comparison of annual benthic microalgal production (gC m^{-2} yr^{-1}) and ratio of annual benthic microalgal to vascular plant net aerial production (BMP/VPP$\times$100%) in different salt marshes of the U.S.

State	Algal prod.	BMP/VPP	Reference
Massachusetts	105	25	Van Raalte et al. (1976*b*)
Delaware	61-99	33	Gallagher and Daiber (1974)
South Carolina	98-234	12-58	Pinckney and Zingmark (1993*b*)
Georgia	200	25	Pomeroy (1959)
Georgia	150	25	Pomeroy et al. (1981)
Mississippi	28-151	10-61	Sullivan and Moncreiff (1988*b*)
Texas	71	8-13	Hall and Fisher (1985)
California	185-341	76-140	Zedler (1980)

Table 2 lists annual rates of benthic microalgal production beneath the canopies of various vascular plant species for four Atlantic, two Gulf Coast, and one Pacific coastal salt marsh. Only the South Carolina study employed oxygen microelectrodes. Hall and Fisher's (1985) study has been included with some reluctance, since their rates are for exposed cyanobacterial mats (primarily *Microcoleus lyngbyaceus*) along a creekbank draining a marsh where the dominant vascular plants were *Spartina patens* and *Distichlis spicata*. As with chl *a* concentrations, no directly comparable European studies are known to us. Moving south from Massachusetts to Georgia, benthic microalgal production increases. Rates for the two northeast Atlantic marshes range from 60 to 105 g C m^{-2} y^{-1}, whereas those for the two southeast Atlantic marshes range from 100 to 235 g C m^{-2} y^{-1}. In those studies where rates were determined beneath both the dwarf (= short) and tall forms of *Spartina alterniflora* (Gallagher and Daiber 1974, Pinckney and Zingmark 1993b), the benthic microalgae were most productive in the former zone (1.3X and 2.4X in Delaware and South Carolina, respectively). The single set of measurements made in the *D. spicata* habitat yielded the lowest value (61 g C m^{-2} y^{-1}) for benthic microalgae in any Atlantic marsh. In the two Gulf Coast marshes, only the benthic microalgae in the sediments beneath the relatively open canopy of the sedge *Scirpus olneyi* possessed an annual production value (151 g C m^{-2} y^{-1}) comparable to their counterparts in southeast Atlantic marshes. The lowest value in Table 2 is for *Juncus roemerianus* (28 g C m^{-2} y^{-1}) in Mississippi, where on average only 7% of light incident at the top of the canopy reaches the mud surface 1.2 m below. In this same marsh, annual benthic microalgal production was twice as high beneath *S. alterniflora* (57 g C m^{-2} y^{-1}; canopy height = 64 cm) and three times higher beneath *D. spicata* (88 g C m^{-2} y^{-1}).

The dominant vascular plants of the southern California marsh studied by Zedler (1980) are all different species than those populating Atlantic and Gulf Coast marshes, and three of the four species are in different genera. Benthic microalgal production is higher (185-340 g C m^{-2} y^{-1}; see Table 2) and vascular plant production is much lower (maximum net aerial primary production = 240-340 g C m^{-2} y^{-1}) in the normally arid conditions and hypersaline soils characterizing marshes of the Tijuana estuary. The highest benthic microalgal production rates were measured in the low marsh beneath the short, open canopy of the succulent *Jaumea carnosa* and the tall, dense canopy of *Spartina foliosa* (341 and 276 g C m^{-2} y^{-1}, respectively). Lower, but still considerable, rates were determined for higher marsh elevations beneath dense mats of the grass *Monanthochloe littoralis* and the open canopy of the succulent *Batis maritima* (253 and 185 g C m^{-2} y^{-1}, respectively). An examination of Table 7 of Colijn and de Jonge (1984) and Table 8 of Pinckney and Zingmark (1993*b*) reveals that annual benthic microalgal production beneath the shading canopies of salt marsh vascular plants are comparable to those measured worldwide for intertidal and shallow subtidal marine sediments. The former authors pointed out that, on a global scale, most annual production values fell within a narrow range of 50 to 200 g C m^{-2} y^{-1}.

In areas of the marsh devoid of vascular plants, benthic microalgal production is mostly at the low end of the scales shown in Table 2. In Delaware, Gallagher and Daiber (1974) estimated annual rates for benthic microalgae resident within a bare creekbank and a salt panne to be 38 and 91 g C m^{-2} y^{-1}, respectively. Despite the fact that the salt panne supported a dense mat of cyanobacteria and the mean annual chlorophyll *a* concentration exceeded 400 mg chl *a* m^{-2}, annual benthic microalgal production was only one-half that of the lowest value reported by Zedler (1980) in southern California. Pinckney and Zingmark (1993b) calculated annual benthic microalgal production in shallow subtidal (depth < 1 m at mean low water), sandflat, and mudflat habitats equal to 56, 93, and 191 g C m^{-2} y^{-1}, respectively, in a South Carolina marsh. Only the benthic microalgae in the dwarf *Spartina alterniflora* habitat possessed a higher annual rate than those in the mudflats of this marsh.

In general, benthic microalgal production tends to peak in late winter or during spring (Gallagher 1971, Van Raalte et al. 1976b, Sullivan and Moncreiff 1988b, Pinckney and Zingmark 1993a). In contrast, Zedler (1980) measured highest rates in summer and lowest in spring, while Pomeroy (1959) found a more or less constant daily rate over a yearly cycle and stated there was a 95% probability that net primary productivity of the benthic microalgae was not less than 90% of their gross primary productivity. The most important point to be made regarding seasonality is that in the Atlantic and Pacific coast marshes a significant portion of benthic microalgal production occurs when the overstory vascular plants are dormant and hence represents the principal source of newly fixed carbon on the marsh. In the Gulf Coast marsh of Sullivan and Moncreiff (1988b) only the canopy of *Scirpus olneyi* dies and collapses during winter, and living stands of the other vascular plant species are present year-round. Except in the *S. olneyi* zone, benthic microalgal production rates are low during fall and winter.

Experimental manipulations carried out in the field have provided insight into the regulation of benthic microalgal production in salt marshes by environmental and biological factors. In a Massachusetts marsh, Van Raalte et al. (1976b) applied urea and sewage sludge containing 10% nitrogen (N) on a long-term basis to sediments populated by *Spartina alterniflora*. Only the highest rate of enrichment (ca. 5 g N m^{-2} 2 wk^{-1}) produced a significant increase in benthic microalgal production. Lower rates of N enrichment (ca. 1.7 g N m^{-2} 2 wk^{-1}) and phosphorus addition (ca. 2.6 g P m^{-2} 2 wk^{-1}) had no effect. Removal of the *S. alterniflora* canopy by clipping also significantly increased benthic microalgal production, but in short-term (summers of 1973 and 1974) experiments where 3 levels of N enrichment were crossed with 3 levels of light the interaction between N and light was not significant. Therefore, the effects of N enrichment were independent of those of light. Darley et al. (1981) carried out short-term (1 to 2 weeks) experiments in the dwarf and tall *S. alterniflora* zones of a Georgia salt marsh where sediment cores received daily enrichment with NH_4Cl at the rate of 8.6 mmole N m^{-2} d^{-1}. Macrograzers (mainly fiddler crabs) were excluded from plots containing the control and enriched cores. Results were essentially identical to those given in the previous section for biomass (as chl *a*) of the benthic microalgae in that both during summer and winter N enrichment significantly increased benthic microalgal production in only the dwarf *S. alterniflora* zone. Exclusion of grazers produced significant increases in benthic microalgal production in both zones with the effect being most pronounced beneath tall *S. alterniflora*. In an ingenious reciprocal transplant experiment involving *unenriched* cores from both vascular plant zones, the 2-5X higher concentrations of exchangeable ammonium in the sediments beneath tall *S. alterniflora* rapidly alleviated the N limitation of transplanted dwarf *S. alterniflora* cores and benthic microalgal production increased by a factor of 8.6X over cores removed and replaced back in the dwarf *S. alterniflora* zone. However, benthic microalgae in cores transplanted from the tall to the dwarf *S. alterniflora* zone experienced a 50% reduction in production due to their rapid depletion of the limited N supplies in the latter zone. Finally, Whitney and Darley (1983) measured summer and winter *in situ* rates of benthic microalgal production in sediments of a bare creekbank and beneath dwarf and intermediate height *S. alterniflora* in the same marsh enriched by Darley et al. (1981). Irradiance reaching the portable incubation chambers was manipulated by placing neutral density screens over each chamber. Maximal rates of benthic microalgal production in the bare creekbank occurred at light levels considerably less than full sunlight. Conversely, the benthic microalgae beneath both height forms of the cord grass were severely limited by light throughout the year in that maximal production rates occurred at irradiances much higher than those actually reaching the sediment surface. Photoinhibition was evident at full sunlight in all three habitats in summer and winter. Darley et al. (1981) reached an opposite conclusion regarding light limitation of benthic microalgae in the dwarf *S. alterniflora* zone of this marsh based on shading experiments conducted in summer and winter without N enrichment. Their results showed the average levels of light reaching the sediment surface were saturating for the production of chl *a* under the prevailing conditions of N limitation that had been previously shown to exist there.

Various authors have inferred the relationships between benthic microalgal production and environmental factors in salt marshes by simple correlation analysis and linear and multiple regression. Of course such analyses only produce hypotheses regarding regulation of benthic microalgal production by various factors which can then be experimentally tested in the field. Nevertheless, such data explorations are of great use and much of what we understand about benthic microalgal distribution, biomass, and primary production comes from the correlative approach. One caveat must be mentioned before proceeding further, however. These standard statistical methods assume that the relationship between the dependent variable and each independent variable is constant (positive or negative with a constant slope) over the entire range of measured values for the latter and thus isolate overall or average effects for each variable. Experience teaches us, however, that this is an oversimplification of real events occurring in natural systems (see Shaffer and Sullivan 1988 for a discussion of differences between events and variables).

Perhaps the most obvious variable that should regulate benthic microalgal production is the concentration of chl *a* in the sediments. Sullivan and Moncreiff (1988b) found a moderate correlation (r^2 = 0.34) between benthic microalgal production and chl *a* only in the *Juncus roemerianus* zone of a Mississippi salt marsh. In contrast, Pinckney and Zingmark (1993a) showed high correlations (r^2 = 0.67-0.73) between these two variables in four of five habitats studied in a South Carolina salt marsh. The highest correlation was in the dwarf *Spartina alterniflora* habitat; this was attributed to use of the microelectrode technique and confining measurements of chl *a* to the upper 2 mm of sediments. Surprisingly, the other studies listed in Table 2 made no attempt to statistically analyze the relationship between benthic microalgal production and biomass. However, Davis and McIntire (1983) and Colijn and de Jonge (1984) found that chl *a* was an excellent predictor of hourly production in intertidal marine sediments devoid of a vascular plant canopy. Light energy reaching the marsh surface is potentially the other "essential" variable for photosynthetic organisms.

Sullivan and Moncreiff (1988b) found light was a poor predictor of benthic microalgal production under all four vascular plant canopies which agrees with results of van Es (1982), Colijn and de Jonge (1984) and Varela and Penas (1985) for intertidal mudflats and sandflats. In contrast, Van Raalte et al. (1976b) demonstrated a highly significant linear relationship between light energy reaching the marsh surface and hourly production. No photoinhibition was observed, even at full sunlight, which is in agreement with the work of Williams (1962) but opposite to that of Whitney and Darley (1983). Williams (1962) wrote that the most important factor controlling benthic microalgal production in salt marshes was light since nutrients (including nitrogen supplies) were adequate and only the extremes of salinity and temperate were likely to be detrimental. These conclusions for light and nutrients were not experimentally tested, however. Van Raalte et al. (1976b) and Sullivan and Moncreiff (1988b) reported little or no correlation existed between temperature and benthic microalgal production, a finding also reported by Davis and McIntire (1983), Colijn and de Jonge (1984), and Varela and Penas (1985).

Sullivan and Moncreiff (1988*b*) carried out multiple regressions for four vascular plant zones using hourly benthic microalgal production as the dependent variable and 8 environmental factors as independent variables. No combination of variables adequately predicted hourly production beneath the *Spartina alterniflora* canopy ($R^2 = 0.39$); however, R^2 values in the remaining three vascular plant zones ranged from 0.70 to 0.86. Overall, the single best predictor of benthic microalgal production was soil moisture, entering first in two equations and second in a third equation. The sign of its partial regression coefficient was constant and benthic microalgal production was predicted to increase as the marsh surface dried out. Pinckney and Zingmark (1991) found that benthic microalgal production in a South Carolina salt marsh was twice as high at low tide than at high tide, which agrees with the statistically inferred importance of soil moisture described above.

It is of considerable interest to calculate the ratio of annual benthic microalgal production to net aerial production of the overstory vascular plant canopy (BMP/VPP) to determine the relative amounts of fixed carbon potentially available to primary consumers. Such values are listed in Table 2 for different salt marshes representing the three coasts of the United States. With the exception of the *Scirpus olneyi* habitat, the two Gulf Coast marshes have the lowest (8 to 13%) BMP/VPP values. This includes the following vascular plants: *Spartina alterniflora, S. patens, Distichlis spicata,* and *Juncus roemerianus*. Annual benthic microalgal production ranged from 28 to 88 g C m^{-2} y^{-1} in these vascular plant zones. Another ''group'' is formed by marshes in Massachusetts, Delaware, and Georgia where BMP/VPP was 25 to 33%, and except for the *D. spicata* habitat in Delaware the vascular plant zones studied were various height forms of *S. alterniflora* in all three states. In South Carolina BMP/VPP for tall *S. alterniflora* was 12%, matching the lower values recorded in the two Gulf Coast marshes; however, BMP/VPP for dwarf *S. alterniflora* (58%) in the former marsh matched that of *Scirpus olneyi* (61%) in Mississippi. The southern California marsh formed a third ''group'' unto itself as the highest BMP/VPP values exist here (range = 76 to 140%). In two of the habitats, annual benthic microalgal production equaled (*Monanthochloe littoralis*) or exceeded (*Jaumea carnosa*) that of the overstory vascular plant canopy. In the remaining habitats, BMP/VPP was 76% and 81% for *Batis maritima* and *Spartina foliosa*, respectively. As discussed above, the large benthic microalgal biomass and reduction in vascular plant production caused by the normally arid conditions and hypersaline soils were cited by Zedler (1980) as reasons for such high BMP/VPP values. In summary, the ranges for benthic microalgal production (28 to 341 g C m^{-2} y^{-1}) and BMP/VPP (10 to 140%) are considerable (Table 2). Almost without exception, whenever benthic microalgal production values are high, so too are those for BMP/VPP.

6. The Role of Benthic Microalgae in Salt Marsh Food Webs

One of the founding principles of salt marsh ecology was that the basis for marsh secondary production was vascular plant production (primarily *Spartina alterniflora*) which had entered a detrital-based food web (Teal 1962, Odum and de la Cruz 1963). However, the utilization of benthic microalgae by numerous salt marsh fauna has been demonstrated in lab and field investigations. Organisms demonstrated to directly ingest benthic microalgae include amphipods (*Talorchestia longicornis*), gastropods (*Ilyanassa obsoleta*), polychaetes (*Nereis* spp.), fiddler crabs (*Uca* spp.), killifish (*Fundulus heteroclitus*), larval shrimp (*Penaeus* spp.) and meiofauna (copepods, nematodes) (Brenner et al. 1976, Wetzel 1977, Kneib et al. 1980, Connor and Teal 1982, Robertson and Newell 1982, Montagna 1984, McTigue and Zimmerman 1991, Weissburg 1992, Carman et al. 1997, Créach et al. 1997). Benthic diatoms were primarily of interest in these studies, with the exception of a study of grazing on cyanobacteria by *T. longicornis* (Brenner et al. 1976). Methods employed in these studies include gut content analysis, growth on algal diets by laboratory-reared organisms, measures of radio-labelled algal uptake, and field measures of the effect of grazing on microalgal biomass. For some organisms, it is evident that benthic microalgae make up a significant portion of the animal's diet, and/or that benthic microalgae can support growth. Organisms for which this has been demonstrated include amphipods (Brenner et al. 1976, Smith et al. 1996), gastropods (Wetzel 1977, Créach et al. 1997), fiddler crabs (Robertson and Newell 1982, Weissburg 1992), polychaetes (Smith et al. 1996) and meiofauna (Carman et al. 1997). The degree to which benthic and epiphytic microalgae support the secondary production of larval and juvenile penaeid shrimp remains unclear (Gleason and Zimmerman 1984, McTigue and Zimmerman 1991), while the ingestion of benthic microalgae by killifish appears to be incidental to the consumption of prey organisms (Kneib et al. 1980). Although the results cited above illuminate the feeding ecology of several marsh herbivores and omnivores, they do not provide an integrated picture of the role of benthic microalgae in salt marsh food webs. This role has been clarified by the application of multiple stable isotope analysis, which has provided a valuable tool for determining the role of benthic microalgae in supporting marsh consumers, including those animals feeding at higher trophic levels.

The following is a brief introduction to the use of stable isotopes in food web analysis, and can be supplemented by reviews of stable isotopes in ecological research by Fry and Sherr (1984), Peterson and Fry (1987), and Lajtha and Michener (1994). The elements most commonly used in multiple stable isotope food web studies are carbon, nitrogen and sulfur, each of which has two or more stable isotopes. The distribution of the rarer, heavier stable isotope of an element varies in the biosphere due to small variations in the rate at which the heavier isotope participates in physical, chemical and biochemical reactions. These variations, which are mass-dependent, are described as isotopic fractionation and have been quantified for numerous enzymatic reactions. For example, the RUBISCO enzyme discriminates against heavier ^{13}C atoms, so that all plant material is depleted in ^{13}C relative to atmospheric CO_2. Moreover, there is a small and predictable isotope fractionation when animals assimilate their food (De Niro and Epstein 1978, 1981, Peterson et al. 1986), so that the isotopic composition of primary producers at the base of an animal's food chain can be deduced from the isotopic composition of the animal. In the discussion that follows, we will observe the following convention for identifying the isotopic composition of samples:

$$\delta X = [(R_{sample}/R_{standard}) - 1] \times 10^3, \quad (1)$$

where $X = {}^{13}C$, ^{15}N, or ^{34}S and $R = {}^{13}C/{}^{12}C$, $^{15}N/{}^{14}N$, or $^{34}S/{}^{32}S$.

The first stable isotope examination of salt marsh biota focused on determining the sources of C to estuarine particulate matter (Haines 1976), and included a discussion of how the isotopic composition of marsh consumers reflected the isotopic composition of primary producers. Food web analysis using this approach was further explored in Haines and Montague (1979), who demonstrated that several consumers, including *Uca* spp. and *Nassarius (Ilyanassa) obsoleta*, had $\delta^{13}C$ values very close to those of benthic microalgae collected from a creekbank. Several other species, in particular the ribbed mussel *Geukensia demissa*, had $\delta^{13}C$ values indicating a mixed diet with a potential significant contribution by benthic microalgae. However, as the $\delta^{13}C$ value for benthic microalgae falls between that of *Spartina alterniflora* and phytoplankton, it is difficult to conclusively determine the contribution of these various primary producers in organisms with intermediate $\delta^{13}C$ values. This difficulty can be partially resolved by adding N and S isotope analysis (Peterson et al. 1986, Peterson and Howarth 1987).

Subsequently, several studies have employed multiple stable isotope (C, N, and S) analysis to determine how marsh primary producers support marsh secondary production. The surprising, and consistent, conclusion has been that vascular plant production, while dominating the standing crop biomass, plays a reduced role in supporting marsh consumers. Further, in some cases it was possible to uniquely identify benthic microalgae as an important component of salt marsh food chains. The first multiple isotope study to suggest an important dietary role for benthic microalgae was from a Georgia marsh dominated by *Spartina alterniflora*. The majority of the 20 consumers analyzed exhibited an isotopic signature consistent with approximately equal contributions of phytoplankton and *S. alterniflora* (Peterson and Howarth 1987). However, these authors noted that, relative to other primary producers, benthic microalgae had intermediate $\delta^{13}C$ values, similar ^{15}N values and unknown ^{34}S values, and may be important contributors to the marsh food web.

Stable isotope values for C, N, and S were determined for benthic (edaphic) microalgae, vascular plants, and over 50 faunal elements from an irregularly flooded *Spartina alterniflora/Juncus roemerianus* marsh in Mississippi and nearby tidal channels (Sullivan and Moncreiff 1990). The majority of the marsh and estuarine consumers sampled had isotopic compositions consistent with a primary input from edaphic algae, and an important contribution from vascular plants could be excluded for many consumers. In the Mississippi study, the combination of $\delta^{13}C$ and $\delta^{34}S$ values proved the most powerful in determining the role of benthic microalgae. In a natural and transplanted *S. alterniflora* marsh in North Carolina, multiple stable isotope analysis (C, N, and S) also provided convincing evidence for the importance of benthic and epiphytic microalgae in the food chains supporting fiddler crabs, snails, and killifish (Currin et al. 1995). In these North Carolina marshes, the combination of C and N isotopes was especially useful in tracking the microalgal contribution, particularly as N_2 fixing cyanobacteria were a significant component of the microalgal community. This paper also reviewed published values for the isotopic composition of estuarine benthic microalgae, and demonstrated that the range of reported values, while large, embrace the range of values reported for many marsh consumers.

Multiple stable isotope analysis has also been used to illuminate the importance of benthic microalgae in California marshes dominated by *Spartina foliosa* or *Salicornia* spp. Kwak and Zedler (1997) examined the food webs in two marshes in southern California using C, N and S isotope analysis. In a marsh dominated by *S. foliosa*, mixing models demonstrated that benthic microalgae were an important component of both invertebrate and fish food chains. In a marsh dominated by *Salicornia virginica*, results from a two-source mixing model suggested that benthic microalgal production supported 20 to 54% of fisheries production. In another southern California marsh dominated by *S. virginica,* Page (1997) demonstrated the importance of benthic microalgal production to a number of marsh consumers, using C and N isotope analysis. In addition to an important role in supporting marsh snails, amphipods and gastropods, Page (1997) presented evidence suggesting that resuspended benthic microalgae are important to filter-feeding bivalves. In southern California marshes, which are marine to hypersaline systems, cyanobacteria as well as diatoms are key components of the microalgal community supporting secondary production (Kwak and Zedler 1997, Page 1997).

Stable C and N isotope analysis has also been used to demonstrate the importance of benthic diatoms in the diets of macroinvertebrates from a European salt marsh. Two species of amphipods, a polychaete and a pulmonate snail were analyzed from low and middle marsh habitats dominated by the vascular plants *Suaeda maritima, Salicornia* sp., *Puccinellia maritima* and *Halimione portulacoides* (Créach et al. 1997). Each of the consumer organisms demonstrated a significant reliance on benthic diatoms, and almost exclusively so for *Ovatella bidentata* and *Corophium volutator*.

In summary, multiple stable isotope analysis has provided an integrated measure of the primary production assimilated by marsh and estuarine consumers. Numerous studies employing this technique have concluded that benthic microalgae are responsible for 50% or more of the C assimilated by consumer organisms. In particular, fiddler crabs, amphipods, snails, and killifish appear to obtain a significant portion of their C from benthic microalgal production. These resident marsh organisms are in turn preyed upon by a variety of transient fish and bird species, and studies that have examined the isotopic composition of fish and birds feeding at higher trophic levels reveal an important role for benthic microalgae in marsh and estuarine food webs (Sullivan and Moncreiff 1990, Deegan and Garritt 1997, Kwak and Zedler 1997).

7. Directions for Future Research

Despite the recent attention accorded the benthic microalgae of salt marshes, thanks mainly to the unexpected results of stable isotope studies, this primary producer group remains somewhat enigmatic to many researchers because of their microscopic size and ''black box'' nature. However, their generally high rates of primary production and its year round availability make them a force to be reckoned with by all who seek to understand the functional roles of salt marsh systems. We therefore consider the following research areas to be potentially fruitful ones.

Few researchers studying various attributes of the vascular plants of a particular salt marsh would not identify each to species. However, this is the usual case for benthic microalgal assemblages within a salt marsh, often because of time constraints but mostly due to the lack of monographic works on the algal groups of shallow coastal habitats. There is great need for such works, particularly for the diatoms and cyanobacteria, if only to standardize the taxonomy of salt marsh microalgae when researchers invest the time to make identifications and cell counts. As discussed above, the relationships between the distribution of the benthic microalgae in space and time and environmental factors are poorly understood and results from different studies may prove contradictory. As suggested to the first author 20 years ago by C. David McIntire (person. commun.), it would be most informative to know how much the variability in species composition and abundance is due to purely stochastic processes (e.g., is it a matter of who gets there first in tidal currents?). If such were the case, then ecologists would know how much of this variability they would ideally have to account for in multivariate statistical approaches.

When biomass or primary production is estimated for the benthic microalgae, a single number (mg chl *a* m^{-2} or mg C m^{-2} hr^{-1}, respectively) is produced leading to the ''black box'' effect. Researchers often write that diatoms or cyanobacteria are the ''dominant'' group in a particular salt marsh habitat but what does this mean in quantitative terms? Hall and Fisher (1985) found that ''diatom-dominated'' areas of a Texas marsh had production rates twice that of cyanobacterial mats. It would be of great value to be able to determine the relative contributions of the different algal groups and even individual species (or at least the more abundant ones) to total assemblage biomass and production. HPLC offers a partial solution to the biomass ''black box'' in that it can only provide information on the relative biomass of a particular algal group in space or time and then only if the measured pigment is not present in other algal groups of the assemblage. In the case of the production ''black box'', track light microscope-autoradiography can determine the relative (but not absolute) production rates of individual taxa (see Coleman and Burkholder 1995 for epiphytic seagrass microalgae) but its application in a sedimentary environment would present great technical problems. When a benthic microalgal assemblage responds to an experimental manipulation (e.g., nutrient enrichment or varying light levels), which species are responding? Cell counts provide important but limited information because cell numbers may not accurately reflect the contribution of a given species to total biomass and production.

The stable isotope approach to investigating the role of benthic microalgae in marsh food webs will benefit from improved resolution in the sampling of the isotopic composition of benthic microalgae, and from studies where quantitative analysis of modelling of the contribution of various primary producers to marsh food chains is performed. Given their occurrence at the chemically dynamic sediment-water interface, it is not surprising that the C, N, and S isotopic composition of benthic microalgae exhibits considerable variation (Currin et al. 1995). As the resolution of mass spectrometry improves, the algal biomass required to obtain an isotopic value has decreased several orders of magnitude from the 1970s. This, in combination with improved methods for separating microalgal biomass from sedimentary material, should provide better spatial and temporal resolution of microalgal isotopic analyses. Future studies may also reveal differences in the assimilation of elements (C, N, and S) from different microalgal food sources. In complex ecosystems such as salt marshes, where there are multiple primary producers, stable isotope analysis can at best provide upper and lower estimates of the possible contribution of various food sources. The combination of stable isotope analysis with other biomarker analyses, in conjunction with improved modelling and statistical analysis, will yield more precise measures of the contribution of benthic microalgae to salt marsh consumers.

Finally, much of what we know concerning the benthic microalgae is biased towards Atlantic coastal marshes and the cord grass *Spartina alterniflora*. More descriptive and experimental studies need to be carried out in different geographical regions (particularly the northwest Pacific coast, continental Europe, and South America) and beneath the canopies of vascular plants other than *S. alterniflora* in order to develop a comprehensive and accurate model of benthic microalgal community structure and functional dynamics.

8. Acknowledgements

The first author would like to thank John L. Gallagher and Franklin C. Daiber for introducing him to the wondrous and challenging complexity of benthic microalgae, and to express his deep appreciation to the late James I. Jones, Director of the Mississippi-Alabama Sea Grant Consortium, for his unwavering personal and financial support.

9. Literature Cited

Admiraal, W. 1977. Tolerance of estuarine benthic diatoms to high concentrations of ammonia, nitrite ion, nitrate ion and orthophosphate. Marine Biology **43**:307-315.

——. 1984. The ecology of estuarine sediment-inhabiting diatoms. Progress in Phycological Research **3**:269-322.

Admiraal, W., H. Peletier and H. Zomer. 1982. Observations and experiments on the population dynamics of epipelic diatoms from an estuarine mudflat. Estuarine, Coastal and Shelf Science **14**:471-487.

Amspoker, M. C. and C. D. McIntire. 1978. Distribution of intertidal diatoms associated with sediments in Yaquina Estuary, Oregon. Journal of Phycology **14**:387-395.

Blair, S. M. 1983. Taxonomic treatment of the *Chaetomorpha* and *Rhizoclonium* species (Cladophorales: Chlorophyta) in New England. Rhodora **85**:175-211.

Blum, J. L. 1968. Salt marsh Spartinas and associated algae. Ecological Monographs **38**:199-221.
Brenner, D., I. Valiela and C. D. Van Raalte. 1976. Grazing by *Talorchestia longicornis* on an algal mat in a New England salt marsh. Journal of Experimental Marine Biology and Ecology **22**:161-169.
Carman, K. R., J. W. Fleeger and S. M. Pomarico. 1997. Response of a benthic food web to hydrocarbon contamination. Limnology and Oceanography **42**:561-571.
Carter, N. 1932. A comparative study of the algal flora of two salt marshes. Part I. Journal of Ecology **20**:341-370.
—— 1933a. A comparative study of the algal flora of two salt marshes. Part II. Journal of Ecology **21**:128-208.
—— 1933b. A comparative study of the algal flora of two salt marshes. Part III. Journal of Ecology **21**:385-403.
Chapman, R. L. 1971. The macroscopic marine algae of Sapelo Island and other sites on the Georgia coast. Bulletin of the Georgia Academy of Science **29**:77-89.
Christian, R. R., K. Bancroft and W. J. Wiebe. 1978. Resistance of the microbial community within salt marsh soils to selected perturbations. Ecology **59**:1200-1210.
Coleman, V. L. and J. M. Burkholder. 1995. Response of microalgal epiphyte communities to nitrate enrichment in an eelgrass (*Zostera marina*) meadow. Journal of Phycology **31**:36-43.
Colijn, F. and V. N. de Jonge. 1984. Primary production of microphytobenthos in the Ems-Dollard estuary. Marine Ecology Progress Series **14**:185-196.
Connor, M.S. and J. M. Teal. 1982. The effect of feeding by mud snails, *Ilyanassa obsoleta* (Say), on the structure and metabolism of a laboratory benthic algal community. Journal of Experimental Marine Biology and Ecology **65**:29-45.
Cook, L. L. and S. A. Whipple. 1982. The distribution of edaphic diatoms along environmental gradients of a Louisiana salt marsh. Journal of Phycology **18**: 64-71.
Créach, V. M. T. Schricke, G. Bertru and A. Mariotti. 1997. Stable isotopes and gut analyses to determine feeding relationships in saltmarsh macroconsumers. Estuarine, Coastal and Shelf Science **44**: 599-611.
Currin, C. A., S. Y. Newell and H. W. Paerl. 1995. The role of benthic microalgae and standing dead *Spartina alterniflora* in salt marsh food webs: implications based on multiple stable isotope analysis. Marine Ecology Progress Series **121**: 99-116.
Currin, C. A. and H. W. Paerl 1998a. Environmental and physiological controls on diel patterns of N_2 fixation in epiphytic cyanobacterial communities. Microbial Ecology **35**:34-45.
—— 1998b. Epiphytic nitrogen fixation associated with standing dead shoots of smooth cordgrass, *Spartina alterniflora*. Estuaries **21**:108-117.
Darley, W. M., C. L. Montague, F. G. Plumley, W. W. Sage and A. T. Psalidas. 1981. Factors limiting edaphic algal biomass and productivity in a Georgia salt marsh. Journal of Phycology **17**:122-128.
Davis, M. W. and C. D. McIntire. 1983. Effects of physical gradients on the production dynamics of sediment-associated algae. Marine Ecology Progress Series **13**:103-114.
Deegan, L. A. and R. H. Garritt. 1997. Evidence for spatial variability in estuarine food webs. Marine Ecology Progress Series **147**:31-47.
De Niro, M. J. and S. Epstein. 1978. Influence of diet on the distribution of carbon isotopes in animals. Geochimica et Cosmochimica Acta **42**:495-506.
—— 1981. Isotopic composition of cellulose from aquatic organisms. Geochimica et Cosmochimica Acta **45**:1885-1894.
Drum, R. W. and E. Webber. 1966. Diatoms from a Massachusetts salt marsh. Botanica Marina **9**:70-77.
Edgar, L. A. and J. D. Pickett-Heaps. 1984. Diatom locomotion. Progress in Phycological Research **3**:47-88.
Estrada, M., I. Valiela and J. M. Teal. 1974. Concentration and distribution of chlorophyll in fertilized plots in a Massachusetts salt marsh. Journal of Experimental Marine Biology and Ecology **14**:47-56.
Fallon, R. D., S. Y. Newell and L. C. Groene. 1985. Phylloplane algae of standing dead *Spartina alterniflora*. Marine Biology **90**:121-127.
Fry, B. and E. B. Sherr. 1984. ^{13}C measurements as indicators of carbon flow in marine and freshwater ecosystems. Contributions in Marine Science **27**:13-47.
Gallagher, J. L. 1971. Algal productivity and some aspects of the ecological physiology of the edaphic communities of Canary Creek tidal marsh. Dissertation, University of Delaware, Newark, Delaware, USA.
Gallagher, J. L. and F. C. Daiber. 1974. Primary production of edaphic algal communities in a Delaware salt marsh. Limnology and Oceanography **19**: 390-395.

Gleason, D. F. and R. J. Zimmerman. 1984. Herbivory potential of postlarval brown shrimp associated with salt marshes. Journal of Experimental Marine Biology and Ecology **84**: 235-246.
Haines, E. B. 1976. Stable carbon isotope ratios in the biota, soils and tidal water of a Georgia salt marsh. Estuarine, Coastal and Marine Science **4:** 609-616.
Haines, E. B. and C. L. Montague. 1979. Food sources of estuarine invertebrates analyzed using $^{13}C/^{12}C$ ratios. Ecology **60:** 48-56.
Hall, S. L. and F. M. Fisher, Jr. 1985. Annual productivity and extracellular release of dissolved organic compounds by the epibenthic algal community of a brackish marsh. Journal of Phycology **21**:277-281.
Jones, R. C. 1980. Productivity of algal epiphytes in a Georgia salt marsh: effect of inundation frequency and implications for total marsh productivity. Estuaries **3**:315-317.
Jonge, V. N. de. 1985. The occurrence of ''epipsammic' diatom populations: a result of interaction between physical sorting of sediment and certain properties of diatom species. Estuarine, Coastal and Shelf Science **21**:607-622.
Jönsson, B., K. Sundbäck and C. Nilsson. 1994. The upright life-form of an epipelic motile diatom: on the behaviour of *Gyrosigma balticum*. European Journal of Phycology **29**:11-15.
Kneib, R. T., A. E. Stiven and E. B. Haines. 1980. Stable carbon isotope ratios in *Fundulus heteroclitus* (L.) muscle tissue and gut contents from a North Carolina *Spartina* marsh. Journal of Experimental Marine Biology and Ecology **46**: 89-98.
Kwak, T. J. and J. B. Zedler. 1997. Food web analysis of southern California coastal wetlands using multiple stable isotopes. Oecologia **110**:262-277.
Lajtha, K. and R. H. Michener. 1994. Stable isotopes in ecology and environmental science. Blackwell Scientific Publications, Boston, Massachusetts, USA.
Leach, J. H. 1970. Epibenthic algal production in an intertidal mudflat. Limnology and Oceanography **15**:514-521.
MacIntyre, H. L., R. J. Geider and D. C. Miller. 1996. Microphytobenthos: the ecological role of the ''secret garden'' of unvegetated, shallow-water marine habitats. I. Distribution, abundance and primary production. Estuaries **19**:186-201.
Maples, R. S. 1982. Bluegreen algae of a coastal salt panne and surrounding angiosperm zones in a Louisiana salt marsh. Northeast Gulf Science **5**:39-43.
McIntire, C. D. 1973. Diatom associations in Yaquina Estuary, Oregon: a multivariate analysis. Journal of Phycology **9**:254-259.
——1978. The distribution of estuarine diatoms along environmental gradients: a canonical correlation. Estuarine and Coastal Marine Science **6**:447-457.
McIntire, C. D. and W. W. Moore. 1977. Marine littoral diatoms: ecological considerations. Pages 333-371 *in* D. Werner, editor. The biology of diatoms. University of California Press, Berkeley, California, USA.
McTigue, T. A. and R. J. Zimmerman. 1991. Carnivory vs. herbivory in juvenile *Penaeus setiferus* (Linnaeus) and *Penaeus aztecus* (Ives). Journal of Experimental Marine Biology and Ecology **151**:1-16.
Miller, D. C., R. J. Geider and H. L. MacIntyre. 1996. Microphytobenthos: the ecological role of the ''secret garden'' of unvegetated, shallow-water marine habitats. II. Role in sediment stability and shallow-water food webs. Estuaries **19**:202-212.
Montagna, P. A. 1984. *In situ* measurement of meiobenthic grazing rates on sediment bacteria and edaphic diatoms. Marine Ecology Progress Series **18**:119-130.
Odum, E. P. and A. A. de la Cruz. 1963. Detritus as a major component of ecosystems. AIBS Bulletin **13:**39-40.
Otte, A. M. and V. J. Bellis. 1985. Edaphic diatoms of a low salinity estuarine marsh system in North Carolina - a comparative floristic study. The Journal of the Elisha Mitchell Scientific Society **10**:116-124.
Page, H. M. 1997. Importance of vascular plant and algal production to macroinvertebrate consumers in a southern California salt marsh. Estuarine, Coastal and Shelf Science **45**:823-834.
Paterson, D. M. 1986 The migratory behaviour of diatom assemblages in a laboratory tidal micro-ecosystem examined by low temperature scanning electron microscopy. Diatom Research **1**:227-239.
Peterson, B. J. and B. Fry. 1987. Stable isotopes in ecosystem studies. Annual Review of Ecology and Systematics **18**:293-320.

Peterson, B. J. and R. W. Howarth. 1987. Sulfur, carbon and nitrogen isotopes used to trace organic matter flow in the salt-marsh estuaries of Sapelo Island, Georgia. Limnology and Oceanography **32**:1195-1213.

Peterson, B. J., R. W. Howarth and R. H. Garritt. 1986. Sulfur and carbon isotopes as tracers of salt-marsh organic matter flow. Ecology **67**:865-874.

Piehler, M. F., C. A. Currin, R. Cassanova and H. W. Paerl. 1998. Development and N_2-fixing activity of the benthic microbial community in transplanted *Spartina alterniflora* marshes in North Carolina. Restoration Ecology **6**:290-296.

Pinckney, J., R. Papa and R. Zingmark. 1994a. Comparison of high-performance liquid chromatographic, spectrophotometric and fluorometric methods for determining chlorophyll *a* concentrations in estuarine sediments. Journal of Microbiological Methods **19**:59-66.

Pinckney, J., Y. Piceno and C. R. Lovell. 1994b. Short-term changes in the vertical distribution of benthic microalgal biomass in intertidal muddy sediments. Diatom Research **9**:143-153.

Pinckney, J. and R. G. Zingmark. 1991. Effects of tidal stage and sun angles on intertidal benthic microalgal productivity. Marine Ecology Progress Series **76**:81-89.

—— 1993a. Biomass and production of benthic microalgal communities in estuarine habitats. Estuaries **16**:887-897.

—— Pinckney, J. and R. G. Zingmark 1993b. Modeling the annual production of intertidal benthic microalgae in estuarine ecosystems. Journal of Phycology **29**:396-407.

Polderman, P. J. 1975. The algal communities of the northeastern part of the saltmarsh "De Mok" on Texel (The Netherlands). Acta Botanica Neerlandica **24**:361-378.

—— 1978. Algae of saltmarshes on the south and southwest coasts of England. British Phycological Journal **13**:235-240.

Polderman, P. J. and R. A. Polderman-Hall. 1980. Algal communities in Scottish saltmarshes. British Phycological Journal **15**:59-71.

Pomeroy, L. R. 1959. Algal productivity in salt marshes of Georgia. Limnology and Oceanography **4**:386-397.

Pomeroy, L., W. Darley, E. Dunn, J. Gallagher, E. Haines and D. Whitney. 1981. Primary production. Pages 39-67 *in* Pomeroy, L. R. and R. G. Wiegert, editors. The ecology of a salt marsh. Springer-Verlag, New York, New York, USA.

Ralph, R. D. 1977. The Myxophyceae of the marshes of southern Delaware. Chesapeake Science **18**:208-221.

Robertson, J. R. and S. Y. Newell. 1982. Experimental studies of particle ingestion by the sand fiddler crab *Uca pugilator* (Bosc). Journal of Experimental Marine Biology and Ecology **52**:1-21.

Round, F. E. 1960. The diatom flora of a salt marsh on the River Dee. New Phytologist **59**:332-348.

Sage, W. W. and M. J. Sullivan. 1978. Distribution of bluegreen algae in a Mississippi Gulf Coast salt marsh. Journal of Phycology **14**:333-337.

Shaffer, G. P. and M. J. Sullivan. 1988. Water column productivity attributable to displaced benthic diatoms in well-mixed shallow estuaries. Journal of Phycology **24**:132-140.

Smith, D., R. G. Hughes and E. J. Cox. 1996. Predation of epipelic diatoms by the amphipod *Corophium volutator* and the polychaete *Nereis diversicolor*. Marine Ecology Progress Series **145**:53-61.

Stowe, W. C. 1980. Vertical distribution of epiphytic *Denticula subtilis* Grunow. Transactions of the American Microscopical Society **99**:323-328.

——. 1982. Diatoms epiphytic on the emergent grass *Spartina alterniflora* in a Louisiana salt marsh. Transactions of the American Microscopical Society **101**:162-173.

Stowe, W. C. and J. G. Gosselink. 1985. Metabolic activity of the epiphytic community associated with *Spartina alterniflora*. Gulf Research Reports **8**:21-25.

Sullivan, M.J. 1975. Diatom communities from a Delaware salt marsh. Journal of Phycology **11**:384-390.

——1976. Long-term effects of manipulating light intensity and nutrient enrichment on the structure of a salt marsh diatom community. Journal of Phycology **12**:205-210.

——1977a. Edaphic diatom communities associated with *Spartina alterniflora* and *S. patens* in New Jersey. Hydrobiologia **52**:207-211.

——1977b. Structural characteristics of a diatom community epiphytic on *Ruppia maritima*. Hydrobiologia **53**:81-86.

—— 1978. Diatom community structure: taxonomic and statistical analyses of a Mississippi salt marsh. Journal of Phycology **14**:468-475.

—— 1981. Effects of canopy removal and nitrogen enrichment on a *Distichlis spicata*-edaphic diatom complex. Estuarine, Coastal and Shelf Science **13**:119-129.

—— 1982a. Similarity of an epiphytic and edaphic diatom community associated with *Spartina alterniflora*. Transactions of the American Microscopical Society **101**:84-90.
—— 1982b. Distribution of edaphic diatoms in a Mississippi salt marsh: a canonical correlation analysis. Journal of Phycology **18**:130-133.
—— 1999. Applied diatom studies in estuaries and shallow coastal environments. Pages 334-351 *in* Stoermer, E. F. and J. P. Smol, editors. The diatoms: applications for the environmental and earth sciences. Cambridge University Press, Cambridge, England.
Sullivan, M.J. and F.C. Daiber. 1975. Light, nitrogen and phosphorus limitation of edaphic algae in a Delaware salt marsh. Journal of Experimental Marine Biology and Ecology **18**:79-88.
Sullivan, M.J. and C.A. Moncreiff 1988a. A multivariate analysis of diatom community structure and distribution in a Mississippi salt marsh. Botanica Marina **31**:93-99.
—— 1988b. Primary production of edaphic algal communities in a Mississippi salt marsh. Journal of Phycology **24**:49-58.
—— 1990. Edaphic algae are an important component of salt marsh food-webs: evidence from multiple stable isotope analyses. Marine Ecology Progress Series **62**:149-159.
Teal, J. M. 1962. Energy flow in the salt marsh ecosystem of Georgia. Ecology **43**:614-624.
Underwood, G. J. C., J. Phillips and K. Saunders. 1998. Distribution of estuarine benthic diatom species along salinity and nutrient gradients. European Journal of Phycology **33**:173-183.
Van Es, F. B. 1982. Community metabolism of intertidal flats in the Ems-Dollard Estuary. Marine Biology **66**:95-108.
Van Raalte, C. D., I. Valiela and J. M. Teal 1976a. The effect of fertilization on the species composition of salt marsh diatoms. Water Research **10**:1-4.
—— 1976b. Production of epibenthic salt marsh algae: light and nutrient limitation. Limnology and Oceanography **21**: 862-872.
Varela, M. and E. Penas. 1985. Primary production of benthic microalgae in an intertidal sand flat of the Ria de Arosa, NW Spain. Marine Ecology Progress Series **25**:111-119.
Webber, E. E. 1967. Bluegreen algae from a Massachusetts salt marsh. Bulletin of the Torrey Botanical Club **94**:99-106.
—— 1968. Systematics and ecology of benthic salt marsh algae at Ipswich, Massachusetts. Dissertation, University of Massachusetts, Amherst, Massachusetts, USA.
Weissburg, M. 1992. Functional analysis of fiddler crab foraging: sex-specific mechanics and constraints in *Uca pugnax* (Smith). Journal of Experimental Marine Biology and Ecology **156**:105-124.
Wetzel, R. L. 1977. Carbon resources of a benthic salt marsh invertebrate *Nassarius obsoletus* Say (Mollusca: Nassariidae). Pages 293-308 *in* M. Wiley, editor. Estuarine processes. Volume 2. Academic Press, New York, New York, USA.
Whitney, D. E. and W. M. Darley. 1979. A method for the determination of chlorophyll *a* in samples containing degradation products. Limnology and Oceanography **24**:183-186.
—— Whitney, D. E. and W. M. Darley. 1983. Effect of light intensity upon salt marsh benthic microalgal photosynthesis. Marine Biology **75**:249-252.
Williams, R. B. 1962. The ecology of diatom populations in a Georgia salt marsh. Thesis, Harvard University, Cambridge, Massachusetts, USA.
—— 1964. Division rates of salt marsh diatoms in relation to salinity and cell size. Ecology **45**:877-880.
—— 1965. Unusual motility of tube-dwelling pennate diatoms. Journal of Phycology **1**:145-146.
Zedler, J. B. 1980. Algal mat productivity: comparisons in a salt marsh. Estuaries **3**:122-131.
—— 1982. Salt marsh algal mat composition: spatial and temporal comparisons. Bulletin of the Southern California Academy of Sciences **81**:41-50.

STRUCTURE AND PRODUCTIVITY OF MICROTIDAL MEDITERRANEAN COASTAL MARSHES

CARLES IBAÑEZ
Departament d'Ecologia, Universitat de Barcelona.
Diagonal 645, 08028 Barcelona, Spain

ANTONI CURCO
Departament de Biologia Vegetal (Botànica), Universitat de Barcelona
Diagonal 645, 08028 Barcelona, Spain

JOHN W. DAY, JR.
Department of Oceanography and Coastal Sciences and
Coastal Ecology Institute, Louisiana State University
Baton Rouge, LA 70803 USA

NARCIS PRAT
Departament d'Ecologia, Universitat de Barcelona
Diagonal 645, 08028 Barcelona, Spain

Abstract

This paper reviews the literature on structure and production of Mediterranean microtidal marshes. Literature on structure and zonation is relatively abundant but there are relatively few studies of coastal wetland primary productivity in the Mediterranean. These tidal marshes are poorly flushed because of the low tidal range and freshwater tidal marshes are rare. Most marshes are found in deltas and fringing coastal lagoons. Recent studies carried out in the Ebre, Po and Rhone deltas show that net primary production (NPP) of marshes is strongly influenced by soil salinity and flooding. The productivity of these marshes is generally low, but there are significant exceptions. Minimum values of NPP of emergent vegetation (below- plus above-ground) were obtained in salt marshes dominated by *Arthrocnemum macrostachyum* (237 g m^{-2} y^{-1}), characterized by low flooding frequency and high salt stress. Maximum values (up to 9685 g m^{-2} y^{-1}) were obtained in fresh marshes dominated by *Cladium mariscus*, with high flooding frequency. In general terms, Mediterranean microtidal marshes have low production due to salt stress and weak tidal flushing. This suggests that there is low export of marsh production to coastal lagoons, bays and open coastal waters.

1. Introduction

In this paper, we review the production ecology of Mediterranean microtidal marshes. In doing so, we include published data from a number of coastal systems as well as unpublished data from the Ebre delta. A comparison of primary production values and factors affecting productivity among Mediterranean climate coastal marshes is also provided, in order to elucidate the importance of climatic factors versus tidal range in the ecology of these systems. Finally, we present and discuss hypotheses about the coupling between coastal waters and coastal wetlands.

1.1 CHARACTERISTICS OF THE MEDITERRANEAN BASIN

In the Mediterranean basin (including the Mediterranean Sea and the Black Sea), both sea level changes and rainfall, 2 fundamental factors affecting coastal marsh productivity, are highly variable in space and time. Continental runoff to the Mediterranean sea is generally low, except in the Adriatic Sea and the Gulf of Lyon (mostly due to the Po and Rhone rivers, respectively). Previously, the Nile River had the highest discharge, but discharge is presently very low due to dam construction and water use (Wahby and Bishara 1981).

This area is microtidal with astronomical tide ranges generally from 20-30 cm. For instance, in the Ebre delta mean and maximum tidal ranges are 16 and 25 cm, respectively (Jiménez 1996). Only near the Strait of Gibraltar and in the northern Adriatic are there higher tidal ranges (about 1 m, Sestini 1992). Maximum meteorological tides are higher than astronomical tides. For instance, monthly maximum surge height due to meteorological tides is about 1 m in the Ebre delta (Jiménez 1996), and water surges up to 2 m have been recorded in Venice lagoon (Sestini 1992). Many Mediterranean coastal marshes are only flooded during these high tides. Minimum sea level is usually recorded in winter or in summer, especially under atmospheric high pressures. Maximum sea level and rainfall occur in fall, with a secondary maximum in spring. In many marshes, there is a hypersaline aquifer few decimeters under the soil surface, and sea level variations may force this saline water towards the surface.

The Mediterranean climate, which occurs between approximately 35-40° N and S, is a transitional type between dry subtropical and temperate zones (Walter 1973). Its distribution includes the Mediterranean basin, the Pacific coast of central and southern California and central Chile, the southern extreme of Africa and SW Australia. It is characterized by dry, hot summers which produce strong hydrologic stress in plants. Winters tend to be moderate and wet.

1.2 MEDITERRANEAN COASTAL MARSHES

We present here some general information on Mediterranean coastal wetlands. The Mediterranean basin is rich in wetlands of great ecological, social and economic value. Yet these important natural assets have been considerably degraded or destroyed, mainly during the 20th century (Skinner and Zalewski 1995). The primary problems affecting coastal wetland conservation in the Mediterranean are reclamation for agriculture, tourism and urban development, sediment starvation due to reservoirs and river dikes, impoundments, overexploitation of natural resources (fisheries, hunting, etc.), eutrophication and chemical pollution. Mediterranean wetlands have been greatly reduced in area and changed by human activities. These changes have taken place since the Greco-Roman times but the greatest changes have been in the 20th century. For example, between 1942 and 1984, more than 30000 ha (about 40 %) of wetlands were lost in the Rhone Delta (Tamisier 1990). Similarly, during this century, coastal wetland loss has been 28% in Tunisia, more than 60% in Spain and Greece, and more than 70% in Italy (Hollis 1992). The largest wetland areas remaining in the Mediterranean occur in the main deltaic areas (Fig. 1 and Table 1): Ebre (Spain), Rhone (France), Po (Italy), Nile (Egypt) and Danube (Romania). The Danube Delta contains a large, mostly undisturbed, wetland area. About 25% of the approximately 5000 km^2 of wetlands have been reclaimed since 1983 (Hollis 1992). There are also important marsh areas in several coastal lagoons such as in the Languedoc region in southwest France and the Venice and Marano lagoons in the north Adriatic (Italy).

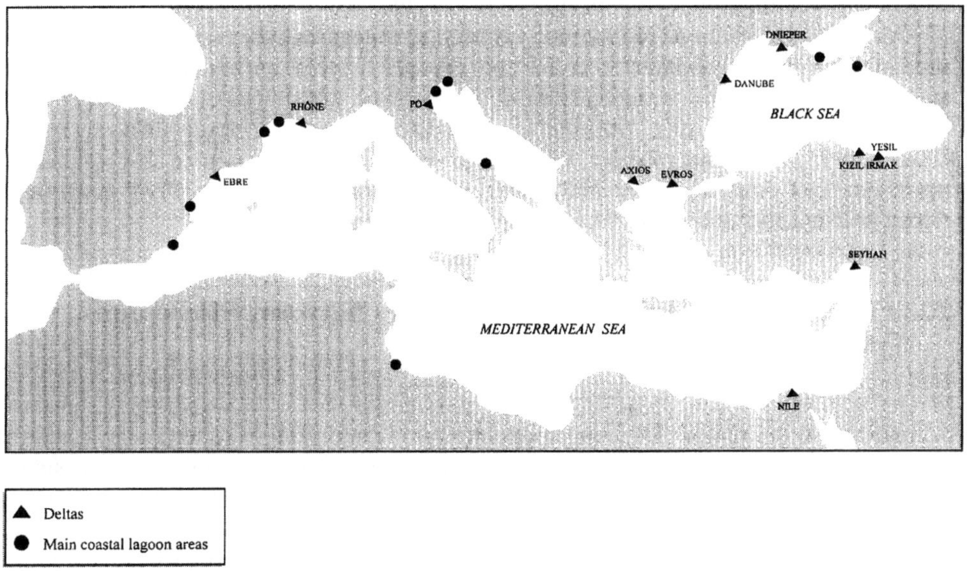

Figure 1. Location of the main deltas (triangles) and coastal lagoons (circles) in the Mediterranean and Black seas. Modified from Pearce and Crivelli (1994).

TABLE 1. Surface area and location of the seven largest Mediterranean and Black Sea deltas.

	Surface area (km^2)	Basin	Country
Po	13398	Mediterranean	Italy
Nile	12512	Mediterranean	Egypt
Danube	4420	Black sea	Romania, Ukraine
Rhone	1736	Mediterranean	France
Seyhan	625	Mediterranean	Turkey
Kizil Irmak	500	Black sea	Turkey
Ebre	330	Mediterranean	Spain

From a descriptive point of view, Mediterranean coastal marsh vegetation is well known, especially due to the studies based in the phytosociological approach of the Zürich-Montpellier School (see Géhu 1984). By contrast, there are few ecological studies, except the research carried out in the 70s in the Rhone delta (Heurteaux 1970, Grouzis et al. 1977, Molinier and Devaux 1978, Corre 1979). In the Rhone delta, one of the most characteristic salt marsh species, *Sarcocornia fruticosa* (formerly *Arthrocnemum fruticosum*), was intensively studied, in terms of structure and primary production (Nichabouri and Corre 1970, Eckardt 1972, Grouzis 1973, Berger et al. 1978, Berger et al. 1979). In other Mediterranean areas, only some general studies on vegetation are found in the literature (Ferrari et al. 1985, García et al. 1993, Shaltout et al. 1995). Recently the first estimates of total net primary production in the Mediterranean were obtained for 3 deltas (Ebre, Rhone and Po) (see Table 4).

Although estimates of productivity in Mediterranean marshes are scarce, literature about factors affecting plant zonation and growth, as well as food web structure, is quite abundant from southern California (Zedler et al. 1980, Zedler 1983, Pearcy and Ustin 1984, Callaway et al. 1990, Pennings and Callaway 1992, Page 1995, Ayala and OLeary 1995, Kwak and Zedler 1997, Page 1997). There are also a few references from southeast Australia (Rea and Ganf 1994, Clarke and Jacoby 1994) and South Africa (Adams and Bate 1994, Naidoo and Rughunanan 1990).

2. From Deltas to Estuaries: Differences between Micro and Macrotidal Coasts

To understand the factors affecting productivity of microtidal Mediterranean estuarine environments, it is important to contrast the differences of these microtidal systems with macrotidal coastal systems (Table 2). Among other microtidal systems are the Baltic Sea and much of the Gulf of Mexico. A major portion of Mediterranean coastal wetlands are located in deltaic areas, which in many senses have opposite features to the classical estuaries. Mediterranean deltas protude into the sea and their estuaries are river-dominated, whereas typical estuaries are coastal identations more dominated by tides. To study Mediterranean estuaries, one must consider the whole deltaic system, comprehensively as a geological, hydrological and ecological unit.

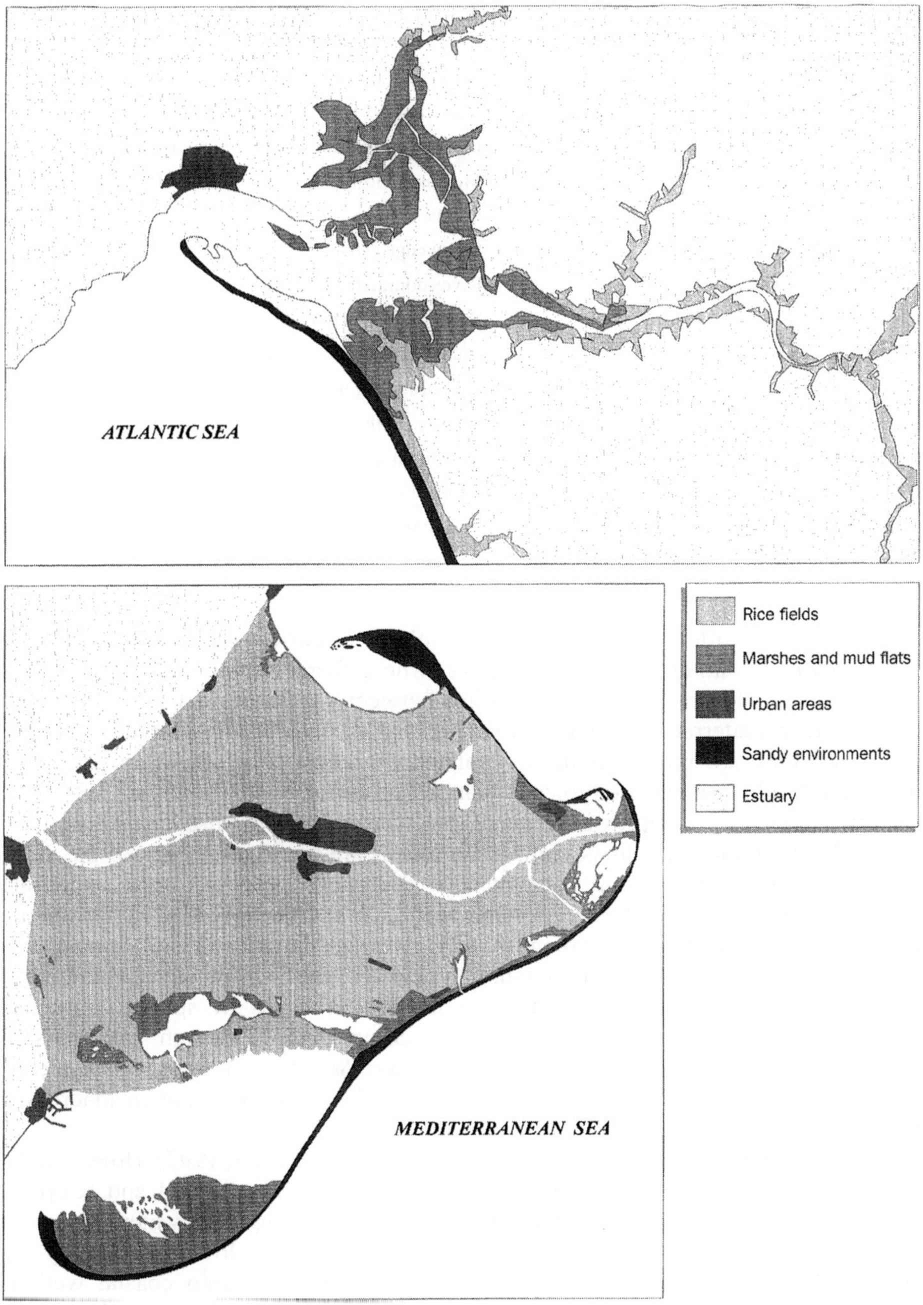

Figure 2. Structural differences between typical macrotidal (Sado estuary, Portugal) and microtidal (Ebre delta, Spain) estuarine areas. Notice that in the Sado estuary marsh areas are located along the inner shore, while in the Ebre delta they are located around the coastal lagoons and along the outer backshore. The Sado estuary is about 50 km long and the Ebre delta is about 30 km long.

TABLE 2. Some differences between macrotidal temperate and microtidal Mediterranean estuarine systems.

Macrotidal temperate systems	Microtidal Mediterranean systems
Tidal range 1-10 m	Tidal range 0.1 - 1 m
High astronomical tides /low meteorological tides	Low astronomical tides/high meteorological tides
High biological productivity	Medium to low biological productivity
Coastal plain estuaries (some deltas)	Deltas, lagoons
Partially and well mixed estuaries	Salt-wedge estuaries
Tide-dominated estuaries	River-dominated estuaries
Extensive tidal marshes	Small and localized tidal marshes
Presence of tidal freshwater marshes	Tidal freshwater marshes rare
Presence of marshes along the shore of the estuary	Marshes along the shore of the estuary rare
Marshes with regular flooding	Marshes with summer drought and salt stress
Maximum productivity from high to middle marsh	Maximum productivity from low to middle marsh
Marsh diversity gradient along the estuary	Marsh diversity gradient across the delta
Plant zonation mainly due to tidal flooding	Plant zonation mainly due to maximum summer salinity
Higher topographic gradient, lower patchiness	Lower topographic gradient, higher patchiness

Tidal marshes often occur in coastal plain estuaries (not present in the Mediterranean) and along the shores of bays, where there is both tidal activity and riverine influence, leading to close coupling between wetlands and adjacent water bodies. In the Mediterranean, coastal marshes are most often not directly riverine influence and they are more lagunal or isolated in nature (Fig. 2). Microtidal river mouths in the Mediterranean are typically salt-wedge estuaries (Ibañez et al. 1997).

Deltaic systems are complex in terms of structure and functioning. The diversity of habitats in Mediterranean deltas is high. River levees are the highest parts of the delta, and under natural conditions, are vegetated by deciduous riparian forests which are flooded only during high discharge. Marshes are fresh, brackish or salty, depending on factors such elevation, inputs of upland runoff, riverine influence or soil drainage. In most cases, there is a clear vegetation zonation mainly related to soil salinity and water regime. Mediterranean coastal marshes are diverse in terms of spatial ecological conditions (and so in richness in vegetation communities), though salt marshes dominated by perennial succulent plants of the genera *Sarcocornia* and *Arthrocnemum* are most abundant. Species of the genus *Spartina*, the most abundant in macrotidal temperate marshes, are rare in the Mediterranean.

Deltas often have high biological productivity (Day et al. 1995, 1997). However, the most characteristic feature of Mediterranean marshes is a high spatial and temporal variability of productivity (Ibañez et al. 1999). Marshes having fresh water inputs can have high productivity, whereas salt marshes usually have low productivity. In summary, heterogeneity is a fundamental feature of Mediterranean coastal wetland environments, both spatially due to the complexity of deltaic and other habitats and temporally due to the occurrence of irregular (and sometimes extreme) pulsing events (strong rainfall and winds, surges, river floods, extreme droughts, etc.).

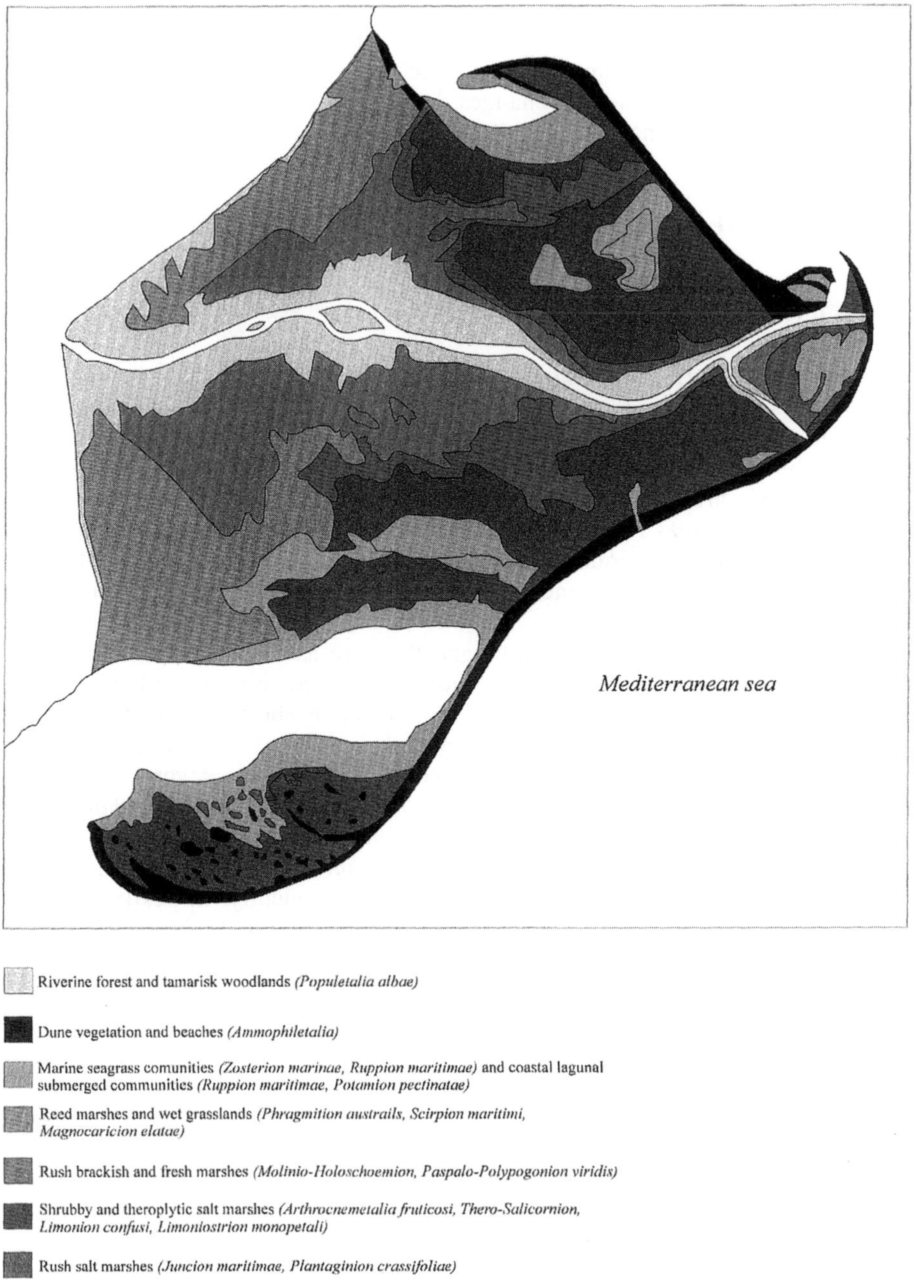

Figure 3. Map of potential vegetation of the Ebre delta (from Curcó et al. 1997). Presently, most of the delta is devoted to rice agriculture (see Fig. 2).

Whereas macrotidal marshes occupy large areas and are characterized by strong and regular tidal fluxes which play a crucial role in ecosystem functioning, microtidal Mediterranean marshes have smaller areas and the role of the weak tide in the ecosystem is less relevant. Sea level changes due to seasonal cycles and storms likely play a more important role than astronomical tides in microtidal marshes.

3. Typology and Productivity of Mediterranean Coastal Marshes

3.1 TYPOLOGY

The variety of Mediterranean coastal marsh-types is high in relation to Arctic and Temperate coastal areas, mainly due to the high variability of salinity and water regimes in relatively small areas, mostly associated with differences in microtopography (for example, the Ebre delta, see Fig. 3). Coastal marshes of the Mediterranean basin are well known from a typological point of view, especially the northern and western areas (see Géhu 1984, Pearce and Crivelli 1994, Ferrari et al. 1985). The types of plant communities are similar around the basin, although there are differences in floristic and environmental conditions associated with geographical and climatic variations (Table 3).

Coastal freshwater marshes are scarce and they are associated with underground freshwater springs in karstic zones or with areas receiving agricultural runoff. Temporary freshwater ponds are dominated by annual or short-lived plants, such as stoneworts (*Chara* sp.). Permanent freshwater habitats (natural wells, old river channels) contain submerged and floating communities with several species of pondweed (*Potamogeton spp, Myiophyllum spicatum*) and water lily (*Nymphaea alba*). In the marshes, emerged vegetation is dominated by cattails (*Typha* sp.) and the common reed (*Phragmites australis*). The spike-rush (*Scirpus holoschoenus*) and some grasses like the water couch (*Paspalum distichum*) also occur frequently. In peatlands with a quite constant water level, the saw sedge (*Cladium mariscus*) and other sedge species like *Carex* sp. can dominate.

TABLE 3. Main wetland-types in the Mediterranean basin, classified according to salinity, structure and dominant species (or genera).

Syntaxonomic syntaxa	Main features	Species
Fresh marshes		
Lemnetea	Floating non-rooted macrophytes, mainly in eutrophic waters	*Lemna* sp.*Riccia* sp., *Azolla* sp.
Utricularietea intermedio-minoris	Floating macrophytes, oligotrophic waters	*Utricularia vulgaris, U. australis*
Charetea	Submerged and annual macrophytes, temporary marshes	*Chara* sp.
Parvopotametalia	Submerged macrophytes, temporary marshes	*Potamogeton* sp.*Ranunculus* sp., *Callitriche* sp.*Elatine* sp.
Magnopotametalia	Submerged macrophytes, permanent marshes	*Potamogeton* sp.*Nymphaea alba, Nuphar luteum*
Phragmitetalia	Emerged and perennial macrophytes, eutrophic waters, permanent marshes	*Phragmites australis, Typha* sp., *Scirpus* sp.
Magnocaricetalia	Emerged and perennial macrophytes, eutrophic waters, permanent marshes	*Cladium mariscus, Carex* sp.
Littorelletea	Emerged and perennial macrophytes, meso-oligotrophic waters, temporary marshes	*Eleocharis* sp.*Juncus* sp.
Isoeto-Nanojuncetea	Emerged and annual macrophytes, oligotrophic waters, temporary marshes	*Isoetes* sp., *Juncus* sp., *Lythrum* sp., *Cyperus* sp.
Molinetalia caeruleae	Emerged and perennial macrophytes, temporary marshes, oligotrophic waters	*Molinia caerulea*
Holoschoenetalia	Emerged and perennial macrophytes, temporary marshes	*Scirpus holoschoenus, Trifolium* sp., *Juncus* sp.,*Paspalum*
Brackish marshes		
Magnopotametalia	Submerged macrophytes, permanent marshes	*Potamogeton pectinatus*
Scirpetalia compacti	Emerged and perennial macrophytes	*Scirpus maritimus, Scirpus littoralis, Phragmites australis*
Salt marshes		
Zosteretea marinae	Submerged macrophytes, marine waters	*Zostera* sp. *Cymodocea nodosa*
Ruppietea	Submerged macrophytes, hypersaline to brackish waters, sometimes temporary	*Ruppia* sp.*Lamprothamnion papulosum*
Spartinetea maritimae	Emerged macrophytes, tidal marshes	*Spartina maritima, Spartina densiflora (introduction)*
Thero-Salicornietea	Annual and succulent halophytes, temporary marshes	*Salicornia, Suaeda*
Arthrocnemetea fruticosi	Perennial and often succulent halophytes, temporary marshes	*Arthrocnemum macostachyum, Sarcocornia fruticosa, S. perennis*
Juncetalia maritimi	Emerged and perennial macrophytes, temporary marshes	*Juncus* sp.*Puccinellia* sp.
Saginetea maritimae	Annual halophytes, temporary and nitrophilous marshes	*Frankenia pulverulenta, Hymenolobus procumbens*

In some coastal areas, former salt and brackish marshes have been transformed into rice fields (about 20 000 ha in the Ebre and Rhone deltas, for instance). Irrigation with river water takes place from April to October, which is in contrast to natural wetlands where winter is the wet period. Rice fields are shallow, temporary, highly productive wetlands. Vegetation can be quite complex, depending on agricultural management, and is dominated by annual species with a pantropical distribution (*Bergia, Ammannia, Lindernia, Echinochloa*, etc.).

Brackish marshes are more widespread than fresh marshes, and have lower species diversity. Temporary ponds are dominated by small submerged macrophytes (such as *Zannichellia* sp. and *Ranunculus* sp.), while permanent aquatic systems (like coastal lagoons) usually are dominated by fennel pondweed (*Potamogeton pectinatus*) and, in more saline waters, by *Ruppia* sp. Freshening of formerly high salinity coastal lagoons by agricultural runoff has led to changes in plant communities. Reed beds, dominated by *Phagmites australis* and occasionally sea club-rush (*Scirpus maritimus*), are now widespread around coastal lagoons. Brackish rush communities occur in slightly higher elevations and contain several species of halophylous rushes (*Juncus maritimus* and *J. acutus*) and grasses (*Aeluropus littoralis* or*Paspalum vaginatum*).

Salt marsh area has been greatly reduced by reclamation and hydrological changes. For example, the reduction has been 60% in the Ebre delta (Curcó et al. 1997) and about 30% in thc Rhonc delta (Tamisier 1990). In the Mediterranean, salt marshes can occur in supra tidal areas, where there is shallow, hypersaline ground water. The elevation range of these salt marshes increases with the intensity of arid conditions and they can become widespread in dry subtropical areas. For instance, they are common in the western Mediterranean and rare in the wetter northern Adriatic. These plant communities are often dominated by non-succulent xerophilous species, which can excrete salts through salt glands, such as *Limonium* sp., *Limoniastrum* sp. or *Frankenia* sp. In areas with high rainfall, xerophilous salt marshes are replaced by halophytic grasslands rich in graminoids, such as *Puccinellia* sp.

Seasonally flooded Mediterranean salt marshes are diverse and show a wide range of water and salinity regimes. In the driest areas, soil salinity can become so high that phanerogamic plants cannot survive and, in this case, they are colonized by algal and microbial mats. When the flooding period does not include the entire summer, annual species of glasswort (*Salicornia* sp.) can form dense stands. However, the more typical and widespread salt marshes are dominated by succulent shrubby species, mainly of the Chenopodiaceae (*Sarcocornia, Arthrocnemum, Suaeda, Halocnemum*). This family also dominates in the high marsh of other Mediterranean-type climate regions (California, South Africa, etc.), and even in some parts of temperate and subtropical regions (Chapman 1977). In most of the northern Mediterranean basin, different *Sarcocornia* and *Arthrocnemum* species occur along the elevation gradient between mean sea level and the highest sea level due to meteorological tides, primarily associated with increasing salinity. In the upper salt marsh, *Arthrocnemum macrostachyum* (=*A. glaucum*) forms sparse communities, sometimes with winter annuals (*Hymenolobus procumbens, Frankenia pulverulenta*, etc.). This community can tolerate strong variations of water level and soil salinity. In the middle marsh, *Sarcocornia fruticosa* forms taller shrublands with a more dense cover. In the low marsh, *Sarcocornia perennis*, a prostrate shrub much rarer than the two other chenopod species, grows at mean sea level, where water and salinity conditions are quite stable during the year. In the southern Mediterranean basin, *A. macrostachyum* is partially replaced by another chenopod more tolerant of a drier climate, *Halocnemum strobilaceum*. Along the northern Adriatic coast, between Marano and Grado lagoons, the low marsh is dominated by *Spartina maritima* due to a high tidal range (Géhu et al. 1984). Salt water coastal lagoons and shallow bays are colonized by seagrass communities, mostly by *Zostera noltii* and *Cymodocea nodosa*.

3.2 VEGETATION ZONATION COMPARISON WITH NON-MEDITERRANEAN AREAS

A number of factors affect the distribution of vascular plant species along the estuarine gradient, including salinity, frequency and duration of inundation, sulfide concentration and substrate composition (Odum 1988). In macrotidal areas, low marshes are flooded daily by sea water, so water and salinity regime of soils are rather independent of climatic conditions. For this reason, the low marsh of macrotidal Mediterranean-type climate coasts has similar vegetation to the European temperate zone and it is composed of *Spartina* communities. The middle and upper marsh are dominated by halophytes, often succulent chenopods, adapted to high soil salinity. Where hypersaline periods are too long, salt flats without perennial vegetation become widespread. In microtidal Mediterranean coasts, low marshes dominated by *Spartina* grasses are practically absent, and glasswort communities can develop from mean sea level to the upper marsh. Several authors showed the importance of summer drought in explaining the distribution of plant communities in these areas (Heurteaux 1970, Callaway et al. 1990, Zedler 1983, Zedler and Beare 1986). Corre (1985) showed that the zonation of salt marsh vegetation across the shore of a temporary salt pond in the Rhone delta was mainly determined by inundation in the lower part, whereas in the higher parts the distribution of soil salinity was the main factor explaining the distribution of plant communities. Salinity, like inundation, has its impact through maximum values and also through seasonal fluctuations.

There are significant differences in zonation of microtidal Mediterranean marshes, macrotidal Mediterranean-climate marshes, and other temperate macrotidal marshes. In macrotidal coasts, factors responsible for marsh plant zonation are essentially similar to those causing zonation in the rocky intertidal zone. At the lower end of a physical gradient (low marsh) the range of a species is limited by its tolerance to physical conditions (e.g., submergence and hypoxia), whereas at the upper end of the gradient (higher in the marsh) a species is excluded by competition. Mediterranean-climate macrotidal salt marshes, however, do not exhibit a simple monotonic gradient of severity of physical factors across marsh elevations, rather there is an interaction between flooding and salinity that creates a band of superior habitat in the middle marsh, where both factors are moderate, a phenomenon not reported elsewhere (Pennings and Callaway 1992).

Peinado et al. (1995) carried out one study comparing vegetation zonation of Mediterranean marshes in Spain and Mediterranean-climate macrotidal marshes in California and Baja California. The marshes in these two areas are similar in terms of taxonomic composition, physiognomy and vegetation zonation. The low marsh is dominated by hydrophytic perennial vegetation: *Spartina* communities in the lowest subzone (which is practically absent in the Mediterranean basin due to low tidal range) and *Sarcocornia* prostrate communities in the highest subzone. In the middle marsh, vegetation is mostly formed by erect species of *Sarcocornia*, and in the upper marsh by more halophytic species of the genus *Arthrocnemum*. Pioneer annual vegetation of

Salicornia species can be found in the bare gaps. Finally, vegetation of the drier upper marsh is made up of halophytic tall rushes (*Juncus* sp.) which form the transitional zone between the marsh and upland vegetation.

In non-Mediterranean macrotidal salt marshes, vascular plants typically are found only in the upper two thirds of the intertidal zone. The lower one third consists of bare mud and, at times, a layer of micro and macro algae. This lack of colonization is primarily a result of high duration of flooding. Exceptions to this general pattern appear to occur where the tidal amplitude is very slight, as along the northern coast of the Gulf of Mexico. Here marsh plants such as *Spartina alterniflora* grow virtually to mean low tide (Odum 1988). This is also true for the Mediterranean coastal wetlands, where normally there are no tidal mud flats between the open water and the marsh.

3.3 BIOMASS AND PRIMARY PRODUCTION

One reason that productivity of salt marshes has been studied so thoroughly is that it is often much higher than other ecosystem types. There is also considerable evidence that salt marsh production forms the basis of important estuarine food chains (Day et al. 1989). However, research on the production ecology of Mediterranean coastal wetlands is quite scarce. One reason in the low level of research in Mediterranean countries, but another reason may be related to perceived low productivity and consequent low commercial exploitation.

There have been numerous studies of primary production in salt marshes but temporal and spatial variability limits generalizations (Odum 1988). Moreover, the use of different methods and high sampling variability make comparisons difficult. Hopkinson et al. (1980) found that different techniques for measuring annual net production of salt marsh plants in Louisiana gave highly variable results. There have been few studies of net primary production (NPP) of coastal Mediterranean marshes; and all of them estimated only the above-ground component during one growing season, using peak standing crop. Recently, a study on NPP of coastal marshes in three Mediterranean deltas (Ebre, Po and Rhone) was carried out. Table 4 compares the NPP values obtained in this study to those of other studies carried out in the Mediterranean and Mediterranean-type marshes. Above-ground values are higher in reed-type brackish marshes (with a maximum of 2989 g m^{-2} y^{-1} in a *Typha angustifolia* marsh), while shrubby salt marshes show lower values (with a minimum of 94 g m^{-2} y^{-1} in an *A. macrostachyum* marsh). Above-ground NPP in reed-type marshes ranges from 452 g m^{-2} y^{-1} in a *Scirpus maritimus* marsh to 2989 g m^{-2} y^{-1} in the *Typha angustifolia* marsh, both in the Rhone delta. In this case, the variation was mainly due to grazing in the *Scirpus maritimus* marsh (the *Typha angustifolia* marsh was protected by an enclosure). Values in the 3 *Phragmites australis* marshes ranged from 824 to 1400 g m^{-2} y^{-1}, the highest value being in the fresher marsh. A *Cladium mariscus* marsh growing in a peatland area had a high above-ground production (1620 g m^{-2} y^{-1}). This marsh had the maximum below-ground NPP (8070 g m^{-2} y^{-1}), whereas the minimum value was for the *Arthrocnemum macrostachyum* salt marsh in the Ebre delta.

There are significant differences in the above and below-ground NPP estimates between *A. macrostachyum* and *Sarcocornia fruticosa* salt marshes. *S. fruticosa* had higher above- (580-1262 g m^{-2} y^{-1}) and below-ground (950-1260 g m^{-2} y^{-1}) NPP than

A. macrostachyum (above-ground 94-294 g m^{-2} y^{-1}, and 50 g m^{-2} y^{-1} for the only below-ground estimate). A *S. fruticosa* salt marsh from the Po delta had relatively low above-ground NPP (683 g m^{-2} y^{-1}); this is likely due to waterlogging by tidal flooding rather than salinity stress. Salt marshes of northern areas (Po and Rhone deltas), showed a strong biomass seasonal pattern, with little or no above-ground biomass during winter, whereas in the southern salt marshes (Ebre delta), there was significant above-ground biomass in winter.

Above-ground NPP of shrubby salt marshes from southern California was similar to equivalent Mediterranean salt marshes (between 215-1200 g m^{-2} y^{-1}). *Sarcocornia pacifica*, the most widespread chenopod in the marshes studied, has 2 different biotypes depending on the position in the salt marsh (low and middle marsh). In the middle marsh *S. pacifica* has an erect form (80-100 cm height), like the Mediterranean *S. fruticosa*, which usually has higher values of NPP. In the low marsh, *S. pacifica* forms a creeping, low-lying shrub (similar to the Mediterranean *S. perennis*), which has low biomass and productivity values. Finally, *Spartina foliosa* marshes in the Californian tidal salt marshes also have a broad range of above-ground NPP (280-1700 g m^{-2} y^{-1}) with the lowest values usually from the low marsh.

Mahall and Park (1976) reported above-ground NPP of 550 to 960 g m^{-2} y^{-1} in *Sarcocornia pacifica* salt marshes of San Francisco Bay (California), with lower values from more saline soils. Mall (1969) reported mean productivity of the same species in another Californian marsh of 722 g m^{-2} y^{-1}, with a low of 276 g m^{-2} y^{-1} on highly saline soils. Jefferies (1972) measured NPP of 867 g m^{-2} y^{-1} for annual *Salicornia* species. Above-ground primary production ranged from 835 to 2316 g m^{-2} y^{-1} in *Batis-Sarcocornia* microtidal marshes from Florida and turnover rates ranged from 1.1 to 5.8 (Rey et al. 1990). Above-ground primary production for similar marshes in California ranged from 300 to 1200 g m^{-2} y^{-1} (Onuf et al. 1978, Eilers 1981, Zedler 1982). Onuf (1987) reported turnover rates of 2.9 and 2.3 for *Batis maritima* and *Sarcocornia pacifica*, respectively, in marshes bordering Mugu Lagoon, California.

Table 5 presents a detailed summary of marsh productivity of the Ebre delta. Soil features (organic matter, C and N content) and water and salinity regimes are more favorable for productivity in the brackish marshes, since in the salt marshes organic matter and nitrogen content are very low, and hypersalinity is present almost all year. Overall, there is an increase of biomass and primary production as salinity decreases. The brackish marshes are dominated by reed-type species which have a pronounced seasonality of above-ground live biomass. Shrubby plants dominate the salt marshes and have a more constant above-ground live biomass during the year, an important part of which is lignified, non-photosynthetic structures. Mean annual values of above-ground live biomass are quite homogeneous in three of the marshes (about 440-470 g m^{-2}), while the *Cladium* marsh (850 g m^{-2}) and the *S. fruticosa* marsh (1660 g m^{-2}) show higher values. Total standing biomass and litter values are high in the two brackish marshes and in the *S. fruticosa* salt marsh. Maximum values of above-ground NPP and turnover occur in brackish marshes (1400 and 1620 g m^{-2} y^{-1}). In the salt marshes, maximum values of above-ground NPP occur in the mixed marsh (840 g m^{-2} y^{-1}), which has an exceptionally high turnover, followed by the *S. fruticosa* marsh (580 g m^{-2} y^{-1}) and the *A. macrostachyum* marsh (350 g m^{-2} y^{-1}). Below-ground material is higher than above-ground one in the brackish marshes, while the above-ground one is greater than the

TABLE 4. Net primary production estimates from different Mediterranean-type climate coastal marshes (brackish and salt marshes). Some estimates from subtropical marshes dominated by *Sarcocornia pacifica* and *Batis maritima* are also included.

	Study area	Location	Species	Above-ground	Below-ground	Method	Reference
Mediterranean	Po delta	Chioggia	*Sarcocornia fruticosa*	683	1260	Smalley	Scarton et al. 1998
	Rhone delta	Palissade	*Sarcocornia fruticosa*	1123		Peak biomass	Ibañez et al. 1999
		Palissade	*Sarcocornia fruticosa*	773		Peak biomass	Rioual et al. *unpublished*
		Digue South	*Sarcocornia fruticosa*	1262		Peak biomass	Ibañez et al. 1999
		La Gacholle 4	*Sarcocornia fruticosa*	1049		Peak biomass	Berger et al. 1978
		La Gacholle 6	*Sarcocornia fruticosa*	1000		Peak biomass	Berger et al. 1978
		Digue North	*Arthocnemum macrostachyum*	187		Peak biomass	Ibañez et al. 1999
		Tour du Valat	*Arthocnemum macrostachyum*	294		Peak biomass	Ibañez et al. 1999
		Bernacles	*Arthocnemum macrostachyum*	94		Peak biomass	Rioual et al. *unpublished*
	Ebre delta	Buda 1	*Arthocnemum macrostachyum*	190	50	Smalley	Curcó et al. *unpublished*
		Buda 2	*A. macrostachyum/S. fruticosa*	840	340	Smalley	Curcó et al. *unpublished*
		Buda 3	*Sarcocornia fruticosa*	580	950	Smalley	Curcó et al. *unpublished*
	Po delta	Barricata	*Phragmites australis*	876	2263	Smalley	Scarton et al. 1998
	Rhone delta	Relongues in	*Typha angustifolia*	2989		Smalley	Ibañez et al. 1999
		Relongues out	*Scirpus maritimus*	452		Smalley	Ibañez et al. 1999
		Palissade	*P. australis/S. maritimus*	824		Smalley	Ibañez et al. 1999
	Ebre delta	Garxal	*Phragmites australis*	1400	3740	Smalley	Curcó et al. *unpublished*
		Vilacoto	*Cladium mariscus*	1620	8070	Smalley	Curcó et al. *unpublished*
California	San Francisco bay	Mare Island	*Sarcocornia pacifica*	815-960		Peak biomass	Mahall & Park 1976
		Petaluma River	*Sarcocornia pacifica*	215-550		Peak biomass	Mahall & Park 1976
	Suisun marsh		*Sarcocornia pacifica*	276-722		?	Mall 1969
	San Francisco bay	Tolay Creek	*Sarcocornia pacifica*	1060-1185		Peak biomass	Cameron 1972
	San Diego bay		*Sarcocornia pacifica*	1500-2500		?	Mudie 1970
	South California	Several locations	*S. pacifica/Batis maritima*	300-1200		?	Zedler 1982
	San Diego bay		*Spartina foliosa*	800		?	Mudie 1970
	San Francisco bay	Petaluma River	*Spartina foliosa*	445-690		Peak biomass	Mahall & Park 1976
		Mare Island	*Spartina foliosa*	280		Peak biomass	Mahall & Park 1976
		Tolay Creek	*Spartina foliosa*	1400-1700		Peak biomass	Cameron 1972
Florida	Indian River lag.	Oslo Road Marsh	*S. pacifica/Batis maritima*	1984-2316		Rey's method	Rey et al. 1990
		North Marsh	*S. pacifica/Batis maritima*	1139-1419		Rey's method	Rey et al. 1990
		IRC12 PD	*S. pacifica/Batis maritima*	937-1413		Rey's method	Rey et al. 1990
		IRC12 UE	*S. pacifica/Batis maritima*	835-1543		Rey's method	Rey et al. 1990

below-ground in the salt marshes. Values of below-ground mean live biomass, total biomass, litter and production are very high in the 2 brackish marshes, specially in the *Cladium* marsh. Below-ground NPP ranges from 3740 to 8070 g m^{-2} y^{-1} in the brackish marshes, and from 50 to 950 g m^{-2} y^{-1} in the salt marshes.

Sarcocornia fruticosa forms dense communities where a large part of the production is used to maintain the permanent, woody structure of the vegetative cover; and old plants are progressively replaced by young ones (Grouzis 1973). Respiration of the lignified stems of *Sarcocornia fruticosa* consumes a considerable part of the energy fixed in photosynthesis, especially during summer drought at high temperatures. This entails, in extreme cases, the elimination of older plants in which the proportion of chlorophyll containing tissues is particularly small (Eckardt 1972).

Berger et al. (1978) carried out a study about the productivity of *Sarcocornia fruticosa* in a salt marsh surrounding a coastal lagoon of the Rhone delta. Above-ground biomass (2300-3700 g m^{-2}) and NPP (1000-1049 g m^{-2} y^{-1}) were high, corresponding to about 402-404 g C m^{-2} y^{-1}. Mineral content in this species is high and variable (200-260 g m^{-2}). Biomass and productivity were higher in less saline study plots, but differences in productivity were small, likely due to the occurrence of higher lignified (i.e., energy-consuming) biomass in the less saline plot. Biomass decreased faster in the more saline plot and by October, biomass in the less saline plot was about 40% higher. The photosynthetic biomass was practically zero during winter, and reached its maximum in July. The authors concluded that above-ground primary production was high during the study period due to high summer rainfall.

Peak production of non-Mediterranean salt marshes is generally higher than that of Mediterranean climates. For example, above-ground primary production of seven coastal marsh species in coastal Louisiana, a microtidal area with high temperature and rainfall, ranged from 1355 to 6043 g m^{-2} y^{-1} (Hopkinson et al. 1978). Above-ground productivity estimates, mostly *Spartina* marshes from the Atlantic Coast of North America, range from 200 to 4000 g m^{-2} y^{-1}, and below-ground estimates from 500 to 6200 g m^{-2} y^{-1} (Turner 1976, Good et al. 1982, Day et al. 1989). Above-ground NPP in freshwater tidal wetlands of the middle Atlantic coast range from 1000 to 4000 g m^{-2} y^{-1}, peak above-ground biomass from 566-2312 g m^{-2}, and below-ground biomass from 500 to over 7100 g m^{-2} (Whigham et al. 1978).

TABLE 5. Net primary production and other parameters from several marshes of the Ebre delta (Curcó, A., C. Ibañez and J.W. Day. 1996, Evaluation of marsh primary production and decomposition, Ebre delta, Spain, *in* Impact of Climatic Change on Northwestern Mediterranean Deltas, Final Report, vol. 1, chapter 6, Universitat Politècnica de Catalunya, Barcelona, Spain). For water level and conductivity the annual mean value and the range (between brackets) are shown. Species composition includes the % of standing biomass in relation the mean standing biomass of the marsh.

	Brackish (1)	Brackish (2)	Salt marsh (1)	Salt marsh (2)	Salt marsh (3)
Conductivity (mS cm^{-1})	2.8 (0.7-6.0)	6.9 (1.1-13.9)	74.2 (37-145)	-	58.9 (23-87)
Water level (cm)	2.3 (-41/22)	-2.1 (-20/26)	-14.2 (-41/10)	-	-11.7 (-31/10)
Number of soil layers	15	15	-	-	10
Sampled thickness (cm)	60	30	-	-	20
Organic matter (%)	38.8	31.8	-	-	5.6
Carbon (mg g^{-1})	235.2	168.3	-	-	50.1
Nitrogen (mg g^{-1})	12.4	7.9	-	-	0.6
Species composition	*Cladium mariscus* (76%) *Phragmites australis* (12%)	*Phragmites australis* (95%) *Scirpus maritimus* (5%)	*Arthrocnemum macrostachyum* (99.2%) *Sarcocornia fruticosa* (0.8%)	*Sarcocornia fruticosa* (62%) *Arthrocnemum macrostachyum* (38%)	*Sarcocornia fruticosa* (100%)
Above-ground					
Mean live biom. (g m^{-2})	850	470	470	440	1660
Total biomass (g m^{-2})	2360	2060	730	950	2800
Litter (g m^{-2})	1180	730	160	350	760
NPP (g m^{-2} y^{-1})	1620	1400	190	840	580
Turnover	1.9	3.0	0.4	1.9	0.3
Below-ground					
Mean live biom. (g m^{-2})	5100	2290	120	170	790
Total biomass (g m^{-2})	7910	3700	140	250	940
Litter (g m^{-2})	2870	1760	70	100	330
NPP (g m^{-2} y^{-1})	8070	3740	50	340	950
Turnover	1.6	1.6	0.4	2.0	1.2
Total					
Mean live biom. (g m^{-2})	5950	2760	590	610	2450
Total biomass (g m^{-2})	10270	5760	870	1200	3740
Litter (g m^{-2})	4050	2490	230	450	1090
NPP (g m^{-2} y^{-1})	9690	5140	240	1180	1530
Turnover	1.6	1.9	0.4	1.9	0.6
Above/belowground ratio					
Mean live biomass	0.17	0.20	3.80	2.59	3.80
Total biomass	0.30	0.55	5.26	3.88	3.00
Litter	0.41	0.41	2.20	3.67	2.28
NPP	0.20	0.38	3.71	2.48	0.61

3.4 FACTORS AFFECTING PRIMARY PRODUCTION

A number of factors affect productivity of Mediterranean coastal marshes. Flooding frequency and duration and soil salinity are perhaps the key factors, but temperature, rainfall, nutrient availability, oxygen levels, sediment type, and drainage are also important. These factors are interrelated and are in turn affected by plant growth (Day et al. 1989). Salt marshes can have significant inter-annual variation in productivity due to variation in the above factors (Teal and Howes 1996). Because Mediterranean marshes experience high variability in the factors affecting productivity, mean values of annual primary production should be based on study periods of several years. Most productivity studies, however, have been for 1 year.

In southern California the growth of *Sarcocornia pacifica* (= *Salicornia virginica*) and *Arthrocnemum subterminalis* is negatively affected by flooding, salinity and competition. However, the relative importance of these factors to the plant varies across the marsh. Thus, the benefit of reduced flooding in the *Arthrocnemum* zone outweighs the disadvantage of higher salinity and consequent lower water potential, and both species grow better in the *Arthrocnemum* zone than in the low *Sarcocornia* zone. In contrast, both species do poorer in the transition zone than in the *Arthrocnemum* zone, even though flooding is greatly reduced, probably because of the high salinity of the transition zone soil. Similarly, although removal of competitors usually increased plant growth dramatically, *Arthrocnemum* plants in the low *Sarcocornia* zone did not have high productivity, presumably because of harsh physical conditions (Pennings and Callaway 1992).

Ground water movement through the marsh significantly affects a wide range of subsurface processes including redox potential, nutrients and toxic compounds such as sulfide, which influence plant distribution, physiological state and primary production (Odum 1988). Evapotranspiration is an important avenue for the vertical flux of water from marsh soils. Replacement of this water occurs via inflow from tidal creek banks, vertical infiltration of flooding tidal water and precipitation, and upland ground water. For expansive, irregularly-flooded coastal wetlands, the critical pathway of pore water exchange is via vertical flux caused by evapotranspiration. This means that subsurface water in these marshes may become stagnant, with high concentrations of toxic substances such as sulfides and greatly reduced redox conditions, resulting in stressed and stunted plants (see Odum 1988). This situation has been observed in Mediterranean salt marshes where lack of tides and low relief (especially in large deltas) cause the stagnation of subsurface waters, which in turn leads to soil hypersalinity via evapotraspiration. Hypersaline, reduced subsurface waters are widespread in Mediterranean deltas. This is the main factor explaining the low productivity of poorly flooded Mediterranean salt marshes.

Another factor which is partly responsible for the high productivity of salt marshes is that many species have the C_4 biochemical pathway of photosynthesis. C_4 plants have, as a group, higher levels of production than most of C_3 plants (Day et al. 1989). The most abundant Mediterranean salt marsh species, the Chenopodiaceae, are C_3 plants.

3.4.1. Tidal Range and Flooding Frequency

Coastal marshes can be flooded either by marine water (meteorological and astronomical tides) or by fresh water (river floods, underground inputs and rainfall). The timing and magnitude of seasonal oscillations in sea level seem to be the critical factors that influence salt-marsh productivity (Morris et al. 1990). However, the way in which tidal range and flooding frequency affect primary production is quite complex. Steever et al. (1976) reported a strong correlation between tidal range and the peak standing crop of *Spartina alterniflora* unrelated to the changes in climatic and edaphic factors. They concluded that data strongly suggested that the energy subsidy provided by tidal action is a significant factor in the standing crop production of *S. alterniflora*. However, production of this species was negatively affected by tides with ranges higher than 3 m, due to physical stress. Moreover, coastal Louisiana has a low tidal range (0.3 m) and a very high productivity (Day et al. 1989). In contrast, Mediterranean salt marshes use to be less productive, likely due to low rainfall during summer leading to salt and water stress.

In a five year study Teal and Howes (1996) found that increasing sea level had a negative effect on biomass of *Spartina alterniflora* in a salt marsh of Massachusetts. They found a relatively low degree of interannual change in biomass and primary production, and concluded that year-to-year changes in production in more frequently flooded salt-marsh areas may be less susceptible to variations in sea level. Conversely, in another 5 y study, Morris et al. (1990) found that annual aboveground productivity of *Spartina alterniflora* in a South Carolina salt marsh varied by a factor of 2 and correlated positively with anomalies in mean sea level during the growing season. They concluded that the effect of sea-level anomalies on the salinity of intertidal sediments probably accounted for the observed changes in primary production. The higher salinity is a result of the combination of lower flooding frequency and higher evapotranspitarion (lower latitude) in relation to the Massachusetts salt marsh.

Increased productivity as a consequence of anomalies in sea level during the growing season may also occur in Mediterranean marshes and is likely to be more important than tidal sea level changes. For instance, in the Ebre delta, the mean astronomical maximum tidal range is only 25 cm, whereas mean maximum monthly sea level varies by 1 m between the minimum in February and the maximum in September (Jiménez 1996). Conversely, the Atlantic coast of North America has a tidal range from 1 to 2 m, whereas seasonal variations of mean sea level are only about 30 cm (Pattullo et al. 1955).

Temporary impounding can cause a decrease in production if rainfall is low and soil salinity increases, but it can also result in increased production if salinity drops because of high precipitation or upland runoff during the impounded period (Zedler et al. 1980). On the other hand, increased flooding may negatively affect production by decreasing sediment oxidation. Waterlogging is a key factor affecting redox potential, which in turn affects the availability of nutrients in the soil to plants (Pennings and Callaway 1992).

3.4.2. Temperature and Rainfall

Solar radiation, temperature and evapotranspiration act together to produce differences in marsh production over a latitudinal gradient. Rainfall is an indirect factor in regulating plant growth, since high sediment salinity results from excess evapotranspiration. In addition, low rainfall leads to lowered new nutrients, either directly or indirectly via upland runoff (Day et al. 1989). Rainfall differences probably are responsible for the differences in productivity between salt marshes from the Gulf of Mexico and the Mediterranean, 2 areas with low tidal range and high solar radiation. Deegan et al. (1986) showed that, within the Gulf of Mexico, areas with little rainfall develop fewer hectares of marsh or mangrove than those with high rainfall, and that in areas with little rain the marsh species tend to be small (salt- and temperature-tolerant) plants such as *Sarcocornia*.

Variations in productivity in Mediterranean terrestrial grasslands are strongly related to variations in rainfall (Figueroa and Davy 1991). In a 13-y study carried out in a salt-marsh in Netherlands (de Leeuw et al. 1990) the authors found that year-to-year variation in peak above-ground biomass of six annual angiosperm communities could be explained by the rainfall deficit during the growing season, while inundation frequency did not contribute to the regression model. These authors suggested that the rainfall deficit may have influenced vegetation production through its impact on soil salinity and soil moisture content, and they concluded that this effect increases with marsh elevation, where soil salinity is determined by the mutually opposing effects of evapotranspiration and precipitation. At tidal elevations below mean high water, fluctuations in soil salinity are strongly related to the salinity of the inundation waters and not to the rainfall deficit. Productivity of coastal Mediterranean marshes seems to be also strongly influenced by rainfall, due to its effect in lowering soil salinity (especially in poorly flooded marshes), so they are more similar to the high marshes in the Dutch study. Other authors (Zedler 1983, Dame and Kenny 1986, Giroux and Bedard 1987) have also attributed year-to-year differences in salt-marsh production to climatic variability.

3.4.3. Salinity

The Mediterranean climate is probably responsible for the control that hypersalinity has on salt marsh productivity. Along the Atlantic and Gulf of Mexico coasts, where rainfall and stream flow are substantial all year, salt marshes are more constant in composition, and vascular plant growth is under much less salinity stress. Although soil salinity is important in wetter climates, it does not have the strong temporal variation seen in semi-arid California (Zedler and Beare 1986). These authors hypothesized that germination and establishment of Mediterranean-climate salt marsh species are limited to the low-salinity gap that follows winter rainfall, and that 2 environmental stresses, hypersaline drought and excessive inundation, limit the expansion and persistence of many plant populations. They observed that in southern California, *Sarcocornia pacifica* expanded within the low marsh during non-flood and non-tidal years, either by natural closing of the mouth bar or following the diking and restriction of tidal flushing to part of the low marsh. These non-tidal conditions are similar to those naturally existing in the Mediterranean sea, where perennial species resistant to hypersaline conditions such as *Sarcocornia fruticosa* and *Arthrocnemum macrostachyum* are the dominant species in coastal marshes.

Zedler (1983) found that a short-term reduction in the salinity of normally hypersaline soils was followed by a 40% increase in the August biomass of *Spartina foliosa* at Tijuana estuary (south California), and at Los Penasquitos lagoon, a longer period of brackish water influence was followed by a 160% increase in August biomass of *Sarcocornia pacifica*. The largest increase in salt marsh biomass occurred in a non-tidal lagoon with a relatively small increase in stream discharge, while tidal marshes underwent lesser changes in biomass following major flooding events. Grouzis (1973) found that maximum growth of *Sarcocornia fruticosa* and *Salicornia emerici* occurred at 10 ‰ Na Cl (Table 6). These species have optimum growth in environments where submersion is long (up to 9 months) and salinity high, usually between 70 and 80 ‰. Grouzis et al. (1977) found that optimal conditions for growth of *Salicornia patula* and *Salicornia brachystachya* (two annual halophytes) in the Rhone delta occurred at mean salinity of 3 ‰, considerably lower than the values observed in other species of the *Salicornia* genus (between 10 and 20 ‰). They also concluded that roots are less sensitive than aerial parts to both deficit and excess of salinity, so at optimal salinity for growth, the shoot to root ratio is maximum. Abdulrahman and Williams (1981) found that maximum growth of *Sarcocornia fruticosa* from a Lybian salt marsh was at 171 mM Na Cl (12.3 ‰) under cool conditions (20/10 °C) and at 342 mM Na Cl (24.6 ‰) under warm conditions (30/15 °C). Similarly, maximal growth of *Salicornia rubra* was at 171-342 mM Na Cl (Tiku 1976), while *Salicornia bigelovii* had maximal growth at 171 mM Na Cl (Webb 1966).

Some halophytes characteristic of extreme salinity conditions are typical of Mediterranean-climate marshes. Usually, the optimal salinity for maximum salt marsh growth is in the range of 100-200 mM (7.2-14.4 ‰), and growth is significantly reduced as salinity increases or decreases. *Salicornia bigelovii* is a succulent annual species that occurs in coastal estuaries and is reported to have maximum growth at about 200 mM Na Cl. The deleterious effects of salinity are thought to result from water stress, ion toxicity, ion imbalance or a combination of these factors (Ayala and O'Leary 1995). These authors also concluded that reduced growth at suboptimal salinity apparently is not due to an insufficient supply of photosyntate to support growth nor is due to less than favorable water relations in the shoots as had been suggested earlier, but it rather seems as if the growth differences may be more closely related to differences in ionic relations.

Rozema (1991) found that in a greenhouse experiment with 17 halophyte species only those from the genera *Salicornia* and *Suaeda* showed an increase in the mean relative growth rate under saline conditions. Chenopodiaceae species like *Atriplex nummularia*, *Suaeda maritima*, *Halimione portulacoides* and *Salicornia dolichostachya* have been found to have maximum growth rates where the external salinity is 50-100 mM Na Cl (3.6-7.2 ‰) (see Rozema 1991). Rozema discussed the suboptimum growth of different Chenopodiaceae species from saline habitats and suggested that unfavorable water relationships at 0 mM Na Cl are implicated in the growth reduction. Salinity is necessary to maintain the turgor pressure potential required for growth.

Salinity greater than 35 ‰ and completely submerged conditions reduced growth of *Sarcocornia perennis*, an important intertidal salt marsh macrophyte occurring in a number of South African estuaries. This species is adapted to a wide range of environmental conditions as it is naturally subjected to flooding by freshwater in the rainy season and is often inundated by tidal sea water (Adams and Bate 1994). This species is more sensitive to submerged conditions than it is to high salinity. Best growth was recorded for freshwater, saturated soil treatments, indicating that *S. perennis* does not necessarily have physiological requirement for salt. From surveys of a number of South African estuaries, the authors found *S. perennis* in salinity ranging from 12 ‰ to 42 ‰.

Naidoo and Rughunanan (1990) found that an increase in salinity from 0 to 300 mM Na Cl (21.6 ‰) stimulated production, increased succulence and shifted resource allocation from roots to shoots in *Sarcocornia natalensis*, a perennial succulent halophyte which frequently occurs as a mat on sandy mud in coastal lagoons and estuaries of South Africa. Growth was optimal at 300 mM and decreased with further increase in salinity. The decrease in total dry mass at 400 and 500 mM Na Cl (28.8 and 36.0 ‰) however, was not associated with a significant reduction in organic dry mass production. The authors suggested that salt tolerance in *S. natalensis* is likely achieved by a delicate balance between ion accumulation, suitable osmotic adjustment, turgor maintenance and growth.

TABLE 6. Salinity values for optimum growth from several halophytes.

Species	Optimum growth (‰ NaCl)	Reference
Sarcocornia fruticosa	10	Grouzis 1973
Sarcocornia fruticosa	12.3 - 24.6	Abdulrahman & Williams 1981
Sarcocornia perennis	0	Adams & Bate 1994
Sarcocornia natalensis	21.6	Naidoo & Rughunanan 1990
Salicornia emerici	10	Grouzis 1973
Salicornia patula	3	Grouzis et al. 1977
Salicornia brachystachya	3	Grouzis et al. 1977
Salicornia rubra	12.3 - 24.6	Tiku 1976
Salicornia bigelovii	14.4	Webb 1966
Salicornia bigelovii	3.6 - 7.2	Ayala & O'Leary 1995
Salicornia dolischostahya	3.6 - 7.2	Rozema 1991
Suaeda maritima	3.6 - 7.2	Rozema 1991
Halimione portulacoides	3.6 - 7.2	Rozema 1991
Atriplex nummularia	3.6 - 7.2	Rozema 1991

3.4.4. Soil aeration and nutrients

The amount of oxygen present in marsh soils is an important factor affecting plant growth. Redox potential of saturated wetland soils affects a variety of processes ranging from the depth to which infauna can penetrate the sediment to the availability of nutrients to plants (Mitsch and Gosselink 1986). Salt marsh sediments have strongly reducing conditions as reflected by low Eh values ranging from -100 to -250 mV. High sulfide concentrations in reduced soils have been shown to be an important factor in reducing salt marsh primary production (Howes et al. 1981). According to these authors, nutrient concentrations and availability are the final growth-limiting factors, but the ability of grasses to use the nutrients is mediated by the extent and degree of sediment oxidation.

Mediterranean salt marsh plants are shallow rooted due to the lack of aeration of deep horizons, with a maximum density between 4 and 20 cm in *Arthrocnemum macrostachyum* and *Sarcocornia fruticosa*, 2 common perennial species (Nichabouri and Corre 1970). These authors showed that maximum root density always corresponded to minimum Na/K ratios in the soil. The cold, wet season (from October to January) is the period of maximum root growth, which is stopped at the end of March, when aerial parts start to growth.

Nitrogen is a key nutrient in coastal ecosystems. In salt marshes, nitrogen plays a critical role in determining the function and structure of the ecosystem. Incresed productivity with nitrogen fertilization has been shown for a number of salt marsh species (Valiela and Teal 1979, Day et al. 1989). Groundwater inputs enhanced the standing crop, above-ground productivity and nitrogen content of *Sarcocornia pacifica* in a southern California salt-marsh (Page 1997). ^{15}N enrichment in *S. virginica* along the tidal marsh boundary, relative to high and middle marsh locations, indicated uptake of groundwater nitrogen (Page 1995).

Studies of nutrient dynamics and limitation in coastal Mediterranean salt marshes are very scarce in the literature, whereas freshwater and brackish marshes have been quite well studied from this point of view, especially in the Rhone delta (see Goltermann 1995).

3.5 CONSUMERS AND SECONDARY PRODUCTION

Direct grazing in salt marshes is generally thought to account for less than 5% of total net primary production (Odum 1988). Important consumers in salt marshes and mangrove swamps include crustaceans such as crabs, amphipods, and caridean shrimp, along with polychaetes, molluscs and adult insects. In tidal freshwater marshes both larval and adult insects play a key role along with oligochaetes and a few crustaceans such as amphipods and caridean shrimp (see Odum 1988). Teal (1962) estimated that the herbivorous insects of a Georgia salt marsh assimilated 4.6% of the net production of *Spartina alterniflora* and concluded that the salt marsh community consisted of 2 trophic pathways, one deriving its energy directly from the living *Spartina* and the other deriving its energy from detritus and algae. Page (1997) found that in Carpinteria salt marsh (southern California) the isotopic composition of macro-invertebrates indicated the incorporation of algae rather than *Sarcocornia pacifica* (=*Salicornia virginica*) biomass or upland sources into the marsh food web.

In a study of the invertebrate community of salt marshes in the Rhone delta, Bigot (1963) reported that the biomass of herbivores and carnivores represented only 0.18 and 0.045%, respectively, of total biomass. Aside from occassional outbreaks of caterpillars of 2 species of microlepidoptera, the primary production of *Sarcocornia fruticosa* and *Arthrocnemum macrostachyum* salt marshes was not used by primary consumers, but by detritus feeders (beetles and springtails) and decomposers (Berger et al. 1979). Detritus feeders (Crustacea, Oniscoidea and Coleoptera) were the most abundant group; herbivorous insects were not abundant and vertebrates did not normally consume these species. Phytoplankton and detrital organic matter, as well as benthic algae are intensively exploited by zooplankton during the irregular and temporary flooding of the salt marsh. During the warm season large populations of Diptera and Coleoptera larvae develop before and after the summer drought. This secondary production is intensively exploited by migrating birds (especially waders). Finally, the organic matter remaining after drying is exploited by terrestrial detritivorous species.

Menéndez and Comín (1990) studied submerged macrophyte consumption by invertebrates in Tancada lagoon, in the Ebre delta. Macroinvertebrate grazing of phytosynthetic biomass was low, but grazing was important in accelerating decomposition of plant material accumulated at the end of the growing season. Important grazers were the crustaceans *Gammarus aequicauda* and *Sphaeroma hookeri*, with a few other less abundant species. The maximum plant biomass consumed was 0.043 % and 0.017 % for *G. aequicauda* feeding on *Ruppia cirrhosa* and *Potamogeton pectinatus*, respectively, and 0.017 % for *S. hookeri* feeding on *P. pectinatus*.

There are no direct estimates of secondary production in coastal Mediterranean wetlands. However, fisheries landings of coastal lagoons is indicative of primary production. Nixon (1982) demonstrated a relationship between the fish catch and the primary production in a number of fresh water and marine systems but that fisheries yield per unit of primary production was 10-20 times higher for marine systems. In coastal systems, primary production was 200-400 g C m^{-2} y^{-1} and fish catch was 50-100 Kg ha^{-1} y^{-1}. In the Ebre delta, phytoplankton production was estimated to be 165 g C m^{-2} y^{-1} in the Tancada lagoon and 480 g C m^{-2} y^{-1} in the Encanyissada lagoon, whereas macrophyte production was in the range 244-467 g C m^{-2} y^{-1} in the Tancada lagoon and zero in the Encanyissada lagoon (Comín et al. 1990). Fisheries yield is at present in the range 40-100 Kg ha^{-1} y^{-1} in the coastal lagoons of the Ebre delta (Fig. 4), but they were much higher in the past before its overexploitation and the deterioration of its water quality. Thus, mean fisheries yield in the period 1966-1976 were in the range 122-388 kg ha^{-1} y^{-1} (Table 7). The decrease in yield has been higher in the Encanyissada lagoon than in the Tancada lagoon, likely due to the disappearance of submerged macrophytes in the first due to eutrophication. The majority of species captured in the coastal lagoons are adapted to live in the coastal waters and they enter the lagoons for feeding and nursery.

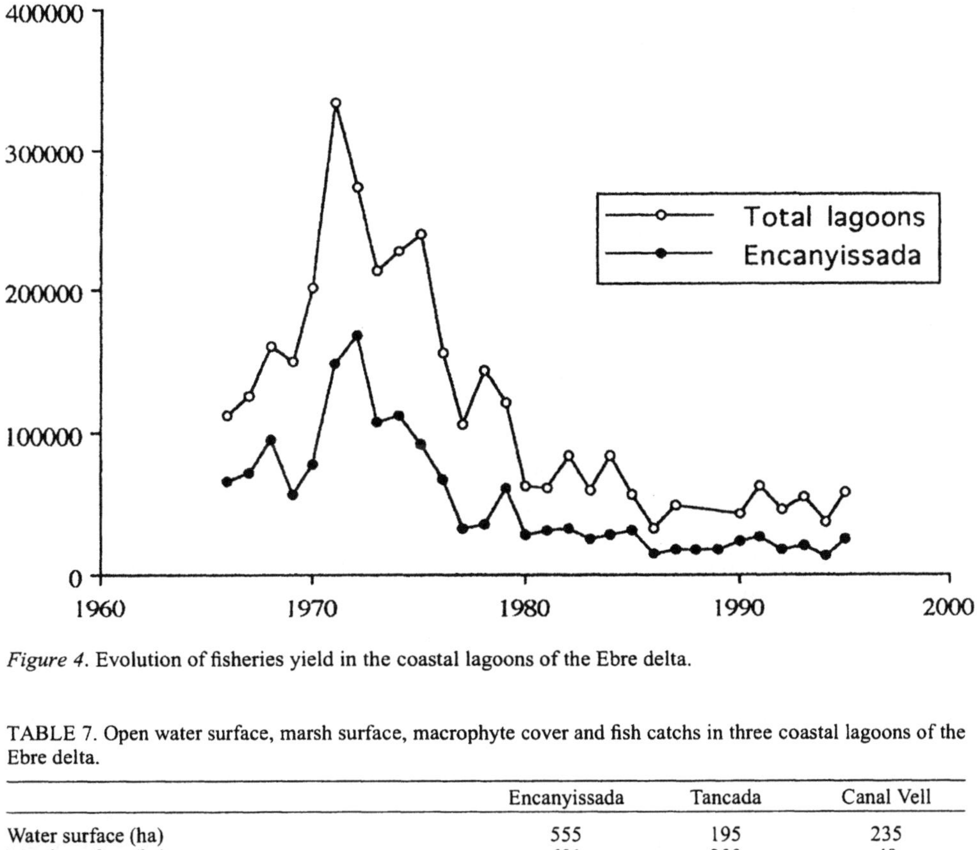

Figure 4. Evolution of fisheries yield in the coastal lagoons of the Ebre delta.

TABLE 7. Open water surface, marsh surface, macrophyte cover and fish catchs in three coastal lagoons of the Ebre delta.

	Encanyissada	Tancada	Canal Vell
Water surface (ha)	555	195	235
Marsh surface (ha)	631	200	40
Marsh/water ratio	1.14	1.02	0.17
Marsh surface (1977) (ha)	643	239	278
Macrophyte cover in 1993 (%)	0.1	80.6	89.5
Macrophyte cover in 1977 (%)	71.6	75.3	100
Fish catchs 1966-76 (kg ha^{-1} y^{-1})	175.3	234	122.4
Fish catchs 1977-87 (kg ha^{-1} y^{-1})	55.1	135.5	49.3
Fish catchs 1989-94 (kg ha^{-1} y^{-1})	39	95.7	44.7
Conductivity range (mS cm^{-1})	4 - 40	8 - 45	3 - 47

4. Some Hypothesis about the Coupling between Production of Coastal Wetlands and Coastal Waters in the Mediterranean

4.1 OUTWELLING VERSUS INWELLING

Salt marshes act primarily as transformers of nutrients and may function as either sinks or sources of nutrients depending upon a variety of factors including the successional age of the marsh, tidal energy, salinity, redox potential, upland and estuarine sources, etc. According to the outwelling hypothesis, tidal inundation of the marsh provides a mechanism for removing large quantities of organic carbon, especially particulate organic carbon, from the marsh into the estuary, and eventually into nearshore coastal regions (see Nixon 1980). This hypothesis was formulated by Odum and Teal in early 60s based on studies carried out in Georgia coastal marshes. The history of the outwelling hypothesis is covered elsewhere in this book, and in this section we want to address this hypothesis from the perspective of Mediterranean coastal marshes.

Based on the ideas of Odum et al. (1979) and Nixon (1980), coastal Mediterranean wetlands likely act as importers of particulate organic carbon, since they are characterized by a low tidal range, low freshwater inputs and nearly closed wetland basins. In the Mediterranean, coastal lagoons play a key role in the sense that they are an intermediate ecosystem between coastal marshes and coastal waters. Presently, many coastal lagoons and bays have artificial freshwater inputs from agriculture and other human activities, but originally they used to be salt water environments with no significant inputs from continental runoff.

In Mediterranean coastal marshes, export of particulate organic matter is likely to be normally low and irregular due to the occurrence of weak tides. Litter accumulation is important in dense salt marshes dominated by *Sarcocornia fruticosa*, brackish marshes dominated by *Phragmites australis* or fresh marshes dominated by *Typha angustifolia* or *Cladium mariscus* (Ibañez et al., 1999). Conversely, litter is very scarce in salt marshes dominated by *Arthrocnemum macrostachyum*, not only because its low productivity but also because low plant cover facilitates its removal by strong winds. Most of the export of material from the marsh to the open water must be in form of dissolved or very fine particulate forms. Mediterranean marshes act as powerful traps of particulate material (organic and inorganic) during resuspension and washover processes caused by strong winds and marine storms (Hensel 1998). Even though materials transport is low during normal tides, strong pulsing events such as river floods and storms can lead to active transformation and import-export of materials. For example, Hensel et al. (1998) showed that a coastal area in the Rhone delta strongly affected by river flow had strong imports of suspended materials which accreted on the marsh surface. Areas of the delta which were isolated from the river and sea by dikes had very low inputs of sediments. Hensel (1998) also showed that this same marsh imported inorganic nutrients and TSS and exported phytoplankton. These facts may lead to the hypothesis that materials transport from Mediterranean marshes is mainly event related, with high imports and exports occurring during river floods, storms and strong winds. These marshes also are active transformers of materials. Since many salt marshes are isolated by dikes and impoundments, the interaction between marshes and

4.2 SUBMERGED AQUATIC VEGETATION, PHYTOPLANKTON, AND BENTHIC ALGAE

In temperate estuaries which are not affected by excessive nutrient input, submerged aquatic vegetation (SAV) can account for more than half of the net primary production. However, due to nutrient inputs and greater overall depth, the organic carbon budget in large temperate estuaries is dominated by plankton production (Stevenson 1988). Stevenson concluded that the coupling of macrophyte production to fisheries is difficult to quantify due to the lack of information such as the palatibility of leaf tissue or the accurate determination of fish biomass in SAV.

Many Mediterranean coastal aquatic ecosystems, like some lagoons of the Ebre delta, have undergone a shift from production dominated by SAV to domination by phytoplankton due to human-induced inputs of nutrient-rich freshwater (Comín et al. 1989). Excessive nutrients leads to higher phytoplankton and epiphyte growth which reduces light levels and leads to SAV decline (Golterman 1995). The loss of SAV was postulated to be the main cause of a drastic decrease in fisheries and waterfowl populations in Encanyissada lagoon (Ebre delta) during the 70s (Comín et al. 1989). During this period, the marsh surface surrounding the lagoon remained almost constant. Fig. 5 shows the evolution of macrophyte cover, and herbivorous waterfowl populations in this lagoon. There is a clear correlation between macrophyte cover and the population of the common coot (*Fulica atra*), an species that feeds directly on macrophytes (Martínez-Vilalta 1995). However, the correlation of macrophyte cover with herbivorous ducks is not so strong, due to the fact that these species also feed on alternative habitats (mainly rice fields). Finally, the correlation of fisheries yield (Fig. 4) and macrophyte cover is not so clear and direct, since most of the species reproduce in the sea and migrate to the lagoons in the post-larval period. However, as mentioned before, the decrease in yield initiated in the 70s (mostly due to overexploitation) was highest in the Encanyissada lagoon, likely due to the loss of SAV. On the other hand, there is no clear relation between fisheries yield and the marsh/open water ratio of the lagoons. Thus, the strong decrease in yield in the Encanyissada lagoon occured with no reduction in marsh surface, while the decrease in yield in the Canal Vell lagoon was lower than in the Encanyissada in despite of an strong reduction in its marsh surface. The Canal Vell lagoon has maintained a high macrophyte cover until the present. However, in the Tancada lagoon, which has maintained both marsh and macrophyte surface, the decrease in yield has been the lowest (see Table 7).

4.3 WHAT CAN BE SAID IN THE END?

It is difficult to make clear quantitative and qualitative links between the productivity of marshes and open coastal waters in the Mediterranean because there is insufficient and often contradictory data. In the Mediterranean, the most productive coastal areas are those influenced by large rivers. Many estuarine ecosystems have high productivity the nearshore zone has been greatly reduced. Thus the support to food chains of coastal lagoons and coastal waters has probably been reduced. However, marsh - lagoon - estuary - open water interactions in the Mediterranean have not been strongly studied and much more information is needed to make general conclusions.

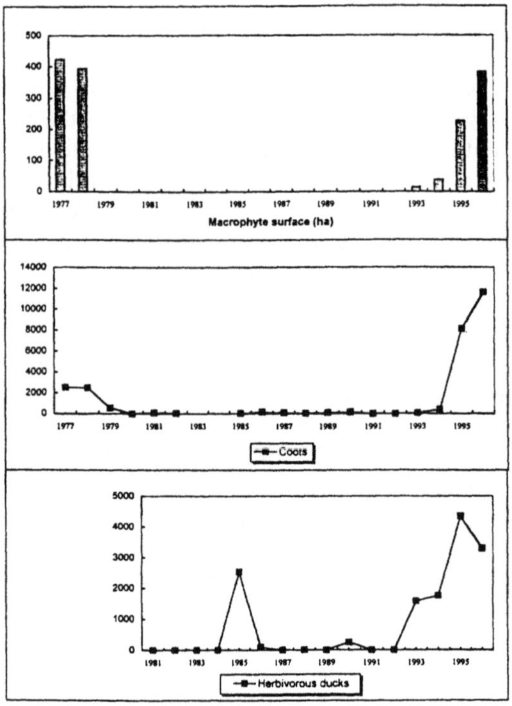

Figure 5. Evolution of macrophyte surface and herbivorous waterfowl (coots and ducks) in the Encanyissada lagoon, Ebre delta (from Martínez-Vilalta 1995).

In southern California, the low area of tidal wetlands and low rates of vascular plant productivity might indicate minimal salt marsh contributions. But the highly productive epibenthic algae which are highly digestible may be an important food source (Kwak and Zedler 1997). These authors found that, despite its dominance in the marsh, *Sarcocornia pacifica* did not play a substantial role in supporting consumers (invertebrates, fishes or birds) in southern California estuaries. They described two complementary food web components in Tijuana Estuary: one of fishes supported primarily by *Spartina*, and another of invertebrates and one bird species (the light-footed clapper rail) utilizing macroalgae as a primary source.

Results from multiple stable isotope research in salt marshes of Georgia (Atlantic Coast) show that *Spartina* and algae are two major sources of organic matter for the fauna of the marshes and estuarine waters (Peterson and Howarth 1987). Multiple isotope research from one Gulf Coast salt marsh suggested that *Spartina* is not an important source of organic matter, and that the food webs are primarily supported by benthic and planktonic algae (Sullivan and Moncreiff 1990). However, Hopkinson and Day (1977) estimated that the contribution of organic matter from *Spartina alterniflora* marshes to adjacent bays equals the amount produced by phytoplankton.

which is due to a number of factors including a shallow well-mixed water column, rapid nutrient cycling, high water temperatures during the growing season, and habitat richness associated to the presence of submerged macrophytes as well as salt marshes. Mediterranean coastal waters are not generally strongly influenced by the export of particulate organic matter from estuarine ecosystems because tidal fluxes are weak but this coupling can be stronger during intense energetic events and as associated with migratory nekton. Finally it seems that part of the problem in determining the mutual influence of the productivity of estuarine and coastal marine ecosystems (from the marsh or the river to the open sea) is the high spatial (structural) and temporal (dynamic) complexity of the whole system, and the difficulty in establishing boundaries when many species (especially nekton) share the different subsystems during their life cycle.

5. Literature Cited

Abdulrahman, F. S. and G. J. Williams. 1981. Temperature and salinity regulation of growth and gas exchange of *Salicornia fruticosa* (L.) L. Oecologia (Berlin) **48**:346-352.

Adams, J. B. and G. C. Bate. 1994. The effect of salinity and inundation on the estuarine macrophyte *Sarcocornia perennis*. Aquatic Botany **47**:341-348.

Ayala, F. and J. W. O'Leary. 1995. Growth and physiology of *Salicornia bigelovii* Torr. at suboptimal salinity. International Journal of Plant Sciences **156**:197-205.

Berger, A., J. J. Corre and G. Heim. 1978. Structure, productivité et régime hydrique de phytocenoses halophiles sous climat méditerranéen. La Terre et la Vie **32**: 241-278.

Berger, A., L. Bigot, A. Champeau, G. Heim and N. Poinsot-Balaguer. 1979. Production primaire et relations trophiques chez les invertebrés des communautés halophiles de Camargue. La Terre et la Vie **2**:175-189.

Bigot, L. 1963. Observations sur les variations de biomasses des principaux groupes d'invertebrés de la 'sansouire' camarguaise. La Terre et la Vie **17**:319-334.

Callaway, R. M., S. Jones, W. R. Ferren and A. Parikh. 1990. Ecology of a mediterranean-climate estuarine wetland at Carpinteria, California: plant distributions and soil salinity in the upper marsh. Canadian Journal of Botany **68**:1139-1146.

Cameron, G. N. 1972. Analysis of insect trophic diversity in two salt marsh communities. Ecology **53**:58-73.

Chapman, V. J. 1977. Wet Coastal Ecosystems. Elsevier, Amsterdam, The Netherlands.

Clarke, P. J. and C. A. Jacoby. 1994. Biomass and above-ground productivity of salt-marsh plants in south-eastern Australia. Australian Journal of Marine and Freshwater Research **45**:1521-1528.

Comín, F., M. Menéndez and J. Lucena. 1989. Proposals for macrophyte restoration in eutrophic coastal lagoons. Hydrobiologia **200/201**:427-436.

Corre, J. J. 1979. Structure des communautés vegetales salées. La Terre et la Vie **2**:105-127.

Corre, J. J. 1985. Environmental structures and variation in coastal vegetation of the Golfe du Lion (France). Vegetatio **61**: 15-22.

Curcó, A., A. Canicio and C. Ibañez. 1997. Mapa d'hàbitats potencials del delta de l'Ebre. Butlletí del Parc Natural del delta de l'Ebre **9**:4-12.

Dame, R. F. and P. D. Kenny. 1986. Variability of *Spartina alterniflora* primary production in the eurihaline North Inlet estuary. Marine Ecology Progress Series **32**: 71-80.

Day, J. W., Jr., C. Hall, W. Kemp and A. Yañez-Arancibia. 1989. Estuarine ecology. Wiley-Interscience, New York, New York, USA.

Day, J. W., Jr., D. Pont, P. F. Hensel and C. Ibañez. 1995. Impacts of sea-level rise on deltas in the Gulf of Mexico and the Mediterranean : the importance of pulsing events to sustainability. Estuaries **18**:636-647.

Day, J. W., Jr., J. F. Martin, L. Cardoch and P. H. Templet. 1997. System functioning as a basis for sustainable management of deltaic ecosystems. Coastal Management **25**: 115-153.

De Leeuw, J., H. Olff and J. P. Bakker. 1990. Year-to-year variation in peak above-ground biomass of six salt-marsh angiosperm communities as related to rainfall deficit and inundation frequency. Aquatic Botany **36**:139-151.

Deegan, L. A., J. W. Day, Jr., J. G. Gosselink, A. Yáñez-Arancibia, G. Soberón and P. Sánchez-Gil. 1986. Relationships among physical characteristics, vegetation distribution and fisheries yield in Gulf of Mexico estuaries. Pages 83-100 *in* Estuarine Variability. Academic Press, New York, New York, USA.

Eckardt, F. E. 1972. Dynamique de l'écosystème, stratégie des végétaux et échanges gazeux: cas des enganes à *Salicornia fruticosa*. Oecologia Plantarum **7**:333-345.

Eilers, H. P. 1981. Production in coastal salt marshes in southern California. U.S. EPA Technical Report EPA-60013-81-023. NTIS, Washington, District of Columbia, USA.

Ferrari, C., R. Gerdol and F. Piccoli. 1985. The halophilous vegetation of the Po Delta (northern Italy). Vegetatio **61**:5-14.

Figueroa, M. E. and A. J. Davy. 1991. Response of Mediterranean grassland species to changing rainfall. Journal of Ecology **79**:925-941.

García, L. V., T. Marañón, A. Moreno and L. Clemente. 1993. Above-ground biomass and species richness in a Mediterranean salt marsh. Journal of Vegetation Science **4**:417-424.

Géhu, J. M. 1984. Mediterranean salt marshes and salt steppes. Pages 166-167 *in* K.S. Dijkema, editor. Salt marshes in Europe. Council of Europe, Nature and Environment Series 30, Strasbourg, France.

Géhu, J. M., A. Scoppola, G. Caniglia, S. Marchiori and J. Géhu-Franck. 1984. Les systèmes végétaux de la côte nord-adriatique italienne. Leur originalité à l'échelle européenne. Colloques phytosociologiques **8**:485-558.

Giroux, J. F. and J. Bedard. 1987. Factors influencing above-ground production of *Scirpus* marshes in the St. Lawrence estuary, Quebec, Canada. Aquatic Botany **29**: 195-204.

Golterman, H.L. 1995. The labyinth of nutrient cycles and buffers in wetlands: results based on research in the Camargue (southern France). Hydrobiologia **315**:39-58.

Good, R., N. Good and B. Frasco. 1982. A review of primary production and decomposition dynamics of the belowground marsh component. Pages 139-157 *in* V.S. Kennedy, editor. Estuarine Comparisons. Academic Press, New York, New York, USA.

Grouzis, M. 1973. Exigences écologiques comparées d'une salicorne vivace et d'une salicorne annuelle: germination et croissance des stades jeunes. Oecologia Plantarum **8** :367-375.

Grouzis, M., G. Heim and A. Berger. 1977. Croissance et accumulation de sels chez deux salicornes annuelles du littoral méditerranéen. Oecologia Plantarum **12**:307-322.

Hensel, P. 1998. Sediment and accretionary dynamics in the Rhone Delta, France. Dissertation, Louisiana State University, Baton Rouge, Louisiana, USA.

Hensel, P., J. W. Day, Jr., D. Pont and J. N. Day. 1998. Short terms sedimentation dynamics in the Rhone Delta, France : The importance of riverine pulsing. Estuaries **21**:52-65.

Heurteaux, P. 1970. Rapports des eaux souterraines avec les sols halomorphes et la végétation en Camargue. La Terre et la Vie **4**:467-510.

Hollis, T. 1992. The causes of wetland loss and degradation in the Mediterranean. Proceedings of IWRB International Symposium, Grado, Italy, 1991.International Waterfowl and Wetlands Research Bureau Special Publication **20**:83-90.

Hopkinson, C. S. and J. W. Day, Jr. 1977. A model of the Barataria Bay salt marsh ecosystem. Pages 235-265 *in* C.A. Hall and J. W. Day, Jr., editors. Modeling in theory and practice. John Wiley and Sons, New York, New York, USA.

Hopkinson, C. S., J. G. Gosselink and R. T. Parrondo. 1978. Aboveground production of seven marsh plant species in coastal Louisiana. Ecology **59**:760-769.

Hopkinson, C. S., J. G. Gosselink and R. T. Parrondo. 1980. Production of coastal Louisiana marsh plants calculated from phenometric techniques. Ecology **61**:1091-1098.

Howes, B. L., R. W. Howarth, J. M. Teal and I. Valiela. 1981. Oxidation-reduction potentials in a salt marsh: spatial patterns and interactions with primary production. Limnology and Oceanography **26**:350-360.

Ibañez, C., D. Pont and N. Prat. 1997. Characterization of the Ebre and Rhone estuaries: A basis for defining and classifying salt-wedge estuaries. Limnology and Oceanography **42**:89-101.

Ibañez, C., J. W. Day, Jr. and D. Pont. 1999. Primary production and decomposition in wetlands of the Rhone delta: possible implications of relative sea-level rise. Journal of Coastal Research **15**:717-731.

Jefferies, R. L. 1972. Aspects of salt-marsh ecology with particular reference to inorganic plant nutrition. Pages 61-85 *in* R. S. Barnes and J. Green, editors. The estuarine environment. Applied Science Publishers, London, England.

Jiménez, J. A. 1996. Evolución costera en el Delta del Ebro. Un proceso a diferentes escalas de tiempo y espacio. Dissertation, University of Barcelona, Spain.

Kwak, T. J. and J. B. Zedler. 1997. Food web analysis of southern California coastal wetlands using multiple stable isotopes. Oecologia **110**:262-277.

Mahall, B. E. and R. B. Park. 1976. The ecotone between *Spartina foliosa* and *Salicornia virginica* in salt marshes of northern San Francisco Bay. I. Biomass and production. Journal of Ecology **64**:421-433.

Mall, R. E. 1969. Soil-water-salt relationships of waterfowl food plants in the Suisun Marsh of California. California Department of Fish and Game, Wildlife Bulletin No. **1**.

Martínez-Vilalta, A. 1995. The rice fields of the Ebro delta. Pages 173-186 *in* C. Morillo and J. L. González, editors. Management of Mediterranean wetlands. Publications MedWet/Ministerio de Medio Ambiente, Madrid, Spain.

Menéndez, M. and F. A. Comín. 1990. Consumption of macrophytes by invertebrates in Tancada lagoon. Scientia Marina **54**:139-144.

Mitsch, W. J. and J. G. Gosselink. 1986. Wetlands. Van Nostrand Reinhold, New York, New York, USA.

Molinier, R. and J. P. Devaux. 1978. Notice explicative de la carte phytosociologique de la Camargue. Revue de Biologie et Écologie méditerranéenne **4**:159-196.

Morris, J. M., B. Kjerfve and J. M. Dean. 1990. Dependence of estuarine productivity on anomalies in mean sea level. Limnology and Oceanography **35**:926-930.

Mudie, P. J. 1970. A survey of the coastal wetland vegetation of San Diego Bay. California Department of Fish and Game Control, Report W26.D25-51.

Naidoo, G. and R. Rughunanan. 1990. Salt tolerance in the succulent, coastal halophyte, *Sarcocornia natalensis*. Journal of Experimental Botany **225**:497-502.

Nichabouri, A. and J. J. Corre. 1970. Comportement de l'appareil radiculaire d'*Arthrocnemum fruticosum* (L.) Moq. et *Arthrocnemum glaucum* (Del.) Ung. Stern. en relation avec les conditions du milieu halomorphe littoral. Oecologia Plantarum **5**:69-86.

Nixon, S. W. 1980. Between coastal marshes and coastal waters: A review of twenty years of speculation and research on the role of salt marshes in estuarine productivity and water chemistry. Pages 437-525 *in* P. Hamilton and K. B. McDonald, editors. Estuarine and wetland processes. Plenum Press, New York, New York, USA.

Nixon, S. W. 1982. Nutrient dynamics, primary production and fisheries yields of lagoons. Oceanologica Acta, Special Volume: 357-371.

Odum, W. E. 1988. Comparative ecology of tidal freshwater and salt marshes. Annual Reviews of Ecology and Systematics **19**:147-76.

Odum, W. E., J. S. Fisher and J. C. Pickral. 1979. Factors controlling the flux of particulate organic carbon from estuarine wetlands. Pages 69-80 *in* R. J. Livingston, editor. ecological processes in coastal and marine systems. Plenum Press, New York, New York, USA.

Onuf, C. P. 1987. The ecology of Mugu Lagoon, California: an estuarine profile. U.S. Fish and Wildlife Service Biological Report **85**(7.15). Washington, District of Columbia, USA.

Onuf, C. P., M. L. Quammen, G. P. Shaffer, C. H. Peterson, J. W. Chapman, J. Cermak and R. W. Holmes. 1978. An analysis of the values of central and southern California coastal wetlands. Pages 189-199 *in* Wetlands functions and values: the state of our understanding. American Water Resources Association, Minneapolis, USA.

Page, H. M. 1995. Variation in the natural abundance of ^{15}N in the halophyte *Salicornia virginica*, associated with groundwater subsidies of nitrogen in a southern California salt-marsh. Oecologia **104**:181-188.

—— 1997. Importance of vascular plant and algal production to macro-invertebrate consumers in a southern California salt marsh. Estuarine, Coastal and Shelf Science **45**:823-834.

Pattullo, J., W. Munk, R. Revelle and E. Strong. 1955. The seasonal oscillation in sea level. Journal of Marine Research **14**:88-156.

Pearce, F. and A. J. Crivelli. 1994. Characteristics of mediterranean wetlands. Publications MedWet/Tour de Valat, Arles, France.

Pearcy, R. W. and S. L. Ustin. 1984. Effects of salinity on growth and photosynthesis of three California tidal marsh species. Oecologia **62**:68-73.

Peinado, M., F. Alcaraz, J. L. Aguirre, J. Delgadillo and J. Alvarez. 1995. Similarity of zonation within Californian – Baja Californian and Mediterranean salt marshes. Southwestern Naturalist **40**:388-405.

Pennings, S. C. and R. M. Callaway. 1992. Salt marsh plant zonation: the relative importance of competition and physical factors. Ecology **73**:681-690.
Peterson, B. J. and R. W. Howarth. 1987. Sulfur, carbon and nitrogen isotopes used to trace organic matter flow in the salt-marsh estuaries of Sapelo Island, Georgia. Limnology and Oceanography **32**:1195-1213.
Rea, N. and G. G. Ganf. 1994. How emergent plants experience water regime in a Mediterranean-type wetland. Aquatic Botany **49**:117-136.
Rey, J. R., J. Shaffer, R. Crossman and D. Tremain. 1990. Above-ground primary production in impounded, ditched and natural *Batis-Salicornia* marshes along the Indian River lagoon, Florida. Wetlands **10**:151-171.
Rozema, J. 1991. Growth, water and ion relationships of halophytic monocotyledonae and dicotyledonae: a unified concept. Aquatic Botany **39**:17-33.
Scarton, F., A. Rismondo and J.W. Day, Jr. 1998. Above- and below-ground production of *Arthrocnemum fruticosum* on a Venice Lagoon saltmarsh. Bollettino del Museo Civico di Storia Naturale di Venezia **48**:237-245.
Sestini, G. 1992. Implications of climatic changes for the Po Delta and Venice Lagoon. Pages 428-494 *in* L. Jeftic, J. D. Milliman and G. Sestini, editors. Climatic change and the mediterranean. UNEP.
Shaltout, K. H., H. F. El Kady and Y. M. Alsodany. 1995. Vegetation analysis of the mediterranean region of the Nile Delta. Vegetatio **116**:73-83.
Skinner, J. and N. Zalewski. 1995. Functions and values of mediterranean wetlands. Publications MedWet/ Tour de Valat, Arles, France.
Steever, E. Z., R. S. Warren and W. A. Wiering. 1976. Tidal energy subsidy and standing crop production of *Spartina alterniflora*. Estuarine, Coastal and Marine Sciences **4**:473-478.
Stevenson, J. C. 1988. Comparative ecology of submersed grass beds in freshwater, estuarine and marine environments. Limnology and Oceanography **33**:867-893.
Sullivan, M. J. and C. A. Moncreiff. 1990. Edaphic algae are an important component of salt marsh food-webs: evidence from multiple stable isotope analyses. Marine Ecology Progress Series **62**:149-159.
Tamisier, A. 1990. The Camargue. Ecological habitats and landscapes. Evolution from 1942 to. 1984. A color map 1:80 000. Association pour les Recherches en Camargue sur la Nature et l'Environment (ARCANE), Arles, France.
Teal, J. M. 1962. Energy flow in the salt marsh ecosystem of Georgia. Ecology **43**: 614-624.
Teal, J. M. and B. L. Howes. 1996. Interannual variability of a salt-marsh ecosystem. Limnology and Oceanography **41**:802-809.
Tiku, B. L. 1976. Effect of salinity on the photosynthesis of the halophytes *Salicornia rubra* and *Distichlis stricta*. Physiology Plantorum **37**:23-28.
Turner, R. E. 1976. Geographic variations in salt marsh macrophyte production: a review. Contributions to Marine Sciences **20**:47-68.
Valiela, I. and J. M. Teal. 1979. The nitrogen budget of a salt marsh ecosystem. Nature **280**:652-656.
Wahby, S. D. and N. F. Bishara. 1981. The effect of the River Nile on Mediterranean Water before and after the construction of the High Dam at Aswan. Pages 311-318 *in* Proceedings of a Review Workshop on River Inputs to Ocean Systems. United Nations, New York, USA.
Walter, H. 1973. Vegetationszonen und Klima. Verlag Eugen Ulmen, Stuttgart, Germany.
Webb, K. L. 1966. Na Cl effect on growth and transpiration in *Salicornia bigelovii*, a salt marsh halophyte. Plant Soil **24**:261-265.
Whigham, D. F., J. McCormick, R. E. Good and R. L. Simpson. 1978. Biomass and primary production in freshwater tidal wetlands of the Middle Atlantic Coast. Pages 3-20 *in* R.E. Good et al., editors. Freshwater wetlands: ecological processes and management potential. Academic Press, New York, New York, USA.
Zedler, J. B. 1982. The ecology of southern California coastal salt marshes: a community profile. U.S. Fish and Wildlife Service, Washington, District of Columbia, USAFWS/OBS-81/54.
——. 1983. Freshwater impacts in normally hypersaline marshes. Estuaries **6**:346-355.
Zedler, J. B., T. Winfield and P. Williams. 1980. Salt marsh productivity with natural and altered tidal circulation. Oecologia **44**:236-240.
Zedler, J. B. and P. A. Beare. 1986. Temporal variability of salt marsh vegetation: the role of low-salinity gaps and environmental stress. Pages 295-306 *in* Estuarine Variability. Academic Press, New York, New York, USA.

DEVELOPMENT AND STRUCTURE OF SALT MARSHES: COMMUNITY PATTERNS IN TIME AND SPACE

A.J. DAVY
School of Biological Sciences
University of East Anglia
Norwich NR4 7TJ, UK

Abstract
The relatively species-rich tidal marshes of western Europe show strong spatial zonation in plant species, communities and ecosystem function that is correlated with elevational and environmental gradients from sea to land. Such spatial zonation has traditionally been interpreted as representing a chronosequence, a classic example of succession, mainly because of the demonstrable dependence of marsh formation and vertical development on sedimentary processes. First I seek to separate zonation from succession. Since the earliest days of salt-marsh ecology, physico-chemical factors related to submergence (elevation) have been invoked to explain patterns of distribution. The dominant halophytes are essentially land plants that occupy physiologically adverse environments by virtue of adaptations to salinity, submergence, hypoxia and tidal scouring. Thus their lower limits on the marsh are likely to be determined directly by physico-chemical tolerances. Generally reduced competitive ability is a trade-off in the evolution of tolerance and so upper limits of species may depend substantially on interactions with other, less tolerant, species. Such interactions may also represent to varying degrees the facilitation, inhibition and tolerance models of succession. A critical review of the role of succession requires a longer perspective. Although auto- and allogenic processes can be demonstrated, many marshes are older than previously appreciated; they originated after the last glaciation and are now in fluctuating equilibrium with their sedimentary environments. Accretion is a function of position in the tidal frame, which in turn depends on eustatic and isostatic changes in sea level. Hence, a significant component of observed vertical accretion can be a response to rising sea level. Furthermore, the genesis or disappearance of certain important lower marsh 'pioneer zone' species may post-date the inception of their marshes.

1. Introduction

The processes that determine the development and structure of tidal salt marsh communities have exercised ecologists for almost a century. Their early interest was probably engaged by the obviously strong influence of the physical environment, both because of the extreme nature of the intertidal environment for plants and also because of the intimate involvement of physiographic processes. As this coincided with the inception of modern ecology, some of its early concepts were developed and deployed in analysing coastal marshes.

Foremost amongst these concepts was succession. Remarkably, in possibly the earliest recorded use of the idea of succession, G.M. Lancisi in 1714 explained in some detail the origins of coastal vegetation near Rome (Pignatti and Ubrizsy Savoia 1989). He recognised the importance of progressive advancement of the shore into the sea, as a result of the accumulation of alluvial sediments from the River Tiber. In descriptions that will be echoed more than once in this paper, Lancisi referred to colonization on small hummocks (*'tumuleto'*) by pioneer plants (*'plantae primigeniae'*) and an explicit time scale for development (*'successio'*) of sea shore vegetation through four successive stages. Nearly two hundred years later, shortly after interest in it had been reawakened, succession was assumed to be the driving force behind another important concept in the analysis of vegetation, the distinctive spatial zonation on European salt marshes (see Oliver 1907, Chapman 1938, Ranwell 1972).

In a tidal system, however, the spatial zonation of vegetation need not necessarily represent the recapitulation of a chronosequence, or changes taking place with time. The tidal cycle itself is sufficient to produce a topographic and environmental gradient from sea to land, and plants with different tolerances of tidal submergence would necessarily occupy different parts of the tidal frame. The physiological requirements of the intertidal salt marsh environment on higher plants are well known (Dainty 1979, Flowers, Hajibagheri and Clipson 1986). These plants have evolved attributes appropriate to an environment that may oscillate between marine and terrestrial as often as twice a day. In the lower parts of the marsh they experience prolonged and regular inundation with seawater, whereas between relatively rare immersions on higher parts, they may experience either hypo- or hypersaline conditions (Jefferies and Davy 1979). To varying degrees, the physiological, morphological and life history characteristics of salt marsh plants are the result of severe selection for tolerance of high ionic concentrations, low water potentials, hypoxic soil conditions, periodic suspension of gas exchange with the atmosphere, and scouring currents of water. These selection pressures may also vary in time and space, depending on complex tidal cycles, the weather, and topography particularly the distribution of pans, creeks and channels (Davy and Smith 1985, 1988, Davy, Noble and Oliver 1990, Noble, Davy and Oliver 1992).

Although both zonation and succession are undoubtedly important features of tidal marsh ecology, the concepts have been consistently confounded. This ambiguity has tended to obscure understanding of the marsh dynamics that underlie the behaviour of everything from their plant populations to their geomorphology. One reason for this is probably that our perspective has been too short: some of what we see today has its origins in the legacy of the last glaciation and some depends on events over a few years, decades or centuries. Another, related, reason may have been insufficient appreciation of the nature of the physical processes involved. The main purpose of this review is to distinguish between spatial zonation and succession, largely using historical evidence; I seek to define the extent to which the former can represent the latter and the conditions that apply to such assumptions. The second aim is to examine evidence for the mechanisms by which succession on tidal marshes proceeds and to illustrate these with recent work.

2. Zonation

Spatial zonation is more or less universal in tidal marshes and it is particularly well developed in the relatively species-rich, minerogenic, marshes of western Europe. Gradients in the abundance of plant species, the composition of their communities and in ecosystem function are inevitably correlated with elevational and environmental gradients between sea and land. The causes of such zonation have attracted much attention over many years and elevational sequences continue to be published for new areas of marsh (e.g., Sánchez, Izco and Medrano 1996). Davy and Costa (1992) have reviewed the large body of literature on environmental and vegetational zonation in salt marshes and so I shall confine myself here to reviewing progress on the essential concepts.

It is important to distinguish between environmental zonation and the corresponding distribution of biota, because the potentially complex interactions between species depend ultimately on their individual responses to the strong abiotic selection pressures. For obvious reasons, environmental zonation is usually defined in terms of elevation relative to the tidal frame. This may take the crude form of 'submergence' marshes (ranging from mean high water neap tide level to mean high water) and 'emergence' marshes (extending from mean high water level to that of the mean high water spring tides) as described by Chapman (1938). Nevertheless, the gradient from land to sea is usually continuous, albeit non-linear, and there may be no clear demarcation between submergence and emergence regions. The all-important gradients of salinity and waterlogging are not simply related to elevation in the tidal frame but depend also on drainage (sediment composition, creek patterns and microtopography), climate (e.g., the seasonal balance between evapotranspiration and precipitation) and geomorpholgy (e.g., position in an estuary or channel). Redox potential, which defines many important aspects of sediment chemistry, generally depends on elevation and drainage. Armstrong et al. (1985) found that the lowest zones were predominantly reducing, except transiently near the surface during neap tides; higher on the marsh, redox potentials were lowered by the spring tides; emergence marsh was predominantly oxidizing, except at the highest spring tides. Nevertheless, waterlogged pans may remain very reducing in upper marsh areas. Intuitive salinity gradients, from seawater concentrations at the lower margin of the marsh and decreasing landward, prevail in some marsh systems but salinity inversions are very common, particularly when evapotranspiration in summer exceeds rainfall (e.g., Jefferies, Davy and Rudmik 1979).

The key issue for vegetational and plant species zonation is how it relates to the underlying environmental gradients. Snow and Vince (1984) discussed a continuum of models:

1. Species are restricted physiologically to different portions of an environmental gradient.

2. Species perform best in different portions of the gradient, but have sufficiently broad tolerances to grow elsewhere, were it not for biotic interactions or inadequate dispersal.

3. All species have the same optimum position on the gradient but are displaced according to their tolerance ranges, competitive ability, susceptibility to herbivory, and dispersal.

Model one is undoubtedly true in an extreme sense. Seagrasses (e.g., *Zostera* sp.) and some algae occupy the lowest zone because they cannot withstand prolonged exposure to the atmosphere. Likewise most higher plant halophytes could not survive transplantation to *Zostera* beds. Given the diversity in structure, physiology, and phylogeny of halophytes, the third model is rather unlikely under most circumstances. To explain most plant zonation we must look to model two. Certain transplantation experiments support this model. For instance, Davy and Smith (1988) showed that closely related populations (or microspecies) of *Salicornia* were both significantly selected against in the alien environment when reciprocally transplanted between upper and lower marsh, even in the absence of interspecific competition; however both types were able to survive for a generation in the alien environment.

In a general sense, it would not be surprising if the lower, seaward distribution limits of most higher plant species were determined primarily by their tolerances of the physico-chemical factors associated with a marine environment, whereas the upper or landward limits of halophytic species might be controlled by a wider variety of mechanisms (as mooted by Pielou and Routledge 1976). Few halophytes are found naturally in non-saline environments and it is generally supposed that a loss of competitive ability is an evolutionary 'trade-off' for energy and resource expenditure on salt tolerance: to oversimplify, halophytes cannot compete on land and more competitive non-halophytes cannot survive even moderate salinity. Assuming that the upper regions are less saline for at least part of the annual cycle, this implies that interactions between plants (and other biotic interactions) may be more important in determining their upper limits on the sea-land gradient, even though the outcome of such interactions is modulated by physical and chemical factors related to elevation. On the other hand, some evidence tends to contradict this general prediction. Interestingly, the established idea that the lower limit of *Salicornia europaea* on tidal mud flats at the seaward margin of saltmarshes is determined by tidal action has been challenged recently by Gerdol and Hughes (1993), who found that this lower limit corresponded with the upper limit of the abundant amphipod *Corophium volutator*, at approximately mean high water neap tide level (MHWNT). Seedlings transplanted below this level were disturbed by the activity of *Corophium* but those in areas treated with insecticide to remove the *Corophium* had a doubled survivorship, similar to that of seedlings transplanted above MHWNT. The populations of *Corophium* of up to about 14,000 m^{-2} were effectively able to prevent seedling establishment. Another recently reported exception is the high-marsh grass *Elymus athericus*, whose lower limit on marshes of the North Sea coast of Germany appears to be limited by competition from the physiognomic dominant, *Atriplex portulacoides* (Bockelmann and Neuhaus 1999).

The literature describing the environmental tolerances of different halophytic species, often at various stages in their life histories, is complex and beyond the scope of this review. Suffice it to say that considerable correlation exists between the traits exhibited by plants and their positions on environmental gradients. More significantly, there is now a great deal of evidence for interactions, both competitive and facilitative, between plant species that affect their distribution on tidal marshes (e.g., Bertness and Yeh 1994, Callaway 1994, 1995, Callaway and King 1996); also, well documented interactions between the plants and vertebrate and invertebrate grazers, and microrganisms have significant implications for distribution (Davy et al. 2000, Ungar 1999). Nevertheless, the nature and outcome of such interactions seem to depend substantially on the particular species present on the environmental gradients of a particular site. Large-scale generalization, however, is still difficult.

In principle vegetational zonation on a salt marsh could represent:

1. Static zonation. The consequence of the ranges of tolerance, on the intertidal environmental gradient, of the individual species whose propagules are able to reach the area, as modified by the negative and positive interactions between those species.
2. Developmental zonation. A faithful recapitulation in space of a contemporary successional series in time, or chronosequence, in an accreting and perhaps prograding marsh.
3. The ghost of succession past. A recapitulation of a historic chronosequence in a mature system that is currently in approximate equilibrium with its sedimentary environment and not prograding or eroding. The effects of tidal inundation could represent 'pulse stabilization' and therefore be inhibiting further successional development, including that to a terrestrial ecosystem.

Chapman (1938, 1939, 1940, 1941), as a result of a wide-ranging series of classic studies, derived elaborate successional diagrams largely on the basis of the spatial distribution of plant communities: British salt marsh communities were represented by separate east, south and west coast seres. Although later Chapman (1960) was careful to distinguish 'static zonation' from 'development zonation', these seres were subsequently compared with similar formulations for marshes in many parts of the world (Chapman 1974) and have implicitly underlain much thinking about salt marsh development ever since.

3. The Development of New Marshes

The driving force of salt marsh development in the minerogenic European marshes is of course the net deposition of fine sediments from sediment-laden tidal waters. Salt marsh may develop in the intertidal zone wherever there is a suitable source of suspended sediment, for instance from river discharge or coastal erosion elsewhere, and some protection from the energy of the sea, afforded perhaps by a barrier island, an estuary, a sheltered embayment or shallow energy-absorbing flats. Mud accumulates high in the intertidal zone because current velocities low enough to allow the settling of fine suspended particles occur only close to high water. This is essentially a physical process and colonization by vegetation begins only when the sediment surface has been raised sufficiently in the tidal frame. Marsh vegetation is generally limited to the zone between mid neap tide level and high water spring tide level (Allen and Pye 1992).

The ecological interpretation of salt marsh structure has been very much coloured by early observations on the physical development of marshes. In certain places, such as the Bouche d'Erquy on the Brittany coast of northern France, tidal marsh could be observed forming on intertidal sand banks over a period of a few years, where sand was being mobilized by the erosive effects of a meandering creek (Oliver 1906, 1907, Carey and Oliver 1918). The processes by which marshes can develop were described in remarkable detail. The key event was colonization of the bank by seedlings of a perennial halophyte *Arthrocnemum perenne* ('*Salicornia radicans*'). These plants trapped water-borne sand and their subsequent lateral growth perpendicular to the flow of the current accreted distinctive pyramidal hummocks, in much the same way as dunes accrete air-borne sand. Interestingly, hummocks forming simultaneously around the annual colonist *Salicornia ramosissima* did not persist through the winter after its senescence. The *Arthrocnemum* hummocks, however, were the fundamental units from which the future relief of the marsh was built up. They were colonized, first in patches, by other typical halophytes, notably *Puccinellia maritima, Suaeda maritima* and *Atriplex portulacoides* (*Halimione portulacoides*). Apparently as a result of this, the hummocks grew and coalesced to form a continuous turfy hummock system, which was added to the general sward of the marsh as the creek continued its meandering. In a remarkable experiment for its time, 1907, it was shown that hummock formation could be initiated by transplanting seedlings of *Arthrocnemum perenne* from elsewhere on the marsh and that the secondary colonization started after about two years (Fig. 1, Carey and Oliver 1918).

Direct evidence of development also came from the British coast. Yapp, Johns and Jones (1917), working on the muddy tidal marshes of the Dovey estuary in Wales, also observed hummock formation by the primary colonizer, in this case *Puccinellia maritima*. As the elevation increased, *Armeria maritima* and other species established themselves; at the same time, lateral extension of the *Puccinellia* caused the hummocks to coalesce and eventually the general level of the marsh surface became more even as a result of reduced sedimentation rates on the higher parts. Work on the north Norfolk coast of eastern England also first provided influential information at about the same time. Blakeney Point is essentially a shingle spit with a hooked end; as the point has extended, successive lateral banks have been formed on its landward side by prolongation of the main axis beyond the current terminal hook. The resulting series of laterals, and the silty salt marshes that have formed in the segments between them, have been aged from historical records and maps. Oliver and Salisbury (1913) were thus able to describe a series of vegetational stages of approximately known age. The earliest stages of colonization consisted of various algae and a scattering of *Salicornia* spp. Next came stands of *Salicornia europaea*, scattered *Aster tripolium*, a greater quantity of *Puccinellia maritima* and a fair abundance of *Limonium vulgare*. *Arthrocnemum perenne*, *Suaeda maritima* and *Atriplex portulacoides* were mostly confined to the edge. Later stages were characterized by the dominance of *Atriplex portulacoides*, spreading centripetally and ousting the pioneers. Similarly, Marsh (1915) reconstructed successional changes in some detail from the known eastward progression of a marsh that had been formed between two ranges of shingle banks at Holme-next-the-Sea, also in Norfolk. Unlike many later workers, Marsh sought extrinsic validation for his propositions. Excavations to reveal the sediment profiles and depths across the marsh were consistent with the interpretation of spatial changes as representing development over time.

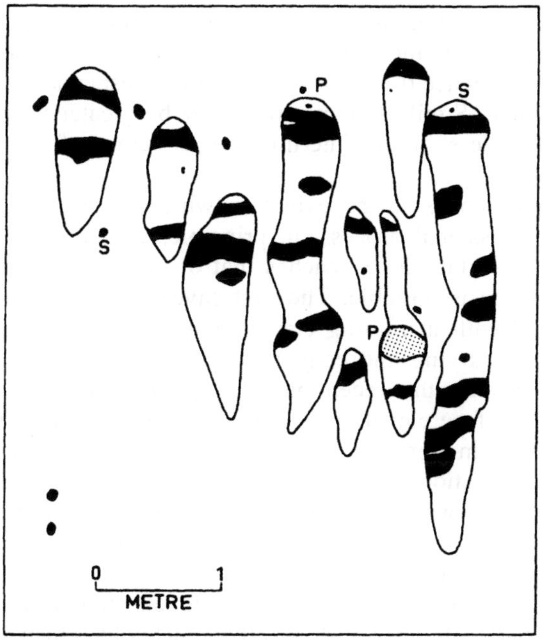

Figure 1. An early transplant experiment. Map drawn in 1909, showing the raised hummocks formed two years after planting seedlings of *Arthrocnemum perenne* (*Salicornia radicans*) on bare sand flats of the Bouche d'Erquy, France, and the beginnings of secondary colonization. Black belts and isolated black dots, *Arthrocnemum perenne*; **P**, colonization by *Puccinellia maritima*; **S**, colonization by *Suaeda maritima*. Redrawn from Carey and Oliver (1918).

These seminal studies were all, in a sense, of changes taking place on a readily observable timescale in response to physiographic perturbations - moving channels and sand banks or newly deposited shingle ridges. They were contemporary with the rediscovery of the idea of ecological succession by Clements (1916) and its more or less universal adoption by ecologists as a framework for analysing vegetational relationships. In consequence, the idea that spatial zonations associated with elevation represented successional change became accepted uncritically, even when marshes were apparently in physiographic equilibrium.

4. Problems with the Successional Interpretation

4.1 ACCRETION, CONSOLIDATION AND SEA LEVEL CHANGE

The impression that spatial zonation necessarily represents a sequence of successional change has been reinforced by measurements of vertical accretion. The technique of placing a marker layer of coloured sand as a base line for accretion measurements appears to have first been used in north Norfolk in 1914 (Carey and Oliver 1918) and related methods have been used extensively ever since (see reviews by Ranwell 1972, Adam 1990, Packham and Willis 1997). The highest rates of accretion are typically found on young or recently colonized areas of marsh; much lower rates obtain on the higher, mature areas because the few tides sufficiently high to reach high elevations, combined with short residence times for settling-out of suspended material, can deposit relatively little sediment. Similarly deposition may be greater near creeks, where the supply is greater, and less in the interfluve areas remote from them (French and Spencer 1993).

This raises a question as to the extent to which surface accretion represents a developmental or successional process. In principle, vertical growth of a salt marsh should cease when its surface level reaches that of the highest astronomical tides, as no more tidal deposition is possible. The first caveat is that surface accretion is not necessarily associated with increasing elevation in the tidal frame. It has long been known that progressive drying and compaction of sediments occurs after deposition (Ranwell 1964). However, the recent novel approach of Cahoon, Reed and Day (1995) in measuring surface elevation relative to a shallow subsurface datum (3-5 m deep) has shown that in some circumstances vertical accretion is a very poor indication of surface elevation change (Cahoon and Lynch 1997). Shallow subsidence resulting from changes in water storage or decomposition of organic material in the sediment profile can result in net lowering of the marsh surface, notwithstanding substantial vertical accretion.

The second caveat relates to changes in sea level relative to land. Mature marsh systems in various parts of the world may be much older than has been generally appreciated by ecologists, their origins having been after the retreat of the last glaciation. Pioneering work by Redfield (1972) reconstructed the postglacial history of an organogenic (peaty) marsh at Barnstable, Massachussetts, USA from radiocarbon-dated peat cores more than 6 m deep and historical evidence. The oldest existing features of this *Spartina alterniflora* marsh are approximately 4000 years old; the protective sand-spit that permitted development of the marsh grew to half its present length in the first thousand years, with ever decreasing rates of extension subsequently. Successive transgressions by a rising sea appear to have allowed sand accumulation at the margin of the marsh and whenever the basement sand surface reached its critical lower elevational limit in the tidal frame, *Spartina alterniflora* became established to initiate peat formation. Pethick (1980) correlated the inception of marshes on bare mud- or sandflats with fluctuations in the Holocene marine transgression. Historical evidence indicates that the colonization of new areas of salt marsh over the last 2000 years has coincided with periods of eustatic sea-level rise. Vertical growth of marshes was very fast in the first 100 years after inception but then it slowed dramatically and, at this point they were considered mature (Pethick 1981).

Upper marsh areas that have persisted for centuries or millennia and still show measurable rates of annual accretion require explanation. A good example is the extensive and ancient area of marsh on the north Norfolk coast of eastern England. Funnell and Pearson (1984, 1989) have investigated the sedimentology and micropalaeontology of radiocarbon dated sediment cores to a depth of 8 m below Ordnance Datum (OD) and up to 8410 years BP. Freshwater peats were initiated between 8410 ± 50 and 4880 ± 60 BP, depending on local conditions, in response to a rising freshwater water-table, itself possibly induced by rising sea level. Marine transgressions, depositing silty sands and muds, followed between 6610 60 and 4630 50 BP. Despite evidence for subsequent periods of regression, a remarkable finding is that the positions of the major channels, tidal flats and marshes appear to have been stabilized in their present positions for at least 4000 years; the present-day distribution of environments is largely determined by structures that were established relatively early in the Holocene or even in the preceding glacial stage. A layer of freshwater peat marks the beginning of the current upper marsh sediments at -1.03 m OD and an age of 2790 40 years BP. As freshwater peats do not form on this coast today below an elevation of about +3.3 m OD, this suggests an average rate of relative sea level rise of 1.5 mm $year^{-1}$ over the last 2800 years, which agrees with the proposed regional rate of crustal subsidence in north Norfolk of 1-2 mm $year^{-1}$. Pye (1992) has reviewed other evidence on different time scales that leads to similar estimates of sea level rise. Marshes reclaimed in the late 17th century are now 0.6 m below the corresponding actively accreting marshes on the seaward side of the embankments, which also points to an increase in relative sea level of about 1.5 mm $year^{-1}$ over the last 400 years. Even the maximum measured vertical accretion rates of 5-8 mm $year^{-1}$ on mature marshes would be consistent with this, after reduction to 25% to allow for changes in bulk density associated with compaction, dewatering, degradation of organic matter and dissolution of calcium carbonate over a period of 50 years (Pye 1992). The point in relation to succession is that this amount of accretion has compensated for the sinking coastline without any overall change in upper marsh elevation relative to the tidal frame. It is possible that strongly zoned vegetation may have remained more or less unchanged for millennia and that measured rates of accretion are at least partly the result of eustatic or isostatic changes in sea-level.

In the Netherlands, Roozen and Westhoff (1985) were the first to draw attention to the fact that the geomorphology of the sand flats on which marshes form could substantially determine the subsequent development of vegetation. They observed changes in the vegetation of permanent quadrats on three transects between 1953 and 1980 on a 4000 ha intertidal marsh that had formed since 1936 on the Frisian Island of Terschelling. The transitions between different community types could be aggregated into four more or less independent series, each of which was characteristic of a different elevation zone on the marsh. There was little evidence of one zone evolving into another: hence the elevation of a zone at the time of colonization strongly influenced the course of further development but the final zonation did not closely reflect previous succession. A combination of elevation and sediment texture must have favoured varying combinations of colonizing species with differing competitive abilities.

On Terschelling only on a small scale could zonation sometimes be interpreted as succession. This idea was developed by de Leeuw et al. (1993) who used aerial photographs and sediment profiles to compare the developmental histories of a barrier island marsh (on Schiermonnikoog, another Frisian Island) and an estuarine marsh (in the Westerschelde). On the barrier island marsh the surface elevation mainly reflected the topography of the initial aolian sand bank rather than the pattern of deposition of saltmarsh sediments: mud accretion was in fact thinnest on the upper parts of the zonation. In contrast, the estuarine marsh conformed to the conventional morphogenetic model with the relief having been formed by sediments deposited in a marsh environment. Both sites showed zoned vegetation but only in the estuarine one could historic succession reasonably be inferred from zonation.

The most recent and detailed analysis of 100 years of intertidal marsh development on Schiermonnikoog (Olff et al. 1997) has confirmed the importance of the underlying beach topography; silt accumulation over the last century has caused a maximum elevational difference of 12 cm, which is rather small in the context of variations in base elevation of over 100 cm along the transects. Simulations show that mean sea level rise of 0.63 mm year $^{-1}$ since 1824 in this area has had a profound effect on the amount and distribution of sedimentation; sediment deposition has been displaced from lower to higher elevations. The abundance and dominance of halophytic species depended strikingly on both the age of the sediment (successional age) and the base elevations of the successions, which were a consequence of larger scale geomorphological features.

4.2 GAINS AND LOSSES OF SPECIES SUBSEQUENT TO MARSH INCEPTION

The third caveat relates to whether the zonation seen in these ancient marshes can even be considered to be 'the ghost of succession past': for instance, are the biota in the lower parts those that would have been the original primary colonizers? An obvious incongruity arises from the position in British marshes of *Spartina anglica*, a species whose origins are now completely elucidated. As is well known, it arose in Southampton Water by hybridization of the native *S. maritima* and the North American introduction *S. alterniflora*, probably in the early 19th century. By 1870 the sterile hybrid (*S. x townsendii*) had doubled its chromosomes to produce the fertile form *S. anglica* (Gray, Marshall and Raybould 1991, Ferris, King and Gray 1997). This was a vigorous colonizer of soft mud at the seaward margin of marshes and in estuaries that promoted rapid accretion; it was widely planted to help reclaim land from the sea in the early part of this century and spread rapidly to become an environmental problem (Oliver 1925). Although in many of the areas where it originally spread *S. anglica* has been for many decades generally in decline (for reasons that are poorly understood), the unwary observer of many marshes would assume it to have been a primary colonist.

Likewise, *Zostera marina*, is seagrass that was classically regarded as an early colonist of low-lying intertidal flats. Its dramatic decline on coasts around the North Atlantic since the 1930s has been conspicuous and attention has recently focused on the recently identified aetiological agent of its 'wasting disease', the slime mould *Labyrinthula zosterae* (Muehlstein 1992), and its possible interactions with coastal eutrophication and climatic change. *Zostera* has been largely replaced in the zonation at different sites by *Spartina anglica*, and more recently by algae such as *Enteromorpha radiata* (den Hartog 1994) and the introduced species *Sargassum muticum* (den Hartog 1997). Where this has happened, the current zonation cannot be an exact recapitulation of succession.

5. The Nature of Succession

Any perturbation to physiographic equilibrium that creates a combination of bare intertidal flats and a supply of suitable suspended sediment will allow the initiation of a genuine succession. The perturbation may be a natural event, such as when an eroding channel changes its course, or a sandbank is thrown up by a violent storm. The best opportunities for ecological investigation often arise from perturbations associated with engineering works (construction of dams, dykes etc.) and other deliberate human interventions. Ecologists have distinguished for a long time between allogenic succession, where vegetational changes are driven by external influences, and autogenic succession in which colonization by plants itself modifies the environment and influences the establishment and performance of future colonists. The distinction between these two extremes is less clear in salt marsh succession. Because tidal marsh development is so dominated by sediment accretion and increasing elevation in the tidal frame, essentially an external physical process, geomorphologists have tended to regard it as an allogenic process. As Gray (1992) picturesquely concluded this implies that 'saltmarsh vegetation is merely the icing on a cake fashioned by physical processes'.

In certain circumstances a predominantly allogenic mechanism can be invoked. Ranwell (1974) chronicled the rapid development of a marsh and its transition to tidal woodland in the sheltered estuary of the River Fal in Cornwall, UK. China clay workings in the catchment probably contributed to the silt load that has been deposited on the extensive mud flats of the estuary. Rapid accretion (some 10 mm y^{-1} vertically) has caused the marsh to extend 800m seawards in a century. The low salinity of the estuarine waters has allowed the tidal woodland community to invade the landward margins of the brackish marsh at about the same pace. The difference in elevation between brackish marsh and tidal woodland can be as little as 0.2 m. Only at this interface do autogenic processes become a factor, because slightly raised tussocks of grasses or mounds formed by ants can be the establishment sites for tree seedlings.

Autogenic mechanisms, however, are typically much more important and their role can begin lower in the tidal frame than the lower margin of the salt marsh. Coles (1979) showed that the accumulation of fine sediments on intertidal mud flats of the Wash, eastern England, was strongly associated with the presence of high densities of benthic microalgae. These were mainly motile epipelic diatoms that produce copious mucous; the presence of mucous on the surface of the mud probably helps to trap fine sediment and migration of algae through the freshly deposited sediment to the surface releases mucous which stabilizes the sediment before it can be remobilized by the ebb tide. Chemical removal of the diatoms suppressed accretion, both on the mud flats and on salt marsh. Conversely, adjacent sand flats also showed little net accretion until the indigenous macroinvertebrate grazers were removed, with a resulting increase in microalgal populations. At slightly higher levels, the first salt marsh colonists, the annual *Salicornia* spp., often arise from seeds trapped in an irregular cover of macroalgae (Costa 1992).

The potential of such colonists in forming raised hummocks that are subsequently colonized by other species has already been referred to. Work on a recent example of such colonization on the coast of southwest Spain has begun to show the complexity of the successional processes involved (Castellanos, Figueroa and Davy 1994). The construction of the Juan Carlos I Dyke was a major engineering project that substantially changed physiographic conditions on the coast of the Gulf of Cadiz. Nearly 15 km in length, the raised dyke carries a road across an extensive area of salt marsh, and projects into the Atlantic Ocean. The result has been the formation of sand-spits to the west of the dyke and enhanced deposition of fine sediments carried down by the rivers on developing marshes to the east. In 1977, continuing construction of the dyke divided a generally uniform, low-lying area of sediment into two lagoons that have developed very different drainage regimes. Developing sand-spits have impeded the drainage of tidal waters from the lagoon on the west of the dyke, such that standing water remains long after the high tide; in contrast, the lagoon to the east drains rapidly into the estuarine channel through a short creek system.

Both lagoons have been colonized by isolated clones of *Spartina maritima*, a rhizomatous perennial grass. These have locally accretion enhanced to form domed hummocks in a process of nucleation. The diameter of the tussocks was highly correlated with the elevation of their sediment surface, the larger tussocks having been the earlier colonists in a generally accreting system (Fig. 2). The sequence of colonization is known from a series of fixed-point photographs (Castellanos 1992). *Spartina* in the interior of the tussocks showed reduced tiller density and vigour. Only in the better-drained lagoon were the central, higher areas of the *Spartina* tussocks invaded by *Arthrocnemum perenne* (*Sarcocornia perennis* (Miller) A.J. Scott ssp. *perennis*), although seed was freely available in both lagoons. There *Arthrocnemum* formed a sprawling, dense canopy and a superficial, relatively impenetrable root system above the rhizomes of *Spartina*; it rapidly suppressed the remaining tillers of *Spartina*, eventually leaving only a fringe around the edge of the hummock (Fig. 3). Areas invaded by *Arthrocnemum* were characterized by a superficial layer (10 cm) of oxidizing sediment ($E_h > +80$ mV). *Spartina*-dominated areas in both lagoons remained highly reducing, even in the surface layers ($E_h < -150$ mV).

The canopy of *Arthrocnemum* caused virtually 100 % light extinction; when it was removed experimentally recovery was rapid from rooted shoots. If the underground parts were also removed, regrowth from the periphery of the plots was also rapid but adventitious rooting only occurred in sediments with positive redox potential. After closure of the encroaching canopy, the light extinction was again higher than 99%. There was evidence of re-establishment of *Spartina* shoots only when *Arthrocnemum* was completely removed from a whole hummock. Subsequent development of the centrifugally growing, domed hummocks has seen coalescence of the oldest ones into larger irregular clumps that are separated only by a network of drainage channels.

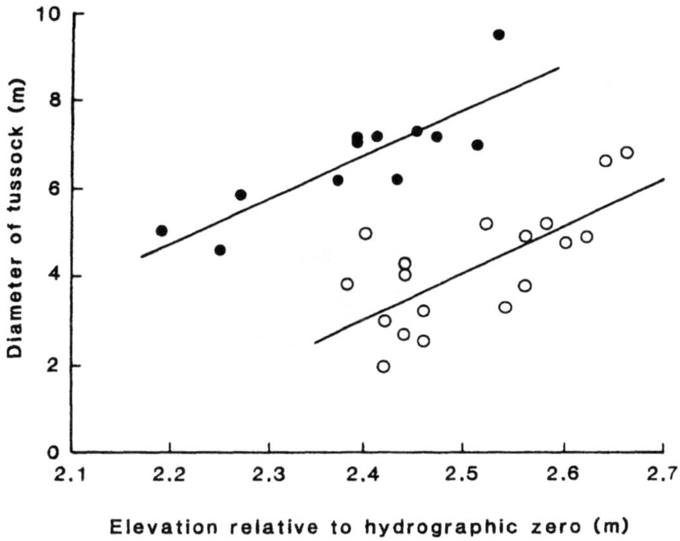

Figure 2. Relationship between the diameter of *Spartina maritima* hummocks (clones) and the elevation of their upper surface relative to Spanish hydrographic zero, under contrasting drainage systems at Odiel Marshes, Huelva, S.W. Spain. Well-drained system at Laguna de Don Claudio (●); Poorly drained, ponded system at Laguna de Ludovico (○). Regression equations: Laguna de Don Claudio, $y = 10.30x - 17.91$, $r^2 = 0.72$, $P< 0.0005$; Laguna de Ludovico, $y = 10.74x - 22.70$, $r^2 = 0.50$, $P<0.005$. Redrawn from Castellanos, Figueroa and Davy (1994).

Connell and Slatyer (1977) proposed a range of widely accepted conceptual models for the species replacements in succession that involved facilitation, inhibition and tolerance. The study of Odiel Marshes adds to the growing weight of evidence that these mechanisms may operate together in complex ways. *Spartina maritima* clearly facilitates the invasion by *Arthrocnemum*, which only becomes established from seed on raised, relatively well drained, oxidising sediments. The interaction is, however, more complicated than this: declining tiller density and moribund tillers of *Spartina* within the hummocks prior to invasion by *Arthrocnemum* are consistent with an inhibition mechanism. The very superior competitive ability of the later successional species *Arthrocnemum*, once established, and the inability of *Spartina* to re-invade suggests that a tolerance mechanism may also operate.

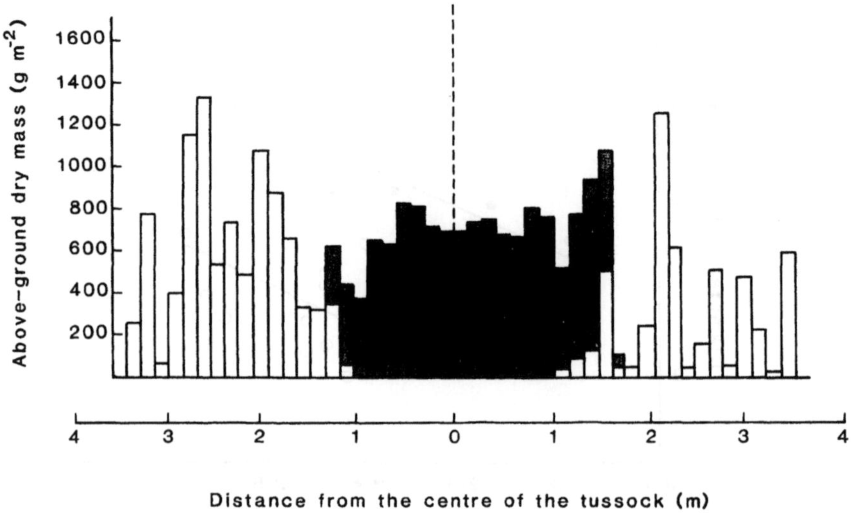

Figure 3. Transect across the diameter of a representative hummock of *Spartina maritima* undergoing invasion by *Arthrocnemum perenne* at the well-drained Laguna de Don Claudio (Odiel marshes, Huelva, S.W. Spain), showing the distribution of above-ground dry mass of *Spartina* (unshaded) and *Arthrocnemum* (shaded) in contiguous 15 x 25 cm quadrats. Redrawn from Castellanos, Figueroa and Davy (1994).

Intraspecific interactions may also change in succession. Certain populations of tillers were examined using conventional demographic techniques (see Silvertown and Lovett Doust 1993) to determine their 'birth' rates (development of juvenile tillers from basal or rhizome buds), rates of maturation of juvenile tillers into adult ones, flowering rates and mortality rates; the dependence of these processes on tiller density was also investigated. Comparison of the tiller dynamics of *Spartina maritima* in these colonizing hummocks with those in a non-successional sward at a similar elevation in the same marsh that had been stable for more than 40 years revealed significant demographic differences between the two populations (Castellanos et al. 1998). Rapid vertical accretion was recorded at the successional site and little net accretion at the non-successional one (Fig. 4); erosion at the seaward boundary of the non-successional site was associated with die-back of *Spartina* but the sward investigated was in the central area, where there was little net change in elevation. As expected, census of permanent quadrats placed near the perimeters of the hummocks at the successional site chronicled moving concentric 'waves' of high tiller density as tussocks expanded. High densities at the start of the study began to decline after one year to low values at the end of the second year (Fig. 5), but they had almost recovered after three years, indicating centripetal as well as centrifugal rhizome growth. The decline represented a combination of reduced numbers of births and increased numbers of deaths. In contrast, tiller densities were substantially higher in the non-successional sward and showed relatively small fluctuations with time; even here, tiller density reflected seasonal and stochastic variations in birth and death rates but the density in June 1990 was very similar to that in June 1988.

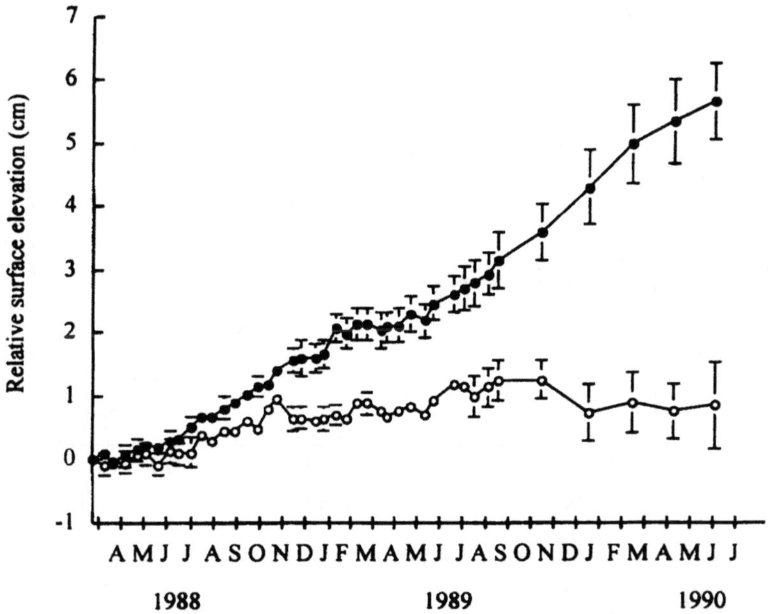

Figure 4. Variations in mean sediment level (SE, n = 4) through accretion and erosion in a coastal salt marsh at Odiel Marshes (Huelva, S.W. Spain) between June 1988 and July 1990: within the *Spartina* stand at the successional site of Laguna de Don Claudio (●); within the middle of the sward at the non-successional site of Marisma de El Muerto (○). Data from Castellanos et al. (1998).

The underlying risk of tiller mortality was similar in the two populations for much of the time but after two years there was increased mortality, mainly associated with flowering, at the successional site; very few tillers flowered in the non-successional sward. This mortality contributed to a shift to a younger age structure in the successional population. Perhaps the most important difference between populations revealed by the study was the presence of density-dependent regulatory processes in only one of them. In the sward population there was evidence for density-dependent mortality of tillers (Castellanos et al. 1998), with e.g., a highly significant relationship between mortality and log. density in juvenile tillers (Fig. 6). Hence, there were apparently compensatory adjustments to subtle variations in density that tended to maintain the relatively high tiller densities observed. No such responses, however, could be found in the population undergoing succession. This lack of density dependence and a relatively low peak density of about 2000 m^{-2} near to the leading edges of the expanding tussocks suggest that tiller placement there was regulated more by physiological mechanisms affecting rhizome growth and bud development in well integrated clones.

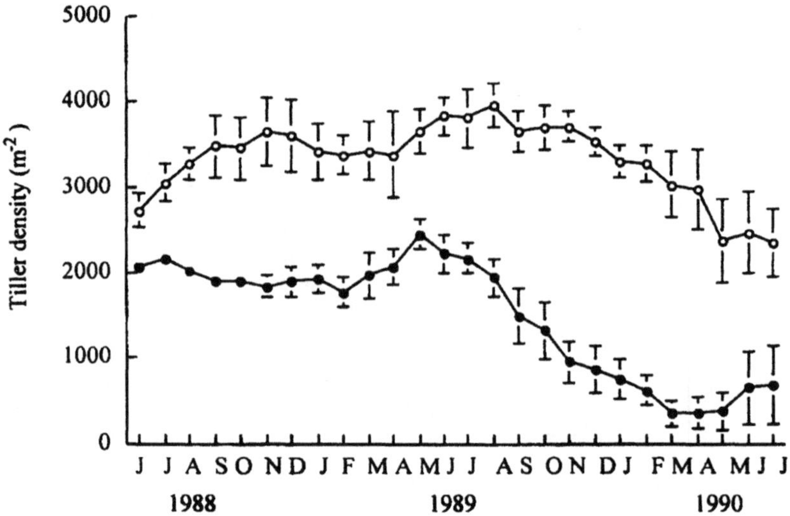

Figure 5. Monthly changes in mean tiller density (SE, n = 4) of *Spartina maritima* in a coastal salt marsh at Odiel Marshes (Huelva, S.W. Spain) between June 1988 and July 1990: colonizing population at the successional site of Laguna de Don Claudio (●); sward population at the non-successional site of Marisma de El Muerto (○). Redrawn from Castellanos et al. (1998).

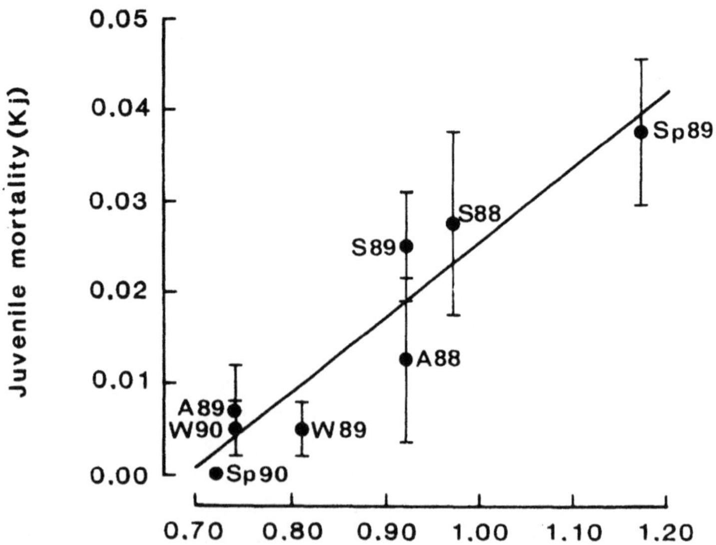

Figure 6. Relationship between juvenile tiller mortality (k_j) and juvenile tiller density at the non-successional sward site (Marisma de El Muerto) in a coastal salt marsh at Odiel Marshes (Huelva, S.W. Spain) between June 1988 and July. 1990. There were no significant effects of density at the contrasting successional site. Data from four permanent plots (each 10 x 10 cm) were seasonally grouped: Summer 1988, S88; Autumn 1988, A88; Winter 1989, W89; Spring 1989, Sp89; Summer 1989, S89; Autumn 1989, A89; Winter 1990, W90; Spring 1990, S90. Regression, $y = 0.08x - 0.06$ ($n = 8$, $r^2 = 0.89$, $P < 0.001$). Redrawn from Castellanos et al. (1998), which presents further details.

6. Conclusion

Experimental manipulations in the field, combined with the techniques of plant population biology and sedimentology, are clearly a key to an integrated understanding of the mechanisms and processes that control the development and maintenance of tidal salt marshes. The behaviour of halophyte populations is inevitably a determinant of local sedimentary environments and vice versa. The detailed information provided by this approach is exemplified in the case study at Odiel Marshes; it will allow us to unravel the complex range of relationships between vegetational zonation and succession that have fascinated coastal ecologists for so long on a larger scale.

Tidal salt marshes have an important role in both conservation and flood defence. The projected rise in global sea level over the next few decades underlines an urgent need to understand all the factors that govern the development, morphology and stability of tidal marshes. We know from responses to environmental perturbations that new marshes can develop relatively rapidly under appropriate conditions and that colonization by algae and pioneer halophytes is important in stabilizing, trapping, dewatering and consolidating sediment; the successional colonization by halophytes also plays a role in continuing marsh development and its protection from erosion. Despite a century of documentary evidence, geomorphologists still seem reluctant to accept the role of vegetational succession as a factor in promoting accretion. Ecologists on the other hand, in confounding succession with spatial zonation, have largely failed to appreciate the timescale over which marshes have persisted and the importance of changes in sea-level relative to land in maintaining them as accreting systems in a tidal frame. Areas such as the Essex coast of eastern England that are no longer considered economically defensible against the sea are likely to undergo 'coastal realignment'. Much of the coastal agricultural land reclaimed from the sea over centuries must be again surrendered to it. However, because of consolidation, chemical changes and oxidization of organic matter, when the dykes are breached, as in the Tollesbury experiment in Essex, the land surface now lies substantially lower in the tidal frame than adjacent areas of lower marsh that support pioneer vegetation. Investigation of marsh development in such circumstances is likely to be mutually rewarding for ecologists and geomorphologists, as well as being of considerable economic importance (Crooks and Turner 1999).

The large areas of mature post-glacial marsh, which show zonation but no current succession, are also of particular interest in the context of accelerating sea-level rise. The slow adjustments that have taken place over the last 3000 years may not be a good indication of future changes. We need to know how the vegetation in different zones will respond and whether accretion will be able to keep pace with the faster rate of sea-level rise. To what extent will the current zones be able to migrate landward? This implies vegetational regression and erosion, rather than succession, especially at what are currently the 'pioneer' zones. Studies of the interactions between species, especially as they are modulated by level in the tidal frame, will be important in predicting likely changes, just as the large-scale experiments imposed upon us will help to clarify and refine the concepts of salt marsh ecology.

7. Literature Cited

Adam, P. 1990. Saltmarsh ecology. Cambridge University Press, Cambridge, England.

Allen, J.R.L. and K. Pye. 1992. Saltmarshes: their nature and importance. Pages 1-18 *in* J.R.L. Allen and K. Pye, editors. Saltmarshes: morphodynamics, conservation and engineering significance. Cambridge University Press, Cambridge, England.

Armstrong, W., P.M. Beckett, S. Lythe and T.J. Gaynard. 1985. Plant zonation and the effects of the spring-neap tidal cycle on soil aeration in a Humber salt marsh. Journal of Ecology **73**: 323-339.

Bertness, M.D. and S.M. Yeh. 1994. Cooperative and competitive interactions in the recruitment of marsh elders. Ecology **75:** 2416-2429.

Bockelmann, A. C. and R. Neuhaus. 1999. Competitive exclusion of *Elymus athericus* from a high-stress habitat in a European salt marsh. Journal of Ecology **87**: 503-513.

Cahoon, D.R. and J.C. Lynch. 1997. Vertical accretion and shallow subsidence in a mangrove forest of southwestern Florida, U.S.A. Mangroves and Saltmarshes **1**: 173-186.

Cahoon, D.R., D.J. Reed and J.W. Day Jr. 1995. Estimating shallow subsidence in microtidal salt marshes of the southeastern United States: Kaye and Barghoorn revisited. Marine Geology **128**: 1-9.

Callaway, R.M. 1994. Facilitative and interfering effects of *Arthrocnemum subterminale* on winter annuals. Ecology **75:** 681-686.

—— 1995. Positive interactions among plants. Botanical Review **61**: 306-349.

Callaway, R.M. and L. King. 1996. Temperature-driven variation in substrate oxygenation and the balance of competition and facilitation. Ecology **77:** 1189-1195.

Carey, A.E. and F.W. Oliver. 1918. Tidal lands: a study of shore problems. Blackie and Son, London, England.

Castellanos, E.M. 1992. Colonización, dinámica poblacional y papel en la sucesión de *Spartina maritima* (Curtis) Fernald en las Marismas del Odiel. Doctoral thesis. Sevilla University, Sevilla, Spain.

Castellanos, E.M., M.E. Figueroa and A.J. Davy. 1994. Nucleation and facilitation in saltmarsh succession: interactions between *Spartina maritima* and *Arthrocnemum perenne.* Journal of Ecology **82**: 239-248.

Castellanos, E.M., C. Heredia, M.E. Figueroa and A.J. Davy. 1998. Tiller dynamics of *Spartina maritima* in successional and non-successional mediterranean salt marsh. Plant Ecology **137**: 213-225.

Chapman, V.J. 1938. Studies in salt-marsh ecology. Sections I-III. Journal of Ecology **26**: 144-179.

—— 1939. Studies in salt-marsh ecology. Sections IV and V. Journal of Ecology **27**: 160-201.

—— 1940. Studies in salt-marsh ecology. Sections VI and VII. Comparison with marshes on the east coast of North America. Journal of Ecology **28**: 118-152.

—— 1941. Studies in salt-marsh ecology. Section VII. Journal of Ecology **28**: 69-82.

—— 1960. The plant ecology of Scolt Head Island. Pages 85-163 *in* J.A. Steers, editor. Scolt Head Island. W. Heffer and Sons, Cambridge, England.

—— 1974 Salt marshes and deserts of the world. 2nd Edition. J. Cramer Verlag Leutershausen, Germany.

Clements, F.E. 1916. Plant succession. An analysis of the development of vegetation. Carnegie Institute of Washington, Publication No. 242, Washington, District of Columbia, USA.

Coles, S.M. 1979. Benthic microalgal populations on intertidal sediments and their role as precursors to salt marsh development. Pages 25-42 *in* R.L. Jefferies and A.J. Davy, editors. Ecological processes in coastal environments. Blackwell Scientific Publications, Oxford, England.

Connell, J.H. and R.O. Slatyer. 1977. Mechanisms of succession in natural communities and their role in community stability and organization. American Naturalist **111**: 1119-1144.

Costa, C.S.B. 1992. Competition and coexistence in the salt marsh annuals *Suaeda maritima* and *Salicornia europaea*. Ph.D. Dissertation, University of East Anglia, Norwich, England.

Crooks, S. and R.K. Turner. 1999. Integrated coastal management: sustaining estuarine natural resources. Advances in Ecological Research **29**: 241-288.

Dainty, J. 1979. The ionic and water relations of plants which adjust to a fluctuating saline environment. Pages 201-209 *in* R.L. Jefferies and A.J. Davy, editors. Ecological processes in coastal environments. Blackwell Scientific Publications, Oxford, England.

Davy, A.J. and C.S.B. Costa. 1992. Development and organization of saltmarsh communities. Pages 157-178 *in* U. Seeliger, editor. Coastal plant communities of Latin America. Academic Press, San Diego, California, USA.

Davy, A.J., C.S.B. Costa, A.R. Yallop, A.M. Proudfoot and M. Mohamed. 2000. Biotic interactions in plant communities of salt marshes. Pages 109-127 *in* B. Sherwood, T.G. Gardiner and T. Harris editors. British Salt marshes. Forrest Text, England.

Davy, A.J., S.M. Noble and R.P. Oliver. 1990. Genetic variation and adaptation to flooding in plants. Aquatic Botany **38**: 91-108.
Davy, A.J. and H. Smith. 1985. Population differentiation in the life-history characteristics of salt-marsh annuals. Vegetatio **61**: 117-125.
—— 1988. Life-history variation and environment. Pages 1-22 *in* A.J. Davy, M.J. Hutchings and A.R. Watkinson, editors. Plant population ecology. Blackwell Scientific Publications, Oxford, England.
French, J.R. and T. Spencer. 1993. Dynamics of sedimentation in a tide-dominated backbarrier salt marsh, Norfolk, UK. Marine Geology **110**: 315-331.
Ferris, C., R.A. King and A.J. Gray. 1997. Molecular evidence for the maternal parentage in the hybrid origin of *Spartina anglica*. Molecular Ecology **6**: 185-187.
Flowers, T.J., M.A. Hajibagheri and N.J.W. Clipson. 1986. Halophytes. Quarterly Review of Biology **61**: 313-337.
Funnell, B.M. and I. Pearson. 1984. A guide to the Holocene geology of North Norfolk. Bulletin of the Geological Society of Norfolk **34**: 123-140.
Funnell, B.M. and I. Pearson. 1989. Holocene sedimentation on the North Norfolk barrier coast in relation to relative sea-level change. Journal of Quaternary Science **4**: 25-36.
Gerdol, V. and R.G. Hughes. 1993. Effect of the amphipod *Corophium volutator* on the colonisation of mud by the halophyte *Salicornia europaea*. Marine Ecology Progress Series **97:** 61-69.
Gray, A.J. 1992. Saltmarsh plant ecology: zonation and succession revisited. Pages 63-79 *in* J.R.L. Allen and K. Pye, editors. Saltmarshes: morphodynamics, conservation and engineering significance. Cambridge University Press, Cambridge, England.
Gray, A.J., D.F. Marshall and A.F. Raybould. 1991. A century of evolution in *Spartina anglica*. Advances in Ecological Research **21**: 1-62.
de Leeuw, J., W. de Munck, H. Olff and J.P. Bakker. 1993. Does zonation reflect the succession of salt-marsh vegetation? A comparison of an estuarine and a coastal bar island marsh in The Netherlands. Acta Botanica Neerlandica **42**: 435-445.
den Hartog, C. 1994. Suffocation of a littoral *Zostera* bed by *Enteromorpha radiata*. Aquatic Botany **47**: 21-28.
—— 1997. Is *Sargassum muticum* a threat to eelgrass beds? Aquatic Botany **58**: 37-41.
Jefferies, R.L., A.J. Davy and T. Rudmik. 1979. The growth strategies of coastal halophytes. Pages 243-268 *in* R.L. Jefferies and A.J. Davy editors. Ecological processes in coastal environments. Blackwell Scientific Publications, Oxford, England.
Marsh, A.S. 1915. The maritime ecology of Holme next the Sea, Norfolk. Journal of Ecology **3**: 65-96.
Muehlstein, L.K. 1992. The host-pathogen interaction in the wasting disease of eelgrass, *Zostera marina*. Canadian Journal of Botany **70**: 2081-2088.
Noble, S.M., A.J. Davy and R.P. Oliver. 1992. Ribosomal DNA variation and population differentiation in *Salicornia* L. New Phytologist **122**: 553-565.
Olff, H., J. de Leeuw, J.P. Bakker, R.J. Platerink, H.J. van Wijnens and W. de Munck. 1997. Vegetation succession and herbivory in a salt marsh: changes induced by sea level rise and silt deposition along an elevational gradient. Journal of Ecology **85**: 799-814.
Oliver, F.W. 1906. The Bouche d'Erquy in. 1906. New Phytologist **5**: 189-195.
—— 1907. The Bouche d'Erquy in. 1907. New Phytologist **6**: 244-252.
Oliver, F.W. and E.J. Salisbury. 1913. The topography and vegetation of the National Trust reserve known as Blakeney Point, Norfolk. Transactions of the Norfolk and Norwich Naturalists' Society **9**: 485-542.
Oliver, F.W. 1925. *Spartina townsendii*; its mode of establishment, economic uses and taxonomic status. Journal of Ecology **13**: 74-91.
Packham, J.R. and A.J. Willis. 1997. Ecology of dunes, salt marsh and shingle. Chapman and Hall, London, England.
Pethick, J.S. 1980. Salt-marsh initiation during the Holocene transgression: the example of the North Norfolk marshes, England. Journal of Biogeography **7**: 1-9.
—— 1981. Long term accretion rates on tidal marshes. Journal of Sedimentary Petrology **51**: 571-577.
Pignatti, S. and A. Ubrizsy Savoia. 1989. Early use of the succession concept by G.M. Lancisi in 1714. Vegetatio **84**: 113-115.
Pielou, E.C. and R.D. Routledge. 1976. Salt marsh vegetation: latitudinal gradients in the zonation pattern. Oecologia **24**: 311-321.
Pye, K. 1992. Saltmarshes on the barrier coastline of North Norfolk, eastern England. Pages 148-178 *in* J.R.L. Allen and K. Pye, editors. Saltmarshes: morphodynamics, conservation and engineering significance. Cambridge University Press, Cambridge, England.

Ranwell, D.S. 1964. *Spartina* marshes in southern England II. Rate and seasonal pattern of sediment accretion. Journal of Ecology **52**: 79-94.
—— 1972. Ecology of salt marshes and sand dunes. Chapman and Hall, London, England.
—— 1974. The salt marsh to tidal woodland transition. Hydrobiological Bulletin (Amsterdam) **8**: 139-151.
Redfield, A. C. 1972. Development of a New England salt marsh. Ecological Monographs **42**: 201-237.
Roozen, A.J.M. and V. Westhoff. 1985. A study on long-term salt-marsh succession using permanent plots. Vegetatio **61**: 23-32.
Sánchez, J.M., J. Izco and M. Medrano. 1996. Relationships between vegetation zonation and altitude in a salt-marsh system in northwest Spain. Journal of Vegetation Science **7**: 695-702.
Silvertown, J.W. and J. Lovett Doust. 1993. Introduction to plant population biology. 3rd Edition. Blackwell Scientific Publications, Oxford, England.
Snow, A.A. and S.W. Vince. 1984. Plant zonation in an Alaskan salt marsh. II. An experimental study of the role of edaphic conditions. Journal of Ecology **72**: 669-684.
Ungar, I.A. 1999. Are biotic factors significant in influencing the distribution of halophytes in saline habitats? Botanical Review **64**: 176-199.
Yapp, R.H., D. Johns and O.T. Jones. 1917. The salt marshes of the Dovey Estuary. Part II. The salt marshes. Journal of Ecology **5**: 65-103.

FATE OF PRODUCTION WITHIN MARSH FOOD WEBS

MICROBIAL SECONDARY PRODUCTION FROM SALT MARSH-GRASS SHOOTS, AND ITS KNOWN AND POTENTIAL FATES

STEVEN Y. NEWELL
Marine Institute
University of Georgia
Sapelo Island, Georgia 31327 USA

DAVID PORTER
Department of Botany
University of Georgia
Athens, Georgia 30602-7271 USA

Abstract
Several lines of evidence (direct microscopy, index biochemicals) point to predominance of eukaryotic decomposers in natural decay of dead shoots of smooth cordgrass (*Spartina alterniflora*). Recent research shows that this is also true for black needlerush (*Juncus roemerianus*). Ascomycetous fungi are the major initial secondary producers based on the dead shoots. There is no overlap between the species of the cordgrass (e.g., *Phaeosphaeria spartinicola*) and needlerush (e.g., *Loratospora aestuarii*) fungal-decay communities. Even when conditions in the marsh are manipulated in directions that would be expected to favor prokaryotes (extra water and nitrogen), the ascomycetes accumulate maximum organic masses in standing-decaying shoots hundreds of times larger than prokaryotic masses. Rates of fungal production are not increased by raising duration of high water availability, probably due to fine-tuned fungal adaptation to periodic dryness, but nitrogen does limit fungal productivity in decaying cordgrass. Content of living-fungal mass can be 10 to 20% of total system (= microbes + remaining plant) mass, depending on nitrogen availability, rates of invertebrate mycophagy, and probably several further factors yet to be determined. Standing crops of living fungi in cordgrass marshes in Georgia (per-m^2 basis) have been calculated to be equal to 3% (summer) to 28% (winter) of living-cordgrass standing crop. This is calculated to be about 50 to 100% of total (non-cyano) bacterial crop; the great bulk of bacterial crop is sedimentary. Fungal productivity per m^2 standing-decaying-cordgrass marsh has been provisionally found to be 10 times greater in winter than in summer (3652 mg per m^2 per day; μ = 0.07 day^{-1}). Total bacterial productivity per m^2 was calculated to be about x2 fungal in summer, and x0.07 fungal in winter. High yields of fungi (on the order of 50%) from cordgrass shoots may be part of the explanation for high rates of animal secondary production in saltmarsh ecosystems. Cordgrass-fungal standing crops and productivities (per unit leaf mass) do not show pronounced variation (in autumn) along a south-north latitudinal gradient from 30° to 44°N. One major known fate of saltmarsh-fungal secondary production is output to shredder gastropods (periwinkles, *Littoraria irrorata*). Other potential substantial fluxes are to

amphipods (especially *Uhlorchestia spartinophila*) and other gastropods (especially *Melampus bidentatus*), and fluxes as sexual propagules (ascospores) and as remnant hyphal wall/sheath mass in fallen, decayed fragments. Key opportunities for saltmarsh-ecological research lie: in learning the details of the life histories of the more important saltmarsh-fungal producers; in determining the biotic and abiotic controls on saltmarsh-fungal productivity; and in investigations of impacts of fungal activities, such as the probable role that saltmarsh ascomycetes have in release of dimethylsulfide to the atmosphere.

1. Marshgrass Shoots as Substrate

Saltmarsh grasses (in common with most other grasses) do not abscise their leaves or aboveground stems (Newell 1993). This is true for both *Spartina alterniflora* (smooth cordgrass) and *Juncus roemerianus* (black needlerush), the two principal primary producers of southern North American saltmarshes (Christian et al. 1990, Newell 1993). Much of the decay of shoots of saltmarsh grasses takes place above the sediment. This is advantageous to the grasses in that it results, after standing decay and partial breakage, in open pathways for gas transport to/from the rhizomes and roots (Arenovski and Howes 1992, Armstrong et al. 1996). It is also possible that standing decay confers advantages upon the grasses through the provision of aerenchymally-transferred high concentrations of CO_2 and H_2O from internal microbial decomposers in dead parts of shoots to living parts of shoots (Newell 1996a). Key characteristics of smooth cordgrass shoots as substrate for microbial decay are its high content of lignocellulose, and its high content of non-lignin cinnamyl phenols (references in Newell 1993, Newell et al. 1996a; see also Bartolomé et al. 1997).

2. Microbial Decomposers

2.1 FUNGI

Mycologists have been aware that ascomycetous fungi (Kingdom Fungi, phylum Ascomycota; Hawksworth et al. 1995) are secondary producers in standing-decaying shoots of smooth cordgrass since the 19th century (Gessner and Kohlmeyer 1976, Kohlmeyer and Volkmann-Kohlmeyer 1991). However, in part due to non-recognition of the absence of abscission of shoot parts (previous section), marsh ecologists did not form partnerships with marsh mycologists, and prokaryotes were designated as the drivers of marshgrass-shoot decomposition in the 1950s (Newell 1993). Using several methodological approaches (direct epifluorescence microscopy, index-biochemical measurements, quantification of sexual-reproductive structures, and transmission electron microscopy [TEM]), it has now been firmly established that it is fungi that pervade and lyse standing shoots of smooth cordgrass (Newell 1993, Newell and Wasowski 1995, Newell 1996a, Newell et al. 1996a). TEM examinations of standing-decaying smooth-cordgrass leaves and stems showed prokaryotic activity (erosion bacteria) only for stems that had collapsed onto the sediment (Newell et al. 1996a), implying, as has experimental research, that prokaryotic decomposers do not have substantial lytic effects upon cordgrass shoots until the shoot parts reach the sediment system (Newell and Palm 1998). Biochemical examination of decaying black needlerush (see next section), along with the discovery of several new species of needlerush ascomycetes (with remarkable ascospore morphologies) by Kohlmeyer and Volkmann-Kohlmeyer (Kohlmeyer et al. 1997, and references therein), strongly suggest that fungal participation in natural decay of black needlerush is as predominant

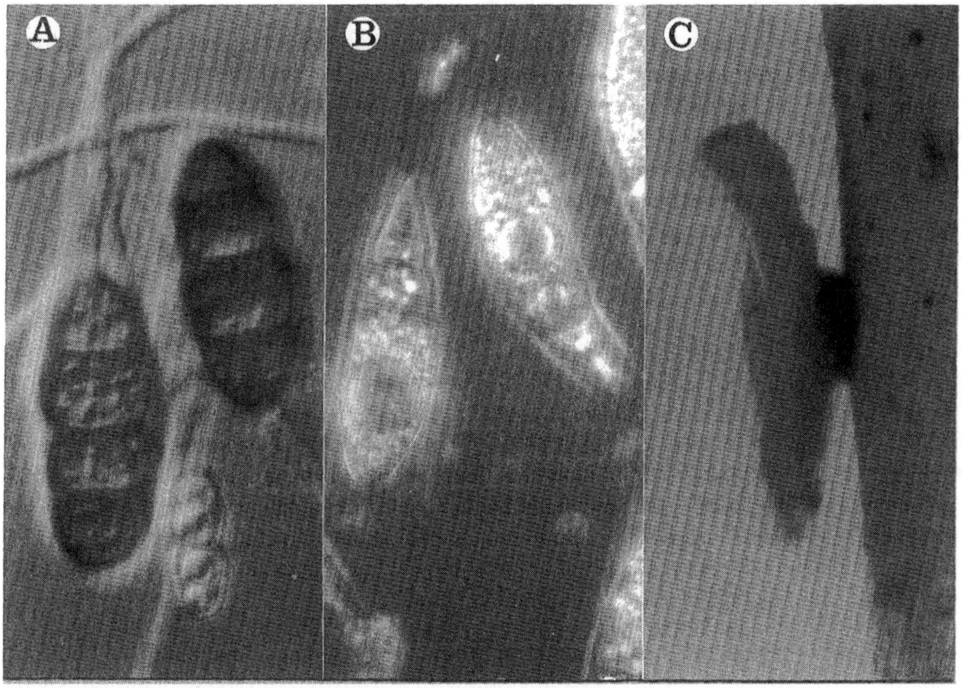

Figure 1. Four common ascomycetes of the smooth-cordgrass (*Spartina alterniflora*) decomposer community. A. Ascospores of two loculoascomycetes that very regularly occur together in leaf blades, *Phaeosphaeria spartinicola* (dark amber-brown, 4-celled, about 11 μm wide) and *Mycosphaerella* sp. 2 (of Kohlmeyer and Kohlmeyer 1979) (colorless, 2-celled, about 7 μm wide). These two species' small (75-160 μm diam) ascomata (the dark brown to black sexual structures embedded just below the abaxial surface) are responsible for the barely visible 'black-peppering' of naturally decaying blades. The ascospores in the photo are germinating to hyphae on ascosporecapture slides (Newell and Wasowski 1995). B. Ascospores of the perithecial ascomycete *Buergenerula spartinae*. The club-shaped, refringent ascospores are about 12 μm wide at their widest point. *B. spartinae* is the most commonascomycete responsible for thorough blackening (melanizing) of large patches on blades, and its large (to 450 μm diam) ascomata can fill a blade aerenchma channel. C. A bright yellow, stalked ascoma of *Lachnum spartinae* in its mushroom-like, natural orientation on a dead shoot. *L. spartinae* is apothecial (the ascoma is shaped like an open saucer, rather than a spheriod with a small opening). The apothecia are about 0.5-2 mm across, and are easily visible to the naked eye in wet, late-autumn rainy periods on older standing-decaying sheaths. See Kohlmeyer and Volkmann-Kohlmeyer (1991), Kendrick (1992), Hawksworth et al. (1995), Alexopoulos et al. (1996), Cantrell et al. (1996), Newel et al. (1996a).

as it is for smooth cordgrass. Preliminary biochemical-index information suggests that decaying parts of shoots of other saltmarsh grasses (*Spartina patens* and *Juncus gerardi*) also contain large (at least 4-5% living-fungal organic mass) standing crops of fungi (Newell and Fell 1992, and Newell unpublished).

Species of hyphomycetes (= nonsexual, mitosporic fungi with simple, hyphal spore-producing structures; see Hawksworth et al. 1995) can be isolated from naturally decaying shoots of smooth cordgrass (references in Gessner 1977). Based on evidence from immunoassay (developed with antibodies against immunogens from a cordgrass ascomycete), most if not all of the living-fungal mass in naturally decaying leaf blades of smooth cordgrass is that of the predominant ascomycetes of the system (Newell and Wasowski 1995). This suggests that the hyphomycetes that can be isolated from decaying cordgrass are present only as inactive propagules or weakly active microcolonies (unless some are unrecognized asexual forms of the cordgrass ascomycetes). This hypothesis has not yet been tested with surface-sterilization techniques (Newell 1996a), in part because application of this technique to smooth-cordgrass leaves is complicated by the undulating adaxial-surface topography of the blades (Anderson 1974).

The result of secondary production by ascomycetes in standing-decaying shoots of smooth cordgrass can be readily detected by direct observation at the stereomicroscope. The sexual-reproductive structures (ascomata) of these fungi are the end result of digestive activity of foregoing mycelium; the presence of dense concentrations of mature ascomata (e.g., for smooth cordgrass in Georgia, USA, = 1000 cm^{-2} blade abaxial surface: Newell and Wasowski 1995) therefore implies substantial supportive mycelial production. One species of ascomycetous cordgrass decomposer (*Lachnum spartinae*) can even be seen on older decaying leaf sheaths with the naked eye (Fig. 1C). Species with smaller ascomata, that are virtually omnipresent in smooth-cordgrass marshes as decomposers of leaf blades, are *Phaeosphaeria spartinicola*, *Mycosphaerella* sp. 2, and *Buergenerula spartinae* (Newell 1993, Newell and Wasowski 1995) (Fig. 1A,B). *Mycosphaerella* sp. 2, for which only a partial, informal description is currently available (Kohlmeyer and Kohlmeyer 1979), was apparently largely overlooked by Newell and Wasowski (1995). Recent seasonal measurements of ascospore expulsion from naturally decaying blades of smooth cordgrass have revealed that *Mycosphaerella* sp. 2, which has smaller ascomata than *P. spartinicola*, is virtually always present along with *P. spartinicola* (Newell unpublished). Additional species that can be common as smooth-cordgrass leaf-blade occupants are: *Phaeosphaeria halima*, *Stagonospora* sp. 2 (of Kohlmeyer and Kohlmeyer 1979; 7-septate conidia; a coelomycete), *Hydropisphaera erubescens* (as *Calonectria* sp. in Newell and Wasowski, 1995; now undergoing description by A. Y. Rossman); in leaf sheaths: *Phaeosphaeria spartinae, Phaeosphaeria neomaritima, Anthostomella* sp. (of Gessner and Kohlmeyer 1976); in naked stems: *Passeriniella obiones* (Newell 1993; see Kohlmeyer and Volkmann-Kohlmeyer 1991).

An amazing diversity of ascomycetes, not including any species of the smooth-cordgrass mycoflora, has recently been formally described from naturally decaying black needlerush (Kohlmeyer et al. 1997 and references therein). Among the most common species appear to be *Loratospora aestuarii, Papulosa amerospora, Aropsiclus junci, Anthostomella poecila, Physalospora citogerminans, Scirrhia annulata, Massarina ricifera*, and *Tremateia halophila*.

2.2 OOMYCOTES

Marine oomycotes are eukaryotic mycelial decomposers that have swimming, biflagellate propagules and lie in the Kingdom Chromista (or Protoctista), phylum Oomycota (Hawksworth et al. 1995, Dick 1997, Fell and Newell 1998). Although oomycotes (species of *Halophytophthora* and *Pythium*) can be isolated from decaying marshgrass, the little evidence presently available suggests that oomycotes are not substantial secondary producers in marshgrass – they appear to direct their activities at leaves that fall into seawater (e.g., mangrove leaves) (Newell 1996a).

2.3 BACTERIA

Although fungal lysis is the most obvious form of alteration of standing-dead shoots of marshgrass (Newell et al. 1996a), it is possible that some species of bacteria of the dead-shoot surfaces interact positively with fungal decomposers, and when fungal-decayed shoot material falls to the sediment surface, it is likely that it moves into a bacterial-lysis system (Moran et al. 1995, González et al. 1997, Newell and Palm 1998). Identifying bacterial species that are active in natural assemblages is not as straightforward as for many species of fungi (e.g., the ascomycetes of marshgrass that form characteristic, unique sexual structures visible under the stereomicroscope) (Fuhrman et al. 1994, Newell 1994, Torsvik et al. 1996, Schut et al. 1997). Moran and Hodson (1990) established that marshgrass lignocellulose can be altered by natural saltmarsh-bacterial assemblages, potentially leading to a substantial contribution of saltmarsh lignin to dissolved humics. González et al. (1996) found that lignin-utilizing bacterial assemblages could be obtained from Georgia saltmarsh waters, and discovered that many numerically abundant species in the lignin-enrichment assemblages were culturable using special techniques. González and Moran (1997) have shown that up to 28% of the bacterioplankton DNA obtained directly from saltmarsh waters of Georgia belongs to the "marine-alpha" group of species (α-3 subclass of the class Proteobacteria, low-nutrient culturable), one of which is *Sagittula stellata*, a lignin-transforming species (González et al. 1997). Chen et al. (1997) have developed a fluorescent-probing, complementary-DNA (cDNA) method for directly detecting *S. stellata* and other bacterial species in natural samples, so at least for some bacterial species, it is now practicable to look for and monitor dynamics of potential combinations of fungal and bacterial partners, or determine which species of bacteria might replace fungal decomposers in fallen shoot material.

3. Lysis of Marshgrass Lignocellulose (LC)

Seventy to seventy-five percent of the organic mass of mature shoots of smooth cordgrass consists of lignocellulose (LC) (Hodson et al. 1984). Early tests of the ability of cordgrass ascomycetes to digest cordgrass LC gave the result that the ascomycetes could not mineralize either major part (polysaccharide or lignin) of the LC faster than 0.09% per day (Newell et al. 1996a). More recent findings indicate that these low rates were likely due to the design of the experiment: testing was done with fine-particulate LC chemically separated from other leaf chemicals and submerged in flasks agitated continuously at 90 rpm. Subsequent experiments with extracted smooth-cordgrass LC (references in Newell et al. 1996a) gave sharply different results: when chemically extracted cordgrass LC was incubated statically, submerged with fungal inocula in a solution of malt and yeast extract, the LC polysaccharide was mineralized at 0.83% per day, and fungal yield efficiency was 40% on cordgrass LC, even if the easily-available carbohydrate was not present.

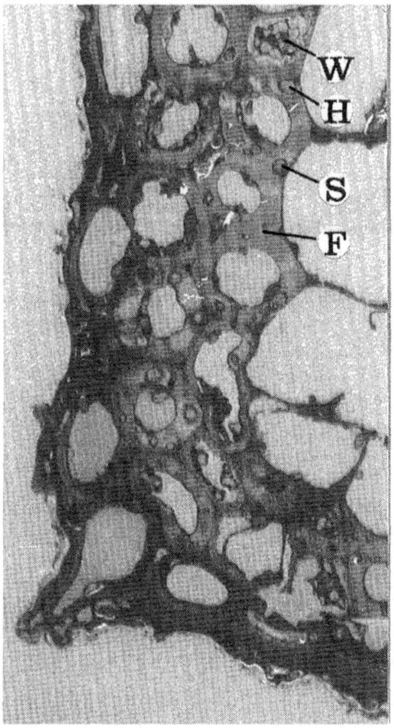

Figure 2. Transmission-electron micrograph of a cross-section of the adaxial surface of a naturally decaying, standing leaf blade of smooth cordgrass (*Spartina alterniflora*). The thick-walled cells (**F**) of the adaxial fiber band show the extent to which the lignocellulosic fiber cells are digested by hyphae (**H**), probably of one or both of *P. spartinicola* and *Mycosphaerella* sp. 2 (Fig. 1A), even on the side of the leaf opposite to that (abaxial) where the ascomata are formed. Both type-1 soft-rot digestion (**S**, wall boring) and white-rot-type digestion (type 2 soft rot) (**W**, lysis from cell lumen outwards) are occuring in this section. See Nilsson et al. (1989). The fiber cells in the photo are about 15 μm across.

The laboratory-experimental result that cordgrass ascomycetes can digest cordgrass LC was challenged by Newell et al. (1996a), through the direct examination of fungal impact on the LC. Transmission electron microscopy was used, with samples from naturally decaying smooth cordgrass shoots. Portions of standing-decaying shoots that contained ascomata of only one of four species of ascomycetes were examined: leaf blades (*P. spartinicola, B. spartinae*), leaf sheaths (*P. spartinae*), and naked stems (*P. obiones*). All four species exhibited clear digestion of the lignocellulose. As pointed out in the preceeding section, it is quite likely that the TEM evidence for *P.*

spartinicola is in fact evidence for the combined activity of *P. spartinicola* and *Mycosphaerella* sp. 2 (of Kohlmeyer and Kohlmeyer 1979). By the time that these two species have formed their ascomata in the leaf blades, the fiber cells (where LC is concentrated) of the blades can be extensively digested, even far from the ascomata (Fig. 2). TEM clearly revealed that cordgrass ascomycetes are vigorous LC lysers, causing destruction not only of the secondary walls of fiber cells, where 60-80% of the lignin of LC resides, but also damaging the middle-lamellar layer where lignin is most concentrated (Fig. 3) (Newell et al. 1996a). It is possible that each of the two types of LC lysis ("type 1 soft rot", digestion from bore holes, Fig. 2; "type 2 soft rot", digestion from the cell lumen outward, Figs. 2 and 3) seen in areas of occupation of decaying blades by *P. spartinicola* and *Mycosphaerella* sp. 2, is caused solely by one or the other of these two species.

There are no data yet available for LC content of black needlerush, nor for LC-lytic

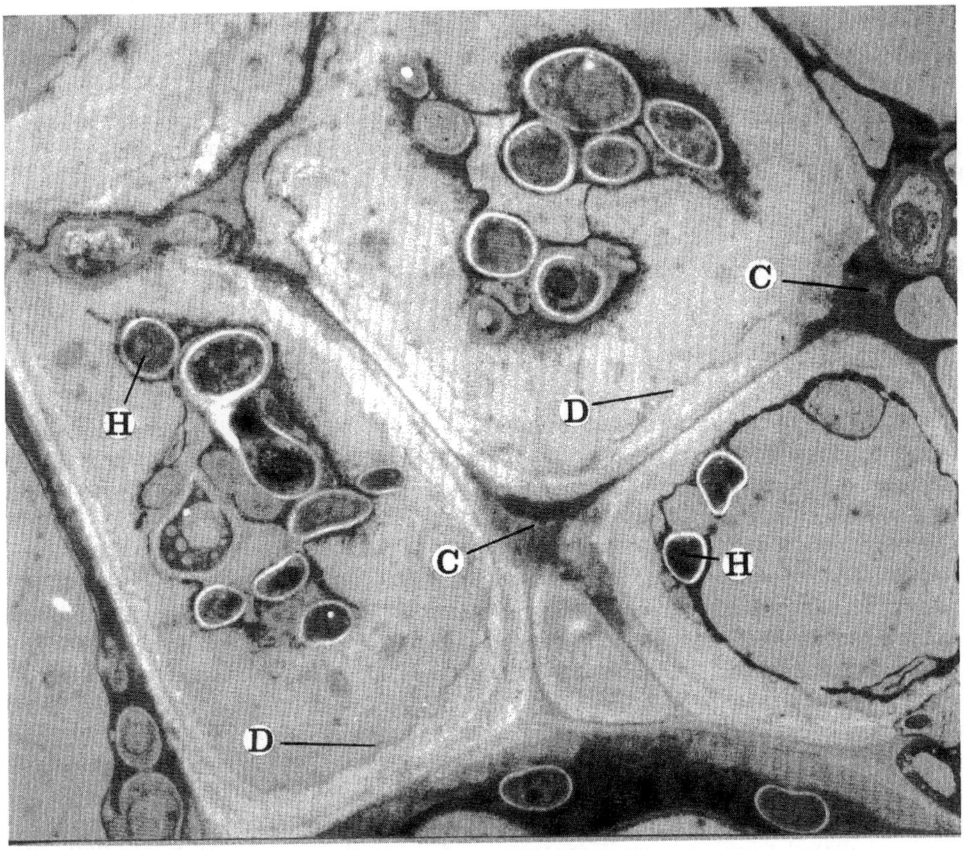

Figure 3. Type 2 soft rot (white-rot like) of fiber cells (see Fig. 2) of smooth cordgrass leaf blades (transmission-electron-microscopic section). Fiber cells are being digested from the cell lumen outwards by hyphae (**H**) within the lumen. Digestion apparently takes place in two phases, as seen by the primary digestive front (**D**) in cells at earlier stages of decay. Note that some hyphea are altering the lignin-rich middle-lamellar layer between cells, even were it is thickest, at the cell corners (**C**). Hyphal sections are about 1.5-3 μm across.

capabilities of needlerush microbes. Since standing decay of the rigid, terete blades of black needlerush can take years, and results in the blades being rendered decolorized and quite fragile (Christian et al. 1990, and Newell unpublished), it is quite likely that the ascomycetes of black needlerush are as capable re LC lysis (though with slower rates) as are the ascomycetes of smooth cordgrass, but this remains to be tested. The invaluable taxonomic work of Kohlmeyer and Volkmann-Kohlmeyer (see preceeding section) has made this potential research direction feasible.

Fiber-cell LC polymeric matrices block entry of large, lytic enzymes, favoring pervasive mycelial decomposers that have matrix-opening power (Newell et al. 1996a, and references therein). But if marshgrass LC is made available through mechanical and chemical intervention (grinding and solvent treatment), saltmarsh bacteria can carry out LC lysis (Benner et al. 1988, Moran and Hodson 1990). Partial fungal destruction of the marshgrass LC framework, which eventually results in decayed-shoot fragmentation, may be equivalent to mechanical/chemical intervention in its provision of degradable particulate LC to bacteria. Thus there may be a loose mutualism between marshgrass ascomycetes and sediment bacterial assemblages, in which the fungi prepare marshgrass-LC particles for efficient flow through a sediment-bacterial processing system (Newell and Palm 1998). Evidence is slight, but both erosion (Newell et al. 1996a) and tunneling bacteria (Newell 1993) are potentially participants in the digestion of fallen, fungal-decayed, marshgrass particles.

4. Standing Crops and Productivities of Cordgrass Microbes

4.1 STANDING CROPS

Standing crops of the three major groups of microbes on or in naturally decaying leaf blades of smooth cordgrass have been measured by the following techniques: green microalga – direct epifluorescence microscopy; bacteria (other than cyanobacteria) – direct epifluorescence microscopy, after pyrophosphate sonication and acridine-orange fluorochroming; fungi – ergosterol analysis as an index of living-fungal mass (see, respectively, Fallon et al. 1985, Newell and Palm 1998, Newell et al. 1996b). The green microalga targeted was *Pseudendoclonium submarinum*, the predominant microautotroph for shoot parts that are not touching or directly adjacent to the sediment (the next-highest group is diatoms, which usually have total biovolumes less than 1/2 those of *P. submarinum*: Newell et al. 1989). Regardless of whether natural conditions of water or leaf-nitrogen availability are manipulated to potentially favor microalgae or bacteria over fungi, fungi accumulate much larger standing crops during cordgrass-leaf decay (Table 1).

The values for fungal crop in Table 1 are for living-fungal (membrane-containing) mass, calculated from ergosterol contents, and using the conversion factor 200 units organic mass per unit ergosterol (Fell and Newell 1998, Gessner and Newell 1997; the 200-units factor is used throughout this communication). As a consequence of the pervasional mode of attack of fungi upon their substrates (''self-extending tubular reactors,'' Aynsley et al. 1990), there is an accumulation of ''fungal necromass''

(Joergensen et al. 1996) in the form of empty hyphal walls during fungal secondary production within decaying substrates. A rough estimate of the ratio of total:living fungal mass in decaying smooth cordgrass is 2:1 (Newell 1996a), so total organic fungal mass may have been as much as 4000 ug per cm^2 leaf abaxial area at the 12-wk point shown in Table 1 ($\approx$ 300 mg organic fungal mass per g decaying system).

TABLE 1. Maximal treatment and contemporaneous-control standing crops (μg organic mass per cm^2 leaf abaxial area) of microbes on and in naturally decaying leaf blades of smooth cordgrass (*Spartina alterniflora*) during a 12-wk field experiment in which water and leaf-nitrogen availability were manipulated (Newell et al. 1996b, Newell and Palm 1998)[H].

	Algal crop	Bacterial crop	Living-fungal crop
Maximal crop (wks)	16(4)	5(4)	2020(12)
Control crop	5	1	1080
Maximizing treatment	Extra H_2O	Extra N & H_2O	Extra N

[H] Shredding periwinkles were nearly absent (Newell 1993, Newell et al. 1996b).

Realization that high levels of accumulation of fungal mass occur within standing-decaying smooth cordgrass leads to the conclusion that the nitrogen of the decay system must be largely captured by the decomposer fungi (Newell 1993, 1996a). Thus, the hypothesis that nitrogen immobilized in naturally decaying shoots is only negligibly microbial and mostly abiotically fixed to shoot humic-like polymers is invalidated for smooth cordgrass (Newell 1993, White and Howes 1994), as it has been for terrestrial straw (Bedrock et al. 1998).

There are no published data for microbial standing crops on/in black needlerush, other than one point in time for fungal mass (40 mg living-fungal organic mass per g organic system mass; Newell and Fell 1992). Data from a multi-season, multiyear study currently underway (Newell unpublished) have revealed that living-fungal standing crop in naturally decaying black needlerush blades can be nearly as high as some contemporaneous values for smooth-cordgrass blades. For example, for data from the summers of 1996 and 1997, for the oldest sampled standing-decaying blades of cordgrass and needlerush, the range for cordgrass for short- to tall-canopy sites averaged 5 to 9% living-fungal of total-system mass; for contemporaneous needlerush blades, the mean value was 5%.

Data from the in-progress multiyear study of the previous paragraph can be used to generate provisional comparative values for types of standing microbial mass in the smooth-cordgrass marsh (Table 2), since the study involves measurement of fungal crop present in the marsh at a given sampling time, regardless of the age beyond senescence of shoot parts, or presence of shredder snails. The estimates in Table 2 demonstrate that shoot-fungal and sediment-bacterial crops are the predominant forms of organoosmotrophic microbial mass in the marsh. Living-fungal mass per unit marsh area approaches one-third that of living cordgrass mass in the winter, and this situation continues into spring, when concentration of fungal mass in decaying shoots is 1.5-fold greater than in winter (data not shown). Note that fungal crop per m^2 marsh was calculated for Table 2 using only values for attached dead-shoot mass. Since fungal content of detached dead parts is unknown, this potential additional fungal material was ignored. For smooth cordgrass, detached dead plant mass has been found to have an annual average of 32% of total dead-shoot mass (26% trapped above the sediment, 6% on the sediment) (Newell et al. 1998). Also excluded from Table 2 is the potential contribution of fungal mass within belowground parts of cordgrass (Mansfield and Bärlocher 1993). See Sullivan/Currin, this volume, for standing-crop information for microalgae.

TABLE 2. Provisionally estimated masses (g dry organic) for living fungi in standing-decaying shoot parts, and for total bacteria, per m^2 smooth-cordgrass marsh, averaged over tall- to short-form stands of cordgrass in three marsh watersheds.

	Summer	Winter
Fungal crop in dead blades	5	23
Fungal crop in dead sheaths + stems	14	29
Sum of fungal crop	19	52
Bacterial crop on dead shoot parts	0.02	0.07
Bacterioplankton, 0.5 m depth flooding tide	0.25	0.15
Bacteria in sediment to 20-cm depth	44	44
Living cordgrass, median	588	183

NOTE: Based on data from an in-progress multiyear analysis of patterns of fungal production in saltmarshes, using a preliminary ratio of fungal concentrations, sheath+stem:blade, of approximately 1:1 (Newell, unpublished), and on: a) data for standing crop of attached dead-shoot material of smooth cordgrass in Sapelo Island, GA marshes (Newell et al. 1998); b) data for bacterial crop on decaying smooth-cordgrass blades (Newell and Palm 1998); c) data for bacterioplankton density in marsh-flooding waters (0.5-m depth in the marshgrass canopy is transient, and primarily a spring-tide occurrence: Newell and Krambeck 1995); d) data for bacterial-cell crop in Sapelo marsh sediments taken in October (Fallon et al. 1983), and used for both seasonal columns in this table (after Rublee 1982); and e) data for standing crop of tall- and short-form living shoots of smooth cordgrass in Sapelo marshes (Dai and Wiegert 1996). Conversion factors: 200 units living-fungal organic mass per unit ergosterol (Fell and Newell 1998, Gessner and Newell 1997); 20 fg C per bacterial cell (Fuhrman 1992); 45% C of organic mass.

4.2 PRODUCTIVITIES

A radioisotopic method for fungal-productivity assessment, congeneric with the thymidine and leucine methods for prokaryotes (Chin-Leo 1997), has recently become available: the acetate-to-ergosterol (Ac→Erg) method (Gessner and Chauvet 1997, Gessner and Newell 1997, Fell and Newell 1998, Suberkropp 1997). The Ac→Erg method is currently being applied as a part of the multiyear analysis of patterns of saltmarsh-fungal production cited in the last two paragraphs of the previous subsection (STANDING CROPS). Comparison of currently available results for cordgrass-fungal productivity with data for prokaryotic productivity from previous work shows that fungal production is the major part of organoosmotrophic production associated with the standing-decaying shoots (Table 3). Note that the fungal-productivity values of Table 3 were obtained at a uniform temperature (20°C, an air temperature near to [within 2°C] or within the usual range of monthly average high-low in all seasons on Sapelo Island [Chalmers 1997]), so the values at seasonal mean temperatures would presumably be higher for summer and lower for winter. A crude and very preliminary value for total annual production of fungi in decaying shoots of smooth cordgrass can be calculated using the mean value for 20°C from Table 3: an annual production of 734 g organic fungal mass per m^2 marsh is obtained (average annual production:biomass of about 20:1). Compare a recent estimate of annual production for smooth-cordgrass shoots at Sapelo – 1313 g dry mass m^2, average for short- and tall-form swards (Dai and Wiegert 1996), implying a yield of fungal mass equivalent to about 50-60% of shoot production. The yield efficiency (fungal mass produced/leaf mass lost) earlier found for fungi in standing-decaying leaves tagged at senescence (no manipulation of nitrogen or water availability) was 56% (Newell et al. 1996b). In considering these high fungal conversion efficiencies, one must keep in mind that some of the fungal mass produced during shoot decay is very likely to be at the expense of dead fungal mass (i.e., some fungal mass is recycled, within and between species; e.g., Boddy and Watkinson 1995, Kerley and Read 1997) (see Chapman and Gray 1986, Fuhrman 1992).

TABLE 3. Provisionally estimated rates of production (mg microbial organic mass per day) and specific growth rates (μ, per day, shown in parentheses) for fungi in standing-decaying shoot parts, and for bacteria, per m^2 smooth-cordgrass marsh, averaged over tall- to short-form stands of cordgrass in three marsh watersheds[H].

	Summer	Winter
Fungal productivity in dead blades	90 (0.018)	1289 (0.056)
Fungal productivity in dead sheaths + stems	279 (0.020)	2363 (0.081)
Sum of fungal productivity	369 (0.019)	3652 (0.070)
Bacterial productivity on dead shoot parts	10 (0.500)	35 (0.500)
Bacterioplankton, 0.5 m depth flooding tide[I]	528 (2.110)	36 (0.240)
Bacteria in sediment to 20-cm depth	750 (0.017)	265 (0.006)

[H] Assuming activity permitted by adequate water availability (tidal contact, rain, and especially dew) in standing-decaying shoots for an average of 12 h per day (Newell et al. 1996b).
[I] This rate applies only to the brief period (< 3 h per diurnal tide) of spring-high-tidal flooding to a 0.5-m depth maximum in the marshgrass sward, during which summer bacterioplanktonic productivity has been found to be enhanced 2-fold (Newell and Krambeck 1995).
NOTE: Based on data from an in-progress multiyear analysis of patterns of fungal production in saltmarshes (Newell unpublished), involving measurement of fungal productivity via the acetate-to-ergosterol (Ac→Erg) method (Fell and Newell 1998, Gessner and Newell 1997), with all Ac→Erg measurements performed at 20C, a temperature common in both summer and winter in Sapelo marshes (Chalmers 1997), and on: a) data for standing crop of attached dead-shoot material of smooth cordgrass in Sapelo Island, GA marshes (Newell et al. 1998); b) data for bacterial μ on non-submerged, wet, decaying smooth-cordgrass blades (Newell and Palm 1998; no seasonality imposed); c) data for bacterioplankton productivity in marsh-flooding waters (Newell and Krambeck 1995); d) data for bacterial-cell production in Sapelo marsh sediments taken in October (O_2-uptake-basis; Fallon et al. 1983; see also Morris and Whiting 1986), with seasonality imposed based on data for muddy sediment in Cammen (1991). Conversion factors: 12.6 μg organic fungal mass per nmol acetate incorporated into ergosterol at 20°C (modified from Newell 1996b, based on new information re HPLC-injection-valve function); 2.0×10^{18} bacterial cells per mole thymidine incorporated into DNA (Chin-Leo 1997).

During the brief periods of tidal partial submergence of marshgrass shoots, the productivities of bacterioplankton in the marsh-flooding water and shoot-decaying fungi can be approximately equivalent (Table 3); this equivalence is extant only during summer. A considerable proportion (half or more?) of the production by bacterioplankton in sward-flooding water is likely to be at the expense of organics dissolving from within the shoots of smooth cordgrass (Kirchman et al. 1984, Coffin et al. 1989, Turner 1993, Hullar et al. 1996; see Newell and Krambeck 1995, who found x2 boosting of bacterioplanktonic per-cell thymidine incorporation as water moved into the marshgrass canopy). See Newell et al. (1988a) and Shia and Ducklow (1997, and references therein) for more information on saltmarsh-bacterioplanktonic productivity.

Measurement of bacterial production in sediments is particularly fraught with methodological problems (Robarts and Zohary 1993), partly due to weak understanding of the bacterial genomes present (e.g., Devereux et al. 1996, Torsvik et al. 1996). However, crudely estimating bacterial production in saltmarsh sediments based on oxygen uptake appears to show that sediment-bacterial production is at least on a par with that of shoot fungi in summer, and it is a major component of marsh-microbial productivity (Table 3). This is not unexpected, if bacteria are the major decomposers of cordgrass rhizome and root mass (Morris and Whiting 1986, Benner et al. 1991, Blum 1993), and of the decayed cordgrass-shoot material (mixtures of fungal hyphae and remnant cordgrass LC: Newell et al. 1996a) that falls to the sediment (Benner et al. 1988, Moran and Hodson 1990, Newell 1993, Sinsabaugh and Findlay 1995, González and Moran 1997, Newell and Palm 1998).

The bacteria on decaying shoots would appear to be negligible participants in saltmarsh-microbial production (Table 3), but it has become clear recently that a common decomposer-bacterial strategy is to build only modest accumulations of biomass on substrates, and to shed cells from solid substrates into the planktonic stage (references in Newell 1996a). When rinsed, naturally-decaying leaf blades of smooth cordgrass were submerged in seawater, they shed from 8 to 149% of their attached bacteria in one hour (Newell and Palm 1998). It is possible that these shed bacterial cells contribute to the high bacterioplanktonic productivity of sward-flooding tidal water (Table 3) (Newell and Palm 1998).

Data from five seasonal samplings of fungal productivity in standing-decaying black needlerush blades have yielded an annual average rate that is equivalent to the rate for blades of smooth cordgrass (about 150 μg organic fungal mass per g decaying-system organic mass per h). The empirical conversion factor from the Ac→Erg method is higher for needlerush than cordgrass by about 40%, but it has been measured for only one species of needlerush ascomycete (17.8 μg organic fungal mass per nmol acetate incorporated into ergosterol; see footnote to Table 3), so the productivity estimate for needlerush is more tenuous. [The conversion factor has been measured for four species of smooth-cordgrass fungi, and is homogeneous among them (12.6 μg organic fungal mass per nmol acetate incorporated into ergosterol; modified from Newell 1996b; see Table 3 footnote).] If it is true that fungal productivities are equivalent between cordgrass and needlerush, then needlerush fungal specific-growth rates are somewhat higher, since living-fungal standing crop is lower in needlerush (STANDING CROP subsection above). This may be a clue that the needlerush conversion factor is too high: needlerush ascomycetes that have been brought into the laboratory have generally slower growth rates in culture than cordgrass ascomycetes (Newell 1996b).

For comparative productivity values for saltmarsh microalgae, see Sullivan/Currin, this volume.

5. Multilatitudinal Information

One problem with the fungal standing-crop and productivity values presented above is that they have all been obtained from Sapelo Island (31°N) (a fact kindly pointed out by various referees over the past decade). In the multi-year study of marshgrass-fungal production (STANDING CROP subsection), we are attempting to broaden our window for collection of data – for autumn sampling, collections are being taken from along the US Atlantic coast, from Maine to eastern Florida, with one site on the Florida Gulf of

Mexico coast (30 to 44°N). No definite north-to-south pattern has been detected for content of living-fungal mass in standing-decaying blades of smooth cordgrass (1997 data are shown in Table 4). The speculation of Newell (1993, p. 317) that more northerly cordgrass marshes might exhibit lower fungal standing crops is not supported by data in Table 4. For rates of fungal production, there may be a hint of lower values for the mid-Atlantic data, but the values for the northernmost site are the highest in the whole 1996-1997 set.

There are published data for living-fungal standing crops in smooth cordgrass marshes at the northern limit of their range (46°N; Samiaji and Bärlocher 1996). The maximum average ergosterol content of whole standing-decaying blades at 46°N was about 100-125 µg per g organic decaying-system mass, suggesting that at the northern end of cordgrass distribution, fungal standing crops are lower (compare values in Table 4). However, Samiaji and Bärlocher point out that in their time-series sampling (30-day intervals after the first month), they may have missed peak standing crops, and that they did find portions of blades that exhibited high ergosterol contents (to 452 µg per g).

TABLE 4. Fungal-mass (µg ergosterol per g organic mass of decaying leaf blade) and productivity (pmol acetate incorporated into ergosterol, per µg ergosterol per h) indices in standing-decaying smooth-cordgrass leaf blades from marsh areas of intermediate canopy height at several latitudes in November, 1997[H].

	Ergosterol	Acetate[I]
44°N (Wells Reserve, ME)	696 (±15)	44.0 (±1.0)
42°N (Waquoit Bay, MA)	300 (±57)	12.7 (±1.5)
37°N (Chesapeake mouth, VA)	583 (±132)	18.7 (±4.5)
31°N (Sapelo Island, GA)	318 (±43)	25.1 (±0.5)
30°N (Anastasia State RA, FL)	561 (±45)	21.9 (±6.2)

[H] Values are means for four blades from low and four from high on shoots, shown with ($\pm\frac{1}{2}$range) in parentheses.
[I] All rates measured at 20°C (see Table 3 footnotes).
NOTE: Based on data from an in-progress multiyear analysis of patterns of fungal production in saltmarshes (Newell unpublished); see footnotes to Tables 2 and 3. See Acknowledgments for names of participants at non-GA/FL sites.

6. Microbial Fates

6.1 MICROBIVORY

Having obtained the first values for fungal production in the saltmarsh, a natural sequel, in addition to better defining these first values, is to turn to the question of the fate of the fungal mass produced. Since fungal yield from naturally decayed smooth cordgrass is high, there is a lot of fungal mass (hundreds of g m^{-2} yr^{-1}) for which to account (PRODUCTIVITIES subsection). Interestingly, the fate of some of the cordgrass-ascomycete production and of the marsh bacterioplankton is the same: to flow to molluscs (periwinkles and mussels, respectively) (Newell and Bärlocher 1993, Newell and Krambeck 1995, McQuaid 1996, Newell/Kreeger, this volume). It may seem odd that periwinkles would be involved in leaf decomposition, but it probably shouldn't, because certain types of snails have been shown repeatedly to fall into the "litter-transformer'' category of Wardle and Lavelle (1997; e.g., Theenhaus and Scheu 1996, Brendelberger 1997, Heller and Abotbol 1997, Slim et al. 1997).

Newell et al. (1989) fortuitously established their plots of tagged smooth-cordgrass leaf blades (for study of natural decay phenomena) in an area of high periwinkle-snail (*Littoraria irrorata*) density (243 m^{-2}). The siting was fortunate, because saltmarsh periwinkles mature enough ($\approx$ 5 mm shell length) to come out of hiding within furled blades are patchily distributed in the marsh (range from near zero to > 400 m^{-2}; Smalley 1959, Newell 1993), and the site chosen enabled the observation that periwinkles were very likely to be effective shredders of standing-decaying leaves (Newell et al. 1989, their Fig. 2). Newell and Bärlocher (1993) subsequently demonstrated experimentally, in microcosms, that *L. irrorata* could indeed effectively shred leaves; they had the potential (with activity enabled 24 h day^{-1}) to ingest 7% of naturally decayed leaves per day, and could digest naturally-decayed blades with an efficiency of 51% (acid-insoluble-ash method; Bärlocher and Newell 1994a). The periwinkles removed living-fungal mass more rapidly (10% per day) than they removed leaf mass, and they can efficiently digest saltmarsh-fungal mycelium (assimilation efficiency $\approx$ 50%; Bebout 1988). The experimental specific rate of fungal removal from leaves, if snails were only active for half the time and were present at 450 m^{-2} (Newell and Bärlocher 1993), would be about 4% day^{-1}, just a little less than the higher measured specific growth rate (Table 3) of cordgrass-blade ascomycetes in the marsh, suggesting that where snails are present in large enough numbers, they have the capability to participate in control of fungal standing crop. When extra nitrogen was made available to the fungal decomposers, through shoot fertilization (see Table 1), it boosted the fungal crop by about x2, but when periwinkles were added (400 m^{-2}), they kept the fungal crop grazed back to control (no extra N) levels (Newell 1993).

Littorinid snails are not the only grazers of saltmarsh-fungal mass. Amphipods are clear suspects, based on their prominence among small smooth-cordgrass-marsh invertebrates (Covi and Kneib 1995), the well-established mycovorous tendencies found for freshwater species (Suberkropp 1992), and experimental evidence that marine amphipods naturally eat cordgrass (Rietsma et al. 1982, Thompson 1984) and can be mycovorous (Boyd 1980). Kneib et al. (1997) tested the capacity of an amphipod (*Ulhorchestia spartinophila*, a prominent inhabitant of smooth-cordgrass shoots) to grow on a diet of senescent or naturally-decayed leaves of cordgrass. Senescent leaf sheaths did not permit reproduction, but decayed leaf parts (including leaf blades that decayed in microcosms after harvest at the senescent stage) all allowed equivalent ecological performance (growth + reproduction), and growth rates equal to those measured for cohorts growing naturally in the marsh. For leaf parts that had clay films removed by rinsing, the highest reproduction ($\approx$ 2 offspring per initial individual per 6 wk), survivorship (84%), and male:female ratio ($\approx$ 1:1) were found for the parts (decayed blades) with the highest living-fungal content. However, results for decayed blades were not statistically significantly different from dead leaf parts with lower fungal content, suggesting that fungal material as food is not an overwhelmingly important characteristic of the dead leaf material eaten. Rather than shred decayed leaf blades when eating them, as do saltmarsh periwinkles, *U. spartinophila* grazes away abaxial surface layers, leaving blades otherwise intact (Kneib et al. 1997, Newell unpublished). This may be a strategy allowing collection of microbes in surface clay films (microalgae, cyano- and other bacteria; Newell 1993) along with fungal mass in shallow layers of leaf cells, including the fungal reproductive structures that are produced just below the abaxial blade surface (Leuchtmann and Newell 1991, Newell et al. 1996a). The capacity of *U. spartinophila* to remove fungal mass from decaying leaves (selectively or non-selectively) has not been determined, but preliminary observations have shown that this amphipod will eat pure fungal mycelium (of *Phaeosphaeria spartinicola* separated from pre-sterilized blades of smooth cordgrass on which it grew [Newell 1996b]; see section Microbial Decomposers) (Newell and Kneib, unpublished). Also, *U. spartinophila* individuals (about 6-mm length) have been observed under the stereomicroscope biting out and swallowing fungal ascomata and associated tissue from abaxial surfaces of naturally decayed *S. alterniflora* blades (Newell and Graça, unpublished).

There are several other invertebrates of cordgrass saltmarshes that are strong candidates as mycovores, based on known feeding habits in other environments (e.g., McGonigle 1997, Wardle and Lavelle 1997), and the fact that these invertebrates utilize cordgrass shoots as primary habitat (e.g., flies, mites, collembolans, enchytraeid polychaetes, nematodes) (Rutledge and Fleeger 1993, Healy and Walters 1994). There is also at least one other gastropod of saltmarshes that is a potential mycovore: *Melampus bidentatus* (eastern melampus, or saltmarsh coffeebean snail; Daiber 1982). It has been established that the very similar mangrove coffeebean snail (*Melampus coffeus*) can shred, ingest, and assimilate leaves of mangroves (Mook 1986, Proffitt et al. 1995). Daiber (1982) lists *M. bidentatus* as a gastropod that will eat leaf material, Thomson (1984) found that bits of saltmarsh *Spartina* constituted more than half of the gut contents of *M. bidentatus*, and Rietsma et al. (1988; see also Spelke et al. 1995) raised *M. bidentatus* on naturally decayed *S. alterniflora* from 5.5 to 6.5 mm shell length in the laboratory (16 wks). We (unpublished data) have observed that: 1) tiny (2-mm shell length) *M. bidentatus* can grow to adult size on a diet of naturally decayed blades of smooth cordgrass; and 2) adult *M. bidentatus* can shred naturally decayed blades to the same extent as can *L. irrorata*. We (unpublished) have found juvenile saltmarsh coffeebean snails on decaying shoots of smooth cordgrass in the central parts of marshes (as opposed to upper marsh edges where they are known to be common; Fell et al. 1982), and since these snails are given to hiding under objects on the sediment or at bases of cordgrass shoots during the day (Hausman 1932, Heard 1982), one wonders whether they might have more pan-marsh impact than one would suspect based on daytime surveys.

Nothing is known regarding interactions of coffeebean snails with cordgrass fungi, but Rietsma et al. (1988) discovered experimentally that older (months) standing-decaying shoots were preferred by *M. bidentatus* over younger (2 wks) dead shoots. Rietsma et al. (1988, and see references therein) concluded that low ferulic acid content was the key to snail preference and greater snail-growth support of older decayed material, but decline in ferulic-acid content was likely to have coincided with increased fungal mass (see Bärlocher and Newell 1994b, Newell et al. 1996a), so it may be that greater presence of fungal material (especially in the damp chambers [= solid-state fermentations; Doelle et al. 1992] used to culture the snails) was an additional factor affecting palatability and nutritional quality.

Two species of saltmarsh invertebrates that deserve a long look as potential important interactors with the natural marshgrass-decay system are the squareback crabs *Armases cinereum* and *Sesarma reticulatum* (Daiber 1982, Pennings et al. 1998). Recent observations of common presence of morphologically characteristic fecal pellets (outer shell of fine, dark-brown particles, inner content of lighter-brown plant fibers) upon standing, naturally decaying and partially shredded leaf blades of smooth cordgrass strongly suggest that squarebacks are cryptic (active principally during darkness?) shredders throughout the marsh canopy (Newell and Graça, unpublished).

Smooth-cordgrass marshes on the Atlantic coasts of South America contain shoot-associated invertebrates (e.g., *Littorina flava, Neritina virginea*, and *Bittium varium* [gastropods], and *Parhyalella whelpleyi* [amphipod]) with the potential to have the same type of feeding (fungal-outflow) niches as *L. irrorata*, *U. spartinophila*, and the other likely mycovores of USA cordgrass marshes (Lana and Guiss 1992).

At the beginning of this subsection, we noted that both saltmarsh-fungal and bacterial output, in the form of eaten microbial mass produced from cordgrass organics, flows partly to molluscs. For the bacterioplankton (see PRODUCTIVITIES subsection), ribbed mussels (*Geukensia demissa*) are a major sink (Newell and Krambeck 1995). Ribbed mussels can filter bacterioplankters directly, but also take bacterial production indirectly, through eating of bacterivorous protozoa (Kemp et al. 1990). Other marsh bivalves can also be sinks for bacterivorous protozoa (Le Gall et al. 1997). See this volume pp 187-220 for more details and references.

6.2 PROPAGULE EXPULSION

A second clear mode of output of fungal material produced in cordgrass shoots takes the form of sexual propagules (ascospores, produced in specialized sexual structures termed ascomata; see Kohlmeyer and Volkmann-Kohlmeyer 1991, Kendrick 1992, Alexopoulos et al. 1996). Suberkropp (1997) calculated that output of conidia of aquatic hyphomycetous fungi (mitosporic fungi; Hawksworth et al. 1995) from decaying leaves in freshwater microcosms was equal to about 50% of total fungal production. Newell and Wasowski (1995) measured ascospore expulsion from naturally decaying leaf blades of smooth cordgrass in late spring, and determined mature ascomatal volume in the decaying blades. They found hourly rates of ascospore release in terms of spore volume that were far lower ($\times 10^{-4}$) than ascomatal volume, implying that the measured rates of ascospore expulsion were conservative. Another indication that the average ascospore-expulsion rates of Newell and Wasowski (1995: 6.5×10^4 m^{-2} marsh h^{-1} of leaf freshwater wetness) are too low is that the rate of expulsion of spores in organic-mass terms, if leaves are wet for 12 h per day (footnotes to Table 3), is only 1-4 $\times 10^{-4}$ (winter-summer) of the total living-fungal crop for blades shown in Table 2. As a part of the multiyear study of saltmarsh-fungal dynamics (STANDING CROP subsection above), ascospore-expulsion rates are being measured seasonally. The duration of incubation of wetted blades was reduced from 168 h (as in Newell and Wasowski 1995) to 72 h, and the average summer (Aug) rate (1996 + 1997) found was 60 spores cm^{-2} h^{-1}, about four-fold greater than Newell and Wasowski's (1995) average spring (Jun) rate. Using even briefer incubations, among other changes (see Newell and Wasowski 1995), seems likely to reveal rates of spore ejection that are higher yet, so we refrain here from trying to fit ascospore output into a budget for flux of fungal mass.

6.3 MELDING INTO THE SEDIMENT SYSTEM

The combined result of fungal deterioration of the lignocellulosic ''skeletal'' structure of marshgrass shoots (Newell et al. 1996a) and invertebrate shredding (Fig. 2 of Newell et al. 1989, Newell and Bärlocher 1993), is that shoot material becomes frangible, allowing fragments to break away under mild physical forcing such as tidal flow (Newell 1993, and unpublished). These fragments probably move to the sediment-surface decay system, where it is likely that bacterial assemblages have the primary impact (BACTERIA subsection above; Sinsabaugh and Findlay 1995, Newell and Palm 1998; Fig. 4 below). In addition to bacterial attack on remnant marshgrass lignocellulose, it is probable that the resident fungal material of the fallen plant fragments is subject to bacterial utilization. Even if most of the fungal material in the decayed fragments is dead and devoid of cytoplasm, much of the chitin/laminaran hyphal-wall mass, and extra-hyphal glucan sheath plus sheath-entrained enzymes, will remain in the fragments (Newell 1993, Gutiérrez et al. 1995, Nicole et al. 1995, Newell et al. 1996a, Barrasa et al. 1998). Svitil et al. (1997) have shown that a marine vibrioid bacterial species produces chitinases that are active upon the chitin *of Phaeosphaeria spartinicola* (strain SAP 93, not identified in

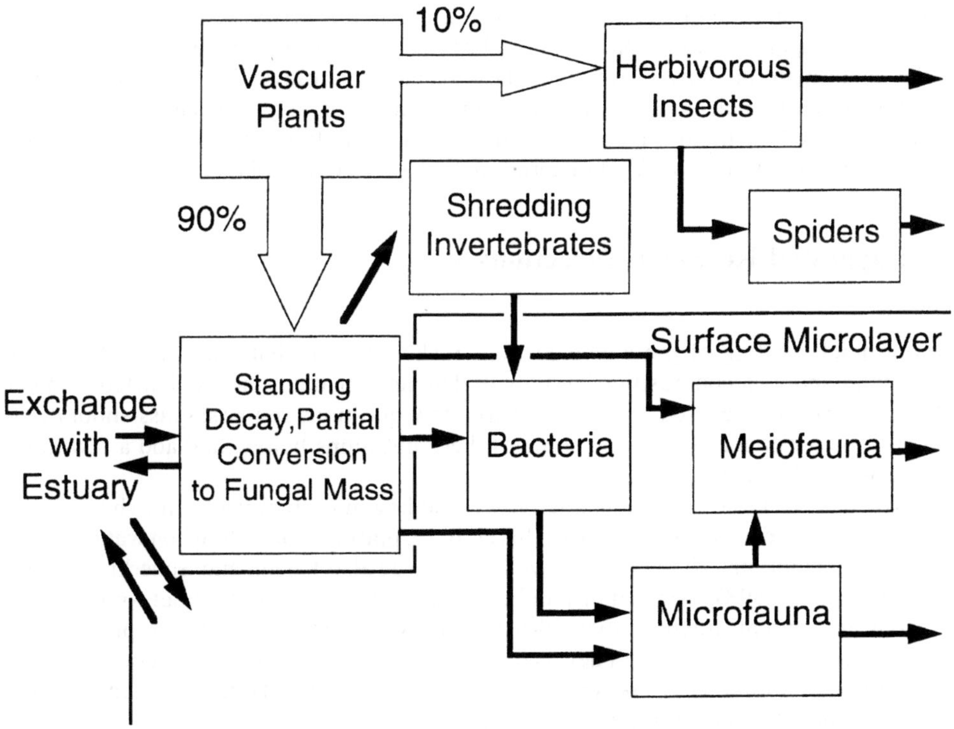

Figure 4. Diagrammatic model of the flow of saltmarsh photosynthate from shoots of marshgrass into the marsh food web. This is a reconfiguration of the upper left corner of the model of Montague and Wiegert (1990). Note that the flow from vascular plants is intended to depict flux of shoot material, not root/rhizome material.

Svitil's publication), a principal fungal decomposer of the standing-dead leaf blades of smooth cordgrass (Microbial Decomposers section above). This is a potentially important finding, since chitins vary greatly in resistance to degradation (Svitil et al. 1997). Bacterial mass produced at the expense of fallen fungal mass could then flow to animals of the sediment system, including meiofauna and filter-feeders, and thereby into the saltmarsh "trophic relay", leading to export of animal mass from the marsh ecosystem (Montagna 1995, Kneib 1997, Newell/Kreeger chapter).

7. Model Reconfiguration

A model for the flow of photosynthate into the saltmarsh food web was published by Montague and Wiegert (1990). Ninety percent of the flow from marshgrass shoots was shown moving into a sediment "surface microlayer" compartment, with the other 10% going to herbivorous insects. I propose, based on the foregoing information on shoot decay and the fate of the naturally-decayed shoots, that the model be reconfigured as depicted in Fig. 4. The main changes involved in the reconfiguration are: 1) the flow as shoots senesce and die moves from shoots to a "standing decay" compartment, where substantial fungal secondary production takes place; 2) following standing decay, part of the shoot-material flow moves into the "surface microlayer" compartment where bacterial, microfaunal, and meiofaunal production occur at its expense; 3) a second avenue followed by the decayed-shoot flux is to a "shredding invertebrates" box; the output of this box moves as fecal material to the "surface microlayer" compartment.

8. Suggested Research Directions

Even a quick reading of the above sections makes it clear that our knowledge of the marshgrass-shoot decomposition system is deficient in many areas. I listed (Newell 1993) and briefly discussed (Newell 1996a) a few questions re marshgrass-shoot decay that are ripe for attack. We revisit some of these questions below, and add a few more, with capsulized discussion.

- Now that we know that fungal activity is the key microbial vector for shoot decay, and that water sufficiency controls this activity, putting it on hold when water content of dead-shoot material is low (Newell 1996a, see also Kuehn and Suberkropp 1998, Kuehn et al. 1998), we need firm information for the duration of shoot wetness: how do water contents fluctuate over 24-h periods; what wetting phenomena predominate (is it in fact, mostly dew?); if dew is the primary wetting phenomenon, does this mean that fungal productivity takes place mostly at night, with adaptation to temperatures cooler than day temperatures?
- Examination of Tables 2 and 3 makes it clear that we need to obtain more definitive information for fungal dynamics (fungal content, fungal productivity, principal species of fungal decomposers) associated with shoot parts other than leaf blades. Data is now being obtained for leaf sheaths (Newell unpublished), but there will

still be a dearth of data for naked stems (one data point: Newell et al. 1988).

- Too few scientists (presently four) are participating in the development of methods for measuring rates of fungal production in nature. More work is needed, for example, on circumscribing the conversion factor (CF) from rate of acetate incorporation into ergosterol, to rate of production of organic fungal mass. Gessner and Newell (1997) cite several empirical CFs ranging from 5.5 to 18.8 μg organic mass per nmol acetate. Recent findings suggest that two high estimates of CF (Newell 1996b) were correct for only one particular analytical-measurement system (wherein nmol acetate incorporated were systematically underestimated), but are high by x1.54 for systems without this peculiarity (see foonote to Table 3; corrected value for smooth cordgrass = 12.6 μg per nmol). Further testing and refinement may reveal that the generally-applicable CF is close to the theoretical value near 7 (Gessner and Newell 1997). Another key methodological problem is the extension of the acetate-to-ergosterol method to decaying solids that are not submerged.
- The work of J. and B. Kohlmeyer (Kohlmeyer et al. 1997, and references therein) has brought us close to complete description of the species of fungi (mostly ascomycetes) that drive the decay of black needlerush (*J. roemerianus*). Many of the key species of smooth-cordgrass fungal decomposers have descriptions in the literature, but some important species do not (e.g., *Mycosphaerella* sp. 1 and 2 of Kohlmeyer and Kohlmeyer 1979; *Anthostomella* sp. of Gessner and Kohlmeyer 1976; two beautiful sporodochial species with tiarosporelloid conidia). We need a full taxonomic treatment of the fungi of the smooth-cordgrass decomposition system to enable accurate ecological research, and at least some initial taxonomic work on fungi of other saltmarsh plants. Note that for some important saltmarsh grasses (e.g., *Distichlis spicata*), only one or two species of the fungal assemblage are known (Kohlmeyer and Kohlmeyer 1979).
- A deplorable gap in our knowledge of the marsh ecosystem lies in our near-total lack of knowledge of the basic life histories of the even the most important species of marshgrass decomposers. For example, Bärlocher and Newell (unpublished) have repeatedly found that the ascospores of *Buergenerula spartinae* (Fig. 1) will not readily germinate. This species forms bacterial-sized (2×5 μm), vibrioid microconidia (spermatia? – Kohlmeyer and Gessner 1976), and it forms hyphopodia on living leaf-sheath surfaces suggesting parasitism or benign endophytic status (Kohlmeyer and Gessner 1976). What is the stimulus required to release ascospores from dormancy, is this species an endophyte or parasite, and how do the microconidia and hyphopodia fit into the life cycle? Why is the ascomatal-production area of *B. spartinae* in standing-decaying blades of smooth cordgrass limited to irregular-sized blackened patches (often no patches at all) (Fig. 1) within the large, nearly whole-blade ascomatal fields of *Phaeosphaeria spartinicola* and *Mycosphaerella* sp. 2 (Fig. 1)? Is the patch-blackening by *B. spartinae* a combat mechanism as part of a battle with blade-fungi for territory (Rayner et al., 1995) (I have seen areas of leaf sheath occupied by *B. spartinae* that were not blackened)?
- "... let us ecologists not neglect to study in greater depth more of the star performers in fungal successions, on which the maintenance of entire ecosystems may depend." (Frankland 1998.)
- The oomycotes (eukaryotic mycelial decomposers that evolved independently from

true fungi) have been found to have low frequencies in smooth-cordgrass samples at the same sites where oomycotic frequencies in fallen deciduous leaves are near 100% (Newell 1996a). However, one genus of oomycotes (*Pythium*) has evinced an association with marshgrasses; *Pythium grandisporangium* was originally described from submerged decaying leaves of *Distichlis spicata* (Fell and Master 1975), and this same marine species has regularly been found in association with shoots of smooth cordgrass (Porter, unpublished). Do oomycotes commonly participate in the decay of saltmarsh grasses, and what are the relationships between oomycotes and fungi in the saltmarsh?

- Methods are now available that enable measurement of biomass dynamics of individual fungal species (e.g., immunoassays: Dewey 1996, DNA assay: Fell and Newell 1998). One could, for example via competitive PCR (Mahuku et al. 1995, Zimmerman and Mannhalter 1996, Edwards et al. 1997), measure the change in mass of one of the important marshgrass-decomposer species (e.g., *Phaeosphaeria spartinicola*) alongside measurements of change in total fungal mass via ergosterol determinations (Newell 1996a). Will it be found that *P. spartinicola* is the predominant species with respect to mass production, or will its nearly everpresent neighbor *Mycosphaerella* sp. 2 (of Kohlmeyer and Kohlmeyer 1979) also be a substantial producer?
- Although microalgae on decaying shoots exhibit low standing crops (Table 1), they are everpresent (especially *Pseudendoclonium submarinum*; Newell 1993). Recent co-culture of the leaf-blade ascomycete *Phaeosphaeria spartinicola* with *P. submarinum* (Newell unpublished) made it apparent that the two enhance each other's growth. What are the physiological details of this interaction (see Honegger 1991, Mouget et al. 1995, Hutchison and Barron 1997), and are there impacts of the interaction upon leaf decay?
- Two saltmarsh invertebrates have recently been added to saltmarsh periwinkles (*Littoraria irrorata*) in the embryonic list of impactors of shoot decay (the amphipod *Uhlorchestia spartinophila* and the gastropod *Melampus bidentatus*; see MICROBIVORY subsection). How many more substantially decay-impacting invertebrates are there in the marsh?
- Fungal secondary production must result in elevated CO_2 and H_2O within the decaying parts of living shoots – can the living shoot take advantage of this intimately-close source of substrate for photosynthesis (Newell 1996a)? Note that cordgrass ascomycetes form hyphal webs in the aerenchymal spaces of standing-decaying shoots (Newell, unpublished); these are likely to be coated with hydrophobins, and efficient gas exchangers (Wessels 1997).
- When oven-dried green smooth-cordgrass shoots are ground to small particles and placed in saltmarsh sediment, they are weak nutritional sources for sediment invertebrates (e.g., Levinton and Stewart 1988: 1-6% C conversion to oligochaete mass). The natural form of cordgrass-shoot input to sediments is as shredder pellets and as remnant lignocellulose plus fungal mass (probably largely in the form of empty hyphal tubes, but possibly including lignocellulolytic enzymes) (Newell 1996a, Newell et al. 1996a). This natural-input form is very unlike the oven-dried green material; e.g., it will have lowered content of cordgrass antifeedant cinnamic acids as a result of fungal activity (Newell 1993, Substrate section above). Do

sediment invertebrates respond differently to natural forms of shoot input than they do to denatured material? To aid in our understanding of the impact of sediment-input of natural particles of marshgrass shoots, we need better data for accumulation of fungal products (e.g., glucosamine, mannoproteins, melanin, hydrophobins) in naturally decaying marshgrass (e.g., Gutiérrez et al. 1995, Newell 1996a, Newell et al. 1996a, Wessels 1997).

- We have discovered that cordgrass ascomycetes are likely to be involved in release of dimethylsulfide from the dimethylsulfoniopropionate (DMSP) produced by smooth cordgrass, which could have atmospheric impact (Bacic et al. 1998). What other effects on flux to/from decaying marshgrass shoots might saltmarsh ascomycetes have (N, P, DOM, etc.)? Are fungi of standing-decaying marshgrass an important part of the marsh nutrient-buffer system (Newell 1993, Newell 1996a)?
- There is evidence for existence of a fungal/bacterial N_2-fixing consortium (Paerl and Pinckney 1996) on/in naturally decaying leaf blades of smooth cordgrass, especially for blades bearing a heavy clay-film layer (Newell et al. 1992, cf. Hill and Patriquin 1992). Could this be *Phaeosphaeria spartinicola* and/or *Mycosphaerella* sp. 2 (of Kohlmeyer and Kohlmeyer 1979) consorting with azospirilla or azoarci (Hill and Patriquin 1992, Hurek et al. 1997)?
- What are the impacts of human perturbations of marshes upon fungal secondary production? One preliminary answer is that saltmarsh ascomycetes can be resilient to toxic impact – at one USEPA Superfund Site (LCP Chemical, Brunswick, GA) where mercury and PCBs were dumped directly into the saltmarsh (current concentrations: Hg and PCBs, tens of μg per g dry sediment, $\geq 1000 \times$ background), standing crops of living-fungal mass (ergosterol basis) in standing-decaying leaves of smooth cordgrass were actually higher than at unpolluted control stations, possibly because of anthropogenic nitrogen inputs (sewage treatment outflows) at the Superfund Site (Newell, unpublished). Considering this fungal resilience, would saltmarsh ascomycetes be good candidates for coastal bioremediation efforts (Newell et al. 1996a)?
- There are indications in the literature that decomposition of grasses can be partially caused by abiotic vectors (e.g., ultraviolet light; e.g., French 1979, p.187, see also Mackay et al. 1994). To what extent might solar radiation enhance marshgrass-lignocellulose degradation by ascomycetes?

9. Acknowledgments

Financial support for much of the research reviewed here was provided by the U.S. National Science Foundation (grants OCE-9521588, 9115642, 8600293, and 8214899, and BSR-8604653) and the U.S. Environmental Protection Agency (NCERQA; R825147-01-0). We thank Wilma Lingle for partnership in the LC-lysis TEM research. Partners in the multilatitudinal research (see Multilatitudinal Information subsection) are: Linda Blum, Rick Crawford, Ting Dai, and Michele Dionne. We thank Darrell Casey for preparation of figures. Contribution 831 of the University of Georgia Marine Institute.

10. Literature Cited

Alexopoulos, C. J., C. W. Mims and M. Blackwell. 1996. Introductory mycology. Wiley, New York, New York, USA.

Anderson, C. E. 1974. A review of structure in several North Carolina saltmarsh plants. Pages 307-344 *in* R. J. Reimold and W. H. Queen, editors. Ecology of halophytes. Academic Press, New York, New York, USA.

Armstrong, J., W. Armstrong, P. M. Beckett, J. E. Halder, S. Lythe, R. Holt and A. Sinclair. 1996. Pathways of aeration and the mechanisms and beneficial effects of humidity- and Venturi-induced convections in *Phragmites australis* (Cav.) Trin. ex Steud. Aquatic Botany **54**:177-197.

Arenovski, A. L. and B. L. Howes. 1992. Lacunal allocation and gas transport capacity in the saltmarsh grass *Spartina alterniflora*. Oecologia **90**:316-322.

Aynsley, M., A. C. Ward and A. R. Wright. 1990. A mathematical model for the growth of mycelial fungi in submerged culture. Biotechnology and Bioengineering **35**:820-830.

Bacic, M. K., S. Y. Newell and D. C. Yoch. 1998. Release of dimethylsulfide from dimethylsulfoniopropionate (DMSP) by plant-associated saltmarsh fungi. Applied and Environmental Microbiology, **64**:1484-1489.

Bärlocher, F. and S. Y. Newell .1994a. Growth of the saltmarsh periwinkle *Littoraria irrorata* on fungal and cordgrass diets. Marine Biology **118**:109-114.

——. 1994b. Phenolics and proteins affecting palatability of *Spartina* leaves to the gastropod *Littoraria irrorata*. PSZNI Marine Ecology **15**:65-75.

Barrasa, J. M., A. Gutiérrez, V. Escaso, F. Guillén, M. J. Martínez and A. T. Martínez. 1998. Electron and fluorescence microscopy of extracellular glucan and aryl-alcohol oxidase during wheat-straw degradation by *Pleurotus eryngii*. Applied and Environmental Microbiology **64**:325-332.

Bartolomé, B., C. B. Faulds, P. A. Kroon, K. Waldron, H. J. Gilbert, G. Hazlewood and G. Williamson. 1997. An *Aspergillus niger* esterase (ferulic acid esterase III) and a recombinant *Pseudomonas fluorescens* subsp. *cellulosa* esterase (XylD) release a 5-5' ferulic dehydrodimer (diferulic acid) from barley and wheat cells. Applied and Environmental Microbiology **63**:208-212.

Bebout, B. M. 1988. The role of marine fungi in the food selection and nutrition of the saltmarsh periwinkle *Littorina irrorata* Say (Gastropoda). MS Thesis, University of North Carolina, Chapel Hill, North Carolina, USA.

Bedrock, C. N., M. V. Cheshire, B. L. Williams, I. Solntseva, S. J. Chapman, J. A. Chudek and B. A. Goodman. 1998. Identification of nitrogenous components of fungal and bacterial origin immobilized in decomposing wheat straw by NMR spectroscopy using ^{15}N CPMAS. Soil Biology and Biochemistry **30**:113-115.

Benner, R., J. Lay, E. K'nees and R. E. Hodson. 1988. Carbon conversion efficiency for bacterial growth on lignocellulose: implications for detritus-based food webs. Limnology and Oceanography **33**:1514-1526.

Benner, R., M. L. Fogel and E. K. Sprague. 1991. Diagenesis of belowground biomass of *Spartina alterniflora* in saltmarsh sediments. Limnology and Oceanography **36**:1358-1374.

Blum, L. K. 1993. *Spartina alterniflora* root dynamics in a Virginia marsh. Marine Ecology Progress Series **102**:169-178.

Boddy, L. and S. C. Watkinson. 1995. Wood decomposition, higher fungi and their role in nutrient redistribution. Canadian Journal of Botany **73**:S1377-S1383.

Boyd, P. E. 1980. Ecology of three arenicolous marine fungi. I. Their role in the diet of the beachhopper *Orchestia grillus*. II. The influence of temperature on the seasonal and geographic distribution of *Asteromyces cruciatus, Sigmoidea marina* and *Varicosporina ramulosa*. MS Thesis, University of North Carolina, Chapel Hill, North Carolina, USA.

Brendelberger, H. 1997. Determination of digestive enzyme kinetics: a new method to define trophic niches in freshwater snails. Oecologia **109**:34-40.

Cammen, L. M. 1991. Annual bacterial production in relation to benthic microalgal production and sediment oxygen uptake in an intertidal sandflat and an intertidal mudflat. Marine Ecology Progress Series **71**:13-25.

Cantrell, S. A., S. Y. Newell and R. T. Hanlin. 1996. A new species of Lachnum on *Spartina alterniflora*. Mycotaxon **57**:479-485.

Chalmers, A. G. 1997. The ecology of the Sapelo Island National Estuarine Research Reserve. Sanctuaries and Reserves Division, Office of Coastal Resource Management, National Oceanic & Atmospheric Administration, Washington, District of Columbia, USA.
Chapman, S. J. and T. R. G. Gray. 1986. Importance of cryptic growth, yield factors and maintenance energy in models of microbial growth in soil. Soil Biology and Biochemistry **18**:1-4.
Chen, F., J. M. González, W. A. Dustman, M. A. Moran and R. E. Hodson. 1997. *In situ* reverse transcription, an approach to characterize genetic diversity and activities of prokaryotes. Applied and Environmental Microbiology **63**:4907-4913.
Chin-Leo, G. 1997. Bacterial secondary productivity. Pages 263-271 *in* M. McInerney, L. Stetzenbach, C.J. Hurst, G. Knudsen and M. Walter, editors. Manual of environmental microbiology. ASM Press, Washington, District of Columbia.
Christian, R. R., W. L. Bryant and M. M. Brinson. 1990. Juncus roemerianus production and decomposition along gradients of salinity and hydroperiod. Marine Ecology Progress Series **68**:137-145.
Coffin, R. B., B. Fry, B. J. Peterson and R. T. Wright. 1989. Carbon isotope composition of estuarine bacteria. Limnology and Oceanography **34**:1305-1310.
Covi, M. P. and R. T. Kneib. 1995. Intertidal distribution, population dynamics and production of the amphipod *Uhlorchestia spartinophila* in a Georgia, USA, saltmarsh. Marine Biology **121**:447-455.
Dai, T. and R. G. Wiegert. 1996. Ramet population dynamics and net aerial primary productivity of *Spartina alterniflora*. Ecology **77**:276-288.
Daiber, F. C. 1982. Animals of the tidal marsh. Van Nostrand Reinhold, New York, New York, USA.
Devereux, R., M. E. Hines and D. A. Stahl. 1996. S cycling: characterization of natural communities of sulfate-reducing bacteria by 16S rRNA sequence comparisons. Microbial Ecology **32**:283-292.
Dewey, F. M. 1996. Production and use of monoclonal antibodies for the detection of fungi. British Crop Protection Council Monograph **65**:85-91.
Dick, M. W. 1997. Fungi, flagella and phylogeny. Mycological Research **101**:385-394.
Doelle, H. W., D. A. Mitchell and C. E. Rolz, editors. 1992. Solid substrate cultivation. Elsevier Applied Science, New York, New York, USA.
Edwards, S. G., A. H. Fitter and J. P. W. Young. 1997. Quantification of an arbuscular mycorrhizal fungus, *Glomus mosseae*, within plant roots by competitive polymerase chain reaction. Mycological Research **101**:1440-1444.
Fallon, R. D., S. Y. Newell and C. S. Hopkinson. 1983. Bacterial production in marine sediments: will cell-specific measures agree with whole-system metabolism? Marine Ecology Progress Series **11**:119-127.
Fell, J. W. and I. M. Master. 1975. Phycomycetes (*Phytophthora* spp. nov. and *Pythium* sp. nov.) associated with degrading mangrove (*Rhizophora mangle*) leaves. Canadian Journal of Botany **53**:2908-2922.
Fell, J. W. and S. Y. Newell. 1998. Biochemical and molecular methods for the study of marine fungi. Pages 259-283 *in* K. Cooksey, editor. Molecular approaches to the study of the ocean. Chapman & Hall, London, England.
Fell, P. E., N. C. Olmstead, E. Carlson, W. Jacob, D. Hitchcock and G. Silber. 1982. Distribution and abundance of macroinvertebrates on certain Connecticut tidal marshes, with emphasis on dominant molluscs. Estuaries **5**:234-239.
Frankland, J. C. 1998. Fungal succession – unravelling the unpredictable. Mycological Research **102**:1-15.
French, N. R. 1979. Perspectives in grassland ecology. Springer-Verlag, New York.
Fuhrman, J. A. 1992. Bacterioplankton roles in cycling of organic matter: the microbial food web. Pages 361-383 *in* P. G. Falkowski and A. D. Woodhead, editors. Primary productivity and biogeochemical cycles in the sea. Plenum Press, New York, New York, USA.
Fuhrman, J. A., S. H. Lee, Y. Masuchi, A. A. Davis and R. M. Wilcox. 1994. Characterization of marine prokaryotic communities via DNA and RNA. Microbial Ecology **28**:133-145.
Gessner, M. O. and E. Chauvet. 1997. Growth and production of aquatic hyphomycetes in decomposing leaf litter. Limnology and Oceanography **42**:496-505.
Gessner, M. O. and S. Y. Newell. 1997. Bulk quantitative methods for the examination of eukaryotic organoosmotrophs in plant litter. Pages 295-308 *in* M. McInerney, L. Stetzenbach, C.J. Hurst, G. Knudsen and M. Walter, editors. Manual of environmental microbiology. ASM Press, Washington, District of Columbia, USA.
Gessner, R. V. 1977. Seasonal occurrence and distribution of fungi associated with *Spartina alterniflora* from a Rhode Island estuary. Mycologia **69**:477-491.
Gessner, R. V. and J. Kohlmeyer. 1976. Geographical distribution and taxonomy of fungi from saltmarsh *Spartina*. Canadian Journal of Botany **54**:2023-2037.

González, J. M., F. Mayer, M. A. Moran, R. E. Hodson and W. B. Whitman. 1997. *Sagittula stellata* gen. nov., sp. nov., a lignin-transforming bacterium from a coastal environment. International Journal of Systematic Bacteriology **47**:773-780.

González, J. M. and M. A. Moran. 1997. Numerical dominance of a group of marine bacteria in the α-subclass of the class Proteobacteria in coastal seawater. Applied and Environmental Microbiology **63**:4237-4242.

González, J. M., W. B. Whitman, R. E. Hodson and M. A. Moran. 1996. Identifying numerically abundant culturable bacteria from complex communities: an example from a lignin enrichment culture. Applied and Environmental Microbiology **62**:4433-4440.

Gutiérrez, A., M. J. Martínez, G. Almendros, A. Prieto, F. J. González-Vila and A. T. Martínez. 1995. Hyphal-sheath polysaccharides in fungal deterioration. Science of the Total Environment **167**:315-328.

Hausman, S. A. 1932. A contribution to the ecology of the saltmarsh snail, *Melampus bidentatus* Say. The Nautilus **66**:541-545.

Hawksworth, D. L., P. M. Kirk, B. C. Sutton and D. N. Pegler. 1995. Ainsworth and Bisby's Dictionary of the Fungi. Eighth edition. CAB International, Wallingford, UK.

Healy, B. and K. Walters. 1994. Oligochaeta in *Spartina* stems: the microdistribution of Enchytraeidae and Tubificidae in a saltmarsh, Sapelo Island, USA. Hydrobiologia **278**:111-123.

Heard, R. W. 1982. Guide to common tidal marsh invertebrates of the northeastern Gulf of Mexico. Mississippi/Alabama Sea Grant Consortium, Mobile, Alabama, USA.

Heller, J. and A. Abotbol. 1997. Litter shredding in a desert oasis by the snail *Melanopsis praemorsa*. Hydrobiologia **344**:65-73.

Hill, N. M. and D. G. Patriquin. 1992. Interactions between fungi and nitrogen-fixing bacteria during decomposition. Pages 783-796 *in* G. C. Carroll and D. T. Wicklow, editors. The fungal community. Second edition. Marcel-Dekker, New York, New York, USA.

Hodson, R. E., R. R. Christian and A. E. Maccubbin. 1984. Lignocellulose and lignin in the saltmarsh grass *Spartina alterniflora*: initial concentrations and short-term, post-depositional changes in detrital matter. Marine Biology **81**:1-7.

Honegger, R. 1991. Functional aspects of the lichen symbiosis. Annual Review of Plant Physiology **42**:553-578.

Hullar, M. A. J., B. Fry, B. J. Peterson and R. T. Wright. 1996. Microbial utilization of estuarine dissolved organic carbon: a stable isotope tracer approach tested by mass balance. Applied and Environmental Microbiology **62**:2489-2493.

Hurek, T., B. Wagner and B. Reinhold-Hurek. 1997. Identification of N_2-fixing plant- and fungus-associated *Azoarcus* species by PCR-based genomic fingerprints. Applied and Environmental Microbiology **63**:4331-4339.

Hutchison, L. J. and G. L. Barron. 1997. Parasitism of algae by lignicolous Basidiomycota and other fungi. Canadian Journal of Botany **75**:1006-1011.

Joergensen, R. G., B. Meyer, A. Roden and B. Wittke. 1996. Microbial activity and biomass in mixture treatments of soil and biogenic municipal refuse compost. Biology and Fertility of Soils **23**:43-49.

Kemp, P. F., S. Y. Newell and C. Krambeck. 1990. Effects of filter-feeding by the ribbed mussel *Geukensia demissa* on the water-column microbiota of a *Spartina alterniflora* saltmarsh. Marine Ecology Progress Series **59**:119-131.

Kendrick, B. 1992. The fifth kingdom. Second edition. Focus Information Group, Newburyport, Massachusetts, USA.

Kerley, S. J. and D. J. Read. 1997. The biology of mycorrhiza in the Ericaceae. XIX. Fungal mycelium as a nitrogen source for the ericoid mycorrhizal fungus *Hymenoscyphus ericae* and its host plants. New Phytologist **136**:691-701.

Kirchman, D. L., B. J. Peterson and D. Juers. 1984. Bacterial growth and tidal variation in bacterial abundance in Great Sippewissett salt marsh. Marine Ecology Progress Series **19**:247-259.

Kneib, R. T. 1997. The role of tidal marshes in the ecology of estuarine nekton. Oceanography and Marine Biology, Annual Review **35**:163-220.

Kneib, R. T., S. Y. Newell and E. T. Hermeno. 1997. Survival, growth and reproduction of the saltmarsh amphipod *Uhlorchestia spartinophila* reared on natural diets of senescent and dead *Spartina alterniflora* leaves. Marine Biology **128**:423-431.

Kohlmeyer, J. and R. V. Gessner. 1976. *Buergenerula spartinae* sp. nov., an Ascomycete from salt marsh cordgrass, *Spartina alterniflora*. Canadian Journal of Botany **54**:1759-1766.

Kohlmeyer, J. and E. Kohlmeyer. 1979. Marine mycology. The higher fungi. Academic Press, New York, New York, USA.
Kohlmeyer, J. and B. Volkmann-Kohlmeyer. 1991. Illustrated key to the filamentous higher marine fungi. Botanica Marina **34**:1-61.
Kohlmeyer, J., B. Volkmann-Kohlmeyer and O. E. Eriksson. 1997. Fungi on *Juncus roemerianus* 9. New obligate and facultative marine Ascomycotina. Botanica Marina **40**:291-300.
Kuehn, K. A. and K. Suberkropp. 1998. Diel fluctuations in rates of CO_2 evolution from standing dead leaf litter of the emergent macrophyte *Juncus effusus* L. Aquatic Microbial Ecology **14**:171-182.
Kuehn, K. A., P. F. Churchill and K. Suberkropp. 1998. Osmoregulatory responses of fungi inhabiting standing-dead litter of the freshwater emergent macrophyte *Juncus effusus*. Applied and Environmental Microbiology **64**:607-612.
Lana, P. C. and C. Guiss. 1992. Macrofauna-plant-biomass interactions in a euhaline saltmarsh in Paranaguá Bay (SE Brazil). Marine Ecology Progress Series **80**:57-64.
Le Gall, S., M. Bel Hassen and P. Le Gall. 1997. Ingestion of a bacterivorous ciliate by the oyster *Crassostrea gigas*: protozoa as a trophic link between picoplankton and benthic suspension feeders. Marine Ecology Progress Series **152**:301-306.
Leuchtmann, A. and S. Y. Newell. 1991. *Phaeosphaeria spartinicola*, a new species on *Spartina*. Mycotaxon **41**:1-7.
Levinton, J. S. and S. Stewart. 1988. Effects of sediment organics, detrital input and temperature on demography, production and body size of a deposit feeder. Marine Ecology Progress Series **49**:259-266.
Mackay, W. P., S. J. Loring, J. C. Zak, S. I. Silva, F. M. Fisher and W. G. Whitford. 1994. Factors affecting loss in mass of creosotebush leaf-litter on the soil surface in the northern Chihuahuan Desert. Southwestern Naturalist **39**:78-82.
Mahuku, G. S., P. H. Goodwin and R. Hall. 1995. A competitive polymerase chain reaction to quantify DNA of *Leptosphaeria maculans* during blackleg development in oilseed rape. Molecular Plant-Microbe Interactions **8**:761-767.
McGonigle, T. P. 1997. Fungivores. Pages 237-248. *in* D. T. Wicklow and B. Söderström, editors. The Mycota. Volume IV. Environmental and microbial relationships. Springer-Verlag, New York, New York, USA.
McQuaid, C. D. 1996. Biology of the gastropod family Littorinidae. II. Role in the ecology of intertidal and shallow marine ecosystems. Oceanography and Marine Biology, Annual Review **34**:263-302.
Montagna, P. A. 1995. Rates of metazoan meiofaunal microbivory: a review. Vie et Milieu 45:1-9.
Montague, C. L. and R. G. Wiegert. 1990. Salt marshes. Pages 481-516 *in* R. L. Myers and J. J. Ewel, editors. Ecosystems of Florida. University of Central Florida Press, Orlando, Florida, USA.
Mook, D. 1986. Absorption efficiencies of the intertidal mangrove dwelling mollusk *Melampus coffeus* Linné and the rocky intertidal mollusk *Acanthopleura granulata* Gmelin. PSZNI Marine Ecology **7**:105-113.
Moran, M. A. and R. E. Hodson. 1990. Contributions of degrading *Spartina alterniflora* lignocellulose to the dissolved organic carbon pool of a saltmarsh. Marine Ecology Progress Series **62**:161-168.
Moran, M. A., L. T. Rutherford and R. E. Hodson. 1995. Evidence for indigenous *Streptomyces* populations in a marine environment determined with a 16S rRNA probe. Applied and Environmental Microbiology **61**:3695-3700.
Morris, J. T. and G. J. Whiting. 1986. Emission of gaseous carbon dioxide from saltmarsh sediments and its relation to other carbon losses. Estuaries **9**:9-19.
Mouget, J. L., A. Dakhama, M. C. Lavoie and J. de la Noüe. 1995. Algal growth enhancement by bacteria: is consumption of photosynthetic oxygen involved? FEMS Microbiology Ecology **18**:35-44.
Newell, S. Y. 1993. Decomposition of shoots of a saltmarsh grass; methodology and dynamics of microbial assemblages. Advances in Microbial Ecology **13**:301-326.
——. 1994. Ecomethodology for organoosmotrophs: prokaryotic unicellular versus eukaryotic mycelial. Microbial Ecology **28**:151-157.
——. 1996a. Established and potential impacts of eukaryotic mycelial decomposers in marine/terrestrial ecotones. Journal of Experimental Marine Biology and Ecology **200**:187-206.
——. 1996b. The [^{14}C]acetate-to-ergosterol method: factors for conversion from acetate incorporated to organic fungal mass synthesized. Soil Biology and Biochemistry **28**:681-683.
Newell, S. Y., R. D. Fallon, B. F. Sherr and E. B. Sherr. 1988a. Mesoscale temporal variation in bacterial standing crop, percent active cells, productivity and output in a saltmarsh tidal river. Verhandlung Internationale Verein Limnologie **23**:1839-1845.

Newell, S.Y. and J.W. Fell. 1992. Ergosterol content of living and submerged, decaying leaves and twigs of red mangrove. Canadian Journal of Microbiology **38**:979-982.
Newell, S. Y. and F. Bärlocher. 1993. Removal of fungal and total organic matter from decaying cordgrass leaves by shredder snails. Journal of Experimental Marine Biology and Ecology **171**:39-49.
Newell, S. Y., C. S. Hopkinson and L. A. Scott. 1992. Patterns of nitrogenase activity (acetylene reduction) associated with standing, decaying shoots of *Spartina alterniflora*. Estuarine, Coastal and Shelf Science **35**:127-140.
Newell, S.Y. and C. Krambeck. 1995. Responses of bacterioplankton to tidal inundations of a saltmarsh in a flume and adjacent mussel enclosures. Journal of Experimental Marine Biology and Ecology **190**:79-95.
Newell, S.Y. and L.A. Palm. 1998. Responses of bacterial assemblages on standing decaying blades of smooth cordgrass to additions of water and nitrogen. International Review of Hydrobiology **83**:115-122.
Newell, S. Y., T. L. Arsuffi and R. D. Fallon. 1988. Fundamental procedures for determining ergosterol content of decaying plant material by liquid chromatography. Applied and Environmental Microbiology **54**:1876-1879.
Newell, S.Y., T.L. Arsuffi and L.A. Palm 1996b. Misting and nitrogen fertilization of shoots of a saltmarsh grass: effects upon fungal decay of leaf blades. Oecologia **108**:495-502.
—— 1998. Seasonal and vertical demography of dead portions of shoots of smooth cordgrass in a south-temperate saltmarsh. Aquatic Botany **60**:325-335.
Newell, S. Y., R. D. Fallon and J. D. Miller. 1989. Decomposition and microbial dynamics for standing, naturally positioned leaves of the saltmarsh grass *Spartina alterniflora*. Marine Biology **101**:471-481.
Newell, S. Y., D. Porter and W. L. Lingle 1996a. Lignocellulolysis by ascomycetes (Fungi) of a saltmarsh grass (smooth cordgrass). Microscopy Research and Technique **33**:32-46.
Newell, S. Y. and J. Wasowski. 1995. Sexual productivity and spring intramarsh distribution of a key saltmarsh microbial secondary producer. Estuaries **18**:241-249.
Nicole, M., H. Chamberland, D. Rioux, X. Xixuan, R. A. Blanchette, J. P. Geiger and G. B. Ouellette. 1995. Wood degradation by *Phellinus noxius*: ultrastructure and cytochemistry. Canadian Journal of Microbiology **41**:253-265.
Nilsson, T., G. Daniel, T. K. Kirk and J. R. Obst. 1989. Chemistry and microscopy of wood decay by some higher ascomycetes. Holzforschung **43**:11-18.
Paerl, H. W. and J. L. Pinckney. 1996. A mini-review of microbial consortia: their roles in aquatic production and biogeochemical cycling. Microbial Ecology **31**:225-247.
Pennings, S. C., T. H. Carefoot, E. L. Siska, M. E. Chase and T. A. Page. 1998. Feeding preferences of a generalist saltmarsh crab: relative importance of multiple plant traits. Ecology **79**:1968-1979.
Proffitt, C. E., K. M. Johns, C. B. Cochrane, D. J. Devlin, T. A. Reynolds, D. L. Payne, S. Jeppesen, D. W. Peel and D. D. Linden. 1995. Field and laboratory experiments of the consumption of mangrove leaf litter by the macrodetritivore *Melampus coffeus* L. (Gastropoda: Pulmonata). Florida Scientist **56**:211-222.
Rayner, A. D. M., G. S. Griffith and A. M. Ainsworth. 1995. Mycelial interconnectedness. Pages 21-40 *in* A. R. Gow and G. M. Gadd, editors. The growing fungus. Chapman & Hall, London, England.
Rietsma, C. S., I. Valiela and R. Buchsbaum. 1988. Detrital chemistry, growth and food choice in the saltmarsh snail (*Melampus bidentatus*). Ecology **69**:261-266.
Rietsma, C. S., I. Valiela and A. Sylvester-Serianni. 1982. Food preferences of dominant saltmarsh herbivores and detritivores. PSZNI Marine Ecology **3**:179-182.
Robarts, R. D. and T. Zohary. 1993. Fact or fiction – bacterial growth rates and production as determined by [methyl-^{3}H]thymidine? Advances in Microbial Ecology **13**:371-425.
Rublee, P. A. 1982. Seasonal distribution of bacteria in salt marsh sediments in North Carolina. Estuarine, Coastal and Shelf Science **15**:67-74.
Rutledge, P. A. and J. W. Fleeger. 1993. Abundance and seasonality of meiofauna, including harpacticoid copepod species, associated with stems of the saltmarsh cord grass, *Spartina alterniflora*. Estuaries **16**:760-768.
Samiaji, J. and F. Bärlocher. 1996. Geratology and decomposition of *Spartina alterniflora* Liosel in a New Brunswick saltmarsh. Journal of Experimental Marine Biology and Ecology **201**:233-252.
Schut, F., R. A. Prins and J. C. Gottschal. 1997. Oligotrophy and pelagic marine bacteria: facts and fiction. Aquatic Microbial Ecology **12**:177-202.

Shia, F. K. and H. W. Ducklow. 1997. Bacterioplankton growth responses to temperature and chlorophyll variations in estuaries measured by thymidine:leucine incorporation ratio. Aquatic Microbial Ecology **13**:151-159.

Sinsabaugh, R.L. and S. Findlay. 1995. Microbial production, enzyme activity, and carbon turnover in surface sediments of the Hudson River estuary. Mircrobial Ecology **30**:127-141.

Slim, F. J., M. A. Hemminga, C. Ochieng, N. T. Jannink, E. Cocheret de la Morinière and G. van der Velde. 1997. Leaf litter removal by the snail *Terebralia palustris* (Linnaeus) and sesarmid crabs in an East African mangrove forest (Gazi Bay, Kenya). Journal of Experimental Marine Biology and Ecology **215**:35-48.

Smalley, A. E. 1959. The role of two invertebrate populations, *Littorina irrorata* and *Orchelimum fidicinium*, in the energy flow of a saltmarsh ecosystem. University Microfilms (59-5687), Ann Arbor, MI, USA.

Spelke, J. A., P. E. Fell and L. L. Helvenston. 1995. Population structure, growth and fecundity of *Melampus bidentatus* (Say) from two regions of a tidal marsh complex in Connecticut. The Nautilus **108**:42-47.

Suberkropp, K. 1992. Interactions with invertebrates. Pages 118-134 *in* F. Bärlocher, editor. The Ecology of aquatic hyphomycetes. Springer-Verlag, New York.

——. 1997. Annual production of leaf-decaying fungi in a woodland stream. Freshwater Biology **38**:169-178.

Svitil, A. L., S. M. Ní Chadhain, J. A. Moore and D. L. Kirchman. 1997. Chitin degradation proteins produced by the marine bacterium *Vibrio harveyi* growing on different forms of chitin. Applied and Environmental Microbiology **63**:408-413.

Theenhaus, A. and S. Scheu. 1996. The influence of slug (*Arion rufus*) mucus and cast material addition on microbial biomass, respiration and nutrient cycling in beech leaf litter. Biology and Fertility of Soils **23**:80-85.

Thompson, L. S. 1984. Comparison of the diets of the tidal marsh snail, *Melampus bidentatus* and the amphipod, *Orchestia grillus*. The Nautilus **98**:44-53.

Torsvik, V., R. Sørheim and J. Goksøyr. 1996. Total bacterial diversity in soil and sediment communities: a review. Journal of Industrial Microbiology **17**:170-178.

Turner, R. E. 1993. Carbon, nitrogen and phosphorus leaching rates from *Spartina alterniflora* Loisel saltmarshes. Marine Ecology Progress Series **92**:135-140.

Wardle, D. A. and P. Lavelle. 1997. Linkages between soil biota, plant litter quality and decomposition. Pages 107-124 *in* G. Cadisch and K. E. Giller, editors. Driven by nature: plant litter quality and decomposition. CAB International, Wallingford, England.

Wessels, J. G. H. 1997. Hydrophobins: proteins that change the nature of the fungal surface. Advances in Microbial Physiology **38**:1-45.

White, D. S. and B. L. Howes. 1994. Nitrogen incorporation into decomposing litter of *Spartina alterniflora*. Limnology and Oceanography **39**:133-140.

Zimmerman, K. and J. W. Mannhalter. 1996. Technical aspects of quantitative competitive PCR. BioTechniques **21**:268-279.

TROPHIC COMPLEXITY BETWEEN PRODUCERS AND INVERTEBRATE CONSUMERS IN SALT MARSHES

DANIEL A. KREEGER
Patrick Center for Environmental Research
Academy of Natural Sciences
1900 Benjamin Franklin Parkway
Philadelphia, PA 19103 USA

ROGER I.E. NEWELL
Horn Point Laboratory
University of Maryland Center for Environmental Science
P.O. Box 775
Cambridge, MD 21613 USA

Abstract
Salt marshes on the Atlantic coast of North America are characterized by having a high biomass of smooth cordgrass, *Spartina alterniflora*. Because of the refractory nature of the lignocellulosic structure of this angiosperm, invertebrates utilize C from these plants with very low efficiency, if at all. This is true for both living cordgrass and post-senescent plant detritus. To balance their C demands, invertebrate consumers living in salt marshes must utilize a wide variety of other resources, including microheterotrophs (bacteria and bacterivorous flagellates) either associated with detritus or free in the water column, fungi colonizing decaying vascular plants, surface-associated algae (e.g., microphytobenthic diatoms and cyanobacteria, epiphytes, surface film algae) and phytoplankton. This high degree of trophic complexity is likely to be an important source of community stability. As an example, we estimate that ribbed mussels, *Geukensia demissa*, in a Delaware marsh must rely on a variety of different food resources since no single food type can meet their nutritional demands for either C or N. To balance their C demands, mussels appear to rely mainly on microheterotrophs, followed by phytoplankton > microphytobenthos > cellulosic detritus. Non-detrital foods are even more important for maintaining positive N balance in *G. demissa*. Previous and emerging evidence from other studies suggests that other important marsh consumers have a similar general diet. Although cordgrass may dominate overall rates of primary production and detritus from cordgrass contributes significantly to secondary production, we challenge the paradigm that salt marshes have a "detritus-based food web." Further research is needed to deduce the importance of microphytobenthos and microheterotrophs as sources of C and N for dominant animal consumers in these marsh systems.

1. Introduction

1.1 DIVERSITY OF PRODUCER/CONSUMER TROPHIC LINKAGES

The intertidal portion of salt marshes contains a wide diversity of food resources for animals that either directly graze on living plants (herbivory) or feed on dead and decaying plant material (detritivory). Autochthonous primary production in marshes can occur via emergent aquatic plants, epiphytic and periphytic films on erect vegetation or other hard surfaces, benthic microalgae and cyanobacteria, surface film algae, macroalgae and phytoplankton. Allochthonous inputs can be derived from the adjoining estuary, and also terrestrial uplands via seasonally pulsed inputs of leaf litter. For each of these materials, there are invertebrate consumers adapted to exploit that resource niche with specialized feeding modes such as skimming and suspension-feeding. The diversity of trophic pathways therefore appears to be high compared with terrestrial ecosystems and other aquatic ecosystems. Our first objective in this chapter will be to qualitatively review the breadth of these types of trophic interactions between dominant marsh producers and dominant invertebrate consumers living in the intertidal (vegetated and bare) zone. We will not attempt to review all invertebrate consumers in tidal marshes since the species composition can differ considerably among marshes and seasons. Rather, we will focus our discussion on the trophic connections among characteristic functional groups, with a bias toward dominant aquatic macroinvertebrates feeding at the base of the food chain.

1.2 QUANTITATIVE IMPORTANCE OF DIFFERENT TROPHIC PATHWAYS

Teal (1962) proposed that energy flow in Georgia salt marshes is predominantly through the decomposer food web. This led to the paradigm that salt marshes have detritus-based food webs (Odum 1980). This persisting view is supported partly because vascular plants in these marshes have some of the highest production rates in the world. Much of this production is known to be decomposed by fungal and microbial processes *in situ* or gets buried or exported from the marsh (e.g., see Newell and Porter, this volume). Dominant marsh plants such as *Spartina alterniflora* are >80% structural lignocellulose (Benner et al. 1984). This material is refractory and consequently is assimilated at low efficiencies (Kreeger et al. 1988, Langdon and Newell, 1990; Charles and Newell 1997) or not at all (Montague, et al. 1981). Teal (1962) observed, however, that a few species of insects were effective at directly grazing cordgrass. Although plant detritus is assimilated poorly by most invertebrates, rates of secondary production and biomass can be very high in these marshes. Results from stable isotope ratio techniques suggest that vascular plants may supply at least part of the carbon requirements of many consumer species (Haines 1979, Haines and Montague 1979, Peterson et al. 1985, 1986, Peterson and Howarth, 1987, Langdon and Newell 1990, Currin et al. 1995). Because of the limited direct utilization of vascular plant carbon, consumers may attain an isotopic signature reflecting a contribution from marsh plants by grazing the detritus-associated microbial community supported by post-senescent plant material. Microbial mediation of vascular plant production represents an *indirect* trophic pathway to metazoan consumers that has long been suspected (e.g., Darnell 1967, Langdon and Newell 1990), but rarely examined directly.

Marsh invertebrates may consume algae in addition to vascular plant material. Sullivan and Moncreiff (1988) used a stable isotope ratio approach in Gulf of Mexico coastal marshes, and found that benthic microalgae appeared to be the dominant resource for macroconsumers. However, results from stable isotope ratio studies with the same or similar consumer species differ among studies and locations. For example, in New England marshes, *Spartina*-derived material has been estimated to contribute up to 80% of the diet of ribbed mussels (Peterson et al. 1985, 1986). The *Spartina* contribution in mid-Atlantic marshes is estimated to be from 30 to 50% (Langdon and Newell, 1990). In Georgia marshes the contribution from vascular plants is reported to be insignificant (Haines 1979, Haines and Montague 1979). These equivocal results may be due to geographic differences in the quantity of angiosperm detritus in the food web, or could result from technical difficulties in discerning isotopic signatures between cordgrass and benthic algae. Both producers derive their inorganic nutrients from the same sources (sediment, water column, and atmosphere). Because the carbon and nitrogen isotopic signatures of cordgrass (Peterson et al. 1985) are similar to those for the microphytobenthos (summarized by Currin et al. 1995 and Newell et al. 1995), it is difficult to distinguish between these two sources of autochthonous production based solely on stable isotope ratios.

Teal (1962) recognized both detrital and algal trophic pathways by observing that a wide variety of marsh invertebrates are ''algae-detritus feeders'' that balance their carbon demands by feeding mainly on dead and decaying plant matter and the associated microbial community. He suggested that detritus-feeders augment their C uptake when detritus supply is low by eating benthic algae. The relative importance of microphytobenthos may be much greater for at least a few of the key marsh invertebrates, such as gastropods (Pace et al. 1979, Lopez and Kofoed 1980, Conner and Edgar 1982, Lopez-Figueroa and Neill 1987), shrimp (R.I.E. Newell and B. Bebout, unpublished data) and mussels (D.A. Kreeger and R.I.E. Newell, unpublished data).

In light of the research that has occurred since Teal's seminal work, in this chapter we will re-examine the paradigm that marshes have detritus-dominated food webs. The biomass of vascular plants can be tremendous in tidal marshes, and since many measures of production are based on quantifying standing stocks of these plants, the dogma in marsh ecology has centered on the fate of this material. There is increasing evidence that rates of primary production by other marsh producers have been underestimated, maybe even grossly underestimated. For example, rates of production by the microphytobenthos can be high even though their biomass is low and inconspicuous compared with emergent vascular plants (see below). Marsh invertebrates may be able to exploit this material with much greater efficiency compared to vascular plants, perhaps compensating for the relatively lower standing stock. We will discuss these relationships in the context of the ''secret garden'' hypothesis (MacIntyre et al. 1996, Miller et al. 1996) which postulates that highly productive microphytobenthos biomass is rapidly turned over by the sustained grazing of consumers.

We will first provide a general review of some of the important trophic links between primary producers and primary consumers in eastern USA salt marshes, defined here as the intertidal zone containing emergent angiosperms. We will then discuss the *sources* and *amounts* of marsh production flowing to the ribbed mussel *Geukensia demissa* (Dillwyn), which can be a dominant invertebrate consumer in these marshes. This species is a good representative of the benthic suspension-feeders, an important functional guild whose biomass can outweigh all other marsh consumers combined, particularly in salt marshes along the Atlantic coast of USA (e.g., Kuenzler 1961a, Lent 1969, Jordan and Valiela 1982, Franz, 1993). Over the last 20 years, data has been generated on the nutrition of *G. demissa* from stable isotope ratio studies (e.g., Peterson et al., 1985, Langdon and Newell 1990), which indirectly trace the ultimate food sources for natural mussel populations. In addition, laboratory feeding experiments have directly examined the routes and conversion efficiencies by which material gets from different producers to this consumer (Kreeger et al. 1988, Kreeger and Newell 1996). Using suspension-feeding mussels as a case study for a resident marsh consumer, our second objective is to examine the relative importance of allochthonous and autochthonous production in the mussel's diet, and the routes by which production flows to this species. We will then discuss the implications of our findings for traditional food web theory in salt marshes.

2. Food Items and Their Nutritional Value

Animals feeding at the base of the food chain in salt marshes are presented with a wide array of food types. These foods may be either living primary producers or constituents of the detritus complex. Photosynthetic organisms that contribute substantially to marsh food webs include vascular plants, epiphytic algae, macroalgae, microphytobenthos (i.e., benthic microalgae and benthic cyanobacteria) and phytoplankton. Most of these primary producers are autochthonous, except for phytoplankton which is likely to be largely allochthonous in origin as it is imported to the marsh with the flood tide. Detritus and detritus-associated organisms that serve as foods for primary consumers include dead producers, fungi, and microheterotrophic bacteria and protists.

In this section we list some of the major foods available to primary consumers in the marsh and briefly compare their food value, which is interpreted to mean their relative availability, digestibility, and intrinsic nutritional value. Of course, the "food value" of these foods will differ markedly among different types of consumers because of their differing abilities to capture, process, digest and assimilate energy and nutrients from the materials, and also because the nutritional requirements of different consumers may vary considerably.

2.1 PRIMARY PRODUCERS

2.1.1 Emergent Vascular Plants

Salt marshes, by definition, are habitats in which emergent vascular plants are abundant and provide a major organizing structure to the ecosystem. In marshes along the Atlantic coast of the USA, the smooth cordgrass *Spartina alterniflora* dominates this plant community, forming an almost ubiquitous stand over the mid and low tidal areas. In high marsh areas near the landward fringe, plant diversity increases; however, less of that production is likely to be available to aquatic macroinvertebrates compared with *S. alterniflora* due to limited duration of tidal inundation.

Annual rates of aboveground net production for *Spartina alterniflora* are among the greatest in the world, ranging generally from 0.5 to 3.5 kg m^{-2} (for reviews, see Pomeroy et al. 1981). Both insects (Smalley 1960, Teal 1962, Marples 1966, Montague et al. 1981) and gastropods (Smalley 1959, Kraeuter and Wolf 1974) are reported to directly graze on this plant biomass. But despite the tremendous standing stock of aboveground and belowground vascular plants, only 5 to 10% of this production flows directly to herbivores (Mann 1972, Heard1982). This is because greater than 80% of the live biomass of *S. alterniflora* is comprised of refractory structural lignocellulose (Benner et al. 1984) that is indigestible by all but a few metazoans. As these plants senesce, the more nutritious, labile components are either resorbed, taken up by fungi that become established in the dead and standing plants (Newell and Porter, this volume), taken up by the microbial community in the water column or sediments (Benner et al. 1984) or leached into the water column (Haines and Hanson 1979, Wilson et al. 1986). Post-senescent standing material is then broken down by physical processes (e.g., ice in northern areas, storms) until it falls to the surface where further decomposition occurs very slowly *in situ* via both physical and microbial processes and by animal feeding activity (e.g., isopods, amphipods), and eventually, it becomes part of the detritus complex (Maccubbin and Hodson 1980, Valiela et al. 1985). Although the abundance of vascular plant material is very high, the digestibility and nutritional value of cordgrass and the resulting post-senescent detritus is therefore regarded as low for animal consumers (Table 1).

2.1.2 Surface-Associated Algae

In addition to vascular plants, the other major group of autochthonous marsh producers are the macroalgae and microalgae that live attached to surfaces (e.g., epiphytes), on the sediment surface (microphytobenthos), or entrained in the water surface film. Compared with research on vascular plants, relatively few workers have studied these algae (Gallagher 1975, D.M. Seliskar, W.L. Carey and J.L. Gallagher, pers. commun.). The contribution of "surface-associated" microalgae such as that comprising the microphytobenthos (benthic diatoms and cyanobacteria) to the overall primary production in salt marshes may have been underestimated (Sullivan and Moncreiff 1988, Cahoon and Cooke 1992, Cahoon et al. 1993, Pinckney and Zingmark 1993,

MacIntyre and Cullen 1995, for review see MacIntyre et al. 1996). Rates of production vary widely among different types of algae, seasons, and marsh systems, and so the relative food values of different forms are difficult to predict. For example, rates of microphytobenthos production have been estimated to vary from less than 10% (Sullivan and Moncreiff 1988) to 20% (Pomeroy 1959), 33% (Gallagher and Daiber 1974), to more than 100% (Zedler 1980) of comparable rates of production by vascular plants living in the same system. This wide disparity likely stems from variability in the benthic microalgal community, but may also result from technical difficulties associated with measuring net rates of production (Pinckney and Zingmark 1993, Pinckney et al. 1994). Photosynthetically active radiation (PAR) available for algal production at the sediment surface is inversely related to cordgrass density, and so PAR will differ considerably throughout the year in northern marshes which have high seasonal variation in canopy density; whereas, in southern marshes, cordgrass grows taller and maintains a high canopy density throughout the year.

The fate of C fixed by these surface-associated microalgae is uncertain, but it seems likely that much of it flows into consumer food webs. For example, the high nutritional value of benthic microalgae for marsh consumers has long been supposed (Teal 1962, Montague et al. 1981), but experimental evidence directly supporting this hypothesis has been slow to develop (Sullivan and Moncreiff 1990, Miller et al. 1996). Unicellular algae have cell diameters between 5 to 40 µm. Hence, as food particles they have a high surface area to volume ratio that facilitates enzymatic digestion once ingested by consumers. Surface-associated unicellular algae are mainly diatoms (Williams 1962, Sullivan 1975) that have siliceous frustules rather than cellulosic cell walls, and compared with vascular plants these cells are more easily disrupted by the physical processes associated with the guts of metazoan consumers. Furthermore, since consumers exploiting these algal resources are unlikely to select for or against specific species, there is likely to be consistency in the nutritional value for consumers since the overall algal community typically has high biodiversity (e.g., up to 60 species per cm^{-2} of marsh surface; Sullivan 1975, Sullivan and Currin, this volume). In summary, areal rates of primary production by surface-associated algae are generally not as high as that for vascular plants such as cordgrass; however, it is much more nutritious and digestible (Table 1), and so we suggest that these algae likely play an important role in salt marsh food webs.

TABLE 1. Conceptual framework for comparing the relative food value of some of the main resources available to dominant primary consumers residing in a typical salt marsh. Approximate relative food values are estimated by integrating the relative availability (i.e., resource abundance), consumer access (i.e., how many major consumers can ingest it), nutritional value (e.g., energy and nutrient content), and digestibility of each resource.

Resource		**Food Value Characteristics**				
Class	**Type**	**Availability**	**Accessibility**	**Nutritional Value**	**Digestibility**	**Relative Food Value**
Primary Producers	Vascular Plants	High	High	Low	Low	Moderately Low
	Surface Associated Algae	Moderate	High	High	High	High
	Phytoplankton	Low	Low[1]	High	High	Low[1]
Detritus Complex	Dead Vascular Plants	High	High	Low	Low	Moderately Low
	Surface-Associated Bacteria	Moderate	High	Moderate	Moderate	Moderate
	Suspended Free Bacteria	Low	Low[1]	Moderate	Moderate	Low[1]
	Heterotrophic Protists	Moderately Low	Low[1]	Moderately High	High	Low[1]
	Fungi	Moderate	Moderately High	Unknown	Unknown	Moderately Low

[1] Except for suspension-feeding consumers, which could derive more from this resource.

2.1.3 Phytoplankton

Primary production by phytoplankton occurs largely in the adjacent estuary; however, it has rarely been directly measured within the water column of the salt marsh when inundated by the tide. Pomeroy et al. (1981) reported that in the Sapelo Island marshes of Georgia USA, *in situ* (i.e., not imported) marsh production by phytoplankton amounted to ~12% that of vascular plants. Phytoplankton typically have cell diameters too small (2 to 15 μm) to be efficiently removed by any consumer type except suspension-feeding animals. Hence, except in marshes dominated by suspension-feeders, phytoplankton are not as important a resource as the microphytobenthos for the overall consumer community. As with surface-associated algae, however, phytoplankton are nutritious for consumers that can access this resource. Most phytoplankton in marshes are either diatoms or dinoflagellates, and both groups are comprised of a high proportion of digestible and metabolizable energy and nutrients. Suspension-feeding animals are well-adapted for capturing large numbers of microalgae, and due to the very high surface area:volume ratio, these algal cells are readily digested and assimilated with efficiencies that are typically 50 to 80% (Webb and Chu 1982, Bayne and Newell 1983). Hence, phytoplankton can represent an important resource for suspension-feeding marsh consumers, especially those residing at the seaward margin where phytoplankton are more abundant (Peterson et al. 1985, Langdon and Newell 1990).

2.2 DETRITUS COMPLEX

For the purposes of this chapter, we consider the detritus complex as including dead and decaying biotic material, together with the associated living community of microbial decomposers. These decomposers consist of fungi and microheterotrophs (bacteria, heterotrophic protists). Omitted from this analysis is dissolved organic material (DOM) associated with detritus breakdown. DOM in marshes can reach high concentrations (Kiel and Kirchman 1991), but little work has been done to study its role as a resource for metazoan consumers in the natural environment (see reviews by Stephens 1985, Newell and Langdon 1996).

2.2.1 Dead Vascular Plants

Most of the organic material associated with the detritus complex in salt marshes is derived from the large inputs of vascular plant material. Although other dead producers and consumers can contribute to detritus, most of this material is rapidly recycled, whereas the recalcitrant lignocellulose from vascular plants degrades slowly.

As with living vascular plants, detritus derived from these producers is difficult for invertebrate consumers to digest and is therefore of low food value. Only a few marsh consumers have been reported to digest cellulose without the aid of external microorganisms (Kreeger et al. 1988, 1990, Langdon and Newell 1990, Charles and Newell 1997), and none have been demonstrated to grow and reproduce when cultured on sterile detritus. Previous studies looking at the nutritional value of detritus have shown direct utilization of this material (e.g., *Capitella capitata*, Tenore and Hanson 1980), but this material was naturally aged and may have contained a nourishing microbial community. As detritus ages, the N content has been shown to increase, and it was long thought that aging therefore led to an increase in the nutritional value of detritus with time (Fenchel 1972, Tenore 1975, 1977, Tenore and Hanson 1980). However, much of this N buildup results from chemical complexation of humic geopolymers that have no nutritional value (Rice 1982). Since the detritus complex is only made available as a resource for metazoan consumers through microheterotrophic intermediaries, its intrinsic resource value for consumers is characterized as being abundant but of poor nutritional value and digestibility (Table 1).

2.2.2 Surface-Associated Bacteria

The nutritional importance of bacteria for aquatic consumers has been questioned for at least 60 years (e.g., see Zobell and Feltham 1938). Bacteria associated with the detritus complex can either be sediment-associated, attached to suspended detrital particles, bound into flocculated aggregates, or be free-living in the water column. Here we consider all but the latter to be "surface-associated" bacteria. In marshes a large proportion of surface-associated bacteria may be cellulolytic, serving to decompose vascular plant detritus (Gallagher et al. 1976, Benner et al. 1984, Newell et al. 1985, Coffin et al. 1989).

By virtue of being associated with sediments or larger particles, bound bacteria are more readily ingested by metazoan consumers such as deposit-feeders and suspension-feeders. Unfortunately, few studies have investigated the abundance of surface-associated bacteria within salt marsh systems. Christian et al. (1981) synthesized a variety of studies using ATP concentrations as indicators of microbial activity and found that 79% of the standing stock of marsh bacteria are associated with sediments. Newell and Porter (this volume) report typical bacterial standing stocks in the sediments to be 44 g bacteria m^{-2} in a Georgia marsh. This represents a substantial food resource for marsh consumers.

Bacteria associated with particles, such as sediment grains, detritus, and flocculated sediments, can be readily ingested by invertebrate consumers, and once ingested, the microbial coating can usually be digested leaving the particulate substrate which gets defecated. This "microbial stripping" idea was first proposed by Newell (1965) who reported that the deposit-feeder *Macoma balthica* derived considerable nutrition from the microbial coating, but egested the organic particles upon which the bacteria were attached. The same idea has been proposed for suspension-feeding invertebrates such as oysters and mussels that can remove bacteria more efficiently from the water column when the bacteria are attached to larger particles. Langdon and Newell (1990) estimated, however, that suspended attached bacteria are not sufficiently abundant to contribute significantly to the nutritional requirements of these animals. Bacteria are N-rich (Fenchel 1982) and are assimilated at efficiencies >50% (Crosby et al. 1990, Langdon and Newell 1990, Werner and Hollibaugh 1993, Kreeger and Newell 1996). For animals such as deposit-feeders, sediment-associated bacteria represent a significant food source (Table 1).

2.2.3 Suspended Free-Living Bacteria

Bacteria free-living in the water column flooding the marsh at high tide can also contribute to marsh food webs. Concentrations of free bacteria are typically ten times more abundant in the intertidal creeks within marshes than in open water areas adjacent to marshes, often exceeding cell densities of 10^7 ml^{-1} (Kreeger 1986, Huang 2000). However, free bacteria are so small (<1 μm diameter) that these high concentrations do not equate with high biomass and resource availability for metazoan consumers. Data of Newell and Porter (this volume) indicate that 0.5 m beneath the marsh water surface free bacteria are present at about 0.2 g biomass m^{-2}, which is low relative to other resources discussed herein. Moreover, the only marsh consumers that potentially have access to unattached bacteria are the suspension-feeders.

There has been a large amount of research into bacteria utilization by bivalves (e.g., Wright et al. 1982, Newell and Field 1983a, b, Seiderer et al. 1984, Muir et al. 1986, Lucas et al. 1987, Mathews et al. 1989, Kemp et al. 1990, Langdon and Newell 1990, Crosby et al. 1990, Werner and Hollibaugh 1993, Newell and Krambeck 1995, Kreeger and Newell 1996). However, only one marsh suspension-feeder (*Geukensia demissa*) has been reported capable of removing these very small cells from suspension (Wright et al.

1982, Riisgård 1988, Langdon and Newell 1990) with an efficiency sufficient to provide substantial nutritional benefit (Kreeger and Newell 1996). Even though free-living bacteria may be quite nutritious and digestible by marsh consumers (see above), we conclude that their direct food value in supplying the bulk nutritional demands for most primary consumers in the marsh is likely to be low (Table 1).

2.2.4 Heterotrophic Protists

In the few studies of the food value of microheterotrophs to marsh consumers, generally only the bacteria have been considered (see above). Yet, measurements of the abundance of heterotrophic flagellates in kelp beds off South Africa, for example, indicate that this group of microheterotrophs can be sufficiently available to represent a significant carbon and nitrogen source for kelp bed consumers (Newell and Field 1983a,b, Linley and Newell 1984).

Although little studied, heterotrophic nanoflagellates and ciliates are increasingly being considered as important constituents of aquatic ecosystems due to their role in remineralization. In systems such as salt marshes where extremely high bacterial abundances are common, bacterivorous protists may also abound (Kemp et al. 1990, Huang 2000). Huang (2000) reported that the typical abundance of heterotrophic nanoflagellates in the water column covering mid-Atlantic marshes at high tide is about 10^3 ml^{-1}, and assuming these cells are similar in C content and nutritional value to autotrophic dinoflagellates (2.5×10^{-8} mg C $cell^{-1}$; D.A. Kreeger, unpublished data), then this would equate to a rather low carbon abundance of about 25 μg C L^{-1}. These microheterotrophic protists have cell diameters of between 3 to 10 μm diameter, a size range readily captured by suspension-feeding marsh consumers (Kemp et al. 1990, Kreeger and Newell 1996). They have also been shown to be readily digested and assimilated by the bivalve, *Geukensia demissa* (Kreeger and Newell, 1996). Therefore, for suspension-feeders, microheterotrophic protists might represent a valuable resource; however for other marsh consumers, low accessibility will restrict their food value.

2.2.5 Fungi

Newell and Porter (this volume) reports that the biomass and productivity of fungi utilizing dead standing marsh angiosperms in a Georgia salt marsh can be considerable. Fungal standing stocks are reported to vary between 3 to 28% of the senescent crop of cordgrass and range from 19 to 52 g biomass m^{-2}, depending on season, which is comparable to sediment bacteria biomass (Newell and Porter, this volume). They also report that rates of production by fungi can exceed that for other microbial decomposers, particularly in winter. These data suggest that fungi are an extremely important constituent of the marsh decomposer community, and at least a few groups of consumers such as gastropods and amphipods derive considerable nutrition from fungal secondary production (Newell and Porter, this volume). However, there are no quantitative data available on how readily metazoan consumers digest this potential resource.

3. Dominant Aquatic Invertebrates and Feeding Modes

Salt marshes support a wide variety of metazoan consumers, both meiofauna (50 to 500 μm) and macrofauna (>500 μm), that reach tremendous abundances. Although the species diversity appears high for certain groups such as insects (e.g., see Rey and McCoy 1997), by comparison to other ecosystems salt marshes have low biodiversity (Pfeiffer and Wiegert 1981). As with the vascular plant community, the community of invertebrate consumers in salt marshes can therefore be characterized as species-poor but biomass-rich (Montague et al. 1981). In this section, we summarize the types of invertebrate primary consumers that are of widespread importance in the ecology of salt marshes, and we provide species names only as examples.

As we will discuss later, many of these animals are omnivores; i.e., they balance their nutrition by using a variety of resources at the base of the food chain such as primary producers and/or constituents of the detritus complex. Larger organisms may feed partly on these materials, but also on meiofauna, and several of these generalists are adapted to feed by facultatively switching to take advantage of temporal or spatial shifts in resource availability. It is therefore difficult to catagorize marsh consumers into discrete feeding guilds. Nevertheless, we organize them here by feeding mode, and we include any consumer that feeds at least partly on producers or detritus—we omit discussion of carnivorous animals that are primarily secondary consumers.

3.1 HERBIVORES

As mentioned previously, only a few species of marsh invertebrates directly consume living plant material as their sole nutritional resource. These are primarily dipteran insects, (Hubbard 1997, Montague et al. 1981, for a more complete review of insect consumers in marshes see Pfeiffer and Wiegert 1981). For example, Smalley (1960) reported that in Georgia marshes two insects, the grasshopper *Orchelimun fidicinium* and the planthopper *Prokelisia marginata*, dominated the functional guild that grazed only on vascular plants. Although insect species diversity in the marsh can be substantial (Rey and McCoy 1997), the biomass of these herbivores is generally low and grazing on living vascular plants has been shown to consume less than 10% of overall plant production (Mann 1972, Heard 1982).

The only other invertebrates that are reported to directly graze on vascular plants are gastropod snails (Haines and Montague 1979) and wharf crabs (Pennings et al. 1998). This conclusion is based on the results of stable isotope studies and so may be equivocal if these species are mainly utilizing surface attached microalgae or fungi colonizing senescent plants. Observational studies have shown that gastropods, such *Melampus bidentatus*, *Littorina irrorata, Ilyanassa obsoleta*, and *Hydrobia ulvae* do feed on epiphytic and epibenthic algae (Wetzel 1976, Montague et al. 1981, Barnes 1989). A variety of reports suggest that these snails preferentially select and digest surface-associated algae, which often comprise the bulk of material in their guts (Pace et al. 1979, Lopez and Kofoed 1980, Conner and Edgar 1982, Lopez-Figueroa and Neill 1987). Recent work by Newell and Bärlocher (1993) and Newell and Porter (this volume) indicate that 3 to 28% of the biomass of senescent cordgrass is composed of fungi and that this fungal component is efficiently utilized by gastropods (Newell and Bärlocher 1993, Newell and Porter, this volume).

3.2 DEPOSIT-FEEDERS

Salt marshes are soft-bottomed habitats with high amounts of organic C in the sediments. This makes them suitable habitats for deposit-feeding invertebrates, which always comprise a large portion of the biomass of the invertebrate community. Deposit-feeders include all sediment-associated meiofauna and many of the macrofauna (Lopez and Levinton 1987). Selective deposit-feeders such as fiddler crabs, some polychaetes, and bivalves are capable of sorting sediments to increase the food value of the ingested ration. Others are non-selective and simply consume large amounts of sediment-associated material, deriving nutrition from parts of the organic material that can be digested and assimilated (Taghon and Jumars 1984, Penry and Jumars 1986), particularly microalgae and microheterotrophs.

Meiofaunal deposit-feeders include nematodes, harpacticoid copepods, amphipods, polychaetes, oligochaetes, turbellarians, ostracods, foraminiferans, gastrotichs, as well as small juveniles of important macrofauna (Coull and Bell 1979, Kruczynski and Ruth 1997). In most marshes, meiofaunal nematodes are the most abundant constituent (for review, see Kruczynski and Ruth 1997). Although spatially patchy in abundance and species composition (Teal and Weiser 1966, Nixon and Oviatt 1973, Bell et al. 1978), meiofauna generally average between 10^6 and 10^7 m^{-2}, and they typically contain 0.5 to 2 g biomass m^{-2} (Nixon and Oviatt 1973, Coull and Bell 1979). As a group, meiofauna can therefore be sufficiently abundant to play an important role in the secondary production of marshes.

Macroinvertebrate deposit-feeders include some of the most important consumers of the marsh ecosystem, such as fiddler crabs, snails, polychaetes, oligochaetes and certain bivalves. Teal (1962) described feeding by crabs, *Uca pugnax, Uca pugilator* and *Armases (=Sesarma) reticulatum* as moderately selective because the animals scoop up sediment, remove large inorganic particles, ingest the remainder, and assimilate about 25% of the ingested ration. Although the feeding mechanisms of other macroinvertebrate deposit feeders vary considerably, they all tend to have a similar strategy for balancing their nutrition; i.e., they process large quantities of sedimentary material but utilize a smaller fraction of that processed through their guts. Crabs can be abundant and are a major consumer of marsh production (Teal 1962, Daiber 1982, Bertness 1987). For example, Daiber (1982) reported the typical abundance of *U. pugnax* in Delaware marshes to range between 80 to 200 crabs m^{-2}. Besides these crabs, other important deposit-feeding invertebrates of USA salt marshes include snails such as *Littorina irrorata* and *Ilyanassa obsoleta* (Montague et al. 1981, Daiber, 1982, Levin et al. 1996), grass shrimp such as *Palaemonetes pugio* (Welsh 1975, Alon and Stancyk 1982), annelids such as *Streblospio* sp., *Capitella* sp., and *Scoloplos fragilis* (Teal 1962; Subrahmanyam et al. 1976, Montague et al. 1981, Levin et al. 1996), bivalves such as clams, *Polymesoda caroliniana*, *Macoma balthica*, and *Cyrenoidea floridana* (Leatham et al. 1976, Heard 1982, Montague et al. 1981, Kruczynski and Ruth 1997), and juvenile blue crabs, *Callinectes sapidus*, may also opportunistically deposit-feed in the marsh (Nixon and Oviatt 1973, Shirley et al. 1990).

3.3 SUSPENSION-FEEDERS

Suspension-feeding invertebrates are adapted to consume large quantities of seston, containing particulate organic material (POM) from a wide diversity of sources, and average POM concentrations range between 1 to 20 mg L^{-1} (Pomeroy and Imberger 1981, Kreeger 1986, Roman and Daiber 1989, Huang 2000). Seston contains phytoplankton, suspended surface-associated algae, bacteria, microheterotrophic protists, detritus, unidentifiable organic aggregates, as well as inorganic particles. As discussed previously, the nutritional value of these food items varies widely, and suspension-feeders possess a wide variety of pre- and post-ingestion sorting mechanisms to maximize exposure of digestive enzymes to suitable substrates (Langdon and Newell 1996).

Although pelagic suspension-feeders such as zooplankton are present in marsh water, and suspension-feeding meiofauna can be epibenthic on stems of *Spartina alterniflora* (Rutledge and Fleeger 1993), it is mainly the larger benthic suspension-feeders that process the bulk of marsh seston (Dame 1996). Important suspension-feeders include oligochaete annelids such as *Manayunkia* (Teal 1962). However, dominant suspension-feeders are often bivalve molluscs such as *Geukensia demissa* (Kuenzler 1961a, Lent 1967, Jordan and Valiela 1982). In the next section, we will discuss how this species has become well adapted to take advantage of the diverse array of foods in the marsh environment enabling it to become abundant and thrive, and we will then consider implications of omnivory for the construction of marsh food webs.

4. A Case Study of Omnivory: The Ribbed Mussel *Geukensia demissa*

Ribbed mussels are common macroconsumers in salt marshes of New England (Fell et al. 1982, Jordan and Valiela 1982, Bertness 1984, Peterson et al. 1985, Franz 1993), the mid-Atlantic (Daiber 1982, Kreeger et al. 1988), southeast (Kuenzler 1961a), and Gulf Coast (Kruczynski and Ruth 1997). In many of the marshes in these areas *Geukensia demissa* thrives, attaining a population biomass that can exceed that of all other marsh metazoans combined (Jordan and Valiela 1982). Indeed, *G. demissa* has been termed a ''keystone species'' due to its domination of animal biomass and rates of secondary production (Kuenzler, 1961a, Lent, 1967, Jordan and Valiela 1982). Ribbed mussels have been shown to be sufficiently abundant in some marshes to collectively filter, perhaps more than once, the entire volume of water overlying the marsh per tidal cycle (Jordan and Valiela 1982). Since much of the material removed from the water column ultimately is deposited as feces and pseudofeces, mussels are important agents in the retention of nutrients within the marsh (Kuenzler 1961b, Jordan and Valiela 1982, Asmus et al. 1995, Dame 1996), which may help fuel characteristically high rates of angiosperm primary production (Bertness 1984, Bertness and Grosholz 1985).

Suspension-feeding bivalves have traditionally been regarded as obtaining the majority of their nutrition from phytoplankton, and this is the case for most species that live in phytoplankton-rich areas such as rocky or sandy shores and estuaries (Bayne and Newell 1983, Dame 1996, Newell and Langdon 1996). Even within marshes, stable isotope data have confirmed that phytoplankton play a role in maintaining the positive carbon balance of bivalves, particularly at the seaward fringes and in southern marshes that have greater *in situ* phytoplankton production (Haines and Montague 1979, Peterson et al. 1985, 1986, Langdon and Newell 1990). However, phytoplankton comprise only a small portion of the organic fraction of the seston of salt marshes (Kreeger 1986, Langdon and Newell 1990, Galvao and Fritz 1991, Huang 2000), particularly in the benthic boundary layer where these bivalves feed (Huang, 2000). In temperate climates phytoplankton availability can vary by more than tenfold during the year (Van Valkenberg et al. 1978, Widdows et al. 1979, Soniat et al. 1984, Berg and Newell 1986, Galvao and Fritz 1991, Huang 2000). Our calculations (see below) suggest phytoplankton abundance is not sufficient to fully meet either the C or N demands of a mid-Atlantic population of *Geukensia demissa*.

The nutritional challenge presented by seston of low organic content within the salt marsh is exacerbated by restricted time available for feeding. Many species of mussels have evolved to live in the intertidal zone, believed to represent a refuge from predation in subtidal locations (Paine 1974, Bertness and Grosholz 1985, West and Williams 1986, Lin 1989, Seed and Suchanek 1992, Stiven and Gardner 1992). The same may be true for *Geukensia demissa* which is restricted to living in the mid- and high intertidal zone (Kuenzler 1961a), and as a consequence can be exposed to air for up to 70% of the time. Despite the severe restriction on feeding time associated with intertidal emersion, Gillmor (1982) demonstrated that among intertidal bivalves, this was the only species that grows better under intertidal conditions than subtidally. Although ribbed mussel productivity appears to be diminished as the elevation and distance from creek bank increases (Franz 1987), ribbed mussels appear to possess an array of sophisticated physiological adaptations to enable them to cope with restricted feeding time in the high intertidal zone. For example, ribbed mussels living high in the intertidal zone exhibit higher feeding rates (Charles and Newell 1997) and assimilate their food with greater efficiency (Kreeger et al. 1990) than those in the mid and low intertidal zone. These adaptations are still not sufficient to fully compensate for reduced time available for feeding, however, as Borrero (1987) reported that high intertidal mussels had both a lower level of gametogenic condition and gametogenesis was delayed compared to low intertidal mussels.

In the sections that follow, we synthesize the growing body of information on the trophic ecology of *Geukensia demissa*, and we then consider the implications of these findings for the general understanding of marsh ecology.

4.1 IMPORTANCE OF DIETARY PHYTOPLANKTON

Evidence for the nutritional importance of phytoplankton for *Geukensia demissa* is partly based on comparisons of the stable isotope signatures of mussel tissues and various primary producers. For example, Peterson et al. (1985) compared isotope ratios for C, N, and S between mussels and producers in a New England marsh. They reported a diet of up to 80% material derived from *Spartina alterniflora*; however, tissue isotopic signatures varied considerably along a seaward to landward gradient and these signatures reflected a greater dietary role of phytoplankton near the seaward margin Langdon and Newell (1990) reported that *G. demissa* living within a Delaware marsh had isotopic signatures reflecting a diet of 30 to 50% non-phytoplankton material. These results are in contrast to the C stable isotope data of Haines and Montague (1979) that indicated that mussels in Georgia derived C mainly from phytoplankton. Kemp et al. (1990) reported that phytoplankton contributed 72% of the C filtered by the mussel population in a Georgia marsh, indicating that phytoplankton C was of much greater importance to the mussel's nutrition than detrital C. However, because Kemp et al. (1990) studied mussels only during the summer, it is unknown whether phytoplankton are as prominent in the diet at other times of the year. These equivocal results could be interpreted as evidence for an increase in the nutritional importance of phytoplankton with decreasing latitude, but recent isotope studies in other southern marshes have shown that consumers might be using suspended benthic microalgae rather than phytoplankton (Sullivan and Moncreiff, 1990, Sullivan and Currin, this volume).

To discern trophic links, stable isotope ratios are best used in combination with direct studies of feeding and digestion processes by consumers (Montague et al. 1981, Newell et al. 1995). In our work summarized here, we have primarily relied on pulse-chase radiotracer techniques to learn more about the mechanisms by which mussels use different producer resources. For example, under simulated natural conditions of tidal exposure, temperature, and diet composition, we have delivered ^{14}C-labeled phytoplankton to freshly collected mussels in the laboratory, and measured mussel filtration rates, ingestion rates, and assimilation efficiencies by quantifying the fate of isotope in their C budget (Kreeger and Newell 1996). To study N balance of mussels, we have also ^{15}N-labeled experimental diets and determined N utilization from experimental diets with the same approach. Isotopically labeled diets are delivered as a small supplement to an otherwise unaltered natural seston diet to minimize shifts in feeding or digestive physiology of mussels. This approach has been repeated at different times of the year and for various dinoflagellate and diatom species characteristic of marsh phytoplankton (Kreeger and Newell 1996, unpublished data).

Results from these experiments have confirmed previous studies showing how readily phytoplankton are ingested and assimilated by bivalve suspension-feeders (see reviews by Webb and Chu 1982, Bayne and Newell 1983, Hawkins and Bayne 1992, Dame 1996, Langdon and Newell 1996). Phytoplankton are efficiently filtered from suspension *by Geukensia demissa* and algal C and N are typically assimilated with efficiencies between 30 and 75%, but the mussels' filtration and assimilation of C from phytoplankton can vary widely during the year (Kreeger and Newell 1996, D.A. Kreeger, R.I.E. Newell, and S. Huang, unpublished data). Seasonal variability in diet utilization may be due to physiological responses to seasonal shifts in nutritional demands and/or the relative abundance of different food resources. In a Delaware salt marsh, Huang (2000) measured seston concentrations of particulate chlorophyll-*a* and reported greater average concentrations in the spring (average during May, 16.4 $\mu g\ L^{-1}$) compared with low mean concentrations in the summer, fall and winter (6.0, 2.8 and 3.9 $\mu g\ L^{-1}$, respectively).

Although phytoplankton availability in this Delaware salt marsh was greatest in the spring, mussel filtration rates and assimilation efficiencies for phytoplankton lagged, being greatest in the summer and fall (D.A. Kreeger, R.I.E. Newell, and S. Huang, unpublished data). By comparing seasonal changes in phytoplankton availability, the mussel's ability to derive nutrition from phytoplankton, and the mussel's C and N demands, we are able to estimate the potential contribution that phytoplankton can provide to satisfy the C demand of a typical adult mussel residing in a Delaware marsh (Fig. 1). In spring, phytoplankton abundance was greatest, but mussel filtration rates for phytoplankton were lowest, and so only 36% of the mussel's C demands could be met by this resource (Fig. 1). Although mussels had higher assimilation rates for phytoplankton C in summer, due to reduced phytoplankton concentrations, we estimated that mussels could satisfy at most 83% of their C needs from phytoplankton during this season. In fall, phytoplankton availability was estimated to slightly exceed mussel C demands. By winter mussels again derived only limited C from phytoplankton (36 %) because phytoplankton concentrations were lower than at other times of the year, and mussels also had the lowest seasonal assimilation efficiency for this resource. This analysis clearly suggests that phytoplankton cannot possibly supply all of the C demands of the mussels in our Delaware study marsh.

4.2 USE OF DETRITAL CELLULOSE

Using the same experimental approach described above for assessing phytoplankton utilization by *Geukensia demissa*, we prepared ^{14}C-labeled, microparticulate cellulosic detritus from radiolabeled *Spartina alterniflora* and fed it to mussels to determine how efficiently mussels can ingest, digest and assimilate C from this resource (Kreeger et al. 1988, 1990). As with the phytoplankton studies, the radiolabeled cellulose was delivered to mussels as only a small portion of an otherwise natural seston diet and under simulated natural conditions in the laboratory. Mussels were found to efficiently filter our experimental diet (2 to 20 μm particle diameter), but mussels generally assimilated C

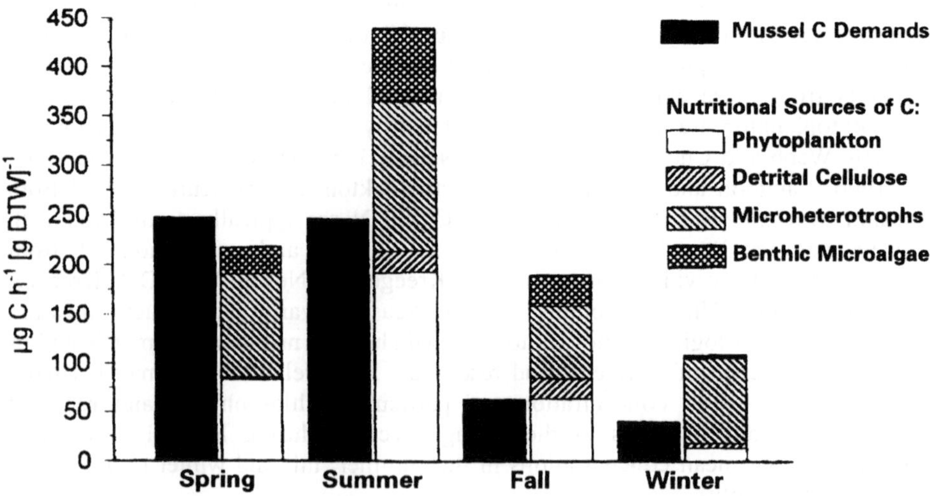

Figure 1. The carbon demand of adult *Geukensia demissa* compared to their potential carbon assimilation rates (units = μg C h^{-1} [g dry tissue weight]$^{-1}$) from phytoplankton, detrital cellulose, microheterotrophs and benthic microalgae during spring, summer, fall and winter, based on data collected in 1995-1996 for mussels in Canary Creek salt marsh, Delaware, USA (D.A. Kreeger, R.I.E. Newell and S. Huang, unpublished data).

from *Spartina* cellulose with low efficiencies of about 15% (Kreeger et al. 1988, 1990). Although this assimilation efficiency is lower than that found for other food types, it is actually higher than might be expected because no other bivalve suspension-feeder has been shown to assimilate unaged (i.e., not colonized by microorganisms) detrital cellulose with more than a few percent efficiency (e.g., <3% for oysters; see Crosby et al. 1989, Langdon and Newell 1990). Since our cellulosic diet was aseptic this suggests that mussels either possess a vigorous gut flora capable of cellulose breakdown or they possess endogenous cellulases.

Subsequent to these studies, we repeated this experiment during different seasons and under different dietary regimes (Charles and Newell 1997, R.I.E. Newell, F. Charles and D.A. Kreeger, unpublished data) and found higher (up to 23%) assimilation efficiencies at certain times of the year. But even considering these new data, we estimate that cellulose in marsh seston can only supply between <1 and 9% of the C demands of *G. demissa,* depending on time of the year (Fig. 1). Carbon contributions from phytoplankton and detrital cellulose (Fig. 1) combined cannot meet the needs of *Geukensia demissa* at any time of the year except fall.

4.3 MICROHETEROTROPHIC MEDIATION OF DETRITUS USE

In addition to direct utilization of cellulosic detritus, mussels may consume detritus derived from *Spartina alterniflora* indirectly by feeding on detritus-associated microheterotrophic intermediaries (Kreeger and Newell 1996). As discussed above, microheterotrophs common in marshes include bacteria associated with surfaces, free bacteria, and heterotrophic protists such as nanoflagellates. Considerable research has been shown that *Geukensia demissa* can remove all of these types of microheterotrophs from the water column (Kreeger et al. 1988, Kemp et al. 1990, Langdon and Newell 1990, Newell and Krambeck 1995, Kreeger and Newell 1996). Compared with other suspension-feeding bivalves, *G. demissa* has an exceptional ability for filtering single-celled bacteria <1 µm in diameter (Wright et al. 1982, Riisgård 1988, Langdon and Newell 1990). Ribbed mussels can remove these small bacteria at rates 33% of those for particles in their ideal size range (Langdon and Newell 1990). Bacteria attached to larger suspended detrital particles are filtered at the same rates as for reference diets such as phytoplankton. Kemp et al. (1990) and Kreeger and Newell (1996) demonstrated that *G. demissa* also readily filters microheterotrophic flagellates (3 to 10 µm) at rates similar to that for phytoplankton.

In contrast to the work on microheterotroph capture, information on microheterotroph digestion and food value has been lacking. Using the same pulse-chase experimental approach described above, we have studied the food value and availability of attached bacteria, free bacteria and heterotrophic protists as potential food sources for *G. demissa* (Kreeger 1986, Langdon and Newell 1990, Kreeger and Newell 1996). Kreeger and Newell (1996) demonstrated that *G. demissa* can digest and assimilate laboratory-cultured strains of bacteria and heterotrophic protists with efficiencies of 42 and 44%, respectively. Furthermore, we found that *G. demissa* fed radiolabeled assemblages of natural microheterotrophs consistently assimilated bacteria and protists with very high efficiency, ranging between 66 and 86% throughout the year. This is consistent with the recorded assimilation efficiency of 52%

for bacteria fed to oysters, *Crassostrea virginica* (Crosby et al. 1990) and 73% for bacteria fed to clams, *Potamocorbula amurensis* (Werner and Hollibaugh 1993). Not only can *G. demissa* use detrital cellulose better than other suspension-feeding bivalves, but it also uses detritus-associated microheterotrophs by both filtering and digesting them efficiently. These data bolster the suggestion that *G. demissa* is uniquely adapted to the food resources present in the high intertidal salt marsh (Langdon and Newell 1990).

Our results on microheterotroph utilization by mussels are contrasted with the annual carbon demands of *Geukensia demissa* in Fig. 1 for different times of the year. This analysis suggests that free bacteria, attached bacteria, and heterotrophic protists can each contribute substantially to the C demands of mussels throughout the year (Fig. 1). During spring and summer, about 50% of the mussel's C demands could be satisfied by bacteria and heterotrophic protists; whereas, in fall and winter these microorganisms could more than fulfill the mussels C requirements. Due to the C shortfall in spring and summer, the combination of free bacteria, attached bacteria, and heterotrophic protists cannot possibly provide all the needed C. At these times phytoplankton imported into the marsh from the adjoining Delaware Bay must supply the balance of the mussels C demands. Our results for a Delaware mussel population are consistent with the conclusions of Kirchman et al. (1984) and Kemp et al. (1990) who reported that microbial C is insufficient to meet the annual C demands of *G. demissa* in Massachusetts and Georgia marshes, respectively.

Since more than half the C demands of *Geukensia demissa* could be met by use of microheterotrophs, we suggest these microorganisms are an important intermediary in the transfer of detrital C derived from vascular plants to a dominant metazoan consumer. Mussels can, in turn, be eaten by a variety of secondary consumers, and they form a conduit of C transport from the microbial community to metazoan food webs. These data demonstrate an unappreciated trophic link in marshes that have high densities of *G. demissa*. Secondly, there may be significant implications for microbial community structure. By feeding on microheterotrophs, mussels may exert considerable top-down grazing pressure, particularly on larger bacterivorous protists which can be filtered more readily than free bacteria (Kemp et al. 1990, Kreeger and Newell 1996).

4.4 THE IMPORTANCE OF SURFACE-ASSOCIATED MICROALGAE

To examine whether *Geukensia demissa* can derive nutrition from benthic microalgae, we isolated numerous algal species from the surface of a Delaware marsh, developed methods for culturing them in the laboratory, radiolabeled them with ^{14}C, and fed them to mussels in pulse-chase feeding experiments as described above for phytoplankton. The different algal species consisted mainly of pennate diatoms having mean cell diameters between 15 and 30 μm, although we included one benthic cyanobacteria in the study. Although phytoplankton have traditionally been considered to be the best diet for suspension-feeding bivalves (Webb and Chu 1982, Bayne and Newell 1983, Hawkins and Bayne 1992, Langdon and Newell 1996), we found that every species of benthic diatom fed to *G. demissa* was filtered, digested and assimilated with efficiencies that equaled or surpassed that for a reference species of phytoplankton (*Isochrysis galbana* clone T-iso) (Webb and Chu 1982). Four of our eight isolates were assimilated with efficiencies between 90 to 95%, and even cyanobacteria were efficiently assimilated (87%).

But what is still uncertain is if these benthic microalgae are available in the water column for suspension-feeders. We now know that pelagic and benthic food webs are closely coupled and should not be treated separately (Threlkeld 1994, Dame 1996, this volume). Bottom currents generated by waves or tides can resuspend individual benthic microalgal cells and even disrupt benthic algal mats (Gallagher 1975, Colijn and Dijkema 1981, Shaffer and Cahoon 1987, de Jonge and van Beusekom 1995). Gallagher (1975) reported that rising tidal waters entrained benthic diatoms from the salt marsh surface, and MacIntyre and Cullen (1996) observed that water column chlorophyll-*a* concentrations were correlated with turbidity presumably due to resuspension of benthic microalgae (see also; Demers et al. 1987, Cloern et al. 1989, Powell et al. 1989, Litaker et al. 1993). MacIntyre and Cullen (1995) estimated that suspended benthic algal cells were responsible for about 25% of total water column primary production in a shallow-water site near Corpus Christi Bay, Texas, and Huang (2000) reported that benthic diatoms comprised 14% of the biovolume of suspended algae in the water column above a bed of *Geukensia demissa* during summer in a Delaware marsh.

Based on benthic microalgae availability (Huang 2000) and our measured efficiencies with which *Geukensia demissa* ingests and assimilates benthic algal biomass, we present the food value of this resource for *G. demissa* in a Delaware marsh (Fig. 1). We estimate that suspended benthic microalgae can contribute between 10% and 53% of the mussels C requirements, depending on time of year (Fig. 1). Even though estimated concentrations of suspended benthic microalgae were greatest in spring (54 μg C L^{-1}), mussels had high C demands and low filtration efficiencies for benthic microalgae at that time. This resulted in a 12% potential contribution of this resource to mussel C balance. The availability of benthic microalgae declined by summer, but they could supply 32% of the mussels C demands in summer because mussels processed that resource with greatest efficiency at that time. In fall, the C resource value of benthic microalgae provided 53% of the C demands of mussels due to high processing rates for benthic microalgae and low C demands by *G. demissa*. These results support the hypothesis that the microphytobenthos can contribute substantially to consumer food webs in salt marshes.

4.5 RELATIVE IMPORTANCE OF DIFFERENT CARBON SOURCES

No single component of marsh seston can supply sufficient C to support the requirements of adult *Geukensia demissa* in the Canary Creek marsh, Delaware, USA. Rather, this species has evolved a diversity of behavioral and physiological processes to derive energy from a wide array of particulate substrates, including phytoplankton, detrital cellulose from vascular plants, microheterotrophic bacteria and protists, and microphytobenthos (microalgae, cyanobacteria). Our calculations based on resource availability, relative food value for mussels, and mussel nutritional requirements indicate that together, these primary C resources can supply 93% of the spring C demands of mussels, and more than 100% at other times of the year (Fig. 1). Suspension-feeding bivalves are able to store energy (Bayne and Newell 1983,

Hawkins and Bayne 1992, Langdon and Newell 1996) and time-optimization of C balance could be important for overcoming a seasonal C deficiency (Hawkins and Bayne 1985). It appears that mussels can potentially assimilate more C than they require over a large part of the year. It is equally possible that during spring, *G. demissa* augments its C uptake by using other potential food resources in the marsh, such as aggregated organic flocs (Alber and Valiela 1994), organic matter in surface-films (D.M. Seliskar, W.L. Carey, and J. L. Gallagher, pers. commun.), dissolved organic material (for review, see Stephens 1985, Newell and Langdon 1996), and perhaps even fragments of the rich fungal community that enter the water column (e.g., see Newell and Porter, this volume). More research is needed to examine the nutritional role that these potential resources can provide for marsh consumers. Nonetheless, our findings clearly suggest that adult *G. demissa* living in our Delaware study marsh are not nutritionally limited by C.

The relative nutritional importance of phytoplankton, microheterotrophs, benthic microalgae, and detrital cellulose varies seasonally (Fig. 2). Microheterotrophs are estimated to be the most valuable C resource for mussels in all seasons except summer when phytoplankton represent the largest C resource (Fig. 2). These results are consistent with the findings of Kemp et al. (1990), who reported that phytoplankton appear to make their greatest contribution to the mussels nutrition in summer, supplying about 50% of the C needs of ribbed mussels in a Georgia marsh. The relative food value of benthic microalgae is sizeable in all seasons except winter, and could represent an essential food source during the spring C shortfall (i.e., when mussel C demands are greatest). Direct use of detrital cellulose can contribute to the mussels C demands throughout the year, but it is of limited value compared to that of the heterotrophic and autotrophic microbial community.

We caution that the data for most of our calculations were collected during the period from August 1995 to November 1996, and the relative importance of these different dietary resources to *Geukensia demissa* at different times of the year will likely vary substantially among years. In addition, our results are representative only for mussels living at mid-tide level on creek banks in the center of a characteristic salt marsh in Delaware. Mussel abundance and population size distribution can vary considerably within and among marshes (Fell et al. 1982, Bertness and Grosholtz 1985, Franz, 1993, 1996). Hence, further research is needed to examine whether the food value of different resources varies spatially along the low to high marsh gradient, the seaward to landward gradient, and perhaps also along a latitudinal gradient. Nevertheless, since *G. demissa* utilizes a wider array of particle types with greater efficiency than typical suspension-feeding bivalves, we conclude that this species is well equipped to tolerate extremes in the availability of any single resource. The adaptations for omnivory possessed by *G. demissa* likely contribute to the ability of this species to thrive in the intertidal zone of salt marshes.

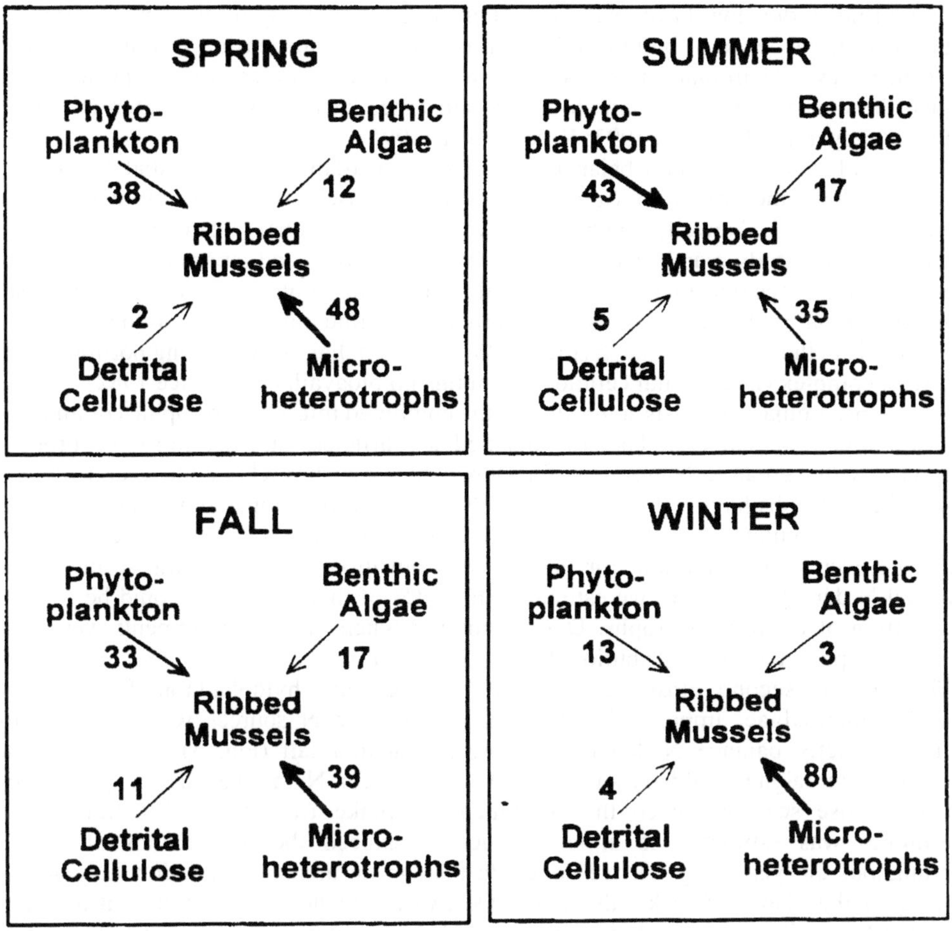

Figure 2. Relative food value of phytoplankton, benthic microalgae, detrital cellulose and microheterotrophs in balancing the carbon demands of *Geukensia demissa* during the spring, summer, fall and winter. Values given are percentages of the total carbon potentially assimilated by mussels during each season (D.A. Kreeger and R.I.E Newell, unpublished data).

4.6 SOURCES OF NITROGEN

Even though the C (=energy) demands of adult *Geukensia demissa* appear to be satisfied throughout most of the year (Fig. 1), it is possible that populations of *G. demissa* are resource limited by N rather than C. For example, in a mass balance study of kelp beds off South Africa, Newell and Field (1983a, b) reported that mussels were more N-limited than C-limited. Nitrogen is certainly a valuable nutrient in salt marshes, as evidenced by the rapid recycling of N that typically characterizes marsh biogeochemistry (Valiela and Teal 1979). In marshes having high densities of *G. demissa*, a significant proportion of the N pool can cycle through the mussel population (Jordan and Valiela 1982), and as yet, we know of no studies that have examined whether the availability of N or other essential nutrients might exert ''bottom-up'' control of these suspension-feeding animals. Organism-level N limitation has been shown to occur in other mussel species even when the diet primarily contains low C:N substrates such as phytoplankton (Kreeger 1993; Kreeger and Langdon 1993; Kreeger et al. 1995).

We calculated the mass balance between the availability of potential N sources (Huang 2000), the ability of *Geukensia demissa* to use different N resources (D.A. Kreeger and R.I.E. Newell, unpublished data), and the seasonal N demands of adult *G. demissa* in our study marsh (Huang, 2000). With these data and reported C:N ratios for the different seston materials, we calculated the potential contribution that phytoplankton, microheterotrophs and microphytobenthic algae supply to the N budget of mussels in the Delaware marsh (Fig. 3). We assumed that the non-living portion of the detritus complex (i.e., detrital cellulose) contains no bioavailable N (Rice 1982).

Our preliminary analysis suggests that the combined microheterotrophic community of water column bacteria, bacteria attached to particles, and heterotrophic protists represents the greatest N resource for ribbed mussels throughout the year (Fig. 3). Microheterotrophs have a C:N ratio of 3.5 (Fenchel 1982), compared with an algal C:N ratio of 6.0 (calculated from Montagnes et al. 1994), and so the N from microheterotrophs may be even more important for the mussel's nutrition than C. During spring when mussel N demands peaked, we estimated that 64% of this N demand could be met by utilization of microheterotrophs. During summer when the N requirements were still high, the potential N contribution from microheterotrophs rose to 98% of demand (Fig. 3). The second most important N resource was phytoplankton, followed by benthic microalgae. Importantly, the sum of all three N sources was estimated to almost exactly balance N demands during spring and fall (Fig. 3), indicating that benthic microalgae could be an essential resource for N (as well as C; see above) during those seasons. Since the combined N uptake from these different N-rich resources was estimated to either balance or exceed the N demands of ribbed throughout the year (Fig. 3), we conclude that on an annual basis adult *G. demissa* living in the Canary Creek salt marsh, Delaware, are not likely to be nutritionally limited by available N (or dietary protein). However, during spring and fall when N resources exactly balanced N demands, the productivity of *G. demissa* populations could be constrained by N availability. If the peak in N demands in the spring results from a high protein requirement associated with gametogenesis, these mussels could be N limited as has been found for other mussel species (Kreeger 1993). Further research is required to discern whether natural populations of *G. demissa* are N-limited.

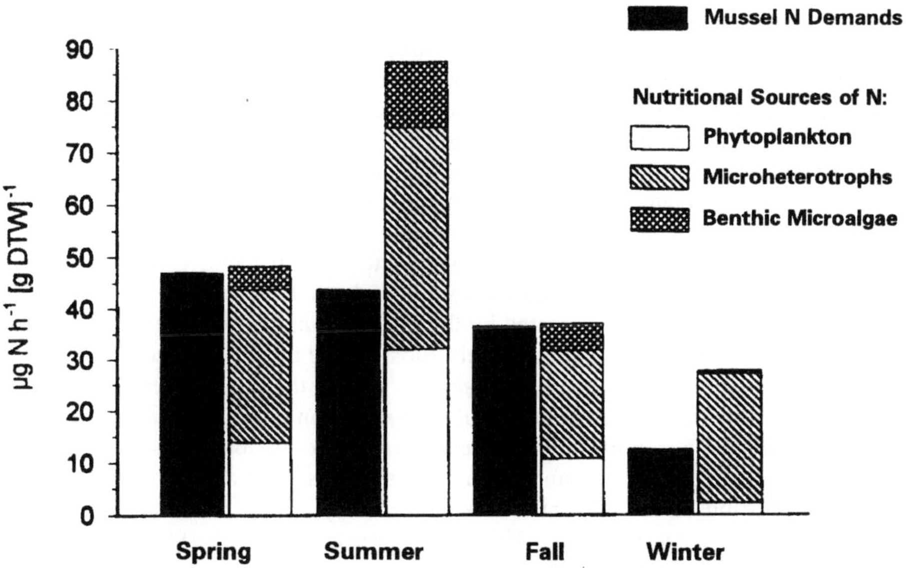

Figure 3. The nitrogen demand of adult *Geukensia demissa* compared to their potential nitrogen assimilation rates (units = $\mu g\ N\ h^{-1}$ [g dry tissue weight]$^{-1}$) from phytoplankton, microheterotrophs and benthic microalgae during spring, summer, fall and winter, based on data collected in 1995-1996 for mussels in Canary Creek salt marsh, Delaware, USA (D.A. Kreeger and R.I.E. Newell, unpublished data).

5. Implications for Marsh Food Webs

Although phytoplankton has been traditionally considered to be the main food resource exploited by suspension feeding bivalves, ribbed mussels may derive a greater proportion of their C and N requirements by the combined use of benthic microalgae, detritus, and microheterotrophs. High rates of secondary production associated with populations of *Geukensia demissa* appear to result from this animal's ability to derive nutrition from non-phytoplankton resources. From an organismal perspective, adaptations for omnivory have permitted this species to penetrate deep into salt marshes where phytoplankton comprise only a small portion of the organic component of the seston (Huang 2000). Phytoplankton production also varies seasonally, but because ribbed mussels are omnivorous, they can continue to balance their nutrition by utilizing these other food resources when phytoplankton become less abundant. The relative contribution of these different food resources to the mussel's carbon requirements during spring in a mid-Atlantic marsh was estimated (D.A. Kreeger and R.I.E. Newell, unpublished data) to be microheterotrophs > phytoplankton >> microphytobenthic algae > > detrital cellulose (Fig. 2). Indeed, phytoplankton constituted the greatest C resource for mussels only during summer; whereas, in winter it represented <13% of the mussels C demands. Our calculations suggest that rather than phytoplankton, microheterotrophs represent the most important C resource for *G. demissa* in a Delaware marsh on an annual basis (Fig. 2).

For the past 30 years, scientists have questioned whether the microbial loop is closed or whether some production from microheterotrophs flows to metazoan consumers (Darnell 1967, Pomeroy 1974). However, little quantitative evidence has been gathered to support the hypothesis that a significant microheterotroph → metazoan pathway exists, and what is known has come mainly from studies in freshwater plankton communities (for review, see Porter, 1996). Deposit-feeding consumers clearly have access to microheterotrophs associated with sediments (Lopez and Levinton 1987) and suspension-feeders have access to microheterotrophs in the water column (Wright et al. 1982, Riisgård 1988, Kemp et al. 1990, Langdon and Newell 1990, Gall et al. 1997). Kreeger and Newell (1996) also demonstrated that both bacteria and heterotrophic nanoflagellates can be efficiently digested by the mussel, *Geukensia demissa*. However, our present analysis is the first to quantitatively substantiate that microheterotrophs may play a significant role in the nutrition of *G. demissa*. Due to the high processing rates for water and seston measured in our studied mussel population (Fig. 2), *G. demissa* might consume a substantial portion of secondary production associated with microheterotrophs in the water column. If this is true, mussels may serve as important elements in structuring the aquatic microbial community in salt marshes having high densities of *G. demissa* (Kemp et al. 1990, Langdon and Newell 1990, Newell and Krambeck 1995, Kreeger and Newell 1996).

The second implication of our analysis is that *Geukensia demissa* is capable of deriving a portion (3 to 17%) of its nutrition by feeding on autochthonous production associated with benthic microalgae (Figs. 1, 3). Although not very abundant in the water column, benthic microalgae are filtered, digested and assimilated more efficiently by *G. demissa* than any other dietary substrate, including phytoplankton. Mobile consumers such as shrimp, snails, and crabs (e.g., the grass shrimp *Palaemonetes pugio*; R.I.E. Newell and B. Bebout, unpublished data) have greater access to microphytobenthic production than sessile suspension-feeders such as the mussel. The idea that benthic microalgae are an important resource for marsh consumers has been around a long time (e.g., see Teal 1962). However, recent evidence suggests that a wide variety of important marsh consumers may rely mainly on benthic microalgae for their nutrition. This supports the "hidden garden" hypothesis (Sullivan and Moncreiff 1990, MacIntyre et al. 1996, Miller et al. 1996) that although rates of benthic algal production are high, grazing by marsh consumers keeps benthic algal standing stocks at a low level even though grazing-associated activity may even facilitate benthic algal production. Hence, we question the paradigm that salt marshes have "detritus-based food webs" (Odum, 1980), considering that the bulk of secondary production by metazoans could actually be linked to primary production by the microphytobenthos rather than through either direct (herbivory) or indirect (detritivory) linkages to primary production by vascular plants.

6. Trophic Complexity and Marsh Community Stability

MacArthur (1955) suggested that community stability is increased whenever each consumer in a low species diversity habitat has a generalized diet. Pimm (1982, 1984) further suggested that communities with low species diversity *must* have a large degree of trophic connectance to remain stable. Relationships between biodiversity, food web structure and community stability continue to be debated, in part, because theoretical treatments suggested that community stability should decrease when consumers have broad diets (e.g., May, 1981). Some ecologists have questioned whether generalist feeders exist widely (e.g., Pimm and Lawton 1978), although increasing evidence demonstrates that omnivores are much more widespread than previously thought (see review by Persson et al. 1996). Experimental tests of these concepts have been lacking since there is basic disagreement among ecologists over what constitutes a food web. Simple conceptual models of marsh food webs show only the presence of trophic linkages between major groups that structure a community. Detailed models depict not only trophic linkages, but also strengths of interactions (e.g., usually with quantitative information on energy flows), and negative and indirect associations (Hastings 1988, Paine 1988).

Nevertheless, it seems appropriate to examine ecological theory in the context of the intertidal zone of the salt marsh because marsh communities are stable when not subjected to major physical perturbations, are well-documented as having low species diversity (Wiegert and Pomeroy 1981), and have functionally dominant invertebrates as omnivores; e.g., ribbed mussels (see above), crabs (Teal 1962, Montague 1980, Bertness 1987), grass shrimp (Welsh 1975), and snails (Curtis and Hurd 1981, Montague et al. 1981, Daiber 1982). Thus, MacArthur (1955) and Pimm (1982) predicted salt marshes would have a high degree of trophic connectance between generalist consumers. Teal (1962) was the first to recognize that salt marshes closely adhere to MacArthur's predictions. Omnivory should promote population stability in low diversity habitats because dominant consumer species are not as prone to periodic oscillations in resource availability; whereas, significant reductions in the food base of specialist feeders would have greater impact. Since the major functional guilds of salt marsh consumers are dominated by omnivores, the entire community is robust to temporal and spatial shifts in resource availability. Theoretical ecology predicts that omnivory will be widespread in ecosystems having abundant resources not significantly affected by consumption (Pimm 1982). This was substantiated by Menge et al. (1996) for a rocky intertidal community. Salt marshes are C-rich environments dominated by vascular plant primary production. If the marsh food web is not detritus-dominated, but fueled by a mix of microautotrophs (microphytobenthos and phytoplankton) microheterotrophs (bacteria and flagellates), then does this prediction of a low ratio of secondary production:primary production still hold for marshes? It does, because direct herbivory of vascular plants accounts for <10% of primary production. How might the microautotroph and microheterotroph → consumer link contribute to community stability if a high proportion of C in these two groups is converted to secondary production? Perhaps the *high* biodiversity typically associated with the microphytobenthos community helps preserve stability by providing a consistent source of production for marsh consumers, no matter the conditions. For example, microphytobenthos are highly diverse with more than 60 species cm^{-2} (Sullivan, 1975); consequently rates of primary production by this community will be robust to the impacts of species-specific stresses. Collectively, these producers would only be affected by major environmental factors such as nutrient and light limitation.

7. Synthesis

Salt marshes are characterized as low species diversity ecosystems having high rates of both primary and secondary production. Most autochthonous marsh production is attributed to vascular plants such as *Spartina alterniflora*, but few organisms are able to directly consume this material. For example, direct utilization of detrital cellulose from vascular plants was estimated to account for <11% of the annual C demand and virtually none of the N demand for a dominant marsh consumer, *Geukensia demissa* (Fig. 2). Because of such low direct utilization of cellulosic detritus there must be other explanations of why C stable isotope ratios of tissues of *G. demissa* reflect a diet comprising 50 to 80% *Spartina*-derived material (Peterson et al. 1985, Langdon and Newell 1990).

Our analysis suggests that *Geukensia demissa* utilizes production derived from vascular plants via microheterotrophic intermediaries. By consuming and digesting microheterotrophs, *G. demissa* may exert some control on the microbial community and represent a trophic link between the microbial loop and higher trophic levels. A complementary explanation is that stable isotope ratios of vascular plants, and microphytobenthos are not sufficiently discrete, even when multiple isotopes are compared, to unequivocally distinguish between these as food sources (see above); i.e, we suggest that the isotopic ratios interpreted previously as resembling *Spartina* might in fact also resemble microphytobenthos.

Only recently has it become widely recognized that microphytobenthos are an important food source for numerous marsh consumers (Sullivan and Moncreiff 1990, MacIntyre et al. 1996). One reason for this oversight is the relatively low biomass of microphytobenthos in salt marshes. This low biomass is not due to low productivity but rather reflects the heavy grazing pressure by primary consumers, a phenomenon termed the ''secret garden'' (MacIntyre et al. 1996, Miller et al. 1996).

We estimate that although mussels can derive little of their nutrition from detrital cellulose, water column microheterotrophs and suspended microphytobenthos appear to provide a large portion of their C and N requirements. The functional importance and prevalence of such omnivory in natural food webs has been overlooked, mainly because of a lack of direct dietary information (Pimm et al. 1991). Although suspension-feeding bivalves are traditionally portrayed as feeding only on phytoplankton, results from our research and that by others over the past ten years clearly demonstrate that *Geukensia demissa* is an omnivore that derives nutrition from a diverse array of non-phytoplankton materials that comprise marsh seston. In many marshes, particularly in the mid-Atlantic region, these mussels can be the keystone consumer and critical agents in key functional processes such as biogeochemical cycling and pelagic-benthic coupling. Hence, models of marsh food webs should be revised accordingly to consider additional trophic linkages between mussels the microheterotrophic community as well as autochthonous marsh producers, rather than simply allochthonous production by phytoplankton. High levels of trophic complexity between primary producers and consumers, such as that shown by *G. demissa* and its different foods, is important for maintaining community stability in the low-diversity, high-productivity salt marsh.

8. Future Directions

Do salt marshes have ''detritus-based food webs?'' To address this question we need to obtain more quantitative information on the strengths of the trophic connections between dominant marsh consumers and producers such as the microphytobenthos. Although stable isotope ratios have been valuable in discerning the relative trophic importance of autochthonous versus allochthonous production in the diet of marsh consumers, this approach can be hampered by poor resolution when used to differentiate trophic links between consumers and different autochthonous producers. Stable isotope ratios also provide little information on whether the trophic source is linked to consumers directly or indirectly. Stable isotope ratio techniques will have greatest power when combined with experimental techniques that can directly trace the path and amounts of resource movement through the food web.

Quantitative data is also needed on the food value of microheterotrophs for most of the important marsh consumers, particularly deposit-feeders. Research is also warranted to examine the relative nutritional importance of fungi, epiphytic algae, surface films, and dissolved organic material for metazoan consumers in the marsh.

In addition to obtaining more qualitative data on the complexity of the marsh food web and more quantitative data on the strength of individual trophic interactions, we also suggest that the role of N should be examined in the context of structuring the marsh community. Bottom-up limitation of secondary production in marshes could result more from N than C (=energy) availability. If so, the feeding behavior of individual consumers might be dictated more by the N content of the food rather than the energy content. If this were true, N might be more important than C in structuring the consumer food web of salt marshes, and trophic models of marshes might do better to use N as a currency rather than energy.

9. Acknowledgments

We thank Shou-Chung Huang for generous access to his data on seston composition and his comments on this manuscript. We are also grateful to two anonymous reviewers who helped improve this text. This work was supported by National Science Foundation Grant No. OCE-9314584 to R.I.E. Newell, D.K. Stoecker and D.A. Kreeger.

10. Literature Cited

Alber, M. and I. Valiela. 1994. Assimilation of organic aggregates by marine mussels. Marine Biology **121**:259-265.

Alon, N.C. and S.E. Stancyk. 1982. Variation in life-history patterns of the grass shrimp *Palaemonetes pugio*, in two South Carolina estuarine systems. Marine Biology **68**:265-276.

Asmus, H., R.N. Asmus and G. Francés Zubillaga. 1995. Do mussel beds intensify the phosphorus exchange between sediment and tidal waters? Ophelia **41**:37-55.

Barnes, R.S.K. 1989. Marine animals of Blakeney Point and Scolt Head Island. Pages 67-75 *in* H. Allison and J.P. Morley, editors. Blakeney Point and Scolt Head Island. National Trust, Norfolk, England.

Bayne, B.L. and R.C. Newell. 1983. Physiological energetics of marine molluscs. Pages 407-515 *in* A.S.M. Saleuddin and K.M Willow, editors. The mollusca. Volume 4, Part 1. Academic Press, New York, New York, USA.

Bell, S.S., M.C. Watzin and B.C. Coull. 1978. Biogenic structure and its effect on the spatial heterogeneity of the meiofauna of a salt marsh. Journal of Experimental Marine Biology and Ecology **35**:99-107.

Benner, R., A.E. MacCubbin and R.E. Hodson. 1984. Preparation, characterization and microbial degradation of specifically radiolabeled [^{14}C] lignocellulose from marine and freshwater macrophytes. Applied Environmental Microbiology **47**:381-389.

Berg, J.A. and R.I.E. Newell. 1986. Temporal and spatial variations in the composition of seston available to the suspension-feeder *Crassostrea virginica*. Estuarine, Coastal and Shelf Science **23**: 375-386.

Bertness, M.D. 1984. Ribbed mussels and *Spartina alterniflora* production in a New England salt marsh. Ecology **65**:1794-1807.

Bertness, M.D. 1987. Fiddler crab regulation of *Spartina alterniflora* production on a New England salt marsh. Ecology **66**:1042-1055.

Bertness, M.D. and E. Grosholz. 1985. Population dynamics of the ribbed mussel, *Geukensia demissa*: the costs and benefits of an aggregated distribution. Oecologia **67**:192-204.

Borrero, F.J. 1987. Tidal height and gametogenesis: reproductive variation among populations of *Geukensia demissa*. Biological Bulletin **173**:160-168.

Cahoon, L.B., G.R. Beretich, Jr., C.J. Thomas and A.M. McDonald. 1993. Benthic microalgal production at Stellwagen Bank, Massachusetts Bay, USA. Marine Ecology Progress Series **102**:179-185.

Cahoon, L.B. and J.E. Cooke. 1992. Benthic microalgal production in Onslow Bay, North Carolina, USA. Marine Ecology Progress Series **84**:185-196.

Charles, F. and R.I.E. Newell. 1997. Digestive physiology of the ribbed mussel *Geukensia demissa* (Dillwyn) held at different tidal heights. Journal of Experimental Marine Biology and Ecology **209**:201-213.

Christian, R.R., R.B. Hansen, J.R. Hall and W.J. Wiebe. 1981. Aerobic microbes and meiofauna. Pages 69-85 *in* L.R. Pomeroy and R.G. Wiegert, editors. The ecology of a salt marsh. Springer-Verlag, New York, New York, USA.

Cloern, J.E., T.M. Powell and L.M. Huzzey. 1989. Spatial and temporal variability in South San Francisco Bay (USA). II. Temporal changes in salinity, resuspended sediments and phytoplankton biomass and productivity over tidal time scales. Estuarine, Coastal and Shelf Science **28**:599-613.

Coffin, R.B., B. Fry, B.J. Peterson and R.T. Wright. 1989. Carbon isotopic compositions of estuarine bacteria. Limnology and Oceanography **34**:1305-1310.

Colijn, F. and K. A. Dijkema. 1981. Species composition of benthic diatoms and distribution of chlorophyll a on an intertidal flat in the Dutch Wadden Sea. Marine Ecology Progress Series **4**:9-21.

Conner, M. S. and R. K. Edgar. 1982. Selective grazing by the mud snail *Ilyanassa obsoleta*. Oecologia **53**:271-275.

Coull, B.C. and S.S. Bell. 1979. Perspectives of marine meiofaunal ecology. Pages 189-216 *in* R.J. Livingston, editor. Ecological processes in coastal and marine systems, Volume 10. Plenum Press, New York, New York, USA.

Crosby, M.P., C.J. Langdon and R.I.E Newell. 1989. The importance of refractory plant material to the carbon budget of the oyster, *Crassostrea virginica* (Gmelin). Marine Biology **100**:343-352.

Crosby, M.P., R.I.E. Newell and C.J. Langdon. 1990. Bacterial mediation in the utilization of carbon and nitrogen from detrital complexes by the oyster, *Crassostrea virginica* (Gmelin). Limnology and Oceanography **35**:625-639.

Currin, C.A., S.Y. Newell and H.W. Paerl. 1995. The role of standing *Spartina alterniflora* and benthic microalgae in salt marsh food webs: considerations based on multiple stable isotope analysis. Marine Ecology Progress Series **121**:99-116.

Curtis, L.A. and L.E. Hurd. 1981. Nutrient procurement strategy of a deposit-feeding estuarine neogastropod, *Ilyanassa obsoleta*. Estuarine, Coastal and Shelf Science **13**:277-285.

Daiber, F. C. 1982. Animals of the tidal marsh. Van Nostrand Reinhold Co., New York, New York, USA.

Dame, R. F. 1996. Ecology of marine bivalves: an ecosystem approach. CRC Press, New York, New York, USA.

Darnell, R. M. 1967. Organic detritus in relation to the estuarine ecosystem. Pages 376-382 *in* G.H. Lauff, editor. Estuaries. AAAS Publ. 83, Washington.

de Jonge, V.N. and J.E.E. van Beusekom. 1995. Wind- and tide-induced resuspension of sediment and microphytobenthos from tidal flats in the Ems estuary. Limnology and Oceanography **40**: 766-778.
Demers, S., J.C. Therriault, E. Bourget and A. Bah. 1987. Resuspension in the shallow sublittoral zone of a macrotidal estuarine environment: wind influence. Limnology and Oceanography **32**: 327-339.
Fell, P.E., N.C. Olmstead, E. Carlson, W. Jacob, D. Hitchcock and G. Silber. 1982. Distribution and abundance of macroinvertebrates on certain Connecticut tidal marshes, with emphasis on dominant molluscs. Estuaries **5**: 234-239.
Fenchel, T. 1972. Aspects of decomposer food chains in marine benthos. Verhandlungen. Deutsche Zoologische Gesellschaft **65**:14-22.
Fenchel, T. 1982. Ecology of heterotrophic microflagellates. I. Some important forms and their functional morphology. Marine Ecology Progress Series **8**:211-223.
Franz, D.R. 1993. Allometry of shell and body weight in relation to shore level in the intertidal bivalve *Geukensia demissa* (Bivalvia: Mytilidae). Journal of Experimental Marine Biology and Ecology **174**:193-207.
Franz, D.R. 1996. Size and age at first reproduction of the ribbed mussel *Geukensia demissa* in relation to shore level in a New York salt marsh. Journal of Experimental Marine Biology and Ecology **205**:1-13.
Franz, D.R. 1997. Resource allocation in the intertidal salt-marsh mussel *Geukensia demissa* in relation to shore level. Estuaries **20**:134-148.
Gallagher, J.L. 1975. The significance of the surface film in salt marsh plankton metabolism. Limnology and Oceanography **29**:120-123.
Gallagher, J.L. and F.C. Daiber. 1974. Primary production of edaphic algal communities in a Delaware salt marsh. Limnology and Oceanography **19**:390-395.
Gallagher, J. L., W. J. Pfeiffer and L. R. Pomeroy. 1976. Leaching and microbial utilization of dissolved organic carbon from leaves of *Spartina alterniflora.* Estuarine, Coastal and Shelf Science **4**:467-471.
Galvao, H.M. and A.T. Fritz. 1991. Microbial trophodynamics in a salt-marsh. Marine Microbial Food Webs **5**:13-26.
Gillmor, R. B. 1982. Assessment of intertidal growth and capacity adaptations in suspension-feeding bivalves. Marine Biology **68**:277-286.
Haines, E. B. 1979. Interactions between Georgia salt marshes and coastal waters: a changing paradigm. Pages 35-46 *in* R.J. Livingston, editor. Ecological processes in coastal and marine systems, Volume 10. Plenum Press, New York, New York, USA.
Haines, E.B. and R.B. Hanson. 1979. Experimental degradation of detritus made from the salt marsh *plants Spartina alterniflora*, *Salicornia virginica* and *Juncus roemerianus*. Journal of Experimental Marine Biology and Ecology **40**: 27-40.
Haines, E.B. and C.L. Montague. 1979. Food sources of estuarine invertebrates analyzed using $^{13}C/^{12}C$ ratios. Ecology **60**:48-56.
Hastings, A. 1988. Food web theory and stability. Ecology **69**: 1665-1668.
Hawkins, A.J.S. and B.L. Bayne. 1985. Seasonal variation in the relative utilization of carbon and nitrogen by the mussel, *Mytilus edulis*: budgets, conversion efficiencies and maintenance requirements. Marine Ecology Progress Series **25**:181-188.
——. 1992. Physiological interrelations and the regulation of production. Pages 171-222 *in* E. Gosling, editor. The Mussel *mytilus*: ecology, physiology, genetics and culture. Elsevier Science Publishers, New York, New York, USA.
Heard, R.W. 1982. Guide to common tidal marsh invertebrates of the Northwestern Gulf of Mexico. MS-AL Sea Grant Consortium Publication No. MASGP-79-004.82 Ocean Springs, Mississippi, USA.
Heydemann, B. 1981. Ecology of the arthropods of the lower salt marsh. Pages 35-57 *in* W.K.R.E. Wingerden and W.J. Wolff, editors. Terrestrial and freshwater fauna of the Wadden sea area, Balkema, Rotterdam, The Netherlands.
Huang, S. C. 2000. A comparison of temporal and spatial variation in nutritional resources and the seasonal nutritional demands of the marsh mussel, *Geukensia demissa*. Dissertation, University of Maryland, College Park, Maryland, USA.
Hubbard, M.D. 1997. Aquatic insects of northern Florida's Gulf Coast tidal marshes. Pages 339-345 *in* C.L. Coultas and Y. Hsieh, editors. Ecology and management of tidal marshes: a model from the Gulf of Mexico. St. Lucie Press, Delray Beach, Florida, USA.
Jordan, T.E. and I. Valiela. 1982. A nitrogen budget of the ribbed mussel, *Geukensia demissa* and its significance in nitrogen flow in a New England salt marsh. Limnology and Oceanography **27**:75-90

Kemp, P.F., S.Y. Newell and C. Krambeck. 1990. Effects of filter-feeding by the ribbed mussel *Geukensia demissa* on the water-column microbiota of a *Spartina alterniflora* saltmarsh. Marine Ecology Progress Series **59**:119-131.

Kiel, R.G. and D.L. Kirchman. 1991. Dissolved combined amino acids in marine waters as determined by a vapor-phase hydrolysis method. Marine Chemistry **33**:243-259.

Kirchman, D., B. Peterson and D. Juers. 1984. Bacterial growth and tidal variation in bacterial abundance in the Great Sippewissett Salt Marsh. Marine Ecology Progress Series **19**:247-259.

Kraeuter, J.N. and P.L. Wolf. 1974. The relationship of marine macroinvertebrates to salt marsh plants. Pages 449-462 *in* R. J. Reimold and W.H. Queen, editors. Ecology of halophytes. Academic Press, New York, New York, USA.

Kreeger, D.A. 1986. Utilization of refractory cellulosic material derived from *Spartina alterniflora* by the ribbed mussel *Geukensia demissa*. Thesis, University of Delaware, Lewes, Delaware, USA.

Kreeger, D.A. 1993. Seasonal patterns in the utilization of dietary protein by the mussel, *Mytilus trossulus*. Marine Ecology Progress Series **95**:215-232.

Kreeger, D. A., A.J.S. Hawkins and B.L. Bayne. 1996. Use of dual-labeled microcapsules to discern the anabolic and catabolic fate of dietary protein in suspension-feeders. Limnology and Oceanography **41**:208-215.

Kreeger, D.A., A.J.S. Hawkins, B.L. Bayne and D.L. Lowe. 1995. Seasonal variation in the relative utilization of dietary carbohydrate, protein-carbon and protein-nitrogen by *Mytilus edulis* (L.) Marine Ecology Progress Series **126**:177-184.

Kreeger, D.A. and C.J. Langdon. 1993. Effect of dietary protein content on growth of juvenile mussels, *Mytilus trossulus* (Gould 1850). Biological Bulletin **185**:123-139.

Kreeger, D.A., C.J. Langdon and R.I.E. Newell. 1988. Utilization of refractory cellulosic carbon derived from *Spartina alterniflora* by the ribbed mussel *Geukensia demissa*. Marine Ecology Progress Series **42**:171-179.

Kreeger, D.A. and R.I.E. Newell. 1996. Ingestion and assimilation of carbon from cellulolytic bacteria and heterotrophic flagellates by the mussels *Geukensia demissa* and *Mytilus edulis* (Bivalvia, Mollusca). Aquatic Microbial Ecology **11**:205-214.

Kreeger, D.A., R.I.E. Newell and C.J. Langdon. 1990. Effect of tidal exposure on utilization of dietary lignocellulose by the ribbed mussel *Geukensia demissa* (Dillwyn) (Mollusca:Bivalvia). Journal of Experimental Marine Biology and Ecology **144**:85-100.

Kruczynski, W.L. and B.F. Ruth. 1997. Fishes and invertebrates. Pages 131-173 in C.L. Coultas and Y. Hsieh, editors. Ecology and management of tidal marshes: a model from the Gulf of Mexico. St. Lucie Press, Delray Beach, Florida, USA.

Kuenzler, E. J. 1961a. Structure and energy flow of a mussel population in a Georgia salt marsh. Limnology and Oceanography **6**:191-204.

——. 1961b. Phosphorus budget of a mussel population. Limnology and Oceanography **6**:400-415.

Langdon, C.J. and R.I.E. Newell. 1990. Utilization of detritus and bacteria as food sources by two bivalve suspension-feeders, the oyster *Crassostrea virginica* and the mussel Geukensia demissa. Marine Ecology Progress Series **58**:299-310.

——. 1996. Digestion and nutrition of larvae and adults. Pages 231-270 *in* V.S. Kennedy, R.I.E. Newell and A. Eble, editors. The eastern oyster, *Crassostrea virginica*. Maryland Sea Grant Publication, College Park, Maryland, USA.

Le Gall, S., M.B. Hassen and P. Le Gall. 1997. Ingestion of a bacterivorous ciliate by the oyster *Crassostrea gigas*: protozoa as a trophic link between picoplankton and benthic suspension-feeders. Marine Ecology Progress Series **152**:301-306.

Leatham, W., P. Kinner and D. Maurer. 1976. Northern range extension of the Florida marsh clam *Cyrenoidea floridana* (Superfamily Cyrenoidacea). Nautilus **90**:93.

Lent, C.M. 1967. Effects and adaptive significance of air-gaping by the ribbed mussel, *Modiolus* (*Arcuatula*) *demissus* (Dillwyn). Dissertation, University of.Delaware, Newark, Delaware, USA.

——. 1969. Adaptations of the ribbed mussel, *Modiolis demissus* (Dillwyn), to the intertidal habitat. American Zoologist **9**:283-292.

Levin, L.A., D. Talley and G. Thayer. 1996. Succession of macrobenthos in a created salt marsh. Marine Ecology Progress Series **141**:67-82.

Lin, J. 1989. Influence of location in the marsh on survivorship of ribbed mussels. Marine Ecology Progress Series **56**:105-110.

Linley, E.A.S. and R.C. Newell. 1984. Estimates of bacterial growth yields based on plant detritus. Bulletin of Marine Science **35**:409-425.

Litaker, W., C.S. Duke, B.E. Kenney and J. Ramus. 1993. Short-term variability and phytoplankton abundance in a shallow tidal estuary. II. Spring and fall. Marine Ecology Progress Series **94**:141-154.
Lopez, G.R. and L.H. Kofoed. 1980. Epipsammic browsing and deposit feeding in mud snails (Hydrobiidae). Journal of Marine Research **38**:585-599.
Lopez, G.R. and J.S. Levinton. 1987. Ecology of deposit feeding animals in marine sediments. Quarterly Review Biology **62**:235-260.
Lopez-Figueroa, F. and F.X. Niell. 1987. Feeding behaviour of Hydrobia ulvae (Pennant) in microcosms. Journal of Experimental Marine Biology and Ecology **114**:153-167.
Lucas, M.I., R.C. Newell, S.E. Shumway, L.J. Seiderer and R. Bally. 1987. Particle clearance and yield in relation to bacterioplankton and suspended particulate availability in estuarine and open coast populations of the mussel *Mytilus edulis*. Marine Ecology Progress Series **36**:215-224.
MacArthur, R.H. 1955. Fluctuations of animal populations and a measure of community stability. Ecology **36**:533-536.
MacCubbin, A.E. and R.E. Hodson. 1980. Mineralization of detrital lignocelluloses by salt marsh sediment microflora. Applied Environmental Microbiology **40**:735-740.
MacIntyre, H.L. and J.J. Cullen. 1995. Fine-scale vertical resolution of chlorophyll and photosynthetic parameters in shallow-water benthos. Marine Ecology Progress Series **122**:227-237.
MacIntyre, H.L. and J.J. Cullen. 1996. Primary production by suspended and benthic microalgae in a turbid estuary: time scales of variability in San Antonio Bay, Texas. Marine Ecology Progress Series **145**:245-268.
MacIntyre, H.L., R.J. Geider and D.C. Miller. 1996. Microphytobenthos: The ecological role of the ''secret garden'' of unvegetated, shallow-water marine habitats. I. Distribution, abundance and primary production. Estuaries. **19**:186-201.
Mann, K,H. 1972. Macrophyte production and detritus food chains in coastal waters. Memorie Istituto Italiano di Idrobiologia 29 Suppl:353-383.
Marples, T.G. 1966. A radionuclide study of arthropod food chains in a *Spartina* salt marsh estuary. Ecology **47**: 270-277.
Matthews, S., M.I. Lucas, J.M.E Stenton-Dozey and A.C. Brown. 1989. Clearance and yield of bacterioplankton and particulates for two suspension-feeding infaunal bivalves, *Donax serra* Roding and *Mactra lilacea* Lam. Journal of Experimental Marine Biology and Ecology **125**:219-234.
May, R.M. 1981. Patterns in multi-species communities. Pages 197-227 *in* R. M. May, editor. Theoretical ecology. Blackwell, Oxford, England.
Menge, B.A., B. Daley and P.A. Wheeler. 1996. Control of interaction strength in marine benthic communities. Pages 258-274 *in* G.A. Polis and K.O. Winemiller, editors. Food webs: integration of patterns and dynamics. Chapman and Hall, New York, New York, USA.
Miller, D.C., R.J. Geider and H.L. MacIntyre. 1996. Microphytobenthos: the ecological role of the ''secret garden'' of unvegetated, shallow-water marine habitats. II. Role in sediment stability and shallow-water food webs. Estuaries **19**:202-212.
Montagnes, D.J.S., J.A. Berges, P.J. Harrison and F.J.R. Taylor. 1994. Estimating carbon, nitrogen, protein and chlorophyll a from volume in marine phytoplankton. Limnology and Oceanography **39**:1044-1060.
Montague, C.L. 1980. A natural history of the temperate western Atlantic fiddler crabs (*Genus Uca*) with reference to their impact on the salt marsh. Contributions in Marine Science **23**:25-55.
Montague, C.L., S.M. Bunker, E.B. Haines, M.L. Pace and R.L. Wetzel. 1981. Aquatic macroconsumers. Pages 69-85 *in* L.R. Pomeroy and R.G. Wiegert, editors. The Ecology of a salt marsh. Springer-Verlag, New York, New York, USA.
Muir, D.G., L.J. Seiderer, C.L. Davis, S.J. Painting and F.T. Robb. 1986. Filtration, lysis and absorption of bacteria by mussels *Choromytilus meridionalis* collected under upwelling and downwelling conditions. South African Journal of Marine Science **4**:169-179.
Newell, R.C. 1965. The role of detritus in the nutrition of two marine deposit-feeders, the prosobranch, *Hydrobia ulvae* and the bivalve, *Macoma balthica*. Proceedings of the Zoological Society of London **144**:25-45.
Newell, R.C. and J.G. Field 1983a. The contribution of bacteria and detritus to carbon and nitrogen flow in a benthic community. Marine Biology Letters **4**:23-36.
—— 1983b. Relative flux of carbon and nitrogen in a kelp-dominated ecosystem. Marine Biology Letters **4**:259-257.

Newell, R.I.E., N. Marshall, A. Sasekumar and V.C. Chong. 1995. Relative importance of benthic microalgae, phytoplankton and mangroves as sources of nutrition for penaeid prawns and other coastal invertebrates from Malaysia. Marine Biology **123**:595-606.

Newell, R.I.E. and C.J. Langdon. 1996. Mechanisms and Physiology of Larval and Adult Feeding. Pages 185-230 *in* V.S. Kennedy, R.I.E. Newell and A. Eble, editors. The Eastern oyster, *Crassostrea virginica.* Maryland Sea Grant Publication, College Park, Maryland, USA.

Newell, S.Y. and F. Bärlocher. 1993. Removal of fungal and total organic matter from decaying cordgrass leaves by shredder snails. Journal of Experimental Marine Biology and Ecology **171**:39-49.

Newell, S.Y., R. D. Fallon, R.M. Cal Rodriguez and L.G. Groene. 1985. Influence of rain, tidal wetting and relative humidity on release of carbon dioxide by standing-dead salt-marsh plants. Oecologia **68**:73-79.

Newell, S.Y. and C. Krambeck. 1995. Responses of bacterioplankton to tidal inundations of a saltmarsh in a flume and adjacent mussel enclosures. Journal of Experimental Marine Biology and Ecology **190**:79-95.

Nixon, S.W. and C.A. Oviatt. 1973. Ecology of a New England salt marsh. Ecological Monographs **43**:463-498.

Odum, E.P. 1980. The status of three ecosystem-level hypotheses regarding salt marsh estuaries: tidal subsidy, outwelling and detritus-based food chains. Pages 485-495 *in* V.S. Kennedy, editor. Estuarine perspectives, Academic Press, New York.

Pace, M.L., S. Shimmel and W.M. Darley. 1979. The effect of grazing by a gastropod, *Nassaius obsoletus*, on the benthic microbial community of a salt marsh mudflat. Estuarine and Coastal Marine Science **9**:121-134.

Paine, R.T. 1974. Intertidal community structure: experimental studies on the relationship between a dominant competitor and its principal predator. Oecologia **15**:93-120.

Paine, R.T. 1988. Food webs: road maps to interactions or grist for theoretical developing problems Ecology **90**:1648-1654.

Pennings, S.C., T.H. Carefoot, E.L. Siska, M.E. Chase and T.A. Page. 1998. Feeding preferences of a generalist salt-marsh crab: relative importance of multiple plant traits. Ecology **79**: 1968-1979.

Penry, D.L. and P.A. Jumars. 1986. Chemical reactor analysis and optimal digestion. BioScience **36**:310-315.

Persson, L., J. Bengtsson, B.A. Menge and M.E. Power. 1996. Productivity and consumer regulation—concepts, patterns and mechanisms. Pages 396-434 *in* G.A. Polis and K.O. Winemiller, editors. Food webs: integration of patterns and dynamics. Chapman and Hall, New York, New York, USA.

Peterson, B.J. and R.W. Howarth. 1987. Sulfur, carbon and nitrogen isotopes used to trace organic matter flow in the salt-marsh estuaries of Sapelo Island. Limnology and Oceanography **32**:1195-1213.

Peterson, B.J., R.W. Howarth and R.H. Garritt. 1985. Multiple stable isotopes used to trace the flow of organic matter in estuarine food webs. Science **227**:1361-1363.

Peterson, B.J., R.W. Howarth and R.H. Garritt. 1986. Sulfur and carbon isotopes as tracers of salt-marsh organic matter flow. Ecology **67**:865-874.

Pfeiffer, W.J. and R.G. Wiegert. 1981. Grazers on Spartina and their predators. Pages 87-112 *in* L.R. Pomeroy and R.G. Wiegert, editors. The ecology of a salt marsh. Springer-Verlag, New York, New York, USA.

Pimm, S.L. 1982. Food webs. Chapman and Hall, New York, New York, USA.

Pimm, S.L. 1984. The complexity and stability of ecosystems. Nature **307**:321-326.

Pimm, S.L. and J.H. Lawton. 1978. On feeding on more than one trophic level. Nature **275**:542-544.

Pimm, S.L., J.H. Lawton and E.G. Cohen. 1991. Food web patterns and their consequences. Nature **350**:669-674.

Pinckney J., R. Papa and R. Zingmark. 1994. Comparison of high-performance liquid chromatographic, spectrophotometric and fluorometric methods for determining chlorophyll a concentrations in estuarine sediments. Journal of Microbiological Methods **19**:59-66.

Pinckney, J. and R.G. Zingmark. 1993. Photophysiological responses of intertidal benthic microalgal communities to *in situ* light environments: methodological considerations. Limnology and Oceanography **38**:1373-1383.

Pomeroy, L.R. 1959. Algal productivity in salt marshes of Georgia. Limnology and Oceanography **4**:386-397.

—— 1974. The ocean's food web, a changing paradigm. Bioscience **24**:499-504.

Pomeroy, L.R. and J. Imberger. 1981. The physical and chemical environment. Pages 21-36 *in* L.R. Pomeroy and R.G. Wiegert, editors. The ecology of a salt marsh. Springer-Verlag, New York, New York, USA.

Pomeroy, L.R., W.M. Darley, E.L. Dunn, J.L. Gallagher, E.B. Haines and D.M. Whitney. 1981. Primary production. Pages 39-67 *in* L.R. Pomeroy and R.G. Wiegert, editors. The ecology of a salt marsh. Springer-Verlag, New York, New York, USA.

Porter, K.G. 1996. Integrating the microbial loop and the classic food chain into a realistic planktonic food web. Pages 51-59 *in* G.A. Polis and K.O. Winemiller, editors. Food webs: integration of patterns and dynamics. Chapman and Hall, New York, New York, USA.

Powell, T.M., J.E. Cloern and L.M. Huzzey. 1989. Spatial and temporal variability in South San Francisco Bay (USA). I. Horizontal distributions of salinity, suspended sediments and phytoplankton biomass and productivity. Estuarine and Coastal Shelf Science **28**:583-597.

Rey, J.R. and E.D. McCoy. 1997. Terrestrial arthropods. Pages 175-208 *in* C.L. Coultas and Y. Hsieh, editors. Ecology and Management of Tidal Marshes: A Model from the Gulf of Mexico. St. Lucie Press, Delray Beach, Florida, USA.

Rice, D.L. 1982. The detritus nitrogen problem: new observations and perspectives from organic geochemistry. Marine Ecology Progress Series **9**:153-162.

Riisgård, H.U. 1988. Efficiency of particle retention and filtration rate in 6 species of northeast American bivalves. Marine Ecology Progress Series **45**:217-223.

Roman, C.T. and F.C. Daiber. 1989. Organic carbon flux through a Delaware Bay salt marsh: tidal exchange, particle size distribution and storms. Marine Ecology Progress Series **54**:149-156.

Rutledge, P.A. and J.W. Fleeger. 1993. Abundance and seasonality of meiofauna, including harpacticoid copepod species, associated with stems of the salt-marsh cord grass, *Spartina alterniflora*. Estuaries **16**:760-768.

Seed, R. and T.H. Suchanek. 1992. Population and community ecology of *Mytilus*. Pages 87-169 *in* E. Gosling, editor. The mussel *Mytilus*: ecology, physiology, genetics and culture. Elsevier Science Publishers, New York, New York, USA.

Seiderer, L.J., C.L. Davis, F.T. Robb and R.C. Newell. 1984. Utilization of bacteria as nitrogen resource by kelp-bed mussel *Choromytilus meridionalis*. Marine Ecology Progress Series **15**:109-116.

Shaffer, G.P. and P. Cahoon. 1987. Extracting information from ecological data containing high spatial and temporal variability: benthic microfloral production. International Journal of Genetics and Systematics **13**:107-123.

Shirley, M.A., A.H. Hines and T.G. Wolcott. 1990. Adaptive significance of habitat selection by molting adult blue crabs *Callinectes sapidus* (Rathburn) within a subestuary of central Chesapeake Bay. Journal of Experimental Marine Biology and Ecology **140**:107-119.

Smalley, A.E. 1959. The role of two invertebrate populations, *Littorina irrorata* and *Orchelimum fidicinium,* in the energy flow of a salt marsh ecosystem. Dissertation, University of Georgia, Athens, Georgia, USA.

Smalley, A.E. 1960. Energy flow of a salt marsh grasshopper population. Ecology **41**:672-677.

Soniat, T.M., S.M. Ray and L.M. Jeffrey. 1984. Components of the seston and possible available food for oysters in Galveston Bay, Texas. Contributions in Marine Science **27**:127-141.

Stephens, G.C. 1985. Uptake of amino acids by marine invertebrates. Pages 280-291 *in* R. Gilles and M. Gilles-Bailien, editors. comparative physiology and biochemistry: current topics and trends. Volume B: Transport, Iono and Osmoregulation. Springer, Berlin, Germany.

Stiven, A.E. and S.A. Gardner. 1992. Population processes in the ribbed mussel, *Geukensia demissa* (Dillwyn), in a North Carolina salt marsh tidal gradient: Spatial pattern, predation, growth and mortality. Journal of Experimental Marine Biology and Ecology **160**:81-102.

Subrahmanyam, C.B., W.L. Kruczynski and S.H. Drake. 1976. Studies on the animal communities of two north Florida salt marshes. Part II. Macroinvertebrate communities. Bulletin of Marine Science **26**:172-195.

Sullivan, M.J. 1975. Diatom communities from a Delaware salt marsh. Journal of Phycology **11**:384-390.

Sullivan, M.J. and C.A. Moncreiff. 1988. Primary production of edaphic algal communities in a Mississippi salt marsh. Journal of Phycology **24**:49-58.

——. 1990. Edaphic algae are an important component of salt marsh food-webs: evidence from multiple stable isotope analyses. Marine Ecology Progress Series **62**:149-159.

Taghon, G.L. and P.A. Jumars. 1984. Variable ingestion rate and its role in optimal foraging behavior of marine deposit-feeders. Ecology **65**:549-558.

Teal, J.M. 1962. Energy flow in the salt marsh ecosystem of Georgia. Ecology **43**:614-624.

Teal, J.M. and W. Wieser. 1966. The distribution and ecology of nematodes in a Georgia salt marsh. Limnology and Oceanography **6**:217-222.

Tenore, K.R. 1975. Detritus utilization by the polychaete, *Capitella capitata*. Journal of Marine Research 33:261-274.

Tenore, K.R. 1977. Growth of the polychaete, *Capitella capitata* cultured in different levels of detritus derived from various sources. Limnology and Oceanography **22**:936-941.

Tenore, K.R. and R.B. Hanson. 1980. Availability of detritus of different types and ages to a polychaete macroconsumer, *Capitella capitata*. Limnology and Oceanography **25**:553-558.

Threlkeld, S.T. 1994. Benthic-pelagic interactions in shallow water columns: an experimentalist's perspective. Hydrobiologia **275/276**: 293-300.

Valiela, I. and J.M. Teal. 1979. The nitrogen budget of a salt marsh ecosystem. Nature **280**:652-656.

Valiela, I., J.M. Teal, S.D. Allen, R. Van Etten, D. Goehringer and S. Volkmann. 1985. Decomposition in salt marsh ecosystems: the phases and major factors affecting the disappearance of above-ground organic matter. Journal of Experimental Marine Biology and Ecology **89**:29-54.

Van Valkenberg, S.D., J.K. Jones and D.R. Heinle. 1978. A comparison by size class and volume of detritus versus phytoplankton in Chesapeake Bay. Estuarine, Coastal and Shelf Science **6**:569-582.

Webb, K.L. and F. L. Chu. 1982. Phytoplankton as a food source for bivalve larvae. Pages 272-291 *in* G.D. Pruder, C.J. Langdon and D.E. Conklin, editors. Second international conference on aquaculture nutrition: biochemical and physiological approaches to shellfish nutrition. Louisiana State University, Baton Rouge, Louisian, USA.

Welsh, B.L. 1975. The role of the grass shrimp, *Palaemonetes pugio*, in the Galveston Bay estuarine system. Contributions in Marine Science **12**:54-79.

Werner, I. and J.T. Hollibaugh. 1993. *Potamocorbula amurensis*: comparison of clearance rates and assimilation efficiencies for phytoplankton and bacterioplankton. Limnology and Oceanography **38**:949-964.

West, D.L. and A.H. Williams. 1986. Predation by *Callinectes sapidus* (Rathbun) within *Spartina alterniflora* (Loisel) marshes. Journal of Experimental Marine Biology and Ecology **100**:75-95.

Wetzel, R.L. 1976. Carbon resources of a benthic salt marsh invertebrate, *Nassarius obsoletus* Say (Mollusca: Nassariidae). Pages 293-308 *in* Estuarine Processes, Volume 2. Academic Press, New York, New York, USA.

Widdows, J., P. Fieth and C.M. Worrall. 1979. Relationships between seston, available food and feeding activity in the common mussel, *Mytilus edulis*. Marine Biology **50**:195-207.

Wiegert, R.G. and L.R. Pomeroy. 1981. The salt marsh ecosystem: a synthesis. Pages 219-230 *in* L.R. Pomeroy and R.G. Wiegert, editors. The Ecology of a Salt Marsh. Springer-Verlag, New York, New York, USA.

Williams, R.B. 1962. The ecology of diatom populations in a Georgia salt marsh. Dissertation, Harvard University, Cambridge, Massachusetts, USA

Williams, P. 1981. Detritus utilization by *Mytilus edulis*. Estuarine, Coastal and Shelf Science **12**:739-746.

Wilson, J.O., R. Buchsbaum, I. Valiela and T. Swain. 1986. Decomposition in salt marsh ecosystems: phenolic dynamics during decay of litter of *Spartina alterniflora*. Marine Ecology Progress Series **29**:177-187.

Wright, R.T., R.B. Coffin, C.P. Ersing and D. Pearson. 1982. Field and laboratory measurements of bivalve filtration of natural marine bacterioplankton. Limnology and Oceanography **27**:91-98.

Zedler, J.B. 1980. Algal mat productivity: comparisons in a salt marsh. Estuaries **3**:122-131.

ZoBell, C.E. and C.B. Feltham. 1938. Bacteria as food for certain marine invertebrates. Journal of Marine Research **1**:312-327

TROPHIC LINKAGES IN MARSHES: ONTOGENETIC CHANGES IN DIET FOR YOUNG-OF-THE-YEAR MUMMICHOG, FUNDULUS HETEROCLITUS

KELLY J. SMITH
GARY L. TAGHON
Institute of Marine and Coastal Sciences
Rutgers, the State University of New Jersey
71 Dudley Rd., New Brunswick, NJ 08901-8521 USA

KENNETH W. ABLE
Marine Field Station
800 Great Bay Blvd., c/o 132 Great Bay Blvd.
Tuckerton, NJ 08087-2004 USA

Abstract

Transfer of salt marsh production in the form of detritus to surrounding coastal and estuarine areas has been an area of interest for a number of years. The common mummichog, *Fundulus heteroclitus* has been proposed as an important link in trophic transfer, however little is known about the role of small young-of-the-year (YOY) in this process. To address this lack of information, YOY were collected from the flooded salt marsh surfaces at a lower Delaware Bay site over 2 summers in order to determine ontogenetic shifts in food habits. Young-of-the-year *F. heteroclitus* were abundant on the marsh surface, and during peak larval recruitment to the marsh, fish densities reached 15 to 30 fish M^{-2}. Multiple cohorts were visible in 1997, indicating up to 4 separate spawning events. Stomach contents for YOY were dominated by detrital-sediment aggregates, harpacticoid copepods, and annelid worms. The food habits changed with size, with the stomach contents of the smallest individuals (6.6 mm < 20.4 mm SL) composed primarily of benthic fauna such as harpacticoid copepods and annelid worms. The stomach contents of larger YOY (20.5 -30.4 mrn SL) shifted to mostly *Spartina* detritus and sediment aggregate. Thus, YOY *F. heteroclitus* consume detritus both directly and indirectly (through consumption of detritivore benthic fauna). Prior research has indicated that the YOY are the dominant fish species and life history stage on the marsh surface in both abundance and biomass, with high measures of secondary production. Thus it is likely that the YOY make a substantial contribution to trophic transfer of production from the marsh surface to the surrounding estuary.

1. Introduction

Most of the primary production of salt marsh macrophytes such as *Spartina is* not grazed directly by marine animals, but enters the coastal estuarine food web via the detrital pathway. The magnitude of the secondary production of salt marshes derived from this detritus, versus other sources of organic matter remains unresolved (Nixon 1980, Boesch and Turner 1984), but assessing this proportion is critical for understanding whether salt marshes function only as structural habitat for upper trophic-level animals like fishes, or supply a substantial part of their diet, either directly or indirectly. Trophic relationships in salt marshes have been examined for a number of geographic areas (Harrington and Harrington 1961, Teal 1962, Nixon and Oviatt 1973, Bell 1980, Nixon 1980, Boesch and Turner 1984, Kneib 1986, Allen et al. 1995, Page 1997). The role of the resident nekton in the actual transfer of marsh production depends upon the amount of energy obtained from the intertidal marsh surface versus subtidal marsh habitats. The concept of ''trophic relay,'' i.e., transfer of marsh surface biological production by resident nekton to the subtidal estuarine habitat, has been recently introduced (Kneib 1997a), and the common mummichog, *Fundulus heteroclitus is* thought to play an important role in this process.

Fundulus heteroclitus is the dominant fish species in salt marshes along the east coast of North America (Bigelow and Schroeder 1953). The flooded marsh surface is an important foraging habitat for this species (Butner and Brattstrorn 1960, Weisberg et al. 1981), and adult *F. heteroclitus* show decreased growth rates when denied access to the intertidal portions of the marsh (Weisberg and Lotrich 1982). In general, *F. heteroclitus* are omnivorous, feeding on small crustaceans, nematodes, polychaetes, insect larvae, algae and detritus (Schmelz 1964, Nixon and Oviatt 1973, Fritz 1974, Baker-Dittus 1978, Kneib and Stiven 1978, Kneib 1986, Allen et al. 1994). Large juvenile and adult *F. heteroclitus* feed opportunistically and show rapid growth rates on a carnivorous diet (Weisberg and Lotrich 1982). When fed a diet of primarily detritus, adult *F. heteroclitus* can not meet basic metabolic demands and lose weight (Prinslow et al. 1974).

Although the trophic relationships of *F. heteroclitus* in salt marshes have been extensively studied, its role is still incompletely understood, in part, because most prior research focused on larger juveniles and adults (but see Kneib 1986). As a result, the trophic significance of this species, either as predator or prey, and its significance as a trophic link in salt marshes needs further study. This is especially true for the young-of-the-year (YOY) fish which have the highest measures of fish production (Meredith and Lotrich 1979) and are the dominant life history stage on the marsh surface Jaylor et al. 1979, Talbot and Able 1984, Kneib 1986, Talbot et al. 1986, Smith 1995). In contrast to the omnivorous diet of larger juveniles and adults, prior research indicates that juveniles (TL < 30 mm) are carnivorous (Kneib and Stiven 1978, Kneib 1986), feeding primarily on small crustaceans. In the laboratory, newly hatched larval *F. heteroclitus* can consume over 100 *Artemia nauplii* daily, and have a conversion efficiency of approximately 35% (Kneib and Parker 1991). Since YOY have a higher conversion efficiency than adults (Prinslow et al. 1974, Kneib and Parker 1991), are the dominant life history stage, and make the greatest contribution to annual production, it is important to understand the trophic role of YOY *F. heteroclitus.*

In this study, we collected YOY *Fundulus heteroclitus* from a salt marsh in lower Delaware Bay over two summers using a quantitative throw trap sampler. We examined diet in terms of dominant food items by biomass (dry weight) and frequency of occurrence in order to describe the feeding habits of these infrequently studied early life history stages, and provide a detailed examination of ontogenetic changes in feeding habits.

2. Materials and Methods

2.1 STUDY SITE

The study area was located in lower Delaware Bay (Fig. 1), along Moore's Beach Road, in Maurice River Township, Cumberland County. This area was formerly a salt hay farm, restored to tidal flow in 1972 and is now composed of extensive stands of tall-form and short-form *Spartina alterniflora,* with patches of *S. patens* at higher elevations. Normal tidal range is about 1.5 m, and during high tide sampling events, the marsh was typically inundated for about 2 h. Salinity and temperature ranged between 19 and 21 ppt, and 21 to 27°C respectively during the summer sampling periods for both years.

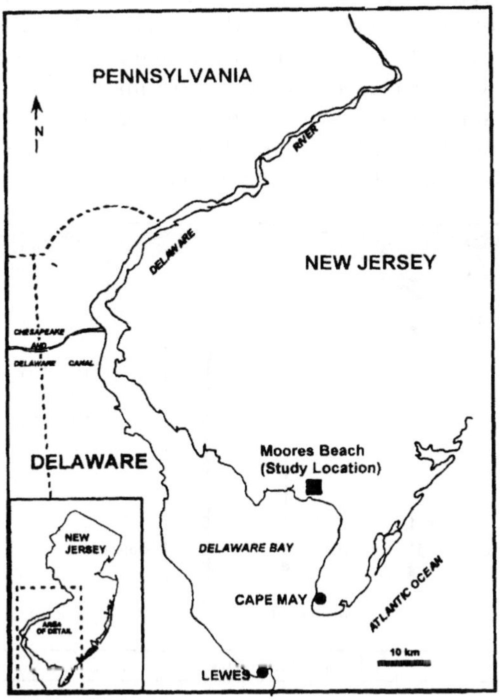

Figure 1. Study area was located in the lower Moores Beach salt marshes in southern New Jersey adjacent to Delaware Bay. Latitude=39°11.54 N, Longitude=74°57.03W.

2.2 FISH COLLECTIONS

Sampling was conducted on flooding tide (once the marsh surface was inundated by at least 5 cm of water), slack high, and ebbing tide (until water receded to below 5 cm depth). Fishes were collected using 0.16 m^2 x 0.45 cm ht cylindrical throw traps. The rough microtopography of the marsh surface precluded using larger samplers, which did not seal adequately along the bottom, and thus allowed fish to escape. Once deployed, the throw traps were cleared with fine-mesh (0.35 mm) dip nets, and sweeps of the net were continued until 5 consecutive sweeps did not collect fish. Dense emergent vegetation was avoided since the sampler would not seal properly, and when vegetation was included in the sampler it was thinned or removed to ensure collection of fishes.

During 1996, three parallel transects were sampled concurrently for each sampling date by three teams, each team following the leading edge of the tide (Fig. 2). Initial samples were collected near an intertidal creek, and as water depth increased, samples were taken at higher elevations on the marsh surface. Water depths sampled ranged from 5 cm to 19 cm, with a median depth of about 13 cm. Distance covered during a transect depended upon the depth and duration of the flooding tide, and typically ranged between 60 and 75 in. During 1996, throw trap sampling was conducted on June 6, June 26, and July 24, and all samples were analyzed for stomach contents. During 1997, a single transect was sampled approximately every two weeks, from late April until late August. Samples collected in 1997 were used to estimate density of fish on the marsh surface, however samples from May 22 and July 7 were deleted because flooding tides did not move far enough onto the marsh surface to allow sufficient replicate sampling. Fishes collected in the field were immediately preserved in a 4% buffered formaldehyde solution in seawater, and transferred to 70% ethanol after one week. All fish were identified based on pigmentation to separate larval and juvenile *F. heteroclitus* from *F. luciae,* and other killifishes (Able and Fahay 1998).

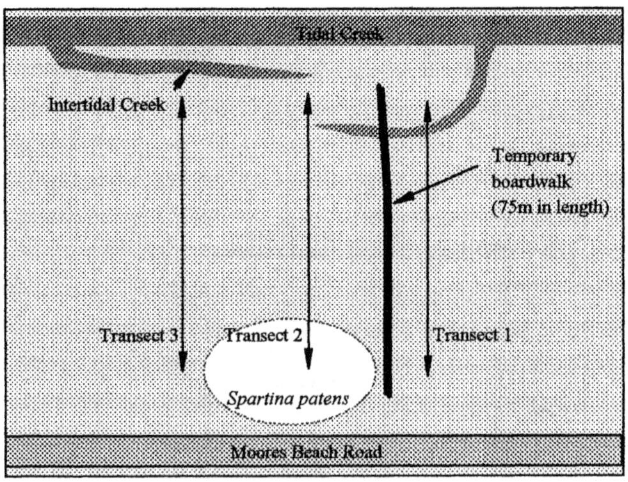

Figure 2. Sampling area diagram of *Spartina alterniflora* dominated marsh within lower Moore's Beach location. Transect locations for 1996 indicated by arrows. Sampling continued along transect 3 in 1997.

2.3 FOOD HABIT ANALYSIS

Fishes were measured to the nearest 0.5 mm standard length, and the entire digestive tract (sections 1, 11, and III, Babkin and Bowie 1928) was cleared of contents. Fish were dried (48 h in a drying oven at 60 ° and dry weights (measured to the nearest 10 μg) of the fish and empty gut were recorded. Gut contents were stained in rose bengal mixed in 70% ethanol prior to identification of contents. Presence/absence of content items for each fish were recorded to generate frequency of occurrence data. Gut content samples were then separated into groups based on throw trap replicate and fish size (5 mm length classes, 5.5 to 30.4 mm standard length), and gut contents of each group were pooled for analysis. Large prey items (amphipods, larval fish, adult insects) were removed, identified, dried, and weights were individually recorded. All items were identified to lowest taxonomic unit, however often prey items were digested beyond identification, and these were lumped into a category ''unidentified animal material'' using a series of 7.6 cm diameter sieves to separate contents into similarly sized fractions (after Carr and Adams 1972). Sieves (meshes 500 μm, 300 μm, 150 μm, and 63 μm) were stacked and shaken gently under running water for 2 to 3 minutes to separate contents adequately. Each fraction was rinsed into a gridded 60 mm diameter petri dish (5 mm grid size) and examined under a stereomicroscope. Percent composition of food items for each fraction was determined by counting grid coverage in the petri dish. Contents were then vacuum filtered through a preweighed filter pad (8 μm pore size), dried in an drying oven at 60 ° for 48 h, and weighed to the nearest 10 μg.

2.4 DATA ANALYSIS

Stomach content items were combined into major categories to reduce the effect of zero values on the analysis. Relative importance of the major categories for each length class group was obtained by plotting the proportion of the stomach contents by weight against proportion by frequency of occurrence (Costello 1990). One-way ANOVA using Bonferroni a posteriori tests were used to test whether frequency of occurrence of major categories of items varied with fish size. Rare items were eliminated from this analysis since most observations were zero. Proportion of the stomach contents by dry weight was arcsine square root transformed and canonical discriminant analysis (SAS version 6.11) was used to examine correlation between specific content items and size classes of fish.

3. Results

3.1 FISH AVAILABILITY

Fundulus heteroclitus were consistently present on the marsh surface during the summer of both years, but abundance varied during the summer. During 1996 sampling, YOY *F. heteroclitus* (n = 175) were collected in 95 throw trap samples over 3 sampling days (June 9, June 26, and July 24). Most of these fish were within a narrow size range, indicating a single cohort (Fig. 3). In 1997, *F. heteroclitus* (n = 338) were taken in 139 throw trap samples over 9 days (Fig. 4). During 1997, the average density of juvenile *F. heteroclitus* on the salt marsh surface was low on April 22, and May 5, prior to recruitment of the larvae. Densities increased by June 6 to 16 (SD 8.6) fish m^{-2}, and peaked in late June at an average density of 29 (SD 7.8) fish m^{-2} (Fig. 5). There is an indication that 2 to 4 cohorts of YOY were present during June and July however, sample sizes were too small for adequate cohort length frequency analysis.

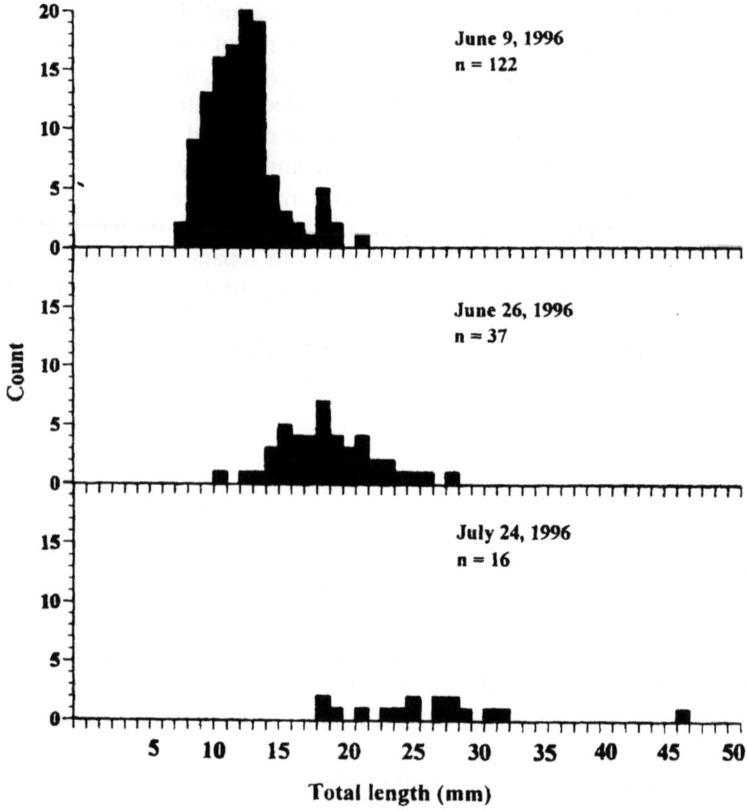

Figure 3. Length frequency of YOY *Fundulus heteroclitus* collected in 1996 using a 0.16 m^2 throw trap sampler.

3.2 COMPOSITION OF STOMACH CONTENTS

A total of 121 YOY *F. heteroclitus* from 1996 were used for food habits analysis (others were larger than YOY or were damaged during collection), producing 38 pooled samples over the sampling period (Table 1). For 1997, a total of 117 fish were selected for analysis from samples collected between early June and late July, yielding 36 pooled samples (Table 1).

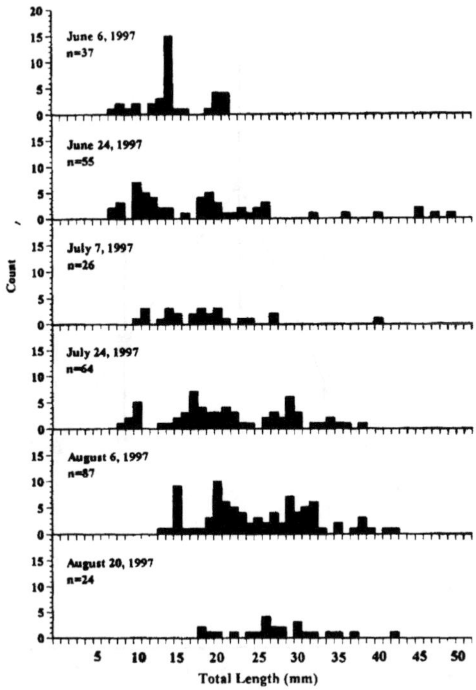

Figure 4. Length frequency of YOY *Fundulus heteroclitus* collected in 1997 using a 0.16 m^2 throw trap sampler.

TABLE 1. Length distributions of fishes analyzed for food habits from 1996 and 1997. Standard length measured to nearest 0.5mm. For each date and length class number of fish included and number of pooled samples (in parentheses) are provided.

	Length Class (SL in mm)				
Date	5.5-10.4	10.5-15.4	15.5-20.4	20.5-25.4	25.5-30.4
6/09/96	45	22	1	0	0
	(5)	(7)	(1)		
6/26/96	1	22	14	2	0
	(1)	(8)	(7)	(2)	
7/24/96	0	2	2	10	0
		(1)	(1)	(5)	
6/24/97	9	11	4	1	2
	(3)	(3)	(2)	(1)	(1)
7/07/97	3	4	3	1	0
	0)	(1)	(1)	(1)	
7/21/97	1	10	10	10	6
	(1)	(2)	(3)	(2)	(2)
8/06/97	0	0	1	15	9
			(1)	(1)	(4)
8/20/97	0	1	6	7	3
		(11)	(3)	(3)	(2)
Total	59	72	41	46	20
	(10)	(23)	(19)	(15)	(6)

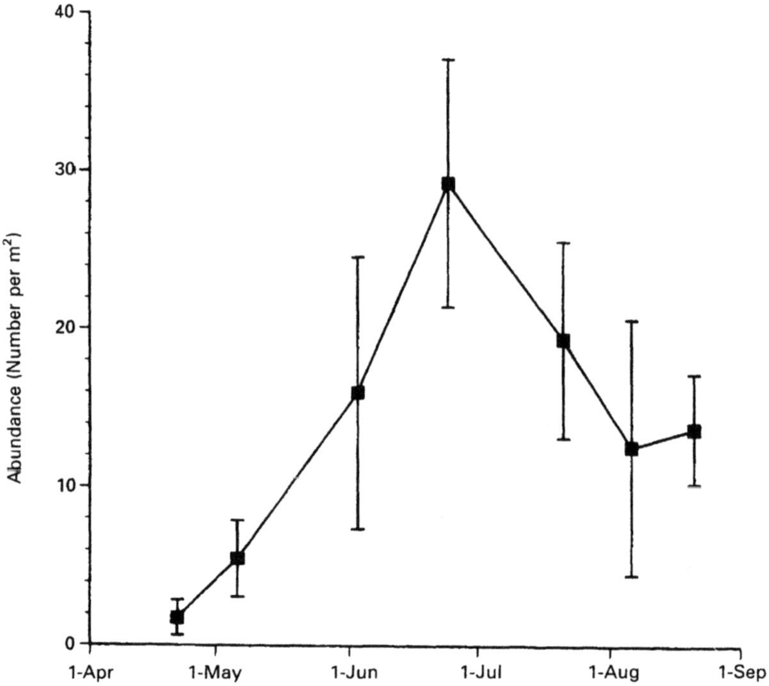

Figure 5. Variation in density (individuals per m^2) of YOY *Fundulus heteroclitus* on the marsh surface during summer of 1997. Density calculated based on samples with a 0.16 m^2 throw trap.

The presence of material in all the guts suggested these fish had been feeding recently during the period of flooding tide. The dominant food category by weight for all size classes was detritus/sediment aggregate (47.5%) which was composed primarily of aggregate of amorphous detritus clumped with sediment, and *Spartina* detritus, but also included worm tubes, and fecal pellets (Table 2). Meiofauna were the most important component by frequency of occurrence (33.1%), and copepods (primarily harpacticoid) dominated the meiofauna, followed by nematodes and foraminiferans. Another important category was the annelid worms (mostly unidentified worm material, with some oligochaetes and ampharetid polychaetes identified), estimated at 5.6% of the contents by weight, and 17.8% of the contents by occurrence. Insects and arachnids combined composed 7.5% of the contents by weight and 16.8% by occurrence. The majority of identified insects belonged to the families *Ceratopogonidae* (punkies), *Chironomidae* (midges), and *Dolichopodidae* (long-legged shore flies), in both larval and adult forms. Crustaceans composed about 10% of the stomach contents in weight and occurrence, and most of those identified were amphipods (gammarid), although several anthurid isopods were also observed. Unidentified animal material was less than 10% of the contents when calculated by weight or by percent occurrence. Finally, less common items were placed into a miscellaneous category for the analysis. These included gastropods, *Fundulus* larvae, anemones, eggs or protist cysts, algae (mostly benthic diatoms), and plantseeds.

TABLE 2. Stomach content composition for all YOY *Fundulus heteroclitus* collected at Moore's Beach salt marsh from 1996 and 1997. Percents at left of each column indicate values for major categories. Percents at the right of each column are for subcategories within major categories.

Diet component		% by weight		% occurrence	
Detritus/aggregate		47.5		11.0	
	aggregate		44.2		5.1
	detritus		3.0		4.9
	worm tube		0.1		0
	fecal pellets		0.1		0.6
Annelida.		5.6			
	unidentified		5.2		14.9
	Oligochaeta		0.2		1.7
	Ampharetidae (Polychaeta)		0.1		0
Arthropoda					
Crustacea		10.0		9.8	
	unidentified		6.0		0.7
	Amphipoda		4.0		8.3
	Isopoda		0.0		0.8
Insecta/Arachnida		7.5		16.8	
Insecta					
	unidentified adults		4.4		6.2
	unidentified larvae and pupae		0.7		4.1
	Ceratopogonidae (larvae)		0.1		0
	Chironomidae (larvae)		0.0		0
	Dolichopodidae (larvae)		1.8		2.8
Arachnida					
	Aranea (spiders)		0.4		1.1
	Acari (mites and ticks)		0.1		2.6
Meiofauna		17.1			
	Nematoda.		0.8		2.4
	Nemertinea		0.0		0
	Foraminiferan		1.3		1.5
	Copepoda		14.6		28.4
	Ostracoda		0.4		0.8
Unidentified animal material		6.4		8.5	
Misc. categories		6.0		3.4	
Mollusca					
	Gastropoda		0.4		0.6
Pisces			0.3		0.4
Cnidaria: anemone, unknown species			0.1		0
egg/cyst		1.2		1.7	
algae: diatom, filamentous algae			3.8		2.0
plant			0.2		

3.3 ONTOGENETIC CHANGES IN FOOD HABITS

As size increased the diet shifted from being largely carnivorous at the smallest sizes to a more omnivorous diet at larger sizes, based on occurrence and weight of major food categories (Fig. 6). Meiofauna, which dominated the stomach contents of fishes within the two smallest size classes (5.5 to 15.4 min SL, Fig. 6) declined in occurrence in fish > 20 mm SL ($F = 31.88$, $p < 0.0001$, Table 3). Although meiofauna still occurred frequently in the contents of fishes ranging from 15.5 to 20.4 mm SL, detritus/aggregate dominated the contents by weight at these sizes (Fig. 6). This food category remained the most important component of the stomach contents for the largest size classes (20.5 to 30.4 min SL, Fig. 6). There was also a significant decline in occurrence of annelid worms in the stomach contents between the smallest fish and those > 15.5 min SL ($F = 17.1$, $p < 0.0001$, Table 3), while relative importance of crustaceans (primarily amphipods) tended to increase with fish size. There was no clear shift in relative importance of insects and spiders in the stomach contents, although they were rare in the smallest size class (SL = 5.5 -10.4 mm).

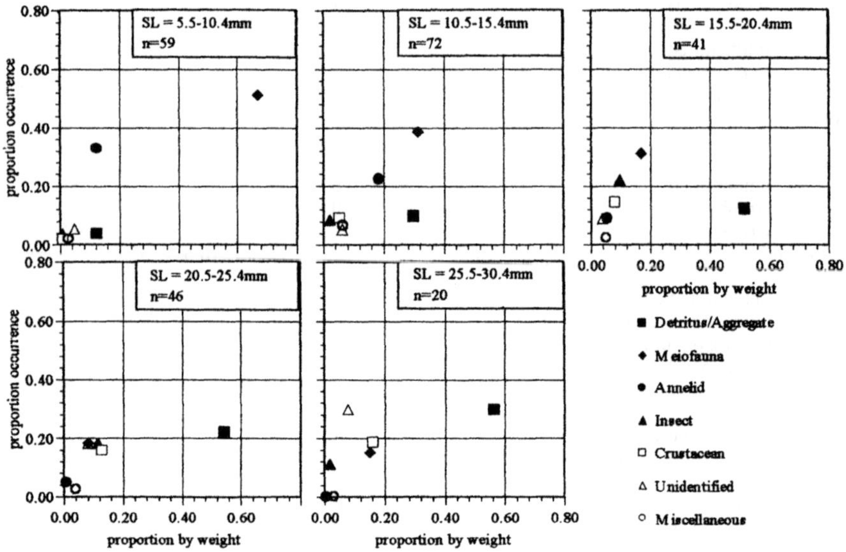

Figure 6. Costello graphs plotting proportion occurrence of major food categories by proportion weight for each length class of YOY *Fundulus heteroclitus*. Data for 1996 and 1997 are combined. See Table 2 for composition of major food groups.

TABLE 3. Ontogenetic changes in *Fundulus heteroclitus* food habits based on one-way analysis of variance on effect of length class on presence of major food categories. Bonferroni comparison of means, rejection level = 0.05. 1 = 5.5-10.4 mm, 2 = 10.5-15.4 mm, 3 = 15.5 - 20.4 mm, 4 = 20.5-25.4 mm, 5:25.5-30.4 mm

Food category	DF	F	P	Present ——→ Absent				
Annelid	4	17.08	0.0000	1	2	3	4	5
Meiofauna	4	31.88	0.0000	1	2	3	4	5
Insect/Arachnid	4	7.52	0.0000	3	4	2	5	1
Crustacean	4	4.18	0.0028	3	4	5	2	1
Detritus/Aggregate	4	5.07	0.0007	5	4	3	2	1
Unidentified animal	4	4.51	0.0017	5	4	3	2	1

Further analysis on specific food items also suggested a significant change in food habits with size. There was a significant effect of size class on mean weight values of the different stomach content components (Pillai's Trace $F_{88,180} = 1.55$, p,< 0.007). Canonical discriminant analysis reduced the data to 4 canonical variates or axes, with over 70% explained by the first canonical variate (Fig. 7, $F_{88,180} = 1.87$, $p < 0.002$).

This axis was positively correlated with copepods and annelids, while sediment aggregate, detritus, amphipods, unidentified crustaceans, insects and spiders were negatively correlated with Canonical I (Table 4). Canonical axis 2 explained approximately 15% of the variation and was positively correlated to food items: sediment aggregate, annelid, diatom, insect, and gastropod, and negatively correlated to detritus, filamentous algae, unidentified animal material, crustacean, and oligochaete (Table 4).

The smaller size classes (5.5-15.4 min SL) tended to have positive values along the 1st canonical axis, indicating that their stomach contents included annelids and meiofauna, and thus were mainly carnivorous. Larger size classes tended to have negative values for the 1st canonical axis, indicating stomach contents composed of detritus, sediment aggregate, amphipod, and insect prey items, thus a more omnivorous diet.

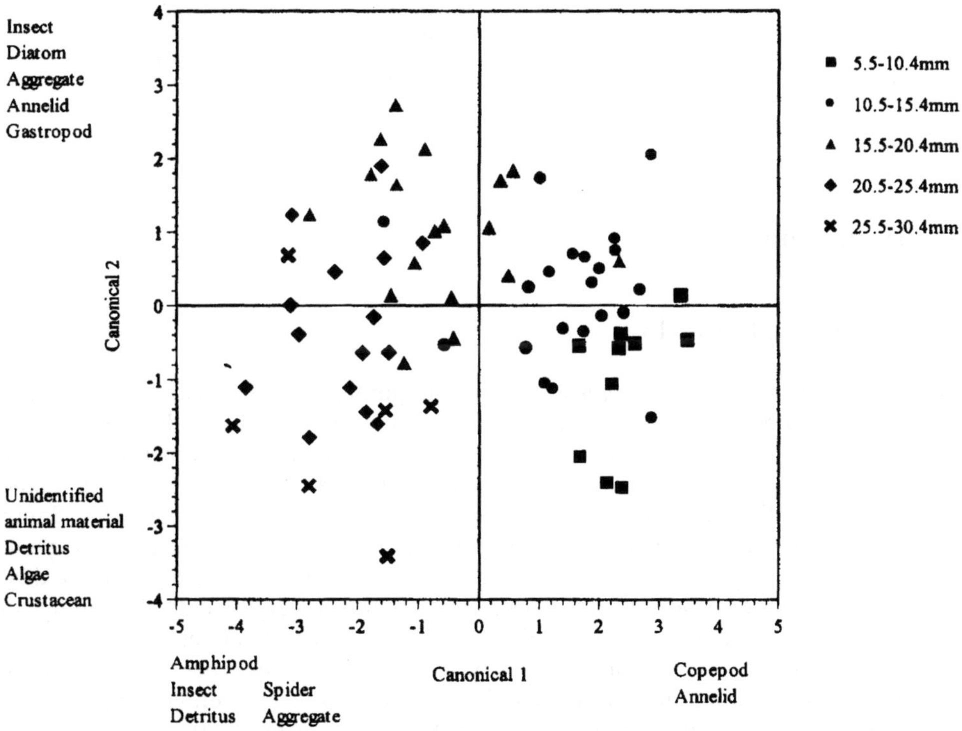

Figure 7 Plot of first two canonical axes for the five length classes of *Fundulus heteroclitus* at the Moores Beach study site. Food items which were significantly correlated with canonical variables listed next to each axis. See Table 4 for correlation coefficients and probability values.

TABLE 4. Analysis of YOY *Fundulus heteroclitus* stomach content composition based on canonical discriminant analysis. Pearson correlation coefficients (p value) are between food item weights and canonical variates 1 and 2.

Food item	Canonical 1	Canonical 2
Aggregate	-0.37 (0.0015)	0.29 (0.0139)
Detritus	-0.32 (0.0065)	-0.25 (0.0371)
Worm tube	0.16 (0.1843)	0.06 (0.6063)
Fecal pellet	-0.21 (0.0808)	-0.11 (0.3487)
Copepod	0.66 (0.0001)	-0.18 (0.1317)
Foraminifera	-0.20 (0.0887)	0.01 (0.9611)
Nematode	0.18 (0.1362)	-0.12 (0.3350)
Ostracod	-0.11 (0.3649)	0.11 (0.3818)
Annelid	0.49 (0.0001)	0.29 (0.0138)
Oligochaete	0.00 (0.9829)	-0.20 (0.0951)
Unknown crustaceans	-0.47 (0.0001)	-0.29 (0.0152)
Amphipod	-0.45 (0.0001)	0.12 (0.3104)
Isopod	0.11 (0.3715)	0.09 (0.4445)
Insect	-0.44 (0.0001)	0.39 (0.0009)
Aranea (spider)	-0.32 (0.0061)	0.02 (0.8956)
Acarina (mite)	0.11 (0.3436)	0.10 (0.4306)
Gastropod	-0.10 (0.4179)	0.26 (0.0259)
Fish	0.12 (0.3212)	0.05 (0.6995)
Egg or cyst	0.13 (0.2911)	-0.22 (0.0626)
Unid. animal material	-0.26 (0.0266)	-0.38 (0.0009)
Diatom	-0.16 (0.1771)	0.38 (0.0011)
Algae	0.21 (0.0754)	-0.24 (0.0414)

4. Discussion

4.1 FISH AVAILABILITY

Young-of-the-year *Fundulus heteroclitus* are ubiquitous on the marsh surface during flooding tides in New Jersey. This pattern has also been observed for adults throughout much of the range of this species including Georgia (Kneib and Wagner 1994), New Jersey (Able et al. 1996), Long Island (Butner and Brattstrom 1960), and Maine (Murphy 1991). Several studies (Kneib 1984a, 1997b, Talbot and Able 1984, Smith 1995) have indicated that small YOY remain on the intertidal marsh surface throughout the tidal cycle, using marsh pools and small depressions as refuges between tidal inundations. Our study indicates that these small YOY are very abundant in the marsh surface areas we sampled during flooding tides, with numbers ranging from 15 to 30 fish m^{-2} during June, a comparable density to Georgia salt marshes (Kneib 1997b). Most of the YOY during this period ranged from 7 to 25 mm TL, falling within the size range for other populations of *F. heteroclitus* in New Jersey (Able 1990). The presence of several cohorts (2-4) in 1997 is not surprising given that *F. heteroclitus* spawn over multiple spring tides throughout the summer (Taylor et al. 1979), thus during part of the summer (until late July), both the more carnivorous small YOY, and the omnivorous larger YOY are feeding on the marsh surface.

4.2 COMPARISON OF FISH FEEDING HABITS WITH AVAILABLE PREY

Our results indicate that benthic macrofauna and meiofauna were important components of the diet of YOY *Fundulus heteroclitus*, especially for the smallest size classes (SL = 5.5 to 25.4 mm). The salt marsh benthic macrofauna at Moore's Beach were dominated by oligochaete worms, sabellid and capitellid polychaetes (Smith and Taghon, pers. obs.). The benthic faunal community at Moore's Beach resembles southern salt marshes (Bell 1979, Kneib 1984b, Levin et al. 1996, Posey et al. 1997) in the general taxa represented, however oligochaetes were far more prevalent in Moore's Beach sediments than in previously published reports on natural salt marshes. Although oligochaetes were extremely abundant in salt marsh sediments, they only compose 5% by dry weight and 20% by occurrence of the diet of YOY *F. heteroclitus,* if we assume most of the unidentified annelid material were oligochaetes. Given the abundance of oligochaetes in the sediments, this may be a valid assumption.

Insects composed about 10% of the diet in YOY F. *heteroclitus,* however relatively little is known about their availability to YOY fishes. In southern New Jersey salt marsh pool fishes feed on aquatic insects (Campbell and Denno 1978), and insect larvae and adult insects play a role in *Fundulus* diets in other systems (Baker-Dittus 1978, Kneib 1986, Allen et al. 1994). Insect larvae were only collected from higher elevation sediments during benthic core sampling taken concurrently at the study site (Smith and Taghon, person. observ.). The most common insect larvae were dipterans from the families *Chironomidae*, *Ceratapogonidae*, and *Dolichopodidae*. These larvae were rare, or not present in lower elevation marsh sediments.

Nematodes are a dominant component of salt marsh meiofauna (Bell 1980), yet rarely occurred in the fishes' diets, even for the smallest size classes of *F. heteroclitus*. Harpacticoid copepods are also common on the marsh surface, although far less abundant than nematodes (Bell 1979, 1980), yet harpacticoid copepods are commonly observed in stomach contents of fishes in this study and others (Alheit and Scheibel 1982, Coull and Wells 1983, Ellis and Coull 1989). Nematodes may be less accessible than harpacticoid copepods because nematodes tend to have a deeper distribution in the sediments (Ellis and Coull 1989). Another possible reason for the observed differences between faunal composition in the marsh sediments and prey composition in the diet is variable digestion rates of different prey items. The disproportionate presence of copepods in the diet may be due to a longer digestion time (Scholz et al. 1991). Diet composition may overestimate the importance of less digestible prey items (Gannon 1976, Singh-Renton and Bromley 1996), which include copepods, crustaceans in general, and insects, which all have carapaces that tend to take longer to digest. Preliminary laboratory studies (Lilly, Smith and Taghon, unpublished data) indicated that YOY *F. heteroclitus* rapidly digest the polychaete *Capitella* sp. I, and identifiable characters are lost within 1 h of ingestion. In contrast, the brine shrimp *Artemia*, which have an exoskeleton, remain recognizable for at least 3 h, and show no appreciable loss of biomass by that point. Because of rapid digestion rates, it is imperative to collect fishes such as *F. heteroclitus* while they are actively feeding, and preserve samples immediately, as we have done in this study. Our technique of sampling fishes during the flooding tide, rather than after the tide subsides, greatly improved our ability to identify stomach contents.

4.3 ONTOGENETIC SHIF S IN FEEDING HABITS

The diet clearly changed with size for YOY *F. heteroclitus.* The smaller fish (5.5 - 20.4 mm SL) ate meiofauna and benthic macrofauna (predominantly annelid worms), while larger YOY (20.5 - 30.4 mm SL) shifted to a diet including mostly detritus and sediment aggregate. Kneib (1986) observed that smaller YOY are more predatory, feeding mostly on copepods, tanaids, and sabellid polychaetes in a Georgia salt marsh. However, he also observed a high incidence of diatoms in the diets of the smallest YOY (5.5 - 9.4 mm). Diatoms were rare for the smallest YOY in our study, which suggests a more carnivorous diet than previously observed. Decreased frequency of occurrence of diatoms in small YOY may indicate these size classes were feeding more selectively than larger YOY. High concentrations of benthic diatoms could be incidentally ingested while fishes were foraging for benthic prey, and our study indicates diatoms were more commonly ingested by larger YOY, along with detritus and aggregate sediments. It is also possible that in Kneib's study (1986) diatoms occurred frequently in the stomach contents, but composed little of the biomass.

Geographic differences in feeding habits of YOY *Fundulus* may also play a role in the observed differences between the studies. If only frequency of occurrence of food items is examined, the importance of detritus and sediment aggregate is underestimated (Fig. 6), while the importance of meiofaunal organisms such as copepods tend to be overestimated. By using both frequency occurrence and weight of diet items, a more reasonable view of diet composition is obtained (Hyslop 1980).

4.4 FUNCTION OF YOY *FUNDULUS HETEROCLITUS* IN SALT MARSHES

Early studies of salt marsh production indicated that emergent plant detritus played an important role in secondary production in surrounding coastal and estuarine habitats (Teal 1962, Nixon and Oviatt 1973). The importance placed on detritus was argued later (Nixon 1980), however it was admitted that the fate of production from the marsh surface is unclear. Our study shows that larger YOY *F. heteroclitus* consume detritus directly, and thus would transport the production directly when they migrate between the marsh surface and surrounding intertidal areas on each tidal stage. Detritus is often considered not nutritional (Prinslow et al. 1974), however stable isotope studies indicate the importance of *Spartina* detritus in marsh food webs (Kneib et al. 1980, Currin et al. 1995), including Delaware Bay marshes (Wainwright et al. in review). Small YOY are primarily predaceous and feed on benthic organisms. These benthic organisms may in turn rely on detritus either directly as a food source, or they feed on the microbial decomposers associated with detritus (Cammen 1980, Lopez and Levinton 1987). These YOY *F. heteroclitus* are the dominant life history stage on the marsh surface in both abundance and biomass (Taylor et al. 1979, Talbot and Able 1984, Talbot et al. 1986, Kneib 1986, Smith 1995). These YOY are potential prey resources for a number of predators, including blue crab, *Callinectes sapidus,* (Kneib 1982), grass shrimp, *Palaemonetes pugio* (Kneib 1987) and summer flounder, *Paralichthyes dentatus,* (Rountree and Able 1992). Since YOY have a high conversion efficiency (Prinslow 1974, Kneib and Parker 1991) and have high measures of secondary production (Meredith and Lotrich 1979), YOY *F. heteroclitus* may contribute the most to trophic transfer of production from the marsh surface to the surrounding estuary.

5. Acknowledgments

Megan Adams, Lori Lilly, and Kimberly Ray were interns who helped with the project, and whose independent studies greatly enriched the research program. The authors would like to thank the many field and laboratory assistants involved in this study, especially Natalie Senyk, who spent many hours dissecting YOY *F. heteroclitus.* Judith McLellan provided assistance in statistical analysis. Funding was provided by the Marsh Estuarine Research Project and Public Service Electric and Gas. This is Rutgers University, Institute of Marine and Coastal Science Contribution No. 99-09.

6. Literature Cited

Able, K. W. 1990. Life history patterns of New Jersey salt marsh killifishes. Bulletin of the New Jersey Academy of Science **35**: 23-30.

Able K. W. and M. P. Fahay. 1998. The first year in the life of estuarine fishes in the Middle Atlantic Bight. Rutgers University Press, New Brunswick, New Jersey, USA.

Able, K. W., D. A. Witting, R. S. McBride, R. A. Rountree and K. J. Smith. 1996. Fishes of polyhaline estuarine shores in Great Bay-Little Egg Harbor, New Jersey: a case study of seasonal and habitat influences. Pages 335-353 *in* K. F. Nordstrom and C. T. Roman, editors. Estuarine shores: evolution, environments and human alterations, John Wiley, Chichester, England.

Alheit, J. and W. Scheibel. 1982. Benthic harpacticoids as a food source for fish. Marine Biology **70**: 141-147.

Allen, D. M., W. S. Johnson and V. Ogbum-Matthews. 1995. Trophic relationships and seasonal utilization of salt-marsh creeks by zooplanktivorous fishes. Environmental Biology of Fishes **42**: 37-50.

Allen, E. A., P. E. Fell, M. A. Peck, J. A. Gieg, C. R. Guthke and M. D. Newkirk. 1994. Gut contents of common mummichogs, *Fundulus heteroclitus* L., in a restored impounded marsh and in natural reference marshes. Estuaries **17**: 462-471.

Babkin, B. P. and D. J. Bowie 1928. The digestive system and its function in *Fundulus heteroclitus.* Biological Bulletin. **54**: 254-277.

Baker-Dittus, A. M. 1978. Foraging patterns of three sympatric killifish. Copeia **1978**: 383-389.

Bell, S. S. 1979. Short- and long-term variation in a high marsh meiofauna community. Estuarine and Coastal Marine Science **9**: 331-350

Bell, S. S. 1980. Meiofauna-macrofauna interactions in a high salt marsh habitat. Ecological Monographs **50**: 487-505.

Bigelow, H. B. and W. C. Schroeder. 1953. Fishes of the Gulf of Maine. Fishery Bulletin of the Fish Wildlife Service **1**: 1-577.

Boesch, D. F. and R. E. Turner. 1984. Dependence of fishery species on salt marshes: the role of food and refuge. Estuaries **7**: 460-468.

Butner, A. and B. Brattstrom. 1960. Local movements in *Menidia* and . *Fundulus.* Copeia **1960**: 139-41

Cammen, L. M. 1980. The significance of microbial carbon in the nutrition of the deposit feeding polychaete *Nereis succinea.* Marine Biology **61**: 920.

Campbell, B. C. and R. F. Denno 1978. The structure of the aquatic insect community associated with intertidal pools on a New Jersey salt marsh. Ecological Entomology **3**: 181-187.

Carr, W. E. S. and C. A. Adams. 1972. Food habits of juvenile marine fishes: evidence of the cleaning habit in the leatherjacket, *Oligoplites saurus,* and the spottail pinfish, *Diplodus holbrooki.* Fishery Bulletin **70**: 1111-1120.

Costello, M. J. 1990. Predator feeding strategy and prey importance: a new graphical analysis. Journal of Fish Biology **36**: 261-263.
Coull, B. C. and J. B. J. Wells. 1983. Refuges from fish predation: experiments with phytal meiofauna from the New Zealand rocky intertidal. Ecology **64**: 1599-1609.
Currin, C. A., S. Y. Newell and H. W. Pearl. 1995. The role of standing dead *Spartina alterniflora* and benthic microalgae in salt marsh food webs: considerations based on multiple stable isotope analysis. Marine Ecology Progress Series **121**:99-116.
Ellis, M. J. and B. C. Coull. 1989. Fish predation on meiobenthos: field experiments with juvenile spot *Leiostomus xanthurus* Lacepede. Journal of Experimental Marine Biology and Ecology **130**: 19-32.
Fritz, E. S. 1974. Total diet comparison in fishes by Spearman rank correlation coefficients. Copeia **1974:** 210-214.
Gannon, J.E. 1976. The effects of differential digestion rates of zooplankton by alewife, *Alosa pseudoharengus,* on determinations of selective feeding. Transactions of the American Fisheries Society: *89-95.*
Harrington, R. W. J. and E. S. Harrington. 1961. Food selection among fishes invading a high subtropical salt marsh: from onset of flooding through the progress of a mosquito brood. Ecology **42**: 646-666.
Hyslop, E. J. 1980. Stomach contents analysis—a review of methods and their application. Journal of Fish Biology **17**: 411-429.
Kneib, R. T. 1982. The effects of predation by wading birds (Ardeidae) and blue crabs *(Callinectes sapidus)* on the population size structure of the common mummichog, *Fundulus heteroclitus.* Estuarine, Coastal and Shelf Science **14**: 159-165.
—— 1984a. Patterns in the utilization of the intertidal salt marsh by larvae and juveniles of *Fundulus heteroclitus* (Linnaeus) and *Fundulus luciae* (Baird). Journal of Experimental Marine Biology and Ecology **83**: 41-51.
—— 1984b. Patterns of invertebrate distribution and abundance in the intertidal salt marsh: causes and questions. Estuaries **7A**: 392-412.
—— 1986. The role of *Fundulus heteroclitus* in salt marsh trophic dynamics. American Zoologist **26**: 259-269.
—— 1987. Predation risk and use of intertidal habitats by young fishes and shrimp. Ecology **68**:379-386
—— 1997a. The role of tidal marshes in the ecology of estuarine nekton. Oceanography and marine Biology an annual review **35**: UCL Press.
—— 1997b. Early life stages of resident nekton in intertidal marshes. Estuaries **20**: 214-230.
Kneib, R. T. and J. H. Parker. 1991. Gross conversion efficiencies of mummichog and spotfin killifish larvae from a Georgia salt marsh. Transactions of the American Fisheries Society **120**: 803-809.
Kneib, R. T. and A. E. Stiven. 1978. Growth, reproduction and feeding of *Fundulus heteroclitus* (L.) on a North Carolina salt marsh. Journal of Experimental Marine Biology and Ecology **31**: 121-140.
Kneib, R. T., A. E. Stiven and E. B. Haines. 1980. Stable carbon isotope ratios in *Fundulus heteroclitus* (L.) muscle tissue and gut contents from a North Carolina *Spartina* marsh. Journal of Experimental Marine Biology and Ecology **46**: 89-98.
Kneib, R. T. and S. L. Wagner. 1994. Nekton use of vegetated marsh habitats at different stages of tidal inundation. Marine Ecology Progress Series **106**: 227-238.
Levin, L. A., D. Talley and G. Thayer. 1996. Succession of macrobenthos in a created salt marsh. Marine Ecology Progress Series **141**: 67-82.
Lopez, G. R. and J. S. Levinton. 1987. Ecology of deposit-feeding animals in marine sediments. Quarterly Review of Biology **62**: 235-260.
Meredith, W. H. and V. A. Lotrich. 1979. Production dynamics of a tidal creek population of *Fundulus heteroclitus* (Linnaeus). Estuarine and Coastal Marine Science **8**: 99-118.
Murphy, S. C. 1991. The ecology of estuarine fishes in southern Maine high salt marshes; access corridors and movement patterns. Thesis. University of Massachusetts, Boston, Massachusetts, USA.
Nixon, S. W. 1980. Between coastal marshes and coastal waters - a review of twenty years of speculation and research on the role of salt marshes in estuarine productivity and water chemistry. Pages 437-525 *in* P. Hamilton and K. B. MacDonald, editors. Estuarine and wetland processes. Plenum Press, New York, New York, USA.
Nixon, S. W. and C. A. Oviatt. 1973. Ecology of a New England salt marsh. Ecological Monographs **43**: 463-498.
Page, H. M. 1997. Importance of vascular plant and algal production to macro-invertebrate consumers in a southern California salt marsh. Estuarine Coastal and Shelf Science **45**: 823-834.

Posey, M. H., T. D. Alphin and C. M. Powell. 1997. Plant and infaunal communities associated with a created marsh. Estuaries **20**:42-47.

Prinslow, T. E., 1. Valiela and J. M. Teal. 1974. The effect of detritus and ration size on the growth of *Fundulus heteroclitus* (L.). Journal of Experimental Marine Biology and Ecology **16**:1 -10.

Rountree, R. A. and K. W. Able. 1992. Foraging habits, growth and temporal patterns of salt-marsh creek habitat use by young-of-the-year summer flounder in New Jersey. Transactions of the American Fisheries Society **121**: 765-776.

Scholz, D. S., L. L. Matthews and R. J. Feller. 1991. Detecting selective digestion of meiobenthic prey by juvenile spot *Leiostomus xanthurus* (Pisces) using immunoassays. Marine Ecology Progress Series **72**: 59-67.

Schmelz, G. W. 1964. A natural history study of the mummichog, *Fundulus heteroclitus* (Linnaeus) in Canary Creek Marsh. Thesis, University of Delaware, Delaware, USA.

Singh-Renton, S. and P. J. Bromley. 1996. Effects of temperature, prey type and prey size on gastric evacuation in small cod and whiting. Journal of Fish Biology **49**: 702-713.

Smith, K. J. 1995. Processes regulating habitat use by salt marsh nekton in a southern New Jersey estuary. Dissertation. Rutgers University, New Brunswick, New Jersey, USA.

Talbot, C. W. and K. W. Able. 1984. Composition and distribution of larval fishes in New Jersey high marshes. Estuaries **7**: 434-443.

Talbot, C. W., K. W. Able and J. K. Shisler. 1986. Fish species composition in New Jersey salt marshes: effects of marsh alterations for mosquito control. Transactions of the American Fisheries Society **115**: 269-278.

Taylor, M. H., G. J. Leach, L. DiMichele, W. M. Levitan and W. F. Jacob. 1979. Lunar spawning cycle in the mummichog, *Fundulus heteroclitus* (Pisces: Cyprinodontidae). Copeia: 291-297.

Teal, J. M. 1962. Energy flow in the salt marsh ecosystem of Georgia. Ecology **43**: 614-624.

Wainright, S. C., M. P. Weinstein, K. W. Able, C. A. Currin (in press). Relative importance of benthic microalgae, phytoplankton and the detritus of smooth cordgrass *(Spartina)* and the common reed *(Phragmites)* to brackish marsh food webs. Marine Ecology Progress Series.

Weisberg, S. B. and V. A. Lotrich. 1982. The importance of an infrequently flooded intertidal marsh surface as an energy source for the mummichog *Fundulus heteroclitus:* an experimental approach. Marine Biology **66**: 307-310.

Weisberg, S. B., R. Whalen and V. A. Lotrich. 1981. Tidal and diurnal influences on food consumption of a salt marsh killifish *Fundulus heteroclitus.* Marine Biology **61**: 243-246.

HABITAT VALUE: FOOD AND/OR REFUGE

FACTORS INFLUENCING HABITAT SELECTION IN FISHES WITH A REVIEW OF MARSH ECOSYSTEMS

J. KEVIN CRAIG
LARRY B. CROWDER
Duke University
Nicholas School of the Environment
Duke Marine Laboratory
135 Duke Marine Lab Rd.
Beaufort, NC 28516-9721 USA

Abstract

We review the general theory regarding habitat selection in fishes and integrate this theory with recent data to evaluate habitat selection by marsh fishes. Models of habitat selection in fishes have evolved rapidly. The earliest models predicted habitat selection based on simple variables like temperature or salinity. Optimal foraging models project habitat selection of individuals based on food availability. Environmental factors and food can be integrated using bioenergetic models to predict the distribution of individuals based on bioenergetic optimization. Habitat selection can be modified by the presence of other individuals including competitors (theory based on ideal free distributions) and predators (theory based on trade-offs of growth vs. predation risk). The most current models use game theory to project the dynamics of habitat selection for both prey species and their mobile predators. In a review of marsh ecosystems, we found that little information is currently available with which to evaluate potential mechanisms underlying patterns of habitat use in these systems. Though marshes are widely considered important for foraging and predator refuge, this function has rarely been measured or critically evaluated. In contrast, measurement has focused on abiotic factors, resulting in a mismatch between factors cited as important and those actually measured. Existing theory on habitat selection in fishes combined with the strong empirical base that has been developed on patterns of habitat use in marsh fishes provides a unique opportunity to begin testing relevant mechanisms underlying distributional patterns. A mechanistic approach is necessary to understand the functional value of habitat to organisms as well as provide a basis for the implementation and evaluation of habitat restoration and management initiatives.

1. Introduction

The distribution of organisms in the environment is influenced by a variety of processes including differential mortality and reproduction, colonization and extinction, passive transport, and habitat selection. Habitat selection is the non-random use of space resulting from the voluntary movements of organisms (Kramer et al. 1997). Habitat selection has both ultimate and proximate components, such as evolved responses to environmental stimuli and behavioral choices among alternative habitats. Organisms potentially respond to a variety of factors when selecting habitats including abiotic conditions, food resources, energetic considerations, inter- and intraspecific competition, and predation risk. Because responses to these factors change over ontogeny as individuals grow and develop, their relative importance in determining habitat selection is dynamic. The integrated response of organisms to their biotic and abiotic environment over their ontogeny presumably results in the selection of habitats that enhance fitness.

Historically the primary value of vegetated marsh habitat to fish has been considered as a carbon source for the production of invertebrate prey that was then utilized by coastal and estuarine-dependent species (Teal 1962, Nixon 1980). This view has recently been questioned, however, and attention has shifted to the direct use of emergent marsh vegetation by fish (Boesch and Turner 1984). A variety of terms have been employed in the literature to describe relationships between fish and vegetated marshes, including habitat ''use,'' ''choice,'' ''selection,'' ''preference,'' and ''utilization.'' Kramer et al. (1997) distinguish habitat *selection*, as defined above, from habitat *use*, in which selection by the organism is not distinguishable from other processes influencing distributional patterns. The difference between *selection* and *use* is partly a function of the investigative methods employed. Correlative studies, such as field surveys, are rarely able to distinguish habitat selection by the organism from other processes. Field surveys or other descriptive methods have been the dominant approach employed to study relationships between fish and marsh habitat (Fig. 1). These approaches can characterize habitat use under a particular set of conditions and provide a basis for developing relevant hypotheses of underlying mechanisms. Descriptive approaches are limited, however, for assessing why fish use marsh habitat. Further, applying results to other similar habitats or forecasting the effects of environmental change is difficult because inferences apply only to the particular location and range of conditions measured. Given recent declines in the amount or quality of aquatic habitat and policy initiatives targeted at managing or restoring habitat (e.g., wetland mitigation, essential fish habitat, NMFS 1996), an understanding of the functional value of habitat to organisms is needed. That is, what fitness benefits does an organism receive by occupying one habitat as opposed to another? Our view is that a mechanistic approach to investigating the relative importance of factors influencing habitat selection is necessary to understand the functional value of habitat to organisms.

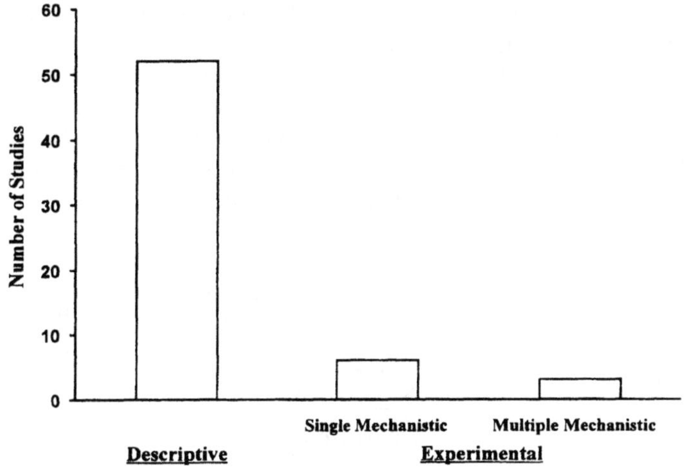

Figure 1. Number of studies employing descriptive approaches, experimental evaluation of a single hypothesis, and experimental evaluation of multiple hypotheses (Appendix I). Multiple factor experimental tests evaluated whether two or more mechanisms (e.g., food and predation risk) influenced habitat use in the same study, though did not necessarily evaluate their simultaneous effects.

Though a mechanistic approach to habitat selection has not been widely employed in marsh ecosystems, this approach has been used in a variety of other aquatic systems. We use examples from this literature to describe the variety of factors that influence habitat selection in fishes. We then review the literature from marsh ecosystems to assess what factors have been considered in these systems. Our primary conclusions are that: 1) processes influencing habitat selection of marsh fishes are probably similar to those in other aquatic systems, but because few experimental manipulations have been performed in marshes the relative importance of particular mechanisms remains obscure and 2) a mismatch exists between the characteristics of marshes cited as important for fish and those that are measured, with foraging and predator refuge widely cited as important but abiotic factors most often measured.

2. Factors Influencing Habitat Selection in Fishes

Defining ''habitat'' from an organisms' perspective is problematic because the appropriate definition depends on the temporal and spatial scale considered as well as the vagility and sensory abilities of the organism. Fish may have preferred microhabitats (e.g., foraging areas) that are occupied on relatively short temporal scales (e.g., daily) within a more general macrohabitat (e.g., lake littoral zone) that is occupied on a seasonal time scale. Some fish are able to sample large areas of potential habitat relatively quickly (e.g., billfish), while others are restricted to a few square meters. We do not explicitly consider the spatial or temporal scale at which habitat selection occurs but focus on processes that are relevant to at least intermediate scales for many fish (tens to hundreds of meters, days to months), such as different tidal creeks or the littoral and vegetated zones of lakes. We categorize factors that influence habitat selection into five general categories: abiotic factors, food resources, bioenergetics, competition, and predation. The role of these factors in determining habitat selection may vary as fish develop and grow (ontogeny) and with other structural characteristics of the environment (e.g., vegetation density and morphology) (Fig. 2). Below we briefly review the effects of each of these processes on habitat selection beginning with single factor hypotheses and building toward more complex models of the interactive effects of multiple factors on habitat selection.

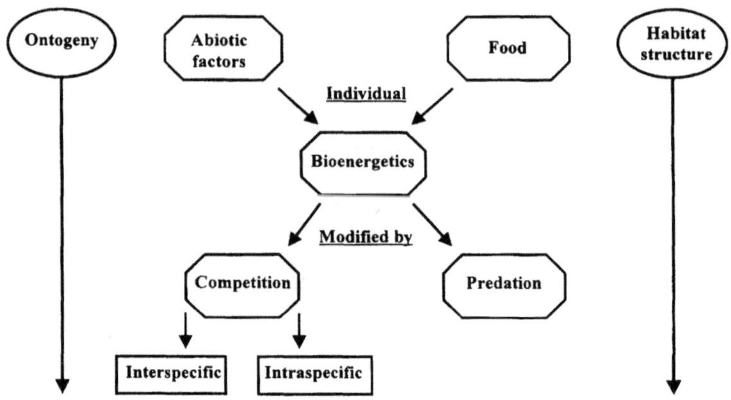

Figure 2. Factors that influence habitat selection in fishes.

2.1 ABIOTIC FACTORS

Field distributions of fish are often correlated with a variety of abiotic variables such as temperature (Neill and Magnuson 1974), salinity (Weisburg 1986), dissolved oxygen (Cech et al. 1990), turbidity (Cyrus and Blaber 1987a,b), sediment type and water velocity (Lough et al. 1989). Abiotic factors can be classified as those that have direct physiological effects on metabolism such as temperature, salinity, and dissolved oxygen, and those that are most likely correlates of these or other factors, such as depth, turbidity, light, and sediment type. Fry's (1971) classification of the environment into ''factors'' based on physiological effects on metabolism is a useful means of identifying mechanisms by which abiotic variables impact habitat selection. Lethal levels set the ultimate limits for habitats that can be occupied. Within these limits fish often have preferences for distinct ranges of abiotic variables (Magnuson et al. 1979). For example, fish generally have narrowly defined thermal preferences (e.g., ~4°C, Fry 1947, Magnuson et al. 1979) and strong capacities to discriminate temperature levels (e.g., 0.03°C, Beitinger and Fitzpatrick 1979). Preferred temperatures are often those that optimize some correlate of fitness such as growth rate (Brett 1971, Magnuson et al. 1979, Crowder and Magnuson 1983). Several abiotic factors may simultaneously influence metabolism via different mechanisms. For example, while temperature defines the limits of metabolism, salinity and dissolved oxygen modify the metabolic rate through regulatory processes such as the maintenance of ionic balance and the rate of supply of oxygen necessary for energy production (Fry 1971, Kirschner 1995). The combined effect of multiple abiotic variables on metabolism are more likely to be important determinants of habitat selection than single abiotic variables. Laboratory studies indicate that fish are able to alter their distribution in response to multiple abiotic variables in a manner that minimizes metabolic costs. For example, preferred temperature often declines as dissolved oxygen levels decrease, suggesting fish reduce metabolic demands for oxygen when levels are low (Bryan et al. 1984, Schurmann et al. 1991, Schurmann and Steffensen 1992).

2.2 FOOD RESOURCES

While abiotic factors may influence distributions through physiological tolerances or effects on the fate of ingested energy, habitats vary in the abundance, availability, and quality of energy sources. Optimal foraging theory is based on the premise that natural selection has resulted in behavioral phenotypes (diet selection, patch choice) that maximize energy intake per unit time of individual foragers (MacArthur and Pianka 1966). In cases where the availability of food resources varies among habitats, changes temporally within a habitat, or with the ontogeny of the fish, optimal foraging theory can be used to predict habitat selection. Models consist of cost-benefit functions in which the energy acquired from a given habitat is a function of the encounter rate with prey and the ability to pursue and capture prey items once encountered, as well as the energetic costs of pursuing, capturing, handling, and assimilating those prey (Werner and Mittlebach 1981, Wright and O'Brien 1984, Osenberg and Mittelbach 1989, Brandt and Mason 1994). For example, in northern glacial lakes the vegetated littoral, unvegetated sediment, and open water pelagic habitats vary in foraging profitability to bluegill sunfish (*Lepomis macrochuris*) over the course of the growing season due to seasonal changes in prey abundance and ontogenetic changes in capture success and handling time by bluegill. Bluegill switch habitats through the season in accord with predictions of an optimal foraging model based on laboratory derived estimates of foraging costs and benefits (Mittelbach 1981, Werner et al. 1983a). Similar models for perch (*Perca fluviatilis*) inhabiting Scandinavian lakes successfully predicted a shift from pelagic to benthic habitats as pelagic resources declined (Persson and Greenberg 1990). While perch exhibited the predicted habitat switch, partial preferences for the pelagic habitat remained and there was a time lag between the predicted and observed shift. Therefore, is some cases these models may be most useful for generating *a priori* qualitative predictions of the relative foraging value of different habitats.

2.3 BIOENERGETICS

Early optimal foraging models incorporated the energetic costs of pursing, capturing, and handling food to predict habitat selection but not metabolic costs associated with variable abiotic conditions. For example, if temperature or other abiotic factors that affect metabolism vary sufficiently among habitats, then the habitat in which net energy gain is maximized after accounting for metabolic costs (bioenergetic net benefit) may be different from that in which foraging rate (gross energy intake) is maximized or temperatures are optimal for growth (Crowder and Magnuson 1983). In this case, bioenergetic models are necessary to translate foraging rates into energy available for growth. Recent foraging models have incorporated energetic costs due to temperature or current velocity to predict habitat selection. For example, stream current velocity determines encounter rates with prey and capture success for drift feeding fish but imposes energetic costs to maintain position. Foraging models incorporating these effects have been accurate predictors of habitat selection by fish in both field (Hughes and Dill 1990, Hill and Grossman 1993, Hughes 1998) and laboratory environments (Tyler and Gilliam 1995). Bioenergetic models that incorporate the effects of food and temperature on growth have also been used at larger spatial scales (i.e., tens to hundreds of kilometers) to predict habitat selection. Coupled foraging and bioenergetic models based on acoustic measures of prey fish biomass and temperature have been used to map pelagic habitats in terms of their growth rate potential, the expected growth rate of a particular predator occupying a particular volume of water (Brandt et al. 1992, Brandt and Kirsch 1993, Goyke and Brandt 1993, Mason et al. 1995). These models, however, have not been rigorously tested to determine if predator distributions are actually consistent with model predictions. While bioenergetic optimization is an intuitively appealing basis for habitat selection of fish, alternative mechanisms are possible. In explicit tests to distinguish whether bluegill sunfish select habitats based on net energy gain (bioenergetic net benefit), maximization of feeding rate (optimal foraging), or optimal temperatures for growth (behavioral thermoregulation), habitat selection in short-term laboratory experiments was most consistent with behavioral thermoregulation (Wildhaber and Crowder 1990).

2.4 COMPETITION

While optimal foraging theory and its extension to bioenergetic optimization can be used to generate predictions of habitat selection by individual foragers, the energetic return to an individual fish foraging in a particular habitat may also depend on the presence of other foragers. Thus, both intra- and interspecific competitive interactions can be important determinants of habitat selection. Intraspecific effects can be addressed with ideal free distribution (IFD) theory which, in its basic form, predicts that at equilibrium the proportion of individuals in a habitat will equal the proportion of resources in that habitat (i.e., resource matching, Fretwell and Lucas 1970, reviewed in Kennedy and Gray 1993). Extensions to the basic model relax various assumptions such as equal competitive abilities among foragers, perfect knowledge of resource distributions, and equal energetic costs among habitats. Experimental tests in simple laboratory environments have yielded support for predictions of the basic ideal free model (Godin and Keenleyside 1984, Milinski 1984, Gillis and Kramer 1987, Abrahams 1989) as well as its variants (Milinski 1994, Grand and Grant 1994, Grand 1997). Many of these studies, however, exhibited anomalous results as well (Kennedy and Gray 1993), and more complicated experimental tests involving multiple habitats have often found the IFD to be a poor predictor of fish distributions (Talbot and Kramer 1986). The few field tests conducted have reported both agreement (Fraser and Sise 1980, Power 1984, Swain and Wade 1993) and departure (Fraser and Sise 1980, Rose and Leggett 1989) from IFD predictions. For example, in years when population size is large, Atlantic cod (*Gadus morhua*) in the Gulf of St. Lawrence expand their geographic range to include marginal habitats in accord with qualitative predictions from IFD theory (Swain and Wade 1993). Further, cod occupy cooler temperatures in years of high abundance suggesting that biotic and abiotic factors interact to determine habitat selection in a manner that reduces density-independent costs (metabolic losses due to high temperature) when density-dependent costs (decreased foraging rate due to intraspecific competition) are high (Swain and Kramer 1995).

Interspecific competitive interactions may impact habitat selection as well. Field removal and enclosure experiments have shown that fish expand the range of habitats occupied when competitors are removed (Larson 1980, Hixon 1980, Schmitt and Holbrook 1986) or shift habitats in response to competition from heterospecifics (Werner and Hall 1976, 1977; Persson 1986, Persson and Greenberg 1990). Competitive interactions may be symmetric such that habitat selection of all competitors is modified by removal of the others (Larson 1980) or asymetric such that only one or a few species are affected (Werner and Hall 1977, Hixon 1980). For example, bluegill sunfish, pumpkinseed (*Lepomis macrochirus*), and green sunfish (*Lepomis cyanellus*) each rank vegetated habitats of lakes as more profitable than unvegetated sediments when foraging alone, but differ in foraging efficiency in vegetation (green>bluegill>pumpkinseed, Werner and Hall 1979). As resources decline through the season and interspecific competition becomes more intense in the preferred vegetated habitat, the species that are less efficient foragers in vegetation shift to alternative habitats first (pumpkinseed then bluegill) while green sunfish remain in the vegetation. Thus, the presence of competitors can qualitatively alter how species rank habitats in terms of foraging profitability.

2.5 PREDATION

Fish not only consider characteristics of habitats that influence net energy gain (e.g., abiotic conditions, food abundance, competitor density) but also those that influence risk of mortality due to predation. The effects of the presence of a predator on habitat selection have been documented in both short-term (<15 days; Lima and Dill 1990, Gotceitas et al. 1995, Jordan et al. 1996, Utne and Bacchi 1997) and long-term experiments (months to >1 year; Tonn and Pazkowski 1992, Persson 1993, Crowder et al. 1994). In some cases, higher predation risk occurs in habitats with higher levels of food resources such that a tradeoff exists. For example, juvenile bluegill (<100mm) do not switch from the vegetated to open water habitat in accord with predictions from simple optimal foraging models when predatory largemouth bass (*Micropterus salmoides*) are present (Werner et al. 1983b, Mittlebach 1984). Similar restrictions on habitat use have been observed in other systems with a range of consequences for subsequent growth (Mittelbach 1988, Persson 1993), encounter rates with other predators (Brabrand and Faafeng 1993), and food web structure (Turner and Mittelbach 1990). Gilliam and Fraser (1987) proposed the quantitative decision rule that fish minimize mortality rate relative to gross foraging rate (μ/f) as an alternative to maximizing gross foraging rate per unit time when selecting among habitats. This model made accurate quantitative predictions of the change in resource density necessary for juvenile creek chubs (*Semotilus atromaculatus*) to shift to a habitat with higher predation risk in short-term behavioral experiments (see Abrahams and Dill 1989 for a similar test). These models and experimental tests, however, assume a fixed, habitat-specific risk of predation. That is, prey move in response to predators in a manner that enhances their fitness (usually to a refuge) but predators do not move in response to the re-distribution of their prey. Three-trophic-level, ideal free distribution models address how fish are distributed among habitats when predators and prey compete intraspecifically within a trophic level, predators move to habitats where prey are more abundant, and prey select habitats based on both the distribution of their resources and predation risk (Hugie and Dill 1994, Sih 1998). These models often make counterintuitive predictions such as predators should aggregate in habitats that contain high densities of their prey's resources more so than high densities of their prey.

2.6 INFLUENCE OF ONTOGENY AND HABITAT STRUCTURE

Above, we have reviewed five classes of factors that influence habitat selection in fishes (abiotic factors, food resources, bioenergetics, competition and risk of predation). The influence of most of these factors on habitat selection scale with body size and habitat structure. Fish exhibit ontogenetic changes in their responses to these factors when they select habitats (Werner and Gilliam 1984, Werner 1988, Stein et al. 1988). Because bioenergetic costs and benefits, as well as competitive ability and risk of predation, scale with body mass, the dynamics of body size play a large role in determining the costs and benefits of selecting a particular habitat (Persson and Crowder 1998). For example, predation risk for juvenile fish frequently decreases as body size increases (Werner and Gilliam 1984, Miller et al. 1988, but see Rice et al. 1997), but vulnerability of larval fish increases to a maximum (as encounter rate increases with larval swimming speed) and then decreases as size increases (Bailey and Houde 1989, Litvak and Leggett 1992).

Habitat structure can alter size-dependent interactions while foraging as well as risk of predation (Vince et al. 1976, Heck and Crowder 1991). For example, Ryer (1987, 1988) showed that both encounter rate and consumption were reduced for pipefish in dense eelgrass, particularly for the larger pipefish. Growth rates of prey fish confined by predators to vegetated refuges can also be reduced by competition within the refuge (Mittelbach 1988). The mechanisms behind decreased predator efficiency in complex habitats can be both decreased encounter rates and decreased capture success once prey have been encountered (Main 1987, Savino and Stein 1989). The interaction of size-dependent factors in structured habitats can lead to a rich variety of complex interactions among predator and prey fishes (see Persson and Crowder 1998).

3. Habitat Use in Marsh Ecosystems

The above review indicates that a variety of factors that potentially interact in complex ways influence habitat selection in fishes. To integrate work conducted in marsh ecosystems with this broader literature we characterized studies of marsh habitats relative to the factors described above (Fig. 2). Our primary goal was to determine what factors have been considered as potential mechanisms underlying patterns of habitat use in marsh fishes and if any conclusions can be drawn regarding the relative importance of different factors. As most studies in marsh ecosystems have been descriptive in nature (Fig. 1), it is difficult to distinguish between habitat use and habitat selection *per se* (Kramer et al. 1997). While our characterization focuses on processes influencing habitat selection, some of the patterns observed may result from other processes, such as differential mortality or colonization. Because most studies in marsh ecosystems have focused on which species are found in what habitat at what time (i.e., habitat use) rather than why fish occupy marsh habitats (i.e., habitat selection), our retrospective analysis may not accurately reflect the context in which some studies were conducted. We argue, however, that the increased emphasis on the functional value of marsh habitats to organisms (Boesch and Turner 1984), as required under recent federal legislation for defining essential fish habitat (1996 Magnuson-Stevens Act, NMFS 1996), necessitates a shift from the characterization of patterns of habitat use to a mechanistic understanding of habitat selection. Our approach facilitates this process.

3.1 SCOPE OF THE REVIEW

What constitutes marsh habitat or its limits is difficult to define. In a recent review tidal marshes were defined as vegetated wetlands periodically flooded by tides and renowned for their high productivity (Kneib 1997). Hackney et al. (1976) distinguished tidal creeks from tidal rivers as streams that drain marsh area, flush to some extent with each tidal cycle, and experience seasonal incursions of saltwater. Rountree and Able (1992a) distinguished five subhabitats of salt marshes as irregularly flooded marsh surface (including marsh pools), regularly flooded intertidal marsh surface, intertidal marsh creeks, subtidal marsh creeks, and bay-marsh fringe habitat. In choosing papers for this review we focused on studies on or in the vicinity of partially or periodically inundated vegetated marsh, including tidal creeks, and marsh pools. Studies conducted in the main portion of estuaries, channels of rivers, or other types of estuarine habitats such as seagrass beds, nonvegetated mudflats or beaches, and shell habitats, were excluded unless considered in relation to marsh habitat. For example, studies addressing the role of submerged aquatic vegetation within marsh creeks as habitat for fish were included (e.g., Rozas and Odum 1987). We did not distinguish between freshwater, brackish, and salt marshes, level of tidal inundation, or the vegetated versus nonvegetated portion of marsh habitats, though some of these distinctions have been previously made (e.g., Odum 1988, Rozas 1995, McIvor and Rozas 1996). In describing the study site, most papers referred specifically to the proximity of the vegetation or characterized the area as a marsh or marsh creek.

Studies addressing the role of marshes as habitat for fish have pursued a wide variety of objectives including microhabitat use (e.g., distribution within *Spartina* spp. patches), macrohabitat comparisons (e.g., *Spartina* spp. marsh versus other habitats), community analyses, relation to environmental factors, and animal movement among or within habitats. We included papers in which the distribution, abundance, or diversity of fish in relation to marsh habitat was a primary focus of the paper. In most cases, spatial or temporal variation in abundance or community composition was the primary focus, but some papers considered marsh habitat in relation to larval recruitment (e.g., Allen and

Barker 1990, Kneib 1997a) or as part of a more general life history study (e.g., Rountree and Able 1996). Papers were omitted when their primary objectives were not directly related to fish use of marsh habitat (e.g., production rate estimates of marsh fish) or focused on large geographic scales (zoogeographic comparisons, Ayvazian et al. 1992). Though a substantial component of the fauna in some systems, invertebrates were not included, partly to limit the number of papers considered but also due to differences in vagility and capacity to survive out of water between fish and invertebrates. We did not include studies whose primary focus was temporary use of marshes for reproduction, diet descriptions (unless part of a broader study of habitat use or specifically related to differential use of habitats for feeding), gear testing and analytic methods, or anthropogenic alteration of habitat. Only full articles published in the primary literature since 1975 were considered.

Our selection process resulted in 61 papers published in 14 journals (Appendix I). The majority of these studies were conducted in or near marshes composed of *Spartina* spp. (82%). 38% of the studies were conducted partially or wholly on the vegetated marsh. Most studies were field surveys (77%) while the remainder conducted field experiments (15%) or used other methods (e.g., telemetry) to investigate fish use of marsh habitat. Approximately one half of the studies (54%) addressed fish communities or several species (generally >5 species). Juvenile stages of transient species and adult stages of resident species were the major life stages considered though larval and adult fish were represented as well. While the studies varied in their objectives, methodology, and temporal and spatial scale, all addressed the abundance, distribution, or diversity of fish in relation to marsh habitat. Most of the studies (51 of 61) were also cited in a recent review on the role of tidal marshes in the ecology of estuarine nekton (Kneib 1997b). Therefore, these studies are probably representative of work that has been conducted on fish use of marsh habitat.

3.2 METHODS OF CHARACTERIZATION

To characterize the selected studies, we determined the number that *mentioned*, *measured*, or *concluded as important* each of the five factors previously discussed (abiotic variables, food, bioenergetics, competition, predation; Appendix II). We did not consider factors related to marsh physiography, such as degree of tidal inundation, bank slope, creek size, or vegetation characteristics (e.g., stem density, height). We did not separate or attempt to determine the relative importance of different abiotic variables or the mechanism by which they influenced habitat use. We considered all factors related to sediments (grain size, organic content, % silt-clay) as a single abiotic factor. A few studies measured sediment organic content, which may be considered a biotic factor, but this did not alter our conclusions.

First, we documented the number of studies that mentioned each of the factors depicted in Fig. 2 as an important factor determining habitat use of marsh fishes. This usually occurred in the *Introduction* of the paper but sometimes in the *Discussion* and interpretation of results. For example, statements referring to marsh habitat as a foraging area and predator refuge were counted as mentioning predation risk and abundant food resources as potential mechanisms underlying habitat use. Second, we documented the number of studies that *measured* each of the factors. A factor had to be measured with the goal of relating it to the abundance, distribution, or diversity of fish in the habitat. Hence cases where abiotic factors were compared among study sites or across seasons at a particular site and related (qualitatively or quantitatively) to spatial or temporal variation in fish community or individual species descriptors were counted, while those measured strictly to describe study sites were not. Third, we determined the number of studies in which a factor was *concluded* to be an important factor underlying habitat use. A factor could not be concluded as important unless it was directly measured. For example, higher abundances at night than during the day did not qualify as measuring predator effects on habitat use although this is one potential interpretation. We did not use statistical significance as a criteria for concluding whether a factor was important because of variation in the use, type, reporting, and interpretation of statistical results. Whether a factor was considered important was based on the interpretation of the authors with the constraint that the factor was measured. For example, though significant correlations between fish abundance and abiotic factors were often obtained, sometimes authors emphasized the small amount of variance explained (abiotic variables not important) while others emphasized the significance of the correlation (abiotic variables important).

3.3 RESULTS

All of the factors previously discussed (Fig. 2) were mentioned in at least some of the studies examined (Fig. 3a). Abundant food resources was most often mentioned as a reason fish occupy marsh habitat (84% of studies). Abiotic factors (69%) and predation refuge (64%) were mentioned in relation to fish use of marsh habitat with nearly equal frequency. About one-third (34%) of the studies mentioned competitive interactions as an important mechanism for observed field distributions of marsh fishes. Nearly all of these studies referred to interspecific competition (20 of 21), while few mentioned intraspecific effects (3 of 21). Bioenergetic considerations were rarely mentioned as an important factor for fish in marsh habitats (16%). When bioenergetics was mentioned authors referred to the effects of tidal currents on energetic costs of position maintenance (Szedlemayer and Able 1993) or acclimation to changing abiotic conditions (Rountree and Able 1992b), or general notions of selecting habitats to maximize net energy gain (Baltz et al. 1993).

The number of studies measuring the various factors that influence habitat selection differed qualitatively from factors mentioned as important (Fig. 3b). Abiotic factors were most often measured (56%), primarily temperature, salinity, and dissolved oxygen but also turbidity, current velocity, and sediment properties. Diet descriptions as a component of larger studies of habitat use or in relation to tidal effects on stomach fullness were sometimes measured (28%). Though often cited as important, few studies have measured the effects of predators on habitat use (7%). Similarly, studies measuring competitive effects were rare (3%). No studies have measured the integrated effects of food and abiotic conditions (bioenergetic net benefit) on habitat use in marsh fishes.

Factors concluded to be important followed the same qualitative pattern as those that were measured (Fig. 3c). Rarely was a factor measured and then considered unimportant to observed distributional patterns in marsh fishes, but there were exceptions. For example, Subrahmanyam and Coultas (1980) discounted significant but weak relationships between seasonal changes in fish abundance and abiotic factors

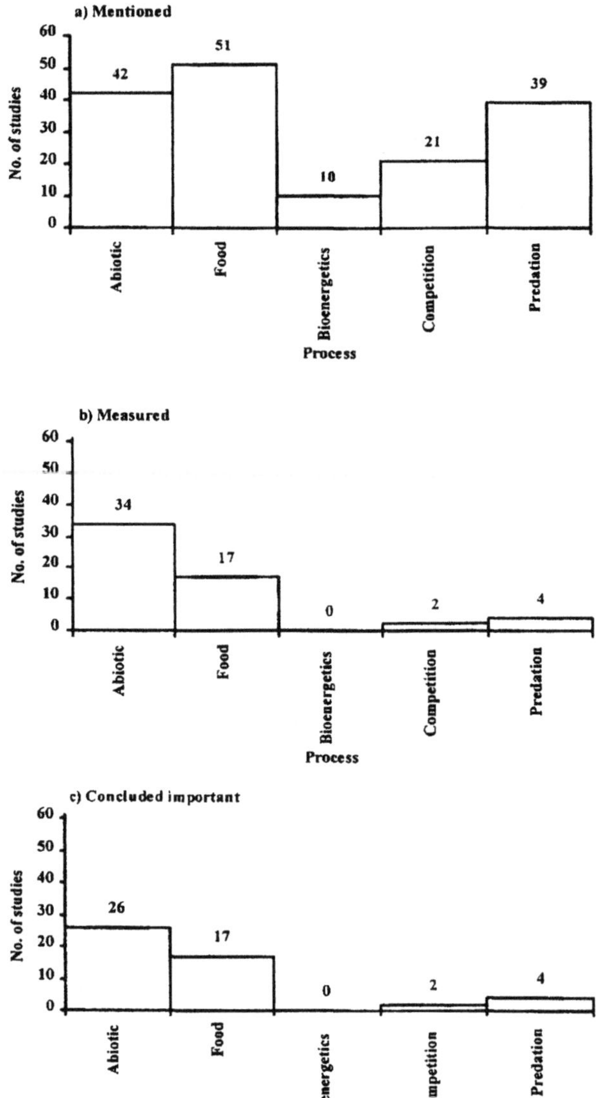

Figure 3. Factors (a) mentioned, (b) measured, and (c) concluded as important in studies of fish use of marsh habitat listed in appendix I (see also appendix II).

(temperature, salinity, dissolved oxygen) as a primary mechanism underlying patterns of habitat use in a north Florida salt marsh, and went on to discuss the adaptive value of staggered patterns in breeding and subsequent recruitment of component species in minimizing interspecific competitive interactions. Similarly, the broad tolerance levels of many estuarine species is often invoked to explain weak correlations between seasonal changes in community structure and abiotic variables (e.g. Cattrijsse et al.

1994). Clearly, a mismatch still exists, however, between the hypotheses emphasized as likely mechanisms driving patterns of habitat use in marsh ecosystems and those which were tested. While abiotic factors have received considerable attention, food and predation risk were mentioned with equal or greater frequency yet remain relatively untested (Fig. 3a-b). Our interpretation is that the high correlation between factors measured and those considered important to habitat use of marsh fishes (Fig. 3b-c) reflects consideration of single factor hypotheses and the relative ease of measuring abiotic variables compared to the difficulty of constructing multiple alternative hypotheses and testing biotic processes.

3.4 WHY DO FISH OCCUPY MARSH HABITATS?

Early studies in marsh ecosystems emphasized the importance of physical factors in driving patterns in the distribution and abundance of marsh fishes. This is evident in statements such as ''physical controls often outweigh biological or geological controls in estuarine ecosystems'' (Reis and Dean 1981) or biotic interactions are ''probably swamped by large fluctuations in the physical environment'' (Allen 1982). Correspondingly, most studies of fish in marsh habitats have measured species abundance or distribution in conjunction with abiotic factors. While significant correlations are often found, whether abiotic factors are really important to habitat selection of marsh fish, or simply correlates of other processes, is unclear. Some abiotic variables, such as dissolved oxygen, certainly impose limits on the distribution of fish if sufficiently extreme. Most fish, however, are tolerant of a wide range of abiotic conditions. This should be particularly true for species that have evolved within the context of fluctuating environmental conditions (marsh residents) or migrate between coastal and estuarine environments (marsh transients). Abiotic variables are likely to influence habitat selection of marsh fishes in more subtle ways via effects on fish energy budgets rather than physiological tolerances. The interpretation that tidal movements of fish in marsh creeks minimize energetic costs of maintaining position in a tidal current (Szedlmayer and Able 1993) or acclimating to tidally induced changes in abiotic conditions (Rountree and Able 1992b) are alternative mechanistic hypotheses of how abiotic factors influence habitat selection that can be experimentally tested.

While diet surveys indicate that fish feed in marsh habitats, we found few studies that compared the foraging value of marshes relative to other available habitat types. Analyses of tidal effects on stomach fullness provide indirect evidence that marsh habitats are preferentially used for feeding. Most studies that have examined tidal effects have found that fish have fuller stomachs on ebb tides than on flood tides (Weisberg et al. 1981, Klepas and Dean 1983, Rozas and LaSalle, 1990, Rountree and Able 1992b; but see Baker-Dittus 1978), suggesting fish move into intertidal creeks and marshes to feed at high tide. Some experimental evidence suggests access to the marsh surface is necessary to achieve growth rates observed in the field (Weisberg and Lotrich 1982). Structural characteristics such as shallow sloping, depositional creek banks (McIvor and Odum 1988) and the presence of submerged aquatic vegetation (Rozas and Odum 1988) may enhance foraging profitability in the subtidal environment and are often correlated with higher fish densities on the adjacent vegetated marsh. These studies suggest marshes may be more profitable foraging areas than other habitats. This conclusion, however, depends critically on the comparative foraging profitability of alternative habitats. Foraging profitability depends not only on the abundance of food in the environment or the presence of food in fish stomachs, but the rate at which fish encounter food, the probability of acquiring it, and the energy expended in doing so. Methods to quantify foraging costs and benefits are well developed (e.g., Mittelbach 1981) and could be used to determine the foraging value of marsh habitats relative to other habitat types.

Abiotic factors or food alone may not be sufficient to explain patterns of habitat use in fishes. This may be particularly true for marsh ecosysystems where habitat availability, as well as physical dynamics that potentially influence abiotic conditions and prey renewal rates, vary over short time scales. Given this variability and suggestions that habitat selection of marsh fish is influenced by energetic considerations (Rountree and Able 1992b, Szedlmayer and Able 1993, Baltz et al. 1993), a bioenergetics approach may be particularly useful in marsh ecosystems. Though fish bioenergetic models have existed for at least 20 years (Kitchell et al. 1977), only recently have they been coupled with foraging models and applied to habitat selection of fish (see *Bioenergetics* above). Prior research in marshes suggests that food resources may be more abundant, but abiotic conditions may impose greater energetic costs than in other habitats. A bioenergetic approach could be employed to determine if this tradeoff actually exists and, if so, whether fish integrate these factors to select the habitat that maximizes net energy gain.

There is little information available to evaluate the role of biotic interactions on habitat selection of marsh fishes. While temporal variation in recruitment to, or emigration from, marsh creeks by transient species has been suggested to reflect the past (Subrahmanyam and Coultas 1980) or current (Rackocinski et al. 1992) effects of interspecific competition, a variety of other explanations for seasonal shifts in species composition can be invoked. The largest discrepancy between factors mentioned as important determinants of habitat use and those measured was for the predation refuge function of marsh habitat. We found only four studies that addressed the effects of predation risk on habitat use though 39 mentioned predation refuge as an important function marshes provide fish. Using manipulative field experiments on the vegetated marsh surface, Kneib (1987) showed that larval killifish (*Fundulus spp.*) had much higher mortality rates when enclosed with adults in artificial subtidal habitats than when adults were excluded. The results suggest that predation risk in addition to food resources and probability of dessication (Kneib and Wagner 1994) are important considerations for young killifish remaining in water filled pits on the marsh surface at low tide rather than subtidal creeks. Tethering experiments suggest that fish occupying marshes with depositional (as opposed to erosional) creek banks or submerged aquatic vegetation (SAV) have lower predation rates, and this (along with more abundant food resources) may be responsible for higher fish densities on adjacent vegetated marshes (McIvor and Odum 1988, Rozas and Odum 1988). Recent observations of high densities of relatively large piscivores in tidal creeks (Rountree and Able 1997) suggests that predation risk in subtidal environments may be substantial for juvenile fish. Risk of predation from birds or mobile invertebrates, however, may be important for species remaining in relatively shallow water on the marsh surface (Kneib 1982, Crowder et al. 1997). Thus, predation risk is likely a complex function of multiple terrestrial and aquatic predators mediated by the tidal cycle. Studies to quantify this risk as a function of predator and prey species and sizes, as well as tidal stage would provide insight on the role of predation in driving tidal movements in and out of vegetated marsh habitats and marsh creeks.

4. Conclusions

While fish are found at relatively higher densities in the vicinity of marsh habitat than unvegetated habitats (Zimmerman et al. 1984, Ayvazian et al. 1992), there is currently little information available to evaluate the relative importance of potential mechanisms driving this pattern. Most studies in marsh systems have used descriptive survey techniques to measure abiotic factors and their relationship to fish use of marsh habitat. Foraging areas and predator refuge, however, are often mentioned as the primary value of marsh habitat to fishes. Clearly, a single factor approach will not be adequate to resolve how the multiple factors that potentially influence habitat selection operate to determine distributional patterns in the field. The relevant question is not which factor is most important, but how do multiple factors interact to determine habitat selection, and how do these interactions change over time. Do fish discriminate among habitats that vary simultaneously with respect to several factors, weigh associated costs and benefits, and select habitats accordingly? A strong empirical basis exists in marsh ecosystems to begin addressing this question (Fig. 4). For example, young fish that remain on the marsh surface or in shallow pools at low tide may experience increased per capita food availability and decreased predation rates from subtidal predators, but increased energetic costs due to extreme abiotic conditions and increased susceptibility to terrestrial predators. Piscivorous fish in subtidal creeks may have increased encounter rates with prey fish leaving the vegetated surface as the tide recedes, but higher energetic costs due to extreme or fluctuating abiotic conditions than would be experience in other habitats. If a single habitat is optimal with respect to energetic gains *and* predation risk for the individual, then competitive interactions within this habitat may be important. Many studies that examine factors influencing habitat selection in fish rely on seasonal changes in habitat profitability (Mittelbach 1981, Persson and Greenberg 1990). Because habitat profitability of tidal marshes varies with the tidal cycle (being zero at low tide for many species), these systems offer a unique opportunity to test alternative behavioral mechanisms of habitat selection on relatively short time scales. We contend that a search for these and other mechanisms of habitat selection is a useful means of understanding the functional value of marsh habitat to aquatic organisms. Ideally, decisions regarding the implementation and evaluation of habitat restoration and management efforts should be based on whether the needs of the organism are satisfied. Without an understanding of the functional value of habitat we can be assured these decision will be based on other criteria.

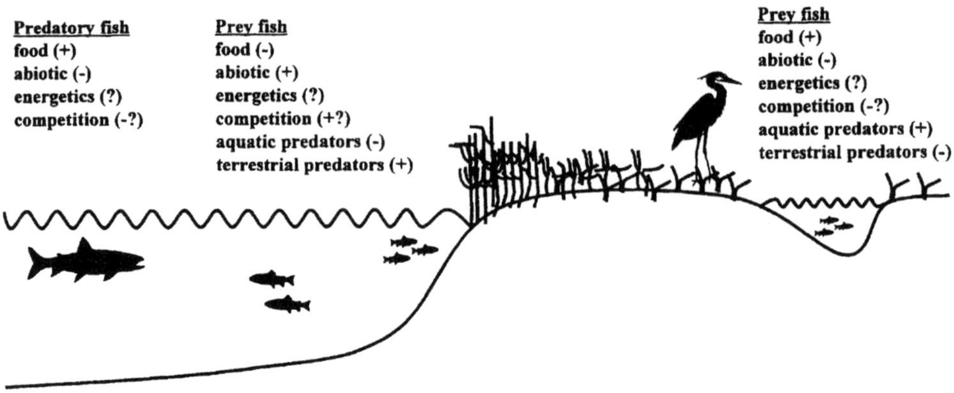

Figure 4. Some potential factors influencing habitat selection of fish in marsh ecosystems. (+) and (-) refer to potential positive or negative effects for that factor relative to the alternative habitats (subtidal vs. marsh surface for prey fish; creek vs. river-estuary for predator fish). Effects are depicted for low tide and would be reversed at high tide.

5. Acknowledgements

We thank Michael Weinstein and co-organizers for the invitation to participate in this symposium. Lisa Eby, Will Figueria and three anonymous reviewers provided helpful comments on the manuscript.

6. Literature Cited

Abrahams, M.V. 1989. Foraging guppies and the ideal free distribution: the influence of information on patch choice. Ethology **82**:116-126.

Abrahams, M.V. and L.M. Dill. 1989. A determination of the energetic equivalence of the risk of predation. Ecology **70**:999-1007.

Allen, L.G. 1982. Seasonal abundance, composition and productivity of the littoral fish assemblage in upper Newport Bay, California. Fishery Bulletin, US **80**:769-790.

Allen, D.M. and D.L. Barker. 1990. Interannual variations in larval fish recruitment to estuarine epibenthic habitats. Marine Ecology Progress Series **63**:113-125.

Ayvazian, S.G., L.A. Deegan and J.T. Finn. 1992. Comparison of habitat use by estuarine fish assemblages in the Acadian and Virginian zoogeographic provinces. Estuaries **15**:368-383.

Bailey, K.M. and E.D. Houde. 1989. Predation on eggs and larvae of marine fishes and the recruitment problem. Advances in Marine Biology **72**:295-301.

Baker-Dittus, A.M. 1978. Foraging patterns of three sympatric killifish. Copeia **1978**:383-389.

Baltz, D.M., C. Rakocinski and J.W. Fleeger. 1993. Microhabitat use by marsh-edge fishes in a Louisiana estuary. Environmental Biology of Fishes **36**:109-126.

Beitinger, T.L. and L.C. Fitzpatrick. 1979. Physiological and ecological correlates of preferred temperature in fish. American Zoologist **19**:319-330.

Boesch, D.F. and R.E. Turner. 1984. Dependence of fishery species on salt marshes: the role of food and refuge. Estuaries **7(4A)**:460-468.

Brabrand, A. and B. Faafeng. 1993. Habitat shift in roach (*Rutilus rutilus*) induced by pikeperch (*Stizostedion lucioperca*) introduction: predation risk versus pelagic behaviour. Oecologia **95**:38-46.

Brandt, S.B. and J. Kirsch. 1993. Spatially explicit models of striped bass growth potential in Chesapeake Bay. Transactions of the American Fisheries Society **122**:845-869.

Brandt, S.B. and D.M. Mason. 1994. Landscape approaches for assessing spatial patterns in fish foraging and growth. Pages 211-240 *in* D.J. Stouder, K.L. Fresh and R.J. Feller, editors. Theory and application of fish feeding ecology, University of South Carolina Press, Columbia, South Carolina, USA.

Brandt, S.B., D.M. Mason and E.V. Patrick. 1992. Spatially-explicit models of fish growth rate. Fisheries **17**:23-33.

Brett, J.R. 1971. Environmental factors and growth. Pages 599-673 *in* W.S. Hoar and D.J. Randall, editors. Fish physiology, Volume Eight. Academic Press, New York, New York, USA.

Bryan, J.D., L.G. Hill and W.H. Neill. 1984. Interdependence of acute temperature preference and respiration in the plains minnow. Transactions of the American Fisheries Society **113**:557-562.

Cattrijsse, A., E.S. Makwaia, H.R. Dankwa, O. Hamerlynck and M.A. Hemminga. 1994. Nekton communities of an intertidal creek of a European estuarine brackish marsh. Marine Ecology Progress Series **109**:195-208.

Cech, J.J., Jr., S.J. Mitchell, D.T. Castleberry and M. McEnroe. 1990. Distribution of California stream fishes: influence of environmental temperature and hypoxia. Environmental Biology of Fishes **29**:95-105.

Crowder, L.B. and J.J. Magnuson. 1983. Cost-benefit analysis of temperature and food resource use: a synthesis with examples from the fishes. Pages 189-221 *in* W.P. Aspey and S.I. Lustick, editors. Behavioral energetics. Ohio State University Press, Columbus, Ohio, USA.

Crowder, L.B., R.A. Wright, K.A. Rose, T.H. Martin and J.A. Rice. 1994. Direct and indirect effects of southern flounder predation on a spot population: Experimental and model analyses. Pages 61-77 *in* D.J. Stouder, K.L. Fresh and R.J. Feller, editors. Theory and application in fish feeding ecology. University of South Carolina Press, Columbia, South Carolina, USA.

Crowder, L.B., D.D. Squires and J.A. Rice. 1997. Nonadditive effects of terrestrial and aquatic predators on juvenile estuarine fish. Ecology **78**:1796-1804.

Cyrus, D.P. and S.J.M. Blaber 1987a. The influence of turbidity on juvenile marine fishes in estuaries. Part 1. Field studies at Lake St. Lucia on the southeastern coast of Africa. Journal of Experimental Marine Biology and Ecology **109**:53-70.

—— 1987b. The influence of turbidity on juvenile marine fishes in estuaries. Part 2. Laboratory studies, comparisons with field data and conclusions. Journal of Experimental Marine Biology and Ecology **109**:71-91.

Fraser, D.F. and T.E. Sise. 1980. Observations on stream minnows in a patchy environment: a test of a theory of habitat distribution. Ecology **61**:790-797.

Fretwell, S.D. and H.L. Lucas. 1970. On territorial behavior and other factors influencing habitat distribution in birds. I. Theoretical development. Acta Biotheoretica **19**:16-36.

Fry, F.E.J. 1971. The effect of environmental factors on the physiology of fish. Pages 1-98 *in* W.S. Hoar and D.J. Randall, editors. Fish physiology, Volume Six. Academic Press, New York, New York, USA.

Gillis, D.M. and D.L. Kramer. 1987. Ideal interference distributions: population density and patch use by zebrafish. Animal Behaviour **35**:1875-1882.

Gilliam, J.F. and D.F. Fraser. 1987. Habitat selection under predation hazard: a test of a model with foraging minnows. Ecology **68**:1856-1862.

Godin, J.J. and M.H.A. Keenleyside. 1984. Foraging on patchily distributed prey by a cichlid fish (Teleostei, Cichlidae): a test of the ideal free distribution theory. Animal Behaviour **32**: 120-131.

Gotceitas, V., S. Fraser and J.A. Brown. 1995. Habitat use by juvenile Atlantic cod (*Gadus morhua*) in the presence of an actively foraging and non-foraging predator. Marine Biology **123**:421-430.

Goyke, A. and S.B. Brandt. 1993. Spatial models of salmonine growth rates in Lake Ontario. Transactions of the American Fisheries Society **122**:870-883.

Grand, T.C. 1997. Foraging site selection by juvenile coho salmon: ideal free distributions of unequal competitors. Animal Behaviour **53**:185-196.

Grand, T.C. and J.W.A. Grant. 1994. Spatial predictability of food influences its monopolization and defense by juvenile convict cichlids. Animal Behaviour **47**:91-100.

Hackney, C.T., W.D. Burbanck and O.P. Hackney. 1976. Biological and physical dynamics of a Georgia tidal creek. Chesapeake Science **17**:271-280.

Heck, K.L., Jr. and L.B. Crowder. 1991. Habitat structure and predator-prey interactions in vegetated aquatic systems. Pages 281-299 *in* S.S. Bell, E.D. McCoy and H.R. Mushinsky, editors. Habitat structure: the physical arrangement of objects in space. Chapman and Hall, London, England.

Hill, J. and G.D. Grossman. 1993. An energetic model of microhabitat use for rainbow trout and rosyside dace. Ecology **74**:685-698.
Hixon, M.A. 1980. Competitive interactions between California reef fishes of the genus *Embiotoca.* Ecology **61**:918-931.
Hughes, N.F. 1998. A model of habitat selection by drift-feeding stream salmonids at different spatial scales. Ecology **70**:281-294.
Hughes, N.F. and L.M. Dill. 1990. Position choice by drift-feeding salmonids: model and test for arctic grayling (*Thymallus arcticus*) in subarctic mountain streams, interior Alaska. Canadian Journal of Fisheries and Aquatic Sciences **47**:2039-2048.
Hugie, D.M. and L.M. Dill. 1994. Fish and game: a game theoretic approach to habitat selection by predators and prey. Journal of Fish Biology **45(supplement A)**:151-169.
Jordan, F., M. Bartolini, C. Nelson, P.E. Patterson and H.L. Soulen. 1996. Risk of predation affects habitat selection by the pinfish (*Lagodon rhomboides*). Journal of Experimental Marine Biology and Ecology **208**:45-56.
Kennedy, M. and R.D. Gray. 1993. Can ecological theory predict the distribution of foraging animals? A critical analysis of experiments on the ideal free distribution. Oikos **68**: 158-166.
Kirschner, L.B. 1995. Energetics of osmoregulation in fresh water vertebrates. The Journal of Experimental Zoology **271**:243-252.
Kitchell, J.F., D.J. Stewart and D. Weininger. 1977. Applications of a bioenergetics model to yellow perch (*Perca flavescens*) and walleye (*Stizostedion vitreum vitreum*). Journal of the Fisheries Research Board of Canada **34**:1922-1935.
Kleypas, J. and J.M. Dean. 1983. Migration and feeding of the predatory fish, *Bairdiella chrysura* Lacepede, in an intertidal creek. Journal of Experimental Marine Biology and Ecology **72**:199-209.
Kneib, R.T. 1982. The effects of predation by wading birds (Ardeidae) and blue crabs (*Callinectes sapidus*) on the population size structure of the common mummichog, *Fundulus heteroclitus*. Estuarine, Coastal and Shelf Science **14**:159-165.
—— 1987. Predation risk and use of intertidal habitats by young fishes and shrimp. Ecology **68**:379-386.
—— 1997a. Early life stages of resident nekton in intertidal marshes. Estuaries **20**:214-230.
—— 1997b. The role of tidal marshes in the ecology of estuarine nekton. Oceanography and Marine Biology: An Annual Review **35**:163-220.
Kneib, R.T. and S.L. Wagner. 1994. Nekton use of vegetated marsh habitats at different stages of tidal inundation. Marine Ecology Progress Series **106**:227-238.
Kramer, D.L., R.W. Rangeley and L.J. Chapman. 1997. Habitat selection: patterns of spatial distribution from behavioural decisions. Pages 37-80 *in* J.J. Godin, editor. Behavioural ecology of teleost fishes. Oxford University Press, New York, New York, USA.
Larson, R.J. 1980. Competition, habitat selection and bathymetric segregation of two rockfish (*Sebastes*) species. Ecology **50**:221-239.
Lima, S.L. and L.M. Dill. 1990. Behavioral decisions made under the risk of predation: a review and prospectus. Canadian Journal of Zoology **68**:619-640.
Litvak, M.K. and W.C. Leggett. 1992. Age and size-selective predation on larval fishes: the bigger-is-better hypothesis revisited. Marine Ecology Progress Series **81**:13-24.
Lough, R., P.C. Valentine, D.C. Potter, P.J. Auditore, G.R. Bolz, J.D. Neilson and R.I. Perry. 1989. Ecology and distribution of juvenile cod and haddock in relation to sediment type and bottom currents on eastern Georges Bank. Marine Ecology Progress Series **56**:1-12.
MacArthur, R.H. and E.R. Pianka. 1966. On optimal use of a patchy environment. American Naturalist **100**:603-609.
Magnuson, J.J., L.B. Crowder and P.A. Medvick. 1979. Temperature as an ecological resource. American Zoologist **19**:331-343.
Main, K.L. 1987. Predator avoidance in seagrass meadows: prey behavior, microhabitat selection and cryptic coloration. Ecology **68**:170-180.
Mason, D.M., A. Goyke and S.B. Brandt. 1995. A spatially explicit bioenergetics measure of habitat quality for adult salmonines: comparison between lakes Michigan and Ontario. Canadian Journal of Fisheries and Aquatic Sciences **52**:1572-1583.
McIvor, C.C. and W.E. Odum. 1988. Food, predation risk and microhabitat selection in a marsh fish assemblage. Ecology **69**:1341-1351.

McIvor, C.C. and L.P. Rozas. 1996. Direct nekton use of intertidal saltmarsh habitat and linkage with adjacent habitats: a review from the southeastern United States. Pages 311-334 *in* K.F. Nordstrom and C.T. Roman, editors. Estuarine shores: evolution, environments and human alterations. John Wiley and Sons Ltd., Chichester, England.
Milinski, M. 1984. Competitive resource sharing: an experimental test of a learning rule for ESSs. Animal Behaviour **32**:233-242.
Milinski, M. 1994. Long-term memory for food patches and implications for ideal free distributions in sticklebacks. Ecology **75**:1150-1156.
Miller, T.J., L.B. Crowder, J.A. Rice and E.A. Marschall. 1988. Larval size and recruitment mechanisms in fishes: toward a conceptual framework. Canadian Journal of Fisheries and Aquatic Sciences **45**:1657-1670.
Mittelbach, G.G. 1981. Foraging efficiency and body size: a study of optimal diet and habitat use by bluegills. Ecology **62**:1370-1386.
—— 1984. Predation and resource partitioning in two sunfishes (Centrarchidae). Ecology **65**:499-513.
—— 1988. Competition among refuging sunfishes and effects of fish density on littoral zone invertebrates. Ecology **69**:614-623.
Neill, W.H. and J.J. Magnuson. 1974. Distributional ecology and behavioral thermoregulation of fishes in relation to heated effluent from a power plant at Lake Monona, Wisconsin. Transactions of the American Fisheries Society **103**:663-710.
Nixon, S.W. 1980. Between coastal marshes and coastal waters-a review of twenty years of speculation and research on the role of salt marshes in estuarine productivity and water chemistry. Pages 437-525 *in* P. Hamilton and K.B. MacDonald, editors. Estuarine and wetland processes. Plenum, New York, New York, USA.
NMFS (National Marine Fisheries Service). 1996. Magnuson Stevens Fishery Conservation and Management Act as Amended through October 11, 1996. U.S. Department of Commerce, National Oceanic and Atmospheric Administration Technical Memorandum NMFS-F/SPO-23, 121 p.
Odum, W.E. 1988. Comparative ecology of tidal freshwater and salt marshes. Annual Review of Ecology and Systematics **19**:147-176.
Osenberg, C.W. and G.G. Mittelbach. 1989. Effects of body size on the predator-prey interaction between pumpkinseed sunfish and gastropods. Ecological Monographs **59**:405-432.
Persson, L. 1986. Effects of reduced interspecific competition on resource utilization in perch (*Perca fluviatilis*). Ecology **67**:355-364.
Persson, L. 1993. Predator-mediated competition in prey refuges: the importance of habitat dependent prey resources. Oikos **68**:12-22.
Persson, L. and L.A. Greenberg. 1990. Juvenile competitive bottlenecks: the perch (*Perca fluviatilis*)-roach (*Rutilus rutilus*) interaction. Ecology **71**:44-56.
Persson, L. and L.B. Crowder. 1998. Fish-habitat interactions mediated via ontogenetic niche shifts. Pages 3-23 *in* E. Jeppesen, Ma. Sondergaard, Mo. Sondergaard and K. Christoffersen, editors. The structuring role of submerged macrophytes in lakes. Ecological Studies, Volume **131**. Springer-Verlag, New York, New York, USA.
Power, M.E. 1984. Habitat quality and the distribution of algae-grazing catfish in a Panamanian stream. Journal of Animal Ecology **53**:357-374.
Rakocinski, C.F., D.M. Baltz and J.W. Fleeger. 1992. Correspondence between environmental gradients and the community structure of marsh-edge fishes in a Louisiana estuary. Marine Ecology Progress Series **80**:135-148.
Reis, R.R. and J.M. Dean. 1981. Temporal variation in the utilization of an intertidal creek by the bay anchovy (*Anchoa mitchilli*). Estuaries **4**:16-23.
Rice, J.A., L.B. Crowder and E.A. Marschall. 1997. Predation on juvenile fishes: dynamic interactions between size-structured predators and prey. Pages 333-356 *in* R.C. Chambers and E.A. Trippel, editors. Early life history and recruitment in fish populations. Chapman & Hall, New York, New York, USA.
Rose, G.A. and W.C. Leggett. 1989. Interactive effects of geophysically-forced sea temperatures and prey abundance on mesoscale coastal distributions of a marine predator, Atlantic cod (*Gadus morhua*). Canadian Journal of Fisheries and Aquatic Sciences. **46**:1904-1913.
Rountree, R.A. and K.W. Able 1992a. Fauna of polyhaline subtidal marsh creeks in southern New Jersey: composition, abundance and biomass. Estuaries **15**:171-185.
—— 1992b. Foraging habits, growth and temporal patterns of salt marsh creek habitat use by young-of-year summer flounder in New Jersey. Transactions of the American Fisheries Society **121**:765-776.

—— 1996. Seasonal abundance, growth and foraging habits of juvenile smooth dogfish, *Mustelus canis*, in a New Jersey estuary. Fishery Bulletin US **94**:522-534.

—— 1997. Nocturnal fish use of New Jersey marsh creek and adjacent bay shoal habitats. Estuarine, Coastal and Shelf Science **44**:703-711.

Rozas, L.P. 1995. Hydroperiod and its influence on nekton use of the salt marsh: a pulsing ecosystem. Estuaries **18**:579-590.

Rozas, L.P. and W.E. Odum. 1987. The role of submerged aquatic vegetation in influencing the abundance of nekton on contiguous tidal fresh-water marshes. Journal of Experimental Marine Biology and Ecology **114**:289-300.

Rozas, L.P. and W.E. Odum. 1988. Occupation of submerged aquatic vegetation by fishes: testing the roles of food and refuge. Oecologia **77**:101-106.

Rozas, L.P. and M.W. LaSalle. 1990. A comparison of the diets of gulf killifish, *Fundulus grandis* Baird and Girard, entering and leaving a Mississippi brackish marsh. Estuaries **13**:332-336.

Ryer, C.H. 1987. Studies on the foraging ecology of pipefish: prey selection and habitat selection. Dissertation, College of William and Mary, Williamsburg, Virginia, USA.

Ryer, C.H. 1988. Pipefish foraging and the effects of altered habitat complexity. Marine Ecology Progress Series **48**:37-45.

Savino, J.F. and R.A. Stein. 1989. Behavioural interactions between fish predators and their prey: effects of plant density. Animal Behaviour **37**:11-32.

Schmitt, R.J. and S.J. Holbrook. 1986. Seasonally fluctuating resources and temporal variability of interspecific competition. Oecologia **69**:1-11.

Schurmann, H. and J.F. Steffensen. 1992. Lethal oxygen levels at different temperatures and the preferred temperature during hypoxia of the Atlantic cod, *Gadus morhua*. Journal of Fish Biology **41**:927-934.

Schurmann, H., J.F. Steffensen and J.P. Lomholt. 1991. The influence of hypoxia on the preferred temperature of rainbow trout *Oncorhynchus mykiss*. Journal of Experimental Biology **157**:75-86.

Sih, A. 1998. Game theory and predator-prey response races. Pages 221-238 *in* L.A. Dugatlin and J.K. Reeve, editors. Game theory and animal behavior. Oxford University Press, New York, New York, USA.

Stein, R.A., S.T. Threlkeld, C.D. Sandgren, W.G. Sprules, L. Persson, W.E. Neill, E.E. Werner and S.I. Dodson. 1988. Size-structured interactions in lake communities. Pages 161-179 *in* S.R. Carpenter, editor. Complex interactions in lake communities. Springer Verlag, Heidelberg, Germany.

Subrahmanyam, C.B. and C.L. Coultas. 1980. Studies on the animal communities in two north Florida salt marshes Part III. Seasonal fluctuations of fish and macroinvertebrates. Bulletin of Marine Science **30**:790-818.

Swain, D.P. and E.J. Wade. 1993. Density-dependent geographic distribution of Atlantic cod (*Gadus morhua*) in the southern Gulf of St. Lawrence. Canadian Journal of Fisheries and Aquatic Science **50**:725-733.

Swain, D.P. and D.L. Kramer. 1995. Annual variation in temperature selection by Atlantic cod (*Gadus morhua*) in the southern Gulf of St. Lawrence and its relation to population size. Marine Ecology Progress Series **116**:11-23.

Szedlmayer, S.T. and K.W. Able. 1993. Ultrasonic telemetry of age-0 summer flounder, *Paralichthys dentatus*, movements in a southern New Jersey estuary. Copeia **1993**:728-736.

Talbot, A.J. and D.L. Kramer. 1986. Effects of food and oxygen availability on habitat selection by guppies in a laboratory environment. Canadian Journal of Zoology **64**:88-93.

Teal, J.M. 1962. Energy flow in a salt marsh ecosystem of Georgia. Ecology **43**:614-624.

Tonn, W.M. and C.A. Paszkowski. 1992. Piscivory and recruitment: Mechanisms structuring prey populations in small lakes. Ecology **73**:951-958.

Turner, A.M. and G.G. Mittelbach. 1990. Predator avoidance and community structure: interactions among piscivores, planktivores and plankton. Ecology **71**:2241-2254.

Tyler, J.A. and J.F. Gilliam. 1995. Ideal free distributions of stream fish: a model and test with minnows *(Rhinicthys atratulus*). Ecology **76**:580-592.

Utne, A.C.W. and B. Bacchi. 1997. The influence of visual and chemical stimuli from cod (*Gadus morhua*) on the distribution of two-spotted goby (*Gobiusculus flavescens*). Sarsia **82**:129-135.

Vince, S.W., I. Valiela, N. Backus and J.M. Teal. 1976. Predation by the salt marsh killifish *Fundulus heteroclitus* (L.) in relation to prey size and habitat structure: consequences for prey distribution and abundance. Journal of Experimental Marine Biology and Ecology **23**:255-266.

Weisberg, S.B. 1986. Competition and coexistence among four estuarine species of *Fundulus*. American Zoologist **26**:249-257.

Weisberg, S.B. and V.A. Lotrich. 1982. The importance of an infrequently flooded intertidal marsh surface as an energy source for the mummichog, *Fundulus heteroclitus*: an experimental approach. Marine Biology **66**:307-310.

Weisberg, S.B., R. Whalen and V.A. Lotrich. 1981. Tidal and diurnal influence on food consumption of a salt marsh killifish, *Fundulus heteroclitus*. Marine Biology **61**:243-246.

Werner, E.E. 1988. Size, scaling and the evolution of complex life cycles. Pages 60-81 *in* B. Ebenman and L. Persson, editors. Size-structured populations: ecology and evolution. Springer Verlag, Heidelberg, Germany.

Werner, E.E. and D.J. Hall. 1976. Niche shifts in sunfishes: experimental evidence and significance. Science **191**:404-406.

Werner, E.E. and D.J. Hall. 1977. Competition and habitat shift in two sunfishes (Centrarchidae). Ecology **58**:869-876.

Werner, E.E. and D.J. Hall. 1979. Foraging efficiency and habitat switching in competing sunfishes. Ecology **60**:256-264.

Werner, E.E. and G.G. Mittelbach. 1981. Optimal foraging: field tests of diet choice and habitat switching. American Zoologist **21**:813-829.

Werner, E.E. and J.F. Gilliam. 1984. The ontogenetic niche and species interactions in size structured populations. Annual Review of Ecology and Systematics **15**:393-425.

Werner, E.E., G.G. Mittelbach, D.J. Hall and J.F. Gilliam 1983a. Experimental tests of optimal habitat use in fish: the role of relative habitat profitability. Ecology **64**:1525-1539.

Werner, E.E., J.F. Gilliam, D.J. Hall and G.G. Mittelbach. 1983b. An experimental test of the effects of predation risk on habitat use in fish. Ecology **64**:1540-1548.

Wildhaber, M.L. and L.B. Crowder. 1990. Testing a bioenergetics-based habitat choice model: bluegill (*Lepomis macrochirus*) responses to food availability and temperature. Canadian Journal of Fisheries and Aquatic Sciences **47**:1664-1671.

Wright, D.I. and W.J. O'Brien. 1984. The development and field test of a tactical model of the planktivorous feeding of the white crappie (*Pomoxis annularis*). Ecological Monographs **54**:65-98.

Zimmerman, R.J. and T.J. Minello. 1984. Densities of *Penaeus aztecus*, *Penaeus setiferus* and other natant macrofauna in a Texas salt marsh. Estuaries. **7**:421-433

Appendix I

Studies included in the review listed in alphabetical order by author's last name.

1. Allen, L.G. 1982. Seasonal abundance, composition and productivity of the littoral fish assemblage in upper Newport Bay, California. Fishery Bulletin, US **80**:769-790.
2. Allen, D.M. and D.L. Barker. 1990. Interannual variations in larval fish recruitment to estuarine epibenthic habitats. Marine Ecology Progress Series **63**:113-125.
3. Allen, R.L. and D.M. Baltz. 1997. Distribution and microhabitat use by flatfishes in a Louisiana estuary. Environmental Biology of Fishes **50**:85-103.
4. Archambault, J.A. and R.J. Feller. 1991. Diel variation in gut fullness of juvenile spot, *Leiostomus xanthurus* (Pisces). Estuaries **14**:94-101.
5. Baker-Dittus, A.M. 1978. Foraging patterns of three sympatric killifish. Copeia **1978:**383-389.
6. Baltz, D.M., C. Rakocinski and J.W. Fleeger. 1993. Microhabitat use by marsh-edge fishes in a Louisiana estuary. Environmental Biology of Fishes **36**:109-126.
7. Bozeman, E.L., Jr. and J.M. Dean. 1980. The abundance of estuarine larval and juvenile fish in a South Carolina Intertidal Creek. Estuaries **3**:89-97.
8. Byrne, D.M. 1978. Life history of the spotfin killifish *Fundulus luciae* (Pisces: Cyprinodontidae), in Fox Creek Marsh, Virginia. Estuaries **1**:211-227.
9. Cain, R.L. and J.M. Dean. 1976. Annual occurrence, abundance and diversity of fish in a South Carolina intertidal creek. Marine Biology **36**:369-379.
10. Cattrijsse, A., E.S. Makwaia, H.R. Dankwa, O. Hamerlynck and M.A. Hemminga. 1994. Nekton communities of an intertidal creek of a European estuarine brackish marsh. Marine Ecology Progress Series **109**:195-208.
11. Feller, R.J., B.C. Coull and B.T. Hentschel. 1990. Meiobenthic copepods: Tracers of where juvenile *Leiostomus xanthurus* (Pisces) feed? Canadian Journal of Fisheries and Aquatic Sciences **47**:1913-1919.
12. Gutierrez-Estrada, J.C., J. Prenda, F. Oliva and C. Fernandez-Delgado. 1998. Distribution and habitat preferences of the introduced mummichog *Fundulus heteroclitus* (Linneaus) in southwestern Spain. Estuarine, Coastal and Shelf Science **46**:827-835.
13. Hackney, C.T., W.D. Burbanck and O.P. Hackney. 1976. Biological and physical dynamics of a Georgia tidal creek. Chesapeake Science **17**:271-280.
14. Halpin, P.M. 1997. Habitat use patterns of the mummichog, *Fundulus heteroclitus*, in New England. I. Intramarsh variation. Estuaries **20**:618-625.
15. Hettler, W.F., Jr. 1989. Nekton use of regularly-flooded saltmarsh cordgrass habitat in North Carolina, USA. Marine Ecology Progress Series. **56**:111-118.
16. Hodson, R.G., J.O. Hackman and C.R. Bennett. 1981. Food habits of young spots in nursery areas of the Cape Fear river estuary, North Carolina. Transactions of the American Fisheries Society **110**:495-501.
17. Irlandi, E.A. and M.K. Crawford. 1997. Habitat linkages: the effect of intertidal saltmarshes and adjacent subtidal habitats on abundance, movement and growth of an estuarine fish. Oecologia **110**:222-230.
18. Kleypas, J. and J.M. Dean. 1983. Migration and feeding of the predatory fish, *Bairdiella chrysura* Lacepede, in an intertidal creek. Journal of Experimental Marine Biology and Ecology **72**:199-209.

19. Kneib, R.T. 1984. Patterns in the utilization of the intertidal salt marsh by larvae and juveniles of *Fundulus heteroclitus* (Linnaeus) and *Fundulus luciae* (Baird). Journal of Experimental Marine Biology and Ecology **83**:41-51.
20. Kneib, R.T. 1987. Predation risk and use of intertidal habitats by young fishes and shrimp. Ecology **68**:379-386.
21. —— 1993. Growth and mortality in successive cohorts of fish larvae within an estuarine nursery. Marine Ecology Progress Series **94**:115-127.
22. Kneib, R.T. and S.L. Wagner. 1994. Nekton use of vegetated marsh habitats at different stages of tidal inundation. Marine Ecology Progress Series **106**:227-238.
23. Kneib, R.T. 1997. Early life stages of resident nekton in intertidal marshes. Estuaries **20**:214-230.
24. Lipcius, R.N. and C.B. Subrahmanyam. 1986. Temporal factors influencing killifish abundance and recruitment in Gulf of Mexico salt marshes. Estuarine, Coastal and Shelf Science **22**:101-114.
25. Lotrich, V.A. 1975. Summer home range and movements of *Fundulus heteroclitus* (Pisces: Cyprinodontidae) in a tidal creek. Ecology **56**:191-198.
26. McIvor, C.C. and W.E. Odum. 1988. Food, predation risk and microhabitat selection in a marsh fish assemblage. Ecology **69**:1341-1351.
27. Miltner, R.J., S.W. Ross and M.H. Posey. 1995. Influence of food and predation on the depth distribution of juvenile spot (*Leiostomus xanthurus*) in tidal nurseries. Canadian Journal of Fisheries and Aquatic Sciences **52**:971-982.
28. Moyle, P.B., R.A. Daniels, B. Herbold and D.M. Baltz. 1986. Patterns in distribution and abundance of a noncoevolved assemblage of estuarine fishes in California. Fishery Bulletin US **84**:105-117.
29. O'Neil, S.P. and M.P. Weinstein. 1987. Feeding habitats of spot, *Leiostomus xanthurus*, in polyhaline versus meso-oligohaline tidal creeks and shoals. Fishery Bulletin, US **85**:785-796.
30. Peterson, G.W. and R.E. Turner. 1994. The value of salt marsh edge vs. interior as a habitat for fish and decapod crustaceans in a Louisiana tidal marsh. Estuaries **17**:235-262.
31. Rakocinski, C.F., D.M. Baltz and J.W. Fleeger. 1992. Correspondence between environmental gradients and the community structure of marsh-edge fishes in a Louisiana estuary. Marine Ecology Progress Series **80**:135-148.
32. Reis, R.R. and J.M. Dean. 1981. Temporal variation in the utilization of an intertidal creek by the bay anchovy (*Anchoa mitchilli*). Estuaries **4**:16-23.
33. Rountree, R.A. and K.W. Able. 1992. Foraging habits, growth and temporal patterns of salt-marsh creek habitat use by young-of-year summer flounder in New Jersey. Transactions of the American Fisheries Society **121**:765-776.
34. —— 1992. Fauna of polyhaline subtidal marsh creeks in southern New Jersey: composition, abundance and biomass. Estuaries **15**:171-185.
35. —— 1993. Diel variation in decapod crustacean and fish assemblages in New Jersey polyhaline marsh creeks. Estuarine, Coastal and Shelf Science **37**:181-201.
36. Rountree, R.A. and K.W. Able. 1996. Seasonal abundance, growth and foraging habits of juvenile smooth dogfish, *Mustelus canis*, in a New Jersey estuary. Fishery Bulletin, US **94**:522-534.

37. —— 1997. Nocturnal fish use of New Jersey marsh creek and adjacent bay shoal habitats. Estuarine, Coastal and Shelf Science **44**:703-711.
38. Rozas, L.P. and Hackney. 1984. Use of oligohaline marshes by fishes and macrofounal crustaceans in North Carolina. Estuaries 7:213-224.
39. Rozas, L.P. and W.E. Odum. 1987. Use of tidal freshwater marshes by fishes and macrofounal crustacean along a marsh stream-order gradient. Estuaries **10**:36-43.
40. —— 1987. Fish and macrocrustacean use of submerged plant beds in tidal freshwater marsh creeks. Marine Ecology Progress Series **38**:101-108.
41. —— 1987. The role of submerged aquatic vegetation in influencing the abundance of nekton on contiguous tidal fresh-water marshes. Journal of Experimental Marine Biology and Ecology **114**:289-300.
42. Rozas, L.P. and W.E. Odum. 1988. Occupation of submerged aquatic vegetation by fishes: testing the roles of food and refuge. Oecologia **77**:101-106.
43. Rozas, L.P., C.C. McIvor and W.E. Odum. 1988. Intertidal rivulets and creekbanks: corridors between tidal creeks and marshes. Marine Ecology Progress Series **47**:303-307.
44. Rozas, L.P. and M.W. LaSalle. 1990. A comparison of the diets of gulf killifish, *Fundulus grandis* Baird and Girard, entering and leaving a Mississippi brackish marsh. Estuaries **13**:332-336.
45. Shenker, J.M. and J.M. Dean. 1979. The utilization of an intertidal salt marsh creek by larval and juvenile fishes: abundance, diversity and temporal variation. Estuaries **2**:154-163.
46. Smith, S.M., J.G. Hoff, S.P. O'Neil and M.P. Weinstein. 1984. Community and trophic organization of nekton utilizing shallow marsh habitats, York River, Virginia. Fishery Bulletin, US **82**:455-467.
47. Smith, K.J. and K.W. Able. 1994. Salt-marsh tide pools as winter refuges for the mummichog, *Fundulus heteroclitus*, in New Jersey. Estuaries **17**:226-234.
48. Sogard, S.M. and K.W. Able. 1991. A comparison of eelgrass sea lettuce, macroalgae and marsh creeks as habitats for epibenthic fishes and decapods. Estuarine, Coastal and Shelf Science **33**:501-519.
49. Subrahmanyam, C.B. and S.H. Drake. 1975. Studies on the animal communities in two north Florida salt marshes. Bulletin of Marine Science **25**:445-465.
50. Subrahmanyam, C.B. and C.L. Coultas. 1980. Studies on the animal communities in two north Florida salt marshes Part III. Seasonal fluctuations of fish and macroinvertebrates. Bulletin of Marine Science **30**:790-818.
51. Szedlmayer, S.T. and K.W. Able. 1993. Ultrasonic telemetry of age-0 summer flounder, *Paralichthys dentatus*, movements in a southern New Jersey estuary. Copeia **1993**:728-736.
52. Szedlmayer, S.T. and K.W. Able. 1996. Patterns of seasonal availability and habitat use by fishes and decapod crustaceans in a southern New Jersey Estuary. Estuaries **19**:697-709.
53. Talbot, C.W. and K.W. Able. 1984. Composition and distribution of larval fishes in New Jersey high marshes. Estuaries 7:434-443.
54. Varnell, L.M., K.J. Havens and C. Hershner. 1995. Daily variability in abundance and population characteristics of tidal salt-marsh fauna. Estuaries **18**:326-334.

55. Weinstein, M.P. 1979. Shallow marsh habitats as primary nurseries for fishes and shellfish, Cape Fear River, North Carolina. Fishery Bulletin, US **77**:339-357.
56. Weinstein, M.P., S.L. Weiss and M.F. Walters. 1980. Multiple determinants of community structure in shallow marsh habitats, Cape Fear River estuary, North Carolina, USA. Marine Biology **58**:227-243.
57. Weinstein, M.P. and H.A. Brooks. 1983. Comparative ecology of nekton residing in a tidal creek and adjacent seagrass meadow: community composition and structure. Marine Ecology Progress Series **12**:15-27.
58. Weisberg, S.B. 1986. Competition and coexistence among four estuarine species of *Fundulus*. American Zoologist **26**:249-257.
59. Weisberg, S.B. and V.A. Lotrich. 1982. The importance of an infrequently flooded intertidal marsh surface as an energy source for the mummichog *Fundulus heteroo-clitus*: an experimental approach. Marine Biology **66**:307-310.
60. Weisberg, S.B., R. Whalen and V.A. Lotrich. 1981. Tidal and diurnal influence of food consumption of a salt marsh killifish *Fundulus heteroclitus*. Marine Biology **61**:243-246.
61. Zimmerman, R.J. and T.J. Minello. 1984. Densities of *Penaeus aztecus*, *Penaeus setiferus* and other natant macrofauna in a Texas salt marsh. Estuaries 7:421-433.

Appendix II. Studies mentioning, measuring, or concluding as important various factors influencing habitat use of marsh fishes. Numbers refer to studies listed in Appendix I.

Abiotic	Food	Bioenergetics	Competition	Predation
		Mentioned		
1,2,3,6,8,9,10,12 13,14,16,19,20, 21,22,24,25,28,29, 30,31,32,33,34 35,38,39,41,46, 47,48,49,50,51, 52,53,54,55,56, 57,60,61	1,3,4,5,6,7,8,9, 10,11,13,14,15 16,17,18,19,20, 21,22,24,25,26, 27,28,29,30,31, 32,33,35,36,38, 39,40,41,42,43, 44,45,46,47,48, 49,51,55,56,58, 59,60,61	3,6,8,25,26,29, 33,35,47,51	1,3,5,6,12,13,14, 17,21,28,30,31, 34,38,46,49,50, 54,55,56,58	1,3,6,7,8,9,10,14 15,16,18,19,20, 21,23,24,26,27, 29,30,31,32,34, 35,37,38,39,40, 41,42,43,45,46, 47,48,54,56,57, 61
		Measured		
1,2,3,6,8,9,10,12 13,19,21,32,33, 24,28,29,31,34, 35,38,39,46,47, 48,49,50,51,52, 53,54,55,56,57, 61	4,5,8,16,18,26, 27,33,36,11,38, 43,44,46,49,59, 60		21,58	20,26,27,42
		Important		
1,3,6,8,9,12,13, 24,28,29,31,32, 33,34,35,38,46, 47,49,51,52,53, 54,55,56,61	4,5,8,16,18,26, 27,33,36,11,38, 43,44,46,49,59, 60		21,58	20,26,27,42

SALT MARSH ECOSCAPES AND PRODUCTION TRANSFERS BY ESTUARINE NEKTON IN THE SOUTHEASTERN UNITED STATES

R. T. KNEIB
The University of Georgia Marine Institute
Sapelo Island, GA 31327 USA

Abstract
Understanding the role of tidal marshes in supporting estuarine nekton populations requires consideration of how different species and life stages use, and depend on, a variety of habitats. The problem might best be viewed from the perspective of a tidal marsh ecoscape, which relates variation in ecological interactions or processes to spatial patterns that emerge when associated marsh habitats are viewed together as a functional unit. Vegetated intertidal habitats, which define the salt marsh and account for most of its areal extent and productivity, are not used directly by most species of estuarine nekton in the southeastern U.S. If they function in the trophic support of these populations, marshes might supply dissolved nutrients to drive primary production in adjacent open waters or they could be a source of passively transported particles (i.e. drift) gathered by nekton from the water column or epibenthos. Alternatively, the few groups of nekton (mostly small marsh resident species) that feed within the marsh vegetation may actively translocate intertidal production horizontally across boundaries within the marsh ecoscape in a type of ''trophic relay''. Transfers to open estuarine waters may occur when material is either excreted in subtidal aquatic refugia at low tide, or accumulated biomass is passed along via predator-prey interactions. Temporal and spatial constraints on mobility and feeding behavior of nekton groups likely limit such production transfers to certain places and times, described as ''shifting interaction zones''. Identifying these interaction zones within the marsh ecoscape is a prerequisite for developing methods and sampling programs that will provide the type of information needed to address long-standing issues involving the role of nekton in the ecology of estuaries and the functional contribution of marshes to estuarine fisheries.

1. Introduction

Support for estuarine fisheries is among the many legendary functions attributed to tidal marshes. It is widely believed that marshes are important sources of food and refuge for nekton, particularly within the context of nurseries for the early life stages of transient marine species (Weinstein 1979, Boesch and Turner 1984). For decades, this piece of ''common knowledge'' has justified not only scientific studies of this habitat, but also encouraged preservation, restoration and creation of coastal marshes. However, as with any legend, it can be difficult to separate the core of truth – which is almost certainly present – from the layers of embellishment that inevitably envelop it over time. When asked for direct evidence of the amount of fishery production associated with a hectare of marsh, or the proportion of annual estuarine nekton biomass attributable to intertidal production, most estuarine ecologists would admit that answers to such questions remain elusive.

There are relatively few estimates of nekton production from tidal marshes, and these have tended to be restricted to the area of aquatic habitat occupied by species at low tide and not the entire intertidal marsh. Some of the first were provided for the mummichog, *Fundulus heteroclitus*, which Valiela et al. (1977) estimated as 16 g DW m^{-2} y^{-1} in Great Sippewissett salt marsh, Massachusetts. It was unclear whether the total area of marsh or only the permanent aquatic habitat was used in calculating this value. Meredith and Lotrich (1979) estimated that production of the same species in Canary Creek, Delaware was 40.7 g wet weight m^{-2} y^{-1} (ca. 10 g DW m^{-2} y^{-1}). This estimate was definitely based on subtidal creek area and not the entire intertidal marsh. Similarly, production of daggerblade grass shrimp (*Palaemonetes pugio*) from marsh creeks and embayments has been reported as ca. 9 -16 g DW m^{-2} y^{-1}(Sikora 1977, Welsh 1975).

A few estimates are also available for species that are only seasonally resident in and around tidal marshes. Allen (1982) reported that the entire fish assemblage from the littoral zone of a tidal marsh in California produced 9.35 g DW m^{-2} y^{-1}. Production of juvenile spot (*Leiostomus xanthurus*), with an average residence time of 86 d in some Virginia tidal marsh creeks, has been estimated at 4.2 - 4.6 g DW m^{-2} (Weinstein et al. 1984). Net export of all nekton (fishes and crustaceans) from two 35.2 hectare impoundments in a Louisiana tidal marsh averaged 21.7 g wet weight (ca. 5.4 g DW) m^{-2} y^{-1} (from Table 1 in Herke et al. 1992). Deegan (1993) calculated that juvenile Gulf menhaden populations emigrating from their nursery habitat in a Louisiana estuary removed 38 g DW m^{-2} y^{-1} in accumulated biomass, accounting for up to 10% of the total primary production of the system. Of course, the portion of that production attributable to tidal marshes could not be determined.

Beyond actual estimates of production from habitats associated with tidal marshes, there is correlative evidence for a positive relationship between fishery yield and primary production in the marine environment, with the greatest yields associated with estuaries (Nixon 1988), particularly those with large areas of intertidal vegetation (Turner 1977). However, for the most part, we have simply accepted that the demonstrably high primary production of intertidal marshes (Mitsch and Gosselink 1993) must eventually translate into enhanced estuarine fishery production (e.g., Nixon 1988).

Estimating production in natural populations is not easy under the best of conditions, but when the populations are mobile and the boundaries of the system are open and ill-defined, the task is formidable indeed. Tidal marshes are located primarily in estuaries at the land-sea boundary, where they are open to fluxes of materials – including organisms (e.g., Dame and Allen 1996) and their potential energy sources – that may be derived from both uplands and the coastal ocean as well as those produced *in situ.* Using evidence from stable isotope analyses, Deegan and Garritt (1997) have argued that estuarine consumer populations tend to utilize locally-produced sources of organic matter. If so, accessibility to food resources in the intertidal zone becomes an important issue in marsh-dominated estuaries. Marshes occur largely above mean low water, and so are potentially available to most aquatic consumers only for a portion of the tidal cycle (see Kneib and Wagner 1994, Rozas 1995, Kneib 1997a). The ability to address questions about production and material transfers involving nekton requires accurate estimates of density, as well as an understanding of spatial and temporal variability in habitat use patterns by mobile aquatic organisms. The complex structure and dynamics of the tidal marsh environment presents a variety of methodological challenges to researchers with respect to sampling designs and gear types needed to provide such information on marsh nekton assemblages (e.g., Kneib 1997a, Rozas and Minello 1997).

Beyond practical issues of sampling, there is a need for a conceptual framework that provides an organized and coordinated strategy from which to formulate and test appropriate hypotheses focusing on the ways nekton function within the marsh/estuary ecosystem. In this paper, I show how structural features of the estuarine marsh, tidal dynamics, nekton behavior, and life histories of different nekton species can be placed in a context that will help influence the way we design research and sampling schemes to address the issue of trophic support from a functional and mechanistic perspective. This contribution builds on a recent review of the literature dealing with the role of tidal marshes in the ecology of nekton (Kneib 1997a).

2. The Marsh Ecoscape

The term ''landscape ecology'' has been used to describe research that deals explicitly with effects of spatial pattern on ecological processes and continues to evolve as a discipline that holds promise to facilitate links between basic and applied ecology (Turner 1989, Farina 1998). However, ''landscape'' connotes a certain spatial scale and human perspective of the environment that tends to consider aquatic and terrestrial elements as physically separate (Kneib 1994). This seems inappropriate when dealing with certain environments, such as tidal marshes, which cycle between terrestrial and aquatic phases. Consequently, I use the term ''ecoscape'' here to convey the same sense of focus on the interplay of spatial pattern and ecological processes, but without a bias toward a particular phase state (i.e., terrestrial or aquatic). When considering movement of intertidal production to the open estuary through nekton assemblages at the ecoscape-level, spatial structure and temporal dynamics are equally important.

2.1 STRUCTURE

Tidal marshes are defined by the presence of distinct vascular plant assemblages that tolerate periodic tidal inundation, but not constant submergence, and are associated largely with estuaries at temperate latitudes (Mitsch and Gosselink 1993). Although the species composition varies regionally, the grasses, herbs and shrubs of marshes are easily distinguishable from subtidal beds of aquatic vegetation (e.g., seagrasses), and the woody plants of tidal freshwater swamps or intertidal mangrove communities of tropical estuaries (Tiner 1993). Thus, two key features defining tidal marshes are elevation relative to the local tidal range and the presence of certain assemblages of vascular plants.

Marshes may be defined by their vegetative cover, but features of the ecoscape that hold or channel water through the system become important when dealing with the nekton assemblages of estuaries because the ability of aquatic organisms to use an area depends on the presence of water. A variety of natural and man-made aquatic features occur within the estuarine system and either border, penetrate, or are embedded within tidal marsh ecoscapes. These include open subtidal bodies of water, channels and ponded aquatic habitats. For purposes of distinguishing the marsh ecoscape *per se* from the estuary as a whole, we can use position within a gradient of tidal inundation to define aquatic habitats embedded within the marsh from those that are adjacent to it. Most would agree that open and permanent subtidal waters of the estuary (e.g., sounds, bays, rivers) or coastal ocean, though fringed by intertidal vegetation, are not tidal marsh. However, intertidal channels and ponded waters, which do not maintain an aquatic connection with the subtidal estuary at low tide, may be considered features embedded within the tidal marsh ecoscape. It may be more difficult to reach agreement on how to categorize subtidal channels associated with marshes. They are commonly referred to as ''marsh creeks,'' but if such channels maintain an uninterrupted aquatic connection with the open estuary at low tide, one could view them as narrow extensions of the subtidal estuary. They could be considered corridors between elements (or habitats) of the larger estuarine ecoscape, or boundaries between the marsh and open estuary. However, being neither intertidal nor supporting the growth of marsh plant assemblages, permanently subtidal creeks cannot be considered part of the marsh proper.

Many shallow, estuarine habitats attractive to nekton often occur adjacent to, or in the general vicinity of, vegetated intertidal marshes. These include intertidal mudflats, oyster reefs and mussel beds, as well as shallow subtidal creek channels and embayments that may support beds of submerged aquatic vegetation. Such habitat features may have important effects on use of intertidal resources by nekton, or on predator-prey interactions contributing to the transfer of marsh production to the open estuary. For example, more small fishes may access the intertidal marsh habitat from the shoal waters on the depositional sections of tidal creeks than along steeper erosional banks (McIvor and Odum 1988). However, predation risk from larger predators may be higher along erosional banks (McIvor and Odum 1988), which suggests that these areas are important ''hot spots'' for the transfer of intertidal production to the subtidal estuary. Beds of aquatic vegetation in subtidal areas adjacent to the marsh may enhance transfer rates of intertidal production to the subtidal estuary by providing low-tide staging areas that concentrate nekton populations, which forage at high tide within or along the edge of intertidal marshes (Rozas and Odum 1987, Irlandi and Crawford 1997).

Other features important to nekton use are embedded within the intertidal marsh ecoscape. For example, small ''rivulets'' that breach creekbank boundaries between subtidal channels and the vegetated intertidal may function as corridors that focus the movements of nekton into and out of the intertidal marsh (Rozas et al. 1988, Hettler 1989). The surface of frequently-flooded intertidal marshes often includes shallow bodies of ponded water that provide habitat for aquatic nekton. Their size may be measured in centimeters (aquatic microhabitats) to hectares (marsh ponds). Aquatic microhabitats (only cm across and mm deep) are important nursery habitats for resident nekton in some marshes (Kneib 1984, Yozzo et al 1994, Kneib 1997b). Larvae and early juvenile stages of some resident nekton derive a measure of protection from predation on the marsh surface (Kneib 1987, 1993) while also maintaining continuous access to the resources of the intertidal habitat (Kneib 1994, 1997b). Larger isolated pools and permanent marsh ponds provide intertidal aquatic habitat for large populations of a few species of adult resident nekton (e.g., Ingólfsson 1994, Smith and Able 1994, Rowe and Dunson 1995). Although these habitats allow aquatic organisms to remain in the intertidal zone during low tide, access to most of the food resources of the vegetated marsh remains restricted to periods of tidal inundation.

There are anthropogenic analogs for many natural aquatic features of tidal marshes (Kneib 1997a). For example, man-made impoundments (e.g., Gilmore et al. 1982, Wenner and Beatty 1988), canals (e.g., Trent et al. 1976, Rozas 1992) and drainage ditches (e.g., Talbot et al. 1986, Bryan et al. 1990) are prominent features of some tidal marshes. Many of these artificial aquatic habitats include steep banks, levees or water control structures that may reduce accessibility to the vegetated marsh at high tide or interfere with the movement of nekton between intertidal and subtidal habitats (e.g., Neill and Turner 1987, Herke 1995). Artificial channels that provide access to infrequently-flooded marsh habitats may serve to entrap nekton populations under suboptimal conditions, causing loss of nekton production (e.g., Poizat and Crivelli 1997).

2.2 NEKTON ASSEMBLAGES OF TIDAL MARSH ESTUARIES

Nekton assemblages associated with tidal marshes tend to be subsets of the species present in the adjacent open estuary and largely comprise fishes and decapod crustaceans (Kneib 1997a). Although these sub-assemblages generally have a lower species richness than the adjacent estuary, they include both marsh resident and marine transient species. All life stages (egg to adult) of resident species are present in the marsh. Although the life cycle of this group can be completed in the marsh, specific life stages (e.g., adults or larvae) may also use other estuarine habitats immediately adjacent to the marsh (e.g., subtidal creeks, seagrass beds). Transient species are represented largely by the juveniles of marine- or estuarine-spawned species that use the shallow waters in and around marshes as nurseries.

2.2.1 Nekton from the vegetated marsh

Tidal inundation of marshes periodically links this productive intertidal environment to the rest of the estuarine ecoscape and provides estuarine nekton with access to potentially important resources. However, relatively few species take advantage of the opportunity. A marsh nekton assemblage from Sapelo Island, Georgia serves as an example (Table 1). Individuals 15 mm total length (TL) were taken at slack high tide in flume weirs (see Kneib 1991), which collect a nearly instantaneous sample from an area of 100 m^2 of flooded marsh surface. A total of 184 flume-weir samples were collected in all months of the year (1989) from low and high relative intertidal elevations at each of two marsh sites – one at the mouth and the other in the upper reaches of the Duplin River. Collections included both day and night as well as spring and neap tides. Smaller individuals (5-15 mm TL) were collected during the same time (1989) from the middle and upper reaches of the Duplin River drainage using simulated aquatic microhabitats (SAMs). SAMs are glass petri dishes (10 cm diameter, 1.7 cm deep) installed within a supporting PVC collar buried so that the top edge is flush with the marsh substratum (see Kneib 1997b for details). Samples of smaller nekton were collected from 10 SAMs installed at each of four marsh locations on 99 days (most samples taken at 3-d intervals).

Resident species, primarily cyprinodontid fishes (*Fundulus* spp.) and caridean shrimps (*Palaemonetes* spp.), dominated the nekton assemblage on the vegetated intertidal marsh (Table 1). Approximately 83% of all specimens $\geq$15 mm in length and practically all of the smaller individuals were classified as marsh residents. Few transient species made extensive use of vegetated intertidal marshes at high tide. The only abundant transient routinely collected in flume weirs on the intertidal marsh around Sapelo Island was the penaeid prawn, *Penaeus setiferus* (Table 1). Unlike residents, which were found year-round, transients were common only during specific seasons. For example, *P. setiferus* was only abundant during July – October. Other transients exhibited considerable interannual variation in abundance. Juveniles of the transient fish *Leiostomus xanthurus* (spot) fit this pattern of occurrence in marshes around Sapelo Island. This species commonly occurred throughout the intertidal marsh at high tide, but it does not appear so from Table 1. Juvenile spot were most abundant in spring (March - May) and, in 1988, composed 42.8% of the total nekton in the flume-weir samples in that season (Kneib, unpublished data). However, in the following year – represented in Table 1 – spot accounted for only 4.6% of the total nekton from the spring flume-weir samples (0.8% on an annual basis).

2.2.2 Nekton from channels and embayments

Choice of sampling location can have a strong influence on the perceived composition of nekton assemblages from tidal marshes (Kneib 1997a). Trawls, seines, block nets, and other gears, usually used outside the vegetated marsh proper (i.e., subtidal or intertidal creeks, bays, canals, etc.), capture a larger subset of estuarine nekton than are found on the marsh proper. Typically, the collections contain some marsh residents, but are usually dominated by transients (e.g., Shenker and Dean 1979, Weinstein 1979, Weinstein and Brooks 1983, Smith et al. 1984, Rountree and Able 1992), including small open water prey species (e.g., engraulid and atherinid fishes), which typically feed in the water column but may be seeking a refuge from predation along the edges of the intertidal marsh (Reis and Dean 1981, Rountree and Able 1993, Allen et al. 1995). Larger predatory species of estuarine nekton also are common in samples from the open water habitats adjacent to the marsh (Kneib 1997a).

TABLE 1. Taxonomic composition and relative abundance (%) of two size classes of nekton from intertidal marshes on Sapelo Island, Georgia USA. Species are identified as either fish (F) or decapod crustacean (C) and are placed in two life history groupings (Resident and Transient) with respect to life-time use of the marsh habitat. Species represented by $\geq$ 0.1% of the total are listed individually. N=total number of individuals in year-round samples of each size class.

Species	(F)ish or (C)rustacean	Marsh life history	5-15 mm, N=31,034	$\geq$ 15 mm, N=28,373
Fundulus heteroclitus	F	Resident	63.4%	33.4%
Palaemonetes pugio	C	Resident	32.8%	32.9%
Penaeus setiferus	C	Transient	—	14.4%
Palaemonetes vulgaris	C	Resident	—	9.0%
Fundulus luciae	F	Resident	3.3%	6.3%
Poecilia latipinna	F	Resident	0.5%	0.8%
Fundulus majalis	F	Transient	—	0.8%
Leiostomus xanthurus	F	Transient	—	0.8%
Mugil cephalus	F	Transient	—	0.3%
Menidia menidia	F	Transient	—	0.3%
Gobionellus boleosoma	F	Resident	—	0.2%
Callinectes sapidus	C	Transient	—	0.2%
Cyprinodon variegatus	F	Resident	—	0.2%
Gobionellus smaragdus	F	Resident	—	0.1%
Brevoortia tyrannus	F	Transient	—	0.1%
12 other species (pooled)	F/C	10 T / 2 R	—	0.2%

2.2.3. *Nekton from the marsh edge*

Most samples of nekton from marshes have come from very near the boundary between the vegetated marsh and the subtidal estuary. When comparisons have been made between the edge and interior marsh habitats at sites both in the Gulf of Mexico (Minello et al. 1994, Peterson and Turner 1994) and on the Atlantic coast of the U.S. (Kneib and Wagner 1994, Kneib 1995) it is clear that many species of transient nekton routinely travel short distances (< 5 m) into the vegetated marsh at high tide, but rarely use a major portion of the intertidal habitat potentially available at high tide. Consequently, species richness and abundance of nekton in samples from the marsh edge often are greater than in those collected further into the interior of the vegetated marsh, particularly at marsh sites adjacent to subtidal water bodies (Hettler 1989, Baltz et al. 1993, Minello et al. 1994, Peterson and Turner 1994).

2.3 DYNAMICS

Changes in the physical and biological environment of estuaries occur across a range of temporal and spatial scales. In marshes bordering the estuary, rapid changes in environmental conditions are common. Many of the important factors affecting nekton over short distances (< 1 km) and over short time intervals (e.g days or weeks) are related to tidal action (e.g., Helfman et al. 1983, Kneib 1994), but seasonal changes induced by regional climatic factors often exert a strong influence over larger areas and on interannual cycles in estuarine nekton assemblages (e.g., Deegan 1990, Knudsen et al. 1996). Behavioral responses of different species to both physical and biological components in the environment can have profound effects on abundance and habitat use patterns (Kneib 1994, 1995).

2.3.1 *Variability in the abiotic environment*

Seasonal changes in salinity, temperature and dissolved oxygen concentrations of estuarine waters often are associated with large-scale movement of nekton populations (e.g., Marotz et al 1990, Peterson and Ross 1991, Rakocinski et al. 1992). These migrations generally lead to relatively predictable changes in the composition of nekton assemblages over periods of months (e.g., Rountree and Able 1992). However, because the components of nekton assemblages are highly mobile, they may also change rapidly over smaller areas in response to local variation in physical environmental conditions. For example, the species richness and stability of nekton assemblages may differ at the mouth and headwaters of tidal marsh creeks because temperature and dissolved oxygen levels tend to be more variable in the shallow headwaters (Hackney et al. 1976). Species composition of nekton assemblages in tidal creeks may vary considerably between day and night (Shenker and Dean 1979, Reis and Dean 1981, Rountree and Able 1993). Most of this variation is associated with marsh transients moving from the open estuary into subtidal or intertidal channels, but not into or out of the intertidal vegetation. Samples collected within the vegetated interior of the flooded marsh habitat do not show strong diel changes in the abundance of the dominant species (Kneib and Wagner 1994). However, mixed semi-diurnal tides are a common feature in many marshes, including those on the U.S. Atlantic coast, and it can be difficult to separate nekton responses to the light-dark cycle *per se* from responses to day-night differences in tidal amplitude.

Tidal flooding is probably the single most important variable controlling nekton access to intertidal marshes and associated estuarine habitats. Opportunities for nekton to use the vegetated marsh are limited to those times and places that provide aquatic habitat. Relationships between the duration of tidal flooding and abundance of the dominant transient (*Penaeus setiferus*) and resident (*Fundulus heteroclitus*) species from flume weir catches on Sapelo Island, Georgia provide an example (Fig. 1). All sampling was at slack high tide during July through October when juvenile *P. setiferus* were abundant. Samples were collected from flume weirs at low (+185 cm above mean low water) and high (+198 cm above MLW) intertidal elevations. There is a significant positive relationship between density and duration of tidal inundation in the transient species, but a negative relationship in the resident species.

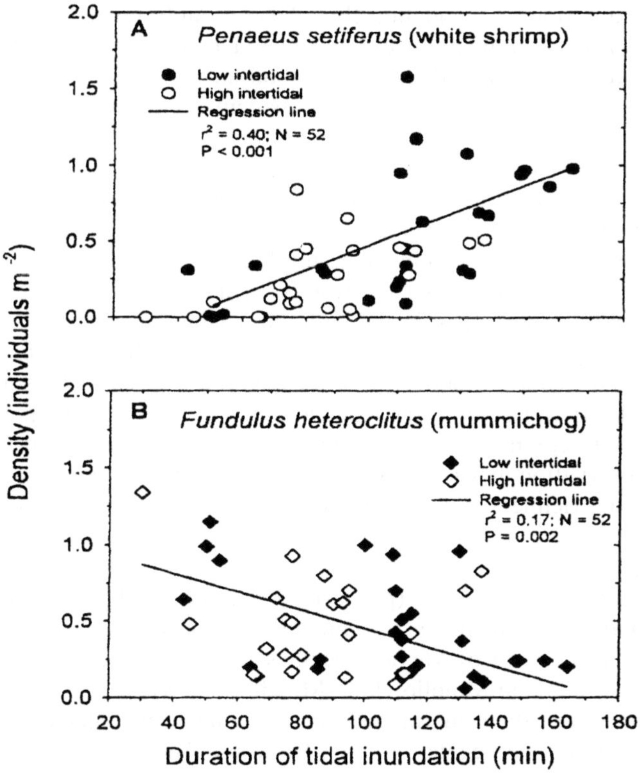

Figure 1. Relationships between duration of tidal inundation and densities of (A) transient juvenile white shrimp, and (B) resident mummichog on the intertidal marsh at slack high tide.

These patterns are likely due to differences in the distance traveled by each species during each tidal cycle. As the tide rises, transients follow it through the intertidal creek system and some, such as juvenile penaeid prawns, eventually reach and traverse the flooded marsh vegetation. The longer the marsh is inundated, the more transients ultimately find their way into the habitat (Fig. 1A). Transient species that venture into the intertidal zone at high tide, rarely remain there at low tide, but follow the receding waters to subtidal aquatic habitats. Having a greater aversion to being stranded in the intertidal zone, the transients tend to leave the flooded marsh earlier than the residents (Kneib and Wagner 1994, Kneib 1995) and so do not take full advantage of the time this habitat is accessible to them.

In contrast, resident species do not travel large distances with each tide, but remain in pools within intertidal creeks or in shallow waters immediately adjacent to the vegetated marsh surface during low tide. These populations enter the vegetated intertidal marsh as soon as it is made accessible by the flooding tide. Densities do not increase with increasing duration of tidal flooding because the entire population enters the habitat very early in the tidal cycle; they are also the last to leave with the receding tide (Kneib and Wagner 1994), thus maximizing the available time in this habitat. The decline in abundance with duration of inundation (Fig. 1B) probably results from dilution of the population. Duration of inundation can be related to the tidal amplitude. Tides that flood the marsh for longer periods may also flood larger areas of the intertidal habitat. This allows more time and area over which resident species can disperse through the available aquatic habitat.

Local tide regimes control temporal opportunities for estuarine nekton to gain access to the vegetated intertidal zone, and regional variation in the frequency and duration of tidal inundation is more or less controlled by either astronomical or meteorological events (Rozas 1995). In some regions (e.g., northern Gulf of Mexico), flooding patterns are largely wind-driven. This usually results in a relatively predictable long-term (seasonal) pattern of flooding associated with the seasonal occurrence of storms, but an unpredictable hydroperiod in the short-term. Consequently, there may be times when the marsh is exposed, or remains flooded, for days at a time. Rozas (1995) suggested that the relative predictability of flooding frequency has little impact on nekton use of vegetated marshes in different regions, but that duration of inundation is important. This was based largely on the observation that samples from marshes along the northern Gulf of Mexico have similar species richness, but greater densities of nekton than those along the Atlantic Coast of the U.S., where tidal inundation occurs with greater regularity and frequency, but each event is of shorter duration (a few hours).

It is the decapod crustaceans (primarily caridean shrimps, penaeid prawns and portunid crabs) rather than the fish component of the estuarine nekton that exhibit higher densities in the Gulf of Mexico marshes relative to those on the U.S. Atlantic coast (e.g.,Zimmerman and Minello 1984, Minello and Webb 1997). Interestingly, it is this same component of the estuarine nekton that tends to be enhanced by the presence of subtidal beds of aquatic vegetation (e.g., Heck and Thoman 1984, Rozas and Odum 1987, Sheridan et al. 1997). It seems that marshes experiencing prolonged periods of inundation may have more in common with seagrass beds than most other intertidal marshes.

Prolonged periods (i.e., weeks or months) of flooding are not conducive to the development and maintenance of intertidal marshes (Chapman 1976). Long-term loss of wetland area due to subsidence or sea level rise (Stumpf and Haines 1998) can occur when sedimentation rates no longer keep pace with rates of submergence. In regions where marshes are associated with large riverine deltas, such as the northern Gulf of Mexico, changes in the directional deposition of sediments from the river mouth may occur in response to either natural or anthropogenic alterations in flows or channel structure. As a consequence, some local areas of marsh may be starved of the sediments required for their maintenance. As these areas subside, they may become much more accessible to estuarine nekton, and can be extremely productive. However, as pointed out by Zimmerman et al. (1991) and Rozas and Reed (1993), this ''supernova-type'' of productivity is likely to be a temporary condition that will decline as the intertidal marsh degrades and is replaced by subtidal open water habitat. Of course, the sediments that are no longer reaching one location are being deposited in another, resulting in the development of new marsh habitat. In such dynamic sedimentary environments, marsh-dependent estuarine nekton production might be expressed as a spatial mosaic that corresponds to the developmental history of local intertidal vegetated environments.

The effects of variation in duration of tidal inundation on production of nekton in tidal marshes is an area that is in need of further study. There is considerable variation in tidal amplitude even within relatively small geographic regions that may affect accessibility to estuarine nekton. For example, average tidal amplitudes along the Atlantic coast of the southeastern U.S. follow a pattern that is related to the contour of the coastline with tides of the highest amplitude associated with the central portion of the bight (Fig. 2). Given a relationship between the frequency and/or duration of tidal inundation and nekton access to intertidal resources, one could hypothesize that, even within the U.S. South Atlantic Bight (North Carolina to Florida), there should be considerable variation in the production of nekton that can be linked to marshes.

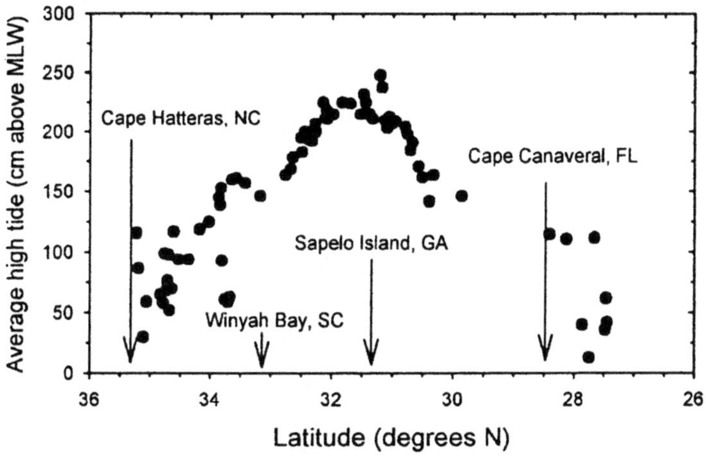

Figure 2. Average heights of tides along the outer coast of the U.S. South Atlantic Bight, Cape Hatteras, North Carolina to Vero Beach, Florida (from Tide.1 software, Micronautics, Inc., Rockport, Maine).

2.3.2 *Variability in the biotic environment*

Variation in the living components of estuarine marsh systems may be even greater than that of the physical environment. Abundance and activity levels of most organisms change over a range of temporal cycles measured in minutes to years and spatial scales of millimeters to many kilometers. Cyclic patterns in abundance of most estuarine nekton species are common and result from intrinsic cycles (e.g., circadian rhythms), temporal patterns in spawning or the collective behavioral responses of individuals in the populations to change in some abiotic or biotic variable.

For example, the timing of reproductive activity (under endocrine control) in a population of the common mummichog (*Fundulus heteroclitus*) from Sapelo Island, Georgia (Kneib 1986a) produced a succession of distinct cohorts of young at approximately 2-wk intervals during the period April to October (Kneib 1997b). Survival in cohorts of young mummichogs on the intertidal marsh surface varied with both the duration of tidal flooding and the level of predation encountered at different sites (Kneib 1993, 1997b). Some cohorts attained high densities (>200 larvae m^{-2}), but also experienced high mortality rates such that entire cohorts virtually disappeared only days after reaching peak abundance. This short-term cycle of abundance was superimposed on a seasonal pattern of reproductive activity (Kneib 1986a) that resulted in a bimodal distribution of young (Kneib 1997b) which was, of course, superimposed upon the annual cycle of spawning. Pulses in abundance are common features of estuarine systems and probably induce similar cycles in the pressure placed on available resources by key species (Kneib 1986b, 1994), and in the movement of production from intertidal marshes (Sardá et al. 1998).

Interactions with other species using marsh habitats may alter the distributions of some nekton in marsh habitats. The seasonal influx of juvenile transients can affect the abundance and distribution of resident nekton populations by either displacing them or increasing mortality rates. For example, Mayer (1985) observed that the seasonal influx of juvenile white shrimp (*Penaeus setiferus*) into Georgia marshes was associated with a decline in the abundance of resident caridean shrimp (*Palaemonetes pugio*). Using a combination of field and laboratory experiments, Kneib and Knowlton (1995) showed that juvenile penaeids did not inflict significant mortality on adults of the resident species, but did increase the mortality of the resident's early life stages.

Several studies (e.g., McIvor and Odum 1988, Ruiz et al. 1993) have indicated that shallow water may provide smaller nektonic organisms a refuge from larger estuarine predators. However, choice of habitats by small resident species may not be based on the physical environment alone (shallow *vs* deeper water), but on the ability of prey to recognize the presence of predator species in their environment and assess the actual mortality risk in each habitat. This was tested experimentally in the laboratory where a common marsh resident prey species – the daggerblade grass shrimp (*Palaemonetes pugio*) – was offered a habitat choice in an 80-l aquarium that was tilted to provide a deep (10 - 15 cm) and shallow (5 - 10 cm) habitat. A plastic barrier (1.0 x 1.5 cm mesh size) divided the deep and shallow sides of the aquarium. Shrimp were small enough (1.5 - 2.0 cm in length; 0.5 cm in width) to pass freely through the mesh. Larger (7.0 - 9.0 cm) fish, which were added to one or the other side of four aquaria, were unable to pass through the mesh divider. The fish used in the experiment were adult mummichogs, *Fundulus heteroclitus*, and juvenile mullet, *Mugil cephalus*. A fifth aquarium received no fish and served as a control for any disturbance due to the presence of fish. Twenty shrimp were added initially to all five aquaria, with each side receiving 10 individuals. The number of shrimp on both sides of each aquarium was recorded at 6-h intervals during each 24-h experimental run. The average (from the four observations in 24 h) proportion of surviving shrimp on the shallow side of each aquarium was the response variable of interest. There were 10 such runs each of the five treatment levels.

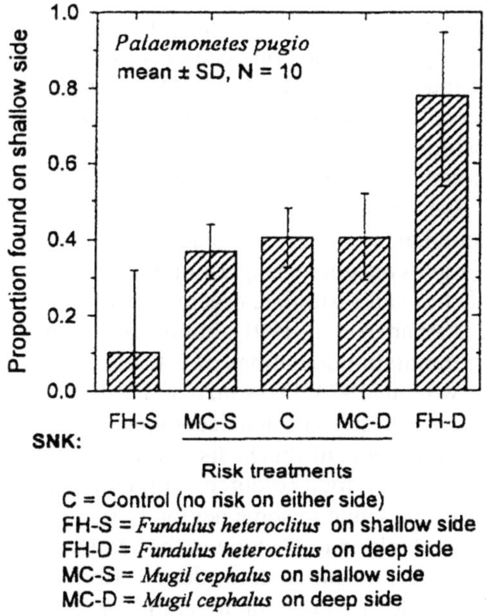

Figure 3. Behavioral response of adult daggerblade grass shrimp (Palaemonetes pugio) to fish species that represent a real risk of mortality (Fundulus heteroclitus) and no risk (Mugil cephalus). Means of treatments sharing a common underline are not significantly different (Student-Newman-Keuls test, α =0.05).

The two principal findings from this experiment were that shrimp showed a slight preference for deeper water in the absence of fish, and when fish were present, shrimp distinguished between species – avoiding mummichogs (the predators) and showing no response to mullet (Fig. 3). The actual daily loss of grass shrimp from aquaria containing mummichogs averaged > 20% compared with < 5% in those with mullet or no fish, showing that mummichogs represented a true mortality risk to the shrimp, while mullet posed no mortal threat. The findings of this simple experiment suggest that the distributions of nekton species are likely affected by behavioral responses to the presence of certain other species and are not simple responses to physical characteristics of the habitat.

3. Food Sources for Marsh Nekton Populations

The source of trophic support for estuarine consumers, in general, and marsh assemblages in particular, has been a key issue at the core of considerable scientific debate for decades (Teal 1962, Odum and de la Cruz 1967, Haines and Montague 1979, Peterson and Howarth 1987, Currin et al. 1995, Deegan and Garritt 1997). Intertidal vascular plants may contribute substantially to total primary production in some estuaries (e.g., Teal 1962, Heymans and Baird 1995), but not in others (e.g., Sullivan and Moncreiff 1990,

Schlacher and Wooldridge 1996, Page 1997). In marsh-dominated estuarine systems there is an abundance of vascular plant production, relatively little of which is consumed as live biomass (Teal 1962, Marinucci 1982, Mann 1988, Mitsch and Gosselink 1993). Much of this vast production enters planktonic or benthic macrofaunal food webs during and after the process of decomposition, which begins while plants are still standing (Newell et al. 1989). Eukaryotic mycelial decomposers can dominate the microbial community of standing-dead leaves of *Spartina alterniflora* (Newell 1996).

Although there is abundant evidence that marsh plant detritus is ingested by nekton, its direct food value to larger consumers is questionable (see Kneib 1997a). Some filter-feeding or herbivorous fishes can assimilate organic material from detrital sources (e.g., Lewis and Peters 1984, D'Avanzo et al. 1991), but the nutritional quality of this material is usually insufficient to maintain populations of estuarine fishes and nektonic decapod crustaceans (e.g., Prinslow et al. 1974, D'Avanzo and Valiela 1990, Deegan et al. 1990, Newell et al. 1995). Leachate from living and decomposing plant materials together with excretions of nektonic organisms contributes dissolved nutrients (Gallagher et al. 1976, Mann 1988), which may enhance food resources in tidal creeks by forming amorphous aggregates of direct nutritional value to some nekton (e.g., D'Avanzo et al. 1991), or by supporting growth of phytoplankton or bacteria in the subtidal water column (e.g., Shiah and Ducklow 1995). However, most marsh resident nekton and many transient species found on the marsh at high tide, tend to exhibit benthic or epibenthic feeding habits and are unlikely to be tapped directly into the trophic dynamics of overlying waters.

Other potentially important sources of nutrition in marsh-estuarine systems are benthic microalgae and cyanobacteria (e.g., Sullivan and Moncreiff 1990, Mallin et al. 1992, Pinckney and Zingmark 1993, Currin et al. 1995, Schmidt and Jónasdóttir 1997). However, except for a few largely herbivorous species, these foods tend to supplement rather than completely support the nutritional requirements of most marsh nekton (e.g., Gleason and Wellington 1988, McTigue and Zimmerman 1991, Schmidt and Jónasdóttir 1997).

Most of the high quality food consumed by marsh nekton is in the form of small benthic and epibenthic invertebrates (Kneib 1997a). It is likely these prey assemblages integrate various sources of primary production (e.g., vascular plant detritus, algae, etc.), re-packaging and enhancing the nutritive value for nekton. Consumer communities of marsh benthic and epibenthic invertebrates graze the microbial community associated with decomposing vegetation (e.g., Newell and Bärlocher 1993) and some, such as the amphipod *Uhlorchestia spartinophila*, have been shown to grow and reproduce on natural diets of standing-dead *Spartina* leaves (Kneib et al. 1997). Harpacticoid copepods – which are very abundant components of intertidal marsh invertebrate assemblages (Coull et al. 1979) and figure prominently in the diets of many marsh-associated nekton (e.g., Bell and Coull 1978, Nelson and Coull 1989, Walters et al. 1996) – are capable of using organic matter from a variety of sources, including vascular plant detritus and algae to support their populations (Guidi 1984, Couch 1989).

After incorporation into biomass of benthic invertebrates, intertidal marsh plant production continues to move across the ecoscape into the open estuary by a mechanism that very likely involves mobile predators, as suggested by results of many recent stable isotope studies (e.g., Deegan and Garritt 1997, Kwak and Zedler 1997,

Paterson and Whitfield 1997). Nekton may have access to this potential food source either as passive invertebrate drift carried by tides from the marsh surface to the subtidal estuary or by actively foraging within the marsh at high tide. Meiofaunal invertebrates, such as harpacticoid copepods and nematodes are commonly suspended and dispersed by tidal action (e.g., Bell and Sherman 1980, Palmer and Brandt 1981, Eskin and Palmer 1985), as are some larger benthic crustaceans, such as tanaids (Mendoza 1982, Kneib 1992) and amphipods (Gilmurray and Daborn 1981). The role and importance of passive invertebrate drift in the trophic dynamics of the marsh-estuary system remain to be determined. However, a large body of evidence supports active foraging by both resident and transient species of nekton in the vegetated intertidal marsh (see Kneib 1997a).

4. Conceptual Models and Hypotheses

4.1 THE TROPHIC RELAY

I recently proposed a conceptual model for the formulation and testing of hypotheses regarding the role of nekton in the transport of production across the marsh ecoscape to the open estuary (Kneib 1997a). Spatial patterns in nekton use of marsh and adjacent estuarine waters by different life stages of resident and transient species can be assembled into a type of map (Fig. 4) that suggests how each species or life stage might play a role in the use and movement of marsh production along corridors and across boundaries within the ecoscape. The young of resident species such as mummichogs and grass shrimp have the most intimate association with the vegetated marsh, remaining in that habitat even during low tide by using the aquatic microhabitats embedded within the fabric of the ecoscape. The cumulative production of this group of young resident species inevitably leaves the vegetated marsh when individuals become too large (by ca. 15 mm total length) to use the aquatic microhabitats and begin to migrate into and out of this habitat with the adults. The young likely are exposed to a higher risk of cannibalism and intraguild predation when they share low-tide aquatic refugia (i.e. intertidal creek pools) with adults (Kneib 1987). Thus, the production incorporated into the biomass of early life stages moves a step closer to the open estuary by being passed to the adult populations through both growth and consumption. Adult residents move into the intertidal marsh with each flood tide and return to intertidal and the shallow portions of subtidal creeks and embayments with the ebb. Material gathered during flood tides is egested or excreted into the waters of creek channels or embayments used by residents as low-tide refugia. Here the distributions of adult marsh residents also overlap with larger juveniles of transient predator species such as weakfish (*Cynoscion regalis*), spotted seatrout (*Cynoscion nebulosus*), flatfishes (*Paralichthys* spp), bluefish (*Pomatomus saltatrix*), dogfish (*Mustelus canis*) and other predatory estuarine nekton, which may not forage directly in the vegetated marsh, but consume resident nekton when the opportunity arises. As these marine transient species grow, they in turn move to deeper waters of the estuary and coastal ocean, completing the final step in the trophic relay of intertidal marsh production to the subtidal estuary.

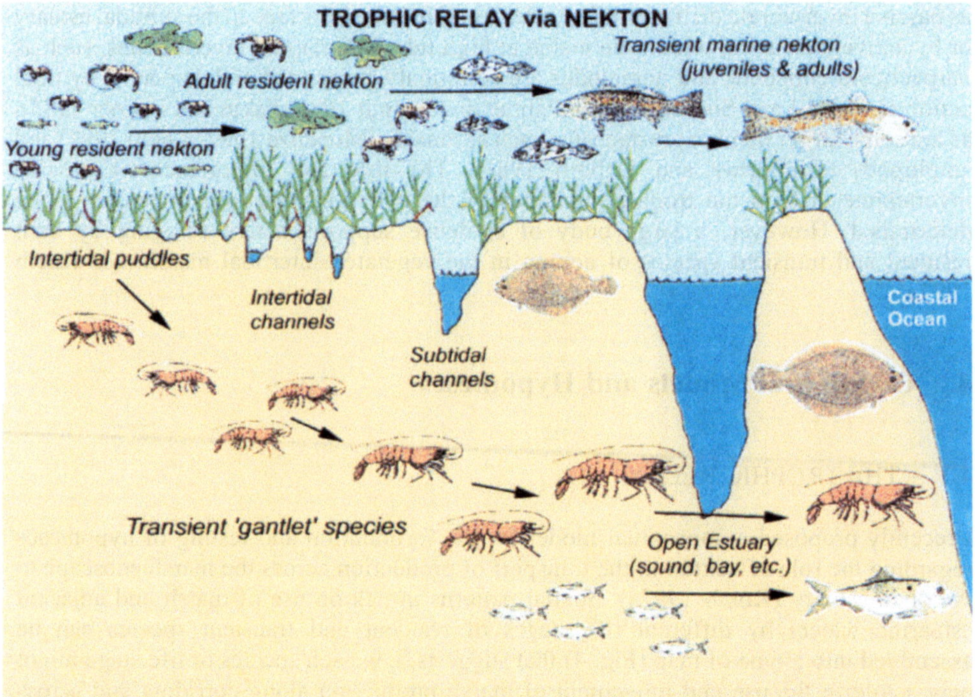

Figure 4. A conceptual model of a trophic relay by which intertidal marsh production may be transferred across the ecoscape to the open estuary through nekton populations. Redrawn from Kneib (1997a).

A separate group of nekton identified as ''transient gantlet species'' (Fig. 4) include those that use estuarine marsh nurseries during a portion of their early life histories and emigrate to deeper waters of the estuary or coastal ocean as adults. For the most part, they are not predatory on other nekton, but are exposed to a changing suite of nektonic predators during their ontogenetic migrations. In effect, they ''run the gantlet'' during their life histories. Some are spawned in the ocean or open waters of the estuary, and immigrate to shallow nursery areas within or adjacent to tidal marshes. Once in the nursery, intertidal marsh production is available to this group either directly or indirectly *via* routes that may bypass marsh resident nekton. Having had time to grow during their journey from spawning grounds in the sea, many of these transient species may enter tidal marsh nurseries at a size similar to that of juvenile or adult residents.

The examples provided here include penaeid prawns, which as juveniles make extensive forays into the flooded vegetated marsh, returning to subtidal habitats on ebbing tides much like adult marsh residents. Like resident species, penaeids pass along intertidal production when they are consumed by predatory nekton. However, they make an additional contribution to the movement of intertidal production when the survivors make large-scale annual migrations to the open estuary or coastal ocean before spawning. When they approach adult size and emigrate, the biomass accumulated by the population from feeding on intertidal production moves out of the estuary. Menhaden (*Brevoortia* spp.) may be included as gantlet species, but being schooling filter-feeders, they tend to remain in open waters and do not forage extensively within the vegetated marsh. Within the tidal marsh creeks, juvenile menhaden are likely to ingest particles that have originated from the intertidal zone, including detrital aggregates, invertebrate drift or planktonic organisms supported by nutrients released through the process of microbial decomposition of marsh-derived detritus.

Atherinids, such as Atlantic silverside (*Menidia menidia*), are not depicted in Fig. 4 but also may be included among the transient gantlet species. The silverside falls into a group of schooling, forage fishes that are typically small (<15 cm adult size), but often abundant in estuaries. It shares some pertinent life history features with both marsh residents and true transients. Atlantic silverside is known to spawn within the intertidal marsh (Middaugh 1981) much like a resident species, but adult populations may seasonally emigrate to deeper waters of the estuary and coastal ocean (Conover and Murawski 1982), like many marine-spawned transients. This ontogenetic shift in habitat use exposes the earliest life stages to predation by both adults of resident species, such as mummichogs (Conover and Kynard 1984), and juveniles of transients such as blue crabs (Middaugh 1981), while juvenile and adult silverside must "run the gantlet" of estuarine and marine predators during their seasonal emigration from estuarine littoral habitats.

4.2 SHIFTING INTERACTION ZONES

Mobile populations of estuarine nekton commonly adjust their distributions to conditions in the estuary (e.g., Friedland et al. 1996). Combining the spatial configuration of physical features in the estuarine ecoscape with the temporal dynamics of tidal action, and superimposing that on what we now know about the life histories, distribution and behavior of nekton in marshes and adjacent habitats provides the basis for developing hypotheses concerning "hot spots" of activity – places and times when trophic transfers involving estuarine nekton are likely to occur. Much of the important activity involved with the direct acquisition of intertidal production by nekton is likely to occur on flooding and high tides, when residents and certain transients take advantage of the expanding foraging opportunities on the marsh surface (Fig. 5A). The intensity of this activity is very likely influenced by the structure of the ecoscape and behavioral responses of nekton to the physical features in their environment. For example, vegetated marshes bordering intertidal creek systems with complex drainage patterns (i.e., numerous small channels and rivulets) or channels with submerged aquatic vegetation – because they offer more low-tide aquatic refugia – may harbor larger populations of resident species (Rozas and Odum 1987, Kneib 1994), which are more likely to be foraging extensively within the marsh vegetation. Transient species that make direct use of marsh resources at high tide might be more abundant in areas nearer subtidal refugia. Alternatively, simple drainage systems – in addition to providing less aquatic area for nekton at low tide – may have steeper and more vertical banks that do not provide easy access to the adjacent marsh surface. This would be true of man-made canals or impoundments that may include bulkheads and other barriers that interfere with the ability of nekton to gain access to intertidal resources in the adjacent vegetated marsh.

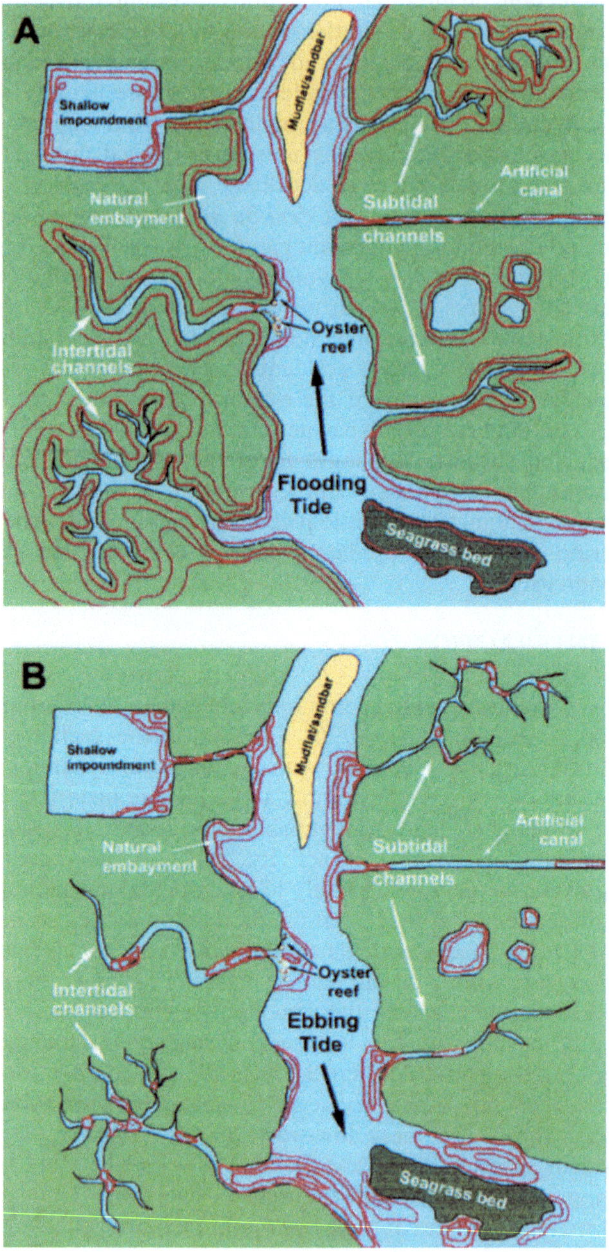

Figure 5. Shifting interaction zones where marsh nekton may be involved in transfers of intertidal production via predator-prey relationships during (A) high and flooding tide stages and (B) ebbing and low tide stages. Red lines are the hypothetical ''hot spots'' of production transfers across the ecoscape at each tidal stage.

When the tide turns and recedes from the marsh (Fig. 5B), the important interaction zones may shift toward habitats occupied by transient nekton feeding either on drift (invertebrates or organic particles) or on smaller nekton that had been foraging on the marsh surface and are returning to low-tide refugia in the channels and embayments (i.e. the trophic relay). Marsh edge habitats, especially near intertidal-subtidal corridors (e.g., rivulets, intertidal creek mouths, etc.), seem to be likely locations for predator-prey interactions involving transient and resident nekton species on ebbing tides. Deeper habitats adjacent to undercut creekbanks also seem ideal locations from which predators could ambush aquatic prey.

Foraging activity of different nekton species usually can be expected to coincide with a particular tidal stage or time of day. Resident marsh fishes, for example, feed largely during daytime high tides (e.g., Butner and Brattstrom 1960, Weisberg et al. 1981, Rozas and LaSalle 1990). Transient species, such as spot (*Leiostomus xanthurus*), which forage extensively within the intertidal marsh vegetation (Feller et al. 1990) feed primarily at high tide, but without a strong preference for day or night (Archambault and Feller 1991). Transient decapod crustaceans, including the penaeid shrimp *Penaeus setiferus*, feed primarily during night-time high tides in marshes around Sapelo Island, Georgia (Mayer 1985). Predatory species of transient nekton often feed on high and ebbing tides, particularly at night (e.g., Helfman et al. 1983, Kleypas and Dean 1983, Rountree and Able 1996). It is important to remember that the behavior of nekton is not a random variable. It will be necessary to take these species-specific differences in feeding activity into account when designing sampling programs and choosing or developing appropriate gears to study the role of mobile fauna in the movement of intertidal production across the estuarine ecoscape.

5. Epilogue

Some may question why it is important to develop the type of mechanistic understanding that I have promoted in this paper. Why not be satisfied simply with knowing that marshes contribute to estuarine production? I would agree that the ability to derive benefits from the use of any system does not require a mechanistic understanding. However, when that system becomes degraded or breaks down and no longer provides the expected benefits, fixing it requires a mechanistic understanding of function. A useful analogy might be found in our ability to maintain a schedule and order in our lives by using timepieces of various types (e.g., sundials, analog and digital clocks) and sizes (e.g., watches to bell towers). We benefit from their use because we grasp the concepts involved in telling time, but how many of us would have the mechanistic understanding to repair any or all of these devices should they fail to function properly? Only an understanding of specific mechanisms can guide a return to proper function.

The propensity of human populations to inhabit lands adjacent to the sea ensures that the loss, protection, creation and restoration of coastal wetlands will continue to be topical environmental issues into the foreseeable future. Attempts to balance the demands of a growing population with protection of the world's coastal environment has led to an increasing emphasis on management of coastal resources. This includes a growing number of efforts to restore perceived losses in the trophic support functions that have long been attributed to intertidal marshes. Given the mixed level of success associated with these efforts (Mitsch and Wilson 1996), it appears that our mechanistic understanding of this system is already seriously lagging behind practical need for such knowledge.

Like many others, I remain a believer in the legendary importance of tidal marshes but recognize that much has been taken on faith. Hopefully, in time, that faith will be confirmed by findings of novel research based on improved sampling methods and testing of hypotheses and new paradigms regarding the ecological role of nekton populations in tidal marsh estuaries. Continuing efforts to improve the quantitative sampling of nekton from intertidal marshes (Kneib 1997a,b, Rozas and Minello 1997) should lead to improved estimates of nekton production from this habitat. Perhaps the concepts of a "trophic relay" and "shifting interaction zones" will contribute to a research focus that ultimately leads to a better mechanistic understanding of production transfers from intertidal marshes to the open estuary.

6. Acknowledgements

Financial support for the original research presented in this contribution was provided by grants from the National Science Foundation (most recently, DEB-96129621), the U.S. Environmental Protection Agency (R825147-01-0), and the Georgia Sea Grant College Program which, together with the Marsh Ecology Research Program, also provided travel support for my attendance at this symposium. The experiment designed to test the behavior of grass shrimp to perceived predation risk was conducted by Leslie Gallagher as part of a student internship sponsored by the Sapelo Foundation. Comments from Jennifer Parker Kneib on an earlier draft improved the manuscript substantially. I am particularly grateful to the symposium organizers for the opportunity to present this paper, as well as their patience during the review process. This paper is contribution number 825 of the University of Georgia Marine Institute.

7. Literature Cited

Allen, D.M., W.S. Johnson and V. Ogburn Matthews. 1995. Trophic relationships and seasonal utilization of salt-marsh creeks by zooplanktivorous fishes. Environmental Biology of Fishes **42**:37-50.

Allen, L.G. 1982. Seasonal abundance, composition and productivity of the littoral fish assemblage in upper Newport Bay, California. Fishery Bulletin **80**:769-790.

Archambault, J.A. and R.J. Feller. 1991. Diel variations in gut fullness of juvenile spot, *Leiostomus xanthurus* (Pisces). Estuaries **14**:94-101.

Baltz, D.M. C. Rakocinski , J.W. Fleeger. 1993. Microhabitat use by marsh-edge fishes in a Louisiana estuary. Environmental Biology of Fishes **36**:109-126.

Bell, S.S. and B.C. Coull. 1978. Field evidence that shrimp predation regulates meiofauna. Oecologia **35**:141-148.

Bell, S.S. and K.M. Sherman. 1980. A field investigation of meiofaunal dispersal: tidal resuspension and implications. Marine Ecology Progress Series **3**:245–249.

Boesch, D.F. and R.E. Turner. 1984. Dependence of fishery species on salt marshes: the role of food and refuge. Estuaries **7**(4A):460-468.

Butner, A. and B.H. Brattstrom. 1960. Local movement in *Menidia* and *Fundulus*. Copeia **1960**:139-141.

Chapman, V.J. 1976. Coastal vegetation, Second edition, Pergamon Press, Oxford, England.

Conover, D.O. and B.E. Kynard. 1984. Field and laboratory observations of spawning periodicity and behavior of a northern population of the Atlantic silverside, *Menidia menidia* (Pisces: *Atherinidae*). Environmental Biology of Fishes **11**:161-171.

Conover, D.O. and S. Murawski. 1982. Offshore winter migration of the Atlantic silverside, *Menidia menidia*. Fishery Bulletin **80**:145-149.

Couch, C.A. 1989. Carbon and nitrogen stable isotopes of meiobenthos and their food resources. Estuarine, Coastal and Shelf Science **28**:433-441.

Coull, B.C., S.S. Bell, A.M. Savory and B.W. Dudley. 1979. Zonation of meiobenthic copepods in a southeastern United States salt marsh. Estuarine, Coastal and Marine Science **9**:181-188.

Currin, C.A., S.Y. Newell and H.W. Paerl. 1995. The role of standing dead *Spartina alterniflora* and benthic microalgae in salt marsh food webs: considerations based on multiple stable isotope analysis. Marine Ecology Progress Series **121**:99-116.

Dame, R.F. and D.M. Allen. 1996. Between estuaries and the sea. Journal of Experimental Marine Biology and Ecology **200**:169-185.

D'Avanzo, C., M. Alber and I. Valiela. 1991. Nitrogen assimilation from amorphous detritus by two coastal consumers. Estuarine, Coastal and Shelf Science **33**:203-209.

D'Avanzo, C. and I. Valiela. 1990. Use of detrital foods and assimilation of nitrogen by coastal detritivores. Estuaries **13**:20-24.

Deegan, L.A. 1990. Effects of estuarine environmental conditions on population dynamics of young-of-the-year gulf menhaden. Marine Ecology Progress Series **68**:195-205.

——. 1993. Nutrient and energy transport between estuaries and coastal marine ecosystems by fish migration. Canadian Journal of Fisheries and Aquatic Sciences **50**:74-79.

Deegan, L.A. and R.H. Garritt. 1997. Evidence for spatial variability in estuarine food webs. Marine Ecology Progress Series **147**:31-47.

Deegan, L.A., B.J. Peterson and R. Portier. 1990. Stable isotopes and cellulase activity as evidence for detritus as a food source for juvenile Gulf menhaden. Estuaries **13**:14–19.

Eskin, R.A. and M.A. Palmer. 1985. Suspension of marine nematodes in a turbulent tidal creek: species patterns. Biological Bulletin **169**:615-623.

Farina, A. 1998. Principles and methods in landscape ecology. Chapman and Hall, London, England.

Feller, R.J., B.C. Coull and B.T. Hentschel. 1990. Meiobenthic copepods: tracers of where juvenile *Leiostomus xanthurus* (Pisces) feed? Canadian Journal of Fisheries and Aquatic Sciences **47**:1913-1919.

Freidland, K.D. D.W. Ahrenholz and J.F. Guthrie. 1996. Formation and seasonal evolution of Atlantic menhaden juvenile nurseries in coastal estuaries. Estuaries **19**:105–114.

Gallagher, J.L., W.J. Pfeiffer and L.R. Pomeroy. 1976. Leaching and microbial utilization of dissolved organic carbon from leaves of *Spartina alterniflora*. Estuarine, Coastal and Marine Science **4**:467-471.

Gilmore, R.G., D.W. Cooke and C.J. Donohoe. 1982. A comparison of the fish populations and habitat in open and closed salt marsh impoundments in east-central Florida. Northeast Gulf Science **5**:25-37.

Gilmurray, M.C. and G.R. Daborn. 1981. Feeding relations of the Atlantic silverside *Menidia menidia* in the Minas Basin, Bay of Fundy. Marine Ecology Progress Series **6**:231-235.

Gleason, D.F. and B.M. Wellington. 1988. Food resources of postlarval brown shrimp (*Penaeus aztecus*) in a Texas salt marsh. Marine Biology **97**:327-337.

Guidi, L.D. 1984. The effect of food composition on ingestion, development and survival of a harpacticoid copepod, *Tisbe cucumariae*. Journal of Experimental Marine Biology and Ecology **84**:101-110.

Hackney, C.T., W.D. Burbanck and O.P. Hackney. 1976. Biological and physical dynamics of a Georgia tidal creek. Chesapeake Science **17**:271-280.

Haines, E.B. and C.L. Montague. 1979. Food sources of estuarine invertebrates analyzed using $^{13}C/^{12}C$ ratios. Ecology **60**:48-56.

Heck, K.L. Jr. and T.A. Thoman. 1984. The nursery role of seagrass meadows in the upper and lower reaches of the Chesapeake Bay. Estuaries **7**:70-92.

Helfman, G.S., D.L. Stoneburner, E.L. Bozeman , P.A. Christian and R. Whalen. 1983. Ultrasonic telemetry of American eel movements in a tidal creek. Transactions of the American Fisheries Society **112**:105-110.

Herke, W.H. 1995. Natural fisheries, marsh management and mariculture: complexity and conflict in Louisiana. Estuaries **18**:10-17.
Herke, W.H., E.E. Knudsen, P.A. Knudsen and B.D. Rogers. 1992. Effects of semi-impoundment of Louisiana marsh on fish and crustacean nursery use and export. North American Journal of Fisheries Management **12**:151-160.
Hettler, W.F. Jr. 1989. Nekton use of regularly-flooded saltmarsh cordgrass habitat in North Carolina, USA. Marine Ecology Progress Series **56**:111-118.
Heymans, J.J. and D. Baird. 1995. Energy flow in the Kromme estuarine ecosystem, St Francis Bay, South Africa. Estuarine, Coastal and Shelf Science **41**:39-59.
Ingólfsson, A. 1994. Species assemblages in saltmarsh ponds in western Iceland in relation to environmental variables. Estuarine, Coastal and Shelf Science **38**:235-248.
Irlandi, E.A. and M.K. Crawford. 1997. Habitat linkages: the effect of intertidal saltmarshes and adjacent subtidal habitats on abundance, movement and growth of an estuarine fish. Oecologia **110**:222-230.
Kleypas, J. and J.M. Dean. 1983. Migration and feeding of the predatory fish, *Bairdiella chrysura* Lacépède, in an intertidal creek. Journal of Experimental Marine Biology and Ecology **72**:199-209.
Kneib, R.T. 1984. Patterns in the utilization of the intertidal salt marsh by larvae and juveniles of *Fundulus heteroclitus* (Linneaus) and *Fundulus luciae* (Baird). Journal of Experimental Marine Biology and Ecology **83**:41-51.
——. 1986a. Size-specific patterns in the reproductive cycle of the killifish, *Fundulus heteroclitus* (Pisces:Fundulidae) from Sapelo Island, Georgia. Copeia **1986**:342-351.
——. 1986b. The role of *Fundulus heteroclitus* in salt marsh trophic dynamics. American Zoologist **26**:259-269.
——. 1987. Predation risk and use of intertidal habitats by young fishes and shrimp. Ecology **68**:379-386.
——. 1991. Flume weir for quantitative collection of nekton from vegetated intertidal habitats. Marine Ecology Progress Series **75**:29-38.
——. 1992. Population dynamics of the tanaid *Hargeria rapax* (Crustacea: Peracarida) in a tidal marsh. Marine Biology **113**:437-445.
——. 1993. Growth and mortality in successive cohorts of fish larvae within an estuarine nursery. Marine Ecology Progress Series **94**:115-127.
——. 1994. Spatial pattern, spatial scale and feeding in fishes. Pages 171-185 *in* D.J. Stouder, K.L. Fresh and R.J. Feller, editors. Theory and application in fish feeding ecology. University of South Carolina Press, Columbia, South Carolina, USA.
——. 1995. Behaviour separates potential and realized effects of decapod crustaceans in salt marsh communities. Journal of Experimental Marine Biology and Ecology **193**:239-256.
——. 1997a. The role of tidal marshes in the ecology of estuarine nekton. Oceanography and Marine Biology: An Annual Review **35**:163-220.
——. 1997b. Early life stages of resident nekton in intertidal marshes. Estuaries **20**:214-230.
Kneib, R.T. and M.K. Knowlton. 1995. Stage-structured interactions between seasonal and permanent residents of an estuarine nekton community. Oecologia **103**:425-434.
Kneib, R.T., S.Y. Newell and E.T. Hermeno. 1997. Survival, growth and reproduction of the salt-marsh amphipod *Uhlorchestia spartinophila* reared on natural diets of senescent and dead *Spartina alterniflora* leaves. Marine Biology **128**:423-431.
Kneib, R.T. and S.L. Wagner. 1994. Nekton use of vegetated marsh habitats at different stages of tidal inundation. Marine Ecology Progress Series **106**:227-238.
Knudsen, E.E., B.D. Rogers, R.F. Paille and W.H. Herke. 1996. Juvenile white shrimp growth, mortality and emigration in weired and unweired Louisiana marsh ponds. North American Journal of Fisheries Management **16**:640-652.
Kwak, T.J. and J.B. Zedler. 1997. Food web analysis of southern California coastal wetlands using multiple stable isotopes. Oecologia **110**:262-277.
Lewis, V.P. and D.S. Peters. 1984. Menhaden – a single step from vascular plant to fishery harvest. Journal of Experimental Marine Biology and Ecology **84**:95-100.
Mallin, M.A., J.M. Burkholder, M.J. Sullivan, W.E. Schaaf and D.S. Peters. 1992. Contributions of benthic microalgae to coastal fishery yield. Transactions of the American Fisheries Society **121**:691-693.
Mann, K.H. 1988. Production and use of detritus in various freshwater, estuarine and coastal marine ecosystems. Limnology and Oceanography **33**:910-930.
Marinucci, A.C. 1982. Trophic importance of *Spartina alterniflora* production and decomposition to the marsh-estuarine ecosystem. Biologial Conservation **22**:35-58.

Marotz, B.L., W.H. Herke and B.D. Rogers. 1990. Movement of Gulf menhaden through three marshland routes in southwestern Louisiana. North American Journal of Fisheries Management **10**:408-417.

Mayer, M.A. 1985. Ecology of juvenile white shrimp, *Penaeus setiferus* Linnaeus, in the salt marsh habitat. Thesis, Georgia Institute of Technology, Atlanta, Georgia, USA.

McIvor, C.C. and W.E. Odum. 1988. Food, predation risk and microhabitat selection in a marsh fish assemblage. Ecology **69**:1341-1351.

McTigue, T.A. and R.J. Zimmerman. 1991. Carnivory vs herbivory in juvenile *Penaeus setiferus* (Linnaeus) and *Penaeus aztecus* (Ives). Journal of Experimental Marine Biology and Ecology **151**:1-6.

Mendoza, J.A. 1982. Some aspects of the autecology of *Leptochelia dubia* (Krøyer, 1842) (Tanaidacea). Crustaceana **43**:225-240.

Meredith, W.H. and V.A. Lotrich. 1979. Production dynamics of a tidal creek population of *Fundulus heteroclitus* (Linnaeus). Estuarine, Coastal and Marine Science **8**:99-118.

Middaugh, D.P. 1981. Reproductive ecology and spawning periodicity of the Atlantic silverside, *Menidia menidia* (Pisces:Atherinidae). Copeia **1981**:766B776.

Minello, T.J. and J.W. Webb, Jr. 1997. Use of natural and created *Spartina alterniflora* salt marshes by fishery species and other aquatic fauna in Galveston Bay, Texas, USA. Marine Ecology Progress Series **151**:165-179.

Minello, T.J., R.J. Zimmerman and R. Medina. 1994. The importance of edge for natant macrofauna in a created salt marsh. Wetlands **14**:184-198.

Mitsch, W.J. and J.G. Gosselink. 1993. Wetlands. Second Edition. Van Nostrand Reinhold, New York, New York, USA.

Mitsch ,W.J. and R.F. Wilson. 1996. Improving success of wetland creation and restoration with know-how, time and self-design. Ecological Applications **61**:77-83.

Neill, C. and R.E. Turner. 1987. Comparison of fish communities in open and plugged backfilled canals in Louisiana coastal marshes. North American Journal of Fisheries Management **7**:57-62.

Nelson, A.L. and B.C. Coull. 1989. Selection of meiobenthic prey by juvenile spot (Pisces): an experimental study. Marine Ecology Progress Series **53**:51-57.

Newell, R.I.E., N. Marshall, A. Sasekumar and V.C. Chong. 1995. Relative importance of benthic microalgae, phytoplankton and mangroves as sources of nutrition for penaeid prawns and other coastal invertebrates from Malaysia. Marine Biology **123**:595-606.

Newell, S.Y. 1996. Established and potential impacts of eukaryotic mycelial decomposers in marine/ terrestrial ecotones. Journal of Experimental Marine Biology and Ecology **200**:187-206.

Newell, S.Y. and F. Bärlocher. 1993. Removal of fungal and total organic matter from decaying cordgrass leaves by shredder snails. Journal of Experimental Marine Biology and Ecology **171**:39-49.

Newell, S.Y., R.D. Fallon and J.D. Miller. 1989. Decomposition and microbial dynamics for standing, naturally positioned leaves of the salt-marsh grass *Spartina alterniflora*. Marine Biology **101**:471-481.

Nixon, S.W. 1988. Physical energy inputs and the comparative ecology of lake and marine ecosystems. Limnology and Oceanography **33**:1005-1025.

O'Bryan, P.D., D.B. Carlson and R.G. Gilmore. 1990. Salt marsh mitigation: an example of the process of balancing mosquito control, natural resource and development interests. Florida Scientist **53**:189-203.

Odum, E.P. and A. de la Cruz. 1967. Particulate organic detritus in a Georgia salt marsh-estuarine ecosystem. Pages 383-388 *in* G.H. Lauff, editor. Estuaries. American Association for the Advancement of Science Publication **83**, Washington, DC, USA.

Page, H.M. 1997. Importance of vascular plant and algal production to macro-invertebrate consumers in a southern California salt marsh. Estuarine, Coastal and Shelf Science **45**:823B834.

Palmer, M.A. and R.R. Brandt. 1981. Tidal variation in sediment densities of marine benthic copepods. Marine Ecology Progress Series **4**:207–212.

Paterson, A.W. and A.K. Whitfield. 1997. A stable carbon isotope study of the food-web in a freshwater-deprived South African estuary, with particular emphasis on the ichthyofauna. Estuarine, Coastal and Shelf Science **45**:705-715.

Peterson, B.J. and R.W. Howarth. 1987. Sulfur, carbon and nitrogen isotopes used to trace organic matter flow in the salt marsh estuaries of Sapelo Island, Georgia. Limnology and Oceanography **32**:1195-1213.

Peterson, G.W. and R.E. Turner. 1994. The value of salt marsh edge vs interior as a habitat for fish and decapod crustaceans in a Louisiana tidal marsh. Estuaries **17**:235-262.

Peterson, M.S. and S.T. Ross. 1991. Dynamics of littoral fishes and decapods along a coastal river-estuarine gradient. Estuarine, Coastal and Shelf Science **33**:467-483.
Pinckney, J. and R.G. Zingmark. 1993. Biomass and production of benthic microalgal communities in estuarine habitats. Estuaries **16**:887-897.
Poizat, G. and A.J. Crivelli. 1997. Use of seasonally flooded marshes by fish in a Mediterranean wetland: timing and demographic consequences. Journal of Fish Biology **51**:106-119.
Prinslow, T.W., I. Valiela and J.M. Teal. 1974. The effect of detritus and ration size on the growth of *Fundulus heteroclitus* (L.). Journal of Experimental Marine Biology and Ecology **16**:1-10.
Rakocinski, C.F., D.M. Baltz, and J.W. Fleeger. 1992. Correspondence between environmental gradients and the community structure of marsh-edge fishes in a Louisiana estuary. Marine Ecology Progress Series **80**:135-148.
Reis, R.R. and J.M. Dean. 1981. Temporal variation in the utilization of an intertidal creeks by the bay anchovy (*Anchoa mitchilli*). Estuaries **4**:16-23.
Rountree, R.A. and K.W. Able 1992 Fauna of subtidal polyhaline marsh creeks in southern New Jersey: composition, abundance and biomass. Estuaries **15**:171-185.
——. 1993. Diel variation in decapod crustacean and fish assemblages in New Jersey polyhaline marsh creeks. Estuarine, Coastal and Shelf Science **37**:181-201.
——. 1996. Seasonal abundance, growth and foraging habits of juvenile smooth dogfish, *Mustelus canis*, in a New Jersey estuary. Fishery Bulletin **94**:522-534.
Rowe, C.L. and W.A. Dunson. 1995. Individual and interactive effects of salinity and initial fish density on a salt marsh assemblage. Marine Ecology Progress Series **128**:271-278.
Rozas, L.P. 1992. Comparison of nekton habitats associated with pipeline canals and natural channels in Louisiana salt marshes. Wetlands **12**:136-146.
——. 1995. Hydroperiod and its influence on nekton use of the salt marsh: a pulsing ecosystem. Estuaries **18**:579-580.
Rozas, L.P. and M.W. LaSalle. 1990. A comparison of the diets of Gulf killifish *Fundulus grandis* Baird and Girard, entering and leaving a Mississippi brackish marsh. Estuaries **13**:332-336.
Rozas, L.P., C.C. McIvor and W.E. Odum. 1988. Intertidal rivulets and creekbanks: corridors between tidal creeks and marshes. Marine Ecology Progress Series **47**:303-307.
Rozas, L.P. and T.J. Minello. 1997. Estimating densities of small fishes and decapod crustaceans in shallow estuarine habitats: a review of sampling design with focus on gear selection. Estuaries **20**:199-213.
Rozas, L.P. and W.E. Odum. 1987. The role of submerged aquatic vegetation in influencing the abundance of nekton on contiguous tidal fresh-water marshes. Journal of Experimental Marine Biology and Ecology **114**:289-300.
Rozas, L.P. and D.J. Reed. 1993. Nekton use of marsh-surface habitats in Louisiana (USA) deltaic salt marshes undergoing submergence. Marine Ecology Progress Series **96**:147-157.
Ruiz, G.M., A.H. Hines and M.H. Posey. 1993. Shallow water as a refuge habitat for fish and crustaceans in non-vegetated estuaries: an example from Chesapeake Bay. Marine Ecology Progress Series **99**:1-16.
Sardá, R., K. Foreman, C.E. Werme and I. Valiela. 1998. The impact of epifaunal predation on the structure of macrofaunal invertebrate communities of tidal saltmarsh creeks. Estuarine, Coastal and Shelf Science **46**:657-669.
Schlacher, T.A. and T.H. Wooldridge. 1996. Origin and trophic importance of detritus – evidence from stable isotopes in the benthos of a small, temperate estuary. Oecologia **106**:382-388.
Schmidt, K. and S.H. Jónasdóttir. 1997. Nutritional quality of two cyanobacteria: how rich is ''poor'' food? Marine Ecology Progress Series **151**:1-10.
Shenker, J.M. and J.M. Dean. 1979. The utilization of an intertidal salt marsh creek by larval and juvenile fishes: abundance, diversity and temporal variation. Estuaries **2**:154-163.
Sheridan, P., G. McMahan, G. Conley, A. Williams and G. Thayer. 1997. Nekton use of macrophyte patches following mortality of turtlegrass, *Thalassia testudinum*, in shallow waters of Florida Bay (Florida, USA). Bulletin of Marine Science **61**:801-820.
Shiah, F-K. and H.W. Ducklow. 1995. Multiscale variability in bacterioplankton abundance, production and specific growth rate in a temperate salt-marsh tidal creek. Limnology and Oceanography **40**:55-66.
Sikora, W.B. 1977. The ecology of *Palaemonetes pugio* in a southeastern salt marsh ecosystem with particular emphasis on production and trophic relationships. Dissertation. University of South Carolina, Columbia, South Carolina, USA.
Smith, K.J. and K.W. Able. 1994. Salt-marsh tide pools as winter refuges for the mummichog, *Fundulus heteroclitus*, in New Jersey. Estuaries **17**(1B):226-234.

Smith, S.M., J.G. Hoff, S.P. O'Neil and M.P. Weinstein. 1984. Community and trophic organization of nekton utilizing shallow marsh habitats, York River, Virginia. Fishery Bulletin **82**:455-467.
Stumpf, R.P. and J.W. Haines. 1998. Variations in tidal level in the Gulf of Mexico and implications for tidal wetlands. Estuarine, Coastal and Shelf Science **46**:165-173.
Sullivan, M.J. and C.A. Moncreiff. 1990. Edaphic algae are an important component of salt marsh food-webs: evidence from multiple stable isotope analyses. Marine Ecology Progress Series **62**:149-159.
Talbot, C.W., K.W. Able and J.K. Shisler. 1986. Fish species composition in New Jersey salt marshes: effects of marsh alterations for mosquito control. Transactions of the American Fisheries Society **115**:269-278.
Teal, J.M. 1962. Energy flow in the salt marsh ecosystem of Georgia. Ecology **43**:614-624.
Tiner, R.W. 1993. Field guide to coastal wetland plants of the southeastern United States. The University of Massachusetts Press, Amherst, Massachusetts, USA.
Trent, L., E.J. Pullen and R. Proctor. 1976. Abundance of macrocrustaceans in a natural marsh and a marsh altered by dredging, bulkheading and filling. Fishery Bulletin **74**:195-200.
Turner, M.G. 1989. Landscape ecology: the effect of pattern on process. Annual Review of Ecology and Systematics **20**:171-197.
Turner, R.E. 1977. Intertidal vegetation and commercial yields of penaeid shrimp. Transactions of the American Fisheries Society **106**:411-416.
Valiela I., J.E. Wright, J.M. Teal and S.B. Volkmann. 1977. Growth, production and energy transformations in the salt-marsh killifish *Fundulus heteroclitus*. Marine Biology **40**:135-144.
Walters, K., E. Jones and L. Etherington. 1996. Experimental studies of predation on metazoans inhabiting *Spartina alterniflora* stems. Journal of Experimental Marine Biology and Ecology **195**:251-265.
Weinstein, M.P. 1979. Shallow marsh habitats as primary nurseries for fishes and shellfish, Cape Fear River, North Carolina. Fishery Bulletin **77**:339-357.
Weinstein, M.P. and H.A. Brooks. 1983. Comparative ecology of nekton residing in a tidal creek and adjacent seagrass meadow: community composition and structure. Marine Ecology Progress Series **12**:15-27.
Weinstein, M.P., L. Scott, S.P. O'Neil, R.C. Siegfried II and S.T. Szedlmayer. 1984. Population dynamics of spot, *Leiostomus xanthurus*, in polyhaline tidal creeks of the York River estuary, Virginia. Estuaries **7**:444-450.
Weisberg, S.B., R. Whalen and V.A. Lotrich. 1981. Tidal and diurnal influence on food consumption of a salt marsh killifish *Fundulus heteroclitus*. Marine Biology **61**:243-246.
Welsh, B.L. 1975. The role of grass shrimp, *Palaemonetes pugio*, in at tidal marsh ecosystem. Ecology **56**:513-530.
Wenner, E.L. and H.R. Beatty. 1988. Macrobenthic communities from wetland impoundments and adjacent open marsh habitats in South Carolina. Estuaries **11**:29-44.
Yozzo, D.J., A. Mannino and D.E. Smith. 1994. Mid-summer abundance of resident sub-adult marsh nekton at the Virginia Coast Reserve. Virginia Journal of Science **45**:21-30.
Zimmerman, R.J. and T.J. Minello. 1984. Densities of *Penaeus aztecus, Penaeus setiferus* and other natant macrofauna in a Texas salt marsh. Estuaries **7**:421-433.
Zimmerman, R.J., T.J. Minello, E.F. Klima and J.M. Nance. 1991. Effects of accelerated sea-level rise on coastal secondary production. Pages 110-124 *in* O. Magoon, W.S. Wilson, H. Converse and L.T. Tobin, editors. Proceedings of the Eighth Symposium on Coastal and Ocean Management. American Society of Coastal Engineers, New York, New York, USA.

SALT MARSH LINKAGES TO PRODUCTIVITY OF PENAEID SHRIMPS AND BLUE CRABS IN THE NORTHERN GULF OF MEXICO

ROGER J. ZIMMERMAN
THOMAS J. MINELLO
LAWRENCE P. ROZAS
NOAA Fisheries
4700 Avenue U
Galveston, TX 77551 USA

Abstract
Secondary production derived from coastal marshes of the northern Gulf of Mexico exceeds that of other regions in the United States and is exemplified by large fishery catches of penaeid shrimps (*Farfantepenaeus aztecus*, *F. duorarum*, and *Litopenaeus setiferus* - 66 % of U.S.) and blue crabs (*Callinectes sapidus* - 25 % of U.S.). We believe that this production arises from coastal wetlands, and is driven by wetland geomorphology and hydrology resulting from the delta building and wetland loss cycles of the Mississippi River. Quantitative surveys document that high densities of shrimps and blue crabs directly use northern Gulf marsh surfaces. Manipulative experiments demonstrate that such marshes provide these fishery species with increased resources for growth and with protective cover to reduce predator-related mortality. Thus, access to the marsh surface is an important component in controlling the link between secondary productivity and coastal wetlands. Marsh access is influenced by tidal flooding patterns, amount of marsh/water edge, and extent of connections between marsh systems and the Gulf. Low-elevation Gulf marshes are flooded nearly continually during some seasons and are extensively fragmented; such characteristics provide maximum access. By contrast, U.S. Atlantic coast marshes have less fragmentation and less flooding. These geomorphic and hydrologic differences coincide with differences in secondary production between the regions, e.g., marsh-derived fishery production is lower on the Atlantic coast. Despite the linkage between coastal wetlands and secondary production, the current rapid loss of wetlands in the Gulf does not appear to be causing a decline, but instead is associated with an increase in fishery productivity. This paradox may be explained by changes in access and habitat function during areal loss of wetlands. Wetland loss is accompanied by increased marsh inundation and fragmentation, expansion of saline zones, and shortened migratory routes. These processes extend the utilization of remaining marsh and support temporary increases in secondary production.

1. Introduction

Shelf fisheries of the southeastern United States are characterized by estuarine dependency and an apparent linkage between coastal wetlands and fishery productivity. The Southeast Region's penaeid shrimp (mainly *Farfantepenaeus aztecus* and *Litopenaeus setiferus*) and blue crab (*Callinectes sapidus*) fisheries are prominent for both their magnitude and their apparent dependency on estuaries as nurseries. A positive correlation has long been recognized between area of coastal wetlands and the shrimp fishery landings (Turner 1977). This relationship suggests a marsh habitat-productivity link, but more direct evidence is required to conclusively document and explain how secondary production of fisheries is derived from coastal marshes.

Evidence for a direct link between coastal marsh habitat and fishery production may come in many forms. In general, the stronger the evidence, the more difficult it is to obtain. Simply documenting the presence of a species in a particular habitat may suggest linkage between this habitat and productivity of the species. Such documentation is relatively easy to obtain using qualitative sampling methods; however, this type of evidence is perhaps the weakest for establishing a habitat-productivity link. More convincing are data showing that a specific habitat, when compared with all other available habitats, contains highest densities of a species. This type of evidence can be obtained in interhabitat comparisons using quantitative sampling methods (Rozas and Minello 1997). Still stronger evidence for a habitat-productivity link requires documentation that a habitat provides one or more production-related necessities (e.g., food, protection, spawning area) for a species. Habitats where species have high survival and growth are most likely to be closely linked with high productivity. The most convincing evidence for a direct link between habitat and productivity would require a comparison of secondary productivity for a species among habitats. This type of analysis is difficult and often not practical. In general, the linkage between habitats and productivity should be demonstrated by direct evidence showing that a habitat both contains higher densities of a species than other habitats and provides essential requirements for survival and growth of that species.

Our objective in this paper is to review the evidence for direct linkages between salt marsh habitat of the northern Gulf of Mexico and productivity of three valuable fishery species, brown shrimp (*Farfantepenaeus aztecus*), white shrimp (*Litopenaeus setiferus*), and blue crab (*Callinectes sapidus*). We examine animal densities, growth, and mortality among marsh and other estuarine habitats. We discuss an emerging paradigm that stimulation of secondary production in northern Gulf fisheries is connected to wetland-loss processes. We explain how northern Gulf marshes may differ from southeastern Atlantic Coast marshes in derivation of secondary production. We examine the productivity responses of several decapod crustacea as affected by differing life history strategies, feeding, and predator avoidance behaviors. Taken as a whole, this body of information provides compelling evidence for a direct but variable connection between estuarine marsh habitat and productivity of brown shrimp, white shrimp, and blue crabs in the northern Gulf of Mexico.

2. Habitat-Related Densities

During their stay in Gulf coast estuaries, small juvenile penaeid shrimps and blue crabs are strongly associated with vegetation structure. Densities of these decapods are generally much more abundant in flooded emergent marsh vegetation or submerged grass beds than over nonvegetated sand or mud bottom. In a synthesis of published papers and unpublished reports on habitat use by nekton (restricted to quantitative enclosure samples), Minello (1999) calculated mean nekton densities in different estuarine habitats of Louisiana and Texas. Habitats examined included salt marsh edge, inner marsh, submerged aquatic vegetation (SAV), and shallow nonvegetated bottom. Overall mean densities were higher in marsh edge habitat than over shallow nonvegetated bottom for all three decapod species. Table 1 lists mean densities of penaeids and blue crabs from published studies that directly compared animal abundance in emergent marsh and over nonvegetated bottom.

TABLE 1. Mean densities (number per m^2) of brown shrimp (*Farfantepenaeus aztecus*), white shrimp (*Litopenaeus setiferus*), and blue crabs (*Callinectes sapidus*) in marsh edge and shallow nonvegetated habitats of Texas and Louisiana. Densities are shown only for seasons when juveniles are abundant in estuaries. A paired-T test was used to compare densities between the two habitats. The data source is cited for each pair of means.

	Marsh edge habitat	Nonvegetated bottom	Difference (Edge-NV)	P value (paired-T)
Brown shrimp (spring-fall)				
Minello and Webb (1997)	12.0	3.3	8.7	
Minello et al. (1991)	9.2	0.4	8.9	
Rozas and Minello (1998)	4.5	0.5	4.0	
Zimmerman and Minello (1984a)	24.9	4.0	20.9	
Zimmerman et al. (1989)	9.6	1.1	8.5	
Zimmerman et al. (1990a)	5.0	1.7	3.3	
Zimmerman et al. (1990b)	5.0	1.2	3.7	
Mean	10.0	1.7	8.3	0.011
White shrimp (summer and fall)				
Minello and Webb (1997)	26.7	6.0	20.7	
Rozas and Minello (1998)	4.2	0.0	4.2	
Zimmerman and Minello (1984a)	26.3	11.7	14.7	
Zimmerman et al. (1989)	4.2	0.3	3.9	
Zimmerman et al. (1990a)	1.6	1.5	0.1	
Zimmerman et al. (1990b)	4.4	0.8	3.6	
Mean	11.2	3.4	7.9	0.061
Blue crab (all seasons)				
Thomas et al. (1990)	6.0	2.5	3.5	
Minello and Webb (1997)	6.4	0.7	5.6	
Minello et al. (1991)	4.6	0.0	4.6	
Rozas and Minello (1998)	10.7	0.3	10.4	
Zimmerman and Minello (1984a)	8.5	1.0	7.5	
Zimmerman et al. (1990a)	4.2	0.4	3.8	
Zimmerman et al. (1990b)	10.8	1.9	8.9	
Mean	7.3	1.0	6.3	< 0.001

Overall mean densities of brown shrimp *F. aztecus* in Texas and Louisiana were estimated at 7.5 m^{-2} (SE 0.7) in *Spartina alterniflora* marsh edge habitat and 1.9 m^{-2} (SE 0.2) on nonvegetated bottom (Minello 1999). Most studies comparing brown shrimp abundance between these habitats reported densities of 3 to 13 m^{-2} in emergent marsh and < 3.5 m^{-2} over nonvegetated bottom (Table 1). The mean density of brown shrimp in SAV was 7.3 m^{-2} (SE 1.0) in Texas and Louisiana (Minello 1999). On the central Texas coast, Rozas and Minello (1998) found that densities of brown shrimp were similar in SAV (5.3 m^{-2}, SE 1.1) and marsh (5.0 m^{-2}, SE 0.8) in fall, but higher in SAV (11.3 m^{-2}, SE 1.3) than marsh (4.5 m^{-2}, SE 0.9) in spring.

White shrimp *L. setiferus* densities in Texas and Louisiana were reported by Minello (1999) to be 5.5 m^{-2} (SE 1.0) in *S. alterniflora* marsh edge compared with 1.2 m^{-2} (SE 0.2) on nonvegetated bottom; and in most comparisons shown in Table 1, white shrimp densities were higher in marsh than over nonvegetated bottom. Densities were similar in the two habitats only in one study (Zimmerman et al. 1990a), where overall white shrimp densities were relatively low (< 2 m^{-2}). Minello (1999) reported low overall densities of white shrimp in SAV (0.5 m^{-2}, SE 0.1), but few studies have directly compared densities in SAV and emergent marsh edge. Zimmerman et al. (1990b) compared marsh edge with *Vallisneria* and *Halodule* habitats, and densities were about five times higher in the marsh. In a study on the central Texas coast, mean densities of white shrimp were twice as high in *Spartina alterniflora* marsh edge as in mixed aquatic beds of *Halodule wrightii* and *Ruppia maritima* (4.2 m^{-2}, SE 0.9 vs 1.9 m^{-2}, SE 1.5), although these differences were not statistically significant (Rozas and Minello 1998).

Young blue crabs *C. sapidus* appear to be closely associated with vegetation in estuaries. In most published reports, densities in marsh habitat are about an order of magnitude greater than on shallow nonvegetated bottom (Table 1). Overall mean densities in Texas and Louisiana were 6.2 m^{-2} (SE 0.7) in *S. alterniflora* marsh edge, 0.9 m^{-2} (SE 0.1) on nonvegetated bottom, and 5.0 m^{-2} (SE 1.2) in SAV (Minello 1999). Where direct measurements were compared between different contiguous vegetated habitats, results are conflicting. Rozas and Minello (1998) found higher mean densities of blue crabs in marsh edge than SAV both in fall (11.1 m^{-2}, SE 0.9 vs 7.5 m^{-2}, SE 1.4) and in spring (10.3 m^{-2}, SE 1.0 vs 2.5 m^{-2}, SE 0.3). In contrast, blue crab densities from *Halodule* beds in Christmas Bay, Texas were higher than in nearby marsh edge in 7 of the 12 months sampled (Thomas et al. 1990).

In most comparisons of nekton densities between salt marsh and nonvegetated bottom, only the marsh edge was sampled. Animal densities along the marsh edge are generally much higher than in marsh located farther from the water-marsh interface (Baltz et al. 1993, Peterson and Turner 1994, Kneib and Wagner 1994, Minello et al. 1994). For example, overall mean densities from the studies examined by Minello (1999) were much higher in marsh-edge than inner-marsh (>5 m from shoreline) habitat (white shrimp: 5.5 m^{-2}, SE 1.0 vs 1.6 m^{-2}, SE 0.9; brown shrimp: 7.5 m^{-2}, SE 0.7 vs 0.4 m^{-2}, SE 0.2; blue crab: 6.2 m^{-2}, SE 0.7 vs 0.5 m^{-2}, SE 0.1). Therefore, nekton densities reported for marsh-edge habitat cannot be extrapolated to the entire marsh surface. However, marsh-edge habitat is extensive in Louisiana and Texas where much of the salt marsh is highly reticulated due to coastal submergence and marsh fragmentation.

3. Marsh-Related Growth

Food is a principal attractant leading to estuarine habitat selection (Boesch and Turner 1984, Kneib 1984, Minello and Zimmerman 1991, McTigue and Zimmerman 1991). Penaeid shrimps generally feed by browsing and digging through surface sediments. Both juvenile white shrimp and brown shrimp are omnivorous and known to eat epiphytic algae, marsh detritus, and animal material in the laboratory (Condrey et al. 1972, Gleason and Zimmerman 1984, McTigue and Zimmerman 1991), but species-specific differences in feeding have been documented. Initial evidence that juvenile brown shrimp and white shrimp feed directly upon marsh infauna was obtained from an unpublished laboratory feeding experiment using marsh sediment cores. Thirty-six cores (10 cm dia. x 5 cm ht.) were collected from a *Spartina alterniflora* marsh on Galveston Island and maintained as microcosms in 25-cm (ht.) PVC sleeves. The cores were held under laboratory conditions of 25 °C, 20‰ salinity, and alternate cycles of 12 h light and 12 h dark. Individual juvenile brown shrimp and white shrimp (28 mm in total length) were placed in 24 cores (12 cores with each species) as treatments. Twelve cores without shrimp served as a control. After 5 d, each core was sieved through a 500-μ screen; and remaining peracarids, annelids, and mollusks were identified and counted. Feeding was quantitatively inferred from differences in numbers of infauna between treatment and control cores. Results of this depletion experiment indicate that marsh annelids (mainly spionids and capitellids) and peracarids (mainly tanaidaceans and amphipods) were readily eaten by the juvenile shrimps. Brown shrimp and white shrimp significantly reduced the numbers of peracarid crustaceans and annelid worms in marsh sediments, and brown shrimp ate significantly more than white shrimp (Fig. 1, ANOVA, $P < 0.05$). Additional experiments reported by McTigue (1993) and McTigue

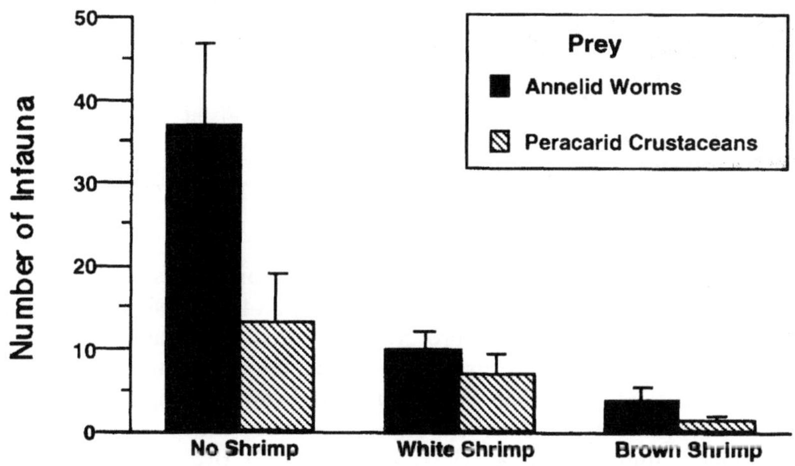

Figure 1. Feeding by brown shrimp (*Farfantepenaeus aztecus*) and white shrimp (*Litopenaeus setiferus*) on annelid worms and peracarid crustaceans in sediments from a subsiding salt marsh on Galveston Island, Texas. Mean number +1 SE of infauna removed by shrimp from 78 cm^2 x 15 cm cores of marsh sediment (n = 12 cores per treatment).

and Zimmerman (1998) confirmed the basic differences in feeding habitats of brown shrimp and white shrimp. Brown shrimp are more effective at removing infauna from sediments and appear to be obligate carnivores that depend on dense numbers of infauna found on the marsh surface. In contrast, white shrimp are less effective at removing infauna from sediments; this species is truly omnivorous and depends more on plant resources than brown shrimp (McTigue 1993).

Figure 2. A newly metamorphosed juvenile white shrimp, *Litopenaeus setiferus*, approximately 23 mm in length, feeding on epiphytic algae attached to smooth cordgrass (*Spartina alterniflora*) culm (photograph by R. Zimmerman).

Growth studies also indicate differences in trophic requirements of brown shrimp and white shrimp. Both species may benefit from feeding on the marsh surface, but brown shrimp productivity appears to be more closely linked with marsh infauna. Brown shrimp will feed on salt marsh detritus and epiphytes, but assimilation was not detected from these treatment diets (Gleason 1986). Juvenile brown shrimp seem to depend on infaunal worms for growth, and densities of these prey organisms are relatively high on the marsh surface (Zimmerman et al. 1991, Whaley 1997). Brown shrimp held in cages showed significantly higher growth rates when they had access to the marsh surface (as high as 1.4 mm d^{-1}) than when they were restricted to subtidal bottom (Zimmerman and Minello 1984b). Simultaneous caging of juvenile white shrimp revealed no difference in growth between marsh and open-water habitats. White shrimp attain better growth on diatoms or epiphytes and some natural animal dietary component (perhaps mysids or copepods) that has yet to be identified (McTigue and Zimmerman 1998). Kneib and Knowlton (1995) and Kneib (1997) suggest that white shrimp may be important predators on early life history stages of daggerblade grass shrimp *Palaemonetes pugio*. The ability of white shrimp to exploit plant resources also is suggested by rapid growth observed in pond studies. In ponds without macrophytes that were fertilized to promote phytoplankton, growth rates were reported for juvenile white shrimp of 2.1 mm d^{-1} (Johnson and Fielding 1956) and 2.3 mm d^{-1} (Wheeler 1968). By comparison, brown shrimp in Wheeler's (1968) experiments grew at 1.0 mm d^{-1} in fertilized ponds and 1.1 mm d^{-1} in unfertilized ponds.

Thomas (1989) tested dietary habits of juvenile blue crabs in natural microcosm cores similar to those described above. Her results demonstrate that blue crab diets are different from those of juvenile shrimps. Post megalopae juvenile blue crabs fed significantly more on epiphytic algae and peracarid crustaceans than on annelid worms. In a caging study conducted on open bay bottom, Minello and Wooten (1993) also found that small juvenile blue crabs (12-17 mm CW) did not appear to feed on infauna. Infaunal densities in this experiment were low, but positive growth of enclosed crabs was measured. Other investigators have documented plant and animal material in guts of blue crabs, and their studies suggest an ontogenetic change in diet to more carnivory as individuals grow (Alexander 1986, Laughlin 1982, Ryer 1987). Rosas et al. (1994) noted that, in general, as blue crabs increase in size, plant matter, sediment, and unidentified animal residues in guts decrease in favor of increases in molluscs and crustaceans. Fitz and Wiegert (1991) found a predominance of feeding on fishes and non-portunid crabs (*Uca* sp., grapsid and xanthid crabs) by large blue crabs in Georgia marshes, and West and Williams (1986) and Schindler et al. (1994) reported that adult blue crabs actively fed upon molluscs such as *Littoraria* sp. on the marsh surface. Ryer (1987) found that blue crab guts contained more food at high tide than at any other period of the tidal cycle and suggested this as evidence that blue crabs foraged in intertidal marshes.

4. Benefits of Marsh-Surface Access to Feeding

A major function of salt-marsh habitat is to serve as a feeding area for opportunistic estuarine species, and there is evidence that this function varies regionally. Historically, salt marshes were thought mainly to contribute to detritus-based food webs by outwelling plant debris into estuaries and coastal areas downstream of marshes (Nixon 1980, Peters and Schaaf 1991). Such indirect use of plant production from Atlantic coast marshes is consistent with relatively high elevations (limiting accessibility for nekton) and large tidal amplitudes (providing energy to transport detritus). But in the northern Gulf of Mexico, direct use of the marsh surface appears to be widespread, fostered by extended tidal flooding associated with low marsh elevations and a narrow tidal range. Greater access to the marsh surface gives young fishery species an opportunity to feed on an abundance of infauna, epiphytic and edaphic algae, and small primary consumers that provide high-quality food necessary for rapid growth. Microalgal trophic pathways have been described from Gulf marshes (Sullivan and

Moncreiff 1990), and the relative importance of algal versus detrital pathways is likely controlled by marsh-surface availability (McIvor and Rozas 1996). Importantly, regional differences in secondary productivity are influenced by differences in opportunistic feeding behavior among estuarine species (Kneib 1995). Differences in productivity and the resulting fishery yields of estuarine-dependent species such as penaeid shrimps and blue crabs are in part due to species-specific abilities to utilize marsh habitat for feeding.

5. Salt Marshes and Mortality of Shrimps and Crabs

Vegetated estuarine habitats also affect productivity of shrimps and crabs by providing cover or refuge and reducing mortality. A major cause of mortality for penaeid shrimps and blue crabs is predation by estuarine fishes (Minello and Zimmerman 1983, Wilson et al. 1987, 1990, Minello et al. 1989, Heck and Coen 1995). Juvenile blue crabs also suffer significant mortality from cannibalism by larger crabs (Orth and van Montfrans 1982, Hines and Ruiz 1995). Mortality due to predators appears to be lower within vegetated estuarine habitats in comparison with nonvegetated bottom. Laboratory experiments have shown that the structure of salt marsh vegetation reduces feeding rates of some estuarine fishes on brown shrimp (Fig. 3) and blue crabs (Minello and Zimmerman 1983, Thomas 1989, Minello et al. 1989). Seagrass structure has also been shown to reduce predation rates on a variety of crustacean prey (Coen et al. 1981, Heck and Thoman 1981, Main 1987) including juvenile blue crabs (Orth and van Montfrons 1982, Orth et al. 1984, Thomas 1989). Predator-induced mortality, however, also depends on the suite of predators present within habitats, and laboratory experiments do not always reflect mortality in the field. Tethering experiments are designed to incorporate differences in trophic webs among habitats in addition to differences in environmental characteristics other than structure. For example, shallow water, that may be associated with some vegetated habitats, has been shown to reduce predation and mortality of blue crabs (Ruiz et al. 1993, Dittel et al. 1995, Hines and Ruiz 1995). Field experiments with tethered blue crabs and brown shrimp prey have shown that mortality is reduced in seagrass and marsh habitats compared with nonvegetated bottom (Heck and Thoman 1981, Wilson et al. 1987, 1990, Minello 1993, Heck et al. 1994). All of these data, therefore, support the hypothesis that vegetated habitats such as salt marshes and seagrass beds reduce predator-related mortality of crustaceans like penaeid shrimps and blue crabs.

The protective value of vegetated habitats varies. Intertidal salt marsh is not always flooded and available for exploitation by shrimps and crabs; thus, regional differences in tidal dynamics can affect the protective value of salt marshes. In the northern Gulf of Mexico, flooding durations during spring and fall are extensive (Rozas and Reed 1993, Minello and Webb 1997). During these seasons, salt marshes may function quite similarly to seagrass in these estuaries (Rozas and Minello 1998). There is also some indication that vegetated habitats with very high densities of plants offer less protective cover, because thick mats of roots and rhizomes prevent burrowing in the substratum (Wilson et al. 1987). Both blue crabs and penaeid shrimps often burrow during the day, and this behavior reduces mortality caused by both predators (Fuss 1964, Fuss and Ogren 1966, Minello et al. 1987) and by temperature extremes (Eldred et al. 1961, Aldrich et al. 1968).

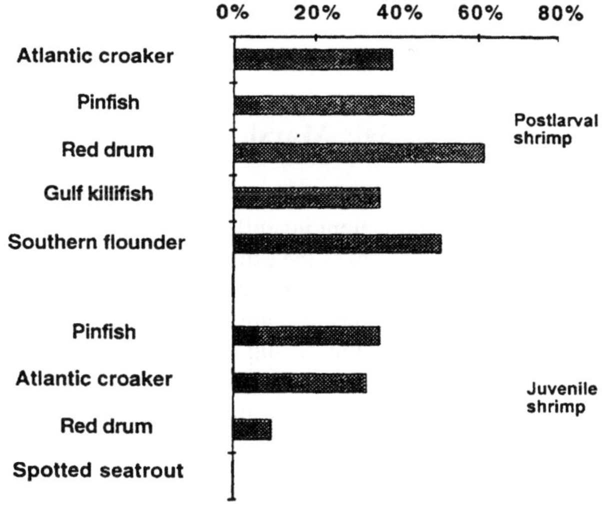

Figure 3. Reduction in feeding rates on brown shrimp due to vegetation. Experiments comparing feeding by various fish predators between bare sand and vegetated habitats were conducted in laboratory microcosms using planted *Spartina alterniflora* and postlarval shrimp (Minello et al. 1989) or simulated vegetation and juvenile shrimp (Minello and Zimmerman 1983).

In addition to predator-related mortality, both penaeid shrimps and blue crabs suffer mass mortality from periodic detrimental physical conditions in the estuary such as freezing weather (Gunter 1941, Gunter and Hildebrand 1951, Dahlberg and Smith 1970) and anoxic water (Gunter 1942, May 1973, Turner and Allen 1982, Turner et al. 1987). Habitats that function to protect shrimps and crabs from predators do not necessarily provide refuge from these sources of mortality. Deep water and an appropriate substratum for burrowing may be important habitat characteristics for reducing mortality from low temperatures (Eldred et al. 1961, Aldrich et al. 1968).

Regional differences in the value of salt-marsh habitats in reducing mortality may be related to differences in tidal dynamics, marsh morphology, trophic structure, and climate. Earlier we discussed the benefits to shrimps and crabs of access to the marsh surface for feeding. The increased marsh access prevalent along the Gulf coast should also provide increased protective benefits. These benefits, however, may be reduced in part by increased predation pressure in Gulf estuaries. Heck and Wilson (1987) and Heck and Coen (1995) found that predation on crabs in vegetated habitats was higher at lower latitudes and higher along the Gulf coast than along the Atlantic coast. The dominant predators may also vary regionally. Fish predators are generally considered the most significant sources of mortality for shrimps and crabs in Gulf estuaries (Minello et al. 1989, Heck and Coen 1995). In Chesapeake Bay, however, Hines and Ruiz (1995) attributed almost all mortality of tethered juvenile blue crabs to cannibalism by larger crabs. In addition, climatic differences between much of the U.S.

Atlantic coast and the Gulf coast may affect the protective value of different habitats. Deep-water habitats with soft substrata may be especially valuable in climates and seasons where temperatures drop to lethal levels for crabs and shrimps. In these situations, seagrasses in deep waters may provide protection from both predators and the physical environment.

6. Seasonal Differences in Salt-Marsh Value

Direct benefits of marsh habitat to transient juveniles of fishery species may depend upon the seasonal timing of larval recruitment to the estuary. Flooding of marsh surfaces in the Gulf varies seasonally, and benefits can be proportionately greater for species that immigrate into the estuary when marshes are most accessible. Postlarval brown shrimp are at their peak abundance during the spring and fall (Baxter and Renfro 1967), coinciding with tidal high water periods that inundate salt marshes extensively (Rozas and Reed 1993, Minello and Webb 1997). Juvenile white shrimp are abundant during fall when marshes are flooded; however, postlarvae mainly recruit in the summer when intermediate water levels persist. Blue crab megalopae recruit into Gulf estuarine habitats in summer and fall (Rabalais et al. 1995), but juveniles overwinter in the estuary. During winter, marshes are relatively inaccessible, and the lowest water levels of the year occur. We suggest that seasonal hydrology affects marsh use and related benefits to production among these three species. On the basis of these seasonal-use patterns, brown shrimp should accrue the most benefit from salt-marsh habitats followed by white shrimp and blue crabs.

Abundances of infaunal prey organisms such as annelid worms and crustaceans in Gulf estuaries are usually more numerous in salt-marsh habitat than on subtidal bottom during most months (Fig. 4). Population levels of infauna vary seasonally, and densities are generally highest in late winter months when predator densities are low (Flint and Younk 1983, Zimmerman et al. 1991, Whaley 1997). This peak in infauna throughout the estuary coincides with the arrival of brown shrimp postlarvae in early spring. Infaunal densities decline to summer low levels, presumably due to predation, by the time white shrimp postlarvae arrive in Gulf estuaries. Blue crab megalopae arrive in the summer and fall, but juveniles overwinter in the estuary and are present in marshes throughout the year (Thomas et al. 1990, Rabalais et al. 1995, Rozas and Minello 1998). Subadult and adult blue crabs also use salt marshes (Thomas et al. 1990, Fitz and Wiegert 1991). This extended period of marsh use by all life-cycle stages for blue crabs contrasts with the seasonally limited use by juvenile penaeid shrimps. Energy derived from foraging in marshes by shrimps is used mainly for growth, while benefits for blue crabs are to growth and reproduction.

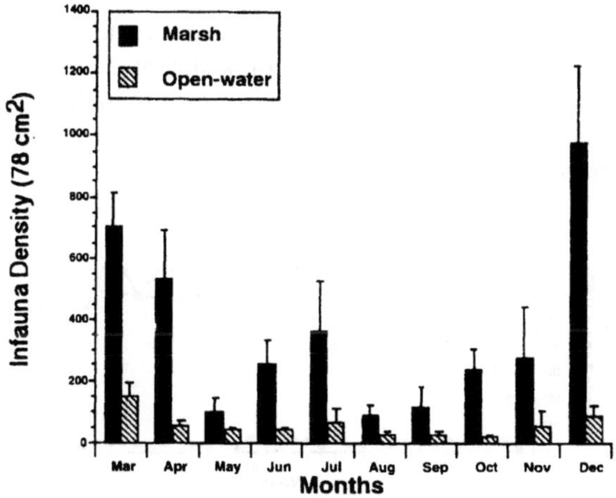

Figure 4. Seasonal densities of infaunal prey (annelid worms and peracarid crustaceans) in a sudsiding marsh and adjacent subtidal habitat from 78 cm^2 x 15 cm cores of sediment taken at Galveston Island in the northern Gulf of Mexico (Zimmerman et al. 1991). Mean and standard error, n = 8 cores in each habitat monthly from 1985 through 1988.

7. Fishery Trends

The fisheries for brown shrimp and white shrimp are the largest crustacean fisheries in the United States, and they are centered in the northern Gulf of Mexico. Because the production of these species appears to depend on coastal wetlands, the high rates of wetland loss in the region are a matter of concern for fishery managers (Condrey and Fuller 1992). The brown shrimp fishery is the largest shrimp fishery in the Gulf of Mexico, and parent stocks of this species have remained relatively stable since 1960 (Eldridge 1988, Nance et al. 1989, Klima et al. 1990, Nance 1993). However, landings of brown shrimp doubled between 1960 and 1991 (Fig. 5) coinciding with increased fishing effort. Using virtual population analyses, the number of shrimp recruits was calculated for each of 31 years of catch data (1960 to 1991) where size composition and fishing effort are known. This analysis showed that recruitment to the brown shrimp fishery increased significantly through this period to historic high levels in 1991, and that increased landings are not solely due to increased fishing effort. The current dominance of brown shrimp in the fishery is recent; white shrimp dominated landings in the Gulf from the late 1800s to 1950 (Condrey and Fuller 1992). The white shrimp fishery in the Gulf is still large, however, and landings more than doubled between 1960 and 1986 (Fig. 5, Nance et al. 1989, Nance 1993). Recruitment of white shrimp has also increased significantly, with the largest increase between 1984 and 1986. Both of these shrimp fisheries have been considered fully exploited at least since the early 1970's, providing further evidence that increases in the Gulf landings of brown shrimp and white shrimp are caused by increases in recruitment. Blue crab landings in the Gulf of Mexico have increased similarly to brown shrimp and white shrimp since 1960

(Fig. 5). No trend in recruitment is known. Another large crustacean fishery, and the third largest shrimp fishery in the Gulf of Mexico, is pink shrimp *F. duorarum*. In contrast to other shrimp species, the principal nurseries for pink shrimp in the Gulf are seagrasses in South Florida and South Texas, and Gulf landings of pink shrimp have not exhibited a steady pattern of increase during the past three decades.

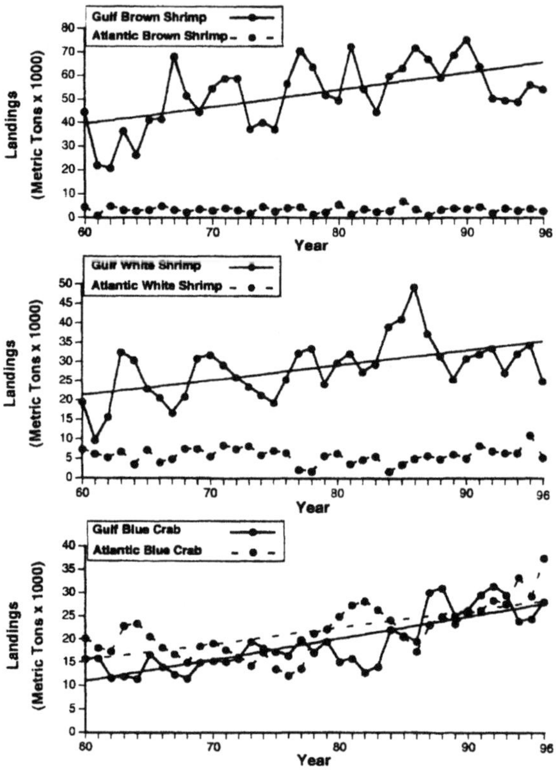

Figure 5. United States southeastern Atlantic and Gulf of Mexico landings of brown shrimp (*Farfantepenaeus aztecus*), white shrimp (*Litopenaeus setiferus*) and blue crab (*Callinectes sapidus*) from 1960 through 1996 (U.S. National Marine Fisheries Service catch statistics data compiled by the Southeast Fisheries Science Center, Miami, Florida). Regression lines denote significant trends.

Fisheries for brown shrimp, white shrimp, and blue crabs of the southeastern U.S. Atlantic coast are associated with salt-marsh dominated estuaries (Weinstein 1979, Wenner and Beatty 1993). However, landings of shrimp per unit area of salt marsh are more than three times higher in the Gulf than the Atlantic (Table 2). In the Atlantic, penaeid shrimp fisheries fluctuate annually like those of the Gulf, but Atlantic landings have not increased significantly since 1960 (Fig. 5), and no trend in recruitment is known. Unlike the Gulf, the southeast Atlantic shrimp fishery has always been dominated by white shrimp. Blue crab landings along the Atlantic have increased similarly to those in the Gulf (Fig. 5). Also, fishery-independent surveys of abundances of juvenile blue crabs appear within the same order of magnitude between the two regions (Heck and Coen

1995). These similarities suggest that regional differences in marsh habitat are not a major factor influencing blue crab production. The northern extension of blue crab landings to the upper mid-Atlantic coast, where penaeid shrimp fisheries are inconsequential, suggests that blue crabs tolerate a wider range of environmental conditions than penaeid shrimps.

TABLE 2. Shrimp fishery landings (kg) per ha of salt marsh in Gulf of Mexico and Atlantic states where brown shrimp (*Farfantepenaeus aztecus*) and white shrimp (*Litopenaeus setiferus*) predominate. The total area of salt marsh (ha), portion of total U.S. salt marsh area (%), total shrimp landings (metric tons, all species combined-*F. aztecus*, *L. setiferus*, and *F. duorarum*), and landings per ha of marsh are given for each state and region.

			1996 Shrimp Landings	
State	Marsh (ha)	% of U.S.	Metric tons	kg/ha
Louisiana	697,223	40.4	36645.6	52.5
Texas	174,862	10.1	28057.6	160.4
Northeast GOM Subtotal	872,085	50.5	64703.2	74.1
South Carolina	148,072	08.6	2433.5	16.4
North Carolina	086,116	05.0	2391.2	27.7
Georgia	078,548	04.5	2637.1	33.5
South Atlantic Subtotal	312,736	18.1	7461.8	23.8
Five State Total	1,184,821	68.6		
Overall U.S. Total	1,728,190			

(Marsh acreage recalculated from: Coastal Wetlands of the United States: an accounting of a valuable natural resource. 1991. U.S. Dept. of Commerce, NOAA/NOS and Dept. of the Interior, U.S. F. W. S. 59 p.; with conversion of acres = 0.4047 hectares)

8. Relationship of Marsh Submergence to Productivity

Increasing yields of brown shrimp and white shrimp over the past 30 years in the northern Gulf of Mexico are correlated with high rates of subsidence and loss of marsh habitat, and there is evidence that wetland-loss processes may have stimulated secondary productivity of these fishery species (Nance et al. 1989, Zimmerman et al. 1991). With high rates of marsh submergence, protection and feeding benefits of nursery habitat are modified for transient juveniles through:

1) Extension of the saline zone inland, providing more salt-marsh nursery area;
2) Lengthened duration of marsh inundation, allowing more time to feed and seek refuge among plant cover;
3) Greater accessibility to marsh habitat from open water due to increased edge; and
4) Shorter migration routes from the sea to inland marshes.

Relationships between submergence and productivity can also be seen by examining annual changes in sea level. Morris et al. (1990) reported that annual growth of *Spartina alterniflora* in South Carolina varies by a factor of two and correlates positively with anomalies in mean sea level. Moreover, commercial landings of shrimp and menhaden of the southeast Atlantic and central Gulf of Mexico are directly correlated with sea level. Childers et al. (1990) noted that relationships between annual water levels and shrimp harvest in the Gulf were curvilinear. Low catches occurred in years of low or high water levels and high catches were in years with intermediate water levels. Low water was attributed to drought. We note that low catch in high water years can be attributed to high rainfall in which lower salinities restrict the area of suitable nursery habitat for young shrimp.

9. Regional Differences

Juvenile shrimps, blue crabs, and other transient marine taxa exhibit regional differences in direct use of the marsh surface, and these differences appear related to regional differences in hydrology and marsh inundation (Zimmerman et al. 1991, Rozas 1995). Northern Gulf salt marshes support densities of penaeid shrimps and blue crabs (Zimmerman and Minello 1984a, Rozas and Reed 1993) that are an order of magnitude greater than densities in East coast marshes (Hettler 1989, Mense and Wenner 1989, Fitz and Wiegert 1991, Kneib 1991, Rozas 1993). We attribute higher densities in Gulf marshes in part to longer inundation times that increase accessibility of subsiding marshes. We also suggest that the differences between Gulf and Atlantic fishery landings, which are largest in brown shrimp, followed by white shrimp and then blue crab, may be attributed to influences of marsh geomorphology and tidal hydrology.

As sea level rises in Gulf of Mexico salt marshes, especially during periods of accelerated rise, marshes are submerged, and habitat characteristics change (Deegan and Thompson 1985, Conner and Day 1987, Wells 1987, DeLaune et al. 1989). The classic configuration of a stable marsh along the Atlantic coast with its dendritic creeks disappears. In the Gulf, the marsh landscape becomes fragmented as interior ponding occurs (Turner and Rao 1990, Turner 1997), and patches of marsh become interspersed within subtidal areas of open water (Fig. 6). This condition creates more edge interface between salt marsh and open water (Browder et al. 1985) resulting in greater direct accessibility of the marsh surface for transient aquatic fauna.

The connections we have outlined between production of fisheries and marsh loss also can be related to characteristics of marsh building and wetland loss cycles of the Mississippi River delta. Rates of sedimentation and subsidence during the aging process of deltaic lobes strongly influence the biological characteristics of marshes (Neill and Deegan 1986, Rejmanek et al. 1987, Reed and Cahoon 1992). For example, recently-formed Atchafalaya delta marshes are dominated by strong riverine inflow, active delta building, and low subsidence rates (Wells 1987, DeLaune et al. 1987). These accreting Atchafalaya marshes, although inundated frequently, may be unavailable to some estuarine consumers due to low salinities; although Castellanos (1997) reports relatively high standing stocks of blue crabs here. Madden et al. (1988) emphasize that secondary production from the building Atchafalaya delta is driven by seasonal input of river-borne nutrients and sediments. By contrast, in an older deltaic system there is little direct river input, and marshes such as those in the lower Barataria Basin are rapidly subsiding and deteriorating (Sasser et al. 1986). These submerging marshes provide additional sources

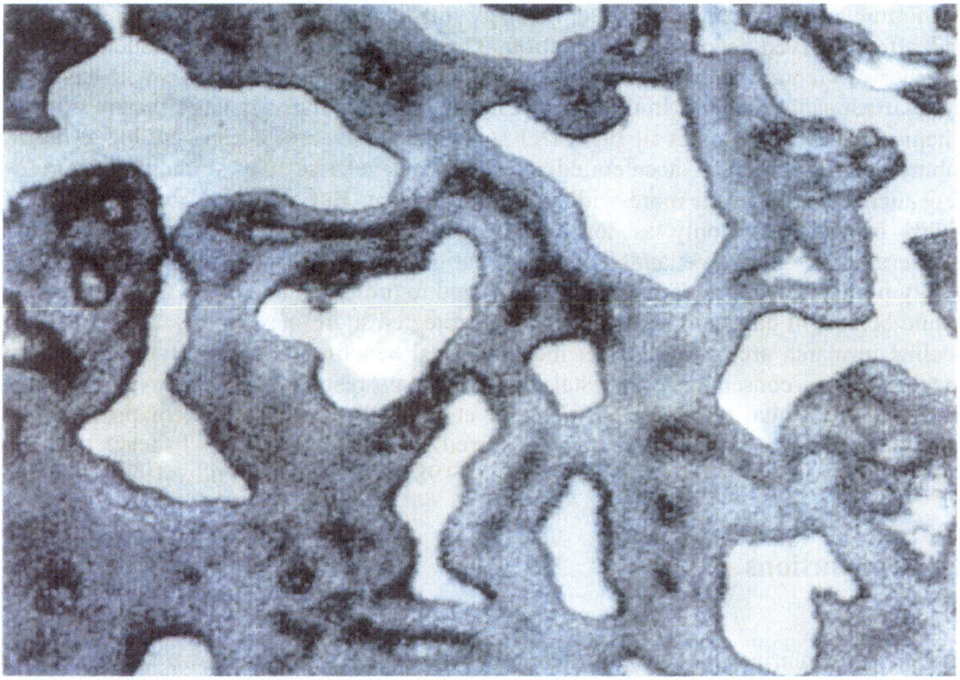

Figure 6. Highly fragmented salt marshes in the northern Gulf of Mexico provide maximum access to the marsh surface for fishery species. Oblique aerial photograph encompassing approximately 3 ha of marsh in Galveston Bay (taken by T. Minello).

of carbon and nitrogen exported to surrounding open waters (Feitjtel et al. 1985, DeLaune et al. 1989). In the Barataria system, marsh utilization by transient marine consumers is favored by higher salinities and organic detritus eroded from old marshes. Deegan and Thompson (1985) reported the mean density of fishes (sampled with otter trawls) to be more than an order of magnitude greater in Barataria Bay (0.32 individuals m^{-2}) than in Atchafalaya Bay (0.02 individuals m^{-2}).

10. Future Trends

The characteristics of drowning marshes, i.e., expansion inland, extended duration of flooding, more edge, and higher erosion rates, may benefit nursery function and enhance fishery production only over the short-term. For example, one model (Browder et al. 1989) suggests that marsh conversion to open water in Barataria Bay will soon reach a point beyond which fisheries will decline due to a reduction in the total amount of marsh area. The implication is that, over the long term, high yields supported by marsh submergence can be maintained only as long as marsh area lost is regenerated elsewhere. Nationally, Dahl and Johnson (1991) reported that areal losses of saline marshes have been more than replaced by encroachment into freshwater wetlands.

Future rates of eustatic sea level rise may further change marsh habitats nationwide and affect derived secondary production. The potential for greater production from marshes of Georgia and the Carolinas rests upon whether submergence will have the same affect as in the Gulf. In the northwestern Gulf, the relatively large marshes of the Chenier plain (Gosselink et al. 1979, DeLaune et al. 1983) are also susceptible to future submergence. Therefore, accelerated rates of sea level rise may stimulate estuarine-dependent fisheries even more widely than at present. But, as noted above, enhanced yields can continue only as long as drowning marshes are replaced by inland progression of saline wetlands. Eventual nationwide losses of total marsh area are highly probable as more barriers are constructed to protect inland areas. Saline marshes would be caught between the rising sea and protected shorelines. In this case, continual decline in marsh area would offset the functional benefits of submergence for fishery species. As a consequence, coastal fisheries may respond to sea-level rise rates predicted by global warming (Armentano et al. 1988) with short-term productivity increases, such as we believe have occurred in the Gulf shrimp fishery, that are unsustainable in the long-term (Browder et al. 1989, Condrey and Fuller 1992).

11. Conclusions

Patterns of estuarine utilization indicate that the productivity of brown shrimp, white shrimp, and blue crabs is linked to salt marshes. Indeed, investigators have amply demonstrated that estuarine wetlands provide the young of these fishery species with an abundant source of food that supports rapid growth, in addition to protective cover that reduces mortality from predators. Correspondingly, the largest area of emergent wetlands, including salt marshes and the largest crustacean fisheries in the U.S. are located in the northern Gulf of Mexico.

The linkages between salt-marsh wetlands and fishery productivity, however, are complex and varied. The importance of salt-marsh availability as nursery habitat has only been recognized fully within the last decade. The availability of coastal marshes to fishery species is determined by tidal flooding patterns, the amount of marsh/water edge, and the extent of connections between interior marsh and the sea. Within the northern Gulf of Mexico, low-elevation marshes are flooded almost continually during some seasons and are extensively fragmented, providing maximum access for young shrimp and blue crabs. By contrast, marshes along the southeastern U.S. Atlantic coast are less inundated and have relatively little marsh/water edge. Densities of transient aquatic species using the marsh surface also differ; the densities in the Gulf are generally an order of magnitude greater than those on the Atlantic coast. We now believe that these differences in wetland availability and degree of use are at least partially responsible for higher production and higher landings of some estuarine-dependent species in the Gulf of Mexico as compared with the U.S. Atlantic.

Overlying the concept of relative wetland value based upon hydrology is modification due to wetland loss. Salt-marsh loss is occurring throughout the southeastern U.S., but the highest rates are in the northern Gulf of Mexico. Because of the proposed linkage between wetlands and fishery production, we might expect estuarine-dependent fisheries to decline as spatial extent of marsh habitat diminishes. In the northern Gulf of Mexico, however, recruitment and landings have increased for brown shrimp and white shrimp over the last 20 to 30 years. By comparison, landings of these species have remained stable along the U.S. Atlantic coast where wetland loss is relatively low. We are left with an interesting paradox — that of increased shrimp fishery production correlated with the loss of nursery habitat. The explanation appears to be related to the process of wetland loss. As the total area of coastal marsh decreases, inundation of existing marshes increases, fragmentation and habitat edge increase, zones of saline and brackish wetlands expand, and connections with the sea are shortened. We believe that the wetland loss process increases the availability and functional value of remaining marsh to transient fishery juveniles, which supports short-term increases in secondary production such as in shrimp. In the long-term, however, these enhanced levels of secondary productivity are not sustainable; continued wetland loss will eventually overtake short-term benefits derived from habitat loss, and future declines in estuarine-dependent shrimp production are unavoidable.

Brown shrimp, white shrimp, and blue crab are opportunistic species. Their productivity does not entirely depend on salt-marsh habitat, because they also occur in estuaries dominated by mangroves and SAV. However, in coastal areas with abundant salt marsh, the productivity of these fishery species appears to depend upon their ability to use the marsh surface directly as determined by hydrographic and geomorphic conditions. This interaction, of productivity and salt marsh habitat, also depends on the life history and behavioral characteristics of the different fishery species. Together, these factors can account for both regional and intraspecific differences in secondary productivity. In the northern Gulf of Mexico, brown shrimp production appears to have benefited the most from this salt marsh relationship, followed by white shrimp and blue crab.

12. Acknowledgements

We would like to thank the Southeast Fisheries Science Center of the National Marine Fisheries Service for supporting research on relationships between fishery species and coastal habitats. The assistance of everyone in the Fishery Ecology Branch (FEB) at the Galveston Laboratory was essential. Members of the FEB were responsible for collecting and analyzing much of the data summarized in this paper.

13. Literature Cited

Aldrich, D. V., C. E. Wood and K. N. Baxter. 1968. An ecological interpretation of low temperature responses in *Penaeus aztecus* and *P. setiferus* postlarvae. Bulletin of Marine Science **18**:61-71.

Alexander, S. K. 1986. Diet of the blue crab, *Callinectes sapidus* Rathbun, from nearshore habitats of Galveston Island, Texas. Texas Journal of Science **38**:85-89.

Armentano, T. V., R. A. Park and C. L. Cloonan. 1988. Impacts on coastal wetlands throughout the United States. Pages 87-149 *in* J. G. Titus, editor. Greenhouse effect, sea level rise and coastal wetlands. EPA-230-05-**86-013**, U.S. EPA, Office of Policy, Planning and Evaluation, Washington, District of Columbia, USA.

Baltz, D. M., C. Rakocinski and J. W. Fleeger. 1993. Microhabitat use by marsh-edge fishes in a Louisiana estuary. Environmental Biology of Fishes **36**:109-126.

Baxter, K. N. and W. C. Renfro. 1967. Seasonal occurrence and size distribution of postlarval brown and white shrimp near Galveston, Texas, with notes on species identification. Fishery Bulletin, U.S. **66**:149-158.

Boesch, D. F. and R. E. Turner. 1984. Dependence of fishery species on salt marshes: the role of food and refuge. Estuaries **7**:460-468.

Browder, J. A., H. A. Bartley and K. S. Davis. 1985. A probabilistic model of the relationship between marshland-water interface and marsh disintegration. Ecological Modeling **29**:245-260.

Browder, J. A., L. N. May, A. Rosenthal, J. G. Gosselink and R. H. Baumann. 1989. Modeling future trends in wetland loss and brown shrimp production in Louisiana using thematic mapper imagery. Remote Sensing of the Environment **28**:45-59.

Castellanos, D. L. 1997. Nekton use of submerged aquatic vegetation, marsh and shallow unvegetated bottom in a Louisiana tidal freshwater ecosystem. Thesis, University of Southwestern Louisiana, Lafayette, Louisiana, USA.

Childers, D. L., J. W. Day and R. A. Muller. 1990. Relating climatological forcing to coastal water levels in Louisiana estuaries and the potential importance of El Nino-Southern Oscillation events. Climate Research **1**:31-42.

Coen, L. D., K. L. Heck and L. G. Abele. 1981. Experiments on competition and predation among shrimps of seagrass meadows. Ecology **62**:1484-1493.

Condrey, R. E. and D. Fuller. 1992. The U.S. Gulf shrimp fishery. Pages 89-119 *in* M. H. Glantz, editor. Climate variability, climate change and fisheries. Cambridge University Press, England.

Condrey, R. E., J. G. Gosselink and H. J. Bennett. 1972. Comparison of the assimilation of different diets by *Penaeus setiferus* and *Penaeus aztecus*. Fishery Bulletin, U.S. **70**:1281-1292.

Conner, W. H. and J. W. Day. 1987. The ecology of Barataria Basin, Louisiana: an estuarine profile. U.S. Fish and Wildlife Service Biological Report **85(7.13)**.

Dahl, T. and C. Johnson. 1991. Wetlands: status and trends in the conterminous United States, mid-1970s to mid-1980s. Report to Congress, U.S. Department of the Interior, Fish and Wildlife Service, Washington, District of Columbia, USA.

Dahlberg, M. D. and F. G. Smith. 1970. Mortality of estuarine animals due to cold in the Georgian coast. Ecology **51**:931-933.

Deegan, L. A. and B. A. Thompson. 1985. The ecology of fish communities in the Mississippi River deltaic plain. Pages 35-56 *in* A. Yanez-Arancibia, editor. Fish community ecology in estuaries and coastal lagoons: towards an ecosystem integration. UNAM Press, Mexico City, Mexico.

DeLaune, R. D., R. H. Baumann and J. G. Gosselink. 1983. Relationships among vertical accretion, coastal submergence and erosion in a Louisiana Gulf coast marsh. Journal of Sedimentary Petrology **53**:147-157.

DeLaune, R. D., T. C. Feijtel and W. H. Patrick. 1989. Nitrogen flows in Louisiana gulf coast salt marsh: spatial considerations. Biogeochemistry **8**:25-37.

DeLaune, R. D., C. J. Smith, W. H. Patrick and H. H. Roberts. 1987. Rejuvenated marsh and bay-bottom accretion on the rapidly subsiding coastal plain of the U.S. Gulf coast: a second order effect of the emerging Atchafalaya Delta. Estuarine, Coastal and Shelf Science **25**:381-389.

Dittel, A. I., A. H. Hines, G. M. Ruiz and K. K. Ruffin. 1995. Effects of shallow water refuge on behavior and density-dependent mortality of juvenile blue crabs in Chesapeake Bay. Bulletin of Marine Science **57**:902-916.

Eldred, B., R. M. Ingle, K. D. Woodburn, R. F. Hutton and H. Jones. 1961. Biological observations on the commercial shrimp, *Penaeus duorarum* Burkenroad, in Florida waters. Florida State Board of Conservation, Professional Paper Series **3**:1-139.

Eldridge, P. J. 1988. The southeast area monitoring and assessment program (SEAMAP): a state-federal-university program for collection, management and dissemination of fishery-independent data and information in the southeastern United States. Marine Fisheries Review **50**:29-39.

Feijtel, T. C., R. D. DeLaune and W. H. Patrick. 1985. Carbon flow in coastal Louisiana. Marine Ecology Progress Series **24**:255-260.

Fitz, H. C. and R. G. Wiegert. 1991. Utilization of the intertidal zone of a salt marsh by the blue crab *Callinectes sapidus* - density, return frequency and feeding habits. Marine Ecology Progress Series **76**:249-260.

Flint, R. W. and J. A. Younk. 1983. Estuarine benthos: long-term community structure variations, Corpus Christi Bay, Texas. Estuaries **6**:126-141.

Fuss, C. M., Jr. 1964. Observations on burrowing behavior of the pink shrimp, *Penaeus duorarum* Burkenroad. Bulletin of Marine Science, Gulf and Caribbean **14**:62-73.

Fuss, C. M. and L. H. Ogren. 1966. Factors affecting activity and burrowing habits of the pink shrimp, *Penaeus duorarum* Burkenroad. Biological Bulletin **130**:170-191.

Gleason, D. F. 1986. Utilization of salt marsh plants by postlarval brown shrimp: carbon assimilation rates and food preferences. Marine Ecology Progress Series **31**:151-158.

Gleason, D. F. and R. J. Zimmerman. 1984. Herbivory potential of postlarval brown shrimp associated with salt marshes. Journal of Experimental Marine Biology and Ecology **84**:235-246.

Gosselink, J. G., C. L. Cordes and J. W. Parsons. 1979. An ecological characterization study of the Chenier Plain coastal ecosystem of Louisiana and Texas, Volumes. 1-3. U.S. Fish and Wildlife Service Report FWS-OBS-**78/9**, 10 & 11.

Gunter, G. 1941. Death of fishes due to cold on the Texas coast, January, 1940. Ecology **22**:203-208.

——. 1942. Offatts Bayou, a locality with recurrent summer mortality of marine organisms. American Midland Naturalist **28**:631-633.

Gunter, G. and H. H. Hildebrand. 1951. Destruction of fishes and other organisms on the South Texas coast by the cold wave of January 28-February 3, 1951. Ecology **32**:731-36.

Heck, K. L., L. Coen, S. G. Morgan and R. K. Zimmer-Faust. 1994. Recruitment and habitat utilization by the blue crab *Callinectes sapidus*: the importance of juvenile nursery habitats to the fishery. Marine Fisheries Initiative (MARFIN) final report to NOAA National Marine Fisheries Service, St. Petersburg, Florida, USA.

Heck, K. L. and L. D. Coen. 1995. Predation and the abundance of juvenile blue crabs: a comparison of selected East and Gulf Coast (USA) studies. Bulletin of Marine Science **57**:877-883.

Heck, K. L. and T. A. Thoman. 1981. Experiments on predator-prey interactions in vegetated aquatic habitats. Journal of Experimental Marine Biology and Ecology **53**:125-134.

Heck, K. L. and K. A. Wilson. 1987. Predation rates on decapod crustaceans in latitudinally separated seagrass communities: a study of spatial and temporal variation using tethering techniques. Journal of Experimental Marine Biology and Ecology **107**:87-100.

Hettler, W. F. 1989. Nekton use of regularly flooded salt marsh cordgrass habitat in North Carolina, USA. Marine Ecology Progress Series **56**:111-118.

Hines, A. H. and G. M. Ruiz. 1995. Temporal variation in juvenile blue crab mortality: nearshore shallows and cannibalism in Chesapeake Bay. Bulletin of Marine Science **57**:884-901.

Johnson, M. C. and J. R. Fielding. 1956. Propagation of the white shrimp, *Penaeus setiferus* (Linn.) in captivity. Tulane Studies in Zoology **4**:175-190.

Klima, E. F., J. M. Nance, E. X. Martinez and T. Leary. 1990. Workshop on definition of shrimp recruitment overfishing. NOAA Technical Memorandum NMFS-SEFC-264.

Kneib, R. T. 1984. Patterns of invertebrate distribution and abundance in the intertidal salt marsh: causes and questions. Estuaries **7**:392-412.

—— 1991. Flume weir for quantitative collection of nekton from vegetated intertidal habitats. Marine Ecology Progress Series **75**:29-38.

—— 1995. Behaviour separates potential and realized effects of decapod crustaceans in salt marsh communities. Journal of Experimental Marine Biology and Ecology **193**:239-256.

—— 1997. The role of tidal marshes in the ecology of estuarine nekton. Oceanography and Marine Biology: an Annual Review **35**:163-220.

Kneib, R. T. and M. K. Knowlton. 1995. Stage-structured interactions between seasonal and permanent residents of an estuarine nekton community. Oecologia **103**:425-434.

Kneib, R. T. and S. L. Wagner. 1994. Nekton use of vegetated marsh habitats at different stages of tidal inundation. Marine Ecology Progress Series **106**:227-238.

Laughlin, R. A. 1982. Feeding habits of the blue crab, *Callinectes sapidus* Rathbun, in the Apalachicola estuary, Florida. Bulletin of Marine Science **32**:807-822.

Madden, C. J., J. W. Day and J. M. Randall. 1988. Freshwater and marine coupling in estuaries of the Mississippi River deltaic plain. Limnology and Oceanography **33**:982-1004.

Main, K. L. 1987. Predator avoidance in seagrass meadows: Prey behavior, microhabitat selection, cryptic coloration. Ecology **68**:170-180.

May, E. B. 1973. Extensive oxygen depletion in Mobile Bay, Alabama. Limnology and Oceanography **18**:353-366.

McIvor, C. and L. P. Rozas. 1996. Direct nekton use of intertidal saltmarsh habitat and linkage with adjacent habitats: a review from the southeastern United States. Pages 311-334 *in* K. F. Nordstrom and C. T. Roman, editors. Estuarine shores: evolution, environments and human alterations. John Wiley & Sons, New York, New York, USA.

McTigue, T. A. 1993. Trophic roles in juvenile *Penaeus aztecus* Ives and *Penaeus setiferus* (Linnaeus) in a Texas salt marsh. Dissertation. Texas A&M University, College Station, Texas, USA.

McTigue, T. A. and R. J. Zimmerman. 1991. Carnivory versus herbivory in juvenile *Penaeus setiferus* (Linnaeus) and *Penaeus aztecus* (Ives). Journal of Experimental Marine Biology and Ecology **15**:1-16.

—— 1998. The use of infauna by juvenile *Penaeus aztecus* Ives and *Penaeus setiferus* (Linnaeus). Estuaries **21**:160-175.

Mense, D. J. and E. L. Wenner. 1989. Distribution and abundance of early life history stages of the blue crab, *Callinectes sapidus*, in tidal marsh creeks near Charleston, South Carolina. Estuaries **12**:157-168.

Minello, T. J. 1993. Chronographic tethering: a technique for measuring prey survival time and testing predation pressure in aquatic habitats. Marine Ecology Progress Series **101**:99-104.

—— 1999. Nekton densities in shallow estuarine habitats of Texas and Louisiana and the identification of Essential Fish Habitat. Pages 43-75 *in* L. Benaka, editor. Fish habitat: essential fish habitat and habitat rehabilitation. American Fisheries Society Symposium **22**.

Minello, T. J. and J. W. Webb, Jr. 1997. Use of natural and created *Spartina alterniflora* salt marshes by fishery species and other aquatic fauna in Galveston Bay, Texas, USA. Marine Ecology Progress Series **151**:165-179.

Minello, T. J., J. W. Webb, R. J. Zimmerman, R. B. Wooten, J. L. Martinez, T. J. Baumer and M. C. Pattillo. 1991. Habitat availability and utilization by benthos and nekton in Hall's Lake and West Galveston Bay, NOAA Technical Memorandum, NMFS-SEFC-**275**.

Minello, T. J. and R. B. Wooten, Jr. 1993. Effects of caging juvenile predators on benthic infaunal populations at experimental open bay disposal areas in Galveston Bay, Texas. Report to Galveston District of the U.S. Army Corps of Engineers. National Marine Fisheries Service, Galveston Laboratory, Galveston, Texas, USA.

Minello, T. J. and R. J. Zimmerman. 1983. Fish predation on juvenile brown shrimp, *Penaeus aztecus* Ives: the effect of simulated *Spartina* structure on predation rates. Journal of Experimental Marine Biology and Ecology **72**:211-231.

—— 1991. The role of estuarine habitats in regulating growth and survival of juvenile penaeid shrimp. Pages 1-16 *in* P. DeLoach, W. J. Dougherty and M. A. Davidson, editors. Frontiers in shrimp research. Elsevier Science Publishers, Amsterdam, The Netherlands.

Minello, T. J., R. J. Zimmerman and E. X. Martinez. 1987. Fish predation on juvenile brown shrimp, *Penaeus aztecus*: effects of turbidity and substratum on predation rates. Fishery Bulletin, U.S. **85**:59-70.

—— 1989. Mortality of young brown shrimp *Penaeus aztecus* in estuarine nurseries. Transactions of the American Fisheries Society **118**:693-708.

Minello, T. J., R. J. Zimmerman and R. Medina. 1994. The importance of edge for natant macrofauna in a created salt marsh. Wetlands **14**:184-198.

Morris, J. T., B. Kjerfve and J. M. Dean. 1990. Dependence of estuarine productivity on anomalies in mean sea level. Limnology and Oceanography **35**:926-930.

Nance, J. M. 1993. Gulf of Mexico shrimp fishery recruitment overfishing definition: workshop 2. NOAA Technical Memorandum, NMFS-SEFSC-**323**.

Nance, J. M., E. F. Klima and T. E. Czapla. 1989. Gulf of Mexico shrimp stock assessment workshop. NOAA Technical Memorandum, NMFS-SEFC-**239**.

Neill, C. and L. A. Deegan. 1986. The effect of the Mississippi River delta lobe development on the habitat composition and diversity of Louisiana coastal wetlands. American Midland Naturalist **116**:296-303.

Nixon, S. W. 1980. Between coastal marshes and coastal waters-a review of twenty years of speculation and research on the role of salt marshes in estuarine productivity and water chemistry. Pages 437-524 *in* P. Hamilton and K. B. Macdonald, editors. Estuarine and wetland processes with emphasis on modeling. Plenum Press, New York, New York, USA.

Orth, R. J., K. L. Heck and J. van Montfrans. 1984. Faunal communities in seagrass beds: a review of the influence of plant structure and prey characteristics on predator-prey relationships. Estuaries **7**:339-350.

Orth, R. J. and J. van Montfrans. 1982. Predator-prey interactions in a *Zostera marina* ecosystem in the lower Chesapeake Bay, Virginia. Pages 81-94 *in* R. J. Orth and J. van Montfrans, editors. Interactions of resident consumers in a temperate estuarine seagrass community: Vaucluse Shores, Virginia, USA. VIMS-SRAMSOE **267**, Gloucester Point, Virginia, USA.

Peters, D. S. and W. E. Schaaf. 1991. Empirical model of the trophic basis for fishery yield in coastal waters of the eastern USA. Transactions of the American Fisheries Society **120**:459-473.

Peterson, G. W. and R. E. Turner. 1994. The value of salt marsh edge vs interior as a habitat for fish and decapod crustaceans in a Louisiana tidal marsh. Estuaries **17**:235-262.

Rabalais, N. N., F. R. Burditt, Jr., L. D. Coen, B. E. Cole, C. Eleuterius, K. L. Heck, Jr., T. A. McTigue, S. G. Morgan, H. M. Perry, F. M. Truesdale, R. K. Zimmer-Faust and R. J. Zimmerman. 1995. Settlement of *Callinectes sapidus* megalopae on artificial collectors in four Gulf of Mexico Estuaries. Bulletin of Marine Science **57**:855-876.

Reed, D. J. and D. R. Cahoon. 1992. The relationship between marsh surface topography, hydroperiod and growth of *Spartina alterniflora* in a deteriorating Louisiana salt marsh. Journal of Coastal Research **8**:77-87.

Rejmanek, M., C. E. Sasser and J. G. Gosselink. 1987. Modeling of vegetation dynamics in the Mississippi River deltaic plain. Vegetatio **69**:133-140.

Rosas, C., E. Lazarochavez and F. Buckleramirez. 1994. Feeding habits and food niche segregation of *Callinectes sapidus*, *C. rathbunae* and *C. similis* in a subtropical coastal lagoon of the Gulf of Mexico. Journal of Crustacean Biology **14**:371-382.

Rozas, L. P. 1993. Nekton use of salt marshes of the Southeast region of the United States. Pages 528-537 *in* O. Magoon, W. S. Wilson, H. Converse and L. T. Tobin, editors. Coastal Zone '93, Volume 2. Proceedings of the 8th symposium on coastal and ocean management, American Society Of Civil Engineers, New York, New York, USA.

—— 1995. Hydroperiod and its influence on nekton use of the salt marsh: a pulsing ecosystem. Estuaries **18**:579-590.

Rozas, L. P. and D. J. Reed. 1993. Nekton use of marsh-surface habitats in Louisiana (USA) deltaic salt marshes undergoing submergence. Marine Ecology Progress Series **96**:147-157.

Rozas, L. P. and T. J. Minello. 1997. Estimating densities of small fishes and decapod crustaceans in shallow estuarine habitats: a review of sampling design with focus on gear selection. Estuaries **20**:199-213.

—— 1998. Nekton use of salt marsh, seagrass and nonvegetated habitats in a South Texas (USA) estuary. Bulletin of Marine Science **63**:481-501.

Ruiz, G. M., A. H. Hines and M. H. Posey. 1993. Shallow water as a refuge habitat for fish and crustaceans in nonvegetated estuaries - an example from Chesapeake Bay. Marine Ecology Progress Series **99**:1-16.

Ryer, C. H. 1987. Temporal patterns of feeding by blue crabs (*Callinectes sapidus*) in a tidal-marsh creek and adjacent seagrass meadow in the lower Chesapeake Bay. Estuaries **10**:136-140.

Sasser, C. E., M. D. Dozier, J. G. Gosselink and J. M. Hill. 1986. Spatial and temporal changes in Louisiana's Barataria Basin marshes, 1945-80. Environmental Management **10**:671-680.

Schindler, D. E., B. M. Johnson, N. A. Mackay, N. Bouwes and J. F. Kitchell. 1994. Crab:snail size-structured interactions and salt marsh predation gradients. Oecologia **97**:49-61.

Sullivan, M. J. and C. A. Moncreiff. 1990. Edaphic algae are an important component of salt marsh food-webs: evidence from multiple stable isotope analyses. Marine Ecology Progress Series **62**:149-159.

Thomas, J. L. 1989. A comparative evaluation of *Halodule wrightii* Aschers, *Spartina alterniflora* Loisel and bare sand as nursery habitats for juvenile *Callinectes sapidus* (Rathbun). Dissertation. Texas A&M University, College Station, Texas, USA.

Thomas, J. L., R. J. Zimmerman and T. J. Minello. 1990. Abundance patterns of juvenile blue crabs (*Callinectes sapidus*) in nursery habitats of two Texas bays. Bulletin of Marine Science **46**:115-125.

Turner, R. E. 1977. Intertidal vegetation and commercial yields of penaeid shrimp. Transactions of the American Fisheries Society **106**:411-416.

—— 1997. Wetland loss in the northern Gulf of Mexico: multiple working hypotheses. Estuaries **20**:1-13.

Turner, R. E. and R. L. Allen. 1982. Bottom water oxygen concentration in the Mississippi River Delta Bight. Contributions in Marine Science **25**:161-172.

Turner, R. E. and Y. S. Rao. 1990. Relationships between wetland fragmentation and recent hydrologic changes in a deltaic coast. Estuaries **13**:272-281.

Turner, R. E., W. W. Schroeder and W. J. Wiseman. 1987. The role of stratification in the deoxygenation of Mobile Bay and adjacent shelf waters. Estuaries **10**:13-19.

Weinstein, M. P. 1979. Shallow marsh habitats as primary nurseries for fishes and shellfish, Cape Fear River, North Carolina. Fishery Bulletin, U.S. **77**:339-358.

Wells, J. T. 1987. Effects of sea-level rise on deltaic sedimentation in South-central Louisiana. Pages 157-166 *in* D. Nummendel, O. H. Pilkey and J. D. Howard, editors. Sea level change and coastal evolution. The Society of Economic Paleontologists and Minerologists, Tulsa, Oklahoma, USA.

Wenner, E. L. and H. R. Beatty. 1993. Utilization of shallow estuarine habitats in South Carolina, USA, by postlarval and juvenile stages of *Penaeus* spp. (Decapoda, Penaeidae). Journal of Crustacean Biology **13**:280-295.

West, D. L. and H. Williams. 1986. Predation by *Callinectes sapidus* (Rathbun) within *Spartina alterniflora* (Loisel) marshes. Journal of Experimental Marine Biology and Ecology **100**:75-95.

Whaley, S. D. 1997. The effects of marsh edge and surface elevation on the distribution of salt marsh infauna and prey availability for nekton predators. Thesis, Texas A&M University, College Station, Texas, USA.

Wheeler, R. S. 1969. Culture of penaeid shrimp in brackish-water ponds, 1966-67. Proceedings of the Southeast Association of Game and Fish Commissioners **22**:387-391.

Wilson, K. A., K. W. Able and K. L. Heck. 1990. Predation rates on juvenile blue crabs in estuarine nursery habitats: evidence for the importance of macroalgae (*Ulva lactuca*). Marine Ecology Progress Series **58**:243-251.

Wilson, K. A., K. L. Heck and K. W. Able. 1987. Juvenile blue crab, *Callinectes sapidus*, survival: an evaluation of eelgrass, *Zostera marina*, as refuge. Fishery Bulletin, U.S. **85**:53-58.

Zimmerman, R. J. and T. J. Minello 1984a. Densities of *Penaeus aztecus*, *P. setiferus* and other natant macrofauna in a Texas salt marsh. Estuaries 7:421-433.

—— 1984b. Fishery habitat requirements: utilization of nursery habitats by juvenile penaeid shrimp in a Gulf of Mexico salt marsh. Pages 371-383 *in* B. J. Copeland, K. Hart, N. Davis and S. Friday, editors. Research for managing the nation's estuaries. UNC-SG-**84-08**, UNC SeaGrant Publications, Raleigh, North Carolina, USA.

Zimmerman, R. J., T. J. Minello, T. J. Baumer and M. C. Castiglione. 1989. Oyster reef as habitat for estuarine macrofauna. NOAA Technical Memorandum, NMFS-SEFC-**249**.

Zimmerman, R. J., T. J. Minello, M. C. Castiglione and D. L. Smith 1990a. The use of *Juncus* and *Spartina* marshes by fisheries species in Lavaca Bay, Texas, with reference to effects of floods. NOAA Technical Memorandum, NMFS-SEFC-**251**.

—— 1990b. Utilization of marsh and associated habitats along a salinity gradient in Galveston Bay. NOAA Technical Memorandum, NMFS-SEFC-**250**.

Zimmerman, R. J., T. J. Minello, E. F. Klima and J. M. Nance. 1991. Effects of accelerated sea-level rise on coastal secondary production. Pages 110-124 *in* H. S. Bolton, editor. Coastal wetlands. American Society of Civil Engineers, New York, New York, USA.

ECOPHYSIOLOGICAL DETERMINANTS OF SECONDARY PRODUCTION IN SALT MARSHES: A SIMULATION STUDY

J. M. MILLER
Zoology Department
North Carolina State University
Raleigh, NC 27695 USA

W. H. NEILL
Department of Wildlife and Fisheries Sciences
Texas A&M University
College Station, TX 77843 USA

K. A. DUCHON
Zoology Department
North Carolina State University
Raleigh, NC 27695 USA

S. W. ROSS
NC NERR
7205 Wrightsville Ave.
Wilmington, NC 28403 USA

Abstract

Variation in the abiotic environment is generally presumed to stress fish in estuarine marshes despite abundant food resources and refuge from predation. Chief among the important variables are dissolved oxygen, temperature, pH and salinity. With new technology for collecting high-resolution abiotic data and with mechanistic models for interpreting these data, it is possible to revisit and refine the conventional paradigm(s) of abiotic stresses and secondary production in salt marshes. With data from National Estuarine Research Reserves (NERR) and our ecophysiological model for juvenile fish, we ask ''what is the relative impact of abiotic factors on growth in different marsh types?'' Based on data from four NERR sites representing a spectrum of marsh hydrotypes and latitudes, we conclude that abiotically-forced variation in growth could explain much of the variation in secondary production in marshes.

1. Introduction

Abiotic variability in both time and space is a prominent feature of estuaries, and is widely recognized as influencing the distribution and abundance of biota. The most obvious factor is salinity, and, owing particularly to ease of measurement, salinity was the focus of many early attempts to explain the distribution of estuarine biota. Indeed, the distributions of many species (or stages of species) are at least loosely correlated with salinity – a few even seem to be restricted to a certain salinity range. Attempts to explain such patterns with laboratory investigations of performance in relation to salinity manipulation have been somewhat less gratifying. The general picture that has emerged is that so-called estuarine species (or stages) are highly tolerant of, and less restricted by, salinity variation. This has led to the general paradigm that estuarine biota are a "tolerant" subset of marine species that thrive in estuaries mainly by escaping competition and predation from their more stenohaline counterparts (e.g., Boesch and Turner 1984, and many others).

Recently, studies of the role of dissolved oxygen on the distribution of estuarine biota (Pihl et al. 1992, Diaz and Rosenberg 1995, Breitburg et al. 1997) have become more prominent with the advent of new technology for recording dissolved oxygen. Still, there seems little appreciation of the fact that many abiotic and biotic factors interact to determine the distribution (and performance) of biota in estuaries. In general, scientists continue to follow reductionist or simplistic pathways to attempt to understand the function of estuaries. Estuaries and their biota are more than the sum of their parts (see Golley 1993 for a reminder). In contrast, some researchers have attempted to construct indices of estuarine "health" (Karr 1981, Deegan et al. 1997, and others). Most often, these indices are without any clear mechanistic basis, and cannot be extended outside the system(s) from which they were derived. While it is widely accepted that biota are 'integrators of their environment', in order to progress in our understanding of estuarine systems, we need to understand the mechanisms of integration.

Not only are the properties of seawater and freshwater mixed at their estuarine interface, but these mixing zones are highly variable, owing to rainfall-, wind-, and tidally-driven flows. Small vertical oscillations in shallow estuaries translate into large horizontal flows. Typically, estuaries represent a dome-shaped gradient of variability, with relatively stable upstream and downstream zones, and an intermediate zone of relatively high variability. And, because most estuaries narrow upstream, horizontal flows are also accompanied by changes in the areal extent of the mixed zone. The distribution of vegetation, and other sessile biota, can often be better explained by the variability rather than the level of abiotic factors (E. Estevez, pers. commun.). Vagile biota often make large excursions up and down the estuary, apparently following preferred isopleths of abiotic factors.

Variability in environment can be expressed in many ways; variance, amplitude, and frequency are but a few. It is not known at present which, if any, are biologically meaningful. In fact, given the usual point-in-time or -space measurements, often *no* measurement of variability is possible; in many other cases, only means are reported. Most researchers probably believe variability is important, but few seem inclined to try to understand (or even express) it. Some of this is to be expected, given the logistical difficulties of adequate spatial or temporal coverage with conventional point-measuring equipment. But the situation is rapidly changing with development of instruments capable of making many high-frequency synoptic measurements and transmitting these to remote locations. Paradoxically, we now seem to be entering a period when we can be overwhelmed with data. But at least now it is possible to ask how many data are enough, and proceed to some practical ''middle ground'' between anecdote and overkill. Hopefully, the biotic responses will be used to determine a useful level of data collection. The same questions face us when we make choices of what type and precision of data to collect, as well as where and when to collect it.

Johnson and Brinton (1963) suggested that oceanic biota are subjected to different kinds of controls at the edges and middle of their ranges. Abiotic factors were suggested to dominate at the edges, whereas biotic factors were more important near the center. MacCall (1990) summarized such ideas into a ''basin theory'', where habitat value decreased toward the margins of distribution. Thus, a habitat was envisioned as a basin with depth proportional to value. Biota were expected to be distributed accordingly, with greater abundances in the deeper (more valuable) regions of the basin. Biota forced to occupy marginal habitat presumably performed worse.

The general idea of greater abiotic constraints at the margins of a species' distribution has been restated many ways. Briggs (1974), for example, pointed out the coincidence of many tropical species' poleward range limits and the 20°C winter isotherm.

If we assume that species' distributions are limited at the margins by inability to tolerate abiotic conditions, most estuarine biota don't fit. There are few data that suggest estuarine species cannot tolerate both marine and oligohaline salinities, for example. We suggest that the fundamental relationship of increasing importance of abiotic factors and decreasing importance of biotic factors towards the edge suggested by Johnson and Brinton (1963), Briggs (1974), MacCall (1990), and many others, including ourselves (Neill et al. 1994) may be, if anything, reversed for estuarine biota. Biotic constraints are probably what define the distribution limits of most ''estuarine'' species, not intolerable abiotic conditions (Miller et al. 1991). Which are more important in the center of their ranges is not clear. But because many estuarine species inhabit the zone of maximum abiotic variability, abiotic factors may both constrain performance and exclude stenotopes (reduce competition and predation). The question then becomes: how relatively important are biotic and abiotic constraints on estuarine biota? This important question cannot be answered until we have a common currency with which to evaluate importance; in this paper we propose both a methodology (our model) and a currency (scope for growth), which can readily be translated into production.

2. Purpose

The purpose of this paper is to explore the use of ecophysiological simulation modeling to properly assess the role of abiotic variability, so that at some future time, the relative, and absolute, importance of biotic and abiotic factors in estuaries can be determined. To illustrate the application of the model, we consider the summer and winter growth performance of juvenile red drum (*Sciaenops ocellata*) in four selected estuaries representing a spectrum of marsh ecotypes.

3. The Model

3.1 MODEL BACKGROUND

In this paper, we emphasize dissolved oxygen (DO), temperature, salinity and pH as abiotic factors, not that these are all that matter, but because: 1) they have been shown to be important for fishes; and 2) data were available. Certainly turbidity (or light) and other abiotic factors, some yet to be discovered, are important to much of the marsh biota. We have chosen juvenile red drum as our model estuarine species because we have access to extensive data from laboratory experiments on its growth relative to abiotic factors.

There is another class of variables, often simply classified as ''structural''. These include depth, bottom topography and substrate, as well as oyster bars, grass beds, and other prominent features of the estuarine landscape. At present, these are not included in our model. In fact, it is not yet clear how to adequately quantify these features in functional terms, much less translate them into a generic biological response. For example, grass blades impede some kinds of predators, but they facilitate others. It is clear that both structural and dynamic factors interact to determine the value of habitat to fishes (Browder and Moore 1981, Edwards 1991). Still, much remains to be done to be able to translate specific cases into more generic hypotheses and mechanistic models.

Several levels of biological organization can be recognized: cellular (metabolic), individual, population, community and ecosystem. Most field studies are at the subpopulation (limited stage or location) level; most laboratory studies are at the individual level. In our model we explicitly consider the metabolic and individual levels, but we suggest our basic approach can be applied at higher levels (Neill et al. 1994).

The generalized production equation, $P = B^{(G-Z)}$ is affected in two distinctly different ways by abiotic factors. Biomass (B) is a function of immigration and settling, which are keyed to structural components of marshes. Growth (G) is sensitive to food limitation and to effects of abiotic controlling, limiting and loading factors; loss of individuals (Z = mortality and emigration) is a function of predators and the action of abiotic lethal and directive factors. This paper focuses on dynamic abiotic factors and their effects on secondary production through the growth term.

Our model output is both at the metabolic and individual levels, with model output of metabolic energy and growth, respectively. Rapid growth is usually interpreted as evidence that fish are healthy and the environment is benign; therefore, we use growth as our measure of performance. The abiotic environment affects many kinds of performance, including growth, at all levels of biological organization; for example, performance analogs at the subpopulation and population levels are production and recruitment, respectively. We speculate that the same model framework can be applied to each. We also consider how environmental effects at one level are transmitted to another.

3.2 MODEL DESCRIPTION

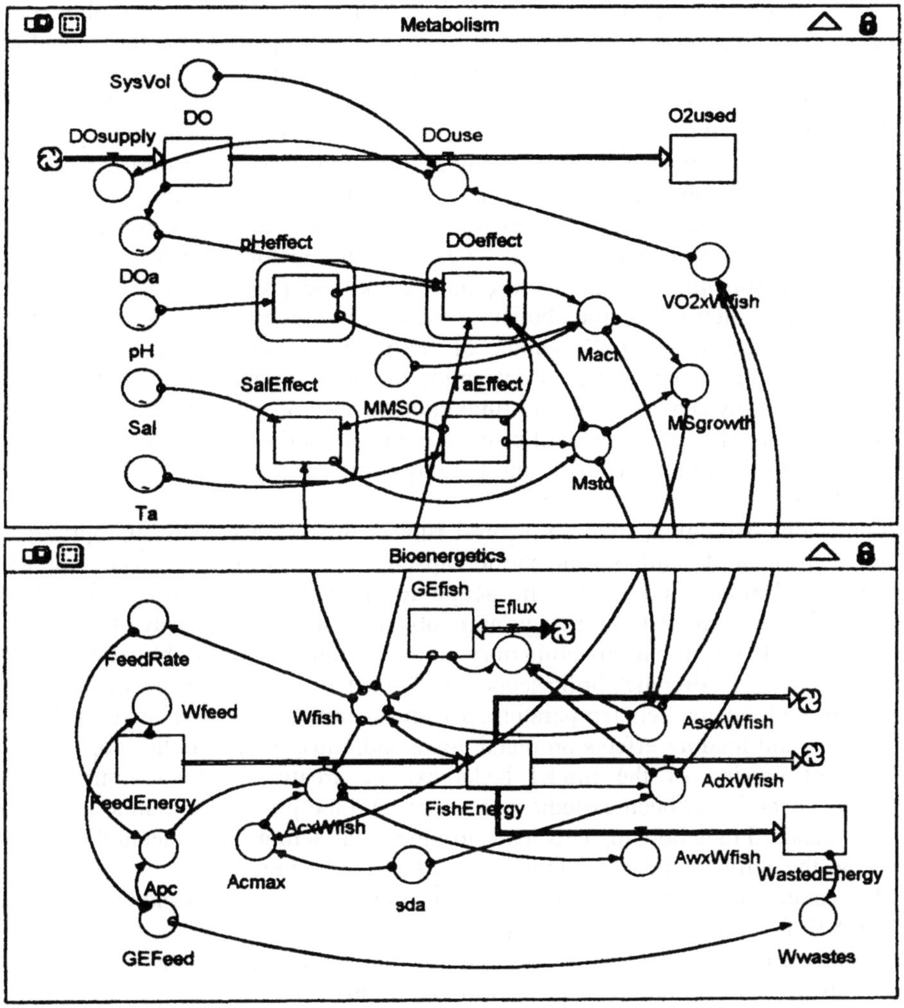

Figure 1. Ecophysiological model for juvenile red drum growth simulation in STELLA II language.

The core model of production-system performance has its mechanistic basis in the ecophysiology of red drum metabolism and growth. The central idea is that the temporal and spatial distribution of abiotic environment within the habitat might best be evaluated as an integral of the physiological and behavioral processes performed by a representative fish living in that habitat. Our ecophysiological model of fish growth incorporates a quantitatively explicit statement of concepts originally formalized by F.E.J. Fry (Fry 1947, 1971) and recently elaborated by Neill and Bryan (1991) and by

Neill et al. (1994). As presently structured, the model simulates fish growth in habitats with varying food, oxygen, temperature, pH, and salinity.

Fry's "physiological classification of environment" and "metabolic scope" concepts were coupled with conventional bioenergetics to provide the theoretical basis for this ecophysiological (oxy-bioenergetic) model. Fry supposed that all environment acts on animal activity through metabolism, and that these metabolic effects can be resolved into those due to five classes of factors: 1) controlling factors like temperature and pH set the inherent pace of metabolism; 2) limiting factors are resources like oxygen and food-energy substrates that, when deficient, restrict maximum, or active, metabolism; 3) masking factors like extreme salinity and parasites load metabolism by increasing obligatory metabolic work; i.e., masking factors increase minimum, or standard, metabolism; 4) lethal factors like toxins and predators kill the animal by completely interdicting its metabolism – and, 5) directive factors like light and photoperiod guide the animal's choice of environment and acclimatory physiology. Jointly, the factors of environment determine the animal's metabolic scope, which is the difference between its active (= maximum aerobic) and standard (= obligatory minimum) metabolic rates; metabolic scope is the animal's capacity to perform useful activities like locomotion, feeding, and the physiological processing of food that leads to growth.

Our ecophysiological model's central rule is this: Fish eat all appropriate food encountered or until available metabolic capacity becomes insufficient to support the processing of more food (metabolic scope for growth is exceeded); the fish then partitions the consumed food energy and substrates in the usual ways (conventional bioenergetics) between various obligatory activities and growth; if obligatory activities cost more than available metabolic scope, the fish enters a state of suspended animation. Time-varying food, oxygen, temperature, pH, and salinity are accommodated as limiting, controlling, and loading effects on metabolism and, thus, on metabolic scope. Because the present version of the model lacks explicit treatment of swimming and its metabolic costs, it has been expedient to adopt "Winberg's rule": routine metabolism is twice standard metabolism. This leads to functional definition of metabolic scope for growth as the active metabolic rate less twice the standard rate.

The conceptual model (Fig. 1) has been customized for red drum and implemented in STELLA© for simulation. Lab experiments with juvenile red drum, mostly at Texas A&M, have enabled functional definition and parameterization of the working model, which has accurately simulated growth of red drum in various aquaculture situations. However, neither the model nor the majority of experiments from which its algorithms and parameters were inferred, has been subjected to rigorous peer evaluation. Moreover, detailed description of the model's structure, and of the various experiments and field trials, obviously are beyond the scope of this paper. Therefore, what follows is offered as a summary sketch of the model and its empirical basis. Predictions of growth are necessarily to be construed as preliminary and tentative, although in recent field tests of the model in TX and FL estuaries, the model predicted growth of juvenile red drum within 8% (our unpublished data).

Physiological responses of red drum to environmental factors and their interactions are consistent with the patterns typical of fishes (Fry 1947, Fry 1971, Brett and Groves 1979, Neill and Bryan 1991). Under the model, optimum temperature is 29°C and lethal (ultimate) temperatures of red drum are about 5 and 35°C—all other factors being optimal. The limiting level of dissolved oxygen (DO) for growth increases with increasing temperature. Again all else being optimum, limiting DO is about 5 mgl^{-1} at temperatures near and above 29°C; limiting DO declines to about 2 mgl^{-1} at 18°C. Thus, the optimum temperature for growth declines when DO or any other limiting factor is at work. Food also limits, at least in nature; whereas, in aquaculture, it is DO that normally limits. Even a maximum daily ration—for juvenile red drum, a mass equivalent to about $0.18*W_{fish}^{-0.27}$ —consisting of naturally energy-dilute food (i.e., gross energy 1,200 $calg^{-1}$) may be limiting at high temperatures and high DOs. pH < ~6.5 induces a pronounced Bohr shift in red-drum hemoglobin's oxygen affinity, thus exacerbating negative metabolic impacts of low DO and high temperature. For the euryhaline red drum, tolerable salinities range from < 1‰ to > 60‰; optimum salinity seems to be near 10‰, which is near the blood iso-osmotic point. We have been unable to establish an expected increase in the optimum salinity with size/age. Red drum, being very adept osmoregulators, exhibit only a 20 to 30% decline in growth performance as salinity diverges from the optimum to extremes of 1 and 45‰. However, there is a dramatic interaction between low salinity and low temperature: at salinities below about 3‰, ultimate lower lethal temperatures of juvenile red drum markedly increase, to values perhaps as high as 15°C. This implies that metabolic costs of osmoregulatory work (the metabolic load imposed by salinity as a masking factor) are greatly exacerbated by low temperature.

Superimposed on these complex steady-state relations are even more complex transient-state dynamics. The present version of the model explicitly accommodates temporal change in environmental temperature and DO, by invoking physiological acclimation in the form of modified (variable rate coefficient) exponential lags in metabolic response. The rate coefficient for thermal acclimation varies from about 0.5 day^{-1} at 18°C, to about 2.0 day^{-1} at 34°C. The rate coefficient for DO acclimation is keyed to standard metabolic rate and ranges typically from $0.1/day^{-1}$ to $0.5/day^{-1}$.

The model has two functional modules, metabolism and bioenergetics (Fig. 1). The metabolism module has a subroutine for each variable of physical-chemical environment. The salinity subroutine computes and returns a standard-metabolism intercept, which is multiplied by a function returned by the temperature subroutine and by $W_{fish}^{-0.2}$, to give standard metabolic rate. The temperature effect on standard metabolism is modeled as the product of steady-state and transient-state components, reflecting both the Ahrennius effect and thermal acclimation. The temperature and pH subroutines both produce outputs that control active metabolism. These controlling effects are modeled as interactions with the limiting effect of DO, as described above. Active metabolic rate also is modeled as weight-dependent, being proportional to $W_{fish}^{-0.22}$. The residual intercept of active metabolism is MMS_o, where MMS is marginal metabolic scope (Neill and Bryan 1991). We interpret MMS_o to represent inherent metabolic efficiency of the fish-environment system, after the effects of temperature, pH, DO, salinity, and fish sizes have been taken into account. MMS offers a practical measure of "environmental quality" from the fish's perspective and is relatively easy to determine, via routine respirometry (Neill and Bryan 1991).

The bioenergetics module reflects conventional "rules of thumb" (Warren and Davis 1967, Brett and Groves 1979), with the addition of a component for metabolic limitation of food intake as suggested by the work of Jobling (1985) and recently embraced by van Dam and Pauly (1995). As mentioned above, our model accommodates metabolic limitation of food intake by setting feeding rate to MIN (rate of food encounter, metabolic scope for growth/sda). Consumed energy is converted to new fish biomass as a residual, after expenditures for standard and routine-activity components of metabolism, specific dynamic action (SDA =0.15*feed energy), and wastes (also 0.15*feed energy) for natural foods. Our modeled fish conserves body form during times of food limitation by reducing caloric density of its tissues.

4. Estuarine Sites

Four National Estuarine Research Reserve (NERR) sites were chosen, mainly on the basis of their geographic locations and availability of summer and winter environmental data. These were Great Bay (GB) in NH, Zeke's Island (ZI) in NC, Weeks Bay (WB) in AL, and Elkhorn Slough (ES) in CA. GB represents an Atlantic cold-temperate estuary, ZI an Atlantic warm-temperate estuary, WB a Gulf warm-temperate estuary, and ES a Pacific warm temperate estuary. We obtained abiotic data from these NERRs to represent a spectrum of actual summer and winter conditions. The 5-day periods used for growth simulations are not considered to be representative; indeed, there is no 5-day period that can represent either summer or winter. They are "typical" in that they were visually judged to be not unusual for that time of year. Likewise, no single site can adequately represent the spatial variability found within any estuary. Rather than trying to represent abiotic conditions, we chose these data to represent an array of real data variability for purposes of demonstrating the model's capability to resolve environmental variability into a biological response.

5. Environmental Variability

To characterize the summer and winter conditions in the different sites, monthly means and ranges for temperature (°C), salinity (‰), DO (mgl^{-1}) and pH were calculated (Table 1). Winter data are not collected at GB, owing to ice cover; so, we used data from April 1996; the winter data from the other three sites are January 1996. Summer data are all August 1996. Plots of the data collected at 30-min intervals are shown in Figures 2-5, along with the 5-day period used in the simulations of growth.

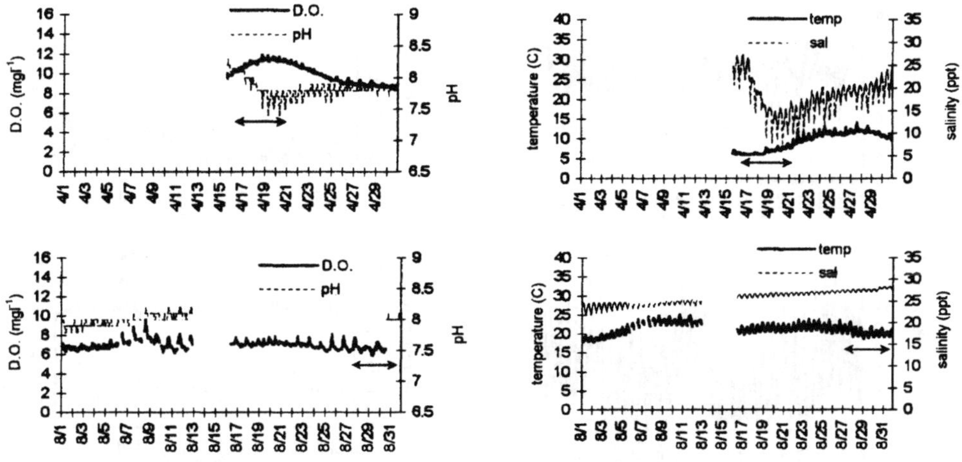

Figure 2 (a + b). Great Bay, NH, NERR site environmental data, August and April 1996. Arrows show periods of simulated growth.

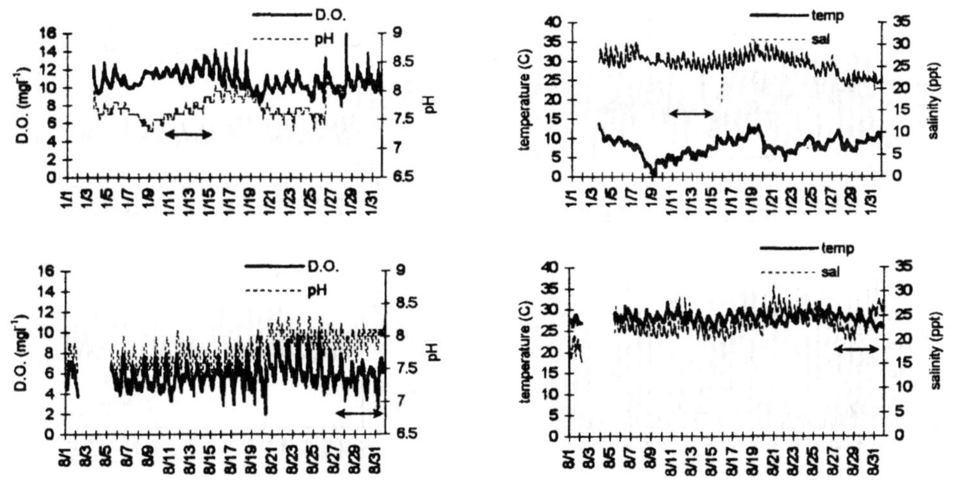

Figure 3 (a + b). Zeke's Island, NC, NERR site environmental data, August and January 1996. Arrows show periods of simulated growth.

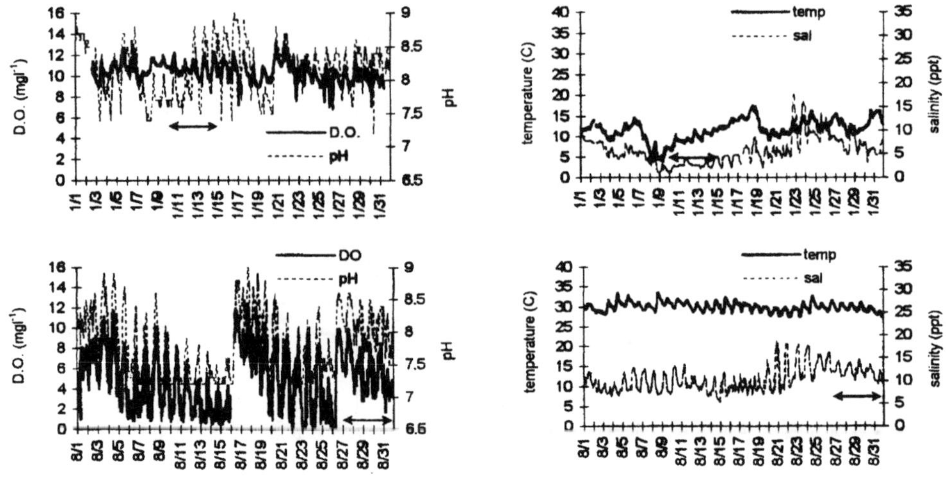

Figure 4 (a + b). Weeks Bay, AL, NERR site environmental data, August and January 1996. Arrows show periods of simulated growth.

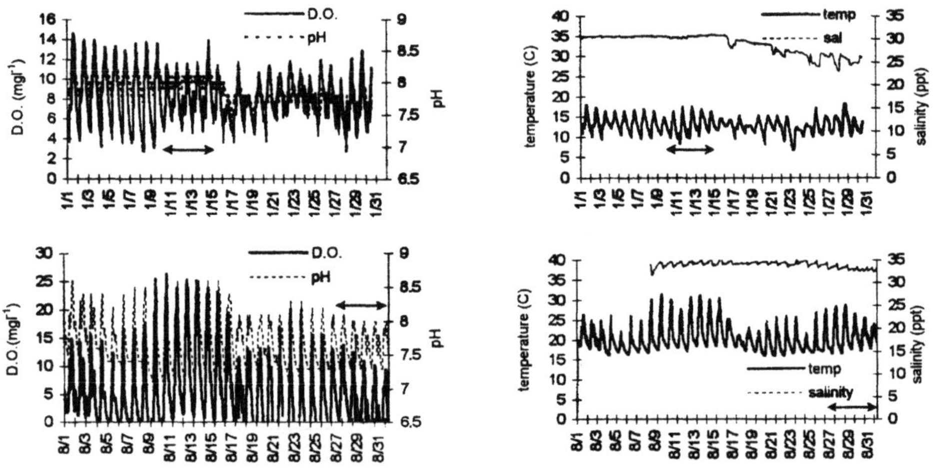

Figure 5 (a + b). Elkhorn Slough, CA, NERR site environmental data, August and January 1996. Arrows show periods of simulated growth.

TABLE 1. Monthly mean and (range) of temperature (°C), salinity (‰), dissolved oxygen (mgl^{-1}) and pH at NERR sites in New Hampshire (GB), North Carolina (ZI), Alabama (WB) and California (ES), 1996.

	TEMP	SAL	DO	pH
SUMMER (Aug 96)				
GB (NH)	21.2(17.8-25.1)	25.9(21.7-28.4)	7.0(5.9-9.6)	8.0(7.8-8.2)
ZI (NC)	28.1(24.6-31.9)	23.4(15.4-31.2)	5.6(1.9-10.0)	7.8(7.2-8.3)
WB(AL)	29.2(26.0-31.4)	4.8(0.5-12.2)	4.7(0.4-10.0)	7.0(5.9-8.8)
ES (CA)	20.5(16.0-31.3)	34.5(31.6-34.5)	6.2(0.0-26.3)	7.7(7.1-8.6)
WINTER (Jan 96)				
GB (NH)*	9.5(6.0-14.2)	18.0(7.7-27.5)	10.2(8.5-11.9)	7.8(7.4-8.3)
ZI (NC)	7.9(-0.3-13.8)	25.7(17.0-30.7)	10.8(7.8-17.2)	7.8(7.3-8.5)
WB(AL)	13.0(7.9-18.6)	2.7(0.2-9.6)	9.8(4.9-13.9)	7.0(6.1-8.6)
ES (CA)	13.0(6.8-18.3)	29.0(23.0-31.1)	8.1(2.7-14.6)	7.9(7.3-8.3)

* April 96 data

Monthly mean summer temperatures (Table 1) ranged from 20.5°C (ES) to 29.2°C (WB) and salinity from 4.8‰ (WB) to 34.5‰ (ES). Mean DO ranged from 4.7 mgl^{-1} (WB) to 10.8 mgl^{-1} (ZI) and pH from 7.0 (WB) to 8.0 (GB). Great Bay (GB) in summer was characterized by relatively low mean temperature (21.2°C), intermediate salinity (25.9‰), intermediate DO (7.0 mgl^{-1}) and relatively high pH (8.0). Zeke's Basin (ZI) had relatively high summer temperature (28.1°C), intermediate salinity (23.4‰), high DO (10.8 mgl^{-1}), and high pH (7.8). Weeks Bay (WB) had the highest mean summer temperature (29.2°C), the lowest salinity (4.8‰), the lowest DO (4.7 mgl^{-1}) and the lowest pH (7.0). Elkhorn Slough (ES) had the lowest mean summer temperature (20.5°C), the highest salinity (34.5‰), an intermediate, but highly variable DO (6.2 mgl^{-1}) and an intermediate pH (7.7).

In winter, the mean temperature ranged from 7.9°C (ZI) to 13.0°C (WB and ES). Winter salinity ranged from 2.7‰ (WB) to 29.0‰ (ES), winter DO from 8.1 mgl^{-1} (ES) to 10.8 mgl^{-1} (ZI), and pH from 7.0 (WB) to 7.9 (ES). GB had intermediate "winter" (April) temperature (9.5°C), salinity was intermediate (18.0‰), DO was high (10.2 mgl^{-1}), and pH was high (7.8). ZI had the lowest mean winter temperature (7.9°C), relatively high salinity (25.7‰), the highest DO (10.8 mgl^{-1}) and a high pH (7.8). WB had (one of the) highest mean winter temperatures (10.8°C), the lowest salinity (2.7‰), intermediate DO (9.8 mgl^{-1}) and the lowest pH (7.0). ES had the (other) highest winter temperatures (10.8°C), the highest salinity (29.0), the lowest DO (8.1 mgl^{-1}) and the highest pH (7.9). Maximum tide range was about 3 m in GB, 1 m in ZI and WB and less than 0.5 m in ES.

6. Model Runs

To investigate the effects of temperature, salinity, DO and pH in the different estuaries in summer and winter, a 5-day period of synoptic data, recorded at 30-min intervals (240 data points for each variable), was selected at each site for each season (Figs. 2-5).

These periods (27-31 August 1996 and 10-14 January 1996, except 16-20 April for GB) were visually selected from graphs of each factor for the entire month to be "typical" in being of intermediate mean and variability. For each of these 5-day periods, the growth (in g wet weight) was simulated for a juvenile red drum with a starting weight of 1 g. These growth increments were expressed as instantaneous daily growth, and used as measures of performance in the different systems. The growth performances of juvenile red drum were simulated in all four systems in both seasons to illustrate: 1) the approach; and 2) the potential importance of different factors in different systems on a single fish type. Indeed, juvenile red drum occur naturally only in ZI and WB.

TABLE 2. Simulated instantaneous daily growth ($\times 100$) of 1g juvenile red drum (*Sciaenops ocellata*) in four estuaries. Dates and sites are same as Table 1.

Site	GB(NH)	ZI(NC)	WB(AL)	ES(CA)
Summer				
Final wt(g)	1.44	1.34	1.27	1.12
G	7.29	5.85	4.78	2.27
Winter				
Final wt(g)	0.98	0.99	0.93	0.97
G	-0.40	-0.20	-1.45	-0.61

Food value was set at 1200 cal g^{-1} to approximate natural food, and avoid food limitation effects. For comparison, growth was also simulated under constant optimum abiotic conditions in addition to actual conditions in the four systems. The constant optima were: temperature 29C, salinity 20‰, DO 8 mgl^{-1}, and pH 7. Simulated instantaneous daily growth of fish under these optimum conditions was 5.40.

Simulated summer instantaneous growth under actual conditions ranged from 2.27 in ES to 7.29 in GB (Table 2). Thus, a 1 g red drum grew 35% better in GB than in optimal conditions, nearly the same as optimal in ZI, 11% less than optimal in WB, and 60% less than optimal in Es.

In winter, a 1 g red drum had negative growth in all four systems: -0.20 in ZI; -0.40 in GB; -0.61 in ES; and -1.45 in WB.

6.1 RELATIVE IMPORTANCE OF FACTORS

To estimate the relative importance of temperature, salinity, DO and pH at different sites in different seasons, we simulated growth while holding one factor constant at its optimum level, while the other factors varied naturally, and compared that growth to growth with all factors varying naturally. The difference between growth when one factor is held at its optimum and growth under natural conditions (expressed as G_{opt} minus G_{act}) is proportional to that factor's effect, thus its importance.

In ZI little difference was found in summer between growth of a 1 g fish under actual conditions ($G_{act} = 5.85$) and that simulated when all factors were held at their optimum levels (Table 3). Temperature had the only effect, and growth under optimum temperature conditions was actually lower by -0.44 (G_{opt} minus G_{act}) than actual, suggesting our optimal temperature of 29°C was too high for our simulated feeding regime (Subsequent experiments and simulations confirmed this, but are beyond the scope of this paper). DO, salinity and pH had no effect. In WB (G_{act} = 4.78), DO was most important (0.47) and none of the other factors had an effect. In GB (G_{act} = 7.29), temperature also had the only effect (-1.89), again suggesting our optimum temperature was to high. In ES (G_{act} = 2.27), DO was most important (3.13), temperature was better than our optimum (-0.73), and neither pH nor salinity had an effect.

TABLE 3. Effects of environmental factors on simulated growth in summer, measured as the difference between growth under actual conditions and growth when factors held at constant optimum levels. Positive responses means growth increased at optima, negative response means decrease. Dates and sites are same as Table 1.

Site	GB(NH)	ZI(NC)	WB(AL)	ES(CA)
Summer, 1g				
Gact	7.29	5.85	4.78	2.27
DO	0	0	0.47	3.13
Temp	-1.89	-0.44	0	-0.73
pH	0.00	0.00	0.06	0.00
Sal	0.00	0.00	0.00	0.00

In winter, a 1 g fish lost weight (negative growth) in all systems (Table 2), Temperature alone depressed growth in all systems in winter, except in WB, where low salinity (2.7‰) further depressed growth.

To summarize, the relative importance of factors differed among the sites, and among seasons. In summer in ZI and GB, temperature was most important; whereas, in WB and ES, DO was most important. In summer, salinity and pH had little effect on growth. The greatest effects were by dissolved oxygen in WB and temperature in GB. In winter, temperature was the only factor to affect growth in all systems, but its effect was exacerbated by low salinity in WB.

6.2 EFFECT OF FISH SIZE

To determine the effect of fish size, we also ran the model in each system in summer and winter for a 10 g juvenile red drum (not shown). The 10 g fish grew more slowly in summer and lost less weight in winter, as expected, and the ranks in the 4 systems for growth of the 10 g fish were the same as those of a 1 g fish.

7. Discussion

The simulation of growth of juvenile red drum in summer and winter in four different estuarine systems showed significant effects of abiotic variables. Fish grew at rates from 40 to 135% of that which would occur under optimal constant abiotic conditions in summer, clearly indicating a strong potential effect of abiotic conditions in the absence of food limitation. In summer, food limitation would exacerbate limiting abiotic effects, especially that of dissolved oxygen. Such effects would not be detectable with typical daytime, or less frequent, measurements, since in all systems daytime DO was high. In fact, in summer when primary production is high, high daytime dissolved oxygen often forecasts low nighttime dissolved oxygen, owing to respiration by primary producers. Therefore, high daytime DO can be quite misleading. Summer nocturnal low tides are often accompanied by water column anoxia, especially in shallow systems dominated by benthic primary production and respiration. Alternatively, nocturnal flood waters can buffer such effects. In many cases, emigration of fish is necessary, and to be expected.

Summer (August 1996) mean temperatures in two systems (ZI = 28.1°C and WB = 29.2°C) were near the optimum for red drum (29°C). In ZI, mean DO (10.8 mgl^{-1}), salinity (23.4‰), and pH (7.8) were also near optimal, and fish growth was near maximal. In WB, the effect of low DO (mean = 4.7 mgl^{-1}) was probably amplified by pH since the range in WB was about three units. In the other systems the pH range was less—1.5 units in ES and less than 1 in ZI and GB. Salinity did not have a substantial effect on growth in any system, even though the salinity range was about 16 and 12‰ in ZI and WB, respectively, and the mean in WB (4.8) was much lower than optimum. In the other two systems (GB and ES), the mean summer temperature was about 8 and 9°C lower than optimum. Suboptimal temperature had a much larger effect in GB than ES, because in ES the effect of DO (6.2 mgl^{-1}) was nearly four times as important (3.13) as that of temperature (0.73). Salinity was near optimal in GB (25.9‰) and, even though high (34.5‰) in ES, it did not have much effect (<0.1).

Fish lost weight in all systems during winter, despite abundant (simulated) food. This may explain why many temperate estuarine fish move oceanward in winter, where temperature is generally higher and more stable.

8. Conclusions

Abiotic differences among estuaries can account for at least a three-fold difference in juvenile fish performance (growth). However, systems differ in the relative importance of individual factors and the actual performance of fish is an integrated response to these factors, the effects of which are not necessarily additive. We have applied an ecophysiological model to assess such joint effects, which, in turn, can specify the time and space scales of necessary measurement by using data selected to simulate less frequent data collection. Only in exceptionally invariate systems will infrequent point measurements suffice to explain growth performance.

Although salinity has been found to be correlated with the distributions of many estuarine fish, it is of little value as an explanation for variable growth rates in different estuaries. Instead, salinity seems to function mainly as a directive factor in estuaries. Among pH, salinity, dissolved oxygen and temperature, the latter two are most important in summer and winter, respectively. The combination of high temperature and low dissolved oxygen in summer is of paramount importance in determining the performance of juvenile fish in estuaries. The same sub-optimal abiotic environment is also likely to reduce mortality by fish predation, since larger, more vagile, fish are expected to avoid such conditions. In general, larger animals are more sensitive to sub-optimal environmental conditions, especially oxygen (van Dam and Pauly 1995). Thus, variation, per se, may provide refuge from predation by larger fish in estuaries.

Owing to rapid changes in abiotic factors, fish inhabiting estuaries are likely to be more or less continuously in the process of acclimation. Acclimating fish have varying preferences and optima. Fish seeking optimal conditions can even be "trapped" in sub-optimal environmental situations where the surrounding environment is even worse (Neill and Galloway 1989), since fish continuously make choices between better and worse conditions. For these reasons, inferences about habitat quality from fish distribution alone can be quite misleading, especially in dynamic systems such as estuaries.

Under low dissolved oxygen, or other sub-optimal, conditions, fish may possess inadequate metabolic scope to process food. Therefore, feeding may be suppressed by low dissolved oxygen (acting as a directive factor) even where abundant food supplies exist (Jobling 1985). It is quite likely that apparent food limitation is, in reality, oxygen limitation. Unfortunately, in most cases, inadequate oxygen data exist to test this hypothesis.

The structural components of habitat cannot yet be evaluated as determinants of performance. Furthermore, many structural features of habitats are responses to the same dynamic environmental conditions that influence fish performance—for example, oyster reefs or submerged vegetation. It may be necessary to manipulate structure in similarly dynamic systems to quantify their importance. It seems to us that a way to evaluate structure is as a loading factor related to stress.

What we have presented is a realistic, internally consistent, model to assess the potential importance of abiotic effects on fish performance—in this case, growth. While the model is incomplete, it does include many of the factors known to affect performance at different levels of biotic organization, including growth at the individual level. What will no doubt emerge from further simulation and tests of the model is the importance of directive factors that strongly influence behavioral and acclimatory responses of fish, and the need for better dynamic quantification of the abiotic environment. Instruments that record the necessary synoptic measurements of important factors, along with ecophysiologically-sound models, offer a new way to evaluate the estuarine environment in terms of fish performance. But it will be field tests of models, such as attempts to restore marshes, which will determine if we have made progress in understanding nature, not more elegant, or even more realistic, models.

In this paper, we outline a new approach to understanding the contribution of growth to variable production in estuaries. Production can be low because productivity (capacity, or scope, for production) is low or because scope is not partitioned into useful activity—in this case, growth. Factors in the system operate either to constrain scope or to partition scope into wasted and useful activity, or both. Fry recognized 50 years ago that environmental factors operated on fish in fundamentally different ways. He also recognized that factors interact, and their effects are additive only within the same factor class. For example, food can be abundant, but unavailable, owing to low oxygen. Oxygen, and other factors, can act both as limiting and directive factors, and it is important to distinguish which. Temperature is often both a directive and controlling factor. Likewise, salinity, which seems to affect the distribution of fish, can act both as a loading factor and as a directive factor. In general, directive responses are poorly understood, but may explain why fish grow poorly, for example, in the presence of abundant food. What we have outlined here is a method of evaluating the importance of environmental factors in terms of scope for performance—in this case, growth. The difference between scope for performance and realized performance will constitute a first estimate of the importance of directive factors. We suggested (Neill et al. 1994) that analogs exist at, and this approach can be applied to the analysis of, higher levels of biotic organization. We also suggested that effects at one level can be translated into performance at higher levels within this scheme (Miller et al. 1997, and Miller 1997).

We are entering a new era where synoptic environmental data are relatively easy to obtain. It is important that we also enter a new era of thinking, where we utilize all our knowledge to begin to build, and field test, mechanistic models of secondary production in salt marshes.

9. Acknowledgement

JMM was supported by Grant NA90AA-D-SGO62 from the National Sea Grant College Program, National Oceanic and Atmospheric Administration, to the North Carolina Sea Grant College Program.

10. Literature Cited

Boesch, D.E. and R.E. Turner. 1984. Dependence of fishery species on salt marshes: the role of food and refuge. Estuaries **7**:460-468.

Breitburg, D.L., T. Loher, C.A. Pacey and A. Gerstein. 1997. Varying effects of low dissolved oxygen on trophic interactions in an estuarine food web. Ecological Monographs. **67**:489-507.

Brett, J.R. and T.D.D. Groves. 1979. Physiological energetics. Pages 279-352 *in* W.S. Hoar and D.J. Randall, editors. Fish physiology. Volume 8. Academic Press, Inc., New York, New York, USA.

Briggs, J. C. 1974. Marine Zoogeography. McGraw Hill, New York, New York, USA

Browder, J.A. and D. Moore. 1981. A new approach to determining the quantitative relationship between fishery production and the flow of fresh water to estuaries. Pages 430-430 *in* R.D. Cross and D.L. Williams, editors. Proceedings—National symposium on freshwater inflow to estuaries 1. U.S. Fisheries and Wildlife Serv. FWS/OBS-81/04.

Deegan, L.A., J.T. Finn, S.G. Ayvazian, C.A. Ryder-Kieffer and J. Buonaccorsi. 1997. Development and validation of an estuarine index of biotic integrity. Estuaries **20**:601-617.

Diaz, R.J. and R. Rosenberg. 1995. Marine benthic hypoxia: a review of its ecological effects and the behavioral response of benthic macrofauna. Oceanography and Marine Biology Annual Review **33**:245-303.

Edwards, R.E. 1991. Nursery habitats of important early-juvenile fishes in the Manatee River estuary system of Tampa Bay. Pages 237-251 *in* S. Treat and P.A. Clark, editors. Proceedings—Tampa Bay area scientific symposium 2. Tampa Bay NEP, Tampa, Florida, USA.

Fry, F.E.J. 1947. Effects of the environment on animal activity. Univeristy of Toronto Studies, Biology Series **55**:1-62.

——. 1971. The effect of environmental factors on the physiology of fish. Pages 1-98 *in* W.S. Hoar and D.J. Randall, editors. Fish physiology. Volume 6. Academic Press, Inc., New York, New York, USA.

Golley, F.J. 1993. A History of the ecosystem concept in ecology. Yale University Press, New Haven, Connecticut, USA.

Jobling, M. 1985. Growth. Pages 213-230 *in* P. Tytler and P. Calow, editors. Fish energetics: new perspectives. Johns Hopkins University Press, Baltimore, Maryland, USA.

Johnson, M.W. and E. Brinton. 1963. Biological species, water masses and currents. Pages 381-414 *in* M. M. Hill, editor. The Sea. Volume 2. J. Wiley and Sons, New York, New York, USA.

Karr, J. R. 1981. Assessment of biotic integrity using fish communities. Fisheries **2**:21-27.

MacCall, A.D. 1990. Dynamic geography of marine fish populations. Washington Sea Grant Program, University of Washington Press, Seattle, Washington, USA.

Miller, J.M., W.H. Neill and K.A. Duchon. 1997. An ecophysiological model for predicting performance of released fish. Bulletin of Natural Resources and the Institute of Aquaculture, Supplement. **3**:87-91.

Miller, J.M. 1997. Opening address of the third flatfish symposium. Journal of Sea Research **37(3/4)**:181-186.

Neill, W.H. and B.J. Galloway. 1989. ''Noise'' in the distributional responses of fish to environment: an exercise in deterministic modeling motivated by the Beaufort Sea experience. Biological Paper of the University of Alaska **24**:123-130.

Neill, W.H. and J.D. Bryan. 1991. Responses of fish to temperature and oxygen and response integration through metabolic scope. Pages 30-57 *in* D.E. Brune and J.R. Tomasso, editors. Aquaculture and water quality: advances in world aquaculture, Volume 3, The World Aquaculture Society, Baton Rouge, Louisiana, USA.

Neill, W.H., J.M. Miller, H.W. Van der Veer and K.O. Winemiller. 1994. Ecophysiology of marine fish recruitment: a conceptual framework for understanding interannual variability. Netherlands Journal of Sea Research. **32**:135-152.

van Dam, A.A. and D. Pauly. 1995. Simulation of the effects of oxygen on food consumption and growth of Nile tilapia, *Oreochromis niloticus* (L.). Aquaculture Research **26**:427-440.

Warren, C.E. and G.E. Davis. 1967. Laboratory studies on the feeding, bioenergetics and growth of fishes. Pages 175-214 *in* S.D. Gerking, editor. The biological basis of freshwater fish production. Blackwell Scientific Publications, Oxford, England.

SALT MARSH ECOSYSTEM SUPPORT OF MARINE TRANSIENT SPECIES

LINDA A. DEEGAN
JEFFREY E. HUGHES
The Ecosystems Center
Marine Biological Laboratory
Woods Hole, MA 02543 USA

RODNEY A. ROUNTREE
University of Massachusetts - Amherst
48 Oregon Road
Mashpee, MA 02649 USA

Abstract

One of the most important reasons stated in legislation for protecting salt marshes is their support of commercially and recreationally important nekton (fish and crustaceans). Yet, there is a surprising level of uncertainty among scientists regarding the role of salt marshes in supporting secondary production. The emphasis has been on "marine transient" species (in earlier literature often referred to as "estuarine dependent") because they have life histories that seem designed to place young-of-the-year or juveniles in marsh habitats and because these species are often of commercial or recreational value. Salt marshes are believed to provide: 1) trophic support resulting in high growth rates, 2) increased survivorship due to lowered mortality, and 3) a suitable physico-chemical environment for development of young fishes. In this paper, we consider the evidence for each of these, with an emphasis on the trophic and survivorship aspects. The seasonally warmer temperatures of estuaries and salt marsh creeks apparently provide a metabolic advantage that supports high growth rates. The influence of marsh-derived organic matter in estuarine food webs is apparent, and its importance to marine transient fishes is supported by dietary, behavioral, and isotopic evidence. The major pathways by which marsh organic matter is transferred to fish are largely indirect, through microbial and invertebrate intermediaries. Invertebrates are the primary link to fish consumers of marsh-associated production, transforming microphytes, organic detritus, and microbial detrital heterotrophs into available biomass. Although most detrital organic carbon entering salt marsh systems, mainly from emergent grasses, is apparently respired by heterotrophs, the support of consumers by marsh plant detritus and microalgae can be equally important. The use of salt marsh detritus in food webs usually occurs in close proximity to the salt marsh indicating that outwelling of salt marsh organic matter offshore is not the dominant way that salt marshes support offshore fisheries. Salt marsh support of offshore fisheries is more probably by direct export of juvenile fish biomass and a trophic relay involving ontogenetic and cyclic migrations of nekton species, rather than export of organic detritus. Understanding the controls on marine transient fish mortality is probably the most problematic and least studied aspect of their ecology. The few estimates of mortality rates of fishes in estuaries are as high as, or higher than, mortality rates of fishes in other marine and freshwater ecosystems. However, because of faster growth rates, fish spend less time in the small stages with the higher mortality rates. Within estuaries, mortality rates for some species, but not all, are lower in marsh creeks compared to more open areas. The value of marshes as refuge habitat is probably due to the interaction of temperature, turbidity, and vegetative structure in restricting the foraging of piscine predators.

1. Introduction

The importance of estuaries and their constituent habitats in the support of coastal commercial and recreational fisheries has been recognized since the early 19th century (Baird 1873). In some parts of the country, coastal species are the major component of commercial and recreational fisheries. According to many accounts, estuaries provide an essential habitat for approximately two-thirds of the commercially and recreationally important species along the east coast of the United States (e.g., Boesch and Turner 1984, Houde and Rutherford 1993). Even where the direct harvesting of estuarine and coastal species is not numerically or economically significant, estuaries and coastal embayments may still be essential for fisheries because they serve as nurseries for the juvenile stages of species harvested offshore or for the prey of commercially important species.

The importance of salt marshes as an essential habitat has been inferred from the high abundance of fishes, particularly young-of-the-year and juveniles, found in these habitats (e.g., Cain and Dean 1976, Weinstein 1979, Weinstein et al. 1980, Haedrich 1983, Weinstein and Brooks 1983). Yet, there is a surprising level of uncertainty about the role of salt marshes in supporting secondary production.

1.1 WHAT SUPPORT DO SALT MARSHES OFFER THAT MAKES THEM ESSENTIAL FISH HABITAT?

The once-stated and oft-repeated hypothesis proposed to explain the widespread use of estuaries by young-of-the-year and juvenile fishes and crustaceans is that these areas provide a refuge from predation in a location where food supplies are abundant and physical factors are suitable (e.g., Joseph 1973, Boesch and Turner 1984, Deegan and Day 1984, Day et al. 1989, Baltz et al. 1993, Miltner et al. 1995). These factors result in high growth rates, low mortality and high abundances of nekton. In this paper, we consider the evidence for each of these, with an emphasis on the trophic support and survivorship aspects. We will examine the growth and mortality of nekton in salt marsh ecosystems in comparison to other habitats and consider evidence for the role of salt marsh in food webs that support nektonic production.

1.2 WHAT ARE MARINE TRANSIENT SPECIES?

Our focus is on the importance of salt marshes to ''marine transient'' species. Fish that use estuaries can do so for all (resident) or only part (transient) of their lives. Transient species spend varying amounts of their lives in other habitats such as the open ocean or in lakes or streams (Fig. 1). Perhaps the most common nektonic life history among the transient species that use estuaries is that of species that spawn in salt water. These marine transient species usually spawn in nearshore coastal waters, then the eggs or larvae move into estuaries where the juveniles spend several weeks to several years before moving offshore to return to adult feeding and spawning grounds. Examples of typical marine transient species are menhaden (*Brevoortia* spp.), mullet (*Mugil* spp.), croaker (*Micropogonias undulatus*), spot (*Leiostomus xanthurus*), and flounder (*Paralichthys* spp. and *Pseudopleuronectes* spp.). Day et al. (1989) list over 55 species of marine transient fish found in estuaries along the Gulf and Atlantic coasts of North America. Although offshore spawning is most common, some species spawn inside estuaries (winter flounder, *Pseudopleuronectes americanus*) or in salt marshes (Atlantic silverside, *Menidia menidia*). Another variation in the marine transient life history is a group of species that are spawned offshore and enter estuaries as juveniles (e.g., bluefish, *Pomatomus saltatrix*). We admit to a ''fish'' bias in our review, although there are many examples of crustaceans (e.g., blue crab, *Callinectes sapidus*; penaeid shrimp, *Penaeus* spp.) that are also marine transients.

One problem with defining salt marsh support of transient species is that many of the behaviors of fishes seem very flexible. All of our observations on estuarine fishes, with few exceptions, point to flexibility in diet and habitat use. There are few species that exhibit behavioral or morphological adaptations designed solely for the use of marshes as a habitat. For example, most fish species use more than a single habitat during their stay in estuaries. They often occupy multiple areas, or use different parts of the estuary sequentially. It is difficult to determine if a salt marsh is ''required'' if the fish spends only a few weeks in that habitat or can apparently use another habitat. These weeks may correspond to a critical period in the life history of the species (e.g., Gulf menhaden, *Brevoortia patronus*, Deegan 1990) or they may occur at a time that does not seem important in determining year-class strength. In addition, the migration pattern of some species varies with latitude and with the size of the estuary. In more northern estuaries, all individuals of some species, such as Atlantic silverside, move offshore for the winter. In southern estuaries, a large portion of the population of the same species may stay in the estuary for most or all of the year. In large estuaries, some species may use the deeper, more saline portions of the estuary as if it were ''offshore''. For example, bay anchovy (*Anchoa mitchilli*) spawns both outside and inside the mouth of Chesapeake Bay, (James Cowan, personal communication) while in smaller, shallower estuaries the entire population moves offshore to spawn. This flexibility in behavior often prompts the question of whether fish would use salt marshes if they had other habitat choices. Thus, answering the question of the importance of marshes to marine transient species becomes not a simple ''Yes'' or ''No'' issue, but requires resolution of the degree of marsh support.

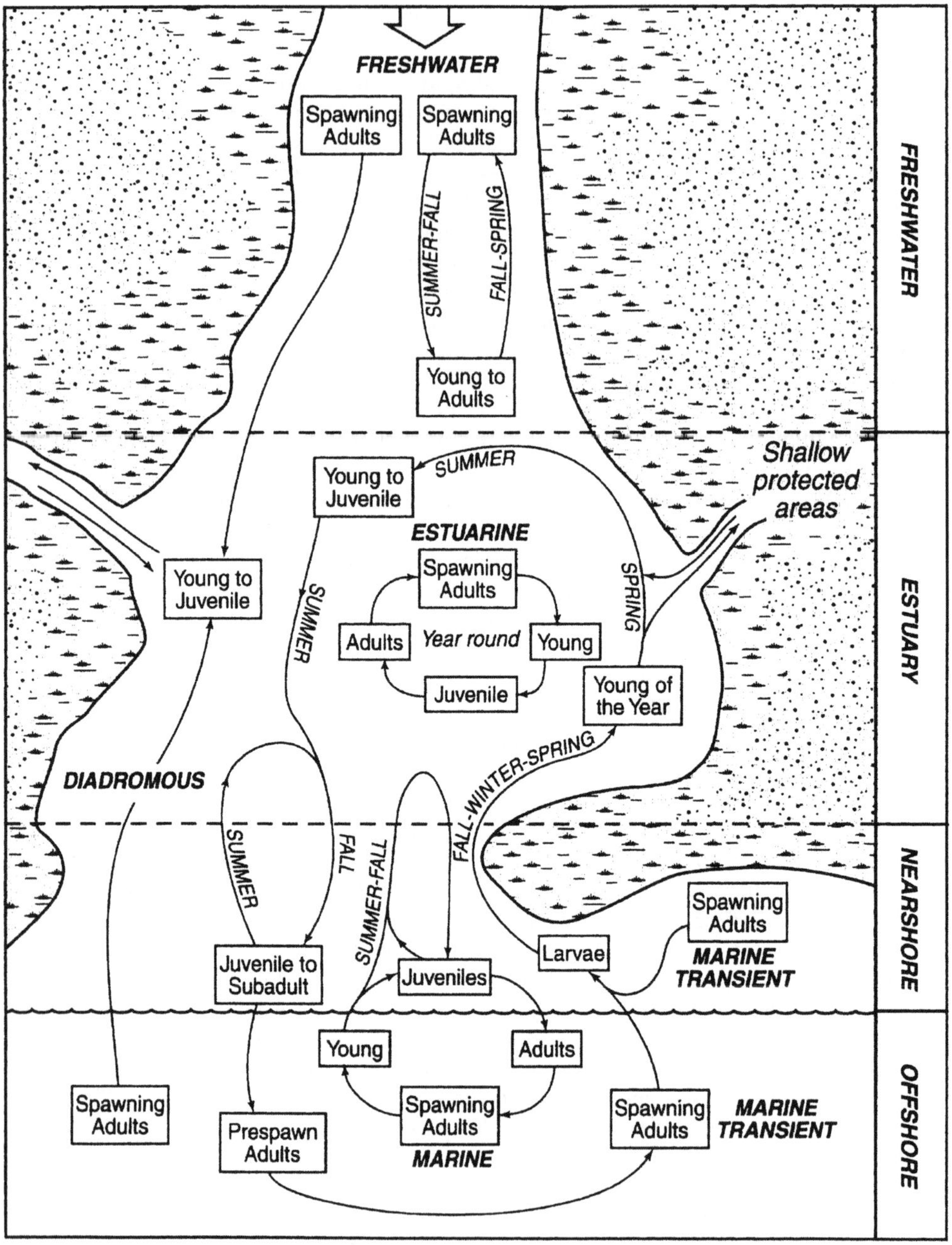

Figure 1. Diagram of temporal and spatial use of estuaries by freshwater, estuarine, marine and estuarine-dependent marine nekton. Adapted from Deegan and Thompson (1985).

1.3 WHAT DO WE MEAN BY SALT MARSH?

One difficulty with defining "salt marsh support" is knowing what different researchers mean by the salt marsh ecosystem. In this paper, we include several contiguous habitats in our definition of a salt marsh ecosystem: the irregularly flooded marsh surface (including vegetated areas and pools or pannes), the regularly flooded intertidal surface, intertidal marsh creeks and small subtidal marsh creeks. We include small subtidal creeks because from the perspective of a fish they are part of the intricate network of channels and pools surrounded by salt marsh vegetation. Our definition differs from Kneib (1997) because we include small subtidal creeks as part of the marsh system. We do not include large open bay areas. These areas are clearly connected to salt marshes by subtidal and intertidal channels and by the movements of marsh-derived materials and animals, but we consider them a different ecosystem because of their hydrology, geomorphology and patterns of productivity. We acknowledge that scale and perspective are somewhat subjective and that small subtidal creeks, large subtidal channels, and the adjacent bay fringed by salt marsh are all connected. When we discuss "estuarine" in general, this includes the larger open bay areas. If we were not certain that the data came from a "salt marsh ecosystem", then we called it "estuarine."

2. Growth and Trophic Support of Nekton

It is universally acknowledged that fish, particularly the young of marine transients, grow rapidly in estuarine environments (e.g., Boesch and Turner 1984, Deegan and Day 1984, Yanez-Arancibia 1985, Day et al. 1989). These high growth rates have been offered as one explanation for the evolution of the marine transient life history (Blaber and Blaber 1980). While growth of young fish in estuaries is indisputably fast, is it faster than we might expect in any other ecosystem?

2.1 COMPARISON OF GROWTH RATES BETWEEN ESTUARIES AND OTHER ECOSYSTEMS

In a recent compilation of growth rates for fish found in estuaries along the mid-Atlantic coast, Able and Fahey (1998) found that the growth rate of marine transient species was substantially higher than that of estuarine resident species in the same estuaries (Fig. 2). This comparison, however, is complicated because growth rates are known to be positively related to adult animal size (Valelia 1995) and the adult size of marine transients is larger than estuarine residents (Able and Fahey 1998). Therefore, based on adult size alone we would expect the growth rates of marine transients to be higher than resident species.

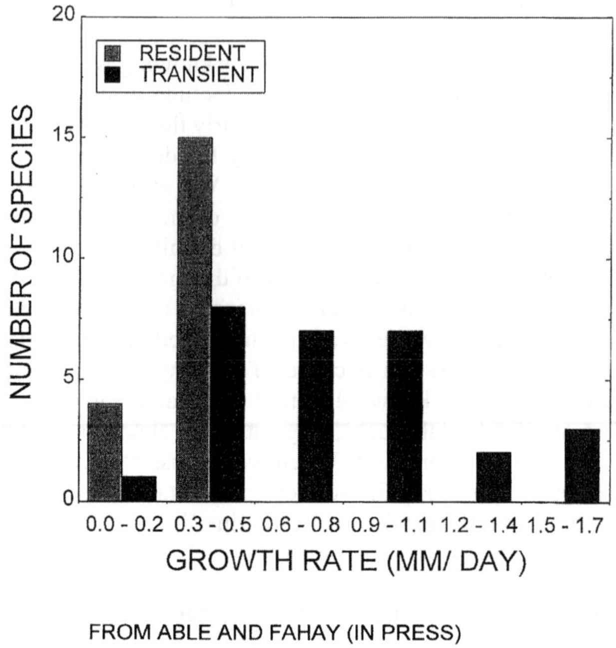

Figure 2. Distribution of growth rates among resident and transient estuarine fishes. From Able and Fahay (1998).

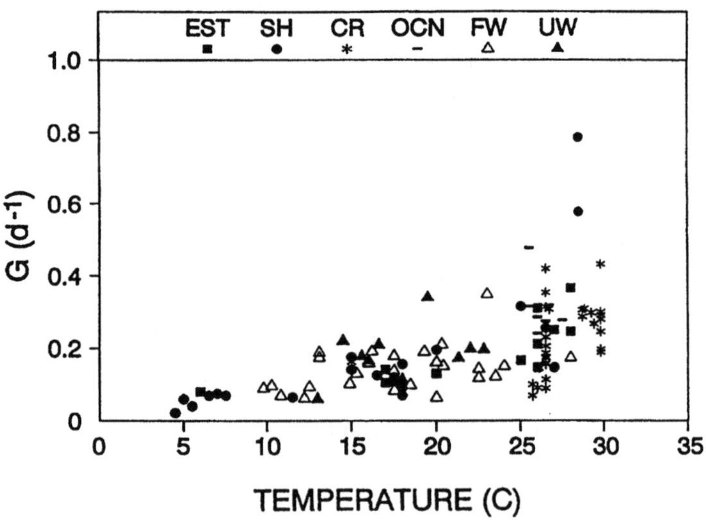

Figure 3. Ecosystem-specific, weight-specific growth coefficient G (d^{-1}) of fish larvae versus water temperature. EST = estuarine, SH = shelf, CR = coral reef, OCN = oceanic, FW = freshwater, UW = upwelling system. From Houde and Zastrow (1993).

In a comparison across many taxa and ecosystems, Houde and Zastrow (1993) found instantaneous growth rates of larval fish were highly correlated with temperature (Fig. 3). The growth rates of larval fish in estuaries, while among the highest measured, were no higher than expected based on the relationship with temperature. In a comparison among habitats within an estuary, Deegan (1990) found higher growth rates of Gulf menhaden correlated with warmer temperatures in marsh creeks compared to open bay areas. This implies that there is a growth advantage to moving into estuarine waters in the spring as these waters are often warmer than the adjacent offshore areas.

2.2 POTENTIAL FOOD SOURCES AT THE BASE OF FOOD WEBS THAT SUPPORT NEKTON

Sustaining high somatic growth in addition to the elevated metabolic demands at high temperatures implies an adequate food supply to leave a sufficient ''scope for growth'' (Hoar et al. 1979, Valelia 1995, Evans 1998). The suggestion that the high primary productivity of salt marsh habitats, in particular intertidal *Spartina* vegetation, supports these high growth rates has been debated for more than three decades (e.g., Darnell 1961, Teal 1962, Nixon 1980, Peters and Schaaf 1981, Pomeroy and Wiegert 1981). We focus first on the processes by which salt marsh-associated organic matter is incorporated into food webs supporting marine transient fish production in estuaries and then discuss the potential for support of more distant areas and coastal fisheries.

The organic matter supporting nekton production in salt marsh estuaries has its origin in several potential sources: 1) marsh macrophytes (e.g., *Spartina* spp.), 2) marine and estuarine phytoplankton, 3) edaphic microphytes, and 4) terrestrial plants of the upland watershed. Although the relative importance of these source materials is only partially resolved, some patterns are clear.

The influence of marsh-derived organic matter in estuarine food webs is apparent, and its importance to marine transient fishes is supported by dietary, behavioral, and isotopic evidence. Stable isotopic analyses (^{13}C, ^{15}N, ^{34}S) demonstrate that marsh-associated organic matter supports nekton in a variety of geographic settings. Research using the multiple stable isotopic content of organic source materials and salt marsh consumers (Peterson et al. 1985, Peterson and Fry 1987, Peterson and Howarth 1987, Currin et al. 1995), and studies of vascular plant diagenesis (Howarth and Teal 1980, Peterson et al. 1980, Benner et al. 1987, Ember et al. 1987, Fogel et al. 1989) have shown that *Spartina* detritus provides important trophic support in salt marsh estuaries. Stable isotopic analyses also indicate that, notwithstanding the immense productivity of salt marsh plants, phytoplankton and benthic algae are also essential in supporting secondary production in the marsh and adjacent estuary (Haines 1977, 1979a, Haines and Montague 1979, Kneib et al. 1980, Hughes and Sherr 1983). Although most detrital organic carbon from emergent grasses either forms peat or is apparently respired by heterotrophs, the support of consumers by marsh plant detritus and microalgae can be equally important (Peterson and Howarth 1987). However, because the relative contribution of micro- and macrophytes to salt marsh estuarine food webs shows considerable geographic variation (Sullivan and Moncreiff 1990), a generalization as to the predominance of either is not possible (Haines 1979a, Mallin et al. 1992 and reply by Schaff and Peters 1992).

A wide variety of nekton have stable isotope values that indicate that salt marsh production is important in the food web. Marsh resident nekton, such as killifish and palaemonetid shrimp generally have a ^{13}C, ^{15}N, and ^{34}S content that reflects feeding in food webs supported by benthic microalgae of the marsh surface (Sullivan and Moncreiff 1990), or *Spartina* detritus (Peterson and Howarth 1987), or both (Hughes and Sherr 1983, Currin et al. 1995, Deegan and Garritt 1997). Similarly, the following benthic- and pelagic- (water column) feeding marine transient nekton from a wide variety of estuaries show a strong connectedness to marsh-based organic matter: menhaden (pelagic), and striped mullet, American eel, blue crab, and penaeid shrimp (benthic) in Georgia salt marsh creeks and tidal rivers (Peterson and Howarth 1987); menhaden in a Louisiana salt marsh embayment (Deegan et al. 1990); menhaden and striped anchovy (pelagic), and spot, silver perch, white perch, and blue crab (benthic) in a Chesapeake Bay tidal marsh creek (Stribling and Cornwell 1997); bluefish, rainbow smelt, and Atlantic silverside (pelagic), and winter flounder, black-spotted stickleback, and green crab (benthic) in a northern Massachusetts salt marsh estuary (Deegan and Garritt 1997).

Although both pelagic and benthic fishes in salt marsh estuaries show a dependence on salt marsh organic matter, in general, benthic-feeding fishes show a greater dependence on salt marsh production than do most fishes that feed in the water column. Analysis of ^{13}C and ^{15}N content of a variety of organisms, including several marine transient fishes, indicates that benthic-feeding organisms show the strongest connection to marsh-based production (Fig. 4) in a northern Massachusetts salt marsh creek (Deegan and Garritt 1997). Most benthic-feeding fishes had ^{13}C values closer to *Spartina* and benthic algae, while pelagic feeders had ^{13}C values closer to phytoplankton.

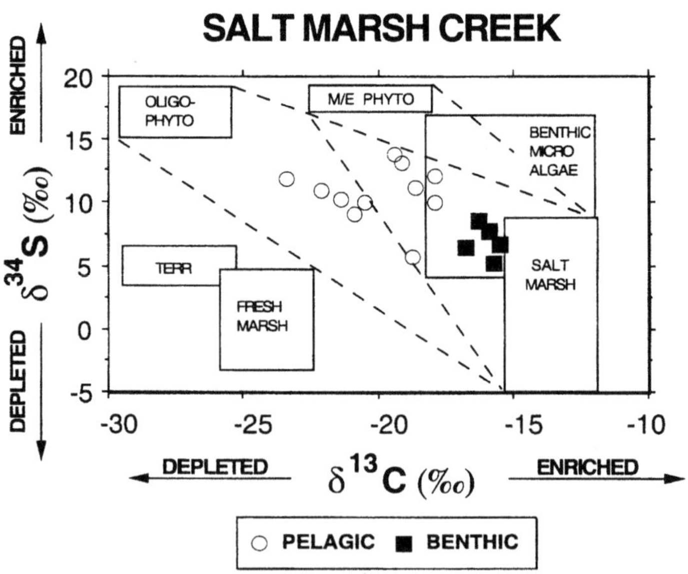

Figure 4. Cross plot of sulfur and carbon stable isotopic values for pelagic and benthic consumer organisms in the salt marsh creek habitat of a northern Massachusetts estuary. Boxes represent typical ranges for candidate organic matter source materials. Abbreviations: "Oligo phyto" is oligohaline phytoplankton, "Terr" is terrestrial plants, and "M/E Phyto" is marine and estuarine phytoplankton. From Deegan and Garritt (1997).

Phytoplankton can be differentiated from marsh primary production using ^{13}C because of distinct sources of dissolved inorganic carbon (Sherr 1982). However, phytoplankton and marsh are potentially linked through the uptake by phytoplankton of remineralized nitrogen from marsh detritus (Thayer 1974, Heinle and Flemer 1976, Valiela et al. 1978, Haines 1979b, Holmes et al. 2000) and the deposition of phytoplankton onto the marsh surface. Because the salt marsh acts as a particle trap and removes phytoplankton and bacteria during marsh flooding (Haines 1979a, Chrzanowski and Spurrier 1987), deposited organic seston should be considered a potential food source for marsh consumers.

The overlap of isotopic values in primary producers sometimes confounds attempts to differentiate organic matter sources at the base of estuarine food webs. One approach to overcoming this limitation is to differentially label one of the primary producers. A recent study applied ^{15}N-nitrate tracer to investigate nitrogen cycling in the food web of the oligohaline tidal reach of a northern Massachusetts estuary (Holmes et al. 2000, Hughes et al. 2000). The addition of ^{15}N-nitrate differentially labeled the phytoplankton, making it isotopically very distinct from marsh production. Uptake of the ^{15}N-nitrate tracer showed that most of the nitrogen assimilated by zooplankton and planktivorous juvenile alewife was based on an oligohaline phytoplankton bloom (Hughes et al. 2000). Many of the epibenthic crustacean consumers (amphipods, white-fingered mud crabs, and palaemonetid shrimp) and juvenile white perch, also depended on phytoplankton for 40 to 70% of their assimilated nitrogen, either through consumption of deposited phytoplanktonic detritus or planktonic copepods. The largely unlabeled detrital material derived from the tidal marsh most likely supported the remainder of their assimilated diets. This detrital trophic base was especially important to the production of benthivorous fishes, including mummichog and white sucker (the numerical and biomass dominants in the reach), and American eel. These results reinforce the idea that estuarine benthic food webs are more closely tied to marsh production than are pelagic food webs (see also Fig. 4).

The importance of terrestrial plant material to consumers in salt marsh food webs has been suggested by some stable isotopic studies (e.g., Hackney and Haines 1980, Stribling and Cornwell 1997). The influence of terrestrial material is expected to decrease with distance from the freshwater source (Cifuentes 1991), and is apparently minimal in salt marsh estuaries with low riverine input (Peterson et al. 1985, Peterson and Howarth 1987, Deegan and Garritt 1997).

2.3 USE OF MARSH FOOD RESOURCES BY NEKTON

Only a few fish species can directly utilize marsh-associated production. Some fishes consume and assimilate algae, including some cyprinodonts and mugilids (Kneib 1997). Vascular plant detritus can comprise a large part of the stomach contents of killifish (Allen et al. 1994), although the importance of this material to the assimilated diet has been questioned (Boesch and Turner 1984, D'Avanzo et al. 1991, Kneib 1997). Gulf menhaden, however, may derive a substantial portion of their assimilated diet from *Spartina* detritus, and contain cellulase-producing microorganisms in their gut (Fig.5, Deegan et al. 1990). Juvenile menhaden captured in salt marsh creeks had ^{13}C and ^{34}S values closer to *Spartina* than did larvae that had fed offshore on zoo- and phytoplankton. Recent research has suggested that amorphous detrital aggregates may be important in the diets of several nektonic species, including palaemonetid shrimp, sheepshead minnow (*Cyprinodon variegatus*), Atlantic menhaden (*B. tyranus*), and two species of mullet (D'Avanzo et al. 1991, Lewis and Peters 1994, Larson and Shanks 1996). The labile dissolved organic matter (DOM) that is the basis for the aggregates is derived in large part from *Spartina* leachate and algae (Turner 1978, Ribelin and Collier 1979, Coffin et al. 1989, Fry et al. 1992, Alber and Valiela 1994, Peterson et al. 1994). The importance of DOM to food webs is potentially large because it is a major component of total organic matter in salt marsh estuaries (Mann 1988, Peterson et al. 1994). However, the degree to which DOM supports fish production, either by detrital aggregates or by way of transfer through microbial heterotrophs and zooplankton (Mann 1988), remains an intriguing but unresolved issue.

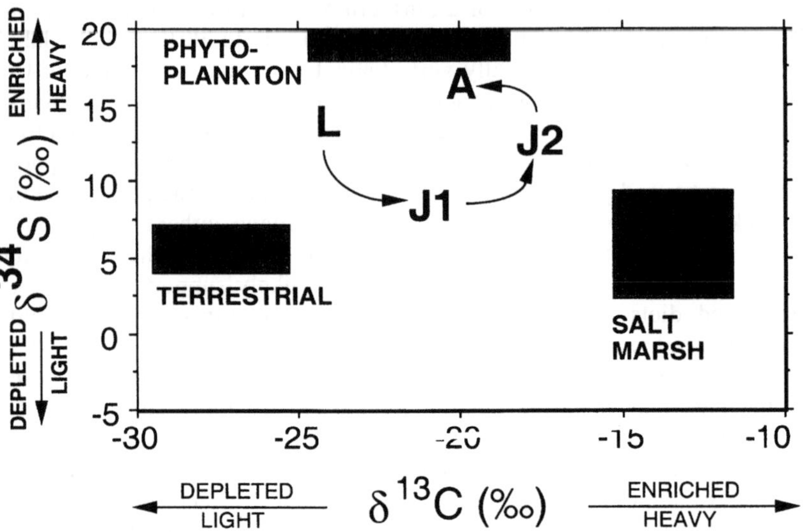

Figure 5. Cross plot of sulfur and carbon stable isotopic values for Gulf menhaden in different life history stages and possible organic matter sources. Boxes represent range of possible values for organic matter sources. L = larvae, J1 = small juveniles in marsh creeks, J2 = larger juveniles of the open bay, L = adult. From Deegan et al. (1990).

The major pathways by which marsh organic matter is transferred to fish are largely indirect, through microbial and invertebrate intermediaries (Tenore et al. 1982, Newell and Langdon 1986). Invertebrates are the primary link to fish consumers of marsh-associated production, transforming microphytes, organic detritus, and microbial detrital heterotrophs into available biomass (Heinle et al. 1977, Tenore et al. 1982, Boesch and Turner 1984, Zagursky and Feller 1985, Kneib 1997). Zooplanktonic, small epibenthic, and shallow-burying invertebrates are abundant in the salt marsh estuary, on the marsh surface, and in adjacent subtidal sediments (Montague et al. 1981). Invertebrates that are important in the diets of juvenile and smaller-sized marsh fishes include polychaetes and oligochaetes, snails, insects and their larvae, and a host of crustaceans - harpacticoid and calanoid copepods, ostracods, mysids, tanaids, amphipods, small crabs (e.g., *Uca* spp.), and palaemonetid shrimp (Feller et al. 1990, Rozas and La Salle 1990, Zimmerman et al. 1990, Cattrijsse et al. 1994, Walters et al. 1996, Gregg and Fleeger 1997, Kneib 1997). These invertebrates can be considered the major primary consumers of marsh production, although "opportunistic omnivory" (Haines 1979) typifies many of their feeding habits (Kneib 1997). Studies of the natural abundance of stable isotopes have demonstrated the trophic linkage of many invertebrate primary consumers to salt marsh organic matter (Haines and Montague 1981, Peterson et al. 1986, Couch 1989, Currin et al. 1995).

It is readily apparent from studies of gut contents that both marsh resident and marine transient nekton feed actively on invertebrate primary consumers during tidal foraging on the marsh surface (reviewed by Kneib 1997). The distribution, abundance, and size composition of invertebrates in the marsh can be correlated with the foraging behavior and abundances of nektonic predators, providing indirect evidence of consumption of marsh prey such as harpacticoid copepods, amphipods, tanaids, and snails (Kneib 1997). Access to the intertidal marsh surface offers increased feeding opportunities for nekton, and promotes growth (Weisberg and Lotrich 1982, Javonillo et al. 1997), although it should be noted that all manipulative field experiments on nekton growth and consumption of marsh prey have been conducted with marsh residents, particularly fundulid fish and palaemonetid shrimp. Caging studies of foraging by nekton have shown significant impacts on marsh invertebrate densities and species composition (Virnstein 1977, Wiltse et al. 1984, Hines et al. 1990). However, fish predator inclusion/exclusion effects on benthos have been equivocal (Kneib 1986, 1988). Nekton behavior interacts with geomorphology and hydrology of estuarine salt marshes to strongly influence the trophic transfer of marsh resources. In a Georgia (USA) salt marsh larger resident nekton range farthest onto the shallow-water vegetated marsh surface during tidal inundation, while smaller nekton remain in deeper water near intertidal creeks (Kneib and Wagner 1994). Accordingly, the effects of nektonic predators on marsh invertebrates should show a spatial dependence based on predator size and species (Kneib and Wagner 1994). In a Louisiana study (Peterson and Turner 1994), the edge of the marsh within a few meters of creek banks was frequented by both marsh resident nekton (e.g., fundulid and gobiid fishes, palaemonetid shrimp) and transient marine nekton (e.g., engraulid fish, penaeid shrimp, portunid crabs). The inner portion of the flooded marsh surface was only of importance to resident species. Marsh edge habitats may be particularly supportive of nekton by providing increased feeding area, elevated prey densities, and protection (Baltz et al. 1993, Rozas and Reed 1993, Miltner et al. 1995).

2.4 THE SPATIAL EXTENT OF MARSH INFLUENCE ON ESTUARINE FOOD WEBS

The influence of salt marsh production at the base of the food web is confined for the most part to the marsh surface and subtidal creeks and embayments immediately adjacent to the marsh. Evidence for a limited spatial scope of "direct" marsh influence in estuarine food webs stems from studies of the composition of estuarine seston and sediments, and from stable isotopic analyses along estuarine gradients. Sherr (1982) found that *Spartina* carbon was an important component of organic matter in intertidal sediments of a Georgia salt marsh, but its contribution was lower in adjacent subtidal sediments, and was scarcely present in seston. An abrupt decline of sedimentary *Spartina* detritus with increasing distance offshore from a salt marsh was reported by Wilson et al. (1985) in a study of subtidal sediments of a Massachusetts coastal embayment. In a Louisiana salt marsh system, the contribution of *Spartina* detritus to subtidal sedimentary organic matter was inversely correlated with the size of the receiving water body (DeLaune and Lindau 1987). Sedimentary ^{13}C content was more similar to *Spartina* in smaller embayments, and more similar to phytoplankton in larger embayments. Cifuentes (1991) concluded that salt marshes exported little organic matter to the Delaware estuary based on the analysis of diagenetic sedimentary components. Similarly localized effects of *Spartina* were observed in sediments of a North Carolina estuarine salt marsh system (Ember et al. 1987) and in western European salt marshes (Hemminga et al. 1996). Bivalve filter-feeders in a Cape Cod salt marsh (Fig. 6, Peterson et al. 1985) and the fauna of a tidal creek in a Georgia estuary (Hughes and Sherr 1983) have ^{13}C contents that suggest highly localized effects of *Spartina* in food webs. In their multiple isotopic study of the biota of a northern Massachusetts salt marsh estuary, Deegan and Garritt (1997) showed that local sources of production were most important to consumers. Fish and invertebrates showed distinct isotopic shifts among habitats that correlated with changes in the isotopic values of marsh plants at the base of the food web. These results echo those of Hughes and Sherr (1983), in which resident and transient nekton in Georgia tidal creeks fed on prey closely related isotopically to local detrital foods.

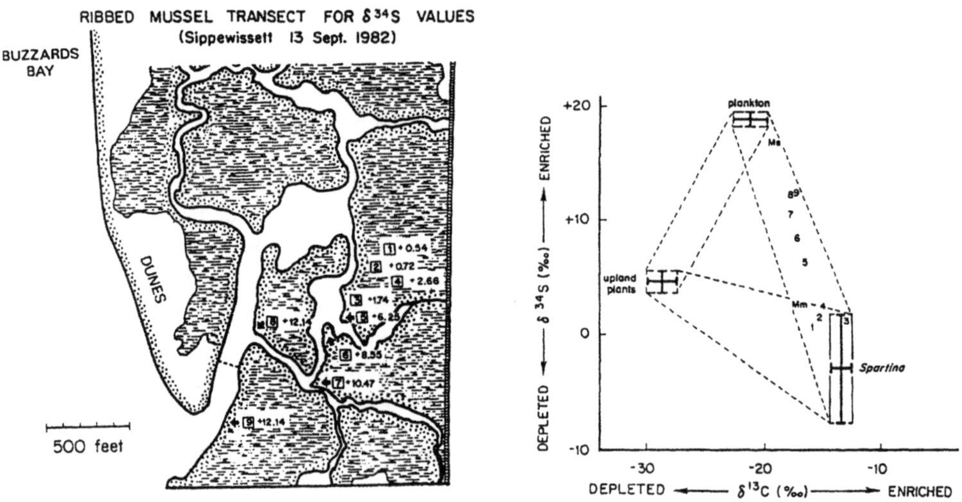

Figure 6. (Left panel) Map of sampling sites and ^{34}S content (per mil) of ribbed mussels (*Geukensia demissa*) in Great Sippewissett Salt Marsh, Massachusetts. (Right panel) Cross plot of sulfur and carbon stable isotopic values for filter-feeding bivalves obtained at corresponding sites in the mapped area. Numbers 1-9 refer to ribbed mussels taken along the transect from the marsh interior (#1) to the adjacent coastal bay (#9). ''Mm'' is the hard clam *Mercenaria mercenaria* from marsh creek sediments. ''Me'' is the blue mussel *Mytilus edulis* from a coastal bay. Boxes in cross plot enclose mean +1 SD of isotopic values for candidate organic matter source materials. From Peterson et al. (1985).

The trophic ''sphere of influence'' of marsh-derived organic matter at the base of estuarine food webs may vary among estuaries, depending in part on the geomorphology of the estuarine basin, the relative magnitude of tidal range and freshwater input (Odum et al. 1979), and the behavior of nekton. The importance of marsh production to nekton production in estuaries depends on the relative area and productivity of marshes compared to other adjacent habitats, which is in part determined by the geological setting (Deegan et al. 1983). In a study of eight Gulf of Mexico estuaries (Deegan et al. 1983), the contribution to total estuarine primary production varied from 16 to 53% for phytoplankton and 0 to 72% for marshes. Undoubtedly the relative importance of phytoplankton and marsh production in the food web also varied.

3. The Marsh as a Refuge From Predation

Reduced risk of mortality is often suggested as a reason for the use of estuaries by young-of-the-year and juvenile fishes (e.g., Joseph 1973, Blaber and Blaber 1980, Boesch and Turner 1984, Deegan & Day 1984). It has been suggested that lower mortality of young nekton in estuarine areas compared to offshore marine areas could be an evolutionary driving force behind the development of complex oceanic-estuarine migration patterns (e.g., Joseph 1973, Blaber and Blaber 1980). We will first examine the evidence for lower mortality rates in estuaries versus offshore areas, and then in salt marshes versus other estuarine habitats. We conclude by discussing the mechanisms governing mortality rates in salt marsh ecosystems.

3.1 COMPARISON OF MORTALITY RATES BETWEEN ESTUARIES AND OTHER ECOSYSTEMS

Comparison of larval mortality rates across a wide variety of marine and freshwater ecosystems does not indicate lower mortality rates in estuarine systems (Fig. 7). In the most comprehensive assessment to date, Houde and Zastrow (1993) found that instantaneous mortality rates of larval fish were correlated with temperature across a wide variety of taxa and ecosystems. Absolute and temperature-adjusted mortality rates for larvae from estuarine systems were high compared to other ecosystems, including the offshore shelf environment. Even though this study includes only a few mortality measurements from estuarine systems (n = 6), the estuarine data are consistent with the general temperature-dependent trend seen across many ecosystems and taxa.

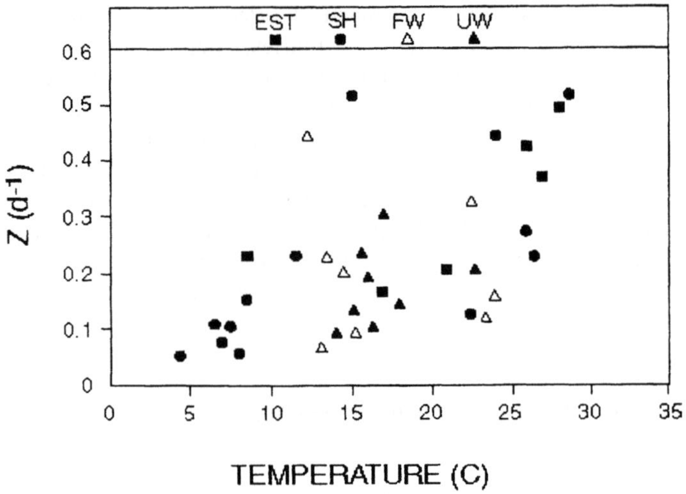

Figure 7. Ecosystem-specific instantaneous mortality coefficient Z (d^{-1}) of fish larvae versus water temperature. EST = estuarine, SH = shelf, FW = freshwater, UW = upwelling system. From Houde and Zastrow (1993).

Estuarine versus offshore comparisons for juvenile nekton are difficult to make because of lack of information on mortality rates in estuaries. Mortality rates for spot (0.02 to 0.05 d^{-1}, Weinstein and Walters 1981, Weinstein 1983), and brown shrimp (*Penaeus aztecus*) (0.04 d^{-1}, Minello et al. 1989) were above the average for marine fishes (0.03 d^{-1}, Gulland 1964), while the mortality rates for Gulf menhaden (0.02 d^{-1}, Deegan 1990) and winter flounder (0.01 d^{-1}, Pearcy 1962) were below (Fig. 8). Mortality estimates for a single species, spot, are both above and below the marine average and show wide year-to-year and geographic differences (Weinstein and Walters 1981, Weinstein 1983). From the limited data available, it is unclear whether mortality rates of juvenile fishes in estuarine systems are in general lower than in offshore ecosystems.

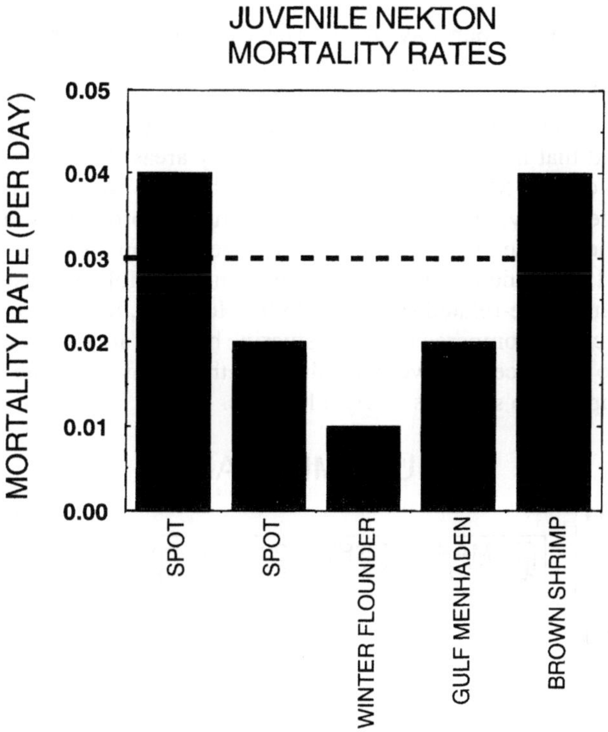

Figure 8. Instantaneous mortality rates of juvenile salt marsh estuarine nekton compared to the average value for juvenile marine fish (dashed line, from Gulland 1964). Spot data at left from Weinstein (1983); other spot data from Weinstein and Walters (1981); winter flounder (Percy 1962); Gulf menhaden (Deegan 1990); brown shrimp (Minello et al. 1989).

The scarcity of data indicates the need for more work on the mortality rates of young-of-the-year and juvenile fishes in estuarine environments. The few data available do not indicate that size-specific mortality rates are lower in estuarine environments compared to offshore environments. However, the higher growth rates of young fish in estuaries may ''overcompensate'' for mortality resulting in net advantage to the fish. Because of the faster growth rates in warmer waters, estuarine fish spend less time in the vulnerable larval stage compared to those in offshore marine areas (Houde and Zastrow 1993). Thus the main advantage of moving into warmer estuarine areas maybe to accelerate growth and decrease the amount of time that fish experience high mortality due to their small size. This may result in a higher net production than if the fish had remained offshore.

3.2 COMPARISON OF MORTALITY RATES AMONG ESTUARINE HABITATS

Within estuaries, salt marshes are suggested to provide protection from predators, resulting in lower mortality rates. For some species there is evidence that mortality rates in marshes may be lower than in open bay areas. Estimates of mortality for Gulf menhaden suggest that, when adjusted for size, fish in marsh creeks may have lower mortality rates than fish in open bay areas. The estimates for mortality (0.0070 to 0.0208 d^{-1}) were the same in the open bay and in marsh creeks (Fig. 9, Deegan 1990), even though fish in the marsh creek were much smaller (20 to 50 mm) than fish in the open bay areas (50 to 100 mm). Because mortality is expected to decrease as fish get larger, we anticipated that mortality rate in the open bay areas should have been much lower than in marsh creeks. Mortality rates in the two habitats were similar despite the difference in fish size, implying that the marsh creek habitat provided some protection from mortality to the smaller fish. Unfortunately, direct comparison of the mortality rates of similar sized menhaden in the two habitats is not possible because use of these habitats by menhaden is size-related (Deegan 1990). However, lower mortality rates of nekton in salt marsh creeks compared to other estuarine habitats are not universal. Wilson et al. (1990) found no evidence for lower mortality of tethered juvenile blue crab (Fig. 10) in marsh creeks compared to seagrass or algal habitats.

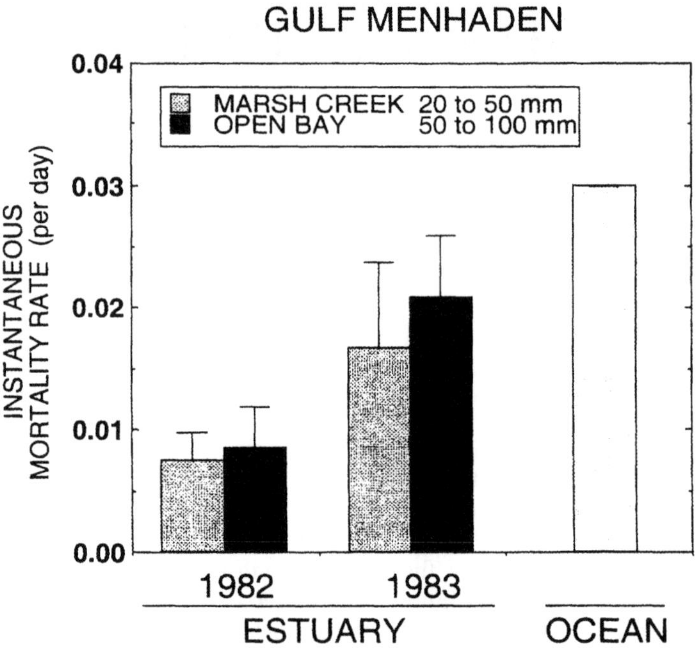

Figure 9. Size-dependent instantaneous mortality rates in two size classes of Gulf menhaden (+ SE) in a Louisiana salt marsh estuary, and the average value for juvenile marine fish ("OCEAN", from Gulland 1964). From Deegan (1990).

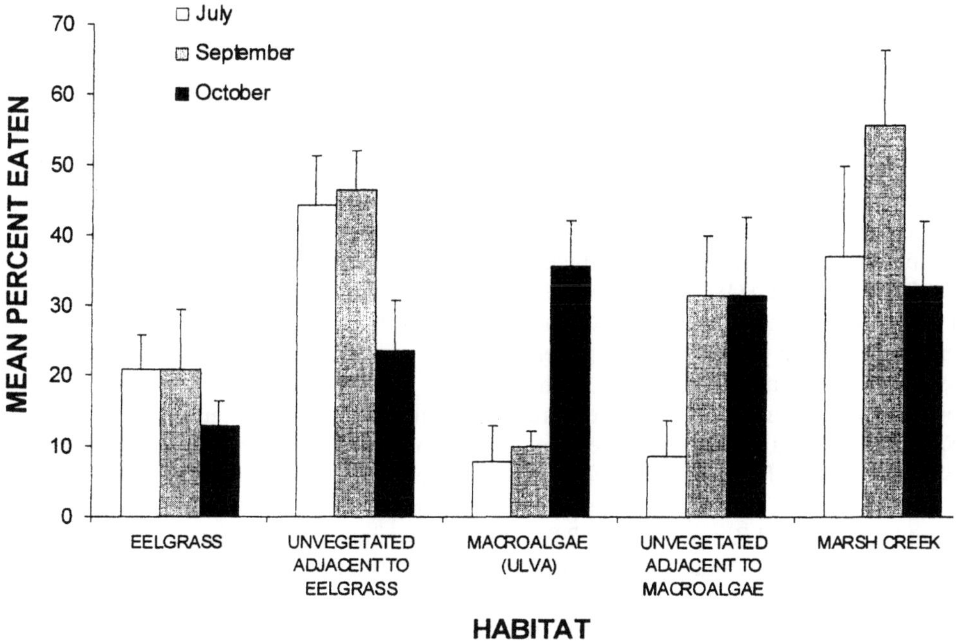

Figure 10. Mean percent (+1 SE) of tethered juvenile blue crabs eaten in eelgrass, adjacent unvegetated substrate, macroalgae (*Ulva lactuca*), and in a marsh creek. From Wilson et al. (1990).

Mortality rates are notoriously difficult to measure and there are very few estimates for estuarine fishes. Because of the paucity of data, it is difficult to make any broad conclusions about mortality rates in salt marsh creeks versus other habitats. The development of new mathematical models to estimate mortality based on mark-and-recapture studies (USFWS, May 1998; http://www.mbr.nbs.gov/software.html) as well as advances in tagging technology should allow progress in estimating population level mortality rates in the near future. Estimating the mortality rates of nekton is clearly an area of research that needs more work.

3.3 MECHANISMS GOVERNING MORTALITY RATES

Predation is an important source of mortality for small nekton. Mortality rates of brown shrimp in marshes declined from ~40% to 3% when predators were excluded, indicating that predators are a major source of mortality for brown shrimp (Minello et al. 1989). Weinstein and Walters (1981) found that mortality of spot was significantly higher in polyhaline creeks than in lower salinity creeks. The difference was attributed to the greater numbers of stenohaline marine predators seasonally occupying higher salinity marshes. Thus, we might expect mortality rates to vary with the attributes of salt marsh creeks. In other habitats, such as submerged aquatic vegetation, complex physical structure affords small fish some measure of protection from larger predators (e.g., Heck and Crowder 1991). It is uncertain if this hypothesis holds for salt marshes because nekton do not live continuously among marsh grass stems, but frequent the marsh and its creeks only during tidal inundation. Marine transient nekton often remain near the edges of salt marshes, even during spring tides, in contrast to marsh resident species such as killifish and palaemonetid shrimp (Murphy 1991, Kneib and Wagner 1994, Peterson and Turner 1994). The vegetative structure available at the marsh edge, the shallow depth, physico-chemical environment and high turbidity of marsh creeks have all been suggested as conveying to nekton protection from predation.

Marsh vegetation near the edges of creeks may provide some protection to transient nekton from predation by large piscine predators. The high densities of fish, penaeid shrimp and portunid crabs along the edges of flooded marsh creeks may be the result of predator avoidance (Weisberg et al. 1981, Murphy 1991, Rountree and Able 1993, Kneib 1997). Most manipulative studies of the effects of marsh vegetation in deterring predation have focused on predatory fish - invertebrate prey interactions (e.g., Vince et al. 1976, Minello and Zimmerman 1983, Minello et al. 1989), and have demonstrated the protective function of vegetated surfaces. In laboratory experiments, simulated *Spartina* vegetative structure reduced predation on brown shrimp by pinfish and Atlantic croaker from 2.5 to 1.5 shrimp d^{-1} (Minello and Zimmerman 1983). Vegetative structure did not, however, affect predation on brown shrimp by red drum or speckled trout (Minello and Zimmerman 1983). After many trials with several predatory species, Minello and Zimmerman found that overall mortality of brown shrimp was reduced by about 40% in marsh edge habitats compared to unvegetated habitat.

The relative importance of the marsh edge as a refuge from predation will vary with the physiography of the marsh (amount of edge), whether the marsh is subsiding or aggrading, and the tidal amplitude. For example, in a northern New England salt marsh that is subsiding relative to sea level rise, the marsh was flooded to a depth used by fishes substantially more often than the typical 25% estimate for New England marshes (Fig. 11; Murphy 1991). Subsiding marshes may provide a better habitat for marine transients than a typical New England salt marsh because fish have more frequent access to the marsh surface. Abundance of penaeid shrimp in some Gulf Coast marshes is greatest in fragmented *Spartina* marshes undergoing submergence, possibly as a result of greater marsh edge habitat (Zimmerman et al. 1984), or because of increased time to forage (Rozas and Reed 1993, Rozas 1995).

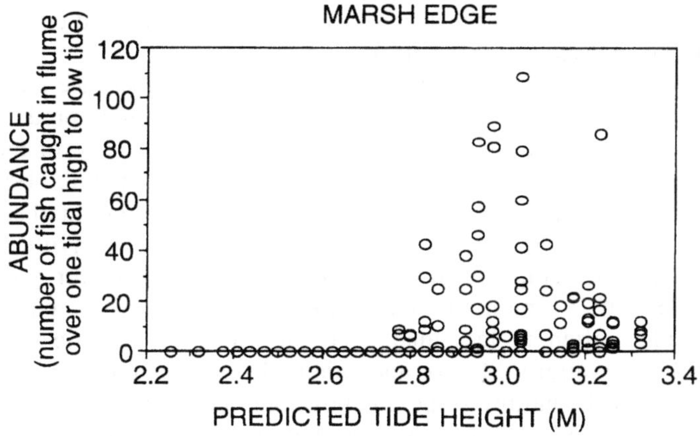

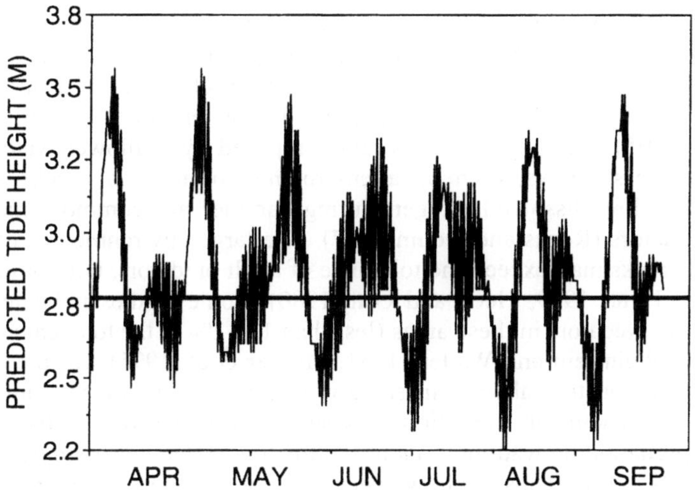

Figure 11. (Top panel) Total abundance of fishes captured in marsh edge flume nets during different tidal heights (m) in two Maine salt marsh estuaries, May-September. 1989. (Bottom panel) Predicted tidal heights (m) in these estuaries,. 1989. Horizontal line indicates the minimum height required to capture fish on the marsh surface. From Murphy (1991).

Occupying a habitat that provides protection from predation, even if for only short periods of time, can substantially increase the number of individuals surviving. For example, consider a nekton cohort with a population size of 10,000 when they enter the salt marsh ecosystem. This species uses the marsh ecosystem for 100 days (e.g., June, July and August) before moving to other habitats. Assume the fish experience a mortality rate of 0.04 d^{-1} in the creek and that by moving up into the vegetated marsh creek edge on high tides their mortality rate decreases by 40% to 0.025 d^{-1} . If the marsh edge is flooded enough to allow the fish access to the vegetated edge 25% of the 100 day period, 1.5 times more individuals will survive by moving into the marsh than by staying in the creek (260 survive with marsh access compared to 180 without marsh access). Obviously, the more time the species can spend in the flooded edge habitat, the greater the number of individuals that will survive.

The distribution of larval and juvenile nekton on the marsh surface and in shallow tidal creeks indicates that shallow-water habitat may also minimize predation. Lower predation pressure was offered as the explanation for the prevalence of molting blue crabs in tidal creeks versus deeper water (Hines et al. 1987). Ruiz et al. (1993) suggest greater predation risk in deep water as a factor causing the concentration of small nekton in shallow bay habitats. Shallow intertidal marsh puddles and small rivulets have been shown to offer a partial refuge from predation and cannibalism for young nekton (Kneib 1987, 1993, 1997). Small fish have often been observed to remain near the water's shallow edge, often in less than 1 cm of water, as the tide creek drained (Deegan, personal observation). McIvor and Odum (1988) found higher abundances of small fish associated with shallow-sloped, depositional creek banks rather than steep-sided, erosional banks. They attributed this distribution to the continuous availability of very shallow areas as the water ebbed in the depositional sites compared to the erosional edge. McIvor and Odum (1988) also found predation on tethered mummichogs was much less off shallow, depositional banks compared to steep-sided, erosional banks.

There are very few studies that examine the depth preferences of fishes, but in one study of hake preying on spot, hake avoided the shallow end of the experimental tank (Miltner et al. 1995). The physical constraints imposed by shallower waters, and the generally more stressful physico-chemical environment of intertidal creeks, where high temperatures and low dissolved oxygen during summer are common, may exclude piscivorous predators (Rozas and Odum 1987). The large environmental variations in shallow marsh creeks may exceed the tolerance of adult predators, but not juvenile fish (Cushing 1975, Hyatt 1979, Heck and Orth 1980). Some studies indicate few large (> 40 cm) piscine predators in these areas (less than 1 to 2% of the total catch; Bozeman and Dean 1980, Weinstein and Walters 1981, Miltner et al. 1995). On the other hand, marsh resident nekton that abound in shallow marsh habitats are tolerant of a wide range of physico-chemical conditions (Kneib 1987). However, fish are more susceptible to predation by wading birds in shallow marsh habitats compared to deeper areas (Kneib 1982, 1987). Thus, shallow water does not provide a perfect refuge from all predators.

Tidal creeks also have high turbidity that may provide protection to small nekton from predators by restricting the vision of predators. Although many senses are involved in feeding, sight is important in successful prey capture for most piscine predators (Nikolsky 1963, Hyatt 1979). The effects of turbidity on predator-prey interactions and feeding success have been researched in freshwater systems (e.g., Abrahams and Kattenfeld 1997), but relatively little work has been done in estuaries (Blaber and Blaber 1980). The effect of turbidity on predation rate depends on the specific predator, prey and even substratum of the habitat (Minello et al. 1987). The turbidity of tidal creeks may confer a survival advantage to juvenile nekton by reducing risk of predation by visual predators (Moore and Moore 1976, Cyrus and Blaber 1987, Hecht and van der Lingen 1992). Turbidity apparently reduced the perceived risk of predation in juvenile chinook salmon (Abrahams and Kattenfeld 1997). However, turbidity did not seem to affect the foraging efficiency of juvenile white perch (Monteleone and Houde 1992) or weakfish (Grecay and Targett 1996), and juvenile weakfish occur abundantly in turbid regions. The role of turbidity in regulating predator/ prey interactions is an area of research that needs much more attention.

The protection from predation for small nekton attributed to salt marshes may derive from a combination of the effects of these four factors (vegetative structure, shallow depth, physico-chemical environment, and turbidity). For example, at high tide the marsh structure may physically interfere with piscivorous fish attacks and at low tide the shallow depth of the water may exclude predators. However, it is important to remember that juvenile stages of some nekton species prey on other juvenile nekton in marsh creeks (Juanes et al. 1993, Miltner et al. 1995) and that large fish have been shown to move into some creeks at night (Rountree 1992). Given that the optimum length of a piscine predator is often about 4 times the length of its prey, most small fish (usually 1 to 10 cm in length) in tidal creeks would be vulnerable to a predator that was only 20 to 40 cm in length.

4. Migration as an Important Linkage among Habitats

One of the original hypotheses of estuarine function was that estuaries produce annual excesses of plant organic matter, some of which is exported seaward as detritus (dissolved or particulate) and supports coastal fisheries (Odum 1968). Current understanding is that detrital outwelling is not a universal phenomenon and that when it occurs the amount transported is often small and relatively refractory (Nixon 1980, Dame 1989). An alternative mechanism for the support of coastal fisheries by salt marshes is by the migration of nekton. Estuarine fish faunas around the world are dominated in numbers and biomass by species that move into the estuary as larvae, accumulate biomass, and then move offshore after attaining a large proportion of their adult size (e.g., Yáñez-Arancibia 1985). Other species make seasonal forays into estuarine systems to feed on the high production of estuarine fishes (e.g., Yáñez-Arancibia 1985). Although many authors have suggested that emigration of fish may export energy from estuaries (Bozeman and Dean 1980, Odum 1980, Weinstein et al. 1980, Wiegert and Pomeroy 1981, Currin et al. 1984, Deegan and Thompson 1985, Zijlstra 1988), the mechanisms of energy transfer and the quantitative estimate of its importance have not been well studied.

The horizontal movement of energy and nutrients from the salt marsh to adjacent habitats and ecosystems via sequential consumption and migration of nekton has been described as a "trophic relay" (Kneib 1997) or the "chain of migration" (Rountree 1992). This concept is similar to Vinogradov's "ladder of migration" hypothesis describing vertical transfer of energy from the photic to abyssal zones of the deep sea (Vinogradov 1953, 1955, Longhurst 1976, Rountree 1992). In much the same way as Vinogradov's ladder rungs link adjacent depth strata, the trophic relay results from a chain of migration that links adjacent habitats (Fig. 12). The first and most important coupling is that between the intertidal marsh and adjacent subtidal habitats. This primary coupling is established by the foraging activities of both permanent and seasonally resident marsh fauna. A secondary coupling is established by the activities of faunal assemblages using subtidal marsh creeks and the marsh fringe and the adjacent estuarine bays. A tertiary link couples estuarine bays with the continental shelf. In some areas auxiliary links might exist between the high marsh and the low intertidal marsh, and also between fully terrestrial habitats and the high marsh. Migration can also link habitats on other spatial scales, such as among freshwater, and upper and lower estuarine habitats.

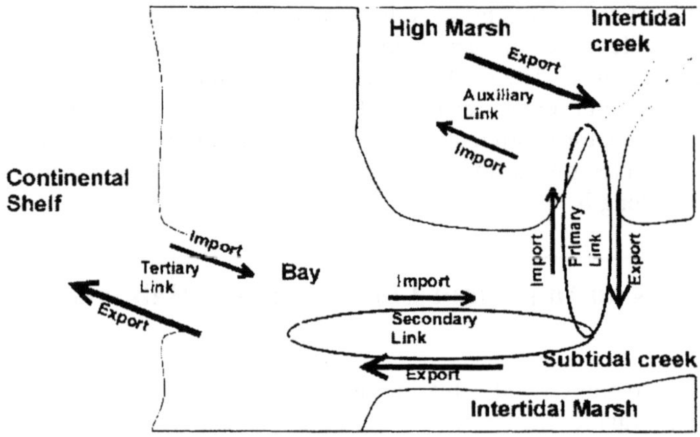

Figure 12. Nekton migration patterns that can result in energy or nutrient exchange among habitats. Trophic coupling between adjacent habitats results from seasonal and ontogenetic migrations, or tidal and diel foraging movements. Three primary couplings are hypothesized: a) a primary link between the intertidal marsh (both marsh surface and intertidal creek habitats) and adjacent subtidal marsh (subtidal creek and shallow subtidal bay habitats along the marsh fringe); b) a coupling between the subtidal marsh and adjacent estuarine bay; and c) a third link between the estuarine bay and continental shelf.

The trophic relay can operate on several different temporal scales (tidal, diel, and seasonal), and through two basic mechanisms: 1) passive diffusion, or 2) migration, either ontogenetic or cyclic foraging migration (Fig. 13). The first and simplest mechanism involves a "diffusion" of energy between overlapping communities through spatially overlapping trophic webs. In natural estuarine systems faunal assemblages are rarely sharply divided, but rather, they tend to grade into each other across habitat clines. It is reasonable to assume then, that these overlapping communities would share overlapping trophic webs. It is difficult, however, to envision how a net one-way export from the marsh could occur by this mechanism. Alternatively, active ontogenetic and cyclic migration patterns result in the direct transfer of energy between habitats through animal movements.

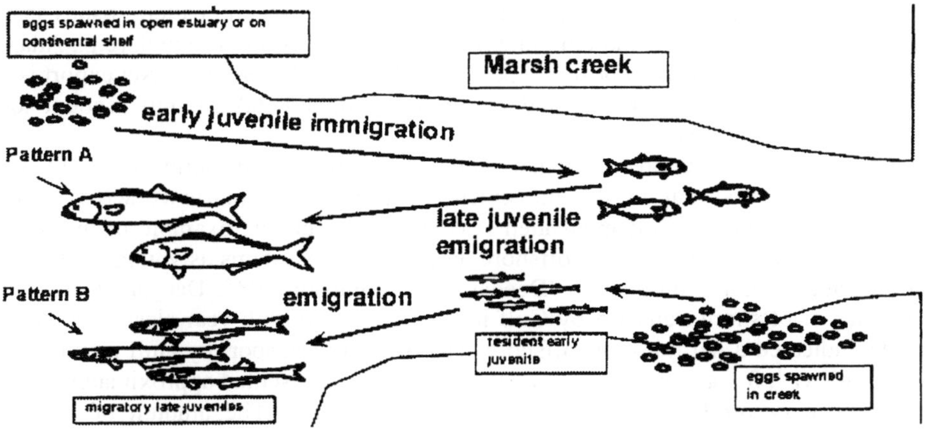

Figure 13. Patterns of trophic transfer of energy, derived from primary production on the marsh surface, through fish and invertebrate communities along an estuarine habitat cline. The simplest mechanism is through a diffusion of energy through the interconnecting food webs of spatially overlapping communities (e.g., estuarine bay, subtidal creek and marsh surface communities). Ontogenetic migration of species between adjacent habitats results in the direct export of energy assimilated in somatic growth. Cyclic foraging migrations of species between habitats also results in direct transfer of energy among the habitats through local consumption, fecal deposition, and mortality.

4.1 ONTOGENETIC MIGRATION

Ontogenetic migration involves successive shifts in the use of adjacent habitats/ ecosystems by different life stages of a given species (Fig. 13), and can occur on several scales: between habitats within the salt marsh, between the salt marsh and estuary and between the estuary and coastal waters.

Numerous species of nekton generally considered true marsh residents, actually are only resident during early larval and juvenile stages, or seasonally during the summer and subsequently move into adjacent subtidal areas with growth or colder temperatures (e.g., mummichog and shore shrimp, *Palaemonetes vulgaris*; Murphy 1991, Rountree 1992, Kneib 1997). Subtidal habitats serve as a low tide refuge for the older juveniles and adults from which they migrate tidally into the intertidal marsh to feed. The primary coupling between the subtidal creek and intertidal marsh is best exemplified by mummichog (Fig. 13). This species is well known to forage on the intertidal marsh during tidal inundation and retreat to subtidal areas with the tide (Baker-Dittus 1978, Weisberg et al 1981, Kneib 1984, 1987, Rountree and Able 1992a, 1993).

Numerous researchers have suggested that substantial biomass may be exported from estuaries as nekton undergo seasonal migrations into coastal waters (e.g., Smith 1966, Welsh 1975, Meredith and Lotrich 1979, Bozeman and Dean 1980, Weinstein and Walters 1981, Conover and Ross 1982, Deegan and Thompson 1985, Vouglitois et al. 1987, Zijlstra 1988, Rountree 1992, Kneib 1997). Two basic patterns of energy exportation through ontogenetic migrations of transient species have been suggested. In one pattern, fishes are spawned either in the open estuary (e.g., winter flounder) or on the continental shelf (e.g., Atlantic menhaden, summer flounder, bluefish, mullet) and recruit to salt marshes during larval or early juvenile stages. In a second pattern, fishes spawn within the intertidal marsh and emigrate offshore after a period of growth (e.g., Atlantic silverside; goby, *Gobiosoma bosc*; striped killifish, *Fundulus majalis*). Net export is the difference between immigrating and emigrating biomass, taking into account local mortality (Deegan 1993).

Deegan (1993) demonstrated that biotic transport by one marine transient, Gulf menhaden, is important in the movement of energy and nutrients across coastal ecosystem boundaries. Gulf menhaden is considered a classic example of the marine transient fishes that are spawned offshore, use marsh ecosystems as young-of-the-year, and depend on marsh production (Deegan and Thompson 1985, Deegan et al. 1990). Estimates of offshore transport by Gulf menhaden varied with year class strength, but always indicated a net transport offshore. The offshore transport by Gulf menhaden of 22 g C m^{-2} yr^{-1}, 3.1 g N m^{-2} yr^{-1}, and 0.9 g P m^{-2} yr^{-1} represented approximately 5-10% of the primary production of inshore coastal Louisiana. The amount of N and P transported by this single species was of the same magnitude as estimates for passive outwelling.

Many other marine transient species also have the potential to transport nutrients and energy. The Atlantic menhaden is the ecological equivalent of Gulf menhaden along the Atlantic coast, with very similar production characteristics. Penaeid shrimp, blue crab, Atlantic croaker, spot and other species have a migration pattern similar to that of Gulf menhaden, suggesting that a net export of energy is likely (Rountree 1992, Deegan 1993). Conover and Ross (1982) estimated that less than 1% of the young-of-the-year population of Atlantic silverside survived the winter to spawn the following spring, suggesting a large one-way export of biomass to the continental shelf. Current work at the Plum Island Sound LTER site in northern Massachusetts is examining the potential of energy and nutrient export by Atlantic silverside (Deegan, Wright and Hughes, in progress). The cumulative transport of 6 to 10 additional species may represent a significant energy source to the offshore ecosystem. Understanding the importance of the translocation of energy and nutrients by the full array of species that use estuaries as juveniles is critical to understanding the role of estuaries in supporting offshore fisheries.

4.2 CYCLIC FORAGING MIGRATIONS

Both tidal and diel cyclic foraging migrations (Miller and Dunn 1980, Rountree 1992, Rountree and Able 1993, 1996, 1997) may be important pathways of energy exchange between the intertidal marsh, subtidal creeks and the adjacent estuary (Fig. 14). The key to this energy transfer is the spatial separation of foraging and refuge habitats due to tides, or diel changes in physical conditions. Energy obtained in the foraging habitat would potentially be transferred to the refuge habitat by two primary mechanisms: 1) through mortality or consumption of the migrant by species resident within the refuge habitat, and 2) through fecal deposition while in the refuge habitat. Energy would be exported from the local linked system through subsequent ontogenetic migration and by predation during migration of other species (i.e., passed along to the next link in the trophic chain).

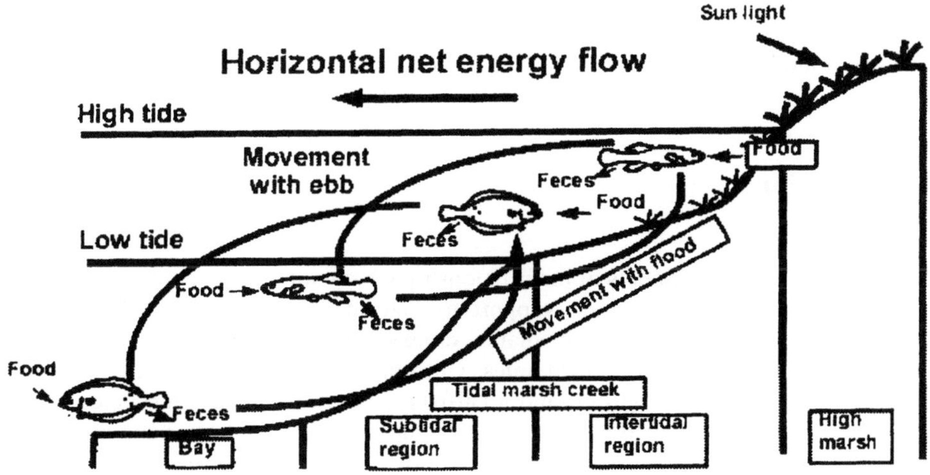

Figure 14. Hypothetical horizontal energy flow among intertidal marsh, subtidal marsh, and estuarine bay habitats through the trophic relay process. Foraging by *Fundulus heteroclitus* in the intertidal marsh results in a transfer of energy (fecal materials and mortality) to the subtidal marsh where the species retreats at low tide, thus establishing a primary coupling between subtidal and intertidal marsh habitats (see text). Predation on *F. heteroclitus* during the tidal foraging activity of *Paralichthys dentatus* results in an export of energy from the subtidal marsh to the adjacent bay establishing a secondary coupling between the bay and intertidal marsh.

Cyclic foraging linkages between the subtidal marsh creeks and adjacent bay habitat are illustrated by summer flounder. Late juveniles of summer flounder undergo regular tidal movements into New Jersey marsh creeks to feed (Rountree and Able 1992b). Individuals captured in gill nets on ebb tide (assumed to be leaving the creeks) had significantly higher gut fullness values than those captured on flood tide. Four marsh species, Atlantic silverside, shore shrimp, sand shrimp (*Crangon septemspinosa)* and mummichogs were the primary food items in summer flounder guts suggesting a close trophic link to the salt marsh. Fecal deposition and mortality while in the bay habitat result in an export of carbon from the subtidal creek to bay. Ultimately, energy derived from subtidal marsh foraging and incorporated into summer flounder biomass, would be exported from the system with seasonal migration of flounder onto the shelf. Hence, this one species is involved in both cyclic (feeding in marsh habitats) and ontogenetic (estuary to offshore) migration pathways of trophic relay, and serves as a vector linking habitats on two different spatial and temporal scales.

The transfer of energy from an area of foraging activity to another area through deposition of fecal material has not been examined in estuaries, although it has been recognized as important for nutrient-poor marine systems such as coral and artificial reefs (e.g., Bray et al. 1981, Meyer et al. 1983, Bray and Miller 1985, Meyer and Schultz 1985a, 1985b, Rountree 1990). It is not clear if this mechanism would be as significant in salt marshes where nutrients are usually high.

It is important to note that the couplings described above assume a two-way transfer rather than a one-way horizontal transfer of energy. We suggest, based on considerations of the balance of growth and mortality that the net transport via trophic coupling and migration is out of the marsh, but mechanisms of import into the marsh must be more fully understood before net export via trophic relay can be confirmed for all migrating species.

5. Conclusions

Although there are many aspects of the relationship we do not yet understand, salt marsh ecosystems apparently do provide support to marine transient fishes. The warmer temperatures of estuaries and salt marsh creeks apparently provide a metabolic advantage that supports high growth rates. Current evidence indicates that estuarine food webs are a mixture of detrital-and algal-based pathways. The importance of salt marsh production to marine transient fishes is supported by dietary, behavioral, and isotopic evidence. Salt marshes support fisheries directly in the case of species that use the habitat as a nursery (e.g., Atlantic and Gulf menhaden, mullet), and in the case of estuarine transients that use many estuarine habitats but derive energy from the salt marsh through trophic relay (e.g., summer flounder, bluefish, striped bass). The salt marsh also indirectly supports fisheries by exporting abundant potential prey species for coastal carnivores (e.g., Atlantic silverside, blue crab and sand shrimp). Unlike nekton, exported detritus is often of low nutritive value and may be rapidly deposited or respired by bacteria without entering the food web. Current evidence suggests that estuarine support for marine fisheries resulting from the direct export of fish biomass and a trophic relay involving ontogenetic and cyclic migrations of nekton species is greater than support via the export of organic detritus. Understanding the controls on marine transient fish mortality is probably the most problematic and least studied aspect of their ecology. The few estimates of mortality rates of fishes in estuaries indicate they are as high or higher than mortality rates of fishes in other marine and freshwater ecosystems. However, the higher growth rates of young fish in estuaries may ''overcompensate'' for mortality resulting in fish spending less time in vulnerable larval stages. This may result in a higher net production than if the fish had remained offshore.

The value of the marsh as a refuge is probably due to the interaction of temperature, turbidity, and vegetative structure in restricting the foraging of predators. We also know that not all marshes provide the same degree of support to marine transients. We expect the importance of marshes to nekton populations to vary with the availability of different organic matter sources, the geomorphology of the estuarine basin, the areal extent and configuration of the marsh, hydrographic features such as frequency and duration of flooding, relative magnitude of tidal range and freshwater input, and behavior of nekton.

At this point in our understanding of the requirements of marine transient fishes, there are several questions we cannot answer. Estuarine areas used by young-of-the-year and juvenile fishes tend to be shallow areas with or adjacent to structure, that have high levels of nutrients, primary production, and invertebrate food. Would the production of these fish be just as high in another habitat that had the same essential features? Could salt marshes, for instance, be replaced by artificial reefs and provide the same benefits to marine transient species? Do habitats with lower secondary production per unit area make an equal or greater contribution to total stock production because of greater areal extent? There are few comparisons of the production of the same species in different habitats within estuaries (e.g., Weinstein and Walters 1981, Weinstein and Brooks 1983) or in other coastal systems (Lenanton 1982) making these questions difficult to answer.

6. Acknowledgments

We thank Sara Wetmore for valuable assistance in the preparation of the manuscript. This work was supported by NSF (NSF-OCE 9214461), the Mellon Foundation, EPA (R825757-01-1) and Saltonstall-Kennedy grants.

7. Literature Cited

Able, K. W. and M. P. Fahay. 1998. The first year in the life of estuarine fishes in the Middle Atlantic Bight, Rutgers University Press, New Brunswick, New Jersey, USA.

Abrahams, M. and M. Kattenfeld. 1997. The role of turbidity as a constraint on predator-prey interactions in aquatic environments. Behavioral Ecology and Sociobiology **40**:169-174.

Alber, M. and I. Valiela. 1994. Production of microbial organic aggregates from macrophyte-derived dissolved organic material. Limnology and Oceanography **39**:37-50.

Allen, E. A., P. E. Fell, M. A. Peck, J. A. Gieg, C. R. Guthke and M. D. Newkirk. 1994. Gut contents of common mummichogs, *Fundulus heteroclitus* L., in a restored impounded marsh and in natural reference marshes. Estuaries **17**:462-471.

Baird, S. F. 1873. Report on the condition of the sea fisheries of the south coast of New England and in 1871 and 1872, Part 1, U. S. Commission of Fish and Fisheries, Washington, District of Columbia, USA.

Baker-Dittus, A. M. 1978. Foraging patterns of three sympatric killifish. Copeia **1978**:383-389.

Baltz, D. M., C. Rakocinski and J. W. Fleeger. 1993. Microhabitat use by marsh-edge fishes in a Louisiana estuary. Environmental Biology of Fishes **36**:109-126.

Benner, R., M. L. Fogel, E. K. Sprague and R. E. Hodson. 1987. Depletion of ^{13}C in lignin and its implications for stable carbon isotope studies. Nature **329**:708-710.

Bigelow, H. B. and W. C. Schroeder. 1953. Fishes of the Gulf of Maine, Fishery Bulletin of the Fish and Wildlife Service 53, U.S. Government Printing Office, Washington, District of Columbia, USA

Blaber, S. J. M. and T. G. Blaber. 1980. Factors affecting the distribution of juvenile estuarine and inshore fish. Journal of Fish Biology **17**:143-162.

Boesch, D. F. and R. E. Turner. 1984. Dependence of fishery species on salt marshes: the role of food and refuge. Estuaries **7**:460-468.

Bozeman, E .L., Jr. and J. M. Dean. 1980. The abundance of estuarine larval and juvenile fish in a South Carolina intertidal creek. Estuaries **3**:89-97.

Bray, R. N., A. C. Miller, and D. G. Geesey. 1981. The fish connection: a trophic link between planktonic and rock reef communities. Science **214**:204-205.

Bray, R. N. and A. C. Miller. 1985. Planktivorous fishes: their potential as nutrient importers to artificial reefs. Bulletin of Marine Science **37**:396

Cain, R. F. and J. M. Dean. 1976. Annual occurrence, abundance and diversity of fish in a South Carolina intertidal creek, Marine Biology **36**:369-379.

Cattrijsse, A., E. S. Makwaia, H. R. Dankwa, O. Hamerlynck and M. A. Hemminga. 1994. Nekton communities of an intertidal creek of a European estuarine brackish marsh. Marine Ecology Progress Series **109**:195-208.

Chrzanowski, T. H. and J. D. Spurrier. 1987. Exchange of microbial biomass between a *Spartina alterniflora* marsh and the adjacent tidal creek. Estuaries **10**:118-125.

Cifuentes, L. A. 1991. Spatial and temporal variations in terrestrially-derived organic matter from sediments of the Delaware Estuary. Estuaries **14**:414-429.
Coffin, R. B., B. Fry, B. J. Peterson and R. T. Wright. 1989. Carbon isotopic compositions of estuarine bacteria. Limnology and Oceanography **34**:1305-1310.
Conover, D. O. and M. R. Ross. 1982. Patterns in seasonal abundance, growth and biomass of the Atlantic silverside, *Menidia menidia*, in a New England estuary. Estuaries **5**:275-286.
Couch, C. A. 1989. Carbon and nitrogen stable isotopes of meiobenthos and their food resources. Estuarine, Coastal and Shelf Science **28**:433-441.
Currin, B. M., J. P. Reed and J. M. Miller. 1984. Growth, production, food consumption and mortality of juvenile spot and croaker: a comparison of tidal and non-tidal nursery areas. Estuaries **7**:451-459.
Currin, C. A., S. Y. Newell and H. W. Paerl. 1995. The role of standing dead *Spartina alterniflora* and benthic microalgae in salt marsh food webs: considerations based on multiple stable isotope analysis. Marine Ecology Progress Series **121**:99-116.
Cushing, D. R. 1975. Marine ecology and fisheries. Cambridge University Press, Cambridge, England.
Cyrus, D. P. and S. J. M. Blaber. 1987. The influence of turbidity on juvenile marine fish in the estuaries of Natal, South Africa. Continental Shelf Research **7**:1411-1416.
D'Avanzo, C., M. Alber and I. Valiela. 1991. Nitrogen assimilation from amorphous detritus by two coastal consumers. Estuarine, Coastal and Shelf Science **33**:203-209.
Dame, R. F. 1989. The importance of *Spartina alterniflora* to Atlantic coast estuaries, Reviews in Aquatic Sciences **1**:639-660.
Darnell, R. M. 1961. Trophic spectrum of an estuarine community, based on studies of Lake Pontchartrain, Louisiana. Ecology **42**:553-568.
Day, J. W. Jr, C. A. S. Hall, W. M. Kemp and A. Yanez-Arancibia. 1989. Estuarine Ecology. John Wiley and Sons, New York, New York, USA.
Deegan, L. A. 1990. Effects of estuarine environmental conditions on population dynamics of young-of-the-year gulf menhaden. Marine Ecology Progress Series **68**:195-205.
—— 1993. Nutrient and energy transport between estuaries and coastal marine ecosystems by fish migration. Canadian Journal of Fisheries and Aquatic Sciences **50**: 74-79.
Deegan, L. A. and B. A. Thompson. 1985. The ecology of fish communities in the Mississippi River deltaic plain. Pages 35-56 *in* A. Yáñez-Arancibia, editor. Fish community ecology in estuaries and coastal lagoons: towards an ecosystems integration. DR (R) UNAM Press, Mexico City, Mexico.
Deegan, L. A. and Day, J. W., Jr. (1984) Estuarine fishery habitat requirements. Pages 315-336 *in* B. J Copeland, K. Hart, N. Davis and S. Friday, editors. Research for managing the nation's estuaries: proceedings of a conference in Raleigh, North Carolina, UNC Sea Grant College Publication UNC-SG-**84-08**, Raleigh, North Carolina, USA.
Deegan, L. A. and R. H. Garritt. 1997. Evidence for spatial variability in estuarine food webs. Marine Ecology Progress Series **147**:31-47.
Deegan, L. A., W. B. Johnson, J. W. Day, Jr., J. G. Gosselink and A. Yáñez-Arancibia. 1983. Relationships between primary productivity and physical characteristics in eight Gulf of Mexico estuaries. Final report to the Office of Marine Pollution, CELFI, Louisiana State University, Baton Rouge, Louisiana, USA.
Deegan, L. A., B. J. Peterson and R. Porter. 1990. Stable isotopes and cellulase activity as evidence for detritus as a food source for juvenile gulf menhaden. Estuaries **13**:14-19.
DeLaune, R. D. and C. W. Lindau. 1987. $\delta^{13}C$ signature of organic carbon in estuarine bottom sediment as an indicator of carbon export from adjacent marshes. Biogeochemistry **4**:225-230.
Ember, L. M., D. F. Williams and J. T. Morris. 1987. Processes that influence carbon isotope variations in salt marsh sediments. Marine Ecology Progress Series **36**:33-42.
Evans, D. 1998. The Physiology of Fishes. CRC Press, Boca Raton, Florida, USA.
Feller, R. J., B. C. Coull and B. T. Hentschel. 1990. Meiobenthic copepods: tracers of where juvenile *Leiostomus xanthurus* (Pisces) feed? Canadian Journal of Fisheries and Aquatic Sciences **47**:1913-1919.
Fogel, M. L., E. K. Sprague, A. P. Gize and R. W. Frey. 1989. Diagenesis of organic matter in Georgia salt marshes. Estuarine, Coastal and Shelf Science **28**:211-230.
Fry, B., M. Hullar, B. J. Peterson, S. Saupe and R. T. Wright. 1992. DOC production in a salt marsh estuary, Archives fur Hydrobioliske Beih. **37**:1-8.
Grecay, P. A. and T. E. Targett. 1996. Effects of turbidity, light level and prey concentration on feeding of juvenile weakfish *Cynoscion regalis*. Marine Ecology Progress Series **131**:11-16.

Gregg, J. C. and J. W. Fleeger Jr. 1997. Importance of emerged and suspended meiofauna to the diet of the darter goby (*Gobionellus boleosoma* Jordan and Gilbert). Journal of Experimental Marine Biology and Ecology **209**:123-142.

Gulland, J. A. 1964. Survival of the youngest stages of fish and its relation to year class strength. ICNAF Special Publication **6**:363-372.

Hackney, C. T. and E. B. Haines. 1980. Stable carbon isotope composition of fauna and organic matter collected in a Mississippi estuary. Estuarine and Coastal Marine Science **10**:703-708.

Haedrich, R. L. 1983. Estuarine fishes. Pages 183-207 *in* B. H. Ketchum, editor. Ecosystems of the world. Elsevier, Amsterdam, The Netherlands.

Haines, E. B. 1977. The origins of detritus in Georgia salt marsh estuaries. Oikos **29**:254- 260.

—— 1979a. Interactions between Georgia salt marshes and coastal waters: a changing paradigm. Pages 35-46 *in* R. J. Livingston, editor. Ecological processes in coastal and marine systems. Plenum Press, New York, New York, USA.

—— 1979b. Nitrogen pools in Georgia coastal waters. Estuaries **2**:34-39.

Haines, E. B. and C. L. Montague. 1979. Food sources of estuarine invertebrates analyzed using $^{13}C/^{12}C$ ratios. Ecology **60**:48-56.

Hecht, T. and C. D. vander Lingen. 1992. Turbidity-induced changes in feeding strategies of fish in estuaries. South African Journal of Zoology **27**:95-107.

Heck, K. L. and L. B. Crowder. 1991. Habitat structure and predator-prey interactions in vegetated aquatic systems. Pages 281-299 *in* S. S. Bell, E. D. McCoy and H. R. Mushinsky, editors. Habitat structure. Chapman and Hall, London, England.

Heck, K.L. and R J. Orth. 1980. Sea grass habitats: the role of habitat complexity, competition and predation in structuring associated fish and motile invertebrate assemblages. Pages 449-464 *in* V.S. Kennedy, editor. Estuarine perspectives. Academic Press, New York, New York, USA.

Heinle, D. R. and D. A. Flemer. 1976. Flows of materials between poorly flooded tidal marshes and an estuary. Marine Biology **35**:359-373.

Heinle, R. R., R. P. Harris, J. F. Ustach and D. A. Flemer. 1977. Detritus as food for estuarine copepods. Marine Biology **40**:341:353.

Hemminga, M. A., A. Cattrijsse and A. Wielemaker. 1996. Bedload and nearbed detritus transport in a tidal saltmarsh creek. Estuarine, Coastal and Shelf Science **42**:55-62.

Hines, A. H., R. N. Lipcius and A. M. Haddon. 1987. Predation dynamics and habitat partitioning by size, sex and molt stage of blue crabs *Callinectes sapidus* in a subestuary of central Chesapeake Bay. Marine Ecology Progress Series **36**:55-64.

Hines, A. H., A. M. Haddon and L. A. Wiechert. 1990. Guild structure and foraging impact of blue crabs and epibenthic fish in a subestuary of Chesapeake Bay. Marine Ecology Progress Series **67**:105-126.

Hoar, W. S., D. J. Randell and J. R. Brett. 1979. Fish physiology, Vol. III. Bioenergetics and growth. Academic Press, New York, New York, USA.

Holmes, R. M., B. J. Peterson, L. A. Deegan, J. E. Hughes and B Fry (2000). Nitrogen biogeochemistry in the oligohaline zone of a New England estuary. Ecology.

Houde, E. and E. Rutherford. 1993. Recent trends in estuarine fisheries: predictions of fish production and yield. Estuaries **16**:161-176.

Houde, E. D. and C. E. Zastrow. 1993. Ecosystem- and taxon-specific dynamic and energetics properties of larval fish assemblages. Bulletin of Marine Science **53**: 290-329.

Howarth, R. W. and J. M. Teal. 1980. Energy flow in a salt marsh ecosystem: the role of reduced inorganic sulfur compounds. American Naturalist **116**:862-872.

Hughes, E. H. and E. B. Sherr. 1983. Subtidal food webs in a Georgia estuary: $\delta^{13}C$ analysis. Journal of Experimental Marine Biology and Ecology **67**:227-242.

Hughes, J. E., L. A. Deegan, B. J. Peterson, R. M. Holmes and B. Fry (2000). Nitrogen flow through the food web in the oligohaline zone of a New England estuary. Ecology **81**:433-452.

Hyatt, K. D. 1979. Feeding strategy. Pages 71-119 *in* W.S. Hoar, D. J. Randell and I.R. Brett, editors. Biogenetics and growth. Volume XI, Academic Press, New York, New York, USA.

Javonillo, R., L. A Deegan, K. Chiaravalle and J. E. Hughes. 1997. The importance of access to salt-marsh surface to short-term growth of *Fundulus heteroclitus* in a New England salt marsh. Biological Bulletin **193**:288-289.

Joseph, E. B. 1973. Analysis of a nursery ground. Pages 118-121 *in* A. L. Pacheco, editor. Proceedings of a workshop on egg, larval and juvenile stages of fish in Atlantic Coast estuaries. Technical Publication No. **1**, National Oceanic and Atmospheric Administration, National Marine Fisheries Service, Mid-Atlantic Coast Fisheries Center, Highlands, New Jersey, USA.

Juanes, F., R. E. Marks, K. A. McKown and D. O. Conover. 1993. Predation by age-0 bluefish on age-0 anodromous fishes in the Hudson River estuary. Transactions of the American Fisheries Society **122**:348-356.

Kneib, R. T. 1982. The effects of predation by wading birds (Areidae) and blue crabs (*Callinectes sapidus*) on the population size structure of the common mummichog, *Fundulus heteroclitus*. Estuarine, Coastal and Shelf Science **14**:159-165.

—— 1984. Patterns of invertebrate distribution and abundance in the intertidal salt marsh: causes and questions. Estuaries 7:392-412.

—— 1986. The role of *Fundulus heteroclitus* in salt marsh trophic dynamics. American Zoologist **26**:259-269.

—— 1987. Predation risk and use of intertidal habitats by young fish and shrimp. Ecology **68**:379-386.

—— 1988. Testing for indirect effects of predation in an intertidal soft-bottom community. Ecology **69**:1795-1805.

—— 1993. Growth and mortality in successive cohorts of fish larvae within an estuarine nursery. Marine Ecology Progress Series **94**:115-127.

—— 1997. The role of tidal marshes in the ecology of estuarine nekton. Pages 163-220 *in* A. D. Ansell, R. N. Gibson and M. Barnes, editors. Oceanography and Marine Biology: An Annual Review. **35**:163-220.

Kneib, R. T., A. E. Stiven and E. B. Haines. 1980. Stable carbon isotope ratios in *Fundulus heteroclitus* (L.) muscle tissue and gut contents from a North Carolina *Spartina* marsh. Journal of Experimental Marine Biology and Ecology **46**:89-98.

Kneib, R. T. and S. L. Wagner. 1994. Nekton use of vegetated marsh habitats at different stages of tidal inundation. Marine Ecology Progress Series **106**:227-238.

Larson, E. T. and A. L. Shanks. 1996. Consumption of marine snow by two species of juvenile mullet and its contribution to their growth. Marine Ecology Progress Series **130**:19-28.

Lenanton, R. C. J. 1982. Alternative nonestuarine nursery habitats for some commercially and recreationally important fish species of southwestern Australia. Australian Journal of Marine and Freshwater Research **33**:881-900.

Lewis, V. P. and D. S. Peters. 1994. Diet of juvenile and adult Atlantic menhaden in estuarine and coastal habitats. Transactions of the American Fisheries Society **123**:803-810.

Longhurst, A. R. 1976. Vertical migration. Pages 116-137 *in* D. H. Cushing and J. J. Walsh, editors. The ecology of the seas. Oxford Blackwell Scientific Publications, London, England.

Mallin, M. A., J. M. Burkholder and M. J. Sullivan. 1992. Contributions of benthic microalgae to coastal fishery yield. Transactions of the American Fisheries Society **121**:691-693.

Mann, K. H. 1988. Production and use of detritus in various freshwater, estuarine and coastal marine ecosystems. Limnology and Oceanography **33**:910-930.

McIvor, C. C. and W. E. Odum. 1988. Food, predation risk and microhabitat selection in a marsh fish assemblage. Ecology **69**:1341-1351.

Meredith, W. H. and V. A. Lotrich. 1979. Production dynamics of a tidal creek population of *Fundulus heteroclitus* (L.). Estuarine, Coastal and Marine Science 7:99-118.

Meyer, J. L. and E. T. Schultz 1985a. Tissue condition and growth rate of corals associated with schooling fish. Limnology and Oceanography **30**:157-166.

—— 1985b. Migrating haemulid fishes as a source of nutrients and organic matter on coral reefs. Limnology and Oceanography **30**:146-156.

Meyer, J. L., E. T. Schultz and G. S. Helfman. 1983. Fish schools: an asset to corals. Science **220**:1047-1049.

Miller, J. M. and M. L. Dunn. 1980. Feeding strategies and patterns of movement in juvenile estuarine fishes. Pages 437-448 *in* V. S. Kennedy, editor. Estuarine perspectives. Academic Press, New York, New York, USA.

Miltner, R. J., S. W. Ross and M. H. Posey. 1995. Influence of food and predation on the depth distribution of juvenile spot (*Leiostomus xanthurus*) in tidal nurseries. Canadian Journal of Fisheries and Aquatic Sciences **52**:971-982.

Minello, T. J. and R. J. Zimmerman. 1983. Fish predation on juvenile brown shrimp, *Penaeus aztecus* Ives: the effect of simulated *Spartina* structure on predation rates. Journal of Experimental Marine Biology and Ecology **72**:211-231.

Minello, T. J., R. J. Zimmerman and E. X. Martinez. 1987. Fish predation on juvenile brown shrimp, *Penaeus aztecus* Ives: effects of turbidity and substratum on predation rates. Fishery Bulletin, US **85**:59-70.

—— 1989. Mortality of young brown shrimp *Penaeus aztecus* in estuarine nurseries. Transactions of the American Fisheries Society **118**:693-708.
Montague, C. L., S. M. Bunker, E. B. Haines, M. L. Pace and R. L. Wetzel. 1981. Aquatic macroconsumers. Pages 69-85 *in* L. R. Pomeroy and R. G. Wiegert, editors. The Ecology of a salt marsh. Springer-Verlag, New York, New York, USA.
Monteleone, D. M. and E. D. Houde. 1992. Vulnerability of striped bass *Morone saxatilis* Waldbaum eggs and larvae to predation by juvenile white perch *Morone americana* Gmelin. Journal of Experimental Marine Biology and Ecology **158**:93-104.
Moore, J. W. and J. A. Moore. 1976. The basis of food selection in flounders, *Platichthys flesus* (L.) in the Severn estuary. Journal of Fish Biology **9**:139-156.
Murphy, S. C. 1991. The ecology of estuarine fishes in southern Maine high salt marshes: access corridors and movement patterns. Thesis, University of Massachusetts, Amherst, Massachusetts, USA.
Newell, R. I. E. and C. J. Langdon. 1986. Digestion and absorption of refractory carbon from the plant *Spartina alterniflora* by the oyster *Crassostrea virginica*. Marine Ecology Progress Series **34**:105-115.
Nikolsky, G.V. 1963. The ecology of fishes, Academic Press, New York, New York, USA, 352 p.
Nixon, S. W. 1980. Between coastal marshes and coastal waters—a review of twenty years of speculation and research on the role of salt marshes in estuarine productivity and water chemistry. Pages 437-525 *in* P. Hamilton and K. B. Macdonald, editors. Estuarine and wetland processes. Plenum Press, New York, New York, USA.
Nixon, S. W. and C. Oviatt. 1973. Ecology of a New England salt marsh. Ecological Monographs **43**:463-498.
Odum, E. P. 1968. A research challenge: evaluating the productivity of coastal and estuarine waters. Proceedings of the 2nd Sea Grant Conference, Graduate School of Oceanography, University of Rhode Island, Kingston, Rhode Island, USA.
—— 1980. The status of three ecosystem level hypotheses regarding salt marshes: tidal subsidy, outwelling and the detritus based food chain. Pages 485-496 *in* V. S. Kennedy, editor. Estuarine perspectives. Academic Press, New York, New York, USA.
Odum, W. E., J. S. Fisher and J. C. Pickral. 1979. Factors controlling the flux of particulate organic carbon from estuarine wetlands. Pages 69-80 *in* R. J. Livingston, editor. Ecological processes in coastal and marine systems. Plenum Press, New York, New York, USA.
Pearcy, W. G. 1962. Ecology of an estuarine populations of winter flounder (*Psuedopleronectes amaericanus*). Bulletin of the Bingham Oceanographic Collection **18**:39-64.
Peters, D. S. and W. E. Schaaf. 1981. Food requirements and sources for juvenile Atlantic menhaden. Transactions of the American Fisheries Society **110**:317-324.
Peterson, B. J. and R. W. Howarth. 1987. Sulfur, carbon and nitrogen isotopes used to trace organic matter flow in the salt-marsh estuaries of Sapelo Island, Georgia. Limnology and Oceanography **32**:1195-1213.
Peterson, B. J. and B. Fry. 1987. Stable isotopes in ecosystem studies. Annual Review of Ecological Systems **18**:293-320.
Peterson, B. J., B. Fry, M. Hullar, S. Saupe and R. Wright. 1994. The distribution and stable carbon isotopic composition of dissolved organic carbon in estuaries. Estuaries **17**:111-121.
Peterson, B. J., R. W. Howarth and R. H. Garritt. 1985. Multiple stable isotope used to trace the flow of organic matter in estuarine food webs. Science **227**:1361-1363.
——. 1986. Sulfur and carbon isotopes as tracers of salt-marsh organic matter flow. Ecology **67**:865-874.
Peterson, B. J., R. W. Howarth, F. Lipschultz and D. Ashendorf. 1980. Salt marsh detritus: an alternative interpretation of stable carbon isotope ratios and the fate of *Spartina alterniflora*. Oikos **34**:1730-177.
Peterson, G. W. and R. E. Turner. 1994. The value of salt marsh edge vs. interior as a habitat for fish and decapod crustaceans in a Louisiana tidal marsh. Estuaries **17**:235-262.
Pomeroy, L. R. and R. G. Wiegert, editors. 1981. The Ecology of a salt marsh. Springer-Verlag Inc., New York, New York, USA.
Ribelin, B. W. and A. W. Collier. 1979. Ecological considerations of detrital aggregates in the salt marsh. Pages 47-68 *in* R. J. Livingston, editor. Ecological processes in coastal and marine systems. Plenum Press, New York, New York, USA.
Rountree, R. A. 1990. Community structure of fishes attracted to shallow water fish aggregation devices off South Carolina, USA. Environmental Biology of Fishes **29**:241-162.

——. 1992. Fish and macroinvertebrate community structure and habitat use patterns in salt marsh creeks of southern New Jersey, with a discussion of marsh carbon export. Dissertation, Rutgers, The State University, New Brunswick, New Jersey, USA.
Rountree, R. A. and K. W. Able 1992a. Fauna of polyhaline subtidal marsh creeks in southern New Jersey: composition, abundance and biomass. Estuaries **15**:171-185.
——. 1992b. Foraging habits, growth and temporal patterns of salt marsh creek habitat use by juvenile summer flounder in New Jersey. Transactions of the American Fisheries Society **121**:765-776.
——. 1993. Diel variation in decapod crustacean and fish assemblages in New Jersey marsh creeks. Estuarine, Coastal and Shelf Science **37**:181-201.
——. 1996. Seasonal abundance, growth and foraging habits of juvenile smooth dogfish, *Mustelus canis*, in a New Jersey estuary. Fishery Bulletin **94**:522-534.
——. 1997 Nocturnal fish use of New Jersey marsh creek and adjacent bay habitats. Estuarine, Coastal and Shelf Science **44**:703-711.
Rozas, L. P. 1995. Hydroperiod and its influence on nekton use of the salt marsh: a pulsing ecosystem. Estuaries **18**:579-590.
Rozas, L. P. and D. J. Reed. 1993. Nekton use of marsh-surface habitats in Louisiana (USA) deltaic salt marshes undergoing submergence. Marine Ecology Progress Series **96**:147:157.
Rozas, L. P. and M. W. LaSalle. 1990. A comparison of the diets of Gulf killifish, *Fundulus grandis* Baird and Girard, entering and leaving a Mississippi brackish marsh. Estuaries **13**:332-336.
Rozas, L. P. and W. E. Odum. 1987. Use of tidal freshwater marshes by fishes and macrofaunal crustaceans along a marsh stream-order gradient. Estuaries **10**:36-43.
Ruiz, G. M., A. H. Hines and M. H. Posey. 1993. Shallow water as a refuge habitat for fish and crustaceans in non-vegetated estuaries: an example from Chesapeake Bay. Marine Ecology Progress Series **99**:1-16.
Schaaf, W. E. and I. S. Peters. 1992. Reply to Mallin et al. (1992). Transactions of the American Fisheries Society **121**:693-695.
Sherr, E. B. 1982. Carbon isotope composition of organic seston and sediments in a Georgia salt marsh estuary. Geochimica et Cosmochimica Acta **46**:1227-1232.
Smith, M. W. 1966. Amount of organic matter lost to a lake by migration of eels. Journal of Fisheries Research Board of Canada **23**:1799-1801.
Sogard, S. M. and K. W. Able. 1991. A comparison of eelgrass, sea lettuce, macroalgae and marsh creeks as habitats for epibenthic fishes and decapods. Estuarine, Coastal and Shelf Science **33**:501-519.
Stribling, J. M. and J. C. Cornwell 1997. Identification of important primary producers in a Chesapeake Bay tidal creek system using stable isotopes of carbon and sulfur. Estuaries **20**:77-85.
Sullivan, M. J. and C. A. Moncreiff. 1990. Edaphic algae are an important component of salt marsh food-webs: evidence from multiple stable isotope analyses. Marine Ecology Progress Series **62**:149-159.
Teal, J. M. 1962. Energy flow in the salt marsh ecosystem of Georgia. Ecology **43**:614-624.
Tenore, K. R., L. Cammen, S. E. G. Findlay and N. Phillips. 1982. Perspectives of research on detritus: do factors controlling the availability of detritus to macroconsumers depend on its source? Journal of Marine Research **40**:473-490.
Thayer, G. W. 1974. Identity and regulation of nutrients limiting phytoplankton production in the shallow estuaries near Beaufort, NC. Oecologia **14**:75-92.
Turner, R. E. 1978. Community plankton respiration in a salt marsh estuary and the importance of macrophytic leachates. Limnology and Oceanography **23**:442-451.
Valiela, I. 1995. Marine ecological processes. Springer-Verlag, New York, New York, USA.
Valiela, I., J. M. Teal, S. Volkmann, D. Shafer and E. J. Carpenter. 1978. Nutrient and particulate fluxes in a salt marsh ecosystem: tidal exchanges and inputs by precipitation and groundwater. Limnology and Oceanography **23**:798-812.
Vince, S., I. Valiela and N. Backus. 1976. Predation by the salt marsh killifish *Fundulus heteroclitus* (L.) in relation to prey size and habitat structure: consequences for prey distribution and abundance. Journal of Experimental Marine Biology and Ecology **23**:255-266.
Vinogradov, M. E. 1953. The role of vertical migration of the zooplankton in the feeding of deep-sea animals. Priroda, Musk **6**:95-96.
——. 1955. Vertical migrations of zooplankton and their importance for the nutrition of abyssal pelagic faunas. Trudy Inst. Okeanol. USSR **13**:71-76.
Virnstein, R .W. 1977. The importance of predation by crabs and fishes on benthic infauna in Chesapeake Bay. Ecology **58**:1199-1217.

Vouglitois, J. J., K. W. Able, R. J. Kurtz and K. A. Tighe. 1987. Life history and population dynamics of the bay anchovy in New Jersey. Transactions of the American Fisheries Society **116**:141-153.
Walters, K., E. Jones and L. Etherington. 1996. Experimental studies of predation on metazoans inhabiting *Spartina alterniflora* stems. Journal of Experimental Marine Biology and Ecology **195**:251-265.
Weinstein, M. P. 1979 Shallow marsh habitats as primary nurseries for fishes and shellfish, Cape Fear River, North Carolina. Fisheries Bulletin **77**:339-357.
—— 1983. Population dynamics of an estuarine dependent fish, *Leiostomus xanthurus*, along a tidal creek-seagrass meadow coenocline. Canadian Journal of Fisheries and Aquatic Sciences **40**:1633-1638.
Weinstein, M. P. and H. A. Brooks. 1983. Comparative ecology of nekton residing in a tidal creek and adjacent seagrass meadow community composition and structure. Marine Ecology Progress Series **12**:15-28.
Weinstein, M. P. and M. F. Walters. 1981. Growth, survival and production in young-of-the-year populations of *Leiostomus xanthurus* Lacepede residing in tidal creeks. Estuaries **4**:185-197.
Weinstein, M. P., S. L. Weiss and M. F. Walters. 1980. Multiple determinants of community structure in shallow marsh habitats, Cape Fear River estuary, North Carolina, USA. Marine Biology **58**:227-243.
Weisberg, S. B. and V. A. Lotrich. 1982. The importance of an infrequently flooded intertidal marsh surface as an energy source for the mummichog *Fundulus heteroclitus*: an experimental approach. Marine Biology **66**:307-310.
Weisberg, S. B., R. Whalen and V. A. Lotrich. 1981. Tidal and diurnal influence on food consumption of a salt marsh killifish, *Fundulus heteroclitus*. Marine Biology **61**:243-246.
Welsh, B. L. 1975. The role of grass shrimp, *Palaemonetes pugio*, in a tidal marsh ecosytem. Ecology **56**:513-530.
Wiegert, R. G. and L. R. Pomeroy. 1981. The salt marsh ecosystem: a synthesis. Pages 219-230 *in* L. R. Pomeroy and R. G. Wiegert, editors. The Ecology of a salt marsh. Springer-Verlag, New York, New York, USA.
Wilson, J. O., I. Valiela and T. Swain. 1985. Sources and concentrations of vascular plant material in sediments of Buzzards Bay, Massachusetts, USA. Marine Biology **90**:12-137.
Wilson, K. A., K. W. Able and K. L. Heck, Jr. 1990. Predation rates on juvenile blue crabs in estuarine nursery habitats: evidence for the importance of macroalgae (*Ulva lactuca).* Marine Ecology Progress Series **58**:243-251.
Wiltse, W. I., K. H. Foreman, J. M. Teal and I. Valiela. 1984. Effects of predators and food resources on the macrobenthos of salt marsh creeks. Journal of Marine Research **42**:923-942.
Yáñez-Arancibia, A., editor. 1985. Fish community ecology in estuaries and coastal lagoons: towards an ecosystem integration. DR (R) UNAM Press, Mexico City, Mexico.
Zagursky, G. and R. J. Feller. 1985. Macrophyte detritus in the winter diet of the estuarine mysid, *Neomysis americana*. Estuaries **8**:355-362.
Zijlstra, J. J. 1988. Fish migration between coastal and offshore areas. Pages 257-272 *in* B. O. Jansson, editor. Coastal-offshore ecosystem interactions. Springer-Verlag, Berlin, Germany.
Zimmerman, R. J., T. J. Minello and G. Zamora, Jr. 1984. Selection of vegetated habitat by brown shrimp, *Penaeus aztecus*, in a Galveston Bay salt marsh. Fishery Bulletin, US **82**:325-336.
Zimmerman, R. J., T. J. Minello, M. C. Castiglione and D. L. Smith. 1990. Utilization of marsh and associated habitats along a salinity gradient in Galveston Bay. NOAA Technical Memorandum NMFS-SEFC-250.

BIOGEOCHEMICAL PROCESSES

BENTHIC-PELAGIC COUPLING IN MARSH-ESTUARINE ECOSYSTEMS

RICHARD F. DAME
ERIC KOEPFLER
LEAH GREGORY
Coastal Carolina University
Conway, SC 29528 USA

Abstract

Active and passive mechanisms utilized by many organisms in marsh-estuarine ecosystems couple the water column to the bottom. These linkages are often engineered by dense populations of plants (marshes) or animals (beds and reefs) that use their organismic structure, i.e., bodies or shells, and functional processes, i.e., water pumping, suspension feeding, etc., to enhance the movement of materials between the two habitats. These adaptations to the benthic boundary layer result in organismically mediated fluxes of materials between the water and the bottom that may dramatically alter either or both habitats. Dense stiff blades of grass dominate the salt marsh component of marsh-estuarine ecosystems. This structure ensures low water flow, low shear velocities, high drag and high roughness at the benthic boundary. These physical factors allow molecular diffusion and sedimentation to dominate exchange mechanisms. Marsh mussels magnify benthic-pelagic coupling by their active pumping and filtration of water. The shells of bivalve beds form a rough benthic surface that enhances turbulent mixing and increases the width of the benthic boundary layer. These beds can remove, via sedimentation (passive) and filtration (active) mechanisms, enormous quantities of suspended materials (phytoplankton, etc.) from the water and release, as a result of metabolism, large amounts of dissolved inorganic substances into tidal currents. In some systems, bivalve beds are equivalent to saltmarshes in processing materials. There have been few studies on benthic-pelagic coupling by marsh-estuarine mudflats. In marsh related mudflats, there is low water flow, shear velocities, drag and smooth surfaces. Both passive and active coupling mechanisms are common to mudflats, but there is little direct information available. The high ratio of bottom surface area to tidal water volume over these flats suggests a great potential for material exchanges. At the system level, coupling processes directly involve marshes and animals in the cycling of major nutrients not only within the shallow tidal marsh-estuarine ecosystem, but also with the adjacent coastal ocean. The magnitude of these system level couplings has only been identified in a few locations, but they are almost always related to high productivity sub-systems, i.e., mussel beds and oyster reefs.

1. Introduction

Where the water column and the bottom meet, these two subsystems are coupled through flows of energy and matter. Rowe (1971) first proposed and others (Hargrave 1973 and Rowe et al. 1975) later showed that benthic secondary production or biomass was related to surface water primary production. Such studies on the benthic organisms feeding on phytoplankton heralded the beginning of investigations into benthic-pelagic (pelagic-benthic to some) coupling. These observations were quickly related to the release of nutrients from the sediments (Davies 1975, Rowe et al. 1975) with the speculation that in shallow systems nutrient regeneration in bottom sediments was sufficient to support adjacent pelagic primary production. In the same time frame, Hammen (1980) was reporting laboratory observations of nitrogen excretion by a number of different estuarine benthic animals, and, from a biogeochemical perspective, and Aller (1978) and Aller and Yingst (1978) were observing the production of inorganic nutrients by benthic animals living on or in estuarine muds. At the ecosystem level, studies on a salt marsh embayment by Nixon et al. (1971 and 1976) strongly implicated mussel beds as major consumers of particulate material and producers of ammonia. Thus, the idea of a two-way exchange of materials between the benthos and the water column was taking shape with ecosystem, sediment biogeochemical, and physiological ecological approaches merging.

The direction and magnitude of the flows across the benthic-pelagic interface can be controlled by both abiotic and biotic factors. In marsh-estuarine ecosystems, strong bi-directional flows of tidal water and dense communities of organisms play important roles in this benthic-pelagic coupling. In some cases, this exchange is passive with only physical principles governing the flows; in other instances, organisms enhance these fluxes through active pumping and bioturbation mechanisms. Because the organisms and functional groups common to marsh-estuarine ecosystems are so abundant, the results of their activities are often observable at the ecosystem level. In this over-view, we will examine the present state of knowledge of material exchanges resulting from benthic-pelagic coupling in marsh-estuarine ecosystems.

2. The Benthic Boundary Layer

In the shallow zones of most marsh-estuarine ecosystems, the bottom is often exposed to bi-directional tidal currents at least once a day. At the scale of millimeters, viscosity or the internal resistance of the water causes the average flow velocity to decrease from its open water magnitude to zero at the bottom boundary (Mann and Lazier 1991). The turbulent motion of the tidal currents is transferred continually to smaller and smaller scales until, at the molecular level, viscosity, acts to resist and smooth out the gradients in velocity and dissipate the energy of motion. When an object on the bottom, i.e., organism, sediment, etc., removes momentum from the moving fluid, that object is said to have created a drag on the fluid (Vogel 1981). The interaction of objects in the water and on the bottom, and of the bottom itself with the moving fluid (water) is important to understanding the ecology of this interface.

2.1 FLOW STRUCTURE AT THE BENTHIC BOUNDARY LAYER

It is the structure of water flow near the bottom that is important to benthic-pelagic coupling in aquatic systems. As the water flows over the bottom, frictional drag (internal resistance to flow) slows the current until the current is zero at the bottom-water column interface. The layer of water between the open water velocity and the bottom is know as the benthic boundary layer (Fig. 1). This boundary layer may exhibit either turbulent or laminar flow characteristics depending on the magnitude of the water currents and the roughness of the bottom. Laminar flows with mainly parallel flow paths, are characteristic of low water velocities and bottoms with smooth textures (mudflats). These laminar flows are not generally well known in natural systems (Nowell and Jumars 1984). In contrast, turbulent flows with chaotic flow paths are typical at higher water velocities and over rough bottoms (oyster reefs). These flows are thought to be a more common feature of natural systems (Nowell and Jumars 1984, Davis and Barmuta 1989, Carling 1992).

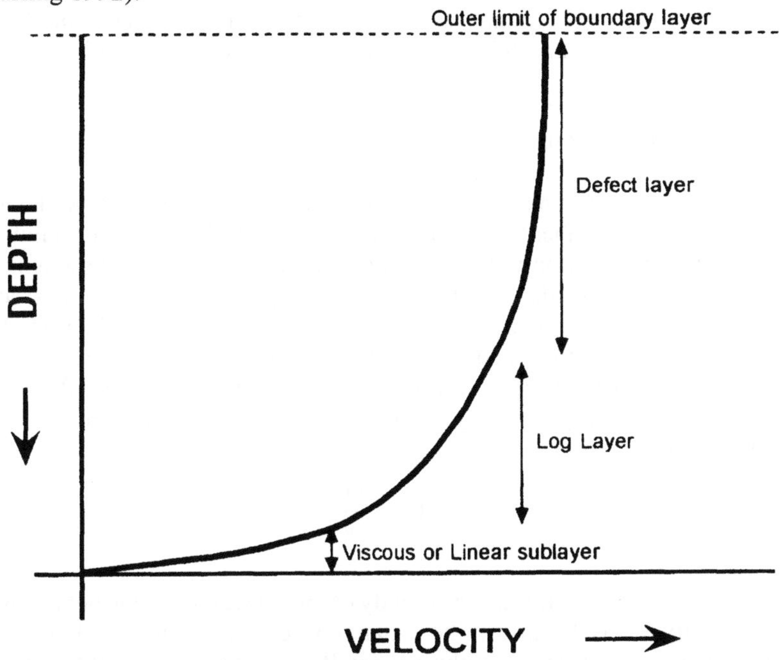

Figure 1. General description of the benthic boundary layer using water velocity profiles. The layer next to the bottom is known as the bed or linear or viscous sublayer. In this layer, water velocities change linearly with depth, laminar flow is evident, and molecular diffusion is important. The middle layer is the log layer. In this region, water velocities change logarithmically with depth and turbulent flow is evident. The outer or defect layer is where the influences of the boundary disappear.

The benthic boundary layer can be divided into three layers (Fig. 1). The outermost layer is called the defect layer and flow behavior is largely independent of bottom roughness. This layer exhibits the open water velocity. The next layer is the log layer where water velocity varies logarithmically with distance above the bottom. These upper layers mix by turbulent diffusion that effectively operates at scales greater than a few millimeters. The layer nearest the bottom is the viscous or linear sublayer where velocity varies linearly with distance above the bottom (Nowell and Jumars 1984, Fréchette et al. 1993). It is within this layer, at spatial scales of millimeters, that molecular diffusion or the slow mixing of molecules by random motion is important (Mann and Lazier 1991). Molecular diffusion is a major mechanism in the transfer of material across the benthic boundary layer.

2.2 ECOLOGICAL PARAMETERS

Historically, mean water flow and discharge were used as descriptors of water flows near the benthic-pelagic interface. However, laboratory studies of flow in this region indicated that these descriptors were only partially adequate (Nowell and Jumars 1984, Davis and Barmuta 1989). The later authors suggested that the principal physical variables that allow the quantitative description of boundary layer flows are: mean free stream velocity and shear velocity; roughness of the bottom; and thickness of the laminar sublayer (Davis and Barmuta 1989). These descriptors will be examined in the next sections.

2.2.1 Average Water Motion

The average water motion in a stream is characterized by two parameters: the Reynolds number and the Froude number (Davis and Barmuta 1989). The Reynolds number (R_e) is a valuable descriptor of not only drag, but also whether the average flow is laminar or turbulent. It is also helpful that the R_e is a scaling parameter that ranges across the spatial scales of living organisms from molecules to ecosystems. Many studies have shown that increasing fluid speed, increasing the size of the object in the flow, increasing the density of the fluid, and decreasing the viscosity of the fluid may shift the character of the flow. The R_e, a dimensionless value, is derived from this combination of parameters.

(1)
$$\mathbf{R_e} = \frac{\mathbf{lU}}{\mathbf{v}}$$

where **l** is the size of the object, **U** is the velocity of the fluid at 0.4 of the depth, and **v** is the kinematic viscosity that is the ratio of the dynamic viscosity and density. For values of R_e < 500 flow is laminar and flow is turbulent for $R_e > 2000$ (Davis and Barmuta 1989). Although turbulent flows are the most common in natural systems, laminar flows are found where water depth is very shallow or velocities are very slow, i.e., environments common to marsh-estuarine ecosystems.

The Froude number (F_r) is a parameter commonly determined for unidirectional gravity induced flows typical of rivers and freshwater streams (Davis and Barmuta 1989). It represents the ratio of inertial forces to gravitational forces and is described as:

(2)
$$\mathbf{F_r} = \frac{\mathbf{U}}{\sqrt{\mathbf{gD}}}$$

where **U** = water velocity at 0.4 of depth, **g** = acceleration due to gravity, and **D** = depth. For conditions where $F_r < 1$ the flow is designated sub-critical or tranquil, where $F_r = 1$ flow is critical, and for $F_r > 1$ flow is super-critical. The later category is characterized by white water. As marsh-estuarine systems are tidal, they cycle through many of these conditions on each tide.

2.2.2 Water Flow Near the Bottom

Water velocities near the bottom are much lower than those in the defect or outer boundary layer (Fig. 1). Shear or frictional stresses are high because of flow interactions with the bottom. Thus, ecologists are interested in a number of parameters that describe conditions in this micro-environment (Davis and Barmuta 1989). Some of these are shear velocity ($\mathbf{U}_*$), roughness Reynolds number (R_{c*}), and relative roughness (**D/k**).

The shear velocity ($\mathbf{U}_*$) describes the flow environment at the boundary between the water column and the bottom (Davis and Barmuta 1989). It is most easily computed from velocity profile measurements as:

(3) $$\mathbf{U}_* = \frac{5.75}{\tan\alpha} \quad \text{and} \quad \tan\alpha = \frac{\mathbf{U}_1 - \mathbf{U}_2}{\log\mathbf{Z}_1 - \log\mathbf{Z}_2}$$

where **U** = measured velocity at depth = **Z**. The shear velocity is generally between 1/10 and 1/30 the mean water velocity.

The roughness Reynolds number (R_{c*}) can be defined as:

(4) $$\mathbf{R}_{e*} = \mathbf{U}_* \frac{\mathbf{k}}{\mathbf{v}}$$

where $\mathbf{U}_*$ = shear velocity, **k** = height of roughness projections, and **v** = kinematic viscosity. At $R_{c*} < 5$ flow is said to be smooth and at $R_{e*} > 70$ flow is rough (Davis and Barmuta 1989). The relative roughness of a given area is simply **D/k**, where **D** is depth and **k** is height of objects on the bottom. Relative roughness determines the flow environment that benthic organisms may experience.

The thickness of the laminar or viscous sublayer, δ' , the region close to the bottom where water flow is entirely laminar, can be calculated from:

(5) $$\delta' = \frac{11.5\mathbf{v}}{\mathbf{U}_*}$$

where $\mathbf{U}_*$ = shear velocity and **v** = kinematic viscosity (Davis and Barmuta 1989). In this layer, also known as the linear layer because within this zone water velocity changes linearly with depth. The thickness of this layer increases with water velocity.

One of the most common types of boundary layer flow patterns found in marsh-estuarine systems is fully developed, turbulent, uniform, steady flow across either smooth or rough bottoms (Nowell and Jumars 1984). These two types are shown in Fig. 2. Even these simple flow environments require the determination of a number of the flow parameters described above in order to understand ecologically meaningful processes. In marsh-estuarine systems, the most meaningful parameters are depth (**D**), mean or average velocity (**U**), velocity profiles, and height of bottom objects (**k**). From these the Reynolds number (R_e), shear velocity (U_*), thickness of the laminar sublayer (δ') and roughness Reynolds number (R_{e*}) can be estimated. With the quantitative parameters of the benthic boundary layer formally defined, the remaining sections will examine specific subsystems within marsh-estuarine ecosystems.

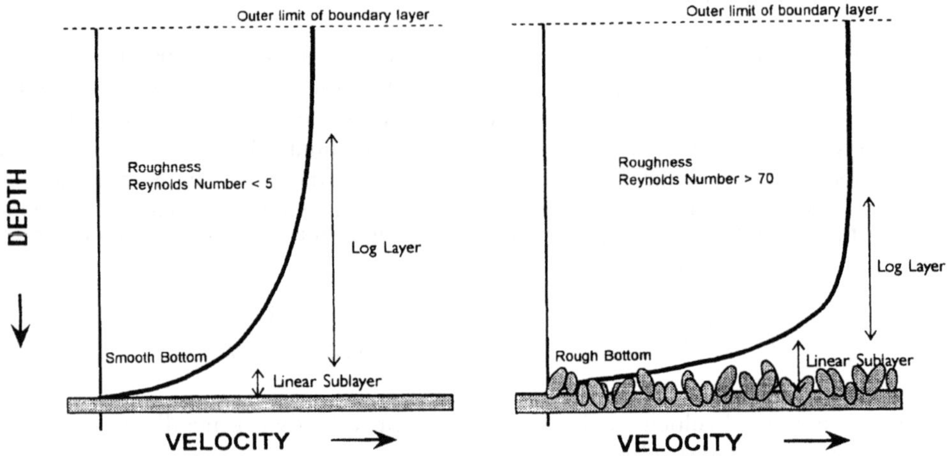

Figure 2. Comparison of water velocity profiles over a smooth bottom and a bottom covered with bivalves. The roughness Reynolds number (R_{e*}) is described in the text. Note the increased thickness of the linear sublayer on the rough bottom system. Profiles after Davis and Barmuta (1989) and Fréchette et al. (1994).

3. Passive and Active Coupling

From an ecological perspective, the coupling between the water column and the bottom can be either passive or active, or both. In passive coupling, the flux of materials between the two environments depends entirely on ambient flow (Wildish and Kristmanson, 1997). Sedimentation and molecular diffusion are processes that fit this category. Active coupling requires that organisms expend energy to move materials between the water column and the bottom. Most active suspension feeders fall into this group. At the organismic level, some barnacles appear to switch between active and passive mechanisms depending on the water velocity (Trager et al. 1990). Similarly, sponges can exhibit active and passive processes simultaneously (Vogel 1974, LaBarbera 1977). At the systems level, marsh-estuarine ecosystems exhibit both active and passive coupling mechanisms. From a systems perspective these mechanisms may be either physically or biologically mediated.

3.1 PASSIVE COUPLING

Passive benthic-pelagic coupling is almost always present in marsh-estuarine systems. In this form, the coupling is governed by the physical interactions of the bottom or organisms with the water column. With passive coupling, there are no active mechanisms, i.e., no energy is expended by organisms in the coupling (Wildish and Kristmanson 1997) to enhance the flux of materials between the bottom and the water. We will examine common passive mechanisms in three subsystems common to marsh-estuarine ecosystem: marshes, bivalve reefs and mudflats.

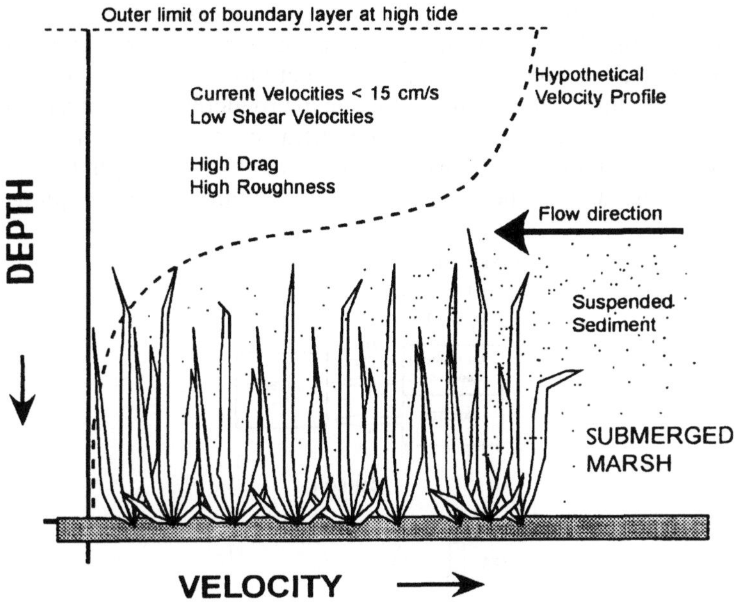

Figure 3. Hypothetical water velocity profile over a salt marsh, showing sediment trapping characteristics. (data from Ke et al. 1994)

3.1.1 Salt Marshes

The structure of salt marshes, usually the most extensive sub-component of marsh-estuarine ecosystems, provides a passive mechanism that couples this system to the tidal waters of the estuary. These systems have often been described as giant sediment traps (Jordan and Valiela 1983, Stevenson et al. 1988) that slow water flow, enhance particle settlement and increase the width of the benthic boundary layer (Fig. 3). They are dominated by vascular plants that are rooted into soft sediments and take most of their nutrients from the substrate. The flat and relatively stiff leaves and stems of grasses like *Spartina alterniflora* act as baffles that create high amounts of drag (high relative roughness), quickly reducing tidal current velocities and dampening waves (Warner 1977, Frey and Basan 1985, Ke et al. 1994). This reduced flow environment enhances sedimentation (Jordan and Valiela 1983, Wolaver and Zieman 1983, Wolaver et al. 1988) to the point that the elevation of many salt marshes is maintained as sea level rises (Stumpf 1983, Reed 1988, Dame 1989).

Sedimentation is not uniform across marshes. In a New England marsh, Jordan and Valiela (1983) observed that sedimentation rates were greatest in the tall *Spartina* zone, slightly less in the muddy marsh creeks, and lowest in the short *Spartina* zone. In addition, these investigators found that fine sediments (4-40 μm) were often resuspended and that within this particular system, resuspension just about offset sedimentation. Depending on location within the shallow marsh-creek system, resuspension is attributed to the shear velocity of tidal currents (Jordan and Valiela 1983), wave action (Anderson 1972), fish feeding (Jordan and Valiela 1983, Feller and Coull 1995), and surface films and rain impaction (Chalmers et al. 1985).

In contrast to reduced water flow increasing sedimentation over salt marshes, this low flow environment may also result in reduced productivity of epiphytes on the blades and stems of the grasses (Jones 1980). This reduction in productivity is the result of the much slower exchange of nutrients via molecular diffusion within the increased viscous boundary layer surrounding the surfaces inhabited by the epiphytes. Thus, some processes are enhanced and others are reduced by benthic-pelagic coupling in the salt marsh.

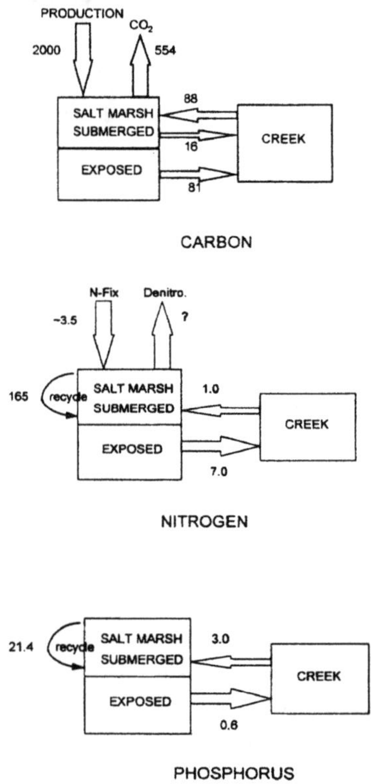

Figure 4. Annual carbon, nitrogen and phosphorus fluxes in the Bly Creek salt marsh at North Inlet, SC. (after Dame et al. 1991)

It is well documented that salt marsh systems are major locations for the recycling of materials (Teal 1962, Day et al. 1973, Valiela and Teal 1979, Pomeroy and Wiegert 1981). For example, Dame et al. (1991) developed input-output budgets for a North Inlet, SC salt marsh utilizing a 140 m (280 m^2) long flume (Fig. 4). These studies showed that, on an annual basis, this submerged salt marsh imported statistically significant quantities of particulate and dissolved materials including inorganic suspended sediments, particulate organic carbon, nitrate + nitrite, particulate phosphorus, orthophosphate, and chlorophyll. Only dissolved organic nitrogen and ammonium showed significant exports. Large quantities of nitrogen and phosphorus were recycled within the marsh. Most of the particulate materials were probably removed passively by sedimentation, but active coupling processes mediated by marsh mussels (Kuenzler 1961, Jordan and Valiela 1982, Bertness 1984) and other filter feeding benthos that probably actively removed some particles are also implied. The removal of dissolved nutrients implies that epiphytes, benthic microalgae and sediments within the marsh (Jones 1980, Pinckney and Zingmark 1993) are actively taking up these materials.

The sediments trapped by the marsh are composed of both organic and inorganic materials. These materials are a potential source of nutrients that help to maintain the high productivity of these grasses and support aerobic and anaerobic decomposition processes within salt marsh sediments. In addition, large amounts of organic material are produced by marsh grass within these systems. A major proportion of these organic materials decompose on and in the marsh sediments (Howarth and Teal 1979, Howarth and Giblin 1983). Because the diffusion of oxygen into wetland soils is four orders of magnitude slower than well drained soils (Gambrell and Patrick 1978), the deeper

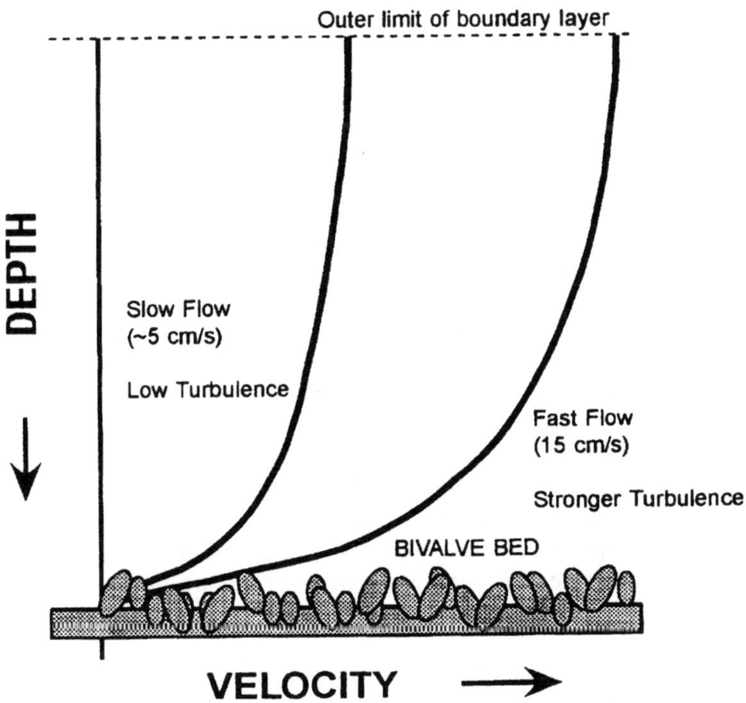

Figure 5. Comparison of water velocity profiles over a mussel bed in a flume at slow and fast flows. (after Fréchette et al. 1994).

layers of marsh soils are usually anaerobic. In this reducing environment, nitrate and sulfate reduction increase, ammonium and phosphate ions increase, reduced iron and hydrogen sulfide increase and organic substrates decrease. In the case of sulfur, Howarth and Teal (1979) and Howarth et al. (1983) calculated that only a small percentage of soluble sulfides may be exported in the marsh pore waters. However, in North Inlet, Dame et al. (1991) found that on an annual basis dissolved organic forms and ammonium were exported from the marsh to the tidal creek. Much of the exported material probably resulted from decomposition processes within the marsh and molecular diffusion (passive coupling) into waters draining the marsh.

Utilizing data from marsh flume studies (Childers 1994) and from a dynamic simulation model (Childers et al. 1993), Childers (1994) presented an argument supporting the idea that tidal subsidies (Odum 1969) control the direction of material fluxes between the water column and the marsh. In essence, Childers (1994) contends that, at tidal ranges below 1 m, net exchange is with the marsh surface and favors export. At tidal ranges above 1 m, the marsh surface favors import of materials with horizontal subsurface and creek bottom advection also becoming important.

In conclusion, the structure of the marsh plants, coupled to tidal flows, supports the passive removal and trapping of materials that can be used by this system to maintain high productivity and elevation as sea level rises. Furthermore, the processes of diffusion and resuspension in ebbing waters on the surface of the marsh can result in the export of materials. Finally, tidal energy may force (pump) horizontal subsurface fluxes to the creeks.

3.1.2. Bivalve Reefs and Beds

It has long been known that the most productive oyster beds were those where the best circulation of water was sustained (Grave 1912). A portion of this circulation can be attributed to the turbulent mixing over oyster reefs that is enhanced by the structural roughness of the shells protruding into the water column (Dame 1996). This mixing is further enhanced by the animals aggregating into dense assemblages or reefs that rise up from the bottom and interact with tidal currents. This view is well supported by a laboratory flume study of a mussel bed (Butman et al. 1994) that showed that, at a flow of 5 cm s^{-1}, the mussel bed roughness significantly enhanced turbulent stress (turbulent mixing) by a factor of three and by a factor of 10 at a flow of 15 cm s^{-1} when compared to a smooth bottom (Figs. 2 and 5). Using constructed subtidal oyster reefs, Lenihan (unpublished) found that flow speed and oyster growth were highest at the top of reefs while sedimentation was highest at the base of these structures. Clearly, elevated water flow and turbulent mixing increases the availability of planktonic food by constantly and rapidly renewing the water near the animals. In turn, this rapid mixing removes inorganic and particulate waste from the bivalves and makes these materials more readily available for recycling through the phytoplankton and the ecosystem.

In a tunnel study of the annual flux of inorganic sediments over an oyster reef, Dame (1987) showed that uptake and release of these materials was almost balanced. From the analysis of individual tidal cycles and a comparison of fluxes over live and dead reefs, he deduced that, at tidal currents above 15 cm s^{-1}, resuspension became a major process.

Thus, the net effect of the interaction of currents and the reef was to keep the reef from being buried in sediments.

Only at very low water velocities (1-2 cm s^{-1}) does it appear that the molecular diffusion in the viscous sublayer dominates the coupling (Dame et al. 1984). In this situation, particle settlement and animal feeding develop gradients of particulate concentrations. Likewise, waste production by organisms also produces concentration gradients of dissolved inorganic materials and particulate organic matter. At low flow rates, molecular diffusion takes place across these established concentration gradients. One byproduct of low flow viscous sublayer formation is the potential development of food limitation due to the animals completely filtering out all potential food material within this section of the boundary layer (Wildish and Kristmanson 1984, Fréchette et al. 1989, Butman et al. 1994). Direct field observations by these and other investigators show that animals on the edge of a bed get the food first and grow the largest when compared to those in the central portions of the bed.

Clearly, shell structure and aggregation of animals interact with tidal flow to influence the processes of sedimentation, resuspension and molecular diffusion. The net result is burial prevention and enhanced material processing. Although flow structure is well known over artificial mussel beds (Butman et al. 1994), the flow environment over natural mussel beds and oyster reefs is not. These natural systems are probably more turbulent because the shells of these bivalves and their beds extend much more into the overflowing water (higher relative roughness). It is also not known how these bivalve beds influence the geomorphological structure of tidal creeks and the associated marshes.

3.1.3 Mudflats

Marsh-estuarine ecosystems are permeated by tidal creeks. Tidal flats are frequently exposed at low tide and reach their greatest extent in the upper reaches of the smallest or ephemeral streams (Peterson and Peterson 1979). There have been very few studies on benthic-pelagic coupling in mudflats associated with marsh-estuarine ecosystems. In a comparative study of water velocity structure and bottom roughness in an English marsh-estuarine system, Ke et al. (1994) found that bottom roughness and drag were much higher over the marsh than over the mudflat (Figs. 2 and 3). In contrast to the marsh that had no flow structure, water flow over the mudflat always exhibited a well-developed logarithmic form (log layer). Passive sedimentation over mudflats probably results from low flow conditions, while passive resuspension of mudflat sediments has been shown to result from water velocities generally exceeding 15 cm s^{-1} (Dame 1987), rain impaction during low tide exposure (Chalmers et al. 1985), ice (Anderson 1983), wave action (Anderson 1972) and wind generated turbulence (Frostick and McCave 1979, Anderson 1980, Kraeuter and Wetzel 1986).

In a New England system, Welsh (1980) found evidence of passive coupling between the mudflat and the tidal water. She determined that her mudflat was autotrophic and consistently removing nutrients through geochemical or biological processes with the sediments and uptake by macroalgae. Welsh (1980) also noted that as the flats become submerged, the macroalgae were suspended in the water column and effectively increased the surface area of the system and the area across which molecular diffusion could take place. Kraeuter and Wetzel (1986) deduced that the various processes taking place at the benthic boundary allowed significant quantities of pore water to be exchanged with the overlying water. They speculated that tidal pumping is the major causative agent in sediment water exchange over their mudflat.

As mudflats are mainly two dimensional (2-D) and they are interacting with the water column that is 3-D, we have hypothesized (Dame et al. 1992) that as the surface to volume ratio increases, i.e., the water level and therefore water volume decreases, so will the passive exchange of materials increase. Thus, with low flow and high surface to volume ratios, material exchange in the mudflat component is probably dominated by molecular diffusion, sedimentation and resuspension. Very little is known about the coupling of these habitats to the water column and, thus, little is known of their magnitudinal importance to the marsh-estuarine system.

3.2 ACTIVE COUPLING

Within each of the systems discussed above, salt marsh, mudflat, and oyster reef, there are organisms that, through their life processes, actively couple the estuarine water column to the bottom. The majority of these active processes focus on the moving or pumping of water by animals as a part of feeding, filtration, respiration or excretion. Because many of these organisms remove particles from suspension as part of feeding, they are often called suspension feeders. Most suspension feeders in marsh-estuarine systems are categorized as active, e.g., many bivalves, bryozoans, sea squirts and segmented worms.

3.2.1 Organismic Mechanisms

Except for a few polychaete species that utilize a muscular piston pump mechanism, most active suspension feeders use a ciliary pump to transport water through their bodies (see Wildish and Kristmanson 1997 for a recent discussion of these). Bivalve suspension feeding appears to be controlled by both external environmental factors and internal physiological conditions. External factors include water velocity, flow direction, suspended particle concentrations, suspended particle quality and water viscosity (Wildish and Saulnier 1993, Jørgensen 1990). Potential physiological controls are gut fullness, and numerous responses of the mantle, gills, and mouth parts to the external factors (Bayne and Newell 1983, Wildish and Saulnier 1993). All of these environmental and physiological factors controlling bivalve suspension feeding have been conceptually linked by Wildish and Saulnier (1993) into an integrated environmental/physiological model.

Both facultative suspension feeders, such as cirripede barnacles, and deposit suspension feeders, such as some bivalves and tube worms, are common to marsh-estuarine systems. Some barnacles appear to switch between active suspension feeding at low velocities and passive suspension feeding at higher velocities (Trager et al. 1990). In contrast, tube building spionid polychaete worms and some tellinid bivalves are deposit feeders at low velocities and switch to suspension feeding at higher velocities (Dauer et al. 1981, Levinton 1991, Taghon and Greene 1992).

From the preceding discussion, we know that water velocity, as well as other external and internal factors, influence the active removal of particles from the water column by benthos. Thus, studies on benthic-pelagic coupling should certainly consider the broad range of flow environments that organisms experience as well as the concentrations of particles in these environments. Although the excretion of inorganic waste by these same animals is obviously related to feeding, almost nothing is known about how the flow environment influences this complimentary component of benthic-pelagic coupling.

3.2.2 Salt Marshes

Although there are numerous snails, worms and crabs in salt marsh sediments (Kraeuter 1976), the most obvious and probably most important biogeochemical active coupler is the marsh mussel, *Geukensia demissa*. In southeastern Atlantic coast salt marshes, marsh mussels are spread throughout the marshes and may form small clumps. In Georgia marshes, Kuenzler (1961) found these animals at maximum densities of about 50 m^{-2} and at elevations that allowed approximately 8 h of submergence per day. By comparison in New England marsh-estuarine systems, Jordan and Valiela (1982) and Bertness (1984) reported denser maximum abundances (800 m^{-2}) of mussels on creekbanks and near the mouths of creeks and attributed the higher densities to longer submergence times of 17 h d^{-1}. In the earliest of these studies, Kuenzler (1961) utilized a combination of field and laboratory methods to estimate that *Geukensia* were capable of filtering a third of the particulate phosphorus suspended in the Georgia marsh water on a given tidal cycle. Most of the filtered material was deposited on the marsh surface as feces and pseudofeces, thus the mussels were an active mechanism for trapping sediments. In a similar study on nitrogen, Jordan and Valiela (1982) found that in a New England salt marsh *Geukensia* filtered a volume of water in excess of the tidal volume of the marsh and annually filtered a quantity of particulate nitrogen 1.8 times that exported by tidal flushing. Of the nitrogen filtered, about half was absorbed into mussel biomass with 55% of that eventually being excreted as ammonia. The importance of the marsh mussel in salt marshes was directly coupled to *Spartina* in field experiments by Bertness (1984). He showed that the presence of *Geukensia* stimulates *Spartina* growth, i.e., grass height, biomass and flowering, and increased both above- and belowground *Spartina* production. These increases were positively correlated with mussel densities and soil nitrogen levels.

The active coupling of marsh mussels within the marsh magnifies material cycling, sedimentation, and *Spartina* growth and production. An interesting question concerning this coupling is, how active coupling mechanisms might influence the long-term stability and geomorphology of the system?

3.2.3 Bivalve Reefs and Beds

Active coupling between the bottom and the water column reaches its zenith with dense assemblages of bivalves in the form of beds or reefs. Usually found in highly productive, shallow estuarine locations, these dense aggregations of suspension feeding animals are often capable of removing a large percentage of the phytoplankton from a system (Officer et al. 1982, Dame 1996). Early energy budget studies on bivalves (Bernard 1974, Dame 1976) were often unbalanced because there was not enough phytoplankton production in the immediate water column to support the existing dense population of bivalves. The solution to this situation was to determine the total volume of water that was available to these animals, e.g., flux studies.

Dame et al. (1980) proposed that oyster reefs in North Inlet, SC were, an important, if not controlling mechanism in the benthic-pelagic coupling within this marsh-estuarine ecosystem. These authors argued that bivalve reefs were capable of translocating and transforming large quantities of matter from the estuarine water column and returning a significant proportion of it to the overlaying waters. Using portable plastic tunnels, this group (Dame et al. 1984, Dame et al. 1989) was able to show that an intertidal oyster reef removed significant amounts of particulate materials from the water while adding significant amounts of dissolved inorganic nutrients (Fig. 6). Applying the concept of turnover, the ratio of throughput to content, Dame et al. (1989 and 1991) were able to show that the oysters in North Inlet grazed more phytoplankton and produced more ammonium than could be accounted for by tidal fluxes. Preliminary studies by Dame and Libes (1983) on creeks with oysters and those with oysters removed directly support these findings. Thus, in this relatively small water volume and short residence time system, the large biomass of suspension feeding bivalves can effectively control phytoplankton concentrations by grazing. They can also influence nutrient cycling by processing large quantities of particulate materials and releasing large amounts of dissolved inorganic nutrients (Dame 1993, Dame et al. 1984).

Bivalves are the preeminent active couplers in many marsh-estuarine systems. Through their activities, they magnify material processing and exchange, and may serve as a mechanism of eutrophication control. They may serve to stabilize these tidally driven systems through their activities (Herman and Scholten 1990).

3.2.4 Mudflats

Suspension and deposit feeding bivalves and worms are common to tidal mud flats. Both forms are capable of moving water into and out of their burrows. For example, the terebellid polychaete worm, *Amphitrite ornata*, is a sedentary, surface deposit feeding animal common to mud flats along the east coast of North America. These worms form permanent, multi-layered, U-shaped tubes or burrows that are ventilated by the peristaltic movements of the animal's body. This activity provides oxygen and carries away waste. Aller and Yingst (1978) found that individual worms transported about

4.5 g d^{-1} of particles while feeding and irrigated their tubes at a rate of 91 ml h^{-1}. They also determined that when irrigation ceased ammonia, phosphate and iron concentrations increased. Some of this increase can be attributed to excretion by the animal, while some is attributed to intense decomposition taking place in the burrow walls. Additional experimental studies by Aller and Yingst (1985) on intertidal mudflats adjacent to marshes in North Inlet, SC found that macrofauna, e.g., *Heteromastus*, *Tellina*, and *Macoma*, had major effects on sedimentary solute transport and increased mudflat production of ammonia by 20 to 30%. Thus, mudflat deposit feeding worms and deposit feeding/suspension feeding bivalves move water, process materials and flush inorganic nutrients into the water column and adjacent sediments. Other more system oriented studies by Carlson et al. (1984) found that active filter feeding by bivalves on a mudflat in Maine did reduce the concentration of phytoplankton in the

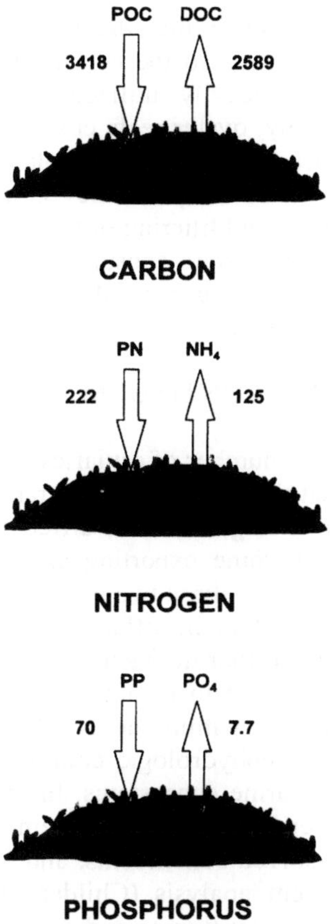

Fig. 6. Annual carbon, nitrogen and phosphorus fluxes for an intertidal oyster reef in South Carolina. (after Dame et al. 1989)

overlying water column. Peterson and Black (1987) studied th bivalve community on a tidal flat and speculated that these animals were depleting suspended food supplies at a system level and , thus, influencing animals at higher elevations. The extent of the mudflat sub-components of the system can equal that of tall *Spartina* (Aller and Yingst 1985), but their system level impact is poorly known.

Bivalves and worms dominate the active coupling processes on mudflats. Their activities magnify material exchanges and processing through the mechanisms of feeding, sedimentation, and resuspension, but the extent of their influence is unknown.

4. System Level Coupling

As noted earlier, there are estuarine ecosystems that apparently cannot supply the phytoplankton necessary to support the benthic suspension feeders present in the system (Smaal and Prins 1993). In North Inlet, the marsh-estuarine system exports most forms of materials on an annual basis to the coastal ocean (Dame et al. 1986), but chlorophyll is imported and is heavily utilized by the various consumer sub-components of the system. Notably, oyster reefs cover only a small percentage of the area of the entire system, but are functionally equivalent to the much more extensive salt marsh (Childers et al. 1993). This equivalency is undoubtedly due to the active coupling exhibited by the pumping and filtering of the oysters. It also gives this estuary the appearance of a food lot with oysters consuming phytoplankton produced in the coastal ocean and releasing inorganic nutrients that are exported to the coastal ocean where they can be used by phytoplankton.

4.1 THE MARSH-ESTUARINE CONTINUUM

Reviews of material transport in a number of estuaries (Odum et al. 1979, Nixon 1980, Dame and Allen 1996) show that the direction of net transport varies dramatically from system to system. Even within a single system, different areas may exhibit different material net flux directions with some exporting all materials, others importing all material, and other properties between these two extremes (Dame et al. 1991). Odum et al. (1979) developed a scheme to explain the differences in particulate fluxes based on the geomorphology of the particular estuarine basin, but his model did not account for dissolved material fluxes. In an effort to explain the flux directions for a variety of particulate and dissolved materials within the North Inlet marsh-estuarine system, Dame et al. (1992) suggested a geohydrologic continuum theory for the spatial and temporal evolution of marsh-estuarine ecosystems. In this scheme, young creeks tend to be net importers of all materials, mid-aged systems tend to be net importers of particulate and exporters of dissolved constituents, and mature components tend to be exporters of everything. A recent analysis (Childers 1994) examined marsh-water column interactions as determined by flumes and generally supported the continuum theory while taking tidal range into account.

At the system level, tidal exchange and benthic-pelagic coupling interact to influence the general pattern of material inwelling or outwelling. Further, the coupling of the marsh-estuarine ecosystem to the coastal ocean suggests the existence of larger scale relationships in system function.

5. Conclusions

Both passive and active benthic-pelagic coupling are important to marsh-estuarine ecosystems. The exact nature of the coupling depends on the specific subsystem and process. All couplings seem to enhance material processing, cycling and exchange. Without benthic-pelagic coupling, marsh-estuarine systems would not be as rich and productive as they are. The causes, processes, effects, and questions raised by this overview of benthic-pelagic coupling in marsh-estuarine ecosystems are summarized in Tables 1 and 2. Clearly both macro and micro flow environments need to be investigated spatially and temporally in all of these environments. Mudflats in marsh-estuarine

TABLE 1. Marsh-estuarine passive benthic-pelagic coupling: causes, processes, effects and questions raised.

Salt Marsh		
	Causes	Tidal flow
		Grass blade structure
		Tidal elevation
	Processes	Sedimentation
		Resuspension
		Molecular diffusion
		Tidal pumping
	Effects	Elevation
		Erosion
		Soil biogeochemistry
		Epiphyte productivity
		Material fluxes
	Questions	Flow structure
		Role of epiphytes
		Resuspension
		Tidal subsidy
Bivalve Beds		
	Causes	Tidal flow
		Shell structure
		Aggregation
	Processes	Sedimentation
		Resuspension
		Molecular diffusion
	Effects	Material input
		Burial prevention
		Enhanced material processing
	Questions	Flow structure over beds
		System structure
Mudflats		
	Causes	Low flow
		High surface to volume ratio
	Processes	Molecular diffusion
		Sedimentation
		Resuspension
	Effects	Material exchange
	Questions	Flow dynamics
		Nutrient dynamics
		Sedimentation/resuspension
		Importance

systems appear to be the least studied, but have the potential to interact passively at the highest rates because of their high surface area to water volume ratios. Studies on marsh-estuarine ecosystems are moving from the mainly descriptive to the explanatory stage and studies that integrate biological and physical factors in an over-arching ecological framework should be profitable.

Table 2. Marsh-estuarine active benthic-pelagic coupling: causes, processes effects, and questions raised.

Salt Marsh		
	Causes	Marsh mussels, etc.
	Processes	Sedimentation
		Pumping
		Feeding
		Excretion
	Effects	Soil biogeochemistry
		Magnifies n & p cycling
		Increases spartina growth & production
	Questions	Influence on marsh morphology
		Long term stability
Bivalve Beds		
	Causes	Bivalves, etc.
	Processes	Sedimentation
		Suspension feeding
		Pumping
		Excretion
	Effects	Magnifies material processing
		Eutrophication control
	Questions	System level importance
Mudflats		
	Causes	Bivalves & worms
	Processes	Sedimentation
		Resuspension
		Deposit feeding
		Suspension feeding
		Material processing
	Effects	Magnifies material exchange
	Questions	Limited data on system level influences

6. Acknowledgements

This work was supported by award No. DEB-950957 from the National Science Foundation. The Coastal Carolina University Library Staff was invaluable in finding requested materials. This is publication No.1178 of the Belle W. Baruch Institute for Marine Biology and Coastal Research.

7. Literature Cited

Aller, R.C. 1978. Experimental studies of changes produced by deposit feeders on pore water, sediment, and overlying water chemistry. American Journal of Science **278**:1185-1234.

Aller, R.C. and J.Y. Yingst. 1978. Biogeochemistry of tube-dwellings: a study of the sedentary polychaeta *Amphitrite ornata* Leidy. Journal of Marine Science **36**:201-254.

Aller, R.C. and J.Y. Yingst. 1985. Effects of the marine deposit-feeders *Heteromastus filiformis* Polychaeta, *Macoma balthica* (Bivalvia), and *Tellina texana* (Bivalvia) on averaged sedimentary solute transport, reaction rates, and microbial distributions. Journal of Marine Science **43**:615-645.
Anderson, F.E. 1972. Resuspension of estuarine sediments by small amplitude waves. Journal of Sedimentary Petrology **42**:602-607.
—— 1980. The variation in suspended sediment and water properties in flood-water front traversing the tidal flat. Estuaries **3**:28-37.
—— 1983. The northern muddy intertidal: a seasonally changing source of suspended sediments to estuarine waters - a review. Canadian Journal of Fisheries and Aquatic Sciences **40**:143-159.
Bayne B.L. and R.C. Newell. 1983. Physiological energetics of marine molluscs. Pages 407-515 *in* K.M. Wilbur, editor. The mollusca, Vol. 4, Academic Press, New York, USA.
Bernard, F.R. 1974. Annual biodeposition and gross energy budget of mature Pacific oyster *Crassostrea gigas*. Journal of the Fisheries Research Board of Canada **31**:185-190.
Bertness, M.D. 1984. Ribbed mussels and *Spartina alterniflora* production in a New England salt marsh. Ecology **65**:1794-1807.
Butman, C.A., M. Fréchette, W.R. Geyer and V.R. Starczak. 1994. Flume experiments on food supply to the blue mussel *Mytilus edulis* L. as a function of boundary-layer flow. Limnology and Oceanography **39**:1755-1768.
Carling, P.A. 1992. The nature of the fluid boundary layer and the selection of parameters for benthic ecology. Freshwater Biology **28**:273-284.
Carlson, D.J., D.W. Townsend, A.L Hilyard and J.A. Eaton. 1984. Effect of an intertidal mudflat on plankton of the overlying water column. Canadian Journal of Fisheries and Aquatic Sciences **4**:1523-1528.
Chalmers, A.G., R.G. Wiegert and P.L. Wolf. 1985. Carbon balance in a salt marsh: interactions of diffusive export, tidal deposition and rainfall-caused erosion. Estuarine, Coastal, and Shelf Science **21**:757-777.
Childers, D.L. 1994. Fifteen years of marsh flumes: a review of marsh-water column interactions in southeastern USA estuaries. Pages 277-293 *in* W.J. Mitsch, editor. Global wetlands: old world and new, Elsevier, New York, USA.
Childers, D.L., H.N. McKellar, R.F. Dame, F. Sklar, and E. Blood. 1993. A dynamic nutrient budget of subsystem interactions in a salt marsh estuary. Estuarine, Coastal, and Shelf Science **36**:105-131.
Dame, R.F. 1976. Energy flow in an intertidal oyster population. Estuarine and Coastal Marine Science **4**:243-253.
—— 1987. The net flux of inorganic sediments by an intertidal oyster reef. Continental Shelf Research **7**:1421-1424.
—— 1989. The importance of *Spartina alterniflora* to Atlantic coast estuaries. Critical Reviews in Aquatic. Sciences **1**:639-660.
—— 1993. The role of bivalve filter feeder material fluxes in estuarine ecosystems. Pages 245-269 *in*. R.F. Dame, editor. Bivalve filter feeders in coastal and estuarine ecosystem processes, Springer-Verlag, Heidelberg, Germany.
—— 1996. Ecology of marine bivalves: an ecosystem approach. CRC Press, Boca Raton, Florida, USA.
Dame, R.F. and D.M. Allen. 1996. Between estuaries and the sea. Journal of Experimental Marine Biology and Ecology **200**:169-185.
Dame, R.F., D.L. Childers and E.T. Koepfler. 1992. A geohydrodynamic continuum theory for the spatial and temporal evolution of marsh-estuarine ecosystems. Netherlands Journal of Sea Research **30**:63-72.
Dame, R.F., T. Chrzanowski, K. Bildstein, B. Kjerfve, H. McKellar, D. Nelson, J. Spurrier, S. Stancyk, H. Stevenson, J. Vernberg and R. Zingmark. 1986. The outwelling hypothesis and North Inlet, South Carolina. Marine Ecology Progress Series **33**:217-229.
Dame, R.F. and S.L. Libes. 1993. Oyster reefs and nutrient retention in tidal creeks. Journal of Experimental Marine Biology and Ecology **171**:251-258.
Dame, R.F., J.D. Spurrier, T. Williams, B. Kjerfve, R. Zingmark, T. Wolaver, T. Chrzanowski, H. McKellar and J. Vernberg. 1991. Annual material processing by a salt marsh-estuarine basin in South Carolina, USA. Marine Ecology Progress Series **71**:153-166.
Dame, R.F., J.D. Spurrier and T.G. Wolaver. 1989. Carbon, nitrogen and phosphorus processing by an oyster reef. Marine Ecology Progress Series **54**:249-256.
Dame, R.F., R.G. Zingmark and E. Haskin. 1984. Oyster reefs as processors of estuarine materials. Journal of Experimental Marine Biology and Ecology **83**:239-247.

Dame, R.F., R.G. Zingmark, H. Stevenson and D. Nelson. 1980. Filter feeder coupling between the water column and benthic subsystems. Pages 520-526 *in* V.S. Kennedy, editor. Estuarine perspectives. Academic Press, New York, New York, USA.

Dauer, D.M., C.A. Maybury and R.M. Ewing. 1981. Feeding behavior and general ecology of several spionid polychaetes from the Chesapeake Bay. Journal of Experimental Marine Biology and Ecology **54**:21-38.

Davies, J.M. 1975. Energy flow through the benthos in a Scottish sea loch. Marine Biology **31**:353-362

Davis, J.A. and L.A. Barmuta. 1989. An ecologically useful classification of mean and near-bed flows in streams and rivers. Freshwater Biology **21**:271-282.

Day, J.W., W.G. Smith, P. Wagner and W. Stowe. 1973. Community structure and carbon budget in a salt marsh and shallow bay estuarine system in Louisiana. Louisiana State University, Baton Rouge, Louisiana, USA, Sea Grant Publ. LSU-SG-72-04.

Feller, R.J. and B.C. Coull. 1995. Non-selective ingestion of meiobenthos by juvenile spot (*Leiostomus xanthurus*)(Pices) and their daily ration. Vie et Milieu **45**:49-59.

Frechétte, M., D. Lefaivre and C.A. Butman. 1993. Bivalve feeding and the benthic boundary layer. Pages 325-367 *in* R.F. Dame, editor. Bivalve filter feeders in estuarine and coastal ecosystem processes, Springer-Verlag, Heidelberg, Germany.

Frey, R.W. and P.B. Basan. 1985. Coastal salt marshes. Pages 225-301 *in* R.A. Davis, editor. Coastal sedimentary environments, Springer-Verlag, New York, New York, USA.

Frostick, L.E. and I.N. McCave. 1979. Seasonal shifts of sediment within an estuary mediated by algal growth. Estuarine and Coastal Marine Science **9**:569-576.

Gambrell, R.P. and W.H. Patrick. 1978. Chemical and microbiological properties of anaerobic soils and sediments. Pages 375-423 *in* D.D. Hook and R.M.M. Crawford, editors. Plant life in anaerobic environments. Ann Arbor Science Publications, Ann Arbor, Michigan, USA.

Grave, C. 1912. A manual of oyster culture in Maryland. Fourth Report Board Shellfish Commissioners. Baltimore, Maryland, USA.

Hammen, C.S. 1980. Marine invertebrates: comparative physiology. University Press of New England, Hanover, New Hampshire, USA.

Hargrave, B.T. 1973. Coupling carbon flow through some pelagic and benthic communities. Journal of the Fisheries Research Board of Canada **30**:1317-1326.

Herman, P.M.J. and H. Scholten. 1990. Can suspension-feeders stabilise estuarine ecosystems? Pages 104-116 *in*. M. Barnes and R.N. Gibson, editors. Trophic relationships in the marine environment. Aberdeen University Press, Aberdeen, Scotland.

Howarth, R.W. and A. Giblin. 1983. Sulfate reduction in the salt marshes at Sapelo Island, Georgia. Limnology and Oceanography **28**:70-82.

Howarth, R.W., A. Giblin, J. Gale, B.J. Peterson and G.W. Luther. 1983. Reduced sulfur compounds in the pore waters of a New England salt marsh. Environmental. Biogeochemistry and Ecology Bulletin Stockholm **35**:135-152.

Howarth, R.W. and J.M. Teal. 1979. Sulfate reduction in a New England salt marsh. Limnology and Oceanography **24**:999-1013.

Jones, R.C. 1980. Productivity of algal epiphytes in a Georgia salt marsh: effect of inundation frequency and implication for total marsh productivity. Estuaries **3**:314-317.

Jordan, T.E. and I. Valiela. 1982. A nitrogen budget of the ribbed mussel, *Geukensia demissa*, and its significance in nitrogen flow in a New England salt marsh. Limnology and Oceanography **27**:75-90.

Jordan, T.E. and I. Valiela. 1983. Sedimentation and resuspension in a New England salt marsh. Hydrobiology **98**:179-184.

Jørgensen, C.B. 1990. Bivalve filter feeding: hydrodynamics, bioenergetics, physiology and ecology. Olsen and Olsen, Fredensborg, Denmark

Ke, X., M.B. Collins and S.E. Poulos. 1994. Velocity structure and sea bed roughness associated with intertidal sand and mud flats and saltmarshes of Wash. U.K. Journal of Coastal Research **10**:702-714.

Kraeuter, J.N. 1976. Biodeposition by salt-marsh invertebrates. Marine Biology **35**:215-223.

Kraeuter, J.N. and R.L. Wetzel. 1986. Surface sediment stabilization-destabilization and suspended sediment cycles on an intertidal mudflat. Pages 203-223 *in*. D.A. Wolfe, editor. Estuarine variability. Academic Press, New York, New York, USA.

Kuenzler, E.J. 1961. Phosphorus budget of a mussel population. Limnology and Oceanography **6**:400-415.

LaBarbera, M. 1977. Brachiopod orientation to water movement. I. Theory, laboratory behavior and field orientations. Paleobiology **3**:270-287.

Levinton, J. 1991. Variable feeding behavior in three species of *Macoma* (Bivalvia: Tellinacea) as a response to water flow and sediment transport. Marine Biology **110**:375-383.

Mann, K.H. and J.R.N. Lazier. 1991. Dynamics of marine ecosystems. Blackwell Scientific Publishers, London, England.

Nixon, S.W. 1980. Between coastal marshes and coastal waters: a review of twenty years of speculation and research on the role of saltmarshes in estuarine productivity and water chemistry. Pages 437-525 *in*. P. Hamilton and K.B. MacDonald, editors. Estuarine and wetland processes. Plenum Press, New York, New York, USA.

Nixon, S.W., C.A. Oviatt, J. Garber and V. Lee. 1976. Diel metabolism and nutrient dynamics in a salt marsh embayment. Ecology **57**:740-750.

Nixon, S.W., C.A. Oviatt, C. Rogers and K. Taylor. 1971. Mass and metabolism of a mussel bed. Oecologia **8**:21-30.

Nowell, A.R.M. and P.A. Jumars. 1984. Flow environments of aquatic benthos. Annual Review of Ecology and Systematics **15**:303-328.

Odum, E.P. 1969. The strategy of ecosystem development. Science **164**:262-269.

Odum, W.E., J.S. Fisher and J.C. Pickral. 1979. Factors controlling the flux of POC from estuarine wetlands. Pages 485-495 *in*. R.J. Livingston, editor. Ecological processes in coastal and marine systems. Academic Press, New York, New York, USA.

Officer, C.B., T.J. Smayda and R. Mann. 1983. Benthic filter feeding: a natural eutrophication control. Marine Ecology Progress Series **9**:203-210.

Peterson, C.H. and R. Black. 1987. Resource depletion by active suspension feeders on tidal flats: influence of local density and tidal elevation. Limnology and Oceanography **32**:143-166.

Peterson, C.H. and N.M. Peterson. 1979. The ecology of intertidal flats of North Carolina: a community profile. U.S. Fish and Wildlife Service, Office of Biological Services. FWS/OBS-**79/39**.

Pinckney, J.L. and R.G. Zingmark. 1993. Modeling the annual production of intertidal benthic microalgae in estuarine ecosystems. Journal of Phycology **29**:396-407.

Pomeroy, L. and R. Wiegert. 1981. The ecology of a salt marsh. Springer-Verlag, New York, USA.

Reed, D.J. 1988. Patterns of sediment deposition in subsiding coastal marshes, Terrbone Bay, Louisiana: the role of winter storms. Estuaries **12**:222-227.

Rowe, G.T. 1971. Benthic biomass and surface productivity. Pages 441-454 *in* J.D. Costlow, editor. Fertility of the sea. Gordon and Breach, New York, New York, USA.

Rowe, G.T., C.H. Clifford, K.L. Smith and P.L. Hamilton. 1975. Benthic nutrient regeneration and its coupling to primary productivity in coastal waters. Nature **255**:215-217.

Smaal, A.C. and T.C. Prins. 1993. The uptake of organic matter and the release of inorganic nutrients by bivalve suspension feeder beds. Pages 271-298 *in* R.F. Dame, editor. Bivalve filter feeders in estuarine and coastal ecosystem processes. Springer-Verlag, Heidelberg, Germany.

Stevenson, J.C., L.G. Ward and M.S. Kearney. 1988. Sediment transport and trapping in marsh systems: implication of tidal flux studies. Marine Geology **880**:37-59.

Stumpf, R.P. 1983. The processes of sedimentation on the surface of a salt marsh. Estuarine and Coastal Marine Science **17**:495-508.

Taghon, G.L. and R.R. Greene. 1992. Utilization of deposited and suspended particulate matter by benthic "interface"' feeders. Limnology and Oceanography **37**:1370-1391.

Teal, J.M. 1962. Energy flow in the salt marsh ecosystem of Georgia. Ecology **43**:614-624.

Trager, G.C., J.S. Hwang and J.R. Strickler. 1990. Barnacle suspension feeding in variable flow. Marine Biology **105**:117-125.

Valiela, I. and J.M. Teal. 1979. The nitrogen budget of a salt marsh ecosystem. Nature **280**:652-656.

Vogel, S. 1974. Current-induced flow through the sponge, *Halichondria*. Biological Bulletin **147**:443-456.

Vogel, S. 1981. Life in moving fluids: the physical biology of flow. Willard Grant Press, Boston, USA.

Warner, G.F. 1977. On the shapes of passive suspension feeders. Pages 567-576 *in* B.F. Keegan, P.O. Ceidigh and P.J.S. Boaden, editors. Biology of benthic organisms. Pergamon Press, Oxford, England.

Welsh, B.L. 1980. Comparative nutrient dynamics of a marsh-mudflat ecosystem. Estuarine and Coastal Marine Science **10**:143-164.

Wildish, D. and D. Kristmanson. 1984. Importance of mussels of the benthic boundary layer. Canadian Journal of Fisheries and Aquatic Sciences **41**:1618-1625.

—— 1997. Benthic suspension feeders and flow. Cambridge University Press, New York, New York, USA.

Wildish, D. and A.M. Saulnier. 1993. The effect of velocity and flow direction on the growth rate of juvenile and adult giant scallops. Journal of Experimental Marine Biology and Ecology **155**:133-143.

Wolaver, T.G., R. F. Dame, J.D. Spurrier and A.B. Miller. 1988. Sediment exchange between a euhaline salt marsh in South Carolina and the adjacent tidal creek. Journal of Coastal Research **4**:17-26.

Wolaver, T.G. and J.C. Zieman. 1983. Effect of water column, sediment, and time over the tidal cycle on the chemical composition of tidal water in a mesohaline marsh. Marine Ecology Progress Series **12**:123-134.

TWENTY MORE YEARS OF MARSH AND ESTUARINE FLUX STUDIES: REVISITING NIXON (1980)

DANIEL L. CHILDERS
Department of Biological Sciences &
Southeast Environmental Research Center
Florida International University
University Park, OE 236
Miami, FL 33199 USA

JOHN W. DAY, JR.
Department of Oceanography and Coastal Sciences
Louisiana State University
Baton Rouge, LA 70803 USA

HENRY N. MCKELLAR, JR.
Department of Environmental Health Sciences and
Marine Science Program
University of South Carolina
Columbia, SC 29208 USA

Abstract

In 1980, Scott Nixon reviewed the role of salt marshes in estuarine and coastal productivity. His review was effectively a progress report on the testing of "The Outwelling Hypothesis" (Odum, 1980). Nixon (1980) signaled a crucial turning point in the direction of estuarine flux studies conducted since then. In this review we revisit Nixon (1980), focusing on research and thinking that has been guided by The Outwelling Concept in the last two decades. Since 1980, estuarine flux studies have been conducted at 41 different sites and presented in over 42 publications. More than a third of these were conducted in Europe, Africa, Australia, or Mexico. Our review of these studies highlighted several important advances. The first was evolution of a conceptual approach that decomposes the estuary-coastal ocean landscape into interacting subsystems (i.e., the coastal ocean, estuarine basins, and marsh). Most post-1980 flux studies have addressed interactions between these individual subsystems, often in an hierarchical sense. Over half of these quantified exchanges between marsh-dominated basins and the greater estuary–generally through a single, well-defined tidal channel. From these data, we found that tidal range, subsystem area, and distance to the ocean together explained 87% of the variability in total organic carbon (TOC) exchanges and 92% of the variability in total suspended solids (TSS) fluxes, with exports occurring at lower tidal ranges, areas, and distances. Tidal range explained 40% of the variability in nitrate + nitrite (NN) exchange (with uptake at ranges below about 1.2 m and export at greater tidal ranges) and 39% of available phosphorus (SRP) flux variation (with export at ranges below about 1.6 m). We were unable to extract similar relationships from whole-estuary exchange studies because so few exist. The geomorphological setting and degree of ecological maturity (analogous to geologic age) of a marsh or basin within an estuary are important controllers of ecological function, thus flux behavior. We applied concepts of community succession and ecosystem development to data from marsh-water column flux studies, and found that slope of flux vs. tidal height relationships was greater for younger marshes compared to all marshes, and much greater for younger marshes compared to older marshes. This change in slope often caused a shift in the inflection point that indicated the tidal range at which export shifted to import, or vice versa. These studies quantified surficial fluxes, though, and a number of post-1980 studies demonstrated the importance of other processes, including subsurface flow, subtidal advection, and the movement of nutrients and organic matter by animals (other chapters in this volume address these processes). Finally, a number of studies showed strong controls on fluxes by exogenous environmental forcing, and we reviewed several studies that used innovative budgeting and modeling of flux dynamics and ecological processes to incorporate these sources of variability. Since 1980 we have learned a great deal more about how estuarine wetlands interact with their estuaries, and of the value of establishing a conceptual framework and system boundaries. Estuarine ecologists have learned a great deal about outwelling as a concept although few flux studies have directly addressed the original Outwelling Hypothesis. We suggest that the question should not be "Is The Outwelling Hypothesis true?" but rather: 1) how are materials being exchanged between different subsystems in estuary-coastal ocean landscapes? 2) what are the mechanisms of this exchange? and 3) how do exogenous forcings control these patterns of exchange? Estuarine scientists are encouraged to view The Outwelling Hypothesis as a conceptual stimulus of ideas and not as a strict statistical hypothesis that must be proven or disproven.

1. Nixon's 1980 Review of Marsh-Estuarine Interactions

In 1980, Scott Nixon published his exhaustive and seminal review of marsh-estuarine interactions. His primary objective with this paper was to summarize 20 years of research, and speculation, into the role of salt marshes in estuarine and coastal productivity. In effect, it was a presentation of progress to date on testing "The Outwelling Hypothesis" (Odum, 1980). His paper was also a critical review of the process by which this idea had, in fact, been scientifically defined and tested. Nixon (1980) represented a milestone in estuarine flux research, as well as a crucial turning point in the very approach and direction of estuarine flux studies conducted since 1980. Notably, our goal is not to prove or disprove the Outwelling Hypothesis. Rather, the goal of this paper is to "revisit" Nixon (1980), focusing on research and thinking that has been guided by the outwelling concept in the last two decades. Our objectives are:

- To summarize the main messages and conclusions of Nixon's (1980) review of marsh-estuarine interactions;
- To present a review of marsh-estuarine flux research that has been conducted in the two decades since Nixon (1980);
- To discuss several important conceptual and methodological advances since Nixon (1980), and;
- To suggest future directions for this line of inquiry, given what we have learned in this time.

1.1 ESTUARINE FLUX RESEARCH FROM 1960 TO 1980

Sapelo Island, Georgia, USA was the center of research and ideas that, in the late 1950s and early 1960s, led to The Outwelling Hypothesis. The road to a formal presentation of this idea appears to have begun with John Teal's 1962 paper published in *Ecology*, and in particular with the last statement in that paper:

> ''...the tides remove 45% of the production before the marsh consumers have a chance to use it and in so doing permit the estuaries to support an abundance of animals.'' (Teal 1962)

E.P. Odum used the term ''outwelling'' for the first time in 1968, when he likened the supply of nutrients and energy to coastal oceans from salt marshes and estuaries to upwelling–which supplies nutrients to these same systems from their deep ocean boundaries. Evidence for this phenomenon came from data presented by Scheleski and Odum (1961), Thomas (1966), Odum and de la Cruz (1967), and Pomeroy et al. (1967). These publications actually contained few data that demonstrated an outwelling of nutrients and organic matter from salt marshes to coastal oceans. Nonetheless, the following quote is representative of the way in which the Outwelling concept became integrated into estuarine research:

> ''The importance of [salt marshes] as ''primary production pumps'' that ''feed'' large areas of adjacent waters has only been recently recognized...'' (Odum, 1968)

This concept of salt marsh and estuarine export quickly became a paradigm, even dogma, in estuarine research. The term ''Outwelling'' was often used as though it was a proven quantitative concept when actually it was first presented as a qualitative idea (Walker 1973). A central theme of Nixon's review (1980) was his concern that the idea of Outwelling gained acceptance by the estuarine research community via emotional acceptance rather than critical, empirical evaluation. To Nixon, this created a serious lack of objectivity that actually threatened scientific credibility. He forcefully argued that marsh-estuarine exchange research must redirect itself to test Outwelling as a hypothesis rather than to substantiate Outwelling as a conclusion (Nixon, 1980). In fact, it is difficult to find the Outwelling concept explicitly presented as a testable hypothesis anywhere in the literature before 1980, when E. P. Odum himself presented it as such (Odum 1980). Interestingly, we conducted a comparative search of Nixon

(1980) and Odum (1980) citations from 1981 through 1997. Nixon's review paper (1980) appears to have been a stronger guiding force in the literature than Odum's paper (1980, Fig. 1). This difference in citations is only one indication of how Nixon (1980) affected the types of estuarine flux work conducted in the last two decades.

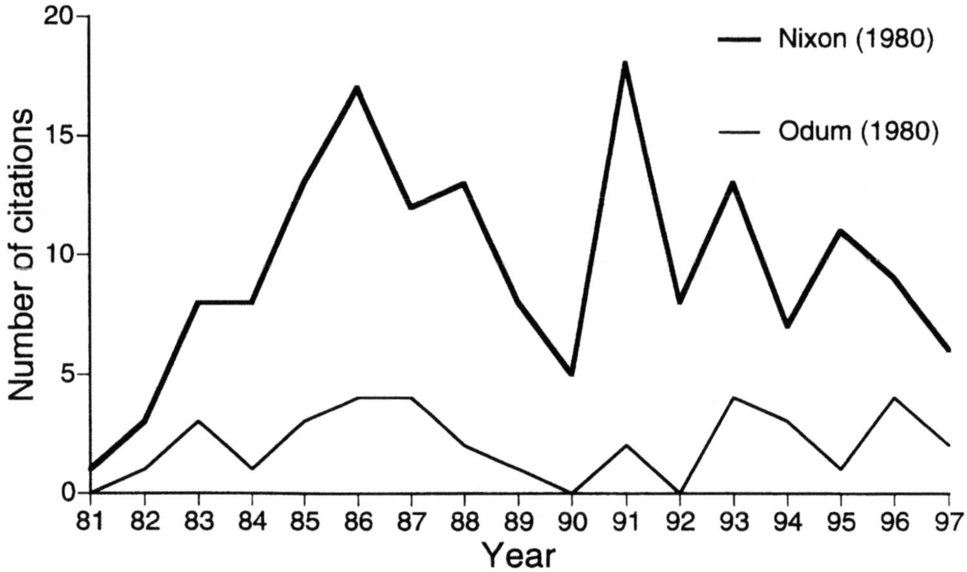

Figure 1. The number of times Nixon (1980) and Odum (1980) were cited each year from 1981 to 1997, as listed in the Science Citation Index.

A number of estuarine flux studies had been conducted by 1980. Many of those directly quantified the exchange of nutrients, organic material, or both, in estuarine ecosystems (Table 1). It seems clear now, in hindsight, that the objective of most—if not all—of these studies was to quantify the Outwelling concept. In reviewing these studies, Nixon also appeared to have this in mind: He described these studies as quantifying ''fluxes between marshes and estuarine waters...'' (Table 8, Nixon 1980), and he presented annual flux estimates from these studies as ''flux...between salt marshes and coastal waters'' (Tables 10, 12, and 14, Nixon 1980). On closer focus, however, only 2 of the 12 estuarine flux datasets available in 1980 actually quantified exchanges between estuaries and the nearshore coastal ocean—the Great Sippewissett Marsh, MA, USA study (Valiela et al. 1978, Valiela and Teal 1979) and the Barataria Basin, LA, USA study (Happ et al. 1977, see Table 1). Of the others, all measured exchange with some intermediate water body and, while two sites were approximately 5 km from the coastal ocean, two others were more than 75 km from the ocean (Table 1). This problem of defining boundaries, terms, and what datasets actually test the Outwelling Hypothesis directly is not trivial. Below, we discuss how this terminology issue helped define the direction of the last two decades of estuarine flux research, and suggest that it remains much more than a simple semantic argument.

TABLE 1: Summary of marsh-estuarine interaction studies summarized in Nixon (1980). "Outwelling Study?" notes whether the flux study in question directly quantified exchanges between an estuarine system and the nearshore coastal ocean. Where studies did not directly quantify outwelling, we show the water body involved in the marsh-estuarine flux that was quantified, and note the approximate distance between the study system and the coastal ocean. We note where Nixon(1980) presented net annual fluxes C, N, or P between a given "salt marsh and the coastal ocean"--see Nixon (1980) for a review of the magnitude and direction of these fluxes.

ESTUARY	OUTWELLING STUDY?	WETLAND TYPE	ADJOINING WATER BODY	DISTANCE TO COASTAL OCEAN	FLUX ESTIMATES PRESENTED? carbon	nitrogen	phosphorus	REFERENCE
Great Sippewissett Marsh, Massachusetts, USA	yes	saltmarsh	Atlantic Ocean	0 km	yes	yes	yes	11,16,17
Barataria Bay, Louisiana, USA	yes	all types	Gulf of Mexico	0 km	yes	no	no	5,11
Block Island, Rhode Island, USA	no	saltmarsh	Block Island Sound	< 5 km	no	no	no	7,11,12
Canary Creek, Delaware, USA	no	saltmarsh	Delaware Bay	≈ 5 km	yes	yes	yes	8,11
Dill Creek, South Carolina, USA	no	saltmarsh	Ashley River	≈15 km	yes	no	no	11,13
Sapelo Island, Georgia, USA	no	saltmarsh	Duplin River	≈ 15 km	no	no	no	2,3,11
Ware Creek, Virginia, USA	no	saltmarsh	York River	≈ 25 km	yes	yes	yes	1,10,11
Carter's Creek, Virginia, USA	no	saltmarsh	York River	≈ 25 km	yes	yes	yes	1,10,11
Providence River, Rhode Island, USA	no	saltmarsh	Providence River	≈ 30 km	no	no	no	7,11,12
Flax Pond, New York, USA	no	saltmarsh	Long Island Sound	≈ 50 km	yes	yes	yes	4,9,1,18,1
Gott's Marsh, Maryland, USA	no	tidal fresh	Patuxent River	≈ 75 km	yes	yes	yes	6,11,15
Hamilton Marsh, New Jersey, USA	no	tidal fresh	Delaware River	> 100 km	no	no	no	11,14

Citations:
1. Axelrad, 1974
2. Haines, 1977
3. Haines et al., 1977
4. Hall et al., 1975
5. Happ et al., 1977
6. Heinle and Flemer, 1976
7. Lee, 1979
8. Lotrich et al., 1979
9. Moll, 1977
10. Moore, 1974
11. Nixon, 1980
12. Oviatt et al., 1977
13. Setttlemyre and Gardner, 1976
14. Simpson et al., 1978
15. Stevenson et al., 1977
16. Valiela et al., 1978
17. Valiela and Teal, 1979
18. Woodwell et al., 1977
19. Woodwell and Whitney, 1977

1.2 TERMINOLOGY VERSUS SEMANTICS

The title of Nixon's review begins with "Between Coastal Marshes and Coastal Waters", and throughout the text he refers to this phenomenon as "marsh-estuarine interactions" and "marsh-estuarine exchanges" (Nixon 1980). He used these terms to review the pool of pre-1980 research driven by the Outwelling concept. Did these terms refer to fluxes between estuaries and the coastal ocean? As we noted above, only 2 of the 12 studies reviewed by Nixon (1980) actually attempted to quantify estuary-coastal ocean exchanges (Table 1). Thus, these two studies were actually directed at the Outwelling concept rather than at marsh-estuarine fluxes. Did these terms refer to fluxes between an intertidal marsh basin and the greater estuary? Eight of the studies reviewed by Nixon (1980), in fact, measured these kinds of fluxes (Table 1). These studies clearly did not directly test the Outwelling concept, but they did address marsh-estuarine flux dynamics. However, these studies did not permit isolation of intertidal marsh exchanges *per se*, because they were conducted in tidal channels draining small marsh basins that also included subtidal benthic communities and uplands inputs. And thus, the terms "marsh-estuarine interactions" and "marsh-estuarine exchanges" might also have referred to direct fluxes between intertidal marshes and their inundating water column. Two of the studies reviewed by Nixon (1980; Block Island and Providence River, RI, USA–Lee, 1979) actually measured these kinds of interactions directly using the marsh flume technique for the first time (Table 1). While this technique does allow separation of intertidal marsh interactions from other estuarine system components, marsh-water column flux studies clearly do not address the central tenet of the Outwelling Hypothesis (in spite of the fact that one of Lee's flumes was in a marsh less than 5 km from the Atlantic Ocean; Table 1).

This ambiguity in terms and boundaries is not a trivial semantic argument. In Fig. 2, we demonstrate the hierarchical nature of these three points of reference. Marsh-water column interactions are inherently part of marsh basin flux dynamics (as are subtidal benthic interactions and uplands inputs), and estuary-coastal ocean flux patterns necessarily integrate marsh and subbasin exchanges throughout the estuary in question. It is not surprising that Nixon had considerable trouble finding patterns in the disparate collection of flux studies with which he had to work. For example, we should expect that the flux behavior of a tidal freshwater marsh basin located on the Patuxent River, MD, USA and about 75 km from the ocean (Gott's Marsh — Heinle and Flemer 1976) would be very different from the flux behavior of a large salt marsh basin located on Delaware Bay about 5 km from the ocean (Canary Creek — Lotrich et al. 1979). Similarly, we should expect that flux behavior measured in a marsh flume at Block Island, Rhode Island, USA (Lee 1979) would be quite different from fluxes measured between the Great Sippewissett Marsh and the coastal ocean (Valiela et al. 1978, Valiela and Teal 1979), in spite of similar intertidal marsh types and the proximity of the former to the ocean. In the last two decades, this hierarchical conceptualization of coastal landscapes has been developed in response to this need to define terminology and identify system boundaries across which fluxes are being measured. It is not clear whether the need to resolve this ambiguity of definitions and boundaries was clear to

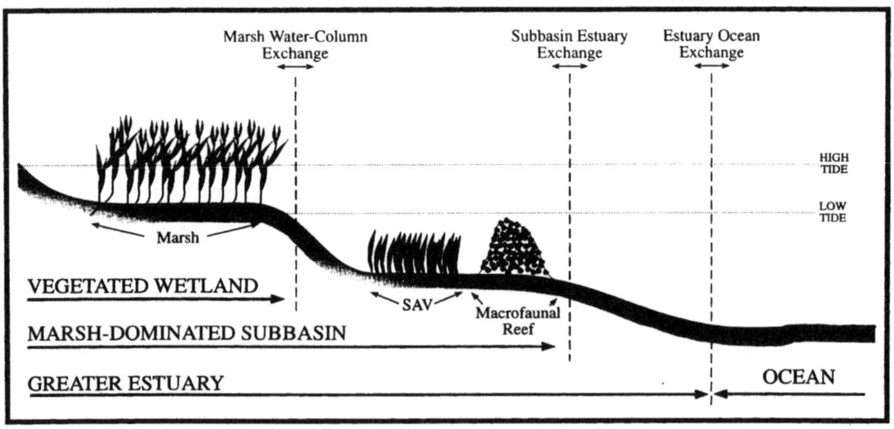

Figure 2. A schematic representation of the estuarine-coastal landscape that includes estuarine subbasins nested within the greater estuary and the vegetated wetland nested within both—in a classic hierarchical sense. This picture forms the basis for our expanded definition of the outwelling concept, in which marsh-water column exchanges affect subbasin-estuary exchanges, which also affect estuary-coastal ocean interactions.

Nixon. Furthermore, we were unable to identify any single publication or group of publications since 1980 that explicitly defined this new hierarchical perspective. Nonetheless, we feel that this conceptual clarification has been a defining feature of the last two decades of estuarine flux work, and we devote considerable time (below) to discussion of this development.

1.3 NIXON'S CONCLUSIONS AND RECOMMENDATIONS

Nixon (1980) did generate some interesting conclusions and generalizations about marsh-estuarine exchanges, in spite of the limitations of the data he was reviewing. As we noted above, he reviewed direct measurements of water flows and constituent fluxes in several clearly defined marsh-estuarine embayments. He also reviewed how estuarine intertidal marshes interact with estuarine and coastal systems by presenting evidence from studies that measured sediment deposition and the associated accumulation of macronutrients. From these data, Nixon concluded that ''tidal marshes'' appeared: 1) to export organic carbon; 2) to transform nitrogen by taking up dissolved oxidized forms (e.g., nitrate and nitrite) and exporting dissolved (e.g., ammonium) and particulate reduced forms (e.g., total nitrogen), and 3) to take up total phosphorus but remineralize and export small amounts of soluble reactive phosphorus. Again, the ambiguity about terminology and system boundaries makes interpretation of these general patterns difficult.

Nixon extended his analysis to the nearshore coastal ocean, and presented a relationship between marsh:open water area and estuarine phytoplankton production. There was no clear pattern here. In fact, when Nixon split the Patuxent River subestuary, MD, USA into Upper River and Lower River subsystems, they effectively spanned the range of area ratios and production (Fig. 12, Nixon 1980). This lack of a clear relationship between the importance of intertidal marsh and open water productivity is also demonstrable in the literature from this time period. Turner et al. (1979) argued that higher nearshore productivity along both Georgia and Louisiana coasts strongly suggested an outwelling phenomenon. At the same time—and also in the Mid-Atlantic Bight—Haines (1975, 1979a,b) used stable isotopic evidence to conclude that high nearshore productivity was supported by autochthonous phytoplankton production stimulated by freshwater nutrient sources and not by salt marsh export. Nixon went one step further, and reviewed data from nearshore coastal ocean fisheries for evidence of an energy subsidy from estuaries and intertidal marshes. The relationship here was also tenuous, largely because few data were available to complement the flux dataset but also because the Chesapeake Bay supported a very large estuarine fishery yet was made up of <10% marsh (Fig. 16, Nixon 1980). Nixon (1980) found a stronger relationship with all other estuarine systems for which data were available. By and large, Nixon concluded that the evidence available to him in 1980 did not support the Outwelling Hypothesis in a convincing way.

Perhaps the most important single contribution of the Nixon (1980) review was controversial, even provocative. Nixon ended by exhorting the estuarine research community to redefine its perspective of salt marshes, and coastal ecosystems in general. ''Are coastal marshes important or not?'' was not the correct question. This overly inflated the importance of subjective valuation and non-empirical emotion to the scientific process. He demonstrated how the Outwelling concept had generated a great deal of enlightening research in the 1960s and 1970s. However, this single issue also tended to stimulate a view in which demonstrating societal importance superceded quantifying ecological function. Nixon urged us, as estuarine scientists, to take more objective and empirical approaches to testing the Outwelling Hypothesis and to quantifying the way intertidal wetlands, estuaries, and coastal ecosystems function and interact. Our review of the last two decades of estuarine flux research suggests that this advice was well heeded.

2. Estuarine Flux Research since 1980

A large number of estuarine flux studies have been conducted in the last two decades. Nixon (1980) reviewed exchange studies conducted at 12 different sites and presented in about 20 publications (Table 1). We have summarized estuarine flux studies, from both tidal marsh and mangrove systems, conducted at 41 different sites and presented in over 42 publications (Tables 2 through 4), recognizing that this list is likely not complete. In addition to the dramatic increase in available datasets since 1980, estuarine flux research took a more international bent during this time. All 12 studies cited in Nixon (1980) were conducted in the US, and of those only one was conducted outside of the Atlantic coast of the US (the Barataria Basin study by Happ et al. 1977, Table 1). In contrast, 14 of the 41 sites we present here—more than a third of the total—were conducted in Europe, Africa, Australia, or Mexico (Tables 2 through 4). Our review of these studies highlights the following important advances, each of which we detail in the following sections:

a) A conceptual approach has evolved that decomposes the estuary-coastal ocean landscape into interacting subsystems, solving the problem of system boundaries to a large extent.
b) Most estuarine flux studies have addressed some specific aspect of interactions between these individual subsystems, often in an hierarchical sense.
c) The geomorphological setting and degree of ecological maturity (analogous to geologic age) of a marsh or basin within an estuary are important controllers of ecological function, thus flux behavior. Concepts of community succession and ecosystem development have been evoked to explain this connection.
d) A number of studies have demonstrated the importance of other processes, which are less obvious than surficial tidal exchange. These include subsurface flow, subtidal advection, and the movement of nutrients and organic matter by animals.
e) A number of studies have shown that fluxes are highly variable in time and are (to various degrees) controlled by exogenous environmental forcing.
f) Several studies have demonstrated the efficacy of innovative budgeting and modeling of flux dynamics and ecological processes.

2.1 A NEW CONCEPTUAL APPROACH: INTERACTING SUBSYSTEMS

There are many problems with generalizing about marsh and estuarine flux behaviors using data from diverse and disparate systems. In sections above, we used the estuarine flux dataset reviewed by Nixon (1980) to demonstrate some of these issues. In fact, many of these problems have been solved with the use of a conceptual approach that divides estuarine ecosystems into interacting subsystems. These subsystems are few and generic enough that they encompass the critical components of virtually all estuarine ecosystems. The subsystems include intertidal marsh, subtidal benthic areas, subtidal and/or intertidal macrofaunal beds (oyster reefs, mussel beds, clam flats, etc.), and adjacent upland systems. In the center of these peripheral subsystems is the water column (Fig. 3). This subsystem effectively links the other subsystems, and mediates and integrates exchanges between all other subsystems in the estuary, as pictured in Fig. 3. The water column directly interacts across the ecosystem boundaries. If the focus is on estuary-coastal ocean interactions, as in studies literally testing the Outwelling Hypothesis, then the extra-boundary system is the nearshore coastal ocean (Fig. 3a). If the focus is on interactions between an estuarine basin and the greater estuary, then the extra-boundary system is the greater estuary (Fig. 3a). If the system of interest is along a river, the conceptual diagram contains upstream and downstream extra-boundary components (Fig. 3b). And finally, if the focus is on marsh-water column interactions, only the marsh and water column subsystems are considered. This approach has not been explicitly employed in all estuarine flux studies, but it has greatly facilitated the synthetic integration of data from different sites and systems, as well as data from different hierarchical levels of the same system. We see this informal standardization of boundaries and system terminology as a major contribution of the last two decades of estuarine flux work. The remainder of our review will emphasize this hierarchical organization scheme.

2.1.1 Studies Quantifying Exchanges Between Estuaries and the Nearshore Coastal Ocean

Thirty years ago, Odum (1968) first presented the idea of estuarine outwelling. By 1980, the idea had stimulated a great deal of novel estuarine research, as well as a measure of controversy and introspection (Nixon 1980). One of the most enlightening findings of our review is that, in fact, very few estuarine flux studies have directly tested the original Outwelling Hypothesis. Of the 12 cited by Nixon (1980), only the Great Sippewissett, MA, USA (Valiela et al. 1978, Valiela and Teal 1979) and Barataria Basin, LA, USA (Happ et al. 1977) studies actually quantified estuary-coastal ocean exchanges (Table 1). And Happ et al. (1977) used indirect methods, rather than direct sampling in a tidal pass. We were able to locate only seven studies that have quantified estuary-coastal ocean exchanges in the last two decades, with two from mangrove-dominated systems (Table 2). The Zwin salt marsh study in the Netherlands did not generate any actual flux estimates — the objective of this work was to quantify how the salt marsh affected the quality of organic matter (Hemminga et al. 1992, Klap 1997). Reed (1988) quantified only suspended sediment flux in the Bridge Creek, England study. Day et al. (1995) have sampled estuary-coastal ocean exchanges between Fourleague Bay and the Gulf

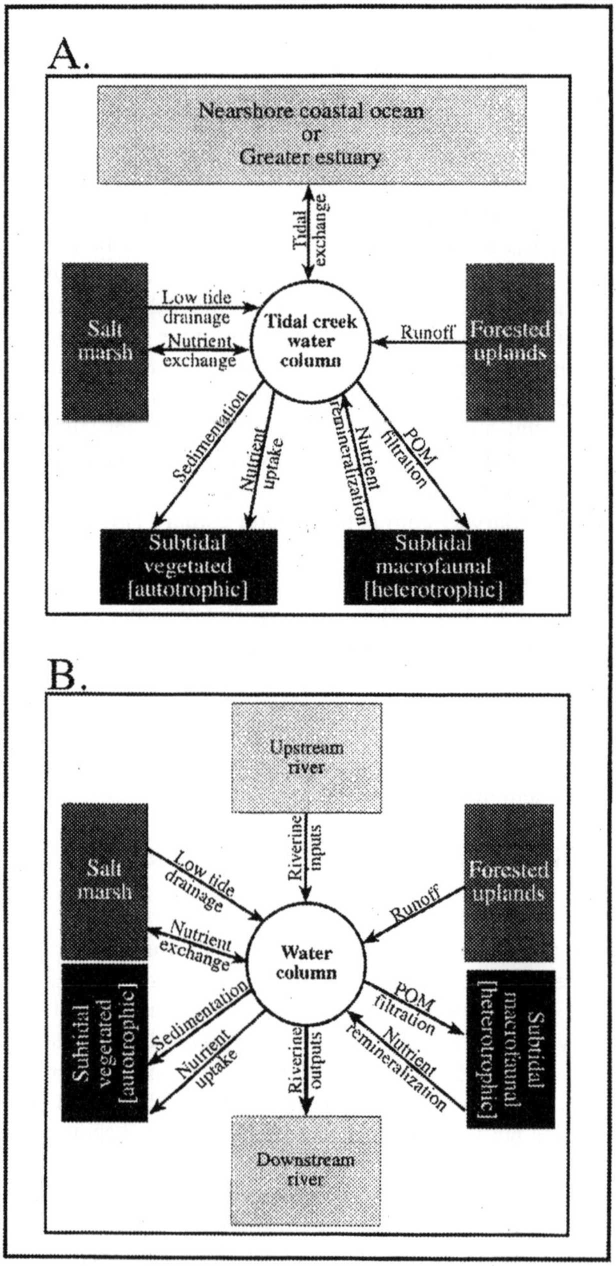

Figure 3. A conceptualization of the estuarine ecosystem as a network of interacting subsystems, with the water column in the central, or ''hub'' position, indicating its importance as integrator of subsystem exchanges (modified from Childers et al. 1993). (a) the general subsystems that characterize a tidal estuary, (b) the general subsystems and interactions that characterize a riverine estuary.

TABLE 2: Summary of whole estuary flux studies published since 1980. Fluxes represent quantification of true estuary-coastal ocean exchanges. In all cases; fluxes are shown from the estuarine reference frame such that negative flux = export from the estuary to the coastal ocean and positive flux = uptake by the estuary. In the classic sense of the definition, negative flux thus represents net outwelling.

SYSTEM	TIDAL RANGE	ESTUARY AREA	VEG.MARSH AREA	NET ANNUAL FLUX (g/m2/yr) NH4	NN	SRP	POC	DOC	TSS	REFS
North Inlet, South Carolina, USA	1.7m	3200 ha	2300 ha	-6.3	-0.9	-1.7	-128	-328	-2200	4
Slufter, Texel, The Netherlands	1.7m	386 ha	?	-0.2	0.27	0.2	19.5	-20.4	347	1, 2
Coral Creek, Australia		500 ha	500 ha	0.15	-0.03	0.13		7.3		3
Fourleague Bay, Lousiana, USA	0.3m	23300 ha	14000 ha	no annual flux estimates were available						5
Huizache-Caimanero Lagoon, Mexico	≈1.5m	17500 ha	?	no annual flux estimates were available						6
Bridge Creek, Essex, England	2.5m	12ha	?	no annual flux estimates were available						9
Zwin saltmarsh, The Netherlands	1.9m	?	?	quantified the quality of POC and DOC only						7,8

Citations:
1. Asjes and Dankers, 1994
2. Boorman et al., 1994
3. Boto and Wellington, 1988
4. Dame et al., 1986
5. Day et al., 1995
6. Espino, G. and M.A.R. Medina, 1993
7. Hemminga et al., 1992
8. Klap, 1997
9. Reed, 1988

of Mexico for a number of years, but did not present any estimates of net annual exchange; nor did Espino and Medina (1993). Flores-Verdugo et al. (1987) estimated that 90% of all mangrove litterfall was exported from the El Verde Lagoon, Mexico, but they made no direct flux measurements. Only the North Inlet, South Carolina, USA (Dame et al. 1986), Slufter, Netherlands (Asjes and Dankers, 1994, Boorman et al. 1994), and Coral Creek, Australia (Boto and Wellington 1988) flux studies thus actually generated data that directly tested the Outwelling Hypothesis. Notably, this was an explicit objective only in the North Inlet Outwelling Study. Thus, after 30 years of estuarine flux studies, only 5 can be used to directly test the original Outwelling Hypothesis. In fact, if we apply some methodological consistency criteria, then only the Sippewissett, North Inlet, Slufter, and Coral Creek studies directly tested this concept that has been central to estuarine research for a generation of scientists. However, these 30 years of estuarine flux studies have generated a wealth of information which contributed to our hierarchically-based model of the outwelling phenomenon (as per Fig. 2) and which expanded our understanding of how estuarine subsystems interact (as per Fig. 3). It is our view that this expanded view of The Outwelling Hypothesis, as demonstrated in Figs. 2 and 3, allows us to discuss the validity of the Outwelling Hypothesis at this point without having to rely solely on research that focused on the estuary-coastal ocean boundary.

There are numerous reasons that very few estuarine flux studies have directly tested the original Outwelling Hypothesis. Sampling water flow and constituent concentrations in tidal channels connecting estuaries to the coastal ocean is the most intuitively direct methodology for such studies. This approach is greatly complicated if the chosen estuary is large, has multiple tidal channels, or is remotely located. Even in the simplest, smallest, and most convenient of systems, this type of research is extremely expensive. Perhaps the most significant limitation to these types of studies, however, is that flux estimates are completely dependent on the precision, accuracy, and error of the water flux measures (Kjerfve et al. 1981, Park and James 1990). Many researchers have avoided these issues by using other techniques to indirectly estimate estuary-coastal ocean interactions. Among these are studies that combine field sampling with modelling (i.e., Vörösmarty and Loder 1994) or depend on complex hydrodynamic (i.e., Rahm and Wulff 1992) or spatial models (i.e., Kjerfve et al. 1991). Other researchers have used budget approaches that involve calculating a mass balance (i.e., Hopkinson, 1988; Boynton et al. 1995) or involve budgets based on nutrient stoichiometry (i.e., Smith et al. 1987, 1991). Still other studies focused on processes in the nearshore coastal ocean (i.e., Hopkinson 1985). These, and the myriad other methods that we did not note here, are all valuable contributions to estuarine science and most provide valuable data relevant to the Outwelling Hypothesis. Reviewing this extensive body of work is beyond the scope of our review, however.

Most estuarine systems appear to show a net export to the coastal ocean. But this export integrates all inputs to the estuary (including riverine and atmospheric inputs) and all processes within the estuary (including the subsystem interactions shown in Fig. 3). In the macrotidal estuaries, such as those of the south Atlantic, USA coast, water flux and estuary-coastal ocean exchange are driven by astronomical tides (see reviews by Childers 1994, Dame 1994). In microtidal systems, such as estuaries in the Gulf of Mexico, however, climatic and riverine forcings play a much more important role in water flux and estuary-coastal exchange. For example, Madden et al. (1988) reported that the maximum water level variation during the winter frontal season in Atchafalaya Bay, Louisiana, USA was 1.2 m compared with a mean astronomical tide of about 0.3 m. A number of workers have shown that low-frequency cold fronts mediated significant estuary-coastal ocean exchanges (Moeller et al. 1993). Smith (1979) showed that even the highest astronomical tides could not effectively flush Aransas and Corpus Christi Bays, Texas, USA and that low frequency local meterological forcings played an important role in bay-shelf exchange. We address the importance of these extra-tidal exogenous forcings in a later section.

2.1.2 *Studies Quantifying Exchanges Between Estuarine Subbasins and the Greater Estuary*

The majority of estuarine flux studies reviewed by Nixon (1980) were actually quantifying exchanges between marsh-dominated basins and the greater estuary—generally through a single, well-defined tidal channel (Table 1). We found that over half of the estuarine flux studies conducted since 1980 also fell into this category (Table 3). Of these, the authors did not publish net annual flux estimates for seven sites. The 14 sites for which we found annual flux estimates encompass a wide range of characteristics, with tidal ranges that vary from 0.3 m to over 6 m and a range of system sizes that spans nearly 4 orders of magnitude (from 0.18 ha to nearly 1000 ha; Table 3). Dissolved inorganic nitrogen flux data, as NH_4^+ and nitrate+nitrite (NN), were available for 10 of 13 sites, some form of organic carbon flux data were available for 11 of 14 sites, and soluble reactive phosphorus (SRP) and total suspended sediment (TSS) flux

TABLE 3: Summary of estuarine subsystem (or subbasin) flux studies published since 1980. In all cases, fluxes are shown from the subsystem or subbasin reference frame such that negative flux = export from the subbasin to the greater estuary and positive flux = uptake by the subbasin.

SUBBASIN	TIDAL RANGE	SUBBASIN AREA	DISTANCE TO COASTAL OCEAN	BASIN TYPE	NET ANNUAL FLUX (g/m2/yr) NH4	NN	SRP	POC	DOC	TSS	REFS
Rhode River low marsh, Maryland, USA	0.3m	0.18 ha	> 100 km	tidal	-1.2	0.84	-0.68	(TOC = -1.8)		100	10
Mira estuary, Portugal	3.4m	1.5 ha	4 km	riverine	0.5	3.8	-0.1		-188	3100	1.8
Rhode River high marsh, Maryland, USA	<0.3m	3 ha	> 100 km	tidal	-0.13	-0.06	-0.29	(TOC = -48.0)		160	10
Crommett Creek, New Hampshire, USA	1m	4.1 ha	20 km	tidal	2.06	0.32	0.56			1166	5
Kariega Estuary Marsh, South Africa	?	4.2 ha	10 km	tidal				3	13		18
Mont St. Michel Bay, France	12m	5.7 ha	15 km	tidal	-4	-31.6		15.9	-74	29970	1.13
Brown's Marsh, South Carolina, USA	0.5 m	8.6 ha	?	riverine	-5.7	8.4			-57.3		14
North River, Massachusetts, USA	≈1m	22.8 ha	?	riverine	1.41	10.6					3
Blackwater estuary, England	4.3m	38.5 ha	6 km	tidal	39	-54	2.2			≈0	1
Bly Creek, South Carolina, USA	1.7m	66 ha	5 km	tidal	-0.54	0.27	-0.3	31	-250		6
Canary Creek, Delaware, USA	1.3m	190 ha	5 km	tidal				-55	-104		16
Coon Creek Marsh, Texas, USA	0.3m	270 ha	25 km	tidal				(TOC = -25.6)			2
Ems-Dollard, The Netherlands	3m	980 ha	50 km	tidal	-1.3	3.9	-0.28	140	-15	14300	7
Salicornia Marsh, Scolt Head Island, England	6.4m	0.88 ha	3 km	tidal				-10.8	-12.7	4740	16
Saeftinge, Westerschelde, The Netherlands	4.8m	2800 ha	50 km	tidal	quantified the quality of POC and DOC only						9,11,12
Sint Annaland, Oosterschelde, The Netherlands	3m	?	40 km	tidal	quantified the quality of POC and DOC only						12
Tomales Bay-Marsh #31, California, USA	≈1m	14 ha	21 km	tidal	no annual flux estimates were available						4
Tomales Bay-Marsh #32, California, USA	≈1m	2.5 ha	25 km	tidal	no annual flux estimates were available						4
Parker River, Massachusetts, USA	1.55m	5100 ha	20 km	riverine	no annual flux estimates were available						20
Branford Harbor-Pawson Marsh, Connecticut, USA	1.8m	?	?	tidal	no annual flux estimates were available						19
Willow Bayou, Lousiana, USA	<0.3m	?	20 km	tidal	no annual flux estimates were available						17

Citations:
1. Boorman et al., 1994
2. Borey et al., 1983
3. Bowden et al., 1991
4. Chambers et al., 1994
5. Daly and Mathieson, 1981
6. Dame et al., 1991
7. Dankers et al., 1984
8. de Bettencourt et al., 1994
9. Hemminga et al., 1993
10. Jordan and Correll, 1991
11. Klap et al., 1996
12. Klap, 1997
13. Lefeuvre et al., 1994
14. McKellar, 1996
15. Murray and Spencer, 1997 (Method 2a)
16. Roman and Daiber, 1989
17. Stern et al., 1991
18. Taylor and Allanson, 1995
19. Welsh, 1980
20. Vörösmarty and Loder, 1994

data were available for 8 of 14 sites. Three continents and three oceans were represented in this flux dataset as well. For all studies, we have also indicated the mean tidal range, the subbasin area (where published), the distance of each subbasin from the coastal ocean (which we estimated when not published), and whether each subsystem was associated with a tidal estuary or a riverine system (Table 3). We used regression analysis, ANOVA, and t-tests to investigate whether any of these variables controlled flux behavior in this

TABLE 4: Summary of regression analysis of the available flux data from subsystem exchange studies listed in Table 3. The independent variables and tidal vs. riverine classifications are as shown in Table 3. All fluxes are in g/m2/yr. Positive flux represents uptake by the subsystem; negative flux represents export (or loss). Positive slope shown below represents a shift from export to uptake; negative slope represents a shift from uptake to export. The I-E switch is the value (in units of the respective independent variable) at which this shift between uptake and export occurred.

Dependent variable	Independent variable(s)	Tidal or Riverine?	model p-value	partial p-value	Regression Parameters			
					r-squared	slope	intercept	I-E switch
A. MULTIPLE REGRESSION MODELS								
1. TOC flux	tidal range	both	(0.073)	0.200	(0.870)	9.580	(-212.9)	NA
	system area	both		0.037		0.234		
	distance	both		0.049		1.896		
2. TSS flux	tidal range	both	(0.031)	0.011	(0.920)	2,910.2	(-7609.8)	NA
	system area	both		0.139		9.527		
	distance	both		0.238		72.9		
B. SIMPLE REGRESSION MODELS								
1. NN flux	tidal range	both	0.051	NA	0.400	-3.661	4.315	1.179
2. SRP flux	tidal range	both	0.132	NA	0.390	0.385	-0.611	1.587
3. SRP flux	tidal range	tidal only	0.096	NA	0.540	0.486	-0.658	1.354
4. TSS flux	tidal range	both	0.009	NA	0.777	2,452.4	-1542.5	0.629
5. POC flux	distance	both	0.036	NA	0.820	3.395	-30.74	9.054
6. POC flux	system area	both	0.099	NA	0.651	0.138	-7.41	53.696
7. TOC flux	system area	both	0.113	NA	0.320	0.175	-77.26	441.486

group of estuarine subbasins. We used 1) mean tidal range, 2) subbasin area, and 3) distance to the ocean as continuous independent variables and estuarine vs. riverine system as a categorical delineator.

There were no significant differentiations in flux dynamics for any constituents based simply on riverine (n=3) versus tidal (n=11) system comparisons. We did find interesting relationships between the continuous descriptor variables and flux in several cases — although, notably, we did not adhere to the strict traditional alpha level of 0.05 (Table 4). Tidal range, subsystem area, and distance to the ocean together explained 87% of the variability in total organic carbon (TOC) flux (n=7) and explained 92% of the variability in TSS flux (n=7). Both relationships were positive, suggesting a pattern of TOC and TSS export at 1) lower tidal ranges, 2) smaller subsystem size, and 3) closer proximity to the ocean (Table 4). Subsystem size played the strongest explanatory role with TOC flux while tidal range played the strongest role with TSS flux. In both cases, simple linear regressions with these independent variables suggested that the switch from export to uptake occurs at about 440 ha in subsystem size for TOC (n=9) and at about 0.63 m in tidal range for TSS (n=7; Table 4) — although the TSS relationship is suspect given that our dataset shows no TSS exports (Table 3). Both subsystem size and distance from the ocean appeared to explain variability in particulate organic carbon (POC) flux as well — 65% and 82% of that variation, respectively (n=5; Table 4). As with TOC, these simple relationships were positive and suggested a switch from POC export to uptake at a system size of about 54 ha and at about 9 km from the ocean —although these regressions are likely somewhat skewed by the large POC export measured at Canary Creek, DE, USA, which is 190 ha in size and has a 1.3 m tidal range (Roman and Daiber, 1989).

Patterns in dissolved nutrient flux appeared to be mainly affected by tidal range. The simple regression of tidal range vs. NN flux suggested that the former explained 40% of the variation in the latter ($p=0.05$; $n=10$). The slope of this relationship was negative, and the switch from NN uptake to export occurred at about 1.2 m mean tidal range (Table 4), although this relationship was heavily weighted by the large exports observed at Mont St. Michel Bay, France and the Blackwater estuary, England—both of which are macrotidal (Table 3). Tidal range explained about 39% of the variation in SRP flux ($n=7$), but in this case the relationship was positive. The switch from export to uptake of SRP occurred at about 1.6 m mean tidal range (Table 4). Interestingly, when we removed the one riverine site with published SRP flux data (Mira estuary, Portugal), the explanatory capabilities of tidal range increased to 54% (Table 4). This is the only relationship that improved when the whole dataset was reduced to only tidal estuarine systems.

We chose to focus on tidal range, subbasin size, and distance to the coast for several reasons. A number of researchers have demonstrated the importance of tidal range as a potential controller of estuarine dynamics (Steever et al. 1975, Odum et al. 1979, Odum 1980, others). Our simple statistical exercise also suggests that tidal range may be an important large-scale physical/environmental control on estuarine subbasin flux behavior. Since virtually all of the studies we list in Table 3 were conducted in a single tidal channel draining an isolated estuarine basin, we assumed that the size of that basin may well affect the types of exchanges measured. In fact, the size of the system did help explain variation in organic carbon fluxes — POC and TOC in particular — such that larger systems tend to take up organic carbon. This phenomenon may be coupled to the geomorphological trapping proposed by Odum et al. (1979) such that the larger the isolated estuarine marsh basin, the greater its trapping ability. Odum et al. (1979) also emphasized the probable effects of estuarine geomorphology and shape on flux characteristics. However, is difficult to represent estuarine geomorphology or shape in a simple numerical fashion, and thus we did not include these parameters in our model.

Finally, we selected distance to the coastal ocean as an estimator of the relative effect of estuarine or upstream versus oceanic influences on subbasin flux behavior. Again, distance was an important explanatory variable for organic carbon fluxes. Perhaps the fraction of "new" water entering an estuarine basin on a given flooding tide is a function of distance from the coast. Nearer the ocean, ebbing waters tend to be flushed and replenished more than farther into an estuary. As a result, the water flooding a basin nearer the ocean will have different concentration characteristics and, for organic carbon, flooding concentrations will nearly always be lower than concentrations on the preceding ebbing tide. This phenomenon will cause greater export of organic carbon in marsh basins closer to the coastal ocean—which is what our analysis indicated.

2.1.3 Studies Quantifying Marsh-Water Column Exchanges

One identifiable advance in the field of estuarine flux measurements since 1980 is the proliferation of marsh flume studies. Lee (1979) first used marsh flumes in two Rhode

Island marshes (Table 1), but these data were not published in the primary literature. Since then, wetland flumes have been used in most types of intertidal marsh, in mangrove wetlands, even on intertidal seagrass banks. Flumes modified with nets for walls have also been used to quantify faunal utilization of wetland habitats (McIvor and Odum 1986, Kneib 1994, Peterson and Turner 1994), and approaches similar to the flume technique have been used in a number of nonwetland settings, including coral reefs (Odum 1956, Rogers 1979, Atkinson and Bilger 1992), oyster reefs (Dame et al. 1984, 1989), and intertidal mussel beds (Dame et al. 1991, Asmus et al. 1992). The 15 wetland-water column flux studies reviewed in Table 5 represent a number of different types of estuaries, including deltaic, river-dominated, and back-barrier wetlands, and a number of different wetland types, including tidal fresh marshes, brackish and saline marshes, and mangroves. While these wetland-water column flux studies span a range of tidal amplitude from <0.1m to 2.5m (Table 5), this range is considerably smaller than that represented by estuarine basin flux studies (Table 3). All wetland-water column flux studies shown in Table 5 used the marsh flume technique. This methodological consistency permits a more rigorous statistical investigation into the causes of flux behavior patterns. In fact, the body of data we present here has been recently reviewed in two separate venues. Childers (1994) presented flume flux data from marshes in the southeastern US estuaries. He found no relationships between patterns of marsh-water column exchange and latitude, although the sites reviewed spanned a very small range of latitudes. He did find a number of relationships between flux behavior and tidal height. Inorganic nutrient and organic matter fluxes were positively related to tidal height, and consistently showed a switch from export to marsh uptake at about 1 m of tidal range (Childers, 1994). Our (similar) analysis of the subbasin flux data showed this same pattern between SRP flux and tidal range, but the opposite for NN (Table 4). Interestingly, Childers (1994) found that TSS flux also switched direction at about 1m of tidal range, but below about 1m marshes tended to take up sediment while they exported sediment at higher tidal ranges. Our subbasin flux analysis shown in Table 4 suggests that marsh basins—which also include subtidal and uplands components — show the opposite relationship between tidal range and TSS flux. In fact, the switch from microtidal TSS export to uptake at higher tidal ranges occurred at about 0.63 m (Table 4).

In a more recent review of wetland-water column interactions, Childers et al. (1999) presented flume flux data from marshes and mangrove forests around the Gulf of Mexico. They used Analysis of Variance to investigate the effects of various wetland characteristics on flux behavior. These characteristics–all of which were bicategorical–included wetland type (marsh or mangrove), geologic setting (terrigenous-clastic or carbonate), geomorphological setting (with the flume located proximal to other wetlands or proximal to open estuarine water), and degree of watershed coupling (intense or minimal). This review found no significant ''treatment'' effects of wetland type or geologic setting on marsh-water column flux. They did find that wetlands proximal to open estuarine water bodies tended to take up organic carbon in particulate form and release it in dissolved form, while wetlands proximal to other wetlands did the opposite. Also, estuarine wetlands associated with intense watershed coupling (analogous to riverine influence) tended to export POC but take up DOC, while organic carbon fluxes were opposite in estuarine wetlands that had minimal watershed coupling

TABLE 5: Summary of wetland-water column flux studies published since 1980 that used the flume technique. See Childers (1994) and Childers et al. (1998) for flux estimates.

ESTUARY	TIDAL RANGE	LATITUDE	ESTUARY TYPE	WETLAND TYPE	WATERSHED COUPLING	GEOMORPHOLOGIC TYPE	RELATIVE GEOLOGIC AGE	REFERENCES
Taylor River dwarf, Florida, USA	<0.1m	25.25	≈riverine	mangrove	minimal	proximal to wetlands	older(?)	6,7
Taylor River fringe, Florida, USA	0.1m	25.20	≈riverine	mangrove	minimal	proximal to wetlands	older(?)	6,7
Barataria Basin, Louisiana, USA	0.25m	29.67	deltaic	brackish marsh	minimal	proximal to open water	older	2,3,5,6
Barataria Basin, Lousiana, USA	0.4m	29.72	deltaic	saltmarsh	minimal	proximal to open water	older	2,3,5,6
Fourleague Bay, Louisiana, USA	0.5m	29.33	deltaic	brackish marsh	intense	proximal to wetlands	younger	2,3,5,6
Boca Chica, Campeche, Mexico	≈0.5m	18.33	riverine	mangrove	intense	proximal to open water	younger	6,9
Rookery Bay basin, Florida, USA	0.55m	26.00	≈barrier island	mangrove	minimal	proximal to wetlands	older(?)	6,10
Rookery Bay fringe, Florida, USA	0.55m	26.00	≈barrier island	mangrove	minimal	proximal to open water	older(?)	6,10
Carter's Creek, Virginia, USA	0.75m	37.25	riverine	saltmarsh	minimal	proximal to open water	older	5,12,13
Estero Pargo, Campeche, Mexico	<1m	18.67	barrier island	mangrove	minimal	proximal to wetlands	older	6,8
North River, Massachusetts, MA	≈1m	42.10	riverine	tidal fresh	intense	proximal to open water	older (?)	1
Bly Creek, South Carolina, USA	1.7m	33.33	barrier island	saltmarsh	minimal	proximal to wetlands	younger	5,11,14,15,16
Cumberland Island I, Georgia, USA	2.5	30.77	barrier island	saltmarsh	minimal	proximal to open water	younger	4,5
Cumberland Island II, Georgia, USA	2.5	30.87	barrier island	saltmarsh	minimal	proximal to wetlands	older	4,5
Cumberland Island III, Georgia, USA	2.5	30.90	barrier island	saltmarsh	minimal	proximal to wetlands	older	4,5

Citations:
1. Bowden, 1986
2. Childers and Day, 1990a
3. Childers and Day, 1990b
4. Childers et al., 1993
5. Childers, 1994
6. Childers et al., 1998
7. Davis and Childers, unpubl.data
8. Rivera-Monroy et al., 1995
9. Rivera-Monroy, unpubl.data
10. Twilley, 1985
11. Whiting et al., 1989
12. Wolaver et al., 1983
13. Wolaver and Zieman, 1983
14. Wolaver and Spurrier, 1988a
15. Wolaver and Spurrier, 1988b
16. Wolaver et al., 1988

(Childers et al. 1999). Perhaps their most interesting finding, however, was that DOC fluxes in marshes were generally orders of magnitude greater than in mangrove wetlands while POC fluxes were of similar magnitude. Rates of carbon fixation in marshes and mangroves are similar while standing crop biomass in mangroves is much greater (Day et al. 1989; Twilley et al. 1992; Mitsch and Gosselink 1993). In fact, much of a marsh's net production goes into biomass that senesces and turns over every winter. Childers et al. (1999) concluded that the greater DOC fluxes from marshes, compared to mangrove wetlands, suggests that marsh productivity is more labile and more readily exchanged with the inundating water column (as per Hopkinson 1992) and most of that exchange is via dissolved organic matter.

The background concentration of nutrients and the related trophic state of the estuary may be a major factor controlling the rates of marsh-water column exchange. Processes of nutrient uptake by tidal marshes (specifically, nitrate uptake rates) may be directly related to the floodwater nutrient concentrations. For example, in the Bly Creek salt marsh, South Carolina, USA, 62-79% of the variability in nitrate (NO_3^-) removal by the marsh was accounted for by seasonal variability in floodwater concentrations (Whiting et al. 1989). Ammonium (NH_4^+) removal by the marsh also exhibited a similar, linear relationship with floodwater NH_4^+ concentrations (r^2=0.40 to 0.67). This concentration-dependent effect was further indicated by comparing results from the Bly Creek study (Whiting et al. 1989), with similar data from a more eutrophic estuary (Cooper River, South Carolina, USA). Both sites were dominated by *Spartina alterniflora* and were characterized by similar tidal ranges (1.4-1.6 m). The Bly Creek marsh removed 0.62 ± 0.13 g NO_3^- -N m^{-2} y^{-1} with a mean annual floodwater concentration of 0.64 ± 0.10mM. In contrast, the Cooper River marsh removed 8.4 g NO_3^- - N m^{-2} yr^{-1} with mean floodwater concentrations of 10.0 ± 1.0 mM (McKellar et al. 1996). A simple extrapolation of these results suggests a linear relationship of about 1 g N m^{-2} y^{-1} uptake per mM of floodwater concentration. This relationship to floodwater concentration provides an important point of comparison among estuaries with variable degrees of nutrient enrichment.

2.1.4 The Importance of Ecological Development and Transgression

Coastal scientists have long recognized that estuaries are transient geologic features. In the contemporary Holocene regime of gradual sea level rise (Milliman and Emery 1968, Redfield 1972, others), estuaries generally change and develop by expanding or contracting either at the seaward margin, following a deltaic model (Coleman and Gagliano 1964, Boesch et al. 1983), or at the landward margin, following a transgressive model (Gardner and Bohn 1980). Another significant advance in the last two decades is the awareness by estuarine ecologists that, as a result of these geomorphological dynamics, estuaries are actually mosaics of differing aged systems. Ecologists have applied theories of community succession (see review by Bertness et al. this volume) and ecosystem development (Odum 1969, Vitousek and Reiners 1975) to this physical age template and thus [attempted to] explain patterns and processes based on relative stages of ecological development (Morris 1988, Hayden et al. 1991, Dame et al. 1992, Hopkinson 1992). In deltaic systems, such as coastal Louisiana, entire basins are at approximately the same stage of ecological development — or deterioration — because they formed as deltaic lobes relatively instantaneously, geologically speaking (Kolb and Van Lopik 1966, Gagliano et al. 1981). Comparing absolute ages and stages of ecological development across non-deltaic estuarine systems is not easy, however. One approach to avoiding this problem is to consider the relative age of different marsh or estuarine systems. In his review of marsh-water column flux studies from southeastern US estuaries, Childers (1994) used this approach. He calculated the relationships between flux and tidal range for younger versus older marshes, compared these to similar relationships for all marshes pooled together, and found a consistent pattern. The slope of the flux vs. tidal height relationship was greater for younger marshes compared to all marshes, and much greater for young marshes compared to older marshes. In many cases, this change in slope associated with stage of marsh development also causes a shift in the inflection point where export changed to import (Childers 1994). Ecosystems in early stages of development should be accumulating biomass and organic matter, and importing nutrients to produce this material, while ecosystems in later stages of development should be at near equilibrium between production and respiration, and show little net nutrient flux (Odum 1969, Vitousek and Reiners 1975, Dame et al. 1992, Hopkinson 1992). Data from Childers (1994) flume flux review demonstrated exactly this pattern. Dame et al. (1991) used this same developmental model to explain why the Bly Creek basin in

North Inlet, South Carolina, USA imported many constituents while the greater North Inlet estuary outwelled most constituents to the Atlantic Ocean.

2.2 OTHER MECHANISMS OF EXCHANGE

Most estuarine flux research prior to 1980 focused on tidally mediated exchanges occurring in the water column. The estuarine flux studies we review here, in Tables 2, 3 and 5, also generally focus on the water column as the medium of interaction (see also Fig. 3). However, a considerable body of work in the last two decades has also investigated other mechanisms by which intertidal wetlands interact with estuaries. Some of these studies directly measured processes such as ''low tide drainage,'' defined as slow drainage of enriched water from the surface of recently exposed marshes (Whiting et al. 1989, Whiting and Childers 1989). Some studies used more indirect techniques, such as stable isotopic signatures, to infer the importance of estuarine marshes and even individual components of those marsh systems, to estuarine and nearshore coastal dynamics (see reviews by Currin et al. and Sullivan et al. this volume). A detailed review of the many alternate mechanisms of marsh-estuarine exchange is beyond the scope of this paper. However, we will briefly touch on two of these mechanisms that we feel are particularly important: exchange via subsurface flow and benthic advection, and exchange via the movement of animal biomass.

2.2.1 Exchanges Via Subsurface Flow and Benthic Advection

Odum (1980) suggested that tidal action caused the irrigation of marsh soils adjacent to tidal creeks, and was thus the primary force behind the flux of nutrient-rich porewater into tidal creeks on ebbing tides. He presented this idea as a testable example of the Tidal Subsidy Hypothesis (Odum et al. 1979). Research in the last two decades has demonstrated the horizontal movement of marsh porewaters towards tidal creeks in a number of different estuarine settings (Agosta 1985, Jordan and Correll 1985a,b, Yelverton and Hackney 1986, Harvey et al. 1987). This process is notably different from groundwater inflow, which is characteristic of peat marshes (such as those in New England, USA estuaries, Millham and Howes 1994, Valiela et al. 1990, 1997) but is not a major process in most other marsh systems.

Marsh porewater enters the tidal creeks in two ways: 1) as ''seepage'' during low tide–in a process analogous to ''low tide drainage'', which we noted above (Agosta 1985, Jordan and Correll 1985a,b, Yelverton and Hackney 1986), or; 2) via the active advection of porewaters into tidal creek bottoms (Whiting and Childers 1989, Howes and Goehringer 1994). Creekbank seepage can only occur at low tide while subtidal advection occurs throughout the tidal cycle. In all cases, these porewaters contained high concentrations of inorganic nutrients and dissolved organic matter and thus appear to represent a significant portion of total marsh-estuary interactions, although the volumes of water are very small compared to surficial tidal volumes. Porewater is replaced primarily by vertical seepage during flood tide inundation of the marsh (Hemond and Fifield 1982, Hemond et al. 1984), and this influx process may be mediated by the macrophyte vegetation itself (Dacey and Howes 1984). In fact, preliminary results of a salt marsh soil dynamics model being developed by J. Morris suggest that this control represents an intricate feedback mechanism that couples marsh processes, flux dynamics, and greater estuarine production (J. Morris, University of South Carolina, person. commun.). The importance of the surface water — porewater — tidal creek advective cycle to estuarine flux dynamics may be determined by tidal range (as per Odum 1980), to the point that it may not be as important in systems with mean tidal ranges much below 1 m (Childers 1994). Based on our review, we argue that this surficial-soil advective cycle couples estuarine marshes to the immediately adjacent tidal creeks in an important way, and may strongly affect estuarine production through its role in the turnover, quality, and net flux of nutrients and, in particular, organic matter.

2.2.2 Mediation of Nutrient and Organic Matter Exchanges by Estuarine Fauna

One of the first indoctrinations of all new estuarine ecology students is the idea that "estuaries are nurseries." A large number of fish, invertebrate, and avian species are either facultative or obligate users of estuarine habitats and productivity, and estuarine-dependent species often support highly productive fisheries (Day et al. 1989, others). A number of studies have demonstrated that estuarine marshes are important habitat for commercially important fish and invertebrates (see review by Zimmerman et al. this volume). Direct measurements of aquatic faunal densities from a range of estuarine wetland settings confirm that these animals utilize the marsh surface for feeding and to escape predation (McIvor and Odum 1986, Rozas 1988, Rozas and Reed 1993, Kneib and Wagner 1994, Peterson and Turner 1994, McIvor and Rozas 1996, review by Kneib this volume). However, we were unable to find any studies in which direct measurements of habitat use by estuarine animals has been integrated with: 1) tidal inundation records (representing habitat availability); 2) growth rates, and 3) tissue nutrient content to generate an actual exchange of carbon and nutrients associated with faunal utilization of marshes. These calculations are actually possible using extant datasets, such as those of McIvor and Odum (1986), Kneib (1994), and Peterson and Turner (1994).

Although most estuaries show a net export of most constituents to the coastal ocean via ebb-dominated water fluxes, larval and juvenile organisms can move against this gradient. A number of Gulf of Mexico studies have shown that the greatest recruitment of larval and juvenile organisms occurs during the strongest cold fronts (Nelson et al. 1977, Shaw et al. 1985, Dagg 1988, Checkley et al. 1988, Goodrich et al. 1989, Rogers et al. 1993). This net movement of organisms into estuaries is behaviorally-mediated. For example, Rogers et al. (1993) showed that postlarval and juvenile brown shrimp tended to move to the bottom when they encounter colder turbid water associated with northerly winds. During prefrontal, southerly winds, the shrimp rise up into the water column at night when they encounter the warmer, high salinity water moving onshore. In this way, the young shrimp aggregated near inlets, facilitating their movement into the estuary. Most of these organisms then moved deep into the marsh for a period of high growth.

Growing organisms require both energy (carbon) and nutrients. Being protein-rich, all fauna require considerably more nitrogen per unit carbon than do primary producers while vertebrate fauna require considerably more nitrogen and phosphorus. Thus, estuarine fauna effectively bioaccumulate nitrogen and phosphorus relative to their food sources. Birds that feed on estuarine fauna, nest in estuaries, or both, are vectors of nutrients and may import, export, or relocate nutrients within an estuarine landscape (Bildstein et al. 1992). Furthermore, any organism that begins its life cycle in an estuary or enters an estuary at a young age, then grows within the estuary (on estuarine productivity), and finally leaves the estuary as an adult is effectively exporting organic carbon but it is also preferentially exporting organic macronutrients. Deegan (1993) quantified this process in an elegant budget of menhaden populations in a Louisiana estuary, and demonstrated that faunal export of carbon and nutrients may be a major component, perhaps even the major component, of estuary-coastal ocean exchange. R. Ulanowicz has used network analysis to suggest that several numerically-dominant species of fish exercise surprising control over the phosphorus budget of the entire Chesapeake Bay (R. Ulanowicz, University of Maryland, person. commun.). Reviews of faunal contributions to estuary-coastal ocean exchanges are reviewed in detail in this volume by Deegan et al. Our review of estuarine flux research has spotlighted the faunal vector issue as an important, albeit difficult to quantify, component of estuary-coastal ocean interactions.

2.2.3 Temporal Variability in Estuarine Fluxes: The Role of Exogenous Forcing

When the Outwelling hypothesis was first formulated, regular daily flushing by the tides was the primary physical mechanism identified for facilitating materials transport (Odum 1968, 1980). For this reason, most estuarine flux studies have emphasized capturing the tidal signal and quantifying tidal water flux (Kjerfve et al. 1981, Park and James 1990). For many flux studies, a common experimental design was to measure flux over 2 to 3 tidal cycles at regular intervals, normally once a month, for a year. Flux patterns measured during individual samplings were extrapolated to the period in which they were made and then summed to obtain annual flux estimates. It has become apparent, however, that major flux events in many coastal systems do not occur during regular astronomical tidal flushing. Rather, other exogenous forcings, such as strong winds, high rainfall, or river flow, can greatly amplify or completely overwhelm daily tidal action. This is especially true for microtidal estuaries, such as those along the northern Gulf of Mexico, where the astronomical tide is about 0.3 m. The role of strong energetic forcings in coastal systems is treated in detail by Day et al. (this volume), but we present an example here to illustrate our point.

Two authors of this paper are currently conducting an estuarine flux study in a southern Everglades mangrove wetland system, Florida, USA. We have been quantifying total nutrient concentrations and salinity daily at the confluence of Taylor River — which drains a sizeable Everglades wetland watershed — and the Florida Bay estuary since April 1996 (D.Childers and J.Day, unpubl. data). Daily TN and TP flux is calculated by combining daily concentrations with water flow data from a U.S. Geological Survey (USGS) gauge at the same location (E. Patino, USGS, unpubl. data). The Taylor River drainage basin receives no exogenous riverine inputs, and interannual variations in rainfall are common. In 1996, precipitation was high throughout the May-November wet season. In 1997, rainfall steadily decreased during the wet season, but remained high into the dry season of 1997-98 because of the El Niño phenomenon (Fig. 4a). It is interesting to note that major variation in monthly rainfall from one year to the next did not have a predictable impact on salinity patterns: Wet season salinity patterns were similar in 1996 and 1997, in spite of very different rainfall patterns, and the 1997-98 El Niño winter prevented the typical dry season increase in salinity seen in previous years (Fig. 4b). Neither precipitation nor salinity patterns appeared to have a direct effect on patterns of daily nutrient flux from the Taylor River watershed into Florida Bay (Figs. 4c, 4d). This is likely because this system is topographically very flat and has little or no astronomical tide, and thus water (and nutrient) exchange between the Florida Bay estuary is likely controlled fairly equally by winds, local precipitation, and drainage from the wetland watershed. Without a clear understanding of all exogenous forces controlling this system, it would be very difficult to explain the patterns in flux behavior we have observed.

Numerous studies have demonstrated how exogenous forcings affect materials flux in estuaries. The majority of sedimentation in estuarine marshes occurs during storms and river floods (Baumann et al. 1984, Roberts et al. 1989, Reed 1989, Childers and Day 1990a, Cahoon et al. 1995, Day et al. 1995, 1998, Hensel et al. 1998). Precipitation

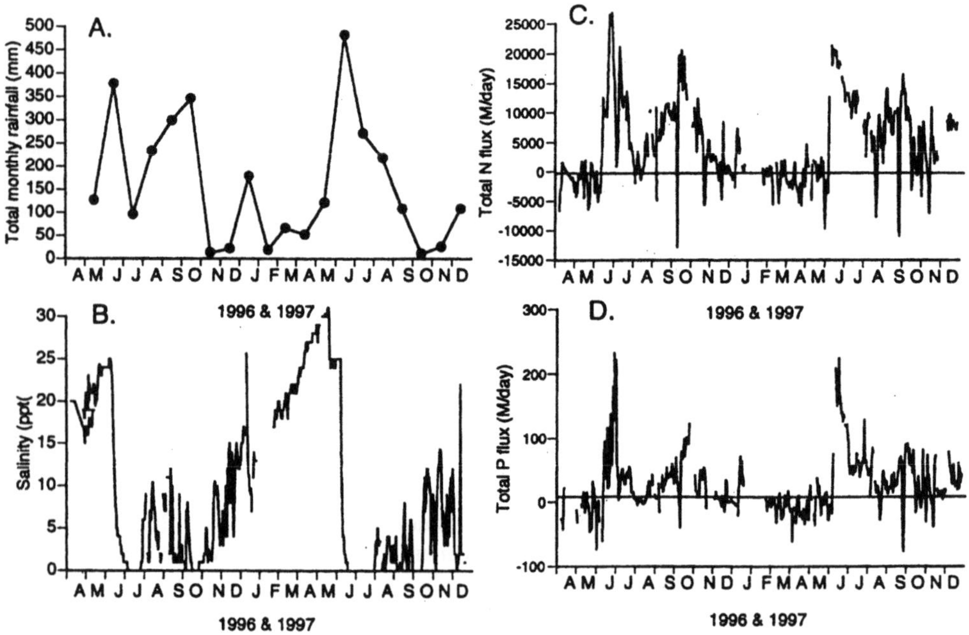

Figure 4. Data from the Taylor River mangrove flux study currently being conducted in the southern Everglades, Florida, USA (Childers and Day, unpubl. data). (a) total monthly precipitation from May 1996 December 1997 measured at the mangrove-freshwater marsh boundary in the Taylor Slough watershed; (b) daily salinity measured in water samples collected from the Taylor River, which drains the Taylor Slough watershed, at its confluence with the Florida Bay estuary; (c) daily total nitrogen flux, in Moles, from the Taylor Slough watershed to Florida Bay, with positive flux = export to the bay and negative flux = uptake by the watershed; (d) daily total phosphorus flux, in Moles, from the Taylor Slough watershed to Florida Bay, with positive flux = export to the bay and negative flux = uptake by the watershed.

events also enhance marsh-water column exchanges of sediment (Settlemeyer and Gardner 1977, Chalmers et al. 1985, Jordan et al. 1986, Whiting et al. 1989). Suspended sediment concentrations may increase by several orders of magnitude in response to winter frontal passages and storm events when compared to fair weather conditions (Caffrey and Day 1986, Leonard et al. 1995, Hensel et al. 1998). Flooding caused by hurricanes and tropical storms may also produce a large influx of sediment to estuaries and estuarine marshes. In the Rappahannock Estuary, tropical storm Agnes (June 1972) increased TSS concentrations up to 25 times normal amounts (Nichols 1977). Schubel and Hirschberg (1978) estimated that the Susquehanna River discharged greater than 30 times its annual input of sediment during 10 days following this same 1972 storm.

2.3 HIERARCHICAL MODELING APPROACHES: THE NORTH INLET, SOUTH CAROLINA EXAMPLE

In the last two decades, modeling has become an accepted component of any integrated systems-level research program, in estuarine science and in ecology in general. In spite of this, we were able to find surprisingly few examples where modeling has been used as a research tool in estuarine flux studies. Several researchers have used spatial models to simulate the movement of water and materials in estuaries (Costanza et al. 1990, Kjerfve et al. 1991, Sklar et al. 1992, Buzzelli et al. 1998). Hydrodynamic numerical simulations have also been used to extend the applicability of field data and to integrate biotic and abiotic processes (Vörösmarty and Loder 1994). We feel that modeling represents a valuable tool for synthesizing field measurements of fluxes and processes, and helping to address questions about how the subsystems that make up estuaries interact with each other and how estuaries interact with the coastal ocean.

Notably, North Inlet, South Carolina, USA was represented at all three hierarchical levels of estuarine flux studies: the Outwelling Study (Table 2), the Bly Creek Basin Study (Table 3), and the Bly Creek Marsh Flume Study (Table 5) were all conducted in this lagoonal, salt marsh-dominated estuary. Dame et al. (1991) used annual budgets of carbon, nitrogen, and phosphorus to compare and contrast the results of these three studies, and to synthesize them into a relatively cohesive picture of the dynamics of this type of estuary. Childers et al. (1993) took this analysis one step further, and developed a dynamic budget of subsystem interactions in the North Inlet estuary. They used a tidal hydrology model to determine the time and area of intertidal habitats inundated every day, and the volumes of water interacting with all habitats. Their model combined this hypsometry + tidal hydrology with the time-series flux data from the Bly Creek and marsh flume studies, and generated expected patterns of nutrient concentrations in a simulated North Inlet tidal creek every day. They validated this temporally dynamic budget output with daily concentrations of nutrients and organic matter in North Inlet tidal creeks. Finally, they integrated the time-series output from the dynamic budget over a year, and compared net surpluses (simulated outwelling) and deficits (simulated import) with actual flux data from the North Inlet Outwelling Study (Dame et al. 1986). With many constituents, the dynamic budget simulated import while the flux data showed export. Childers et al. (1993) attributed this to the fact that they had used flux data from a marsh basin which is geologically young and ecologically immature (Gardner and Bohn 1980). By extrapolating these fluxes to the entire North Inlet estuary, the dynamic budget thus simulated the entire estuary as an immature and developing system when it is actually a mosaic of different aged marsh basins. We note here that this result adds further credence to the concept that relatively ''young'' estuarine marshes and basins seem to follow the basic pattern of biomass and nutrient sequestration while ''mature'' systems approximate steady state in both, as set forth by Odum (1969) and Vitousek and Reiners (1975).

3. Summary

3.1 RECOMMENDATIONS FOR FUTURE FLUX RESEARCH

Nixon (1980) began his summary by pointing out that, while a summary should be brief and simple, the issues he reviewed and answers he presented were neither. After nearly 20 more years of estuarine flux research, an extensive body of literature has developed and, like Nixon, we are also unable to be either simple or brief. Unlike Nixon, though, we find ourselves considerably closer to substantive answers. Intensive and expensive estuarine flux studies are an excellent tool for addressing questions and issues that are specific to a given system, or perhaps to a given coastal region. We have learned a great deal more about how estuarine wetlands interact with their estuaries, and of the value of establishing a conceptual framework and system boundaries. The most important message of this review may be that estuarine ecologists have learned a great deal about outwelling as a concept in spite of the fact that few flux studies have directly addressed the original Outwelling Hypothesis. Most estuarine flux studies, taken as a body of work, fit well into our broader conceptualization of outwelling. We thus suggest that the question should not be ''Is The Outwelling Hypothesis true?'' Rather, the questions are 1) how are materials being exchanged between the different subsystems that make up the estuary-coastal ocean landscape? 2) what are the mechanisms of this exchange? and 3) how do exogenous forcings control these patterns of exchange?

No review of a large body of research would be complete without some recommendations for avenues of future research. Investigations into issues of organic matter quality, lability, and utility are likely to be a particularly fruitful avenue for future estuarine interactions work. Stable isotopic techniques will continue to generate more accurate, albeit indirect, measures of the source of organic matter and nutrients (see review by Currin et al. this volume). Identifying and quantifying the various sources of organic matter in estuarine marshes is also an important component of this line of inquiry (see review by Sullivan this volume). We should continue to focus on how decompositional processes on the marsh surface itself — including those mediated by fungi, bacteria, and meiofauna — affect the quality of organic matter supplied to the estuary (see review by S. Newell this volume). A group of estuarine ecologists have done some interesting work quantifying how [dissolved and particulate] organic matter changes in chemical composition, molecular size and content, and degree of lability as it interacts with estuarine marshes over the course of a tidal cycle (Hemminga et al. 1992, 1993, Klap et al. 1996). This research has included direct measures of bacterial utilization of organic matter taken at various stages of the tide (thus after varying periods of time exposed to the marsh surface) as a measure of material quality and lability (Klap 1997). We feel that a great deal can be learned by combining this direction of work with recent advances in microbial ecology. We should also focus on temporal variability in flux patterns — at a range of ecologically important time scales — and how exogenous forcings affect and even control this variability (see review by Day et al. this volume). The importance of these environmental controls is further supported by the statistical relationships that we found between estuarine fluxes (at the marsh and basin levels) and general geomorphological and physical characteristics of the estuary in question (including tidal range, basin area, distance to the ocean, and proximity of a marsh to large estuarine water bodies).

In the last two decades the estuarine scientific community has come to realize that these systems are landscape mosaics. The pieces are of different geologic ages, successional stages, and degrees of ecosystem development. The pattern of this developmental mosaic is often predictable: In deltaic estuaries, whole basins are generally of similar age and stage in the deltaic cycle, as they were formed during the same river switching event. In transgressive back-barrier systems, the entire estuary is migrating landward in response to sea level rise, setting up an upland to ocean developmental gradient. It is important that we explicitly note this context in future studies, for several reasons: 1) the relative age of a particular study site has important bearing on most ecological processes in a classical developmental sense, and this is a critical component of data interpretation; 2) the developmental context is important when comparing patterns or processes measured at several different sites within the same estuary, or across estuaries, and; 3) since estuaries are clearly dynamic systems, knowing the developmental context of systems is critical to interpreting and managing for anthropogenic influences. For instance, many transgressive back-barrier salt marsh estuaries are now bounded on their landward margins by development (seawalls, subdivisions, roads, shopping centers, etc.). This development effectively prevents further landward migration of these systems, which will gradually eliminate the ocean<–>land developmental gradient and ultimately eliminate the age mosaic of the system, leaving an estuary that is basically all mature marsh.

It is particularly intriguing to us to consider estuarine animals as conveyors of high-quality organic matter and [relatively] concentrated nutrients. We suggest that a great deal may be learned about the estuarine dynamics if flux studies incorporate a faunal component and faunal studies incorporate a biogeochemical component. We recognize that this is not a trivial suggestion. During the 1960s, a sizable rift developed between ecologists who focus on organisms and populations, and ecologists who focus on ecosystems. To some extent, the underpinnings of this schism still exist today, though it is subtle. It may not be an easy collaboration at first, but we strongly feel that it is important. Many species of estuarine fauna are dependent on marshes for habitat, marsh function is a critical component of habitat quality, subsystem interactions are critical to marsh processes, and (it appears that) estuarine fauna are an important component of subsystem interactions. The connections are clear from the perspective of the tidal creek; we need only clarify these connections in our laboratories and meeting rooms.

3.2 CONCLUSIONS

Nixon ended his 1980 review of marsh-estuarine interactions by reminding us that the question to be answered is not "are coastal salt marshes important?" After two more decades of marsh flux research, we conclude here that the question also is not "have we accepted or rejected the Outwelling Hypothesis?" This idea has a rich history, and has clearly stimulated a great deal of debate, research, and thought. We will argue that, to that end, the Outwelling Hypothesis has made a very important heuristic contribution to estuarine science. However, neither Nixon (1980) nor our current review generated any substantive proof or disproof of the Outwelling Hypothesis. In the nearly 40 years since Teal's 1962 paper, only four flux studies have generated datasets that directly tested this hypothesis as it was originally presented. We began this review with an expansion of the Outwelling Hypothesis that encompassed the connectivity and hierarchical nature of estuarine subsystems and subbasins. In this way, we were able to review the last two decades of estuarine flux research from the perspective of outwelling as a concept, not viewing it as a strict hypothesis to be tested and interpreted in its original form.

Fig. 5 is an example of how futile it would be to attempt any statement supporting or refuting outwelling as a testable scientific hypothesis. In this LANDSAT image of the South Carolina, USA coast, three estuaries are show. The uppermost is North Inlet, which is arguably the best studied estuary in terms of estuarine interactions. As we noted before, it is a lagoonal, marsh-dominated back-barrier system with negligible freshwater inflows. Just south of North Inlet is Winyah Bay. This estuary has only fringing marshes and receives large freshwater inputs from the Pee Dee and Waccamaw Rivers. Just south of Winyah Bay is the Santee Delta, which is marsh-dominated (as is North Inlet) but also receives large freshwater inflows, from the Santee River (as does Winyah Bay). It seems obvious even from a satellite that all three of these very different estuaries have plumes extending from their inlets into the nearshore coastal ocean. But are they exporting nutrients, or organic matter, or faunal biomass? Are they supporting coastal ocean productivity? In fact, we will conclude by encouraging estuarine scientists to view the Outwelling Hypothesis as a conceptual stimulus of ideas and not as a strict statistical hypothesis that must be proven or disproven. And finally, at the very least Nixon (1980) implored us to think rather than feel. Our review of the last 20 years of estuarine flux research demonstrates that our research community did, in fact, think.

4. Acknowledgements

The idea for this paper has been ruminating among the authors for nearly 10 years, and ideas for our approach came from many discussions with many colleagues, far more than we can mention here. We received productive comments and ideas from many members of the FIU Wetland Ecosystems Ecology Lab, from colleagues at the April 1998 conference, from Scott Nixon, and from one anonymous reviewer. Portions of this research were supported by research grants from the South Florida Water Management District to the first two authors. This paper is Contribution No. 125 to the Southeast Environmental Research Center Publication Series.

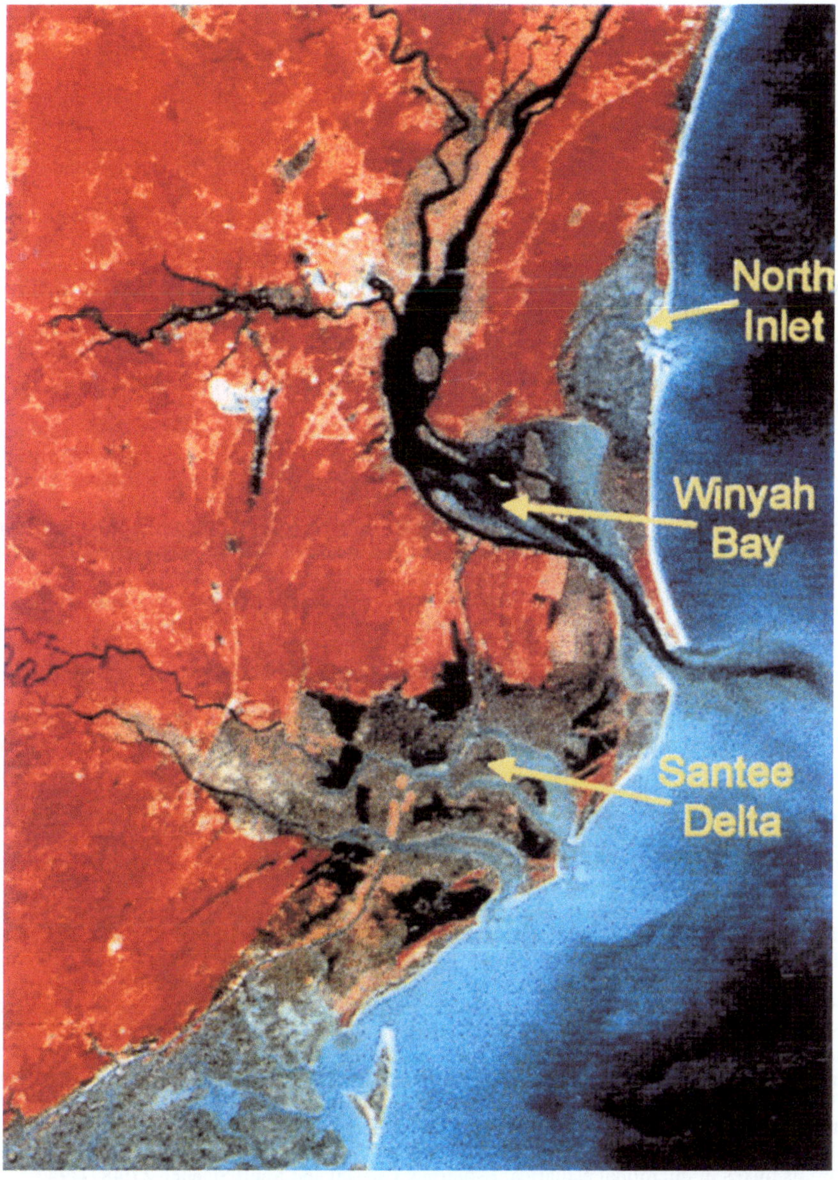

Figure 5. A portion of a LANDSAT image of the South Carolina, USA coast, showing the North Inlet estuary, the Winyah Bay estuary, and the Santee Delta estuary (where shown). In spite of dramatically different marsh:open water ratios and degrees of freshwater/riverine influence across these three diverse estuaries, an export plume is clear at the mouth of all three. That three so different systems, located in such proximity to each other, all appear to be exporting material to the nearshore coastal ocean seems to suggest that it would be very difficult to disprove the Outwelling Hypothesis as originally presented. Alternatively, that it would take extensive and expensive sampling to determine how each of these different systems is, in fact, influencing nearshore coastal productivity also suggests that it may be best to find other avenues for investigating the concept of outwelling.

5. Literature Cited

Agosta, K. 1985. The effect of tidally induced changes in the creekbank water table on pore water chemistry. Estuarine, Coastal and Shelf Science **21**: 389-100.

Asjes, J. and N.M.J.A. Dankers. 1994. Variations in tidal exchange processes between a Dutch salt marsh, the Slufter and the North Sea. Pages 201-204 *in* W. Mitsch, editor. Global wetlands. Elsevier Publishing, Amsterdam, The Netherlands.

Asmus, H., R.M. Asmus, T.C. Prins, N. Dankers, G. Frances, B. Maaß and K. Reise. 1992. Benthic-pelagic flux rates on mussel beds — tunnel and tidal flume methodology compared. Helgolaender Meeresuntersuchungen **46**: 341-361.

Atkinson, M. J. and R.W. Bilger. 1992. Effects of water velocity on phosphate uptake in coral reef-flat communities. Limnology and Oceanography **37**: 273-279.

Baumann, R. H., J. W. Day, Jr. and C. A. Miller. 1984. Mississippi deltaic wetland survival: sedimentation versus coastal submergence. Science **224**:1093-1095.

Bildstein, K.L., E. Blood and P. Frederick. 1992. The relative importance of biotic and abiotic vectors in nutrient transport. Estuaries **15**:147-157.

Boorman, L.A., J. Hazelden, P.J. Loveland, J.G. Wells and J.E. Levasseur. 1994. Comparative relationships between primary productivity and organic and nutrient fluxes in four European salt marshes. Pages 181-199 *in* W. Mitsch, editor. Global wetlands. Elsevier Publishing, Amsterdam, The Netherlands.

Borey, R.B., P.A. Harcombe and F.M. Fisher. 1983. Water and organic carbon fluxes from an irregularly flooded brackish marsh on the upper Texas coast, USA. Estuarine, Coastal and Shelf Science **16**:379-402.

Boto, K.G. and J.T. Wellington. 1988. Seasonal variations in concentrations and fluxes of dissolved organic and inorganic materials in a tropical, tidally-dominated, mangrove waterway. Marine Ecology Progress Series **50**:151-160.

Bowden, W.B. 1986. Nitrification, nitrate reduction and nitrogen immobilization in a tidal freshwater marsh sediment. Ecology **67**:88-99.

Bowden, W. B., C.J. Vörösmarty, J.T. Morris, B.J. Peterson, J.E. Hobbie, P.A. Steudler and B. Moore III. 1991. Transport and processing of nitrogen in a tidal freshwater wetland. Water Resources Research **27**: 389-408.

Boynton, W.R., J.H. Garber, R. Summers and W.M. Kemp. 1995. Inputs, transformations and transport of nitrogen and phosphorus in Chesapeake Bay and selected tributaries. Estuaries **18**:285-314.

Buzzelli, C. P., R.L. Wetzel and M.B. Meyers,. 1998. Dynamic simulation of littoral zone habitats in the lower Chesapeake Bay. II. Seagrass habitat primary production and water quality relationships. Estuaries **21**:673-689.

Caffrey, J. M. and J. W. Day,. 1986. Control of the variability of nutrients and suspended sediments in a gulf coast estuary by climatic forcing and spring discharge of the Atchafalaya river. Estuaries **9**:295-300.

Cahoon, D. R., D. J. Reed and J. W. Day, Jr. 1995. Estimating shallow subsidence in microtidal salt marshes of the southeastern United States: Kaye and Barghoorn revisited. Marine Geology **128**:1-9.

Chalmers, A.G., R.G. Wiegert and P. Wolf. 1985. Carbon balance in a saltmarsh: interactions of diffusive export, tidal deposition and rainfall-caused erosion. Estuarine, Coastal and Shelf Science **21**:757-771.

Chambers, R. M., S.V. Smith and J.T. Hollibaugh. 1994. An ecosystem-level context for tidal exchange studies in salt marshes of Tomales Bay, CA, USA. Pages 265-276 *in* W. Mitsch, editor. Global Wetlands. Elsevier Publishing, Amsterdam, The Netherlands.

Childers, D.L. and J. W. Day, Jr.,. 1988. A flow-through flume technique for quantifying nutrient and materials fluxes in microtidal estuaries. Estuarine, Coastal and Shelf Science **27**:483-494.

Childers, D.L. and J. W. Day, Jr., 1990a. Marsh-water column interactions in two Louisiana estuaries. I. Sediment dynamics. Estuaries **13**:393-403.

—— 1990b. Marsh-water column interactions in two Louisiana estuaries. II. Nutrient dynamics. Estuaries **13**:404-417.

—— 1991. The dilution and loss of wetland function associated with conversion to open water. Wetlands Ecology and Management **3**:163-171.

Childers, D.L., H.N. McKellar, Jr., R. Dame, F. Sklar and E. Blood. 1993. A dynamic nutrient budget of subsystem interactions in a salt marsh estuary. Estuarine, Coastal and Shelf Science **36**:105-131.

Childers, D.L., S. Cofer-Shabica and L. Nakashima. 1993. Spatial and temporal variability in marsh-water column interactions in a Georgia salt marsh. Marine Ecology Progress Series **95**:25-38.
Childers, D.L. 1994. Fifteen years of marsh flumes - A review of marsh-water column interactions in Southeastern USA estuaries. Pages 277-294 *in* W.Mitsch, editor. Global wetlands. Elsevier Publishing, Amsterdam, The Netherlands.
Childers, D.L., S. Davis, V. Rivera-Monroy and R.R. Twilley. 1999. Wetland-water column interactions and the biogeochemistry of estuary-watershed coupling around the Gulf of Mexico. Pages 211-236 *in* T.S. Bianchi, J. Pennock and R. Twilley, editors. Biogeochemistry of Gulf of Mexico estuaries. John Wiley & Sons, New York, New York, USA.
Dacey, J.W.H. and B.L. Howes. 1984. Water uptake by roots controls water table movement and sediment oxidation in short *Spartina* marsh. Science **224**:487-489.
Daly, M.A. and A.C. Mathieson. 1981. Nutrient fluxes within a small north temperate salt marsh. Marine Biology **61**:337-344.
Dame, R. F., R.G. Zingmark and E. Haskin. 1984. Oyster reefs as processors of estuarine materials. Journal of Experimental Marine Biology and Ecology **83**: 239-247.
Dame, R. F., T. H. Chrzanowski, K. Bildstein, B. Kjerfve, H. McKellar, D. Nelson, J. Spurrier, S. Stancyk, H. Stevenson, J. Vernberg and R. Zingmark. 1986. The outwelling hypothesis and North Inlet, SC. Marine Ecology Progress Series **33**: 217-229.
Dame, R. F., J.D. Spurrier and T.G. Wolaver. 1989. Carbon, nitrogen and phosphorus processing by an oyster reef. Marine Ecology Progress Series **54**: 249-256.
Dame, R.F., N. Dankers, T. Prins, H. Jongsma and A. Smaal. 1991. The influence of mussel beds on nutrients in the western Wadden Sea and eastern Scheldt estuaries. Estuaries **14**:130-138.
Dame, R. F., J.D. Spurrier, T.M. Williams, B. Kjerfve, R.G. Zingmark, T.G. Wolaver, T.H. Chrzanowski, H.N. McKellar and F.J. Vernberg. 1991. Annual material processing by a salt marsh-estuarine basin in South Carolina. Marine Ecology Progress Series **72**: 153.
Dame, R. F., D.L. Childers and E.T. Koepfler. 1992. A geohydrologic continuum theory for the spatial and temporal evolution of marsh-estuarine ecosystems. Netherlands Journal of Sea Research **30**:63-72.
Dame, R. F. and L.R. Gardner. 1993. Nutrient processing and the development of tidal creek ecosystems. Marine Chemistry **43**: 175-183.
Dame, R. F.,. 1994. The net flux of materials between marsh-estuarine systems and the sea. The Atlantic coast of the United States. Pages 295-302 *in* W.Mitsch, editor. Global wetlands. Elsevier Publishing, Amsterdam, The Netherlands.
Dankers, N. M. Binsbergen, K. Zegers, R. Laane and M.R. van der Loeff. 1984. Transportation of water, particulate and dissolved organic and inorganic matter between a salt marsh and the Ems-Dollard estuary, the Netherlands. Estuarine, Coastal and Shelf Science **19**: 143-165.
Day, J.W. Jr., C.J. Madden, R.R. Twilley, R.F. Shaw, B.A. McKee, M.J. Dagg, D.L. Childers, R.C. Raynie and L.J. Rouse. 1995. The influence of Atchafalaya River discharge on Fourleague Bay, Louisiana (USA). Pages 151-160 *in* K.R. Dyer and R.J. Orth, editors. Changes in fluxes in estuaries, ECSA 22/ ERF Symposium Series, Olsen and Olsen Publishers, Fredensborg, Denmark.
Day, J.W. Jr., F. Scarton, A. Rismondo and D. Are. 1999. Rapid deterioration of a salt marsh in Venice Lagoon, Italy. Journal of Coastal Research **14**:583-590.
Deegan, L. A. 1993. Nutrient and energy transport between estuaries and coastal marine ecosystems by fish migration. Canadian Journal of Fisheries and Aquatic Science **50**:74-79.
Espino, G. and M.A. R. Medina. 1993. Nutrient exchange between subtropical lagoons and the marine environment. Estuaries **16**:273-279.
Flores-Verdugo, F.J., J.W. Day, Jr. and R. Briseño-Dueñas. 1987. Structure, litterfall, decomposition and detritus dynamics of mangroves in a Mexican coastal lagoon with an ephemeral inlet. Marine Ecology Progress Series **35**:83-90.
Haines, E.B. 1975. Nutrient inputs to the coastal zone: the Georgia and South Carolina shelf. Pages 303-322 *in* L. E. Cronin, editor. Estuarine research. Volume 1. Academic Press, New York, New York, USA.
Haines, E.B. 1979. Interactions between Georgia salt marshes and coastal waters: a changing paradigm. Pages 35-46 *in* R. J. Livingston, editor. Ecological processes in coastal and marine systems. Plenum Press, New York, New York, USA.
Haines, E.B. 1979. Nitrogen pools in Georgia coastal waters. Estuaries **2**:34-39.
Harvey, J.W., P.F. Germann and W.E. Odum. 1987. Geomorphological control of subsurface hydrology in the creekbank zone of tidal marshes. Estuarine, Coastal and Shelf Science **25**:677-691.

Hemminga, M.A., V.A. Klap, J. van Soelen and J. de Leeuw. 1992. Shifts in seston characteristics after inundation of a European salt marsh. Limnology and Oceanography **37**:1559-1564.
Hemminga, M.A., V.A. Klap, J. van Soelen and J.J. Boon. 1993. Effect of salt marsh inundation on estuarine particulate organic matter characteristics. Marine Ecology Progress Series **99**:153-161.
Hemond, H.F. and J.L. Fifield. 1982. Subsurface flow in salt marsh peat: a model and field study. Limnology and Oceanography **27**:126-136.
Hemond, H.F., W.K. Nuttle, R.W. Burke and K.D. Stolzenbach. 1984. Surface infiltration in salt marshes: theory, measurement and biogeochemical implications. Water Research **20**:591-600.
Hensel, P., J. W. Day, Jr., D. Pont and J. N. Day. 1998. Short-term sedimentation dynamics in the Rhone River delta, France: the importance of riverine pulsing. Estuaries **21**: 1 (52-65).
Hopkinson, C.S. 1985. Shallow-water benthic and pelagic metabolism: evidence of heterotrophy in the nearshore Georgia Bight. Marine Biology **87**:19-32.
—— 1988. Patterns of organic carbon exchange between coastal ecosystems — the mass balance approach in salt marsh ecosystems. Pages 122-154 *in* B.O. Jansson, editor. Lecture notes on coastal and estuarine studies **22**: 122-154.
—— 1992. A comparison of ecosystem dynamics in freshwater wetlands. Estuaries **15**: 549-562.
Howes, B.L. and D.D. Goehringer. 1994. Porewater drainage and dissolved organic carbon and nutrient losses through the intertidal creekbanks of a New England salt marsh. Marine Ecology Progress Series **114**: 289-301.
Hyakumachi, M. 1989. Methods for determination of carbon loss from fungal propagules incubated in soil. Biochemistry **21**: 567-571.
Jordan, T. E. and D.L. Correll. 1985. Nutrient chemistry and hydrology of interstitial water in brackish tidal marshes of Chesapeake Bay. Estuarine, Coastal and Shelf Science **21**:45.
Jordan, T. E. and D.L. Correll. 1991. Continuous automated sampling of tidal exchanges of nutrients by brackish marshes. Estuarine, Coastal and Shelf Science. **32**: 527-545.
Jordan, T. E., J. W. Pierce and D. L. Correll. 1986. Flux of particulate matter in the tidal marshes and subtidal shallows of the Rhode River Estuary. Estuaries **9**:310-319.
Kjerfve, B. L., H. Stevenson, J.A. Proehl, T.H. Chrzanowski and W.M. Kitchens. 1981. Estimation of material fluxes in an estuarine cross section: a critical analysis of spatial measurement density and errors. Limnology and Oceanography **26**:324-335.
Kjerfve, B., L.B. Miranda and E. Wolanski. 1991. Modeling water circulation in an estuary and intertidal salt marsh system. Netherlands Journal of Sea Research **28**:141-147.
Klap, V.A. 1997. Biogeochemical aspects of salt marsh exchange processes in the SW Netherlands. Dissertation and Publ. No. 2288, Netherlands Institute of Ecology, Centre Estuarine and Coastal Ecology.
Kneib, R. T. and S.L. Wagner. 1994. Nekton use of vegetated marsh habitats at different stages of tidal inundation. Marine Ecology Progress Series **106**: 227-238.
Lee, V. 1979. Net nitrogen flux between the emergent marsh and tidal waters. Thesis, University of Rhode Island, Kingston, Rhode Island, USA.
Lefeuvre, J.C., G. Bertru, F. Burel, L. Brient, V. Creach, Y. Gueunne, J. Levasseur, A. Mariotti, A. Radureau, C. Retiere, B. Savoure and O. Troccaz. 1994. Comparative studies on salt marsh processes: Mont Saint Michel Bay, a multi-disciplinary study. Pages 215-234 *in* W. Mitsch, editor. Global wetlands. Elsevier Publishing, Amsterdam, The Netherlands.
Leonard, L. A., A. C. Hine, M.E. Luther, R. P. Stumpf and E. E. Wright. 1995. Sediment transport processes in a west-central Florida open marine marsh tidal creek: the role of tides and extra-tropical storms. Estuarine, Coastal and Shelf Science **41**:225-248.
Madden, C.J., J.W. Day, Jr. and J. Randall. 1988. Freshwater and marine coupling in estuaries of the Mississippi River deltaic plain. Limnology and Oceanography **33**:982-1004.
McIvor, C. C. and W.E. Odum. 1986. The flume net: a quantitative method for sampling fishes and macrocrustaceans on tidal marsh surfaces. Estuaries **9**:208.
McIvor, C. C. and L.P. Rozas. 1996. Direct nekton use of intertidal saltmarsh habitat and linkage with adjacent habitats: a review from the southeastern United States. Pages 311-331 *in* Nordstrom, K.F. and C. T. Roman, editors. Estuarine shores: evolution, environments and human alterations. John Wiley & Sons Ltd., New York, New York, USA.
McKellar, H., A. Douglas, A. Smith, T. Munnerlyn and R. Rao. 1996. Tidal wetlands help control water quality. Harbor News (Sept./Oct.), Charleston Harbor Project, South Carolina Department of Health and Environmental Control.

Millham, N. P. and B.L. Howes. 1994. Freshwater flow into a coastal embayment: groundwater and surface water inputs. Limnology and Oceanography **39**: 1928-1944.

Minello, T. J., R.J. Zimmerman and R. Medina. 1995. The importance of edge for natant macrofauna in a created salt marsh. Wetlands **14**: 184-198.

Moeller, C. C., O.K. Huh, H.H. Roberts, L.E. Gumley and W.P. Menzel. 1993. Response of Louisiana coastal environments to a cold front passage. Journal of Coastal Research **9**:434-447.

Murray, A. L. and T. Spencer. 1997. On the wisdom of calculating annual material budgets in tidal wetlands. Marine Ecology Progress Series **150**:207-216.

Nichols, M. M. 1977. Response and recovery of an estuary following a river flood. Journal of Sedimentary Petrology **47**:1171-1186.

Nixon, S.W. 1980. Between coastal marshes and coastal waters A review of twenty years of speculation and research on the role of salt marshes in estuarine productivity and water chemistry. Pages 437-525 *in* P. Hamilton and K. B. MacDonald, editors. Estuarine and wetland processes. Plenum Press, New York, New York, USA.

Odum, E.P. and A.A. de la Cruz. 1967. Particulate organic detritus in a Georgia salt marsh-estuarine ecosystem. Pages 383-388 *in* G.H. Lauff, editor. Estuaries, AAAS Publication No. 83.

Odum, E.P. 1968. A research challenge: evaluating the productivity of coastal and estuarine water. Proceedings of the Second Sea Grant Conference, University of Rhode Island, Kingston, Rhode Island, USA.

—— 1969. The strategy of ecosystem development. Science **164**:262-270.

—— 1980. The status of three ecosystem-level hypotheses regarding salt marsh estuaries: tidal subsidy, outwelling and detritus-based food chains. Pages 485-495 *in* V.S. Kennedy, editor. Estuarine perspectives. Academic Press, New York, New York, USA.

Odum, W. E., J. S. Fisher and J.C. Pickral. 1979. Factors controlling the flux of particulate organic carbon from estuarine wetlands. Pages 69-80 *in* R. J. Livingston, editor. Ecological processes in coastal and marine systems. Plenum Press, New York, New York, USA.

Oviatt, C.A., S.W. Nixon and J. Garber. 1977. Variation and evaluation of coastal salt marshes. Environmental Management **1**:201-211.

Park, J. K. and A. James. 1990. Mass flux estimation and mass transport mechanism in estuaries. Limnology and Oceanography **35**: 1301-1313.

Peterson, G. W. and R. E. Turner. 1994. The value of salt marsh edge vs. interior as a habitat for fish and decapod crustaceans in a Louisiana tidal marsh. Estuaries **17**:235-262.

Pomeroy, L. R., R.E. Johannes, E. P. Odum and B. Roffman. 1967. The phosphorus and zinc cycles and productivity of a salt marsh. Pages 412-430 *in* D. J. Nelson and F.C. Evans, editors. Symposium on Radioecology.

Rahm, L. and F. Wulff. 1992. Using parallel salinity and temperature profiles for calculations of estuarine fluxes with reference to the Baltic proper. Netherlands Journal of Sea Research **29**:281-289.

Reed, D.J. 1988. Sediment dynamics and deposition in a retreating coastal salt marsh. Estuarine, Coastal and Shelf Science **26**:67-79.

Reed, D. J. 1989. Patterns of sediment deposition in subsiding coastal salt marshes: the role of winter storms. Estuaries **12**:222-227.

Rivera-Monroy, V. H., J.W. Day, R.R. Twilley, F. Vera-Herrera and C. Coronado-Molina. 1995. Flux of nitrogen and sediment in a fringe mangrove forest in Terminos Lagoon, Mexico. Estuarine, Coastal and Shelf Science **40**: 139-160.

Roberts, H.H., O.K. Huh, S.A. Hsu, L.J. Rouse and D.A. Rickman. 1989. Winter storm impacts on the Chenier Plain coast of southwestern Louisiana. Transactions of Gulf Coast Association Geological Society **29**:515-522.

Rogers, C. S. 1979. Productivity of San Cristobal Reef, Puerto Rico. Limnology and Oceanography **24**:342-349.

Roman, C.T. and F.C. Daiber. 1989. Organic carbon flux through a Delaware Bay salt marsh: tidal exchanges, particle size distribution and storms. Marine Ecology Progress Series **54**:149-156.

Rozas, L. P., C.C. McIvor and W.E. Odum. 1988. Intertidal rivulets and creekbanks: corridors between tidal creeks and marshes. Marine Ecology Progress Series **47**: 303-307.

Rozas, L. P. and D.J. Reed. 1993. Nekton use of marsh-surface habitats in Louisiana deltaic salt marshes undergoing submergence. Marine Ecology Progress Series **96**: 147-157.

Scheleski, C.L. and E.P. Odum. 1961. Mechanisms maintaining high productivity in Georgia estuaries. Proceedings of Gulf Caribbean Fisheries Institute **14**:5-80.

Schubel, J. R. and D.J. Hirschberg. 1978. Estuarine graveyards, climatic change and the importance of the estuarine environment. Pages 285-303 *in* M. L. Wiley, editor. Estuarine Interactions, Academic Press, New York, New York, USA.

Settlemyre, J. L. and L.R. Gardner. 1977. Suspended sediment flux through a salt marsh drainage basin Estuarine, Coastal and Shelf Science **5**:653-663.

Sklar, F.H., R. Costanza, D.L. Childers, E.B. DeBellevue, M.S. Jacobsen, T. Maxwell and M.L. White. 1992. Developments in regional scale simulation and analysis: case studies from coastal wetland ecosystems. Pages 47-62 *in* R.B.Singh, editor. Environmental monitoring–application of remote sensing and GIS. Geocarto Int. Centre, Hong Kong.

Smith, N. P. 1979. Tidal dynamics and low-frequency exchanges in the Aransas Pass, Texas. Estuaries **2**:218-227.

Smith, S.V., W.J. Wiebe, J.T. Hollibaugh, S.J. Dollar, S.W. Hager, B.E. Cole, G.W. Tribble and P.A. Wheeler. 1987. Stoichiometry of C, N, P and Si fluxes in a temperate-climate embayment. Journal of Marine Research **45**:427-460.

Smith, S.V., J. T. Hollibaugh, S.J. Dollar and S. Vink. 1991. Tomales Bay metabolism: C-N-P stoichiometry and ecosystem heterotrophy at the land-sea interface. Estuarine, Coastal and Shelf Science **33**:223-257.

Spurrier, J. D. and B. J. Kjerfve. 1988. Estimating the net flux of nutrients between a saltmarsh and a tidal creek. Estuaries **11**:10-14.

Stern, M. K., J.W. Day, Jr. and K.G. Teague. 1991. Nutrient transport in a riverine-influenced, tidal freshwater bayou in Louisiana. Estuaries **14**: 382-394.

Stevenson, J.C., D.R. Heinle, D.A. Flemer, R.J. Small, R.A. Rowland and J.F. Ustach. 1977. Nutrient exchange between brackish water marshes and the estuary. Pages 219-240 *in* Estuarine Processes, Volume 2. Academic Press, New York, New York, USA.

Taylor, D. I. and B. R. Allanson. 1995. Organic carbon fluxes between an high marsh and estuary and the inapplicability of the outwelling hypothesis. Marine Ecology Progress Series **120**:263-270.

Twilley, R. R. 1985. The exchange of organic carbon in basin mangrove forests in a SW Florida estuary. Estuarine, Coastal and Shelf Science **20**: 543.

Valiela, I. J. Costa, K. Foreman, J.M. Teal, B. Howes and D. Aubrey. 1990. Transport of groundwater-borne nutrients from watersheds and their effects on coastal waters. Biogeochemistry **10**: 177-197.

Valiela, I., G. Collins, J. Kremer, K. Lajtha, M. Geist, B. Seely, J. Brawley and C.H. Sham. 1997. Nitrogen loading from coastal watersheds to receiving estuaries: new method and application. Ecological Applications **7**:358-380.

Vitousek, P.M. and W.A. Reiners. 1975. Ecosystem succession and nutrient retention: a hypothesis. Bioscience **25**: 376-381.

Vörösmarty, C.J. and T.C. Loder, III. 1994. Spring-neap tidal contrasts and nutrient dynamics in a marsh-dominated estuary. Estuaries **17**:537-551.

Walker, R.A. 1973. Wetlands preservation and management on Chesapeake Bay: The role of science in natural resource policy. Journal of Coastal Zone Management **1**:75-101.

Welsh, B.L. 1980. Comparative nutrient dynamics of a marsh-mudflat ecosystem. Estuarine, Coastal and Shelf Science **10**:143-164.

Wetzel, R. G. 1992. Wetlands as metabolic gates. Journal of Great Lakes Research **18**:529-532.

Whiting, G. and D.L. Childers. 1989. Subtidal advective water as a potentially important nutrient input to southeastern U.S.A. salt marsh estuaries. Estuarine, Coastal and Shelf Science **28**:417-431.

Whiting, G. J., H.N. McKellar, Jr., J.D. Spurrier and T.G. Wolaver. 1989. Nitrogen exchange between a vegetated saltmarsh and the adjoining creek. Limnology and Oceanography **34**:463-473.

Wolaver, T. G., R.L. Wetzel, J.C. Zieman and K.L. Webb. 1980. Nutrient interactions between saltmarsh, mudflats and estuarine water. Page 123 *in* V.S. Kennedy, editor. Estuarine perspectives. Academic Press, New York, New York, USA.

Wolaver, T. G., R.L. Wetzel, J.C. Zieman and K.L. Webb,. 1983. Tidal exchange of nitrogen and phosphorus between a mesohaline vegetated marsh and the surrounding estuary in the lower Chesapeake Bay. Estuarine, Coastal and Shelf Science **16**:321-332.

Wolaver, T. G., G.J. Whiting, B. Kjerfve, J.D. Spurrier, H.N. McKellar, R.F. Dame, T. H. Chrzanowski, R. Zingmark and T. Williams. 1985. The flume design — a methodology for evaluating material fluxes between a vegetated saltmarsh and the adjacent tidal creek. Journal Experimental Marine Biology Ecology **91**:281-291.

Wolaver, T. G. and J. D. Spurrier. 1988. The exchange of phosphorus between a euhaline vegetated marsh and the adjacent tidal creek. Estuarine, Coastal and Shelf Science **26**: 203-214.

Wolaver, T. G., R. F. Dame, J. D. Spurrier and A.B. Miller. 1988. Sediment exchange between a euhaline saltmarsh in South Carolina and the adjacent tidal creek. Journal Coastal Research **4**: 17-26.

Yelverton, G.F. and C. T. Hackney. 1986. Flux of dissolved organic carbon and pore water through the substrate of a *Spartina alterniflora* marsh in North Carolina. Estuarine, Coastal and Shelf Science **22**:255-267.

Yoder, J.A., P.G. Verity, S.S. Bishop and F.E. Hoge 1993. Phytoplankton chl-a, primary production and nutrient distributions across a coastal frontal zone off Georgia, USA. Continental Shelf Research **13**:131-141.

THE ROLE OF OLIGOHALINE MARSHES IN ESTUARINE NUTRIENT CYCLING

JENNIFER Z. MERRILL
JEFFREY C. CORNWELL
University of Maryland Center for Environmental Science
Horn Point Laboratory
Cambridge, MD 21613 USA

Abstract

Oligohaline marshes, poised at the land-sea margin, often occur where the estuary is most enriched in inorganic particles and nutrients. Although light can limit the production of planktonic communities, high nutrient concentrations and regular tidal inundation results in highly productive macrophyte and algal communities. Despite potentially important water quality values, relatively few detailed studies of N and P cycling in oligohaline marshes are evident in the literature. Because of the temporal variability in marsh flux studies, the net annual retention of N and P is best assessed by measurement of N and P burial in the sediment. In the Chesapeake Bay and other estuaries and subestuaries, high rates of tidal marsh N and P burial indicate an important water quality function. A recent study shows the marshes of a Chesapeake Bay tributary retain a large portion of nitrogen and phosphorus entering the river from above the fall line. The marshes trap 35% of the nitrogen and 81% of the phosphorus which would otherwise be recycled, exported, or buried in the subtidal sediments of the estuary. Although there are few studies, high nitrate supply rates, potentially high nitrification rates, and high rates of sediment metabolism can result in high rates of denitrification. More complete studies of tidal marsh nutrient cycling, particularly nitrogen cycling, are needed for a better understanding of the importance of these tidal freshwater marshes to estuarine nutrient balances. Alternative methodologies for denitrification measurement are needed for more accurate measurements, and more attention needs to be paid to scaling individual measurements to whole marsh ecosystems. A new method for the measurement of net $N_{2(g)}$ exchange was applied to a Chesapeake Bay tributary to develop an annual estimate of net denitrification in the marsh sediments. Denitrification rates were ~ 60 μmol N m^{-2} h^{-1} with high seasonal variability. Annual calculations were made based on a loose correlation to annual ambient nitrate concentrations. This preliminary calculation suggests that an additional 10% of the fall line nitrogen may be removed by such marsh systems. More measurements of net $N_{2(g)}$ exchange and computer simulation models are required to determine the net removal of fall line nitrogen by the upper estuarine marshes.

1. Introduction

Despite the widely-held belief that wetland systems are important nutrient sinks, surprisingly few studies have attempted to quantify their overall role in estuarine ecosystems. While the net influence of marshes on the nutrient chemistry has been studied intensively at a modest number of sites (Heinle and Flemer 1976, Stevenson et al. 1977, Woodwell et al. 1979, Dame et al. 1986), the literature regarding the influence of tidal wetlands on water quality improvement in anthropogenically-stressed estuaries is remarkably small. Recent efforts to improve global nitrogen budget estimates have pointed toward an inherent value of marshes in removing nitrogen through denitrification (Howarth et al. 1996, Vitousek et al. 1997), but relatively little marsh data was available to support these estimates. Permanent net losses of nitrogen and phosphorus to sediment burial in tidal marsh systems are available for relatively few systems.

Tidal freshwater and oligohaline marshes, often located immediately seaward of the fall-line at the head of tide, exchange materials with adjacent tidal waters. The high nutrient, contaminant, and sediment concentrations often associated with these upper estuarine systems make these tidal marshes potentially important in mitigating anthropogenic stress in the more saline regions of the estuary. Odum's (1988) review of tidal freshwater marsh ecology identifies the major ecological features of these systems and points out the lack of study of nutrient cycles in these systems. Insights provided by Odum's review continue to guide the development of more detailed understanding of these systems.

The objective of this chapter is to examine the role of low salinity tidal wetlands in estuarine nutrient cycles. Our focus is on processes leading to a long-term or permanent loss of N and P from active N and P cycling in the estuary; specifically, we will examine the rates and importance of denitrification and sediment N and P burial (Fig. 1).

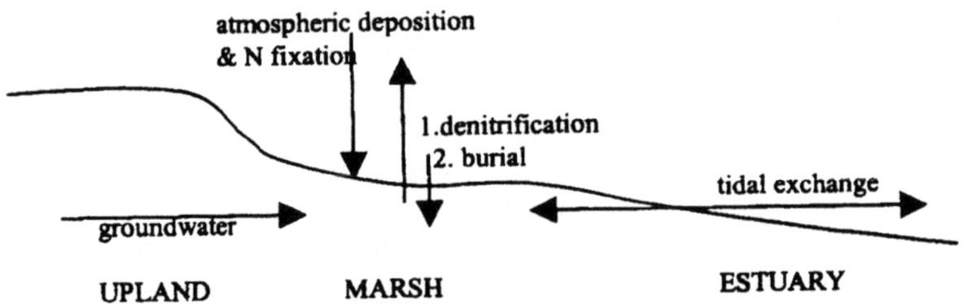

Figure 1. Schematic of tidal marsh nitrogen cycling. Burial and denitrification are discussed in this review.

2. Tidal Freshwater and Oligohaline Habitats

Tidal freshwater marshes are defined as those marshes which see salinities ranging between 0.5 and 5.0 psu (Odum 1988). These two marsh types form a continuum at a point in the estuary where they may provide disproportional benefits to the water quality of the estuary relative to their area. Tidal freshwater marshes occur in a region of focused tidal energy, nutrient inputs, and human development. Enhanced productivity from tidal energy subsidies is common, occurring without the constraints on plant diversity imposed by high salinities (Odum 1988). As a result, marsh vegetation is very diverse and highly productive. Rates of primary production are generally between 1 and 3 kg biomass m^{-2} y^{-1} (Mitsch and Gosselink 1993).

Human development is frequently concentrated near the fall line of riparian watersheds as a result of the high-energy fall line and a limit to oceanic shipping. High development density causes increased nutrient inputs, either as localized sources from industry and wastewater treatment plants and from agricultural or other non-point sources (Fig. 2). The proximity of some tidal freshwater marshes to these developed areas may increase their potential for interception of nutrient loads before they can impact the lower estuary. Tidal freshwater marshes are located above the turbidity maxima of estuarine systems. These zones are caused by a combination of estuarine circulation patterns and a response to increased salinity. Flocculation of colloidal particles occurs and results in the increase in total suspended solids. These particulates are potential sediment sources for tidal freshwater marshes.

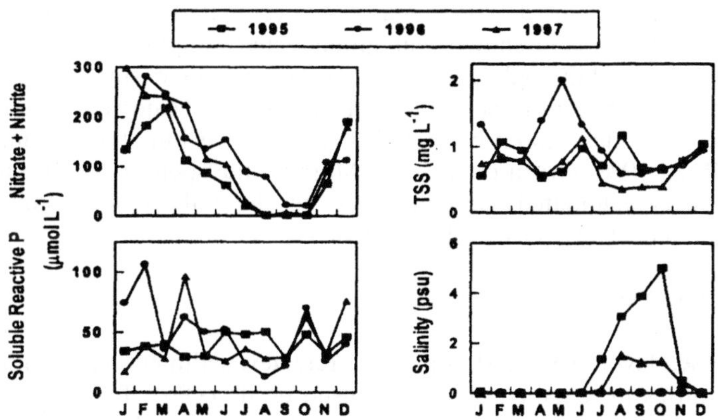

Figure 2. Selected water quality parameters for three years in the tidal freshwater/oligohaline Choptank River subestuary (Chesapeake Bay Program Station ET 5.1; http://www.chesapeakebay.net/). This station is located in a part of the estuary that is characterized by high biomass and areal coverage of tidal freshwater plant species. The strong seasonality of salinity and nitrate concentrations are characteristic of other coastal plain subestuaries; interannual salinity differences are driven by differences in freshwater inputs.

Biogeochemical processes are strongly affected by changes in salinity, with soil anaerobic metabolism in low salinity systems generally dominated by methanogenesis (Capone and Kiene 1988). With increasing salinity and higher concentrations of sulfate in marshes, a switch from methanogenesis to sulfate reduction is observed (Bartlett et al. 1987). Changes in salinity can have important biogeochemical consequences for processes ranging from P recycling (Caraco et al. 1989) to ammonium adsorption (Seitzinger et al. 1991); for the most part, the salinity dependence of these processes has not been examined in marshes.

3. A Review of Methods

3.1 FLUX STUDIES

The spatially and temporally dynamic nature of marshes can obscure findings and lead to interesting, but often inconclusive, results. One of the most popular approaches to determine small temporal scale estuarine-marsh interactions has been to measure tidal exchanges of nutrients. Numerous studies have attempted to draw inferences about marsh-estuarine interactions by measuring nutrient concentrations of flooding and draining marsh water over tidal cycles (Heinle and Flemer 1976, Roberts and Pierce 1976, Stevenson et al. 1977, Woodwell et al. 1979, Jordan et al. 1983, Simpson et al. 1983, Dame et al. 1986, Chambers et al. 1992, Childers et al. 1993). However, tidal exchange studies do not always provide an adequate means for estimating marsh-estuary nutrient budgets. Uncertainty in the hydrologic budget allows for calculated errors large enough to obscure any possible exchanges. Despite significant improvements to the basic technique of tidal exchange, such as the use of flumes, none of these studies has measured or even reasonably estimated groundwater inputs. Staver and Brinsfield (1996) have shown that groundwater supplies of nitrogen to coastal marshes can be an additional large input. Tidal exchange studies also rely on the extrapolation of discrete time points to annual budgets. Major events such as storms and even irregular tides cannot be included, yet may be important to annual material processing (Reed 1989, Orson et al. 1990, Murray and Spencer 1997). Intensive long-term (multi-year) automated sampling may be the only way to develop realistic estimates of tidal exchanges using this approach.

3.2 GEOCHRONOLOGY

In the face of sea level rise, marshes must accrete new sediment on a continual basis for plant communities to remain viable (Stevenson et al. 1986). This necessary burial of particulates generally entails the burial of both inorganic and organic sediments. The deep burial of N primarily as organic N and of P as both organic and inorganic P represent losses to the surface waters. The decomposition of organic matter in near surface soils is ubiquitous in marshes, and nutrient burial estimates based on near-surface sediment horizons may lead to erroneous burial estimates.

Measuring vertical accretion of marshes over longer time scales may be the simplest approach to estimating nutrient retention. Artificial markers such as brick dust and feldspar have been used effectively for the determination of sediment accretion (Richard 1978), but the method requires repeated measures and considerable temporal resolution to reliably estimate long-term burial. Radiotracers utilized for the measurement of accretion include ^{137}Cs derived from bomb testing and ^{210}Pb dating. The early 1960s ^{137}Cs peak has been used for reliable measures of sediment accretion in many locations (DeLaune et al. 1989, Nyman et al. 1993), but concerns about post-depositional mobility (Davis et al. 1984, Comans et al. 1989), non-steady state deposition (Milan et al. 1995) and mixing (Sharma et al. 1987) can make it difficult to apply in some circumstances. Atmospherically-derived ^{210}Pb has been used to successfully determine marsh accretion in many marsh systems (Armentano and Woodwell 1975, Sharma et al. 1987, Bricker-Urso et al. 1989, Orson et al. 1990, Wang and Benoit 1992). Analysis of the naturally occurring ^{210}Pb isotope has allowed researchers to determine the age of recently deposited sediments (Appleby and Oldfield 1978). Detection of five half-lives of ^{210}Pb ($t_{1/2}$ = 22.3 years), allows the application of this technique to sediments deposited over the previous 100 years. Bioturbation (Sharma et al. 1987) and potential mobility with transition metals (Anderson et al. 1987, Benoit and Hemond 1990) may affect the ^{210}Pb vertical profiles in some circumstances. Pollen analysis of European plant species (i.e., ragweed:oak ratio) has been used to estimate deposition in tidal fresh water marshes (Kahn and Brush 1994), but the technique relies on historical records to recreate time periods of European colonization and limits resolution. To calculate the rates of nutrient burial, the sediment burial rate (g m^{-2} y^{-1}) is multiplied by the core's nutrient concentration (mg g^{-1}) to estimate nitrogen or phosphorus burial (Fig. 3).

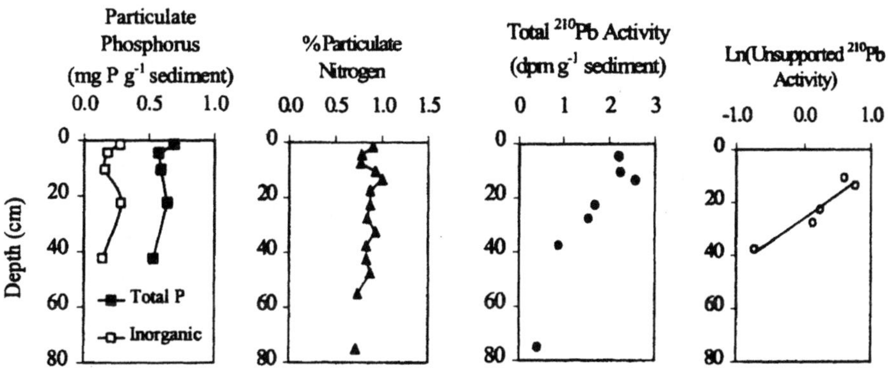

Figure 3. Sediment profiles from an oligohaline marsh on the Choptank River, Maryland (Merrill and Cornwell *unpublished*).

4. Losses of Nitrogen and Phosphorus: Short-term vs. Long-term

4.1 SHORT-TERM: DENITRIFICATION

Denitrification, the microbial reduction of oxidized NO_3^- or NO_2^- to N_2, represents a sink term in the ecosystem nitrogen budget of marshes (Kaplan et al. 1979). The NO_3^- that is generally the nitrogen source for denitrification may be derived from overlying water or from nitrification of ammonium in the sediment. High rates of NH_4^+ production in marsh soils occur through the microbially-mediated coupling of reduced organic matter decomposition with the utilization of a sequence (O_2, NO_3, Fe(III), SO_4) of electron acceptors (Howarth and Hobbie 1982). Diffusion of NH_4^+ to aerobic sediment horizons leads to microbial conversion of NH_4^+ to NO_3^-, with subsequent reduction to

N_2O and eventually N_2 via denitrification (Jenkins and Kemp 1984). Microbial communities capable of denitrification are believed to be ubiquitous and require 1) labile organic matter as a substrate, 2) nitrate for reduction, and 3) hypoxic or anoxic conditions. Organic matter is abundant in marsh sediment (>12% versus <4% in most subtidal sediments) and should not be a limiting factor of denitrification in marsh sediments, although measurements of dissolved organic carbon in natural marsh sediment have not been reported.

In marsh sediments, supplies of remineralized ammonium and the large redox boundary surface area in the rhizosphere may lead to high rate of coupled nitrification-denitrification (Reddy et al. 1989). While this high dependence on coupled nitrification-denitrification, as opposed to nitrate uptake, may exist in some marsh sediments, low rates of nitrification have been reported in salt marsh sediments (DeLaune and Smith 1987). However, nitrification did not appear to limit denitrification in a recent study by Risgaard-Petersen and Jensen (1997) who report an increase in denitrification and an increased reliance on the coupling of nitrification and denitrification in sediments inhabited by submerged aquatic plants. Published denitrification rates in tidal fresh and oligohaline marshes are rare.

The measurement of denitrification has been one of the major challenges in biogeochemistry (Table 1). High concentrations of N_2 dissolved in water present a major difficulty in measurement of denitrification and there have been a number of reviews on the problems with denitrification methodologies (Seitzinger et al. 1993, Cornwell et al. 1999). The acetylene inhibition technique (Chan and Knowles 1979) is still used (Urban et al. 1988) even though major drawbacks have been identified (Seitzinger et al. 1993). Another approach is the use of ^{15}N as a tracer of nitrification and denitrification (Nishio et al. 1982, Jenkins and Kemp 1984, Bowden 1986, Nielsen 1992), but the heterogeneity of labeled ^{15}N mixing with the nitrogen pools of interest can be of concern (Middelburg et al. 1996). Seitzinger et al. (1984) developed a direct measurement of denitrification by observing the change in N_2 over time, but long incubations were necessary to minimize the flux of background N_2. This approach has been successfully applied to wetland soils (Seitzinger 1994).

TABLE 1. Comparisons of methods used to measure denitrification.

Technique	Approach	Limitations	Example References
acetylene inhibition	Blocks N_2O to N_2, measure N_2O accumulation	Incomplete block; blocks nitrification	Sorensen 1978, Chan and Knowles 1979
^{15}N	Label NO_3^- or NH_4^+, follow incorporation in N_2	Potential enhancement by NO_3^- additions	Nielsen 1992
Direct N_2	Purge system of N_2, measure N_2 increase	Degassing step, long-term incubation, controls	Seitzinger 1994
Direct N_2	Time course change in N_2 measured by gas chromatography	High background, gas disequilibria from changing temperatures or from CH_4 ebullition	Devol 1991, LaMontagne and Valiela 1995
N_2/Ar ratio change	Measure time-course changes in gas ratio	Gas disequilibria from changing temperatures or from CH_4 ebullition	Kaplan et al. 1979, Kana et al. 1994; 1998

Direct measurement techniques of N_2 flux using high precision gas chromatography techniques has been used successfully for aquatic sediments, generally with a precision of

0.5%. Recently, Kana et al. (1994) adapted a membrane inlet mass spectrometer to the precise (<0.05%) measurement of N_2: Ar gas ratios; we have successfully applied this technique to tidal marsh sediments by using core incubations (Merrill unpublished). Limitations to the technique include a need for tight temperature control during incubation and potential gas disequilibria from bubbles of O_2 or CH_4 produced during incubation.

4.2 LONG-TERM: BURIAL OF NITROGEN AND PHOSPHORUS

Particulate deposition and organic matter retention in marsh sediment is a sink for nitrogen and phosphorus from the surrounding ecosystem. Sediments in marshes accrete vertically, incorporating organic and mineral matter, and retaining nutrients within this matrix. Reported rates of vertical accretion in coastal marshes range between 0.2 (Harrison and Bloom 1974) and 1.35 cm y^{-1} (DeLaune et al. 1981). Nixon (1980) estimates nutrient burial may range from 5 to 20 g N m^{-2} y^{-1} and 0.05 to 0.3 g P m^{-2} y^{-1}.

Limited examples of nitrogen and phosphorus burial studies are found in North American estuaries, including Chesapeake Bay (Table 2). Nutrient burial literature is scattered, with the majority of the work completed within inland freshwater systems (Johnston 1991), Louisiana (Hatton et al. 1982) and North Carolina salt marshes (Craft et al. 1988). Louisiana marshes bury large amounts of nitrogen and phosphorus, as well as a large portion of the carbon fixed by photosynthesis within the marsh (DeLaune et al. 1981, Hatton et al. 1982). These salt marshes retain a large amount of nitrogen, up to 21 g of nitrogen per square meter of marsh annually (DeLaune et al. 1981). However, the long-term potential for nutrient retention remains unclear if these marshes cannot keep pace with rapid rates of sea level rise on the Mississippi Delta. Kahn and Brush (1994) have used pollen analysis to estimate nutrient burial rates at two sites in Jug Bay, on the Patuxent River. High and low marsh profiles were compared, and higher nutrient accumulation rates in the high marsh are attributed to a slower rate of decomposition and more efficient retention of plant material. A review of nutrient retention in primarily non-tidal freshwater marshes has been published by Johnston (1991).

5. Recent Research

5.1 DENITRIFICATION RATES

Denitrification has been investigated in a variety of aquatic systems, with the majority of studies having been conducted in subtidal environments. Marsh denitrification rates are limited primarily to inland freshwater and salt marsh systems, but there is reason to believe that a gradient of denitrification rates may be observed with position in the estuary. Organic matter supply and lability (Odum 1988) coupled with sulfate availability may affect the ability of marsh sediments to remove fixed nitrogen. Sulfate can reduce rates of nitrification (Joye and Hollibaugh 1995) which may limit denitrification in brackish and salt marshes. In an early study of Great Sippewissett salt

TABLE 2. A comparison of nutrient burial rates in marsh systems of varying salinity. All studies were based on calculations of burial by measurements of sediment deposition and nutrient concentration.

	Tracer	N burial $g\ m^{-2}\ y^{-1}$	P burial $g\ m^{-2}\ y^{-1}$	Author(s)
Salt Marsh				
North Carolina	^{137}Cs	1.3-4.1	-	Craft et al. 1993
Oligohaline/Mesohaline				
Louisiana	^{137}Cs	21	-	DeLaune et al. 1981
North Carolina	^{137}Cs	6.9-10.3	-	Craft et al. 1993
Choptank River, MD	^{210}Pb	19.2-27.1	0.18-1.96	Merrill and Cornwell unpublished
Monie Bay, MD	^{210}Pb	13.6	0.01-1.30	Merrill and Cornwell unpublished
Tidal Freshwater				
Patuxent River, MD	^{210}Pb	23.4	3.54	Merrill and Cornwell unpublished
Otter Point Creek, MD	^{210}Pb	2.74-11.7	0.47-2.09	Merrill and Cornwell unpublished
Tivoli Bays, NY	^{210}Pb	2.37-13.3	0.667-3.06	Merrill unpublished
Nontidal Freshwater				
Wisconsin	^{137}Cs	12.8	2.6	Johnston et al. 1984
Average organic soils	various	14.6	1.46	Johnston 1991
Average inorganic soils	various	1.6	0.26	Johnston 1991
Florida everglades	^{210}Pb	14.1	0.66	Craft and Richardson 1993

marsh, Kaplan et al. (1979) report denitrification rates of 3 to 21 g N $m^{-2}\ y^{-1}$. The Nixon and Lee (1986) review reported rates of denitrification for various marsh types, yet tidal freshwater wetlands have no representative studies. A study by Groszkowski (1995) suggests that the tidal freshwater marshes of the Patuxent River may be capable of denitrification rates higher than that of subtidal sediment (Fig. 4). Rates nearing 100 μmol N $m^{-2}\ h^{-1}$ were measured, approximately four times the measured rate of nearby subtidal sediments (Kemp et al. 1990). However, Bowden et al. (1991) measured only small rates of denitrification in a tidal freshwater wetland using the acetylene-block technique. Clearly, denitrification estimates for tidal freshwater marsh systems are needed to determine their capacity for removing fixed nitrogen from aquatic systems.

The authors have completed a more detailed study of denitrification in the tidal freshwater marshes of the Patuxent River. Core incubation experiments were conducted on cores collected from four locations in the marshes of the Jug Bay National Estuarine Research Reserve during the spring, summer and fall. The technique was adapted from Kana et al. (1997). Cores were kept in gas-tight tubes with lids fitted with magnetic stir bars with 15 cm of overlying water. Water samples were removed over a period of twelve hours and analyzed for oxygen, nitrogen ($N_{2(g)}$) and argon using the membrane-inlet mass spectrometer (Kana et al. 1994). Declines

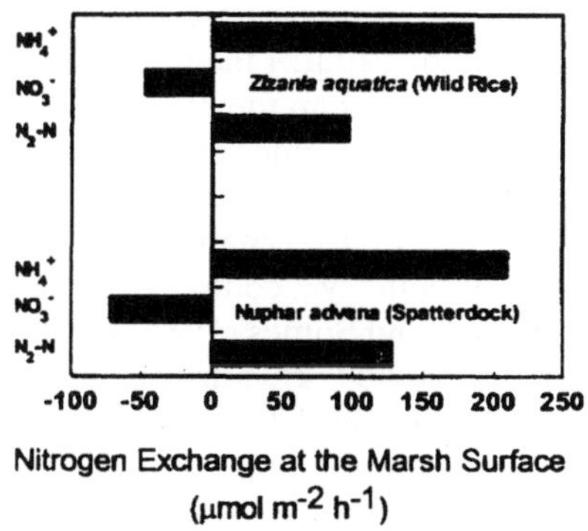

Figure 4. Nitrogen flux rates given as μmol N m^{-2} h^{-1} measured during core incubation experiments in the summer of 1995. Negative fluxes are into the marsh sediment. Cores were collected from the Maryland National Estuarine Research Reserve System (NERRS) at Jug Bay (from Groszkowski 1995).

in $N_{2(g)}$ over time relative to argon, assumed to be a conservative tracer were measured and used to calculate net denitrification (Fig. 5).

Denitrification occurred during the spring experiments at rates up to 60 μmol N m^{-2} h^{-1}, comparable to that measured by Jenkins and Kemp (1984) in local subtidal sediments (Fig. 6). Summer experiments found denitrification in some cores, but not in all replicates. Fall experiments measured net nitrogen fixation is occurring at rates up to 28 μmol N m^{-2} h^{-1}. Nitrate supplements increased net denitrification in spring experiments, suggesting nitrate limitation during spring sampling. It is likely River water collected during sampling at low tide was already depleted in nitrate as shown by Swarth and Peters (1993). Ambient water used in the experiments commonly contained only 20 μM NO3, while Patuxent River water reaches nitrate concentrations of 120 μM during the spring (Magnien et al. 1992).

Nitrate concentration was the most closely correlated of all measured factors to net denitrification, based on data from the Patuxent River and a similar study in Hudson River tidal freshwater marshes. Typical seasonal Patuxent River nitrate concentrations

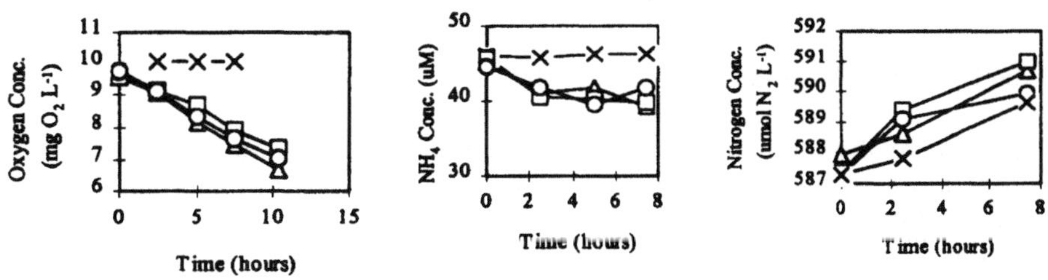

Figure 5. Time course incubations of intact tidal freshwater marsh cores collected from the low marsh of Jug Bay NERRS during the spring of 1997. Incubations were run at field temperature, 14°C. Nitrogen gas was measured using the Membrane Inlet Mass Spectrometer (Kana et al. 1994). Data presented are triplicates cores from the low marsh site; 'x'' designates the blank which contained only River water.

were used to generate a first estimate of the annual removal of fixed nitrogen by the tidal freshwater marshes of an ecosystem. Spring nitrate concentrations are the highest of the year, and at a nitrate concentration of 120 μM the tidal freshwater marshes may remove as much as 178 μmol N m^{-2} h^{-1}, higher than the rates reported by Groszkowski (1995). Nitrate concentrations drop during the summer and fall and likely limit denitrification. Assuming denitrification rates drop to half of those found in the spring, that each season consists of 90 days, and the marsh is only capable for 12 hours each day due to tidal cycling, approximately 360 mmol N m^{-2} y^{-1} could be removed. Using tidal fresh marsh acreage estimates from McCormick and Somes (1982) for the Patuxent River, the marsh may remove the equivalent of 100,800 kg nitrogen as $N_{2(g)}$. An admittedly loose calculation, this example provides a useful format for comparison to the surrounding ecosystem. If this estimate is a reasonable approximation, then these marshes remove approximately 10% of nitrogen entering the Patuxent River from above the fall line (Boynton et al. 1995), a region of heavy human development. Clearly denitrification needs to be more closely examined in the context of landscape and ecosystem function.

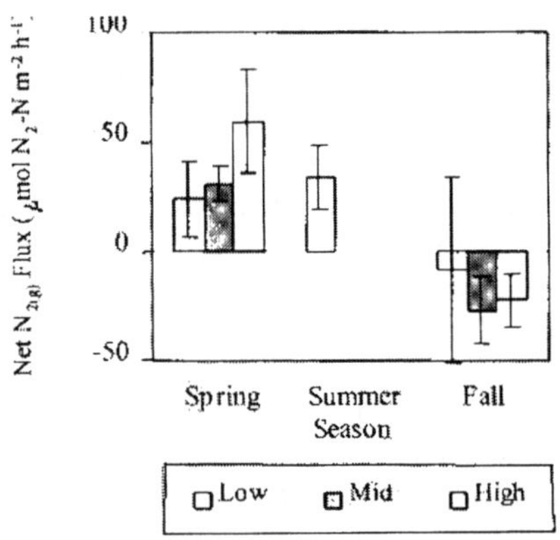

Figure 6. Marsh sediment net denitrification measured using membrane inlet mass spectrometry technology. Three elevations within the marsh are shown. Error bars from triplicate experimental cores are shown. Missing summer data is the result of one of three replicates producing a net $N_{2(g)}$ flux. Subtidal sediment cores were included in the experiment but exhibited no net flux of $N_{2(g)}$. All experiments were conducted at field temperatures, 14, 23, and 15°C, respectively.

5.2 NITROGEN AND PHOSPHORUS BURIAL RATES

A recent study by the authors gathered data from four marshes of the Chesapeake Bay in order to measure sedimentation, nitrogen and phosphorus to calculate nutrient burial. One meter sediment cores were collected from the Patuxent River, Otter Point Creek, near Baltimore, the Choptank River and from Monie Bay, at the mouth of the Wicomico River (Fig. 7). In all, thirty-eight cores were used for the determination of sediment accretion and nutrient burial. Cores were dated using ^{210}Pb geochronology. Nitrogen analysis was carried out using a Control Instruments CHN analyzer with a precision better than 5%. Phosphorus (total and inorganic) was analyzed following

(Aspila et al. 1976) using the molybdenum blue technique of Parsons et al. (1984). Repeated measurements were within 5%. Nitrogen burial rates were calculated using average N concentration to 100 cm following DeLaune et al. (1981). Phosphorus burial was estimated using average total phosphorus concentration found between 20 and 100 cm depth to avoid inflated estimates due to enhanced surface concentrations which could result from post-diagenetic mobility with iron. The nutrient concentrations were multiplied by the mass accretion (g m^{-2} y^{-1}) of sediment as determined by a linear regression of the ^{210}Pb data on the cumulative bulk density at each site. This calculation corrects for sediment compaction to the depth of the limit of the ^{210}Pb detection (in most cases, approximately 40 cm).

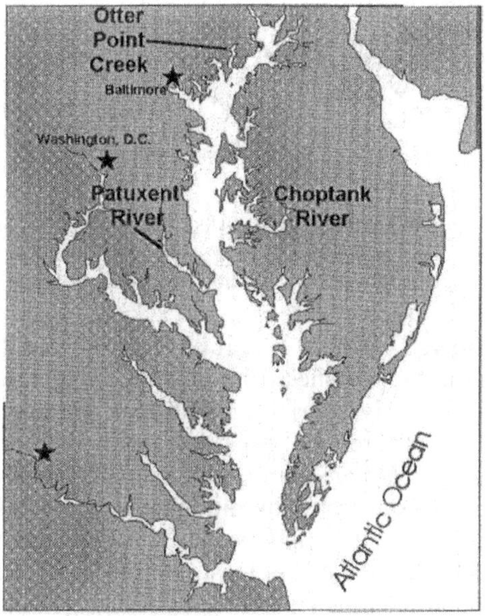

Figure 7. Map of sampled marshes in the Chesapeake Bay

Twenty-five cores collected from the Patuxent River tidal fresh and oligohaline marshes were used to compare marsh nutrient burial to the nutrient inputs to the River as presented by Boynton et al. (1995). Salt marsh sediment accretion rates decline with distance from tidal water source (DeLaune et al. 1981, Bricker-Urso et al. 1989) while vegetation associations are controlled by various mechanisms associated with flooding regime (Simpson et al. 1983, Odum et al. 1984). Sediment core data and nutrient burial calculations were extrapolated to the marsh surface area using acreage estimates of specific vegetation communities and associations reported by McCormick and Somes (1982). Sampled communities accounted for 79% of the reported vegetation coverage. For the remaining 21% the Patuxent River marsh average nitrogen and phosphorus burial rates were used.

Sedimentation rates in the Patuxent River marshes were highly variable ranging from 25 mm y^{-1} to 1.1 mm y^{-1} (Fig. 8) and were generally higher than those found in the other three marshes. Patterns of sedimentation distributed by vegetation community support earlier work showing higher rates of accretion close to the tidal river. Rapidly-accreting *Nuphar*-dominated associations were always located at the river's edge, which was true also for *Spartina cynosuroides*-dominated associations with few exceptions.

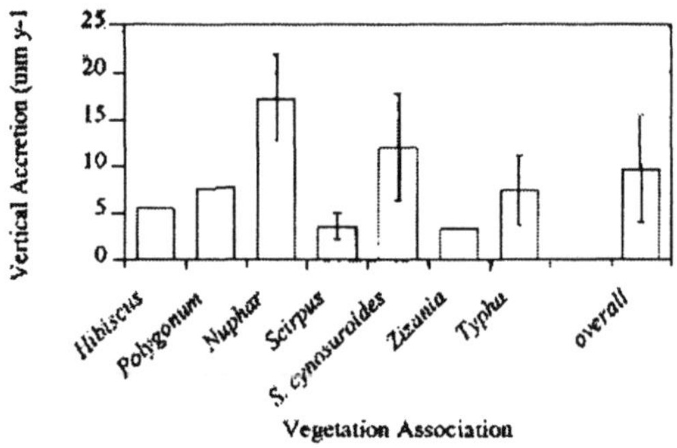

Figure 8. Sedimentation in the Patuxent River marshes presented by vegetation community or association named for the dominant species. Error bars represent standard error of the calculated average from all cores collected in the specified vegetation type. Data without error bars indicates data from one core was used.

Patuxent River marsh nutrient burial also varied greatly, and again, rates were generally higher than in the marshes from this and other studies (Table 2). Phosphorus burial rates were similarly variable. Nutrient burial, while calculated using sedimentation rate, did not remain consistent with the same relative rates between vegetation communities. *Nuphar*-dominated locations, while by far the most rapidly accreting, had only moderate rates of nitrogen burial, while phosphorus was buried much more rapidly than at the other sites. The opposite was true for the *Scirpus*-dominated sites, which were located at least 30 m from the River. At these locations the high sediment N:P supported much higher rates of nitrogen burial relative to phosphorus. The origin of sedimentary matter, phosphorus-laden mineral inputs in the *Nuphar* associations at the edge of the River, and organic matter retention by the interior marshes, clearly drives the ability of a marsh to retain sediment nutrients. The Patuxent River marsh study included this nuance in the estimate of marsh system nutrient retention by basing the final ecosystem calculations on vegetation community acreage.

Tidal fresh and oligohaline marshes of the Patuxent River retain 418,000 kg total nitrogen and 71,000 kg total phosphorus annually. This upper region of the Patuxent receives approximately 1,182,000 kg nitrogen and 88,000 kg phosphorus over the same time period from various sources, including but not limited to 43 million gallons daily of sewage effluent from eight treatment plants. Local marshes may be retaining 35% of this incoming nitrogen and 81% of the incoming phosphorus. Phosphorus has a high affinity for particulate material and the River receives high loads of sediment. This may account for the disparity between the two nutrients. Tidal fresh and oligohaline marshes in the Patuxent River are providing a necessary nutrient sink for much of the incoming anthropogenic pollution.

5.3 THE IMPORTANCE OF CHESAPEAKE BAY LOW SALINITY MARSHES

Chesapeake Bay tidal marshes cover approximately 1.7 million acres, yet the importance of these marshes as nutrient sinks within the estuary remains largely unassessed. Marsh-estuarine nutrient studies in the Chesapeake have focused on creek flux studies (Heinle and Flemer 1976, Stevenson et al. 1977, Jordan et al. 1983) which have not lead to a synthesized picture of marshes within the Bay. Other studies such as Kahn and Brush (1994) incorporate long term data but have been very limited in scope and cannot be applied to the Chesapeake as a whole. The first step to understanding the role of the tidal marshes in permanent nutrient retention is to find their rate of sedimentation. Long-term nutrient retention in tidal marshes along the Chesapeake Bay has not been determined. Nixon's (1980) review of marsh ecology inspired studies of nutrient deposition areas such as the Louisiana delta, but similar research on Chesapeake Bay tidal marshes is nonexistent.

Our studies of Chesapeake tidal freshwater and oligohaline marshes (Table 2) generally show high rates of nitrogen and phosphorus burial. In four different systems similar, if slightly more modest, results to those for the Patuxent River show these marshes to be critical for water quality maintenance. A considerable range in nutrient burial rate may be found in each system, with creek bank nitrogen and phosphorus retention rates being higher than rates measured in tidally-limited areas of the marsh. Vegetation coverage appears to be a reasonable approximation for ecosystem calculations.

With new seasonal estimates of marsh denitrification for a Patuxent River marsh a first attempt at including tidal freshwater marshes in an ecosystem nutrient budget has been completed. The Groszkowski (1995) summer study showed high rates of denitrification, and work by the authors showed gaseous nitrogen flux is a major biogeochemical process at the surface of the marsh. Although denitrification has received much attention as a microbial pathway for nitrogen transformation, salt marsh studies are limited, and tidal fresh marshes are just beginning. In order to establish how the upper estuarine marshes relate to the larger ecosystem more work must be done to develop our understanding of denitrification in the landscape.

The highest net denitrification rates measured, when extrapolated to the whole marsh system, are much less substantial than nitrogen burial rates. More detailed studies are required for a better understanding of nitrogen cycling in these tidal fresh and oligohaline marshes.

6. Conclusions and Recommendations

Little has been added to the understanding of seasonal cycling of nitrogen and phosphorus in tidal freshwater marshes since the work of Bowden (1986, Bowden et al. 1991). Recent studies suggest that natural tidal freshwater marshes are important to the water quality of estuarine ecosystems. Burial of nutrients in the marsh sediments represents a measurable loss of both nitrogen and phosphorus. Tidal freshwater marshes are situated near areas of high suspended solids in the estuarine water column, allowing them to trap and preserve large quantities of nitrogen and phosphorus. The marshes are located in regions that are anthropogenically impacted, leading to increased concentrations of fixed nitrogen. New studies suggest that this increased supply of nitrogen may fuel high rates of denitrification in the marsh sediments. This variability may be critical in determining the role these marshes play in the estuarine nutrient budgets. Furthermore, the local inputs of nitrogen and phosphorus and the generally unassessed process of nitrogen fixation may have important effects on the assessment of the water quality function of these marshes.

Multiple terms within conceptual models of nitrogen and phosphorus cycling in tidal freshwater marshes remain unmeasured. Unlike coastal salt marshes, these systems remain virtually unstudied, causing generalizations about their functioning and potential importance to be largely unsubstantiated. Investigations into seasonal and annual nutrient processing cycles need to be completed. Emphasis should be placed on the location of these marshes relative to the estuarine watershed and nutrient supply.

Care must be used in the development of nutrient budgets for tidal freshwater marshes. Patterns following the course of the river in these regions suggest zones of heavy and light deposition. Clear depositional patterns are observed in transects from the riverbank to the back marsh areas (Merrill and Cornwell unpublished, Kahn and Brush 1994, Bricker-Urso et al. 1989), further complicating the development of system budgets. Spatial coverage needs to be emphasized by completing nutrient burial surveys distributed across the surface of these tidal marshes.

7. Literature Cited

Anderson, R. F., S. L. Schiff and R. H. Hesslein. 1987. Determining sediment accumulation and mixing rates using ^{210}Pb, ^{137}Cs and other tracers: problems due to post-depositional mobility or coring artifacts. Canadian Journal of Fisheries and Aquatic Sciences **44**:231-250.

Appleby, P. G. and F. Oldfield. 1978. The calculation of lead-210 dates assuming a constant rate of supply of unsupported ^{210}Pb to the sediment. Catena **5**:1-8.

Armentano, T. V. and G. M. Woodwell. 1975. Sedimentation rates in a Long Island marsh determined by ^{210}Pb dating. Limnology and Oceanography **20**:452-456.

Bartlett, K. B., D. S. Bartlett, R. C. Harriss and D. I. Sebacher. 1987. Methane emissions along a salt marsh salinity gradient. Biogeochemistry **4**:183-202.

Benoit, G. and H. F. Hemond. 1990. ^{210}Po and ^{210}Pb remobilization from lake sediments in relation to iron and manganese cycling. Environmental Science and Technology **8**:1224-1134.

Bowden, W. B. 1986 Nitrification, nitrate reduction and nitrogen immobilization in a tidal freshwater sediment. Ecology **67(1)**:88-99.

Bowden, W. B., C. J. Vorosmarty, J. T. Morris, B. J. Peterson, J. E. Hobbie, P. A. Steudler and B. Moore III. 1991. Transport and processing of nitrogen in a tidal freshwater wetland. Water Resources Research **27**:389-408.

Boynton, W. R., J. H. Garber, R. Summers and W. M. Kemp. 1995. Inputs, transformations and transport of nitrogen and phosphorus in Chesapeake Bay and selected tributaries. Estuaries **18 (1B)**: 285-314.

Bricker-Urso, S., S. W. Nixon, J. K. Cochran, D. J. Hirschberg and C. Hunt. 1989. Accretion rates and sediment accumulation in Rhode Island salt marshes. Estuaries **12**:300-317.

Capone, D. G. and R. P. Kiene. 1988. Comparison of microbial dynamics in freshwater and marine environments: Contrasts in anaerobic carbon metabolism. Limnology and Oceanography **33**:725-749.

Caraco, N. F., J. J. Cole and G. E. Likens. 1989. Evidence for sulphate-controlled phosphorus release from sediments of aquatic systems. Nature **341**:316-318.

Chambers, R. M., J. W. Harvey and W. E. Odum. 1992. Ammonium and phosphate dynamics in a Virginia salt marsh. Estuaries **15**:349-359.

Chan, Y.-K. and R. Knowles. 1979. Measurement of denitrification in two freshwater sediments by an *in situ* acetylene inhibition method. Applied and Environmental Microbiology **37**:1067-1072.

Childers, D. L., S. Cofer-Shabica and L. Nakashima. 1993. Spatial and temporal variability in marsh-water column interactions in a southeastern USA salt marsh estuary. Marine Ecology Progress Series **95**:25-38.

Comans, R. N. J., J. J. Middelburg, J. Zonderhuis, J. R. W. Woittiez, G. J. DeLange, H. A. Das and C. H. Van Der Weijden. 1989. Mobilization of radiocesium in pore water of lake sediments. Nature **339**:367-369.

Cornwell, J.C., W.M. Kemp and T.M. Kana. 1999. Denitrification in coastal ecosystems: methods, environmental controls and ecosystem level controls, a review. Aquatic Ecology **33**: in press.

Craft, C. B., S. W. Broome and E. D. Seneca. 1988. Nitrogen, phosphorus and organic carbon pools in natural and transplanted marsh soils. Estuaries **11**:272-280.

Craft, C. B. and C. J. Richardson. 1993. Peat accretion and N, P and organic C accumulation in nutrient-enriched and unenriched Everglades peatlands. Ecological Applications **3**:446-458.

Craft, C. B., E. D. Seneca and S. W. Broome. 1993. Vertical accretion in microtidal regularly and irregularly flooded estuarine marshes. Estuarine and Coastal Shelf Science **37**:371-386.

Dame, R., T. Chrzanowski, K. Bildstein, B. Kjerfve, H. McKellar, D. Nelson, J. Spurrier, S. Stancyk, H. Stevenson, J. Vernberg and R. Zingmark. 1986. The outwelling hypothesis and North Inlet, South Carolina. Marine Ecology Progress Series **33**:217-229.

Davis, R. B., C. T. Hess, S. A. Norton, D. W. Hanson, K. D. Hoagland and D. S. Anderson. 1984. ^{137}Cs and ^{210}Pb dating of sediments from soft-water lakes in New England (USA) and Scandinavia, a failure of ^{137}Cs dating. Chemical Geology **44**:151-185.

DeLaune, R. D., C. N. Reddy and W. H. Patrick. 1981. Accumulation of plant nutrients and heavy metals through sedimentation processes and accretion in a Louisiana salt marsh. Estuaries **4**:328-334.

DeLaune, R. D. and C. J. Smith. 1987. Simultaneous determination of nitrification and nitrate reduction in sediment-water columns by nitrate-15 dilution. Journal of Environmental Quality **16**:227-230.

DeLaune, R. D., J. H. Whitcomb, W. H. Patrick Jr., J. H. Pardue and S. R. Pezeshki. 1989. Accretion and canal impacts in a rapidly subsiding wetland: ^{137}Cs and ^{210}Pb techniques. Estuaries **12**:247-259.

Devol, A. H. 1991. Direct measurement of nitrogen gas fluxes from continental shelf sediments. Nature **349**:319-321.

Groszkowski, K. 1995. Denitrification in a tidal freshwater marsh. Thesis, Harvard University, Cambridge, Massachusetts, USA.

Harrison, E. Z. and A. L. Bloom. 1977. Sedimentation rates on tidal salt marshes in Connecticut. Journal of Sedimentary Petrology **47**:1484-1490.

Hatton, R. S., W. H. Patrick Jr. and R. D. DeLaune. 1982. Sedimentation, nutrient accumulation and early diagenesis in Louisiana Barataria Basin coastal marshes. Pages 255-267 *in* V. S. Kennedy, editor. Estuarine Comparisons. Academic Press, New York, New York, USA.

Heinle, D. R. and D. A. Flemer. 1976. Flows of materials between poorly flooded tidal marshes and an estuary. Marine Biology **35**:359-373.

Howarth, R. W., G. Billen, D. Swaney, A. Townsend, N. Joworski, K. Lajtha, J. A. Downing, R. Elmgren, N. Caraco, T. Jordan, F. Berendse, J. Freney, V. Kudeyarov, P. Murdoch and Z. Zhao-Liang. 1996. Regional nitrogen budgets and riverine N and P fluxes for the drainages to the North Atlantic Ocean: natural and human influences. Biogeochemistry **35**:75-139.

Howarth, R. W. and J. E. Hobbie. 1982. The regulation of decomposition and heterotrophic microbial activity in salt marsh soils: a review. Pages 183-207 *in* V. S. Kennedy, editor. Estuarine comparisons. Academic Press, New York, New York, USA.

Jenkins, M. C. and W. M. Kemp. 1984. The coupling of nitrification and denitrification in two estuarine sediments. Limnology and Oceanography **29**:609-619.

Johnston, C. A. 1991. Sediment and nutrient retention by freshwater wetlands: effects on surface water quality. Critical Reviews in Environmental Control. **21**:491-565.

Johnston, C. A., G. D. Bubenzer, G. B. Lee, F. W. Madison and J. R. McHenry. 1984. Nutrient trapping by sediment deposition in a seasonally flooded lakeside wetland. Journal of Environmental Quality **13**:283-290.

Jordan, T. E., D. L. Correll and D. F. Whigham. 1983. Nutrient flux in the Rhode River: tidal exchange of nutrients by brackish marshes. Estuarine and Coastal Shelf Science **17**:651-667.
Joye, S. B. and J. T. Hollibaugh. 1995. Influence of sulfide inhibition on nitrification on nitrogen regeneration in sediments. Science **270**:623-625.
LaMontagne, M. G. and I. Valiela. 1995. Denitrification measured by a direct N_2 flux method in sediments of Waquoit Bay, MA. Biogeochemistry **31**:63-83
Kahn, H. and G. S. Brush. 1994. Nutrient and metal accumulation in a freshwater tidal marsh. Estuaries **17**:345-360.
Kana, T. M., C. Darkangelo, D. M. Hunt, J. B. Oldham, G. E. Bennett and J. C. Cornwell. 1994. Membrane inlet mass spectrometer for rapid high-precision determination of N_2, O_2 and Ar in environmental water samples. Analytical Chemistry **66**:4166-4170.
Kana, T. M., M. B. Sullivan, J. C. Cornwell and K. M. Groszkowski. 1998. Denitrification in estuarine sediments determined by membrane inlet mass spectrometry. Limnology and Oceanography **43**:334-339.
Kaplan, W., I. Valiela and J. M. Teal. 1979. Denitrification in a salt marsh ecosystem. Limnology and Oceanography **24**:726-734.
Kemp, W. M., P. Sampou, J. Caffrey, M. Mayer, K. Henriksen and W. R. Boynton. 1990. Ammonium recycling versus denitrification in Chesapeake Bay sediments. Limnology and Oceanography **35**:1545-1563.
Magnien, R. E., R. M. Summers and K. G. Sellner. 1992. External nutrient sources, internal nutrient pools and phytoplankton production in the Chesapeake Bay. Estuaries **15**:497-516.
Middelburg, J. J., K. Soetaert, P. M. J. Herman and C. H. R. Heip. 1996. Denitrification in marine sediments: a model study. Global Biogeochemical Cycles **10**:661-673.
Milan, C. S., E. M. Swenson, R. E. Turner and J. M. Lee1995. Assessment of the ^{137}Cs method for estimating sediment accumulation rates: Louisiana salt marshes. Journal of Coastal Research **11(2)**:296-307.
Mitsch, W. J. and J. G. Gosselink. 1993. Wetlands. Second edition. Van Nostrand Reinhold, New York, New York, USA.
Murray, A. L. and T. Spencer. 1997. On the wisdom of calculating annual material budgets in tidal wetlands. Marine Ecology Progress Series **150**:207-216.
Nielsen, L. P. 1992. Denitrification in sediment determined from nitrogen isotope pairing. Microbial Ecology **86**:357-362.
Nishio, T., I. Koike and A. Hattori. 1982. Denitrification, nitrate reduction and oxygen consumption in coastal and estuarine sediments. Applied and Environmental Microbiology **43**:648-653.
Nixon, S. W. 1980. Between coastal marshes and coastal waters. A review of twenty years of speculation and research on the role of salt marshes in estuarine productivity and water chemistry. Pages 437-525 *in* P. Hamilton and K. B. MacDonald, editors. Estuarine and wetland processes with emphasis on modeling. Plenum Publishing Company, New York, New York, USA.
Nixon, S. W. and V. Lee. 1986. Wetlands and water quality. United States Army Corps of Engineers, Washington, District of Columbia, USA.
Nyman, J. A., R. D. DeLaune, H. H. Roberts and W. H. J. Patrick. 1993. Relationship between vegetation and soil formation in a rapidly submerging coastal marsh. Marine Ecology Progress Series **96**:269-279.
Odum, W. E. 1988. Comparative ecology of tidal freshwater and salt marshes. Annual Review of Ecology and Systematics **19**:147-176.
Orson, R. A., R. L. Simpson and R. E. Good. 1990. Rates of sediment accumulation in a tidal freshwater marsh. Journal of Sedimentary Petrology **60**:859-869.
Reddy, K. R., W. H. J. Patrick and C. W. Lindau. 1989. Nitrification-denitrification at the plant root-sediment interface in wetlands. Limnology and Oceanography **34**:1004-1013.
Reed, D. J. 1989. Patterns of sediment deposition in subsiding coastal salt marshes, Terrebonne Bay, Louisiana: The role of winter storms. Estuaries **12**:222-227.
Richard, G. A. 1978. Seasonal and environmental variations in sediment accretion in a Long Island salt marsh. Estuaries **1**:29-35.
Risgaard-Petersen, N. and K. Jensen. 1997. Nitrification and denitrification in the rhizosphere of the aquatic macrophyte *Lobelia dortmanna* L. Limnology and Oceanography **42**:529-537.
Roberts, W. P. and J. W. Pierce. 1976. Deposition in upper Patuxent estuary, Maryland, 1968-1969. Estuarine and Coastal Marine Science **4**:267-280.

Seitzinger, S. P. 1994. Linkages between organic matter mineralization and denitrification in eight riparian wetlands. Biogeochemistry **25**:19-39.

Seitzinger, S. P., W. S. Gardner and A. K. Spratt. 1991. The effect of salinity on ammonium sorption in aquatic sediments: Implications for benthic nutrient recycling. Estuaries **14**:167-174.

Seitzinger, S. P., L. P. Nielsen, J. Caffrey and P. B. Christensen. 1993. Denitrification measurements in aquatic sediments: A comparison of three methods. Biogeochemistry **23**:147-167.

Seitzinger, S. P., S. W. Nixon and M. E. Q. Pilson. 1984. Denitrification and nitrous oxide production in a coastal marine sediment. Limnology and Oceanography **29**:73-83.

Sharma, P., L. R. Gardner, W. S. Moore and M. S. Bollinger. 1987. Sedimentation and bioturbation in a salt marsh as revealed by ^{210}Pb, ^{137}Cs and ^{7}Be studies. Limnology and Oceanography **32**:313-326.

Simpson, R. L., R. E. Good, R. Walker and B. R. Frasco. 1983. The role of Delaware River freshwater tidal wetlands in the retention of nutrients and heavy metals. Journal of Environmental Quality **12**:41-48.

Sorensen, J. 1978. Capacity for denitrification and reduction of nitrate to ammonia in a coastal marine sediment. Applied and Environmental Microbiology **35**:301-305.

Staver, K. W. and R. B. Brinsfield. 1996. Seepage of groundwater nitrate from a riparian agroecosystem into the Wye River Estuary. Estuaries **19**:359-370.

Stevenson, J. C., D. R. Heinle, D. A. Flemer, R. J. Small, R. A. Rowland and J. F. Ustach. 1977. Nutrient exchanges between brackish water marshes and the estuary. Pages 219-240 *in* M. Wiley, editor. Estuarine Processes. Academic Press, New York, New York, USA.

Stevenson, J. C., L. G. Ward and M. S. Kearney. 1986. Vertical accretion in marshes with varying rates of sea level rise. Pages 241-259 *in* D. A. Wolfe, editor. Estuarine variability. Academic Press, New York, New York, USA.

Urban, N. R., S. J. Eisenreich and S. E. Bayley. 1988. The relative importance of denitrification and nitrate assimilation in midcontinental bogs. Limnology and Oceanography **33**:1611-1617.

Vitousek, P. M., J. D. Aber, R. W. Howarth, G. E. Likens, P. A. Matson, D. W. Schindler, W. H. Schlesinger and D. Tilman. 1997. Human alteration of the global nitrogen cycle: sources and consequences. Ecological Applications **7**:737-750.

Wang, X. and G. Benoit. 1992. Chronological variations in concentrations of heavy metals in sediments of the Tivoli South Bay: a study using ^{210}Pb dating methodology. Final report to the Tibor T. Polgar Fellowship Program, Hudson River Foundation, Inc., New York, New York, USA.

Woodwell, G. M., C. A. S. Hall, D. E. Whitney and R. A. Houghton. 1979. The Flax Pond ecosystem study: exchanges of inorganic nitrogen between an estuarine marsh and Long Island Sound. Ecology **60**:695-702.

MOLECULAR TOOLS FOR STUDYING BIOGEOCHEMICAL CYCLING IN SALT MARSHES

Molecular Studies of Bacteria

LEE KERKHOF
DAVID J. SCALA
Institute of Marine and Coastal Sciences
71 Dudley Road, Cook College,
Rutgers University
New Brunswick, NJ 08901-8521USA

Abstract

Understanding biogeochemical processes in salt marshes will help elucidate their role as essential habitats. Since microbial activity accounts for nearly all of the biogeochemical cycling that occurs in the marsh environment, monitoring bacteria and their activity is fundamental to assessing marshes as sites for biogeochemical change. In the past, this has been accomplished using approaches that estimated the average response of the entire population of micro-organisms. These studies have proven very useful for computing overall fluxes and secondary production. However, questions of diversity, population dynamics, microbial ecology, and the role of specific bacteria responsible for a biogeochemical transformation have been difficult to approach using the traditional, bulk rate techniques. The recent revolution in biochemical methods has allowed microbiologists to now identify specific groups of bacteria in a natural sample. This is done by targeting specific macromolecules in the bacterial cells such as fatty acids, proteins, and nucleic acids to characterize the various microbial members of or community independent of the other bacteria and eukaryotes present in the sample. Such studies have begun to provide information on the variety, distribution, and gene regulation of particular bacteria responsible for a given biogeochemical process. Although a comprehensive overview of molecular techniques will not be feasible in this chapter, we shall discuss some principles of applying biochemical analysis to complex microbial communities. It is hoped the data obtained from molecular studies in marsh habitats in the future will lead to a better understanding of the linkages between the structure and the function of the microbial communities that mediate biogeochemical cycling in the environment.

1. Introduction

Marshes are sites of intense microbial activity affecting decomposition and biogeochemical cycling (Carpenter 1983, Goldhaber 1974, Van Es 1982). For example, the importance of bacteria as consumers of organic carbon in marine food webs has long been recognized (Cho 1988, Fuhrman 1980, Pomeroy 1974). This understanding has been based on classical methodologies which measure a change (typically uptake or evolution of an isotopic label) and average that change over larger areas and the entire population of bacteria to estimate an overall biogeochemical rate. These traditional approaches have generally considered the bacterial community at the sample site as a "black box" and monitored the flow of material inside and out. However, the classic methods have been unable to assess the role of different species within the complex microbial assemblage, or determine the relative contribution of any particular microbial species to biogeochemical cycling under different conditions.

At present, there is a need to understand what is going on inside the "black box" since shifts within the bacterial community will impact biogeochemical cycling in the biosphere. Historically, major changes in the bacterial community have had significant impacts on conditions for life on Earth. For example, the modern atmosphere is a direct result of the evolution of oxygenic photosynthesis by early cyanobacteria billions of years ago (Atlas 1993) The dominance of N_2 in this oxic environment illustrates the importance of present-day populations of denitrifying bacteria. In order to understand and predict how various microbial populations will affect biogeochemical cycling in marshes, some means of addressing microbial population dynamics needs to be developed.

This chapter will focus on the development of tools and techniques to approach the issues of which bacteria are inside the "black box", which particular bacterial species is most numerous, who is changing and growing the fastest, and which traits (or genes) are being expressed in an environmental sample. Much of the research discussed here stems from work done in other parts of the marine environment and awaits application to salt marshes. This research approach is extremely important because, in regions undergoing active biogeochemical transformations (such as salt marshes), the most abundant and the fastest growing micro-organisms must be playing significant roles.

2. Definition of Molecular Methods

Bacteria generally lack defining morphological characteristics. Therefore, efforts within the last decade have focused on identifying specific molecules present within the bacteria to differentiate the various members of the microbial community. A recent example would include the direct analysis of the nucleic acids present in microbial biomass (for review, Head et al. 1998, Hugenholtz et al., 1998, Amann et al. 1995, Torsvik et al. 1996, Woese 1987). For our purposes, "molecular methods" will be defined as any procedure that tracks a specific cellular constituent that can differentiate the various players within the microbial population. These biomarkers can include lipids, proteins, and nucleic acids. (Although some components may be unique to various groups, biomarkers that cannot differentiate the various members of a group will not be considered here.) Each approach and methodology will be explored in this review chapter and particular emphasis will be placed on both the capability of distinguishing specific bacterial species and the ability to monitor the changes in abundance of these bacteria. Furthermore, a number of excellent reviews exist for many of the biogeochemical processes discussed in this chapter and these references are indicated in the text. The reader is encouraged to seek out these reviews as well.

Finally, the quantitative nature of any methods will be discussed since this information will be crucial for understanding whether the composition of the microbial community can affect biogeochemical rates. In this chapter we envision three possible microbial-based mechanisms which could lead to differences in biogeochemical transformation rates between areas or times in the marsh environment:

1) A change in the types (species) of bacteria catalyzing a biogeochemical change.

2) A change in the numbers of a few species of bacteria catalyzing a biogeochemical change.

3) A change in the velocity of the biochemical reactions associated with a biogeochemical change without an increase in species or numbers of the particular bacteria (null hypothesis).

At present, these hypotheses are untested. However, molecular methods will be able to provide a means to address specific microbial population biodiversity/dynamics. This information should lead to an understanding of the role of microbial-based mechanisms in affecting biogeochemical rates between regions of the marsh environment. The ability to monitor bacterial population changes may also provide an early warning system to monitor environmental degradation or restoration far sooner than larger organisms with longer doubling times. Clearly, understanding the fate of micro-organisms in a variety of ecosystems has many important implications in managing marshes and the global ecosystem.

2.1 LIPID-BASED ANALYSES

All living organisms possess cell membranes composed of phospholipids. The structure and abundance of these phospholipids has been used to glean information on the microbial population (for reviews, see White 1993, White 1994, White et al. 1996). The individual fatty acids can differ in length, the presence/absence and position of double bonds, cis/trans conformations, methyl groups, iso/anti-iso branching, or cyclopropyl moieties. Additional cellular lipids/biomarkers including steroids, sphingolipids, lipopolysaccharides, glycolipids, di/tri glycerides, and poly b hydroxyalkanoates (PHA) can also be used as signature biomarkers to identify the microbial communities. Profiles are generally examined using principal component analysis and the mole percent of signature phospholipids can quantitatively track groups of micro-organisms (White 1988).

A wide variety of habitats have been studied using phospholipid fatty acid (PLFA) analysis, including marine and freshwater sediments (Boon et al. 1996, Klieft et al. 1997, Rajendran et al. 1997, Rajendran et al. 1994), soils (Baath et al. 1998), bioreactors (Herrmann and Shann, 1997), and the deep subsurface (Ringelberg et al. 1997). It has been possible using these techniques to differentiate between microeukaryotes, gram negative bacteria, gram positive, sulfate reducers, and methane oxidizers (for reviews, White 1988, White et al. 1996) For example, Findlay (Findlay et al. 1990a,b) monitored the effects of predation and disturbance on microbial populations in coastal sediments. They found reductions in microbial biomass and enhancement of activity after physical or biotic translocation of sediments. Additionally, the methods could distinguish between the aerobic bacteria and the anaerobic bacterial response to these disturbances. (Sundh et al. 1997) studied 2 boreal peatlands and found site and depth accounting for most of the variability seen in PLFA profiles rather than sampling time during the summer. (Rajendran et al. 1994) detected significant regional differences in microbial populations in Osaka Bay sediments believed to result from strong tidal fronts and pollutant loads.

Although a number of researchers have used the PLFA methods to profile various microbial populations, the levels of resolution cannot easily differentiate the various members of the major groups For example, a commonly used biomarker for sulfate reducing bacteria (SRBs) is 10me16:0 (by convention, indicating a 16 carbon, unsaturated fatty acid chain with a methyl side group on carbon -10). However, a comparison of fatty acid profiles and 16S rRNA sequence data indicates only 2 of 25 sulfate reducers contain this biomarker (Kohring et al. 1994). The remaining bacteria in the SRB group have a wide range of mole percent of other biomarkers. An exception to this lack of resolution for specific bacteria by PLFA analysis is the methanotrophic bacteria. Which fall into 2 clusters, (Type I and II) which can be differentiated by their unique mono-unsaturated 16 and 18 carbon fatty acids (Bowman 1991, Hanson and Hanson 1996).

Recently, an important development using PLFA analysis and isotopic labelling of substrates was described which provides a window on activity currently unavailable with other methods. (Roslev et al. 1998) were able to generate substrate specific radio-assays by following [^{14}-C] phenanthrene and [^{14}-C] methane incorporation into [^{14}-C] PLFA using soil, activated sludge, and a coastal marine sediment sample. A similar study using [^{14}-C] acetate and [^{14}-C] methane by Boschker et al. (1998) in aquatic sediments demonstrated the gram-positive micro-organism, *Desulfotomaculum acetoxidans,* incorporated more label rather than the gram-negative genus *Desulfobacter*. Likewise, Type I methanotrophs were found to incorporate more label than the Type II methanotrophs. These studies emphasize the complementary approach of using lipid-based approaches with nucleic acid techniques which can provide more information on the various member of the major groups (see below).

2.2 PROTEIN BASED ANALYSES

Proteins are another cellular macromolecule which can be used to identify specific bacteria in a complex environment. Each bacterial protein may have a unique sequence of amino acids or 3-dimensional structure that can be unambiguously recognized using antisera generated against the bacterial antigens. Furthermore, many universities and commercial companies are now equipped to provide antibodies for bacteria that have been obtained in culture. The immuno approach for marine samples is exemplified by Ward and Carlucci (1985), who developed immunofluorescent assays for nitrifying bacteria in seawater. Polyclonal antibodies were obtained which recognized ammonium oxidizing and nitrite oxidizing groups. Depth profiles of nitrifying bacteria have been generated for the Southern California Bight, the Washington coast, and the Peruvian upwelling system (Ward 1984, Ward et al. 1989, Ward and Zafiriou, 1988). Additionally, Ward and Cockcroft (1993) used a species specific immunofluorescence assay to detect the presence and abundance of *Pseudomonas stutzeri*, the traditional laboratory marine denitrifier, in the water column of Monterey Bay and in microbial mats from Tomales Bay. This *P. stutzeri* antisera was recently used to develop a species specific productivity assay combining radio labelled thymidine and leucine incorporation with cell sorting on magnetic beads (Bard and Ward 1997) It was found that *Pseudomonas stutzeri* represented <0.1% of the population yet accounted for 1-3% of the radiolabel assimilation. Another immuno approach for studying bacteria used antibodies to the nitrogenase protein (Currin et al. 1990). In this investigation, both single cells and heterocysts in chain forming bacteria which possessed nitrogenase could be identified. Paerl et al. (1989) used this technique on *Trichodesmium* colonies and demonstrated nearly all cells in a colony contained nitrogenase and no localization of nitrogenase within the cells.

An unfortunate drawback to the immuno approach is the necessity for the microorganism to be in culture. Furthermore, many antibody preparations experience problems in specificity with some sera detecting bacteria at the group level but not identifying bacteria at the species level and other sera detecting at the strain level but not identifying at the species level. Finally, the sensitivity of antibody based approaches is surpassed by nucleic acid based methods for detecting small numbers of target molecules.

2.3 NUCLEIC ACIDS BASED ANALYSIS

Bacterial evolution based on sequence variation of the 16S ribosomal RNA genes (Woese 1987) has profoundly influenced our understanding of bacterial diversity and microbial ecology in the natural environment (Head 1998, Torsvik 1996). It is now possible to study microbial assemblages by analyzing nucleic acids directly extracted from the naturally occurring bacteria, revealing a myriad of undescribed bacterial lineages (DeLong et al. 1993, Devereux and Mundfrom 1994, Fuhrman et al. 1993, Giovannoni et al. 1990, Schmidt et al. 1991, Torsvik et al. 1996). Furthermore, this 16S rRNA analysis indicates many bacteria in the environment have yet to be cultured (Suzuki et al. 1997).

In order to identify or characterise these uncultured bacteria both 16S rRNA based and functional gene approaches has been implemented. In some cases, the bacteria responsible for a particular biogeochemical transformation form a coherent phylogenetic cluster for example, sulfate reducers in the δ subdivision of the Proteobacteria (for review, see Devereux et al. 1996a) and small subunit ribosomal based probes can be used to follow the ecology of these micro-organisms. Unfortunately, not all important biogeochemical processes are mediated by bacteria which form distinct phylogenetic groups, i.e., the 16S rRNA phylogenies do not always provide information on metabolic capabilities of the micro-organisms. Nitrogen fixation (for review, see Zehr and Capone 1996) and denitrification (for review, see Zumft 1997) are cases in point with members across many eubacterial phyla.

3. Biogeochemical Pathways using Nucleic Acid Analysis

In the following sections, we shall describe the molecular tools being developed to monitor important biogeochemical processes occurring in the salt marsh. Both phylogenetic and functional tools will be addressed.

3.1 CARBON FIXATION

Carbon fixation (via RuBisCO, see below) is an ancient process which is shared by both eubacteria and eukaryotes. There is an amazing diversity of photosynthetic organisms which includes cyanobacteria, prochlorophytes, cryptophytes, diatoms and higher plants (Pichard et al. 1993) Due to this diversity of photosynthetic organisms, 16S rRNA-based approaches will only target specific primary producers such as cyanobacteria.

Phytoplankton in marine systems play a key role in the global carbon cycle through the fixation of carbon dioxide into organic carbon. The primary pathway for photosynthetic carbon fixation in the oceans is the Calvin cycle (Paul 1996). Ribulose-1,5-bisphosphate carboxylase/oxygenase (RuBisCO) catalyzes the first step of this cycle. There are two natural forms of RuBisCO: form I which is composed of equal amounts (usually 8) of large and small subunits and form II which is composed only of two large subunits (Pichard et al. 1997, Stein et al. 1990). The *rbcL* and *rbcS* genes encode the large and small subunits, respectively.

Stein et al. (1990) cloned the RuBisCO gene from a sulfur-oxidizing, chemoautotrophic endosymbiont of a deep-sea, hydrothermal vent gastropod. They designed PCR primers specific to *rbcL* and *rbcS* to amplify and sequence these genes from the endosymbiont. The amino acid sequence of *rbcL* revealed several regions of nearly total conservation, among the six sequences (from various organisms) examined. There was less correspondence found in the amino acid sequence of *rbcS*. Therefore, *rbcL* has subsequently been used in molecular examinations of carbon fixation.

(Paul et al. 1990) designed 2 sets of *rbcL*-specific PCR primers to detect *rbcL* sequences in dissolved and particulate DNA from aquatic samples. Several algal isolates were also examined with the *rbcL* primers. (Pichard and Paul 1991) used an *rbcL* probe to detect RuBisCO expression in *Synechococcus* cultures and natural phytoplankton assemblages. This was the first report of measuring gene expression in natural populations via mRNA isolation and probing. Additional studies of RuBisCO gene expression in natural phytoplankton communities using the *rbcL* marker have been conducted (Pichard et al. 1993, Xu and Tabita 1996, Pichard et al. 1997) investigated the diversity of *rbcL* in natural phytoplankton assemblages from the Gulf of Mexico. Paul (1996) presented an overview of the current status and future directions of carbon fixation research using *rbcL*.

3.2 METHANE OXIDATION

Methane oxidizing bacteria, or methanotrophs, are a unique group of bacteria which utilize methane as their sole carbon and energy source (Hanson and Hanson 1996). They belong to a larger physiological group of bacteria termed methylotrophs, which are aerobic bacteria that utilize one-carbon compounds more reduced than CO_2 as sources of carbon and energy (Brusseau et al. 1994).

Methanotrophs have been intensely studied because of their wide distribution in many natural habitats, including terrestrial, freshwater and marine environments, as well as for their important role in carbon cycling. Furthermore, methanotrophic bacteria are capable of degrading many environmental pollutants, including trichloroethylene and vinyl chloride (Brusseau et al. 1994). Finally, methanotrophs may be the largest biological sink for methane in aerobic soils (McDonald and Murrell 1997a). Methane is an important greenhouse gas, and its atmospheric concentration, until recently, has been increasing by about 1% per year (McDonald and Murrell 1997a). A generalized pathway for methane oxidation is:

$$CH_4 \;\; --> CH_3OH \;\; --> CH_2O \;\; --> \text{Cellular Material}$$

The first step in this pathway, the oxidation of methane to methanol is catalyzed by methane mono-oxygenase. There are 2 forms of this enzyme, a particulate, membrane-bound form (pMMO) and a soluble, cytoplasmic form (sMMO) (Hanson and Hanson 1996). The second step in this pathway, the oxidation of methanol to formaldehyde is catalyzed by methanol dehydrogenase (McDonald et al. 1995),

Methanotrophs are classified by their mode of formaldehyde assimilation. Type I methanotrophs incorporate formaldehyde into cellular material via the ribulose monophosphate (RuMP) pathway. Type II methanotrophs assimilate formaldehyde using a serine pathway (Hanson and Hanson 1996).

Both 16S rDNA and functional gene nucleic acid-based approaches have been used to study the microbial ecology of methanotrophic bacteria in natural environments. Brusseau et al. (1994) sequenced 15 small-subunit rRNA genes from methylotrophic bacteria. Five new deoxyoligonucleotide probes were designed, synthesized, and labelled. These probes were hybridized to RNA extracted from representative methylotrophs and to RNA purified from soils enriched for methanotrophs. The probes were able to differentiate between Type I and II methanotrophs, and could also distinguish between methylotrophs that utilize methane and those that cannot.

Other studies utilizing a 16S rDNA based approach to studying methanotrophs include Holmes et al. (1995b), who examined marine samples using group-specific probes (i.e., *Methylobacter*, *Methylococcus*, *Methylomonas*, and *Methylosinus*), and McDonald et al. (1996), who applied the same group-specific probes to study methanotrophs in blanket bog peat. Holmes et al. (1996) used 16S gene probes to detect and identify marine methanotrophs.

Researchers have targeted several functional genes specific to methanotrophs for molecular ecology studies. McDonald et al. (1995), designed PCR primers specific to several genes of the soluble methane monooxygenase (sMMO) gene cluster. These sMMO specific primers (for genes *mmoX*, *Y*, *Z*, *B*, and *C*) were based on alignment of the sMMO sequences of *Methylococcus capsulatus* and *Methylosinus trichosporium*. PCR primers specific to the mxaF gene (the gene that codes for the large subunit of methanol dehydrogenase) were designed based on the MDH sequence of *Paracoccus denitrificans*. These primers were successfully applied to environmental samples from various aquatic and terrestrial sites. All primers yielded the appropriately sized PCR product, however the *mmoZ* primers were found to amplify non-specifically.

McDonald and Murrell (1997a) applied the *mxaF* PCR primers to blanket bog peat sediments and found that the high level of conservation of the *mxaF* sequences would preclude use of this gene for any group- or genus-specific studies. However, these primers would be very efficient for detecting the presence of methylotrophs in environmental samples.

Miguez et al. (1997) report the development of PCR primers specific to the *mmoX* and *mmoY* genes of the sMMO gene cluster, based on the sequence of *Methylococcus capsulatus* (Bath). The *mmoY*1-*Y*2 primer set was found to be specific to *M. capsulatus* only. The *mmoX*1-*X*2 primer set did amplify the appropriately sized fragment from the environmental sites tested.

Components of pMMO have also been targeted as methanotroph- specific probes. pMMO is encoded by the *pmoA* and *pmoB* genes. PCR primers specific to *pmoA* have been developed (Holmes et al. 1996, McDonald and Murrell 1997b). McDonald and Murrell (1997b) used the *pmoA*189/*A*682 primer set to amplify a fragment of the *pmoA* gene from blanket peat bog samples and from enrichment cultures. Holmes et al. (1996) amplified *pmoA* fragments from enrichment cultures using the same primer set. In addition, they used a labelled fragment of the *pmoB* gene to probe enrichment cultures for the presence of methanotroph specific DNA.

3.3 METHANOGENESIS

Methanogens are a large and diverse group of bacteria that produce methane as a by-product of their metabolism. They are strict anaerobes and found entirely within the domain *Archaea* (Hales et al. 1996). A typical pathway for methanogenesis is provided below, however it should be noted that methanogens can also metabolize other simple compounds, such as formate or methanol.

$$CO_2 + H_2 \; --> CH_4$$

Given the recent rate of increase of atmospheric methane concentration by about 1% per year, methanogenesis is a process of considerable interest (McDonald and Murrell 1997a).

Methanogens possess several unique coenzymes necessary for methane synthesis. Coenzyme M is used in the final step of methanogenesis, where a methyl moiety attached to it is reduced to methane, in a reaction catalyzed by methyl coenzyme M reductase (MCR) (Hales et al. 1996). The α subunit of MCR is encoded by the *mcrA* gene.

Hales et al. (1996) identified methanogen-specific DNA in blanket bog peat samples using both 16S rRNA- and *mcrA*-specific PCR primers. Methanogen-specific DNA was typically only detected in the deepest (anaerobic) sections of the peat cores. The 16S rRNA analysis yielded 8 methanogen variants, while the *mcrA* analysis produced only 2 sequence variants. As the *mcr* gene database increases, the *mcrA* primers should become more useful for targeting methanogens in natural environments.

3.4 NITROGEN FIXATION

Nitrogen fixation is the reduction of dinitrogen gas to ammonia and is performed solely by prokaryotes under both aerobic and anaerobic conditions. The ability to fix nitrogen is found in a diverse group of micro-organisms, which includes representatives from the cyanobacteria, the α and γ subclasses of the Proteobacteria, the *Clostridia*, and the *Archaea* (Zehr et al. 1995). Nitrogen fixers are found as both free-livers and symbionts. Given the phylogenetic diversity of nitrogen fixing micro-organisms, molecular based studies of nitrogen fixation have relied upon the use of a functional gene.

A generalized pathway for nitrogen fixation is provided below:

$$N_2 \; --> NH_3 \; --> \text{Cellular Material}$$

The reduction of nitrogen gas to ammonia is catalyzed by the nitrogenase enzyme, which is composed of 2 multisubunit proteins. The MoFe subunit is encoded by *nifD* and *nifK* genes, while the Fe subunit is coded for by the *nifH* gene (Ben-Porath and Zehr 1994). The *nifH* gene encodes component II of nitrogenase, dinitrogenase reductase (Zehr et al. 1995). To date, *nifH* has been the only gene to be used in molecular studies.

Kirshstein et al. (1993, 1991) utilized PCR primers specific to *nifH* to detect the presence of nitrogen fixing organisms in seawater samples. Ben-Porath and Zehr (1994) used *nifH* specific PCR primers to characterize cyanobacterial *nifH* genes and found sufficient variation within the DNA sequence to distinguish broad taxonomic groups of nitrogen fixers. Zehr et al. (1995) studied the diversity of *nifH* genes in a marine cyanobacterial mat and found a diverse group of *nifH* sequences belonging largely to heterotrophic nitrogen fixing organisms. Wyman et al. (1996) used a *nifH* probe to measure the abundance of *nifH* mRNA in natural populations of a nonheterocystous marine cyanobacterium. Zehr and Capone (1996) provided a rationale for the use of molecular techniques to study nitrogen fixation. They also highlighted the strengths and weakness of the *nifH*-based approach, and described several case studies that demonstrate the contribution of this approach to our understanding of nitrogen fixation in marine environments.

3.5 NITRIFICATION

Nitrification is the successive oxidation of ammonium to nitrate and is carried out by a group of gram-negative, obligate aerobic chemolithotrophs (Voytek and Ward 1995) The generalized nitrification pathway is listed below:

$$NH_4^+ \, --> NO_2^- \, --> NO_3^-$$

Ammonium oxidizers perform the first step in the nitrification pathway, while the oxidation of nitrite to nitrate is carried out by the nitrite oxidizers. Ammonium oxidation and nitrite oxidation are carried out by distinct groups of bacteria (Ward 1992). The oxidation of reduced nitrogen compounds often leads to the loss of biologically available nitrogen from an ecosystem, either via gaseous intermediates or via coupled nitrification/denitrification (Ward 1996).

Voytek and Ward (1995) reported the development of PCR primers specific to the 16S rRNA genes of ammonium oxidizers belonging to the beta-subclass of the *Proteobacteria*. These primers were used to amplify nitrifier rDNA fragments from water samples taken at Lake Bonney, Antarctica and the Southern California Bight. Ward et al. (1997) developed 16S rRNA gene-specific PCR primers to detect specific subsets of the total ammonium-oxidizer population. These primers were applied to environmental sites, including 3 lakes in Germany. Ward (1996) reviewed the current status of nitrifier PCR primers being applied in aquatic environments, while Utaker and Nes (1998) have evaluated thirty published 16S rRNA gene PCR primers which target various groups of ammonium oxidizing bacteria.

Ammonia oxidizers cluster into closely interrelated phylogenetic groups, allowing physiological inferences to be made from ribosomal RNA based phylogenies (Ward 1996). However, a more powerful probe for nitrification would be based on a functional gene unique to the process. Researchers have targeted the ammonia monooxygenase (*amoA*) gene, which codes for the key enzyme in ammonia oxidation. *amoA* is found in all ammonium-oxidizing bacteria and the level of conservation of *amoA* gene sequences allows for fine-scale differentiation of ammonium-oxidizing bacteria, making it an ideal target for molecular detection (Rotthauwe et al. 1995).

Rotthauwe et al. (1997) designed PCR primers specific to the *amoA* gene, which were used to amplify *amoA* fragments from a variety of natural habitats. While the primers did detect *amoA* in these environments, it did not amplify the homologous region of particulate methane monooxygenase (pMMO, a key enzyme in methane oxidation), which shares a close evolutionary relationship with *amoA* (Holmes et al. 1995a). Sinigalliano et al. (1995) designed *amoA* specific PCR primers which they used to amplify *amoA* fragments from a seawater sample.

3.6 DENITRIFICATION

Denitrification is the reductive respiration of nitrate or nitrite to N2 or N2O, and is carried out by a diverse group of bacteria generally under anaerobic conditions. The denitrification pathways is:

$$NO_3^- \; -- > NO_2^- \; -- > NO \; -- > N_2O \; -- > N_2$$

Denitrifiers have been isolated from the *Proteobacteria* (α, β, γ), the *Cytophaga/ Flavobacter*, and the Gram positive bacterial phyla and the *Archaea* (Zumft et al. 1992). Many of the isolated denitrifiers are representatives of the *Pseudomonas*, *Alteromonas*, *Thiomicrospira*, or *Erythrobacteria* groups within the *Proteobacteria*. However, if most marine bacteria have not been cultured (Suzuki et al. 1997), then the model micro-organisms for studying denitrification (*Paracoccus denitrificans*,

Pseudomonas stutzeri Zobell, and *Alcaligenes eutrophus*) may not be the most active or abundant denitrifiers in the marsh environment.

Approaches using 16S rRNA genes will not identify many denitrifiers due to their phylogenetic diversity. However, 2 approaches have been implemented to monitor those bacteria capable of denitrification in marsh sediment samples by analyzing for the genes in the denitrification pathway. The 2 genes specific to denitrifying bacteria that have been studied are the nitrite reductase gene (*nir*) (Ward 1995) and the nitrous oxide reductase (*nosZ*) gene (Scala and Kerkhof 1998). Dissimilatory nitrite reduction is the first step in the denitrification pathway which is unique to denitrifiers. (Dissimilatory nitrate reduction can be accomplished by a variety of bacteria not capable of the remaining step in the denitrification pathway.) Nitrous oxide reduction is the final step in the denitrification pathway and represents loss of biologically available N from the ecosystem.

Ward et al. (1993) compared the ability of antibody and DNA probes to detect the presence of nitrite reductase (*nir*) in water samples collected off the California coast (in the Santa Barbara Channel). They concluded that the DNA probe had broader use as a functional probe in environmental samples, while the anitbody probe was strain- or species-specific. Ward (1995) compared the use of the *nir* probe versus *nir* PCR primers for detecting denitrifying isolates. Hybridization with the *nir* probe was generally more successful than amplification with the *nir* PCR primers. Finally, Ward (1996) reviews the current state of *nir* as a probe for denitrifiers in aquatic environments.

The *nosZ* gene has been used by Scala and Kerkhof (1998) to investigate diversity of denitrifiers in sediment communities from the Mid Atlantic Bight. Fig. 1 demonstrates the denitrifiers present in the environment are not the cultured micro-organisms on which much of our understanding is based.

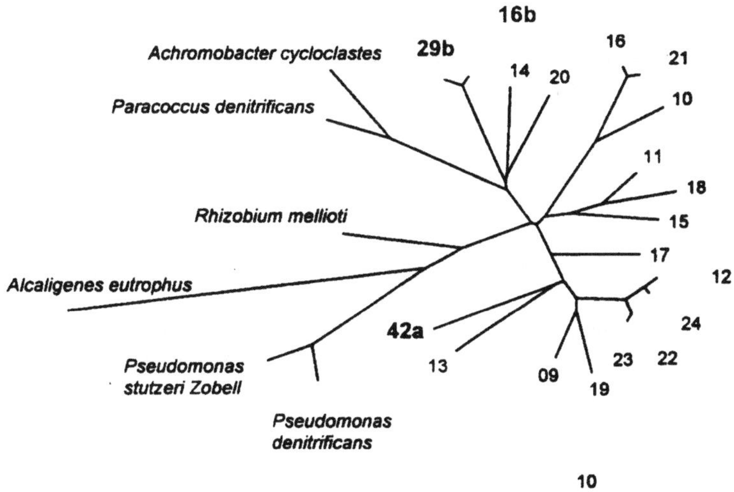

Figure 1. Phylogenetic tree reconstructed for denitrifying bacteria (*nosZ* genes) in coastal sediments off New Jersey. The numbers indicate the various *nosZ* genes present in the clonal library. The tree is based on alignment of ~ 500 bp of the *nosZ* gene.

3.7 SULFUR OXIDATION

Sulfur oxidation is the oxidation of reduced or partially reduced sulfur compounds to sulfuric acid. A composite pathway is presented below:

$$H_2S \; -- > S_0 \; -- > S_2O_3^{-2} \; -- > SO_3^{-2} \; -- > SO_4^{-2}$$

It is carried out by a physiologically and phylogenetically diverse group of bacteria, under aerobic or anaerobic conditions (Karl 1995) Sulfur oxidizers obtain their energy from the oxidation of these reduced sulfur compounds. Some can fix carbon dioxide via the Calvin cycle, i.e., they are chemolithoautotrophs (Nelson 1995) while others utilize organic substrates, i.e., they are chemolithoheterotrophs (Karl 1995). Sulfide oxidation can be carried out by a wide range of aerobic chemotrophs (e.g., *Thiobacillus*), as well as by anaerobic phototrophs (e.g., *Chromatium*) (Prescott 1993). Sulfur oxidizing bacteria are represented in the β, γ and ε divisions of the *Proteobacteria*, and in the *Archaea*.

Much research has been conducted on the sulfur bacteria inhabiting hydrothermal vent environments. At these systems, the predominant sulfur bacteria may be facultative chemoautotrophs (Jannasch and Mottl 1985). There are also many examples of symbioses between sulfur-oxidizing chemoautotrophs and marine invertebrates at hydrothermal vents (Cary et al. 1993b, Krueger et al. 1996).

The work of (Stein et al. 1990) has been summarized above (Section 3.1). These researchers cloned the RuBisCO gene from a sulfur-oxidizing, chemoautotrophic endosymbiont of a deep-sea, hydrothermal vent gastropod. Other investigations of sulfur-oxidizing endosymbionts, utilizing the RuBisCO gene as a symbiont-specific marker, include Krueger et al. (1996).

Cary et al. (1993a) used group-specific 16S rRNA probes to detect sulfur-oxidizing endosymbionts in 3 vent organisms. Cary and Giovannoni (1993) investigated inheritance of endosymbiotic bacteria in vent clams using a 16S rRNA probe. Gros et al. (1996) used endosymbiont-specific 16S rRNA PCR primers to investigate environmental transmission of the sulfur-oxidizing endosymbiont in a tropical marine bivalve. Muyzer et al. (1995) and Brinkhoff and Muyzer (1997) examined the phylogenetic relationships and diversity of sulfur-oxidizing bacteria (esp. *Thiomicrospira*) using denaturing gradient gel electrophoresis (DGGE) of 16S rDNA fragments.

3.8 SULFATE REDUCTION

Sulfate reduction is an anaerobic process in which the bacteria respire sulfate when most other terminal electron acceptors have been exhausted. The pathway is:

$$SO_4^{-2} \; -- > SO_3^{-2} \; -- > S_2O_3^{-2} \; -- > S_0 \; -- > H_2S$$

Gram positive, Gram negative, and *Archael* sulfate reducing bacteria (SRBs) have been isolated from the environment (for review, see Devereux et al. 1996a). Mainly, approaches using 16S rRNA have been implemented to monitor SRB bacteria since many of the bacteria form a coherent phylogenetic cluster in the δ subdivision of the *Proteobacteria* (Devereux et al. 1989). These probes have been used to monitor the population (Amann et al. 1992), activity (Poulsen et al. 1993), and isolation (Kane et al. 1993) of SRBs from anaerobic bioreactors. Abundance measurements of SRBs in depth profiles of coastal sediments from Florida have been done by hybridization of various 16S based probes to total RNA (Devereux et al. 1996b). Additionally, functional genes specific to SRBs bacteria have been cloned and sequenced. The dissimilatory sulfite reductase gene (*dsvA*) from *Desulfovibrio vulgaris* was aligned with the homologous gene from an Archaea, *Archaeoglobus fulgidus* (Karkhoffschweizer et al. 1995). Dissimilatory sulfite reduction is the second step in the sulfate reduction pathway and is unique to SRBs. PCR primers were designed using these 2 sequences to *dsvA* which would yield a 1400 bp fragment. These primers were capable of amplifying the *dsvA* gene in 12 other SRB's tested.

4. Quantitative Methods of Analysis for Nucleic Acids

4.1 TRADITIONAL CLONING AND SEQUENCING

Currently, the simplest way to identify bacteria is by PCR amplification of target genes and traditional cloning and sequencing. The method involves amplification, creation of a clonal library in *E. coli*, and screening of the clones to assess which particular target genes are present before sequence analysis. A schematic diagram illustrating the traditional method is shown in Fig. 2. The most common use of this technique involves analysis of small subunit (16S) ribosomal RNA genes. Although 16S rRNA gene characterization is now becoming routine in many labs, potential biases to the traditional clone and sequence approach exist. Primarily, DNA extraction procedures can miss entire groups that are difficult to lyse such as gram-positive organisms. Additionally, large amounts of template DNA, high cycle numbers in the PCR amplification, and large amounts of transforming DNA are used to maximize the amount of colonies obtained during cloning.

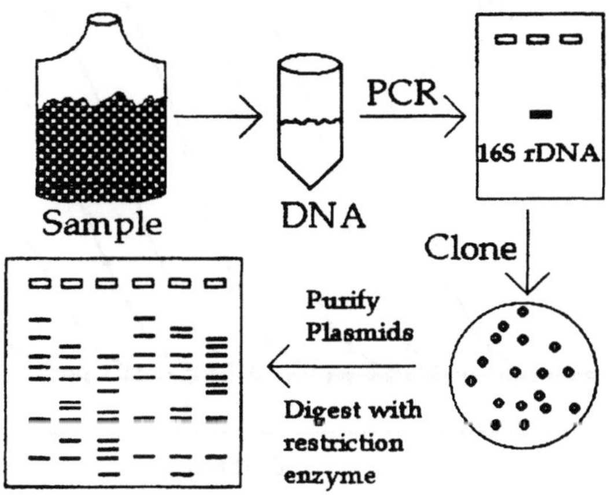

Figure 2. Schematic of the clone and sequence approach to analyzing target genes (16S rDNA in this case) from environmental samples

However, this strategy will confound any attempts at quantitative analysis due to PCR (Alard et al. 1993, Suzuki and Giovannoni 1996) or cloning biases resulting from asymptotic transformation of *E. coli* at DNA masses >10 ng for hexamine cobalt or rubidium chloride treatment or >300 ng for electroporation (see below; (Dower et al. 1988, Hanahan 1983). It is recommended that investigators routinely use minimal template concentration (<10 ng of genomic DNA), low cycle numbers (20 to 25), and low transforming DNA (<6 ng) to create clonal libraries.

4.2 QUANTITATIVE PCR

Numerous examples regarding quantitative PCR have been reported (Alard P et al. 1993, Becker-Andre and Hahlbrock 1989, Gilliland et al. 1990, Wang et al. 1989). Most techniques utilize low cycle numbers (<25) and there is a reasonable correspondence between the amount of a particular template added to a reaction and the amount of product obtained over a broad range of target molecule concentrations (10^1-10^8 copies, see Fig. 1). For *Escherichia coli*, a single copy gene would yield a target molecule concentration of 10^5 if 1 ng of genomic DNA were used for amplification. Studies on ribosomal RNA gene dosage in marine bacteria indicate 6 ± 2 operons per genome, only slightly less than *E. coli* (Kerkhof and Speck 1997). Therefore, genomic DNA extracted from environmental samples used in a quantitative PCR reaction should be at template concentrations approaching 1 ng and amplified using cycles numbers less than 25.

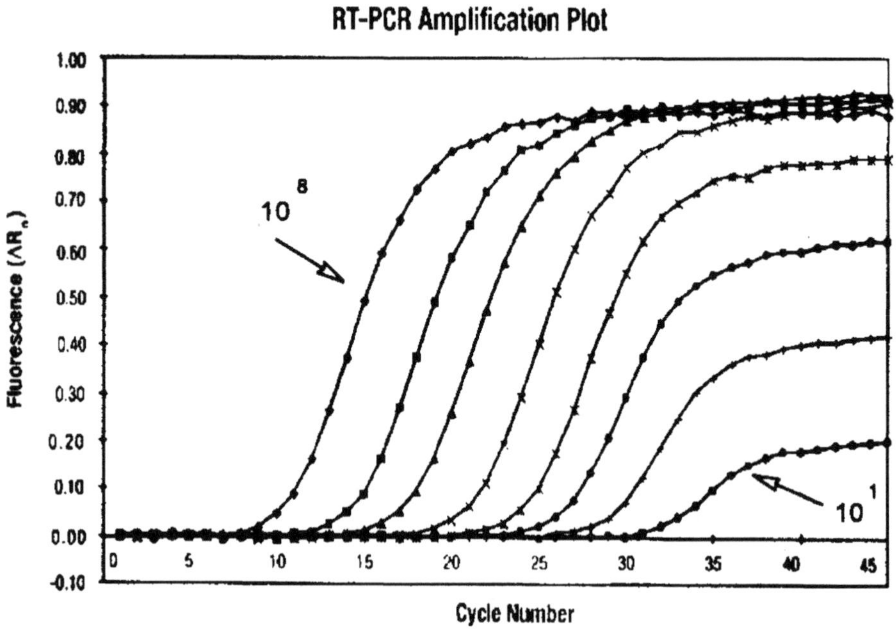

Figure 3. Quantitation of RT PCR product. Data from Perkin-Elmer (http://www2.perkin-elmer.com:80/ab/about/pcr/sds/wpfig2.html)

A consequence of quantitative PCR using the 16S rRNA primers is that the relative abundance of any particular species will be biased toward those bacteria possessing numerous 16S rRNA genes (Farrelly et al. 1995). Thus, it may not be possible using this methodology to distinguish between those bacteria with 10 copies of rRNA genes that have doubled from those bacteria with two copies of rRNA genes that have increased by a factor of 10.

4.3 QUANTITATIVE TRANSFORMATION

A second concern in creating 16S rRNA clonal libraries is that the cells used for transformations can become saturated. When DNA concentrations less than 1 ng are used to transform hexamine cobalt treated cells a linear response to DNA concentration is seen (Hanahan 1983)). Concentrations above 3.5 ng lead to saturation of this type of competent cell [calcium chloride treated cells saturate at lower DNA concentrations (<500 pg, data not shown)]. In contrast, a linear response to DNA concentration is seen with electro-competent cells transformed with up to 300 ng of DNA (Dower et al. 1988) By transforming with lower amounts of ligated plasmid (e.g., 1 ng), it should be possible to remain in the linear range of the curves shown below and obtain a sufficient number of transformants to select numerous recombinant clones at random.

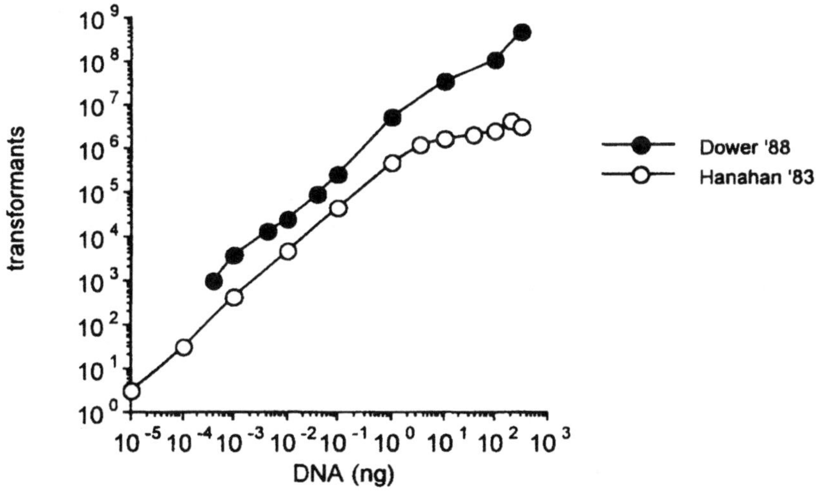

Figure 4. Graph showing number of transformant obtained vs. amount of DNA used in the transformation. The symbols refer to electro-competent cells (l) and hexamine cobalt treated cells (m)

5. Nucleic Acid Profiling Methods

Two recently developed methodologies that are replacing the traditional clone and sequence approach are described below.

5.1 DENATURING GRADIENT GEL ELECTROPHORESIS (DGGE)

Traditional gel electrophoresis techniques, such as agarose gel electrophoresis, separate DNA fragments based upon differences in size (i.e., molecular weight). Denaturing gradient gel electrophoresis (DGGE) is an electrophoresis technique in which DNA fragments of the same size, but differing sequence, can be physically separated (Myers et al. 1987). This technique relies upon the fact that changes in sequence between 2 DNA fragments affect the molecules' melting temperature, due to the sequence differences themselves as well as by stacking interactions (Myers 1988). As the DNA molecules progress through the gel, and into regions of an increasing denaturant, they will melt at their particular T_m and therefore cease to migrate into the gel (Fischer and Lerman 1983). The final result is that DNA molecules of the same size are now separated and can be individually visualized as bands within the gel. These bands may be excised and used for subsequent enzymatic analyses or DNA sequencing.

DGGE is useful in the analysis of genetic diversity in complex microbial populations from natural samples (Ferris et al. 1996, Muyzer et al. 1993, Wawer and Muyzer 1995) DGGE profiles generated from natural samples have been used to distinguish individual constituents, identify novel genotypes and to monitor changes in the composition of the community (Murray et al. 1996). Muyzer and Smalla (1998) review the current status of the application of DGGE to microbial ecological studies.

5.2 TERMINAL RESTRICTION LENGTH POLYMORPHISM ANALYSIS (TRFLP)

Although these other methods are becoming routine in many labs, they are very laborious. However, a rapid target gene fingerprinting method has been recently reported which utilizes fluorescent end labeling of target genes (PCR product) and screening by terminal restriction length polymorphism (TRFLP) (Avaniss-Aghajani et al. 1994, 1996, Clement et al. 1998, Liu et al. 1997, Phelps et al. 1998). A schematic diagram of the methodology is shown in Fig. 5. The principal involves sorting of target genes based on the position of a restriction enzyme site with respect to a terminal label. The current generation of automated, fluorescent sequencers (Perkin-Elmer/ABI) have the capability and software to detect these terminally labelled restriction fragments and present the data in a format akin to a chromatogram. Each peak in the fingerprint represents target gene(s) of different sizes after restriction enzyme digestion.

The TRFLP technology significantly decreases sample processing time. Yet, it must be kept in mind that a single TRFLP analysis will not be able to resolve all possible target genes in an amplification. No restriction enzyme exists which is capable of that degree of resolution. However, setting up digests requires a fraction of the time necessary for traditional cloning/sequencing or DGGE. The crux of the TRFLP technique is judicious selection of restriction enzymes.

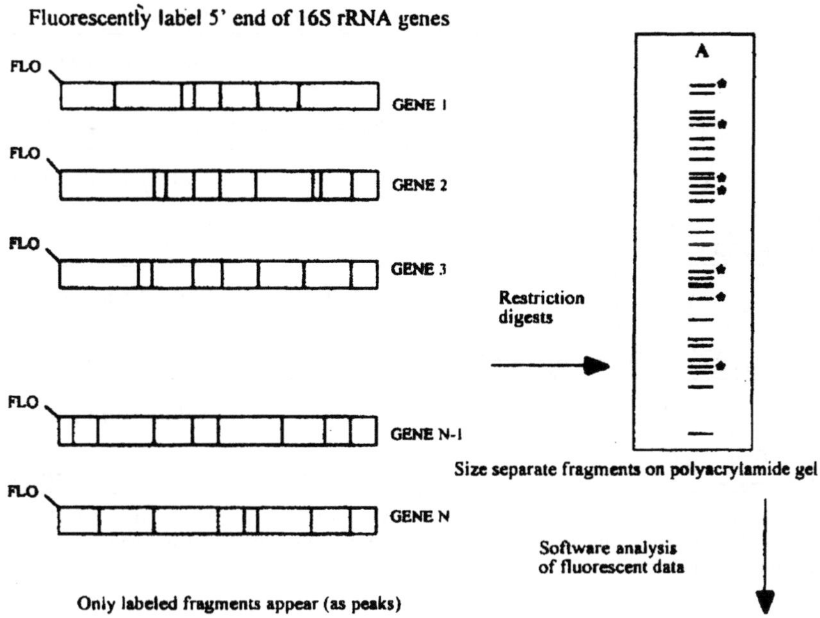

Figure 5. Schematic of terminal restriction length polymorphism (TRFLP) analysis

Furthermore, we are not bound to a single enzyme for the analysis. The odds that two bacteria will contain different target genes, yet contain two to three identical restriction sites with respect to a terminal label are very small. The simplest way to increase resolution is to perform more than one diagnostic digest and search for only those clones that contain the specific fragments one is looking for if one insists on species-specific identification at the onset. These additional digests will double the number of runs needed to fingerprint an environmental sample, yet it still represents over a 90% reduction in effort compared with the other approaches.

Finally, the greatest advantage of the TRFLP assay is to identify fingerprint bands of the target genes that are different between samples. Whether the band that appears in a sample of interest contains more than one gene is not nearly as important as the fact that a clear difference between bacterial populations can be demonstrated and the members ultimately identified. Given the high throughput capabilities of the current generation of automated sequencers, it may turn out to be more cost effective to simply identify all clones bearing a single diagnostic restriction site from a single enzyme fingerprint and sequence them all. An example of identification of unique target genes using salt marsh samples is presented in Fig. 6. Here we are characterizing denitrifier populations in various habitats in a New Jersey salt marsh using the *nosZ* gene (Scala unpubl. data). Unique *nosZ* genes can be seen for subtidal, tall-form *Spartina*, short-form *Spartina*, and *Salicornia* dominated parts of the marsh.

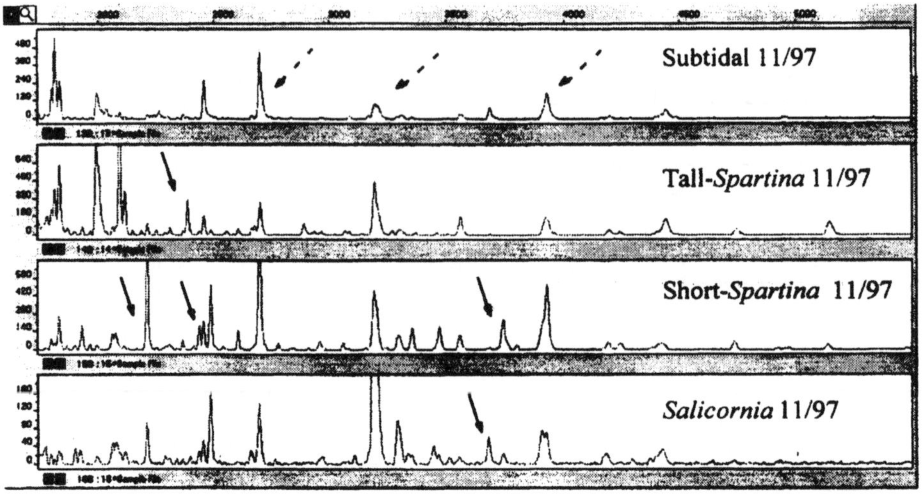

Figure 6. Comparison of denitrifying bacteria (*nosZ* genes) at different sites at Tuckerton, NJ [HinP I Digest]. Date and areas of the marsh sampled are indicated. The arrows indicate *nosZ* genes which are unique to the different habitats (Scala unpubl. data).

6. Activity measurements using molecular tools

6.1 CAN FUNCTIONAL GENE TRANSCRIPTION PREDICT ACTIVITY AND BIOGEOCHEMICAL RATES?

A simple way to ascertain whether a biogeochemical process is occurring involves determining if the functional genes (like *nosZ*) are being transcribed. Ultimately, we would like to know if a particular mRNA relates to the overall biogeochemical rate so we can map large geographic areas and determine spatial patterns in various processes. Unfortunately, it is possible the absolute levels of target mRNA may not correlate with a biogeochemical rate since transcript level is not always a good predictor of protein produced, protein level doesn't equate with activity, and activity is what determines the biogeochemical rate. However, it may be that in samples with a higher biogeochemical rate, more members of the microbial community responsible for the transformation will be transcribing the functional gene. Thus, both abundance and complexity of the target gene mRNA response may relate to the overall biogeochemical transformation rate.

Finally, if molecular techniques based on target gene expression are to be routinely used in natural samples, a clear understanding of the mRNA half life will be necessary for interpretation of field measurements. This information is crucial since an mRNA half-life of 5 min would not lead to a practical assay for environmental samples. For example, *nosZ* transcript analysis in *P. denitrificans* indicates a half-life of 3 h (Baumann et al. 1996) in the laboratory. But, we are unsure of the half-life behavior of *nosZ* mRNA from indigenous denitrifiers since they are not the model bacteria we have in culture (see Fig. 1).

6.2 ACTIVITY BY RIBOSOMAL RNA CONTENT

The relationship between ribosomal RNA content and growth rate has been well established for bacteria (Neidhardt and Magasanik 1960, Rosset et al. 1966, Schaechter et al. 1958). Later work extended the range of growth rates exhibiting a relationship between rRNA content and growth and presented evidence for a global relationship (at least among *Proteobacteria*) between RNA/DNA ratio and growth rate under steady-state conditions at doubling times from 0.3-60 hours [RNA:DNA = $1.65 + 6.01\ \lambda^{0.73}$, $r^2 = 0.85$] (Kemp 1995, Kerkhof and Ward 1993). This global data set comes RNA and growth rate measurements collected over 40 years using 6 different methodologies to analyze nucleic acid concentrations (from colorimetric methods to *in situ* hybridization). The authors of these studies (Kemp et al. 1993, Kerkhof and Ward 1993, Poulsen et al. 1993) found the predictability between RNA content and growth to be quite high even at low growth rates ($r^2 > 0.92$) and the global relationship should be applicable to most 16S rRNA genes currently identified with biogeochemical process in salt marshes since the micro-organisms are mostly *Proteobacteria*.

Therefore, if one can measure the amount of a specific 16S rRNA subunit and it's associated genomic DNA in marsh sample, it will be possible to estimate the growth-rate of that particular bacterium if one assumes:

1) a global relationship between rRNA content and growth-rate
2) the bacteria being monitored are in a steady-state of growth.

However, it may be that many bacteria in the marine environment are not in steady-state (Morita 1982) and other molecules such as pre-16S mRNA can predict growth rate (Cangelosi and Brabant 1997) on a species-specific basis in natural samples during non-steady state growth. Finally, the fact that some micro-organisms are growing may not necessarily indicate a biogeochemical transformation is occurring if the bacterial are capable of alternate metabolism. For example, many denitrifiers are facultative anaerobes and will happily respire oxygen if available. These bacteria could be actively growing in an environment where denitrification is not occurring.

6.3 QUANTITATIVE ANALYSIS OF SPECIFIC BACTERIA

Determining the number of a particular bacterium present within a sample when all you know is the 16S rDNA sequence from that micro-organism can be a daunting task. A number of methods are currently in use to monitor changes in abundance of specific micro-organisms present in natural samples including specific hybridization to rRNA or rDNA (DeLong 1992, Gordon and Giovannoni 1996) or *in situ* hybridization with fluorescently labelled oligo-nucleotide probes (for review, see Amann et al. 1995). An alternative method would involve the use of a modification of the most probable number (MPN) method to assess changes in abundance of specific bacteria in space and time. In classical MPN, serial dilutions of samples are grown on selective media to ascertain the numbers of a bacterium present in the original sample. Recently, replacement of the classical growth step with a PCR detection has been implemented (Degrange and Bardin 1995). This modification circumvents the need for a culturing step and has detection limits in the 10 to 100 cells range. The only requirement is a vigorous extraction method and species-specific primers for PCR amplification. The MPN/PCR method may be preferable for quantitating specific bacteria since the other methods either do not provide absolute numbers or have not been shown to reliably work in the marine environment on a group or species-specific level.

7. Conclusion

In this chapter, methods and approaches are presented which have been designed to address the response of a single species of bacterium in a natural sample independent of the other bacteria and eukaryotes present in that sample. This research is important since the mechanisms which are controlling microbial diversity and the population dynamics of bacteria in complex environments, such as salt marshes, remain largely unknown. In the future, these molecular methods will provide an avenue to address the fundamental questions in marine microbiology and shed light on the microbial processes occurring in salt marshes.

8. Literature Cited

Alard P, O. Lantz, M. Sebagh, C.F. Calvo, D. Weill, G. Chavanel, A. Senik and B. Charpentier. 1993. A versatile ELISA-PCR assay for mRNA quantitation from a few cells. Biotechniques **15**: 730-7.

Amann, R.I., W. Ludwig and K.H. Schleifer. 1995. Phylogenetic identification and *in situ* detection of individual microbial cells without cultivation. Microbiological Reviews **59**:143-169.

Amann, R.I., J. Stromley, R. Devereux, R. Key and D.A. Stahl. 1992. Molecular and microscopic identification of sulfate-reducing bacteria in multispecies biofilms. Applied & Environmental Microbiology **58**:614-623.

Atlas, R.M. and R. Bartha. 1993. Microbial ecology: fundamentals and applications. The Benjamin/ Cummings Publishing Company, Redwood City, California, USA.

Avaniss-Aghajani, E., K. Jones, D. Chapman and C. Brunk. 1994. A molecular technique for identification of bacteria using small subunit ribosomal RNA sequences. Biotechniques **17**:144-6.

Avaniss-Aghajani, E., K. Jones, A. Holtzman, T. Aronson, N. Glover, M. Boian, S. Froman and C.F. Brunk. 1996. Molecular technique for rapid identification of Mycobacteria. Journal of Clinical Microbiology **34**:98-102.

Baath, E., M. Diazravina, A. Frostegard and C.D. Campbell. 1998. Effect of metal-rich sludge amendments on the soil microbial community. Applied & Environmental Microbiology **64**: 238-245.

Bard, D.G. and B.B. Ward. 1997. A species-specific bacterial productivity method using immunomagnetic separation and radiotracer experiments. Journal of Microbiological Methods **28**: 207-219.

Baumann, B., M. Snozzi, A.J.B. Zehnder and J.R. Vandermeer. 1996. Dynamics of denitrification activity of *Paracoccus denitrificans* in continuous culture during aerobic-anaerobic changes. Journal of Bacteriology **178**: 4367-4374.

Becker-Andre, M. and K. Hahlbrock. 1989. Absolute mRNA quantification using the polymerase chain reaction (PCR): a novel approach by a PCR aided transcript titration assay (PATTY). Nucleic Acids Research **17**: 9437-9446.

Ben-Porath, J. and J.P. Zehr. 1994. Detection and characterization of Cyanobacterial *nifH* genes. Applied and Environmental Microbiology **60**: 880-887.

Boon, P.I., P. Virtue and P.D. Nichols. 1996. Microbial consortia in wetland sediments — a biomarker analysis of the effects of hydrological regime, vegetation and season on benthic microbes. Marine and Freshwater Research **47**: 27-41.

Boschker, H.T.S., S.C. Nold, P. Wellsbury, D. Bos, W. de Graaf, R. Pel, R.J. Parkes and T.E. Cappenberg. 1998. Direct linking of microbial populations to specific biogeochemical processes by ^{13}C-labelling of biomarkers, Nature **392**: 801-805.
Bowman J.P., P.E. Nichols, L.I. Sly. 1991. Phospholipid fatty acid and lipopolysaccharide fatty acid signature lipids in methane-utilizing bacteria. FEMS Microbiology and Ecology **85**: 15-22.
Brinkhoff, T. and G. Muyzer. 1997. Increased species diversity and extended habitat range of sulfur-oxidizing *Thiomicrospira spp.* Applied and Environmental Microbiology **63**: 3789-3796.
Brusseau, G.A., E.S. Bulygina and R.S. Hanson. 1994. Phylogenetic analysis and development of probes for differentiating methylotrophic bacteria. Applied and Environmental Microbiology **60**: 626-636.
Cangelosi, G.A. and W.H. Brabant. 1997. Depletion of pre-16s rRNA in starved *escherichia coli* cells. Journal of Bacteriology **179**:4457-4463.
Carpenter, E. 1983. Nitrogen in the marine environment. Academic Press, London, England.
Cary, S.C. and S.J. Giovannoni. 1993. Transovarial inheritance of endosymbiotic bacteria in clams inhabiting deep-sea hydrothermal vents and cold seeps. Proceedings of the National Academy of Sciences, USA **90**: 5695-5699.
Cary, S.C., W. Warren, E. Anderson and S.J.Giovannoni 1993a. Identification and localization of bacterial endosymbionts in hydrothermal vent taxa with symbiont-specific polymerase chain reaction amplification and *in situ* hybridization techniques. Molecular Marine Biology and Biotechnology **2**:51-62.
——. 1993b. Identification and localization of bacterial endosymbionts in hydrothermal vent taxa with symbiont-specific polymerase chain reaction amplification and *in situ* hybridization techniques, Molecular Marine Biology and Biotechnology **2**: 51-62.
Cho, B.C. and F.Azam. 1988. Major role of bacteria in biochemical fluxes in the ocean's interior. Nature **332**: 441-443.
Clement, B.G., L.E. Kehl, K.L. Debord and C.L. Kitts. 1998. Terminal restriction fragment patterns (TRFPS), a rapid, PCR-based method for the comparison of complex bacterial communities. Journal of Microbiological Methods **31**:135-142.
Currin, C.A., H.W. Paerl, G.K. Suba and R.S. Alberte. 1990. Immunofluorescence detection and characterization of N_2-fixing microorganisms from aquatic environments. Limnology and Oceanography **35**: 59-71.
Degrange, V. and R. Bardin. 1995. Detection and counting of *Nitrobacter* populations in soil by PCR. Applied and Environmental Microbiology **61**: 2093-2098.
DeLong, E.F. 1992. *Archaea* in coastal marine environments, Proceedings of the National Academy of Sciences, USA **89**: 5685-5689.
DeLong, E.F., R.B. Frankel and D.A. Bazylinski. 1993. Multiple evolutionary origins of magnetotaxis in bacteria. Science **259**: 803-806.
Devereux, R., M.E. Hines and D.A. Stahl 1996a. S cycling — characterization of natural communities of sulfate-reducing bacteria by 16s rRNA sequence comparisons. Microbial Ecology **32**: 283-292.
Devereux, R. and G.W. Mundfrom. 1994. A phylogenetic tree of 16s rRNA sequences from sulfate-reducing bacteria in a sandy marine sediment. Applied and Environmental Microbiology **60**: 3437-3439.
Devereux, R., M.R Winfrey, J. Winfrey and D.A. Stahl 1996b. Depth profile of sulfate-reducing bacterial ribosomal RNA and mercury methylation in an estuarine sediment. FEMS Microbiology and Ecology **20**: 23-31.
Dower, W., J. F. Miller and C. W. Ragsdale. 1988. High efficiency transformation of *E. coli* by high voltage electroporation. Nucleic Acids Research **16**: 6127-45.
Farrelly, V., F.A. Rainey and E. Stackebrandt. 1995. Effect of genome size and *rrn* gene copy number on PCR amplification of 16s rRNA genes from a mixture of bacterial species. Applied and Environmental Microbiology **61**: 2798-2801.
Ferris, M.J., G. Muyzer and D.M. Ward. 1996. Denaturing gradient gel electrophoresis profiles of 16s rRNA-defined populations inhabiting a hot spring microbial mat community. Applied and Environmental Microbiology **62**: 340-346.
Findlay, R.H., M.B. Trexler, J.B. Guckert and D.C. White 1990a. Laboratory study of disturbance in marine sediments: response of a microbial community. Marine Ecology Progress Series **62**:121-134.
Findlay, R.H., M.B. Trexler and D.C. White 1990b. Response of a benthic microbial community to biotic disturbance. Marine Ecology Progress Series **62**:135-148.

Fischer, S.G. and L.S. Lerman. 1983. DNA fragments differing by single base-pair substitutions are separated in denaturing gradient gels: correspondence with melting theory. Proceedings of the National Academy of Sciences, USA **80**:1579-1583.

Fuhrman, J.A., K. McCallum and A.A. Davis. 1993. Phylogenetic diversity of subsurface marine microbial communities from the Atlantic and Pacific oceans. Applied and Environmental Microbiology **59**: 1294-1302.

Fuhrman, J. and F. Azam. 1980. Bacterioplankton secondary production estimates for coastal waters of British Columbia, Antarctica and California. Applied Environmental Microbiology **39**: 1085-1091.

Gilliland, G., S. Perrin, K. Blanchard and H.F. Bunn. 1990. Analysis of cytokine mRNA and DNA: detection and quantitation by competitive polymerase chain reaction. Proceedings of the National Academy of Sciences, USA **87**: 2725-2729.

Giovannoni, S.J., T.B. Britschgi, C.L. Moyer and K.G. Field. 1990. Genetic diversity in Sargasso Sea bacterioplankton. Nature **345**: 60-63.

Goldhaber, M.a.K., I. 1974 The Sulfur cycle. Pages 569-655 *in* E. Goldberg, editor. The Sea. John Wiley and Sons, New York, New York, USA.

Gordon, D.A. and S. J. Giovannoni. 1996. Detection of stratified microbial populations related to *Chlorobium* and *Fibrobacter* species in the Atlantic and Pacific oceans. Applied and Environmental Microbiology **62**: 1171-1177.

Gros, O., A. Darrasse, P. Durand, L. Frenkiel and M. Moueza. 1996. Environmental transmission of a sulfur-oxidizing bacterial gill endosymbiont in the tropical *Lucinid* bivalve *Codakia orbicularis*. Applied and Environmental Microbiology **62**:2324-2330.

Hales, B.A., C. Edwards, D.A. Ritchie, G. Hall, R.W. Pickup and J.R. Saunders. 1996. Isolation and identification of methanogen-specific DNA from blanket bog peat by PCR amplification and sequence analysis. Applied and Environmental Microbiology **62**:668-675.

Hanahan, D. 1983. Studies on transformation of *Escherichia coli* with plasmids. Journal of Molecular Biology **166**: 557-80.

Hanson, R.S. and T.E. Hanson. 1996. Methanotrophic bacteria. Microbiological Reviews **60**: 439-471.

Head, I. M., J. R. Saunders and R. W. Pickup. 1998. Microbial evolution, diveristy and ecology — a decade of ribosomal RNA analysis of uncultivated microorganisms. Microbial Ecology **35**:1-21.

Herrmann, R.F. and J.F. Shann. 1997. Microbial community changes during the composting of municipal solid waste. Microbial Ecology **33**: 78-85.

Holmes, A.J., A. Costello, M.E Lidstrom and J.C. Murrell 1995a. Evidence that particulate methane monooxygenase and ammonia monooxygenase may be evolutionarily related. FEMS Microbiology Letters **132**: 203-208.

Holmes, A.J., N.J.P. Owens and J.C. Murrell 1995b. Detection of novel marine methanotrophs using phylogenetic and functional gene probes after methane enrichment. Microbiology **141**: 1947-1955.

—— 1996. Molecular analysis of enrichment cultures of marine methane oxidising bacteria. Journal of Experimental Marine Biology and Ecology **203**: 27-38.

Hugenholtz, P., B. M. Goebel and N. R. Pace. 1998. Impact of culture-independent studies on the emerging phylogenetic view of bacterial diversity. Journal of Bacteriology **180**:4765-4774.

Jannasch, H.W. and M.J. Mottl. 1985. Geomicrobiology of deep-sea hydrothermal vents. Science **229**: 717-725.

Kane, M.D., L.K. Poulsen and D.A. Stahl. 1993. Monitoring the enrichment and isolation of sulfate-reducing bacteria by using oligonucleotide hybridization probes designed from environmentally derived 16s rRNA sequences. Applied and Environmental Microbiology **59**: 682-686.

Karkhoffschweizer, R.R., D.P.W. Huber and G. Voordouw. 1995. Conservation of the genes for dissimilatory sulfite reductase from *Desulfovibrio vulgaris* and *Archaeoglobus fulgidus* allows their detection by PCR. Applied and Environmental Microbiology **61**: 290-296.

Karl, D.M. 1995. Ecology of free-living, hydrothermal vent microbial communities. Pages 35-124 *in* D. M. Karl, editor. The microbiology of deep-sea hydrothermal vents. CRC Press, Boca Raton, Florida, USA.

Kemp, P.F. 1995. Can we estimate bacterial growth rates from ribosomal RNA content? Pages 279-302 *in* I. Joint, editor. Molecular ecology of aquatic microbes. Springer-Verlag, Berlin, Germany.

Kemp, P.F., S. Lee and J. LaRoche. 1993. Estimating the growth rate of slowly growing marine bacteria from RNA content. Applied and Environmental Microbiology **59**: 2594-2601.

Kerkhof, L. and Speck, M. 1997. Ribosomal RNA gene dosage in marine bacteria, Molecular Marine Biology and Biotechnology **6**: 260-267.

Kerkhof, L. and B.B. Ward. 1993. Comparison of nucleic acid hybridization and fluorometry for measurement of the relationship between RNA/DNA ratio and growth rate in a marine bacterium. Applied and Environmental Microbiology **59**: 1303-1309.

Kirshtein, J.D., Paerl, H.W. and J. Zehr. 1991. Amplification, cloning and sequencing of a *nifH* segment from aquatic microorganisms and natural communities, Applied & Environmental Microbiology **57**: 2645-2650.

Kirshstein, J.D., J.P. Zehr and H.W. Paerl. 1993. Determination of N_2 fixation potential in the marine environment: application of the polymerase chain reaction. Marine Ecology Progress Series **95**: 305-309.

Kohring, L.L., D.B. Ringelberg, R. Devereux, D.A. Stahl, M.W. Mittelman and D.C. White. 1994. Comparison of phylogenetic relationships based on phospholipid fatty acid profiles and ribosomal RNA sequence similarities among dissimilatory sulfate-reducing bacteria. FEMS Microbiology Letters **119**: 303-308.

Krueger, D.M., R.G. Gustafson and C.M. Cavanaugh. 1996. Vertical transmission of chemoautotrophic symbionts in the bivalve *Solemya velum* (*Bivalvia, Protobranchia*). Biological Bulletin **190**:195-202.

Liu, W.T., T.L. Marsh, H. Cheng and L.J. Forney. 1997. Characterization of microbial diversity by determining terminal restriction fragment length polymorphisms of genes encoding 16s rRNA. Applied and Environmental Microbiology **63**: 4516-4522.

McDonald, I.R., G.H. Hall, R.W. Pickup and J.C. Murrell. 1996. Methane oxidation potential and preliminary analysis of methanotrophs in blanket bog peat using molecular ecology techniques. FEMS Microbiology and Ecology **21**: 197-211.

McDonald, I.R., E.M. Kenna and J.C. Murrell. 1995. Detection of methanotrophic bacteria in environmental samples with the PCR. Applied and Environmental Microbiology **61**: 116-121.

McDonald, I.R. and J.C. Murrell 1997a. The methanol dehydrogenase structural gene *mxaF* and its use as a functional gene probe for methanotrophs and methylotrophs. Applied and Environmental Microbiology **63**: 3218-3224.

——. 1997b. The particulate methane monooxygenase gene *pmoA* and its use as a functional gene probe for methanotrophs. FEMS Microbiology Letters **156**: 205-210.

Miguez, C.B., D. Bourque, J.A. Sealy, C.W. Greer and D. Groleau. 1997. Detection and isolation of methanotrophic bacteria possessing soluble methane monooxygenase (*smmo*) genes using the polymerase chain reaction (PCR). Microbial Ecology **33**: 21-31.

Morita, R.Y. 1982. Starvation-survival of heterotrophs in the marine environment. Pages 171-198 *in* K. Marshall, editor. Advances in Microbial Ecology. Plenum, New York, New York, USA.

Murray, A.E., J.T. Hollibaugh and C. Orrego. 1996. Phylogenetic compositions of bacterioplankton from two California estuaries compared by denaturing gradient gel electrophoresis of 16s rDNA fragments. Applied and Environmental Microbiology **62**: 2676-2680.

Muyzer, G., E.C. De Waal and A.G. Uitterlinden. 1993. Profiling of complex microbial populations by denaturing gradient gel electrophoresis analysis of polymerase chain reaction-amplified genes coding for 16S rRNA. Applied and Environmental Microbiology **59**: 695-700.

Muyzer, G., A. Teske, C.O. Wirsen and H.W. Jannasch. 1995. Phylogenetic relationships of *Thiomicrospira* species and their identification in deep-sea hydrothermal vent samples by denaturing gradient gel electrophoresis of 16S rDNA fragments. Archives of Microbiology **164**:165-172.

Myers, R.M., V. Sheffield and D.R. Cox. 1988. Detection of single base changes in DNA: ribonuclease cleavage and denaturing gradient gel electrophoresis. Pages 000-000 *in* K. Davies, editor. Genomic Analysis: A practical approach.

Myers, R.M., T. Maniatis and L.S. Lerman. 1987. Detection and localization of single base changes by denaturing gradient gel electrophoresis. Methods in Enzymology **155**: 501-27.

Neidhardt, F.C. and B. Magasanik. 1960. Studies on the role of ribonucleic acid in the growth of bacteria. Biochimica Biophysica Acta **42**: 99-116.

Nelson, D.C.and C.R. Fischer. 1995. Chemoautotrophic and methanotrophic endosymbiotic bacteria at deep-sea vents and seeps. Pages 125-000 *in* D. M. Karl, editor. The microbiology of deep-sea hydrothermal vents. CRC Press, Boca Raton, Florida, USA.

Paerl, H.W., J.C. Priscu and D.L. Brawner. 1989. Immunochemical localization of nitrogenase in marine Trichodesmium aggregates: relationship to N_2 fixation potential. Applied and Environmental Microbiology **55**: 2965-2975.

Paul, J.H. 1996. Carbon cycling — molecular regulation of photosynthetic carbon fixation. Microbial Ecology **32**: 231-245.

Paul, J.H., L. Cazares and J. Thurmond. 1990. Amplification of the *rbcL* gene from dissolved and particulate DNA from aquatic environments. Applied and Environmental Microbiology **56**: 1963-1966.

Phelps, C. D., L. J. Kerkhof and L. Y. Young. 1998. Molecular characterization of a sulfate-reducing consortium which mineralizes benzene. FEMS Microbiology and Ecology **27**:269-279.

Pichard, S.L., L. Campbell and J.H. Paul. 1997. Diversity of the ribulose bisphosphate carboxylase/oxygenase form I gene (*rbcl*) in natural phytoplankton communities. Applied and Environmental Microbiology **63**: 3600-3606.

Pichard, S.L., M.E. Frischer and J.H. Paul. 1993. Ribulaose bisphosphate carboxylase gene expression in subtropical marine phytoplankton populations. Marine Ecology Progress Series **101**: 55-65.

Pichard, S.L. and J.H. Paul. 1991. Detection of gene expression in genetically engineered microorganisms and natural phytoplankton populations in the marine environment by mRNA analysis. Applied and Environmental Microbiology **57**: 1721-1727.

Pomeroy, L.R. 1974. The ocean's food web, a changing paradigm. Bioscience **24**: 499.

Poulsen, L.K., G. Ballard and D.A. Stahl. 1993. Use of rRNA fluorescence *in situ* hybridization for measuring the activity of single cells in young and established biofilms. Applied and Environmental Microbiology **59**:1354-1360.

Prescott, L.M., J.P. Harley and D.A. Klein. 1993. Microbiology. Wm. C. Brown Publishers, Dubuque, Iowa, USA.

Rajendran, N., O. Matsuda, R. Rajendran and Y. Urushigawa. 1997. Comparative description of microbial community structure in surface sediments of eutrophic bays. Marine Pollution Bulletin **34**: 26-33.

Rajendran, N., O. Matsuda, Y. Urushigawa and U. Simidu. 1994. Characterization of microbial community structure in the surface sediment of Osaka Bay, Japan, by phospholipid fatty acid analysis. Applied and Environmental Microbiology **60**: 248-257.

Ringelberg, D.B., S. Sutton and D.C. White. 1997. Biomass, bioactivity and biodiversity—microbial ecology of the deep subsurface—analysis of ester-linked phospholipid fatty acids. FEMS Microbiology Reviews **20**: 371-377.

Roslev, P., N. Iversen and K. Henriksen. 1998. Direct fingerprinting of metabolically active bacteria in environmental samples by substrate specific radiolabelling and lipid analysis. Journal of Microbiological Methods **31**: 99-111.

Rosset, R., J. Julien and R. Monier. 1966. Ribonucleic acid composition of bacteria as a function of growth rate. Journal of Molecular Biology **18**: 308-320.

Rotthauwe, J.H., W. Deboer and W. Liesack. 1995. Comparative analysis of gene sequences encoding ammonia monooxygenase of *Nitrosospira* sp ahb-1 and *Nitrosolobus multiformis* c-71. FEMS Microbiology Letters **133**:131-135.

Rotthauwe, J.H., K.P. Witzel and W. Liesack. 1997. The ammonia monooxygenase structural gene *amoA* as a functional marker—molecular fine-scale analysis of natural ammonia-oxidizing populations. Applied and Environmental Microbiology **63**: 4704-4712.

Scala, D. and L. Kerkhof. 1998. Three nitrous oxide reductase (*nosz*) genes from marine sediments and development of PCR primers for identification of denitrifiers. FEMS Microbiology Letters **162**: 61-68.

Schaechter, M., O. Maalow and N.O. Kjelgaard. 1958. Depencency on medium and temperature of cell size and chemical composition during balanced growth of *Salmonella typhimurium.* Journal of General Microbiology **19**: 592-606.

Schmidt, T.M., D.F. DeLong and N.R. Pace. 1991. Analysis of a marine picoplankton community by 16S rRNA gene cloning and sequencing. Journal of Bacteriology **173**: 4371-4378.

Sinigalliano, C.D., D.N. Kuhn and R.D. Jones. 1995. Amplification of the *amoA* gene from diverse species of ammonium-oxidizing bacteria and from an indigenous bacterial population from seawater. Applied and Environmental Microbiology **61**: 2702-2706.

Stein, J.L., M. Haygood and H. Felbeck. 1990. Nucleotide sequence and expression of a deep-sea ribulose-1,5-bisphosphate carboxylase gene cloned from a chemoautotrophic bacterial endosymbiont. Proceedings of the National Academy of Sciences, USA **87**: 8850-8854.

Sundh, I., M. Nilsson and P. Borga. 1997. Variation in microbial community structure in two boreal peatlands as determined by analysis of phospholipid fatty acid profiles. Applied and Environmental Microbiology **63**: 1476-1482.

Suzuki, M.T. and S.J. Giovannoni. 1996. Bias caused by template annealing in the amplification of mixtures of 16S rRNA genes by PCR. Applied and Environmental Microbiology **62**: 625-630.

Suzuki, M.T., M.S. Rappe, Z.W. Haimberger, H. Winfield, N. Adair, J. Stroebel and S.J. Giovannoni. 1997. Bacterial diversity among small-subunit rRNA gene clones and cellular isolates from the same seawater sample. Applied and Environmental Microbiology **63**: 983-989.

Torsvik, V., R. Sorheim and J. Goksoyr. 1996. Total bacterial diversity in soil and sediment communities: A review. Journal of Industrial Microbiology **17**: 170-178.

Utaker, J.B. and I.F. Nes. 1998. A qualitative evaluation of the published oligonucleotides specific for the 16S rRNA gene sequences of the ammonia-oxidizing bacteria. Systematic and Applied Microbiology **21**: 72-88.

Van Es, F. and L. Meyer-Reil. 1982. Biomass and metabolic activity of heterotrophic marine bacteria. Pages 111-169 *in* K. Marshall, editor. Advances in Microbial Ecology. Plenum, New York, New York, USA.

Voytek, M.A. and B.B. Ward. 1995. Detection of ammonium-oxidizing bacteria of the beta-subclass of the class *Proteobacteria* in aquatic samples with the PCR. Applied and Environmental Microbiology **61**: 1444-1450.

Wang, A.M., M.V. Doyle and D.F. Mark. 1989. Quantitation of mRNA by the polymerase chain reaction, Proceedings of the National Academy of Sciences, USA **86**: 9717-9721.

Ward, B.B. 1984. Combined autoradiography and immunofluorescence for estimation of single cell activity by ammonium-oxidizing bacteria. Limnology and Oceanography **29**:402-410.

—— 1995. Diversity of culturable denitrifying bacteria. Limits of rDNA RFLP analysis and probes for the functional gene, nitrite reductase. Archives of Microbiology **163**: 167-175.

—— 1996. Nitrification and denitrification: probing the nitrogen cycle in aquatic environments. Microbial Ecology **32**: 247-261.

Ward, B.B. and A.F. Carlucci. 1985. Marine ammonia- and nitrite-oxidizing bacteria: serological diversity determined by immunofluorescence in culture and in the environment. Applied and Environmental Microbiology **50**: 194-201.

Ward, B.B. and A.R. Cockcroft. 1993. Immunofluorescence detection of the denitrifying strain *Pseudomonas stutzeri* (ATCC 14405) in seawater and intertidal sediment environments. Microbial Ecology **25**: 233-246.

Ward, B.B., A.R. Cockcroft and K.A. Kilpatrick. 1993. Antibody and DNA probes for detection of nitrite reductase in seawater. Journal of General Microbiology **139**: 2285-2293.

Ward, B.B., H.E. Glover and F. Lipschultz. 1989. Chemoautotrophic activity and nitrification in the oxygen minimum zone off Peru. Deep-sea Research **36**: 1031-1057.

Ward, B.B., M.A. Voytek and K. P. Witzel. 1997. Phylogenetic diversity of natural populations of ammonia oxidizers investigated by specific PCR amplification. Microbial Ecology **33**: 87-96.

Ward, B.B. and O.C. Zafiriou. 1988. Nitrification and nitric oxide in the oxygen minimum of the eastern tropical north Pacific. Deep-Sea Research **35**: 1127-1142.

Wawer, C. and G. Muyzer. 1995. Genetic diversity of *Desulfovibrio* spp in environmental samples analyzed by denaturing gradient gel electrophoresis of [*nifE*] hydrogenase gene fragments, Applied and Environmental Microbiology **61**: 2203-2210.

White, D.C. 1988. Validation of quantitative analysis for microbial biomass, community structure and metabolic activity. Advances in Limnology **31**: 1-18.

White, D.C. 1993. *In situ* measurement of microbial biomass, community structure and nutritional status. Transactions of the Royal Society of London **344**: 59-67.

White, D.C. 1994. Is there anything else you need to understand about the microbiota that cannot be derived from analysis of nucleic acids? Microbial Ecology **28**: 163-166.

White, D.C., J.O. Stair and D.B. Ringelberg. 1996. Quantitative comparisons of in situ microbial biodiversity by signature biomarker analysis. Journal of Industrial Microbiology and Biotechnology **17**: 3-4.

Woese, C.R. 1987. Bacterial evolution. Microbiological Review **51**: 221-271.

Wyman, M., J.P. Zehr and D.G. Capone. 1996. Temporal variability in nitrogenase gene expression in natural populations of the marine cyanobacterium *Trichodesmium thiebautii*. Applied and Environmental Microbiology **62**: 1073-1075.

Xu, H.H. and F.R. Tabita. 1996. Ribulose-1,5-bisphosphate carboxylase/oxygenase gene expression and diversity of Lake Erie planktonic microorganisms, Applied and Environmental Microbiology **62**:1913-1921.

Zehr, J.P. and D.G. Capone. 1996. Problems and promises of assaying the genetic potential for nitrogen fixation in the marine environment, Microbial Ecology **32**:263-281.

Zehr, J.P., M. Mellon, S. Braun, W. Litaker, T. Steppe and H.W. Paerl. 1995. Diversity of heterotrophic nitrogen fixation genes in a marine cyanobacterial mat. Applied and Environmental Microbiology **61**: 2527-2532.

Zumft, W.G. 1997. Cell biology and molecular basis of denitrification. Microbiology and Molecular Biology Reviews **61**: 533-616.

Zumft, W.G., A. Dreusch, S. Lochelt, H. Cuypers, B. Friedrich and B. Schneider. 1992. Derived amino acid sequences of the nosZ gene (respiratory N_2O reductase) from *Alcaligenes eutrophus*, *Pseudomonas aeruginosa* and *Pseudomonas stutzeri* reveal potential copper-binding residues. Implications for the CuA site of N_2O reductase and cytochrome-c oxidase. European Journal of Biochemistry **208**: 31-40.

NITROGEN AND VEGETATION DYNAMICS IN EUROPEAN SALT MARSHES

JELTE ROZEMA
PETER LEENDERTSE*
Department of Ecology and Ecotoxicology
Vrije Universiteit Amsterdam
The Netherlands
**(now at: Centre for Agriculture and Environment, Utrecht, The Netherlands)*

JAN BAKKER
HARM VAN WIJNEN
Dept. of Plant Ecology, University of Groningen, The Netherlands

. . . and having little or no wind, and a smooth sea, the sun shining upon it, the sight was, as I thought, the most delightful that ever I saw.
—Daniel Defoe, 1719, *Robinson Crusoe*, "I go to sea."

Abstract
In this paper we discuss the role of nitrogen in the vegation dynamics of European salt marshes. An overview is presented of various fluxes of nitrogen to and from salt marsh ecosystems. Our primary view is on European salt marshes. A nitrogen budget constructed for salt marsh ecosystems in the Wadden Sea area shows that N-influx rates via floodwater (50-200 kg N ha^{-1} y^{-1}) and via atmospheric deposition (30 kg N ha^{-1} y^{-1}), the rate of N assimilation (95-230 kg N ha^{-1} y^{-1}) and the rate of mineralization (28-225 kg N ha^{-1} y^{-1}) vary and fluctuate in space and time but are of the same order of magnitude. These N-influx and efflux rates may differ markedly between salt marshes and within salt marshes. Within salt marshes the in and effluxes vary along elevational gradients with varying flooding frequencies and sedimentation rates and along successional gradients. Salt marshes are productive and dynamic ecosystems with variable and extreme environmental conditions. Rather than acting as a source or sink for nitrogen the transformations of organic and inorganic nitrogen tend to characterize salt marsh ecosystems. The cycling of N through a sediment-plant-sea water-atmosphere salt marsh ecosystem is also discussed. In addition, the impact of coastal salt marshes in reducing levels of inorganic N from sea water is assessed. ^{15}N- stable isotope studies indicate that, despite the significant influxes of N via atmospheric deposition and seawater flooding to the salt marsh ecosystems, most inorganic N is taken up via the root system and not through foliar uptake. There is strong evidence that availability of nitrogen limits plant growth and vegetation succession in salt marshes. It is hypothesized that increased input of nitrogen via floodwaters (polluted rivers and estuaries) and atmospheric deposition (agriculture, industry, traffic) has affected the development of salt marsh vegetation. In particular the increasing dominance of the

grass *Elymus athericus* during the last decades is related to the increased supply of nutrients to salt marshes. The impact of eutrophication via seawater and atmospheric deposition is discussed in relation to changes in species composition of various vegetation zones and successional stages in grazed and ungrazed salt marshes. Analysing changes in species composition over periods of years and decades, longer term changes of other environmental factors such as sedimentation, flooding frequency and flooding duration and nitrogen content of the salt marsh soil are also surveyed. The increased occurrence of *Elymus athericus* in West European salt marshes may relate to eutrophication. Alternatively, the dominance of the tall and rapidly growing *Elymus athericus* is linked with the ''naturally'' enhanced N-soil content and increased rate of N-mineralization in late successional stages of salt marshes.

1. Nitrogen and Vegetation Dynamics in Salt Marshes: Introduction

The salt marsh environment is inherently dynamic. Both spatially and temporally, salt marsh vegetation zones reflect different frequencies of sea water flooding and flooding duration (Beeftink and Rozema 1988, Bertness 1991). Salt marsh species of different zones may differ both in salt and flooding tolerance (Rozema et al. 1985). In addition differences in competitive ability determine the position and range of species in vegetation zonation (Rozema et al. 1988, Scholten and Rozema 1990, Davy and Costa 1992). The canopy of the lower marsh is often rather open (with a leaf area index (LAI) <1), while upper marsh zones, may have a more closed canopy (with a LAI > 1). Competitive interference between *Spartina anglica* and *Puccinellia maritima*, both of the lower marsh, relates to different time of emergence of shoots in the spring and the acquisition of solar radiation (Scholten et al. 1987, Rozema et al. 1988).

Seawater flooding also leads to deposition of small silt (clay) and organic (dead plant and animal material) particles. Over longer periods, sedimentation may lead to a higher elevation (and reduced sea water flooding), increased soil organic matter and an altered nutrient status of the soil (Beeftink and Rozema 1988). It is possible that long term successional vegetation changes in salt marshes do not only relate to differential tolerance of salt marsh species to salinity and flooding, but also to various abilities to acquire and assimilate photosynthetically active radiation and nutrients. Of course, long term vegetation changes may also depend on the management and the occurrence of grazing by small herbivore such as geese, rodents or by larger herbivores such as deer or cattle (Bakker 1989, Olff 1992). Grazing by small herbivores may be rather natural on salt marshes, but grazing by larger herbivores (ruminants) only rarely. Grazing by cattle, sheep or horses on salt marshes often relates to farm practices or to nature management (Bakker 1993).

Although all of these biotic and biotic factors may govern vegetation production,the focus of this paper is on the role of nitrogen in vegetation changes in salt marshes. We observed that the tall salt marsh grass *Elymus athericus* (sea couch) of the upper marsh increased in many ungrazed areas of salt marshes of the Dutch Wadden Sea during the last decades (Leendertse et al. 1997a). This may be according to natural vegetation succession (Beeftink and Rozema 1988). The tall grass *E. athericus* occurs in the drift-line zone and is a strong competitor capable of outshading smaller plant species such as *Festuca rubra* (Beeftink 1977, Scholten et al. 1987, Adam 1990). The efficient uptake and assimilation of nitrogen by *Elymus athericus* may help to explain the dominance of this grass in upper salt marsh zones (Olff 1992). Alternatively, we discuss the hypothesis that the increasing dominance of *E. athericus* is the result of eutrophication of coastal seawater and increased levels of atmospheric nitrogen deposition (Rozema and Leendertse 1991).

These possibilities are: 1). correlative evidence from long-term salt marsh vegetation changes. 2). the use of possible differences in the natural abundance of the stable isotope ^{15}N in the ecosystem compartments for studies of N-fluxes through the salt marsh ecosystem. 3). nitrogen uptake experiments in the field and greenhouse. Uptake of ^{15}N depleted ammonium nitrate (0.01 % ^{15}N), by the roots and leaves of *Elymus athericus* (and *Spartina anglica*) from flooding seawater, nutrient solution and salt marsh sediment, both in greenhouse and field studies, are measured and evaluated.

1.1 POOLS AND FLUXES OF NITROGEN IN SALT MARSH ECOSYSTEMS

Every part of a salt marsh is under the influence of sea water. Flooding with sea water leads to an input of inorganic and organic substances into the marsh. The composition of sea water (Rozema et al. 1985) includes elements with a less variable concentration (Na^+, Cl) and elements with more variable concentrations (e.g., NO_3^-, NH_4^+, PO_{43-}, SO_4^{2-}). The sea water concentration of the latter ions may relate to various natural and anthropogenic sources such as the load supplied by (polluted) rivers and estuaries as well as from atmospheric deposition. The anthropogenic sources of nutrient elements in seawater and in atmospheric deposition are of industrial, agricultural or domestic origin. There has been a 5-10 times increase in the supply of phosphate to the Wadden Sea (van der Veer et al. 1989, Janssen 1993) and doubling of inorganic nitrogen concentrations of Wadden sea water over the period 1950-1990 (Janssen 1993) The increased phosphate content of the Wadden Sea is held responsible for increased phytoplankton production (de Jonge 1990). Reduced discharges during the last few years, into rivers and estuaries have led to reduced phosphate in the Wadden Sea, but no such decreases have been observed for nitrogen so far (Janssen 1993).

We discuss influxes and effluxes and transformations of nitrogen in salt marshes more quantitatively, in order to compare the input of nitrogen via polluted flood water and through atmospheric deposition (Fig. 1).

Relatively few data on the N-content of salt marsh compartments have been published. In addition, estimates on the size of the N-pool do not always indicate the availability of nitrogen to plants. Recently, data on the N-pool in salt marsh soil on the Wadden Island of Schiermonnikoog have become available (see Tables 1 and 6, van Wijnen et al. 1998a,b). The N-pool data in Table 1 indicate increasing soil nitrogen content with salt marsh age from 10 to 100 years. The increasing size of the N-pool in salt marsh soil with increasing age of the salt marsh indicates that the salt marsh acts as a sink for N.

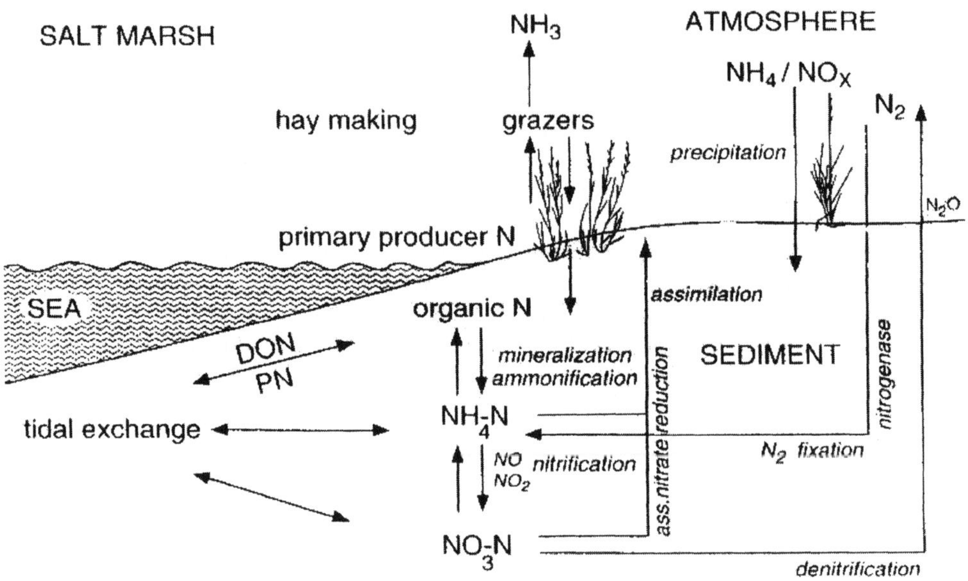

Figure 1. Nitrogen dynamics of a salt marsh ecosystem. DON = Dissolved Organic Nitrogen. PN = Particulate Nitrogen. Ass. nitrate reduction = Assimilatory nitrate reduction. The influxes and effluxes and conversions of nitrogen are further quantified in Tables 1-7.

2. A Nitrogen Budget of Salt Marshes of the Wadden Sea

The Dutch Wadden Sea is a shallow coastal sea between the mainland of Friesland an Groningen and a chain of barrier islands (Fig. 2). Supply of river water is via the IJsselmeer and via the Ems-Dollard estuary, which has led to an increase of P and N. Also nutrients enter the Wadden Sea with tidal exchange from the North Sea. The nutrient increase in the Wadden Sea is held responsible for increased phytoplankton and algal growth such as of *Ulva lactuca* between 1980 and 1990 (de Jonge 1990).

The data of N influx and N efflux rates summarized here (Table 1) are based on detailed literature studies and field and laboratory measurements by Leendertse et al. (1993a,b). For nitrate, salt marshes appear to act as a sink. In many studies, a net annual import of NO_3- into the salt marsh predominates (Table 2). This may be linked with denitrification (Table 1) and nitrate uptake (assimilation) by the salt marsh plants. An occassional net efflux of NO_3- from marshes may relate to local conditions such as ground-water flow from the upland. For ammonium, a net efflux from salt marshes is more common (Tables 2 and 3). This is attributed to a high rate of mineralization (ammonification) (Fig. 1).

TABLE 1. Nitrogen budget for salt marshes in the Dutch Wadden Sea area.
+ = import, - = export units in kg N $ha^{-1}y^{-1}$.

	Process/Pool kg N ha^{-1}	kg N $ha^{-1}y^{-1}$
Influx	Tidal exchange	+80/+470 [a]
	Atmospheric deposition	+30 [b]
	Nitrogen fixation	+ 6 [c]
Nitrogen pool	young salt marsh (10 yr)	600 [g, h]
	old salt marsh (100 yr)	2300/2900 [g, h]
Internal cycling	Assimilation	160 (95-230) [d]
	Mineralization	120 (30-225) [d]
	young salt marsh(10 yr)	69 [g, h]
	old salt marsh (100 yr)	127 [g, h]
Efflux	Denitrification	-8[e]
	Grazing	0 (-10/-70) [f]
	Hay making	0 (-15/-13) [f]

a. Table 3, including sedimentation. b. Bakker et al. (1993), Koerselman et al. (1992).
c. Table 5, Abd. Aziz and Nedwell (1986). d. Table 6. e. Abd. Aziz and Nedwell 1986b. f. Bakker et al. (1993) g. van Wijnen et al 1998a, h. van Wijnen et al. 1998b.

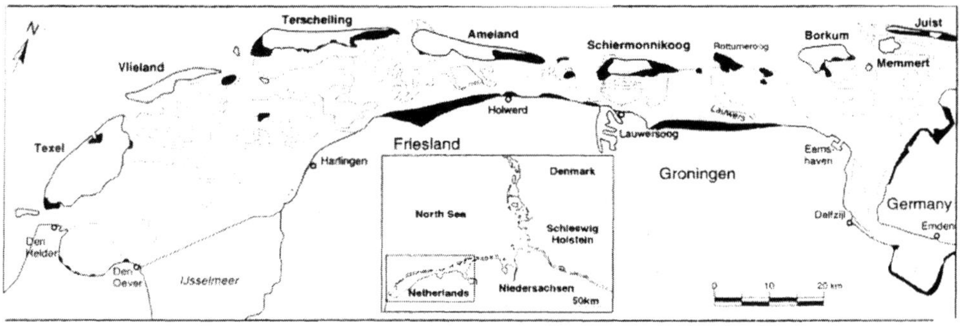

Figure 2. Salt marshes (marked black) at the barrier islands and at the mainland of the Dutch Waddensea area as considered in this paper. The longterm studies (1953-1990) of salt marsh vegetation zonation (Leendertse et al. 1997) were performed on the island of Terschelling.

TABLE 2. Net nitrogen exchange by tide in a number of salt marshes (kg N ha-1yr^{-1}). Values for amonium, nitrate, dissolved organic nitrogen (DON), particulate nitrogen (PN) and total nitrogen are given, + = import; - = export. (1) = Dankers *et al.* (1984). (2) = Asjes and Dankers (1993), (3) = Abd. Aziz and Nedwell (1986b), (4) = Woodwell *et al.* (1979), (5) = Valiela and Teal (1979), (6) = Daly and Mathieson (1981), (7) = Wolaver *et al.* 1983), (8) = Jordan *et al.* (1983), (9) = Heinle and Flemer (1976), (10) =Dame *et al.* (1986), (11) = Whiting *et al.* (1987), (12) = Whiting *et al.* (1989).

Study site	Dominant plant species	Nitrogen form NO_3^-	NH_4^+	DON	PN	Total
Ems-Dollard (NL)(1)	*Puccinellia maritima*	+40	-13			
Slufter (NL)(2)	*Limonium vulgare*	+14[a]	-3[a]			
Colne Point (UK)(3)	*Puccinellia maritima*	+27	-2	-43	+11	-7
Flax Pond (USA)(4)	*Spartina alterniflora*	+12	-21			
GSM* (USA)(5)	*Spartina alterniflora*	-17	-19	-44	-30	-111
Grommet Creek (USA)(6)	*Spartina alterniflora*	+3	+21			
Carter Creek (USA)(7)	*Spartina alterniflora*	+8	+19	+48	+241	+316
Rhode River (USA)(8)	*Typha angustifolia*	+1	+1	-49	+25	-4
	Spartina patens	+3	-14	-16	+35	+9
Gott's marsh (USA)(9)	*Spartina cynosuroides*	+10	-4	-21	-3	-37
North Inlet (USA)(10)	*Spartina alterniflora*	-9	-63			-427
North Inlet (USA)(11)	*Spartina alterniflora*	-6	-47			
North Inlet (USA)(12)	*Spartina alterniflora*	+61	+27	-87	+9	-45

a. Mean of four methods of calculation. * GSM = Great Sippewisset Marsh

TABLE 3. Net nitrogen exchange by tide in Wadden Sea salt marshes (kg N ha^{-1} yr^{-1}). Values for ammonium, nitrate, dissolved organic nitrogen (DON), particulate nitrogen (PN) and total nitrogen are given += import; - = export. The mean nitrogen exchange is also given. (1) = Dankers *et al.* (1984), (2) = Asjes and Dankers (1993), (3) = Abd. Aziz and Nedwell (1986b), (4) = Table 4.

Study site	Dominant plant species	Nitrogen form				
		NO_3^-	NH_4^+	DON	PN	Total
Ems-Dollard (NL)(1)	*Puccinellia maritima*	+40	-13			
Slufter (NL)(2)	*Limonium vulgare*	+14	-3			
Colne Point (UK)(3)	*Puccinellia maritima*	+27[a]	-2	-43	+11[a]	-7[a]
Island marshes ((NL)(4)					+112	
Mainland marshes ((NL)(4)					+500	
Wadden Sea marshes mean of (1) and (2)		+27	-8	-50	+112/ +500	+81/ +469

a. Authors give +31.0 (NO_3^-), +12.4 (PN) and -2.2 (total). Their calculations needed a slight modification.

TABLE 4. Estimations of import, export and net exchange of particulate nitrogen by tide in Wadden Sea salt marshes (kg N ha^{-1} yr^{-1}) + = import. Average values. Minimum and maximum values are given in brackets.

Salt marsh	Import PN (sedimentation)	Export PN (dead plant material)	Net exchange PN
Island	+122 (+43/+200) [a]	-10 (-5/-15) [b]	+112 (+28/+195)
Mainland	+525 (+300/+750) [c]	-10 (-5/-15) [b]	+515 (+285/+735)

a. Estimation by Olff (1992) and Leendertse et al. (1999a)
b. Estimation based on data from Wolff et al. (1979) and Dankers et al. (1984)
c. Estimation based on data from Dankers et al (1984) and De Glopper (1988)

TABLE 5. Mean nitrogen fixation in a number of salt marshes (kg N $ha^{-1}yr^{-1}$). (1) = Abd. Aziz and Nedwell (1986b), (2) = Jones (1974), (3) = Valiela and Teal (1979), (4) = Hanson (1977), (5) = Casselman *et al.* (1981), (6) = Delaune and Patrick (1990). The mean nitrogen fixation in *Puccinellia maritima* is calculated as representative for the Wadden Sea salt marshes.

Study site	Vegetation	Mean
Colne Point (UK)(1)	*Puccinellia maritima*	6[a]
Bank End (UK)(2)	*Spartina anglica*	171
Bank End (UK)(2)	*Puccinellia maritima*	51
Bank End (UK)(2)	*Salicornia dolichostachya*	28
GSM (USA)(3)	*Spartina alterniflora*	68
Sapelo Island (USA)(4)	*Spartina alterniflora*	150
Barataria Basin (USA)(5)	*Spartina alterniflora* (inland)	45
Barataria Basin (USA)(5)	*Spartina alterniflora* (streamside)	153
Louisiana (USA)(6)	*Spartina patens*	68
Wadden Sea marshes	mean of *Puccinellia zone* (Colne point)	6[b]

a. mean value for the whole salt marsh including creeks.
b. mean value for the salt marh zone with *Puccinellia.*

Input of nitrogen via tidal water of the Wadden Sea and by atmospheric deposition to the salt marshes often equals or exceeds the assessed rate of nitrogen mineralization (decomposition of organic matter) and the rate of nitrogen assimilation by salt marsh vegetation (Fig. 2, Tables 1 and 3). The influx of nitrogen into the marsh by nitrogen fixation is low relative to the above influx and internal cycling rates. Output of nitrogen by denitrification slightly exceeds the rate of nitrogen fixation. However, few published data on (relevant) denitrification measurements are available.

Results of nitrogen in- and efflux studies of Wadden Sea salt marshes (Tables 1, 2, 3, 4) suggest an import of N based on the import of NO_3^- and particulate nitrogen (Table 3).

The input and output of particulate nitrogen for Wadden Sea salt marshes has been calculated based on the sedimentation rate, nitrogen concentrations of sediment and the bulk density of the sediment (Table 4) (cf. Delaune et al. 1979) and the efflux rate of dead plant material (Leendertse et al. 1993). The calculations indicate a large input of particulate nitrogen, particularly for mainland salt marshes (Fig. 2) associated with high sedimentation rate (cf. Dijkema et al. 1990, Olff 1992, Bakker et al. 1997, Leendertse et al. 1997a).

TABLE 6. Nitrogen mineralization and assimilation (the total of NO_3^--N) and NH_4^+ -N) in salt marshes in the Wadden Sea, Ostsee and United Kingdom (kg N ha^{-1} yr^{-1}). Henriksen and Jensen (1979)(1), Zimmek (1975)(2), Scherfose (1987)(3), Abd. Aziz and Nedwell (1979)(4), Abd. Aziz and Nedwell (1986a)(5). Mean values of the studies are calculated as representative for the Wadden Sea salt marshes.

Area	Dominant plant species	Mineralization	Assimilation
Skallingen (DK)(1)	*Halimione portulacoides*	110	120
Spiekeroog (BRD)(3)[a]	*Juncus maritimus*	103	
	Artemisia maritima	142	
	Halimione portulacoides	118	
	Plantago maritima	105	
	Puccinellia maritima	78	
	Spartina anglica	44	
	Salicornia europaea	64	
Heilighafen (BRD)(2)[b]	*Festuca rubra*	77	
	Festuca rubra	178	
	Puccinellia maritima	28	
	Puccinellia maritima	41	95
Colne Point (UK)(4)	*Puccinellia maritima/Spartina anglica*	178	193
	Halimione portulacoides	116	
Colne Point (UK)(5)	*Puccinellia maritima/Spartina anglica*	225	
	Puccinellia/Halimione	185	229
Wadden Sea marshes The Netherlands	average values of above studies	120	160
Schiermonnikoog[c, d]	*Festuca rubra* 10 yr old salt marsh	25	
	Elymus athericus 100 yr old salt marsh	70/127	

a. The values of Scherfose (1987) represented nitrogen mineralization between June and October. The nitrogen mineralization in this period is estimated to be 65-80% of the annual mineralization. (Henriksen and Jensen 1979; Abd. Aziz and Nedwell, 1986a). Therefore the values of Scherfose were corrected by adding 27.5% for the non-measured period to obtain the annual nitrogen mineralization.
b. Average values of grazed and non-grazed salt marsh areas.
c., d. van Wijnen et al 1998a, and van Wijnen et al. 1998b.

This nitrogen budget of salt marsh ecosystems in the Dutch Wadden Sea makes it plausible that an enhanced influx of nitrogen by polluted tidal water and by atmospheric deposition as occurred from 1950-1990 may affect plant growth and salt-marsh vegetation development (Rozema and Leendertse 1991; Leendertse et al. 1993a).

Previous studies (e.g., Tyler 1967, Gallagher 1975, Haines 1979, Morris 1988, Kiehl et al. 1997) indicate that plant available inorganic nitrogen often limits growth of salt marsh plants. Experiments by Valiela (1984) showed that the biomass production of dominant salt marsh species increased with nitrogen fertilization, indicating that not salinity and flooding, but the availability of nitrogen, was limiting growth of *Spartina alterniflora.* Later, we discuss the role of nitrogen as a factor limiting plant growth in salt marshes in some more detail.

The nitrogen presented here (Table 1) applies to the 8500 ha of salt marshes of the Dutch Wadden area (Fig. 2). The range in values of N- fluxes varies markedly both between salt marshes and within salt marshes. The values depend on many factors such as the geomorphology, tidal regime, velocity of tidal currents, slope of the marshes, the volume and rate of supply of river water (Whiting et al. 1987).

In addition to the above dynamics, the salt marsh environment varies markedly both in space along gradients and in time along successional lines (Ranwell 1972, Huiskes et al. 1987, Beeftink and Rozema 1988, Adam 1990, Bakker et al. 1997). This implies differences in flooding frequency and duration, sedimentation rate and rates of mineralization and denitrification for various vegetation zones. Often, methodology differs between studieswith the number of measurements within one year varying between 4 (Abd Aziz and Nedwell 1986b) and 52 (Woodwell et al. 1979). Reliable estimates of both spatial and seasonal variation require a high frequency of sampling.

The above sources of variation, help to explain the large range of assessed influxes and effluxes of N in the literature. Of course, this variability of nitrogen fluxes and pools will obstruct generalizations of the N-budget of salt marshes. Also, it will onstruct the detection of possible effects of (anthropogenic) nitrogen pollution on vegetation composition in salt marshes. Before further analysing the relationship between nitrogen and vegetation dynamics, we discuss some N influxes and effluxes of salt marshes in more detail.

2.1 NITROGEN FIXATION IN SALT MARSHES

To our knowledge, nitrogen fixation has not been measured in the Dutch Wadden Sea salt marshes. Data of field measurements in some UK and USA salt marshes reveal a range from 6-171 kg N $ha^{-1}y^{-1}$ (Table 5). Also more generally, published field measurements of nitrogen fixation in salt marsh ecosystems are scarce. The value presented in Table 1 is from Abd Azizz and Nedwell (1986b) for the *Puccinellia* zone (lower marsh) at the Colne Point salt marsh on the East coast of the United Kingdom, indicated as (1) in Table 5). These salt marshes are comparable to the Dutch Wadden Sea salt marshes. Nitrogen fixation in salt marshes result from activity of nitrogenase in N_2 fixing bacteria (Casselman et al. 1981) or in cyanobacteria (Jones 1974). Nitrogen fixation rates in USA salt marshes and those of Jones (1974) in the UK have been assessed in the range 45-153 kg N $ha^{-1}y^{-1}$ (1974). These values exceed the N_2 fixation rate of 6 kg N ha^{-1} y^{-1} for Colne Point, but have been criticized. It can not be ruled out that in the field studies of Jones (1974) disturbance of salt marsh soil by sampling has led to overestimates.

Soil concentrations of ammonium exceeding 20 mg kg^{-1} dry soil will inhibit nitrogenase. Ammonium concentrations in Wadden Sea salt marsh soil are in the range of 20-100 mg N kg^{-1} dry soil. Therefore, the rate of nitrogen fixation in Wadden Sea salt marshes is estimated to be low: 6 kg N ha^{-1} y^{-1} (Abd Aziz and Nedwell 1986b). This value is close to the measurement of nitrogen fixation in a cyanobacterial mat in the Wadden Sea (Stal et al. 1984).

2.2 ATMOSPHERIC DEPOSITION OF NITROGEN

In addition to the import of nitrogen by flood water, the N influx by atmospheric deposition is significant (30 kg N ha^{-1} y^{-1} reported for coastal areas by Koerselman 1992) and this influx is of the same order of magnitude as the import of NO_3^- by tidal exchange (Tables 1, 2, 3). Atmospheric deposition of nitrogen in coastal foredunes of the Netherlands (0.5 km from the North Sea) was estimated at 28.8 kg N ha^{-1} y^{-1} and 27.4 kg N ha^{-1} y^{-1} for inner dunes, 2.0 km from the sea (ten Harkel 1998). In Europe, a major fraction (80%) of the NH_3 emissions is from livestock waste (Morris 1991). For a North American salt marsh, bulk (atmospheric) deposition of nitrogen is estimated at 7.6 kg N ha^{-1} y^{-1} (Valiela and Teal 1979). More generally, nitrogen deposition and NH_4^+ deposition in particular in areas of (Western) Europe appear to exceed that in North America.

2.3 NITROGEN MINERALIZATION AND ASSIMILATION

N-mineralization represents the biological breakdown of organic matter to NH_4^+ (ammonification) and to NO_3^- (nitrification) (Fig. 1). In practice, the rate N-mineralization is often assessed in salt marsh sites by incubating tubes or bottles in the salt marsh soil after removal or cutting of roots and aerial plant parts (to prevent uptake of NO_3^- and NH_4^+). At the end of an incubation period the soil content of $NH_4^+$and NO_3^- is then measured (e.g., Rozema et al. 1977, van Wijnen et al. 1998b)

Rates of mineralization and assimilation for USA salt marshes are in the range 100-400 kg N ha^{-1} y^{-1} (Teal 1986) and exceed values published for European salt marshes (28-225 kg N ha^{-1} y^{-1}) measured in different vegetation zones (Table 6). In addition, there is significant variation in the mineralization rate between the various vegetation zones. This may relate to differences in flooding frequency, sedimentation rate, sediment characteristics and chemical composition of the salt marsh plant litter (Hemminga and Buth 1991) and climatological conditions. Apparently, the rate of litter decomposition in the drift line in the upper marsh zone, where large amounts of dead plant and animal material accumulate, may differ from mineralization at the middle and lower marsh. For *Halimione portulacoides*, the values measured on three European salt marshes are similar (94, 110 and 116 kg N ha^{-1} y^{-1}). For mineralization in the lower marsh (*Puccinellia maritima*) large differences have been reported (28-225 kg N ha^{-1} y^{-1}). For the overall nitrogen budget (Table 1), average values of the mineralization and assimilation rates reported in Table 6 were taken. It appears that mineralization and assimilation rates in salt marshes are similar (Henriksen and Jensen 1979, Abd Aziz and Nedwell 1986a).

In recent papers van Wijnen et al. (1998a,b) assessed net nitrogen mineralization rates along a successional gradient of an island salt marsh in the Netherlands. The mineralization rates estimated by van Wijnen et al. (1998a,b) are in the range of values indicated in Table 1. Interestingly, it was found that nitrogen mineralization was higher in ungrazed salt marshes, areas compared with grazed sites. In ungrazed salt marsh areas, nitrogen mineralization increased with increasing age of the salt marsh (10-100 years).

Enhanced nitrogen mineralization in late succesional salt marsh stages was associated with an increased nitrogen content of the salt marsh soil (Table 1). The dominance of the tall growing salt marsh grass *Elymus athericus* in late succesional stages is related to the (naturally) high rate of mineralization and nitrification in the salt marsh soil (van Wijnen et al 1998a,b).

2.4 DENITRIFICATION IN SALT MARSHES

Published values for the efflux of N by denitrification in European salt marshes are scanty. The rate of denitrification measured at Colne point, 8 kg N ha^{-1} y^{-1} (Abd Aziz and Nedwell (1986b) is low compared to values of denitrification in salt marshes in the USA: 70 kg N ha^{-1} y^{-1} (Smith et al. 1983), 120 kg N ha^{-1} y^{-1} (Haines et al. 1977) and 140 kg N ha^{-1} y^{-1} (Kaplan et al. 1979). For inland fresh water wetland ecosystems, estimates of the output of N via denitrification are higher than in salt marshes. Values of up to 3500 kg N ha^{-1} y^{-1} have been published for fresh water marshes (Table 1) (see Rozema and Verhoef 1996, Vymazal 1998). There are no published data of *in situ* denitrification for salt marshes of the Wadden Sea.

3. Nitrogen Removal from Flood Water in Salt Marsh Ecosystems

The ability of wetlands to remove inorganic nitrogen from the surface water relates to denitrification in the sediment. Freshwater wetlands are used to treat waste water Verhoeven 1990, Rozema and Verhoef 1996, Vymazal et al. 1998). Denitrification in marshes is based on microbial activity in the sediment converting nitrate (NO_3^-) to nitrous oxide (N_2O) and nitrogen gas (N_2)(Fig. 1). In the anaerobic marsh sediment inorganic nitrogen occurs as NH_4^+ and NO_3^-. The radial oxygen loss from roots of (salt) marsh plants will stimulate the conversion of NH_4^+ and NO_3^-. Diffusion of NO_3^- ions to anaerobic sites in the (salt) marsh sediment will promote denitrification to N_2O and N_2 (Reddy and Patrick 1984).

In principle, the microbial process of denitrication in salt is expected to be similar to that of fresh water wetlands (Valiela et al. 1973, Rozema and Verhoef 1996). A more recent survey is presented by Vymazal et al. 1998). We regard it unlikely that (varying) salinity will markedly affect the rate of denitrification in salt marshes.

3.1 NITROGEN REMOVAL EFFICIENCY

The nitrogen removal efficiency of a wetland can be expressed as the percentage reduction of the nitrogen concentration of the surface water during passage through the wetland (eco)system. Nitrogen removal via denitrification in wetlands depends on various environmental factors (Nichols 1983, Patrick 1990, 1998, Verhoeven 1990, Verhoeven and van Oorschot 1990, Vymazal):

1. Nitrogen load of the surface water
2. Retention time of flood water

3. Flooding frequency
4. Aeration of the (anaerobic) salt marsh sediment by oxygen loss from plant roots
5. Temperature

In man-made (fresh water) wetlands, planted for example with *Phragmites australis*, the above environmental factors may be manipulated such that the nitrogen removal efficiency is high. In lower parts of salt marshes the retention time of flood(sea)water is about 6 h. This is low compared with values of days and weeks in inland freshwater wetlands.

There is now broad experience with the use of freshwater wetlands for nutrient removal of surface water, particularly in remote areas where sewerage is absent (Vymazal et al. 1998). Based on the limited published data on denitrification for salt marsh wetlands it is difficult as yet to compare the capacity of nitrogen removal by salt marshes with freshwater wetlands.

In a research project for the Dutch Waterways authorities we have theoretically and experimentally explored the possibility of removing nutrients from flood water by salt marshes (Leendertse et al. 1993b). Experimental salt marsh systems were capable of removing inorganic nitrogen (and phosphorus) from overlying water. The removal efficiency of nitrogen (and phosphorus) improved from increasing retention time of the flood water (up to 30 days). The removal efficiency of nitrogen (and phosphorus) in the fall was lower than in the summer, probably due to lower temperature reducing microbial denitrification and reduced plant activity (senescence).

We conclude from the above that apart from salinity, no essential differences exist between salt water and fresh water wetlands as regards the process of denitrification. Therefore, we expect in principle no differences in the capacity to remove nutrients from floodwater. Lower values of denitrification in salt marshes may relate to a lower retention time of the flood water.

In addition to nutrient (nitrogen) removal from flood water by denitrification, nutrients may be removed from the surface water due to deposition of silt and organic particles, adsorption of NH_4^+ to organic matter and the cation exchange complex of the soil, and nutrient uptake by the marsh vegetation. We refer the reader to (Nichols 1983, Patrick 1990, Verhoeven and van Oorschot 1990, and Vymazal 1998) and will not further discuss these aspects here.

3.2 NITROGEN REMOVAL BY GRAZING AND HAY-MAKING

Removal of nitrogen by grazing by cattle or birds (geese) does not represent a high efflux of nitrogen rom the salt marsh. Since most of the nitrogen will return to the salt marsh soil, nitrogen is not removed from the salt marsh. Grazing may rather accelerate the internal cycle of ammonification (mineralization) and assimilation of nitrogen (cf. van de Koppel 1997, van Wijnen et al 1998a,b).

Haymaking may represent a significant efflux of N (16-131 kg N ha^{-1} y^{-1}) but is not a common management practice for salt marshes in the Wadden Sea area (Bakker et al. 1993). Volatilization of ammonia, which may accompany grazing activity of herbivores, is regarded only an insignificant efflux of N in salt marshes (Valiela and Teal 1979, De Laune et al. 1989).

3.3 SALT MARSHES ACT AS SINKS AND SOURCES AND AS TRANSFORMERS OF NITROGEN

The net exchange of NO_3^-, NH_4^+, dissolved organic nitrogen (DON) and particulate nitrogen (PN) for various European and USA salt marshes has been reviewed and compared in Tables 1 through 6. The net yearly import to the salt marsh of NO_3^-N ranges from 3 to 40 kg NO_3^-N ha^{-1} y^{-1} (Table 2). For ammonium, a net export rate on a yearly basis is from 2 to 61 kg NH_4^+-N ha^{-1} y^{-1} is common. (Table 3). This net NH_4^+ export rate is explained by mineralization (ammonification) of organic matter in the salt marsh soil. There is a net yearly export of dissolved organic matter (16-87 kg N ha^{-1} y^{-1}, and an import of particulate nitrogen (9-241 kg N ha^{-1} y^{-1}) (Table 3).

This demonstrates that salt marshes are not merely sinks *or* sources of nitrogen. Salt marshes may act as sinks *and* sources of nitrogen dependent on the time scale regarded and on what nitrogen ion is considered.

In a majority of studies (Table 2), there is a net export of (total) nitrogen from the salt marsh via tidal fluxes to the coastal water. This indicates that these salt marshes should be regarded sources of nitrogen to the coastal water. However, the net efflux of N is generally small, relative to the nitrogen mineralization assimilation rates (Table 6). Some authors therefore conclude that salt marshes act rather as transformers of nitrogen than as sinks or sources (Valiela and Teal 1979, Abd Azziz and Nedwell (1986). The increased N-content of salt marsh soil (related to a thickening clay layer) with increasing age of the salt marsh (van Wijnen et al 1998a,b) indicates that a (growing) salt marsh acts a a sink for N. Rather than emphasizing the source/sink *or* transformer function of marshes, marshes in general may act as both a source/sink *as well as* a transformer of elements.

4. Sedimentation, Nitrogen Dynamics and Succession of Salt Marsh Vegetation

Salt marshes are flooded with seawater and dependent on the frequency and duration of flooding and velocity of tidal currents, silt or larger particles (e.g., PN, POC, clays), as well as particulate organic matter, including particulate nitrogen (PN, Table 4) are deposited.

Sedimentation rates for the Wadden Sea island salt marshes range from 1-5 mm y^{-1} with an estimated import of 43-200 kg N ha^{-1} y^{-1}cf. Tables 4 and 7). Sedimentation at the mainland marshes is higher (10-20 mm y^{-1}) (Dijkema et al. 1990), which implies an import of 300-750 kg N ha^{-1} y^{-1} (cf. Table 4). In Table 4, the import of N by sedimentation has been compared with the export of N in dead plant material with tidal exchange. Based on *this* balance salt marshes in the Dutch Wadden Sea act as a sink for (particulate) nitrogen and sedimentation represents a major input of nitrogen into salt marshes. Also, when the data for net nitrogen exchange in Dutch Wadden Sea salt marshes for NO_3^-, NH_4^+, DON and PN are combined (Table 3), it is concluded that the relatively high sedimentation rates (6-20 mm) implies that these salt marshes act as a sink for nitrogen.

4.1 HAS EUTROPHICATED FLOOD WATER AND INCREASED ATMOSPHERIC DEPOSITION AFFECTED VEGETATION CHANGES IN SALT MARSHES?

The doubling of inorganic nitrogen in Wadden Sea water over the period 1950-1990 (van der Veer et al. 1989, Janssen 1993, Rozema and Verhoef 1996) and the increase of atmospheric N deposition (Rozema et al. 1983, Koerselman et al. 1992, Bakker et al. 1993, ten Harkel 1998) as a result of agricultural (livestock) and industrial activities may have affected plant growth and vegetation development in salt marshes. In marine aquatic ecosystems phosphorus is often limiting primary production and growth of aquatic plants. It has been demonstrated that nitrogen is limiting plant growth in ''terrestrial'' salt marsh vegetations (Teal 1986, Jefferies and Perkins 1977, Kiehl et al. 1997). Teal (1986) indicated that increased nitrogen was associated with enhanced biomass production, the rate of litter decomposition, the abundance of herbivores and detrivores and with a reduced number of plant species in the salt marsh vegetations. This may also imply that (some) salt marsh plants will respond to enhanced nitrogen supply and, if plant species differ in this response, the species composition of salt marshes may change under eutrophication. Kiehl et al. (1997) confirmed the inorganic nitrogen is limiting plant growth in temperate salt marshes. Plant responses to nitrogen addition were more pronounced in the lower marsh than in the upper marsh. These authors further indicated that the responsiveness of individual plant species to N fertilization may depend on the growth rate, plant growth form and plant morphology. We hypothesize that the increasing dominance of the salt marsh grass *Elymus athericus* in various long-term studies of vegetation succession (Bakker 1989, Leendertse et al. 1997a) may relate to the above indicated eutrophication of Wadden Sea water. The capacity of *E. athericus* to rapidly and efficiently assimilate inorganic nitrogen contributes to its strong competitive ability in of the upper marsh zone where it reduces or replaces *Festucua rubra* under conditions of nutrient enrichment (cf. Roozen and Westhoff 1985, Olff 1992, Lenssen et al. 1993, Olff et al. 1997).

In the following discussion, we consider the increasing dominance of *Elymus athericus* as occurring in long-term vegetation studies both at the island salt marshes and the mainland marshes (Bakker 1989, Leendertse et al. 1993).

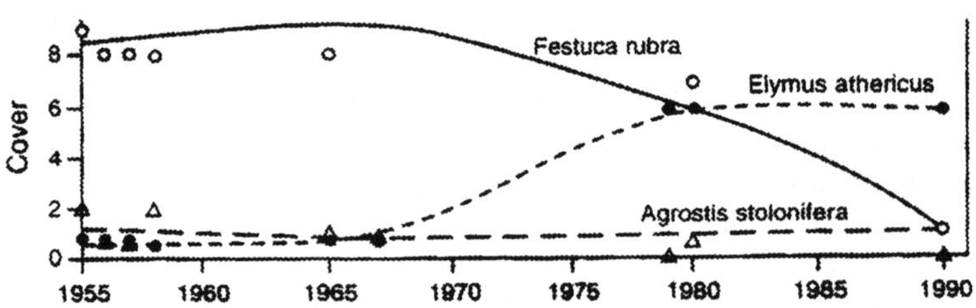

Figure 3. Successional change of cover of dominant salt marsh plant species of the uppermarsh in a permanent plot of transect 10 between 1955 and 1990. Note the increasing dominance of *Elymus athericus* and decrease of *Festuca rubra*. After Leendertse et al. (1997).

4.2 LONG-TERM (1953-1990) VEGETATION STUDIES AND INCREASING DOMINANCE OF *ELYMUS ATHERICUS*: NATURAL SUCCESSION OR RESPONSE TO EUTROPHICATION?

Between 1953 and 1990, permanent plots along transects from the lower to the upper marsh at the Boschplaat (Fig. 2) have been analysed and the total number of salt marsh species decreased between 1953 and 1990. There was a marked increase of the grass *Elymus athericus* in the upper marsh vegetation leading to dominance between 1980 and 1990 (Fig. 3). The increasing dominance of *E. athericus* may relate to a high nitrogen mineralization rate as a result of decomposition of dead organic material deposited in the *E. athericus* vegetation zone, the flood mark in the upper marsh (Ranwell 1972, Beeftink 1977, Rozema et al. 1985, Adam 1990). A large increase in occurrence and vegetation cover by *E. athericus* on a salt marsh at Schiermonnikoog during 1970-1990 may be related to an increase of the nitrogen pool in the salt marsh soil as a result of continuous sedimentation (Olff 1992, Olff et al. 1997). Alternatively, increased input during this period of both organic and inorganic nitrogen from flood water and atmospheric deposition (Rozema and Leendertse, Leendertse et al. 1997a) may have lead to the steady and gradual increase of *E. athericus* in permanent plots on Terschelling (Leendertse et al. 1997a). This is further supported by the observation that *E. athericus* increased markedly at the mainland salt marshes during about the same period (1975-1992) (Leendertse et al. 1993a). However, cattle grazing was also reduced during this period at the mainland marshes. Since it is known that cattle grazing reduces the abundance of *E. athericus* in salt marshes, the increase of *E. athericus* at the mainland marshes may be attributed to both eutrophication and reduced cattle grazing. For the plots with increasing dominance of *E. athericus* there was a small increase in elevation as a result of silt accretion (81 cm silt layer) with an accretion rate of about 5 mm y^{-1}. An expected reduction in flooding frequency with increased silt accretion has not occurred. The number of sea water floodings was 41 per year in 1990 compared with 21 per year for 1953 (Table 7). This is, in part, attributed to simultaneous (relative) sea level rise and is discussed in more detail by Dijkema et al. (1990) and Bakker et al. (1997). This relative sea level rise relates to the local exploitation of earth gas in the Wadden Sea.

The above long-term vegetation changes and increasing dominance of *E. athericus* may be the result of natural successional changes relating to silt accretion and an increase of the N pool in the salt marsh soil. It can not be ruled out however that the increase of *E. athericus*, at least in part, relates to eutrophication of flood water and the atmosphere. From field studies with various abiotic and biotic factors changing simultaneously during longer periods, it will be difficult to demonstrate causal relationships.

TABLE 7. Dominant salt marsh plant species, elevation, thickness of the silt (clay) layer and number of seawater floodings in 1953 and 1990 in plots of transect 10 and 18 (see Roozen and Westhoff 1985), where the salt marsh grass *Elymus athericus* became dominant. Average values and standard deviation of three replications.

Parameter	1953	1990
Transect 10		
Dominant species	*Festuca rubra*	*Elymus athericus*
Elevation (cm + NAP)	151±10	156±11
Silt layer (cm)	0±0	8±1
Number of floodings	21±9	41±9
Transect 18		
Dominant plant species	*Festuca rubra*	*Elymus athericus*
Elevation (cm + NAP)	137±5	145±7
Silt layer (cm)	0±0	4±1
Number of floodings	34±7	53±10

5. Nitrogen as a Factor Limiting Plant Growth in Salt Marshes

It was hypothesized that the marked increase of the grass *E. athericus* may partly be due to the enhanced input of nitrogen (and phosphorus) by eutrophicated flood water and atmospheric deposition. One assumption in this regard is that nitrogen is limiting growth of *E. athericus* in salt marshes. Addition of nitrogen has led to increased biomass production in many salt marsh in North America (Gallagher 1975, Chalmers 1979, Valiela 1984, Teal 1986, Morris 1988) and in Europe (Tyler 1967, Jeffries and Perkins 1977, Rozema 1978, Beeftink 1982, Kiehl et al. 1997) indicating that not salinity and flooding but the availability of nitrogen was limiting growth salt marsh plants. In a North American salt marsh, the standing crop of *Spartina alteriflora* increased from 1000 g/m^2 to 1300 g/m^2 by nitrogen fertilization. Phosphorus alone did not increase biomass production in most studies (Teal 1986).

5.1 TRACING AND QUANTIFYING N FLUXES IN SALT MARSHES: APPLICATION OF ^{15}N STABLE ISOTOPE ANALYSIS

The nitrogen budget presented (Table 1) shows large temporal variation of N fluxes and differences between salt marshes from different areas. In view of the large variations, fluctuations and many uncertainties regarding the data, the construction of a nitrogen budget for a salt marsh is a time-consuming task. In traditional studies, large scale and replicated sampling of salt marsh soil, flood water and above ground and below ground plant biomass is necessary to reliably quantify the exchange of nitrogen between salt marsh plants, sediment and flood water. In such studies it is difficult to make a distinction between the effects of natural nutrient supply (from for example mineralization) and that by anthropogenic sources (polluted flood water and atmospheric deposition). In some recent studies the stable isotope ^{15}N has been applied to trace various N-sources in the natural environment (Moraghan 1993, White and House 1994). Here we briefly indicate possibilities of the application of the stable isotope ^{15}N in N flux and N uptake studies.

The abundance of the stable isotope ^{15}N relative to ^{14}N is expressed as the mass ratio (R) $^{15}N/^{14}N$ relative to this ratio under standard conditions, that is, the ratio $^{15}N/^{14}N$ of atmospheric N_2 with an abundance of ^{15}N of 0.3663% (Ehleringer and Rundel 1989)

The $\delta^{15}N$ is defined as follows:

$\delta^{15}N = 1000\ (R_{sample}/R_{standard})\ \text{-}1$

where R = $^{15}N/^{14}N$ ratio.

The ^{15}N abundance of atmospheric N_2 ($\delta^{15}N$ = 0‰) is globally uniform and generally lower than in other N-pools such as soil and plant material. For soils on average a value of 0.3699% ($\delta^{15}N$ = + 9‰) is reported (Shearer and Kohl 1989). N-containing chemicals and fertilizers are often synthesized on the basis of atmospheric N_2. Therefore, the ^{15}N abundance of such compounds almost equals that of atmospheric N_2. The analysis of the stable isotope ^{15}N may be applied in two ways in studies of N-fluxes in salt marshes, briefly discussed below.

5.2 EVIDENCE FOR SALT MARSH POLLUTION WITH FERTILIZERS?- DISTRIBUTION OF ABUNDANCE OF ^{15}N IN SALT MARSH SEDIMENT PROFILES

Industrially synthesized fertilizer has a low ^{15}N abundance, for ammonium nitrate this value is 0.3660 % ^{15}N. Salt marshes are regularly flooded with sea water which may contain nitrogen from agricultural, domestic and industrial sources (with a reduced abundance of ^{15}N). Usually there is steady and gradual sedimentation of silt (Beeftink and Rozema 1988, Adam 1990, Dijkema et al. 1990, Leendertse et al 1997). This implies that deeper sediment layers reflect silt deposited in an (earlier) period with generally less agricultural and industrial pollution. We sampled sediment profiles from salt marshes in Portugal (Pancas), The Netherlands (Oosterkwelder, Schiermonnikoog) and Germany (Westerhefer), as part of a salt marsh research project funded by the EU ''disturbance of European saltmarsh ecosystems: the impact of environmental pollution (eutrophication) in relation to sedimentation patterns (SEES)''. An analysis was made of the $\delta^{15}N$ values in the profiles of (0-50 cm) of salt marsh soil from the Netherlands, Germany and Portugal

The mean natural abundance of ^{15}N was almost similar in the three salt marshes and the $\delta^{15}N$ varied between 9.6. (Dutch salt marsh) and 11.7‰ (German salt marsh). These values are close to the mean value of 9.92‰ reported for a number of surface soils in the USA as reported by Shearer and Kohl (1989) and are higher than the $\delta^{15}N$ value of atmospheric nitrogen (0‰) There was a small decrease in the natural abundances of ^{15}N with increasing depth (from 0 to 50 cm). This does, however, not confirm the expectation that deposition of fertilizer (with low $\delta^{15}N$ values) at the surface of the salt marsh soil profile would lead to increasing $\delta^{15}N$ values with increasing depth. Perhaps the decrease of $\delta^{15}N$ values with depth relates to increasing particle size. Shearer and Kohl (1989) found fine clay to have relatively high $\delta^{15}N$ values Deeper layers of the salt marsh soil at Schiermonnikoog contain more sand than surface layers. This observation may also explain the decrease of total N in the salt marsh soil. The total nitrogen concentration of the salt marsh sediment decreased with depth. In the 100-year old Oosterkwelder salt marsh at Schiermonnikoog, the total nitrogen content (g N/kg soil) of the soil was 6.6 ± 0.4 in the 0-10 cm layer, 3.3 ± 0.2. in the 10-15 cm layer and 0.1 ± 0.2 in the 15-30 cm layer (cf. Leendertse et al. 1997b).

The perspective of the use of ^{15}N in similar retrospective studies is still promising (Lindau et al. 1989, Shearer and Kohl 1989, Handley and Raven 1992). For example, Paerl and Fogel (1994) showed that the $\delta^{15}N$ values of phytoplankton decreased from +4.5‰ to 2‰, when rainwater with a $\delta^{15}N_4^+$ of -4‰ was added in bioassays with coast water. Thus, they showed that the phytoplankton absorbed inorganic N from the rain water.

5.3 APPLICATION OF ^{15}N STABLE ISOTOPE ANALYSIS IN UPTAKE EXPERIMENTS WITH NH_4NO_3 DEPLETED IN ^{15}N

When $^{15}NH_4$ $^{15}N\,O_3$, depleted in ^{15}N is applied as nutrient or fertilizer (abundance of ^{15}N in this $^{15}NH_4$ $^{15}NO_3$is 0.01 %), uptake of ^{15}N depleted ammonium nitrate by plants will reduce the $^{15}N/^{14}N$ ratio in plant material. Of ^{15}N depleted NH_4NO_3 added to salt marsh soil 6 to 8% was taken up by *Spartina anglica*, the recovery of the ^{15}N label in the soil was 37-44 %. For *Elymus athericus* the percentage ^{15}N label found in the plant was 1 to 6 % and 57-86% of the ^{15}N label was found in the soil (cf. Leendertse et al. 1997b, Leendertse et al. 1998). The remaining part of the ^{15}N added may have left the soil by denitrification (White and Howes 1994). In other experiments we demonstrated that application of ^{15}N depleted NH_4NO_3 to flood water led to a significant reduction of ^{15}N percentages in leaves, stems and in the roots of *E. athericus* an of *S. anglica* at a concentration of 5 mg NH_4NO_3 - N/l in the floodwater. Further calculations revealed that only a minor fraction (1-2%) of ^{15}N added to floodwater enters the plant by foliar uptake. When ^{15}N was added to artificial rain applied (as a spray) to the leaves of *E. athericus* and *S. anglica* no foliar uptake could be demonstrated. This illustrates that stable isotope studies represent a promising tool in tracing and quantifying nutrient fluxes between ecosystem compartments and nutrient uptake routes by plants (Handley and Raven 1992).

We admit however that neither the nitrogen budget, the longterm vegetation studies, nor the more experimental N flux and N uptake studies deliver a direct proof for the assumed link between the eutrophication and increasing dominance of *Elymus athericus* in island and mainland salt marshes.

6. Nitrogen and Vegetation Dynamics in Salt Marshes: A Perspective

Salt marshes represent an inherently dynamic environment in terms of sea water flooding, exchange of nutrients, sedimentation and vegetation development. Rather than acting as sinks or sources for nutrients, the high rates of transformation from organic nitrogen to inorganic nitrogen forms and *vice versa* during mineralization and assimilation represent a characteristic process of salt marshes. Accretion of silt and subsequent increased elevation and nutrient accumulation govern natural succession of salt marsh vegetation. Enhanced nutrient supply from polluted rivers and estuaries and from atmospheric deposition may affect the natural vegetation changes. The increasing dominance of *Elymus athericus,* in salt marshes of the Wadden Sea area known for its efficient nitrogen uptake and assimilation system, may at least in part, reflect a response of the species composition of the salt marsh to eutrophication.

In tracing and quantifying N fluxes in salt marshes, the application of ^{15}N stable isotope analysis is promising. ^{15}N stable isotope analysis can be used for N flux studies in salt marshes in several ways: based on variation in the natural abundance of ^{15}N in salt marsh ecosystem compartments and as tracer using chemical compounds enriched or depleted in ^{15}N. We briefly discussed the use of ^{15}N as a tracer in uptake experiments with NH_4NO_3 depleted in ^{15}N.

There is substantial experimental evidence that nitrogen is limiting plant growth and vegetation development in salt marshes. The nitrogen budget presented here and the analysis of in- and effluxes of nitrogen reveal large spatial an temporal variation. We have indicated that "natural" and/or anthropogenic changes in the nitrogen pools and fluxes underlie vegetation changes in salt marshes. In the dynamic salt marsh environment it will remain difficult to separate consequences of natural processes and human influences.

7. Acknowledgements

Rik Zoomer is acknowledged for his assistance with ammonium and nitrate analysis. We thank Rob Broekman for his skillful technical and chemical assistance. We are indebted to Mr. G.W. Valkenburg (AB-DLO) for the ^{15}N analyses. Part of the research presented here was financed by the EC Environment and Climate Programme (contract EV5V CT 93 0265). We are grateful to Staatsbosbeheer for permission to work on the salt marsh De Boschplaat and the Ministerie van Verkeer en Waterstaat, DG Rijkswaterstaat, RIKZ for data of the tides and for financial support. Karin Uyldert is gratefully acknowledged for the word processing. An anonymous reviewer is acknowledged for constructive comments. The authors acknowledge the organizers of the "Marsh Conference," Dr. Michael Weinstein and Dr. Dan Kreeger in particular, for their stimulating and integrating activity and their editorial improvement of the manuscript.

8. Literature Cited

Abd. Aziz, S.A. and D.B. Nedwell. 1979. Microbial nitrogen transformations in the salt marsh environment. Pages 385-398 *in* R.L. Jefferies and A.J. Davy, editors. Ecological processes in coastal environments. Blackwell, Oxford, England.

Abd. Aziz, S.A. and D.B. Nedwell 1986a. The nitrogen cycle of an East coast, U.K. saltmarsh: I. Nitrogen assimilation during primary production; detrital mineralization. Estuarine Coastal and Shelf Science **22**: 559-575.

—— 1986b. The nitrogen cycle of an east coast, U.K. saltmarsh: II. Nitrogen fixation, nitrification, denitrification, tidal exchange. Estuarine Coastal and Shelf Science **22**: 689-704.

Adam, P. 1990. Salt marsh ecology. Cambridge University Press, Cambridge, England.

Asjes, J. and N. M. J. A. Dankers. 1993. Variations in tidal exchange processes between a Dutch salt marsh, the Slufter and the North Sea. Proc. 4th International Wetland Conference. Columbus, Ohio, USA.

Bakker, J.P. 1989. Nature management by grazing and cutting. Kluwer, Dordrecht, the Netherlands

Bakker, J.P., J. de Leeuw, K.S. Dijkema, P.C. Leendertse, H.H.T. Prins and J. Rozema. 1993. Salt marshes along the coast of the Netherlands. Hydrobiologia **265**: 73-95.

Bakker, J.P., P. Esselink, R. van der Wal and K.S. Dijkema. 1997. Options for restoration and management of coastal salt marshes in Europe. Pages 286-322 *in* K.M. Urbanska, N.R.Webb and P.J. Edwards, editors. Resoration ecology and sustainable development. Cambridge University Press, Cambridge, England.

Beeftink, W.G. 1977. Salt marshes. Pages 93-121 *in* R.S.K. Barnes, editors. The coastline. Wiley, New York, New York, USA.

Beeftink, W.G.and J. Rozema 1988. The nature and functioning of salt marshes. Pages 59-87 *in* W. Salomons, B.L. Bayne, E.K. Duursma and U. Förstner, editors. Pollution of the North Sea: an assessment. Springer-Verlag, Berlin, Germany.

Beeftink, W.G., J. Rozema and A.H.L. Huiskes 1985. Ecology of coastal vegetation. Junk, The Hague, the Netherlands.

Bertness, M.D. 1991. Zonation of *Spartina patens* and *Spartina alterniflora* in a New England salt marsh. Ecology **72**: 138-148.

Casselman, M.E., W.H. Patrick Jr. and R.D. Delaune. 1981. Nitrogen fixation in a Gulf Coast salt marsh. Soil Science Society America, Journal **45**: 51-56.

Daly, M.A. and A.C. Mathieson. 1981. Nutrient fluxes within a small north temperate salt marsh. Marine Biology **61**: 337-344.

Dame, R., T. Chrzanowski, K. Bildstein, B. Kjerfve, H. McKellar, D. Nelson, J. Spurrier, S. Stancyk, H. Stevenson, J. Vernberg and R. Zingmark,. 1986. The outwelling hypothesis and North Inlet, South Carolina. Marine Ecology Progress Series **33**: 217-229.

Dankers, N., M. Binsbergen, K. Zegers, R. Laane and M. Rutgers van der Loeff. 1984. Transportation of water, particulate and dissolved organic and inorganic matter between a salt marsh and the Ems-Dollard Estuary, the Netherlands. Estuarine Coastal and Shelf Science **19**: 143-165.

Davy, A.J. and C. S. B. Costa 1992. Development and organization of salt marsh communities. Pages 157-178 *in* U. Seeliger, editor. Coastal plant communities of Latin America. Academic Press, San Diego, California, USA.

Defoe, D. 1719. Robinson Crusoe. Published in 1994 as Penguin Books, London, England.

De Glopper, R.J. 1988. De bodem van buitendijkse gronden in Nederland. Rijkswaterstaat Directie Flevoland, Lelystad.

De Jonge, V.N. 1990. Response of the Dutch Wadden Sea ecosystem to phosphorus discharges from the River Rhine. Hydrobiologia **195**: 49-62.

Delaune, R.D., T.C. Feijtel and W.H. Patrick, Jr. 1989. Nitrogen flows in Louisiana Gulf Coast salt marsh: spatial considerations. Biogeochemistry **8**: 25-37.

Delaune, R.D. and W.H. Patrick, Jr. 1990. Nitrogen cycling in Louisiana Gulf Coast brackish marshes. Hydrobiologia **199**: 73-79.

Dijkema, K.S., J.H. Bossinade, P. Bouwsema and R.J. De Glopper 1990. Salt marshes in the Netherlands Wadden Sea: rising high-tide levels and accretion enhancement. Pages 173-188 *in* J.J. Beukema, W.J. Wolff and J.J.W. Brouns, editors. Expected effects of climatic change on marine coastal ecosystems. Kluwer, Dordecht, the Netherlands.

Ehleringer, J.R. and P.W. Rundel. 1989. Stable isotopes: history, units and instrumentation *in* P.W. Rundel, J.R. Ehleringer and K.A. Nagy, editors. Stable isotopes in ecological research. Ecological studies. **68**:

Gallagher, J.L. 1975. Effect of an ammonium nitrate pulse on the growth and elemental composition of natural stands of *Spartina alterniflora* and *Juncus roemerianus*. American Journal of Botany **62**: 644-648.

Haines, E.B., A. Chalmers, R. Hanson and B. Sherr 1977. Nitrogen pools and fluxes in a Georgia salt marsh. Pages 241-254 *in* V.S. Kennedy, editor. Estuarine perspectives. Academic Press, New York, New York, USA.

Haines, E.B. 1979. Growth dynamics of cordgrass, *Spartina alterniflora* Loisel., on control and sewage fertilized plots in a Georgia salt marsh. Estuaries **2**: 50-53.

Handley, L.L. and J.A. Raven. 1992. The use of natural abundance of nitrogen isotopes in plant physiology and ecology. *Plant Cell Environt* **15**: 965-985.

Hanson, R.B. 1977. Comparison of nitrogen fixation activity in tall and short *Spartina alterniflora* salt marsh soil. Applications of Environmental Microbiology **33**: 596-602.

Harkel, M.J. ten. 1998. Nutrient pools and fluxes in dry coastal grasslands. Thesis. University of Amsterdam, The Netherlands.

Heinle, D.R. and D.A. Flemer. 1976. Flows of materials between poorly flooded tidal marshes and an estuary. Marine Biology **35**: 359-373.

Henriksen, K. and A. Jensen. 1979. Nitrogen mineralization in a salt marsh ecosystem dominated by *Halimione portulacoides*. Pages 373-384 *in* R.L. Jefferies and A.J. Davy, editors. Ecological processes in coastal environments. Blackwell, Oxford, England.

Huiskes, A.H.L., C.W.P.M. Blom and J. Rozema 1987. Vegetation Between Land and Sea. Junk, Dordrecht, The Netherlands.

Janssen, G.M. 1993. De eutrofiëring van de Noordzee en Waddenzee, een tussenbalans. Zien we al resultaten van het saneringsbeleid? *H_2O* **26**: 86-91

Jefferies, R.L. and N. Perkins, N. 1977. The effects on the vegetation of the additions of inorganic nutrients to salt marsh soils at Stiffkey, Norfolk. Journal of Ecology **65**: 867-882.

Jones, K. 1974. Nitrogen fixation in a salt marsh. Journal of Ecology **62**: 553-565.

Jordan, T.E., D.L. Correll and D.F. Whigham. 1983. Nutrient flux in the Rhode River: tidal exchange of nutrients by brackish marshes. Estuarine, Coastal and Shelf Science **17**: 651-667.

Kaplan, W., I. Valiela and J.M. Teal. 1979. Denitrification in a salt marsh ecosystem. Limnology and Oceanography **24**: 726-734.

Kiehl, K., P. Esselink and J.P.Bakker. 1997. Nutrient limitation and plant species composition in temperate salt marshes. Oecologia **111**: 325-330

Koerselman, W., H. Doing, L.H.W.T. Geelen, G.M. Leltz and E.M. Nijssen 1992. De voedselrijkdom van duinvalleien in relatie tot standplaatscondities en externe inputs. KIWA N.V., Nieuwegein. SWE 91.037.

Koppel, J. van de. 1997. Trophic interactions in gradients of primary production. Thesis, University of Groningen, The Netherlands.

Leendertse, P.C., C.C. Karman and J. Rozema 1993b. Helofytenfilters langs de Afsluitdijk -een haalbaarheidsstudie- VU-Oecologie & Oecotoxicologie/ RWS-DGW, Amsterdam, The Netherlands

Leendertse, P.C., H.J. van Wijnen, P. Cartaxana, R. Neuhaus, J.P. Bakker and J. Rozema. 1998. Natural abundance of ^{15}N nitrogen in salt marsh soils. Submitted.

Leendertse, P.C., A.J.M. Roozen and J. Rozema 1997a. Long-term changes (1953-1990) in the salt marsh vegetation at the Boschplaat on Terschelling in relation to sedimentation and flooding. Plant Ecology **132**: 49-58.

Leendertse, P.C., J. Rozema and A. Andrea 1997b. Effects of nitrogen addition on the growth of the salt marsh grass *Elymus athericus*. Journal of Coastal Conservation **3**: 35-40.

Leendertse, P.C., J. Rozema, G.M. Janssen and J.H. Bossinade 1993a. Kwelders en waterkwaliteit. Een verkenning in het Waddengebied. Landschap **10**: 3-15.

Leendertse, P.C., J. Rozema, R. Spaargaren and N.H. Weller. 1998. The use of ^{15}N depleted ammonium nitrate to measure nitrogen uptake of two salt marsh grasses. Submitted.

Lenssen, G.M., J. Lamers, M. Stroetenga and J.Rozema. 1993. Interactive effects of atmospheric CO_2 enrichment, salinity and flooding on the growth of C_3 (*Elymus athericus*) and C_4 (*Spartina anglica)* salt marsh species. *Vegetatio* **104/105**: 379-390

Lindau, C.W., R.D. Delaune, W.H. Patrick Jr. and E.N. Lambremont. 1989. Assessment of stable nitrogen isotopes in fingerprinting surface water inorganic nitrogen sources. Water Air Soil Pollution **48**: 489-496.

Moraghan, J.T. 1993. Loss and assimilation of ^{15}N -nitrate added to a North Dakota cattail mash. Aquatic Botany **46**: 225-234.

Morris, J.T., 1988. Pathways and controls of the carbon cycle in salt marshes. Pages 496-510 *in* D.D. Hook, W.H. McKee Jr., H.K. Smith, J Gregory, V.G. Burnell, Jr., M.R. DeVoe, R.E. Sojka, S. Gilbert, R. Banks, L.H. Stolzy, C. Brooks, T.D. Matthews and T.H. Shear, editors. The ecology and management of wetlands. Croom Helm, London, England.

Morris, J.T. 1991. Effects of nitrogen loading on wetland ecosystems with particular reference to atmospheric deposition. Annual Review of Ecology Systematics **22**: 257-279

Nichols, D.S.,. 1983. Capacity of natural wetlands to remove nutrients from wastewater. Journal of Water Pollution, Control Federation **55**: 495-505.

Olff, H. 1992. On the mechanisms of vegetation succession. Thesis, RUG, Groningen, the Netherlands.

Olff, H., J. de Leeuw, J.P. Bakker, R.J. Platerink, H.J. van Wijnen and W. de Munck. 1997. Vegetation succession and herbivory in a salt marsh: changes induced by sea level rise and salt deposition along an elevational gradient. Journal of Ecology **85**: 799-814.

Patrick, W.H, Jr. 1990. Microbial reactions of nitrogen and phosphorus in wetlands. The Utrecht Plant Ecology News Report **11**: 52-63.

Pearl, H.W. and M.L.Fogel. 1994. Isotopic characterization of atmospheric ntrogen inputs as sources on enhanced primary production in coastal Atlantic Ocean water. Marine Biology **119**: 635-645.

Ranwell, D.S. 1972. Ecology of Salt Marshes and Sand Dunes. Chapman and Hall, London, England.

Reddy, K.R. and W.H. Patrick, Jr. 1984. Nitrogen transformations and loss in flooded soils and sediments. CRC Crit. Review Environmental Control **13**: 273-309.

Roozen, A.J.M. and V. Westhoff. 1985. A study on long-term salt-marsh succession using permanent plots. Vegetatio **61**: 23-32.

Rozema, J. and J.A.C. Verkleij. 1991. Ecological Responses to Environmental Stresses. Kluwer Academic Publishing, Dordrecht, The Netherlands.

Rozema, J. 1996. Biology of halophytes. Pages 17-30 *in* R. Choukr-Allah, C.V. Malcolm and A. Hamdy, editors. Halophytes and biosaline agriculture. Marcel Dekker, New York, New York, USA.

Rozema, J. and H.A. Verhoef. 1996. Leerboek Toegepaste Oecologie. Free University Press, Amsterdam, The Netherlands.

Rozema, J. and P.C. Leendertse 1991. Natural and man-made environmental stresses in coastal wetlands. Pages 92-101 *in* J. Rozema and J.A.C. Verkleij, editors. Ecological responses to environmental stresses. Kluwer Academic Publishing, Dordrecht, The Netherlands.

Rozema,J., H.J.M.Nelissen, M. van der Kroft and W.H.O. Ernst. 1977. Nitrogen mineralization in sandy salt marsh soils of the Netherlands. Z. Pflanzenernährung and Bodenkd. **140**: 707-717.

Rozema,J. Y. van Manen, H. Vugts and A. Leusink. 1983. The ecological relevance of air-borne salt in the distribution of three *Elytrigia* species. Acta Botanica Neerlandaise **32**: 447-456.

Rozema, J., M.C. Th. Scholten, P.A. Blaauw and J. van Diggelen. 1988. Distribution limits and physiological tolerances with particular reference to the salt marsh environment. Pages 137-164 *in* A.J. Davy, M.J. Hutchings and A.R. Watkinson, editors. Plant population ecology, Blackwell, Oxford, England.

Rozema, J., P. Bijwaard, G. Prast and R. Broekman 1985a. Ecophysiological adaptions of coastal halophytes from foredunes and salt marshes. Vegetatio **62**: 449-521.

Scherfose, V. 1987. Pflanzensoziologische und ökologische Untersuchungen in Salzrasen der Nordseeinsel Spiekeroog. II. Bodenchemische Untersuchungen. Stickstoff-Netto-Mineralisation und Salzbelastung. Tuexenia **7**: 173-198.

Scholten, M., P.A. Blaauw, M. Stroetenga and J. Rozema 1987. The impact of competitive interactions on the growth and distributin of plant species in salt marshes. Pages 270-281 *in* A.H.L. Huiskes, C.W.P.M. Blom and J. Rozema, editors. Vegetation between land and sea. Junk, Dordrecht, The Netherlands.

Scholten, M.C. Th. and J. Rozema. Pages 39-49. 1990. The competitive ability of *Spartina anglica* on Dutch salt marshes. *In* A.J Gray and P.E.M. Bernham, editors. *Spartina anglica*: a research review. ITE Research Publication No.2.

Shearer, G. and D.H. Kohl, 1989. Estimates of nitrogen fixation in ecosystems: the need for and basis of the ^{15}N natural abundance method. Pages 342-374 *in* P.W. Rundel, J.R. Ehrlinger and K.A. Nogy, editors. Stable isotopes in ecological research. Ecological studies **68**: Springer Verlag, Berlin, Germany.

Smith, C.J., R.D. Delaune and W.H. Patrick, Jr. 1983. Nitrous oxide emission from Gulf Coast wetlands. Geochimica Cosmochimica Acta **47**: 1805-1814.

Stal, L.J., S. Grossberger and W.E. Krumbein. 1984. Nitrogen fixation associated with the cyanobacterial mat of a marine laminated microbial ecosystem. Marine Biology **82**: 217-224.

Teal, J.M. 1986. The ecology of regularly flooded salt marshes of New England: a community profile. U.S. Fish Wildlife Services Biology Report **85** (7.4).
Tyler, G.,. 1967. On the effect of phosphorus and nitrogen, supplied to baltic shore-meadow vegetation. Botaniska Notiser **120**: 433-447.
Valiela, I. and J.M. Teal. 1974. Nutrient limitation in salt marsh vegetation. Pages 547-563 *in* R.J. Reimold and W.H. Queen, editors. Ecology of halophytes. Academic Press, London, England.
Valiela, I. and J.M. Teal. 1979. The nitrogen budget of a salt marsh ecosystem. Nature **280**: 652-656.
Valiela, I.,. 1984. Marine Ecological Processes. Springer Verlag, New York, New York, USA.
Valiela, I., J.M. Teal and W. Sass. 1973. Nutrient retention in salt marsh plots experimentally fertilized with sewage sludge. Estuarine Coastal and Marine Science **1**: 261-269.
Valiela, I., S. Vince and J.M. Teal 1976. Assimilation of sewage by wetlands. Pages 234-253 *in* M. Wiley, editor. Estuarine processes. Vol. I. Academic Press, New York, New York, USA.
Veen, A. van der, A.P. Grootjans, J. de Jong and J. Rozema. 1997. Reconstruction of an interrupted primary beach plain, using a geographical information system. Journal of Coastal Conservation **3**: 70-78.
Veer, H.W. van der, W. Van Raaphorst and M.J.N. Bergman. 1989. Eutrophication of the Dutch Wadden Sea: external nutrient loadings of the Marsdiep and Vliestroom Basin. Helgolander Meeresuntersuchingen **43**: 501-515.
Verhoeven, J.T.A. 1990. Nutrient dynamics of freshwater wetlands, with special reference to enrichment. The Utrecht Plant Ecology News Report **11**: 5-21.
White, D.S. and B.L. Howes. 1994. Long-term ^{15}N-nitrogen limitation in the vegetated sediments of a New England salt marsh. Limnology and Oceanography **39**: 167-175.
Whiting, G.J., H.N. McKellar Jr., B. Kjerfve and J.D. Spurrier. 1987. Nitrogen exchange between a southeastern USA salt marsh ecosystem and the coastal ocean. Marine Biology **95**:173-182.
Whiting, G.J., H.N. McKellar Jr., J.D. Spurrier and T.G. Wolaver. 1989. Nitrogen exchange between a portion of vegetated salt marsh and the adjoining creek. Limnology and Oceanography **34**: 463-473.
Wijnen, H.J. van and J.P. Bakker. 1997. Nitrogen accumulation and plant species replacement in three salt marsh systems in the Wadden Sea. Journal of Coastal Conservation **3**: 19-26.
Wijnen, H.J. van, R. van de Wal and J.P. Bakker. 1998. The impact of small herbivores on soil net nitrogen mineralization rate: consequences for salt marsh succession. Oecologia (submitted)
Wijnen, H.J. van, A.A.H. Hazekamp, E.C. van Hooff and J.P.Bakker. 1998. Net nitrogen mineralization rates along a succesional gradient. Mangroves and Salt Marshes (submitted)
Wolaver, T.G., J.C. Zieman, R. Wetzel and K.L. Webb. 1983. Tidal exchange of nitrogen and phosphorus between a mesohaline vegetated marsh and the surrounding estuary in the lower Chesapeake Bay. Estuarine Coastal and Shelf Science **16**: 321-332.
Wolff, W. J., M.J. Van Eeden and E. Lammens. 1979. Primary production and import of particulate matter on a salt marsh in the Netherlands. Netherlands Journal of Sea Research **13**: 242-255.
Woodwell, G.M., C.A.S. Hall, D.E. Whitney and R.A. Houghton. 1979. The Flax Pond ecosystem study: exchanges of inorganic nitrogen between an estuarine marsh and Long Island Sound. Ecology **60**: 695-702.
Zimmek, G.E. 1975 Die Mineralstickstoff-Versorgung einiger Salzrasen Gesellschaften des Graswarders vor Heiligenhafen/Ostsee. Thesis, Göttingen, The Netherlands.

MODELLING NUTRIENT AND ENERGY FLUX

A STABLE ISOTOPE MODEL APPROACH TO ESTIMATING THE CONTRIBUTION OF ORGANIC MATTER FROM MARSHES TO ESTUARIES

PETER M. ELDRIDGE
Western Ecology Division
U.S. EPA
2111 SE Marine Science Dr.
Newport, Oregon 97365 USA

LUIS A. CIFUENTES
Department of Oceanography
Texas A&M University
College Station, Texas 77843 USA

Abstract
The impact of marsh carbon export (outwelling) on estuarine metabolism has been debated for the last three decades. Much of this controversy stems from interpretations of stable isotope data. Although the outwelling of marsh carbon to the estuaries can be substantial, the stable isotope signal ($\delta^{13}C$) of marsh material is generally not detected except in sediments and infauna at marsh fringes. However, most of these studies focus on the $\delta^{13}C$ of either particulate organic carbon (POC) or the $\delta^{13}C$ of estuarine organisms that depend on POC, even though this carbon pool may be the least likely to provide a marsh signal. A series of simple models are developed to show how marsh outwelling affects the $\delta^{13}C$ of POC and the other major estuarine carbon pools, i.e., dissolved organic carbon (DOC), and dissolved inorganic carbon (DIC). The models show that a marsh $\delta^{13}C$ signal will only be detected in estuarine POC or DIC when the marsh area is substantially larger than the estuary. However, because estuarine *in-situ* DOC production is only a fraction of POC production, our mixing models suggest that a marsh signal should be found in estuarine DOC, even when the marsh areas is smaller that the estuarine area. Finally, a transport model, incorporating a simplified bathymetry and hydrology for the Parker River, MA, is used to back calculate marsh outwelling from the estuarine $\delta^{13}C$-DOC. The model estimate of marsh DOC outwelling is consistent with other estimates, and suggests that our parameterization of estuarine transport and degradation processes that regulate DOC isotope ratios is probably also correct.

1. Introduction

Isotope geochemistry has been used extensively to describe sources and sinks of organic material in estuaries (see Lajtha and Michener 1994, Peterson and Howarth 1987, Fry and Sherr 1984). However, because our knowledge of the mechanisms by which transport processes and estuarine foodwebs effect stable isotope ratios is incomplete, results of naturally occurring isotope studies in salt-marsh estuaries are often puzzling. Early on, Haines (1976, 1977) showed that stable carbon isotope ratios ($\delta^{13}C$) of consumers, seston and sediments from Georgia salt-marsh estuaries did not exhibit the distinct isotopic signal associated with marsh primary production (see review of Fry and Sherr 1984). This result was unexpected considering outwelling from marshes was long thought to contribute significantly to estuarine food webs (Teal 1962). Later on in his review of marsh studies, Nixon (1980) provided evidence for the outwelling hypothesis, when he showed carbon (C) export rates from marshes were typically greater than 100 g-C m^{-2} y^{-1}. Finally, Peterson et al. (1985) confirmed, with natural abundance of two stable isotopes, that marsh outwelling of organic material was assimilated in an estuary. They found that in the Great Sippewissett Marsh, shellfish have $\delta^{13}C$ and stable sulfur isotope ratios ($\delta^{34}S$) intermediate between *Spartina* and phytoplankton and argued that marsh primary production contributed to their diet (Fig. 1).

Similar results were found in salt-marsh estuaries near Sapelo Island, Georgia (Peterson and Howarth 1987). However, the issue of estuarine organism assimilation of marsh material has now been further confounded by the presence of microalgal mats that have $\delta^{13}C$ and $\delta^{34}S$ values intermediate between *Spartina* and phytoplankton (e.g., Currin et al. 1995). Moreover, most studies documenting assimilation of marsh organic matter by organisms have been conducted in close proximity to the marsh, and therefore, can not be extrapolated to the whole estuary. In spite of extensive data sets and the use of multiple isotope tracers, our present description of carbon outwelling based on isotope analyses is still ambiguous after 20 years of effort.

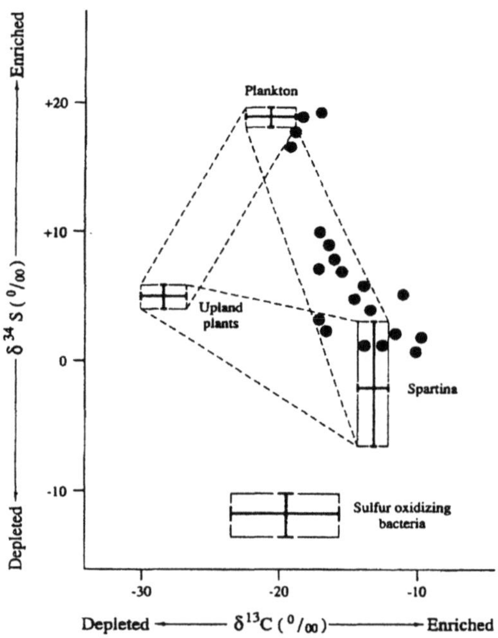

Figure 1. Peterson et al. (1985) demonstrated that shellfish in Great Sippewissett Marsh have $\delta^{13}C$ and $\delta^{34}S$ isotope rations intermediate between *Spartina* and phytoplankton suggesting that marsh primary production contributes to their diet.

1.1 POC AND DOC EXCHANGE

If significant export of marsh organic carbon does occur, its distinct $\delta^{13}C$ signal must be traceable to one or more carbon pools (e. g., particulate organic carbon, dissolved organic carbon, dissolved inorganic carbon, organisms) in the estuary. Sediments and organisms rarely have $\delta^{13}C$ values expected for marsh detritus (-12 to -16‰; see review of Fry and Sherr 1984), except in inter-tidal sediments. Similarly, Coffin et al. (1994) compilation of $\delta^{13}C$ data for particulate organic carbon (POC) from estuaries throughout North America displayed no values within this range (Fig. 2), suggesting a minor role for marsh POC in estuarine foodwebs. The other large carbon pools in estuaries in which we might expect to find a marsh signal are the dissolve inorganic carbon (DIC) and the dissolved organic carbon (DOC).

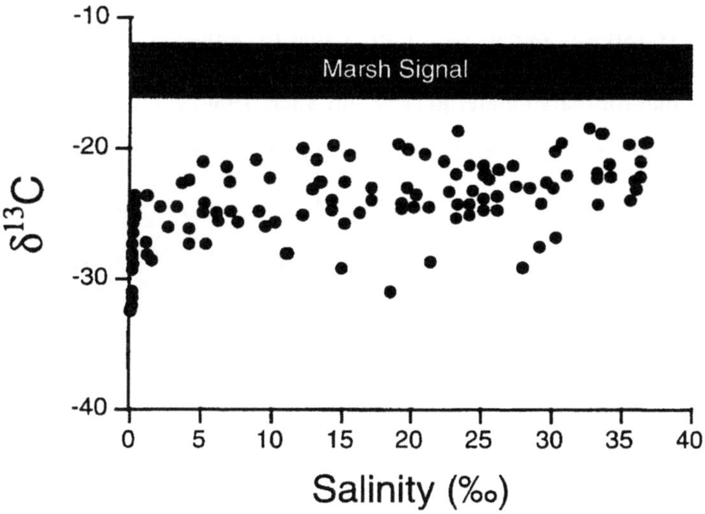

Figure 2. A compilation of $\delta^{13}C$ values for POC plotted against salinity (revised from Coffin et al., 1994). There are no values near the range (-12 to -16%) expected for marsh particulate detritus indicating that outwelling of marsh POC has little influence on the isotopic composition of estuarine POC..

Although DIC isotope measurements vary extensively in estuaries (see Fig. 10.4 in Coffin et al. 1994), this variation can not be linked to marsh export processes without further analysis. The substantial marsh organic carbon outwelling in the form of DOC reported in Nixon's (1980) review would suggest that we might find a distinct $\delta^{13}C$ marsh signal in estuarine DOC pools. Dissolved organic carbon is not as prone to sedimentation as is POC. Furthermore, comparatively less DOC is produced during phytoplankton production (Lignell, 1990). Thus, the relative contribution of marsh derived DOC to the total estuary DOC is likely to be far greater than marsh POC to the estuarine POC pool. Below, we use a simple box model to compare the influence of marsh POC and DOC outwelling on isotope ratios in estuaries.

The contribution of phytoplankton and marsh-derived ^{13}C to estuarine POC and

DOC can be represented as a function of relevant fluxes and isotope ratios (EQs 1 & 2):

$$\delta^{13}C_{POC} = \frac{P_f\delta^{13}C_P(1-\phi) + ME_rM_f\delta^{13}C_M}{P_f(1-\phi) + ME_rM_f} \tag{1}$$

$$\delta^{13}C_{DOC} = \frac{P_f\delta^{13}C_P\phi + ME_rM_f\delta^{13}C_M}{P_f\phi + ME_rM_f} \tag{2}$$

where P_f is primary phytoplankton production (50 to 500 gC m^{-2} y^{-1}; Heip et al. 1995), $\delta^{13}C_p$ (-22; typical value from Fry and Sherr 1984) is the isotope ratio of the phytoplankton, ϕ (0.1; (we use 0.2 later) Lignell 1990) is the fraction of primary production released as DOC, ME_r is the ratio of marsh to estuary area (0.1 to 5.0; see Nixon, 1980), M_f (50 g-C m^{-2} y^{-1}; a middle value for either POC or DOC according to Nixon, 1980) is the marsh outwelling of POC or DOC, and $\delta^{13}C_M$ ($-13^0/_{00}$) is an intermediate value for marsh material (Fry and Sherr 1984, Michener and Schell, 1994).

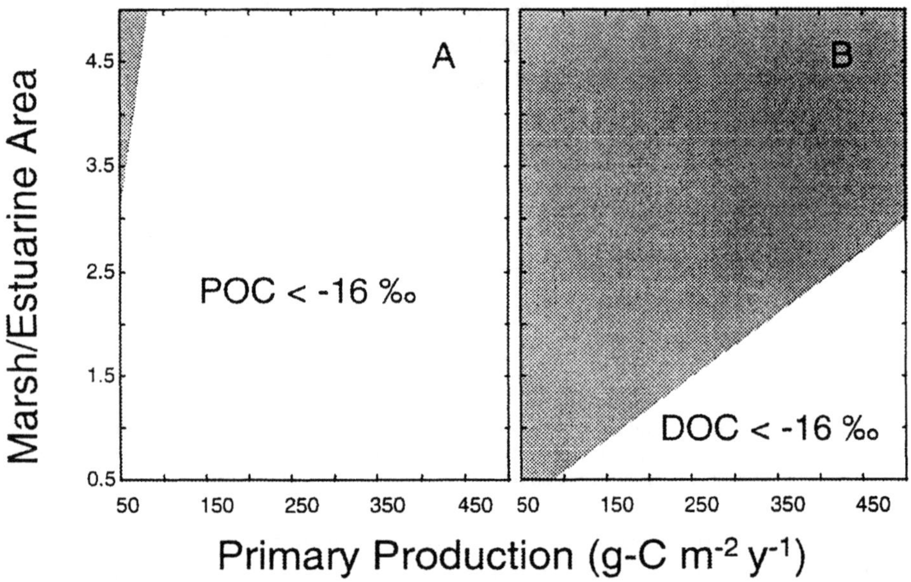

Figure 3. Results of a simple model showing the dependence of estuarine: A) POC and B) DOC isotope ratio on marsh outwelling and phytoplankton primary production. Marsh outwelling for both POC and DOC is a function of marsh:estuary area (assumes an export rate of 50 g-C m^{-2} y^{-1}). The shaded area indicates regions where the marsh signature dominates i.e. $\delta^{13}C_{POC\ or\ DOC} > -16$.

Simplistically, significant contributions from marshes to either estuarine POC or DOC pools should result in estuarine $\delta^{13}C$ of $-16^0/_{00}$ or greater. The model shows that equal fluxes of marsh POC (Fig. 3A) and DOC (Fig. 3B) to the estuary, when mixed with phytoplankton sources, result in different isotope ratios. Even at low levels of primary production (< 100 g-C m^{-2} y^{-1}), substantial marsh outwelling (i.e., high marsh to estuarine area) is required before a marsh signal appears in the estuarine POC (see shaded area in Fig. 3A). In fact, at typical primary production rates (e.g. 200 to 300 g-C m^{-2} y^{-1}) marsh values are not predicted even in the presence of exceptionally high marsh export. The opposite result is observed in the DOC analysis. In contrast to POC at typical primary production values a marsh DOC signal is seen in the $\delta^{13}C$ of estuarine DOC at marsh/estuarine areas as low as 0.8. An obvious conclusion of the analysis is that energy relationships based on natural isotope ratios are better established with DOC than with POC.

1.2 CARBON CYCLING BETWEEN MARSH AND ESTUARY

Models of carbon cycling and transport can be constrained by combining concentration and stable isotope measurements. This combination must correctly predict both variables in a system (e. g., Wigley et al. 1978) and should integrate over estuarine time-scales. However, a simple reliance on DOC concentration and isotope data does not encompass the full set of carbon transformations (Fig. 4) within an estuary and may not ultimately provide evidence of marsh influence on estuary energy flow when it actually occurs. For example, the absence of a marsh signal in a $\delta^{13}C$ measurement of DOC may denote that it was not transported to that point in the estuary, or alternately, the DOC was metabolised to dissolved inorganic carbon (DIC). Because both DIC and DOC are recycled on time-scales of the estuarine residence time (Cifuentes and Eldridge 1998), any analysis that purports to describe marsh/estuarine interactions quantitatively must incorporate a reasonably full description of carbon dynamics (Fig. 4). Since the $\delta^{13}C$ of DIC in estuaries is a balance between respiration, production, mixing, and evasion and invasion of CO_2 from the atmosphere, this description must also include estuarine hydrodynamics, the carbonate system, and air-sea exchange.

Recent improvements in accuracy and costs of isotope analysis now make it feasible to make fine spatial and temporal scale measurements of stable carbon isotope ratios. Furthermore, the refinement of gas exchange and circulation models provides a quantitative means of analysing this data.

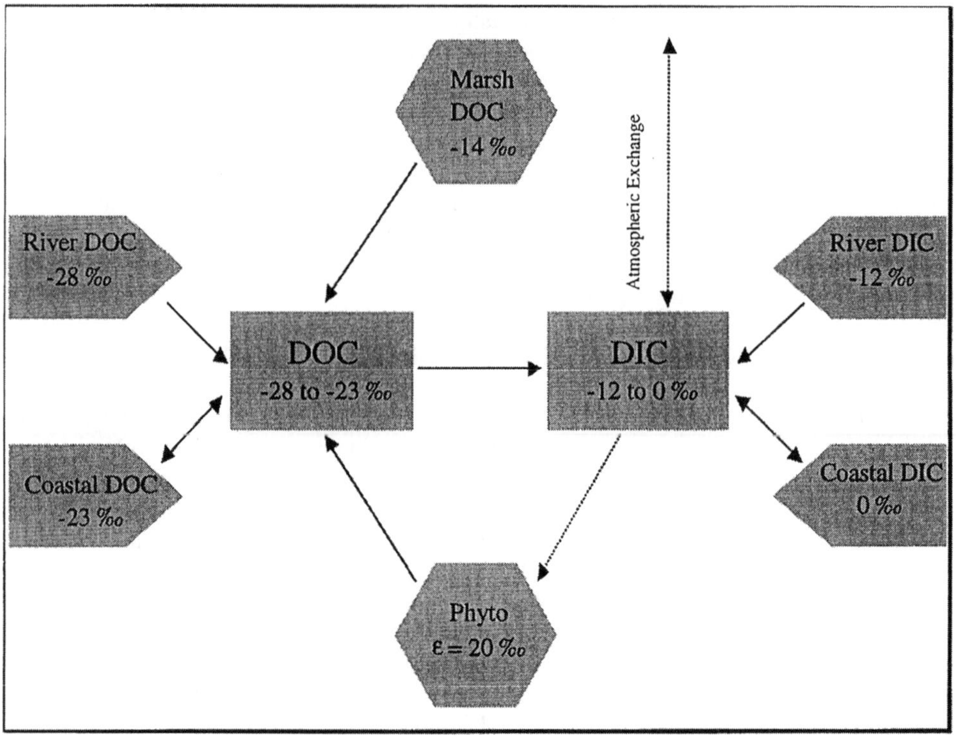

Figure 4. Conceptual isotope model of marsh:estuary DOC and DIC interactions. Coastal and riverine end-members of DOC (-28 and -23‰) and DIC (-12 and ‰) are isotopically distinct and define the conservative concentration vs. salinity gradient within the estuary. Marsh DOC is the only other allochthonous DOC source in the estuary while atmospheric CO_2 represents either a sink or source of DIC to the estuary. Autochthonous sinks and sources include primary production and respiration. There is a 20‰ isotope effect in the model due to selective uptake of ^{13}C-depleted DIC by phytoplankton and due to atmospheric exchange indicated by dotted lines.

1.3 MODELLING MARSH OUTWELLING

In this chapter, we describe a simple model that incorporates important features of DOC and DIC dynamics (e. g., marsh outwelling, production, loss, and transport) in estuaries. In the following sections the model is used first to predict the influence of marsh DOC outwelling on isotope ratios of DOC and DIC in the estuary. Then through iterations of the model, we show how variations in hydraulic residence time alter the effect of marsh outwelling and primary production on DOC and DIC concentrations and isotope ratios. Finally, the model provides an estimate of marsh outwelling from concentration and isotope measurements of estuarine carbon pools.

2 Model Description

The isotope model included two DOC constituents (labile and refractory), DIC, their ^{13}C analogs, and salt. We defined riverine and coastal end members, allochthonous and autochthonous sources and sinks of DOC and DIC, and then solved for their concentrations along the axis of the estuary with a transport equation. The end member concentrations for the ^{13}C isotope were derived from the riverine and coastal DOC and DIC concentrations and their isotope ratio (EQs 3 & 4):

$$rY = \left(\frac{\delta^{13}Y}{1000} + 1\right)pdb \tag{3}$$

$$^{13}Y = rY[Y] \tag{4}$$

where rY was the ^{13}C:^{12}C, [Y] is DOC or DIC concentration, and *pdb* was the absolute ^{13}C:^{12}C of Pee Dee Belemnite (Craig, 1953).

2.1 TRANSPORT MODEL

We used a time-dependent, finite-difference model with constant geometry (i.e., constant width and depth) to solve the transport equations. A system of these equations was formulated using the general form of (e. g., EQ 5):

$$\frac{\partial[Y]}{\partial t} = \frac{\partial\left(D(x)\frac{\partial[Y]}{\partial x}\right)}{\partial x} - u\frac{\partial[Y]}{\partial x} + \frac{F(x)}{A} - R_i \tag{5}$$

where Y was either salinity, refractory (r) or labile (l) DOC, and DIC or their ^{13}C concentration (i.e., ^{13}C-DOC_r, ^{13}C-DOC_l, and ^{13}C-DIC), A was cross-sectional area (m^2), D was the eddy diffusion coefficient ($m^2\ s^{-1}$), u was velocity ($m\ s^{-1}$) and $F_i(x)$ was the DOC input (mmol-C $m^{-1}\ s^{-1}$) along the estuary with i referring to either marsh- or phytoplankton-derived DOC. Transport in the model had the units of mmol-C $m^{-3}\ s^{-1}$ and the units of concentration were mmol-C m^{-3}. Because the ^{13}C-abundance in DOC is only about 1 %, we assumed [DOC]$\approx$[^{12}C-DOC] without a significant decrease in accuracy. The DOC flux was distributed along the estuary according to a Gaussian probability distribution (EQ 6):

$$P(j, \mu, \sigma) = \frac{1}{(2\pi\sigma^2)^{\frac{1}{2}}}e^{-(j-\mu)^2}/_{2\sigma^2}, (1 \geq j \geq np), \tag{6}$$

where j was the index for each gridpoint, np was the number of gridpoints, l was the gridpoint in the center of the estuary and σ^2 dictated the number of gridpoints over which the flux occurred. In turn, the flux of DOC at each gridpoint was (EQ 7):

$$F_i(x) = \frac{P(i, \mu, \delta) F_T}{\Delta x} \tag{7}$$

where F_T was the total flux of marsh derived DOC (mol s^{-1}) and Δx was the length between gridpoints. Finally, the loss term R_i (mol m^{-3} s^{-1}) was based on a first order decay model (EQ 8):

$$R_i = -\frac{d[DOC_i]}{dt} = k_1[DOC_i, \tag{8}$$

where k was the first-order decay constant and i referred to either labile or refractory DOC. Model parameters are shown in Table 1. Concentration and isotopic boundary conditions were provided at the riverine and coastal mixing end members. The above equations were solved with a finite-difference scheme (see Boudreau 1986).

Once the system of equations was solved we calculate the $\delta^{13}C$ along the estuarine axis from the ^{12}C and ^{13}C model results (EQ 9):

$$\delta^{13}C = \left(\frac{^{13}C/^{12}C_{sa}}{^{13}C/^{12}C_{std}} - 1 \right) \times 1000, \tag{9}$$

where subscript *sa* is the sample ratio and *std* is the standard *pdb* ratio.

TABLE 1. Physical and biological parameters used in the model. The generic model used a 1:1 marsh/estuary area and a 6 m depth. The range of marsh/estuary area and depth were used in the contour model shown in Fig. 9.

Physical Parameter	Value	Biological Parameter	Value
Length	50 km	Marsh DOC Flux	50 gC m^2 yr^{-1}
Width	5 km	Phyto Productivity	100 to 250 gC m^2 yr^{-1}
Depth	2 - 10 m	Phyto DOC release	20 %
Marsh:Estuary Area	0.5 to 5	Marsh DOC (% Refractory)	60 %
Eddy Dispersion	100 m^2 s^{-1}	Labile DOC reaction rate (k_l)	36 yr^{-1}
River Inflow	150 m^3 s^{-1}	Refractory DOC reaction rate (k_r)	0.365 yr^{-1}
D/z	0.005 cm^1 s^{-1}	pH	7.5 to 8.2
T°C	20		

2.2 BOUNDARY CONDITIONS

Flux conditions for all constituents were imposed at the river boundary using the loading (concentration x flow) as the first term in equation 10:

$$u[Y] - D\frac{\delta[Y]}{\delta x} = F_r, \tag{10}$$

where F_r was the flux at the riverine end member. The river loading term was set to zero for salinity. Concentration boundary conditions were used at the coastal end member of the analysis for all the model constituents.

2.3 GAS EXCHANGE MODEL

The concentration of DIC at any point within the estuary was a function of riverine and coastal end-member mixing, the amount added by respiration (R) and lost to primary production (P), and that lost or gained from the atmosphere (EQs 11 & 12).

$$\frac{\partial DIC}{\partial t} = transport - P + R + F_{CO_2} \tag{11}$$

$$F_{CO_2} = -\frac{D}{z}\frac{[pCO_2]K_0(t,s) \times 1000 - [H_2CO_3]}{h \times 100} \tag{12}$$

The atmosphere pCO_2 concentration was 333 ppm (Whitfield and Turner, 1986). D/z was in cm s^{-1} and depth (h) was meters. Multiplication by 1000 and division by 100 provides units of mmol C m^{-2} s^{-1}. The thermodynamic constants including $K_0(t,s)$ used in equation 12 were calculated from temperature-salinity dependent algorithms (Whitfield and Turner, 1986). To determine the distribution of the DIC species (H_2CO_3*, HCO_3^-, and CO_3^{2-}) along the salinity gradient we assumed that alkalinity was a conservative property and that the estuary acts as a semi-closed system (Whitfield and Turner, 1986). Hence, there was a limited exchange of pCO_2 from the atmosphere. The end-member alkalinities were calculated from pH(NBS) and [DIC] using a closed system equilibrium model (Stumm and Morgan, 1981). Parallel equations for the ^{13}C-DIC species (e. g., EQs. 6 -7) were used in the model with an additional term of ε=-0.9 for an isotope fractionation for differential diffusion through the atmospheric/water layer and a second ε=-1.1 ‰ isotope effect due to hydration of pCO_2 (Mook et al., 1974).

3. Results

The model included fluxes of DOC to the estuary from both marsh outwelling and phytoplankton release. Although the model can place these fluxes anywhere along the estuary, we chose, for illustrative purposes, to place maximal marsh outwelling and phytoplankton release at the center of the 50 km estuary (Fig. 5). The flux of marsh DOC was set at 50 g-C m^{-2} y^{-1}, an intermediate value from Nixon (1980), and assumed to be composed of 60 % refractory carbon (Table 1). In turn, phytoplankton production was set at 100 and 250 g-C m^{-2} y^{-1} in the winter and summer, respectively. We set the labile and refractory DOC decay rates at 36 yr^{-1} and 1 yr^{-1} (Cifuentes and Eldridge, 1998). Finally, 20% of the phytoplankton production was released as labile DOC, (e. g., Lignell 1990). As shown in Fig. 5, the DOC flux into the estuary was dominated by marsh outwelling at the maximum marsh:estuary area (5:1), whereas marsh and phytoplankton fluxes were nearly equivalent at the minimum marsh:estuary area (0.5:1).

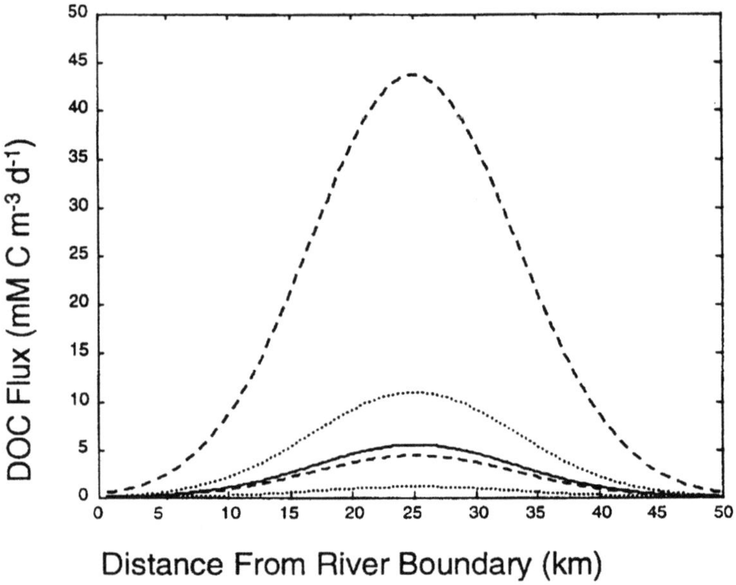

Figure 5. The outwelling of refractory and labile DOC had a Gaussian probability density function, which was maximal at the center of the 50 km estuary and extended over its length. At the 5:1 marsh:estuary area, marsh outwelling was by far the greatest source of refractory (upper dashed line) and labile DOC (upper dotted line). Phytoplankton-derived DOC (solid line) providing only a small portion of the total DOC. When the marsh to estuary area was only 0.5 to 1, refractory (lower dashed line) and labile (lower dotted line) were smaller than the input of phytoplankton-derived DOC.

Estuarine residence time was manipulated in the model by changing the depth of the estuary (width is always constant in this model). Changing freshwater inflow or dispersion would also have altered residence time, but this would have changed the input rate of allochthonous DOC. A constant input rate allows us to test the effect of marsh outwelling independently of residence time. The changes caused by deepening the estuary, however, influenced the size of the salinity zone (i.e., where salinity > 0). This produced a conundrum for our analysis. Initially, we numerically defined the spatial scale of the estuary with a finite difference scheme. However, the physics of the estuary changed the length of the salinity zone (e. g., Fig. 6), and, therefore, re-defined the dimensions of the estuary. This put some of the salinity zone above the riverine boundary set by the model. Because most of our results were referenced to salinity, it was important that we encompass most of the estuary within the model's riverine and coastal boundaries. By constraining the estuarine depth in all our analyses between 2 and 10 m, we were able to provide a reasonable compromise through which the salinity profile was generally within the bounds of the finite difference scheme, yet encompassed a reasonable range of residence times (Fig. 6).

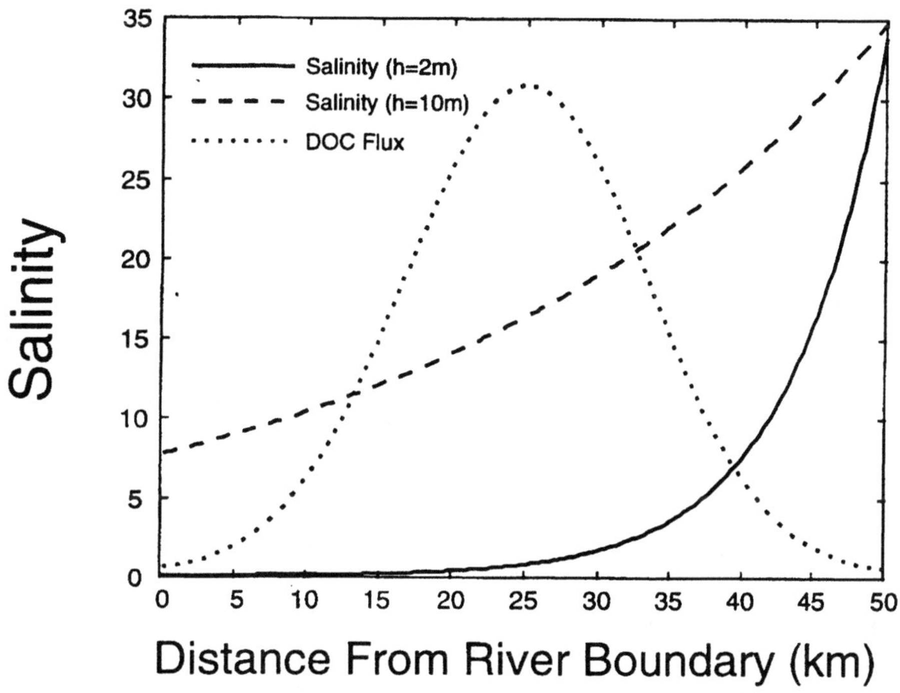

Figure 6. Changes in estuary depth influenced the extent of the salinity gradient. The analyses herein were constrained approximately to the 50 km basin by modelling depths (h) between 2 and 10 meters. With a constant eddy diffusion coefficient, the salinity gradient extended further up river in the deep estuary than in a shallow estuary causing the maximum marsh outwelling occurred at widely varying salinities in these simulations.

3.1 SALINITY PROPERTY ANALYSES

As an example, we show results for an estuary with a basin depth of 6.0 m (hydraulic residence time of about 0.3 y), a 1:1 marsh:estuary area, and a D/z of 0.005 cm s^{-1} (typical of calm conditions; Whitfield & Turner 1986). To demonstrate the influence that phytoplankton and/or marsh outwelling had on estuarine DIC and DOC concentrations and $\delta^{13}C$, we compared all model runs to a conservative case (i.e., only considers mixing of end-members).

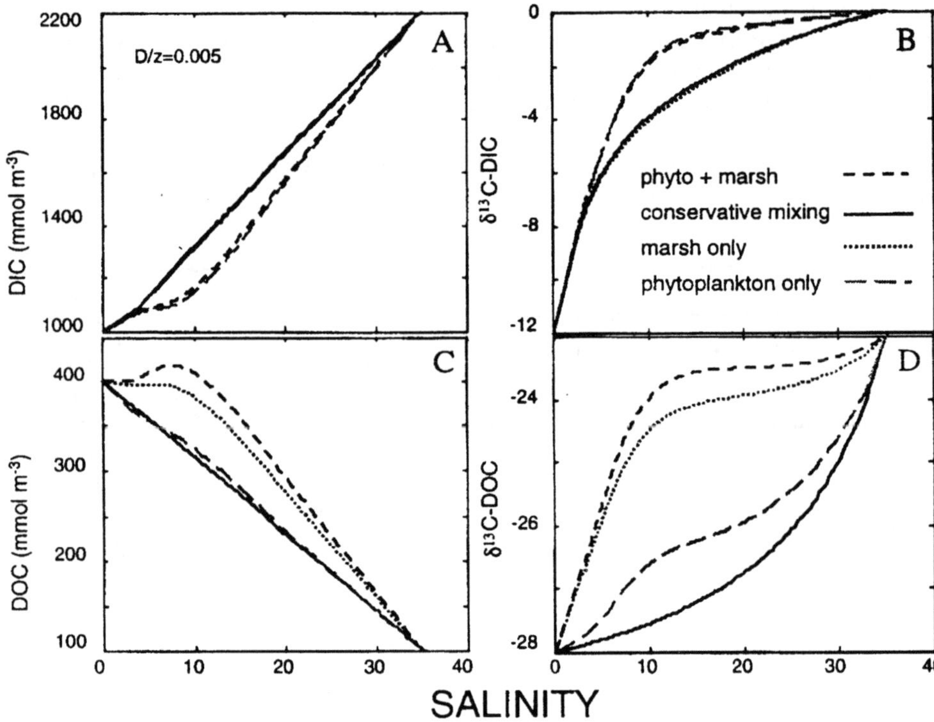

Figure 7. Model simulation with estuary depth of 6 m, equal marsh:estuary area, and D/z of 0.005. A. Shows [DIC]-salinity diagram. B. δ^{13}C-DIC-salinity diagram. C. [DOC]-salinity diagram. D. δ^{13}C-DOC-salinity diagram. Results of simulations that included marsh DIC outwelling (δ^{13}C= -14) equivalent to the DOC were nearly identical to those shown here.

For this analysis, we developed three cases describing marsh interactions with an estuary: 1) phytoplankton only (F_R); 2) marsh only (F_M) and; 3) combined marsh and phytoplankton ($F_{M,R}$) (Fig. 7). For both the (F_R) and ($F_{M,R}$) analyses uptake by phytoplankton depleted the [DIC] relative to the conservative analysis (Fig. 7A). Although this was expected, the actual depression in DIC concentration was greater than reported in the literature (e. g., Spiker and Schemel 1979, Fogel et al., 1992). Additional DIC sources such as benthic respiration, not considered in the model, could have accounted for this difference. The model may also have underestimated the DIC exchange with the atmosphere. Many large estuaries may have a higher D/z than we used (0.005 cm s^{-1}) or may even behave like an open system with respect to the atmosphere when conditions such as whitecaps are present (Whitfield and Turner 1986).

The F_M simulation did not produce enough respiration from outwelled organic carbon to alter the [DIC] salinity profile (Fig. 7A). Owing to the large size of the estuarine DIC pool, the moderate marsh outwelling had little effect on DIC dynamics. When we added DIC outwelling equivalent to the DOC outwelling the results remained the same. This was analogous to the minimal effect, discussed earlier, that marsh POC outwelling had on estuarine POC. The DIC isotope model showed a strong response to phytoplankton primary productivity with up to a $3^0/_{00}$ difference from the conservative case, but again with little difference between the F_M and conservative case (Fig. 7B). Because much of the marsh DOC is refractory, the resulting flux to the metabolizable DOC pool in the estuary was small relative to primary production and hence the enhancement of estuarine respiration due to marsh outwelling was also minimal. Based on these results we might expect that in estuaries in which *in-situ* and marsh sources alone provide most of the DOC, bacterial production will be low relative to primary production. Indeed, Heip et al. (1997) provide examples where the depth integrated ratio of bacteria to primary production (B:P) is low (e. g., 0.2 and 0.35 for the Chesapeake and Delaware Bays respectively). In contrast, the Hudson and Westerschelde, Netherlands provide examples of estuaries where allochthonous sources augment labile DOC pools to the point that they can support bacterial production levels significantly greater than primary production (e. g., B:P— 4.2 and 6.0 respectively).

In contrast to the DIC analyses, marsh outwelling had a precipitous effect on both [DOC] and isotope ratios. There was an enhancement of nearly 100 mmol m^{-3} in the [DOC] salinity profile while $\delta^{13}C$ increased by up to 4 ‰ (Fig. 7C and D). DOC release by phytoplankton caused less than a 20 mmol m^{-3} change in concentration and about a 1‰ change in the $\delta^{13}C$ relative to the other cases.

3.2 MODEL RESPONSE TO ESTUARINE RESIDENCE TIME AND MARSH OUTWELLING

It was not clear from the above analyses (Fig. 7) how the DIC and DOC concentration and isotope model results will change with respect to variations in marsh DOC outwelling and estuarine mixing times. To address these question we produced a matrix from 900 runs of the model for which we calculated the mean concentration difference as a percentage change (*mean* ΔDIC or ΔDOC) and the maximum difference in $\delta^{13}C$ (*max* Δ^{13}C-DIC or Δ^{13}C-DOC) from the conservative case (Figs. 8 and 9). We assembled simulations for both summer (primary production of 250 g-C m^{-2} y^{-1}) and winter (primary production of 100 g-C m^{-2} y^{-1}) conditions.

The analyses shown (Fig 8 and 9) include only marsh DOC outwelling. DIC outwelling may also occur, since much of the marsh photosynthesis is supported by atmospheric CO_2 while degradation of leaf litter occurs to some extent in the water. Simulations with and without DIC marsh outwelling were similar except at the extremes of marsh/estuarine ratio and are discussed below.

When only DOC marsh outwelling in considered, we found that over the range of residence times tested, the DOC concentration and isotope models essentially responded only to changes in marsh outwelling, exhibiting a nearly linear relationship between outwelling and both *mean* ΔDOC and *max* Δ^{13}C-DOC (Figs. 8 B and D; 9 B and D). The *max* Δ^{13}C-DIC decrease with residence time in response to reductions in the volume weighed rate of primary production (i. e., increased water depth with a constant rate of primary production). This occurred for both the summer and winter analysis albeit with smaller *max* Δ^{13}C-DIC's in the winter (Figs. 8 C and 9C). The *mean* Δ[DIC] predicted for both the summer and winter analysis were too small to be meaningful in the context of any actual measurement we could make (Figs. 8A and 9A).

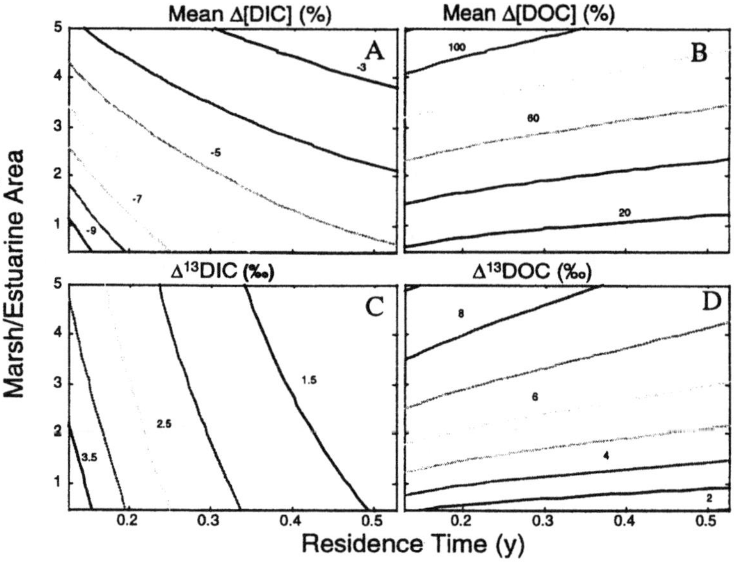

Figure 8. Contour plots showing the relationship between residence time on mean Δ[DIC] and Δ[DOC] concentration and maximum δ^{13}C-DIC and δ^{13}C-DOC within estuaries during the summer. Residence time was altered in this analysis by varying the estuary depth from 2 to 10 m. Primary production was set at 250 gC m^{-2} y^{-1}.

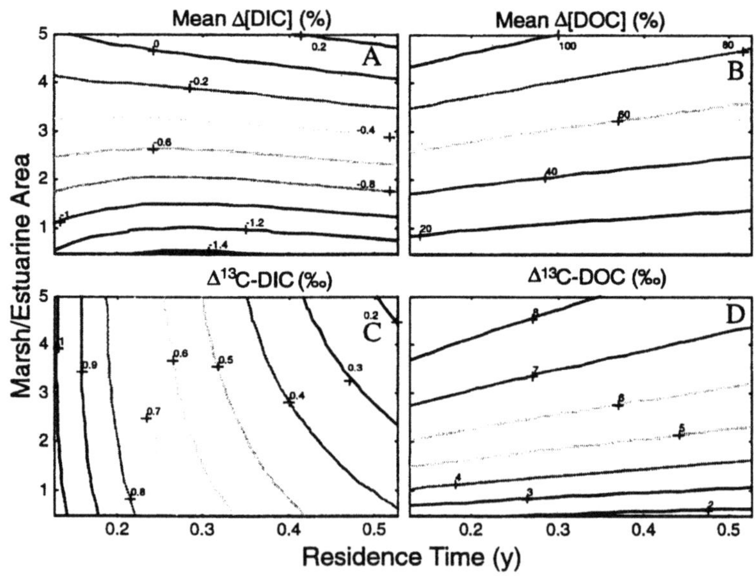

Figure 9. Contour plots as shown in Fig. 8, but with primary production set at 100 gC m^{-2} y^{-1} to simulate winter conditions

The inclusion of marsh DIC outwelling ($\delta^{13}C$= 14‰) resulted in enriched $\delta^{13}C$-DIC and increased [DIC] only at high marsh/estuary ratios and long residence times (results not shown). This DIC concentration and isotope effect diminished rapidly with increased atmospheric exchange (i.e., D/Z > 0.005). Our analyses suggest that the combination of both marsh DIC outwelling, if it occurs, and enhanced estuarine metabolism due to marsh DOC outwelling could result in a 1 to 3‰ enrichment of the estuarine DIC.

3.3 MARSH OUTWELLING BACK CALCULATED FROM ESTUARINE DOC DYNAMICS

In order to explain non-conservative [DOC]-salinity and $\delta^{13}C$-DOC-salinity profiles from several eastern estuaries, Peterson et al. (1994) suggested that phytoplankton and/or marsh fluxes were significant in some systems, whereas a dynamic cycle of DOC input and removal occurred in others. These authors described the correspondence between freshwater flushing and observed trends qualitatively, but did not explain how incorporation of mixing dynamics could be used to put constraints on the magnitude of sources and sinks within the estuary. Below, their data from the Parker River estuaries were re-examined with our DOC model. Admittedly, this analysis was based on a simplistic physical description (depth- and width-averaged) of these systems and assumed steady-state. However, the purpose of this exercise was to demonstrate how DOC-salinity and $\delta^{13}C$-DOC-salinity data, coupled with an adequate physical model, could be used to 1) estimate the size of sources and/or sinks, and 2) identify their origins.

TABLE 2. Parameters used in the model runs for the Parker river estuary.

System	L* (km)	Width† (m)	Depth‡ (m)	R* (m^3 s^{-1})	F_R ‡‡ (moles s^{-1})	Pe	$Area_M$/* $Area_E$
Parker	24.2	170	1.5	0.92	0.65	0.2	1.7

*Peterson et al. (1994)
†(Area (high water) - Area (low water))/2/L using data of Peterson et al. (1994)
‡Assumed value
‡‡Equivalent to R x [DOC] at S = 0

The Parker River estuary has a relatively low river inflow and substantial marsh area:estuarine area (Table 2). According to Peterson et al. (1994), this estuary receives significant DOC input from marshes. Their DOC-salinity (Fig. 10A) and $\delta^{13}C$-DOC-salinity (Fig. 10B) diagrams supported this hypothesis. With these data, we calculated a *mean* ΔDOC of +18.6% and a *max* $\Delta^{13}C$-DOC of +3.6$^0/_{00}$. The question is—what influx would explain these changes? Because the tidal range is substantial (mean range = 2.6 m; Wright and Coffin 1984) in the region, we assumed an intermediate eddy diffusion (Dyer 1974) coefficient of 100 m^2 s^{-1} in the model. The dependence between the amount of added DOC to the Parker River estuary and *mean* ΔDOC is shown in Fig. 11A. As expected, this relationship was linear whereas the response of *max* $\Delta^{13}C$-DOC was curvilinear (Fig. 11B). Based on Fig. 11A, about 1.95 molesC s^{-2}, or 3 times the riverine loading (Table 2), were needed to attain a *mean* ΔDOC of 18.6%. If this DOC originated from adjacent marshes, it would represent an annual influx of about 30 gC m^{-2} marsh, which is well within the range of values, 8 to 140 gC m^{-2} marsh y^{-1}, reported by Nixon (1980). Alternatively, phytoplankton release may have been the major DOC source. However, primary productivity of about 1300 gC m^{-2} y^{-1}, assuming a DOC release of 15 % would be required to produce a flux of 1.95 moles s^{-1}. This productivity is much higher than reported for these types of estuaries (Heip et al. 1995). Another argument against phytoplankton DOC release was found in the $\delta^{13}C$ data. As shown in Fig. 11B, even more DOC release (>>1.95 moles s^{-1}) by primary producers ($\delta^{13}C$= -20) would not lead to the *max* $\Delta DO^{13}C$ measured in the Parker River estuary. In contrast, the *max* $\Delta^{13}C$-DOC predicted by 1.95 molesC s^{-1} of marsh DOC ($\delta^{13}C$ = -14) was similar to the measured value (Fig. 11B). Thus, we agree with Peterson et al.'s (1994) qualitative assessment that marsh inputs explained non-conservative relationships of DOC and $\delta^{13}C$-DOC in the Parker River estuary. However, by analysing their data in the context of estuarine mixing, we obtained a quantitative estimate of the exchange rate (30 gC m^{-2} y^{-1}) that led to the observed concentration (Fig. 10A) and isotopic changes (Fig. 10B).

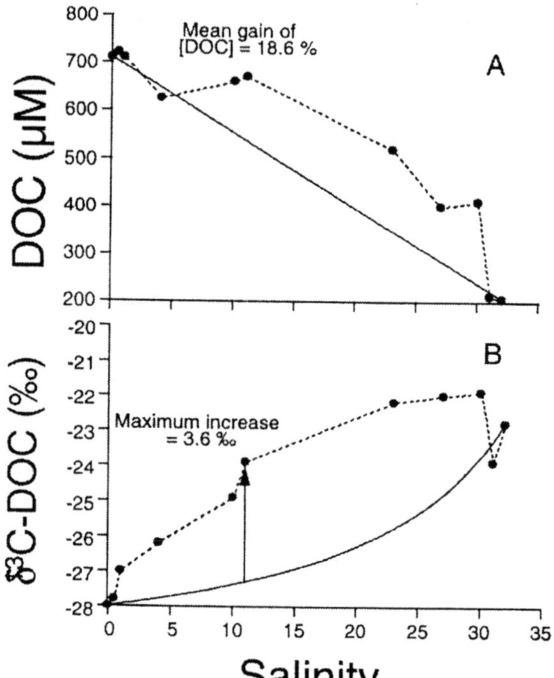

Figure 10. [DOC] versus salinity for (A) the Parker River estuary. $\delta^{13}C$-DOC versus salinity for (B) the Parker River estuary. Conservative mixing lines (solid) are included. Redrawn from Peterson et al. (1994).

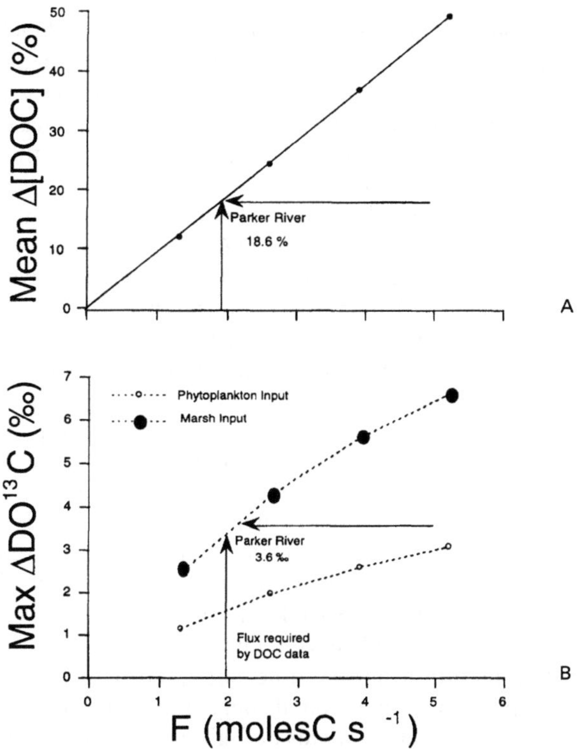

Figure 11. Model results for (A) *mean* ΔDOC and (B) *max* $\Delta DO^{13}C$ versus flux (moles s^{-1}) in the Parker River estuary. Values derived from the data of Peterson et al. (1994) for the Parker River estuary are included (refer to Fig. 10).

4. Conclusions

We have developed evidence through models and other data suggesting that measurable impact of marsh outwelling on estuarine carbon cycling can be best ascertained through analysis of [DOC] and $\delta^{13}C$-DOC. Analysis of other carbon pools provide useful information about autochthonous production within the estuary and may shed light on the effect of riverine loadings of carbon. The models showed, however, that only exceptional levels of marsh outwelling will produce a marsh signal in the estuarine POC. Nixon (1980) using a similar line of reasoning suggested this might be the case even with a mixture of 30 to 40% marsh material. This was not the case for DOC. Because phytoplankton generally release only about 10-20 % of primary production as DOC, their relative contribution of this pool is often smaller than marsh outwelling. Our simple box model approach suggests that a marsh signal would be found in the $\delta^{13}C$-DOC isotope ratio. This is not to say that marsh outwelling does not have an impact on the suspended POC pools, but instead that it will be difficult to ascertain the effect of marsh outwelling on this pool using $\delta^{13}C$-POC and concentration measurements. Our subsequent analyses, therefore, focused on developing a more refined DOC and isotope transport model including, primary production, marginal fluxes from marsh, and loading from the riverine and coastal boundaries. DOC dynamics are of course tightly linked to the CO_2 system through respiration and recycling. We assumed that if marsh DOC outwelling was reactive, a marsh signal might be seen in the DIC pool. The results of our analysis strongly suggest that this was not the case. The large size of the estuarine DIC pool caused this model to be insensitive to marsh outwelling. On the other hand the DIC isotope model was sensitive to primary productivity. Results suggest that up to a 3‰ difference from the conservative case may occur and that the analysis should be sensitive to spatial as well as temporal changes in production. Both the DOC isotope and concentration models were sensitive indicators of marsh outwelling. The combined models thus provide a nice description of both phytoplankton and marsh outwelling impacts on carbon cycling in an estuary.

In a third analysis we attempted to define how changes in the marsh to estuarine area ratio and the estuarine hydraulic residence time affected carbon pool size and isotope ratios. As might be expected from the earlier phases of the study, marsh/estuarine area substantially altered [DOC] and $\delta^{13}C$-DOC. [DIC] was not sensitive to changes in either marsh/estuarine area or residence time although $\delta^{13}C$-DIC decreased with longer residence time. This result may be partially an artefact of the way we increase residence time by deepening the estuary and hence reducing volume specific primary productivity. Other processes that modify residence time, such as mixing or riverine flow, could also effectively reduce primary production through nutrient limitation. This issue is still equivocal since primary production can be light as well as nutrient limited (Cifuentes and Eldridge submitted, Heip et al. 1995).

In the final analysis we back calculate marsh DOC outwelling based on changes in estuarine [DOC] and $\delta^{13}C$-DOC (data from Peterson et al., 1994). The resulting estimates were in the range predicted by Nixon (1980) and suggested that this simplistic model was a reasonable abstraction of a real marsh-estuary system.

5. Acknowledgments

This work was done before Peter Eldridge joined the U.S. Environmental Protection Agency. It has been subjected to the Agency's peer and administrative review, and it had been approved for publication as an EPA document. Mention of trade names or commercial products does not constitute endorsement or recommendation for use. We would like to thank three anonymous reviewers for many helpful suggestions to this manuscript.

6. Literature Cited

Boudreau, B. P. 1986. Mathematics of tracer mixing in sediments: I. spatially-dependent, diffusive mixing. American Journal of Science **286**: 161-198.

Cifuentes, L. A. and P.M. Eldridge. 1998. A mass- and isotope-balance model of DOC mixing in estuaries. Limnology and Oceanography **43**: 1872-1882.

Cifuentes, L. A. and P. M. Eldridge (submitted). How and where is net heterotrophy possible in estuaries? Limnology and Oceanography.

Coffin, R. B., L. A. Cifuentes, and P. M. Eldridge. 1994. The use of stable isotopes to study microbial processes in estuaries. Pages 222-239 *in* K. Lajtha and R. H. Michener, editors. Stable isotopes in ecology. Scientific Publications, London, England.

Craig, H. 1953. The Geochemistry of the stable carbon isotopes. Geochemica et Cosmochimica Acta. **3**: 53-92.

Currin, C. A., S. Y. Newell, H. W. Paerl. 1995. The role of standing dead *Spartina alternifora* and benthic microalgae in salt marsh food webs: considerations based on multiple stable isotope analysis. Marine Ecology Progress Series **121**: 99-116.

Dyer, K. R. 1974. The salt balance in stratified estuaries. Estuarine, Coastal and Marine Science **2**: 273-281.

Fogel, M. L., L. A. Cifuentes, D. J. Velinsky and J. H. Sharp. 1992. The relationship of carbon availability in estuarine phytoplankton to isotopic composition. Marine Ecology Progress Series **82**: 291-300.

Fry, B. and E.B. Sherr. 1984. $\delta^{13}C$ measurements as indicators of carbon flow in marine and freshwater ecosystems. Contributions in Marine Science **27**: 13-47.

Haines, E. B. 1976. Relationship between the stable carbon isotope composition of fiddler crabs, plants, and soils in a salt marsh, Limnology and Oceanography **21**: 880-883.

—— 1977. The origins of detritus in Georgia salt marsh estuaries. Oikos **29**: 254-260.

Heip, C. H. R., N. K. Goosen, P. M. Herman, J. J. Kromkamp, J. J. Middleburg, and K. Soetaert. 1995. Production and consumption of biological particles in temperate tidal estuaries. Oceanography and marine biology: an annual review **33:** 1- 149.

Lignell, R. 1990. Excretion of organic carbon by phytoplankton: its relationship to algal biomass, primary productivity and bacterial secondary productivity in the Baltic Sea. Marine Ecology Progress Series **68**:85-99.

Michener, R. H. and D. M. Schell. 1994. Stable isotope ratios as tracers in marine aquatic food webs. Pages 138-157 *in* K. Lajtha and R. H. Michener, editors. Stable isotopes in ecology. Blackwell Scientific Publications, London, England.

Mook, W. G., J. C. Bommerson, and W. H. Staverman. 1974. Carbon isotope fractionations between dissolved bicarbonate and gasous carbon dioxide. Earth Planetary Science Letters **22**: 169-176.

Nixon, S. 1980 Between coastal marshes and coastal waters - a review of twenty years of speculation and research on the role of salt marshes in estuarine productivity and water chemistry. Pages 437-525 *in* P. Hamilton and K. B. MacDonald, editors. Estuarine and wetland processes with emphasis on modeling. Plenum Publishing Corp., New York, New York, USA.

Peterson, B. J., R. W. Howarth, and R.H. Garitt. 1985. Multiple stable isotopes used to trace the flow of organic matter in estuarine food webs. Science **227**: 1361-1363.

Peterson, B. J. and R.W. Howarth. 1987. Sulfur, carbon and nitrogen isotopes used to trace organic matter flow in the salt-marsh estuaries of Sapelo Island, Georgia. Limnology and Oceaography **32**:1195-1213.

Peterson, B. J., B. Fry, M. Hullar, S. Saupe, and R.Wright. 1994. The distribution and stable carbon isotopic composition of dissolved organic carbon in estuaries. Estuaries **18**:111-121.

Spiker, E. C. and L. E. Schemel. 1979. Distribution and stable-isotope composition of carbon in San Francisco Bay. San Francisco Bay: the urbanized estuary. Pacific Division of the American Association for the Advancement of Science, San Francisco, California, USA.

Stumm W. and J. Morgan. 1981. Aquatic Chemistry: an introduction emphasizing chemical equilibria in natural waters, 2nd Edition. John Wiley & Sons New York, New York, USA.

Teal, J. M. 1962. Energy flow in the salt marsh ecosystem of Georgia. Ecology **43**: 614-624.

Whitfield, M. and D. R. Turner. 1986. The carbon dioxide system in estuaries — an inorganic perspective. The Science of Total Environment. **49**: 235-255.

Wigley, T.M.L., L.N. Plummer, and F. J. Pearson, Jr. 1978. Mass transfer and carbon isotope evolution in natural water systems. Geochimica Cosmochimica Acta **42**:1117-1139.

Wright, R. T. and R. B. Coffin. 1983. Planktonic bacteria in estuaries and coastal waters of northern Massachusetts: spatial and temporal distribution. Marine Ecology Progress Series **11**:205-216.

TYPES OF SALT MARSH EDGE AND EXPORT OF TROPHIC ENERGY FROM MARSHES TO DEEPER HABITATS

G. CICCHETTI
U.S. E.P.A
Atlantic Ecology Division
27 Tarzwell Drive
Narragansett, Rhode Island 02882 USA

R. J. DIAZ
School of Marine Science
Virginia Institute of Marine Science
Gloucester Point, Virginia 23062 USA

Abstract

We quantified nekton and estimated trophic export at salt marshes with both erosional and depositional edges at the Goodwin Islands (York River, Virginia, USA). At depositional-edge marshes, we examined trophic flows through quantitative sampling with 1.75 m^2 drop rings, and through gut content analyses of captured nekton. Six habitats were sampled on a transect from the marsh surface to the unvegetated subtidal. Consumption of animal prey by nekton, and export of trophic energy via transient nekton, was estimated for each habitat and for the entire marsh transect. The marsh edge was the habitat where we estimated the greatest contribution to export per m^2. An estimated 28.0 g dry weight of animal tissue was removed as prey per m^2 of depositional marsh edge into the open estuary over 150 days, primarily by blue crabs *Callinectes sapidus*. When the entire marsh was examined, however, marsh interior areas provided most of the trophic support for resident and transient species due to the greater area of the interior. When we considered the entire tidal cycle, the unvegetated intertidal area was also very productive, and contributed substantially to these trophic pathways. The blue crab *Callinectes sapidus* was the biomass dominant and probably the most important predator in all habitats. In a separate study, erosional-edge marshes facing open bays were examined with an 80 m^2 enclosure net that sampled the marsh edge and the adjacent unvegetated area. For these marshes we report a high biomass of larger transient species. Blue crabs were the biomass dominant at every sampling date (mean from June to September 1996, 0.31 inds m^{-2} and 2.81 g dry weight m^{-2}). A high biomass of transient fish species was also seen (mean from June - September, 0.90 inds m^{-2} and 1.25 g dry weight m^{-2}). We suggest this high biomass implies that these species receive an appreciable benefit from this habitat. The high biomass of predators also suggests the potential for export from erosional-edge marsh areas. Although the gear used to examine erosional and depositional marsh edges was clearly different, both types of edge saw considerable use by transient species. We therefore conclude that marshes with both erosional and depositional edges can export significant biomass to deeper water ecosystems as the consumption of marsh secondary production by transient nekton.

1. Introduction

The extent to which export from the salt marsh surface supports nearby subtidal ecosystems has been a topic of recent interest. Primary production on the surface of salt marshes is certainly very high, and has been estimated at 1150 gC m^{-2} yr^{-1} at the Goodwin Islands (York River, VA) near where our study took place (Buzzelli 1996). In general, marsh primary production must be processed by decomposers and detritivorous invertebrates before it is available to marsh nekton (Kneib 1997). Larger predatory invertebrates and fishes may then consume these decomposers and detritivores. Some of these predators are transient and leave the marsh system on seasonal or tidal time scales, exporting with them the energy obtained at the marsh. It is this form of export which we consider in this paper.

We divide nekton into two groups, marsh residents and marsh transients, in order to examine export of invertebrate biomass by predatory nekton. We define marsh residents as those species that spend their entire juvenile and adult stages within salt marshes and the immediately adjacent shallow water areas. We define marsh transients as species that spend only a portion of their life history within these areas. Marsh resident nekton such as fundulids and palaemonid shrimp are well adapted to remain on the marsh surface and in shallow water refugia (Kneib 1997) and may not export significant marsh energy into deeper water via their own migration (Currin et al. 1984). Marsh resident nekton may, however, contribute importantly to other ecosystems when eaten by predators from the deeper ecosystem (Kneib and Wagner 1994). Kneib (1997) extends this idea as a ''trophic relay'' where predation moves energy from shallow to deep waters through suites of progressively larger nekton. The full trophic relay was not examined in our study due to a lack of proper data on resident nekton as prey. Primarily, we investigated the pathway of trophic export from the marsh to the estuary via consumption of invertebrates on the marsh surface by transient predators. Export occurs as these predators leave the marsh system for the larger estuary.

Kneib (1997) suggested that emigration of transient species from marsh habitats holds a large potential to move production from the marsh surface into deeper estuarine and oceanic waters. Indeed, Deegan (1993) estimated that average export of gulf menhaden (*Brevoortia patronus*) out of a Louisiana estuary was 38 gdw (grams dry weight) m^{-2} yr^{-1} of fish biomass, calculated per area of marsh habitat. This represented about 5 to 10% of the primary productivity of the area (Deegan 1993) and shows not just an enormous abundance and biomass of fish, but also a tremendous export from the shallow marsh and bay to the coastal ocean. The Atlantic silverside, *Menidia menidia*, (which we consider a marsh transient) may also be a very important species in the transfer of energy from marsh and estuarine areas into the coastal ocean. This species spawns in shallow estuarine waters including the marsh surface (Fay et al. 1983). Silversides are preyed upon by a variety of predators within estuaries (Fay et al. 1983). Silversides also undertake a winter migration into the coastal ocean where they are further consumed by oceanic fishes, and experience considerable mortality (Fay et al. 1983). Fitz and Wiegert (1991) hypothesized that blue crabs, *Callinectes sapidus*, may also function as vectors of carbon transport from the marsh surface, though densities of crabs in their study were relatively small (40 to 50 individuals per hectare). Kneib (1982) suggested that blue crab predation on mummichogs may constitute a significant export of marsh production into deeper water. A growing number of studies are showing the importance of a variety of marsh transient species to the export of energy from marshes into deeper waters.

Trophic export from marshes has been relatively well studied in tidal creeks. Weinstein et al. (1984) estimated that production of spot (*Leiostomus xanthurus*) in polyhaline marsh tidal creeks of the York River (Virginia) was 4.6 gdw m^{-2} over 90 days. Seasonal export of this production from marsh creeks to deeper waters occurs as these fish migrate out of the creeks at the end of the season. In a different study, Weinstein et al. (1980) estimated that marine transient biomass present in tidal creeks at the end of marsh residency was about 1.51 gdw m^{-2} of low tide habitat. Kleypas and Dean (1983) suggested that silver perch feeding on palaemonids in intertidal creek areas transfer energy from marshes into the deeper estuary. Allen et al. (1995) documented a contribution of exported biomass in the form of marsh creek zooplanktivorous fishes. A direct connection to the marsh surface was suggested based on the presence of *Spartina*-consuming leafhoppers in guts of rough and Atlantic silversides. Bozeman and Dean (1980) showed considerable use of tidal creeks by larval fishes of 16 species, and suggested that the export of biomass from these habitats as fishes matured over a season may be a valuable contribution. Rountree and Able (1992a) found a tremendous biomass of nekton moving out of marsh creek areas with each tidal cycle. This tremendous preponderance of evidence leaves little doubt that salt marsh tidal creeks are very important conduits for the export of marsh-derived production.

Trophic export from the marsh to the estuary has been less well studied on the marsh surface and at open embayment sites than it has at marsh creeks. Marsh areas facing open water have been termed bay-marsh fringes by Rountree and Able (1992a). Our study was conducted at narrow strips of marsh facing open water; no creek habitats were sampled. We feel these areas represent an important baseline marsh system that, paradoxically, has not been well studied for export of marsh energy.

We carried out two separate projects to examine export at open embayment marshes: a depositional edge study in 1995, and an erosional edge study in 1996 (Fig. 1). Evidence suggests that erosional and depositional marsh edges are used differently by nekton. McIvor and Odum (1988) used flume nets to show that, in tidal freshwater creeks, depositional marsh edges were characterized by higher abundances of small fishes than were erosional marsh edges. While presence of SAV may also have played a role in these processes, experimentation showed greater infaunal food availability at depositional sites and higher levels of piscivorous predation at erosional sites (McIvor and Odum 1988). Hettler (1989) carried out a flume net study in a polyhaline creek system and suggested that gradually sloping depositional marshes offer a shallow water refuge from predation for small fishes. Furthermore, Hettler (1989) suggested that piscivores forage more effectively in deeper erosional-edge areas.

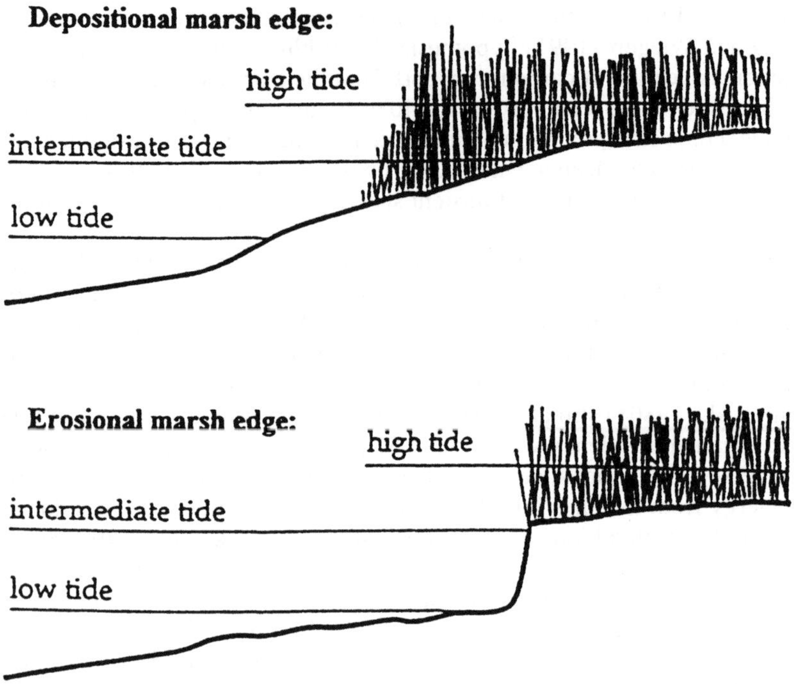

Figure 1. Depositional vs. erosional marsh edges. A depositional-edge marsh area was studied in 1995. Nearby erosional-edge marsh areas were studied using different gear in 1996.

Both of these studies indicate different patterns of fish use between erosional and depositional edges in marsh creeks, linked also to stream order in Hettler's study. Note, however, that Rozas (1992) found no significant differences in predation on tethered *Fundulus grandis* along different types of edge in Louisiana salt marsh channels. Rozas suggested that the difference in edge profile between sites might not have been sufficient (due to subsequent edge slumping) to cause significant differences in predation. In general, the literature suggests that nekton use of marsh edge habitat is linked to edge type. Our work examines marshes with both erosional and depositional edges.

2. Methods

2.1 SITE DESCRIPTION

The Goodwin Islands are a small group of uninhabited islands located at the mouth of the York River in lower Chesapeake Bay, Virginia. These islands are maintained by the Chesapeake Bay National Estuarine Research Reserve in Virginia. Tides are astronomically forced, and the marsh areas flood regularly. The mean tidal amplitude was 0.69 m during our study. Salinities ranged from 13 to 22 ‰, with a mean of 18 ‰ for the time period of our project (K. A. Moore, unpublished data). Summertime temperatures in the shallow embayments that surround this island can reach 30°C, and the shallow marsh areas sampled in this project occasionally experienced even higher temperatures. The sites selected for this work were narrow fringe marshes bordering a small embayment, and were exposed to moderate wave energy. Work was done in one embayment with a depositional marsh edge, and at several nearby areas with erosional marsh edges. The depositional marsh surface experienced a mean horizontal flooding distance of 16 m during the time period of the study, and a mean horizontal flooding distance of 23 m over the spring tide days when sampling took place. Erosional marsh areas flooded for shorter horizontal distances due to their higher elevation at the marsh-unvegetated interface, but this was not quantified. Vegetation at the marsh edge was tall form or short form *Spartina alterniflora* grading into short form *S. alterniflora* at a distance of no more than 2 m from the marsh edge. Short form *S. alterniflora* was replaced by a *S. patens/Distichlis spicata* mixture in marsh interior areas. No levee was present at the marsh edge. The gross morphology of this marsh is a type that is fairly common in this geographic region.

2.2 NEKTON SAMPLING: DEPOSITIONAL-EDGE MARSHES

Sampling for natant macrofauna at the depositional-edge marsh was carried out from June through October 1995 using 1.75 m^2 drop rings and throw rings. Habitats sampled at high tide included the marsh interior (3 to 15 m from the marsh edge), the band of marsh from 1 to 3 m from the edge (henceforth referred to as the ''marsh fringe''), the marsh edge itself (with the drop ring half on the marsh and half in the unvegetated area), and the unvegetated sand/mud area within 10 m of the marsh. Sampled low tide habitats included the 0 to 10 cm deep unvegetated shallows within 2 m of the water's edge, and the slightly deeper (10 to 25 cm of water) unvegetated shallows within 6 m of the water's edge.

The stratified random design for drop sampling thus considered one depositional marsh as a sampling area, with six habitats (four high tide, two low tide) as strata within this area. The sampling design was randomized spatially within the marsh area by dividing the area into sections, each of which included the six habitat strata. Sections and habitats (strata) were then selected for sampling using a random numbers table. Sampling was repeated every two weeks at daytime spring tides to minimize tidal and diurnal variation. Replication for the two week interval was achieved by sampling on three nearly consecutive days, with each of the six habitats (strata) sampled once each day. This was collapsed for analysis, and data are reported monthly, with each month represented by five drop samples per habitat. The marsh fringe habitat was not sampled in June.

In use, the 1.75 m^2 cylindrical sheet metal drop trap was deployed from a 3 m boom in front of a small boat. The trap was sufficiently heavy (80 kg) to cut through thick marsh vegetation and create a stable and effective seal with the sediment. A lighter (24 kg) shallow-water model of this ring was used as a throw trap where water depths excluded a boat; it also had a diameter of 1.48 m but was effective only in short form *Spartina alterniflora* or in unvegetated habitats because it lacked the weight to cut through heavily vegetated tall form *S. alterniflora*. Drop rings and throw rings were pounded into the sediment as necessary to ensure an effective seal. Rings were dropped or thrown in a different random location each time.

In order to empty these traps, a hinged rotating clearing device was inserted into the ring. This clearing device consisted of two halves connected at a vertical hinge. Each half had a width equal to the radius of the drop ring. One half acted as a stationary bag-like cod end (2 mm mesh) that sealed to both the drop ring and the substrate, and provided a perceived refuge for nekton to enter. The other half rotated on the vertical hinge in the center of the drop ring, traveling around the entire inner periphery of the ring. This rotating section pressed a rubber seal against the inside of the drop ring, and scraped the substrate with rake teeth spaced 6 mm apart. In use, the rake teeth were forced down into the substrate. As this rotating section raked the entire area of the drop ring, mobile creatures were scraped into the stationary bag-like cod end until the movable half was pressed tightly into the stationary half, trapping all creatures in the mesh bag and preventing escape. The entire clearing device was then lifted from the drop ring in this closed position, and all organisms were removed from the mesh cod end. This clearing device performed well in both unvegetated and vegetated habitats, removed samples rapidly, and could be worked through all the types of vegetation encountered at these sites. The sampled marsh was fairly flat, with few hummocks or pits that would have made sampling less effective. Results of gear testing for removal efficiencies are shown in Table 1. This gear worked well for our purposes in all sampled habitats.

All captured fishes and crustaceans were preserved immediately in the field using liquid nitrogen, stored in an ultracold freezer, then identified, enumerated, and measured in the lab. Weights for *Palaemonetes* shrimp and for blue crabs $\leq$30 mm were estimated with length-weight regressions; all other captured individuals were weighed directly as wet weight. Wet weights were converted mathematically to dry weights using information from Cummins and Wuycheck (1971) and Nixon and Oviatt (1973).

Gut contents were examined quantitatively for all captured fishes > 20 mm, all blue crabs > 30 mm, and for subsamples of *Palaemonetes*, subsamples of fishes $\leq$20 mm, and subsamples of blue crabs $\leq$30 mm. The contents of the stomach were examined for fishes having a distinct stomach. The contents of Section I of the gut (Babkin and Bowie 1928) were examined for fundulids, and the contents of the cardiac stomach ("gastric mill") were examined for crustaceans. The materials in these gut sections were removed and placed on a microscope slide. Percent composition by volume of all items in the gut section was then estimated using a grid on the stage of a dissecting microscope (Odum 1970). Percent volumes were converted directly into estimates of percent composition by weight, using the assumption that volumes of items in the gut are roughly proportional to weights (as in Swedberg and Walburg 1970).

A survey of the marsh surface determined elevations in July 1996 at 133 locations (total) on 11 transects at the sampling site. Tidal heights were recorded on each sampling day between June and October 1995 at the site. NOAA tide gauge data collected at Gloucester Point (10 km distant) was then adjusted to fit the tidal height data recorded at the site (n = 58, r-squared = 0.95). The adjusted tidal signal was referenced to the marsh elevation survey, and was used to calculate inundation times for the sampled habitats. All collected data were evaluated using computer modeling software as described below.

TABLE 1. Estimates of removal efficiences (R.E.) and confidence intervals (C.I.) for the device used to clear drop ring samples. In serial resampling (*), large numbers of individuals were placed in the dropring, and the clearing device was employed three times in succession. Statistics were derived using the Moran-Zippin method for estimating closed populations in repeated sampling without replacement, as described in Youngs and Robson 1978.

Group	**Size (TL or CW)**	**Habitat**	**Trials**	**Inds**	**R.E.**	**95% C.I.**	**Method**
fundulids, cyprinodontids	28 - 102 mm	*S. alterniflora*	8	106	84%	72 - 97%	mark-recapture
fundulids, cyprinodontids	28 - 102 mm	seagrass	5	80	93%	83 - 100%	mark-recapture
fundulids, cyprinodontids	28 - 102 mm	unvegetated	4	80	99%	95 - 100%	mark-recapture
Callinectes sapidus	50 - 100 mm	*S. alterniflora*	7	9	86%	63 - 100%	mark-recapture
Callinectes sapidus	3 - 30 mm	*S. alterniflora*	6	52	16%	4 - 29%	mark-recapture
Callinectes sapidus	3 - 30 mm	seagrass	4	31	39%	24 - 54%	mark-recapture
palaemonids	15 - 45 mm	*S. alterniflora*	3 x 3	2703	78%	63 - 93%	serial resample*
palaemonids	15 - 45 mm	unvegetated	3 x 3	428	72%	53 - 92%	serial resample*

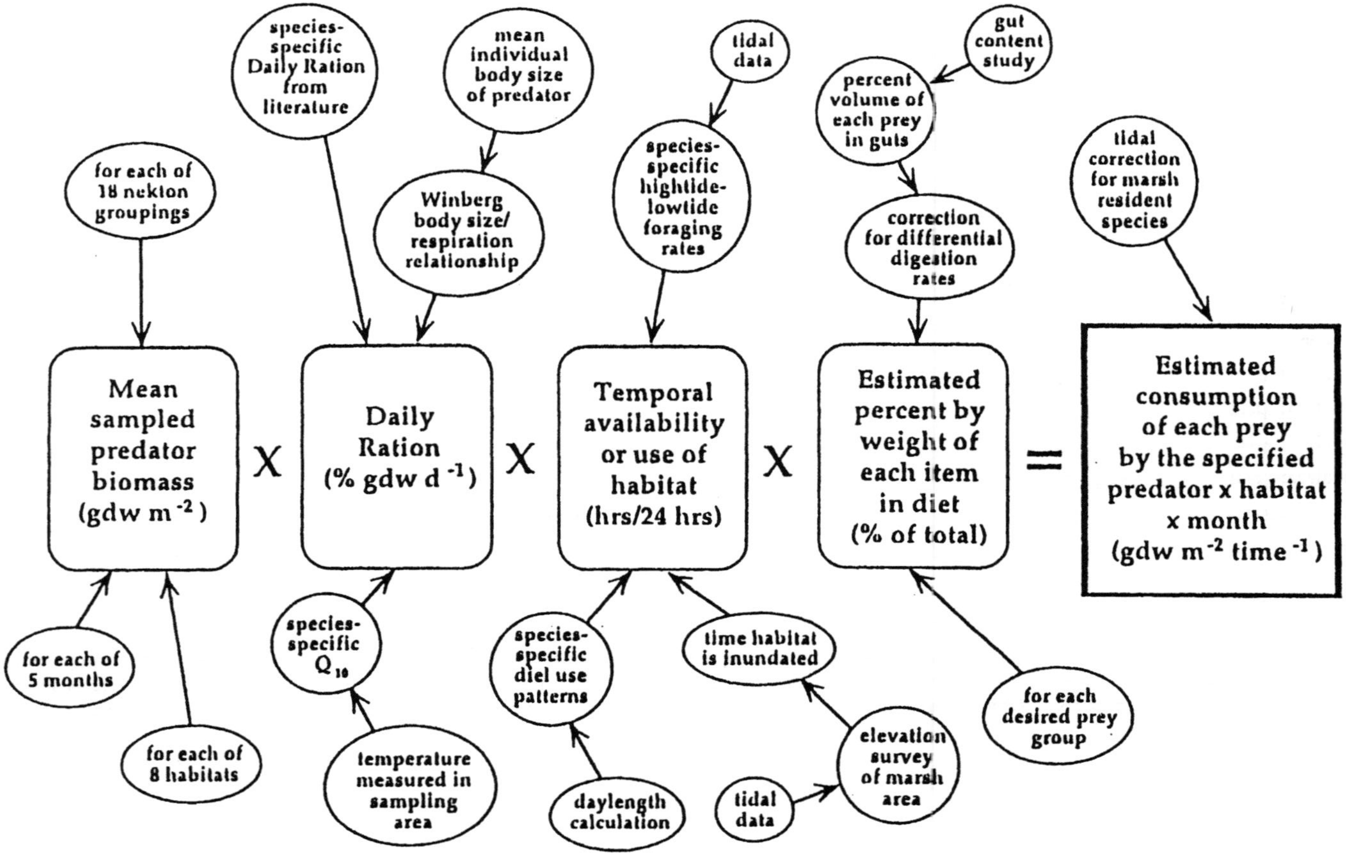

Figure 2. Diagram of model calculations and design. The calculation is repeated for each of the possible predator x prey x habitat x month combinations to provide monthly values for each. The monthly values are summed to yield total consumption over the five months of the study.

2.3 CONSUMPTION CALCULATION: DEPOSITIONAL-EDGE MARSHES

A dynamic calculation model was constructed to evaluate flows of energy from invertebrates to nekton in shallow water habitats at this depositional open-embayment marsh site. The model actually is very simple and in reality does only one thing: it estimates the consumption by sampled predators of each available prey in each quantified habitat. Fig. 2 shows a diagram of the calculations that are carried out to achieve this. The calculations of the model synthesize our data into a trophic description, and provide insight as to ecosystem connections. The model does not simulate or predict over time; it uses quantitative sampling data, gut content data, tide data, and bioenergetic information from the literature to calculate the consumption estimated to have occurred during the time period of sampling.

The model was written using the software package Madonna-PPC, version 5.8 (Macey et al. 1996). Fig. 2 shows the conceptual basis for the calculations of this model. These calculations were based on a daily ration for each predator, the sampled biomass of the predators, the percent by weight made up by each prey item in the diet, and the time allowed by tidal inundation for feeding in each habitat. Daily rations were taken from the literature (Table 2). These rations were adjusted to daily temperature using Q_{10} values from the literature. Rations for fishes were adjusted to body size based on Winberg's (1956) summary that the rate of metabolism in fishes is generally proportional to body weight raised to the power -0.2. Rations for decapods were similarly adjusted to body weight, using an exponent of -0.29 (Laird and Haefner 1976). Model forcing functions were temperature (measured at the site), tidal height (from adjusted NOAA tide data), and daylength (using an algorithm for the region from the literature). These forcing functions were then applied to the calculation of species-specific consumption. Output of the model for each of the 18 nekton species/size class groups for each of the 5 month x 6 habitat combinations was gdw consumed of each dietary item per m^2 per day. The unedited equations of the model and more detailed explanations are available in Cicchetti (1998).

2.3.1. Corrections for Differential Digestion

The differential digestion of prey items is always a concern in gut studies. Several authors have found that a variety of fish species digest small particles between 1.5 and 2.0 times faster than they digest larger and harder food items (Kionka and Windell 1972, Swenson and Smith 1973, Lankford and Targett 1997). The mean value of these three studies, 1.8, was used as a correction factor to account for slower digestion of large hard food items in all fishes. The literature also suggests that rates of digestion in fishes may be similar between different types of small prey items. (Kionka and Windell 1972, Weisberg et al. 1981). In particular, Kionka and Windell (1972) found that smaller chitinous parts were passed through the guts of rainbow trout at about the same rate as "digestible organic matter" and that large size of chitinous particles played a bigger role in impeding progress than did hardness alone. The correction for differential digestion in fishes, then, was to divide percents of gut content for large, hard-shelled prey such as crabs and mussels by 1.8 relative to other prey so as to approximate the actual percent of diet. Following this correction, all percents of gut content were then readjusted by the factor necessary to bring the total back to 100%.

Digestion in blue crabs (*Callinectes sapidus*) differs from digestion in fishes, in part due to the toothed, muscular cardiac stomach (gastric mill) of these crustaceans. Custer (1985) studied differential digestion in blue crabs and found no significant differences in rates of digestion of mussels (eaten with shell), shrimp, or whole fish for small and medium crabs, mean carapace widths 35 and 51 mm. These size classes (and smaller) made up the vast majority of crab numbers in our study. Custer noted that crabs regurgitated mussel shells within 2 to 6 hours of feeding, and that this accounted for the unexpected rapid removal of these hard items from the stomach. Custer also noted that, for small and medium sized crabs, soft flesh was always cleared within two hours while harder parts typically remained in the gut for closer to 6 hours. Based on this, we infer that digestion of soft organic tissue in crabs was perhaps three times faster than digestion of shell, carapace, and bone. Consequently, a correction was made for digestion of very soft prey (annelids, etc.) in crabs but not for digestion of any hard bodied creatures. It was assumed that soft unshelled organisms disappear from crab guts three times faster than do shelled organisms or the general matrix of refractory material that typified crab gut contents and this correction was applied. Based on Custer, no correction was made for differences in rates of digestion between any hard-shelled organisms and the general refractory matrix of crab stomach content. Correction factors were applied in the same way that corrections for fishes were applied.

Palaemonid shrimp also consume invertebrates (Sikora 1977, Morgan 1980, Kneib 1985). No information could be found dealing with differential digestion in these crustaceans, although McTigue and Feller (1989) suggest that differential digestion probably takes place in penaeid shrimp. The palaemonid shrimp we studied did not show a tremendous variety of food size and hardness in their diets. Most gut material was decaying organic detritus, at times mixed with algae or other vegetative matter. Common prey items were nereids and insect larvae. No hard parts of larger prey were seen. Based on the uniform consistency of the gut content of these animals, and on the lack of information in the literature, no correction factor for differential digestion in palaemonid shrimp was applied.

2.3.2. Tidal Correction for Marsh Resident Species

Marsh resident fishes live in habitats that expand and contract with the level of tide. As tide levels fall from slack high water, densities of marsh resident fishes increase per m^2 of still-available marsh surface. Drop samples taken near the marsh edge when the tide had fallen significantly from slack high generally contained extremely high abundances of marsh resident nekton (no samples taken at such tides were included in the data set used in this paper). This tidal concentration of marsh resident populations complicates the analysis of nekton consumption per m^2 of marsh.

Unfortunately, little quantitative information exists to describe differences in feeding on the marsh surface within the length of time that the marsh is inundated. It is not known to what extent nekton are actively feeding when they are compressed by tides into edge habitats. If an assumption is made that each individual feeds at the same rate at all times that any marsh surface is covered by water, then consumption in habitats near the edge is calculated to be unrealistically high due to this tidal population compression. It seems likely from this type of calculation that feeding cannot be uniform through the entire tidal cycle as residents are concentrated by falling tides. The data of Kneib and Wagner (1994) may suggest that larger marsh resident nekton feed more actively and efficiently in high marsh habitats when these habitats are flooded near slack high tide, but this is not described quantitatively in the literature.

Some mathematical adjustment to consumption must be made to account for movement of animals with the tide. Residents captured in the high marsh were assumed to move with the tides, and consumption was corrected evenly on all habitats using the correction factor which produced the result that each animal consumed an entire daily ration each 24 hours. This empirically-derived correction factor was remarkably similar for all species of fundulid at about 3.0, implying that fishes foraged three times more efficiently in the high marsh habitat, and that low marsh habitats experienced a three-fold increase in consumption as high marsh fishes were compressed into edge areas at early and late stages of high tide. This approach minimizes assumptions, maintains calculated daily rations, does not hide any information, and allows the presentation of results that are derived as much as possible from actual sampling. Note that these corrections were empirically derived for this marsh area as an adjustment to daily ration and will not apply to other marshes with different tidal regimes, differing availability of marsh surface refugia, and different edge-to-interior ratios.

The assumption that marsh resident fishes remain on the marsh surface for the entire period of inundation does not hold for marsh transient species, which are not captive to the marsh area within a tidal cycle. Unlike residents, transients may leave the marsh surface entirely rather than concentrate in edge habitats as the tide recedes. No tidal correction factors are applied to transient species; they are evaluated as a "snapshot" quantified at slack high tide. Since we define export as removal of prey by transients, our export estimates do not include any corrections for tidal habitat compression.

2.3.3. Corrections for Removal Efficiency

Results of the model for depositional-edge marshes have been corrected for removal efficiencies of the clearing device used to empty the drop rings. These efficiencies ranged from 16% for very small blue crabs in *Spartina alterniflora* habitat to 99% for cyprinodontids and fundulids in unvegetated habitats (Table 1). Efficiencies estimated for marsh resident fishes were applied to all fishes. Efficiencies for large crabs were assumed to be 86% in all habitats, though large crabs were tested for efficiency only in *Spartina alterniflora* habitat (Table 1). Juvenile crabs were not tested for removal efficiency in unvegetated habitat. This efficiency is assumed to be higher than the 39% and 16% estimates for juvenile crabs in seagrass and marsh habitats (Table 1). Because of this uncertainty, a range of values was generated from the model based on the assumption that removal efficiency for small crabs in unvegetated habitats was between 40% and 100%. All model calculations involving juvenile crabs in unvegetated habitats incorporate this range of estimates. In this way, model results for depositional-edge marshes described in this paper were corrected for the removal efficiency of the gear. This provides a more accurate picture of energy flow in the sampled habitats. Note, however, that none of the results reported for fishes and crabs at erosional-edge sites were corrected for gear efficiency.

2.3.4. *Assumptions and Validation*

It is important to bear in mind that any model involves certain assumptions, which must be considered as model results are evaluated. The following are among the assumptions and caveats incorporated into this model:

1. Populations are adequately described by the sampling results, as modified by gear correction factors for removal efficiency.

2. The same nekton access the marsh surface at neap tides (or at intermediate tides) and at sampled spring tides, the only difference being the amount of time available for nekton to access a habitat at each tide, due to different durations of flooding.

3. The gut contents of predators reflects feeding in the habitat where captured. This may not be true for individual predators. The assumption becomes more accurate on larger sampling scales, when means of gut information from many predators are considered.

4. The proportion of each prey item in the diet of each predator species over the 5-month period of the study can be described by a single value that represents the mean of all months.

5. Corrections for differential digestion, gear removal efficiency, and tidal population compression (as discussed above) are applied appropriately.

In spite of these assumptions, we feel that model output can be interpreted in a meaningful and informative way. The strength of this model is in the calculation of total consumption over the 150-day sampling period. Comparisons between prey type and habitat type within one month are more subject to error. For this reason, results are presented as total values for the entire five months of the study.

This is not a predictive or simulation model, but rather a descriptive model that estimates consumption of sampled predators using daily rations, Q_{10} values, and other parameters from the literature. Many of the assumptions of the model have to do with how marsh resident fishes partition their daily ration with regard to habitats, tidal cycles, and diel cycles. The major conclusions of the model can be traced back to applying daily rations to a sampled biomass of predators. This is the foundation of our model; it is our opinion that this premise is well established in the scientific literature. We did not feel that a separate study to ascertain whether predators consumed the daily rations suggested in the literature was necessary as validation. We did, however, conduct separate validation sampling to evaluate the model on the basis of spatial and temporal variability. Drop ring sampling was conducted at three different areas at the Goodwin Islands in 1996. Results from the area upon which the model was based fell between results from the two comparison areas, and no significant differences were found between the modeled area and either of the two comparison areas. A significant difference in total animal biomass was found, however, between the two comparison areas themselves. Another validation drop ring study compared the modeled marsh in

1995 and in 1996. This study found very similar total biomass between the two years, and found no significant differences in any of the abundant species except for small *Callinectes sapidus*, which were much less abundant in 1996. Both the area comparison and the year-to-year study are described in detail in Cicchetti (1998). Large year-to-year differences in *C. sapidus* recruitment have also been noted by Orth and van Montfrans (1987). We conclude from these investigations that our depositional-edge marsh model provides a robust description of these populations at the Goodwin Islands, with the caveat that *C. sapidus* abundances are subject to year-to-year recruitment variation.

2.4 NEKTON SAMPLING: EROSIONAL-EDGE MARSHES

Sampling areas were defined as stretches of erosional marsh edge at least 50 m long, separated from the next closest sampling area by at least 200 m, and at least 200 m from the nearest tidal creek. A total of ten areas on the southeast facing side of the Goodwin Islands met these criteria. Three of these ten areas were selected with a random numbers table in each of 4 months (June through September, 1996) for sampling. Each of the three areas selected in a month was sampled once, and all samples for the month were collected within a four-day period of spring tides. Thus, 3 samples were taken each month, and were treated as replicates for that month. All samples were collected during daylight. Sampling took place when the tide level was at the top of the erosional marsh edge. The exact sampling site (20 m of marsh edge) to be enclosed by the gear within a sampling area was determined using a random numbers table. To sample, a 30 m x 1.2 m heavily weighted net (6.3 mm mesh) with a large foamcore floatline was deployed from a small (1.5 m) catamaran equipped with a net spool. The catamaran and net were used to quantitatively enclose nekton on the erosional salt marsh edge and in the adjacent unvegetated intertidal/subtidal area. The net was stealthily deployed by quickly pulling the catamaran from the marsh edge out into the open water in an arc and back to the marsh edge using a 5 mm line. A 20 to 25 m section of erosional marsh edge and about 80 m^2 of adjacent unvegetated mud or sand was quietly enclosed in about 20 seconds. Sampled water depths ranged from 55 to 120 cm. Fishes and crabs were removed by closing in the net while raking animals along the marsh edge into a 1.2 m x 1.2 m x 2 m mesh box. Following this, a hinged door on the mesh box was raised to seal in the last section of the deployed net and the trapped nekton, then the entire mesh box was lifted from the water. This method worked well to capture larger fishes and crustaceans on erosional marsh edges.

3. Results

3.1 DEPOSITIONAL-EDGE MARSHES

A total of 30 species and 4048 individual animals were captured in 138 drop samples taken in six habitats between June and October 1995. Data are reported here as means over the five months of the study (or, in the case of consumption, as the estimate of total biomass consumed during these five months, 150 days). This provides a summary view of trophic export and allows a better examination of overall differences between habitats. Fig. 3 shows nekton abundance and biomass as well as total estimated consumption of all animal prey in each habitat over five months. This figure does not include organic detritus, vegetative matter, sediment, or other inorganic materials in reporting consumption; in every habitat this non-animal consumption was greater than consumption of animal tissue. Our interest, however, was in estimating the quantity of high quality marsh invertebrate production removed as export. Note that consumption in each habitat is influenced by the amount of time available for nekton to feed in that habitat, as determined by tidal and marsh elevation information. Nekton abundance, biomass, and consumption in the unvegetated habitat adjacent to the marsh in Fig. 3 is calculated as a tidally driven composite of information collected at high and low tide sampling.

Nekton boxes in this figure are arranged in order of decreasing biomass in each habitat. Each nekton box gives the mean abundance and biomass of that predator at high tide, corrected for gear removal efficiency (Table 1), over the five months of the study (23 to 25 drop samples per habitat). Species composition of the marsh resident group shown in Fig. 3 by percent of biomass (all habitats) includes: *Fundulus heteroclitus* (69%); *F. majalis* (23%); *Lucania parva* (7%); and *Cyprinodon variegatus* (1%). *Palaemonetes* shrimp are composed of 98% *P. pugio* in these habitats. Species composition of the marine transient box is described in the figure caption.

Fig. 3 shows that *Callinectes sapidus* makes up the largest fraction of biomass in each habitat (overall 62%); this species is also responsible for the largest part of consumption estimated in each habitat (overall 45 to 50%). This high biomass of crabs is also seen in certain other marsh surface studies (for example Rozas and Reed 1993). However, the biomass of crabs seen in our study at the marsh edge (mean 4.3 g m^{-2} dry weight even when uncorrected for gear efficiency, equivalent to 17 g m^{-2} wet weight) is very high in comparison to what has been reported in other studies. Many previous studies have not quantified crabs at the edge itself. Densities of other marsh fauna shown in Fig. 3 are within the ranges reported for other studies (Kneib 1997).

The question of trophic export from the depositional-edge marsh is addressed by dividing the biota into two categories, marsh residents and marsh transients. Export is defined as marsh secondary production (animal tissue) consumed by the sampled marsh transient nekton and subsequently removed from the marsh system. Fig. 3 thus only considers export by transient marsh fauna such as blue crabs and small transient fishes as they migrate on seasonal or shorter time scales into deeper waters. Fig. 3 does not account for export as resident species such as mummichogs are consumed by larger unsampled aquatic or avian predators. While larger fishes were effectively sampled at erosional marsh edges using the catamaran gear described above, they may not have been effectively sampled at depositional marsh edges. The scale of the drop ring gear that was used at depositional edges was probably not appropriate to the capture of these larger organisms.

Fig. 3 shows that, on a per m^2 basis, the marsh edge is the region of highest biomass at slack high tide, and is estimated to be the region of greatest foraging activity and trophic export per m^2. However, the marsh edge represents the smallest area of any sampled habitat. The proportional contribution of edge to the overall marsh dynamic is dependent on the area of marsh edge relative to the area of the entire marsh surface.

This ratio at our site can be seen in Fig. 4, which shows an averaged 1 m wide by 23 m long transect through the sampled Goodwin Islands site. The horizontal extent of marsh shown (23 m) is the mean flooding distance on spring tide sampling dates, so that the area of marsh shown in the figure is consistent with densities determined at sampling. The marsh flooded to an average distance of 16 m at mean high tide. Differences between spring and neap tide consumption are accounted for in Fig. 3 and Fig. 4, because the model includes a dynamic duration-of-flooding calculation based on tidal data.

Fig. 4 shows consumption of all animal prey by all predators for a 1 m wide strip of marsh habitat, as well as export of animal prey via blue crabs and transient fishes into the open estuary. This figure estimates a total of 82.4 gdw of animal tissue exported by marsh transient predation over five months per meter-wide transect of marsh, equal to 23 m^2 of area at spring high tide or 16 m^2 at mean high tide. This number is small in comparison to the primary production or standing plant biomass of the marsh, and it is also small in comparison to export by clupeids in some marsh systems (Deegan 1993). Nonetheless, this 82.4 gdw represents a clear biomass contribution to deeper water ecosystems. The sampled unvegetated intertidal area contributes another 80 to 117 gdw to export per 10 m x 1 m (Fig. 4), which includes feeding at high tide by species such as *Callinectes sapidus* and *Symphurus plagiusa*. Much of the contribution of this unvegetated area stems from consumption of small nereids (especially *Laeonereis culveri*) which were the biomass dominant group in this habitat during the year of the study (R. J. Diaz, unpublished data). In fact, infaunal standing stocks were greater in this unvegetated area than they were on the marsh edge or in the marsh interior (R. J. Diaz, unpublished data). The unvegetated intertidal area was estimated to be very productive at this site.

A high biomass of the transient crab *Callinectes sapidus* was documented at depositional sites; these crabs were the major exporters of animal tissue in our study. Few larger transient fish species were caught during the depositional edge study. It is unclear if these fishes were not present, or if the drop ring gear did not enclose enough area to capture larger, low-density fishes, or selected against them in some other way. A gear designed specifically to capture these fishes was used at erosional edge sites.

3.2 EROSIONAL-EDGE MARSHES

We found a high abundance and biomass of transient fishes and crabs on erosional marsh edges adjacent to the open bay using our catamaran-deployed enclosure net. A total of 919 marsh transient fishes and 331 blue crabs were captured from a total sampled area of 1084 m^2. All of these animals are visitors to the marsh system. We infer that these organisms can export substantial amounts of energy from these marshes. We investigated this area at the time that the flooding or ebbing tide had just reached the level of the marsh surface. The gear enclosed on average 3.7 m^2 of unvegetated area for every linear meter of marsh edge sampled. We report our data in Fig. 5 per square meter of sampled area, and per linear meter of marsh edge. We believe that the highest density of piscivores and crabs occurs at the erosional edge itself. The majority of crabs we observed while sampling were located on the marsh edge rather than in the adjacent unvegetated area, which was also enclosed by the sampling gear. A remote-recording underwater video camera study done at these sites in October 1997 showed that the vast majority of prey fishes (*Menidia menidia* and *Lucania parva*) were using the edge

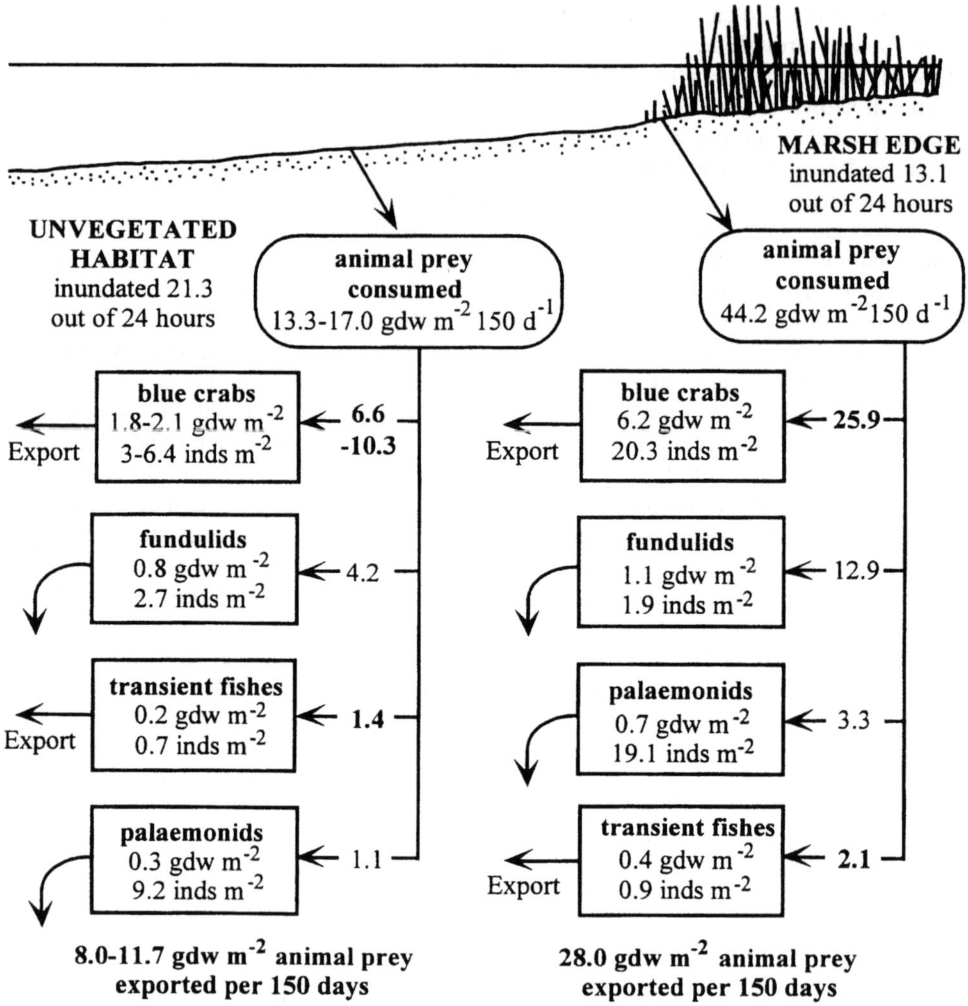

Figure 3. Energy flows from marsh and unvegetated habitats. Boxes show biomass (grams dry weight) and abundance per square meter of habitat at high tide, corrected for removal efficiency of the gear. Arrows show dynamically calculated flows in total grams dry weight of animal prey consumed per square meter over the entire tidal cycle during a 150 day sampling period, June - October 1995. Consumption exported via predation by marsh transients is shown in bold with a left-pointing arrow. Consumption by residents is assumed not to leave the marsh (note that this is not entirely true) and is shown with a downward pointing arrow. Ranges for blue crabs in unvegetated habitat provide high and low estimates of gear removal efficiency. Corrections for tidal habitat compression (see text) are applied to marsh residents only. Each marsh habitat represents means of 23 to 25 drop ring samples. The unvegetated habitat shows tidally averaged abundance, biomass, and consumption, created as a composite of 72 high and low tide samples. Marsh fringe was not sampled in June, and represents 20 drop samples.

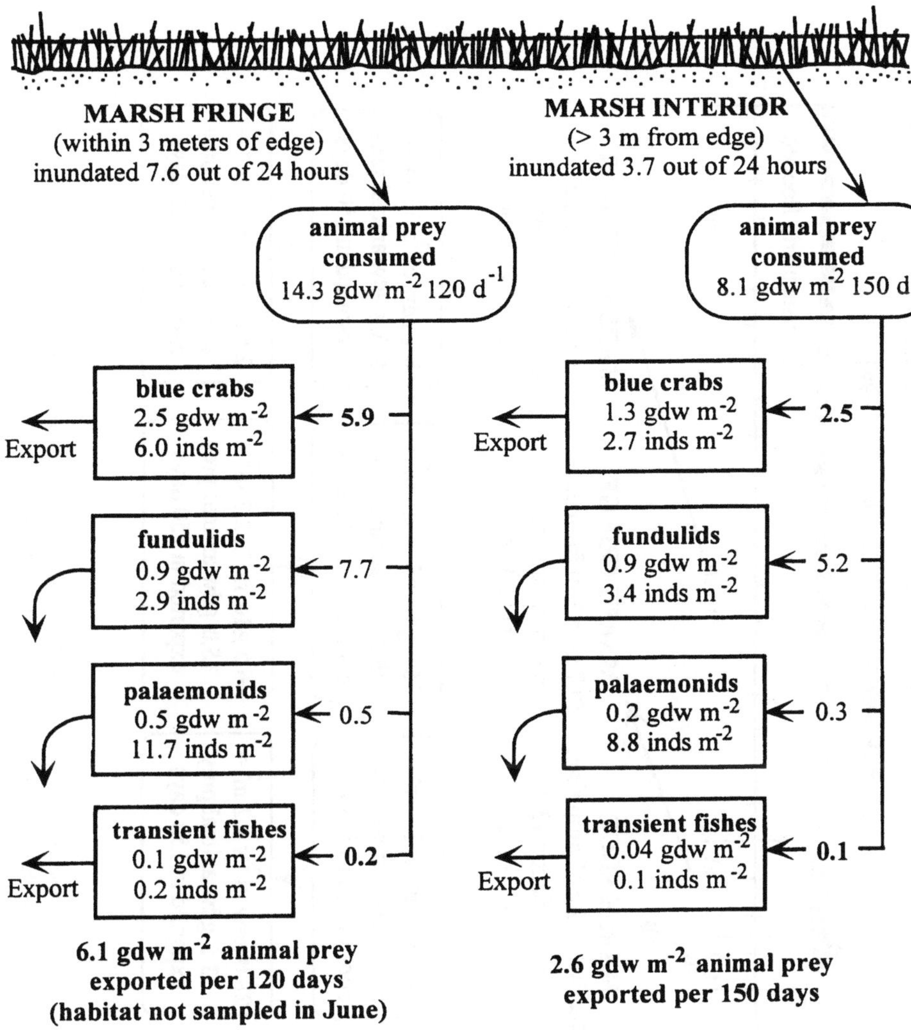

Figure 3, continued. Energy flows from marsh and unvegetated habitats. Marsh transient fish species composition, percent by weight: Unvegetated habitat: 54% *Mugil cephalus;* 22% *Menidia menidia;* 12% *Leiostomus xanthurus;* 11% *Symphurus plagiusa;* 2% *Gobiosoma bosc.* Marsh Edge: 30% *M. menidia;* 28% *Cynoscion nebulosus;* 21% *Bairdiella chrysoura;* 19% gobies (three species); 2% three other species. Marsh fringe: 100% *M. menidia.* Marsh interior: 77% *B. chrysoura;* 23% *M. menidia.* Species composition of other groups is described in the text. Approximate dry weight to wet weight conversions: for fishes, ww=5 x dw; for crustaceans, ww=4 x dw.

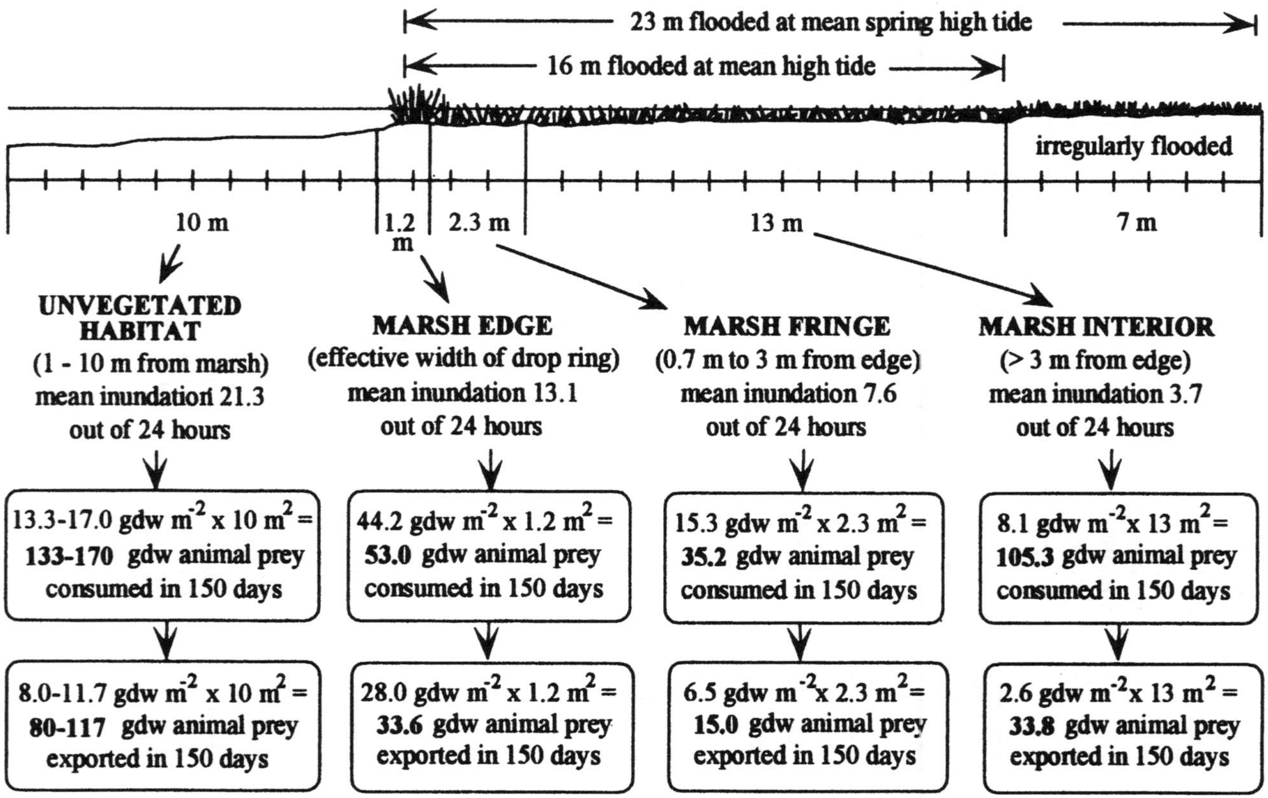

Figure 4. Estimated energy flows on 1 m wide mean transect of sampled marsh. Scale ticks below marsh profile are meters. Consumption and export estimates per square meter are taken from Figure 3. Marsh fringe habitat was not sampled in June but flow estimates here are extrapolated to provide values per 150 day time period.

TABLE 2. Daily rations of predatory nekton (percent of body weight consumed per day), taken from the literature. In calculating consumption, these rations were adjusted to temperature and animal body size as discussed in the text.

Species/group	Daily Ration	Details/Reference
F. heteroclitus	12.89	for 58 - 63 mm fish in summer based on lab studies at 20 degrees C, calculated by Weisberg and Lotrich (1982) using data in Weisberg et al. (1981)
L. xanthurus	7.7	for larger juveniles at 22 degrees C, from Brooks et al. (1981)
Large C. sapidus	7.9	for 60 - 130 mm CW crabs at 25 - 32 degrees C, calculated from data in Ryer (1987) using the method of Elliott and Perrson (1978) based on finding of exponential evacuation in Custer (1985)
Small C. sapidus	14.2	extrapolated from above to a 1 g wet weight (gww) crab using metabolism/size relationship in Laird and Haefner (1976)
P. pugio	21	for 6 - 52 mg dry weight shrimp at 31.9 degrees C, calculated by Morgan (1980) using data in Wood (1967)
Other fundulids	12.89	value for *F. heteroclitus* used as above
Other sciaenids	7.7	value for *L. xanthurus* used as above
Other palaemonics	21	value for *P. pugio* used as above
Other animals	12	for 1 gww organism at 20 degrees C, mean of all above values converted to 1 gww at 20 degrees C (this category makes up 3% of total biomass)

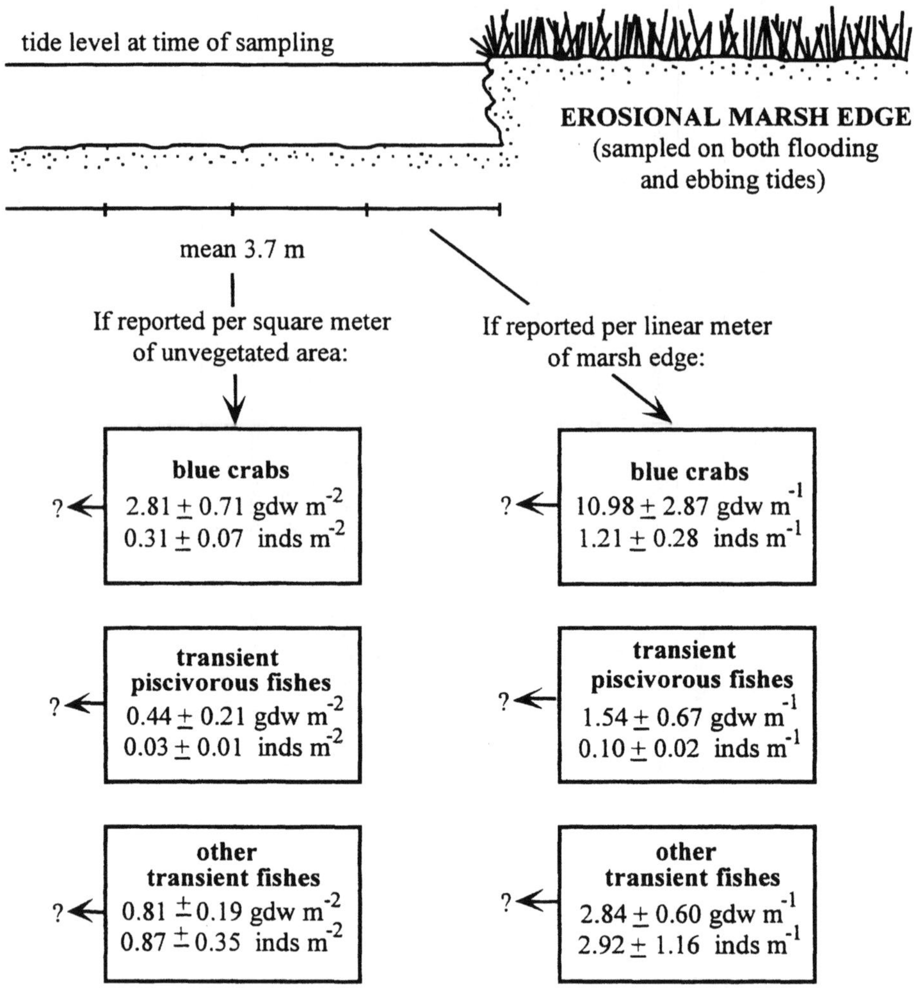

Figure 5. Nekton use of bay-exposed erosional salt marsh edges, means of 12 samples ± SE. Values are not corrected for gear removal efficiency. Question marks denote a lack of precise information about consumption rates and tropic export; first-order estimates are provided in the text. Species composition is also described in the text. Mean carapace width of blue crabs, 64.8 mm. Approximate dry weight to wet weight conversions: for fishes, ww=5 x dw; for crustaceans, ww=4 x dw.

itself and not the unvegetated area. We have no information about the distribution of larger fishes within the area sampled, but we suggest that piscivores may have been closely associated with prey species at the marsh edge.

Preliminary mark-recapture gear testing suggests a removal efficiency of about 50% for fishes and crabs, and reported numbers are probably underestimates for this reason. Because of small sample size in these mark-recapture tests ($n = 29$) we did not correct our reported numbers at erosional edges for gear efficiency. Quantifications of larger fishes and crabs are rare in these habitats, and few studies could be found for comparative purposes. Nonetheless, the biomass numbers we report (even uncorrected for gear efficiency) are high for larger fishes and crabs in any habitat in this region of lower Chesapeake Bay.

As can be seen in Fig. 5, biomass of blue crabs, *Callinectes sapidus*, is greater than of transient fish species. A mean of 0.31 crabs m^{-2} or 2.81 gdw m^{-2} was caught; 0.90 inds per m^2 or 1.25 gdw m^{-2} made up the catch of transient fishes. A total of 23 marsh transient fish species were captured, and the biomass dominant in the fish community was spot, *Leiostomus xanthurus* (27% of all fishes by biomass). These abundance and biomass figures almost certainly underestimate the true values, since numbers are uncorrected for gear efficiency and since reporting per m^2 includes considerable unvegetated habitat that we suspect was less utilized than marsh edge itself.

Marsh transient fishes were broken down into piscivorous fishes and non-piscivorous fishes (Fig. 5). A total of 29 piscivorous fishes were captured; by number of individuals the top three species were: bluefish (*Pomatomus saltatrix*) 26%; summer flounder (*Paralichthys dentatus*) 26%; and inshore lizardfish (*Synodus foetens*) 13%. By percent of piscivore biomass, this group included: bluefish 58%; summer flounder 14%; inshore lizardfish 10%; striped bass (*Morone saxatilis*) 9%; 6 other species 9%. A total of 890 non-piscivorous transient fishes were captured; by number of individuals the top three species were: Atlantic silverside (*Menidia menidia*) 49%; bay anchovy (*Anchoa mitchelli*) 28%; and spot (*Leiostomus xanthurus)* 16%. This non-piscivorous transient group included (by percent of biomass) spot 42%; Atlantic silverside 30%; gizzard shad (*Dorosoma cepedianum*) 14%; bay anchovy 7%; southern kingfish (*Menticirrhus americanus*) 2%; Atlantic croaker (*Micropogonias undulatus*) 2%; 6 other species 3%. A large number of transient species were found to be using this habitat.

With the exception of crabs, most of these species may not ever move onto the marsh surface; they nonetheless are using and exporting marsh production if they feed at the marsh edge. These species are what Rozas (1993) refers to as peripheral species and Peterson and Turner (1994) refer to as the marsh subtidal group. Unfortunately it is not possible for us to estimate consumption and trophic export for these fishes with the same precision as our depositional marsh estimates. More data on tidal use, more gut content information, and more flood vs. ebb sampling would be required for precise analyses. But, the high biomass of these predators in these areas seems to imply a trophic connection to marsh habitat. Since the erosional marsh edge is intertidal at these sites, fishes and large crabs are migrating with the tides into this habitat as it becomes available. Indeed, the erosional marsh edge seems to be a habitat of choice for many transient species at certain stages of tide. High animal densities, as we report from these areas, are generally taken as indicative of preferred habitat and of high habitat quality (Rozas and Minello 1998). A high density of transient species also implies a significant export from these areas.

In order to derive a crude estimation of this export from the marsh edge, we can consider only blue crabs and piscivorous fishes. We omit spot and other non-piscivorous transient fishes from this exercise since they may well have been feeding in the unvegetated habitat, not specifically on the marsh edge. The mean abundance and biomass of larger piscivorous fish species between June and September 1996 was 0.03 ± 0.01 (SE) inds m^{-2} and 0.44 ± 0.21 (SE) gdw m^{-2}. Gut examinations of piscivores showed evidence of feeding on mummichogs and palaemonids, but guts of many piscivores were empty (possibly due to vomiting at capture) and the gut content results were inconclusive. Assuming a 6% daily ration at 20°C, 75% of the diet being animal prey, 6 hours to feed at the marsh per 24 hour period, and a Q_{10} of 2.0, then 1.2 gdw m^{-2} of animal prey would have been consumed by these piscivores from June through September (120 days) in the areas sampled. If we use the same set of assumptions for crabs, but with 40% of the diet being animal prey (based on crab gut information from depositional areas, this study) then 3.9 gdw m^{-2} of animal prey would have been consumed per 120 days. The value of 5.1 gdw m^{-2} (or 18.87 gdw m^{-1} if reported per linear meter of marsh edge) is offered as a first order estimate of total predation over 120 days from June through September by piscivorous fishes plus crabs at an erosional edge site. This should be considered an underestimate since these numbers are not corrected for gear efficiency, which is suspected to be around 50%. Also, the consumption and export by non-piscivorous transient fishes would add to this estimate. Predation and export by larger transient piscivores does occur in marsh habitats, but is very difficult to quantify. Further study is needed to better clarify the link between these larger predators and erosional marsh edge habitat.

4. Discussion

Our study represents one of the first attempts in this geographic area to quantify the export of living animal tissue from the marsh surface. Trophic export from the depositional-edge marsh we studied made a significant contribution to deeper waters, with an estimated 82.4 gdw of animal matter being removed per 1 m x 16 m section of marsh over five months (See Fig. 4). Our results indicate a significant contribution of the marsh surface to deeper waters. This is true both in the quantity of the export and in the efficiency with which this energy is removed, given the limited time that the marsh surface is flooded. It is also clear from our data that a comparatively abundant and diverse group of transient fishes is selecting erosional marsh edges during certain stages of the tide. This use of erosional edge by larger fishes has been suggested within marsh creeks (McIvor and Odum 1988, Hettler 1989, Rozas 1992) but not for marshes facing open water, and is rarely quantified per m^2 as we do here. Creeks are clearly important pathways for larger transient fishes and crustaceans (Weinstein and Brooks 1983, Weinstein et al. 1984, Rozas and Odum 1987, Rulifson 1991, Rountree and Able 1992a,b). Use of marsh edge that is not in a creek by larger species is relatively unstudied, however. We suggest that bay-exposed erosional and depositional marshes also provide a pathway of energy flow from the marsh surface into deeper water via larger transient fishes and crustaceans.

It is important to bear in mind that our conclusions are based on a total of 138 drop samples at one depositional-edge area, and on a total of 12 erosional edge samples from a single island. It is clear from previous work that many factors affect nekton use of marsh edge. Variation in nekton use of marshes occurs along gradients of salinity (Weinstein et al. 1980, Rakocinski et al. 1992), substrate type (Weinstein et al. 1980), and depth (Rakocinski et al. 1992). This must be taken into account in any analysis. Our study is not meant to describe all marshes, but rather to provide one example of export processes from a single relatively pristine area. We have no reason to believe that our Goodwin Islands marsh represents a particularly unusual area. Abundances of marsh resident nekton that we report were within the ranges described in a review by Kneib (1997). In fact, marshes in other geographic locations may experience much higher animal abundances than we report here. Rozas (1993), in a review of published studies, concluded that densities of nekton on Gulf Coast marshes are at least an order of magnitude greater than on Atlantic Coast marshes. While densities as reported by Rozas (1993) may not correlate directly to consumption, production, or export, the tenfold difference in animal densities suggests that marshes from the Gulf Coast and certain other geographic regions may contribute at least as much production to deeper waters as does our marsh.

In interpreting these results, it should be noted that the depositional edge study is not directly comparable to the erosional edge study. The drop ring gear used in the depositional study, while very effective in capturing smaller fishes, did not capture larger fishes. It is unclear whether this was due to the absence of larger fishes at depositional marsh edges, or was due to sampling limitations brought about by the small size of the gear relative to these larger fishes. A gear capable of sampling much larger areas was developed specifically to capture larger transient fishes in a quantitative way. Unfortunately, this gear relied on the erosional marsh edge as a natural barricade to prevent escape of the enclosed fishes onto the marsh surface. The gear could not used on the depositional-edge marshes. Nor could drop rings be deployed at the steep erosional areas where the larger gear was used. For this reason, and because the two studies were conducted in different years, a direct comparison between erosional and depositional marshes is not possible. Nonetheless, conclusions can be drawn from our work at each type of marsh.

4.1 THE VALUE OF MARSH EDGE IN EXPORT TO DEEPER SYSTEMS

Many previous studies have shown that the edges of tidal marshes support a higher biomass and diversity of fishes and crustaceans than do areas in the marsh interior (Minello and Zimmerman 1992, Baltz et al. 1993, Minello et al. 1994, Peterson and Turner 1994). Rozas (1993), in a review of published papers, concluded that marsh edge was selected for by estuarine transients, including species of commercial value. Our study adds to this bulk of evidence, and expands the geographic regions in which these studies have been conducted. Much of the previous work has been in Gulf and South Atlantic marshes, which function differently than Mid-Atlantic marshes, in part because of the very different tidal regimes in each area (Rozas 1993, Kneib 1997). Our study examines a marsh transect in more spatial and temporal detail than have many previous works, quantifying nekton in four habitats at high tide and two habitats at low tide over a five month period. We have used our results to examine certain aspects of consumption and trophic export along this transect, which has not previously been done. This consumption and export analysis lends further support to the notion that edges are extremely valuable areas for salt marsh function.

It has been suggested that marshes with more edge habitat per unit area should support higher densities of estuarine nekton per total area, and that created marshes should attempt to maximize marsh edge by incorporating reticulation into design of constructed marshes (Minello and Zimmerman 1992, Peterson and Turner 1994). Our study supports these ideas. We also suggest that planning of constructed marshes should not underestimate the value of intertidal unvegetated areas. The adjacent unvegetated intertidal at our site was quite productive with regard to infauna (R. J. Diaz, unpublished data). Nekton biomass at these areas was high at low tide, and gut content work on captured nekton showed extensive feeding on infauna. Other unvegetated intertidal areas can also be quite productive. Diaz et al. (1982) found a very high annual average infaunal biomass of 236 g m^{-2} (wet weight) in a Virginia intertidal mud area, which was higher than values quantified by Diaz et al. (1982) in subtidal sand and seagrass habitats. Sarda et al. (1995) reported high infaunal biomass and production in the sandy organic unvegetated areas of a Massachusetts salt marsh. Buzzelli (1996) estimated that production of sediment microalgae at the Goodwin Islands was higher in the unvegetated intertidal than in seagrass or marsh surface habitats. Consequently, if a productive intertidal unvegetated area were replaced with a featureless high elevation marsh, the net trophic benefit to the deeper aquatic ecosystem might not be as planned. However, since marsh edge seems to be utilized to a greater extent than does either unvegetated intertidal or marsh interior, creation of an extensive marsh edge system should lead to a net improvement of ecosystem function as trophic support for deeper habitats.

4.2 THE VALUE OF MARSHES

Our study reinforces the generally-held belief that marshes contribute importantly to estuarine function. We examined marshes facing open water, where trophic export has been less well studied. Blue crab use of depositional marsh edges was seen to be a particularly important mechanism for movement of trophic energy off the marsh surface. The erosional marsh edge was also used extensively by blue crabs, and was selected for by a high abundance and diversity of transient fishes. The marsh surface provides primary productivity that fuels much of marsh secondary production, both as invertebrate biomass and as marsh resident fish biomass. The unvegetated area adjacent to the marsh was heavily used by nekton at low tide, and also contributed substantially to export from these shallow systems. The marsh surface, depositional marsh edge, erosional marsh edge, and adjacent unvegetated intertidal flat are linked ecosystems, in that resident and transient marsh nekton migrate regularly with the tides from one habitat to another and feed opportunistically in each. We believe that these intertidal habitats function to provide an important source of energy that is exported to deeper water ecosystems via predation.

5. Acknowledgments

Many people were invaluable in the design and completion of these studies, especially Jennifer Cicchetti, Ken Able, Jim Bauer, Randy Cutter, Beth Hinchey, Mo Lynch, Janet Nestlerode, Willy Reay, Linda Schaffner, Steve Schimmel, Treda Smith, Jacques van Montfrans, Eric Wooden, and Dave Yozzo. This work was supported by the Chesapeake Bay National Estuarine Research Reserve in Virginia, the William J. Hargis Award, the John and Marilyn Zeigler Award, and by VIMS Graduate Student Association minigrants. The erosional edge study was conducted under an award from the Sanctuaries and Reserves Division, Office of Ocean and Coastal Resource Management, National Ocean Service, National Oceanic and Atmospheric Administration. This is contribution number NHEERL-NAR-2156 of the U.S. EPA National Health and Environmental Effects Research Laboratory, Atlantic Ecology Division.

6. References

Allen, D. M., W. S. Johnson and V. Ogburn-Matthews. 1995. Trophic relationships and seasonal utilization of salt-marsh creeks by zooplanktivorous fishes. Environmental Biology of Fishes **42**:47-50.

Babkin, B. P. and D. J. Bowie. 1928. The digestive system and its function in *Fundulus heteroclitus*. Biological Bulletin **54**:254-277.

Baltz, D. M., C. Rakocinski and J. W. Fleeger. 1993. Microhabitat use by marsh-edge fishes in a Louisiana estuary. Environmental Biology of Fishes **36**:109-126.

Bozeman, E. A. Jr. and J. M. Dean. 1980. The abundance of estuarine larval and juvenile fish in a South Carolina intertidal creek. Estuaries **3**:89-97.

Brooks, H. A., J. V. Merriner, C. E. Meyers, J. E. Olney, G. W. Boehlert, J. V. Lascara, A. D. Estes and T. A. Munroe. 1981. Higher level consumer interactions. Volume IV *in* R. J. Orth and J. van Montfrans, editors. Structural and functional aspects of the biology of submerged aquatic macrophyte communities in the lower Chesapeake Bay. Special Report Number 267 in Applied Marine Science and Ocean Engineering, Chesapeake Bay Program, U.S. E.PA.

Buzzelli, C. P. 1996. Integrative analysis of ecosystem processes in the littoral zone of lower Chesapeake Bay: a modeling study of the Goodwin Islands National Estuarine Research Reserve. Doctoral Dissertation, The College of William and Mary, Williamsburg, Virginia, USA.

Cicchetti, G. 1998. Habitat use, secondary production and trophic export by salt marsh nekton in shallow waters. Doctoral Dissertation, The College of William and Mary, Williamsburg, Virginia, USA.

Cummins, K. W. and J. C. Wuycheck. 1971. Caloric equivalents for investigations in ecological energetics. Mitteilung Internationale Vereinigung fur Theoretische und Angewandte Limnologie No. **18**.

Currin, B. M., J. P. Reed and J. M. Miller. 1984. Growth, production, food consumption and mortality of juvenile spot and croaker: a comparison of tidal and nontidal nursery areas. Estuaries **7**:451-459.

Custer, K. J. 1985. Gut clearance rates of three prey species of the blue crab, *Callinectes sapidus* Rathbun. Masters Thesis, The Florida State University, Tallahassee, Florida, USA.

Deegan, L. A. 1993. Nutrient and energy transport between estuaries and coastal marine ecosystems by fish migration. Canadian Journal of Fisheries and Aquatic Sciences **50**:74-79.

Diaz, R. J., G. Markwith, R. J. Orth, W. Rizzo and R. Wetzel. 1982. Examination of tidal flats: Volume 1, Research Report. Final Report, Office of Research, Federal Highway Administration. Contract No. DOT-FH-11-9360.

Elliott, J. M. and L. Persson. 1978. The estimation of daily rates of food consumption for fish. Journal of Animal Ecology **47**:977-991.

Fay, C. W., R. J. Neves and G. B. Pardue. 1983. Species profiles: life histories and environmental requirements of coastal fishes and invertebrates (Mid-Atlantic). Atlantic silverside. U. S. Fish and Wildlife Service Biological Report FWS/OBS-82/11.10.

Fitz, H. C. and R. G. Wiegert. 1991. Utilization of the intertidal zone of a salt marsh by the blue crab *Callinectes sapidus*: density, return frequency and feeding habits. Marine Ecology Progress Series **76**:249-260.

Hettler, W. F., Jr. 1989. Nekton use of regularly-flooded saltmarsh cordgrass habitat in North Carolina, USA. Marine Ecology Progress Series **56**:111-118.

Kionka, B. C. and J. T. Windell. 1972. Differential movement of digestible and indigestible food fractions in rainbow trout, *Salmo gairdneri*. Transactions of the American Fisheries Society **101**:112-115.

Kleypas, J. and J. M. Dean. 1983. Migration and feeding of the predatory fish, *Bairdiella chrysoura* Lacepede, in an intertidal creek. Journal of Experimental Marine Biology and Ecology **72**:199-209.

Kneib, R. T. 1982. The effects of predation by wading birds (Ardeidae) and blue crabs (*Callinectes sapidus*) on the population size structure of the common mummichog, *Fundulus heteroclitus*. Estuarine, Coastal and Shelf Science **14**:159-165.

—— 1985. Predation and disturbance by the grass shrimp, *Palaemonetes pugio* Holthuis, in soft-substratum benthic invertebrate assemblages. Journal of Experimental Marine Biology and Ecology **93**:91-102.

—— 1997. The role of tidal marshes in the ecology of estuarine nekton. Oceanography and Marine Biology: An Annual Review **35**:163-220.

Kneib, R. T. and S. L. Wagner. 1994. Nekton use of vegetated marsh habitats at different stages of tidal inundation. Marine Ecology Progress Series **106**:227-238.

Laird, C. E. and P. A. Haefner, Jr. 1976. Effects of intrinsic and environmental factors on oxygen consumption in the blue crab, *Callinectes sapidus* Rathbun. Journal of Experimental Marine Biology and Ecology **22**:171-178.

Lankford, T. E., Jr. and T. E. Targett. 1997. Selective predation by juvenile weakfish: post-consumptive constraints on energy maximization and growth. Ecology **78**:1049-1061.

Macey, R., G. Oster and T. Zahnley. 1996. Madonna-PPC, version 5.8. YouSeeSoftware, Berkeley CA.

McIvor, C. C. and W. E. Odum. 1988. Food, predation risk and microhabitat selection in a marsh fish assemblage. Ecology **69**:1341-1351.

McTigue, T. A. and R. J. Feller. 1989. Feeding of juvenile white shrimp *Penaeus setiferus*: periodic or continuous? Marine Ecology Progress Series **52**:227-233.

Minello, T. J. and R. J. Zimmerman. 1992. Utilization of natural and transplanted Texas salt marshes by fish and decapod crustaceans. Marine Ecology Progress Series **90**:273-285.

Minello, T. J., R. J. Zimmerman and R. E. Medina. 1994. The importance of edge for natant macrofauna in a created salt marsh. Wetlands **14**:184-198.

Morgan, M. D. 1980. Grazing and predation of the grass shrimp, *Palaemonetes pugio*. Limnology and Oceanography **25**:896-902.

Nixon, S. W. and C. A. Oviatt. 1973. Ecology of a New England salt marsh. Ecological Monographs **43**:463-498.

Odum, W. E. 1970. Pathways of energy flow in a South Florida estuary. Dissertation, The University of Miami, Miami, Florida, USA.

Orth, R. J. and J. van Montfrans. 1987. Utilization of a seagrass meadow and tidal marsh creek by blue crabs *Callinectes sapidus*. I. Seasonal and annual variations in abundance with emphasis on post-settlement juveniles. Marine Ecology Progress Series **41**:283-294.

Peterson, G. W. and R. E. Turner. 1994. The value of salt marsh edge vs. interior as a habitat for fish and decapod crustaceans in a Louisiana tidal marsh. Estuaries **17**:235-262.

Rakocinski, C. F., D. M. Baltz and J. W. Fleeger. 1992. Correspondence between environmental gradients and the community structure of marsh-edge fishes in a Louisiana estuary. Marine Ecology Progress Series **80**:135-148.

Rountree, R. A. and K. A. Able 1992a. Fauna of polyhaline subtidal marsh creeks in Southern New Jersey: composition, abundance and biomass. Estuaries **15**:171-185.

—— 1992b. Foraging habits, growth and temporal patterns of salt-marsh creek habitat use by young-of-year summer flounder in New Jersey. Transactions of the American Fisheries Society **121**:765-776.

Rozas, L. P. 1992. Comparison of nekton habitats associated with pipeline canals and natural channels in Louisiana salt marshes. Wetlands **12**:136-146.

—— 1993. Nekton use of salt marshes of the Southeast region of the United States. Coastal Zone '93, Proceedings, 8th Symposium on Coastal and Ocean Management.

Rozas, L. P. and T. J. Minello. 1998. Nekton use of salt marsh, seagrass and nonvegetated habitats in a south Texas (USA) estuary. Bulletin of Marine Science **63**:481-501.

Rozas, L. P. and W. E. Odum. 1987. Use of tidal freshwater marshes by fishes and macrofaunal crustaceans along a marsh stream-order gradient. Estuaries **10**:36-43.

Rozas, L. P. and D. J. Reed. 1993. Nekton use of marsh-surface habitats in Louisiana (USA) deltaic salt marshes undergoing submergence. Marine Ecology Progress Series **96**:147-157.

Rulifson, R. A. 1991. Finfish utilization of man-initiated and adjacent natural creeks of South Creek Estuary, North Carolina using multiple gear types. Estuaries **14**:447-464.

Ryer, C. H. 1987. Temporal patterns of feeding by blue crabs (*Callinectes sapidus*) in a tidal-marsh creek and adjacent seagrass meadow in the lower Chesapeake Bay. Estuaries **10**:136-140.

Sarda, R., K. Foreman and I. Valiela. 1995. Macroinfauna of a Southern New England salt marsh: seasonal dynamics and production. Marine Biology **121**:431-445.

Shenker, J. M. and J. M. Dean. 1979. The utilization of an intertidal salt marsh creek by larval and juvenile fishes: abundance, diversity and temporal variation. Estuaries **2**:154-163.

Sikora, W. B. 1977. The ecology of *Palaemonetes pugio* in a southeastern salt marsh system with particular emphasis on production and trophic relationships. Dissertation, The University of South Carolina, Columbia, South Carolina, USA.

Swedberg, D. V. and C. H. Walburg. 1970. Spawning and early life history of the freshwater drum in Lewis and Clark Lake, Missouri River. Transactions of the American Fisheries Society **99**:560-570.

Swenson, W. A and L. L. Smith, Jr. 1973. Gastric digestion, food consumption, feeding periodicity and food conversion efficiency in walleye (*Stizostedion vitreum vitreum*). Journal Fisheries Research Board of Canada **30**:1327-1336.

Weinstein, M. P. 1979. Shallow marsh habitats as primary nurseries for fishes and shellfish, Cape Fear River, North Carolina. Fisheries Bulletin **77**:339-357.

Weinstein, M. P. and H. A. Brooks. 1983. Comparative ecology of nekton residing in a tidal creek and adjacent seagrass meadow: community composition and structure. Marine Ecology Progress Series **12**:15-27.

Weinstein, M. P., L. Scott, S. P. O'Neil, R. C. Siegfried II and S. T. Szedlmayer. 1984. Population dynamics of spot (*Leiostomus xanthurus*) in polyhaline tidal creeks of the York River Estuary, Virginia. Estuaries **7**:444-450.

Weinstein, M. P., S. C. Weiss and M.F. Walters. 1980. Multiple determinants of community structure in shallow marsh habitats, Cape Fear River Estuary, North Carolina, USA. Marine Biology **58**:227-243.

Weisberg, S. B. and V. A. Lotrich. 1982. Ingestion, egestion, excretion, growth and conversion efficiency for the mummichog, *Fundulus heteroclitus* (L.). Journal of Experimental Marine Biology and Ecology **62**:237-249.

Weisberg, S. B., R. Whalen and V. A. Lotrich. 1981. Tidal and diurnal influence on food consumption of a salt marsh killifish *Fundulus heteroclitus*. Marine Biology **61**:243-246.

Winberg, G. G. 1956. Rate of metabolism and food requirements of fishes. Fisheries Research Board of Canada Translation Series No. 194.

Wood, C. E. 1967. Physioecology of the grass shrimp, *Palaemonetes pugio*, in the Galveston Bay estuarine system. Contributions in Marine Science **12**:54-79.

Youngs, W. D. and D. S. Robson. 1978. Estimation of population number and mortality rates. Pages 137-164 *in* T. Bagenal, editor. Methods for assessment of fish production in fresh waters. Blackwell Scientific Publications, Oxford, England.

SILICON IS THE LINK BETWEEN TIDAL MARSHES AND ESTUARINE FISHERIES: A NEW PARADIGM

COURTNEY T. HACKNEY
LAWRENCE B. CAHOON
CHRISTIAN PREZIOSI
AMY NORRIS
University of North Carolina at Wilmington
Wilmington, NC 28403 USA.

Abstract

Tidal marshes accumulate large quantities of biogenic silica which accumulates in sediments, in plants and in porewaters. Higher salinity tidal marshes lower in the estuary contain higher levels of biogenic silica than those in the upper portion. Vascular plants, *Spartina alterniflora* and *Juncus roemerianus,* contain approximately 0.5% (by weight) biogenic silica, which does not vary along the estuarine gradient. The tidal marsh is an efficient biogenic silica trap, concentrating biogenic silica and dissolved silicate above levels found in nearby estuarine waters. High dissolved silicate levels provide ample silicon for benthic diatoms. We hypothesize that diatoms sequester silicate from surface waters supplied by tides and freshwater runoff. These benthic diatoms are eventually consumed and deposited into marsh sediments by macrofauna where their tests are dissolved into the marsh porewater. Vascular plants remove silicate from porewater and deposit amorphous silica into cell walls where it remains until liberated by decomposition into surface water where it can again be taken up by diatoms. Drainage of porewater containing high concentrations of dissolved silicate from marsh sediments is very slow, so flux of dissolved silicate from marsh sediments is uncoupled from the flux of surface water. Most porewater is released when tides are near low or rising resulting in the retention of silicate in the marsh/creek system. High concentrations of biogenic silica in tidal marshes are necessary for maximum benthic diatom production which in turn is necessary for high secondary production of commercial fish and crustaceans. Created tidal wetlands may require many years to concentrate biogenic silica sufficient to maintain high secondary production via the benthic diatom grazing foodchain.

1. Introduction

Recent studies on tidal marshes, diatoms, and silicon in the Cape Fear River estuary in North Carolina have prompted a reevaluation of the role tidal marshes play in the production of commercial and non-commercial marine species and the commercial fishing industry they support. This reevaluation has led to an alternative paradigm on the role of tidal marshes in secondary production and provided questions and testable hypotheses as to the value of created tidal wetlands in supporting secondary production.

1.1 HISTORICAL PERSPECTIVE

Tidal marsh plants were thought to be the energetic source for most secondary consumers through the detrital food chain (Teal 1962, Darnell 1967a,b, Odum and de la Cruz 1967). This original concept was challenged when analyses of stable carbon isotopes failed to find characteristic signatures of *Spartina alterniflora* (salt marsh cordgrass) in most estuarine fish and invertebrates (Haines 1977, 1979). Instead, carbon signatures resembled those of fishes offshore that primarily depend on phytoplankton production for their energy. The lack of a single energy source responsible for high secondary production in estuaries, coupled with conflicting and confusing data on flux of carbon from tidal marshes (Heald 1969, Day et al. 1973, Moore 1974, Shisler 1975, Settlemeyre and Gardner 1972, Axelrad 1974, Heinle and Flemer 1976, Hackney 1977, Woodwell et al. 1977, Happ et al. 1977) led to an assumption that there was no general model that related tidal marshes to secondary production. This implied that the energetics of each estuary were to some degree unique and dependent upon either detritus from marsh plants, algal production within estuaries, inputs of organic material from rivers, or some combination of these three energetic sources (Odum et al. 1979, Howarth and Teal 1979, 1980). Currently most estuarine scientists accept the argument that microalgae (benthic and planktonic) are the most important energetic source for estuarine food chains (Peterson and Howarth 1987, Sullivan and Moncreiff 1990) and consider the value of vascular plant production secondary in the energetic equation.

1.2 TIDAL MARSHES AS HABITAT

Other investigators argue a direct link between vascular plants in tidal marshes and secondary production, not just on an energetic basis, but through use of the tidal marsh as a structural habitat where feeding on microalgae can occur (Turner 1977, Boesch and Turner 1984). Much of the current emphasis of regulatory agencies on the restoration and creation of tidal marsh is based on this direct relationship between tidal marshes and the production of economically valuable fish and invertebrates. Thus, the clear establishment of a direct relationship is more than just an intellectual exercise. We offer here an alternative explanation of the direct relationship of tidal marsh production with secondary production, consistent with published data, but based on the ability of tidal marsh vascular plants to sequester and enhance internal recycling of both biogenic and aqueous silicate within the tidal marsh/creek system, thus stimulating production of estuarine diatoms.

1.3 DIATOMS AND SILICON

Are estuarine microalgae silicon limited in the intertidal habitat? In unvegetated intertidal habitat benthic diatoms can limit planktonic diatom production in the overlying water column through silicate uptake during at least half of the year (Sigmon and Cahoon 1997). Benthic microalgae are extremely rich in silicon (14.3:1 Si:Chlorophyll *a*) compared to their planktonic counterparts (2.8:1) because life amid sediment particles and in turbulent flow at the marsh surface requires armor, while planktonic forms must be only slightly silicified because of silica's density. Although silicate availability may limit diatom production in the water column, it is not thought to be a limiting nutrient for benthic microalgae because intertidal sediments contain high levels of dissolved silicate in substrate porewaters.

Diatoms on the marsh surface have several different sources of silicate available to them. First they are exposed to a continual supply of silicate from adjacent surface waters, albeit low in concentration, as water flows past them during flood stages of the tide. Diatoms on coral reefs where silicate concentrations are always low (<10 M) have an infinite supply if enough water moves past them (Adey 1987) and are capable of removing silicate at these low concentrations. A second source of silicate in marsh systems is fluxed directly from sediments due to the hydraulic head between the marsh surface and the lower surface of the non-vegetated, marsh edge. Diatoms intercept this highly concentrated silicate that would otherwise diffuse from sediments into the water column (Sigmon and Cahoon 1997). The time lag that results from typical tidal marsh sediments that have low hydraulic conductivity leads to the third mechanism that provides dissolved silicate to diatoms. A study on the flux of water through marsh sediments (Yelverton and Hackney 1986) demonstrated that most water enters the marsh porewater as vertical flux down into the marsh substrate at high tide and leaves as lateral flux usually through creek edges (Fig. 1). Flux of porewater is asynchronous with surface water flux adding small quantities of highly concentrated dissolved silicate to the floodwater that will cover the marsh and be available to diatoms.

Silicate concentrations of 15 μM, typical of estuarine waters, are not limiting for most planktonic diatoms (Hecky and Kilham 1988). As already noted, however,

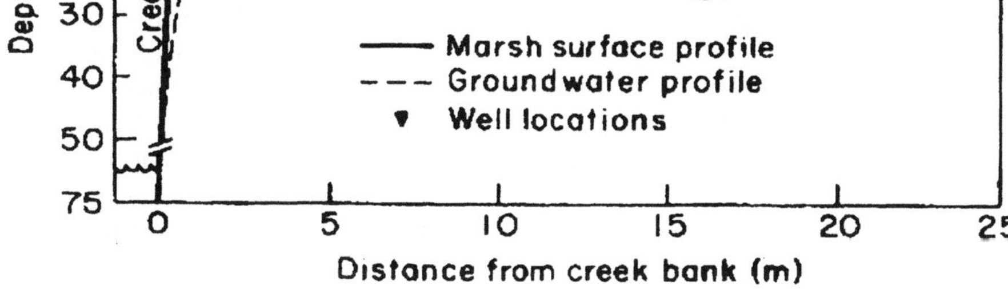

Figure 1. Cross sectional depiction of the maximum depression of pore water which occurs during each tidal cycle, and transport direction (arrows). Most water flux from the marsh occurs along the creek bank, but a small portion exits in depressions 25 cm below maximum marsh elevation on the marsh surface (from Yelverton and Hackney 1986). Inverted triangles on the marsh surface represent monitor piesometers.

benthic diatoms are more robust and require more silicon for their tests. We do not know their silicate requirements. Heins (1995) found that silicate additions increased growth of benthic diatoms in one North Carolina marsh subjected to increased phosphorus and nitrogen inputs, suggesting a silicate limitation. Low silicate concentrations may also alter the taxonomic composition of the benthic algal community as it does the planktonic community (Hecky and Kilham 1988) by favoring species that do not require silica, e.g., dinoflagellates.

1.4 VASCULAR PLANTS AND SILICA

For most of the upper intertidal habitat, however, vascular plants are a potential competitor of benthic diatoms for dissolved silica. Norris and Hackney (1999) found the two dominant vascular plants of western Atlantic tidal marshes (*S. alterniflora* and *Juncus roemerianus*) highly siliceous (around 0.5% silica by weight). These plants are capable of absorbing and retaining large quantities of dissolved silicate during the growing season. However, vascular plants likely obtain their dissolved silicate from sediments where roots are located (usually within the top 30 cm of sediment). Spatial separation of vascular plant roots and diatoms, and the very high concentrations of biogenic (available) silica within tidal marshes, make it unlikely that silicate is limiting to either primary producer. As an example, one marsh in the Cape Fear River estuary contained 89 g $m^{-2}$of biogenic silica in plants and sediments (Norris and Hackney, 1999 (Fig. 2), in addition to porewater silicate concentrations as high as 687 µM.

1.5 SILICATE AND TIDAL MARSHES

Are high biogenic silica levels common in all marshes? Do vascular plants in other marshes also take up large quantities of silicate? Numerous studies have demonstrated dissolved silicate concentration gradients of surface waters in estuaries due to conservative mixing of silicate-rich river water with ocean waters containing low silicate concentrations (e.g., Bien et al. 1958, Stephens and Oppenheimer 1972, Liss and Pointon 1973, Boyle et al. 1974, Calvert 1983, Blanchard 1988, Mallin et al 1997). If this relationship follows in tidal marshes, then the low salinity marsh studied by Norris and Hackney (1999) should contain more silicate and biogenic silica than more saline marshes.

However, silicate content of porewater and sediments do not exhibit an inverse relationship with salinity (Fig. 3) (unpublished). Instead, sediments in saline marshes contained high concentrations of biogenic silica and porewaters high concentrations of silicate (unpublished). Vascular plants from marshes along the same estuarine salinity gradient exhibited essentially the same concentrations of biogenic silica in their tissues even though less was available in sediments of lower salinity marshes suggesting an active uptake mechanism.

If diatoms are the foundation of estuarine and salt marsh food chains as stable isotope studies suggest (Peterson and Howarth 1989, Sullivan and Moncreiff 1990), then these marshes can supply a large quantity of silicon, an essential nutrient, to benthic diatoms on the marsh surface. The processes outlined here are consistent with studies demonstrating that the Cape Fear tidal marshes provide a substantial energetic base (Cammen et al. 1982) and support a large recreational and commercial fishery.

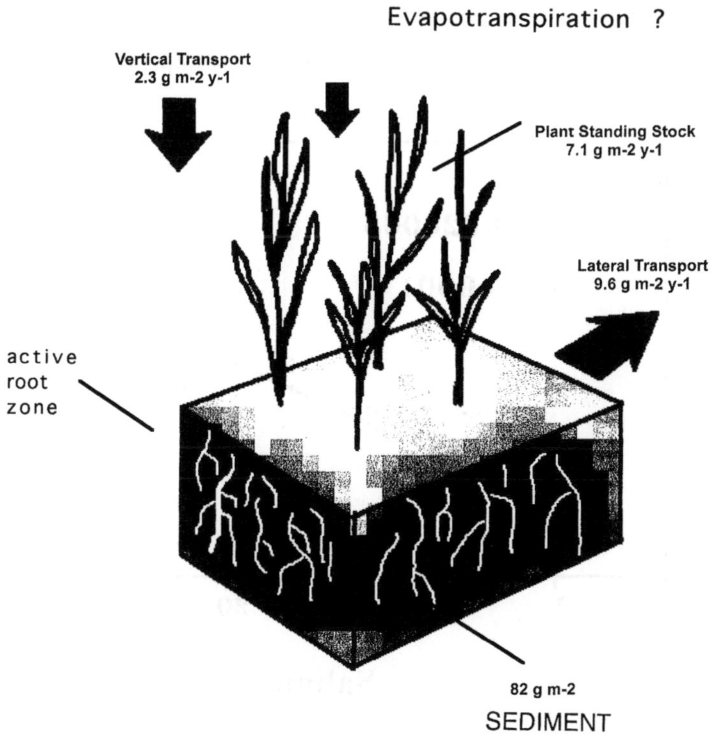

Figure 2. Section of mesohaline tidal marsh in the Cape Fear estuary showing accumulation of biogenic silica from each year's production of vascular plants and the biogenic silica pool (sediments and pore water) in the 30 cm root zone per mean square meter of marsh (unpublished). Annual flux of biogenic silica was obtained using concentrations of pore water and surface water and annual flux rates of water (Yelverton and Hackney 1986). The relationship to evapotranspiration is not known as some of the water which enters pore space at high tide is likely lost through evapotranspiration, not to lateral flux. This diagram assumes that all water that enters is lost as lateral transport.

2. Tidal Marshes as a Silicon Pump

What is the source of biogenic silica and how is it available to surface dwelling diatoms? We hypothesize that the tidal marsh and its vascular plants concentrate bioavailable silicon and make it available to diatoms through the following process. Each tide brings dissolved silicate onto the marsh as well as particulate matter containing biogenic silica (silica tests, plant debris, etc.). Dissolved silicate is either removed by benthic diatoms or added to the pore water. Diatoms turn over rapidly on the marsh and are consumed by a variety of macrofauna that deposit feces and pseudofeces laden with diatom tests in sediments where tests dissolve. Vascular plants take up silicate from the silicate rich porewater and transport it to leaves and stems where it is incorporated into tissue and remains sequestered until the plant decomposes. Decomposition begins on dead plants while they still stand and continues when leaves and stems fall to the marsh surface. There, silicate is available to the highly siliceous benthic diatoms.

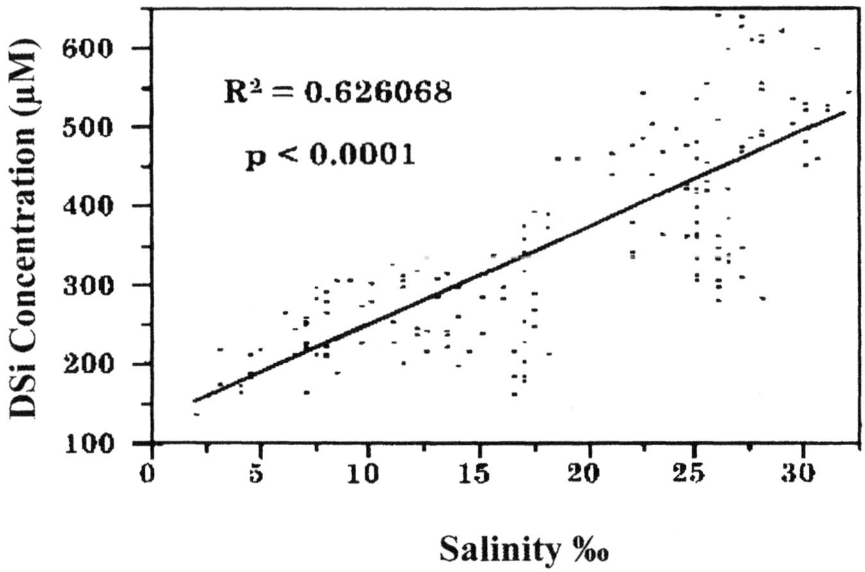

Figure 3. Dissolved silicate (mean for all seasons) versus salinity for pore water at three tidal marshes along a salinity gradient in the Cape Fear estuary (unpublished). d.f. = 152. DSi concentration (µM) = 129.377 + salinity (‰).

In addition to this internal cycling between aqueous phases (surface water and porewater) and amorphous phases (biogenic silica of diatom frustules and plants), there is another mechanism that retains silica within the marsh creek system. Porewater dissolved silicate not taken up by vascular plants flows laterally through marsh sediments (Fig. 1). The marsh operates as a leaky dam (Yelverton and Hackney 1986), releasing porewater slowly as the tide falls (Fig. 4). Instead of dissolved silicate exiting sediments and being carried from the marsh creek system, it is retained because most of this water does not leave marsh sediments until the tide has already begun to rise. As a result, dissolved silicate concentrates and is highly conserved within the marsh/creek system (Fig. 5).

The apparent relationship between salinity and the concentration of dissolved silicate in porewater and biogenic silica in sediments may be an artifact of the position of tidal marshes in the estuary. Sixteen percent of biogenic silica leaving rivers is in particulate form (Conley 1997) and the presence of biogenic silica may relate to greater transport of particulate matter into marshes lower in the estuary. High water clarity in marshes lower in the estuary may fuel higher diatom production and consequently lower loss rates of silicate from the marsh.

3. Relevance to Created Wetlands

Not all tidal marshes may be equivalent energy sources for secondary production, especially if the time required to accumulate biogenic silica is significant or if accumulation rates differ with salinity or position in the estuary. This is especially true for created tidal wetlands. The first created tidal marsh we examined near Morehead City, North Carolina (5 years old and fed only by tidal flux) contained an average of 55.5 $\pm$ 9 μM dissolved silica in porewaters. Created tidal marshes with upland drainage originating from deposited dredge material contained more dissolved silicate (204 $\pm$ 29 μM), whereas natural tidal marshes reported here ranged from 130 to 650 μM dissolved silicate. Design specifications can lead to marshes that contain the same structure as natural marshes, but lack the important diatom component because silicate is limited. Created tidal wetlands could be augmented with forms of diatomaceous earth or other biogenic silica additions, e.g., marsh wrack, to simulate older marshes. Similarly, sediment type used, and/or tidal elevation may enhance silicate conservation, thus enhancing the value of created tidal marshes.

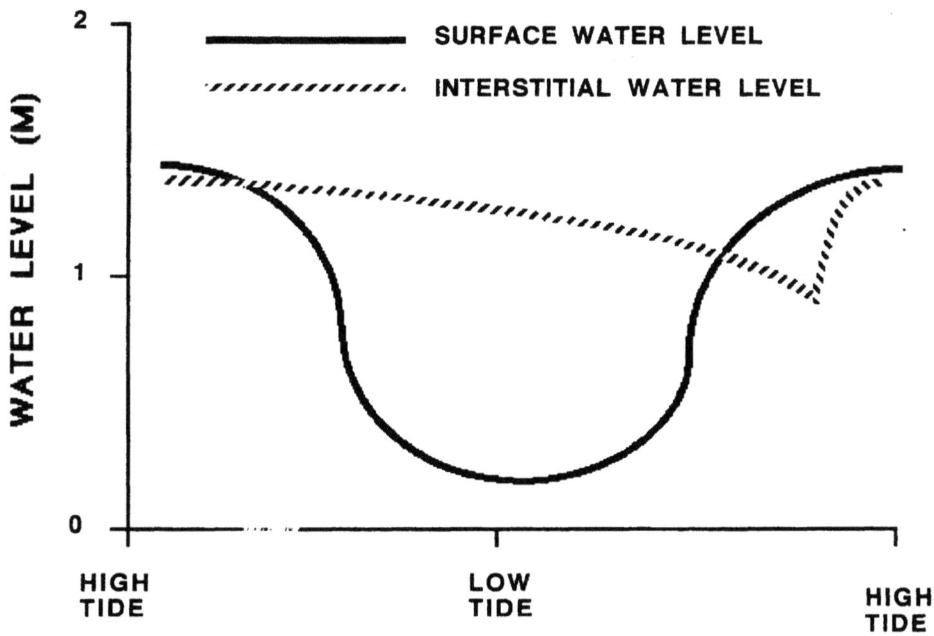

Figure 4. Interstitial water level in marsh sediments relative to surface water in adjacent creeks during each tidal cycle showing the lag in pore water loss. Pore water (including dissolved silicate) draining from the marsh is largely retained in the marsh/creek system. This figure is based on unpublished data from sandy substrate and fine substrate marshes in and near the Cape Fear estuary. The depth of porewater depression is dependent on substrate type and tidal amplitude.

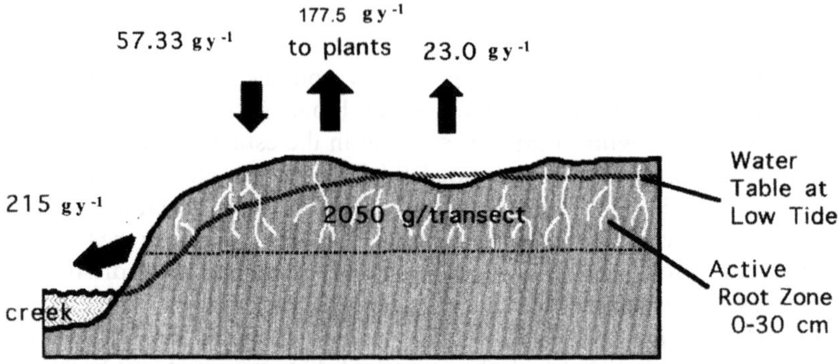

Figure 5. Microscale flux of biogenic silica along a one meter wide by 25 m long transect using flow rates from Yelverton and Hackney (1986) and pore water silicate concentrations (unpublished). Dissolved silicate is released to the creek (215 g y^{-1}) and in surface depressions (23 g y^{-1}). This will likely reenter the marsh at high tide. Silicate leaves sediment via vertical flux when lateral flow of pore water intersects depressions on the marsh surface. Dissolved silicate enters the marsh soil as the tide floods the marsh refilling pore space created through evapotranspiration and marsh drainage (57 g y^{-1}). Plants remove (177 g y^{-1}) from the soil and incorporate it in their cell walls.

4. Conclusion

The hypothesis we are proposing for a link between secondary production of tidal marshes and vascular plant production does not preclude the importance of tidal marsh as structural habitat or of vascular plant detritus as an energetic source for some species. If correct, our paradigm may explain differences found in energetic sources (algal versus vascular plant) for secondary production. It also questions the functional value of created tidal marshes if biogenic silica and silicate accumulation in marshes requires a long time. Finally, it suggests more saline marshes (or those lower in the estuary) are more important for secondary production of commercially important species dependent on benthic diatom production.

5. Literature Cited

Adey, W.H. 1987. Food production in low-nutrient seas. Bioscience **37**:340-348.

Axelrad, D.M. 1974. Nutrient flux through the salt marsh. Dissertation. The College of William and Mary, Williamsburg, Virginia, USA.

Bien, G.S., D.E. Contois and W.H. Thomas. 1958. The removal of soluble silica from freshwater entering the sea. Geochemica et Cosmochimica Acta **14**:35-54.

Blanchard, S.F. 1988. Dissolved silica in the tidal Potomac River and Estuary, 1979-1981 water years. U.S. Geological Survey Water Supply Paper 2234-H.

Boesch, D.F. and R.E. Turner. 1984. Dependence of fishery species on salt marshes: The role of food and refuge. Estuaries **7**:460-468.

Boyle, E., R. Collier, A. Dengler, A. Edmond and R. Stallard. 1974. On chemical mass balance in estuaries. Geochemica et Cosmochimica Acta **38**:1719-1728.
Calvert, S.E. 1983. Sedimentary geochemistry of silicon. Pages 143-186 *in* S.R. Aston, editor. Silicon geochemistry and biogeochemistry. Academic Press, London, England.
Cammen, L.M., U. Blum, E.D. Seneca and L.M. Stroud. 1982. Energy flow in a North Carolina salt marsh: a synthesis of experimental and published data. Association of Southeastern Biologists Bulletin **29**:111-134.
Conley, D.J. 1997. Riverine contribution of biogenic silica to the ocean silica budget. Limnology and Oceanography **42**:774-777.
Darnell, R.M. 1967a. The organic detritus problem. Pages 374-375 *in* G.H. Lauff, editor. Estuaries. AAAS, Washington, District of Columbia, USA.
——. 1967b. Organic detritus in relation to the estuarine ecosystem. Pages 376-382 *in* G.H. Lauff, editor. Estuaries. AAAS, Washington, District of Columbia, USA.
Day, J.W., Jr., W.G. Smith, R.R. Wagner and W.C. Stowe. 1973. Community structure and carbon budget of a salt marsh and shallow bay estuarine system in Louisiana. **LSU-SG-7204**, Center for Wetland Resources, Louisiana State University, Baton Rouge, Louisiana, USA.
Hackney, C.T. 1977. Energy flow in a tidal creek draining an irregularly flooded *Juncus* marsh. Dissertation. Mississippi State University, Starkeville, Mississippi, USA.
Haines, E.B. 1977. The origin of detritus in Georgia salt marshes. Oikos **29**:254-260.
——. 1979. Interactions between Georgia salt marshes and coastal waters: A changing paradigm. Pages 35-46 *in* R.J. Livingston, editor. Ecological processes in coastal and marine systems. Plenum Press, New York, New York, USA.
Happ, G., J.F. Gosselink and J.W. Day, Jr. 1977. The seasonal distribution of organic carbon in a Louisiana estuary. Estuarine and Coastal Marine Science **5**:695-705.
Heald, E.J. 1969. The production of organic detritus in a South Florida estuary. Dissertation. University of Miami, Miami, Florida, USA.
Hecky, R.F. and P. Kilham. 1988. Nutrient limitation of phytoplankton in freshwater and marine environments: A review of recent evidence on the effects of enrichment. Limnology and Oceanography **33**:796-822.
Heinle, D.R. and D.E. Flemer. 1976. Flows of material between poorly flooded tidal marshes and an estuary. Marine Biology **35**:359-373.
Heins, C. 1995. Phytoplankton and benthic microalgal responses to nutrient enrichment in two North Carolina tidal creeks. Thesis, University of North Carolina at Wilmington, Wilmington, North Carolina, USA.
Howarth, R.W. and J.M. Teal. 1979. Sulfate reduction in a New England salt marsh. Limnology and Oceanography **24**:999-1013.
—— 1980. Energy flow in a salt marsh ecosystem: Role of reduced inorganic sulfur compounds. American Naturalist **116**:861-872.
Liss, P.S. and M.J. Pointon. 1973. Removal of dissolved boron and silicon during estuarine mixing of sea and river waters. Geochemica et Cosmochimica Acta **37**:1493-1498.
Mallin, M.A., L.B. Cahoon, M.R. McIver, D.C. Parsons and G.C. Shank. 1997. Nutrient limitation and eutrophication potential in the Cape Fear and New River estuaries. WRRI REPORT **313**.
Moore, K.A. 1974. Carbon transport in two York River, Virginia tidal marshes. Thesis, University of Virginia, Charlottesville, Virginia. USA.
Norris, A.R. and C.T. Hackney (1999). Silica content of vascular plants and porewater in a mesohaline marsh in North Carolina. Estuarine, Coastal and Shelf Science **49**: 597-607.
Odum, E.P. and A.A. de la Cruz. 1967. Particulate organic detritus in a Georgia salt marsh. Pages 383-388 *in* G.H. Lauff, editor. Estuaries. AAAS, Washington, District of Columbia, USA.
Odum, W.E. , J.S. Fisher and J.C. Pickral. 1979. Factors controlling the flux of particulate organic carbon from estuarine wetlands. Pages 69-82 *in* R.J. Livingston, editor. Ecological Processes in Coastal and Marine Systems. Plenum Press, New York, New York, USA.
Peterson, B.J. and R.W. Howarth. 1987. Sulfur, carbon and nitrogen isotopes used to trace organic matter flow in the salt-marsh estuaries of Sapelo Island, Georgia. Limnology and Oceanography **32**: 1195-1213.
Settlemeyre, J.L. and L.R. Gardner. 1977. Suspended sediment flux through a salt marsh drainage basin. Estuarine and Coastal Marine Science **5**: 653-664.

Shisler, J.K. 1975. Movement of organic carbon in natural and mosquito ditched salt marshes. Dissertation, Rutgers University, New Brunswick, New Jersey, USA.

Sigmon, D.E. and L.B. Cahoon. 1997. Comparative effects of benthic microalgae and phytoplankton on dissolved silica fluxes. Aquatic Microbial Ecology **12**:275-284.

Stephens, C.F. and C.H. Oppenheimer. 1972. Silica contents in the northwestern Florida Gulf coast. Contributions to Marine Science **16**:99-108.

Sullivan, M.J. and C.A. Moncreiff. 1990. Edaphic algae are an important component of salt marsh food webs: evidence from multiple stable isotope analyses. Marine Ecology Progress Series **62**:149-159.

Teal, J.M. 1962. Energy flow in a salt marsh ecosystem of Georgia. Ecology **43**:614-624.

Turner, R.E. 1977. Intertidal vegetation and commercial yields of penaeid shrimp. Transactions of the American Fisheries Society **106**:411-416.

Weinstein, M.P. 1979. Shallow marsh habitats as primary nurseries for fish and shellfish, Cape Fear River, North Carolina. Fisheries Bulletin **77**:339-357.

Woodwell, G.M., D.E. Whitney and D.W. Juers. 1977. The Flax Pond ecosystem study: exchanges of carbon in water between a salt marsh and Long Island Sound. Limnology and Oceanography **22**:833-838.

Yelverton, G.F. and C.T. Hackney. 1986. Flux of dissolved organic carbon and pore water through the substrate of a *Spartina alterniflora* marsh in North Carolina. Estuarine, Coastal and Shelf Science **22**:255-267.

TIDAL MARSH RESTORATION: FACT OR FICTION?

SELF-DESIGN APPLIED TO COASTAL RESTORATION

An Application of Ecological Engineering

WILLIAM J. MITSCH
School of Natural Resources
The Ohio State University
Columbus, OH 43210 USA

1. Introduction

Coastlines around the world have been altered and heavily managed for centuries. They have been the favorite location for human settlement because of proximity to the sea and because estuaries are among the most productive aquatic ecosystems in the world. We are now realizing the importance of these coastal areas for the ecosystem services that they provide while also recognizing that many coastlines have seriously deteriorated due to pollution, drainage and filling, over-development, and resource over-harvesting. Costanza et al. (1997) valued estuarine systems as among the highest values per unit area on any ecosystems on earth (Table 1). For all of these reasons, there is a renewed effort to restore our coastlines of estuaries, coastal marshes, and embayments, and the watersheds that drain to them.

Ecological engineering as the practice and self-design as the theoretical concept may offer the framework in which coastal restoration can take place on a large scale around the world. This paper introduces the concept of ecological engineering, contrasting it with the more familiar term ecosystem restoration. It then focuses on several attempts at large-scale coastal restoration projects that have been undertaken in the USA, describing the scale at which the projects are being developed and the general approaches that are being used. Finally, the paper points out practices that pass the self-design litmus test and those that do not.

1.1 ECOLOGICAL ENGINEERING AND ECOSYSTEM RESTORATION

Ecological engineering is defined as "the design of sustainable ecosystems that integrate human society with its natural environment for the benefit of both" (Mitsch 1996, 1998). Ecological engineering provides an alternative approach to solving environmental problems while utilizing nature's energy flow (Odum 1971, Mitsch 1993). The design of sustainable ecosystems protects and enhances biodiversity, as illustrated by Aldo Leopold's concept that the first rule of a tinkerer is to not throw away any of the parts. It also focuses on using and enhancing the so-called "nature's services" (Costanza et al. 1997) that ecosystems provide.

The related field of ecosystem restoration has been defined as "the return of an ecosystem to a close approximation of its condition prior to disturbance" (NRC

Committee on Restoration of Aquatic Ecosystems, 1992). We restore ecosystems to protect and enhance biodiversity, to provide a less costly and more sustainable landscape, to prevent irreversible deterioration, and to restore ''nature's services'' on the planet.

The fields of ecological engineering and ecosystem restoration are simply versions of the same general approach, approximating a spectrum (Fig. 1) where human involvement can range from minimal as in the case of restoring a prairie with light seeding and/or burning to the construction of artificially contained ecosystems such as those found at the experimental Biosphere 2 (Nelson et al. 1993, Marino and Odum 1999).

With large scale restoration projects already underway and many more planned, obvious questions arise. Do we have enough resources to carry out all of these projects? What kind of maintenance commitments will these projects require? What are the probabilities of their long-term success? How do we measure this success?

TABLE 1. Estimated unit values of ecosystems (from Costanza et al. 1997)

Ecosystem	Unit Value $\$ ha^{-1} yr^{-1}$
Estuaries	22,832
Wetlands	14,785
Lakes/Rivers	8,498
Forest	969
Grasslands	232

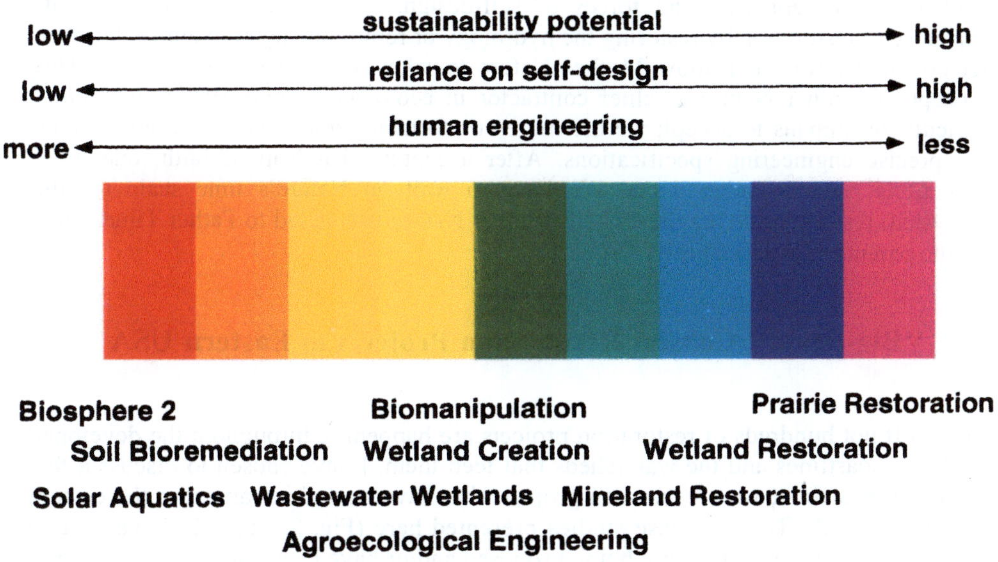

Figure 1. The spectrum of ecological engineering, which ranges from simple restoration to the construction of ecosystems in enclosures (from Mitsch 1998).

1.2 SELF-DESIGN—WORKING WITH MOTHER NATURE AND FATHER TIME

Self-design may offer the theoretical framework by which coastal restoration projects can be designed to minimize the resource costs and maximize the probability of success. Self-design is the process in ecosystem development whereby natural processes contribute to species introduction and selection. It is a manifestation of the concept of self-organization, described by Odum (1989) for microcosms and newly created ecosystems as follows: "after the first period of competitive colonization, the species prevailing are those that reinforce other species through nutrient cycles, aids to reproduction, control of spatial diversity, population regulation, and other means." Self-design relies on the self-organizing ability of ecosystems. In self-design, the presence and survival of species due to their and their propagule's continuous introduction is the essence of the successional and functional development of an ecosystem (Mitsch et al. 1998).

In the context of coastal ecosystem restoration and creation, self-design means that if an ecosystem is open to allow "seeding," through human or natural means, of enough species' propagules, the system itself will optimize its design by selecting for the assemblage of plants, microbes, and animals that is best adapted for existing conditions. "Seeding" here refers to the introduction of any protist, plant, invertebrate, and vertebrate and/or their propagules. Seeding of propagules can, of course, occur through biotic, atmospheric, and, most importantly in coastal restoration, hydrologic inputs. The opposite approach of restoration whereby specific organisms are introduced and expected to survive is more akin to gardening and zoo-keeping.

The concepts of ecological engineering and self-design described above as ones which could partially be understood by the epithet "Mother Nature and Father Time." Mother Nature represents the forces of self-design. Start an ecosystem toward its trajectory of restoration by restoring the hydrology or reintroducing propagules, but then get out of the way and allow Mother Nature to take over and sort things out. This concept of Mother Nature as chief contractor in ecological engineering is inherently difficult for humans to accept; we are more comfortable with certainty, safety factors, and precise engineering specifications. After accepting that leap of faith, one more element is needed--Father Time. We have to wait on Nature's time scale for the ecosystem restoration to manifest itself. Mother Nature is wedded to Father Time. They are the parents of Sustainability.

2. "Big-ticket" Coastal Restoration Projects in Eastern USA

Dozens if not hundreds of restoration projects are happening throughout the developed world on coastlines and the watersheds that feed them. I have chosen to discuss a few of the large-scale coastal restoration projects that have started in earnest in the eastern half of the USA. The four case studies presented here (Fig. 2, Table 2) have several features in common: 1) they are being proposed and/or have been carried out on a large (>50 km^2) spatial scale; 2) the restoration in all cases is simple in concept but enormously difficult in practice: 3) the time scale for improvements to manifest themselves ranges

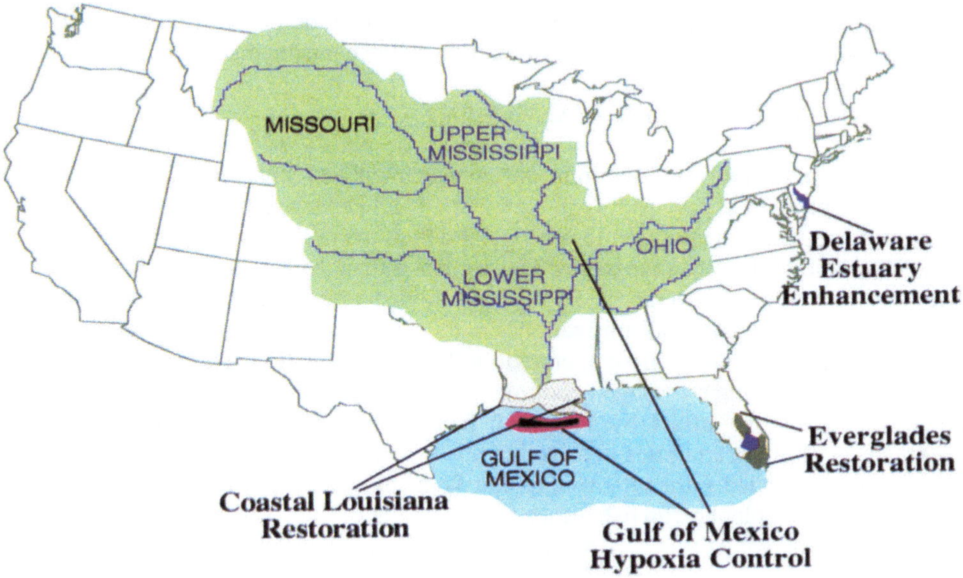

Figure 2. Location of four major restoration projects discussed in this paper.

TABLE 2. Large-scale restoration projects discussed in this paper

Project	scale, km^2	approaches
Delaware Estuary enhancement	58	- salt marsh reflooding - *Phragmites* control - hydrologic modifications in marshes
Everglades restoration		
Kissimmee River restoration	~100 (restored wetlands)	- reinstall meanders in river
Everglades nutrient removal	160 (constructed wetlands) ? (protected wetlands)	- slow the spread of *Typha* in Everglades by wetland construction for P removal
Coastal Louisiana restoration	300	- artificial diversion of Mississippi River into delta - marsh management through dikes - barrier island restoration - plantings - dredge spoil disposal
Gulf of Mexico hypoxia control	17,000 (Gulf of Mexico) 40,000*(Misssissippi River Basin)	- modifying agricultural practices - tertiary treatment of point sources - riparian buffers, wetlands - stream and delta restoration - atmospheric controls of NO_X

* estimated area in 2,950,000 km^2 Mississippi River watershed that would need to be restored to wetlands and riparian zones to control nitrate-nitrogen, the primary cause of the hypoxia.

from several years to decades; 4) upstream watershed restoration is often as much a part of the restoration as is any direct manipulation of the coastline itself.

2.1 DELAWARE ESTUARY ENHANCEMENT

This complex project involves the restoration, enhancement, and conservation of 58 km^2 of coastal salt marshes on Delaware Bay in New Jersey and Delaware in northeastern USA (Fig. 3). With its size and attention to ecological detail, overall this is one of the largest wetland restoration projects in the world. This estuary enhancement, being carried out by Public Service Electric and Gas (PSE&G) with advice from a team of scientists and consultants, was undertaken as mitigation for the potential impacts of once-through cooling from a nuclear power plant operated by PSE&G on the Bay. The reasoning was that the impact of once-through cooling on fin fish, through entrainment and impingement, could be offset by increased fisheries production from restored salt marshes. Because of uncertainties involved in this kind of ecological trading, the area of restoration was estimated as the salt marshes that would be necessary to compensate for the impacts of the power plant on fin fish, times a safety factor of four. There are three distinct approaches that are being utilized in this project to restore the Delaware Bay coastline:

1. Reintroduce flooding. The most important type of restoration involves the reintroduction of tidal inundation to about 18 km^2 of diked salt hay farms. Many marshes along Delaware Bay have been isolated by dikes from the bay, sometimes for centuries, and put into the commercial production of "salt-hay" (*Spartina patens*). Restoration is being accomplished by excavating breaches in the dikes that were built to protect the salt hay farms from tidal inundation and, in most cases, connecting these new inlets to a system of recreated tidal creeks and existing canal systems.

2. Re-excavate tidal marshes. Additional restoration involves enhancing drainage by re-excavating higher order tidal creeks in salt marshes, thereby increasing tidal circulation in existing marshes. This is particularly important in marshes that were formerly diked, as the isolation from the sea has led to the filling of former tidal creeks.

3. Reduce *Phragmites* domination. In yet another set of restoration projects in Delaware and New Jersey, restoration involves the reduction in cover of the aggressive and invasive *Phragmites australis* in 21 km^2 of non-impounded coastal wetlands. Alternatives that are being investigated include hydrological modifications such as channel excavation, breaching remnant dikes, and microtopographic changes, mowing, planting and herbicide application (Gary Bickel, person. commun.). The most common method that is currently being used is herbicide spraying followed by controlled burning and selected hydromodification (Weinstein et al. 1997).

Typical goals for the marsh restoration along Delaware Bay include a high percent cover of desirable vegetation such as *Spartina alterniflora*, a relatively low percent open water, and the absence of the invasive reedgrass *Phragmites australis* (Table 3). The success of this coastal restoration project, subject to a combination of legal, hydrologic,

TABLE 3. Endpoints for salt marsh restoration in Delaware Bay (from Weinstein et al. 1997)

1. Open water ≤ 20% of total marsh area
2. *Phragmites* coverage ≤ 4% of total marsh area
3. 'Desirable vegetation' cover ≥ 76% of total marsh area

as well as ecological constraints, is being estimated through comparison of restored sites to natural reference marshes on variables such as those listed in Table 3.

Early results of this project are generally encouraging. In those marshes where tidal exchange was restored to formerly diked salt hay farms (sites Dennis Township, Maurice River, and Commercial Township in Fig. 3; approximately 1600 ha total), the reestablishment of *Spartina alterniflora* and other favorable vegetation has been rapid and extensive. At Dennis Township, approximately 64% of the site is already dominated by *Spartina alterniflora* after only two growing seasons. Tidal restoration was completed at the Maurice River site in early 1988 and major revegetation by *Spartina alterniflora* and some *Salicornia* has already occurred. At the third and largest

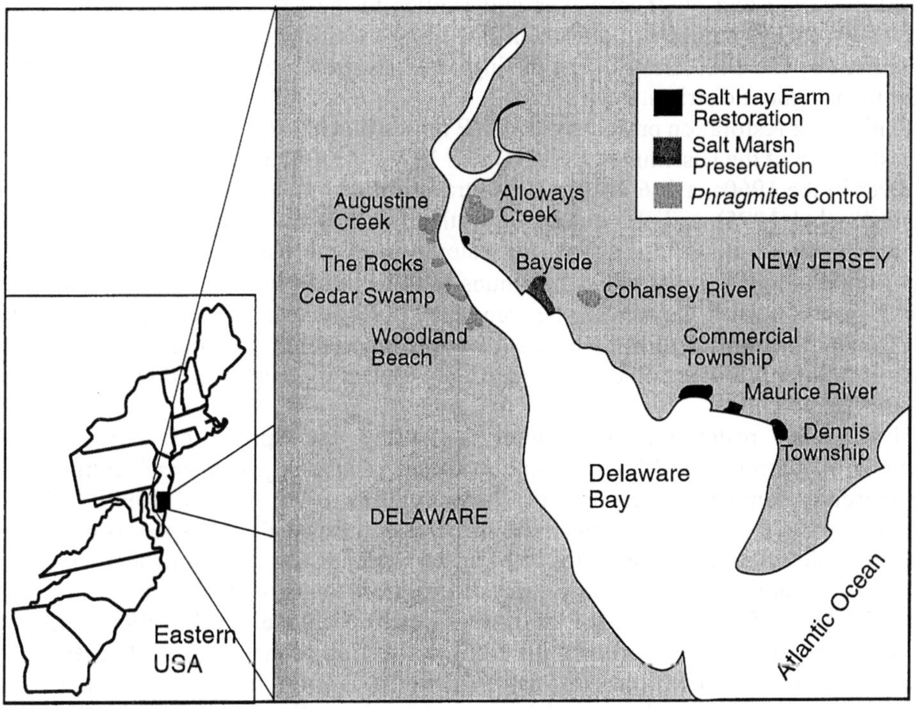

Figure 3. Delaware Bay estuary enhancement project, showing reflooded salt hay farms and *Phragmites* control sites (Mitsch and Gosselink 2000; copyright 2000 John Wiley&Sons, reprinted with permission)

salt hay farm restoration site, Commercial Township, revegetation is occurring rapidly from the bayside eastward at this recently tidally restored site.

For the restoration of marshes being carried out through spraying and burning of *Phragmites* of 1600 ha of marshes at the Cohansey River and Alloway Creek sites (Fig. 3), results are not as promising as those at the hydrologic restoration sites. Burning followed by spraying has not eliminated *Phragmites* at any given site, so repeated treatment is necessary. On these sites, alternative methods to spraying are being investigated for the control of *Phragmites* including mowing and rhizome cutting, grading of dikes where *Phragmites* tends to dominate, selective planting, and microtopographic modifications of the marsh surface.

2.2 EVERGLADES RESTORATION

The restoration of the Florida Everglades, the largest wetland area in the United States, is not one but several separate initiatives being carried out in the 46,000 km^2 KOE (Kissimmee-Okeechobee-Everglades) region in the southern half of Florida (Fig. 2). Problems in the Everglades have developed because of 1) excessive nutrient loading to Lake Okeechobee (Reddy and Flaig 1995) and to the Everglades itself (Wu et al. 1997), 2) loss and fragmentation of habitat caused by urban and agricultural development, 3) spread of *Typha* and other invasives and exotics to the Everglades, replacing native vegetation, and 4) hydrologic alteration due to an extensive canal system and straight rivers built by the U.S. Army Corps of Engineers and others and maintained by several water management districts.

One major restoration project in the KOE region that has received a lot of attention is the restoration of the Kissimmee River (Dahm 1995). As a result of the channelization of the river in the 1960s, a 166-km river was transformed into a 90-km long, 100-m wide canal (Koebel 1995) and the extent of wetlands along the river decreased by 65% (Table 4; Toth et al. 1995, 1998). The restoration of the Kissimmee River will be a major undertaking to reintroduce the sinuosity to the artificially straightened river. The river restoration work, expected to be completed in stages over the next several decades, will return some portion of lost wetland habitat to the riparian zone and will also provide sinks for nutrients that are otherwise cause increased eutrophication in downstream Lake Okeechobee.

Everglades restoration also involves halting the spread of cattail (*Typha domingensis)* through the low-nutrient sawgrass (*Cladium jamaicense*) communities that presently dominate the Everglades (Newman et al. 1996, Wu et al. 1997). Since the main causes of the spread of *Typha* are nutrients and especially phosphorus emanating from agricultural areas in the basin, 160 km^2 of wetlands, called Stormwater Treatment Areas (STAs) are planned for phosphorus control from the agricultural area. A prototype of the STAs, a 15 km^2 site called the Everglades Nutrient Removal (ENR) project has operated since mid-1994. Early results from this testing of the concept of creating and restoring wetlands to protect downstream wetlands from nutrient enrichment are encouraging. Experimental wetland basins are consistently decreasing phosphorus to background levels (T. Fontaine, person. commun.).

2.3 COASTAL LOUISIANA RESTORATION

Louisiana is one of the most wetland-rich regions of the world with 36,000 km^2 of marshes, swamps and shallow lakes. Yet Louisiana is suffering a rate of coastal wetland loss of 66 - 100 km^2/yr due to natural (land subsidence) and human causes such as river levee construction, oil and gas exploration, urban development, sediment diversion, and possibly climate change. Several approaches are currently being investigated in a number of projects in Louisiana to bring back these lost wetlands or at least slow the rate of loss. These approaches include artificial diversion of Mississippi River into the surrounding landscape, marsh management through dikes, barrier island restoration, plantings, and dredge spoil disposal (Turner and Boyer 1997). The Coastal Wetlands Planning, Protection and Restoration Act passed by U.S. Congress in 1990 led to the Louisiana Coastal Wetland Restoration Plan which was completed in 1993. That plan now has led to the spending of approximately $40 million per year. Eighty CWPPRA projects have been started or selected and all projects are expected to restore 295 km^2 over the 20 years (LCWCR Task Force, 1997).

TABLE 4. Wetland losses due to channelization of the Kissimmee River (from Toth et al. 1995)

Wetland Type	Pre-channelization km^2	Post channelization km^2	percent change
Marsh	88.9	12.4	-86
Wet Prairie	41.3	21.3	-48
Shrub-Scrub Wetland	20.7	10.0	-51
Forested Wetland	1.5	2.4	+62
Other	5.3	9.2	+72
Total	157.7	55.3	-65

2.4 GULF OF MEXICO HYPOXIA CONTROL

A hypoxic zone has developed off the shore of Louisiana in the Gulf of Mexico where hypolimnetic waters with dissolved oxygen less than 2 mg/L O_2 now extend over an area of 16,000 to 18,000 km^2 (Rabalais et al. 1996, 1998). The generally accepted theory is that nitrogen, particularly nitrate-nitrogen, is the most probable cause; 80% of that input is from 3,000,000 km^2 Mississippi River basin (41% of lower-48 states of the USA). The control of this hypoxia is important in the Gulf of Mexico because the continental shelf fishery in the Gulf is approximately 25% of the US total.

A number of approaches are being considered for controlling nitrogen flow into the Gulf; many of them involve large-scale modifications of land-use practices in Midwestern USA. Among the options being considered are modifying agricultural practices, e.g., reduce fertilizer use or alternate cropping techniques; tertiary treatment (biological, chemical, physical) of point sources; landscape restoration, e.g., riparian buffers and wetland creation, to control nonpoint source pollution from farmland; stream and delta restoration; and atmospheric controls of NO_x. The approach that appears to have the highest probability of success with a minimum impact on farming in the Midwestern USA is landscape restoration. Mitsch et al. (1999) have suggested that wetlands and riparian zones on the order of 10 million ha would be necessary to provide enough denitrification to substantially reduce the nitrogen that enters the Gulf of Mexico. Restoring the Gulf of Mexico requires restoring the Mississippi River Basin.

3. Self-Design and Coastal Restoration

There are many practices in the above projects that appear to embrace the concept of self-design while there are some practices that tend to conflict with that approach (Table 5). Some of the current coastal practices that use self-design include the following:

1. *Openings for tidal exchange.* Humans have generally developed coastlines and riverline by separating the seas and rivers from land. Dikes and levees along bodies of water are more the norm than the exception in developed parts of the world. Ecological restoration that reestablishes hydrologic connections of coastal ecosystems to adjacent bodies of water, such as the breaching of dikes around salt marshes, is a good example of the application of self-design. In Eastern USA, the reintroduction of tidal exchange has brought in *Spartina alterniflora* seeds in a remarkably rapid fashion and has further selected for salt-tolerant organisms. Marshes open to the bay for only a few months are already well along to developing a good cover of *Spartina* and conduits for feeding and spawning of fin fish.

2. *Increasing stream density and meanders.* The Kissimmee River restoration project north of the Everglades in Florida, while inland rather than coastal, illustrates the principle of increasing the fractal dimensions or ''edges'' of streams and rivers to maximize contact between the river and its shoreline and increase the spatial heterogeneity of the aquatic environment. Likewise, increasing the stream density of tidal creeks in marsh restoration accelerates the distribution of propagules while allowing better connectivity between aquatic organisms and the rich marsh habitat.

TABLE 5. Coastal 'restoration' practices that encourage or discourage ecosystem self-design

Encourages self-design
- opening for tidal exchange
- increasing stream density and meanders
- connecting to upstream watersheds
- restoring upstream aquatic systems
Discourages self-design
- construction of dikes and levees
- supporting monospecific goals
- spraying with herbicides
- over-management in general

3. *Connecting to upstream watersheds.* In many cases our rivers have been engineered to carry their loads of rich sediments out to sea with little if any connectivity to the shallow wetlands, backwaters and embayments along the way. Coastlines, particularly where large rivers discharge to the sea, need to be restored to allow flood flows to spread across the low-lying landscape, capturing the fertile sediments and species propagules rather than shunting them to burial in the sea. The hypoxia in the Gulf of Mexico is a direct result of not only the upstream fertilizers but also the disconnect between the Mississippi River and its delta.

4. *Restoring upstream aquatic ecosystems.* The restoration of watersheds upstream of coastal systems does not appear, on first glance, to be coastal restoration. But the restoration of rivers and wetlands upstream of coastal areas may be a more effective way of enhancing coastal systems than almost any approach. A restored watershed leads to an increased pollution assimilation capability and a higher probability of cleaner river water reaching the coastal areas.

Coastal restoration practices that discourage self-design include the following:

1. *Construction dikes and levees.* Dikes and levees to protect humans from flooding is often the right thing to do. Building dikes to restore ecosystems is generally the wrong thing to do. Additionally, it commits managers to an ultimately unsustainable system of continued maintenance.

2. *Supporting monospecific goals.* Restoration to introduce and/or support only one or two desired species, e.g., a sport fish or selected endangered species, misses the point. Niches open and close with remarkable frequency in dynamic environments such as coastlines so species often come and go with these changes. A better approach is to have a suite of desirable species as a goal and to design the restoration project accordingly.

3. *Spraying with herbicides.* Herbicides are often used with other management practices, such as burning, to remove undesirable species from heavily managed ecosystems. In coastal areas of Eastern USA, *Phragmites* spread is frequently "controlled" through this means. Arrested succession practices such as this one are ultimately doomed to failure unless the forcing functions, e.g., hydrology, are changed at the same time that spraying and burning occur. These practices do not take advantage of nature's ability to self-design but impose a heavy degree of management and subsequent use of resources to maintain these artificial systems.

4. *Over-managing in general.* Humans have a general tendency to over-manage any ecological restoration that they undertake, usually in the name of "ensuring success." There is nothing inherently wrong with managing ecosystems but if we wish to return a salt hay farm to a salt marsh or a straight canal to a meandering river, the very idea of restoration means returning the landscape from a heavily managed and maintained landscape to one that Nature generally manages.

4. Literature Cited

Costanza, R., R. d'Arge, R. de Groot, S. Farber, M. Grasso, B. Hannon, K. Limburg, S. Naeem, R.V. O'Neill, J. Paruelo, R.G. Raskin, P. Sutton and M. van den Belt. 1997. The value of the world's ecosystem services and natural capital. Nature **387**:253-260.

Dahm, C.N., editor. 1995. Kissimmee River Restoration. Special issue of Restoration Ecology **3**: 145-238.

Louisiana Coastal Wetlands Conservation and Restoration Task Force. 1997. Louisiana Coastal Wetland Restoration Projects: The 1997 Evaluation Report to the U.S. Congress. Louisiana Department of Natural Resources, Baton Rouge, Louisiana, USA.

Marino, B. and H.T. Odum, editors,. 1999. Biosphere 2: Introduction and research progress. Special Issue of Ecological Engineering **13** 1-356.

Mitsch, W.J. 1993. Ecological engineering—a cooperative role with the planetary life-support systems. Environmental Science and Technology **27**:438-445.

——. 1996. Ecological engineering: A new paradigm for engineers and ecologists. 1996. Pages 111-128 *in* P.C. Schulze, editor. Engineering within ecological constraints. National Academy Press, Washington, District of Columbia, USA.

——. 1998. Ecological engineering–the seven-year itch. Ecological Engineering **10**:119-138.

Mitsch, W.J., X. Wu, R.W. Nairn, R.E. Weihe, N. Wang, R. Deal and C.E. Boucher. 1998. Creating and restoring wetlands–a whole-ecosystem experiment in self-design. BioScience **48**:1019-1030.

Mitsch, W.J., J.W. Day, Jr., J.W. Gilliam, P. M. Groffman, D. L. Hey, G.W. Randall and N. Wang. 1999. Reducing nutrient loads, especially nitrate-nitrogen, to surface water, groundwater and the Gulf of Mexico. Final Report to NOAA, Coastal Program, Silver Spring, Maryland, USA.

National Research Council Committee on the Restoration of Aquatic Systems. 1992. Restoration of Aquatic Ecosystems. National Academy Press, Washington, District of Columbia, USA.

Nelson, M., T.L. Burgess, A. Alling, N. Alvarez-Romo, W.F. Dempster, R.J. Walford and J.P. Allen. 1993. Using a closed ecological system to study Earth's biosphere initial results from Biosphere 2. BioScience **43**: 225-236.

Newman, S., J.B. Grace and J.W. Kooebel. 1996. The effects of nutrients and hydroperiod on mixtures of *Typha domingensis*, *Cladium jamaicense* and *Eleocharis interstincta*: Implications for Everglades restoration. Ecological Applications **6**:774-783.

Odum, H.T. 1971. Environment, power and society. J. Wiley, New York, New York, USA.

Odum, H.T. 1989. Ecological engineering and self-organization. Pages 79-101 *in* W.J. Mitsch and S.E. Jorgensen, editors. Ecological engineering: An introduction to ecotechnology. J. Wiley, New York, New York, USA.

Rabalais, N.N., W.J. Wiseman, R.E. Turner, B.K. Sengupta and Q. Dortch. 1996. Nutrient changes in the Mississippi River and system responses on the adjacent continental shelf. Estuaries **19**:386-407.

Rabalais, N.N., R.E. Turner, W.J. Wiseman and Q, Dortch. 1998. Consequences of the 1993 Mississippi River flood in the Gulf of Mexico. Regulated Rivers **14**:161-177.

Reddy, K.R. and E.G. Flaig, editors. 1995. Phosphorus dynamics in the Lake Okeechobee watershed, Florida. Special Issue of Ecological Engineering **5**:127-414.

Toth, L.A., D.A. Arrington, M.A. Brady and D.A. Muszick. 1995. Conceptual evaluation of factors potentially affecting restoration of habitat structure within the channelized Kissimme River ecosystem. Restoration Ecology **3**: 160-180.

Toth, L.A., S.L. Melvin, D.A. Arrington and J. Chamberlain. 1998. Hydrologic manipulations of the channelized Kissimmee River. BioScience **48**:757-764.

Turner, R.E. and M.E. Boyer. 1997. Mississippi River diversions, coastal wetland restoration/creation and an economy of scale. Ecological Engineering **8**:117-128.

Weinstein, M.P., J.H. Balletto, J.M. Teal and D.F. Ludwig. 1997. Success criteria and adaptive management for a large-scale wetland restoration project. Wetlands Ecology and Management **4**:111-1127.

Wu, Y, F.H. Sklar and K. Rutchey. 1997. Analysis and simulations of fragmentation patterns in the Everglades. Ecological Applications **7**:268-276.

FUNCTIONAL EQUIVALENCY OF RESTORED AND NATURAL SALT MARSHES

JOY B. ZEDLER
ROBERTO LINDIG-CISNEROS
Arboretum and Botany Department
University of Wisconsin - Madison
1207 Seminole Highway
Madison, WI 53711 USA

Abstract

Although achieving functional equivalency of restored and natural ecosystems is a desirable restoration goal, direct assessments of function are rare, as are data supporting the use of indicators of function. The issue is how structural attributes can be used to assess ecosystem functioning. A functional equivalency index mixed both structural and functional measures, and an often-cited model of ecosystem degradation and restoration depicted a straight-line relationship between the two variables. Several points need clarification, not only for tidal wetlands but for restoration ecology generally. It is unrealistic to expect linear relationships among structure, function, and time; it is also inappropriate to assume, for natural and restored ecosystems, that equivalent structure means equivalent function. For example, using plant biomass to compare primary productivity rates assumes equal biomass:productivity relationships among sites; restored and natural wetlands are less likely to be similar in grazing, decomposition, and export rates than are two natural sites. Likewise, using soil organic matter (OM) and total Kjeldahl nitrogen (TKN) to indicate nutrient availability may be less appropriate for restored and natural marsh comparisons than for two natural sites. We recommend that comparisons of structural attributes be labeled structural equivalency measures, thereby avoiding misconceptions. Useful structural measures for assessing tidal wetland restoration are: soil texture; soil OM, soil nutrients, vegetation structure (height distributions); invertebrate and fish populations (especially fish size distributions), and topographic complexity.

1. Introduction

An ideal restored ecosystem would be functionally equivalent to one or more natural reference ecosystems. It would support the natural biodiversity of the region, resist invasion by alien species, and be self-sustaining. Such pronouncements about goals are much more easily made than assessments of progress, i.e., determining when a site has attained functional equivalency and predicting when it might.

What functions are essential for an ecosystem to sustain itself? Given that functions are difficult to measure, what structural attributes can be used to predict how restored and natural ecosystems function? The terms ''structure'' (a condition at one point in time) and "function" (a process that occurs over time, i.e., a rate; Fig. 1) are often used interchangeably in restoration ecology. That is, similarities in the structure of restored and natural habitats are used to indicate functional equivalency. Is this appropriate? The related concept of biological integrity also involves mixing structural and functional attributes of natural ecosystems to provide an overall index of health (Karr and Dudley 1981). The common mixing of structural and functional attributes suggests the need for a literature review of current usage. In this paper, we review the terms used in the functional assessment of tidal wetlands, evaluate the relationship between structural and functional attributes, and use a southern California case study to suggest several salt marsh attributes that are useful in comparing ecosystems.

Table 1. Papers that make explicit reference to functional equivalency.

Source	Use of the term	Attributes measured
LaSalle and Landin 1991	"The present study addresses the issue of functional value of the developed marsh through an evaluation of selected biotic components of two varying-aged zones at the site." "The various-aged zones...seemed to represent well established and functionally viable habitats."	Plant height, density, cover, biomass; abundance of benthic macrofauna, fish and shellfish abundance, fish gut contents in North Carolina salt marshes established on dredge spoil
Moy and Levin 1991	"... few programs have addressed the functional ecological equivalence of man made marshes and their natural counterparts. This study addresses structural and functional interactions in a man-made and two natural marshes."	Substrate characteristics and marsh use by organisms of two trophic levels were studied in Dills Creek, North Carolina
Minello and Zimmerman 1992	"Comparisons of functional equivalency between natural and transplanted salt marshes require adequate information on how salt marshes actually function for fish and decapod crustaceans."	Comparison of habitat use by fish and decapod crustaceans among three transplanted and three natural marshes
Scatolini and Zedler 1996	"This study of the San Diego Bay mitigation site indicates that constructed and natural salt marshes were far from being structurally or functionally equivalent. Vegetation cover and sediment type were different for the two study sites." "The invertebrate assemblage in the 4-year-old marsh was not equivalent to that of the natural marsh."	Epibenthic invertebrates in restored and natural salt marshes

Simenstad and Thom 1996	"If we knew whether, and which, wetland attributes would follow linear, asymptotic, logistic, or other development patterns, we could theoretically predict the functional equivalency and contribute extensively to the future predictive power of wetland restoration"	16 attributes, including topography, sediments, and vegetation; water chemistry and temp.; survival, distribution, and growth of planted and naturally recruited emergent plants; benthic and planktonic invertebrate composition and standing stock; fish and bird species occurrence and density
Gibson et al. 1994	"Constructed salt marshes may fail to achieve functional equivalency with natural marshes if their nutrient status limits ecosystem development"	Responses of *Spartina foliosa* to different soil amendments in a constructed salt marsh
Knott et al. 1997	"The recovery of functional equivalency of the excavations, compared with natural marshes, was not evaluated during our study, but we speculate that their functioning may equilibrate over time"	Density, biomass and spatial coverage of *Spartina alternifolia;* assessed recovery of the macrofaunal community in the backfilled marshes
Miller and Simenstad 1997	"Short term otolith growth rates, residence times, and forage of two species of juveniles salmon (*Oncorhynchus* spp.) were compared ....to assess the functional equivalency of the created slough in providing suitable rearing habitat."	Short-term otolith growth rates, residence times and forage

2. Functional Equivalency and Biotic Integrity

Our review of the functional equivalency literature drew on an earlier search for peer-reviewed evaluations of salt marsh restoration (Zedler and Callaway, in press), which identified 26 peer-reviewed papers. Eight papers that compared restored and natural salt marshes explicitly mention functional equivalency (Table 1), and an additional 10 implied functional comparisons (Table 2). Additional literature on biological integrity (Table 3) focused on aquatic habitats, of which one study concerned estuaries (Deegan et al. 1997). In most comparisons of restored and natural salt marshes, structural measures were used as surrogates for function (e.g., LaSalle et al. 1991, Minello and Zimmerman 1992, Gibson et al. 1994, Scatolini and Zedler 1996, Simenstad and Thom 1996), although two studies estimated fish growth rates (Moy and Levin 1991, Miller and Simenstad 1997) and one estimated nitrogen-fixation (see below).

A "functional equivalency index" (PERL 1990, Zedler and Langis 1991) included 11 attributes measured in both constructed and natural tidal wetlands of Sweetwater Marsh National Wildlife Refuge (32°10'N, 117°10'W), within San Diego Bay. In this case (Fig. 2), the objective was to compare the best areas of the constructed marsh with average conditions in the remnant of natural marsh that was not damaged by highway construction. Measurements were chosen as indicators of three functions important for the target species, the endangered light-footed clapper rail (*Rallus longirostris levipes):* cordgrass *(Spartina foliosa)* productivity, nutrient dynamics, and food chain support.

TABLE 2. Papers that make implicit reference to functional equivalency.

Source	Concepts	System
Seneca et al. 1976	"Five main functions are usually attributed to natural salt-marshes: (1) primary production, (2) nutrient cycling, (3) substrate stabilization, (4) habitat formation, and (5) offering of nursery areas for a wide range of fishes and other biota."	*Spartina alternifolia* marsh in North Carolina
Sheffler et al. 1990	"The extensive historical losses and alterations of wetland habitats in this region have undoubtedly diminished the function by many estuarine environments as temporary residence areas that once contributed to increase survivals of juvenile salmon."	Juvenile salmon in a restored estuarine wetland, in the Puyallup river estuary in Tacoma, Washington
Craft et al. 1991	"The large differences in soil organic matter and Fe and Mn content between upland and wetland soils suggest that upland soils will not provide the same nutrient cycling capabilities as natural wetland soils. For this reason, we compared soil pore water nutrient chemistry of a created marsh established on an upland site and a nearby natural marsh"	Studies on water chemistry in natural and created wetlands on the Palmico River estuary, North Carolina
Frenkel and Morlan 1991	"First, we assess restoration by examining changes in a limited number of wetland characteristics including: plant species, plant communities, elevation of the site, the role of salinity and soil texture, width and depth of creek cross-sections, and estimated above ground net primary production."	Several parameters were monitored in a restored salt marsh in the Salmon River, Oregon
Langis et al. 1991	By comparing nitrogen levels in one constructed and one natural marsh the functioning of both systems is compared. "Exchanging an acre of natural wetland results in a net loss in wetland function."	Restored (excavated from dredge spoil) and natural salt marshes of San Diego Bay

Chamberlain and Barnhart 1993	"The objective of this study was to determine whether the restored marsh provided adequate mitigation. Fish abundance, species diversity, and use (spawning, rearing, duration of residence) were compared among the mitigation site, a nearby undisturbed marsh, channels adjacent to marshes, and the Woodley Island Marina"	Use of habitat by fish of a mitigation salt marsh in Humboldt Bay, California
Peck et al. 1994	"The presence, distribution, and abundance of various macro invertebrates on the restored impounded marsh in comparison to the marshes below the impoundment dike and to a nearby unimpounded valley marsh indicates that restoration is in an advanced phase."	Macroinvertebrates in restored impounded valley and unimpounded valley marshes in Stonington, Connecticut
Sacco et al. 1994	"Given the high degree of variability inherent in natural marshes and their corresponding infauna, it is possible that the infaunal community of an artificially established marsh may eventually develop into one resembling that of a natural marsh, in some, if not all attributes."	Infauna sampled from an artificial and a natural marsh
Havens et al. 1995	By comparing the natural and the constructed wetland, the authors make several recommendations to make wetland compensation projects more successful. They reach to the conclusion that some functions reach levels similar to those of natural wetlands and others do not (such as carbon accumulation).	A restored tidal marsh and two natural marshes in coastal Virginia
Haltiner et al. 1997	"Perhaps rather than expecting newly restored marshes to quickly replicate the functions and values of natural sites (which have evolved over thousands of years), it is more important to assess whether or not we have created the appropriate "template" to allow the restores site to develop the desired characteristics over an appropriate time frame."	Restored (excavated from dredge spoil) tidal wetlands in California

Nine of the attributes were structural features (measurements representing the condition of the system at one point in time); only two (nitrogen-fixation rates in the rhizosphere and at the soil surface) were functional measures. On this basis, the constructed marsh was considered to be < 60% functionally equivalent to the natural marsh. At 5 years of age (counting from year of planting), 10 of the attributes assessed had lower means in the constructed marsh. Only nitrogen-fixation in the rhizosphere was greater. Various explanations of the patterns indicate interrelationships among the 11 attributes; they are far from independent indicators of function. For example, the higher rate of nitrogen-fixation in the rhizosphere of the constructed marsh may have been related to the low concentration of soil inorganic nitrogen. Also, the low soil organic

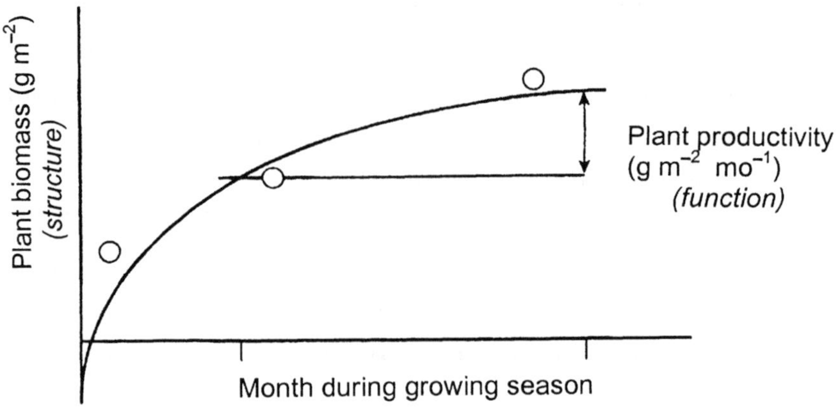

Figure 1. Simple model relating structure (condition at one point in time) with function (change over time), using biomass and productivity as examples.

matter concentrations helped explain why nitrogen-fixation rates were quite low in both the natural and constructed marshes, as shown experimentally by Zalejko (1989). Subsequent experimental work (Gibson et al. 1994, Boyer and Zedler 1998) showed that soil nitrogen concentrations determined cordgrass height.

Indices of functional equivalency and biotic integrity are needed to evaluate the progress of restoration efforts, and it is tempting to measure structure in restored and natural sites as indicators of their functioning. We suggest the need for caution, however, because the structural measures may have different relationships between such sites. We provide two examples from the San Diego Bay study: First, plant biomass is often harvested to provide an estimate of plant productivity (Fig. 1), especially in tidal marshes (Mitsch and Gosselink 1993). Changes in biomass underestimate net productivity, because there are losses (due to grazing, decomposition, and export) between samples, and these losses may well differ for restored and natural sites. At San Diego Bay, epibenthic invertebrate populations, which include grazers and decomposers, differed two-fold, and with very different ratios for many species (Scatolini and Zedler 1996). Overall, the most abundant species was a fly larva (genus *Pericoma*), but snails and crabs were more abundant in the constructed marsh, where less of the substrate was vegetated. The effects of snails and crabs on plant biomass are unknown. Also suggesting differences in grazing, we witnessed a scale insect population irruption only in the constructed marsh, where much higher consumption rates were likely (Boyer and Zedler 1996). In general, losses of plant biomass due to grazing were probably different in the restored and natural sites, and the relationship between total stem length (a nondestructive measure of biomass) and primary productivity was also probably different. Second, soil OM and soil nutrient concentrations (e.g., TKN) are used to indicate soil development and nutrient supply (cf. Craft et al. 1988 and Langis et al. 1991); however, supplies and losses may differ for restored and natural sites. At San Diego Bay, the constructed marsh had high decomposition rates and rapid leaching, as determined from litter bag data for OM and N concentrations over time (Gibson et al. 1994); but because comparable measures were not made in reference wetlands, we cannot compare aboveground decomposition rates directly. But it may be the below ground decomposition and leaching rates that differ most. Sandy soils could have much greater influx and efflux of nitrogen than clayey soils, making it difficult to compare nutrient supply rates using concentration data.

The above examples suggest caution in using structural attributes to estimate function, especially to compare the functioning of restored and natural wetlands. Both consider structure:function relationships that are quite direct. What about other relationships that would be useful for assessing how well restoration sites function?

Table 3. Papers concerning biotic integrity.

Author	Concepts	System
Karr and Dudley 1981	"the capability of supporting and maintaining a balanced, integrated, adaptive community of organisms having a species composition and functional organization comparable to that of natural habitat of the region."	Black Creek watershed, Indiana
Kerans and Karr 1994	"The complexity of biological systems and the diverse influences of human society require a broadly based, multimetric approach that reflects important aspects of stream biology from individual to assemblage levels and responds to the effects of human society in detectable ways."	Invertebrate data from rivers in the Tennessee Valley
Hoiland et al. 1994	"The biotic integrity of impacted sites continues to lag behind reference sites, probably due to poor habitat structure and metals leaching from mine tailings."	The Coeur d'Alene River Drainage of northern Idaho
Minns et al. 1994	"The fish-related uses represent the majority of the intrinsic ecosystems properties, and thus, fish ought to be useful in selecting targets and criteria for assessing, restoring, and managing aquatic ecosystems."	The littoral zone of the Great Lakes
Shields et al. 1995	"extension of the concepts underlying the IBI to lakes, reservoirs, wetlands, and coastal environments is underway."	The Mississippi River alluvial plain

Roth et al. 1996	"We used the IBI based on the species richness and composition of stream fish assemblages as a biological assessment tool."	The River Raisin
Hugueny et al. 1996	"The biotic integrity of an ecosystem can be defined as its capacity to maintain a community with a species richness, composition and functional organization comparable to that in similar ecosystems that are undisturbed by human activity.	The Konkoure river, Republic if Guinea
Deegan et al. 1997	"Our purpose was to develop an Estuarine Biotic Integrity Index (EBI) to evaluate the ecosystem status of estuaries consistent with Karr's approach. The EBI must reflect the ecological processes that support higher trophic levels (fishes) in estuaries."	Two estuaries in Waquoit Bay and Buttermilk Bay in Massachusetts

Community structure (species composition, size distributions of woody taxa, ratios of perennials:annuals) would be very useful if it could indicate rates and directions of compositional change (succession). Unfortunately, most restoration sites are young and are not well sampled for such structural attributes (Zedler and Callaway, in press). Likewise, structural attributes of one trophic level (e.g., primary producer biomass) would be useful if they could predict the functioning of another trophic level (e.g., herbivore productivity). For tidal wetlands, there may be such a surplus of primary productivity, and such an open system for its distribution, that herbivory differences between restored and natural sites would have to be very large before they would be detectable.

Additional suggestions for using vegetation structure to indicate ecosystem function come from terrestrial ecology, where there is considerable interest in exploring how the number of plant species influences other components of the ecosystem, e.g., efficient nutrient cycling and wildlife support (Naeem et al. 1994, Tilman et al. 1996, Tilman and Downing 1994). Recent experiments (Chapin et al. 1998) that vary species richness and assess function have documented predictable patterns of net primary production (Fig. 3A) and drought resistance (Fig. 3B) measured as the log of the ratio of plant biomass during and before drought. Another study (Fig. 3C) suggests a more complex pattern between productivity and successional stage, however. None of these data suggest straight-line relationships between structure and function. Nor does any study explore the relationship for restored versus natural sites. It seems premature to assume that function is readily predicted from structure.

Summary data sets	%
Height of vascular plants	65
Biomass of vascular plants	42
Foliar nitrogen concentration	84
Epibenthic invertebrate numbers	36
Epibenthic invertebrate species lists	78
Nitrogen fixation (surface cm)	51
Nitrogen fixation (rhizosphere)	110
Organic matter content	51
Sediment nitrogen (inorganic N)	45
Sediment nitrogen (TKN)	52
Pore-water nitrogen (inorganic N)	17
Average of comparisons	≈ 57%

Figure 2. An index of functional equivalency of four constructed islands in relation to the natural cordgrass (*Spartina foliosa*) marsh at Paradise Creek. Each percentage represents a value from the constructed marsh divided by the value from natural marsh times 100. This index compares soils, nutrients, plants, and epibenthos, holding intertidal elevation constant. Note that the constructed marsh sampling stations were in areas with the most vigorous cordgrass; hence, the index is higher than it might be if based on a random sample of the four islands. The average (<60%) is best interpreted as a maximum value, because the constructed marsh areas with little or no cordgrass were not included. Modified from PERL 1990 and Restoration & Management Notes 9(1):21-25, copyright 1991, reprinted by permission of The University of Wisconsin Press.

3. Is It Widely Assumed that Structure Measures Function?

The conceptual framework of restoration ecology rests on a model first published by Bradshaw in 1984 (Fig. 4). The pre-disturbance ecosystem is depicted as the reference system for restoration, with the restoration process moving the degraded system toward that target. The axes represent structure (X = species and complexity) and function (Y = biomass and nutrient content, both of which are structural attributes). This model, which was derived from a simpler one without axes (Magnuson et al. 1980), has been reprinted repeatedly, with minor rewording (Bradshaw 1987, 1995, Meffe and Carroll 1997, Dobson et al. 1997), suggesting broad acceptance of the concept that function increases linearly as structure increases (Bradshaw 1984).

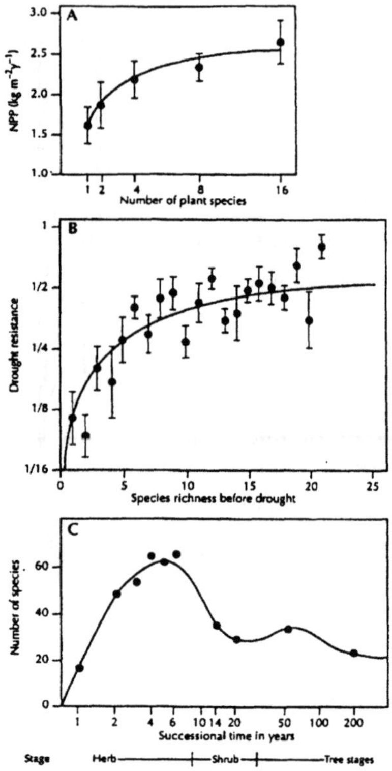

Figure 3. Changes in ecosystem attributes as species composition becomes more complicated, either in space (3A-B, from experiments with different numbers of species) or time (3C). Figures were redrawn from Chapin et al. (1998: 3A-B) and Whittaker (1975: 3C).

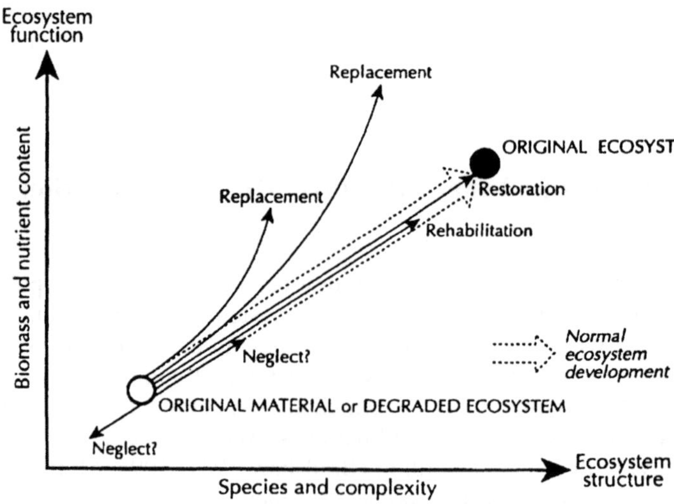

Figure 4. Conceptual model of pre-disturbance and restored ecosystems. Redrawn from Bradshaw (1987).

Two main assumptions of Bradshaw's model are that a) the relationship between structure and function is linear, and b) structural and functional attributes become more complex through time (succession). Details and supportive data are lacking in the papers where this model is presented (Bradshaw 1984, 1987, 1995 and Dobson et. al. 1997). The recent data in Fig. 3A-B fail to support a linear relationship between structure and function. Nor does the pattern in Fig. 3C indicate a simple relationship between function and successional stage, i.e., function is not linear with time. Hence, we prefer a model without axes that lacks straight-line connections between the restoration site and the reference system (Fig. 5).

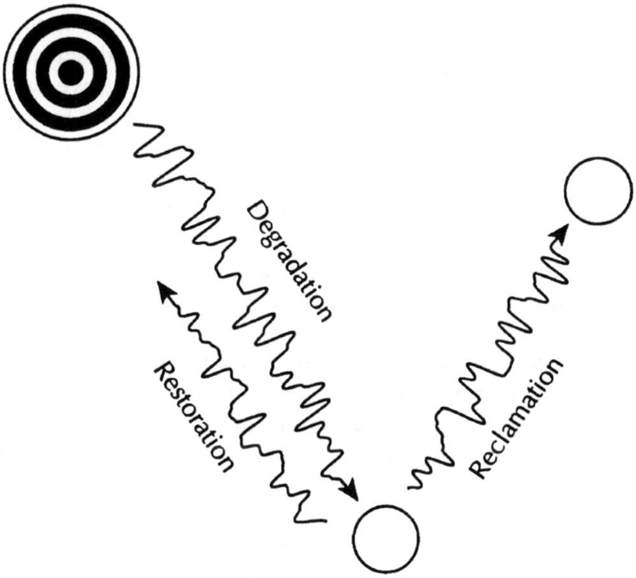

Figure 5. A simple model for pre-disturbance (the target) and restored ecosystems. Outcomes that differ substantially from the pre-disturbance condition are not considered restoration, but replacement, reclamation or rehabilitation.

Determining whether or not structure and function are as tightly linked as suggested by the functional equivalency index, the biotic integrity index, and the Bradshaw model is important to restoration efforts. It would be highly desirable to have one-time measures of structure that could predict function, and also to have structural attributes that, measured through time, could predict the long-term development of a restoration site. This is tantamount to predicting the course of a motion picture (function analog) by examining one or more frames (structure analog)–only more difficult, because the actors, scenes, and script can all change without notice. Hence, we suggest that structural measures not be assumed to indicate ecosystem function, recognizing that there are few straight-line relationships, either between structure and function or between structure and time or function and time (Fig. 3). ''Structural equivalency'' would be a reasonable term for comparing the structure of paired ecosystems (restored and natural), and ''biotic integrity'' remains a good term for describing a site relative to a broader standard. If we avoid the reference to function where direct measures are lacking, we reduce chances of misunderstanding.

4. What Structural Attributes Best Indicate the Status or Biotic Integrity of Restored Salt Marshes?

4.1 VEGETATION STRUCTURE

Species composition, stem density, percent cover, and biomass (or the nondestructive measure, total stem length) are probably the most widespread descriptors of salt marsh structure. Where habitat use by wildlife is a restoration target, we would add canopy architecture (stem height distributions for grasses and/or layering for species for which individuals are hard to identify). Zedler (1993) proposed a specific height standard for assessing the suitability of cordgrass for nesting by clapper rails, namely, the presence of more than 30 stems over 90 cm tall m^{-2} (and at least 100 stems m^{-2}). Where too few plants were tall in natural salt marshes, clapper rails failed to nest (Zedler 1993). Tall vegetation is thought to camouflage nests, especially during high tides when the nest floats and when eggs and chicks are more vulnerable to aerial predators (raptors), and also to anchor the nest, which might float away if the canopy were too short. A tall canopy is also an important high-tide refuge for predatory beetles (*Coleomegilla fucilabris*), which help keep populations of scale insects (*Heliaspis spartina*) in check (Boyer and Zedler 1996), thereby sustaining the tall canopies. Whereas earlier descriptors of vegetation grew out of ecological studies, the use of cordgrass height distributions grew out of the need to assess the suitability of restored marshes for clapper rail nesting.

The relative abundance of exotic and native species is also of great interest in comparing the structure of restored and natural wetlands. We suggest comparing both the species lists and a measure of abundance to determine the status of exotic plant invasions. If possible, the proportions of natural and restored sites that are invaded by exotics should also be compared.

4.2 SOIL ATTRIBUTES: TEXTURE, NUTRIENTS, AND ORGANIC MATTER

Long-term studies at a San Diego Bay mitigation site have demonstrated that soil texture and nitrogen supplies are related to height distributions of cordgrass. Coarse dredge spoils were found to be depauperate in nitrogen (Langis et al. 1991), as well as leaky following nitrogen addition (Gibson et al. 1994). Nitrogen supply rates that produce tall canopies were determined experimentally: the coarse dredge spoil substrate produced the most tall stems with biweekly additions of urea from March through August (Boyer and Zedler 1998). We have not yet determined standards for soil texture (minimum % clay) and soil nitrogen pools (minimum TKN) that can produce sufficient tall stems without nitrogen fertilization, but this would be a useful criterion to include in a biotic integrity index.

Although nitrogen additions produced tall canopies, the nutrient-subsidized cordgrass was not considered to be self-sustaining; hence, the mitigation requirements were not met (PERL 1997). The unfertilized plants remained short for 13 years after planting, with no indication of improvement over time (PERL 1997). In North Carolina (Craft et al. 1988) and Texas (Lindau and Hosner 1981), restored tidal marshes have also been shown to lack the organic matter and nutrient content of reference ecosystems. Webb and Newling (1985) found high aboveground biomass but low root biomass in constructed marshes of Galveston Bay; however, they attributed the differences from natural marshes to environmental variation and young age, rather than to marsh origin. Like soil texture, the soil nutrient levels that can sustain tall plant canopies need to be determined and used in assessing salt marsh restoration projects.

Although most coastal wetland restoration projects have focused on the lower intertidal elevations, high marsh restoration is also of concern in southern California. Soils in the high marsh experience tidal and rainfall regimes that interact to create a long, dry and hypersaline period during the warmest months of the year (August-September). Low OM and coarse textured soils retain little moisture and rapidly develop salt crusts. Planting vegetation during the winter is also chancy, as rainfall is unpredictable in both timing and amount. Hence, it is difficult to establish high marsh vegetation on constructed marsh substrates that lack sufficient organic matter; in fact, plots where kelp compost was added supported visibly higher plant growth than unamended plots (PERL 1997). Freshwater irrigation and organic amendments show promise for restoring higher marsh vegetation (PERL 1997). We recommend that soil texture and nutrient levels necessary to support high marsh vegetation be further explored.

4.3 INVERTEBRATE AND FISH ASSEMBLAGES

Constructed salt marshes are readily occupied by native invertebrates (Fell et al. 1991, LaSalle et al. 1991, Scatolini and Zedler 1996) and fish species (Rulifson 1991, Simenstad and Thom 1996, Havens et. al. 1995, Minello and Zimmerman 1992). In North Carolina, Sacco et. al. (1994) found similar patterns for marsh invertebrates; that species lists were similar, but densities differed. Likewise, Scatolini and Zedler (1996) found 75% similarity between epibenthic invertebrate species lists but only one-third as many individuals in the constructed marsh compared to the reference site. Because presence of a species does not appear to be very useful in distinguishing restored and natural marshes, abundance and some measure of functioning is preferred. Moy and Levin (1991) linked differences in fish diets to differences in invertebrate macrofauna of the marsh surface, which in turn were linked to less soil organic matter. Nursery and food-chain-support functions are also of interest.

The marshes constructed at San Diego Bay have mostly large tidal channels (no small creeks), although the list of fish species is similar to that of more complex tidal creek networks (Williams and Zedler 1999). What appears to be lacking is the nursery function for California killifish (*Fundulus parvipinnus*) that is provided by smaller tidal creeks (Desmond et al. 1999). We recommend assessing size distributions of key populations.

Feeding relationships are more difficult to document. A multiple-stable-isotope study of the food web of natural marsh remnants (Kwak and Zedler 1997) showed that the food chain leading to fish had a strong signature from salt marsh producers. This led to subsequent study of the route by which fishes obtain their food–marsh export of food to the channels versus fishes moving into the marsh to feed. Recent work (Johnson 1999) documented 5 species feeding in the marshes and their utilization of marsh foods (including spiders and insects) that are available to fishes only when high spring tides allow access to the marshes. The stable isotope (Kwak and Zedler 1997) and feeding (Johnson 1999) analyses are the first southern California studies to demonstrate that wetland fishes use the foods produced by the salt marsh. It follows that comparison of salt marsh habitat structure, i.e., the complexity of tidal creek networks and potential for animals to move between the marsh and channels should be assessed.

4.4 COMPLEX TOPOGRAPHY

In tidal marshes of Galveston Bay, habitat complexity is proving to be a good indicator of faunal use. Minello and Zimmerman (1992) added tidal channels to restored marshes and found increased abundances of fishes and invertebrates of commercial value. Zedler et al. (1999) have shown that tidal creek networks are important to the distribution of native plant species in natural wetlands of San Quintín Bay; they speculate that restored marshes would support more plant species if complex creek networks were included in restoration sites.

Two simple, but effective, measures of habitat complexity are creek density (the length of creeks per area of wetland) and the relative distribution of creeks of different order (first, second, third, etc., Desmond et al. in press). At Tijuana Estuary, a natural reference site in southern California, creek density is 0.007 in the least disturbed tidal arm, and about 45% of the length of channels is contributed by first-order creeks (*ibid.*).

Because the composition of tidal marshes is so closely linked to small changes in elevation, rates of erosion and accumulation should also be assessed. Uni-directional changes will suggest major losses of constructed habitats in the long term (e.g., filling-in of deepwater habitat constructed for fish, Simenstad and Thom 1996).

5. Are Restored Tidal Marshes Likely to Become Structurally Equivalent to Natural Reference Sites?

In a recent analysis of long-term data from the San Diego Bay marshes, Zedler and Callaway (in press) found little evidence that the constructed marsh would ever achieve the restoration target, namely providing nesting habitat (tall cordgrass) for an endangered bird. Three of the four indicators failed to show "progress," and the fourth (soil TKN) suggested it would be a long time to equivalency. Based on the 10-year sampling period, TKN was projected to achieve levels in the reference site in 20+ years. The curve for OM leveled off at about 75% of levels in the reference marsh. The essential vegetation attributes (cordgrass total stem length and the number of tall stems, i.e., those >90 cm) were erratic and appear to be declining over time. Nitrogen limitation is responsible for the failure of the constructed marsh to produce tall vegetation (Boyer and Zedler 1998); while additional factors, not assessed over time, probably contribute to the continuing decline in cordgrass heights. We know that sedimentation has raised the topography in some locations (Haltiner et al. 1997), and this favors highly competitive species of *Salicornia (S. virginica, S. bigelovii)*. We predict that large areas of the marsh will convert from cordgrass to other vegetation as time passes. Cordgrass is likely to remain the dominant near tidal creeks.

Simenstad and Thom (1996) followed 16 attributes in a long-term study of a brackish marsh and embayment and found three that reached asymptotes, and three measures of bird use that showed progress. Other attributes were erratic or declining over time. They concluded that most attributes were too variable to decide if the site would achieve functional equivalency with natural wetlands in the region. The marsh portion of the restoration project was still developing at seven years of age, with naturally-recruiting vegetation advancing onto the mudflat as sediment accreted. Although *Carex lyngbei* was planted over about 30% of the intertidal area, few of the transplants persisted. Some 56 vascular plant species colonized on their own, however, and the authors (*ibid.*) predicted continued conversion of mudflat to marsh. As reference data for brackish marsh were lacking, they had difficulty deciding if the vegetation was on a trajectory toward a natural target.

Too few salt marsh restoration sites have been studied long enough to draw generalizations about the potential for achieving structural or functional equivalency. More attributes need to be measured over longer periods of time (National Research Council 1992, Zedler and Callaway in press).

6. Conclusions

Functional equivalency is often stated explicitly or implied in the peer-reviewed literature concerning salt marsh restoration. However, the functions (rates, processes) are usually assumed to be indicated by structural attributes (one-time measures of condition).

Bradshaw's (1984) model of ecosystem degradation and restoration presents a linear relationship for structure and function. However, we found no support for linear relationships between structure and function or between function and time (successional stage). A restoration model without axes is preferable, given the current state of understanding.

Indices of functional equivalency and biotic integrity both pool structural and functional attributes. "Structural equivalency" would be a reasonable term for comparing the structure of paired ecosystems (restored and natural), and "biotic integrity" remains a good term for describing a site relative to a broader standard.

Useful indicators of salt marsh structure (and in some cases function) include: the composition, percent cover, biomass, density, and vertical structure of the vegetation; three soil attributes (texture, nitrogen pools, and organic matter concentrations); fish and invertebrate abundance and measures of their growth or size distributions; and topographic complexity (especially tidal creek density and the relative total length of creek orders).

Future research should more carefully consider the relationships between structure and function. Understanding how to restore tidal marshes to structural *and* functional equivalency with reference wetlands requires that we study more restoration sites, conduct longer-term studies, and assess more attributes at each location.

7. Acknowledgments

We thank the conference organizers for inviting this presentation. Work on this paper was supported in part by funding from the National Science Foundation (DEB-9619875) and Earth Island Institute. We thank Kandis Elliot, UW Botany Department, Senior Artist, for preparing the graphics and Kimberly Hamblin Hart for help with the final manuscript.

8. Literature Cited

Boyer, K. E. and J. B. Zedler. 1996. Damage to cordgrass by scale insects in a constructed salt marsh: effects of nitrogen additions. Estuaries **19**:1-12.

—— 1998. Effects of nitrogen additions on the vertical structure of a constructed cordgrass marsh. Ecological Applications **8**:692-705.

Bradshaw, A. D. 1984. Ecological principles and land reclamation practice. Landscape Planning **11**:35-48.

—— 1987. The reclamation of derelict land and the ecology of ecosystems. Pages 53-74 *in* W. R. Jordan, M.R. Gilpin and J. D. Aber, editors. Restoration ecology: a synthetic approach to ecological research. Cambridge University Press, Cambridge, England.

—— 1995. Alternative end points for reclamation. Pages 165-186 *in* J. Cairns, editor. Rehabilitating damaged ecosystems. Second edition. Lewis Publishers, Boca Raton, Florida, USA.

Chamberlain, R. H. and R. A. Barnhart. 1993. Early use by fish of a mitigation salt marsh, Humboldt Bay, California. Estuaries **16**: 769-783.

Chapin, F. S., O. E. Sala, I. C. Burke, J. P. Grime, D. U. Hooper, W. K. Lauenroth, A. Lombard, H. A. Mooney, A. R. Mosier, S. Naeem, S. W. Pacala, J. Roy, W. L. Steffen and D. Tilman. 1998. Ecosystem consequences of changing biodiversity: experimental evidence and research agenda for the future. BioScience **48**:45-52.

Craft, C. B., E. D. Seneca and S. W. Broome. 1991. Porewater chemistry of natural and created marsh soils. Journal of Experimental Marine Biology and Ecology **152**:187-200.

Craft, C. B., S. W. Broome and E. D. Seneca. 1988. Nitrogen, phosphorus and organic carbon pools in natural and transplanted marsh soils. Estuaries 11:272-280.

Deegan, L. A., J. T. Finn and J. Bounaccorsi. 1997. Development and validation of an estuarine biotic integrity index. Estuaries **20**:601-617.

Desmond, J., J. B. Zedler and G. D. Williams. 1999. Fish use of tidal creek habitats in two southern California salt marshes. Ecological Engineering **14**:233-252.

Dobson, A. P., A. D. Bradshaw and A. J. M. Baker. 1997. Hopes for the future: restoration ecology and conservation biology. Science **227**:515-522.

Fell, P. E., K. A. Murphy, M. A. Peck and M. L. Recchia. 1991. Re-establishment of *Melampus bidentatus* (Say) and other macroinvertebrates on a restored impounded tidal marsh: comparison of populations above and below the impoundment dike. Journal of Experimental Marine Biology and Ecology **152**:33-48.

Frenkel, R. E. and J. C. Morlan. 1991. Can we restore our salt marshes? Lessons from the Salmon River, Oregon. The Northwest Environmental Journal **7**:119-135.

Gibson, K. D., J. B. Zedler and R. Langis. 1994. Limited response of *Spartina foliosa* to soil amendments in constructed salt marshes. Ecological Applications **4**: 757-767.

Haltiner, J., J. B. Zedler, K. E. Boyer, G. D. Williams and J. C. Callaway. 1997. Influence of physical processes on the design, functioning and evolution of restored tidal wetlands in California (USA). Wetlands Ecology and Management **4**: 73-91.

Havens, K. J., L. M. Varnell and J. G. Bradshaw. 1995. An assessment of ecological conditions in a constructed tidal marsh and two natural reference tidal marshes in coastal Virginia. Ecological Engineering **4**:117-141.
Hoiland, W. K., F. W. Rabe and R. C. Biggam. 1994. Recovery of macroinvertebrate communities from metal pollution in the South Fork and mainstream of the Coeur d'Alene River, Idaho. Water Environment Research **66**:84-88.
Hugueny, B., S. Camara, B. Samoura and M. Magassouba. 1996. Applying an index of biotic integrity based on fish assemblages in a West African river. Hydrobiologia **331**:71-78.
Johnson, J. M. 1999. Fish use of a southern California salt marsh. Thesis, San Diego State University. San Diego, California, USA.
Karr, J. R. and D. R. Dudley. 1981. Ecological perspective on water quality goals. Environmental Management **5**:55-68.
Kerans, B. L. and J. R. Karr. 1994. A benthic index of biotic integrity (B-IBI) for rivers of the Tennessee Valley. Ecological Applications **4**:768-785.
Knott, D. M., E. L. Wenner and P. H. Wendt. 1997. Effects of pipeline construction on the vegetation and macrofauna of two South Carolina, USA, salt marshes. Wetlands **17**:65-81.
Kwak, T. J. and J. B. Zedler. 1997. Food web analysis of southern California coastal wetlands using multiple stable isotopes. Oecologia **110**:262-277.
Langis, R., M. Zalejko and J. B. Zedler. 1991. Nitrogen assessments in a constructed and a natural salt marsh of San Diego Bay, California. Ecological Applications **1**:40-51.
LaSalle, M. W., M. C. Landin and G. S. Jerre. 1991. Evaluation of the flora and fauna of a *Spartina alterniflora* marsh established on dredged material in Winyah Bay, South Carolina. Wetlands **11**:191-208.
Lindau, C. W. and L. R. Hosner. 1981. Substrate characterization of an experimental marsh and three natural marshes. Soil Science Society of America Journal **45**:1171-1176.
Magnuson, J. J., H. A. Reiger, W. J. Christie and W. C. Sonzogni. 1980. To rehabilitate and restore Great Lakes ecosystems. Pages 95-112 *in* J. Cairns, editor. The recovery process in damaged ecosystems. Ann Arbor Science, Ann Arbor, Michigan, USA.
Meffe, G. K., C. R. Carroll and Contributors 1997. Principles of conservation biology, second edition. Sinauer Associates, Inc. Sunderland, Massachusetts, USA.
Miller, J. A. and C. A. Simenstad. 1997. A comparative assessment of natural and created estuarine slough as rearing habitat for juvenile chinook and coho salmon. Estuaries **20**:792-806.
Minello, T. J. and R. J. Zimmerman. 1992. Utilization of natural and transplanted Texas salt marshes by fish and decapod crustaceans. Marine Ecology Progress Series **90**:273-285.
Minns, C. K., V. W. Cairns, R. G. Randall and J. E. Moore. 1994. An index of biotic integrity (IBI) for fish assemblages in the littoral zone of Great Lakes' areas of concern. Canadian Journal of Fisheries and Aquatic Systems **51**:1804-1822.
Mitsch, W. J. and J. G. Gosselink. 1993. Wetlands, second edition. Van Nostrand Reinhold, New York, New York, USA.
Moy, L. D. and L. A. Levin. 1991. Are *Spartina* marshes a replaceable resource? A functional approach to the evaluation of marsh creation efforts. Estuaries **14**:1-16.
Naeem, S., L. J. Thompson, S. P. Lawler, J. H. Lawton and R. M. Woodfin. 1994. Declining biodiversity can alter the performance of ecosystems. Nature **368**:734-737.
National Research Council. 1992. Restoration of aquatic ecosystems. National Academy Press, Washington, District of Columbia, USA.
Pacific Estuarine Research Laboratory (PERL). 1990. A manual for assessing restored and natural coastal wetlands with examples from southern California. California Sea Grant Report No. T-CSGCP-**021**. La Jolla, California, USA.
—— 1997. The status of constructed wetlands at Sweetwater Marsh National Wildlife Refuge. Annual report to the California Department of Transportation, District 11, San Diego, California, USA.
Peck, M. A., P. E. Fell, E. A. Allen, J. A. Gieg, C. R. Guthke and M. D. Newkirk. 1994. Evaluation of a tidal marsh restoration: comparison of selected macroinvertebrate populations on a restored impounded valley marsh and an unimpounded valley marsh within the same salt marsh system in Connecticut, USA. Environmental Management **18**:283-293.
Roth, N. E., J. D. Allen and D. L. Erickson. 1996. Landscape influences on stream biotic integrity assessed at multiple spatial scales. Landscape Ecology **11**:141-156.
Rulifson, R. A. 1991. Finfish utilization of man-initiated and adjacent natural creeks of South Creek Estuary, North Carolina, using multiple gear types. Estuaries **14**:447-464.

Sacco, J. N., E. D. Seneca and T. R. Wentworth. 1994. Infaunal community development of artificially established salt marshes in North Carolina. Estuaries **17**:489-500.

Scatolini, S. R. and J. B. Zedler. 1996. Epibenthic invertebrates of natural and constructed marshes of San Diego Bay. Wetlands **16**:24-37.

Seneca, E. D., S. W. Broome, W. W. Woodhouse, Jr., L. M. Cammen and J. T. Lyon, III. 1976. Establishing *Spartina alterniflora* marsh in North Carolina. Environmental Conservation **3**:185-188.

Shields, F. D., S. S. Knight and C. M. Cooper. 1995. Use of the index of biotic integrity to assess physical habitat degradation in warm water streams. Hydrobiologia **312**:191-208.

Shreffler, D. K., C. A. Simenstad and R. M. Thom. 1990. Temporary residence by juvenile salmon in a restored estuarine wetland. Canadian Journal of Fisheries and Aquatic Sciences **47**:2079-2984.

Simenstad, C. A. and R. M. Thom. 1996. Functional equivalency trajectories of the restored Gog-Le-Hi-Te estuarine wetland. Ecological Applications **6**:38-56.

Tilman, D. and J. A. Downing. 1994. Biodiversity and stability in grasslands. Nature **367**:363-365.

Tilman, D., D. Wedin and J. Knops. 1996. Productivity and sustainability influenced by biodiversity in grassland ecosystems. Nature **379**:718-720.

Webb, J. W. and C. J. Newling. 1985. Comparison of natural and man-made salt marshes in Galveston Bay complex, Texas. Wetlands **4**:75-86.

Whittaker, R. H. 1975. Communities and ecosystems, second edition. Macmillan Publishing Company, Inc., New York, New York, USA.

Williams, G.D. and J.B. Zedler. 1999. Fish assemblage composition in constructed and natural tidal marshes of San Diego Bay: Relative influence of channel morphology and restoration history. Estuaries **22**:702-716.

Zalejko, M. K. 1989. Nitrogen fixation in a natural and a constructed southern California salt marsh. Thesis, San Diego State University, San Diego, California, USA.

Zedler, J. B. 1993. Canopy architecture of natural and planted cordgrass marshes: selecting habitat evaluation criteria. Ecological Applications **3**:123-138.

Zedler, J.B. and J.C. Callaway. In press. Evalutating the progress of engineered tidal wetlands. Ecological Engineering.

Zedler, J. B., J. C. Callaway, J. Desmond, G. Vivian-Smith, G. Williams, G. Sullivan, A. Brewster and B. Bradshaw. 1999. Californian salt marsh vegetation: an improved model of spatial pattern. Ecosystems **2**:19-35.

Zedler, J. B. and R. Langis. 1991. Comparisons of constructed and natural salt marshes of San Diego Bay. Restoration & Management Notes **9**:21-25.

Zedler, J. B. and C. S. Nordby. 1986. The ecology of Tijuana Estuary: an estuarine profile. U.S. Fish Wild. Serv. Biol. Rep. **85** (7.5).

Zedler, J. B. and A. Powell. 1993. Problems in managing coastal wetlands: complexities, compromises and concerns. Oceanus **36**:19-28.

ORGANIC AND INORGANIC CONTRIBUTIONS TO VERTICAL ACCRETION IN SALT MARSH SEDIMENTS

R. E. TURNER
E. M. SWENSON
C. S. MILAN
Coastal Ecology Institute,
Louisiana State University
Baton Rouge, LA 70803 USA

Abstract
The contribution of organic and inorganic constituents to recent vertical accretion rates (since 1963/4) was estimated for 141 salt marshes ranging from New England to the Gulf of Mexico. The range of vertical accretion and inorganic accumulation rates were 0.09 to 1.78 cm y^{-1}, and 0.01 to 0.53 g cm y^{-1}, respectively. The volume of the accumulated organic and inorganic in all salt marshes averaged 3.8 and 4.9%, respectively, of the total, which is relatively low among soil types. The remaining soil volume is water and air. There was a direct relationship between vertical accretion and organic accumulation that explained 59% of the variation for all samples combined. In contrast, the bulk density is strongly and directly related to inorganic content, but not the vertical accretion rate. A multiple regression equation describing the vertical accumulation as a function of mineral and organic accumulation suggests that organic accumulation is five times more important than inorganic accumulation for East coast salt marshes ($n = 19$; weight basis), but that inorganic content is a statistically-insignificant factor for Gulf of Mexico salt marshes ($n = 122$), or for all salt marshes examined ($n = 141$). A simple linear regression showed that a 1 cm rise in salt marsh elevation was composed of 10.9 g of organic matter. A threshold level of 0.02 g organic matter y^{-1} accumulation can continue without inorganic accumulation. It appears that it is the accumulation of organic matter that controls inorganic accumulation in established marshes, not the reverse. These results document the dominant role of below ground plant material in maintaining salt marshes once they are established. When wetland hydrology is altered, it is the organic soil constituents that are affected (through oxidation or plant growth below ground), thus explaining salt marsh conversion to open water through indirect changes in hydrology. Salt marsh management and restoration efforts would do well to keep in mind the plant's health, especially belowground, if the long-term and effective strategies are to be successfully implemented. The biological components, not the geological components, appear to control the fate of established salt marshes.

> "There is no other case in nature, save in the coral reefs, where the adjustment of organic relations to physical conditions is seen in such a beautiful way as the balance between the growing marshes and the tidal streams by which they are at once nourished and worn away." (Shaler, 1886).

"My belief is that the present low sea marsh does not owe its origin to any silt depositing river, but is simply an accumulation of vegetable matter." (C. W. Thomas, of the Louisiana Land and Exploration Company, in a letter about the Louisiana coast; cited in Nesbitt, 1885).

1. Introduction

Coastal salt marshes have accumulated enough material *in situ* to survive for the last 7,000 years in the macrotidal environments of the Bay of Fundy and the microtidal regimes of the Gulf of Mexico (Redfield and Rubin 1962, Coleman and Smith 1964). This accumulation, consisting of both organic and inorganic material, has varied among salt marshes and over centuries as the underlying strata rises or sinks, or sea level changes, or because of the variability in the source materials and quantity, plant growth and belowground decomposition, and re-suspension. Whether the net accumulation rate equals the relative sea level rise is of particular interest now because the present sea level rise (about 0.23 cm y^{-1}; Gornitz et al. 1982) is projected to double, or more, in the next century, thus equaling the relative sea level rise of 4,000 years BP, and perhaps threatening coastal wetland stability (e.g., Titus 1991).

The accumulated volume in the upper 2 m of a salt marsh contains water, air space, live and dead plant organics, and inorganic materials. Some of the surface accretion volume is eventually reduced through compaction, organic oxidation, water loss, and shrinkage. The annual belowground additions of live plant roots also undergo decomposition. When organic material is lost, then the soil's ability to hold water volume and sustain pore space is eventually diminished. Most of these volumetric changes apparently take place within the first several hundred years of accumulation. The subsidence of wetland soils in Louisiana after 200 years, for example, is relatively very small compared to the first 100 years, or even negligible (Turner 1991). These volumetric changes have the potential to be substantial and rapid. DeLaune et al. (1994), for example, described how the soil surface of a brackish marsh (dominated by *Spartina patens)* fell 15 cm within 2 years after the above-surface plant died.

Inorganic materials are essential to the creation of salt marshes. *Spartina alterniflora,* the dominant emergent macrophyte of US salt marshes, has a limited tolerance that varies in proportion to the tidal range (McKee and Patrick 1988) and will not become established if flooding is too great. The forefront of expanding salt marshes is usually mostly composed of inorganic materials. However, once *S. altemiflora* is rooted and spreads, then it forms a physical structure that traps some material (Wang 1997, Stumpf 1983), resists erosion and, importantly, adds organic material on the surface and belowground. In other words, the maintenance of salt marshes involves an additional constellation of forces than those found only during their formation.

The relative significance of the inorganic and organic materials to net salt marsh sedimentation and maintenance is unclear and may be quite dependent on tidal range, the underlying substrata, or climate. Callaway et al. (1997) demonstrated a strong statistical relationship between the vertical accretion rate and organic accumulation in 5 Gulf of Mexico coastal wetlands, and Nyman et al. (1993) discussed the volumetric significance of organic materials in 15 salt and brackish marshes in one south Louisiana watershed. DeLaune et al. (1979), however, suggested that there was a minimum inorganic content necessary in salt marshes. Many of Louisiana's coastal wetland restoration efforts involve re-routing sediments to degraded marshes (Turner 1997) under the assumption that their regression was caused by inorganic sediment deprivation (Day and Templet 1989, Anon 1995).

We measured the accretion rate and content in 55 salt marshes from the northern Gulf of Mexico and 3 in Rhode Island, and compared the results to literature values, where appropriate. These data were used to establish the relative contribution of organic and inorganic material to the vertical accumulation rate. We thought that a fundamental relationship would be exposed if we could find a pattern in these data, despite the natural variability, and also because of the technical issues related to sampling.

2. Methods

2.1 SAMPLE SITES

Salt marsh sites were sampled in 1991 and 1993 in each of three Louisiana coastal watersheds: Terrebonne, Barataria and St. Bernard (Fig. 1). The resulting distribution of sites is shown in Fig. 1. Additional samples were taken in coastal Texas in 1993, and from upper Narragansett Bay (Palmer and Barrington Rivers, Rhode Island) in 1994. Data from others are from the literature. The Texas and Louisiana sites were selected prior to field sampling using various maps indicating that salt marsh vegetation was present (the procedures are described in Turner et al. 1995).

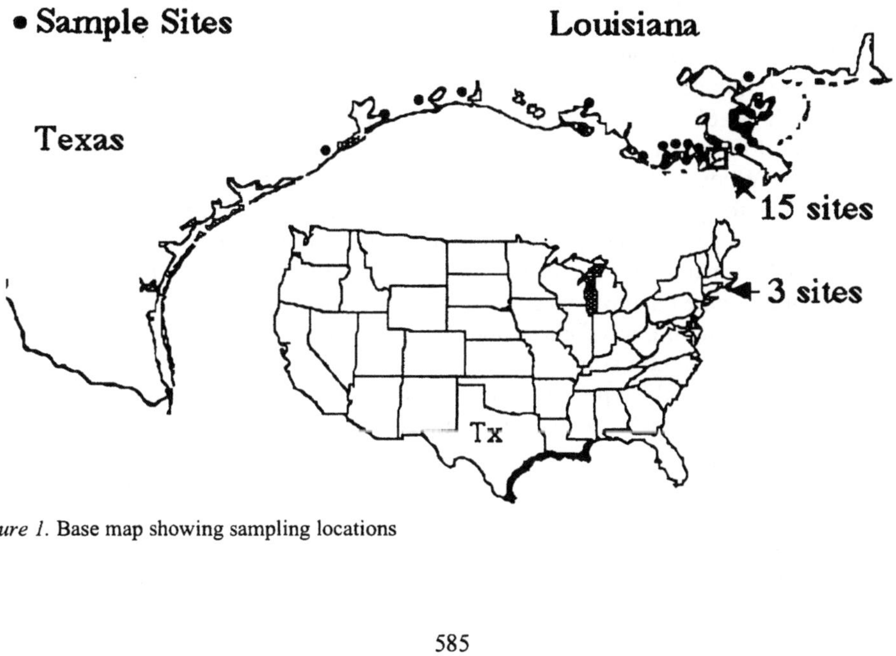

Figure 1. Base map showing sampling locations

2.2 SEDIMENT CORE COLLECTION

Salt marshes were sampled at 50 m inland to ensure that edge effects did not influence the data. A single core was taken from most sites, and triplicate cores taken from 10 sites. We inserted (twisted) a 12 cm diameter stainless steel core tube into the sediment between plants to within 3 cm of the marsh surface and were careful not to compress the core. The depth of the core surface inside and outside the tube was recorded during core collection to determine if there were any sediment compaction. The cores were initially refrigerated upon return to the laboratory and later frozen. The frozen core was extruded from the core tube by placing the sample in a core defroster until the edges thawed sufficiently to allow the core to be pushed out of the tube. The core was extruded into a pre-labeled plastic bag and returned to the freezer until sectioning.

2.3 SEDIMENT CORE ANALYSIS

The distance from the top of the tube to the frozen mud surface was re-measured before the core was extruded to document compaction or expansion during storage. The average total compaction post-collection was 17% of the total core length, of which an average 4.7‰ was due to freezing. No core was used with a compaction greater than 20‰. Whatever compaction occurred was assumed to be homogeneously distributed throughout the core. The extruded frozen cores were sectioned into 1 cm increments using a band saw. The blade thickness (1 mm) was included in the increment measurement to ensure that the depth to the top of each consecutive section was always a 1 cm interval from the top of the previous section. The thickness of every fifth section was measured with a digital micrometer to ensure overall accuracy of the technique. Using this approach we can cut sections with a mean thickness of 0.98 ± 0.2 cm (mean $\pm$ two standard deviations). The material lost during cutting was therefore known, and subsequent calculations of constituent concentrations were adjusted by this amount to estimate the actual *in situ* amount in the core sections. Cut sections were placed into weighed and pre-labeled dishes. The dishes with the core sections were then weighed, dried for 24 h at 60 °C, and then weighed again. This procedure allows for the determination of both wet and dry bulk densities. The dried sections were homogenized with a Thomas-Wiley Mill equipped with a #20 mesh screen. A sub-sample of 1 to 2 grams was taken for percent organic content determination. The percent organic matter for all sections was determined by weighing the sample in a clean pre-weighed crucible, ashing at 550 °C for one hour, and then re-weighing after cooling (APHA 1992). The remaining homogenized sample was placed in a weighed and pre-labeled 26 cm^3 petri dish with a tight fitting lid, reweighed, and then analyzed for ^{137}Cs.

Samples were counted using a Princeton Gamma-Tech 60 mm diameter hyperpure germanium (HPGe) ''N'' type coaxial detector. The detector was interfaced to an EG&G Ortec 92X® spectrum master integrated gamma-spectroscopy system. Data was acquired using EG&G's Maestro II® software and analyzed with EG&G's Minigam® software. Samples were analyzed continuously using a Gamma Products sample changer and G3200® software that interfaces the sample changer with the EG&G software.

The detector was calibrated for the respective geometry and sample matrix using a certified mixed standard (Amersham Corporation; QCY.44). The calibration was checked monthly. Samples were counted for 4 h, which yielded a counting error of 10-20% for samples around the 1963/64 peak. Results for ^{137}Cs are expressed as pCi/g (dry weight). Detection limit for ^{137}Cs, depending on sample weight, was ≈ pCi g^{-1} dry $wt.^{-1}$ with a counting error of ≈ 45%. Longer counting times would have yielded better counting error statistics, but no alteration in accumulation rates was obtained by replicate analyses of samples around the 1963/64 peak from 12 cores.

Literature values for accretion rates and constituents are from Rhode Island (Bricker et al. 1989), Long Island (McCaffrey and Thomson 1980, Armentano and Woodwell 1975), North Carolina (Craft et al. 1993), South Carolina (Vogel et al. 1996), Florida (Callaway et al. 1997), Mississippi (Callaway et al. 1997), Louisiana (DeLaune et al. 1989, Nyman et al. 1993), and Texas (Callaway et al. 1997).

The accumulated organic and inorganic material (since 1963/4) was converted to a volumetric basis assuming 1.2 and 2.4 g cm^{-3}, respectively, by relying on the literature values shown in Table 1. Simple linear and multiple regression statistics used Statview II® Version 1.04 (Abacus Concepts, Inc., 1984 Bonita Ave., Berkeley, California 94704).

3. Results

Appendix 1 includes the data for accretion rate and accumulation rates for all data collected in Louisiana and Texas as part of this study. Estimates of bulk density measurments are illustrated in Table 1.

TABLE 1. Estimates of bulk density measurements in salt marsh soils.

Parameter	Source
A. Organic Matter (g cm^{-3})	
1.14	Callaway et al. 1997 (citing Smith 1943 and DeLaune et al. 1983)
1.34	Craft et al. 1993
1.24	Craft et al. 1993
B. Mineral Matter (g cm^{-3})	
2.61	Callaway, et al. 1997, (citing Smith 1943 and DeLaune et al. 1983)
2.4	Craft et al. 1993
2.5	Craft et al. 1993

The vertical accretion rate for all salt marshes (n = 141) is directly related to the accumulation rate of organic matter, but not inorganic matter (Fig. 2). Table 2 has the results of a multiple regression analysis of the two variables, separated by geographic areas or, in the case of large data sets, investigators. The average vertical accretion for all sites was an approximate 10.9 cm y^{-1} accretion for each 1 g cm^{-3} y^{-1} of organic matter. Individual station groups showed similar relationships, but there was more scatter in the slopes. The slope for all Long Island Sound salt marshes (n = 12) had a negative relationship between these two variables in this multiple regression analysis, and was in contrast to the other regional groupings for Rhode Island, the Mid-Atlantic, and any grouping for the Gulf of Mexico. The larger data set for the entire east coast (n = 19) had a positive and significant relationship between both inorganic and organic accumulation rate and the dependent value, vertical accretion rate. There was a statistically-significant relationship between organic accumulation and accretion rate for the Gulf of Mexico data set (n = 122), but inorganic accumulation was not statistically-significant in the multiple regression analysis. Nine of the 16 regional groupings had either a negative slope or statistically-insignificant slope for inorganic accumulation in the multiple regression analysis. There were five groupings that had a positive and statistically-significant slope between inorganic and organic accumulation and accretion rate, and, in each case, the organic accumulation had a larger slope than the inorganic accumulation (comparison based on a weight per volume basis). Each of these five groupings were located on the east coast.

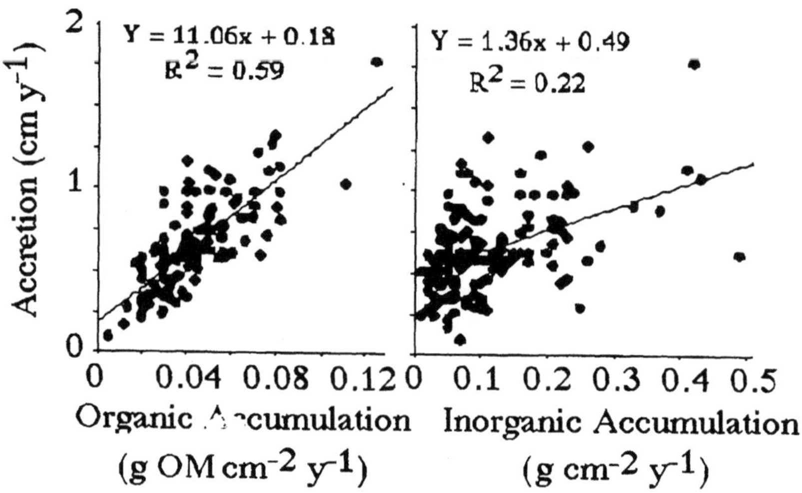

Figure 2. The relationship between the vertical accretion rate and the accumulated organic (left panel) and inorganic material (right panel). The data are for post 1963/4 accumulations.

The slope of the linear relationship between organic accumulation rate and vertical accretion was less variable than in the multiple regression analysis (Table 3). The vertical accretion rate increased 1 cm for each 10 g cm^{-2} accretion of organic matter, and varied between 9 to 11.1 g organic cm^{-2} in the three geographic regions (Long Island Sound, Mid Atlantic and Gulf of Mexico). There was no statistically-significant difference in these slopes among these three regions.

The volume of organic and inorganic matter averaged for all cores was 3.8% (range = 0.4 to 10.4%) and 4.9% (range 0.4 to 20.4%), respectively, of the volume in the individual cores. The volume of each is therefore a minor component of the total volume, which otherwise consists of water, but also some air spaces trapped in plant tissues and soil.

The average soil bulk density in the post 1963/4 horizon was strongly and directly related to the inorganic content and inversely (curvilinearly) related to the % organic content (Fig. 3). Gosselink et al. (1984) found a similar result for surface samples from Louisiana salt marshes.

TABLE 2. Results from a multiple regression analysis of the independent value, vertical accretion rate ($cm^{-2}\ y^{-1}$), and the dependent values organic and inorganic accumulation ($g\ cm^{-3}y^{-1}$). Adj. R^2 is the adjusted Coefficient of Determination Symbols: * and ** = significant at the 95% and 99% confidence level, respectively.

Location	n	Adj. r^2	Slope Organic	Slope Inorganic	Accretion Rate Range ($cm\ y^{-1}$)
Rhode Island (Bricker et al. 1989)	6	0.96**	4.67*	1.37**	0.24-0.60
Rhode Island (this study)	3	0.97**	3**	2**	0.21-0.34
All Rhode Island	9	0.97**	4.83**	1.41**	0.21-0.60
Long Island (Armentano and Woodwell 1975; McCaffrey and Thomson 1980)	3	1.0**	-26.7**	8.67**	0.35-0.63
All Long Island Sound	12	0.79**	n. s.	1.83**	0.21-0.63
North Carolina (Craft et al. 1993)	4	0.97**	13.9**	n. s.	0.09-0.36
South Carolina (Vogel et al. 1996)	3	1.0 **	16.9**	-0.27**	0.16-0.35
Mid-Atlantic	7	0.95**	12.5**	0.8*	0.09-0.36
All East coast	19	0.79**	5.82**	1.15**	0.09-0.63
Texas and Florida (Callaway et al. 1997)	5	0.92**	17.0**	n.s.	0.27-0.62
Terrebonne (LA) (Nyman et al. 1993)	15	0.81**	12.9**	n.s.	0.55-1.78
Louisiana (DeLaune et al. 1989)	47	0.49**	13.7**	n.s.	0.22-1.17
Louisiana and Texas (this study)	55	0.66**	7.98**	n.s.	0.28-1.14
All Gulf of Mexico except DeLaune et al. 1989	75	0.67**	10.9**	n.s.	0.22-1.78
All Gulf of Mexico	122	0.56**	10.2**	n.s.	0.1-1.78
All data	141	0.59**	10.9**	n.s.	0.09-1.78

TABLE 3. Simple linear regression statistics for the relationship between the dependent variable vertical accretion rate ($cm\ y^{-1}$) and the independent variable accumulation of organic matter ($g\ cm^{-2}y^{-1}$). All data are for the depth interval above the 1963/4 horizon. The average inorganic accumulation rate ($g\ cm^{-2}y^{-1}$) is included in the last column.

	n	Slope	Std. Error	Adj. r^2	Average inorganic rate ($g\ cm^{-2}\ y^{-1}$)
Long Island Sound	12	7.9	2.11	0.54	0.097
Mid Atlantic	7	11.8	1.95	0.86	0.066
East Coast	19	7.9	1.21	0.70	0.085
Gulf of Mexico	122	10.2	1.00	0.56	0.12
All	141	11.06	0.78	0.59	0.12

The relationship between the annual accumulation of organic and inorganic material for the post 1963/4 period is shown in Fig. 4. A polynomial fit of the data is shown (Adjusted r^2 = 0.40). The relationship between the 2 variables is positive, as expected. However, it is also curvilinear at organic accumulation values above 0.06 g organic y^{-1}, which is equivalent to about 0. 8 cm y^{-1} vertical accretion (Fig. 2). The Y intercept for this plot is positive (at about 0.02 g organic y^{-1}) and at a point in Fig. 2 where there is 0.2 cm y^{-1} vertical accretion. This result supports, but does not prove, the hypothesis that there is a minimum amount of organic matter (not inorganic matter) that must occur for the salt marsh to survive a relative sea level rise of 0.2 cm y^{-1}. That is to say, the accumulation of organic matter appears to control inorganic accumulation, not the reverse, and, there is a threshold level of 0.02 g organic y^{-1} accumulation that can survive without inorganic accumulation.

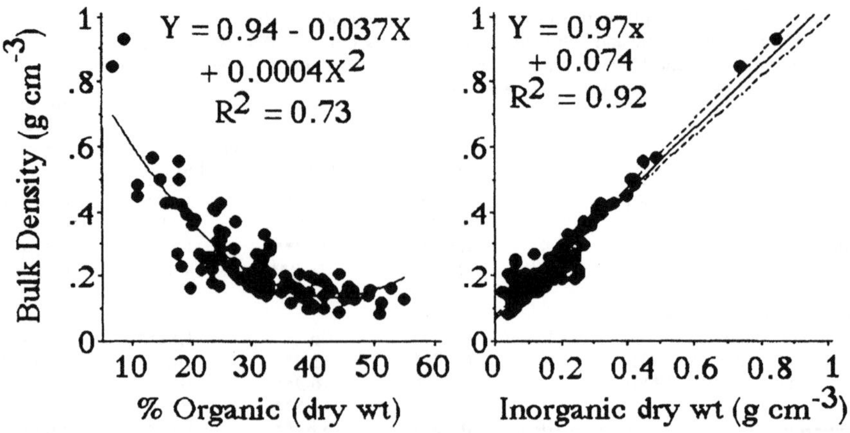

Figure 3. The relationship between bulk density and both inorganic and organic content. A polynomial regression is plotted for the left panel and a linear regression of the data for the right panel.

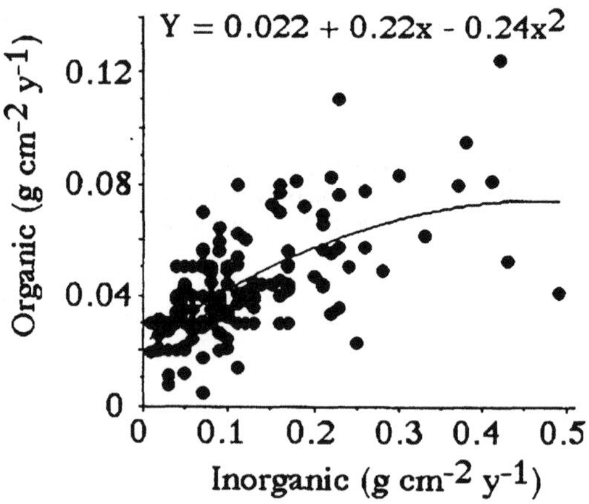

Figure 4. The relationship between the annual accumulation of organic and inorganic material accumulating after 1963/4. A polynomial fit of the data is shown. (Adjusted $r^2 = 0.40$)

4. Discussion

These robust results demonstrate the dominant influence of organic matter accumulation on salt marsh vertical accretion. The inorganic material controls the density of material accumulating, but not the vertical accretion rate. It appears that inorganic materials may play an important role in vertical accretion rates on the east coast (but not on the GOM coast), although these are of lesser importance than organic accumulation (on a weight basis).

While there were no obvious relationship between tidal range or climate and the importance of organic or inorganic content to accretion rate that was uncovered during this analysis, we do not mean to imply that such relationships do not exist. We acknowledge that measuring errors are introduced for a variety of sampling and methodological reasons. The relationships described here might be improved upon with further details on the soil types or consistent interpretation of the dating techniques. Some investigators, for example, use 1963 as the ^{137}Cs peak, and others use the monthly values which had peaks in both 1963 and 1964. Milan et al. (1995) estimated that there was more ^{137}Cs deposited in Louisiana sediments in 1964 compared to 1963, for example. Other considerations that may influence the variability among locations include the thickness of soil segments sampled, inorganic particle size distributions, and core compaction.

Salt marsh soils are built over mostly inorganic materials accumulating before plant colonization. It is a perhaps beguiling transfer of perspectives to assume that they are like the soils in nearby upland environments, and to consider them as a mostly geochemical environment with only some slight biological influences. Once vegetation is established, the dynamics of soil growth change dramatically. The plant roots extend more than 1 meter below ground in many salt marshes. These roots and the associated microbial community affects the material accumulating over 100 y (assuming an accretion rate of 1 cm y^{-1}. Less than 5% of the salt marsh soil volume is inorganic material. The rest of this volume is composed of organic material and water. The below ground environment remains in a dynamic state and can be easily affected by hydrologic events. The results of some analyses of freshwater peats might therefore be useful to put these results in perspective. Hobbs (1986) said of peat soils: ''Undoubtedly the most striking characteristic of peat is its ability to hold water. Five metres of fibrous peat may contain 4.7 in of water and as little as 300 mm of solid plant matter and yet possess a significant shear strength and . . . behave in many respects like normally consolidated clay.'' What he is saying is that the organic material is the matrix within which the soil volume, which is mostly water and therefore neutrally-buoyant, is formed. His comment suggests a ratio of 15.7, which compares favorably to those in Table 3. Eggelsmann (1976) showed how quickly this volume may be lost in peats, which might be at a rates of 1 to 2 cm y^{-1} for peats from northern latitudes, and 10 cm in the Gulf of Mexico (Okey 1915).

Volumetric changes in organics, perhaps resulting from an altered hydrology, could be caused indirectly by the reduction in net accumulation below ground (e.g., from sulfide toxicity), increased decomposition rates (from increased drying), or both. These hydrologic affects have been documented many times and can be subtle (e.g., Tate 1979). For example, the indirect effects of semi-impounding a Louisiana salt marsh appears to have caused increased drying periods and the duration of flooding (Swenson and Turner 1987). The freshwater swamp at the famous Holmes Post (England) is still sinking after 100+ y of drainage (Fillenhain 1963). The below ground losses of organic material would be permanent, but the inorganic material would remain as the surface sinks (unless it eroded, of course).

Increasing inorganic sediments to a marsh would not, apparently, have the same effect as the equal volumetric increases in below ground organic material, whose volumetric leverage is about 10:1. These results therefore have some bearing on restoration efforts. Massive additions of inorganic material to a degraded salt marsh can restore or create a salt marsh. But such an approach will be a brute force action whose long-term success may remain dependent on the below ground accumulation of organics. In the long run, perhaps over 100 y, the organic materials will eventually be needed to replace this volumetric addition that is subject to sea level rise and subsidence. The long-term perspective for restoration must consider the plant health and its contribution to salt marsh stability. If the plant dies for lack of adequate hydrologic conditions, then the marsh will not survive without considerable additional human engineering and maintenance.

Finally, if this analysis is substantially accurate, then it forms a basis (once again) for recognizing that our knowledge of salt marsh ecosystems remains shallow and inadequate for some management. If the role of the plants and organic matter could have been recognized earlier, when the role of inorganic material was so well accepted, then perhaps some of Louisiana's coastal marshes (and others) would not have been lost (see Turner 1997 for a discussion of the conflicting hypotheses leading to the dramatic wetland losses on the Louisiana coast). A slow recognition of the subtle effects of changing estuarine hydrology may have inexorably delayed the application of useful management measures, to the detriment of wetland conservation and restoration. Humility and inquisitiveness mixed with quantitative analyses have a major role to play, in this regard, if given a chance.

5. Acknowledgments

This research was supported by the Environmental Monitoring and Assessment Program (EMAP) administered by the Environmental Protection Agency's (EPA) Office of Research and Development, Dr. K. Summer, Contract Officer. We thank the following for assistance in the field sampling: A. Bass, I. Hesse, G. Holm, D. Justic', J. M. Lee, S. Nesbitt, T. A. Oswald, and G. W. Peterson. We thank Dr. J. Callaway and two anonymous reviewers for helpful comments of an earlier manuscript.

6. Literature Cited

Anon. 1995. A white paper: the state of Louisiana's policy for coastal restoration activities. Signed by the Governor, the Secretary and Assistant Secretary of the Louisiana Department of Natural Resources and the Executive Assistant of the Governor's Office of Coastal Activities. Dated 24 April,. 1995.

APHA 1992. Standard methods for the examination of water and wastewater, 18th editon. American Public Health Association, Inc. New York, New York, USA.

Armentano, T. V. and G. M Woodwell. 1975. Sedimentation rate in a Long Island marsh determined by ^{210}Pb dating. Limnology and Oceanography **74**:755-774.

Bricker-Urso, S., S. W. Nixon, J. K. Cochran, D. J. Hirschberg and C. Hunt. 1989. Accretion rates and sediment accumulation in Rhode Island salt marshes. Estuaries **12**:30-317.

Callaway, J. C., R. D. DeLaune and W. H. Patrick, Jr. 1997. Sediment accretion rates from four coastal wetlands along the Gulf of Mexico. Journal of Coastal Research **13**:181-191.

Coleman, J. M. and W. G. Smith. 1964. Late recent rise of sea level. Geological Society of American Annual Bulletin **75**:833-840.

Craft, C. B., E. D. Seneca and S. W. Broome. 1993. Vertical accretion in microtidal regularly and irregularly flooded estuarine marshes. Estuarine, Coastal and Shelf Science. **27**:371-386.

Day, J. W. Jr. and P. H. Templet. 1989. Consequences of sea level rise: implications from the Mississippi Delta. Coastal Management **17**:241-257.

DeLaune, R. D., R. J. Buresh and W. H. Patrick, Jr. 1979. Relationship of soil properties to standing crop biomass of *Spartina altemiflora* in a Louisiana marsh. Estuarine and Coastal Marine Science **8**:477-487.

DeLaune, R. D., R. H. Baumann and J. G. Gosselink. 1983. Relationships among vertical accretion, coastal submergence and erosion in a Louisiana Gulf Coast marsh. Journal Sedimentary Petrology **53**:147-157.

DeLaune, R. D., J. H. Whitcomb, W. H. Patrick, Jr., J. H Pardue and S. R. Pezeski. 1989. Accretion and canal impacts in a rapidly subsiding wetland. I. ^{137}Cs and ^{210}Pb techniques. Estuaries **12**:247-259.

DeLaune, R. D., J. A. Nyman and W. H. Patrick, Jr. 1994. Peat collapse, ponding and wetland loss in a rapidly submerging coastal marsh. Journal of Coastal Research **10**: 1021-1030.

Eggelsmann, R. 1976. Peat consumption under influence of climate, soil condition and utilization. Proceeding International Peat Congress. Vol. 1, Poznan, Poland, pp. 233-247.

Fillenham, L. F. 1963. Holmes Fen Post. Geographical Journal **129**:502-503.

Gornitz, V., S. Lebedeff and J. Hansen. 1982. Global sea-level trend in the past century. Science **215**:1611-1614.

Gosselink, J. G., R. Hatton and C. S. Hopkinson. 1984. Relationship of organic carbon and mineral content to bulk density in Louisiana marsh soils. Soil Science **137**:177-180.

Hobbs, N. B. 1986. Mire morphology and the properties and behaviour of some British and foreign peats. Quarterly Journal of Engineering Geology (London) **19**:7-80.

McCaffrey, R. J. and J. Thomson. 1980. A record of the accumulation of sediment and trace metals in a Connecticut salt marsh. Pages 165-236 *in* B. Saltzman, editor. Advances in geophysics, estuarine physics and chemistry. Studies in Long Island Sound, Vol. 22. Academic Press, New York, New York, USA.

McKee, K. L. and W. H. Patrick, Jr. 1988. The relationship of smooth cordgrass *(Spartina alterniflora)* to tidal datums: a review. Estuaries **11**: 143-15 1.

Milan, C. S., E. M. Swenson, R. E. Turner and J. M. Lee. 1995. Accumulation rates estimated from ^{137}CS activity: Variability in Louisiana salt marshes. Journal of Coastal Research **11**:296-307.

Nesbitt, D. M. 1885. Tide marshes of the United States. Department of Agriculture Miscellaneous Special Report No. 7. United States Department of Agriculture, Washington, District of Columbia, USA.

Nyman, J. A., R. D. DeLaune, H. H. Roberts and W. H. Patrick, Jr. 1993. Relationship between vegetation and soil formation in a rapidly submerging coastal marsh. Marine Ecology Progress Series **96**:269-279.

Okey, C. W. 1918. The subsidence of muck and peat soils in southern Louisiana and Florida. American Society of Civil Engineers **82**:396-422.

Redfield, A. C. and M. Rubin. 1962. The age of salt marsh peat and its relation to recent changes in Barnstable Harbor, Massachusetts. Proceedings of the National Academy of Sciences **48**:1728-1735.

Shaler, N. S. 1886. Preliminary report on seacoast swamps of the Eastern United States. U.S. Geological Survey 6th Annual Report, U.S. Geological Survey, Washington, District of Columbia, USA.

Smith, W. 1943. Density of soil solids and their genetic relations. Soil Science **56**:263-272.

Stumpf, R. P. 1983. The process of sedimentation on the surface of a salt marsh. Estuarine, Coastal and Shelf Science **17**:495-508.

Swenson, E. M. and R. E. Turner. 1987. Spoil banks: effects on coastal marsh water level regime. Estuarine, Coastal and Shelf Science **24**:599-609.

Tate, R. L. 1979. Effect of flooding on microbial activities in organic soils: carbon metabolism. Soil Science **128**:267-273.

Titus, J. G. 1991. Greenhouse effect and sea level rise: the cost of holding back the sea. Coastal Management **19**:171-204.

Turner, R. E. 1991. Tide gage records, water level rise and subsidence in the northern Gulf of Mexico. Estuaries **14**:139-147.

—— 1997. Wetland loss in the northern Gulf of Mexico: multiple working hypotheses. Estuaries **20**:1-13.

Turner, R. E. and R. R. Lewis, 111. 1996. Hydrologic restoration of coastal wetlands. Wetlands Ecology and Management **4**:65-72.

Turner, R. E., E. M. Swenson and J. K. Summers. 1995. Coastal Wetlands Indicator Study: EMAP Estuaries Louisiana Province - 199 1. US EPA, Office of Research and Development, Environmental Research Laboratory, Gulf Breeze, Florida, USA. EPA/620/R-**95/005**.

Vogel, R. L., B. Kjerfve and L. R. Gardner 1996, Inorganic sediment budget for the North Inlet salt marsh, South Carolina. Mangroves and Salt Marshes **1**:23-35.
Wang, F. C. 1997. Dynamics of intertidal marshes near shallow estuaries in Louisiana. Wetlands Ecology and Management **5**:131-143.

Appendix 1. Data collected as part of this study. Accumulation rates are for the period since 1963/4.

Sample	Location	Year collected	ID	Vertical Accumulation cm y^{-1}	Organic Accumulation g cm^{-2} y^{-1}	Inorganic Accumulation g cm^{-2} y^{-1}
1	Texas	1993	3	0.29	0.02	0.25
2	Texas	1993	4	1.14	0.08	0.41
3	Texas	1993	5	0.35	0.03	0.06
4	Louisiana	1993	6	0.33	0.03	0.10
5	Louisiana	1993	7	0.61	0.06	0.07
6	Louisiana	1993	8	1.04	0.11	0.23
7	Louisiana	1993	10	0.67	0.05	0.28
8	Louisiana	1993	13	0.53	0.03	0.07
9	Louisiana	1993	14	0.54	0.02	0.07
10	Louisiana	1993	15	0.62	0.04	0.12
11	Louisiana	1993	16	0.64	0.04	0.05
12	Louisiana	1993	17	0.72	0.06	0.07
13	Louisiana	1993	18	0.61	0.03	0.02
14	Louisiana	1993	19	0.64	0.03	0.09
15	Louisiana	1991	bh1a-h	0.42	0.03	0.04
16	Louisiana	1991	bh2an	0.57	0.03	0.06
17	Louisiana	1991	bh2bn	0.82	0.06	0.09
18	Louisiana	1991	bh2cn	0.63	0.05	0.07
19	Louisiana	1991	bh3an	0.88	0.05	0.07
20	Louisiana	1991	bh4al	0.38	0.03	0.06
21	Louisiana	1991	bh5an	0.86	0.05	0.17
22	Louisiana	1991	bh6al	0.28	0.02	0.10
23	Louisiana	1991	bl1-ah	0.74	0.05	0.05
24	Louisiana	1991	bi1-bh	0.60	0.04	0.05
25	Louisiana	1991	bi1-ch	0.62	0.04	0.04
26	Louisiana	1991	bi2-an	0.43	0.03	0.09
27	Louisiana	1991	b13a-n	0.92	0.06	0.33
28	Louisiana	1991	b14a-l	0.98	0.08	0.18
29	Louisiana	1991	b15a-l	0.42	0.02	0.06
30	Louisiana	1991	b16a-l	0.90	0.07	0.07
31	Louisiana	1991	sh1a-l	0.58	0.06	0.26
32	Louisiana	1991	sh2a-n	0.57	0.04	0.17
33	Louisiana	1991	sh3a-h	0.72	0.08	0.23
34	Louisiana	1991	sh41-h	0.65	0.04	0.13
35	Louisiana	1991	sh5a-h	0.47	0.04	0.23
36	Louisiana	1991	sh6b-h	0.68	0.07	0.21
37	Louisiana	1991	sh6c-h	0.81	0.08	0.22
38	Louisiana	1991	s11a-h	0.30	0.02	0.03
39	Louisiana	1991	s12a-l	0.52	0.04	0.13
40	Louisiana	1991	s12b-l	0.58	0.05	0.11
41	Louisiana	1991	s12c-l	0.53	0.04	0.12
42	Louisiana	1991	s14a-n	0.61	0.06	0.12
43	Louisiana	1993	th1a-n	0.83	0.07	0.21
44	Louisiana	1993	th2a-n	0.89	0.08	0.37
45	Louisiana	1993	th2b-n	0.63	0.05	0.20
46	Louisiana	1993	th2c-n	0.57	0.04	0.17
47	Louisiana	1993	th3a-h	0.62	0.05	0.08
48	Louisiana	1993	th4a-h	0.77	0.06	0.22
49	Louisiana	1993	th5a-h	0.58	0.03	0.10
50	Louisiana	1993	th6a-h	0.79	0.06	0.17
51	Louisiana	1993	t11a-i	0.68	0.05	0.10
52	Louisiana	1993	t13a-n	0.70	0.06	0.09
53	Louisiana	1993	t14a-i	1.11	0.08	0.16
54	Louisiana	1993	t15a-i	1.08	0.05	0.43
55	Louisiana	1993	t16b-i	0.76	0.05	0.08
56	Rhode Island	1994	NB	0.26	0.03	0.06
57	Rhode Island	1994	NB	0.34	0.04	0.08
58	Rhode Island	1994	NB 3	0.21	0.02	0.05

LANDSCAPE STRUCTURE AND SCALE CONSTRAINTS ON RESTORING ESTUARINE WETLANDS FOR PACIFIC COAST JUVENILE FISHES

CHARLES A. SIMENSTAD
W. GREGORY HOOD
School of Fisheries, University of Washington
Seattle, WA 98195-5020 USA

RONALD M. THOM
Battelle Pacific Northwest Laboratories, Marine Science Laboratory
Sequim, WA 98382 USA

DAVID A. LEVY
Hatfield Consultants, Ltd.
201-1571 Bellevue Avenue
West Vancouver, BC Canada V7V 1A6

DANIEL L. BOTTOM
NOAA-NMFS/NWFSC
Hatfield Marine Science Center
2030 SE Marine Science Drive
Newport, OR 97365

Abstract

Juveniles of many Pacific Northwest coastal fishes and particularly anadromous species, utilise coastal marshes as ''nursery'' habitats, predicating the assumption that restoration of marsh sites will promote increased fish survival and production. However, species such as anadromous salmonids have evolved life history strategies that to various degrees depend upon the structure and scale of the estuarine landscape rather than habitat sites *per se*. Examples include: 1) migration of juvenile salmon among interconnected wetlands along estuarine gradient, 2) their access to dendritic tidal channels, and 3) extended residence in tidal freshwater sloughs. Unfortunately, estuarine habitat restoration is seldom designed or implemented with landscape structure and scale in mind, ignoring important landscape attributes and processes such as habitat matrix heterogeneity, dendritic tidal channel complexity, allometric relationships of estuarine sloughs, and disturbance frequency and intensity. In this analysis, we draw on several estuarine wetland mitigation and restoration sites in the Pacific Northwest to explore the effect of estuarine landscape structure and scale on their effectiveness for protecting and rehabilitating coastal fisheries resources. We argue that basing restoration solely on site-specific criteria may be significantly inhibiting our ability to re-establish estuarine support function for fisheries resources. To significantly recover the function of juvenile fish migration and survival in coastal ecosystems, future marsh restoration must be conceptualized, designed, constructed and assessed taking into account estuarine landscape structure and scale.

1. Introduction

1.1 RATIONALE BEHIND RESTORATION OF PACIFIC NORTHWEST ESTUARINE ECOSYSTEMS

A broad generalization of ecological and fisheries literature is that tidal marshes support high secondary production and contribute significantly to the early life history of important fisheries species (Weinstein 1979, Boesch and Turner 1984, Adam 1990), albeit not without some controversy (Nixon 1980, Lenanton and Potter 1987). This argument has been applied equally to coastal marshes of the Pacific Northwest (Thom 1987). A particular paradigm of "estuarine dependence" has been applied to Pacific salmon (Healey 1982, Levings et al.1989), wherein marshes are considered to be one of the more requisite habitats for rearing of small juvenile salmon ("fry"), in particular chinook (*Oncorhynchus tshawytscha*) and chum (*O. keta*) (Levy and Northcote, 1982, Simenstad et al. 1982). If salmon and other important Pacific Northwest fishes are to some degree dependent upon tidal wetlands, the historic loss of between 50% and >75% of the region's estuarine wetlands (Simenstad et al. 1982, Levings and Thom 1994) has likely contributed to recent declines in fish production. This may be particularly relevant with anadromous species such as salmon. With rapidly declining Pacific salmon populations, and some approaching extinction (Nehlsen et al. 1991, National Research Council 1996), conservation and restoration of tidal marshes are not being considered as important aspects required for salmon recovery.

However, the success of tidal wetland conservation and restoration actions is predicated upon the fundamental assumption that the production of important fish and shellfish resources such as salmon will significantly increase (Simenstad and Thom 1994, Simenstad and Cordell 2000). Yet, other than inference from estuarine-scale experiments (e.g., Levings et al. 1989), evidence of survival benefits derived specifically from natural marsh residence is meagre. Despite the concerted effort dedicated to studies of the "fisheries habitat value" of restoring or creating estuarine habitats, the results, while often persuasive, have hardly been conclusive (Shreffler et al. 1990, 1992, Chamberlain and Barnhart 1993, Miller and Simenstad 1997).

This conundrum raises the question of whether we are examining tidal marsh structure and processes at scales appropriate to the processes that account for fish production. We herein propose that to resolve the contribution of tidal marshes to fisheries production requires understanding both the temporal and spatial scale of the dynamic interactions between the complex estuarine landscape ("ecoscape" of Kneib, this volume) and the ontogenetic "scales" of fishes that transcend and reside in it. If the fish must integrate a heterogeneous array of landscape elements, our anthropomorphic designation of an "essential habitat" is ignoring the larger-scale function of the proverbial "forest" while myopically focused on the "trees." Similarly, we suggest that the importance of variation in within-habitat microstructure is seldom taken into account when recognizing the dependence of larval and juvenile fishes on shallow coastal habitats.

We consequently argue that our ability to restore such estuarine ecosystem functions as juvenile fish rearing is contingent upon the application of such a landscape perspective to restoration planning, design and assessment. To some degree, we substantiate arguments and landscape criteria provided earlier by Lewis (1992) and Shreffler and Thom (1993) in applying them specifically to juvenile fish production in tidal marshes. We adopt the National Research Council's (1992) perspective that "Restoration is different from habitat creation, reclamation, and rehabilitation – it is a holistic process not achieved through the isolated manipulation of individual elements." Our approach to restoration of tidal marsh and estuarine landscape structure explicitly implies conservation and restoration of the ecosystem processes that promote and sustain marshes and their related elements.

1.2 TIDAL WETLAND HABITATS

Tidal marshes generally do not dominate Pacific Northwest estuarine landscapes as they do in relatively old tidal marsh systems of the mid- and southern Atlantic and Gulf estuaries. Because of relatively recent (~10-15,000 y BP) glaciation, isostatic rebound (producing relative sea level fall), large-scale subsidence associated with massive subduction zone earthquakes (the most recent in 1700), and extensive, sediment rich freshwater discharges, much of estuarine intertidal area remains unvegetated (Simenstad et al. 1997b). For example, among 17 coastal estuaries of Oregon, an average of 37% (±1 S.D. = 13.6%) of the total intertidal area is mud and sand flats, as compared to 34% (±19.2%) tidal marsh (Cortright et al. 1987). As a consequence, fish fauna in this region has coevolved with the development of a complex landscape of unvegetated as well as vegetated marsh elements.

Pacific Northwest tidal marshes vary in vegetation structure and geomorphology along the estuarine and oligohaline gradient as a function of freshwater discharge, regional climate and geology, and semidiurnally-mixed tides that can range over more than 3 m (Seliskar and Gallagher 1983, Simenstad et al. 1997b). Seasonal variation in freshwater discharge can strongly influence their structure. Although distributed as a continuum, they can be generally classified from oligohaline to euhaline salinity zones as: 1) freshwater tidal, 2) brackish, 3) estuarine, and 4) marine, only the latter three are generally classified under the colloquial definition of salt marsh in this region. Certain vegetation assemblages, as distinguished by low or high tidal elevations, often characterise each zone (Table 1).

Although the Pacific Northwest's tidal marshes are generally small in area, they have a relatively diverse assemblage of plant species. The typical freshwater-seawater gradient in marsh assemblages involves *Carex lyngbyei*-dominated tidal freshwater marshes that fringe rivers and streams, high (above MHHW) brackish marshes dominated by grasses, and low (below MHHW and above ~1.8 m above MLLW) salt marshes in which *Salicornia virginica* and *Distichlis spicata* dominate. *Typha latifolia* is often found in

1 These estimates do not take account estuarine area historically developed, which would generally lead to a considerable underestimate of natural tidal marsh area because development (e.g., diking and filling) typically focused on high tidal marsh systems.

the upper reaches of tidal-freshwater systems. Salt marshes typically grade by tidal elevation from *Deschampsia cespitosa*-dominated assemblages at lower elevations. As an early coloniser, *Triglochin maritimum* often occurs in patches on mudflats at the leading edge immediately seaward of contiguous low salt marsh, although it is also found in integrated into the low marsh assemblage. Contrary to our oversimplification of these assemblages, it is not uncommon to find exceedingly complex marshes composed of integrated marine, estuarine and brackish assemblages within one narrow system due to spatially heterogeneous distributions of elevation, freshwater influence, soil conditions and disturbance (e.g., Levings and Moody 1976, Berg et al. 1980, Ewing 1983). In addition to the horizontal gradient associated with salinity distribution, there are often strong vertical gradients in tidal marsh communities that reflect both the frequency and duration of tidal flooding but also vertical salinity variation in estuaries with strong variation in seasonal freshwater outflow and water column stratification, substrate and disturbance regimes (Frenkel 1981, Frenkel et al.1981).

TABLE 1. General tidal marsh communities in Pacific Northwest estuaries (adapted from Seliskar and Gallagher 1983 and Weinmann et al. 1984)

Estuarine Zone	Relative Vertical Elevation	Dominant Vegetation	Characteristic Geomorphic Features
Tidal Fresh	High	*Typha latifolia* *Carex obnupta* *Scirpus acutus*	Substrate composed of clay and peat; often large, sinuous channels (sloughs) with extensive large woody debris; associated with forested (Sitka spruce, red alder) swamps and shrub-scrub wetlands and considerable overhanging vegetation
	Low	*Carex lyngbyei* *Lilaeopsis occidentalis*	Often occur in unique, monotypic terraces or ''benches;'' substrate is fine, unconsolidated silt
Brackish	High	*Agostris alba* *Juncus balticus* *Eleocharis palustris* *Scirpus maritimus* *Hordeum brachyantherum*	Extremely complex, dendritic tidal channel systems; all but highest order channels dewater during Spring low tides
	Low	*Carex lyngbyei*	Adjoining narrow, moderate gradient mudflats
Estuarine	High	*Potentilla pacifica* *Deschampsia cespitosa* *Carex lyngbyei* *Plantago maritima* *Atriplex patula*	Broad marsh plain; tidal channels often wide
	Low	*Troglochin maritimum* *Distichlis spicata* *Jaumea carnosa* *Salicornia virginica* *Spergularia marina* *Scirpus americanus*	Often adjoining or prograding over extensive, low gradient mudflats
Marine	High	*Elymus mollis* *Juncus lesueurii*	Coarser substrates, e.g., sand and gravel; dunes and steep-gradient beaches
	Low	*Salicornia virginica*	

Considerable variation in tidal marsh community structure and distribution, as well as ecological function, is also associated with the type of estuary, within the underlying context of geologic setting and history, and on-going tectonic activity (Simenstad et al. 1997b). Pacific Northwest estuaries are predominantly drowned river valley estuaries, but fjord types are not uncommon in the northern sections of the region and some of the southern estuaries take on lagoon estuary characteristics. Many of the coastal estuaries are the termini of short, steep watersheds confined to the coastal zone that are characterized by low mean annual freshwater discharge. During particularly low summer flows, sand-gravel spits often close off the estuary entrance. In these systems, the associated tidal marshes tend to be small peripheral or "pocket" (between rocky outcrops) estuarine marshes. Larger watersheds, epitomized by the 660,000-km^2 Columbia River basin, have extensive tidal floodplains that penetrate coastal mountain ranges. Due to the higher freshwater flow influence and associated fine sediment transport and deposition, estuaries of these systems have extensive tidal-freshwater and brackish marshes, shrub-scrub and forested swamp complexes intertwined by tidal slough (little to no freshwater input) networks. In the Columbia River estuary, these complexes dominate extensive islands. Many of these estuaries, such as the Columbia River, Grays Harbor and Willapa Bay, Washington and Tillamook Bay, Oregon, have multiple watersheds and contributing rivers that form extensive estuaries feeding the larger estuarine bays. Other coastal systems, such as Yaquina Bay, the Umpqua River estuary and Coos Bay, Oregon, have small peripheral drainages to the primary river estuary that generate comparatively little freshwater input but promote peripheral marshes along these lateral sloughs. The most complex tidal landscapes occur where peripheral flooded river valleys intersect extensive fjords that form the large inland seas of Puget Sound and the straits of Georgia and Juan de Fuca. Because of the abrupt bathymetry associated with the fjords and the often steep watershed gradients (e.g., eastern and northern margins of Olympic Mountains), the river valley estuaries tend to be comparatively short and built out over low-gradient intertidal deltas.

Tidal channels and sloughs are dominant features of tidal marshes. This is particularly the case in the Pacific Northwest's macrotidal estuaries, where complex, dendritic tidal channels characterise brackish and estuarine marshes and slough channel networks are typical of the tidal-freshwater floodplains of large river estuaries. Fish access to marshes occurs primarily through tidal channels during cycles of tidal pulsing (Rozas 1995). In tidal-freshwater and brackish marshes where natural levees tend to form along the channel margin of the marshes, fish access via sheet flow across the face of the marsh may be inhibited. Access is constrained by channel depth, depending upon the stage of the tide and tidal month. Only the higher order tidal channels (e.g., 4th or 5th and greater) tend to be deeper than MLLW elevations. Thus, most channels dewater during spring low tides, except when high freshwater discharge significantly enhances the maximum flood tidal elevation, but retain water during neap tides.

Therefore, on spring tides fish can access all marsh surfaces but must evacuate the marsh and much of the (upper) tidal channel complex by the end of ebb tide unless

2 Dendritic tidal channels can be described like streams, where first order streams are the samllest unbranched headwater triburary channels, second order channels are those formed by the confluence of two first order channels, third order channels are formed by the confluence of two second order channels, etc.

they are small enough to occupy the remnant channel pools. Conversely, the lower high tide elevations during neap tides limits the access to the surface of higher elevation marshes even though the fish are not forced out of the marsh system. For instance, if we generalise low marshes of Grays Harbor (Chehalis River estuary, Fig. 1) to occur between 1.8 m Mean Lower Low Water (MLLW) to Mean Higher High Water (MHHW) and high marshes from Mean Higher High Water (MHHW) to Extreme Higher High Water (EHHW), then during the period of juvenile fish occupation (February-October) of tidal marshes in this estuary: 1) most tidal channels never dewater during the neap tides but dewater 15 hr (4.7%) of a spring tide series; 2) the low marsh is always flooded once a tidal cycle (100% flooding frequency); and 3) the high marsh is never inundated on neap tides and only for 17 hr (4.4%) on a spring tide series. Furthermore, the period when the channels dewater and high marshes flood is only during the stronger spring tide cycle once a month, which averages only six or seven days.

Not all functions for the marshes apply uniformly, either. If all the juvenile fishes rearing in marshes feed visually, as we suspect, then flood tides that occur at night do not necessarily serve equivalent function for foraging as those during daylight, although there may be no difference in other functions (physiological adaptation, refugia). In fact, most of the spring higher high tides that inundate high marshes between February and October occur at night, while the extreme low tides occur during daylight.

2. Fish Utilization of Pacific Northwest Estuaries

Compared to the plethora of prior and emerging literature on tidal marsh fishes of the Atlantic and Gulf coasts of North America (encapsulated in Kneib 1997), descriptions of juvenile fishes specifically found using tidal marsh ecosystems in the Pacific Northwest are comparatively rare. This is especially the case if we apply rigorous criteria to consider only the results of methods that sample fish accessing the marsh plain or dendritic tidal channels, as most estuarine fish sampling data derives from beach seining or trawling in shoreline or neritic sites. Furthermore, multiple gear/ habitat research on tidal marsh fish ecology in the other regions has explored their fine-grain distribution, feeding and behaviour in a variety of marsh microhabitats such as vegetated and unvegetated marsh plains, salt ponds and pans, and tidal channels of different order, size, and depth (e.g., Halpin1997). In contrast, quantitative data from the Pacific Northwest originates almost exclusively from trap (fyke) or plankton nets deployed in tidal channels in seven estuaries (Fig. 1).

Experimental gillnets have recently been deployed to semiqualitatively sample fish along mudflat or tidal channel/mudflat and marsh vegetation ecotones in non-indigenous (i.e., *Spartina alterniflora*) and restoring marshes (e.g., Cordell et al. 1998). However, because of the preliminary or incomplete nature of these results, we have not incorporated these data except for the results reported herein for the Chehalis River estuary.

2.1 FISH COMMUNITIES

Twenty-seven species typify fish assemblages documented in the Pacific Northwest tidal marshes (Table 2). Two thirds of the species occur as larvae or juveniles. While only four species are considered resident species, seven are seasonal transients (occupying marshes for weeks to months during particular life history stage and/or seasonal riverflow period), and 16 are ephemeral transients (rapidly passing into/out of or through the estuary). Thus, 41% of the assemblage occupies tidal marshes for a significant period of their life history. The anadromous salmonids (*Oncorhynchus* spp.) and smelts (Osmeridae) comprise almost a third of the assemblage. Eight species are common in four or more of the seven estuaries. Most species are found throughout the region, but six (pink and chum salmon, longfin smelt, eulachon, redside shiner, snake prickleback) are relatively confined to the northern, and five (topsmelt, jacksmelt, walleye and white surfperch,

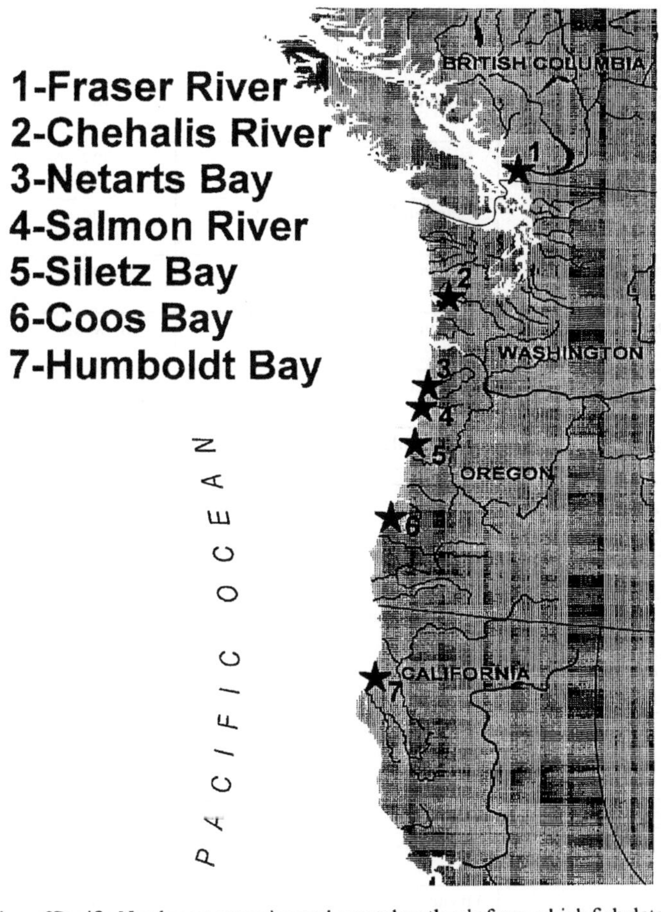

Figure 1. Location of Pacific Northwest estuarine and coastal wetlands from which fish data were derived: (1) Fraser River, BC, (2) = Chehalis River, WA, (3) Netarts Bay, (4) Salmon River, OR, (5) Siletz River, OR, (6) Coos Bay, OR, and (7) Humboldt Bay, CA.

arrow goby) to the more southern estuaries. However, this list does not represent uniform sampling and should not be considered a reliable index of zoogeographic distribution of marsh fishes. For instance, sampling that specifically targets planktonic fish larvae and post-larval juveniles has only been applied to tidal marshes from Humboldt Bay (Chamberlain and Barnhart 1993) and the brackish-tidal freshwater ecotone has only been sampled in a few estuaries.

TABLE 2 Characteristics of fish species commonly found in estuarine and coastal wetlands of the Pacific Northwest (expanded from original synthesis by Seliskar and Gallagher 1983). Nomenclature and phylogenetic organization based on Robins et al. 1980. References-Locations: FrR = Fraser River estuary, Levy et al. 1979, Levy and Northcote 1981; ChR = Chehalis River estuary, Simenstad et al. 1992, 1993, 1997a; NeB = Netarts Bay, Higley and Holton 1981; SaR = Salmon River estuary, D. Bottom, ODFW, unpubl., SeR = Siletz River, Higley and Holton 1981; CoB = Coos Bay, Hoffnagle et al. 1976; HuB = Humboldt Bay, Chamberlain and Barnhardt 1993.

Species	Life History Stage	Duration of Residence[1]	Wetland Habitat2	Reference-Location
Anadromous species				
Oncorhynchus gorbuscha, pink salmon	juvenile	t	mc	FrR
Oncorhynchus keta, chum salmon	juvenile	s	mc	FrR, ChR, NeB, SeR
Oncorhynchus kisutch, coho salmon	juvenile	s	mc	FrR, ChR, SaR
Oncorhynchus mykiss, steelhead	juvenile	t	mc	ChR
Oncorhynchus nerka, sockeye salmon	juvenile	t	mc	FrR
Oncorhynchus tshawytscha, chinook salmon	juvenile	s	tfws, mc	FrR, ChR, NeB, SaR, SeR
Spirinchus thaleichthys, longfin smelt	juvenile	t	mc	FrR, ChR
Thaleichthys pacificus, eulachon	juvenile	t	mc?	FrR
Freshwater species				
Mylocheilus caurinus, Peamouth chub	juvenile	r	tfws, mp	FrR, ChR
Catostomus macrocheilus, largescale sucker	juvenile-young adult	s	tfws, mc	ChR
Gasterosteus aculeatus, threespine stickleback3	juvenile-adult	r	tfws, mc, mp	FrR, ChR, NeB, SaR, SeR, HuB
Cottus asper, prickly sculpin	juvenile-young adult	r	mc	FrR, ChR, SaR, HuB
Estuarine/Marine species				
Clupea harengus pallasi, Pacific herring	larvae-juvenile	t	mc	FrR, HuB
Engraulis mordax, northern anchovy	larvae-juvenile	t	mc	HuB, NeB, SeR

Species	Life History Stage	Duration of Residence[1]	Wetland Habitat2	Reference-Location
Hypomesus pretiosus, surf smelt	juvenile	t	mc	FrR, ChR, NeB, SeR, HuB
Richardsonius balteatus, redside shiner	juvenile-adult	s	tfws, mc	FrR, ChR
Atherinops affinis, topsmelt	larvae-juvenile	t	mc	HuB
Atherinops californiensis, jacksmelt	larvae	t	mc	HuB
Cymatogaster aggregata, shiner perch	juvenile-adult	s	tfws, mc	FrR, ChR, NeB, SaR, SeR, CoB, HuB
Hyperprosopon argenteum, walleye surfperch	adult	t	mc	HuB
Phanerodon furcatus, white surfperch	adult	t	mv	HuB
Clevelandia ios, arrow goby	larvae-juvenile	t	mc	HuB
Lepidogobius lepidus, bay goby	larvae-juvenile	t	mc	HuB
Ammodytes hexapterus, Pacific sand lance	juvenile	t	mc	SaR
Leptocottus armatus, Pacific staghorn sculpin	juvenile-young adult	r	mc, mp	FrR, ChR, NeB, SaR, SeR, CoB, HuB
Lumpenus sagitta, Snake prickleback	juvenile-adult	t	mc	ChR
Platichthys stellatus, Starry flounder	juvenile	s	tfws, mc	FrR, ChR, NeB, SeR

2.2 ONTOGENETIC AND OTHER PATTERNS OF HABITAT UTILISATION

That anadromous fishes and other seasonal transients occupy tidal marshes for extended periods during their early life history implies strong selective pressures to maximise growth and minimise predation (Crowder and Craig this volume). This association may include specific marsh zones or features that are particularly important foraging sites or refuge. Juvenile Pacific salmon exemplify the importance of these habitats because their emigration from freshwater habitats into a saline environment carries with it physiological and ecological challenges. At this transition, they must adjust osmoregulatory (gills, intestine, kidneys, bladder) and other metabolic processes to hyperosmotic waters, shift from essentially non-evasive prey (e.g., drift and aquatic insects) to progressively more evasive prey, and new suites of predators. While many species such as juvenile chum salmon are capable of rapidly making these transitions, or avoid them, others must use specific estuarine zones in which to ''stage'' their metabolism and behaviour as they encounter these physicochemical and ecological changes during their seaward migration along the estuarine gradient. Variability in fish response to estuarine gradient, prey-predator assemblages, and intense forcing events such as flooding is conceptualized in five movement patterns among tidal marshes (Fig. 2):

1. anadromous punctuated migration; directed movement of anadromous species through the estuary, although the rate of movement and the time of residence in different reaches of the estuary may be quite variable
2. overwintering; anadromous species that during high winter river flows will drop down from watersheds into the tidal freshwater reaches of estuaries but do not necessarily migrate seaward
3. tidal/event; movements regulated principally by tidal or events (e.g., floods, storm and snowmelt freshets)
4. opportunistic marine; marine fishes that venture into estuaries with circulation of saline water masses
5. marine rearing; marine fishes that utilize estuaries for nursery habitats

Anadromous salmon and smelts demonstrate the prevailing punctuated migration pattern, wherein their juveniles occupy tidal-freshwater to marine marshes on their migration to the ocean during a period from a few days to several months. The pattern is characterised as punctuated because some species and life history stages appear to stage at specific junctures in the estuarine landscape. Anadromous species such as coho salmon that reside in freshwater for the first 9-12 months of their early life history may actually overwinter in tidal-freshwater floodplain marshes during high river discharge periods. Juvenile fishes penetrating from the ocean can appear episodically during short-term, opportunistic forays into saline and estuarine marshes as well as occupy estuarine habitats for extended periods.

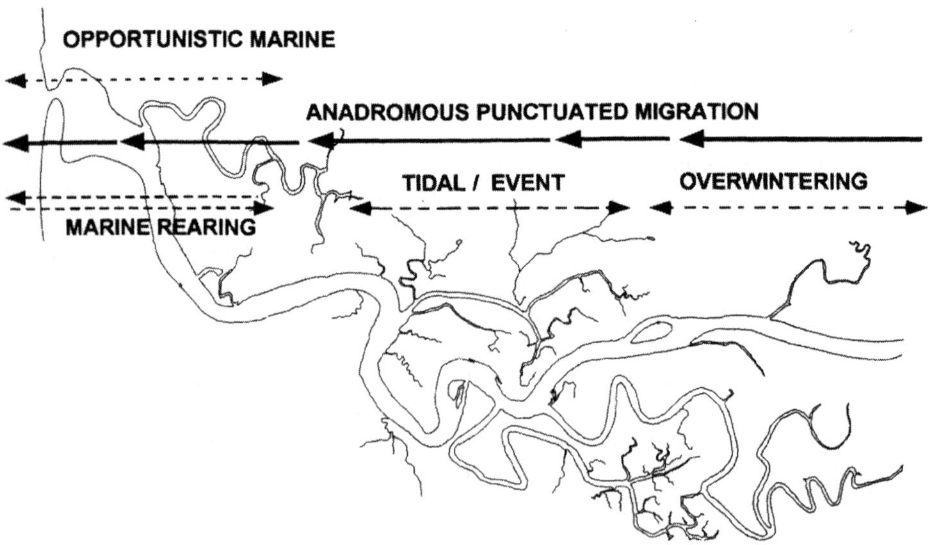

Figure 2. Patterns of juvenile fish movement and migration through estuarine habitats of the Pacific Northwest.

Anadromous species migration is often cited as the essence of estuarine dependence, but utilisation of estuaries as nursery areas by marine species should not be discounted. While we know of no specific data on the marine species we have identified to occupy tidal marshes, Gunderson et al. (1990) and others have demonstrated that juvenile fish such as the English sole, *Pleuronectes vetulus*, along the Washington coast extensively use estuaries as a nursery during their first year before completing their life history off shore. The importance of this brief estuarine life history stage to overall growth and mortality is evidenced by the fact that year-class strength is strongly influenced by variability in estuarine conditions (Shi et al. 1995, 1997).

2.3 SPATIAL AND TEMPORAL SCALES OF INTERACTION

Juvenile fish occur in Pacific Northwest tidal marshes over multiple, sometimes hierarchical temporal and spatial scales (Table 3), similar to nekton migration frequencies and scales identified by Kneib (1997, this volume). In most cases, different species display distinct patterns and extents of occurrence in marshes. For instance, some species such as post-larval and juvenile eulachon may appear throughout the spectrum of tidal marshes but only during the brief ontogenetic period of their seaward transport or migration in late spring and early summer. In contrast, juvenile shiner perch may occupy brackish marshes for months.

However, marsh use may also vary among different life history types of the same species. Juvenile chinook salmon may display perhaps the most variable suite of genetic and tactical strategies that includes tidal marsh residence (Healey 1991), from "ocean-type" fry (juveniles ~30-50 mm FL) that can spend weeks to months in tidal-freshwater and brackish marshes, while larger fingerlings (juveniles ~50-100 mm FL) less frequently range into estuarine marsh channels, and "stream-type" smolts (juveniles ~100-150 mm FL) that may not use tidal marshes at all as they migrate rapidly through the estuary. While it may be tempting to argue that fish size can explain the fidelity for shallow-water environs like marshes, such a simple explanation does not apply to salmon. For example, pink and chum salmon fry of the same size can readily be found in tidal marshes at the same time as chinook fry. Despite co-occurrence at the same life history stage (but not necessarily the same physiological or behavioral state), pink salmon fry are essentially transients in tidal marshes, while chum fry remain at the most two weeks and "ocean-type" chinook may reside for up to three months in estuaries, a significant portion of which occurs in tidal marshes (Reimers 1973, Levy et al. 1979, Levy and Northcote 1981, 1982, Shreffler et al. 1990). Mark and recapture experiments confirm that chinook fry have the highest degree of fidelity to particular tidal channels, followed in decreasing fidelity by chum and pink salmon (Levy et al. 1979). Studies of juvenile salmon occurrence in natural and restoring tidal-freshwater/brackish channels in the Squamish River estuary (Ryall and Levings 1978), Chehalis River estuary (Simenstad et al. 1992, 1993, 1997a), and the Salmon River estuary (D. Bottom, unpublished) also indicate that subyearling coho fry occupy channels in tidal-freshwater and brackish marshes, which may be a relatively underestimated life history pattern for coho salmon (Miller and Simenstad 1997).

TABLE 3. Spatial and temporal scales of juvenile fish occurrence in Pacific Northwest tidal marshes

Spatial	Temporal
Intramarsh	Tidal
• Within marsh; m's to 10's of m	• Tidal cycles; hours
Intralandscape	Diel/Diurnal
• Between and among landscape elements, e.g., channel to marsh surface; 10's to 100's of m	• Diel cycles; within days
Estuarine Gradient	Climatic Event
• Transitions between estuarine zones, or with zones during tidal excursion and retreat; km's	• High river discharge associated with storm events, most intense during winter/spring rain-on-snow events; days
Estuarine	Neap/spring Tidal
• Estuarine migration; 10's of km	• Strong/weak neap/spring tidally monthly cycles; ~weeks
Land-Margin	Life History State Duration
• Freshwater-estuary and estuary-ocean transitions; up to 100's of km	• Physiological, growth, and behavioural stanzas; weeks to months
	Seasonal-Annual
	• Nursery, maturation and reproduction stages; months to years

2.3.1 Short-term, fine-scale dynamics

Due to the strong semi-diurnal cycling of the tides, juvenile fish must within their energetic capabilities continually reposition themselves to maximise access to prey resources, and minimise vulnerability to predation and unfavourable salinity and temperature regimes. This involves both response to tidal flooding and exposure of intertidal wetlands but also natural displacement of salinity zones. While peripheral shallow water habitats may serve as refugia during ebb tide periods, changes in salinity distributions along the estuarine gradient may encompass km's depending upon river flow and stage in tidal monthly cycles. The later into their life history, with increasing energetic capability, the more likely the fish are capable of seeking out marshes that are specific landscape patches or corridors, allowing them to avoid displacement. For this reason, it is difficult to interpret whether the occurrence of fish larvae in tidal channels has a functional linkage to marshes or is simply a consequence of tidal transport.

In the case of juvenile salmon, penetration into tidal marsh systems can be extensive and prolonged. Levy et al. (1979) and Levy and Northcote (1981, 1982) investigations of juvenile salmon and other fishes utilising tidal marshes of the lower Fraser River estuary, British Columbia, Canada, effectively illustrate these patterns (Fig. 3). When abundant in the marshes until mid-June, densities of all three species of juvenile salmon are usually highest at the entrance of the channels but densities also increase with increasing distance into the marsh along the length of the channel (Fig. 3b). Chinook salmon fry demonstrate the highest densities deeper in the marsh, but pink and chum fry still show an increasing density gradient with tidal channel distance into the marsh. Other species tend to illustrate different patterns: 1) highest densities deep within the marsh, as with threespine stickleback, suggest marsh residency; 2) decreasing density with increasing distance

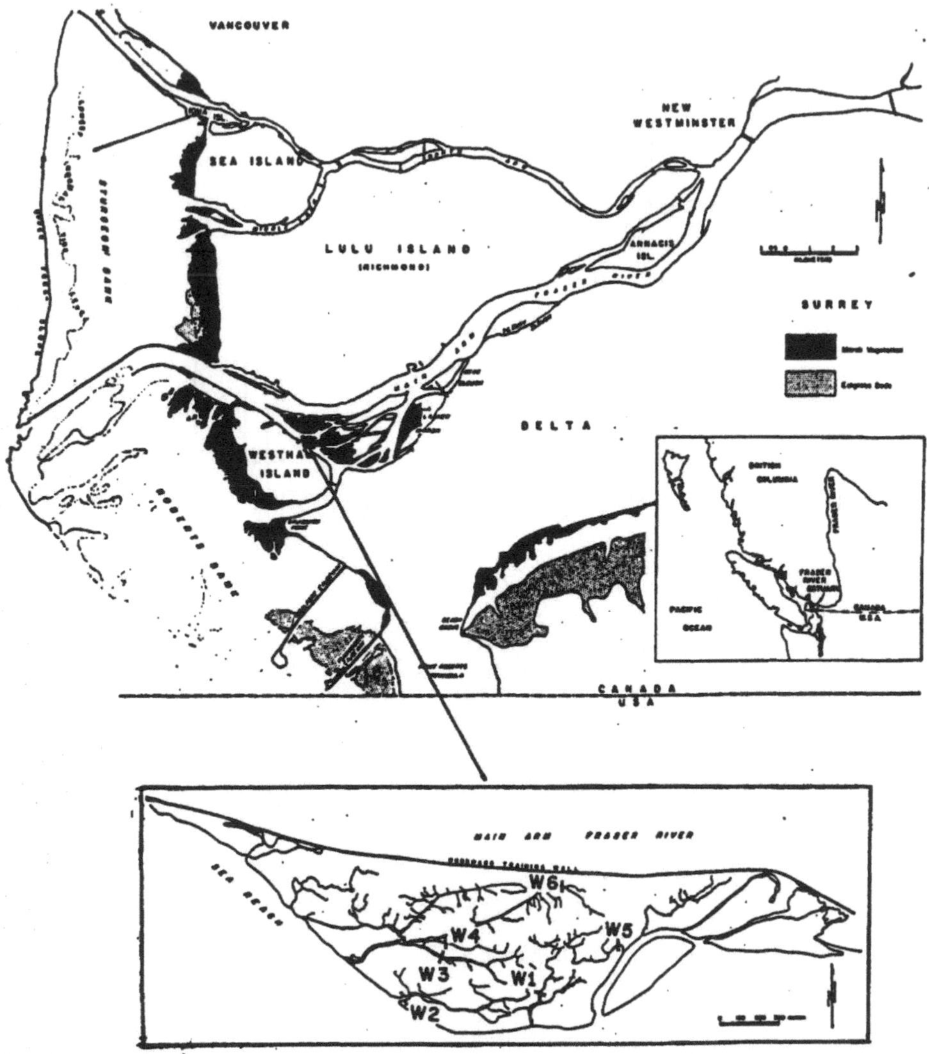

a

Figure 3. Fish densities in tidal channels of Woodward Island marshes in lower Fraser River estuary, British Columbia, Canada (a) as a function of position along "Main TC" tidal channel gradient in early May-late June 1978 (b). Adapted from Levy et al. (1979).

into the marsh, as with Pacific staghorn sculpin, implies penetration into the marsh with the flooding tide; and, 3) relatively uniform distribution, as with longfin smelt, suggests random movement throughout the marsh. Essentially the same patterns were observed in a different Woodward Island channel in a subsequent year (Levy and Northcote 1981). It may be important to note that the apparently strong association of fish with lower order channels may not reflect the same mechanism as described by Rozas and Odum (1987) because submerged aquatic vegetation is generally not present in these Pacific Northwest

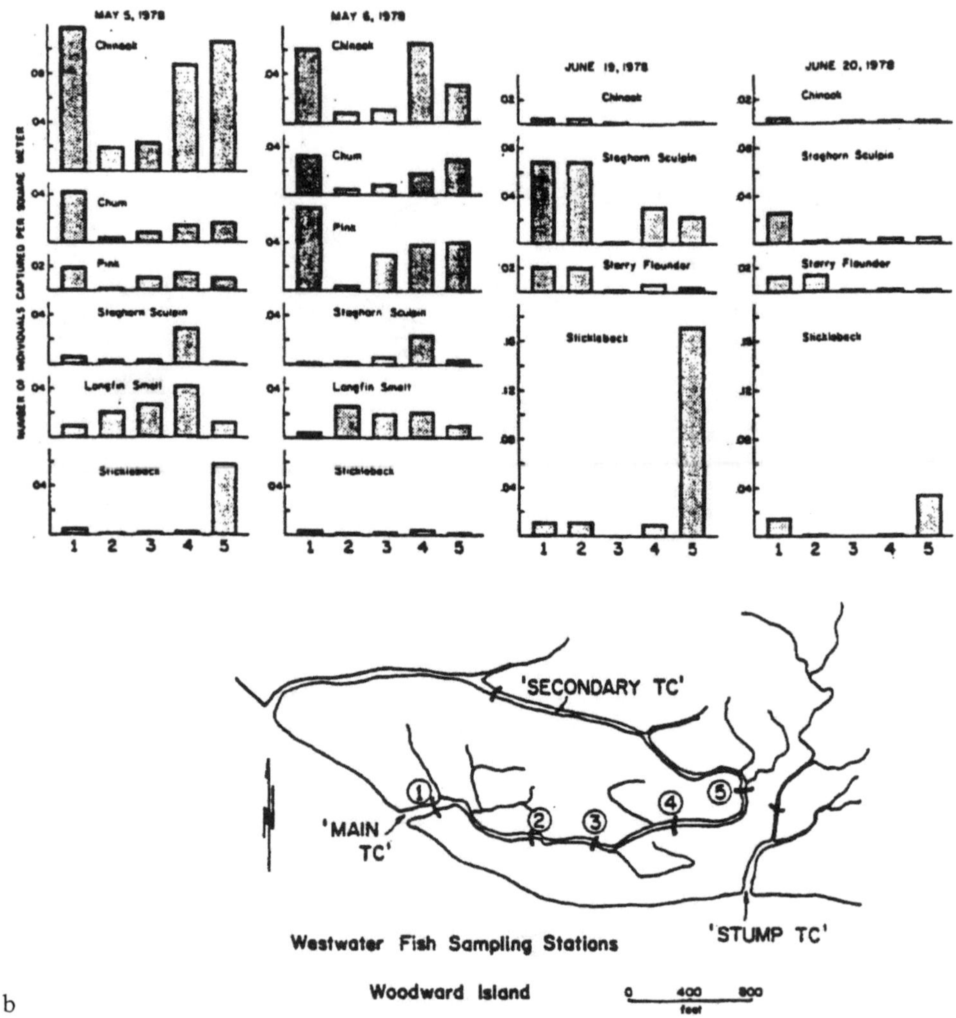

b

Figure 3 (continued).

tidal channel and marsh systems except for eelgrass (*Zostera marina*) in subtidal, high order tidal channels. In this region, lower water velocities and closer proximity to prey resources may provide some of the explanation for the evident association of many fishes with lower order tidal channels.

Most fish emigrate from channels with ebbing tides (Fig. 4), although this likely varies with stage of the tidal month depending upon whether or not tidal channels entirely dewater. Pink fry, accompanied by some chum fry, emigrated from the Lower Fraser River marsh channel during the first two to three hours of the ebbing tide. Almost all the chinook fry were not caught until the very end of the ebb tide, indicating both the concentration of their distribution deep in the marsh and perhaps the behavioural fidelity to the channel. In trapnet sampling of a natural and a created tidal-freshwater/ brackish slough in the Chehalis River estuary in 1995, Simenstad et al. (1997a) did not

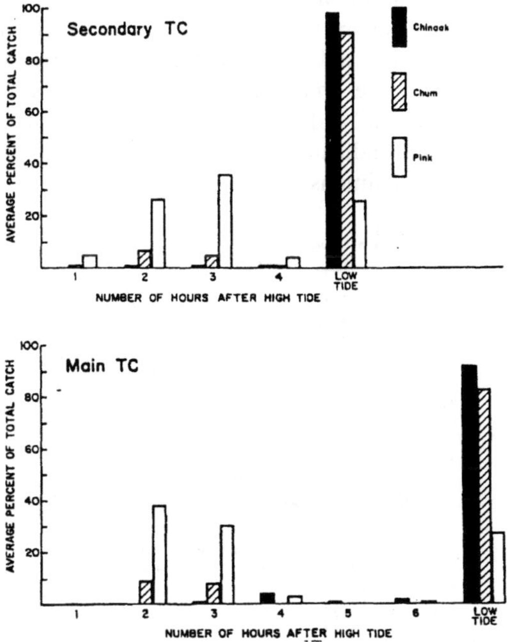

Figure 4. Average percent of juvenile salmon population caught as they emigrated from the "Secondary TC" and "Main TC" tidal channels of Woodward Island marshes in lower Fraser River estuary, British Columbia, Canada, from early May to late June 1978; see Fig. 3a for channel locations. Adapted from Levy et al. (1979).

find a consistent trend in fish emigration, although densities often peaked 0.5 to 1.5 hr before the end of the ebb tide. They found threespine stickleback to be particularly recalcitrant in moving out of the sloughs.

The fidelity to tidal marsh channels by juvenile chinook salmon may be explainable in part by trophic affiliations with prey resources uniquely derived from the marsh. Diet composition (percent volume, occurrence and frequency) of chinook from three locations along the main tidal channel ('Main TC') of Woodward Island sampled by Levy et al. (1979) in 1978 indicated that chironomid dipteran fly pupae were the principal prey (Fig. 5). Chironomid adults and the estuarine gammarid amphipod *Anisogammarus* (*pugettensis*?) became more prominent components of the diet deeper in the marsh. Juvenile chum and pink salmon also consumed chironomid pupae and adult dipterans, but their diet tended to be dominated more by benthic/epibenthic harpacticoid copepods, while dipteran larvae were more prominent components of starry flounder and threespine stickleback diets, and *Anisogammarus*, the isopod *Gnorimosphaeroma* (*oregonensis*?) and the mysid *Neomysis* (*mercedis*?) comprised the primary prey of Pacific staghorn sculpin and longfin smelt (Levy et al. 1979, Levy and Northcote 1981, 1982). Thus, the highest fidelity to the tidal marsh was demonstrated by fish that were foraging primarily on marsh fauna. Alternate prey that were utilized by other fish in the tidal channel/marsh system were more characteristic of channel, mudflat and adjoining shallow water communities.

Marsh associated insects dominated the diet of juvenile chum, chinook and coho salmon in the Chehalis River estuary tidal-freshwater and brackish sloughs (Simenstad et al. 1991, 1992, 1997a, Miller and Simenstad 1997). While chironomid and other dipteran larvae and adults were important components (as well as mysids and amphipods), adult aphids typically dominated their diets. Further sampling of the potential sources of the aphids indicated that they were being washed off the fringing *Carex lyngbyei* sedge marshes (Simenstad et al. 1997a).

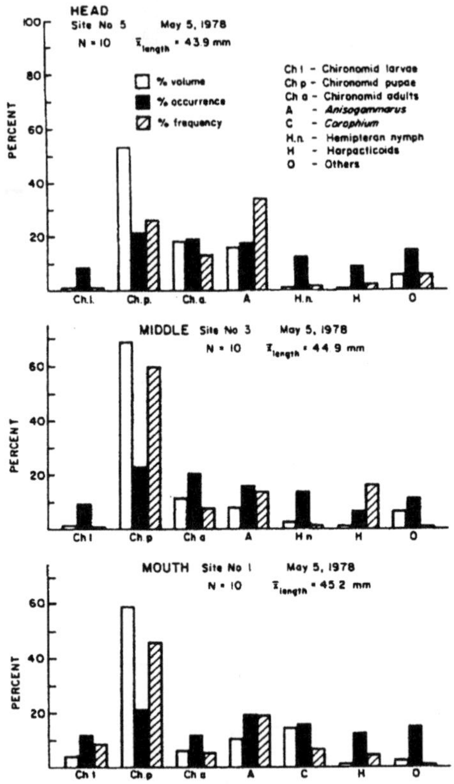

Figure 5. Diet composition of juvenile chinook salmon at three locations in the "Main TC" tidal channel of Woodward Island marshes in lower Fraser River estuary, British Columbia, Canada, in early May 1978, see Fig. 3b for channel locations. Adapted from Levy et al. (1979).

Diet data from the two Oregon coastal estuaries sampled by Higley and Holton (1981) show somewhat different diet compositions, however. While larval and adult dipterans were still prominent in the diet of juvenile chum salmon, epibenthic crustaceans such as amphipods (*Corophium* spp., *Anisogammarus* sp.), cumaceans (*Hemileucon* sp.) and tanaids were more prevalent in the diets of Pacific staghorn sculpin, shiner perch, surf smelt and threespine stickleback. This may reflect prey differences as a function of marsh structure (i.e., brackish marshes in the Fraser and Chehalis rivers estuaries vs. estuarine marshes in Oregon) or differences in marsh use by the fish (e.g., more access of the marsh plain).

While tidal cycles appear to regulate prey availability and fish feeding activity, diurnal and diel processes can also contribute to the availability of prey in marsh channels. Tschaplinski (1987) documented that terrestrial invertebrates trapped at the air-water interface in the Carnation Creek estuary reached a maximum during dusk, when juvenile coho fed intensively upon them.

There are few data that indicate how fish specifically use marsh features, and how such behaviour may vary among species and life history stages. Most fish sampling has been focused on tidal channels and tidal-freshwater sloughs, but only three species (peamouth chub, threespine stickleback, Pacific staghorn sculpin) have been documented to use the marsh plain. Higley and Holton's (1981) studies in two coastal Oregon estuaries is one of the few that has investigated different marsh features in this region. They found that while only two species (threespine stickleback, Pacific staghorn sculpin) were abundant in the tidal channels, two additional species (surf smelt, chinook salmon) were abundant in the low marsh plain and two other species (shiner perch, starry flounder) occurred uniquely there. Only threespine stickleback and Pacific staghorn sculpin were found in marsh pans. Because fish sampling has so seldom been applied discretely to marsh features, it is conceivable that other species may utilise marsh features more than this list suggests.

Results from variable mesh gillnet sampling along marsh ecotones between the tidal channel and *Carex lyngbyei* vegetation in the tidal-freshwater and brackish sloughs of the Chehalis River estuary offers further insight into the interface between fish and the marsh community (Fig. 6). Juvenile peamouth chub, shiner perch and chinook salmon were captured along the edge and in the small, first-order channels ('rivulets') of the marsh but not actually within the marsh sedge vegetation. Even when the tide rose over the marsh vegetation, fish were captured along the top but not within the sedge canopy. These distributions imply that the fish are seeking prey at the marsh edge or as potential prey organisms are washed off and out of the vegetation. Divers observing juvenile salmon behaviour in marshes of the Salmon River estuary, Oregon, have also described finding juvenile chinook oriented to the marsh vegetation edge and within extremely small rivulet channels. This importance of the marsh edge and rivulets further echoes previous evidence of nekton use of marshes in other regions (Baltz et al. 1993, Rozas et al. 1988, Rozas 1992, Minello et al. 1994, Kneib 1997).

Given the dynamic tidal- and event-driven variability in estuarine salinities, temperatures, and water velocities, it is intriguing that some fishes will actually occupy tidal marshes for periods as long as three months. Climatic events in the form of short-term storm driven increases in freshwater inflow to estuaries have the biggest potential to disrupt fish use of tidal marshes. Normal late winter and spring mean flows of 0.01 to 0.1 $m^3sec^{-1}km^{-2}$ watershed area can within 24 hr increase to ~1 $m^3sec^{-1}km^{-2}$ in normal storm events and over 1-2 $m^3sec^{-1}km^{-2}$ during rapid snowmelt events (Wissmar, 1998). In small estuaries, this could result in compression of estuarine salinity levels into a narrow zone at the mouth of the estuary, although bathymetry and estuarine circulation could dampen the effect (e.g., increased stratification would permit greater saltwater intrusion near bottom). Anadromous fishes such as juvenile salmon may be less vulnerable because of their occupation of tidal-freshwater and brackish conditions, and their residence in extensive channel and slough systems that may offer refuge from high current flows. Storm events can actually flush juvenile salmon from freshwater into the

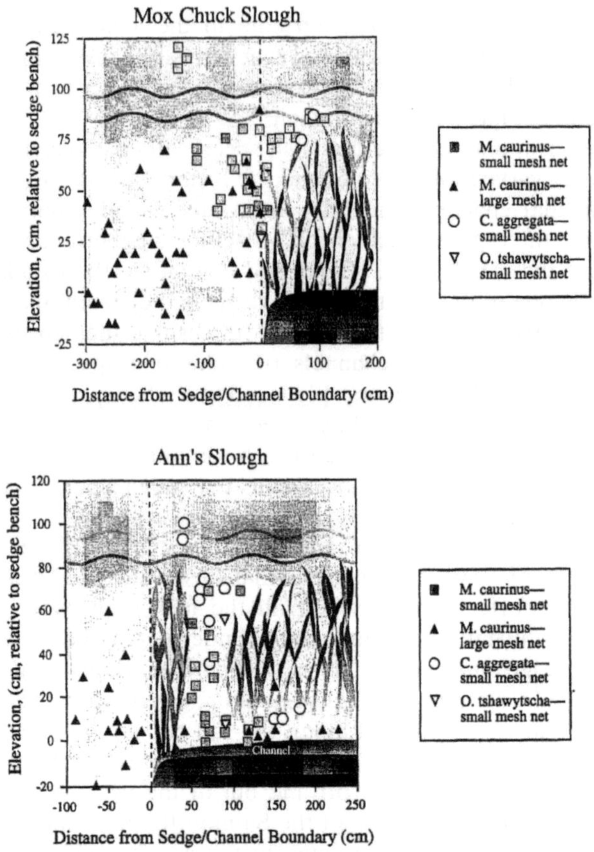

Figure 6. Capture locations of juvenile peamouth chub (*M. caruinus*), shiner perch (*C. aggregata*) and chinook salmon (*O. tshawytscha*) in variable mesh gill nets relative to *Carex lyngbyei* sedge marsh edges and tidal channels in two tidal-freshwater and brackish sloughs of the Chehalis River estuary. ''Channel'' label in vegetation indicates first order, rivulet channel running parallel into sedge marsh; the gill net was located within this channel rather than the vegetation.

estuary and increasing rearing populations, as demonstrated by Tschamlinski (1982) for the Carnation Creek estuary, Vancouver Island, British Columbia. However, estuarine and marine species responding to decreasing salinity would have to emigrate downstream from marshes normally within their range of tolerance.

It is likely, although entirely speculative, that the ecological consequences of these changes will depend to a large degree upon the landscape structure within the range of fish movement. Large-scale distributions of marsh patches and corridors along the estuarine gradient should logically provide viable pathways of movement as well as prey resources and refugia. Conversely, significant gaps in natural landscape structure such as expanses of armoured or developed shorelines could be detrimental.

2.3.2 Mid-term, mid-scale dynamics

Mid-scale patterns in fish access and utilisation of tidal marshes involve tidal monthly variation in flooding regimes, salinity structure and ontogenetic changes in fish and their movement patterns. Depending upon estuarine circulation characteristics, which are influenced by bathymetry and river flow, the differences between spring and neap tides can influence the location of salinity zones as well as the frequency and duration of marsh flooding. Many of the larger Pacific Northwest estuaries can alternate between mixed and stratified circulation structure during spring and neap tide regimes, respectively, resulting in differences in the length of salinity ranges: narrow under mixed conditions but potentially long under stratified conditions (although the mesohaline interface may be quite thin). As mentioned earlier, during spring tide series this corresponds to increased access to high elevation marshes at high tides but being forced our of tidal channel systems at low tides, while during neap tide series continuous occupation of tidal channels is possible but high elevation marsh communities are inaccessible.

The extent of fish utilization of tidal marshes over these mid-term and mid-scale variations in estuarine conditions is contingent upon life history timing and change. While juvenile salmon may be found in estuaries throughout the year, extensive use of tidal marshes as fry tend to be concentrated between February and July. Their appearances are pulsed in accordance with the timing of adult migrations the previous fall and winter, winter and spring water temperatures and flows, the outcome of intraspecific competition and interspecific interactions in freshwater, and physiological cues. Thus, characteristics of immigration to the estuary are set by both genetic and tactical influences, particularly in the case for the complex of ''ocean-type'' and ''stream-type'' life history patterns of chinook salmon (Healey 1991).

2.3.3 Long-term, coarse-scale dynamics

Long-term tidal marsh use involves species that are capable of residency over long periods, and capable of integrating or compensating higher frequency variations in tidal flooding and water physicochemistry. Extended marsh utilization is exemplified by coho salmon fry overwintering in tidal-freshwater marshes and sloughs and by chinook fry rearing in brackish and estuarine marshes during oceanward migrations in spring and early summer. Patterns of juvenile salmon occurrence in the Fraser River estuary, Woodward Island marshes (Fig. 7, Levy and Northcote 1981, 1982) illustrate how the more transitory (short-residence time) chum salmon fry pass continually through the system (especially a ~4-week peak in 1979) between March and June, while chinook fry rear for several months between April and July. Differences between juvenile chum and chinook salmon in the level of rearing and residence in marshes was evidenced by the relative changes in size of fish over their occupation of marshes. Chum fry mean length varied little (within 5 mm) between March and May, reflecting rapid transit of fry through the Woodward Island marshes, while chinook fry mean length increased progressively from ~40 mm FL after late April, suggesting growth of rearing individuals to ~60-70 mm FL by July.

Species that have adapted to occupy a relatively confined zone of tidal marshes, and complete much or all of their life cycle within it, include peamouth chub, threespine stickleback, prickly sculpin and Pacific staghorn sculpin (Table 2). In the Woodward Island marshes, threespine stickleback and staghorn sculpin are present in the system at low densities until May, when recruitment (primarily young-of-year) occurs and rearing continues through July (Fig. 7). Many of the species that occupy tidal marshes on season scales also move into other elements and regions of the estuarine landscape for continued rearing. For instance, with increasing growth starry flounder move into deeper channels and subtidal habitats. Unfortunately, there have not been any studies that have examined landscape-scale transitions in estuarine fish fauna in this region.

3. Role of Landscape Structure and Processes in Coastal Fish Production

We have presented evidence on fish occurrence and apparent volitional residence in tidal marshes that implies the marshes contribute significantly to nekton production. However, we have not established the mechanisms whereby production is realised, or how marshes in the estuarine landscape influence the level of production. Simple occurrence in the marsh does not necessarily explain the evolution of fish life history traits that promote marsh use. In this respect, it is important to discern the relative importance of the capacity of the marshes to support fish, the opportunity of fish to access that capacity,

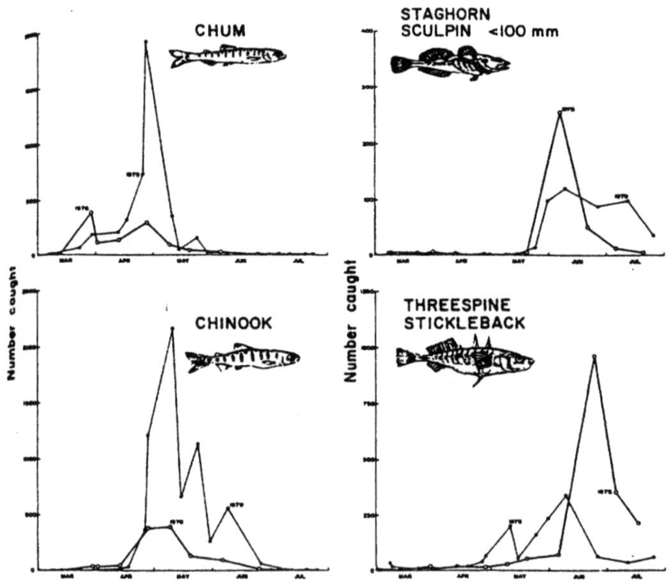

Figure 7. Total abundance of juvenile chum and chinook salmon, juvenile (<100 mm TL) Pacific staghorn sculpins, and threespine sticklebacks in "Stump TC" tidal channel (see Fig. 3b) in 1978 and 1979. Adapted from Levy and Northcote (1981).

and the product of the two in terms of the realised function in the form of increased survival due to high growth rates, reduction in predation rates and maximal fish health (Simenstad and Cordell, 2000).

3.1 TIDAL MARSH STRUCTURE

The landscape structure of tidal marshes directly influences the opportunity of juvenile fishes to seek prey resources and refugia. Although meagre, evidence from the Pacific Northwest suggests that most access occurs through the tidal channels and sloughs, and marsh edge is the primary ecotone at which foraging and refugia from predation are realised. Tidal channel systems, and the development of vegetated marsh around them reflect both the geologic history (see 1.2) and the contemporary energy regime. Tidal channel geomorphology (order, size, density and sinuosity) is regulated primarily by processes that dissipate tidal energy, but tidal prism and freshwater inflow are also factors (Pethick 1992, Coats et al. 1995). There appear to be no universal, first-principle laws describing the structure of tidal channel systems (Coats et al. 1995). Therefore, Coats et al. (1995) recommend development of empirical relationships for fundamental metrics that describe tidal channel morphometry and related marsh landscape characteristics, such as patch size and shape, connectivity and edge:patch area. In a study of 12 diverse marshes around San Francisco Bay, Coats et al. (1995) described a number of morphometric relationships that potentially relate fish responses to channel order. Higher order, more complex tidal channels are associated with greater marsh drainage area. There are relatively linear semi-log relationships between tidal channel system order and the number of channels and average sinuous length of channels (Fig. 8). Thus, length of channel edge and drainage density that may prompt juvenile fish production in tidal marshes is regulated by the processes that determine tidal channel system order. Landscape elements within tidal marshes may also be important but essentially unknown at this time. Heterogeneity of topography, vegetation patch structure, and the occurrence, distribution and size of pans on the marsh plain may also influence fish use if the tidal channel network provides access. In addition to tidal channel morphometry, these landscape attributes all likely contribute to the "trophic relay" processes of Kneib (1997, this volume) that facilitates the active translocations of intertidal production horizontally across marsh ecotones.

Large woody debris (LWD) can also represent potentially important structural features in tidal marshes. Logs and rootwads that have drifted into and settled in marshes during extreme high tides can disturb marsh vegetation, thus creating ponds, forming new channels, increasing edge length, as well as promoting sedimentation and providing an energy source and substrate for marsh consumer organisms (Maser and Sedell 1994).

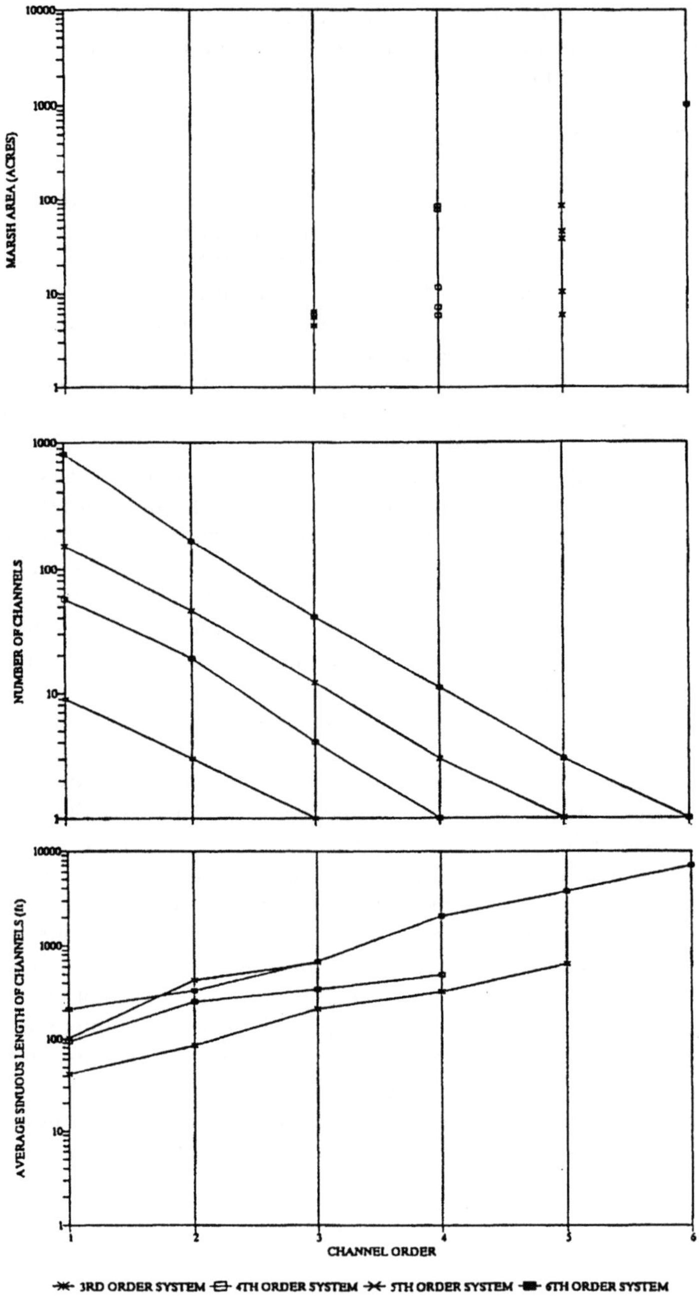

Figure 8. Empirical relationships between tidal channel system order and marsh area (top), number of channels (middle) and average sinuous length of channels from 12 marshes in San Francisco Bay. Adapted from Coats et al. (1995).

3.2 MIGRATION CORRIDORS

Certain types of marsh vegetation or topographic features with tidal marshes also provide corridors, prey resources and refugia along migratory pathways. Particularly important marsh corridors are the tidal-freshwater and brackish *Carex lynbyei* marshes that fringe distributary channels and sloughs in tidal floodplains and deltas. These marshes are flooded for approximately half of most tidal days and provide corridors during downstream or tidal movements along shore. Some of the *C. lygbyei* marshes are continuous terraces but they can also form discontinuous but closely distributed patches where the sedge has colonised block slumps of marsh soil along the channel edge. Shallow-water distributary channels, mud- and sandflats and eelgrass may also provide important corridors bridging marshes.

3.3 PREY RESOURCE CONCENTRATION

Tidal marshes obviously produce abundant, diverse invertebrate prey resources for foraging juvenile fish but fish access to the source of prey is often limited. Marsh topography significantly enhances fish production by trapping, concentrating, and transporting invertebrates from the marsh surface into rivulets and tidal channels where the prey become more available to consumption by fishes. These ebb tide processes would likely increase the efficiency of fish foraging at the nodes of the rivulets and the deeper, high order channels along the marsh edge. However, we have not found any literature that documents such fine-resolution foraging behaviour.

Tidal cycling also concentrates terrestrial aerial insects that are not actually contributed by the marsh, but by surrounding coniferous and deciduous vegetation (Tschaplinski 1987). On-going research by one of the authors (W. G. Hood, unpublished Ph.D. dissertation) has determined experimentally that the residence time of surface drift is directly related to the distance up-channel from the mouths of tidal-freshwater and brackish sloughs in the tidal floodplain of the Chehalis River. This implies that retention of drift insects, and their availability by fishes foraging in the sloughs, increases as a function of slough system morphometry as well as the characteristics of the vegetation patches that are arrayed along the length of the slough.

3.4 PRIMARY-SECONDARY PRODUCTION LINKAGES

Tidal marshes are both direct contributors to estuarine detritus-based foodwebs as well as sites of extensive detritus trapping and decomposition. Patterns of production, senescence and decomposition of vascular plant material can be highly specific to the vegetation community, such that different marsh types produce detrital matter over different seasonal schedules. For example, Thom (1981) determined that in Grays Harbor, Washington, the mean standing stock of live *Carex lyngbyei* sedge peaked in June to August, and the dead plant matter in November, high marsh plant assemblages demonstrated peak live standing stock in August and maximal dead biomass the next April; patterns in the low estuarine marsh and tidal-freshwater marshes were similar to the sedge marsh. Depending upon tidal elevation and vegetation structure, which affects shading of the benthos, different marsh types may also support different levels and timing of benthic microalgae production. Import and export of floating detritus will occur primarily as a function of the tidal channel system. But, detritus trapping and retention is also influenced by marsh structure, including both marsh elevations and retention structures such as erect emergent vegetation and LWD.

Riverine import are also a significant source of organic matter to estuarine food webs (Thom 1987) and can provide as much as 80-87% of the total estuarine organic matter budget (e.g., Grays Harbor and Columbia River estuary, respectively, Simenstad et al.1997b). An unknown quantity of this is coarse particular, floating debris that can be deposited in tidal marshes, but also the considerable fine particular organic matter that is suspended can also settle out in marshes. Trapping, retention and decomposition of these contributions also contribute to the support of tidal marsh fishes on different time and space scales.

The consequence of disjunct, overlapping patterns of primary organic matter availability to marsh primary consumers is that some fish prey resources are often predicated on the deposition and decomposition of organic matter. A unique aspect of prey such as aphids, which are sucking insects, and other direct herbivores such as benthic amphipods and copepods, is that they are directly related to live plant production. Many other primary consumers that constitute important fish prey resources are detritivores and are more dependent upon the patterns of detritus accumulation and decomposition. For instance, in the Fraser River estuary (Woodward Island), *Carex lyngbyei* reaches peak aboveground standing stock in June and July, senesces from July to November, and gradually becomes incorporated into accreting sediment by June, or is exported by ''detritus rafting'' in February-April (Kistritz and Yesaki 1979, Kistritz et al. 1983). The nutrient (N, P) level of the detritus increases through the winter, and maximum nutritional value of *C. lynbyei* detritus occurs in late winter to early spring, coincident with peak production cycles of detritivores that are fed upon by juvenile fishes recruiting to the marshes (chironomids fed upon by juvenile salmon). This is but one example of the disparate primary production and consumption cycles that takes place in tidal marshes and estuarine landscapes, but serves to illustrate that organic matter transfer to consumers, especially detritivores, can be both directly (production) and indirectly (decomposition) related to landscape structure.

3.5 ROLE OF NATURAL DISTURBANCE REGIMES

Estuarine landscapes are both the expression and source of natural disturbance that can shape juvenile fish use. While marshes will tend to be more stable than elements in the landscape that are more vulnerable to river and tidal erosion and sedimentation events (e.g., mud and sand flats), large-scale flooding and storm events can modify marsh structure. This may be particularly evident in exposed marine and estuarine marshes, that are vulnerable to erosion from wind-driven waves, and marshes within confined tidal floodplains, where channel meander processes can account for significant realignment of distributary channels. As a result of the moderate to heavy flood and storm disturbance frequency and intensity in the Pacific Northwest, tidal marshes are relatively heterogeneous, even within a single marsh plain.

4. Planning and Designing Restoration with Landscape in Mind

Restoration of tidal marshes in the Pacific Northwest has to date been primarily an opportunistic endeavour. Except for a few estuaries (e.g., the Snohomish River, Washington) where landscape-scale planning is emerging, the design and location of tidal marsh restoration sites, and the relationship among restoration sites and natural marshes, has been dictated more by legal constraints, jurisdictional agendas and site availability and cost than by ecological criteria. Obviously, extensively modified and urban estuaries offer severe constraints to restoration *per se* (Shreffler and Thom 1993), and rehabilitation and reallocation (enhancement) are likely the primary or only options (Aronson and Le Floc'h, 1996). In these cases, site design and site selection considerations such as the influence of pollution and contaminated sediments, control of exotics, alterations to freshwater and tidal flows and proximity of donor populations can become critical constraints. However, even in less-impacted estuaries, it should be evident that restoration of tidal marshes may not contribute significantly to the production and rehabilitation of fisheries resources without considering landscape structure. Certain fishes have evolved to utilise different marshes and other, interacting estuarine landscape elements as a dynamic continuum. Planning and design of tidal marsh restoration, as well as functional assessment of restoration's contribution to fish production, must include criteria for all relevant spatial and temporal scales that encompass this continuum. This initially requires basic information at two levels: 1) empirical information about marsh structure that is derived from local or regional tidal marshes and captures sources of tidal, climatic, geologic and other site-specific factors [see below for Coats et al. (1995) example of marsh morphometric relationships]; and 2) analysis of the existing and historic landscape structure of the estuary.

4.1 MARSH STRUCTURE

Marsh structure directly influences both the extent (opportunity, access) and utility (realized function) of juvenile fish use. It is necessary to understand not only what structural attributes enhance fish production but also how different fish life history stages and patterns respond to different structural attributes and what ecosystem processes sustain them.

For example, the Coats et al. (1995) empirical relationships between marsh area and tidal channel system morphometry (see Fig. 8) suggest scientific criteria for decisions about restoring tidal marshes to enhance juvenile fish production. If maximizing capacity for rearing juvenile fish such as salmon is paramount, a large, complex tidal channel system may be most desirable in order to facilitate access deep into the marsh, maximize marsh edge for foraging and retention of drift insects, and generate the highest possible drainage order system (e.g., increase likelihood that system wouldn't entirely dewater). Alternatively, distributing many, smaller drainage area marshes may not generate equally complex tidal channel systems but would offer a broader distribution of accessible marshes that might be required to compensate for neap-spring tide variability or climatic events. This was effectively illustrated in Coats et al. (1995) hypothetical illustration of one 40-ac tidal marsh, the area of which could sustain a 4th

order system in San Francisco Bay, as compared to four 10-ac marsh systems that could only maintain 3rd order tidal channel systems (Fig. 9).

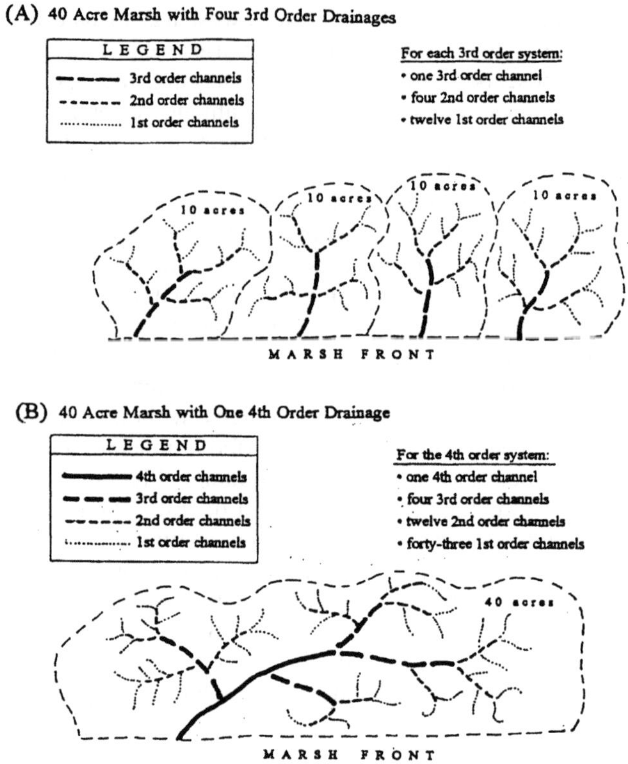

Figure 9. Hypothetical trade-off between alternative tidal channel drainage patterns in a 40-ac restored marsh, wherein four 10-ac drainages are created (A) or one 40-ac drainage. Adapted from Coats et al. (1995).

The total sinuous length that would be generated by the 4th order system would be approximately 5,314 m, while the total length in a 3rd order system would be 1,329 m, or 5,318 m for the total over four systems. Therefore, while the quantitative difference in marsh:channel edge may not be significantly different between the two configurations because the total tidal prisms do not differ, the qualitative difference of the increased channel order versus access to four systems instead of one over the same area may have substantial implications for juvenile fish rearing. The quantitative differences in other channel morphometry metrics such as bifurcation ratio and sinuosity may offer similar trade-off implications. In example illustrated in Fig. 9, the bifurcation ratios in the 40-ac marsh would be 3.58 for 1st order, 3.0 for 2nd order and 4.0 for 3rd order, as compared to 3.0 for 1st order and 4.0 for 2nd order for each 10-ac marsh. 1st order bifurcation ratio may be particularly important because they may indicate ecological "hot spots" (i.e., "interaction zones" of Kneib, this volume) where transport of prey off the marsh plain is intercepted by fish in 2nd order channels.

Considerable more studies of the association between marsh morphometry and fish production are required to develop the information required to determine whether such metrics are even meaningful. For instance, there are many factors that influence the ecological consequences of patch size and implications to restoration (Shreffler and Thom 1993), and no real empirical data with which to determine ''optimal'' size criteria.

4.2 LANDSCAPE STRUCTURE

We have argued that the composition, orientation, and linkages among tidal marshes and other estuarine landscape elements are fundamental to effective rearing of juvenile fishes in Pacific Northwest estuaries. Furthermore, landscape organisation should not be viewed as steady state, but should be examined over the scales of spatial and temporal variability that influence the opportunity of fish to seek tidal marshes under other than average conditions. Landscape structure can also be used to decrease uncertainity in the potential contribution of marsh restoration to the recovery of specific fish populations, such as depressed Pacific salmon in this region. In initial explorations of juvenile fish responses to restoring tidal marshes in the Salmon River estuary, several of the authors (D. Bottom, C. Simenstad) are documenting the occurrence, relative density and diet of juvenile salmon in restoring marshes of different ages (dikes breached in 1978, 1987 and 1996), with associated changes in marsh channel morphometry, and landscape positions (the 1996 site is located in the brackish region of the estuary, while the 1978 and 1987 sites are located in more estuarine regions). Tidal channel trapnet catches of chinook and coho fry in 1997 (Fig. 10) indicate that, except for one sample from the 1996 restoration site, chinook fry were not utilising the restoring marshes at the same density that they were an undiked control (located in the estuarine region of the estuary). However, densities of coho fry were often higher in the restoring marshes than the control marshes. Diet analyses are available only for chinook fry at this time, but they indicated somewhat comparable diets dominated by marsh insects, with highest prey taxa diversity in the control and 1996 restoring marsh and lowest diversity in the 1978 marsh, which after 20 years has developed into a comparatively monotypic, *Carex lyngbyei*-dominated low marsh assemblage. Consistent with results from the Chehalis River slough studies, aphids were particularly prominent prey of chinook fry in the 1978 restoring marsh. The apparent functionality of these restoring systems indicates that recovery of some processes, such as insect production, can be quite rapid and beneficial to foraging fish, if not exactly synonymous with natural, mature marshes. Because these marshes typically subside after being diked, it may also suggest that more frequent access to lower marsh surfaces may to some degree compensate for the lack of fully developed prey resources, e.g., there is a trade-off between accessibility (higher in restoring marshes) and capacity (higher in natural marshes?). The abundance of juvenile salmon in the brackish 1996 restoring marsh also implies that there is an additional benefit, particularly to coho fry, for marshes situated in this region of the estuarine gradient.

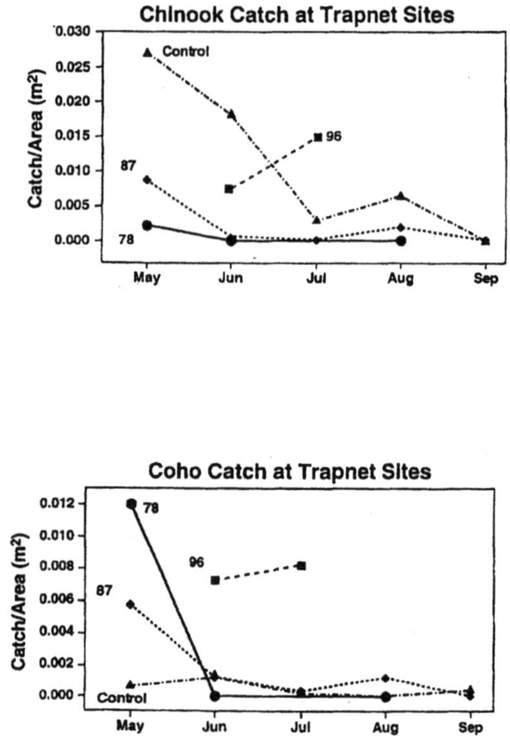

Figure 10. Densities of juvenile (fry) chinook and coho salmon in tidal channels of tidal marshes restored in 1978, 1987 and 1996 compared to an unaltered control marsh in the Salmon River estuary, Oregon. Data from D. Bottom, unpublished.

5. Summary and Recommendations

It was our intent to provide persuasive evidence that the functionality of Pacific Northwest tidal marshes for juvenile fish is to a large degree dependent upon landscape structure and scale, and that the function of marshes varies with fish ontogeny over a broad spectrum of space and time scales. We acknowledge that the evidence is as speculative as it is suggestive. However, it does pose some intriguing avenues of inquiry into factors important to understanding, managing and restoring tidal marshes for fish production that recommends: 1) incorporation of landscape structure and dynamics into restoration planning; 2) consideration of differences in fish use under both mean and extreme conditions; 3) development of functional relationships between intramarsh and intralandscape structure and fish utilisation; 4) recognition that different estuarine landscapes and elements such as tidal marshes function differently under stochastic or cyclic variability; and, 5) emphasising diversity of structure and process will also optimise the expression of fish life history diversity, which is of particular importance to the recovery of species like Pacific salmon.

5.1 NEED TO BREAK BEYOND HABITAT/SITE-SPECIFIC MANAGEMENT AND RESTORATION

The strict concept of habitat as the ''locality, site and particular type of local environment occupied by an organism'' (Lincoln et al. 1985), especially applied as ''critical'' or ''essential'' habitat of economically important fish and wildlife, is inappropriate when applied to juvenile fishes in tidal marshes. Both the rapidly changing physiological and ecological requirements of fish and the dynamics of estuarine environments imply that juvenile fish have evolved to occupy a landscape mosaic in which tidal marshes are but one element. To discount or sacrifice the role of other landscape elements along the continuum of use is likely to result in negative feedback to the functional of tidal marshes to support fishes. Thus, managing and restoring tidal marsh ''habitats'' for fish imposes a significant level of risk that landscape-scale functions will be sacrificed. A landscape perspective also suggests that the application of ''mitigation banking'' to tidal marsh restoration also does not justify or compensate for continued loss of estuarine landscapes. Whether intended or not, the operative outcome of mitigation banking is often to simplify and concentrate restored or created marshes irrespective of their ecological contribution to the estuarine landscape, e.g., based on opportunity of ownership or economic incentives. Landscape criteria for restoration of tidal marshes should apply equally to the processes involved in mitigation banking.

We suggest the following as approaches for incorporating landscape structure in tidal marsh restoration:

1. Use natural landscape templates that are specific to the estuary and local region to guide restoration;
2. Emphasize corridors and other linkages among marshes and other tidal landscape elements that facilitate physiological, foraging and refuge requirements of different fish species and life history stages;
3. Incorporate landscape elements and a mosaic that maintain a natural diversity of primary producers and detritus sources; and,
4. Promote landscape structure that accommodates fish responses to climatic variability and natural disturbance regimes.

5.2 ANADROMOUS SPECIES IN PARTICULAR NEED CONSIDERATION AT MULTIPLE, HIERARCHICAL SCALES

Fish life histories present the most viable models over which to examine scales of ecosystem interactions. This is particularly the case with anadromous fishes because the diversity of life history types and the variability in tidal marsh interactions associated with changing ontogeny. There is accumulating evidence, some presented in this paper, that estuarine residence, growth and survival of anadromous fishes relates to intramarsh attributes. This suggests that intramarsh performance is integrated with performance at landscape to estuary scales, while we know little about such feedback loops. But, the functionality of tidal marshes for anadromous species likely depends just as much on opportunity as capacity. Landscape connectivity and other patch or landscape attributes should be considered equally important as population productivity and ecosystem capacity when addressing the recovery of Pacific salmon (Mobrand et al. 1997). Volitional migration patterns set by the ontogenetic programs of juvenile salmon, as well as more reactive movements in response to estuarine conditions (e.g., salinity changes), spans the tidal-freshwaterestuarine gradient over tidal to seasonal time scales. As much as tidal marsh restoration should address historic loss of capacity, it also needs to reinstate access to landscape elements and corridors that support punctuated migration, overwintering, and tidal/event movement patterns. In this respect, restoration should not focus *a priori* on particular types of tidal marshes but instead be grounded on the historic landscape template that influenced evolution of anadromous species and metapopulations in that system. Restoration of lost and fragmented landscape elements and corridors should be a primary goal.

5.3 IMPORTANCE OF UNDERSTANDING NATURAL LANDSCAPE-SCALE STRUCTURE AND PROCESSES

Neither tidal marsh attributes important to juvenile fish nor the relationship between the attributes and fish production have been investigated to any degree in Pacific Northwest estuaries. Simenstad et al. (1991) provided fish prey attributes that were characteristic of distinct estuarine habitats but could not identify data on the habitats' physical attributes that contributed to the support of fish and wildlife. While we have suggested potentially important attributes at the intramarsh scale, such as marsh:channel edge, empirical data on this and larger landscape scales are rare. Advancing scientific information about tidal marsh use by fish and application to restoration ecology will require considerable more dedicated research, especially at the larger scales. We suggest that research addressing the following questions would significantly advance the science and ecotechnology of tidal marsh restoration:

1. What are the functional relationships between tidal channel system morphometrics and fish residence time, consumption rate, and survival?
2. What are the food web and bioenergetic implications of fish foraging in tidal marsh communities of different estuarine zones, tidal elevations and states of restoration?
3. Do marshes provide fish refugia from predation, e.g., what natural attributes of tidal marshes reduce the vulnerability of fish to predation?
4. What is the significance of ecological "hot spots" in marsh structure, such as 1st:2nd order channel bifurcation points, that may be particularly important in transferring marsh secondary production to higher trophic levels? and,
5. What determines the affinity of resident and short- and long-term transient fish toward particular marsh sites and marsh attributes?

Finally, in advocating the development of functional relationships between tidal marsh structure and fish use we do not necessarily imply that they can or should be deployed directly as construction criteria for created marshes. The morphometric structures of marshes and the landscapes within which they are imbedded are the result of extremely complex processes that are at present difficult, if not impossible, to predict using existing hydrological or sedimentological first principles. Attempts to short-circuit natural processes and accelerate the development rate of important attributes such as tidal channel morphometry or soil structure have generally not been successful, and have often delayed the natural evolution of the restoration site (Simenstad and Thom 1996). We recommend that these functional relationships are best applied to guide 'self-design'' (Mitsch and Wilson 1996) of restoring marshes and reduce the uncertainty of alternative trajectories (Aronson and Le Floc'h 1996) while taking advantage of the natural processes of ecosystem succession and disturbance that shape the role of tidal marshes within dynamic and complex estuarine landscapes (Done and Reichelt 1998).

6. Acknowledgments

We gratefully acknowledge the support of Washington and Oregon Sea Grant programs for both participation in the International Marsh Symposium and preparation of the manuscript. Maryland Sea Grant also provided support for C. Simenstad to attend the Symposium. The research upon which this synthesis is based would not have been possible without our eminent colleagues, Jeff Cordell, David Shreffler and Jessica Miller. Two anonymous reviewers provided exceedingly constructive comments that we broadly adopted to improve the manuscript.

7. Literature Cited

Able, K., D. Nemerson, P. Light and R. Bush (this volume). Response of fishes to restoration of salt hay farms in Delaware Bay, Chap. *in* M. P. Weinstein and D. A. Kreeger, editors. Concepts and controversies in tidal marsh ecology, Kluwer Academic Publishing, Dordrecht, The Netherlands.

Adam, P. 1990. Salt Marshes. Cambridge University Press, Cambridge, England.

Aronson, J. and E. Le Floc'h. 1996. Vital landscape attributes: missing tools for restoration ecology. Restoration Ecology **4**:377-387.

Baltz, D. M., C. Rakocindski and J. W. Fleeger. 1993. Microhabitat use by marsh edge fishes in a Louisiana estuary. Environmental Biology of Fishes **36**: 109-126.

Burg, M. E., D. R. Tripp and E. S. Rosenberg. 1980. Plant associations and primary productivity of the Nisqually salt marsh on southern Puget Sound, Washington. Northwest Science **54**: 222-236.

Boesch, D. F. and R. E. Turner. 1984. Dependence of fishery species on salt marshes: the role of food and refuge. Estuaries **7**: 460-468.

Chamberlain, R. H. and R. A. Barnhart. 1993. Early use by fish of a mitigation salt marsh, Humboldt Bay, California. Estuaries **16**:769-783.

Coats, R. N., P. B. Williams, C. K. Cuffe, J. B. Zedler, D. Reed, S. M. Waltry and J. S. Noller. 1995. Design guidelines for tidal channels in coastal wetlands, Report 934, Phillip Williams & Associates, San Francisco, California, USA.

Cordell, J. R., H. Higgins, C. Tanner and J. K. Aitkin. 1998. Biological status of fish and invertebrate assemblages in a breached-dike wetland site at Spencer Island, Washington, FRI-UW-9805, Fisheries Research Institute, School of Fisheries, University of Washington, Seattle, Washington, USA.

Cortright, R., J. Weber and R. Bailey. 1987. The Oregon estuary plan book. Oregon Department of Land Conservation and Development, Salem, Oregon, USA.

Crowder, L. and J. K. Craig (this volume). Factors influencing habitat choice in fish: the utility of the salt marsh habitat. *in* M. P. Weinstein and D. A. Kreeger, editors. Concepts and controversies in tidal marsh ecology. Kluwer Academic Publishing, Dordrecht, The Netherlands.

Done, T. J. and R. E. Reichelt. 1998. Integrated coastal zone and fisheries ecosystem management: generic goals and performance indices. Ecological Applications **8**: S110-S118.

Ewing, K. 1983. Environmental controls in Pacific Northwest intertidal marsh plant communities. Canadian Journal of Botany **61**: 1105-1116.
Frenkel, R. E. 1981. Phytosociological classification of Pacific Northwest coastal salt marshes. Estuaries **4**:260-261.
Frenkel, R. E., H. P Eilers and C. A. Jefferson. 1981. Oregon coastal salt marsh upper limits and tidal datums. Estuaries **4**:198-205.
Gunderson, D. R., D. A. Armstrong, Y. B Shi and R. A. McConnaughey. 1990. Patterns of estuarine use by juvenile English sole (*Parophrys vetulus*) and Dungeness crab (*Cancer magister*). Estuaries **13**:59-71.
Halpin, P. M. 1997. Habitat use patterns of the mummichog, *Fundulus heteroclitus*, in New England. 1. Intramarsh variation. Estuaries **20**:618-625.
Healey, M. C. 1982. Juvenile Pacific salmon in estuaries: the life support system. Pages 315-341 *in* V. S. Kennedy, editor. Estuarine comparisons. Academic Press, New York, New York, USA.
Healey, M. C. 1991. Life history of chinook salmon (*Oncorhynchus tshawytscha*). Pages 311-393 *in* C. Groot and L. Margolis, editors. Pacific salmon life histories. UBC Press, Vancouver, British Columbia, Canada.
Higley, D. L. and R. L. Holton. 1981. A study of the invertebrates and fishes of salt marshes in two Oregon estuaries. Final Report to US Army Corps of Engineers, School of Oceanography, Oregon State University, Corvallis, Oregon, USA.
Hofnagle, J., R. Ashley, B. Cherrick, M. Gant, R. Hall, C. Magwire, M. Martin, J. Schrag, L. Stunz, L. Vanderzanden and B. Van Ness. 1976. A comparative study of salt marshes in the Coos Bay estuary, A National Science Foundation Student Oriented Study, University of Oregon, Eugene, Oregon, USA.
Kistritz, R. U. and I. Yesaki. (1979). Primary production, detritus flux and nutrient cycling in a sedge marsh, Fraser River estuary. Technical Report 17, Westwater Research Center, University of British Columbia, Vancouver, British Columbia, Canada.
Kistritz, R. U., K. J. Hall and I. Yesaki. 1983. Productivity, detritus flux and nutrient cycling in a *Carex lyngbyei* tidal marsh. Estuaries **6**: 227-236.
Kneib, R. T. 1997. The role of tidal marshes in the ecology of estuarine nekton. Oceanography and Marine Biology, Annual Review **35**: 163-220.
—— (this volume). Salt marsh ecoscapes and production of estuarine nekton along the southeastern coast of the U.S. *in* M. P. Weinstein and D. A. Kreeger, editors. Concepts and controversies in tidal marsh ecology, Kluwer Academic Publishing, Dordrecht, The Netherlands.
Lenanton, R. and I. Potter. 1987. Contribution of estuaries to commercial fisheries in temperate Western Australia and the concept of estuarine dependence. Estuaries **10**: 28-35.
Levings, C. D. and R. M. Thom. 1994. Habitat changes in Georgia Basin: implications for resource management and restoration. Pages 330-351 *in* R. C. H. Wilson, R. J. Beamish, F. Aitkens and J. Bell, editors. Review of the marine environment and biota of strait of Georgia, Puget Sound and Juan de Fuca strait. Proceedings of the British Columbia/Washington Symposium on the Marine Environment, Canadian Technical Report, 1948, Fisheries and Aquatic Sciences.
Levings, C. D., C. D. McAllister, J. S. Macdonald, T. J. Brown, M. S. Kotyk and B. A. Kask. 1989. Chinook salmon (*Oncorhynchus tshawytscha*) and estuarine habitat: a transfer experiment can help evaluate estuarine dependency. Pages 116-122 *in* C. D. Levings, L. B. Holtby and M. A. Henderson, editors. Proceedings of the national workshop on effects of habitat alteration on salmonid stocks. Canadian Special Publications in Fisheries and Aquatic Sciences **105**.
Levings, C. D. and A. I. Moody. 1976. Studies of intertidal vascular plants, especially sedge (*Carex lyngbyei*), on the disrupted Squamish River delta. British Columbia Technical Report 606, Fisheries and Marine Service, Environment Canada, Ottawa, Canada.
Levy, D. A. and T. G. Northcote. 1982. Juvenile salmon residency in a marsh area of the Fraser River estuary. Canadian Journal of Fisheries and Aquatic Sciences **39**: 270-276.
Levy, D. A., T. G. Northcote and G. J. Birch. 1979. Juvenile salmon utilization of tidal channels in the Fraser River estuary, British Columbia. Technical Report 23, Westwater Research Center, University of British Columbia, Vancouver, British Columbia, Canada.
Levy, D. A. and T. G. Northcote. 1981. The distribution and abundance of juvenile salmon in marsh habitats of the Fraser River estuary. Techical Report 25, Westwater Research Center, University of British Columbia, Vancouver, British Columbia, Canada.
Lewis, R. R., III. 1992. Coastal ecosystems. Restoration and Management Notes **10**: 18-20.

Lincoln, R. J., G. A. Boxshall and P. F. Clark. (1985) A Dictionary of Ecology, Evolution and Systematics, Cambridge University Press, Cambridge, England.
Maser, C. and J. R. Sedell. 1994. From the Forest to the Sea: The Ecology of Wood in Streams, Rivers, Estuaries and Oceans. St. Lucie Press, Delray Beach, Florida, USA.
Miller, J. A. and C. A. Simenstad. 1997. A comparative assessment of a natural and created estuarine slough as rearing habitat for juvenile chinook and coho salmon, Estuaries **20**: 792-806.
Mitsch, W. J. and R. F. Wilson. 1996. Improving the success of wetland creation and restoration with know-how, time and self-design. Ecological Applications **6**: 77-83.
Minello, T. J., R. J. Zimmerman and R. Medina. 1994. The importance of edge for natant macrofauna in a created salt marsh. Wetlands **14**: 184-198.
Mobrand, L. E., J. A. Lichatowich, L. C. Lestelle and T. S. Vogel. 1997. An approach to describing ecosystem performance "through the eyes of salmon." Canadian Journal of Fisheries and Aquatic Sciences **54**:2964-2973.
National Research Council. 1992. Restoration of Aquatic Ecosystems: Science, Technology and Public Policy, National Academic Press, Washington, District of Columbia, USA.
National Research Council. 1996. Upstream: salmon and society in the Pacific Northwest. National Academic Press, Washington, District of Columbia, USA.
Nehlsen, W., J. E. Williams and J. A. Lichatowich. 1991. Pacific salmon at the crossroads: stocks at risk from California, Oregon, Idaho and Washington. Fisheries **16**: 4-21.
Nixon, S. W. 1980. Between coastal marshes and coastal waters—a review of twenty years of speculation and research on the role of salt marshes in estuarine productivity and water chemistry. Pages 437-525 *in* P. Hamilton and K. B. Macdonald, editors. Estuarine and wetlands processes with emphasis on modeling. Plenum Press, New York, New York, USA.
Northcote, T. G., N. T. Johnston and K. Tsumura. 1979. Feeding relationships and food web structure of lower Fraser River fishes. Technical Report 16, University of British Columbia, Westwater Research Center, Vancouver, British Columbia, Canada.
Peterson, G. W. and R. E. Turner,. 1994. The value of saltmarsh edge vs. interior as a habitat for fish and decapod crustaceans in a Louisiana tidal marsh. Estuaries **17**: 235-262.
Pethick, J. S. 1992. Saltmarsh geomorphology. Pages 41-62 *in* J. R. L. Allen and K. Pye, editors. Saltmarshes: morphodynamics, conservation and engineering significance. Cambridge University Press, Cambridge, England.
Reimers, P. E. 1973. The length of residence of juvenile fall chinook salmon in the Sixes River, Oregon. Oregon Fish Game Commissions Research **4**:1-43.
Robins, C. R., R. M. Bailey, C. E. Bond, J. R. Brooker, E. A. Lachner, R. N. Lea and W. B. Scott. 1980. A List of Common and Scientific Names from the United States and Canada. Special Publication 12, American Fisheries Society, fourth edition, Bethesda, Maryland, USA.
Rozas, L. P. 1992. Comparison of nekton habitats associated with pipeline canals and natural canals in Louisiana marshes. Wetlands **12**: 136-146.
—— 1995. Hydroperiod and its influence on nekton use of the salt marsh: a pulsing ecosystem. Estuaries **18**: 579-590.
Rozas, L. P. and W. E. Odum. 1987. Use of tidal freshwater marshes by fishes and macrofaunal crustaceans along a marsh stream-order gradient. Estuaries **10**: 36-43.
Ryall, R. and C. D. Levings. 1987. Juvenile salmon utilization of rejuvenated tidal channels in the Squamish estuary, British Columbia, Canadian Manuscript, Report Fisheries and Aquatic Sciences 1904, Department of Fisheries and Oceans, West Vancouver, British Columbia, Canada.
Seliskar, D. M. and J. L. Gallagher. 1983. The ecology of tidal marshes of the Pacific Northwest coast: a community profile. FWS/OBS-**82/32**, U.S. Fish and Wildlife Services, Biological Service Program, Washington, District of Columbia, USA.
Shi, Y., Gunderson, D. R. and D. A. Armstrong. 1995. Population dynamics of 0^+ English sole, *Pleuronectes vetulus*, in estuaries and nearshore areas off Washington. Pages 343-365 *in* Proceedings of the international symposium on North Pacific flatfish. Alaska Sea Grant College Program, Fairbanks, Alaska, USA.
Shi, Y., Gunderson, D. R. and P. J. Sullivan. 1997. Growth and survival of 0^+ English sole, *Pleuronectes vetulus*, in estuaries and adjacent nearshore waters off Washington. Fishery Bulletin, US **95**: 161-173.
Shreffler, D. K. and R. M. Thom. 1993. Restoration of urban estuaries: new approaches for site location and design, Report prepared for Washington State Department of Natural Resources, Battelle Pacific Northwest Lab, Richland, Washington, USA.

Shreffler, D. K., C. A. Simenstad and R. M. Thom. 1990. Temporary residence by juvenile salmon of a restored estuarine wetland. Canadian Journal of Fisheries and Aquatic Sciences **47**: 2079-2084.

—— 1992. Juvenile salmon foraging in a restored estuarine wetland. Estuaries **15**: 204-213.

Simenstad, C. A. 1983. The ecology of estuarine channels of the Pacific Northwest: a community profile. FWS/OBS **83/05**, U.S. Fish and Wildlife Services, Biological Service Program, Washington, District of Columbia, USA.

Simenstad, C. A. and J. R. Cordell (in review). Ecological assessment criteria for restoring anadromous salmonids'' habitat in Pacific Northwest estuaries. Ecological Engineering.

Simenstad, C. A. and R. M. Thom. 1992. Restoring wetland habitats in urbanized Pacific Northwest estuaries. Pages 423-472 *in* G. W. Thayer, editor. Restoring the nation's marine environment. Maryland Sea Grant, College Park, Maryland, USA.

Simenstad, C. A., J. R. Cordell, R. C., Wissmar, K. L. Fresh, S. Schroder, M. Carr and M. Berg. 1988. Assemblage structure, microhabitat distribution and food web linkages of epibenthic crustaceans in Padilla Bay National Estuarine Research Reserve, Washington, NOAA Tech. Rep. Ser. OCRM/ MEMD, FRI-UW-**8813**, Fisheries Research Institute, University of Washington, Seattle, Washington, USA.

Simenstad, C. A., J. R. Cordell, W. G. Hood, J. A. Miller and R. M. Thom. 1992. Ecological status of a created estuarine slough in the Chehalis River estuary: report of monitoring in created and natural estuarine sloughs, January-December. 1991. FRI-UW-**9206**. Fisheries Research Institute, University of Washinton, Seattle, Washington, USA.

Simenstad, C. A., J. R. Cordell, J. A. Miller, W. G. Hood and R. M. Thom. 1993. Ecological status of a created estuarine slough in the Chehalis River estuary: assessment of created and natural estuarine sloughs, January-December 1992, FRI-UW-**9305**. Fisheries Research Institute, University of Washington, Seattle, Washington, USA.

Simenstad, C. A., J. R. Cordell, W. G. Hood, B. E. Feist and R. M. Thom 1997a. Ecological status of a created estuarine slough in the Chehalis River estuary: assessment of created and natural estuarine sloughs, January-December 1995, FRI-UW-**9621**, Fisheries Research Institute, University of Washington, Seattle, Washington, USA.

Simenstad, C. A., M. Dethier, C. Levings and D. Hay, 1997b. The land-margin interface of coastal temperate rain forest ecosytems: shaping the nature of coastal interactions. Pages 149-187 *in* P. Schoonmaker, B. von Hagen and E. Wolf, editors. The rain forests of home: profile of a North American bioregion. Ecotrust/Interain Pacific, Portland, Oregon, USA and Island Press, Covelo, California, USA.

Simenstad, C. A., K. L. Fresh and E. O. Salo. 1982. The role of Puget Sound and Washington coastal estuaries in the life history of Pacific salmon: an unappreciated function. Pages 343-364 *in* V. S. Kennedy, editor. Estuarine comparisons. Academic Press, New York, New York, USA.

Simenstad, C. A., C. D. Tanner, R. M. Thom and L. Conquest. 1991. Estuarine habitat assessment protocol. EPA 910/9-**91-037**, U.S. Environmental Protection Agency, Region 10, Seattle, Washington, USA.

Tschaplinski, P. J. 1982. Aspects of the population biology of estuary-reared and stream-reared juvenile coho salmon in Carnation Creek: a summary of current research. Pages 289-307 *in* G. F. Hartman, editor. Proceedings of the Carnation Creek workshop, ten-year review, Malaspina College, Nanaimo, British Columbia, Canada.

Tschaplinski, P. J. 1987. The use of estuaries as rearing habitats by juvenile coho salmon. Pages 123-142 *in* T. W. Chamberlin, editor. Proceedings of the workshop: applying 15 Years of Carnation Creek results. Carnation Creek Steering Committee, Pacific Biology Station, Nanaimo, British Columbia, Canada.

Thom, R. M. 1981. Primary productivity and carbon input to Grays Harbor estuary, Washington, Grays Harbor. Chehalis River improvement to navigation. Environmental Studies, U.S. Army Corps Engineers, Seattle District, Seattle, Washington, USA.

Thom, R. M. 1987. The biological importance of estuaries. The Northwest Environmental Journal **3**: 21-42.

Weinmann, F., M. Boulé, K. Brunner, J. Malek and V.Yoshino. 1984. Wetland plants of the Pacific Northwest. U.S. Army Corps Engineers, Seattle Distict, Seattle, Washington, USA.

Weinstein, M. P. 1979. Shallow marsh habitats as primary nurseries for fish and shellfish, Cape Fear River North Carolina, Fishey Bulletin, US. **77**: 339-357.

Wissmar, R.C. 1998 Catchment hydrology and status of chum salmon (*Oncorhynchus keta*) in Washington State. 535-541 in K. Kovar, U. Tappeiner, N. R. G. Cragn (eds). Headwater Control IV: Hydrology, Water Resources and Ecology in Headwaters, UNESCO International Hydrology Program IHP-V, 1996-2001, IAHS Press, Wallingford, Oxfordshire, England.

ECOLOGICAL ENGINEERING OF RESTORED MARSHES

THE ROLE OF PULSING EVENTS IN THE FUNCTIONING OF COASTAL BARRIERS AND WETLANDS: IMPLICATIONS FOR HUMAN IMPACT, MANAGEMENT AND THE RESPONSE TO SEA LEVEL RISE

JOHN W. DAY, JR.
Department of Oceanography and Coastal Sciences and
Coastal Ecology Institute
Louisiana State University
Baton Rouge, LA 70803 USA

NORBERT P. PSUTY
Institute of Marine and Coastal Sciences
Rutgers University
P.O. Box 231
New Brunswick, NJ 08903 USA

BRIAN C. PEREZ
Department of Oceanography and Coastal Sciences and
Coastal Ecology Institute
Louisiana State University
Baton Rouge, LA 70803 USA

Abstract

Coastal wetland and barrier systems exist in a dynamic equilibrium, in both horizontal and vertical planes, between forces which lead to their establishment and maintenance and forces which lead to deterioration. Rising sea level will affect both of these systems. The functioning of coastal barriers and coastal wetlands is affected by energetic forcings which serve to enhance productivity, increase materials fluxes, and affect the morphology and evolution of these systems. These forcings occur over a hierarchy of different spatial and temporal scales and include waves and daily tides, frontal passages and other frequent storms, normal river floods, strong storms, great river floods, and switching of river channels. Human impact over the past century has diminished these forcings at all temporal and spatial scales. Sustainable management should include the re-integration of these forcings into coastal management.

1. Introduction

Coastal wetlands and coastal barriers exist in a dynamic equilibrium, in both horizontal and vertical planes, between forces which lead to their establishment and maintenance and forces which lead to deterioration. In the vertical plane, one of the most important processes affecting these systems over the past century is rising sea level. If wetlands are to survive rising water levels, they must be able to accrete at a rate such that surface elevation gain is sufficient to offset the rate of water level rise (Cahoon et al. 1995b). A number of studies have shown that coastal marshes are able to accrete at a rate equal to the historical rate of eustatic sea level rise (1-2 mm y^{-1}, Gornitz et al. 1982) and survive for long periods of time (Redfield 1972, McCaffey and Thompson 1980, Orson et al. 1987). There are reports, however, that sea level rise is leading to wetland loss in a number of coastal areas including North Carolina (Hackney and Cleary 1987), New York (Clark 1986), South Carolina (Kana et al. 1986), Maryland (Stevenson et al. 1988), Louisiana (Salinas et al. 1986, Conner and Day 1989), the Po delta (Sestini 1992), and the Nile delta (Stanley 1988). Coastal barriers have likewise been shown to have existed through time and space as a variably-shifting mass of sand, encroaching upon the continental margin as sea level was rising and adequate quantities of sediment were maintained (Shepard et al. 1960, Riggs and Cleary 1993).

It is likely that the rate of sea level rise will accelerate over the coming 100 years by 4-6 mm y^{-1} (Raper et al. 1996, Gornitz 1995). In addition, subsidence due to a number of factors, has caused relative sea level rise (RSLR) to be much greater than the eustatic rate in a number of coastal systems. For example, in the Mississippi delta, RSLR is 10-12 mm y^{-1}, primarily due to regional subsidence (Penland and Ramsey 1990) and for the Nile Delta, the rate of subsidence is as high as 5 mm y^{-1} (Stanley 1988). The background rate of geologic subsidence in Venice Lagoon during the 20th century has been 1.2-1.5 mm y^{-1} (Pirazzoli 1987, Carbognin et al. 1996) resulting in a RSLR between 2.4 and 3.0 mm y^{-1} (Albani et al. 1983, Rusconi et al. 1993). In the mid-Atlantic coast, sea level rise is much higher than the global rate (Stevenson et al. 1988). Marshes, however, can survive such high rates of RSLR if there is sufficient sediment input and in situ organic soil formation. In the Mississippi Delta, for example, tidal creek streamside levee marshes and those near sources of riverine sediments are able to accrete vertically at rates higher than RSLR while back marshes generally have accretion rates less than RSLR (Baumann et al. 1984, Hatton et al. 1983). In Venice Lagoon, however, practically all riverine sediment input has been stopped (Bendoricchio et al. 1993) and there is a strong net loss of sediments from the lagoon (Bettinetti et al. 1995). As a result, salt marsh area in the lagoon has fallen from about 12000 ha at the beginning of the century to about 4000 ha at present due to reclamation, erosion, pollution, and natural and human-induced subsidence (Favero et al. 1988, Runca et al. 1993).

Barrier islands are usually considered to have concentrated sediment that had been supplied from offshore during the latest Holocene sea-level rise (Swift 1975, Nummedal 1983, Walker and Coleman 1987). However, that supply was finite and most barrier islands in the world are currently experiencing a negative sediment budget as sand is transferred to sediment sinks offshore or in deep estuaries (Bird 1993). An accelerated rate of sea-level rise will exacerbate the negative sediment budget on the barriers and will contribute to a general attenuation of the form, both in width and height. (Bruun 1988, Dubois 1990, McBride and Byrnes 1997)

In this paper, we attempt to integrate coastal system functioning with human impacts and management and show how this affects system response to global change. We first present a conceptual framework of coastal system functioning that illustrates how energetic pulsing events on different spatial and temporal scales are critical to the maintenance of these systems. We then show that many detrimental human impacts are a result of reduction or elimination of these pulsing events. Finally, we develop a comprehensive theory of coastal management that shows that system functioning should form the basis for management and, specifically, that integration of pulsing events into management is necessary for sustainable management, especially in the face of sea level rise.

2. A Spatial-Temporal Hierarchy of Natural Subsidies

The functioning of coastal barriers and coastal wetlands is affected by energetic forcings which serve to enhance productivity, increase materials fluxes, and affect the morphology and evolution of these systems. This type of pulsing applies to many natural systems, but it is especially important for coastal ecosystems (e.g., Estuaries, Volume 18, Blum 1995). Eugene Odum (1980) recognized this early on when he described estuaries as tidally subsidized, fluctuating water level ecosystems. But the tide is not the only energy subsidy to coastal systems. Energetic forcings occur over a hierarchy of different spatial and temporal scales (Day et al. 1995, 1997). These energetic events range from waves and daily tides to switching of river channels in deltas which occur on the order of every 1000 years, and include frontal passages and other frequent storms, normal river floods, strong storms, and great river floods (Table 1). The primary importance of the infrequent events such as channel switching, great river floods and very strong storms such as hurricanes is in sediment delivery to wetlands and coastal barriers and in major spatial changes in geomorphology. The more frequent events such as annual river floods, seasonal storms such as frontal passages and tidal exchange are also important in maintaining salinity gradients, delivering nutrients and regulating biological processes. In the remainder of this section, we will discuss the role of energetic pulsing in maintaining coastal wetland and coastal barriers.

TABLE 1. Temporal scale of pulsing events in deltaic systems. Modified from J. W. Day et al. (1995).

Event	Timescale	Impact
River Switching	1000 y	Deltaic lobe formation. Net advance of deltaic landmass.
Major river floods	50-100 y	Channel switching initiation. Crevass splay formation deposition.
Major storms	5-20 y	Major deposition. Enhanced production.
Average river floods	3. Annual	Enhanced deposition. Freshening (lower salinity) Nutrient imput. Enhanced 1st and 2nd production.
Normal storm events (frontal passage)	Weekly	Enhanced deposition. Organism transport. Net material transport.
Tides	Daily	Drainage/marsh production. Low net transport.

2.1 WAVES

Strong wave energy can lead to the relatively rapid erosion of wetlands (Day et al. 1998). Thus a major importance of coastal beach/dune systems is to protect estuarine wetlands from the full force of oceanic waves. Waves, however, are the primary agent in maintaining and modifying the barrier islands. The basic sediment budget of any barrier island is the product of the along-shore and cross-shore exchange of sediment that is the product of fluid flows driven by the incident wave energy. The alongshore gradient in wave energy largely explains the conditions of erosion or accretion, with increasing energy levels mobilizing more sediment than is entéring coastal compartments and resulting in a deficit sediment supply and subsequent erosion or inland displacement of the shoreline. The opposite is true for a decreasing alongshore energy gradient.

2.2 TIDES

The daily rise and fall of tides leads to higher biological production, enhanced interaction between wetlands and adjacent water bodies, allows drainage of wetland sediments (Howes et al. 1986), and permits fish to use the surface of the marsh for feeding and protection during periods of high tide (Zimmerman and Minello 1984). Because of this, E.P. Odum (1980) called estuaries ''tidally subsidized, fluctuating water level ecosystems.''

The vertical elevation growth range of wetland vegetation has been correlated to tidal range (McKee and Patrick 1988), therefore, coastal wetlands in areas of low tide range are more sensitive to sea level rise. For example, the Gulf of Mexico (GOM) and the Mediterranean generally have low tide ranges on the order of 20-40 cm and wetlands in these areas exist within a relatively narrow elevation range. This sensitivity is important in considering the potential effects of increasing water levels because in the absence of vertical accretion, the elevation of these wetlands will quickly fall below the acceptable growth range.

Tides have several functions on barriers islands. Initially, barriers do not exist in macrotidal locations, and there tend to be more continuous barriers in microtidal seas (Hayes 1979). Tides extend the vertical range of the barriers in proportion to the tidal range, with a tremendous amount of sand in storage within the intertidal zone. Tides tend to add to the character of the beach by constructing a high tide berm flat slightly above high water level, related to spring tide elevations, and a migrating low-tide berm that expands and contracts during the tidal cycle. There is a suggestion that incremental accumulation on narrow barrier forms may have a temporal pattern associated with the 18.6-year tidal epoch (Orford et al. 1992). An important component of tides is the creation of tidal currents at inlets in the barriers and the opportunities for transfer of sediment into the bay or lagoon and the export of materials from the estuary. Inlets are a highly dynamic assemblage of sediment and form and produce features that transfer sediments on the oceanside as well as on the bayside and offer submarine topography that affects wave refraction and tidal channel alignment that interacts with the barrier island form and topography (Hayes 1979, FitzGerald 1982, Boothroyd 1985, Fenster and Dolan 1996).

2.3 FRONTAL STORMS

Winter cold fronts and other similar storm events have a profound effect on processes governing coastal marsh sustainability. High wind speeds associated with cold fronts cause resuspension of bottom sediments from coastal bays and the nearshore coastal ocean and the onshore winds preceeding frontal passage lead to flooding of coastal marshes and sediment deposition on the marsh surface (Baumann et al. 1984, Reed 1989, Fig. 1). In addition to higher suspended sediment concentrations as a result of storms, tidal hydrodynamics are also altered (Leonard et al. 1995, Hsu and Blanchard 1993, Moeller et al. 1993, Smith 1979). In the northern Gulf coast, for example, the astronomical tidal amplitude averages approximately 0.35m. However, water level changes in excess of 1.2 m have been recorded in Atchafalaya Bay by Kemp et al. (1980) in response to a cold front event. These storms lead to enhanced flushing of coastal bays and strong net movements of water. For example, Perez et al. (2000) calculated that approximately 56% of the volume of Fourleague Bay, LA was exported to the Gulf of Mexico as a result of 45 hours of strong northerly winds following the passage of a cold front. In addition to large exports of water, suspended sediment concentrations in excess of 1500 mg L-1 and ebb-directed suspended sediment fluxes near 1100 kg s-1 were measured in response to strong frontal events (Fig. 2). Frontal passages and other seasonal storms are especially important in enhancing sedimentation in coastal marshes in areas of low tidal range (Baumann et al. 1984, Reed 1989, Roberts et al. 1989, Cahoon et al. 1995a, Day et al. 1994, 1995, Hensel et al. 1998).

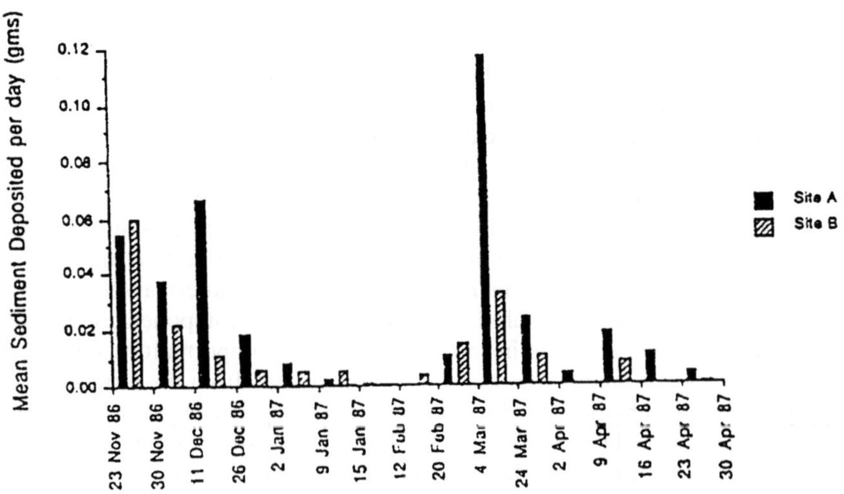

Figure 1. Mean sediment deposition for individual monitoring periods. Cold fronts with a strong set-up occurred during the 11-26 December 1986 and 4-24 March 1987 periods. Modified from Reed 1989.

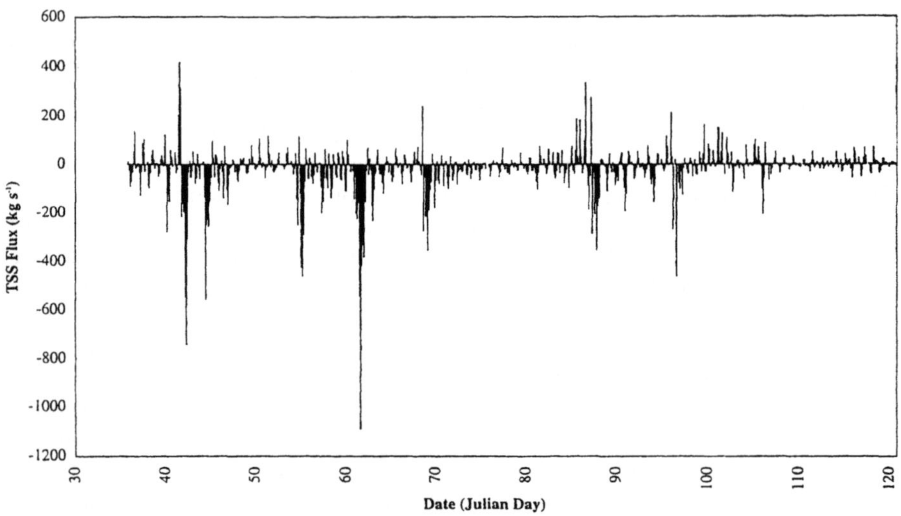

Figure 2. Total suspended sediment (TSS) flux from Feb. 1, 1994 to Apr. 30, 1994 in Oyster Bayou, LA. Positive values are fluxes into the bay (flood-directed) and negative values are fluxes to the Gulf of Mexico (ebb-directed). Extreme negative fluxes (days 42, 44, 54, 61, 69, 87, 96) are associated with strong cold front passages. Modified from Perez et al. (2000).

Precipitation events also enhance marsh sediment import/export processes. Heavy rainfall events at low tide have been shown to cause strong export of sediments from the marsh surface (Chalmers et al. 1985, Childers and Day 1990a). Heavy rains during low tide can increase the sediment loads in small streams by an order of magnitude and lead to high sedimentation on the marsh surface (Settlemyer and Gardner 1975). In the Rhode River Estuary, Jordan et al. (1986) reported that 32% of the total suspended matter discharge occurred during just 3 out of 156 weeks of measurement and that the large discharges were produced by strong precipitation events. The majority of the sediment discharged during storms remained in tidal creeks or bordering marshes due to rapid settling of suspended material.

Erosion of natural channels and tidal flats can also lead to high TSS concentrations and account for a significant portion of the sediment advected onto adjacent marshes, leading to a more stable wetland. Settlemyer and Gardner (1977) found that winter sediment transport in Dill Creek, S.C. was nearly balanced except during tidal cycles that occurred after high winds generated by storms. Tidal flats were eroded and significant sedimentation resulted from these ephemeral events which are "primarily responsible for supplying the growth requirements of marshes"' in the study area (Settlemyer and Gardner 1977).

Suspended sediment concentrations can increase by several orders of magnitude in response to winter frontal passages and storm events when compared to fair weather conditions (Caffrey and Day 1986, Leonard et al. 1995, Hensel et al. 1998, Perez et al. 2000, Fig. 3). Enhanced water levels and wind generated turbulence in conjunction with increased TSS concentrations provides a mechanism for sediment transport into coastal marshes which leads to high levels of sediment deposition. In the Holland Glade marsh, Delaware, Stumpf (1983) stated that with "sufficiently deep flood waters, wind stress will greatly increase the suspended load into the marsh"' because the wind leads to higher suspended sediment concentrations and the storm floods may distribute sediment more evenly across the marsh surface, even into the back marshes. In fact, two storms, in October and November 1980, each deposited approximately 2 mm of sediment onto the marsh surface in the Holland Glade marsh, suggesting that 1-2 major

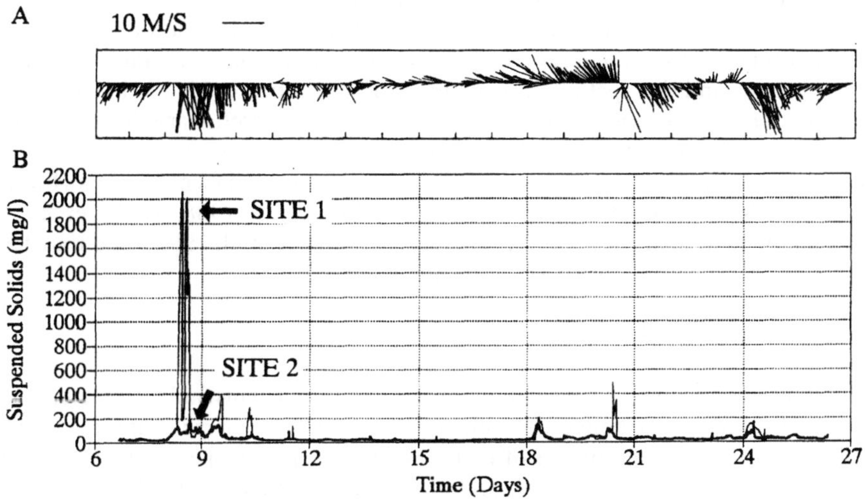

Figure 3. (A) Wind vectors at Grande Isle, Louisiana, November 6-27, 1991. (B) Suspended sediment concentration (C_s) at 2 sites in Lake Barre, Louisiana. Modified from Murray et al. (1993).

storms per year can supply practically all of the annual sedimentation of 5 mm y^{-1} (Stumpf 1983). The importance of the combination of forces in enhancing marsh accretion was also recognized by Leonard et al. (1995) who reported that maximum sediment transport into a marsh on the west coast of Florida occurred when resuspension of sediments coincided with high current velocities thus allowing the sediment to remain in suspension and be advected onto the marsh surface.

Currents generated by frontal passages are also important in transporting organisms and organic matter into and out of estuaries (Rogers et al. 1993, Incze and Roman 1983). Rogers et al. (1993) reported peak concentrations of brown shrimp post larvae catch during the post-frontal return period of a major cold front in the northwestern Gulf of Mexico. The authors attributed this to a postlarval response to environmental cues such as temperature, pressure, and salinity. As the cold front and associated north winds generate colder temperatures, fresher water, and strong ebb-dominated discharges, the postlarval brown shrimp mediate their depth in the water column and wait to take advantage of strong flooddirected currents characteristic of the post-frontal return, thus maximizing recruitment into the estuary (Rogers et al. 1993).

Geomorphological responses on barrier island beaches are often characterized as seasonal or cyclic, referring to the summer/winter comparison on a longer time scale and the passage of fronts on the shorter time scale (Nordstrom 1980, Hallermeier 1981). The inference is that the beach-dune system goes through a period of morphological erosion and recovery at different time scales of months to several years, but eventually returning to a morphological equilibrium (Thom and Hall 1991). There may be a net loss or gain of sediment, but the dimensions of the form are similar, albeit desplaced slightly spatially because of the sediment budget (Nakashima 1988). Coastal dune development and dune mobility also have been related to the periodic displacements associated with storm events (Edelman 1972, Psuty 1988, 1992, Sherman and Bauer 1993). Roman et al. (1997) showed that accretion in a Massachusetts salt marsh was related to proximity to a barrier inlet and strong storms.

Barrier island systems also show periodic form and spatial variation related to frontal storm occurrence. Part of the response is the sequential transgressive transfer of sediment onto the barrier and the spatial shift of the barrier (Forbes et al. 1991, Dubois 1995). Transfers along the barrier serve to create recurrent accretionary forms (Shaw and Forbes 1987, Ollerhead and Davidson-Arnott 1993, Psuty 1994) as well as to interact with inlet dynamics and the periodic exchanges between adjacent barriers (FitzGerald et al. 1984, Boothroyd 1985, Oertel 1988, Orford et al. 1988, Hicks et al. 1999); transfers on the bay margin on to the barrier (Armbruster et al. 1995), and transfers that relate to the entire barrier form (McBride and Byrnes 1997).

2.4 RIVER FLOODS

Annual river floods supply a yearly input of sediments, nutrients and freshwater to coastal systems. The sediments delivered include both fine and coarse materials which can contribute to the sediment budget of marshes and beaches, respectively. The regularity of these pulses results in an annual, and predictable, reduction of salinity and input of nutrients. It has been demonstrated that river discharge is related to estuarine primary production (Boynton et al. 1982) and fisheries (Sutcliffe 1972). The biota within deltaic systems have adapted to this seasonality, and are therefore dependent upon their regular occurrence (Day et al. 1989).

River floods can carry very high suspended sediment loads. For example, concentrations up to 1.6 g l^{-1} were measured in a Rhone River flood in November 1992 (Fig. 4). Riverine flooding of coastal marshes causes not only increased accretion but also enhances primary production (Nyman et al. 1990) which leads to higher rates of organic soil formation. Management approaches should therefore be designed to increase the input of flood waters to wetlands. In areas of lower elevation, inundation frequency and duration are important factors influencing vegetation growth and distribution (Swenson and Sasser 1992). In Willow Bayou, LA, Stern et al. (1991) calculated that tidal forcing was dominant only during periods of high mean sea level and low river flow. They also determined that the Atchafalaya River was the source of nitrate, soluble reactive phosphorous, and TSS to the marsh system (Stern et al. 1991).

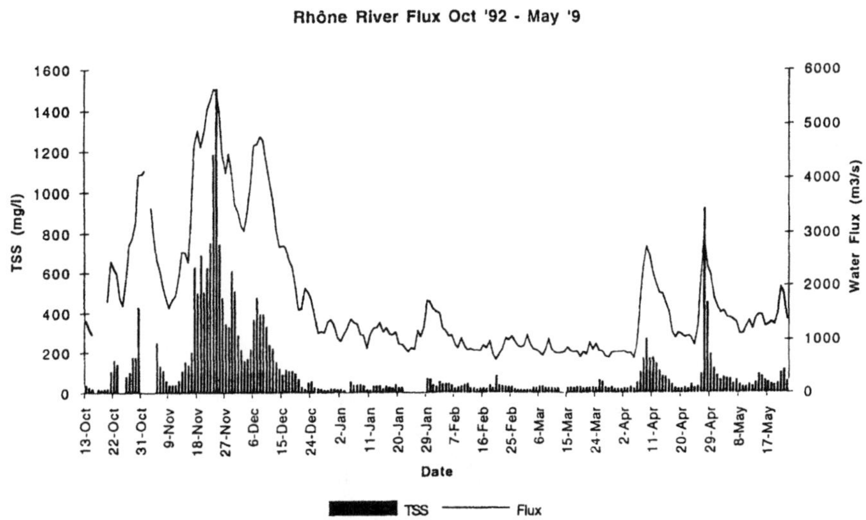

Figure 4. Total suspended sediment (TSS) concentrations and water flux in the Rhone River from Oct. 1992 to May 1993. Modified from Hensel et al. (1998).

High nutrient and suspended sediment concentrations in river water is vital to the sustainability of coastal marsh systems. The Mississippi deltaic plain has a high rate of subsidence and Baumann et al. (1984) showed that marshes near the Atchafalaya had high sediment deposition rates during the period of high river discharge. Estuaries receiving little to no direct river input, such as the Barataria Basin in Louisiana, are dependent more on organic production and resuspension and relocation of bay sediment for marsh building processes (Madden et al. 1988). Coastal systems take up nutrients during periods of high riverine input and release nutrients via remineralization during low discharge (Kemp and Boynton 1984, Childers and Day 1990b).

2.5 MAJOR STORMS

Large storms such as hurricanes and typhoons, occurring every 10 to 20 years, are pulsing mechanisms which supply large sediment inputs to coastal systems. Storm events resuspend large quantities of bottom sediments of coastal bays and the nearshore coastal ocean and deposit them on coastal wetlands. Strong storms breach barrier islands but they also mobilize large volumes of sand from offshore and move it in front of beaches where it is transported to barrier islands by normal waves and winds. For example, Baumann et al. (1984) reported that two tropical storms were responsible for 40% of total accretion over a five-year period in salt marshes in the Mississippi delta. Similarly, Cahoon et al. (1995a) reported that during the passage of Hurricane Andrew in 1992, short term sedimentation rates in Mississippi delta marshes were between 3-8 g m^{-2} d^{-1} as compared to rates generally less than 0.5 g m^{-2} d^{-1} during non-storm periods (Fig. 5). Longer term accretion as measured by marker horizons was 2-12 times higher than non-hurricane periods. (A special issue of the Journal of Coastal Research addressed the impact of Hurricane Andrew on coastal ecosystems of Florida and Louisiana, Stone and Finkl 1995). Rejmanek et al. (1988) reported that more than 2 cm of material was deposited in a tidal freshwater Phragmites community in the Mississippi delta during the passage of Hurricane Danny in 1985.

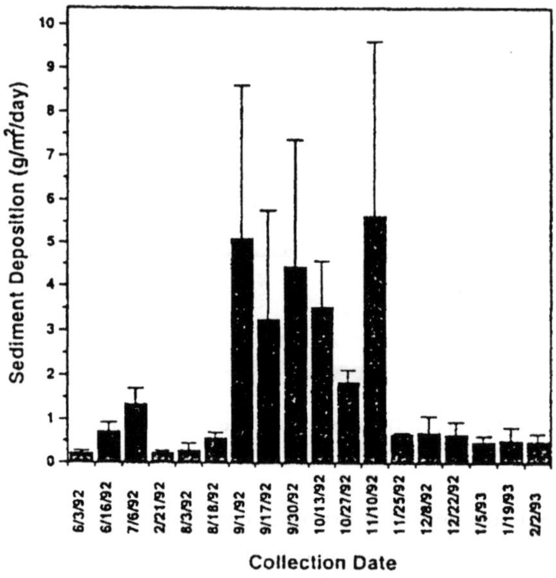

Figure 5. Sediment deposition at Bayou Blue immediately before, during and after Hurricane Andrew, measured by marsh surface sediment traps. Means 1 SE, n=3. Hurricane Andrew made landfall on August 26, 1992. Modified from Cahoon et al. (1995).

Flooding caused by hurricanes and tropical storms can produce a large influx of sediment to coastal systems. In the Rappahannock Estuary, tropical storm Agnes increased TSS concentrations up to 25 times over normal amounts (Nichols 1977). Over 110,000 tons of sediment were transported into the estuary in 16 days, 90% of which was trapped (Nichols 1977). Thus, the high sediment load trapped as a result of Tropical Storm Agnes supplied the estuary with nearly 100,000 tons of sediment which has the potential of being reworked, resuspended, and distributed into the adjacent marshes. Similarly, Schubel (1974, 1976) estimated that the Susquehanna River discharged greater than 30 times the annual input of sediment during 10 days following Tropical Storm Agnes, most of which settled in upper Chesapeake Bay. In fact, nearly half of the sediment deposition in northern Chesapeake Bay since 1900 occurred as a result of Tropical Storm Agnes and a large flood in 1936 (Schubel and Hirshberg 1978).

Major storms have been documented to have impacts on coastal systems even when there is not a direct incursion on a specific area. For instance, in 1996 Tropical Storm Josephine propagated from the western Gulf of Mexico across the northern gulf coast into central Florida. The storm did not directly impact the coast of Louisiana, however, water levels in the Atchafalaya Delta were elevated for a period of a week and salinities increased from 0 to 6 (Holm 1998). Thus, an indirect effect of Tropical Storm Josephine was to override the local wind forcing and tidal regime in the Delta (Holm 1998).

The morphological changes in barrier islands are decidedly stepwise in occurrence. This non-linear nature of beach change is recognized as the core of barrier island dynamics (Terwindt and Battjes 1991, Chappell 1983, Phillips 1992). Conceptually, the minor events produce changes (oscillations) that are masked or removed by the year to year or shorter conditions. However, on a less frequent basis, a major storm will cause a displacement that establishes a new position of the shoreline that will oscillate in that place until the next major storm. The shoreline responds stepwise to an adjustment that extends seaward to the depths of sediment exchange and landward into the foredune or through washover into the adjacent bay (McGowen and Scott 1975, Nummedal et al. 1980, Balsillie 1986, Morton 1988, Nakashima 1989, Bush 1996). Major displacements of the coastal foredune and major episodes of beack-dune sediment exchange, or major exports to the washover fan or tidal deltas are part of this non-linear episodic alteration of the barrier island (Psuty 1990, Stone et al. 1993, Stone et al. 1997).

2.6 MAJOR RIVER FLOODS

Major river floods occur a few times a century. These floods deliver great amounts of sediment to coastal systems. When conditions are right for channel switching in a delta, the major shift in flow between channels normally occurs during great river floods. These processes are exemplified for the Atchafalaya delta in the great flood of 1973 on the Mississippi River (Belt 1975). Peak discharge for the 1973 flood was 64,051 m^3s^{-1} ; compared to a peak discharge of 66,345 m^3s^{-1} for the great 1927 flood (U.S. Army Corps of Engineers 1987). For several decades prior to the 1973 flood, Atchafalaya Bay filled with fine grained sediments. In 1973, large amounts of coarse sediments were mobilized and the Atchafalaya delta became sub-aerial for the first time (van Heerden and Roberts 1980). It is mainly during floods such as 1973 that current velocities and bedload are large enough for course-grain material to reach the new delta lobe and provide a foundation upon which to build land (Roberts et al. 1980). The flood al.most undermined the control structure at Old River which prevents the Atchafalaya from capturing the Mississippi. If the control structure were not in place, the major portion of the Mississippi would probably have been captured by the Atchafalaya. While every major river flood does not result in delta switching, levees are often breached and large amounts of sediments contribute to the delta plain via overbank flooding at crevasses (Kesel 1988). In the Ebro delta in Spain, the last major switch in the position of the river mouth occurred during the large flood in 1937 (Ibañez et al. 1996, 1997). The effects of such events are clearly evident in areas affected by flood waters. In 1993-94, there were two 50 and 90 year floods on the Rhone River (Fig. 6). Massive flooding of the upper delta occurred as the levee along the Petit Rh6ne broke in several locations during each flood. In sites affected by the floods, there was accretion up to 24 mm (Hensel et al. 1998). Accretion in impounded habitats not impacted by the river was very low showing that these habitats were largely uncoupled from riverine processes (Fig. 7).

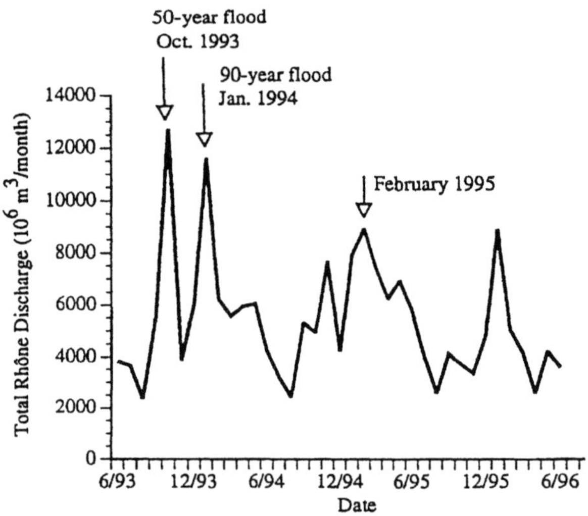

Figure 6. Total discharge of the Rhone River from June 1993 to June 1996. Arrows indicate major floods during this time period. Modified from Hensel (1998).

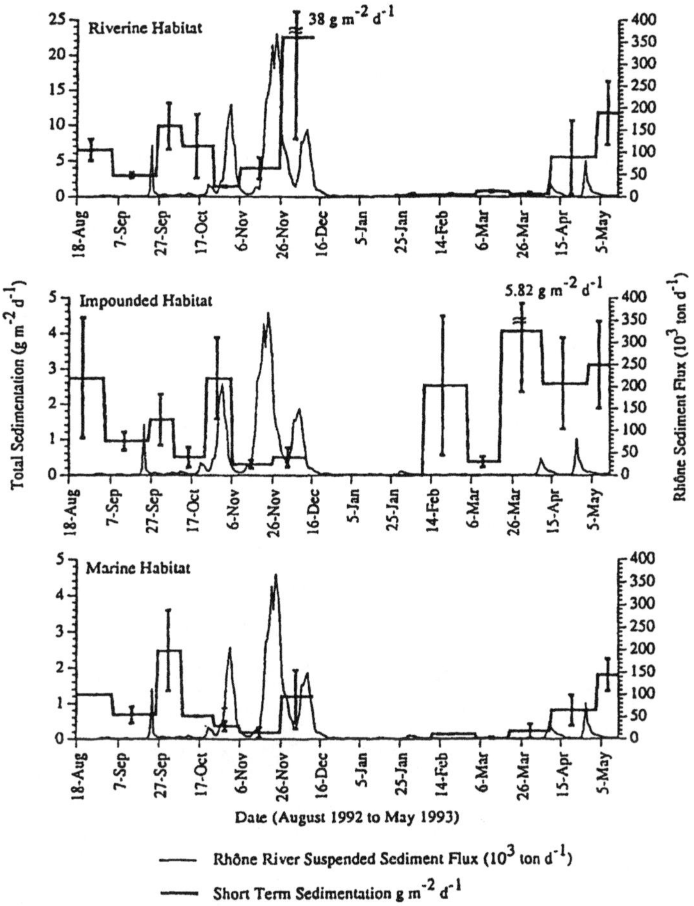

Figure 7. Total (inorganic+organic) short-term sedimentation in the Rhone Delta, averaged over sites within a given habitat type per sampling period. Error bars represent 1 standard error. High variability characterizes all habitat types. Riverine habitat regularly received more sediment than either marine or impounded sites. Peak sedimentation recovered on Dec. 12 was related to a semi-annual flood event in the Rhone River. High deposition at an impounded site within the April 9 sample was related to a wind-induced resuspension of autochthonous sediments. Modified from Hensel et al. 1998.

2.7 DELTAIC LOBE SWITCHING

The major growth cycles of deltas take place through the formation of new delta lobes. A series of overlapping deltaic lobes is an efficient way to distribute sediments and continually build land over the entire coastal plain. Evidence of major changes in the route to the sea (which, in the Mississippi delta, occur approximately every 1000 years and affect 1000s of km^2, Roberts et al. 1980) has been documented for many deltaic systems (Fig. 8, Coleman and Wright 1971, Wells and Coleman 1984, Van Andel 1967,

Todd and Eliassen 1938, Freeman 1928, Ibañez et al. 1997, Tornqvist et al. 1996, Kazmi 1984, Stanley and Warner 1993). Channel switching occurs as the existing channel lengthens and the slope decreases and the channel becomes less efficient. Eventually, the height of the river bed is raised (Freeman 1928, Roberts 1997) and the upstream levee is breached permanently in favor of a more hydraulically efficient, shorter route to the sea. This process is pulse dependent as the breaching of the levee usually takes place during large flood events. River flow is never confined to one channel, but generally the primary channel receives on the order of 80% of total discharge with the remainder divided among older distributaries (Gagliano and Van Beek 1973), thus insuring efficient dispersal of sediments over the entire deltaic plain. Barrier islands of deltas are nourished by sands delivered to the near shore coastal area and undergo a cycle of formation and deterioration which follows river switching (Penland et al. 1988).

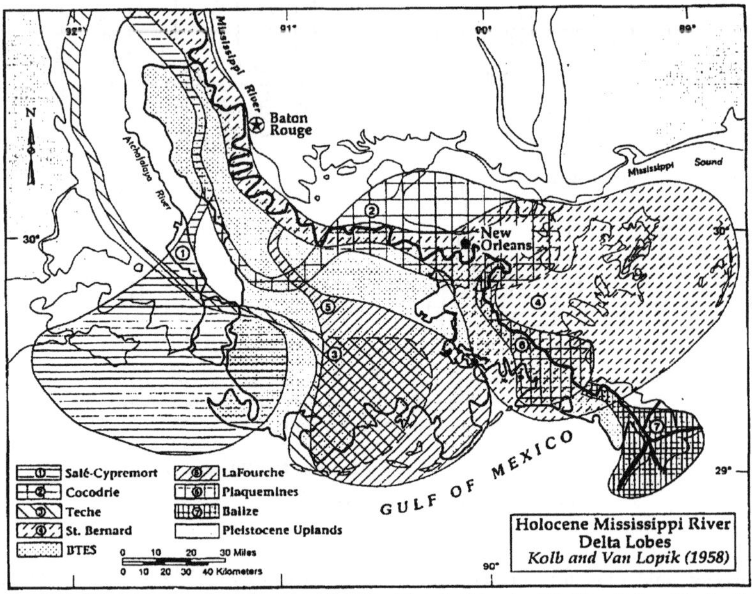

Major delta lobes that make up the active delta of the Mississippi River (modified from Kolb and Van Lopik 1958).

Figure 8. Major delta lobes that make up the active delta of the Mississippi River. Modified from Kolb and Van Lopik (1958).

3. Human Impact in Coastal Systems and Management for Sustainability

Human activities have had a pervasive impact on coastal systems. These impacts most often are classified and discussed in terms of the types of impacts which occur. Thus, for example, there are discussions of water quality deterioration in terms of eutrophication and toxic materials; physical alterations such as jetties, groins, dredging, channelization, and filling; habitat loss; heat pollution and entrainment by electricity generating stations; declines in fishery populations; and introduction of exotic species (e.g., Day et al. 1989). Many solutions have been proposed to deal with these individual impacts. But from a comprehensive, holistic point of view, human activity has reduced the inputs of energy and materials at all spatial and temporal scales.

In order to deal with the problems of coastal wetlands and coastal barriers, especially within the context of rising water levels, comprehensive management is needed because these problems cannot be solved in a piecemeal way. It is the often unorganized, fragmented way that these areas have been managed in the past which have reduced the energy pulses which sustain them and has given rise to the problems which exist today. Management must take into consideration not only the coastal system itself but also the drainage basin.

Many important rivers have been dammed, which has reduced floods and resulted in a reduction in the amount of freshwater and sediments reaching coastal systems. The amount of sediment carried in Nile, Indus, and Ebro has been reduced by over 95%; for the Po and Mississippi the reduction is about 70% and for the Rhone, the reduction is greater than 50% (Stanley and Warner 1993, Milliman et al. 1984, Varela et al. 1983, Kessel 1988, 1989, Sestini 1992, L'Homer 1992, Day and Templet 1989). Kesel et al. (1992) calculated that about 25% of sediment transported into the lower Mississippi River below Red River in the second half of the 19th century was trapped within the delta. This percentage is greatly reduced at present due to levees. Reduction of freshwater inflow can lead to salinity intrusion and, in arid and semi-arid areas, to hypersalinity, which in turn can lead to wetland vegetation death. In the Indus delta, for example, more than 99% of the original quarter million ha of mangroves have disappeared, primarily because of hypersalinity (Snedaker 1984). Hypersalinity and increased waterlogging due to lack of sedimentation is leading to wetland deterioration in the Rhone delta as well (Hensel et al. 1998). For sustainable management to take place, there will likely have to be some degree of mobilization of sediments which are now trapped in reservoirs (Wasp et al. 1977). A great need is for engineering methods to accomplish this and then to move these sediments to the coastal zone.

At the longest temporal scale and the broadest spatial scale, riverine channel switching and the development of new deltaic lobes has been stopped or greatly reduced for many deltaic systems. This has been done using water control structures, closure of minor distributaries, and construction of dikes. In the Mississippi delta, for example, there are currently two functioning distributaries—the lower Mississippi river, and the Atchafalaya river which carries about one third of the flow of the Mississippi. There were at least four other distributaries that carried significant flows at the beginning of European colonization but these have all been closed. Crevasse splays have also been largely eliminated. As a result the accelerated deterioration of barrier islands of the Mississippi delta has been shown to be due to the reduction in sand delivery to the coastal zone (Penland et al. 1988)

3.1 COASTAL WETLANDS

Human activities in the coastal zone have had a pervasive impact on coastal wetlands. Within many coastal systems, dikes have isolated large areas from riverine input and many rivers are channelized and diked so that they discharge directly to the sea. River dikes prevent changes in the course of the lower river, the development of crevass splays, and input of riverine freshwater, sediments and nutrients during river floods (Kesel 1988, 1989). Sea dikes and canals with their associated spoil banks inhibit water movement into marshes and the deposition of sediments during pulsing events such as coastal storms and frontal passages (Swenson and Turner 1987, Reed 1992, Hensel et al. 1998).

In order to offset the effects of dikes, diversions of river water into coastal areas should be considered in order to enhance accretion and to maintain high productivity, wetland habitat, and low salinity areas. Such freshwater diversions are now being carried out in the Mississippi delta (Day et al. 1997). Large scale diversions are already carried out in many coastal areas of the world for irrigation purposes and these could be incorporated into an overall management plan for salinity, sediments and nutrients. Combating salinity intrusion is often attempted by the use of barriers that lead to closed systems. Salinity management using freshwater to form a buffer against saltwater intrusion allows the coastal systems to remain open to some extent thus allowing the movement of fishery species that use brackish water and wetlands as important habitat. This also maintains important energetic pulses originating from the sea such as storms.

Within many coastal systems, hydrology has been altered through such activities as stream channelization, dredging, canal construction, and impoundments. These activities have often led to degradation of habitat. In the Mississippi delta, for example, wetland loss rates have been related to canal densities (Scaif et al. 1983). Impoundments consisting of a system of dikes and water control structures have been shown to reduce tidal exchange and the influx of suspended sediments, lower accretion rates, lower productivity and reduce the movement of migratory fishes (Reed 1992, Rogers et al. 1992, Cahoon 1994, Boumans and Day 1994, Hensel et al. 1998). In a study of impoundment marsh-management in two Louisiana marshes, Boumans and Day (1994) and Cahoon (1994) reported higher deposition in unmanaged wetlands (Fig. 9). The water control structures greatly reduced water exchange and sediment input to the managed areas. During Hurricane Gilbert, the dikes at one of the managed areas were low and were overtopped leading to deposition as high as the control area. In the other managed area, the dikes were higher and were not overtopped by the hurricane. This led to significantly lower deposition rates leading to decreased accretion (Boumans and Day 1994). The authors concluded that the ''exclusion of hurricane flood waters from marsh management areas block an important mechanism for sediment delivery''' (Boumans and Day 1994).

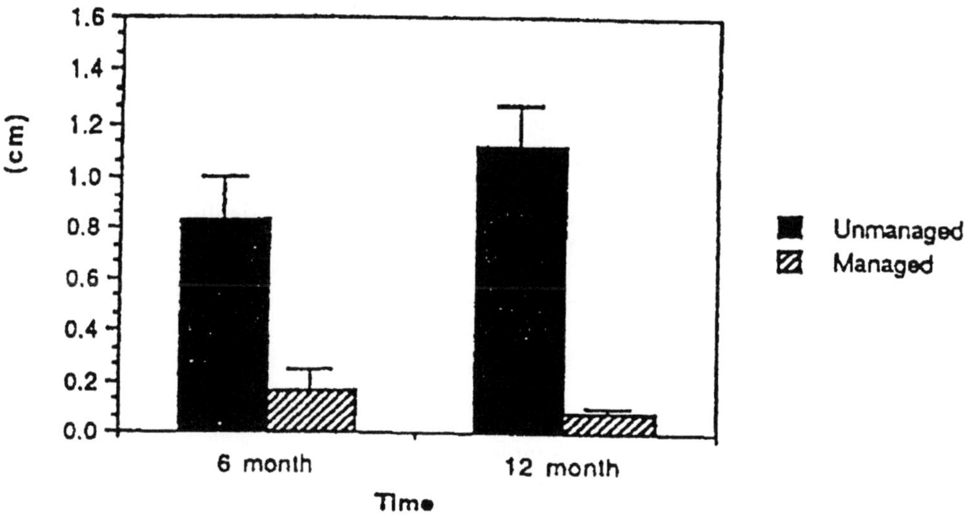

Figure 9. Vertical accretion as measured by marker horizons in two brackish marshes in coastal Louisiana. One of the sites was an impounded (managed) marsh and the other was a nearby nonimpounded area (unmanaged). The accretion rate in the nonimpounded site is approximately equal to local relative sea-level rise and thus is geomorphically sustainable. Modified from Cahoon (1994).

Sediment management should include plans for movement of sediments into coastal systems both from the river and the sea, reducing impediments to sediment movement within the coastal system as well as utilization of dredged sediments whenever possible. Important sources of sediments are river water, resuspended sediments from coastal lakes and bays and those transported from the nearshore zone. For example, much of the sediments deposited on the surface of coastal marshes in the Mississippi delta are resuspended from bay bottoms or transported from the nearshore area (Baumann et al. 1984, Reed 1989, Cahoon et al. 1995a). The work of resuspending and transporting these sediments is done by natural forces of wind, waves and tidal currents. Brush fence baffles have been used in the Dutch Wadden Sea and in the Mississippi delta to encourage settling of suspended sediments and inhibit resuspension (Schoot and de Jong 1982, Boumans et al. 1997, Fig. 10). This raises the elevation of the sediment surface allowing revegetation to occur. Along the north coast of the Netherlands, thousands of ha of new wetlands have been created using sediment fences. Dredge spoil should also be used to create habitat whenever possible.

Reclamation of coastal wetlands and shallow water bodies for agricultural, urban and industrial development is widespread and is, in essence, the complete elimination of the energy subsidies which maintain coastal wetlands. Practically all wetlands of the Nile, Ebro, Po, Rhine, Sacramento and many other deltas have been reclaimed (Stanley 1988, Stanley and Warner 1993, Ibañez et al. 1997, Sestini 1992, Knights 1979, Newmarch 1981, Weir 1950) while in others, such as the Mississippi and Rhone (Harrison and Kollmorgen 1948, Jeftic et al. 1992, Day and Templet 1989, R. Day et al. 1990, Tamisier 1990, Corre 1992, Day et al. 1995), large portions have been reclaimed. Similar reclamation has taken place in many non-deltaic coastal wetlands.

Reclamation and drainage almost always leads to high rates of subsidence due to soil oxidation so that the reclaimed land sinks below sea level and must be put permanently under pump (Kazmann and Heath 1968, Knights 1979, Wagner and Durabb 1976). Removal of subsurface fluids can greatly increase the rate of subsidence. In the Po delta, for example, extraction of high water content natural gas led to total subsidence of 2-4 m (Sestini 1992). In the Mississippi delta, wetlands have been impounded and reclaimed for urban and agricultural purposes. Most of the agricultural reclamations have failed and are now shallow ponds (Day et al. 1990).

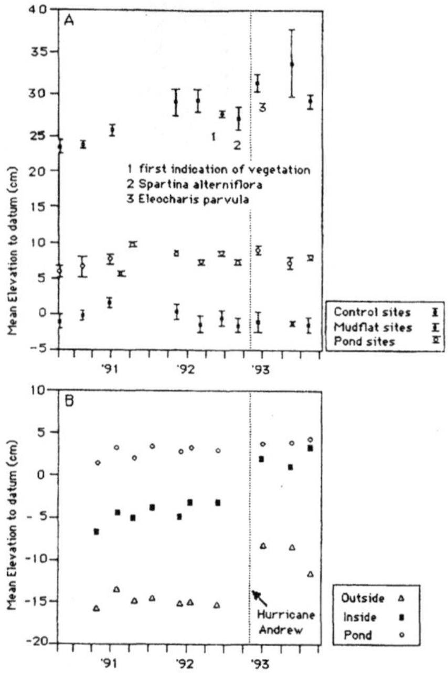

Figure 10. Marsh surface elevations measured by the sedimentation-erosion table (SET) at the LaBranche (A) and Leeville areas (B), showing Hurricane Andrew and points at which different vegetation types came in at LaBranche. Put elevations on common datum asssuming LaBranche marsh at +30 cm NGVD. Modified from Boumans et al. (1997).

3.2 COASTAL BARRIERS

Despite the specter of a decreasing availability of sediment in the system and a rising sea level that is slowly drowning the barrier islands, there is an increasing scale of attempts to "stabilize"' the shoreline and "protect"' the barriers from further change. In many areas, the long term presence of construction on barrier islands is witnessing the final phases of the development syndrome that, in many places, has lowered all of the high dunes, flattened the barrier to facilitate construction, and pushed the sand into back barrier wetlands to foster lagoon housing. The natural buffers have all been usurped in the massive sand mobilization process associated with development. There is often no high ground remaining. There is no setback from the water's edge. There is little or no recognition that the natural system has been undergoing dimensional change and that mass of the barrier island is diminishing.

In some older developed areas on barrier islands, the ocean front is composed of bulkheads and seawalls that separate the infrastructure from the water because the natural sand buffer has been eroded and there is no way to dissipate the incident wave energy (Kraus and Pilky 1988). Structures at inlets and lengthy groins have interfered with the alongshore transport processes and have added downdrift sediment starvation to the problems of natural sediment deficits(Reynolds 1987, Fenster and Dolan 1996). Coastal communities that have occupied their locations for most of this century are seeing the direct effects of sea level rise. Storm surge is raising the local water levels to the point that water is coming from the bay through the storm drains and into the streets. Roads that were constructed near water level at some time in the past, now flood during spring tide. The barrier islands are being drowned by a rising sea level and they are being reduced in dimension by long-term episodic sediment loss.

The Federal Emergency Management Agency (FEMA) has initiated a program to reduce the magnitude of people and things at risk in disaster-prone areas, such as the coastal zone. It has introduced a mitigation strategy to assist in lowering the losses in low-lying areas and to encourage modification in behavior toward combatting the natural forces (FEMA 1995). FEMA has stated that its mission is to reduce the insurance payouts by 50% by the year 2010. Policies are evolving that are moving away from beach nourishment as the only approach to counter the changing shoreline. Funds are becoming available for buyout. There is an encouragement to rebuild the dunes to provide for a return of the natural buffer. Although there may be limited space in some areas to accomodate dune builidng if the development is adjacent to the shoreline.

Management on developed barrier islands should look to the improvement of public safety in the short term and in the long term to management in keeping with the episodic nature of island dynamics. Areas of high hazard should be considered for buyout and a change of land use. This is especially appropriate in a post-disaster scenario when there is an opportunity to make changes in the land use in the process of rebuilding. There should be an emphasis away from beach protection (a classic misnomer) to the concept of coastal hazard management (Psuty et al. 1996, Burby 1998) in which the objective is to reduce the magnitude of the people and infrastructure at risk in a very hazardous environment and whose risk will escalate into the future.

3.3 OTHER ASPECTS OF MANAGEMENT

In many coastal areas, there have been great changes in hydrology leading to alterations in the fresh/saline water balance and changes in the way water flows (Day and Templet 1989, Day et al. 1995, Ibañez et al. 1997). In order to maintain and restore wetlands, as well as improve water quality, there needs to be better management of hydrology, including both the amount and timing of water flowing into coastal systems as well as the pathways of flow within the systems.

Channelization and construction of canals has led to such hydrological changes as more rapid flushing of some water bodies, isolation of wetlands behind spoil deposits, and salt water intrusion (Swenson and Turner 1987, Boumans and Day 1994). As mentioned above, impoundments consisting of a system of dikes and water control structures have been widely constructed in coastal wetland and lagoon areas. Studies have shown that these impoundments can reduce the influx of suspended sediments, lower accretion rates, lower wetland productivity, and reduce the movement of migratory marine fishes. Careful planning of canal construction and development of impoundments is necessary if the negative impacts are to be avoided. Care should be taken not to isolate wetlands so that tidal action is maintained.

An integral part of any management plan is the conservation and restoration of natural habitat, especially wetlands. The proper design of systems to deliver freshwater, sediments and nutrients to deltaic areas will enhance the conservation and productivity of natural habitat. Proper planning will also ensure that there are a diversity of fresh, brackish and saline habitats including wetlands, submerged vegetation and open water. This will lead to an enhancement of fisheries and wildlife.

4. Holistic Management

Management actions should be taken as part of a holistic strategy which aims to reintegrate the natural subsidies into coastal functioning. A number of elements of such an approach have been proposed including sand nourishment of beaches, construction of salt marshes and tidal flats with dredge material, vegetative plantings, reintroduction of river inflow to deltas for sediments, salinity management and to enhance productivity, and use of wetlands to reduce nutrient levels. Pethick (1993), Day and Templet (1989), and Templet and Meyer-Arendt (1988) have proposed such holistic management approaches for the southeastern coast of the UK and for the Mississippi delta. Management should anticipate future change, especially accelerated sea level since coastal wetlands are so sensitive to water level changes. A reintegration of the natural energy pulses into coastal area management does not mean that humans cannot continue to thrive in these areas. It only means doing some things differently. Navigation, flood control, agriculture, and urban development can all exist in a sustainable coastal system. But approaches which reduce subsidies from energetic forcings must be changed. What has to change is the large-scale alteration of coastal hydrology and attempts to stabilize the dynamics of coastal systems. In conclusion, system functioning should form the basis for the sustainable managment of coastal systems.

5. Acknowledgements

We acknowledge the input and comments of a number of colleagues including Don Cahoon, Denise Reed, Paul Kemp, Joe Suhayda, Carles Ibanez, Didier Pont, and Paul Templet. This work was supported by the Louisiana and New Jersey Sea Grant College Programs Institutional support was also provided by the Department of Oceanography and Coastal Sciences and the Coastal Ecology Institute at LSU and the Institute of Marine and Coastal Sciences at Rutgers University.

6. Literature Cited

Albani, A.D., V. Favero and R. Serandrei Barbero. 1983. Apparatiintertidali della laguna di Venezia. Afti del convegno ''Laguna, fiumi, lidi; cinque secoli di gestione delle acque nelle Venezie''. Venice, 10-12 June. 1983.

Armbruster, C.K., G.W. Stone and J.P. Xu. 1995. Episodic atmospheric forcing and bayside foreshore erosion: Santa Rosa Island Florida. Gulf Coast Association of Geological Societies, Transactions **45**: 31-38.

Balsillie, J.H. 1986. Beach and storm erosion due to extreme event impact. Shore and Beach **54**: 22-36.

Baumann, R. H., J. W. Day Jr. and C. A. Miller. 1984. Mississippi deltaic wetland survival: sedimentation versus coastal submergence. Science **224**: 1093-1095.

Belt, C.B. 1975. The 1973 flood and man's constriction of the Mississippi River. Science. **189**: 681-684.

Bendoricchio, G., M. Di Luzio, P. Baschieri and A. Capodaglio. 1993. Diffuse pollution in the lagoon of Venice. Water, Science and Technology **28**: 69-78.

Bettinetti, A., F. Mattarolo and P. Silva. 1995. Reconstruction of saltmarshes in the Venice Lagoon. Pages 921-935 *in* E. Ozhan, editor. Proceedings– MEDCOAST 94 Conference, 22-27 October 1995, Tarragona, Spain.

Bird, E.C.F. 1993. Submerging Coasts: the effects of a rising sea level on coastal environments. John Wiley and Sons, New York, New York, USA.

Blum, L. K. 1995. Papers from William E. Odum Memorial Symposium. Estuaries **18**: 545-659.

Boothroyd, J. C. 1985. Tidal inlets and tidal deltas. Pages 445-532 *in* R.A. Davis, editor. Coastal sedimentary environments, 2nd edition, Springer-Verlag, New York, New York, USA.

Boumans, R.M.J., J.W. Day, Jr., G. P. Kemp and K. Kilgen. 1997. The effect of intertidal sediment fences on wetland surface elevation, wave energy and vegetation establishment in two Louisiana coastal marshes. Ecological Engineering **9**: 37-50.

Boumans, R. M. and J.W. Day, Jr. 1994. Effects of two Louisiana marsh management plans on water and materials flux and short-term sedimentation. Wetlands **14**:247-261.

Boyton, W., M. Kemp and C. Keefe. 1982. A comparative analysis of nutrients and other factors influencing estuarine phytoplankton production. Pages 69-90 *in* V. Kennedy, editor. Estuarine Comparisons. Academic Press, New York, New York, USA.

Bruun, P. 1988. The Bruun Rule or erosion by sea level rise: A discussion of large-scale two-and three-dimensional usages. Journal of Coastal Research **4**: 627-648.

Burby, R.J., editor,. 1998. Cooperating with nature: confronting Natural Hazards with LandUse Planning for Sustainable Communities. John Henry Press, Washington, D.C., USA.

Bush, D.M., R .S. Young, C.A. Webb and E.R. Thieler. 1996. Soundside impacts of a northward tracking tropical cyclone: Hurricane Emily (31Aug93), Cape Hatteras area, North Carolina. Journal of Coastal Research **12**: 229-239.

Caffrey, J. M., J.W. Day, Jr. 1986. Control of the variability of nutrients and suspended sediments in a Gulf Coast estuary by climatic forcing and spring discharge of the Atchafalaya River. Estuaries, **9**(4A): 295-300.

Cahoon, D. 1994. Recent accretion in two managed marsh impoundments in coastal Louisiana. Ecological Applications **4**: 166-176.

Cahoon, D., D. Reed, J.W. Day, Jr., G. Steyer, R. Boumans, J. Lynch, D. McNally and N. Latif. 1995a. The influence of hurricane Andrew on sediment distribution in louisiana coastal marshes. Journal of Coastal Research **21**: 280-294.

Cahoon, D. R., D.J. Reed and J.W. Day, Jr. 1995b. Estimating shallow subsidence in microtidal salt marshes of the southeastern United States: Kaye and Barghoorn revisited. Marine Geology **128**:1-9.

Carbognin, L., G. Gambolati, F. Marabini, G. Taroni, P. Teatini and L. Tosi. 1996. Analisi del procecsso di subsidenza nell'area Veneziana e sua simulazione con un modello tridimensionale non linear. Ministero dell'Universita e della Ricerca Scientifica e Tecnologica. Progetto Sistema Lagunare Veneziano.

Chalmers, A.G., R.G. Wiegert and P. Wolf. 1985. Carbon balance in a salt marsh: interactions of diffusive export, tidal deposition and rainfall-caused erosion. Estuarine, Coastal and Shelf Science **21**: 757-771.

Chappell, J. 1983. Thresholds and lags in geomorphological changes. Australian Geographer **15**: 357-66.

Childers, D.L. and J.W. Day, Jr. 1990a. Marsh-water column interactions in two Louisiana estuaries. 1. Sediment dynamics. Estuaries **13**:393-403.

—— 990b. Marsh-water column interactions in two Louisiana estuaries: 11. Nutrient dynamics. Estuaries, **13**: 404-417.

Clark, J. S. 1986. Coastal forest tree populations in a changing environment, southeastern Long Island, New York. Ecological Monographs **56**: 259-277.

Coleman, J.M. and Wright, L.D. 1971. Analysis of major river systems and their deltas: procedures and rationale, with two examples. Tech. Rept. 95. Coastal Studies Institute, Louisiana State University, Baton Rouge, Louisiana, USA.

Conner, W.H. and J.W. Day, Jr. 1989. Response of coastal wetland forests to human and natural changes in the environment with emphasis on hydrology. Pages 34-43 *in* D.D. Hook and R. Lea, editors. Proceedings: The forested wetlands of the Southern United States, Orlando, FL. Gen. Tech. Rep. SE-50. Asheville, NC: U.S. Department of Agriculture, Forest Service, Southeastern Forest Experiment Station.

Corre, J.J. 1992. Implications des changements climatiques etude de cas: le Golfe du Lion. Pages 328-427 *in* L. Jeftic, J. Milliman and G. Sestini, editors. Climatic change and the Mediterranean. Edward Arnold, London, England.

Day, J.W., Jr., F. Scarton, A. Rismondo and D. Are. 1998. Rapid deterioration of a salt marsh in Venice Lagoon, Italy. Journal of Coastal Research **14**: 583-590.

Day, J.W., Jr., J.F. Martin, L. Cardoch and P.H. Templet. 1997. System functioning as a basis for sustainable management of deltaic ecosystems. Coastal Management **25**: 115-153.

Day, J. W., Jr., C.J. Madden, R.R.Twilley, R.F. Shaw, B.A. McKee, M.J. Dagg, D.L. Childers, R.C. Raynie and L.J Rouse. 1994. The influence of Atchafalaya River discharge on Fourleague Bay, Louisiana (USA). Pages 151-160 *in* K. Dyer and R. Orth, editors. Changes in fluxes in estuaries. Olsen and Olsen, New York, New York, USA.

Day, J.W., Jr., D. Pont, P. Hensel and C. Ibanez. 1995. Impacts of sea level rise on deltas in the coastal zone of Louisiana, USA: historical trends. Environmental Management **14**:229-240.

Day, J., Jr., C. Hall, M. Kemp and A. Yanez-Arancibia. 1989. Estuarine Ecology. Wiley-Interscience, New York, New York, USA.

Day, J. W., Jr. and P.H. Templet. 1989. Consequences of sea level rise: implications from the Mississippi Delta. Coastal Management **17**: 241-257.

Day, R., R. Holz and J.W. Day, Jr. 1990. an inventory of wetland impoundments in the coastal zone of Louisiana, USA: historical trends. Environmental Management **14**:229-240.

Dubois, R.N. 1990. Barrier-beach erosion and rising sea level. Geology **18**: 150-152.

—— 995. The transgressive barrier model: An alternative to two-dimensional volume balance models. Journal of Coastal Research 11: 1272-1286.

Edelman,T. 1972. Dune erosion during storm conditions. Pages 1305-1312. Proceedings of the 13th Conference on Coastal Engineering. American Society of Coastal Engineers.

Favero, V., R. Paarolini and M. Scattolin. 1988. Morofologia Storica della Laguna di Venezia. Arsenale Editrice, Venice.

FEMA. 1995. National mitigation strategy: partnerships for building safer communities. Washington, District of Columbia, USA.

Fenster, M. and R. Dolan. 1996. Assessing the impact of tidal inlets on adjacent barrier island shorelines. Journal of Coastal Research **12**: 294-310.

FitzGerald, D. M. 1982. Sediment bypassing at mixed energy tidal inlets, Pages 1094-1118. Proceedings,18th Coastal Engineering Conference, American Society of Civil Engineers.

FitzGerald, D.M., S. Penland and D. Nummedal. 1984. Control of barrier island shape by inlet sediment bypassing: East Frisian Islands, West Germany. Marine Geology **60**: 355-376.

Forbes, D.L., R.B. Taylor, J.D. Orford, R.W.G. Carter and J. Shaw. 1991. Gravel barrier migration and overstepping. Marine Geology **97**: 305-313.

Freeman, J.R. 1928. Flood control on the River Po in Italy. American Society of Civil Engineers Proceedings **54**: 957-992.

Gagliano, S. M. and J.L. VanBeek. 1973. Environmental management in the Mississippi delta system. Gulf Coast Association Geological Societies, Transactions **23**: 203-209.

Gornitz, V., S. Lebedeff and J. Hansen. 1982. Global sea level trend in the past century. Science **215**: 1611-1614.

Gornitz, V. 1995. Sea-Level rise: a review of recent, past and near-future trends. Earth Surface Processes and Landforms **20**: 7-20.

Hackney, C. T. and W.J. Cleary. 1987. Saltmarsh loss in southeastern North Carolina lagoons: importance of sea level rise and inlet dredging. Journal of Coastal Research **3**: 9397.

Hallermeier, R.J. 1981. A profile zornation for seasonal sand beaches from wave climate. Coastal Engineering **4**: 253-277.

Hatton, R. S., R.D. DeLaune and W.H. Patrick, Jr. 1983. Sedimentation, accretion and subsidence in marshes of Barataria Basin, Louisiana. Limnology and Oceanography **28**:494-502.

Hayes, M.O. 1979. Barrier island morphology as a function of tidal and wave regime. Pages 1-27 *in* S. Leatherman, editor. Barrier Islands. Academic Press, New York, New York, USA.

Harrison, R.W. and W.M. Kollmorgen. 1948. Drainage reclamation in the coastal marshlands of the Mississippi River delta system. Louisiana Historical Quarterly **30**: 654-709.

Hensel, P. 1998. Material fluxes across different scales of time and space in the Rhone River delta, France. Dissertation, Louisiana State University, Baton Rouge, Louisiana, USA.

Hensel, P., J.W. Day, Jr., D. Pont and J.N. Day. 1998. Short-term sedimentation dynamics in the Rhone River delta, France: The importance of riverine pulsing. Estuaries **21:** 52-65.

Hicks, B.M., T.M. Hume, A. Swales and M.O. Green. 1999. Magnitudes, spatial extent time scales and causes of shoreline change adjacent to an ebb tidal delta, Katikati Inlet, New Zealand. Journal of Coastal Research **15**: 220-240.

Holm, G. 1998. A comparison of the Atchafalaya and Wax Lake Deltas: Salinity Incursions and the Loss of *Sagitaria latifolia Willd* (Broadleaf Arrowhead). Thesis, Louisiana State University, Baton Rouge, Lousiana, USA.

Howes, B., J. Dacy and D. Goehringer. 1986. Factors controlling the growth form of *Spartina alterniflora*: feedbacks between above-ground production, sediment oxidation, nitrogen and salinity. Journal of Ecology **74**: 881-898.

Hsu, S.-A. and B.W. Blanchard. 1993. Meteorological forcing on Louisiana wetlands. Pages 230-242 in 8th Symposium on Coastal and Ocean Management, New Orleans, Louisiana, USA.

Ibanez, C., N. Prat and A. Canicio. 1996. Changes in the hydrology and sediment transport produced by large dams on the lower Ebro river and its estuary. Regulated Rivers. **12**: 51-62.

Ibanez, C., A. Canicio, J. W. Day Jr. and A. Curco. 1997. Morphologic development, relative sea level rise and sustainable management of water and sediment in the Ebre delta, Spain. Journal of Coastal Conservation **3**: 191-202.

Incze, M. L. and M.R. Roman. 1983. Carbon Production and Export from Biscayne Bay, Florida. II. Episodic export of organic carbon. Estuarine, Coastal and Shelf Science 61-73.

Jeftic, L., J. Milliman and G. Sestini, editors. 1992. Climatic change and the mediterranean. Edward Arnold, London.

Jordan, T. E., J.W. Pierce and D.L. Correll. 1986. Flux of particulate matter in the tidal marshes and subtidal shallows of the Rhode River estuary. Estuaries **9**: 310-319.

Kana, T. W., B.J. Baca and M.L. Williams. 1986. Potential impacts of sea level rise on wetlands around Charleston, South Carolina. U. S. Environmental Protection Agency, EPA 230-10-**85-014**.

Kazmann, R. and M. Heath. 1968. Land subsidence related to ground-water offtake in the New Orleans area. Gulf Coast Association Geological Societies. Transactions **18**: 108-113.

Kazmi, A. 1984. Geology of the Indus delta. Pages 71-84 *in* B.H. Haq and J.D. Milliman, editors. Marine geology and oceanography of Arabian Sea and coastal Pakistan. Van Nostrand Reinhold Co., New York, New York, USA.

Kemp, G.P., J.T. Wells and I. Li. van Heerden. 1980. Frontal passages affect delta development in Louisiana. Coastal Oceanography and Climatological News **3**: 4-5.

Kemp, W. M. and W.R. Boynton. 1984. Spatial and temporal coupling of nutrient inputs to estuarine primary production: the role of particulate transport and decomposition. Bulletin of Marine Science, **35**: 522-535.

Kesel, R. H. 1988. The decline in the suspended load of the lower Mississippi River and its influence on adjacent wetlands. Environmental and Geological Water Sciences **11**: 271-281.

Kesel, R.H. 1989. The role of the Mississippi River in wetland loss in southeastern Louisiana, U.S.A. Environmental and Geological Water Sciences **13**: 183-193.

Kesel, R.H., E. Yodis and D. McCraw. 1992. An approximation of the sediment budget of the lower Mississippi river prior to major human modification. Earth Surface Processes and Landforms **17**: 711-722.

Knights, B. 1979. Reclamation in the Netherlands. Pages 66-81 *in* B. Knights and A. Phillipps, editors. Estuarine and coastal land reclamation and water storage. Saxon House, London, England.

Kolb, C.R. and J.R. Van Lopik. 1958. Geology of the Mississippi River deltaic plain, southeastern Louisiana. Vicksburg, Mississippi: U.S. Army Corps of Engineers, Waterways Experiment Station. Technical report **2**.

Kraus, N.C. and O.H. Pilkey, editors. 1988. Effects of seawalls on the beach. Journal of Coastal Research Special Issue **4**.

L'Homer, A. 1992. Sea-level changes and impacts on the Rhone coastal lowlands. Pages 136-152 *in* A. Tooley, M. and S. Jelgersma, editors. Impacts of sea-level rise on European coastal lowlands. Blackwell Publishers, Oxford, England.

Leonard, L. A., A.C. Hine, M.E. Luther, R.P. Stumpf and E.E. Wright. 1995. Sediment transport processes in a west-central Florida open marine marsh tidal creek; the role of tides and extra-tropical storms. Estuarine, Coastal and Shelf Science **41**:225-248.

Madden, C. J., J.W. Day Jr. and J.M. Randall. 1988. Freshwater and marine coupling in estuaries of the Mississippi River deltaic plain. Limnology and Oceanography **33**: 982-1004.

McBride, R.A. and M.R. Byrnes. 1997. Regional variations in shore response along barrier island systems of the Mississippi River delta plain: historical change and future prediction. Journal of Coastal Research **13**: 628-655.

McCaffrey, R. J. and J. Thomson. 1980. A record of the accumulation of sediment and trace metals in a Connecticut salt marsh. Pages 165-236 *in* B. Saltzman, editor. Estuarine physics and chemistry: studies in Long Island Sound. Academic Press: New York, New York, USA.

McGowen, J.H. and A.J. Scott. 1975. Hurricanes as geologic agents on the Texas coast. Pages 23-46 *in* L.E. Cronin, editor. Estuarine Research, Vol 2. Academic Press: New York, New York, USA.

McKee, K. and W. Patrick. 1988. The relationship of smooth cordgrass *(Spartina alterniflora)* to tidal datums: a review. Estuaries **11**: 143-151.

Milliman, J.D., G.S. Quraishee and M.A. Beg. 1984. Sediment discharge from the Indus River to the ocean: Past, present and future. Pages 65-70 *in* B.H. Haq and J.D. Milliman, editors. Marine geology and oceanography of Arabian Sea and coastal Pakistan. Van Nostrand Reinhold Co., New York, New York, USA.

Moeller, C. C.,O.K. Huh, H.H. Roberts, L.E. Gumley and W.P. Menzel. 1993. Response of Louisiana coastal environments to a cold front passage. Journal of Coastal Research **9**: 434-447.

Morton, R.A. 1988. Nearshore responses to great storms. Pages 7-22 *in* H.E. Clifton, editor. Sedimentological consequences of convulsive geologic events. Geological Society of America Special Paper **229**.

Nakashima, L.D. 1988. Short-term changes in beach morphology on the Louisiana coast. Gulf Coast Association of Geological Societies. Transactions **38**: 323-329.

Nakashima, L.D. 1989. Shoreline responses to Hurricane Bonnie in southwestern Louisiana. Journal of Coastal Research **5**: 127-136.

National Research Council. 1995. Beach Nourishment and Protection. National Academy Press. Washington, District of Columbia, USA.

Newmarch, G. 1981. Subsidence of organic soils: Sacramento-San Joaquin Delta. California Geology **34**:135-141.

Nichols, M. M. 1977. Response and recovery of an estuary following a river flood. Journal of Sedimentary Petrology **47**: 1171-1186.

Nordstrom, K.F. 1980. Cyclic and seasonal beach responce: a comparison of oceanside and bayside beaches. Physical Geography **1**:177-196.

Nummedal, D. 1983. Barrier islands. Pages 77-121 *in* P.D. Kamar, editor. Handbook of Coastal Processes and Erosion. CRC Press, Boca Raton, Florida, USA.

Nummedal, D., S. Penland, R. Gerdes, W. Schramm, J. Kahn and H. Roberts. 1980. Geologic responce to hurricane impact on low-profile Gulf Coast Barriers. Gulf Coast Association of Geological Societies, Transactions **30**: 183-197.

Nyman, J., R. Chabreck and R. Linscombe. 1990. Effects of weir management on marsh loss, Marsh Island, Louisiana, USA. Environmental Management **14**: 809-814.

Odum, E.P. 1980. The status of three ecosystem-level hypotheses regarding salt marsh estuaries: tidal subsidy, outwelling and detrital-based food chains. Pages 485-495 *in* V. Kennedy, editor. Estuarine Perspectives. Academic Press, New York, New York, USA.

Oertel, G.F. 1988. Processes of sediment exchange between tidal inlets, ebb deltas and barrier islands. Pages 297-318 *in* D.G. Aubrey and L. Weishar, editors. Hydrodynamics and sediment dynamics of tidal inlets. Springer-Verlag, New York, New York, USA.

Ollerhead, J. and R.G.D. Davidson-Arnott. 1993. Controls on sandy spit dune evolution: A comparison of Buctouche Spit, New Brunswick, Canada and Long Pint Spit, Ontario, Canada. Pages 151-153 *in* J. List, editor. Large Scale Coastal Behavior '93. U.S. Geological Survey Open File Report 93-381.

Orford, J.D., R.W.G. Carter, D.L. Forbes and R.B. Taylor. 1988. Overwash occurrence consequent on morphodynamic changes following lagoon outlet closure on a coarse slastic barrier. Earth Surfaces Processes and Landforms **13**: 27-35.

Orford, J.D., A.C. Hinton, R.W.G. Carter and S.C. Jennings. 1992. A tidal link between sea-level rise and coastal response of a gravel-dominated barrier: Story Head, Nova Scotia. Pages 71-79 *in* Woodworth, P., D.T. Pugh, J. De Ronde, R.G. Warrick and J. Hannah, editors. Sea-level changes: determination and effects. Geophysical Monograph series, Vol.69, American Geophysical Union, Washington, District of Columbia, USA.

Orson, R., R. Warren and W. Niering. 1987. Development of a tidal marsh in a New England river valley. Estuaries **10**: 20-27.

Penland, S., R. Boyd and J.R. Sutter. 1988. The transgressive depositional systems of the Mississippi delta plain: A model for barrier shoreline and shelf sand development. Journal of Sedimentary Petrology **58**: 932-949.

Penland, S. and K. Ramsey. 1990. Relative sea-level rise in Louisiana and the Gulf of Mexico, 1908-1988. Journal of Coastal Research **6**: 323-342.

Perez, B.C., J.W. Day, L.J. Rouse, R.F. Shaw and M. Wang. 1999. Influence of Atchafalaya river discharge and winter frontal passage on suspended sediment concentration and flux in Fourleague bay, Louisiana. Estuarine Coastal and Shelf Science **50**:271-290.

Pethick, J.S. 1993. Shoreline adjustments and coastal management: physical and biological processes under accelerated sea-level rise. The Geographical Journal **159**:162-168.

Phillips, J. D. 1992. Nonlinear dynamical systems in geomorphology: revolution or evolution. Geomorphology **5**: 219-29.

Pilkey, 0. H. and Dixon. 1998. The Corps and the shore. Island Press, Covelo, California.

Pirazzoli, P. 1987. Recent sea-level changes and related engineering problems in the Lagoon of Venice, Italy. Progress in Oceanography **18**: 323-346.

Psuty, N.P., editor. 1988. Dune/Beach interaction. Journal of Coastal Research Special Issue No. 3.

—— 992. Spatial variation in coastal foredune development. Pages 3-13 *in* R.W.G. Carter, T.G. F. Curtis and M.J. Sheehy-Skeffington, editors. Coastal dunes: geomorphology, ecology and management for conservation. The Hague: Balkema.

—— 994. Spatial/temporal sequencing in coastal foredune development. Littoral '94, Eurocoast Proceedings, Lisbon, Spain.

Psuty, N.P., M. DeLuca, M. Grace, J. McDonnell, S. Pata and E. Spence. 1996. coastal hazard management report: New Jersey's shoreline future: preparing for tomorrow, 2 vols. plus appendices (draft). Institute of Marine and Coastal Sciences, Rutgers University, New Brunswick, New Jersey, USA.

Psuty, N.P. 1990. Foredune mobility and stability, Fire Island, New York. Pages 159-176 *in* K.F. Nordstrom, N.P. Psuty and R.W.G. Carter, editors. Coastal dunes: form and process. John Wiley, Interscience, London, England.

Raper, S., T. Wigley and R. Warrick. 1996. Global sea-level rise: past and future. Pages 11-45 *in* J. Milliman and B. Haq, editors. Sea-Level Rise and Coastal Subsidence. Kluwer Academic Publishers, Dordrecht, The Netherlands.

Redfield, A. C. 1972. Development of a New England salt marsh. Ecological Monographs **42**: 201-237.

Reed, D. J. 1989. Patterns of sediment deposition in subsiding coastal salt marshes: the role of winter storms. Estuaries **12**: 222-227.

Reed, D. 1992. Effect of weirs on sediment deposition in Louisiana coastal marshes. Environmental Management **16**: 55-65.

Rejmanek, M., C.E. Sasser and G.W. Peterson. 1988. Hurricane-induced sediment depostition in a Gulf Coast marsh. Estuarine, Coastal and Shelf Science **27**: 217-222.

Reynolds, W.J. 1987. Coastal structures and long term shore migration. Coastal Zone '87. Pages 414-426. Proceedings of the American Society of Civil Engineers, New York, New York, USA.

Riggs, S. R. and W.J. Cleary. 1993. Influence of inherited geological framework upon barrier island morphology and shoreface dynamics. Pages 173-176 *in* J. H. List, editor. Large-scale coastal behavior. U.S. Geological Survey Open File Report 93-381.

Roberts, H. 1997. Dynamic changes of the Holocene Mississippi River delta plain: the delta cycle. Journal of Coastal Research **13**:605-627.

Roberts, H. H., R.D. Adams and R.H.W. Cunningham. 1980. Evolution of the sand dominant subaerial phase, Atchafalaya delta, Louisiana. American Association Petroleum Geology Bulletin **64**: 264-279.

Roberts, H.H., O.K. Huh, S.A. Hsu, L.J. Rouse, D.A. Rickman. 1989. Winter storm impacts on the Chenier Plain coast of Southwestern Louisiana. Transactions: Gulf Coast Association of Geological Societies **29**: 515-522.

Rogers, D., B. Rogers and W. Herke. 1992. Effects of a marsh management plan on fishery communities in Louisiana. Wetlands **12**: 53-62.

Rogers, B. D., R.F. Shaw, W.H. Herke and H. Blanchet. 1993. Recruitment of postlarval and juvenile brown shrimp (*Penaeus aztecus* Ives) from offshore to estuarine waters of the Northwestern Gulf of Mexico. Estuarine, Coastal and Shelf Science **36**: 377-394.

Runca, E., A. Bernstein, L. Postma and G. DiSilvio. 1993. Control of macroalgae blooms in the Lagoon of Venice. Pages 421-435 *in* Proceedings, MEDCOAST 93 Conference, November 2-5, 1993, Antalya, Turkey.

Rusconi, A., M. Ferla and M. Filippi. 1993. Tidal observations in the Venice lagoon: the variations in sea level observed in the last 120 years. Proceedings of Sea Change 93, The Hague, Netherlands, April 19-23,. 1993.

Salinas, L. M., R.D. DeLaune and W.H. Patrick Jr. 1986. Changes occurring along a rapidly submerging coastal area: Louisiana, USA. Journal of Coastal Research **2**: 269-284.

Scaife, W., R.E. Turner and R. Costanza. 1983. Coastal Louisiana recent land loss and canal impacts. Environmental Management **7**: 433-442.

Schoot, P. M. and J.E.A. de Jong. 1982. Sedimentatie en erosie metingen met behulp van de Sedi-Eros-tafel (Set). Rijkswaterstaat. Notitie. DDMI-**82**.401.

Schubel, J.R. 1974. Effects of tropical storm Agnes on the suspended solids of northern Chesapeake Bay, in suspended solids in water. Pages 113-132 *in* R. Gibbs, editor. Plenum Marine Science, Volume 4. Plenum Press, New York, New York, USA.

—— 976. Effects of Agnes on the suspended sediment of the Chesapeake Bay and contiguous shelf waters. Pages 179-200 *in* J. Davis, B. Laird and J.R. Schubel, editors. The Effects of tropical storm Agnes on the Chesapeake Bay estuarine system. The Johns Hopkins University Press, Baltimore, Maryland, USA.

Schubel, J. R. and D.J. Hirschberg. 1978. Estuarine graveyards, climatic change and the importance of the estuarine environment. Estuarine Interactions 285-303.

Sestini, G. 1992. Implications of climatic changes for the Po delta and Venice lagoon *in* L. Jeftic, J. Milliman and G. Sestini, editors. Climatic change and the mediterranean. Edward Arnold, London, England.

Settlemyre, J. L. and L.R. Gardner. 1975. Chemical and sediment budgets for a small tidal creek, Charleston Harbor, S.C. (Water Resources Research Insititute Report No. 57). Clemson University, South Carolina, USA.

—— 977. Suspended sediment flux through a salt marsh drainage basin. Estuarine and Coastal Marine Science, **5**: 653-663.

Shaw, J. and D. L. Forbes. 1987. Coastal barrier and beach-ridge sedimentation in Newfoundland. Pages 437-454 *in* Proceedings Canadian Coastal Conference, National Research Council Canada, Ottawa.

Shepard, F. P., F.B. Phleger and T.H. van Andel. 1960. Recent sediments, Northwest Gulf of Mexico. American Association of Petroleum Geologists, Tulsa, Oklahoma, USA.

Sherman, D.J. and B.O. Bauer. 1993. Dynamics of beach-dune systems. Progress in Physical Geography **17**: 413-447.

Smith, N. P. 1979. Tidal dynamics and low-frequency exchanges in the Aransas Pass, Texas. Estuaries **2**: 218-227.

Snedaker, S. 1984. Mangroves: A summary of knowledge with emphasis on Pakistan. Pages 255-262 *in* B.H. Haq and J.D. Milliman, editors. Marine Geology and Oceanography of Arabian Sea and Coastal Pakistan. Van Nostrand Reinhold Co., New York, New York, USA.

Stanley, D. 1988. Subsidence in the northeastern Nile delta: rapid rates, possible causes and consequences. Science **240**: 497-500.

Stanley, D. J. and A.G. Warner. 1993. Nile Delta: Geological Evolution and Human Impact. Science **260**: 628-634.

Stern, M. K., J.W. Day, Jr. and K.G. Teague. 1991. Nutrient transport in a riverine influenced, tidal freshwater bayou in Louisiana. Estuaries **14**: 382-394.

Stevenson, J.C., L. Ward and M. Kearney. 1988. Sediment transport and trapping in marsh systems: implications of tidal flux studies. Marine Geology **80**: 37-59.

Stone, G.W., J.M. Grymes, K.D. Robbins, S.G. Underwood, G.D. Steyer and R.A. Muller. 1993. A chronological overview of climatological and hydrological aspects associated with Hurricane Andrew and its morphological effects along the Louisiana coast, USA. Shore and Beach **61**: 2-12.

Stone, G., J.M. Grymes, J.R. Dingler and D.A. Pepper. 1997. Overview and significance of hurricanes on the Louisiana Coast, USA. Journal of Coastal Research **13**: 656-669.

Stone, G. W. and C.W. Finkl. 1995. Journal of Coastal Research. The Coastal Education and Research Foundation.

Stumpf, R. P. 1983. The Process of sedimentation on the surface of a salt marsh. Estuarine, Coastal and Shelf Science **17**: 495-508.

Sutcliffe, W. 1972. Some relations of land drainage, nutrients, particulate material and fish catch in two eastern Canadian bays. Journal of Fisheries and Research Board Canada **29**:357-362.

Swenson, E. M. and C.E. Sasser. 1992. Water level fluctuations in the Atchafalaya Delta, Louisiana: tidal forcing versus river forcing. Pages 191-208 *in* D. Prandle, editor. Coastal and estuarine studies. American Geophysical Union, Washington, District of Columbia, USA.

Swenson, E., R. Turner. 1987. Spoil banks — Effects on a coastal marsh water-level regime. Estuarine, Coastal and Shelf Science **24**:599-609.

Swift, D.J.P. 1975. Barrier island genesis: evidence from the central Atlantic shelf, eastern USA. Sedimentary Geology **14**:1-43.

Tamisier, A. 1990. Gamargue: Milieux et Paysages, Evolution de 1942 a. 1984. Association Pour les Recherches en Camargue Sur la Nature et I'Environnement, Montpellier, France.

Templet, P. and K. Meyer Arendt. 1988. Louisiana wetland loss: a regional water management approach to the problem. Environmental Management **12**: 181-192.

Terwindt, J. H. J. and J.A. Battjes. 1991. Research on large-scale coastal behavior. Pages 1975-1983 *in* Proceedings, 2nd International Conference on Coastal Engineering. American Society of Civil Engineers, New York, New York, USA.

Thom, B.G. and W. Hall. 1991. Behavior of beach profiles during accretion and erosion dominated periods. Earth Surface Processes and Landforms **16**: 113-127.

Todd, O.J. and S. Eliassen. 1938. The Yellow River problem. American Society of Civil Engineers Proceedings **64**: 1921-1991.

Tornqvist, T., T. Kidder, W. Autin, K. van der Borg, A. de Jong, C. Klerks, E. Snijders, J. Storms, R. van Dam and M. Wiemann. 1996. A revised chronology for Mississippi River subdeltas. Science **273**: 1693-1696.

U.S. Army Corps of Engineers. 1987. Water Central Manual. New Orleans District, New Orleans, Louisiana, USA.

Van Andel, T.H. 1967. The Orinoco Delta. Journal of Sedimentary Petrology. **37**: 297-310.

van Heerden, I.L. and H.H. Roberts. 1980. The Atchafalaya delta: rapid progradation along a traditionally retreating coast (South-Central Louisiana). Z. Geomorph. N.F., 188-201.

Varela, J., A. Gallardo and A. Lopez De Velasco. 1983. Retencion de solidos por los embalses de Mequinenza y Ribarroja. Efectos sobre los aportes al Delta del Ebro, Pages 203-219 *in* M. Marino, editor. Sistema Integrado del Ebro. Madrid, Spain.

Wagner, F. W. and E. J. Durabb. 1976. The Sinking city. Environment **18**: 32-39.

Walker, H.J. and J.M. Coleman. 1987. Atlantic and Gulf coastal province. Pages 51 -110 *in* W.L. Graf, editor. Geomorphic Systems of North America, Centennial Special Volume 2. Geological Society of America, Boulder, Colorado, USA.

Wasp, E., J. Kenny and R. Gandhi. 1977. Solid-liquid flow slurry pipeline transportation. Trans. Tech. Publications, Clausthal, Germany.

Weir, W. 1950. Subsidence of peat lands of the Sacramento-San Joaquin Delta, California. Hilgardia **20**: 37-56.

Wells, J. and J. Coleman. 1984. Pages 85-100 *in* B.H. Haq and J.D. Milliman, editors. Marine Geology and Oceanography of Arabian Sea and Coastal Pakistan. Van Nostrand Reinhold Co., New York, New York, USA.

Zimmerman, R. and T. Minello. 1984. Densities of Penaeus aztecus, Penaeus setiferus and other natant macfauna in a Texas salt marsh. Estuaries **7**: 421-433.

INFLUENCES OF VEGETATION AND ABIOTIC ENVIRONMENTAL FACTORS ON SALT MARSH INVERTEBRATES

LISA A. LEVIN
THERESA S. TALLEY
Marine Life Research Group - 0218
Scripps Institution of Oceanography
La Jolla, CA 92093-0218 USA

Abstract
Sediment-dwelling fauna are a ubiquitous component of salt marshes yet we have limited understanding of their roles in marsh functioning and of the environmental conditions that control their distributions and abundances. This paper examines the influence of vegetation (presence, type, density, and biomass) and other environmental variables (marsh age, sediment and porewater properties, elevation, flow, oxygen, and biogenic structures) on salt marsh macrofauna and meiofauna. We review studies from a variety of geographical locations and include new information from systems with adjacent natural and restored sites in southern California. The influence of environmental factors on faunal assemblages varies with marsh system, factor intensity or concentration, taxon studied, and with other interacting factors present. We hypothesize a hierarchy of environmental variables in which abiotic properties such as marsh age, elevation and salinity act over large space and time scales, and are most likely to influence the presence or absence of species. Sediment properties (organic matter and particle size) and vegetation presence or type act on intermediate scales affecting macrofaunal abundance and composition. Plant biomass, culms and biogenic structures generated by fauna are patchy and act on small scales, often interacting with flow, to affect distribution and abundance patterns. Resolution of these processes in salt marshes should improve our understanding of controls on invertebrate communities and will ultimately aid the conservation and restoration of salt marsh habitat.

1. Introduction

Sediment-dwelling fauna (infauna) are a ubiquitous and numerically abundant component of salt marshes. Although many infauna are recognized as diet items for crustaceans and fishes, rather little is known about their ecological roles within salt marshes, and what controls their distribution and abundance. The last major review of invertebrate community patterns in salt marshes was prepared by Kneib (1984) over 15 years ago. Since that time, increased salt marsh restoration and creation activities have generated a growing interest in this component of marsh systems, and have stimulated investigations of infaunal ecology. Created marshes often have a wider range of vegetation biomass, particle size, organic matter content, and other environmental variables, compared to natural marshes, especially when these variables are manipulated experimentally. By utilizing this variation it is possible to distinguish the influences of different environmental factors, and to better understand their roles in structuring infaunal marsh assemblages.

Most environmental parameters do not act independently of one another in salt marshes. For example, elevation, salinity and marsh age influence the presence, type and density of vascular plant vegetation. Similarly, the presence of vegetation or coarseness of sediments may influence the intensity of effects of elevation or marsh age on macrofauna. The importance of biogenic structures, such as burrows or tubes, almost certainly varies with elevation, organic loading (and thus redox) and vegetation properties. A more thorough understanding is needed of how environmental factors interact with one another to produce observed patterns. Facilitative interactions may become prevalent among marsh plants (Bertness and Callaway 1994) and rocky intertidal invertebrates (Stephens and Bertness 1991) under physically harsh conditions. Limited evidence reviewed below, in which plants stabilize and oxygenate sediments and biogenic structures create habitat, supports facilitation of invertebrate marsh faunas by plants and animals under conditions of physical stress. Complex interactions also are likely to occur between abiotic properties, vegetation and biotic processes such as predation and competition (e.g., Vince et al. 1976).

This paper will review current knowledge of the influence of vascular marsh vegetation, and environmental parameters such as elevation, flow, oxygen, soil properties and marsh age, on salt marsh macrofauna (defined by various authors as animals retained on either a 300- or 500-μm mesh) and meiofauna (defined as animals passing through a 300-μm mesh but retained on either a 63- or 42-μm mesh). Emphasis will be placed on environmental properties which are often manipulated (or could be) during habitat restoration. Within the context of this review, new information will be presented for macrofauna from natural and restored salt marshes in southern California. In these investigations, environmental differences across created and natural marsh systems are employed to investigate controls on infaunal community composition and abundance.

No attempt is made here to review the large body of literature addressing direct animal interactions such as predation or competition; this would require another paper. Similarly, we do not address effects of anthropogenic influences such as contaminants or introduced species, although both are becoming more important in salt marshes with time. We acknowledge the significance of these processes and suggest that a better understanding of marsh invertebrate dynamics requires a knowledge of both biotic and abiotic as well as natural and anthropogenic influences.

2. Materials and Methods

We studied 5 Pacific coast salt marshes located along a 200-km stretch of coastline in southern California. The marshes were in Tijuana River Estuary (32°34′N, 117°7′W), San Diego Bay (Paradise Creek Marsh, 32°38′N, 117°6′W), Mission Bay (Northern

Wildlife Preserve, 32°47′N, 117°14′W), Upper Newport Bay (33°37′N, 117°53′W), and Anaheim Bay (33°44′N, 118°5′W). All embayments except for Tijuana Estuary contain natural and restored salt marsh areas. Restored *Salicornia virginica/Salicornia bigelovii* marshes and nearby natural counterparts were sampled during February 1995 in San Diego (10-y old), Upper Newport Bay (6-y old) and Anaheim Bay (5-y old) and in Mission Bay in April 1997 (16-mo old). Restored *Spartina foliosa* marshes and natural counterparts were sampled in San Diego Bay (10-y old) during February 1995, and in Mission Bay during April 1997 (16-mo old). Natural *S. foliosa* and *Salicornia* spp. zones were sampled in Mission Bay during February 1995 and in Tijuana Estuary during August 1996.

The restored marsh site within Mission Bay, Crown Point Mitigation Site (CPMS), had an experimental design consisting of six 25 x 12 m blocks bordered along one edge by the main creek. Within each block, 8 treatment plots (5 x 2 m) were established and each contained one of eight treatments: (1) peat, (2) alfalfa, (3) kelp or (4) milorganite amendments all rototilled and with *S. foliosa* transplants, (5) *S. foliosa* transplants with rototilling, (6) rototilling only (no amendment or transplants), (7) *S. foliosa* transplants only (no amendment or rototilling) or (8) no manipulation. Those plots with *S. foliosa* transplants received 21 plants per plot. The organic amendments consisted of 2000 g m^{-2} of material rototilled in to the top 6 cm of sediment in March 1996. Only data from the milorganite (a sewage-based fertilizer) and non-amended treatments will be discussed within this review.

Within each marsh, except the restored Mission Bay marsh, sampling sites were selected along a transect line at 5-m intervals, with random positioning of 0.25 m^2 quadrats as discussed in Levin et al. (1998). In the created Mission Bay system (CPMS) quadrats were located within each 5 x 2 m plot (n=6 plots per treatment). Estimates of percent cover of open space and vascular plant species, as well as plant densities, were recorded within each quadrat. Cores (18 cm^2x 6 cm deep) were collected within the quadrat for analyses of sediment parameters (percent sand and percent organic matter content) and infauna (1 core each). Sediment salinities were measured by extruding porewater from the upper 4 cm of sediment with a syringe through a Whatman filter No. 2 on to a Leica hand-held salinity refractometer. In Tijuana Estuary, the lower 10 cm of *S. foliosa* and *S. virginica* culms were cut from each quadrat, preserved and processed (n=5 samples per site) as discussed below for infaunal samples.

Additionally in Mission Bay during April 1997, benthic microalgal biomass and tidal elevations were measured. Benthic microalgal biomass was estimated by measuring the chlorophyll a concentration (spectrophotometric method) of a plug of surface sediment (1 cm diameter x 2 mm deep). Elevations were obtained at each sampling site using a surveyor's transit and expressed in terms of height above mean lower low water (MLLW).

Infaunal cores were processed as described in Levin et al. (1998). Briefly, they were preserved intact in 10% buffered formalin and seawater. Animals retained on a 300-μm mesh were sorted and identified. Below-ground plant material (live and dead) was sorted from infaunal cores and dried to estimate below-ground biomass (g dw). Sediment cores were frozen at -20°C until analyzed. Organic matter content (%) was determined by calculating the amount of loss after combustion (> 500°C, overnight) of dried sediment. Particle size, expressed as weight percent sand content, was determined by wet sieving the sediment sample through a 63-μm mesh, drying both fractions (>63-μm = sand, <63-μm = silt and clay) at 65°C, overnight and weighing the dried sediment.

Comparisons of macrofaunal (density, species richness, composition) and sediment properties (organic matter, sand, below-ground biomass, salinity) across marsh zones or marsh types were made using ANOVA's or *t*-tests (JMP Statistical software). Relationships between environmental and macrofaunal properties were explored within embayments across marsh zones (*S. foliosa* vs. *Salicornia* spp.) (Mission Bay, February 1995 and Tijuana Estuary, August 1996) and across marsh type (i.e., natural and restored; February 1995 and April 1997) using forward stepwise multiple regressions (Statistica statistical software). All numeric data were $\log_{10}$ (x+1) transformed and percent data were arcsin square root transformed prior to statistical analyses.

The remainder of the paper is a review in which results of the southern California marsh sampling described above have been integrated into appropriate sections.

3. Review and Results

3.1 MARSH INFAUNA

Invertebrates inhabiting salt marsh sediments are generally hardy species that tolerate diurnal and seasonal fluctuations in salinity, temperature, oxygen, inundation and other environmental factors (Kneib 1984). Species diversity in salt marsh sediments is usually low. For macrofauna, typically most of the individuals present at any single site belong to only a few major taxa (e.g., Weigert and Pomeroy 1982, Rader 1984, Frid and James 1989, Lana and Guiss 1991). Macrofaunal groups commonly encountered in salt marshes include oligochaetes, polychaetes, crustaceans, molluscs and insects (Table 1, Kneib 1984). Numerically dominant saltmarsh meiofauna are nematodes, harpacticoid copepods, turbellarians, ostracods and foraminiferans (Bell 1983, Watzin 1983, Kneib 1984). Studies characterizing the faunal assemblages of salt marshes typically examine densities of animals rather than biomass. While small invertebrates (e.g., oligochaetes or peracarid crustaceans) often may be the numerically dominant macrofauna in salt marshes, larger, less abundant invertebrates such as nereid polychaetes, mussels, crabs and gastropods may contribute much more biomass (Table 1). In *Spartina alterniflora* marshes of North Carolina, 41% of annual secondary productivity was attributed to large, mostly epifaunal invertebrates (fiddler crabs, gastropods and mussels); the remainder was accounted for by small macrofauna (polychaetes, isopods and insects) and meiofauna (Cammen et al. 1982).

TABLE 1. Relative contribution (%) of each macrofaunal taxon to total number of individuals and to total biomass in *Spartina* marshes. Values were calculated from taxonomic lists given in the literature and from this study. Studies focused on macro-infauna, so large epifauna were not collected.

Reference	Jackson et al. 1985		Cammen 1976		This study	
Location	England		North Carolina (Snow's Cut)		California (Mission Bay)	
Core size (diameter x depth)	7.6 x 20 cm		9.5 x 13 cm		4.8 x 6 cm	
Mesh size	0.5 mm		1 mm		0.3 mm	
Dominant Vegetation	*Spartina anglica*		*Spartina alterniflora*		*Spartina foliosa*	
	Percentage of Total					
Taxon	Density	Biomass	Density	Biomass	Density	Biomass
Nereidae	32.9	73.6	3.7	45.3	0	0
Other Polychaeta	1.1	0.1	0.2	1.1	1.2	12.3
Oligochaeta	37.8	4.9	0	0	63.9	30
Peracarid Crustacea	8.7	3.6	95.5	32.3	1.3	8.6
Decapoda	0.9	4.8	0.1	3.2	0	0
Gastropoda	4.6	8.9	0	0	3.3	15.2
Bivalvia	0.6	0.8	0.3	17.9	0	0
Insecta & Arachnida	10.5	3.1	0	0	28.2	11.1
Larvae	1.6	1.2	0	0	0	0
Nemertea	0	0	0	0	1.7	6.6
Other	0.6	0.3	0.1	0.4	0.3	16.4

3.2 MARSH AGE

Over the past decade, increased mitigation efforts have stimulated monitoring of macrofauna in relatively young created marshes (Table 2). Most of these studies have compared single created systems to one or more natural marsh counterparts (Cammen 1976a, Moy and Levin 1991, Havens 1995, Scatolini and Zedler 1996, Levin et al. 1996, Toomey 1997). In a created marsh, it is possible to evaluate faunal succession over time (marsh age). Sacco et al. (1987) reported significant changes in macrofaunal density and species composition in the Snow's Cut *S. alterniflora* marsh, North Carolina from 2 to 15 years of age. Ten-fold increases in density, reductions in proportional representation by amphipods and insects, and increased importance of polychaetes after 15 years, indicated a faunal trajectory converging with the natural system. Zones of differing age (4 and 8 years) within a South Carolina *S. alterniflora* marsh were found to exhibit macrofaunal density differences that appear to reflect time since establishment, although diversity was similar (La Salle et al. 1991). In particular, the presence of large-bodied molluscs within the older system was thought to be age-related. Varying macrofaunal compositions and densities were observed in *S. alterniflora* and *Schoenoplectus robustus* marshes of three different ages (ca. 15-, 11- and 4-y old), with intermediate densities of several of the taxa found in the intermediate-aged marsh (Posey et al. 1997). Evidence of macrofaunal succession was

TABLE 2. Summary of marsh age influence on macrofaunal populations and communities in salt marshes. Only those studies comparing created marshes of different ages or a single marsh at different times are included. S.a. = *Spartina alterniflora*, S.f. = *Spartina foliosa*, S.b. = *Salicornia bigelovii*, S.v. = *Salicornia virginica*, S.r. = *Schoenoplectus robustus*.

MARSH TYPE/ LOCATION	TAXON STUDIED	INFLUENCE ON FAUNA	REFERENCE
S.a./ North Carolina	Macrofauna	Major changes in density & species composition in a created marsh between 2 & 15 yr. age, resembled natural system after 15 years.	Sacco et al. 1987
S.a./ South Carolina	Macrofauna	4- & 8-yr. old marshes exhibited different faunal densities but similar diversity.	La Salle et al. 1991
S.a./ North Carolina	Macrofauna	Comparison of 7 created/natural marsh pairs indicated marsh age was less important in determining similarity than environmental factors.	Sacco et al. 1994
S.f./California	Macrofauna	Epibenthic macrofaunal densities were 2-3x lower, species richness was similar and taxonomic compositions differed (including higher exotic species abundances) in a 4-yr. old constructed vs. a natural marsh.	Scatolini and Zedler 1996
S.a./ North Carolina	Macrofauna	Marsh age can affect likelihood of colonization by taxa with different reproductive modes. A 4-yr. old marsh didn't attain species composition of a nearby natural marsh.	Levin et al. 1996
S.a. & S.r./ North Carolina	Macrofauna	Infaunal composition and abundances differed between sites of different ages with most of the differences occurring between the oldest (15-yr. old) and youngest (4-yr. old) sites.	Posey et al. 1997
S.v. & S.b./ California	Macrofauna	Macrofaunal densities were higher in created marshes (5-10 yr. old) than adjacent natural marshes, but species richness was similar. Early successional assemblages were largely insects and naidid oligochaetes; later successional stages were mainly tubificid and enchytraeid oligochaetes.	Talley and Levin 1999

observed by Talley and Levin (1999) in 4 created *S. virginica* and *S. bigelovii* marshes in southern California (16-mo to 10-y). Insects and naidid oligochaetes dominated the younger marshes, whereas tubificid and enchytraeid oligochaetes exhibited higher densities in the older marshes and in adjacent natural marshes.

Generally, during the first 6 years, created marsh macrofaunal communities develop from a sparse, low-diversity assemblage dominated by opportunists, to those with densities more comparable to natural systems. However, studies in North Carolina (Cammen 1976a,b, Moy and Levin 1991, Sacco et al. 1994, Levin et al. 1996, Posey et al. 1997, Toomey 1997) and in California (Scatolini and Zedler 1996, Talley and Levin 1999), suggest that assemblages often do not develop the functional attributes of natural marshes within this period, despite rapid growth of marsh vegetation. Sacco et al. (1994) compared 7 created-natural marsh pairs in North Carolina ranging in age from 1 to 17 years. He found that age of the created marsh had little to do with the extent to which densities and functional group composition resembled those of natural counterparts. Instead, environmental factors such as physical disturbance, as well as developmental age of the natural marsh, were more important. It is likely that marsh-age effects, when they are observed within created marshes, are associated with temporal changes in many of the environmental factors discussed below, including above- and below-ground plant biomass, benthic microalgal biomass, and soil organic and particle size properties, as well as with time available for colonization. This latter property may be of particular importance for species with direct development and limited dispersal capability (Levin et al. 1996, Talley and Levin 1999). Even within natural marshes, age affects soil organic matter accumulation (Friedman and DeWitt 1978) and elevation (Redfield 1972), which in turn influence vegetation success (Bertness 1988). Thus, marsh age may underlie some of the patterns discussed later in this review, even though age was not examined explicitly.

3.3 VEGETATION INFLUENCE

Vascular halophytes, specifically marsh grasses and succulents, are usually the dominant biological component of salt marshes, in terms of biomass, physical structure and elemental cycling (Adam 1990, Mitsch and Gosselink 1993). Marsh plants generate distinct zonation along elevation or salinity gradients (Adam 1990), provide extensive above and below-ground structure (Bertness 1984, Lana and Guiss 1992), modify flow (Leonard and Luther 1995), increase rates of sediment accretion (Zipperer 1996), and alter soil oxygenation and moisture (Howes et al. 1986, De La Cruz et al. 1989), organic content and texture, light penetration, evaporation and salinity (Valiela et al. 1984, Morris 1989, Bertness and Hacker 1994, Netto and Lana 1997a). A number of species, including *S. alterniflora* and *Spartina anglica,* have invaded previously unvegetated tidal flat habitats following their introduction to non-native sites (Meixner 1983, Jackson 1985, Callaway and Josselyn 1992, Zipperer 1996). In some cases, vascular marsh plants provide habitat for endangered species (Massey et. al. 1984, Zedler 1993). Thus, it is not surprising that vascular marsh plant effects have been the focus of numerous macrofaunal studies (Table 3).

Investigations have addressed: 1) faunal differences between vegetated and unvegetated sediments, 2) effects of different vascular plant species, 3) the influence of above- and below-ground plant biomass, 4) the influence of plant culms, 5) direct utilization of plant structure as habitat, and 6) the importance of vascular plant biomass or detritus as food. Most of the information about these processes has been gathered within *Spartina* marshes of the northern hemisphere. Where possible, we have tried to discuss influences of other marsh plants as well. Vegetation effects on marsh invertebrates are summarized in Table 3 and discussed below.

3.3.1 Vegetated vs. Non-Vegetated Sediments

Despite the well-entrenched paradigm developed for seagrasses, that the presence of vegetation enhances infaunal densities via increased survivorship, food, and substrate stability (Orth 1977, Orth et al. 1991), no consistent results have emerged for fauna in salt marshes. Unvegetated habitats in salt marshes consist of either channels, channel banks, bare pools, or adjacent tidal flats. Within both *S. anglica* and *S. foliosa* marshes, investigators have reported greater diversity or density and different species composition of macrofauna in unvegetated pools (Frid and James 1989) and mudflats (Jackson 1985, Frid and James 1989, Levin et al. 1998). However, in SE Brazil, Lana and Guiss (1991) observed higher density and species richness, as well as greater species persistence in vegetated *S. alterniflora* habitat relative to nearby unvegetated areas. In their study, principal component analysis separated species assemblages primarily by the presence or absence of plant canopies. Similarly, mobile crustaceans, including the shrimps *Palaemonetes pugio, Penaeus aztecus* and the blue crab *Callinectes sapidis* were found to be more abundant within vegetated portions of a Texas salt marsh (Zimmerman and Minello 1984). Mudflat patches invaded by *S. alterniflora* in Willapa Bay, Washington were shown by Zipperer (1996) to exhibit higher macrofaunal densities than unvegetated sites in April, but lower densities in August. An increase in subsurface-deposit feeding taxa such as *Capitella*, was noted in older meadows of *S. alterniflora.*

With yet a different result, studies of infaunal succession by Levin et al. (1996) in North Carolina revealed no difference in colonization rates or community composition between *Spartina*-vegetated and unvegetated plots. The small size of the experimental units (2 x 7 m) and the fact that unvegetated plots were surrounded by *S. alterniflora,* may have been responsible. In Mission Bay, macrofaunal density (t_{22}=0.16, P=0.88) and species richness (t_{22}=0.01, P=0.99) exhibited no differences over 16 months between created-marsh plots originally planted with *S. foliosa* and plots left unvegetated. However, *Spartina* survivorship was patchy and environmental conditions (i.e., grain size, organic matter content) were similar in the vegetated and unvegetated areas. In a comparison of channel vs. marsh effects on natant macrofauna, Minello et al. (1994) observed no influence of vegetation presence on benthic diversity or on abundances of fiddler or juvenile blue crabs.

One general pattern to emerge in many of the studies cited above is that, independent of overall macrofaunal trends, oligochaetes, especially Enchytraeidae, form a larger fraction of the total infauna in vegetated than unvegetated sediments (Frid and James 1989, Minello et al. 1994, Levin et al. 1997a, 1998). The extent to which physiological growth requirements, physiological tolerances, competitive interactions or other factors are responsible, remains to be determined.

The diverse findings, in which vascular vegetation can enhance, inhibit, or not influence macrofauna, may be related to the range of conditions studied and varying methodologies. We hypothesize that in physically stressful settings, vascular plants can enhance macrofaunal communities by reducing high salinity or evaporation through shading, or by oxygenating soils. As has been documented in seagrass systems (e.g., Heck and Orth 1980), where predation pressure is intense the plants also may inhibit predator access to macrofauna (Vaughn and Fisher 1988, Lee and Kneib 1994). However, in mature marshes, the presence of vegetation may be associated with light reduction (inhibiting growth of epibenthic algal food sources), rhizomes that inhibit burrowing, and detritus buildup that lowers redox potential. All of these factors may hinder macrofaunal activities relative to unvegetated sediments. These ideas merit testing by experimentation, much as Bertness and co-workers have done for plant-plant interactions in marshes (Bertness and Callaway 1994, Bertness and Hacker 1994).

3.3.2 Vegetation Type or Zone

The majority of salt marsh macrofaunal studies have been carried out in *S. alterniflora* systems. Although many marshes contain distinct vascular-plant zones, only a few investigations have examined the influence of vegetation type on macrofauna. Lana et al. (1997) compared the polychaetes of a *S. alterniflora* marsh to those of 3 mangrove habitats (*Rhizophora mangle, Laguncularia racemosa* and *Avicennia schaueriana*) along cross-elevation transects at seven stations within Paranagua Bay, SE Brazil. They found that abundance and species composition were relatively unaffected by vegetation zone, and that salinity and energy gradients were of greater importance in distinguishing assemblages. Substantial community differences were reported by Levin et al. (1998) between macrofauna of Pacific *S. foliosa* marshes and Atlantic *S. alterniflora* marshes, with proportionally more oligochaetes, especially enchytraeids, in the Pacific, and proportionally more polychaetes and tubificid oligochaetes in Atlantic marshes. The macrofaunal assemblage in an unvegetated Pacific mudflat was more similar at the generic level to that of the Atlantic than Pacific *Spartina* marshes (Levin et al. 1998). This pattern may reflect the fact that the Atlantic S. *alterniflora* extends lower into the intertidal zone than does the Pacific *S. foliosa.* Capehart and Hackney (1989) examined distribution of the clam *Polymesoda caroliniana* in three tidal marshes that varied in plant composition (*S. alterniflora, S. cynosuroides* and *Juncus roemerianus*) but not elevation. They attributed higher clam densities in the *Juncus* marsh to reduced root density relative to the *Spartina* habitats. The development of an infaunal community through time was found by Posey et al. (1997) to correspond to the replacement of *S. alterniflora* by *S. robustus*.

In poorly flushed embayments of southern California, *S. virginica* replaces *S. foliosa* at low tide levels. Construction of roads and railroads and diversion of freshwater inflow has dramatically altered lagoon circulation in southern California. *Spartina foliosa*, which once exhibited extensive coverage, but which requires regular tidal flushing, now thrives in only 5 of the 18 embayments south of Los Angeles. Invertebrate communities and food webs differ in *Spartina*- and *Salicornia*- dominated embayments of southern California (Nordby and Zedler 1991, Kwak and Zedler 1997). A detailed comparison of macrofauna in adjacent *S. foliosa* and *Salicornia* (*S. bigelovii and S. virginica)* zones within several southern California marshes revealed varying taxonomic composition with vegetation type (Levin et al. 1997a, Talley and Levin 1999). In Tijuana estuary, *S. foliosa*-vegetated sediments generally supported greater densities of polychaetes (*Streblopsio benedicti*, *Polydora nuchalis* and *Capitella* spp.) while the *Salicornia*-vegetated sediments, which sometimes occurred at higher elevations, had more gastropods *(Assiminea californica*), isopods and tubificid oligochaetes (*Monopylephorus rubroniveus* and *Tubificoides fraseri*). Both zones contained high densities of enchytraeid oligochaetes and scale insects (Coccidae). In Mission Bay, California, macrofauna were surveyed in natural *Spartina foliosa* and *Salicornia* spp. zones at similar elevations (*Spartina* 1.57 - 1.70m, *Salicornia* 17.8 - 1.80m above MLLW). The *Spartina* zone had significantly higher densities of tubificid oligochaetes (t_{18}=5.8, $p \leq 0.0001$); the *Salicornia* zone supported higher densities of naidid oligochaetes, insects, and peracarid crustaceans (t_{18}>3.8, $p \leq 0.001$).

Created *Salicornia* marshes (6 to 10-yr old) were shown by Talley and Levin (1999) to exhibit assemblages intermediate between those of natural *Salicornia* marshes and natural *S. foliosa* marshes. Macrofaunal succession within created *Salicornia* marshes of southern California was suggested to mirror that which occurs naturally over longer time periods as the *S. foliosa* zone develops into a *Salicornia* zone.

To date, assessments of plant-type effects on invertebrates have been correlative. Mechanistic assessments of the influence of vascular plant type on macrofauna will probably require an experimental approach. Variations in above- and below-ground structural and chemical properties of different plant species or growth forms (e.g., grasses vs. succulents) are certain to influence light attenuation, soil characteristics, and epifauna in ways that alter infauna. Experimental manipulation of plant biomass and structure, or creation of mimic structures could be used to test relevant hypotheses.

3.3.3 Plant Biomass and Density

A number of investigations have examined the influence of above- and below-ground plant biomass on macrofaunal communities (Table 3). Detailed studies by Lana and Guiss (1992) in *S. alterniflora* habitat of Paranagua Bay revealed that total macrofaunal density and densities of dominant taxa such as the polychaetes *Isolda pulchella* and *Nereis oligohalina* were positively correlated with live below-ground biomass (dry wt), but not with above-ground biomass. These authors suggested that plant material is used mainly as a refuge or physical support, rather than as food. They found total numbers of epifaunal taxa, and densities of the gastropod *Littorina flava*, were negatively correlated with above-ground biomass. Lana and Guiss (1992) suggested that dense above-ground canopy decreases light availability and subsequent algal development on sediment and plants, thus inhibiting epifaunal grazers. At their study site, plant biomass had no influence on species richness. However, positive correlations between *S. alterniflora* density and plant height with densities of the gastropod *Neritina virginea* were observed by Bonnett et al. (1994) in the same region. In a related study, Lana and Guiss (1991) attributed seasonal increases in macrofaunal density in Paranagua Bay to temporal variation in plant litter availability (as well as to lowered predation pressure).

Marsh invertebrates exhibit both positive and negative relationships with *S. alterniflora* stem density. A positive correlation between stem density and *Geukensia demissa* density has been documented in *S. alterniflora* marshes (Bertness 1984, West

TABLE 3. Summary of vascular vegetation influence on benthic invertebrate populations and communities in salt marshes.
S.a. = *Spartina alterniflora*, *S. an.* = *Spartina anglica*, *S.f.* = *Spartina foliosa*, *S.c.* = *Spartina cynosuroides*, *J.r.* = *Juncus roemerianus*
S.b. = *Salicornia bigelovii*, *S.v.* = *Salicornia virginica*, *S.e.* = *Salicornia europea*, *S.r.* = *Schoenoplectus robustus*

VEGETATION TYPE/ LOCATION	TAXON STUDIED	INFLUENCE ON FAUNA	REFERENCE
ENVIRONMENTAL AGENT: VASCULAR VEGETATION (PRESENCE/ABSENCE)			
S.a./ North Carolina, USA	Macrofauna	No differences in succession of vegetated vs unvegetated plots in a created marsh.	Levin et al. 1996
S.a./ Galveston, Texas, USA	Natant macrofauna	Juvenile blue crabs & fiddler crabs, and macro-benthic diversity unaffected by channel presence. Oligochaetes usually in vegetated bottom only.	Minello et al. 1994
S.a./ Texas, USA	Natant macrofauna	*Palaemonetes pugio*, *P. aztecus* and *Callinectes sapidus* more abundant in vegetated than non-vegetated portions of the marsh.	Zimmerman & Minello 1984
S.a./ Washington, USA	Macrofauna	Comparison with unvegetated tidal flat revealed higher macrofaunal densities in *S.a.* patches in April and lower densities in August. Increased densities of subsurface- deposit feeders occurred in vegetated patches.	Zipperer 1996
S.a./ SE Brazil	Macrofauna	Higher density & species richness, greater temporal stability in vegetated than unvegetated areas.	Lana & Guiss 1991
S. an../ Suffolk, England	Macrofauna	Differing species composition in vegetated vs mudflat habitat.	Jackson 1985
S. an., *Salicornia* spp./ Norfolk, England	Macrofauna	Macrofauna more diverse in unvegetated marsh pools than vegetated areas, especially *Nereis diversicolor* and *Nephtys hombergi*. Lower diversity in marsh than sand, mudflats. Oligochaetes more abundant in vegetated sediments. *Hydrobia ulvae* density enhanced by vegetation.	Frid & James 1989

TABLE 3 (Continued)

VEGETATION TYPE/ LOCATION	TAXON STUDIED	INFLUENCE ON FAUNA	REFERENCE
S.f./ Mission Bay, CA, USA	Macrofauna	Higher density & species richness on mudflat than salt marsh.	Levin et al. 1998
S.v./ mesocosms, CA, USA	Macrofauna	Experimental mowing (unvegetated treatments) of epiphytic/benthic S.v. yielded higher densities of benthic fauna & lower densities of epiphytic fauna.	de Szalay et al. 1996
ENVIRONMENTAL AGENT: VASCULAR VEGETATION (PLANT TYPE / ZONE)			
S.a., Mangroves/ SE Brazil	Macrofauna	Little effect on species composition or abundance of marsh vs mangrove zones.	Lana et al. 1997
S.a. & *S.r.*/ North Carolina, USA	Macrofauna	Found relationship between vegetation development and infaunal community development.	Posey et al. 1997
S.a., *S.c.* & *J.r.*/ North Carolina, USA	*Polymesoda caroliniana*	Higher clam densities in the *J.r.* than *Spartina* marshes.	Capehart & Hackney 1989
S.f./ California vs *S.a.*/ North Carolina,	Macrofauna	Considerable taxonomic difference associated with *Spartina* species/ocean. Enchytraeid oligochaetes dominated in the Pacific, polychaetes & tubificid oligochaetes in the Atlantic.	Levin et al. 1998
S.f., *S.v.* & *S.b.*/ Mission & San Diego Bays, CA,USA	Macrofauna	Varying taxonomic composition with vegetation types. Higher tubificid oligochaete densities in *S.f.*, higher insect, naidid oligochaetes and peracarid densities in *Salicornia.*	This study Talley and Levin 1999
S.f. & *S.v.*/ Tijuana Estuary, CA, USA	Macrofauna	Varying taxonomic composition with vegetation types. Higher density of polychaetes and insects in *S.f.*, higher density of gastropods, naidid and enchytraeid oligochaetes & peracarids in *S.v.*	Levin et al. 1997
S.v. & *S.r.*/ N. California, USA	Macrofauna	Higher density and greater trophic diversity in *S.v.* epiphytic/ benthic than *S.r.* marshes. Comparable species richness.	de Szalay & Resh 1996

TABLE 3 (Continued)

VEGETATION TYPE/ LOCATION	TAXON STUDIED	INFLUENCE ON FAUNA	REFERENCE
ENVIRONMENTAL AGENT: VEGETATION ABOVE-GROUND BIOMASS & DENSITY			
S.a./ Rhode Island, USA	*Geukensia demissa*	Facultative mutualism between *S.a.* and *G. demissa. S.a.* production increases mussel growth, and mussel increases *S.a.* production.	Bertness 1984
S.a./ Rhode Island, USA	*Uca pugnax*	Facultative mutualism between *S.a.* and *U. pugnax. S.a.* cover facilitates crab burrowing through substrate stabilization, and crab increases *S.a.* production through aeration of the sediment.	Bertness & Miller 1984 Bertness 1985
S.a./ North Carolina, USA	*Geukensia demissa*	Mortality of the mussel decreased with distance from the marsh edge probably due to protection from predators that were impeded by the dense grasses towards the interior of the marsh.	Lin 1989
S.a., *S.c.*, & *J.r.*/ North Carolina, USA	*Polymesoda caroliniana*	Positive relationship between stem density and clam density	Capehart & Hackney 1989
S.a./ Alabama, USA	*Geukensia demissa*	Positive correlation between stem density and mussel density	West & Williams 1986
S.a./ Washington, USA	Macrofauna	Shoot density was positively correlated with *Capitella* density and negatively correlated with densities of corophiid amphipods and cirratulid polychaetes.	Zipperer 1996
S.a./ SE Brazil	*Neritina virginea*	Positive correlation of gastropod density with *Spartina* biomass, density and height.	Bonnet et al. 1994
S.v. & *S.b.*/ Georgia, USA	*Uca pugnax* *Uca pugilator*	Positive relationship between crab burrows and plant cover due to increased use of plants by crabs for predator avoidance. No effect of crabs on plants observed.	Nomann & Pennings 1998

TABLE 3 (Continued)

VEGETATION TYPE/ LOCATION	TAXON STUDIED	INFLUENCE ON FAUNA	REFERENCE
S.v. & *S.b.*/ Mission Bay, CA, USA (natural & 16-mo old restored)	Macrofauna	Negative correlation of % cover of all plants with enchytraeid oligochaetes.	This study
S.v. & *S.b.*/ So. California Bays, USA (natural & 5-10 yr old restored	Macrofauna	Positive correlation of % *S.b.* cover with species richness, density of macrofauna, oligochaetes, insects, polychaetes, nemerteans and turbellarians. Negative correlation of % *S.v.* cover with species richness, density of macrofauna, enchytraeid oligochaetes, molluscs; positive correlation with polychaete density	This study
S.f. & *S.b.*/ Mission Bay, CA, USA (natural & 16-mo old restored)	Macrofauna	Density of *S.f.* was positively correlated with density of naidid oligochaetes and negatively correlated with insect density. % *S.b.* cover was positively correlated with nemertean and turbellarian density and negatively correlated with tubificid oligochaete density.	This study
ENVIRONMENTAL AGENT: VEGETATION BELOW-GROUND BIOMASS			
S.a./ Maryland Rhode Island, USA	*Uca pugnax*	Root mat inhibits burrowing of crabs.	Ringold 1979 Bertness 1985
S.a./ South Carolina, USA	Meiofauna	Positive association of living root biomass with nematode density.	Osenga & Coull 1983
S.a./ North Carolina, USA	Macrofauna	Linear positive correlation of macrofaunal density with root volume.	Rader 1984

TABLE 3 (Continued)

VEGETATION TYPE/ LOCATION	TAXON STUDIED	INFLUENCE ON FAUNA	REFERENCE
S.a., *S.c.*, & *J.r.*/ North Carolina, USA	*Polymesoda caroliniana*	Negative relationship between root/rhizome biomass and density.	Capehart & Hackney 1989
S.a./ Washington, USA	Macrofauna	Below-ground biomass was positively correlated with *Capitella* and dipteran larvae density and negatively correlated with cirratulid polychaete density.	Zipperer 1996
S.a./ SE Brazil	Macrofauna	Total abundance & density of 2 polychaete spp. positively correlated with live below-ground biomass.	Lana & Guiss 1992
S.a./ SE Brazil	Macrofauna	Abundances correlated with below-ground and dead above-ground *Spartina* biomass, vertical distributions were unaffected.	Netto & Lana 1997b
S.f. & *S.b.*/ Mission Bay, CA, USA (natural & 5-10 yr old restored)	Macrofauna	Positive correlation of below-ground plant biomass with density of macrofauna, species richness and number of naidid oligochaetes.	This study
S.v. & *S.b.*/ So. CA, Bays, USA (natural & 5-10 yr old restored)	Macrofauna	Positive correlation of below-ground plant biomass with density of peracarid crustaceans & molluscs; negative correlation with density of macrofauna, enchytraeid oligochaetes & polychaetes.	This study
ENVIRONMENTAL AGENT: PROXIMITY TO CULMS			
S.a./ Rhode Island, USA	*Uca pugnax*	Burrow density and longevity in soft sediments higher near *S.f.* stems than adjacent bare areas.	Bertness 1985
S.a./ South Carolina, USA	Meiofauna	No effects of proximity to *S.a* culms or root biomass.	Bell et al. 1978

TABLE 3 (Continued)

VEGETATION TYPE/ LOCATION	TAXON STUDIED	INFLUENCE ON FAUNA	REFERENCE
S.a./ North Carolina, USA	Macrofauna	3-9 x higher juvenile & adult densities near culms. Greatest effects in mid and high marsh.	Rader 1984
S.f./ South California, USA	Macro/ meiofauna	No effects of *S.a.* culms on macrofaunal colonization.	Levin et al. 1996
S.f./ Mission Bay, CA, USA	*Cerithidea californica*	Retention of transplanted snails within 10 cm of *S.f.* plugs.	T. Talley, unpublished
ENVIRONMENTAL AGENT: PLANTS AS HABITAT			
S.a./ Georgia, USA	Oligochaeta	Densities higher in *S.a.* leaf sheaths at the base of plants than in sediment, roots, surface debris. Oligochaete distribution affected by stage of sheath decay. Some species specialize on stem/sheath habitat.	Healy & Walters 1994
S.a./ Louisiana, USA	Meiofauna	Stems support high densities, often greater than in surrounding sediments.	Rutledge & Fleeger 1993
S.a./ Maryland, USA	*Gammarus palustris*	Amphipods live on stems or at culm base.	Van Dolah 1978
S.a./ eastern England	Macrofauna	13 taxa associated with *S.a.* canopy specifically; they account for only 2% of secondary production.	Jackson et al. 1985
S.v./ Mesocosm, N. California, USA	Macrofauna	22 taxa were associated with *S.v.* stems and insects were numerically dominant (70 to 90%).	de Szalay et al. 1996
S.v. & *S.f.*/ Tijuana Estuary, California, USA	Macro/ meiofauna	Bottom 10 cm of *S.v.* stems inhabited by naidid and enchytraeid oligochaetes, turbellaria and mites. Bottom 10 cm of *S.f.* stems inhabited by dolichopodid insect larvae, enchytraeid oligochaetes and turbellarians.	Levin et al. 1997

TABLE 3 (Continued)

VEGETATION TYPE/ LOCATION	TAXON STUDIED	INFLUENCE ON FAUNA	REFERENCE
S.r. & *S.v.*/ California, USA	Macrofauna	Herbivores (all insects) were numerically more important in epiphytic than benthic habitats probably due to the presence of epiphytic algae.	de Szalay & Resh 1996
ENVIRONMENTAL AGENT: TROPHIC INFLUENCE OF PLANTS			
S.a.	Large macrofauna	Stable isotopic evidence for *S.a.* in diets.	Haines 1976 Haines & Montague 1979 Peterson et al. 1985; 1986 Peterson & Howarth 1987 Currin et al. 1995 Deegan & Garrit 1997
S.a./ LAB	*Orchestisa grillus*	Amphipods utilize microbes while *S.a.* organic debris is poorly digested.	Lopez et al. 1977
S.a./ LAB	Meiofauna	Detrital additions to lab microcosms did not affect meiofaunal biomass.	Schwinghammer & Kepkay 1987
S.a./ LAB	*Crassostrea virginica*	Carbon from bacteria had a higher (52.5%) assimilation efficiency than *S.a.* and bacteria combined (10.3%) or *S.a.* alone (2.7%).	Crosby et al. 1990
S.a./ LAB	*Littoraria irrorata*	Snails utilize dead leaves as an energy source and prefer those colonized by fungi or bacteria.	Barlocher & Newell 1994a;b
S.a./ LAB	*Geukensia demissa*, *Mytilus edulis*	Microheterotrophic organisms (e.g., flagellates and bacteria) contribute to the carbon requirements of mussels.	Kreeger & Newell 1996

TABLE 3 (Continued)

VEGETATION TYPE/ LOCATION	TAXON STUDIED	INFLUENCE ON FAUNA	REFERENCE
S.a./ LAB	*Uhlorchestia spartinophila*	Amphipods utilize senescent blades and sheaths as an energy source. Food value is related to decomposition stage.	Kneib et al. 1997
S.a./ LAB *S. an.* & *S.e.*/ LAB	*Nereis virens*, *Nereis diversicolor*	Macroalgae has higher values of assimilation and shorter digestion time than vascular plants in juveniles.	Olivier et al. 1996, 1997
S.a./ LAB Delaware, USA	*Geukensia demissa*	Mussels utilize *S.a.* detritus as an energy source.	Kreeger et al. 1988 Langdon & Newell 1990
S.a./ Chesapeake Bay, USA	*Crassostrea virginica*	Crude fiber of *S.a.* can potentially contribute up to 20% of the oysters' diet if fiber concentrations in the seston are high.	Crosby et al. 1989
S.f./ California, USA	Lg. Macrofauna	Stable isotopic evidence for *S.f.* in diets.	Kwak & Zedler 1997
S.v./ California, USA	Amphipods Isopods	Stable isotopic evidence for *S.v.* in diets.	Page 1997

and Williams 1986). *Spartina* stems provide critical attachment surfaces *for G. demissa* (Bertness 1984, Bertness and Grosholtz 1985) while the mussels stimulate *S. alterniflora* production (Bertness 1984). Mortality of *G. demissa* was seen to decrease with distance from the marsh edge into the interior of the marsh where marsh grasses became dense (Lin 1989). Increased plant density probably impeded predators, allowing higher survival of the mussels (Lin 1989). Where *S. alterniflora* has invaded Pacific coast tidal flats, stem density is positively correlated with densities of the polychaete *Capitella* sp., and negatively correlated with amphipod (*Corophium* spp.) and cirratulid (*Aphelochaeta* sp.) densities (Zipperer 1996).

Increased plant cover appears to facilitate burrowing by fiddler crabs (Bertness and Miller 1984, Nomann and Pennings 1998). In the lower marsh *(S. alterniflora)*, this effect is attributed to substrate stabilization by plants (Bertness and Miller 1984, Bertness 1985). High-marsh experiments in Georgia suggest that reduced salinity and temperature from shading and refuge from predators may be the mechanisms underlying vascular plant facilitation of *Uca* spp. in hypersaline habitats (Nomann and Pennings 1998).

Root biomass is of interest because roots occupy a significant fraction of the space in marsh sediments. Bell et al. (1978) observed no or negative correlations between root biomass and meiofauna in an *S. alterniflora* marsh in South Carolina. However, nearby, Osenga and Coull (1983) observed a positive association of living root biomass and nematode abundance, although no correlation with total or dead root biomass. A positive correlation between *S. alterniflora* root volume and macrofaunal abundance was reported by Rader (1984). Netto and Lana (1997) found below-ground and dead above-ground biomass to account for much variation in macrofaunal abundance. However, in their system *S. alterniflora* had a pioneering role and most detritus was from nearby mangroves. Below-ground biomass of *S. alterniflora* was positively correlated with densities of capitellid polychaetes and dipteran larvae in an invaded tidal flat, accounting for 69% and 28% of the variance, respectively (Zipperer 1996). Micro-oxygenation of sediments by *Spartina* roots (Teal and Weiser 1966) has been proposed by many of these authors to contribute to the positive associations between below-ground plant biomass and infauna. Root exudates might stimulate microbial growth and enhance food supply for nematodes (Osenga and Coull 1983).

Inhibition of fauna by *S. alterniflora* roots also occurs. Root mat hinders burrowing by fiddler crabs. An inverse relationship between *Uca pugnax* densities and habitat root mat development was observed by Ringold (1979). Zipperer (1996) reported a negative correlation between below-ground biomass and density of a surface-feeding cirratulid polychaete (r^2 = 0.49). Similarly, a negative relationship was observed between root and rhizome biomass and density of the Carolina marsh clam (*Polymesoda caroliniana*) (Capehart and Hackney 1989). There was however, a positive relationship between marsh plant stem density and this clam, which might have resulted from reduced predation in more dense vegetation. Experiments conducted by Lee and Kneib (1994) document the potential of *S. alterniflora* culms to reduce predation by the xanthid crab, *Eurytium limosum* on another xanthid crab, *Panopeus herbstii*.

In a comparison of natural and created (16-mo old) *S. foliosa* marshes in Mission Bay, California, below-ground biomass (living and dead) was positively correlated with macrofaunal density in the *Spartina* zone and with species richness and naidid oligochaete density in the *Salicornia* zone (Table 4). In older restored (5-10-yrs old)

TABLE 4. Multiple regression analyses of macrofaunal and environmental properties sampled from the *Salicornia* spp. (*S. bigelovii* and *S. virginica*) and the *Spartina foliosa* habitats of the Crown Point Mitigation Site (16-mo old) and adjacent Northern Wildlife Preserve (natural) in Mission Bay, California during April 1997. Independent variables analyzed include elevation (cm above MLLW), % organic matter content (OM), % sand, salinity (ppt), chlorophyll a concentration (mg/L) (Chl a), below-ground biomass (g dw) (Bgb), % cover of open space (absence of plants), and density of vascular plants (all plants in *Salicornia* habitat; and *Spartina foliosa* [*S.f.*] and *Salicornia bigelovii* [*S.b.*] in the *Spartina* habitat).

Criterion for variables included in each model was $p \leq 0.10$, and for significance of each model the criterion was $P \leq 0.05$.

Dependent variable (No. per $18 cm^2$)	Independent variable:	Elev (cm)	OM (%)	Sand (%)	Salinity (ppt)	Chl a (mg/L)	Bgb (g dw)	Open (%)	Plant density	*S.b.* density	*S.f.* density	R^2 or r^2	P
Macrofauna	*Salicornia*		+	+								0.87	$\leq 1x10^{-7}$
	Spartina	-		-			+					0.63	$\leq 1x10^{-4}$
Species	*Salicornia*						+					0.31	0.01
	Spartina							-				0.33	$\leq 3x10^{-3}$
Oligochaeta	*Salicornia*		+	+					-			0.95	$\leq 1x10^{-7}$
	Spartina		+	+		+						0.86	$\leq 1x10^{-7}$
Tubificidae	*Salicornia*		+									0.27	0.02
	Spartina		+							-		0.85	$\leq 1x10^{-7}$

TABLE 4 (Continued)

Dependent variable (No. per $18cm^2$)	Independent variable:	Elev (cm)	OM (%)	Sand (%)	Salinity (ppt)	Chl a (mg/L)	Bgb (g dw)	Open (%)	Plant density	*S.b.* density	*S.f.* density	R^2 or r^2	P
Enchytraeidae	*Salicornia*		+	+					-			0.94	$\leq 1x10^{-7}$
	Spartina		+	+		+						0.83	$\leq 1x10^{-7}$
Naididae	*Salicornia*												N.S.
	Spartina		-			+					+	0.70	$\leq 1x10^{-5}$
Insecta	*Salicornia*					+						0.21	0.05
	Spartina		+	-	+						-	0.65	$\leq 2x10^{-4}$
Polychaeta	*Salicornia*		+									0.41	$\leq 3x10^{-3}$
	Spartina	-			-							0.30	0.02
Nemertea &	*Salicornia*			-								0.29	0.02
Turbellaria	*Spartina*									+		0.24	0.01
Peracarid	*Salicornia*												N.S.
crustacea	*Spartina*				-	-						0.26	0.03

TABLE 5. Multiple regression of macrofaunal properties against environmental variables in natural and restored *Salicornia* spp. (*S. bigelovii* & *S. virginica*) marshes in three bays in southern California. Independent variables include % organic matter content (OM), % sand content, salinity (ppt), below-ground biomass (g dw) (Bgb), % cover of open space (absence of plants), % cover of *Salicornia bigelovii* (*S.b.*), and % cover of *Salicornia virginica* (*S.v.*). Created marsh ages are 10, 6 and 5 yrs. old for San Diego, Upper Newport and Anaheim Bays, respectively. Criterion for variables included in each model was $p \leq 0.10$, and for significance of each model the criterion was $P \leq 0.05$.

Dependent variable (No. per $18cm^2$)	Independent variable:	OM (%)	Sand (%)	Salinity (ppt)	Bgb (g dw)	Open (%)	S.b. (%)	S.v. (%)	R^2 or r^2	P
Macrofauna	San Diego						+		0.20	0.05
	Upper Newport			+			+		0.84	$1x10^{-5}$
	Anaheim				-			-	0.63	$2x10^{-4}$
Species	San Diego					-	+		0.36	0.02
	Upper Newport							-	0.24	0.03
	Anaheim									N.S.
Oligochaeta	San Diego						+		0.20	0.05
	Upper Newport			+			+		0.81	$1x10^{-6}$
	Anaheim							-	0.27	0.02
Tubificidae	San Diego					-	+		0.46	$5x10^{-3}$
	Upper Newport		+						0.54	$2x10^{-4}$
	Anaheim									N.S.
Enchytraeidae	San Diego						+		0.24	0.03
	Upper Newport			+					0.46	$1x10^{-3}$
	Anaheim	+			-	-		-	0.80	$4x10^{-4}$
Naididae	San Diego									N.S.
	Upper Newport						+		0.74	$1x10^{-6}$
	Anaheim	+				+			0.40	0.01

TABLE 5 (Continued)

Dependent variable (No. per 18cm^2)	Independent variable:	OM (%)	Sand (%)	Salinity (ppt)	Bgb (g dw)	Open (%)	S.b. (%)	S.v. (%)	R^2 or r^2	P
Insecta	San Diego									N.S.
	Upper Newport	-	-				+		0.81	$5x10^{-6}$
	Anaheim	-	+						0.68	$6x10^{-5}$
Polychaeta	San Diego									N.S.
	Upper Newport			-	-		+	+	0.71	$5x10^{-4}$
	Anaheim					+			0.31	0.01
Nemertea &	San Diego						+		0.23	0.03
Turbellaria	Upper Newport						+		0.19	0.05
	Anaheim									N.S.
Peracarid	San Diego									N.S.
crustacea	Upper Newport									N.S.
	Anaheim					-		-	0.30	0.05
Mollusca	San Diego									N.S.
	Upper Newport				+				0.29	0.01
	Anaheim	-		-		-	-	-	0.86	$2x10^{-5}$

and natural *Salicornia* marshes, below-ground biomass was negatively correlated with numbers of macrofauna, enchytraeids, and polychaetes, and positively correlated with mollusc density (Table 5). Above-ground plant densities in Mission Bay were positively correlated with naidid oligochaete and nemertean/turbellarian densities, but negatively correlated with tubificid oligochaete and insect densities in the *Spartina* zone. In the *Salicornia* zone, plant density was negatively associated with oligochaete density. There was no relationship between total macrofaunal density or species richness and plant density in either vegetation zone (Table 4). In the older restored and natural *Salicornia* marshes, species richness, and density of total macrofauna, oligochaetes, insects, polychaetes, peracarid crustaceans, nemerteans and turbellarians, all were positively associated with percent cover of *S. bigelovii*, a pioneering species. There was a negative association of percent cover of *S. virginica* with macrofaunal species richness, and with densities of macrofauna, oligochaetes, nemerteans and turbellarians. Molluscs were negatively associated with cover of both *Salicornia* species (Table 5).

It appears that the influence of above- and below-ground vascular plant biomass or density varies with the type and amount of vegetation present, and with the faunal taxon examined. We hypothesize that positive biomass or density effects (e.g., oxygenation by roots, predator reduction, shade, attachment substrate) are experienced by infauna up to some density threshold and that beyond that plant effects become inhibitory. Small infauna will almost certainly experience different effects of below- or above-ground biomass than large burrowing taxa such as fiddler or grapsid crabs. In young, created marshes where there often is initially little above or below-ground plant material, plant effects are more likely to be beneficial to benthic invertebrates. In mature systems, dense vegetation may reduce algal productivity and dense root mat will hinder infaunal burrowing and feeding activities (Ringold 1979, Bertness 1985, Brenchley 1982).

3.3.4 Proximity to Plant Culms

Investigations on small spatial scales have sometimes revealed increased densities of macrofauna with increasing proximity to *Spartina* culms, often despite decreased sediment volume in the sample (Van Dolah 1978, Rader 1984). However, the absence of this effect also has been observed for meiofauna (Bell et al. 1978) and macrofauna (Rader 1984, Levin et al. 1996) in *S. alterniflora* marshes.

The marsh mussel, *G. demissa,* attaches with byssus threads to *S. alterniflora* rhizomes (Bertness 1984) and in California, where the mussel has been introduced, to *S. foliosa* rhizomes (personal observation). The fiddler crab, *Uca pugnax*, also lives in contact with the bases of *S. alterniflora* culms and roots. The rhizomes apparently stabilize sediments and provide a refuge from predators for associated fauna (Bertness 1984, 1985). *Spartina townsendii*, an exotic species in German tidal marshes, provides substrate for the mussel *Mytilus edulis* (Meixner 1983). In a 3-month old restored marsh in Mission Bay, southern California, 139 California hornsnails (*Cerithidea californica*) were introduced with *S. foliosa* transplanted plugs. After two months, 96% of these snails were still found within 10 cm or less of the transplant culms (Talley et al. unpublished data), probably due to the shade and moisture provided by the *Spartina* relative to the barren areas between transplants.

3.3.5 Above-ground Plant Structure as Habitat

Where investigators have looked, they have found invertebrates utilizing plant culms, sheaths, blades and detritus as habitat. Jackson et al. (1985) reported 13 species associated with *S. anglica* canopy. However, these contributed < 2% of the overall annual production and assimilation of macrofauna in this marsh in eastern England. Only one of the species, a sap feeding insect, *Philaenus sparmanus*, was thought to feed on live *Spartina*. High densities of meiofauna were found on *S. alterniflora* stems in Louisiana by Rutledge and Fleeger (1993); these densities often were greater than in surrounding sediments. Algal cover on the *S. alterniflora* stems was positively correlated with densities of mites, amphipods and isopods, but negatively correlated with harpacticoid copepod densities (Rutledge and Fleeger 1993). The presence of epiphytic algae on plant stems is thought to explain why herbivores were numerically more important in epiphytic than benthic habitats in *S. virginica* and *S. robustus* marshes in California (de Szalay and Resh 1996). Insects were found to be the dominant taxon in epiphytic habitats of *S. virginica* and *S. robustus* marshes (de Szalay and Resh 1996) and in *S. virginica* meso- and macrocosms (de Szalay et al. 1996) in northern California. In Georgia *S. alterniflora* marshes, Healy and Walters (1994) observed much higher oligochaete densities in leaf sheaths at the base of *S. alterniflora* plants than in surrounding sediments, root and surface debris. Oligochaete densities were influenced by position in the marsh, height on stems and stage of sheath decay. A number of species appeared to be highly specialized for sheath and stem habitat. Healy and Walters (1994) suggest that the stem habitat may be more important in marshes in the southeastern US than in more northern marshes because the *Spartina* plants are taller and thicker in the south. Analyses of vascular plant culms and stems in the Tijuana River Estuary revealed a sparse fauna, with most species present also occurring in surrounding sediments. *S. foliosa* culms (lower 10 cm) were inhabited mainly by dolichopodid insects, turbellarians and enchytraeid oligochaetes; *S. virginica* stems were inhabited by naidid and enchytraeid oligochaetes, turbellaria and mites (Levin et al. 1997a, this study). The overall importance of above-ground portions of plant stems and leaves as invertebrate habitat appears to vary regionally, and possibly with vascular plant species. There are presently too few investigations to generalize further.

3.3.6 Vegetation as Food

Long-standing interest in nutrient cycling within salt marshes, and in the fate of marsh production (see Nixon 1980), has led to studies of invertebrate consumption of marsh plants. These investigations have involved both direct experimentation in the laboratory and indirect approaches such as stable isotopic techniques. *Spartina* (detritus) may serve as a food source for the ribbed mussel *Geukensia demissa* (Kreeger et al. 1988, Lin 1989, Langdon and Newell 1990) and the oyster *Crassostrea virginica* (Crosby et al. 1989). Direct utilization of senescent sheaths and blades of *S. alterniflora* as an energy source has been observed for the talitrid amphipods *Orchestia grillus* (Lopez et al. 1977) and *Uhlorchestia spartinophila* (Kneib et al. 1997), and for the gastropod, *Littoraria irrorata* (Barlocher and Newell 1994a, b). Value of this food source appears related to stage of decomposition and level of fungal biomass. Schwinghammer and Kepkay (1987) however, found that additions of *S. alterniflora* detritus to microcosms had no effect on meiofaunal biomass. They suggested that any food benefits were countered by detrital inhibition of algal growth. Olivier et al. (1996, 1997) found that juveniles of *Nereis virens* and *N. diversicolor* exhibited higher rates of assimilation and shorter digestion times when fed macroalgae than when fed vascular plants (including *S. alterniflora*, *Zostera marina*; *S. anglica* and *Salicornia europea*).

For many invertebrate species, microheterotrophic decomposition of marsh plants by bacteria, protists, or fungus enhances nutritional value. *Geukensia demissa* (Kreeger and Newell 1996) and *C. virginica* (Crosby et al. 1990), for example, benefit far more from detritus-associated microbes than from the detritus itself.

It was early interest in *Spartina* fates that stimulated development of stable isotopic methods to evaluate food chains. The approach was based on the fact that C, N and S stable isotopic signatures are passed from primary producer to consumer tissues with little modification. Although the early studies implicated *Spartina* as an important energy source for animals within salt marshes (e.g., Haines 1976, Haines and Montague 1979, Peterson et al. 1985, 1986, Peterson and Howarth 1987), the results were not definitive because primary producers such as algae were not included in the analyses. More recent studies (Langdon and Newell 1990, Currin et al. 1995, Page 1997, Deegan and Garritt 1997, Kwak and Zedler 1997) confirm that vascular plants such as *Spartina,* and to a lesser extent *Salicornia*, are used by some invertebrate species, but suggest that macro and microalgae may be of generally greater importance (Carman et al. 1997, Currin et al. unpublished data). Stable isotope studies in marshes of California (Page 1997, Currin et al. unpublished) and in Massachusetts (Peterson et al. 1985, Deegan and Garritt 1997) demonstrate that the spatial location of individuals within a marsh can affect the relative importance of vascular plants in their diets.

3.4 ABIOTIC INFLUENCE

A variety of abiotic factors exert major influence on the composition, distribution, and standing stock of invertebrates in salt marshes. These include organic matter within sediments, particle grain size, elevation, flow regime, and salinity and oxygen within porewaters. Table 6 summarizes investigations into the effects of these factors on marsh infauna.

3.4.1 Salinity

Salinity gradients are common within bays and salinity differences often occur among embayments. Large-scale shifts in estuarine communities are documented along salinity gradients, but few investigators have specifically addressed the influence of surfacewater or porewater salinity on salt marsh macrofauna (Table 6). Insects and oligochaetes dominate assemblages in the fresh to brackish heads of estuaries while polychaetes dominate those towards the marine (more saline) parts of the estuary (McLusky et al. 1993, Ysebaert et al. 1993). In salt marsh tidal pools there is a negative relationship between chironomid larval density and interstitial salinities, and a positive relationship between polychaete density and overlying water salinities (Ward and FitzGerald 1983). Nordby and Zedler (1991) examined channel infauna of two

TABLE 6. Summary of abiotic environmental factors reported to influence macrofaunal populations and communities in salt marshes. *S.a.* = *Spartina alterniflora*, *S. an.* = *Spartina anglica*, *S.f.* = *Spartina foliosa*, *S.c.* = *Spartina cynosuroides*, *J.r.* = *Juncus roemerianus*, *S.b.* = *Salicornia bigelovii*, *S.v.* = *Salicornia virginica*, *S.r.* = *Schoenoplectus robustus*, M= Mollusc

VEGETATION TYPE/ LOCATION	TAXON STUDIED	INFLUENCE ON FAUNA	REFERENCE
ENVIRONMENTAL AGENT: SALINITY			
S.f./ Tijuana Estuary, CA, USA	Macrofauna	Flooding lowered salinity and reduced abundances and species richness; population structure was skewed to young animals, dominance by species with early maturity. Bivalve size structure was sensitive.	Nordby & Zedler 1991
S.f. & *S.b.*/ Mission Bay, CA, USA (natural & 16-mo old restored)	Macrofauna	Salinity was positively associated with density of insects and negatively associated with density of polychaetes and peracarid crustaceans.	This study
S.v./ Los Penasquitos Lagoon, CA, USA	Macrofauna	Flooding and reduced salinity selected for species tolerant to salinity shock and low dissolved O_2 levels.	Nordby & Zedler 1991
ENVIRONMENTAL AGENT: ELEVATION			
S.a./ North Carolina, USA	*Geukensia demissa*	Mortality of mussels increased with decreased tidal elevation due to increased exposure to predators that feed while inundated (e.g., crabs).	Lin 1989

TABLE 6 (Continued)

VEGETATION TYPE/ LOCATION	TAXON STUDIED	INFLUENCE ON FAUNA	REFERENCE
S.a./ North Carolina, USA	*Uca* spp.	Fiddler crab recruit and adult distribution varies with elevation.	O'Connor 1993
S.a./ North Carolina, USA Created Marsh	Macrofauna	More rapid recovery of macrofauna at lower elevations (0-4 yrs) in a created marsh.	Levin et al. 1996
S.a./ North Carolina, USA Created Marsh	Macrofauna	Positive correlation of *Capitella* & oligochaete densities with elevation in a created marsh but negative correlation of oligochaete, *Tharyx*, *Gemma gemma* and ostracod densities in a natural marsh.	Toomey 1997
S.a./ North Carolina, USA Created & Natural marsh	Macrofauna	Macrofaunal and *Streblospio benedicti* densities increased with decreasing elevations. Oligochaete & *Capitella* densities were highest at low & mid elevation.	Moy & Levin 1991
S.a./ Texas, USA	Natant macrofauna	Higher daggerblade grass shrimp & brown shrimp densities at low elevations with 24 cm but not 14 cm difference.	Minello et al. 1994
S.a./ SE Brazil	Macrofauna	Species composition and abundance changes with elevation, partly related to plant cover.	Netto & Lana 1997
S.f./ Mission Bay, CA, USA (natural & 16-mo old restored)	Macrofauna	Higher densities of total macrofauna and polychaetes at lower elevation.	This study

TABLE 6 (Continued)

VEGETATION TYPE/ LOCATION	TAXON STUDIED	INFLUENCE ON FAUNA	REFERENCE
ENVIRONMENTAL AGENT: OXYGEN			
S.a./ Louisiana, USA	Nematode	Nematode densities decline with depth in anaerobic sediments.	Sikora & Sikora 1982
S.a./ Lab, USA	Macrofauna	*Streblospio benedicti* and *Capitella sp.* I exhibited differences in sensitivity of growth rates to low oxygen concentrations associated with organic amendments.	Bridges et al. 1994 Levin et al. 1996b
ENVIRONMENTAL AGENT: HYDRODYNAMICS			
S.a./ Lab, USA	Meiofauna	*Spartina* culms (structure) interact with flow to influence resuspension of copepods. Nematodes unaffected.	Palmer 1986
S.a./ North Carolina, USA	Meiofauna	Use of burrow mimics and natural burrows indicate meiofauna are passively entrained in fiddler crab burrows.	DePatra & Levin 1989
S.a. & Mudflat/ South Carolina, USA	Meiofauna	Tidal currents may resuspend salt marsh meiofauna, yielding lower infaunal density at ebbing tide than low tide.	Palmer & Brandt 1981 Palmer & Gust 1985
S.a. & Mudflat/ Louisiana, USA	Meiofauna	Aggregations of harpacticoid copepods occur in topographic depressions, at least in part due to reduced shear stress and passive deposition.	Sun et al. 1993 Sun & Fleeger 1994
S.a. & Mudflat/ Louisiana, USA	Macrofauna	Epibenthic species are more heavily influenced by tidal flows than burrowing species.	Sun & Fleeger 1994

TABLE 6 (Continued)

VEGETATION TYPE/ LOCATION	TAXON STUDIED	INFLUENCE ON FAUNA	REFERENCE
S.a./ Louisiana, USA	Meiofauna	Sediment densities of nematodes, copepod nauplii and 3 copepod species were reduced during ebbing tidal flow relative to low tide.	Fleeger et al. 1984
ENVIRONMENTAL AGENT: SOIL GRAIN SIZE			
S.a./ North Carolina, USA	Macrofauna	Positive association of *Capitella* with % sand Positive association of *S. benedicti* with % silt.	Toomey 1997
S.a./ Washington, USA	Macrofauna	Increasing grain size is correlated with greater species richness, *Capitella* and dipteran larval density.	Zipperer 1996
S.v. & *S.b.*/ So. California Bays, USA (natural & 5-10-yr old restored)	Macrofauna	Percent sand content was positively associated with density of tubificid oligochaetes.	This study
S.f./ So. California Bays, USA	Macrofauna	59% of variance in assemblage composition & density explained largely by % sand & organic matter (1st principal axis in CCA).	Levin et al. 1998
S.f., *S.v.* & *S.b.*/ Mission Bay, USA (natural & 16-mo old restored)	Macrofauna	Densities of total macrofauna and enchytraeid oligochaetes are negatively associated with sand in the *S.f.* zone and positively associated with sand in the *Salicornia* zone. Insects in the *S.f.* zone, and nemerteans & turbellarians in the *Salicornia* zone are negatively associated with sand.	This study
S.f./ SE Brazil	*Lucina pectinata*	Positively associated with finer grained, organic-rich sediments.	Lana & Guiss 1991

TABLE 6 (Continued)

VEGETATION TYPE/ LOCATION	TAXON STUDIED	INFLUENCE ON FAUNA	REFERENCE
ENVIRONMENTAL AGENT: SOIL ORGANIC MATTER			
S.a./ Massachusetts, USA Channels	Macrofauna	Sewage sludge fertilization led to 3x increase in 2° production, increases in density of *Capitella, Paranais litoralis, Streblospio benedicti* and other species. Effects were largest in the Spring.	Sarda et al. 1992 & 1996
S.a./ North Carolina, USA	Oligochaetes Crustaceans Insects	Marsh receiving sewage effluent contained higher densities of *Monopylephorus rubroniveus, Talorchestia longicornis* and *Sminthurides aquatica* var. *levanderi* than a nearby marsh that did not.	McMahan 1972
S.a./ North Carolina, USA	Macrofauna	In a created marsh: Negative associations of oligochaete density w/ sediment TOC and OM content. Negative associations of amphipod density w/ water soluble organic C & TOC. In a natural marsh: Positive associations of *S. benedicti* and ostracod density with sediment TOC, of *G. gemma* & gastropod density with sediment TOM; negative correlations of *Tharyx* density with WSOC.	Toomey 1997
S.a./ North Carolina, USA	Macrofauna	Initial amendments with peat, alfalfa & straw inhibited community development for 1st 6 months but had little effect later.	Levin et al. 1997b

TABLE 6 (Continued)

VEGETATION TYPE/ LOCATION	TAXON STUDIED	INFLUENCE ON FAUNA	REFERENCE
S.a./ SE Brazil	Macrofauna	*Lucina pectinata* positively associated with sediment organic content.	Lana & Guiss 1991
S.f./ So. California Bays, USA	Macrofauna	Organic content is main component of 1st principal axis explaining community variation via CCA.	Levin et al. 1998
S.f. & *S.b.*/ Mission Bay, USA (natural & 16-mo old restored)	Macrofauna	Density of macrofauna was higher in a planted, sewage-amended than planted, control treatment. Organic matter positively associated with density of tubificid and enchytraeid oligochaetes, and insects; negatively associated with density of naidid oligochaetes.	This study
S.v. & *S.b.*/ Mission Bay, USA (natural & 16-mo old restored)	Macrofauna	Sediment organic matter positively associated with densities of macrofauna, tubificid and enchytraeid oligochaetes and polychaetes.	This study
S.v. & *S.b.*/ So. California Bays, USA (natural & 5-10-yr old restored)	Macrofauna	Sediment organic matter positively associated with densities of naidid and enchytraeid oligochaetes, negatively associated with densities of insects and molluscs.	This study

southern California bays subject to hypersalinity from mouth closure, followed by flooding. They found that reduced salinities led to decreased abundances and species richness in the Tijuana R. Estuary, with population structure skewed to young animals, and dominance by species with early maturity. Bivalve species composition was especially sensitive. In Los Peñasquitos Lagoon, species present were more tolerant to salinity shock and low dissolved oxygen levels. In the *S. foliosa* marsh in Mission Bay, soil salinity was positively correlated with insect density, but negatively correlated with densities of polychaetes and peracarid crustaceans (Table 4). The low, more moist elevations had salinities similar to seawater while the upper, fairly dry elevations reached salinities of up to 80. In *Salicornia* marshes of other southern California bays, salinity was positively associated with densities of macrofauna and enchytraeid oligochaetes, but negatively associated with polychaete and mollusc density (Table 5).

Spatial and temporal fluctuations in salinity tend to be greater in restored compared to natural marshes, probably due to the lack of ground cover (vascular plants and algae) which may reduce evaporation of water from and penetration of rainwater into the sediments. In southern California, salinity fluctuations are largest in lagoons with restricted or intermittent tidal flushing (Levin et al. unpublished data). Such trends may be more pronounced in arid marshes where freshwater inputs are highly seasonal. Although marsh taxa are often exposed to wide fluctuations in salinity, physiological tolerances at the lowest and highest ends of the salinity spectrum may regulate community composition and diversity.

3.4.2 Elevation

Elevation exerts a strong influence on macrofaunal density and species composition in *Spartina* marshes (Table 6). Significant shifts in species composition and density of invertebrates occurred along an elevation transect which spanned marsh and tidal flat habitat in Paranagua Bay, Brazil (Netto and Lana 1997b). Densities of total macrofauna and *S. benedicti* increased with decreasing tidal height in a study of three elevation levels within a natural and created marsh in North Carolina (Moy and Levin 1991). There were fewer oligochaetes, *Capitella* spp. and meiofauna at the uppermost elevation in these *S. alterniflora* marshes. Macrofauna recovered more rapidly at lower than higher elevations during the first four years in another North Carolina created marsh (Levin et al. 1996). Minello et al. (1994) reported higher densities of daggerblade grass shrimp and brown shrimp at low elevations in a Galveston Bay marsh, when elevation differences were 24 cm, but not when they were 14 cm. Despite the potential benefits of living at lower tidal elevations (decreased desiccation stress and increased feeding time), mortality of the mussel *G. demissa* was greater at lower tidal elevations due to increased exposure to predators that feed while inundated (e.g., crabs) (Lin 1989). In studies of three elevation zones in the Tijuana River Estuary, Levin et al. (1997a) reported a general trend towards decreasing importance of polychaetes and increasing importance of oligochaetes and insects as one moves from low (unvegetated) to higher tidal levels (*S. foliosa* then *S. virginica*) within the estuarine system. Within the created *S. foliosa* marsh in Mission Bay, California, we found an inverse relationship between elevation and total macrofaunal or polychaete densities (Table 4). Elevation-related zonation of fiddler crabs (*Uca* spp.) occurs in the Gulf of Mexico (Mouton and Felder 1996) and Atlantic Ocean

(O'Connor 1993). *Uca longisignalis* increases burrow depth with distance from the shoreline into the marsh (increasing elevation) (Mouton and Felder 1996).

Elevation gradients are associated with changes in evaporation, soil salinity, and plant cover, as well as with inundation period, which for many taxa indicates feeding time and susceptibility to aquatic and aerial predators. Although we suspect physiological salinity and temperature tolerances are involved in many of the elevation effects reported above, definitive tests have yet to be conducted.

3.4.3 Oxygen

Oxygen availability can affect macrofauna on both small (cm to m) and large (10s of m to km) scales. Areas of salt marsh which are well flushed by tides (e.g., along marsh edges) may have higher soil oxygen concentrations and lower concentrations of reduced compounds (e.g., sulfides) than areas which are not well flushed (Steever et al. 1976, Odum 1980). On a smaller scale, sediments may be oxygenated by animal burrowing activities (Montague 1982, Bertness 1985) or plant roots and rhizomes (Moorehead and Reddy 1988). Organic enrichment on large (e.g., pollution), medium (e.g., algal blooms, soil amendment treatments) and small (e.g., decaying plant material) spatial scales also may reduce oxygen availability to infauna due to increased microbial activity. Macrofaunal abundances and species richness are reduced in anoxic relative to oxygenated areas. Nilsson and Rosenberg (1994) demonstrated experimentally that numbers of species and macrofaunal abundance in soft sediments declined when exposed to severe and moderate hypoxic conditions in the overlying water (0.5-1.0 mg O_2/l) compared with normoxic conditions (>8.0 mg O_2/l). Two opportunistic salt marsh polychaetes, *S. benedicti* and *Capitella* sp. I, exhibited differences in sensitivity of individual and population growth rates to low oxygen concentrations associated with experimental organic amendments (Bridges et al. 1994, Levin et al. 1996b). Similar low oxygen concentrations have been shown to affect the feeding behavior of *S. benedicti*. LT_{50} under anoxic conditions for *S. benedicti* was 1.8 days (Llanso 1991); LT_{50} for *Capitella* sp. was seven days (Warren 1977).

Oxygen is known to penetrate only the first few millimeters of flooded wetland sediment surface (Patrick and Delaune 1972, Gambrell and Patrick 1978), except in crab burrows. This may explain the near-surface distributions of most macrofauna within salt marshes. In most salt marshes, infaunal density and biomass are concentrated in the upper sediments (McCann and Levin 1989, Palacio et al. 1991, Levin et al. 1998). The depth at which the macrofauna are found declines with increased organic enrichment in intertidal sandy sediments. Nematode densities and redox values decreased with increased depth in the anaerobic sediments of a Louisiana salt marsh (Sikora and Sikora 1982). Nematodes are adapted to anaerobic conditions and may be an important link between anaerobic and aerobic energy cycles. Although oxygen availability within sediments appears to influence the distribution and growth of infaunal species, there are no definitive studies examining the extent to which oxygen availability affects species composition (presence) or density in salt marshes. However, recent research by Woodin et al. (1998, person. commun.) indicates that sulfide, ammonium and oxygen concentrations in porewaters of near surface sediments influence habitat selection by settling larvae.

3.4.4 Hydrodynamics

Hydrodynamic factors such as drainage pattern, flow speed, and turbulence intensity are strongly influence by the presence and density of vascular plant vegetation in wetlands (Fonseca et al. 1982, Pethick et al. 1990, Leonard and Luther 1995). Flow energy may be an order of magnitude lower on the vegetated marsh surface than in channels or tidal flats; turbulent energy decreases exponentially with increasing distance from the creek edge (Leonard and Luther 1995). These flow parameters control resuspension, transport and deposition of sediment particles, detritus, larvae, and small benthic invertebrates (e.g., Eckman 1983, 1990). They also relate directly to variations in sediment particle size, organic matter content, geochemical properties, stability or vegetation patterns (Collins et al. 1987, Leopold et al. 1993, Streever and Genders 1997), which are shown elsewhere in this chapter to influence benthic marsh invertebrates.

Tidal currents may resuspend salt marsh meiofauna (Palmer and Gust 1985) yielding lower infaunal densities at ebbing tide than low tide in *S. alterniflora* marshes (Palmer and Brandt 1981, Fleeger et al. 1984). Meiofaunal susceptibility to hydrodynamic effects differs with functional group. Epibenthic species are heavily influenced by tidal flows; burrowing species much less so (Sun and Fleeger 1994). Aggregations of meiofauna, especially harpacticoid copepods, occur in topographic depressions on the marsh surface, at least in part due to reduced shear stress and passive deposition (DePatra and Levin 1989, Sun et al. 1993).

Although flow regime is known to affect the flux of food particles and larvae of macrofauna and megafauna in seagrass beds (Peterson et al. 1984, Thistle et al. 1984, Eckman 1987, 1990) and tidal flats (Eckman 1983), there are only a few studies of animal-flow interactions in salt marshes. As found on small scales for animal tubes, seagrass shoots or mimics, vegetation-induced variation in bottom stress should influence faunal feeding environments, chemical cues associated with larval settlement or prey detection (Weissburg and Zimmer-Faust 1991, 1993, 1994, Moore et al. 1994, Turner et al. 1994), microbial and algal activity (Thistle et al. 1984, Eckman et al. 1985) and substrate stability (Eckman and Nowell 1984). On larger scales, drainage patterns influence the chemistry of marsh sediments (Portnoy and Giblin 1997), including acidity, redox and metal mobility. These factors will in turn affect the composition and density of marsh invertebrates. Episodic storm events can remove marsh substrate, uproot plants and animals, or bury them under excessive sediment deposits.

3.4.5 Organic Matter and Grain Size

Food availability is widely recognized as a major structuring agent of infaunal communities (Pearson and Rosenberg 1978, 1987). Organic matter within sediments may act as a direct food source or may stimulate microbial production, which is then consumed by infauna (Fenchel 1970, Lopez and Levinton 1978). Organic matter and its lability also affect chemical oxidation gradients and the environment experienced by marsh infauna. Sediment organic content is negatively correlated with grain size; finer-grained sediments typically contain more organic matter than coarser sediments due to increased surface area and a hydrodynamic regime that promotes deposition.

Sediment grain size is believed to regulate the distribution of 3 *Uca* species in the NW Atlantic; *Uca* mouthparts appear adapted for different particle sizes (Miller 1961). Mesocosm experiments, in which soil clay content was varied, revealed distinct preferences by the fiddler crabs *Uca spinicarpa* and *Uca longisignalis* for high and low clay content, respectively (Mouton and Felder 1996). However, Ringold (1979) found particle size was uncorrelated with crab distributions in Bell Creek Marsh in North Carolina.

Differences in sediment organic matter content and grain size, and presumably food availability for macrofauna, are commonly observed between created salt marshes and their nearby natural counterparts (e.g., Lindau and Hossner 1981, Shisler and Charette 1984, Craft et al. 1988, Langis et al. 1991, Moy and Levin 1991, Levin et al. 1996, Scatolini and Zeddler 1996, Talley and Levin 1999). A number of studies which compared macrofaunal assemblages in restored and natural marshes have implicated reduced organic matter content of sediments and/or coarse grain size as contributing to reduced densities of total macrofauna or selected taxa in the created systems (e.g., Moy and Levin 1991, Sacco et al. 1994, Levin et al. 1996). However, these conclusions were not based on experimental analyses or even direct correlations between faunal and sediment properties in discreet samples.

Several investigators have carried out correlative or multivariate studies of organic matter and particle size relationships to macrofaunal parameters across stations, marsh microhabitats or treatments (created vs. natural, amended vs. unamended) (Table 6). Lana and Guiss (1991) reported positive relationships of macrofaunal abundance with increased organic content and smaller grain size in Brazilian salt marshes, and cited these factors as among the most important structuring agents. Canonical correspondence analyses of macrofaunal assemblages in *S. foliosa* marshes of southern California yielded a principal axis based primarily on organic matter and percent sand content that explained 59% of the variance in composition and abundance (Levin et al. 1998). Within patches of invasive *S. alterniflora* on a northwest Pacific tidal flat, Zipperer (1996) found positive correlations of grain size with macrofaunal species richness and with densities of *Capitella* and dipteran larvae. In each case grain size accounted for about 50% of the faunal variation.

Toomey (1997) found that density of the polychaete *Capitella* spp. was positively associated with percent silt (particles 2-50 μm) in a created North Carolina, *S. alterniflora* marsh but with percent sand (particles 50 μm-2 mm) in the natural system, where grain size was much finer. *Streblospio benedicti* densities were negatively correlated with percent silt in the natural marsh. An extensive evaluation of different forms of organic matter, and their relationships to macroinvertebrate abundances was conducted in this *S. alterniflora* system (Toomey 1997). Within the created marsh there were negative associations of oligochaete density with total carbon and total organic matter, of amphipod density with water soluble organic C and total organic carbon, and of density of the bivalve *Gemma gemma* with water soluble organic carbon. Within the natural marsh there were mainly positive associations of macrofaunal density with organic matter. *Streblospio benedicti* and ostracod densities increased with organic carbon content; *G. gemma* and gastropod densities increased with total organic matter content. The processes underlying the association of different invertebrate taxa with varying forms of organic matter are unknown, but merit investigation.

The associations of vegetation and soil parameters with macrofauna in *Salicornia* and *S. foliosa* marshes of southern California were examined by multiple regression across natural and restored systems (Tables 4 and 5) and within systems (Talley and Levin 1999). Both studies demonstrated a widespread positive relationship between sediment organic matter and macrofaunal density. In Mission Bay, organic content of sediments in *Salicornia* habitat (both created and natural) was positively correlated with densities of macrofauna, oligochaetes and polychaetes; densities of oligochaetes and insects were positively correlated with percent organic matter content in the *Spartina* habitat (both created and natural) (Table 4). Species richness and all other taxa exhibited no relationship to organic matter content. In *Salicornia* habitat (created and natural) in other southern California bays, % soil organic content was positively correlated with numbers of naidid and enchytraeid oligochaetes but was negatively correlated with numbers of insects and molluscs (Table 5). When natural and created *Salicornia* marshes were examined separately, positive relationships between sediment organic content and macrofaunal species richness, total density, and densities of most of the major taxa were evident in both marsh types (Talley and Levin 1999). However, in the natural marshes the faunal relationships were with combustible organic matter, whereas in the restored marshes these relationships were observed primarily with below-ground biomass (a measure of large organic particles including roots and detritus) (Talley and Levin 1999).

Grain size effects were relatively minor for these habitats. In Mission Bay an increasing proportion of sand had a positive influence on faunal densities in the *Salicornia* zone and a negative influence on faunal abundance in the *Spartina* zone (Table 4). Benthic microalgal biomass in surface sediments (measured as chlorophyll a concentration) was positively correlated with oligochaete density and negatively correlated with peracarid crustacean density in the *Spartina* zone of Mission Bay (Table 4). In the *Salicornia* zone, chlorophyll a was positively correlated with insect densities (Table 4).

Several programs have artificially amended sediments with different forms of organic matter and examined infaunal responses (Table 6). A comparison of two *S. alterniflora* marshes by Mc Mahan (1972) revealed a sewage-exposed marsh to have higher densities of the oligochaete *Monopylephorus rubroniveus* and the amphipod *Talorchestia longicornis*, and lower densities of the collembolid *Sminthurides aquatica* var. *levanderi* than an unexposed marsh. Long-term fertilization of creekbed sediments with sewage-sludge based fertilizer within the Great Sippewissett *S. alterniflora* marsh in Massachusetts led to a 40% increase in sediment organic matter content and a lowering of the C:N ratio (Sarda et al. 1992). There were corresponding increases in density and secondary production of numerous invertebrates, including the polychaetes *Capitella* sp., *S. benedicti, Polydora cornuta*, and *Amphicteis gunneri*, the coelenterate *Nematostella victensis*, and the oligochaetes *Paranais litoralis* and *Monopylephorus evertus* (the dominants in the enriched area) (Sarda et al. 1996). Seasonal variations in population responses led the authors to hypothesize that food availability influences selected infaunal populations during spring and early summer, but that later in the year, when predators are abundant, there is top-down control of macrofauna. Amendments of peat, alfalfa, and straw to created *S. alterniflora* marsh sediments in the Newport R., North Carolina, led to initial inhibition of macrofauna, probably due to development of anoxic conditions. However, after 6 months, assemblages in amended sediments did not differ from those in unamended sediments (Levin et al. 1997b).

A one-time amendment of milorganite to created marsh sediments in Mission Bay, CA led to increased macrofaunal densities 9 and 16 months later, but no significant change in composition or diversity of the assemblage (Levin et al., unpublished data). Notably elevated $\delta^{15}N$ signatures of oligochaetes and insects from these treatments relative to surrounding sediments suggests that the milorganite was directly ingested or that the amended N was incorporated into invertebrates via bacteria (Levin et al. unpublished data). Milorganite, in addition to being rich in N is also rich in heavy metals. Addition of essential metals to created-marsh sediments may have enhanced infaunal recovery.

The extent to which soil organic matter promotes or inhibits invertebrate assemblages will depend on its lability, size, chemical structure, and absolute concentration. We hypothesize that when soil organics are at low levels (as in many created marsh sediments), additions may be beneficial. However, at high concentrations, organic matter promotes loss of oxygen and a reduced chemical environment that may inhibit infauna.

3.5 FAUNAL BIOGENIC STRUCTURES

Structures made by animals, in particular tubes, semi-permanent burrows, shells and pseudofeces, are frequently present within marsh sediments. Such structures are known to alter sediment chemistry, particularly oxidation- reduction processes (Aller 1982), and metal accumulation (Doyle and Otte 1997), and to modify flow in the boundary layer (Nowell and Jumars 1984). Sometimes these structures form mechanical barriers, inhibiting burrowing (Woodin 1976, Brenchley 1982) or predator access to prey (Lee and Kneib 1994). Although effects of biogenic structures are well documented in unvegetated sediments in both shallow (Reise 1985) and deep water (Levin 1991), comparable investigations in salt marshes are more limited.

Much of the focus in salt marshes has been on effects of plant culms (discussed earlier) or crab burrows. Fiddler crab burrows were observed by Bell et al. (1978) to enhance nematode densities but inhibit copepods in a *S. alterniflora* marsh. The deep distribution of some foraminifera, to 30 cm, was attributed to the burrowing of crabs in a Georgia salt marsh (Goldstein et al. 1995). Observations and experiments conducted by DePatra and Levin (1989) suggested that increased meiofaunal densities within fiddler crab burrows in *S. alterniflora* marshes may be in part a passive process associated with entrainment of tidally suspended animals. However, active habitat selection may be involved as well. Burrow linings provide a better oxygenated, more moist, finer-grained habitat, as well as refuge from predators (Reise 1981, 1985, DePatra and Levin 1989).

Experimentally elevated densities of the tube-builder *Manayunkia aestuarina* were shown to inhibit some nematode and copepod species in a South Carolina *S. alterniflora* marsh, although the tubes had a positive effect on oncholaimid nematodes (Bell 1983). Often the distribution, size structure, or sex ratios of prey species can be affected by association with protective biogenic structures (Lee and Kneib 1994). Oyster shells, which are common in some low *S. alterniflora* marshes, provide substrate for some species and refuge for taxa such as mussels, from decapod predators (Lee and Kneib 1994). Many of these interactions, and the positive plant-animal interactions reviewed in Table 3, typify a class of biotic effects termed facilitation. Facilitative interactions among marsh invertebrates, especially those mediated by biogenic structures, are likely to be as ecologically significant as those involving marsh plants (Montague 1982, Bertness 1984, 1985, Bertness and Callaway 1994).

4. Conclusions

Several decades of research indicate that numerous environmental factors influence the density, distribution, composition and diversity of marsh invertebrates. The influence of environmental factors appears to vary with marsh system, factor intensity (effect on presence or abundance), taxon studied, and with other interacting factors. However, it is likely that the relative importance of environmental parameters is hierarchical, with certain factors acting over larger scales or with greater intensity than others. In Fig. 1 we propose a hypothetical scheme in which abiotic properties such as marsh age, elevation and salinity are most likely to influence the presence or absence of species over large space and time scales. Flow regime, oxygen concentrations, soil properties such as organic content and particle size, as well as the presence or type of vegetation are hypothesized to act on more moderate spatial scales, with effects on both species composition and abundance. Factors such as above and below-ground plant biomass, and the presence or density of culms and faunally generated biogenic structures contribute to small-scale patchiness in species abundance patterns, but in most cases do not appear to determine the presence or absence of species. Pearson and Rosenberg (1987) presented a similar organizational scheme for soft-bottom macrofauna, in which physical factors have spatially greater influence over evolutionary time scales, and biotic processes create patchiness over shorter periods on spatial scales of mm to 10s of cm. Resolution of the complex interactions among abiotic and biotic factors in salt marshes, and of the scales on which they act, should improve our understanding of invertebrate communities and ultimately aid the conservation and restoration of salt marsh ecosystems.

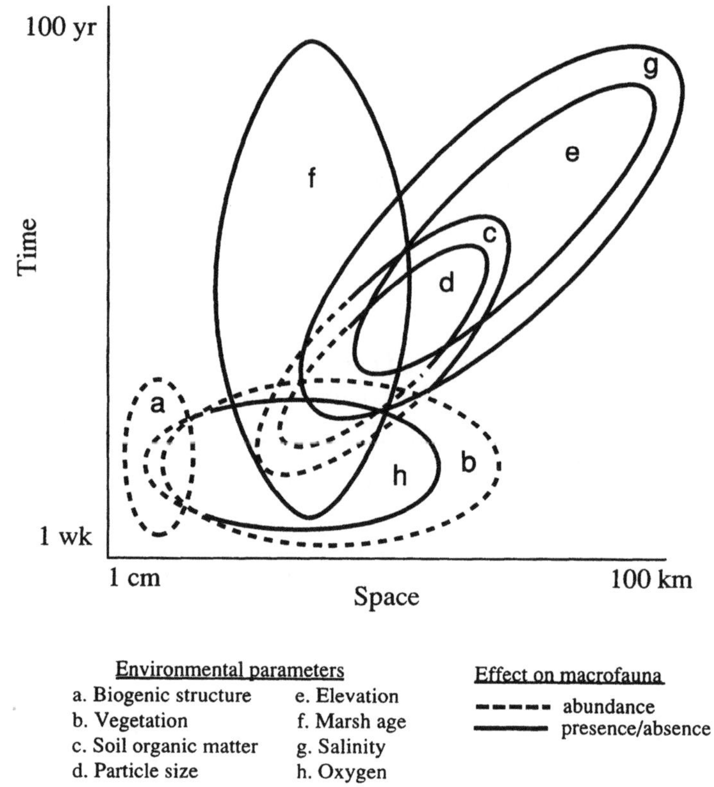

Figure 1. Schematic diagram of the heirarchy of environmental parameters acting on marsh macrofaunal presence or abundances over a range of spatial and temporal scales.

5. ACKNOWLEDGEMENTS

We thank the many people who have aided us with the southern California field sampling and sample processing, including D. Talley, A. Larson, D. James, J. Bernd, L. McConnico, L. Foote, A. McCray, A. Robles, A. Jones, C. Martin, M. Sigala, M. Tambakuchi, P. Alsop, M. Woo, K. Stern, S. Ross, K. Stanfield, D. Hennan, J. Ellis, C. Currin and J. Scope. We thank C. Martin for assistance with polychaete identifications, M. Milligan and B. Healy for help with oligochaete identifications, B. Isham for help with insect identifications, and L. McConnico for assistance with manuscript preparation. Helpful comments on the manuscript were provided by D. Talley, J. Crooks, D. Kreeger, and three anonymous reviewers. The research was funded by grants from the Ellen Browning Scripps Foundation and the National Oceanic and Atmospheric Administration: Sanctuaries and Reserves Division (NA670R0237), California Sea Grant College System and the California State Resources Agency (NA36RG0537 R/ CZ-125 and NA66RG0477 R/CZ-140). The views expressed herein are those of the authors and do not necessarily reflect the views of NOAA or any of its subagencies.

6. Literature Cited

Adam, P. 1990. Saltmarsh Ecology. Cambridge University Press, Cambridge, England.

Aller, R.C. 1982. The effects of macrobenthos on chemical properties of marine sediment and overlying water. Pages 53-102 *in* P.L. McCall and M.J.S. Tevesz, editors. Animal-sediment relations. Plenum Press, New York, New York, USA.

Barlocher, F. and S.Y. Newell 1994*a*. Phenolics and proteins affecting palatability of *Spartina* leaves to the gastropod *Littoraria irrorata*. Marine Ecology **15**: 65-75.

——. 1994*b*. Growth of salt marsh periwinkle *Littoraria irrorata* on fungal and cordgrass diets. Marine Biology **118**: 109-114.

Bell, S.S. 1983. An experimental study of the relationship between below-ground structure and meiofaunal taxa. Marine Biology **76**: 33-39.

Bell, S.S., M.C. Watzin and B.C. Coull. 1978. Biogenic structure and its effects on the spatial heterogeneity of meiofauna in a salt marsh. Journal of Experimental Marine Biology and Ecology **35**: 99-107.

Bertness, M.D. 1984. Ribbed mussels and *Spartina alterniflora* production in a New England salt marsh. Ecology **65**: 1794-1807.

——. 1985. Fiddler crab regulation of *Spartina alterniflora* production on a New England saltmarsh. Ecology **66**: 1042-1055.

——. 1988. Peat accumulation and the success of marsh plants. Ecology **69**: 703-713.

Bertness, M.D. and R. Callaway. 1994. Positive interactions in communities. Trends in Ecology and Evolution **9**: 191-193.

Bertness, M.D. and E. Grosholz. 1985. Population dynamics of the ribbed mussel, *Geukensia demissa*: the costs and benefits of an aggregated distribution. Oecologia **67**: 192-204.

Bertness, M.D. and S.D. Hacker. 1994. Physical stress and positive associations among marsh plants. American Naturalist **144**: 363-372.

Bertness, M.D. and T. Miller. 1984. The distribution dynamics of *Uca pugnax* (Smith) burrows in a New England salt marsh. Journal of Experimental Marine Biology and Ecology **83**: 211-237.

Bonnet, B.R.P., P.C. Lana and C. Guiss. 1994. Influência de gramínea *Spartina alterniflora* sobre a distribuição e densidade de *Neritina virginea* (Gastropoda: Neritidae) em Marismas da Baia de Paranaguá (Paraná, Brasil). Neritica Curitiba **8**: 99-108.

Brenchley, G.A. 1982. Mechanisms of spatial competition in marine soft-bottom communities. Journal of Experimental Marine Biology and Ecology **60**: 17-33.

Bridges, T.S., L.A. Levin, D. Cabrera and G. Plaia. 1994. Effects of sediment amended with sewage, algae, or hydrocarbons on growth and reproduction in two opportunistic polychaetes. Journal of Experimental Marine Biology and Ecology **177**: 99-119.

Callaway, J.C. and M.N. Josselyn. 1992. The introduction and spread of smooth cordgrass (*Spartina alterniflora*) in southern San Francisco Bay. Estuaries **15**: 218-26.

Cammen, L.M. 1976a. Abundance and production of macroinvertebrates from natural and artificially established salt marshes in North Carolina. American Midland Naturalist **96**: 244-253.

——. 1976b. Macroinvertebrate colonization of *Spartina* marshes artificially established on dredge spoil. Estuarine, Coastal and Marine Science **4**: 357-372.

Cammen, L.M., U. Blum, E.D. Seneca and L.M. Stroud. 1982. Energy flow in a North Carolina salt marsh: a synthesis of experimental and published data. ASB Bulletin **29**: 111-134.

Capehart, A.A. and C. Hackney. 1989. The potential role of roots and rhizomes in structuring salt marsh benthic communities. Estuaries **12**: 119-122.

Carman, K.R., J.W. Fleeger and S. Pomarico. 1997. Response of a benthic food web to hydrocarbon contamination. Limnology and Oceanography **42**: 561-571.

Collins, L.M., J.N. Collins and L.B. Leopold. 1987. Geomorphic processes of an estuarine marsh: preliminary results and hypotheses. International Geomorphology Part I: 1049-1072.

Craft, C.B., S.W. Broome and E.D. Seneca. 1988. Nitrogen, phosphorus and organic carbon pools in natural and transplanted marsh soils. Estuaries **11**: 272-280.

Crosby, M.P., C.J. Langdon and R.I.E. Newell. 1989. Importance of refractory plant material to the carbon budget of the oyster *Crassostrea virginica*. Marine Biology, Berlin **100**: 343-352.

Crosby, M.P., R.I.E. Newell and C.J. Langdon. 1990. Bacterial mediation in the utilization of carbon and nitrogen form detrital complexes by *Crassostrea virginica*. Limnology and Oceanography **35**: 625-639.

Currin, C.A., S.Y. Newell and H.W. Paerl. 1995. The role of benthic microalgae and standing dead *Spartina alterniflora* in salt marsh food webs: implications based on multiple stable isotope analysis. Marine Ecology Progress Series **121**: 99-116.

Deegan, L.A. and R.H. Garritt. 1997. Evidence for spatial variability in estuarine food webs. Marine Ecology Progress Series **147**: 31-47.

De la Cruz, A., C. T. Hackney and N. Bhardwaj. 1989. Temporal and spatial patterns of redox potential (Eh) in three tidalmarsh communities. Wetlands **9**: 181-190.

DePatra, K.D. and L.A. Levin. 1989. Evidence of the passive deposition of meiofauna into fiddler crab burrows. Journal of Experimental Marine Biology and Ecology **125**: 173-192.

De Szalay, F.A., D.P. Batzer and V.H. Resh. 1996. Mesocosm and macrocosm experiments to examine effects of mowing emergent vegetation on wetland invertebrates. Environmental Entomology **25**: 303-309.

De Szalay, F.A. and V.H. Resh. 1996. Spatial and temporal variability of trophic relationships among aquatic macroinvertebrates in a seasonal marsh. Wetlands **16**: 458-466.

Doyle, M.O. and M.L. Otte. 1997. Organism-induced accumulation of iron, zinc and arsenic in wetland soils. Environmental Pollution **96**: 1-11.

Eckman, J.E. 1983. Hydrodynamic processes affecting benthic recruitment. Limnology and Oceanography **28**: 241-257.

——. 1985. Flow disruption by an animal-tube mimic affects sediment bacterial colonization. Journal of Marine Research **43**: 419-435.

——. 1987. The role of hydrodynamics in recruitment, growth and survival of *Argopecten irradians* (L.) and *Anomia simplex* (D'Orbigny) within eelgrass meadows. Journal of Experimental Marine Biology and Ecology **106**: 165-191.

——. 1990. A model of passive settlement by planktonic larvae onto bottoms of differing roughness. Limnology and Oceanography **35**: 887-901.

Eckman, J.E. and A.R.M. Nowell. 1984. Boundary skin friction and sediment transport about an animal-tube mimic. Sedimentology **31**: 851-862.

Fenchal, T. 1970. Studies on the decomposition of organic detritus derived from turtle grass, *Thalassia testudinum*. Limnology and Oceanography **15**: 14-20.

Fleeger, J.W., G.T. Chandler, G.R. Fitzhugh and F.E. Phillips. 1984. Effects of tidal currents on meiofauna densities in vegetated salt marsh sediments. Marine Ecology Progress Series **19**: 49-53.

Fonseca, M.S., J.S. Fisher, J.C. Zieman and G.W. Thayer. 1982. Influence of the seagrass, *Zostera marina* L., on current flow. Estuarine Coastal and Shelf Science **15**: 351-364.

Frid, C. and R. James. 1989. The marine invertebrate fauna of a British coastal salt marsh. Holarctic Ecology **12**: 9-15.

Friedman, R.M. and C.B. DeWitt. 1978. Wetlands as carbon and nutrient reservoirs: a spatial, historical and societal perspective. Pages 175-185 *in* P.E. Greeson, J.R. Clark and J.E. Clark, editors. Wetland functions and values: the state of our understanding. American Water Resources Association, Minneapolis, Minnesota, USA.

Gambrell, R.P. and W.H. Patrick. 1978. Chemical and microbiological properties of anaerobic soils and sediments. Pages 375-423 *in* D.D. Hook and R.M.M. Crawford, editors. Plant life in anaerobic environment. Ann Arbor Scientific Publishing, Inc., Ann Arbor, Michigan, USA.

Goldstein, S.T., G.T. Watkins and R.M. Kuhn. 1995. Microhabitats of salt marsh foraminifera: St. Catherines Island, Georgia, USA. Marine Micropaleontology **26**: 17-29.

Haines, E.B. 1976. Relation between the stable carbon isotope composition of fiddler crabs, plants and soils in a salt marsh. Limnology and Oceanography **21**: 880-883.

Haines, E.B. and C.L. Montague. 1979. Food sources of estuarine invertebrates analyzed using $^{13}C/^{12}C$ ratios. Ecology **60**: 48-56.

Havens, K.J. 1995. An assessment of ecological conditions in a constructed tidal marsh and two natural reference tidal marshes in coastal Virginia. Ecological Engineering **4**: 117-141.

Healy, B. and K. Walters. 1994. Oligochaeta in *Spartina* stems: the microdistribution of Enchytraeidae and Tubificidae in a salt marsh, Sapelo Island, USA. Hydrobiologia **278**: 111-123.

Heck, K.L. and R.J. Orth. 1980. Seagrass habitats: the roles of habitat complexity, competition and predation in structuring associated fish and motile macroinvertebrate assemblages. Pages 449-464 *in* V.S. Kennedy, editor. Estuarine Perspectives. Academic Press, Inc. New York, New York, USA.

Howes, B.L., J.W.H. Dacey and D.D. Goehringer. 1986. Factors controlling the growth form of *Spartina alterniflora*: feedbacks between above-ground production, sediment oxidation, nitrogen and salinity. Journal of Ecology **74**: 881-898.

Jackson, D. 1985. Invertebrate populations associated with *Spartina anglica* salt-marsh and adjacent intertidal mud flats. Estuarine Brackishwater Scjence Association Bulletin **40**: 8-14.

Jackson, D., C.F. Mason and S.P. Long. 1985. Macro-invertebrate populations and production on a salt-marsh in east England dominated by *Spartina anglica.* Oecologia **65**: 406-411.

Kneib, R.T. 1984. Patterns of invertebrate distribution and abundance in the intertidal salt marsh: causes and questions. Estuaries **7**: 392-412.

Kneib, R.T., S.Y. Newell and E.T. Hermeno. 1997. Survival, growth and reproduction of the salt-marsh amphipod *Uhlorchestia spartinophila* reared on natural diets of senescent and dead *Spartina alterniflora* leaves. Marine Biology **128**: 423-431.

Kreeger, D.A., C.J. Langdon and R.I.E. Newell. 1988. Utilization of refractory cellulosic carbon derived from *Spartina alterniflora* by the ribbed mussel *Geukensia demissa.* Marine Ecology Progress Series **42**: 171-179.

Kreeger, D.A. and R.I.E. Newell. 1996. Ingestion and assimilation of carbon from cellulolytic bacteria and heterotrophic flagellates by the mussel *Geukensia demissa* and *Mytilus edulis* (Bivalvia, Mollusca). Aquatic Microbial Ecology **11**: 205-214.

Kwak, T.J. and J.B. Zedler. 1997. Food web analysis of southern California coastal wetlands using multiple stable isotopes. Oecologia **110**: 262-277

Lana, P.C., E.C.G. Couto and M.V.O. Almeida. 1997. Polychaete distribution and abundance in intertidal flats of Paranagua Bay (SE Brazil). Bulletin of Marine Science **60**: 433-442.

Lana, P. da Cunha and C. Guiss. 1991. Influence of *Spartina alterniflora* on structure and temporal variability of macrobenthic associations in a tidal flat of Paranagua Bay (southeastern Brazil). Marine Ecology Progress Series **73**: 231-244.

——. 1992. Macrofauna-plant-biomass interactions in a salt marsh in Paranagua Bay (SE Brazil). Marine Ecology Progress Series **80**: 57-64.

Langdon, C.J. and R.I.E. Newell. 1990. Utilization of detritus and bacteria as food sources by two bivalve suspension-feeders, the oyster *Crassostrea virginica* and the mussel *Geukensia demissa.* Marine Ecology Progress Series **58**: 299-310.

Langis, R., M. Zalejko and J.B. Zedler. 1991. Nitrogen assessments in a constructed and a natural salt marsh of San Diego Bay, California. Ecological Applications **1**: 40-51.

La Salle, M.W., M.C. Landin and J.G. Sims. 1991. Evaluation of the flora and fauna of a *Spartina alterniflora* marsh established on dredged material in Winyah Bay, South Carolina. Wetlands **11**: 191-209.

Lee, S.Y. and R.T. Kneib. 1994. Effects of biogenic structure on prey consumption by the xanthid crabs *Eurytium limosum* and *Panopeus herbstii* in a salt marsh. Marine Ecology Progress Series **104**: 39-47.

Leonard, L.A. and M.E. Luther. 1995. Flow dynamics in tidal marsh canopies. Limnology and Oceanography **40**: 1474-1484.

Leopold, L.B., J.N. Collins and L.M. Collins. 1993. Hydrology of some tidal channels in estuarine marshland near San Francisco. Catena **20**: 469-493.

Levin, L.A. 1991. Interactions between metazoans and large, agglutinating protozoans: Implications for the community structure of deep-sea benthos. American Zoologist **31**: 886-900.

Levin, L.A., D.M. Talley and G. Thayer. 1996. Macrobenthic succession in a created salt marsh. Marine Ecology Progress Series **141**: 67-82.

Levin, L.A., T.S. Talley, A.A. Larson and A. Jones 1997a. Faunal composition in the Tijuana River estuarine intertidal habitats and the role of life histories in faunal recovery of southern California restored wetlands. Final Report for the National Estuarine Research Reserve System, NOAA Award No. NA 670R0237, August. 1997. 41pp.

Levin, L.A., D. Talley, T. Talley, A. Larson, A. Jones, G. Thayer, C. Currin and C. Lund 1997b. Restoration of *Spartina* marsh function: an infaunal perspective. Pages 190-194 *in* K.B. Macdonald and F. Weinmann, editors. Wetland and riparian restoration: Taking a broader view. U.S. Environmental Protection Agency, Seattle, Washington, USA.

Levin, L.A., T.S. Talley and J. Hewitt. 1998. Macrobenthos of *Spartina foliosa* (Pacific cordgrass) salt marshes in southern California: Community structure and comparison to a Pacific mudflat and a *Spartina alterniflora* (Atlantic smooth cordgrass) marsh. Estuaries **21**: 129-144.

Lin, J. 1989. Importance of location in the salt marsh and clump size on growth of ribbed mussels. Journal of Experimental Marine Biology and Ecology **128**: 75-86.

Lindau, C.W. and L.R. Hossner. 1981. Substrate characterization of an experimental marsh and three natural marshes. Soil Science Society of America Journal **45**: 1171-1176.

Llanso, R.J. 1991. Tolerance of low dissolved oxygen and hydrogen sulfide by the polychaete *Streblospio benedicti* (Webster). Journal of Experimental Marine Biology and Ecology **153**: 165-178.

Lopez, G.R. and J.S. Levinton. 1978. Ecology of deposit feeding animals in marine sediment. Quarterly Review of Biology **62**: 235-259.

Lopez, G., J.S. Levinton and L.B. Slobodkin. 1977. The effect of grazing by the detritivore, *Orchestia grillus*, on *Spartina* litter and its associated microbial community. Oecologia **30**: 111-127.

Massey, B., R. Zembal and P.D. Jorgenson. 1984. Nesting habitat of the light-footed clapper rail in southern California. Journal of Field Ornithology **55**: 67-80.

McCann, L.D. and L.A. Levin. 1989. Oligochaete influence on settlement, growth and reproduction in a surface-deposit-feeding polychaete. Journal of Experimental Marine Biology and Ecology **131**: 233-253.

McLusky, D.S., S.C. Hull and M. Elliot. 1993. Variations in the intertidal and subtidal macrofauna and sediments along a salinity gradient in the upper Forth Estuary. Netherlands Journal of Aquatic Ecology **27**: 101-109.

McMahan, E. 1972. Relative abundance of three marsh-floor organisms in a sewage-affected marsh and a sewage-free marsh. Journal of the Elisha Mitchell Society **88**: 61-65.

Meixner, R. 1983. The introduced rice grass *Spartina townsendii* as a substrate for *Mytilus edulis*. Archiv fur Fischereiwissenschaft **33**: 151-159.

Miller, D.C. 1961. The feeding mechanisms of fiddler crabs with ecological considerations of feeding adaptations. Zoologica, N.Y. **46**: 89-100.

Minello, T.J., R.J. Zimmerman and R. Medina. 1994. The importance of edge for natant macrofauna in a created salt marsh. Wetlands **14**: 184-198.

Mitsch, W.J. and J.G. Gosselink. 1993. Wetlands. Second Edition. Van Nostrand and Reinhold, New York, USA. 722 pp.

Montague, C.L. 1982. The influence of fiddler crab burrows and burrowing on metabolic processes in salt marsh sediments. Pages 283-301 *in* V.S. Kennedy, editor. Estuarine Comparisons. Academic Press, New York, New York, USA.

Moore, P.A., M. Weissburg, J.M. Parrish, R.K. Zimmer-Faust and G.A. Gerhardt. 1994. Spatial distribution of odors in simulated benthic boundary layer flows. Journal of Chemical Ecology **20**: 255-279.

Moorehead, K.K. and K.R. Reddy. 1988. Oxygen transport through selected aquatic macrophytes. Journal of Environmental Quality **17**: 138-142.

Morris, J.T. 1989. Modeling light distribution within the canopy of the marsh grass *Spartina alterniflora* as a function of biomass and solar angle. Agriculture and Forest Methodology **46**: 349-361.

Mouton, E.C. and D.L. Felder. 1996. Burrow distributions and population estimates for the fiddler crabs *Uca spinicarpa* and *Uca longisignalis* in a Gulf of Mexico salt marsh. Estuaries **19**: 51-61.

Moy, L.D. and L.A. Levin. 1991. Are *Spartina* marshes a renewable resource? A functional approach to the evaluation of marsh creation efforts. Estuaries **14**: 1-16.

Netto, S.A. and P.C. Lana 1997a. Influence of *Spartina alterniflora* on superficial sediment characteristics of tidal flats in Paranagua Bay (South-eastern Brazil). Estuarine, Coastal and Shelf Science **44**: 641-648.

——. 1997b. Intertidal zonation of benthic macrofauna in a subtropical salt marsh and nearby unvegetated flat (S. E. Brazil). Hydrobiologia **353**: 171-180.

Nilsson, H.C. and R. Rosenberg. 1994. Hypoxic response of two marine benthic communities. Marine Ecology Progress Series **115**: 209-217.

Nixon, S.W. 1980. Between coastal marshes and coastal waters — A review of twenty years of speculation and research on the role of salt marshes. Pages 437-525 *in* P. Hamilton and K.B. MacDonald, editors. Estuarine Productivity and Water Chemistry. Plenum Publishing Corporation, New York, New York, USA.

Nomann, B.E. and S.C. Pennings. 1998. Fiddler crab-vegetation interactions in hypersaline habitats. Journal of Experimental Marine Biology and Ecology **225**: 53-68.

Nordby, C.S. and J.B. Zedler. 1991. Responses of fish and macrobenthic assemblages to hydrologic disturbances in Tijuana Estuary and Los Peñasquitos Lagoon, California. Estuaries **14**: 80-93.

Nowell, A.R.M. and P.A. Jumars. 1984. Flow environments of aquatic benthos. Annual Review of Ecology and Systematics **15**: 303-328.

O'Connor, N.J. 1993. Settlement and recruitment of the fiddler crabs *Uca pugnax* and *U. pugilator* in a North Carolina, USA, salt marsh. Marine Ecology Progress Series **93**: 227-234.

Odum, E.P. 1980. The status of three ecosystem-level hypotheses regarding salt marsh estuaries: tidal subsidy, outwelling and detritus-based food chains. Pages 485-495 *in* V.S. Kennedy, editors. Estuarine Perspectives. Academic Press, New York, New York, USA.
Olivier, M., G. Desrosiers, A. Caron and C. Retiere. 1996. Juvenile growth of the polychaete *Nereis virens* feeding on a range of marine vascular and macroalgal plant sources. Marine Biology (Berlin) **125**: 693-699.
Olivier, M., G. Desrosiers, A. Caron, C. Retiere and A. Caillou. 1997. Juvenile growth of *Nereis diversicolor* (O. F. Muller) feeding on a range of marine vascular and macroalgal plant sources under experimental conditions. Journal of Marine Biology and Ecology **208**: 1-12.
Orth, R.J. 1977. The importance of sediment stability in seagrass communities. Pages 281-300 *in* B. Coull, editor. Ecology of Marine Benthos. University of South Carolina Press, Columbia, South Carolina, USA.
Orth, R.J., K.L. Heck and R.J. Diaz. 1991. Littoral and intertidal systems in the mid-Atlantic coast of the United States. Pages 193-209 *in* A.C. Mathieson and P.H. Nienhuis, editors. Intertidal and littoral ecosystems. Elsevier, Amsterdam.
Osenga, G.A. and B.C. Coull. 1983. *Spartina alterniflora* Loisel root structure and meiofaunal abundance. Journal of Experimental Marine Biology and Ecology **67**: 221-225.
Page, H.M. 1997. Importance of vascular plant and algal production to macro-invertebrate consumers in a southern California salt marsh. Estuarine Coastal and Shelf Science **45**: 823-324.
Palacio, J., M. Lastra and J. Mora. 1991. Vertical distribution of the intertidal macrofauna in the Lourizan Inlet (Ria de Pontevedra, Galicia). Thalassas **9**: 49-62.
Palmer, M.A. and R.R. Brandt. 1981. Tidal variation in sediment densities of marine benthic copepods. Marine Ecology Progress Series **4**: 207-212.
Palmer, M.A. and G. Gust. 1985. Dispersal of meiofauna in a turbulent tidal creek. Journal of Marine Research **43**: 179-210.
Patrick, W.H. Jr. and R.D. Delaune. 1972. Characterization of the oxidized and reduced zones in flooded soil. Soil Science Society of American Proceedings **36**: 573-576.
Pearson, T.H. and R. Rosenberg. 1978. Macrobenthic succession in relation to organic enrichment and pollution of the marine environment. Oceanography and Marine Biology Annual Review **16**: 229-311.
——. 1987. Feast and famine: structuring factors in marine benthic communities. Pages 373-395 *in* J.H.R. Gee and P.S. Giller, editors. Organization of communities: past and present. Blackwell Scientific Press, Oxford, United Kingdom.
Peterson, B.J. and R.W. Howarth. 1987. Sulfur, carbon and nitrogen isotopes used to trace organic matter flow in the salt-marsh estuaries of Sapelo Island, Georgia. Limnology and Oceanography **32**: 1195-1213.
Peterson, B.J., R.W. Howarth and R H. Garritt. 1985. Multiple stable isotopes used to trace the flow of organic matter in estuarine food webs. Science **227**: 1361-1363.
Peterson, B. J., R. W. Howarth and R. H. Garritt. 1986. Sulfur and carbon isotopes as tracers of salt-marsh organic matter flow. Ecology **67**: 865-874.
Peterson, C.H., H.C. Summerson and P.B. Duncan. 1984. The influence of seagrass cover on population structure and individual growth rate of a suspension feeding bivalve: *Mercenaria mercenaria*. Journal of Marine Research **42**: 123-138.
Pethick, J.S., D. Legget and L. Husain. 1990. Boundary layers under salt marsh vegetation developed in tidal currents. Pages 113-123 *in* J.B. Thornes, editor. Vegetation and Erosion. Wiley, New York, USA.
Portnoy, J.W. and A. E. Giblin. 1997. Effects of historic tidal restrictions on salt marsh sediment chemistry. Biogeochemistry **36**: 275-303.
Posey, M., T. Alphin and C. Powell. 1997. Plant and infaunal communities associated with a created marsh. Estuaries **20**: 42-47.
Rader, D.N. 1984. Salt marsh benthic invertebrates: small-scale patterns of distribution and abundance. Estuaries **7**: 413-420.
Redfield, A.C. 1972. Development of a New England salt marsh. Ecological Monographs **42**: 201-237.
Reise, K. 1981. High abundance of small zoobenthos around biogenic structures in tidal sediments of the Wadden Sea. Helgol Meeresunters **32**: 413-425.
——. 1985. Tidal Flat Ecology. Springer Verlag, Berlin, pp. 1-191.
Ringold, P. 1979. Burrowing, root mat density and the distribution of fiddler crabs in the eastern United States. Journal of Experimental Marine Biology and Ecology **36**: 11-21.

Rutledge, P.A. and J.W. Fleeger. 1993. Abundance and seasonality of meiofauna, including harpacticoid copepod species, associated with stems of the salt-marsh cord grass, *Spartina alterniflora*. Estuaries **16**: 760-768.

Sacco, J.N., B.L. Booker and E.D. Seneca. 1987. Comparison of the macrofaunal communities of a human initiated salt marsh at two and fifteen years of age. Pages 282-285 *in* J. Zelazny and S. Feierabend, editors. Proceedings from the Increasing Our Wetland Resources Conference. National Wildlife Federation, Washington, District of Columbia, USA.

Sacco, J.N., E.D. Seneca and T.R. Wentworth. 1994. Infaunal community development of artificially established salt marshes in North Carolina. Estuaries **17**: 489-500.

Sarda, R., K. Foreman and I. Valiela. 1992. Controls of benthic invertebrate populations and production of salt marsh tidal creeks: experimental enrichment and short- and long-term effects. Pages 85-91 *in* G. Colombo et al., editors. Marine eutrophication and population dynamics. Olsen and Olsen, Feredensborg.

Sarda, R., I. Valiela and K. Foreman. 1996. Decadal shifts in a salt marsh macroinfaunal community in response to sustained long-term experimental nutrient enrichment. Journal of Experimental Marine Biology and Ecology **205**: 63-81.

Scatolini, S.R. and J.B. Zedler. 1996. Epibenthic invertebrates of natural and constructed marshes of San Diego Bay. Wetlands **16**: 24-37.

Schwinghamer, P. and P. Kepkay. 1987. Effects of experimental enrichment with *Spartina* detritus on sediment community biomass and metabolism. Biological Oceanography **4**: 289-322.

Shisler, J.K. and D.J. Charette. 1984. Evaluation of artificial salt marshes in New Jersey. New Jersey Agricultural Experimental Station Publication No. **P-40502-01-84**. Mosquito Research and Control, Cook College, Rutgers University, New Brunswick, New Jersey, USA.

Sikora, W.B. and J.P. Sikora. 1982. Ecological implications of the vertical distribution of meiofauna in salt marsh sediments. Pages 269-282 *in* V.S. Kennedy, editor. Estuarine comparisons. Academic Press, New York, New York, USA.

Steever, E.Z., R.S. Warren and W.A. Niering. 1976. Tidal energy subsidy and standing crop production of *Spartina alterniflora*. Estuarine Coastal Marine Science **4**: 473-478.

Stephens, E.G. and M.D. Bertness. 1991. Mussel facilitation of barnacle survival in a sheltered bay habitat. Journal of Experimental Marine Biology and Ecology **145**: 33-48.

Streever, W.J. and A.J. Genders. 1997. Effect of improved tidal flushing and competitive interactions at the boundary between salt marsh and pasture. Estuaries **20**: 807-818.

Sun, B. and J.W. Fleeger. 1994. Field experiments on the colonization of meiofauna into sediment depressions. Marine Ecology Progress Series **110**: 167-175.

Sun, B., J.W. Fleeger and R.S. Carney. 1993. Sediment microtopography and the small- scale spatial distribution of meiofauna. Journal of Experimental Marine Biology and Ecology **167**: 73-90.

Talley, T.S. and L.A. Levin (1999). Macrofaunal succession and community structure in *Salicornia* marshes of southern California. Estuarine, Coastal and Shelf Science **49**: 713-731.

Teal, J.M. and J.W. Wieser. 1966. The distribution and ecology of nematodes in a Georgia salt marsh. Limnology and Oceanography **11**: 217-222.

Thistle, D., A. Reidenauer, R.H. Findlay and R. Waldo. 1984. An experimental investigation of enhanced harpacticoid (Copepoda) abundances around isolated seagrass shoots. Oecologia **63**: 295-299.

Toomey, W.A., Jr. 1997. The effect of soil organic matter quantity on salt marsh functioning and infaunal community structure. Thesis. North Carolina State University, Raleigh, North Carolina, USA.

Turner, E.J., R.K. Zimmer-Faust, M.A. Palmer, M. Luckenbach and N.D. Pentcheff. 1994. Settlement of oyster (*Crassostrea virginica*) larvae: effects of water flow and a water-soluble chemical cue. Limnology and Oceanography **39**: 1579-1593.

Valiela, I., J. Wilson, R. Buchsbaum, C. Rietsma, D. Bryant, K, Foreman and J. Teal. 1984. Importance of chemical composition of salt marsh litter on decay rates and feeding by detritivores. Bulletin of Marine Science **35**: 261-269.

Van Dolah, R.F. 1978. Factors regulating the distribution and population dynamics of the amphipod *Gammarus palustris* in an intertidal salt marsh community. Ecological Monographs **48**: 191-217.

Vaughn, C.C. and F.M. Fisher. 1988. Vertical migration as a refuge from predation in intertidal marsh snails: a field test. Journal of Experimental Marine Biology and Ecology **123**: 163-176.

Vince, S., I. Valiela, N. Backus and J. Teal. 1976. Predation by the salt marsh killifish *Fundulus heteroclitus* (L.) in relation to prey size and habitat structure: Consequences for prey distribution and abundance. Journal of Experimental Marine Biology and Ecology **23**: 255-266.

Ward, G. and G.J. FitzGerald. 1983. Macrobenthic abundance and distribution in tidal pools of a Quebec salt marsh. Canadian Journal of Zoology **61**: 1071-1085.

Warren, T.M. 1977. The ecology of *Capitella capitata* in British waters. Journal of Marine Biological Association, U. K. **57**: 151-159.

Watzin, M.C. 1983. The effects of meiofauna on settling macrofauna: meiofauna may structure macrofaunal communities. Oecologia **59**: 163-166.

Weigert, R.G. and L.R. Pomeroy. 1982. The salt marsh ecosystem: a synthesis. Ecological Studies **38**: 219-230.

Weissburg, M. and R.K. Zimmer-Faust. 1991. Effects of chemoreception and hydrodynamics in mediating predation. American Zoologist **31**: 103A.

——. 1993. Life and death in moving fluids: hydrodynamic effects on chemosensory-mediated predation. Ecology **74**: 1428-1443.

——. 1994. Odor plumes and how blue crabs use them in finding prey. Journal of Experimental Biology **197**: 349-375.

West, D.L. and A.H. Williams. 1986. Predation by *Callinectes sapidus* (Rathbun) within *Spartina alterniflora* (Loisel) marshes. Journal of Experimental Marine Biology and Ecology **100**: 75-95.

Woodin, S.A. 1976. Adult-larval interactions in dense infaunal assemblages: Patterns of abundance. Journal of Marine Research **34**: 25-41.

Woodin, S.A., R.L. Marinelli and S.M. Lindsay. 1998. Process-specific cues for recruitment in sedimentary environments: geochemical signals? Journal of Marine Research **58**: 535-558.

Ysebaert, T., P. Meire, D. Maes and J. Buijs. 1993. The benthic macrofauna along the estuarine gradient of the Schelde Estuary. Netherlands Journal of Aquatic Ecology **27**: 327-341.

Zedler, J.B. 1993. Canopy architecture of natural and planted cordgrass marshes: Selecting habitat evaluation criteria. Ecological Applications **3**: 123-138.

Zimmerman, R.J. and T.J. Minello. 1984. Fishery habitat requirements: utilization of nursery habitats by juvenile penaeid shrimp in a Gulf of Mexico salt marsh. Pages 371-383 *in* B.J. Copeland, K. Hart and N. Davis, editors. Research for Managing the Nation's Estuaries. Proceedings of a Conference, Raleigh, North Carolina, USA.

Zipperer, V.T. 1996. Ecological effects of the introduced cordgrass, *Spartina alterniflora,* on the benthic community structure of Willapa Bay, Washington. Thesis. University of Washington, Seattle, Washington, USA.

THE HEALTH AND LONG TERM STABILITY OF NATURAL AND RESTORED MARSHES IN CHESAPEAKE BAY

J. C. STEVENSON
J.E. ROOTH
University of Maryland Center for Environmental Sciences
Horn Point Laboratory, P.O. Box 775
Cambridge, MD 21613 USA

M.S. KEARNEY
Department of Geography
University of Maryland
College Park MD 20742 USA

K.L. SUNDBERG
University of Maryland Center for Environmental Sciences
Horn Point Laboratory, P.O. Box 775
Cambridge, MD 21613 USA

Abstract
Recent evidence in Chesapeake Bay suggests that the majority of the tidal marsh acreage has been negatively affected by sea-level rise in recent years. The ultimate impact, total marsh loss, centers on Blackwater National Wildlife Refuge (NWR) which has encountered particularly high rates of land subsidence in the twentieth century. Analysis using digitized photography has revealed that marshes are being lost more rapidly in the northern than southern sections of Blackwater. The former is closest to the center of a large cone of depression in the most important underlying aquifer in the region. The groundwater withdrawals at Cambridge correspond to a rapid rise in sea level which appears to be two to three times the present global rate of 1 to 2 mm y^{-1}. The declining health of the marshes, reflected in reduced productivity, canopy thinning, channel enlargement, rotten spots and salt pans as well as ultimate conversion to mudflat and open water, can be tracked using satellite imagery. Although not as dramatic as marsh conversion to open water at Blackwater NWR (over 50% during the 20th century), other marshes on the lower Eastern Shore of Maryland show clear signs of incipient change. Past attempts at restoring Blackwater marshes have not been successful due in part to the combination of excessive grazing by muskrats and nutria as well as anthropogenic influences (reduced diel tidal amplitude because of road building, increasing salinity because of canals, and possibly large-scale burning). Ultimately, restoration efforts depend on the maintenance of groundwater pressure and/or on supplementation of the system with sediment from other sources to keep them abreast of rising sea level. Restoration efforts will depend not only on controlling groundwater withdrawals, but possibly in revitalizing existing marshes by promoting rhizosphere oxygenation. Where

these strategies are not practical, more innovative approaches may be necessary such as the use of highly productive species (e.g., *Phragmites australis*) that appear to be more efficient in promoting sedimentation and long term accretion than other marsh species.

1. Introduction

The Chesapeake Bay has one of the largest concentrations of tidal marsh along the East Coast with an estimated 124,000 ha (Table 1) and is only exceeded by South Carolina (Stevenson et al. 1985a). Chesapeake Bay also appears to be the focus of tidal marsh loss along this coastline. Although the documented conversion at Blackwater (20 ha y^{-1}; Table 2) is dwarfed by that in Louisiana (5600 ha y^{-1}; Barras et al. 1994), it remains the greatest source of wetland loss in Maryland. Ironically, most resource managers in Chesapeake Bay have concentrated restoration efforts on non-tidal wetlands where net change has been essentially zero (and possibly even positive) over the last decade. While State and Federal agencies continue to focus their efforts on non-tidal wetlands, tidal marshes are being lost at a rapid rate. The purpose of this paper is to examine the causes of tidal marsh declines in Chesapeake Bay and suggest how present management policies have contributed to the problem. We also outline strategies that could be implemented to slow and possibly arrest marsh loss in various environments. Finally, we attempt to draw comparisons to other marsh systems in the mid-Atlantic region and place them in a global context.

TABLE 1. Estimates of aerial coverage (hecatares) of functional categories of tidal marshes in the Chesapeake Bay (1 acre= 0.4 ha)

MARSH TYPE	Maryland	Virginia	Both States
Coastal High Salinity	0	7,163	7,163
Submerged Upland *	46,662	0	46,662
Brackish Estuarine	22,355	21,597	43,952
Tidal Freshwater **	10,345	16,000	26,345
TOTAL TIDAL	79,362	447,630	124,122

* From Darmody (1975)
**From Odum et al. (1984)

TABLE 2. Marsh conversion to open water at Blackwater NWR as determined using ArcInfo.

	Area (in ha)			
	1938	1980	difference	rate (ha y^{-1})
Marsh	2,524	1,796	-728	-17.3
Water	1,173	1,999	+816	+19.4

2. Background

The reticence that marsh ecologists have in directing public attention to the plight of tidal marshes under high sea-level scenarios may seem puzzling, but lies partially in the general belief that present losses are driven by underlying geological factors which are not easily controlled. Most evidence points to the fact that global sea-level has risen slightly over 120 meters since the onset of the Holocene Period when the glaciers began to melt 17-18,000 y BP (Curray 1965, Fairbanks 1989, Shroeder et al. 1995). Although a simple calculation would suggest that overall sea-level rise averaged 7 mm y^{-1} during the Holocene, most of the rise occurred before 4000 y BP, after which sea-level change was slight (Rampino and Sanders 1981). Kearney (1996) found that sea-level rose about only about 0.56 mm y^{-1} during the last thousand years in the Chesapeake. This deceleration reinforces Kearney and Stevenson's (1991) suggestion that sea-level rise was low during the ''Little Ice Age'' but beginning about 1850 accelerated, with a sharp inflection between 1900 and 1920.

This upward trend tracks the increasing output of global industrial gas emissions in the mid- to late- 1800s (Fleming 1998). Although once largely ignored by the scientific community, Arrhenius' suggestion in 1895 that industrial gas emissions could cause global greenhouse warming has now been shown to be remarkably accurate. At least on the time scale of millennia, Stauffer et al. (1998) have shown a close correspondence between temperature changes and CO_2 concentrations. Most recent estimates of global sea-level rise lie between 1 and 2 mm y^{-1}, and although complicated by astronomical factors generally reflect the global increase in temperatures of the last century (Douglas 1991). While this recent increase may be alarming in many ways, the reality is that the majority of marshes along the Atlantic seaboard appear to be able to accrete fast enough to keep abreast of present global sea-level rise (Stevenson et al. 1986). However, where local land subsidence produces high rates of relative sea-level (RSL) rise, marsh losses can increase dramatically.

Not all the impacts of rise in CO_2 are negative. In fact there are indications that C_3 photosynthetic pathway species, such as *Scirpus americanus* (formerly *olneyi*), may benefit from increased CO_2 in terms of overall productivity (Curtis et al. 1989a). Furthermore since overall litter decomposition appears to be little affected even by doubling CO_2 levels, the end result may be an increase in peat production, helping to offset the sea-level problem (Curtis et al. 1989b). However, marsh loss is complicated and involves much more than CO_2 increase in the atmosphere and eustatic rise of the global oceans. Increasing temperatures associated with global warming may increase anoxia in the root zone that can have serious consequences for the energy budget of marsh plants (Mendelssohn et al. 1981). Decomposition rates can also increase with rhizosphere temperatures resulting in less peat production. Two other important factors in long-term marsh survival are subsidence of the land and loss of sediment inputs which help to build marshes against sea-level rise. Stevenson et al. (1988) have discussed the importance of ebb and flood dominated tides in determining whether a marsh will export or trap sediments. They found that many tidal systems appear to have ebb-dominated tidal dynamics which exacerbates particulate losses. Sediment losses translate into long-term health problems for marshes when they are in areas where RSL rise is high.

The question remains: How do we manage these systems? In order to approach this issue we have chosen to use our experience in mid-Chesapeake Bay where marsh loss has been documented over the past century. The quantification of sea-level rates of change usually begins with data produced at major tide gauges around the world where long records can be obtained. Because sea-level does not rise monotonically, relatively long records are needed to ascertain changes. Douglas (1995) concluded that at minimum, a 50-year record is necessary to derive trends. In an earlier review, Stevenson et al. (1986) observed that RSL was rising from 2-5 mm y^{-1} in Chesapeake Bay. This rate is higher than any other region along the eastern seaboard and is due in part to the fact that this region lies south of the terminal line of the last glaciation which extends in an irregular line from southern New York City (Staten Island) to Long Island NY to Harrisburg PA.

The position of the ice sheet north of Chesapeake Bay had important consequences for the present landscape. One aspect is that the disruption of vegetative cover caused large dust storms which deposited loess on the upper eastern shore of present Chesapeake Bay (Foss et al. 1978). Many Chesapeake Bay marshes developed over this silt-clay layer as sea-level rose during the Holocene Period (Darmody and Foss 1979). Also at maximum glaciation (15,000 to 20,000 y BP), the land north under the glaciers was compressed due to the weight of ice, while the land south of the line bulged upwards (Nerem et al. 1998). At the beginning of the Holocene, when the glaciers began to recede, the underlying terrain began to rebound. The land at the immediate edge of the glaciers was a terminal moraine which consisted of boulders and gravels brought with the ice and deposited as the ice melted. Beyond the advance of the glaciers, the terrain was elevated in an extensive fore-bulge, which collapsed as the glaciers receded. The Chesapeake and Delaware Bays fall within the region of the fore-bulge collapse and consequently sea-level rise rates are relatively high in this region of the mid-Atlantic (Nerem et al. 1998).

While it appears that many marshes along the Atlantic seacoast are capable of keeping up with current rates of sea-level rise, there are several long term implications of sea level which impact the basic accretionary potential and detrital particulate budget of marshes. No studies we have seen of marsh function take into account the sea-level variability in their studies which might be attributed to sea-level rise. Also very little has been investigated concerning the consequences of fluid withdrawal from underlying strata and the importance of maintaining water levels at the surface. With over half the population of the U.S. now living on the 10% of the land defined as coastal, and populations projected to increase 10 to 15% by 2010 (Culliton et al. 1990), this looms as an important issue for marsh maintenance as well as restoration. Ironically, even though the Louisiana Delta has had massive fluid withdrawals over the last 100 years, virtually no one has projected impacts of that on deteriorating marsh systems on the surface. Our goal is to focus on this issue and provide suggestions on future policy options. The long-term health of these key natural resources depends on steps that need to be taken on a variety of scales from local to regional and ultimately global.

3. Marsh Acreage and Health in Chesapeake Bay

Traditional methods of assessing marsh condition or ''health'' have relied on delineating changes in the surface coverage of marsh versus open water in a given area over time. Although this is a reasonable approach to infer temporal changes in marsh coverage, it does not give much information concerning the functional vitality ''health'' of the remaining marsh. This can lead to erroneous interpretations of marsh contributions to the ecology of local waters. For example, instead of acting as a habitat for commercially important species of the estuary (crabs, shrimp, anadromous fish, etc.), anoxic shallows of declining marshes often are inhospitable in late spring and summer when organisms depend on them most. Moreover, as with all change detection procedures, the length of temporal coverage is crucial to the accuracy of end-point assessments; rate evaluations based on short-term data can be unduly biased by occasional extreme events. This last point can be difficult to overcome, given the unequal availability of historical aerial photographs.

An alternate approach for assessing marsh vulnerability and potential loss involves the examination of the physical characteristics (primarily the arrangement, type and number of open water areas within a marsh and, secondarily, characteristics of the biomass) of a marsh surface. In Maryland much of the marshes on the Western Shore have luxuriant sediment supplies which have promoted marsh expansion since settlement (Froomer 1980, Stevenson and Kearney 1996). These marshes can be used as ''controls'' for determining healthy marsh signatures. This approach has the advantage of not being limited by the length of the aerial photographic record. In fact, the method can produce useable information on marsh condition with only aerial photographs from one time period, particularly if it is the most recent aerial photography for the area. Temporal change detection investigations can also be made if photography from a number of years is available. In either case, the results show not a general estimate of the amount of marsh versus open water, but the actual estimates of the surface condition present within the marsh coverage.

Marsh submergence typically occurs in several distinct stages (Kearney et al. 1988). In stable coastal marshes, large areas of open water (interior ponds) are absent, and the number of tidal creeks relatively few. As marsh loss initiates, tidal creeks increase in number and widen, and small interior ponds form. Finally, interior ponds enlarge and eventually coalesce; extensive strings of coalescing interior ponds mark an advanced stage of marsh loss. Also, concurrent lateral erosion of the shoreline of submerging marshes occurs, particularly if they are contiguous to large embayments with long fetches where wave action is strong during storm events.

Kearney et al. (1988) developed a marsh surface condition index (MSCI) to classify marshes in the Nanticoke River estuary according their vulnerability to sea-level rise. This classification recognized five categories of marsh surface condition, ranging from intact ''healthy'' marshes to degraded marshes that had become open water. Assessing whether a marsh belongs in one category or another is based on the presence, relative location, and number of key attributes that are generally identifiable on high quality aerial photographs (preferably large-scale color photography). Although several of the intermediate categories employ tonal or texture characteristics, this method basically relies on gross pattern recognition, and therefore determining the absolute size or position of the individual features is not critical. It thus avoids one of the major pitfalls of conventional change detection approaches that inherently rely on a high degree of planimetric precision: inaccurate registration of base points between different generations of images.

4. Changes in the Nanticoke Estuary Using the MSCI, 1938-1985

Marsh deterioration and loss has been occurring in the Nanticoke Estuary since at least the early decades of this century (Kearney et al. 1988). By 1938, almost 40% of the submerged upland marshes had undergone moderate to complete deterioration (essentially open water). Marshes situated in the meanders of the brackish and fresh reaches of the estuary had either declined moderately in surface quality (brackish meander marshes) or declined negligibly (fresh meander marshes). By 1985, deterioration of marshes was widespread, especially in the lower estuary where submerged upland marshes predominate. These data illustrate the chief advantage of MSCI analysis over a more conventional aerial photographic study of marsh loss that simply reports areal loss. Aerial losses alone for submerged marshes in the Nanticoke between 1938-1985 were 807 ha, whereas a total of 2690 ha of this marsh type had become moderately to severely degraded over the same time period. The latter figure is more realistic in assessing the loss of functional marshes in the estuary, especially in terms of its potential impact on the overall ecology of the system.

Reasons for this rapid decline in the Nanticoke marshes are tied to the low rates of vertical accretion in most of the marshes. Kearney et al. (1988) showed that vertical accretion rates in the large submerged upland marshes of the estuary are below the relative rate of sea-level rise for the middle Chesapeake Bay. In fact, rates of vertical accretion of all marshes in the system generally decrease down estuary. Only in the upper, sediment-trap portion of the estuary do vertical accretion rates equal or exceed the relative rate of sea-level rise. Not surprisingly, these marshes are among the most stable in the system, showing little or no evidence of the classic patterns of marsh loss documented here and elsewhere in Chesapeake Bay.

5. Remote Sensing of the Health of Chesapeake Bay Marshes

Current approaches to modeling marsh coverages using Thematic Mapper Imagery involve decomposing the scene pixels into basic spectral components of vegetation, water, and earth materials or soil. At its simplest, such methods assume that each pixel is a linear combination of these components that may be approximated by the additive sum of the reflectance of all the surface covers in the parcel, weighted by the fraction of the parcel occupied by each surface cover. Three end members with three spectral bands may be represented as:

$$rw1fw + rv1fv + rs1fs = R1$$
$$rw2fw + rv2fv + rs2fs = R2$$
$$rw3fw + rv3fv + rs3fs = R3$$

where w, v and s represent water, vegetation, and soil, respectively; f represents the proportion of the pixel covered by a particular cover type; and r represents the reflectance of the various surface cover types in the respective bands. An additional assumption is that $fw + fv + fs = 1$.

In this study the end members were selected from pixels in the actual images used for the development of the mixture model. Field investigations had shown these areas to be good representatives of marsh vegetation (encompassing various types) and marsh soil (again, encompassing a range of types). Pixels to be used for end members were selected from open ocean areas on the images where the spectral characteristics of water would not be confounded by the presence of suspended inorganic or organic materials, a distinct possibility if pixels for the water end members had been selected from the Chesapeake Bay. The Landsat Thematic Mapper image in false color infra-red (Fig. 1) and the processed MSCI (Fig. 2) image for the Wingate, MD, USGS quadrangle illustrate the utility of this technique. The MSCI image shows healthier marshes at the mouth of the Blackwater River (northeast corner of the image) and moderately-to heavily-degraded marshes in the interior of the peninsula and along the western shore of Fishing Bay (eastern portion of the image).

First model results for sections of the Maryland (essentially the middle and upper Bay) and Virginia (essentially the lower Bay) parts of the Chesapeake Bay are shown in Fig. 3. Although the model is being refined, the data indicate that most Bay marshes are to some degree degraded. Marshes in the middle and upper Bay range fairly evenly between all surface condition classes, with moderately degraded marshes being the most common. This relatively even spread across "health" classes occurs despite the

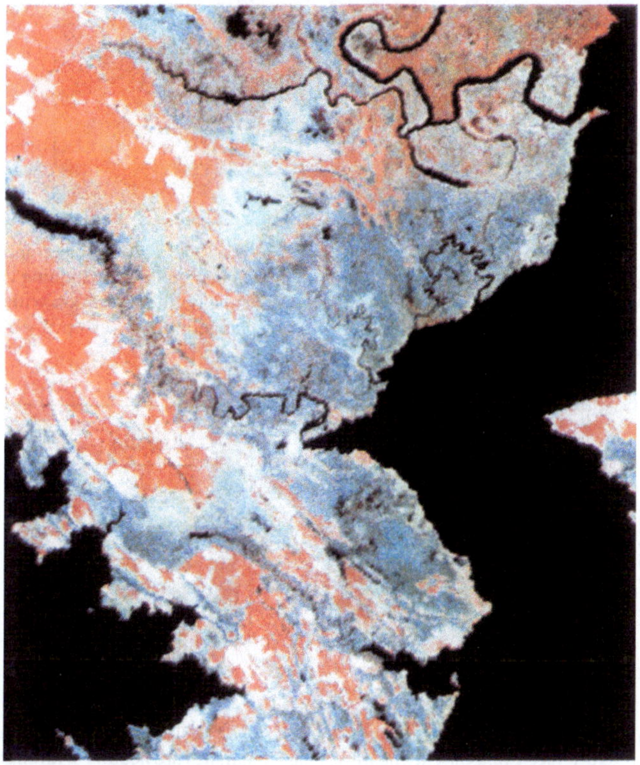

Figure 1. Thematic Mapper Image (TMI) in false color infra-red of the Wingate, MD U.S.G.S. Quadrangle. Black = open water; blue-grey = wetlands; and red = upland.

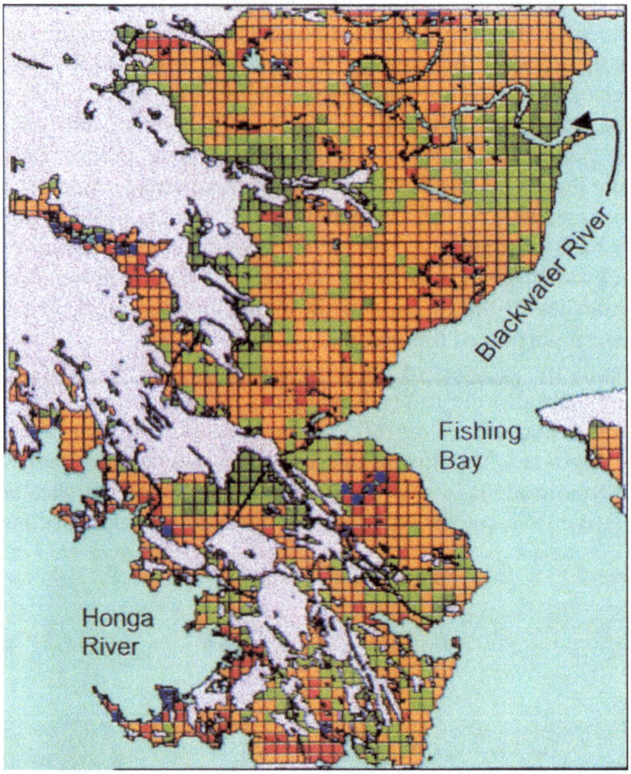

Figure 2. MSCI image derived from a National Wetlands Inventory (U.S.G.S 1:24000) map and the TMI of Fig. 1. Dark green represents healthy marsh; light green is slightly degraded marsh with thinning canopies; orange indicates moderate deterioration with stunted plants; red indicates heavy deterioration with frequent rotten spots; and dark blue represents complete deterioration (void of plants, but not yet open water). Non-marsh areas are colored pinkish-grey.

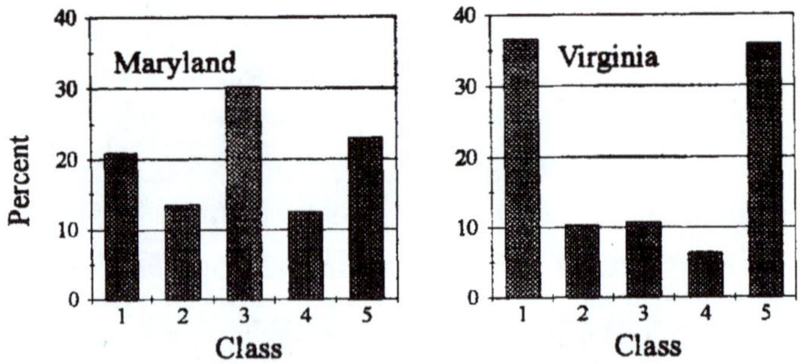

Figure 3. Distribution of health classes from marshes in the Maryland (~79,000 ha) and Virginia (~45,000 ha) portions of Chesapeake Bay. Class 1 marshes are healthy, Class 2 marshes are healthy with slight deterioration, Class 3 marshes exhibit slight to moderate deterioration, Class 4 marshes are moderately to heavily deteriorated, and Class 5 marshes exhibit heavy to complete deterioration.

occurrence of several large areas of severely to completely degraded marshes along the lower Eastern Shore of Maryland (e.g., the Blackwater National Wildlife Refuge and other areas in South Dorchester County). This reflects subsidence patterns on the Eastern Shore (see next section) and the large number of nominally ''healthy'' marshes in the upper Chesapeake Bay where suspended sediment fluxes are high from the Susquehanna River.

By comparison, marshes in the Virginia, or the lower Bay, fall into predominantly two surface condition classes, either ''healthy'' or completely degraded, with more marshes falling into the latter class. Relatively few marshes are characterized by the model in the intervening classes of slightly to severely degraded marsh. Inspection of the geographic distribution of surface condition classes in the lower Bay suggests that this apparent bimodality results from the bias imparted by several large severely to completely degraded marshes to the smaller total area of marshes in the lower Bay. Without these marshes included, the surface condition of lower Bay marshes is relatively ''healthy''.

6. Fluid Withdrawal and Subsidence Drives Localized Marsh Loss

Groundwater is increasingly becoming recognized as an important issue in the Chesapeake Bay and its watershed. Although groundwater quality is obviously a critical component in the nutrient budgets in coastal systems, other aspects are also important. The fragility of the underlying substrate was demonstrated in the 1990s when farmland and highways collapsed near Interstate 70 between Washington D. C. and Frederick, MD. A recent investigation of this event (Boyer 1997) suggests that the collapsing land and portions of highways are located over subterranean sinkholes which lie in a dendritic pattern from an earlier drainage system. This dramatically underscores how groundwater is often key in understanding the stability of the land surface. Although comparatively little studied along the Eastern Seaboard, the effects of aquifer pumping and deformation of the overlying land surface is well documented in California and other western states (Poland and Davis 1969). For example, the region adjacent to the South San Francisco Bay Estuary has subsided several meters over the last century when large scale irrigation was initiated around San Jose (Poland 1972). Davis (1985) concluded that land subsidence is common in many areas along the Gulf and Atlantic Coasts of the U.S. where aquifer withdrawals are significant. Texas has also experienced subsidence problems in Houston and Galveston. Zimmerman (this volume) has linked marsh decline along the Texas coast with excessive subsidence due to groundwater withdrawals.

A recent study (Rule 1995) of the aquifers in the mid-Chesapeake Bay region (Fig. 4) suggests that aquifer pumping is linked to underlying subsidence. This in part accounts for the excessive rates of sea-level change noted by Hohldal and Morrison (1974). Aquatards (sedimentary layers with low water permeability) between the aquifers are particularly susceptible to compression when fluid is withdrawn because of their high clay content (Johnson and Morris 1962). In addition, old peat layers can compress as they dry out and may also be susceptible to oxidization by high nitrate groundwater.

Rule (1995) found that the average compaction for Dorchester County was 8.5 cm and primarily due to compressibility of the Nanjemoy formation (Fig. 5). This compaction was due to a large cone of depression centered at Cambridge, where cannery operations have withdrawn large amounts of water primarily from the Piney Point aquifer for oyster packing beginning early in the twentieth century (Jones 1902). Presently, the cone of depression extends far under the Blackwater NWR (Fig. 6) and is spreading outwards due to increasing demands by domestic use and by farmers for irrigation.

A more detailed model of groundwater movement in response to current withdrawal is necessary for precise estimates of compaction actually at Blackwater, however it is obviously an important component in the high land subsidence rates at Cambridge reported by Nerem et al. (1998). Although our analysis of the tide gauge rate at Cambridge is substantially less (3.5 mm y^{-1} from 1943-1996) than Nerem et al. reported (7.9 mm y^{-1}), increasing groundwater withdrawal could turn out to be a decisive factor confronting marsh restoration on the eastern shore of Chesapeake Bay. In addition to the deep aquifer withdrawals, there is also the problem of surficial supplies of groundwater which traditionally had supplied marshes with freshwater.

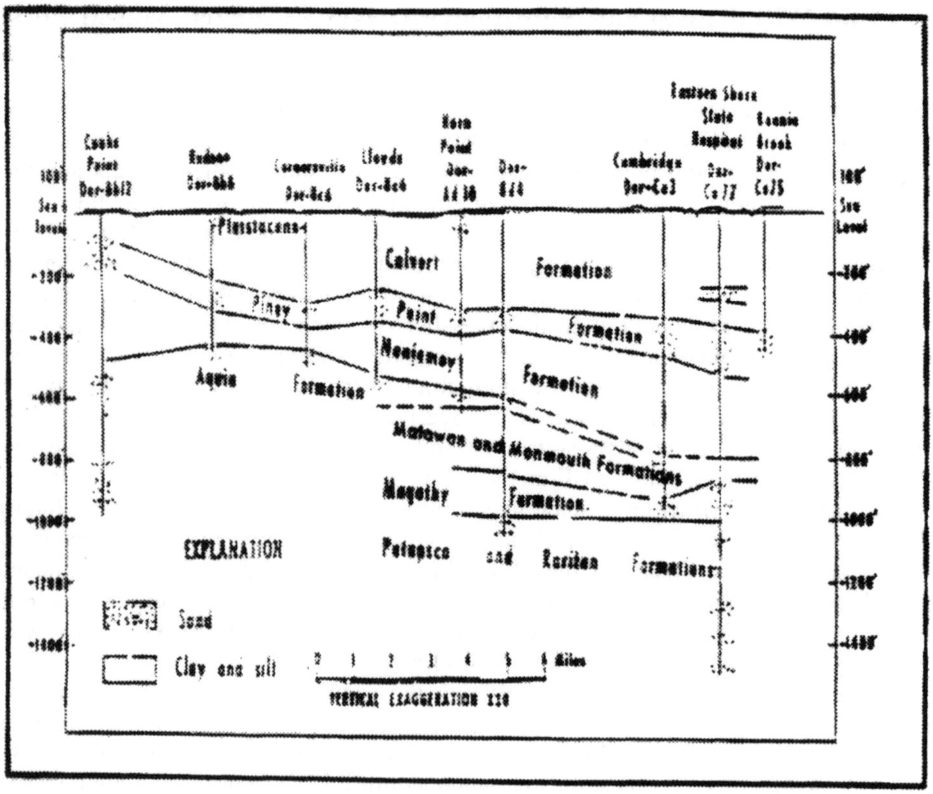

Figure 4. Generalized stratigraphy coastal plain sediments in Dorchester County, Maryland. Observation wells indicated on the surface range from Cooks Point (mouth of the Choptank River) to Bonnie Brook, 2 miles east of the city of Cambridge. (From: Mack et al. 1971).

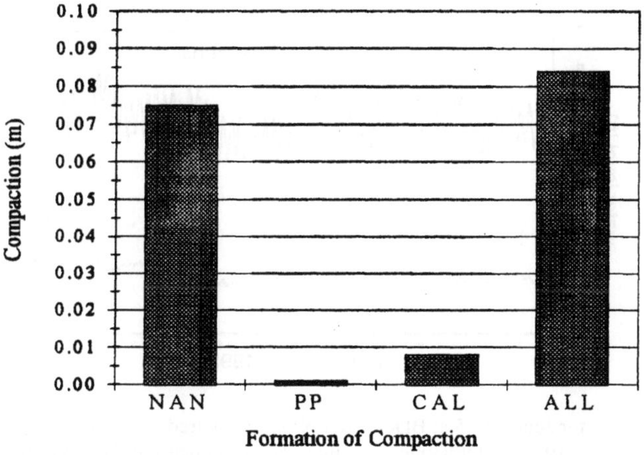

Figure 5. Formation compaction, portion of total Dorchester County, Maryland. PP is Piney Point formation. NAN is Nanjemoy formation. CAL is Calvert formation. ALL is the total compaction of the formation. Drawdown is in Piney Point formation.

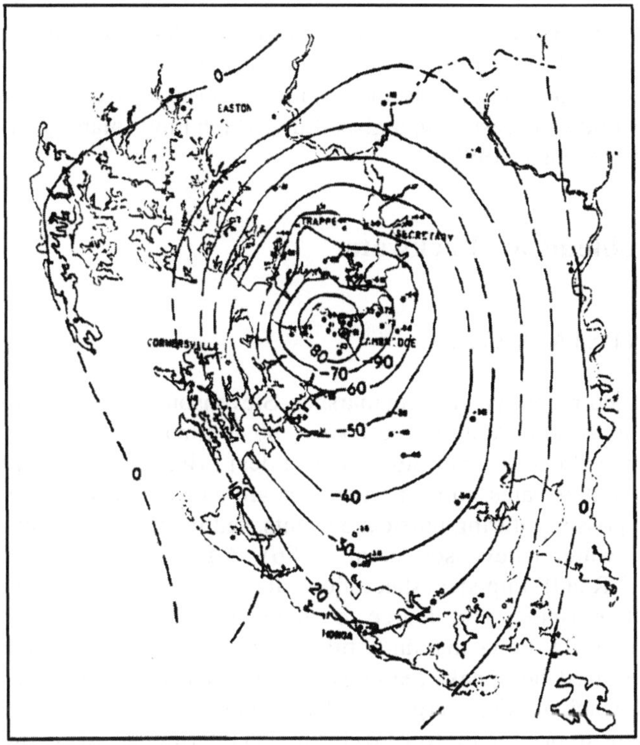

Figure 6. Potentiometric surface of the Piney Point aquifer centered at Cambridge, Maryland, May, 1976. Contours are in feet. Datum is mean sea level. In 1976, nearly all of Dorchester County, except for the southern-most reaches, was affected. (From: Rule 1995).

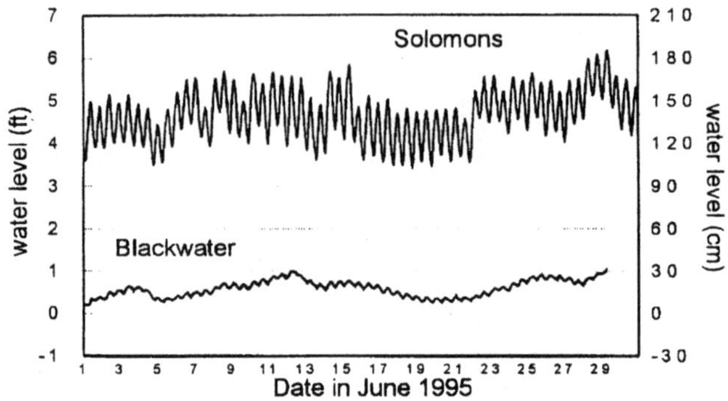

Figure 7. Tidal gauge data for June, 1995 at Blackwater NWR compared to mid-Bay at Solomons Island (30 km west of Blackwater). The Blackwater gauge is located 250 m southeast of the old refuge headquarters, on the Little Blackwater River. This is about 10.5 km upstream of Shorter's Wharf Bridge, which limits tidal exchange.

Shallow well observations suggest that there is now less groundwater available to wet the fresh and brackish portions of marshes (Stevenson, pers. obs.). Not only can this cause compaction of clays, sands and organic material of marsh sediments (Meade 1966), but less groundwater results in high sulfides which limits plant growth (Koch et al. 1990; Dewar and Stevenson 1998).

7. Other Problems at Blackwater

7.1 SEDIMENT STARVATION

In their review of sediment dynamics in marshes, Stevenson et al. (1988) found that most tidal systems along the mid-Atlantic coast were ebb dominated (i.e. out-going tides have higher maximum velocities than in-coming tides). Because sediment transport is regulated by water flow rates, more particulates are carried out of these systems on a net basis. Except possibly during hurricane events, sediments are not deposited by tidal flooding at Blackwater (Stevenson et al. 1986). Additionally, an 8 km long road (Shorter's Wharf Road) impedes sheet flow across the marsh surface and further reduces sediment transport. The degree to which tides are dampened is remarkable. Tide gauge data (Fig. 7) shows that daily tidal range is only about 6 cm in the center of the Refuge, while mid-Chesapeake readings at Solomons are in the range of 30 cm.

Because of the lack of tidal energy and of allocthonous sediment input, peat becomes increasingly organic as one moves towards the marsh interior and away from the levees of the tidal creeks. Another problem with restricting tidal flow is that water movement is accelerated in the channel beneath Shorter's Wharf Bridge (on the Blackwater River, approx. 15 km from the mouth). This causes a jet which is again ebb dominated and results in tremendous amounts of sediment being transported downstream and out of the Blackwater system. Researchers found that up to 720,000 metric tons of sediment were lost in one year from the Blackwater NWR upstream from Shorter's Wharf Road (Stevenson et al. 1985b). Even though there is considerable amount of farmland surrounding the Little Blackwater River, the sediment transport in agricultural runoff is not enough to balance that leaving the main channel of the system.

7.2 GRAZING BY ANIMALS AND WATERFOWL

Grazers can denude an area completely, leaving it exposed to the erosive forces of wind and water. At Blackwater, muskrats caused damage called ''eat-outs'' in the 1940s (Dozier et al. 1948). The nutria population has been increasing especially after the cold winters of 1977-79, when they were almost eliminated because of frostbite. One of the nutria's favorite foods is *Scirpus americanus* (Olney's Three-Square), a dominant plant in some parts of the refuge. Nutria and muskrats cause damage by eating the grass, by using the grass for shelter, and by digging up the marsh to get to the roots. In their study of fire and herbivory on brackish species in Louisiana, Ford and Grace (1998) found that all the dominant plant species they studied had higher biomass in plots where grazing was eliminated with exclosures.

7.3 BURNING MARSHES

It appears that the marshlands of Maryland's eastern shore have been burned periodically since settlement in the 1650s. Fire is advantageous to trappers because it clears out undesirable shrubs from the marsh and promotes the growth of *Scirpus americanus* (Chabreck 1982). The central part of Blackwater was a large fur farm before it became a National Wildlife Refuge and was burned routinely to encourage fur production (Dozier et al. 1948). Burning is now an integral part of the management practices carried out by the U.S. Fish and Wildlife Service at Blackwater. Although at first justified by trapping interests, it is now seen as a way to reduce the potential destructive power of massive fires and is encouraged by policy of the Department of the Interior (which emerged as a result of the fires at Yellowstone in 1988).

Studies in 1979 and 1980 at Blackwater (Pendleton and Stevenson 1983) indicated that burned plots had significantly greater culm density and above-ground biomass of *Scirpus americanus* than unburned plots (Table 3). A follow up study in 1981 corroborated the finding that plot above-ground and below-ground biomass was stimulated by burning (Table 4). A further clipping treatment suggested that the stimulation of growth was not only due to reduction of shading but also to increasing oxidation of the upper rhizosphere. The ratio of NO_3 to NH_4 was much lower in the unburned controls than burned plots (and clipped plots). Although these studies suggested positive short term effects of burning, Pendleton and Stevenson (1983) further estimated that from 380 to 460 g m^{-2} were removed during annual burns at blackwater. This translates into 2 to 3 mm of peat accretion (assuming dry bulk density of 0.14 g cm^{-3} measured in sediments at Blackwater) and appears to cover the amount needed to combat eustatic sea-level rise. Furthermore, recent studies in Louisiana suggest that unlike *Scirpus americanus*, biomass of other marsh species including *Spartina patens* is significantly reduced by burning and grazing (Ford and Grace 1998).

TABLE 3. Mean standing crops of *Scirpus olneyi* and other marsh vegetation (gdw m^{-2}) on burned (N=24) and unburned (N=12) tracts at Blackwater NWR, July 1980. All means are presented ± 1 S.E. Statistical significances are based on log-transformed data. From Pendleton and Stevenson 1983.

Vegetation Attribute	Gdw m^{-2}		
	Burned	Unburned	Significance
Living *S. olneyi*	550.6 ± 29.1	330.3 ± 69.9	P<0.05
Dead *S. olneyi*	337.4 ± 37.8	487.4 ± 50.7	P<0.05
Other plant biomass	56.5 ± 10.9	23.1 ± 9.4	NS
Total biomass	944.4 ± 47.6	840.8 ± 92.0	NS
S. olneyi culm density	745 ± 40	450 ± 78	P<0.05
Biomass/culm	0.8 ± 0.0	0.6 ± 0.1	NS

TABLE 4. Comparison of burned and hand-cleared (clipped) plots to unburned plots at Blackwater NWR during the 1981 growing season. All means are presented ± 1 Standard Deviation. From Pendleton and Stevenson 1983.

Variable	Treatment		
	Burned	Clipped	Unburned
Live culm density (m^{-2})	846 ± 167	784 ± 226	460 ± 31
Average culm height (cm)	72.6 ± 2.1	55.8 ± 2.1	77.6 ± 2.5
Live belowground biomass (mg dry wt cm^{-3})	31.3 ± 2.1	18.4 ± 1.1	14.4 ± 2.3

Pendleton and Stevenson (1983) concluded that "the Refuge's practice of burning each selected area once or twice every three years may be the best available compromise..." between long term peat accumulation and short term biomass production. However they did not take into account the possibility that more productive species would most likely replace *Scirpus americanus* if fire was eliminated from the system. Recently, Blackwater has been burned more frequently than Pendleton and Stevenson (1983) originally suggested. Fire frequency appears to be the most important factor in encouraging the growth of Olney's Three-Square (Ford and Grace 1998) which draws muskrats and nutria to the Refuge. Thus although not completely clear cut, it would appear that the present fire management program at Blackwater Refuge (and carried out in other managed marsh areas) has contributed to marsh losses. We are initiating a study to measure sedimentation and accretion at Blackwater burn vs. non-burn sites in fall of 1998. This study should clarify some of the issues mentioned above.

7.4 INCREASING SALINITY

Historically, the Blackwater River and Parsons Creek (near Taylor's Island) were not connected. In the early 1800's, Stewart's Canal was constructed to connect the two rivers in order to facilitate transport of lumber out of Moneystump Swamp. With a direct connection between the saltier Little Choptank River and the fresh headwaters of the Blackwater River, salt water intruded into the freshwater marshes and killed several of the less salt-tolerant plant species (Stevenson, personal observation). In the last decade, salinities have been especially high in the upper Blackwater River, killing species such as water lilies (*Nymphea odorata*). Decaying plant material increases anoxia (lack of oxygen in the water) which can be harmful to fish.

8. Past and Present Restoration Efforts at Blackwater

8.1 ELEVATION PLOTS

Small scale (>1m^2) elevation plots (Pendleton and Stevenson 1983) indicated significant improvement in marsh production when marsh plants were higher in the tidal spectrum at Blackwater. Using this experiment as a prototype, in the early 1980's, the U.S. Fish and Wildlife Service attempted to re-establish acre-sized plots in the center of the Refuge. The perimeters of the plots were rimmed with straw bales which were spiked into the mudflats to hold the sediments. Sediment from the channels of the marsh was dredged with a small ''Mudcat'' dredge. *Spartina alterniflora*, *Spartina patens*, *Scirpus americanus* and *Distichlis spicata* were sprigged into the sediment using armies of volunteers. However, the organic sediments compacted, producing low spots that did not support plant growth. Even where compaction was not significant, growth was not luxuriant most likely because of oxidation of pyrite in the sediment, producing low pH. Ice in successive winters completely scoured the plots and eventually even the straw bales disappeared leaving no trace of the restoration project. Calculations suggested that the restoration attempts were in the range of $20,000 per acre (and had failed). However, what limited further refinement of the technique was the lack of clean sandy sediment in local creeks to support further plots. The highly organic sediment now in the creeks is difficult for plant colonization because of poor aeration properties.

A series of snow fences were established in 1979-1980. Despite failures at most locations to attract sediment, occasionally marshes were restored. However, ice eventually eliminated the snow fencing. In the early 1990s, following the example of researchers in Louisiana, Christmas trees were collected from the public and placed in a 100-meter barrier in a location pre-determined by the USFWS to catch incoming sediment from the Little Blackwater River. Although the trees had much more staying power than the snow fencing, little sediment actually accumulated and no marshes formed on either side of the fencing. Again the sediment at Blackwater is very fine grained and flocculent and not the best substrate for *Spartina* spp. growth under low tidal ranges. Observations at Monie Bay suggest that sediments along the shoreline are much sandier and support luxuriant populations of *Spartina alterniflora* (Kearney et al. 1994).

8.2 GRAZING PLOTS

Although much of the preferred foods of muskrat and nutria at Blackwater have disappeared, the animals still exist in significant densities, and are thought to contribute to the continuing problems with marsh destruction. Although Pendleton and Stevenson (1983) reported no significant difference in plant growth when grazers were excluded from experimental plots in each of two experiments, the plots were perhaps too small (0.25 m^2 in 1979 and 1.0 m^2 in 1980) to realistically evaluate grazing impacts of fairly large animals. Larger grazing plots, 30 m x 30 m exclosures, are now being used to test the effects of grazing by nutria.

While the plots of the Pendleton and Stevenson (1983) study were designed to examine the effects of grazing, an interesting experimental artifact has been observed. These plots still exist, and in most cases, the area inside the plots is noticeably more productive than the adjacent areas outside (Rooth, personal observation). Exclusion of grazers over the long term (18 years) is probably a factor in the current success of these plots, but it is not the only factor. The fine mesh on the exclosures slows currents and causes more sediment deposition as water flows through the plots, which most likely caused the visible elevation of the marsh surface within the plots, and subsequently enhanced production. These observations led to the use of large-mesh enclosures in the current experiments.

9. Other Restoration Strategies

9.1 STOP BURNING

Although this is a controversial issue, not burning the marsh may allow the marsh to deposit organic material which now literally goes up in smoke. It could also allow the marsh to progress to a more mature successional stage with species less palatable to nutria. These plants include Black Needle Rush *(Juncus roemarianus)*, High Tide Bush or Groundsel Tree (*Baccharis halimifolia*) and Marsh Elder (*Iva frutescens*). One of the reasons trappers favor burning marshes on the Eastern Shore is to keep the shrubs out. However these are the very species which might be beneficial not only to help build marsh sediments but also for attenuating nutrient inputs.

9.2 CONTROL GRAZERS

Trapping and hunting muskrats, nutria, and waterfowl that destroy marsh grasses may help the marsh combat sea-level rise. Nutria in particular are so large that they can eat large quantities of marsh grasses. Because nutria are not indigenous to the Chesapeake, some wildlife biologists believe that with concerted effort, they could be eradicated completely, as they have been in England (Gosling and Baker 1989).

9.3 ARTIFICIALLY NOURISH MARSHES WITH SEDIMENT

In Louisiana, resource managers have sprayed sediments onto marshes from barges in the fight against sea-level rise and subsidence. Follow-up studies suggest that even when relatively crudely applied, sediment supplements to drowning marshes helped maintain their health (Cahoon and Cowan 1988). This technology uses a high-pressure spray which can jet a liquefied slurry 250 feet into the marsh. In addition to adding sediments to the marsh surface, this may be an acceptable alternative for disposal of dredge spoil from channel dredging projects, providing that the sediment supply consists of a large percentage of sand and not organic material (as in the Blackwater Creeks). Care must also be taken to spread the sediment as evenly as possibly so large mounds do not develop which are not intertidal.

9.4 INCREASE TIDAL ACTION

Constructing more culverts under roads such as Shorter's Wharf Road would increase tidal action across the marsh surface. E. P. Odum (1971,1980) first articulated the hypothesis that tidal action acted as an energy subsidy which accounts for the high productivity of Georgia salt marsh systems. Steever et al. (1976) later corroborated this hypothesis by demonstrating a high correlation between peak standing biomass of *Spartina alterniflora* and tidal range at several locations on Long Island Sound. Furthermore Morris et al. (1990) showed that annual variations in productivity were directly related to sea-level flooding frequencies. The greater the flooding frequency (and the less the interstitial salinities), the higher the rate of production. Besides benefiting marshes because of increased productivity, increased flooding can also lead to more sediment deposition on the interior marshes, which are now sediment starved (Stevenson et al. 1985b). Another benefit is that increased tidal circulation reduces anoxic conditions in the marsh. Thus increased circulation around Shorter's Wharf Road would have two major benefits for Blackwater NWR. Installation of culverts is one of the cheaper ''quick-fixes'' which might help slow down marsh loss considerably.

9.5 REDUCE SALT WATER INTRUSION INTO FRESH WATER MARSHES

The USFWS has proposed blocking Stewart's Canal which would prevent intrusions of high salinity water from the Little Choptank River into the freshwaters at the head of the Blackwater River. This would be beneficial to the traditional freshwater species (including riparian trees) in this region and prevent die off events. However, there may be trade-offs (both ecological and sociological) in this approach. First, blocking the canal reduces tidal action which could create even more anoxic conditions in the upper Blackwater River. Furthermore salt-tolerant species such as *Spartina alterniflora* which have now have been re-established along the higher salinity sections of Stewart's Canal would be displaced once again. Also, a water control structure would block navigation and thus far has been opposed by surrounding landowners and regulatory agencies. There are also serious doubts about whether a water control structure could withstand hurricane force winds, the effect a sudden breach would have on the Blackwater system, and whether public funds could be found quickly enough to re-establish a structure which has so little local support from the local population. Although water control structures have been used in Louisiana for marsh management (Cahoon 1994, Cahoon and Groat 1990), recent quantification of their effectiveness in the Barataria basin indicates that plant production is not increased (Johnson and Foote 1997). In fact, biomass of *Scirpus americanus* was significantly lower in areas where water control structures have been installed. Finally, the question remains whether sealing just one entrance to the Upper Blackwater is enough to stop intrusions. Presently, there is also high salinity water entering the Upper Blackwater from the upper parts of World's End Creek, a tributary of the Honga River.

9.6 RE-ALLOCATE GROUNDWATER WITHDRAWALS

Decreasing the amount of groundwater withdrawn from the aquifer would reduce underlying compaction and reduce relative sea-level rise. This technique has apparently been successful in reducing subsidence in the Venice Lagoon in Italy (Stevenson et al., in press). In Venice, an 80 km long aqueduct was constructed to transport water from mountain lakes to the city. Cambridge may be able to exploit desalination technology (reverse-osmosis filtration) to convert Choptank River water into safe drinking water. Currently there is a debate concerning the future allocation of water from aquifers such as the Piney Point aquifer. We need to begin to consider environmental impacts of water withdrawal in areas where marsh loss (and land subsidence) is prevalent.

9.7 REDUCE GREENHOUSE GAS EMISSIONS

Unless we plan on writing off large amounts of land area in Maryland (e.g. Dorchester, Somerset and Wicomico and Worcester Counties) over the next couple centuries, a more aggressive policy of reducing greenhouse gases needs to be formulated. This could include increased gasoline and power plant taxes on the state and/or federal levels to help compensate for losses which will be incurred by riparian property owners as global sea-levels rise. On other issues it has been shown that when pollution is taxed, there is often a favorable market adjustment toward conservation. Ultimately, the U.S. needs to support international efforts to bring greenhouse gas emissions under control.

9.8 DO NOTHING

If we decide to do nothing, the extensive marshland that is left in Dorchester County will become open water embayments. Unfortunately, the amount of organic material and turbidity severely limits the growth of submersed aquatic plants, and the embayment is not at present suitable habitat for fish and crabs. Measurements show that because of the excessive amount of organic debris in the ponds from degraded marsh plants, the area becomes sporadically anoxic (Fig. 8). It could take decades to centuries for these embayments to become productive environments.

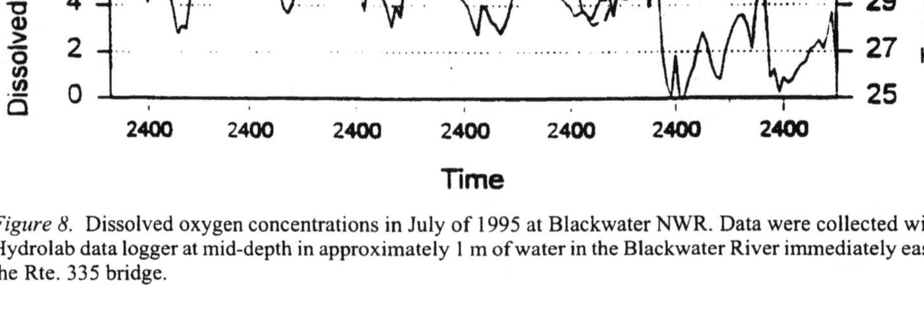

Figure 8. Dissolved oxygen concentrations in July of 1995 at Blackwater NWR. Data were collected with a Hydrolab data logger at mid-depth in approximately 1 m of water in the Blackwater River immediately east of the Rte. 335 bridge.

10. A Novel Approach

One of the least expensive and possibly most effective approaches in controlling loss in brackish zones of eroding marshes may be to selectively use highly productive species to slow and possibly even arrest deterioration of marshes. *Phragmites australis* is a perennial grass reaching heights between 1.5 and 4 m in the Chesapeake region (Brown and Brown 1984) and is an active colonizer of wet areas found everywhere throughout the world except Antarctica (Tucker 1990). *Phragmites australis* appears to be native to the east coast as it was mentioned by Thomas Jefferson in 1787 (Jefferson 1982) and has been found in peat dated 3000 years old (Niering and Warren 1977, Metzler and Rozsa 1987, Orson et al. 1987). However this is a very polymorphic species having different ploidy levels. In fact, Chrysler (1910) shows a photo of a stand of *P. australis* along the Patuxent River with culms appearing to have more numerous branching of leaves than stands of invasive *P. australis*, and perhaps the photo captured the east coast's native genotype. It is possible that many of the biotypes of *P. australis* currently on the east coast of the United States were imported from Europe since they first were reported around the ports of Baltimore, Philadelphia and New York (Chrysler 1910, Tucker 1990, Besitka 1996).

Rapid expansion of monotypic stands of *P. australis* has occurred over the last 40 years in many Atlantic coast wetlands (Marks et al. 1994, Rice and Stevenson 1996). For example, nearly one third of Delaware's coastal wetlands are now occupied by *P. australis* (Jones and Lehman 1987, Hellings and Gallagher 1992). In the Chesapeake Bay region, Rice and Stevenson (1996) found increased *P. australis* expansion in all six tidal freshwater and brackish marshes examined, and the intrinsic rate of increase was greater in the more recently colonized brackish systems. Older, more established stands may reach an equilibrium point where the rate of increase levels off before *P. australis* can occupy more than 50% of the marsh (Rice et al., in prep).

Although the appearance of this species is recognized as detrimental to the marsh ecosystem in the U.S. (i.e., reducing vegetative biodiversity, animal habitat and food for waterfowl; Marks et al. 1994), the *P. australis* communities of Europe are considered very important components in the wetland landscape (Ostendorp 1989, 1993). They are valued for pollution control (high nutrient uptake), as well as habitat and food for mammals and birds. Most importantly for marshes undergoing erosion, *P. australis* rhizomes anchor deeply into the sediment and thus help stabilize shorelines. In comparison with a variety of wetland shrubs and plants, *P. australis* affords superior bank protection and erosion control (Ostendorp 1993). For instance, *P. australis* reduces erosion by entrapping sediment within the vegetation to a much greater extent than observed for *Scirpus lacustris* in an experimental wave tank, resulting in overall reduction of wave loading on the coastal edge (Coops et al. 1996). Additionally, *P. australis* communities reduce resuspension of fine material (Takeda and Kurihara 1988) and increase retention of sand particles at high stem densities (Knutson 1988).

The effectiveness of *P. australis*' ability to reduce erosion is largely due to the high above ground productivity and slow breakdown and decomposition of senesced plant material. Culms remain standing from 1 to 4 years after dying (Graneli 1989). Consequently, large quantities of litter accumulate on the marsh surface, and enhanced organic deposition often occurs if the litter is not readily exported from the marsh. This phenomenon is observed at a subsiding marsh along Little Creek in the Monie Bay National Estuarine Research Reserve, and nearby at Little Deal Island, where active coastal erosion is occurring (Rooth and Stevenson 1998). In this ongoing study, *P. australis* communities are exhibiting a pattern of higher total and organic deposition when compared to nearby *Spartina* sp. communities (Fig. 9). Fig. 9b shows a clear trend of higher total deposition at Little Creek in the *P. australis* community versus *Spartina* spp. at all four stations representing the coastal edge to the marsh interior. The evidence at Little Deal Island is not as definitive due to higher litter export, but suggests that at three of the four stations sampled the deposition is higher in *P. australis* (Fig. 9a). Storm events cause an accumulation of approximately 80% mineral sediment that can be as much as three times higher in the *P. australis* community than in the *Spartina* sp. Therefore, in the Chesapeake Bay, *P. australis* communities are potentially counteracting the effects of rising sea level (both from subsidence and anthropogenic causes) by increasing both organic and mineral deposition on the marsh surface.

These findings suggest that because *P. australis* enhances overall sediment

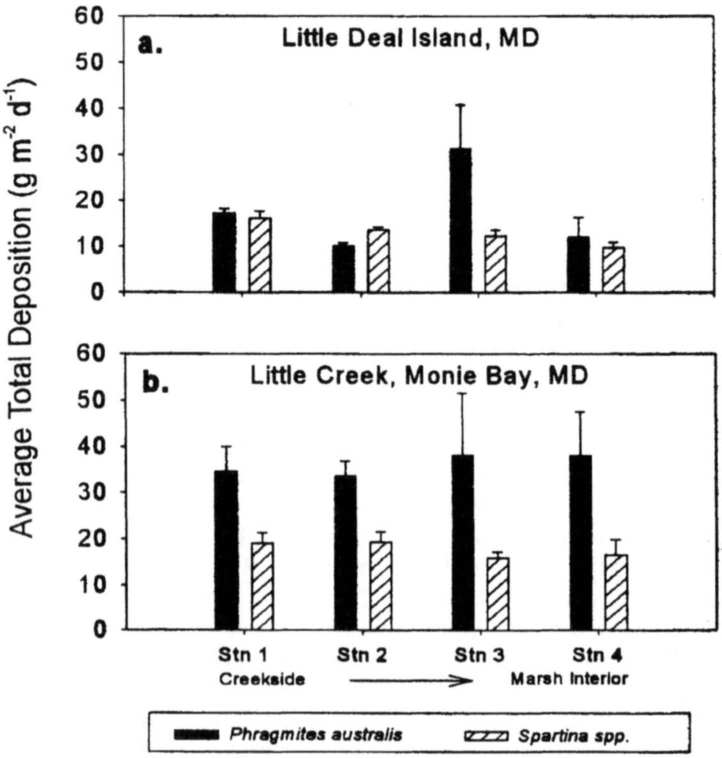

Figure 9. Average total deposition in Phragmites australis and Spartina sp. at (a) Little Deal Island, MD (erosional marsh) and (b) Little Creek, Monie, MD (subsiding marsh). For both sites we used 6 replicate filter paper plates (Reed 1989) at each of 4 stations, with Station 1 at the marsh edge (creekside) and Stations 2 to 4 at intervals into the marsh interior of 3, 10, and 17 m, respectively. We sampled on 6 occasions at Little Deal Island (n=36) between July and November 1997; 4 times at Little Creek (n=24) in August and September 1997.

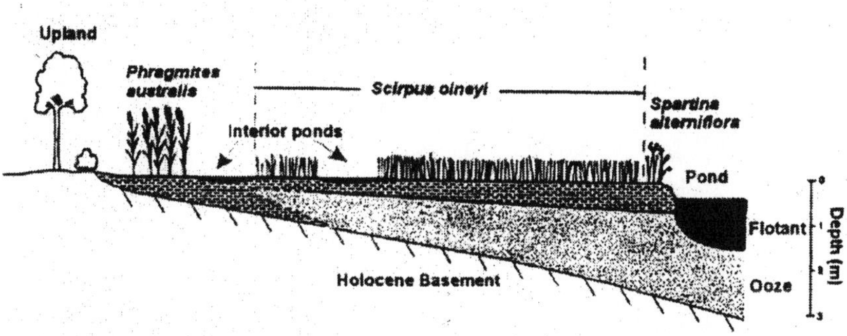

Figure 10. Generalized profile of a submerged upland marsh at Blackwater NWR showing the establishment of P. australis at the landward edge where incipient submergence is occurring. Without enhanced peat deposition such as that provided by productive macrophytes, this area often deteriorates first to rotten spots, and eventually to enlarging ponds which are unproductive due to periodic anoxia and turbidity. As has been observed at Blackwater NWR, the ponds coalesce, culminating in complete marsh loss.

deposition, current management practices at Blackwater should incorporate planting and maintaining *P. australis* communities. In addition to increasing the vertical accretion potential of the marsh, *P. australis* can reduce soil moisture if growing along a terrestrial edge (Rooth unpublished data) via extremely high rates of evapotranspiration (Ostendorp 1993) and creates a greater depth to watertable (Windham 1995). Therefore, we are implying that autogenic succession in the *P. australis* community acts to increase marsh elevation by altering the substrate over time, and this inevitably reduces marsh degradation. It has also been demonstrated that when interspersed with open water and available food sources, *P. australis* is a suitable habitat for waterfowl (Cross and Fleming 1989). Although the planting of an invasive grass in key areas of wetland loss appears unorthodox, the alternative is continued loss of extensive areas of marsh. *P. australis* may not be the preferential vegetative type in a waterfowl sanctuary, but it may be one way of stabilizing the land interface (Fig. 10) that otherwise is particularly vulnerable to sea-level rise because of sediment starvation and compaction.

11. Conclusions

Sea-level rise and subsidence need to be incorporated into tidal marsh restoration projects for long-term success. Although few places along the mid-Atlantic coast face a combination of environmental factors that make marsh health and restoration as tenuous a proposition as at Blackwater NWR, recent assessment of the health of Chesapeake marshes suggests widespread problems (Fig. 11). Similar problems have emerged in analyzing marsh health on the Delaware estuary (Kearney et al., unpub. data). It is likely that the 4050 ha restoration project now being carried out by Public Service Electric and Gas (PSE&G) in New Jersey may also be impacted by ''sea-level rise.'' Historically, these marshes were largely salt hay farms but increasing submergence in the 1920s and 1930s forced farmers to install seaward dikes to reduce

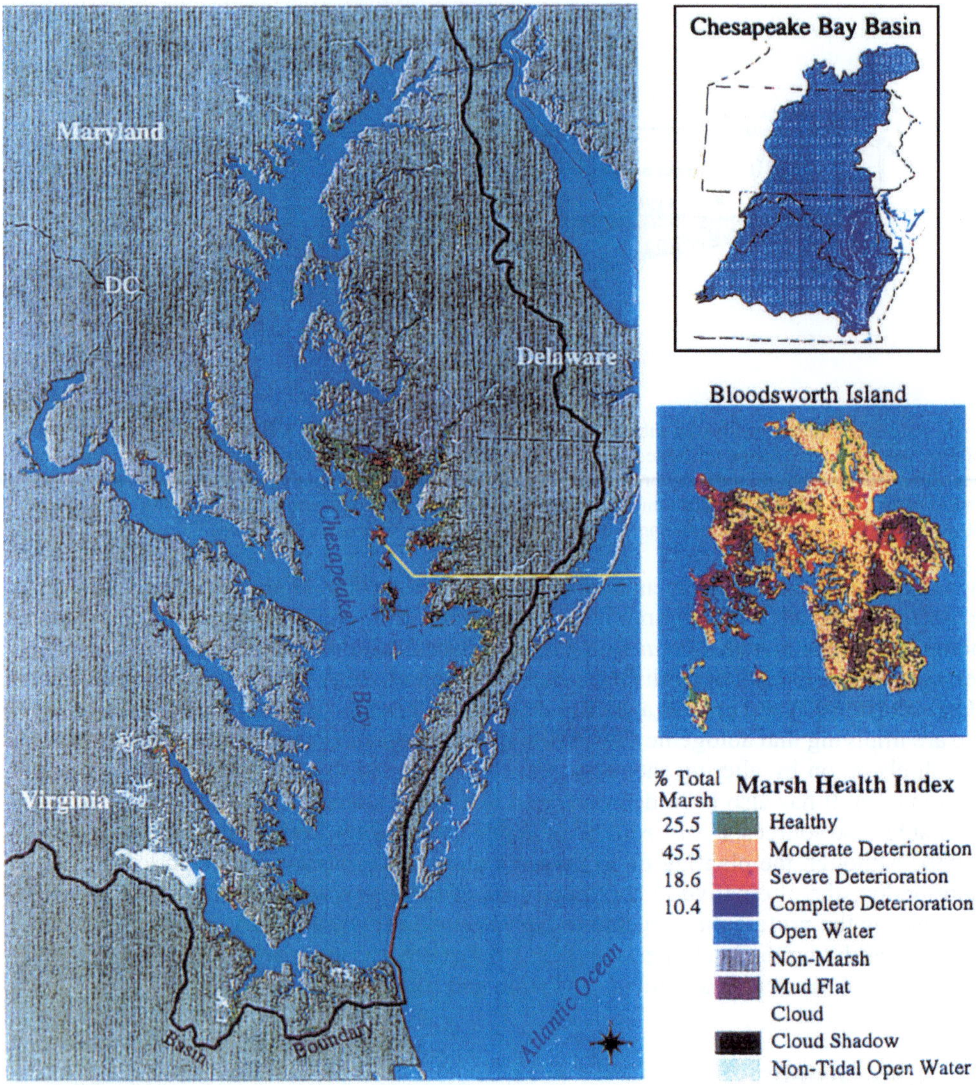

Figure 11. Health of Chesapeake Bay marshes in 1993.

the excessive tides. This resulted in a much drier substrate that caused oxidation of peats, compaction of the substrate and an ultimate shift in species composition towards *Phragmites australis* in drier areas after World War II. Where the dikes impounded sea-water, hypersaline conditions persisted and caused localized invasion of highly salt tolerant species such as *Suaeda maritima*. Over the last three years, PSE&G has had short term success in restoring *Spartina alterniflora* by opening up the seaward dikes to tidal action allowing sediment from the Delaware Bay to re-enter the systems via flood dominated hydraulics.

However, longer-term prospects for the PSE&G project depend on solving the underlying problems of sea-level rise and submergence which drove the original salt-hay farmers to create the seaward dikes in the first place. It is clear from analysis of old maps and plats of the area that property lines of the hay farms extended much further into Delaware Bay and are now under water. One of the most problematic aspects for the future of the PSE&G project may lie in the installation of landward dikes to prevent tidal flooding into contiguous landowners' properties during hurricane and other storm events. This barrier effectively prevents any avenue of inward migration for these marshes as sea-levels continue to rise. Unless global warming, sea-level rise and local subsidence are slowed, the ultimate viability of the PSE&G marshes is in doubt. It is slightly ironic that the reason for the large marsh project is a nuclear facility. It does not contribute significantly to the CO_2 problem, yet its large mitigation project suffers from the output of competing fossil fuel plants which have not had to carry out any mitigative projects for the environmental damage they are causing to coastal wetlands.

Blackwater NWR and the PSE&G project are just two examples of many tidal marsh systems throughout the world which face daunting prospects because of complications involving apparent sea-level rise in coastal systems. Perhaps the most publicized groundwater withdrawal problems and resulting land subsidence are in the Venice Lagoon, which was once largely marsh (Gambolati et al. 1974, Stevenson et al. 1999). Over the last several decades the Italian government has attempted to reduce subsidence at Venice by severely curtailing groundwater withdrawals at the industrial centers of Mestre and Marghera next to the lagoon. However, recent attempts to arrest or even slow marsh losses at Punta Cane have been very difficult because of wave energy during storms and other anthropogenic effects that have altered this coastal lagoon's sediment balance towards long-term erosion (Day et al. 1998). Many of the hydraulic modifications (including river diversions, dike building along the edges, ship channel dredging and inlet stabilization) have combined to make the Venice lagoon into an ebb dominated system which debauches sediments needed for marsh accretion into the Adriatic Sea (Gatto and Caribognin 1981). As in Blackwater NWR and the Delaware River Estuary, the trade-off in the Venice Lagoon is not a loss in marshes for a gain submersed aquatic vegetation because of the turbidity induced when organic sediments are eroded into the shallows. Instead, phytoplankton and benthic algae increase, causing increased anoxia and ultimately less desirable habitat for a variety of estuary-dependent species.

Solving many of these complex environmental problems in impacted coastal systems will take the concerted effort of numerous regulatory agencies. This only underscores Norton's (1991) general observation that the narrow field of resource management (where one worries primarily about the health of a single resource) must be expanded into environmental management (where the comprehensive health of the ecosystem is considered). Only then can these complex issues concerning tidal marsh health and restoration be adequately addressed. Although there is obviously a large economic cost to society in reducing greenhouse gas emissions and in controlling land subsidence, the costs of drowning coastal shorelines (including private properties and marshes) and of global warming in general are also large (Wigley et al. 1996). Even when increased plant growth in the higher northern latitudes (Myneni et al. 1997) is factored in, we predict that it will become increasingly obvious that greenhouse gas emissions need to be curtailed in order to prevent massive resource losses. Of course controlling atmospheric inputs will not solve local groundwater withdrawal problems and these will obviously have to be addressed by regional governmental bodies. Since many groundwater issues cross state boundaries, the federal government should take a leadership role in assessing where the subsidence hot-spots are (using various satellite technologies) and in applying already developed 3-D models to predict where subsidence is likely to result from groundwater withdrawals. If these efforts are successful, it may obviate the need for ecologists to quarrel over the establishment of the much-scorned *Phragmites australis* in tidal wetlands facing high rates of RSL rise.

12. Acknowledgements

We would like to express appreciation for past collaboration and discussions with William Boicourt, Jeff Cornwell, John Day, Bruce Douglas, Edward Pendleton, Denise Reed and Larry Ward. This paper is a synthesis of results from several research projects funded primarily from the National Oceanic and Atmospheric Administration (NOAA), C-Cap via Maryland Sea-Grant and NASA with ancillary support from NOAA - National Estuarine Research Reserve Program, and Coastal Zone Program (via the Maryland Department of Natural Resources) as well as the U.S. Fish and Wildlife Service and U.S. Environmental Protection Agency, Multiscale Experimental Ecosystem Research Center (MEERC). This is contribution #3163 from the University of Maryland, Center for Environmental Sciences.

13. Literature Cited

Barras, J.A., P.E. Bourgeois, L.R. Handley. 1994. Land loss in coastal Louisiana 1956-1990. National Biological Survey, National Wetlands Research Center, Open File Report **94-01**.

Besitka, M.A. 1996. An ecological and historical study of *Phragmites australis* along the Atlantic Coast. Thesis. Drexel University, Philadelphia, Pennsylvania, USA.

Boyer, B.W. 1997. Sinkholes, soils, fractures and drainage: Interstate 70 near Frederick, Maryland. Environmental and Engineering Geoscience **3**:469-485.

Brown, M.L. and R. G. Brown. 1984. Herbaceous Plants of Maryland. Port City Press, Baltimore, Maryland, USA.

Cahoon, D.R. 1994. Recent accretion in two managed marsh impoundments in coastal Louisiana. Ecological Applications **4**:166-176.

Cahoon, D.R. and J.H. Cowan, Jr. 1988. Environmental impacts and regulatory policy implications of spray disposal of dredged material in Louisiana wetlands. Coastal Management **16**:341-362.

Cahoon, D.R. and C.G. Groat, editors,. 1990. A study of marsh management practice in coastal Louisiana. Volume I: Executive summary. Final report submitted to Minerals Management Service, New Orleans, Contract No. 14-12-0001-30410, OCS Study/MMS **90-0075**.

Chabreck, R. H. 1982. Effects of burn date and regrowth rate of *Scirpus olneyi* and *Spartina patens*. Proceedings of the Annual Conference of Southeastern Association of Fish and Wildlife Agencies **35**:201-210.

Chrysler, M. A. 1910. The ecological plant geography of Maryland, coastal zone, western shore district. Pages 149-193 *in* F. Shreve, M.A. Chrysler, F. H. Blodgett and F. W. Besley, editors. The plant life of Maryland. The Johns Hopkins Press, Baltimore, Maryland, USA.

Coops, H., N. Geilen, H. J. Verheij, R. Boeters, G. van der Velde. 1996. Interactions between waves, bank erosion and emergent vegetation: an experimental study in a wave tank. Aquatic Botany **53**:187-198.

Cross, D.H. and K. L. Fleming. 1989. Control of *Phragmites* or common reed. U.S. Fish and Wildlife Service Report CO.RWS/OIT-**13.4.12**: 1-5.

Culliton, TJ., M.A. Warren, L.R. Goodspeed, D.G. Remer, C.M. Blackwell and J.J. McDonough. 1990. The second report of a coastal trends series: 50 years of population change along the nation's coasts, 19602010. Strategic Assessment Branch, National Oceanic and Atmospheric Administration, Rockville, Maryland, USA.
Curray, J. R. 1965. Late quaternary history, continental shelves of the United States. Pages 723-735 *in* H. E. Wright and D. G. Frey, editors, The Quaternary of the United States. Princeton University Press, Princeton , New Jersey, USA.
Curtis, P. B., B. G. Drake, P. W. Leadley, W. J. Arp and D. F. Whigham 1989a. Growth and senescence in plant communities exposed to elevated CO_2 concentrations on an estuarine marsh. Oecologia **78**:20-26.
Curtis, P. B. and B.G. Drake and D.F. Whigham 1989b. Nitrogen and carbon dynamics in C_3 and C_4 estuarine marsh plants grown under elevated CO_2 *in situ*. Oecologia **78**:297-301.
Darmody, R. G. and J. E. Foss. 1979. Soil-landscape relationships of the tidal marshes of Maryland. Soil Science of America Journal **43**: 534-541.
Darmody, R.G. 1975. Reconnaissance survey of the tidal marsh soils of Maryland. Thesis, University of Maryland, College Park, Maryland, USA.
Davis, G. H. 1985. Land subsidence and sea-level rise on the Atlantic Coastal Plain of the United States. Environmental Geology and Water Science **10**:67-80.
Day, J. W. Jr., F. Scarton, A. Rismondo and D. Are. 1998. Rapid deterioration of a salt marsh in Venice Lagoon, Italy. Journal of Coastal Research **14**: 583-590.
Dewar, J.L. and J.C. Stevenson. 1998. High nitrate groundwater effects on a sandy fringe marsh mesocosm. Presentation, concepts and controversies in tidal Marsh ecology, Vineland, New Jersey, USA, April. 1998.
Douglas, B.C. 1991. Global sea-level rise. Journal of Geophysical Research **96**:6981-6982.
Douglas, B.C. 1995. Global sea-level change: determination and interpretation. U.S. National Representative of the International Union Geodetics and Geophysics. 1991-1994. Review of Geophysics **33**:1425-1432.
Dozier H. L., M. H. Markley and L. M. Lewllyn. 1948. Muskrat investigations on the Blackwater National Wildlife Refuge, Maryland, 19411945. Journal of Wildlife Management **12**:177-190.
Fairbanks, R. G. 1989. A 17,000-year glacio-eustatic sea-level record: Influence of glacial melting rates on the Younger Dryas event and deep-ocean circulation. Nature **342**:637-642.
Fleming, J. E. 1998. Historical Perspectives on Climate Change, Oxford University Press, New York, New York, USA.
Ford, M. and J. B. Grace. 1998. The interactive effects of fire and herbivory on a coastal marsh in Louisiana. Wetlands **18**:1-9.
Foss, J. E., D. S. Fanning, F. P. Miller and D. P. Wagner. 1978. Loess deposits of the eastern shore of Maryland. Soil Science Society of America Journal **42**: 329-334.
Froomer, N. L. 1980. Sea-level changes in the Chesapeake Bay during historic times. Marine Geology **36**: 289-305.
Gambolati, G., P. Gatto and R. A. Freeze. 1974. Predictive simulation of subsidence of Venice. Science **183**:849-851.
Gatto P. and L. Carbognin. 1981. The lagoon of Venice: natural environmental trend and man-induced modification. Hydrological Science Bulletin **26**: 379-391.
Gosling, L.M. and S.J. Baker. 1989. The eradication of muskrats and coypus from Britain. Biological Journal of the Linnean Society **38**:39-51.
Graneli, W. 1989. Influence of standing litter on shoot production in reed, *Phragmites australis* (Cav.) Trin. Ex. Steudel. Aquatic Botany **35**:99-109.
Hellings, S. E and J. L. Gallagher. 1992. The effects of salinity and flooding on *Phragmites australis*. Journal of Applied Ecology **29**: 41-49.
Holdahl, S. R. and N. L. Morrison. 1974. Regional investigations of vertical crustal movements in the U. S., using precise relevelings and mareograph data. Tectonophysics **23**:373-390.
Jefferson, T. 1982. Notes on the State of Virginia. W.Peden, editor, The University of North Carolina Press, Chapel Hill, North Carolina, USA.
Johnson, A. I. and D. A. Morris. 1962. Relation of volumetric shrinkage to clay content of sediments from the San Joaquin Valley, California. U. S. Geological Survey Professional Paper **450B**, 59 pp.
Johnson, L.A. and A.L. Foote. 1997. Vertebrate herbivory in managed coastal wetlands: a manipulative experiment. Aquatic Botany **59**:17-32.
Jones, E. 1902. History of Dorchester County. Williams and Wilkins Co., Baltimore Maryland, USA.

Jones, W. L. and W. C. Lehman. 1987. *Phragmites* control and revegetation following aerial applications of glyphosate in Delaware. Pages 184-199 *in* W.R. Whitman and W. H. Meredith, editors. Waterfowl and wetlands symposium, Dover, Delaware, USA.
Kearney, M.S. 1996. Sea-level change during the last thousand years in Chesapeake Bay. Journal of Coastal Research, **12**:977-983.
Kearney, M.S., R. E. Grace and J. C. Stevenson. 1988. Marsh loss in the Nanticoke Estuary, Chesapeake Bay. Geographical Review **78**:205-220.
Kearney, M S. and J. C. Stevenson. 1991. Island land loss and marsh vertical accretion rate evidence for historical sea-level changes in Chesapeake Bay. Journal of Coastal Research **7**:403-415.
Kearney, M.S., J. C. Stevenson and L. G. Ward. 1994. Spatial and temporal changes in marsh vertical accretion rates at Monie Bay: implications for sea-level rise. Journal of Coastal Research **10**: 1010-1020.
Knutson, P. L. 1988. Wave damping in *Spartina alterniflora* marshes. Wetlands **2**:87-104.
Koch, M.S., I.A. Mendelssohn and K.L. McKee. 1990. Mechanism for the hydrogen sulfide-induced growth limitation in wetland macrophytes. Limnology and Oceanography **35**:399-408.
Mack, F.K., W.E. Webb and R.A. Gardner. 1971. Water resources of Dorchester and Talbot Counties, Maryland, with special emphasis on the ground-water potential of the Cambridge and Easton areas. Report of Investigations No. **17**, Maryland Geological Survey.
Marks, M., B. Lapin and J. Randall. 1994. *Phragmites australis* (*P. communis*): threats, management and monitoring. Natural Areas Journal **14**:285-294.
Meade, R. H. 1966. Factors influencing the early stages of the compaction of clays and sands—a review. Journal of Sedimentary Petrology. **36**: 1085-1101.
Mendelssohn, I. A., K. L. McKee and W. H. Patrick. 1981. Oxygen deficiency in *Spartina alterniflora* roots: metabolic adaptation to anoxia. Science **214**:214-439-441.
Metzler, K. and R. Rozsa. 1987. Additional notes on the tidal wetlands of the Connecticut River. Newsletter of the Connecticut Botanical Society **15**:1-6.
Morris, J. T. B. Kjerfve and J. M. Dean. 1990. Dependence of estuarine productivity on anomalies in mean sea-level. Limnology and Oceanography **24**:960-966.
Myneni, R. B., C. D. Keeling, C. J. Tucker, G. Asar and R. R. Newmani. 1997. Increased plant growth in the northern high latitudes from 1981 to. 1991. Nature **386**:698-702.
Nerem, R.S., T.M. van Dam and M.S. Schenewerk. 1998. Chesapeake Bay subsidence monitored as wetlands loss continues. EOS Transactions, American Geophysical Union **79**:149,156-157.
Niering, W.A. and R.S. Warren. 1977. Our dynamic tidal marshes: vegetation changes as revealed by peat analysis. Connecticut Arboretum Bulletin **12**.
Norton, B. G. 1991. Ecological health and resource management Pages 102-117 *in* R. Costanza, editor. Ecological economics. Columbia University Press, New York, New York, USA.
Odum, E. P. 1971. Fundamentals of Ecology. W.B. Saunders Co., Philadelphia, Pennsylvania, USA.
——. 1980. The status of three ecosystem-level hypotheses regarding salt marsh estuaries: tidal subsidy, outwelling and detritus-based food chains. Pages 485-495 *in* V. S. Kennedy, editor. Estuarine perspectives, Academic Press New York, New York, USA.
Odum, W.E., T.J. Smith III, J.K. Hoover and C.C. McIvor. 1984. The ecology of tidal freshwater marshes of the United States. Biological Services Program (FWS/OBS-**83/17**). Slidell, Louisiana: United States Department of Interior, Fish and Wildlife Service.
Orson, R.A., W.A. Niering and R.S. Warren. 1987. The development of a New England river valley tidal marsh. Estuaries **10**:20-27.
Ostendorp, W. 1989. Die-back of reeds in Europe - a critical review of the literature. Aquatic Botany **35**:5-26.
——. 1993. Reed bed characteristics and significance of reeds in landscape ecology. Pages 149-161 *in* W. Ostendorp and P. Krumscheid-Plankert, editors. Lakeshore deterioration and restoration works in central Europe. Gustav Fischer, New York, New York, USA.
Pendleton, E. and J. C. Stevenson. 1983. Investigations of marsh losses at Blackwater Refuge. Final Report to U.S. Fish and Wildlife Service and Tidewater Administration of Maryland Department of Natural Resources, Annapolis, Maryland, USA.
Poland, J. F. 1972. Land subsidence in the western United States due to groundwater overdraft. Water Resources Bulletin **8**:118-131.
Poland J. F. and G. H. Davis. 1969. Land subsidence due to withdrawal of fluids. Pages 187-269 *in* D. J. Varness and G. Kiersch, editors. Reviews in engineering geology, Volume 2, Geological Society of America, Boulder, Colorado, USA.

Rampino, M. E. and J. Sanders. 1981. Episodic growth of Holocene tidal marshes in the northeastern United States: a possible indicator of eustatic sea-level fluctuations. Geology **9**: 63-67.

Reed, D.J. 1989. Patterns of sediment deposition in subsiding coastal salt marshes, Terrebone Bay, Louisiana: the role of winter storms. Estuaries **12**: 222-227.

Rice, D. and J. C. Stevenson. 1996. The distribution and expansion rate of *Phragmites australis* in six marshes in Chesapeake Bay. Pages 467-476 *in* C.A. Hallam, J. M. Salisbury, K. L. Lanfer and W.A. Battaglin, editors. Proceedings of the American Water Resource Association annual (AWRA) symposium, GIS and Water Resources, AWRA, Herndon, Virginia, USA.

Rooth, J.E. and J.C. Stevenson. 1998. Vertical accretion in *Phragmites australis* and *Spartina* spp. communities in mid-Atlantic marshes: implications for coastal areas threatened by rising sea-level. Presentation, concepts and controversies in tidal marsh ecology, Vineland, New Jersey, USA, April. 1998.

Rule, T.C. 1995. Groundwater withdrawal, land subsidence and marsh loss in the Chesapeake Bay region of Maryland. Thesis, University of Maryland, College Park, Maryland, USA.

Schroeder, W. W. and A. Shultz and O. H. Pilkey. 1995. Late Quaternary oyster shells and sea-level history, Inner Shelf, northeast Gulf of Mexico. Journal of Coastal Research **11**:664-674.

Stauffer, B. , T. Blunier, A. Dällenbach, A. Indermühle, J. Schwander, T. F. Stocker, T. Tshumi, J. Chappellaz, D. Raynaud, C.U. Hammer and H. B. Clausen. 1998. Atmospheric CO_2 concentration and millennial-scale climate changes during the last glacial period. Nature **392**:59-62.

Steever, E. Z., R. S. Warren and W. A. Neiring. 1976. Tidal energy subsidy and standing crop production of *Spartina alterniflora*. Estuarine, Coastal and Marine Science **4**:473-478.

Stevenson, J. C., L.G. Ward, M.S. Kearney and T.E. Jordan1985a. Sedimentary processes and sea-level rise in tidal marsh systems of Chesapeake Bay, Pages 37-62. *in* H.A. Groman et al., editors. Wetlands of the Chesapeake. Environmental Law Institute, Washington, District of Columbia, USA.

Stevenson, J. C., M.S. Kearney and E.C. Pendleton. 1985b. Sedimentation and erosion in a Chesapeake Bay brackish marsh system. Marine Geology **67**:213-235.

Stevenson, J. C., L.G. Ward and M.S. Kearney. 1986. Vertical accretion in marshes with varying rates of sea-level rise. Pages 241-259 *in* D. Wolfe, editor. Estuarine variability. Academic Press, New York, New York, USA.

Stevenson, J. C., L.G. Ward and M.S. Kearney. 1988. Sediment transport and trapping in marsh systems: implications of tidal flux studies. Marine Geology **80**:37-59.

Stevenson, J. C. and M. S. Kearney. 1996. Shoreline dynamics on the windward and leeward shores of a large temperate estuary. Pages 233-259 *in* K. F. Nordstrom and C. T. Roman, editors. Estuarine shores: evolution, environments and human alterations. John Wiley & Sons, Chichester, England.

Stevenson,J. C., I. J. Marusic, B. Ozretic, A. Marson, G. Cecconii and M. S. Kearney. 1999. Shallow water and shoreline ecosystems of the Chesapeake Bay compared to the Northern Adriatic Sea: transformation of habitat at the land-sea margin. Pages 29-80 *in* T.C. Malone, A. Malej, L.W. Harding, N. Smodlaka and R.E. Turner, editors. Ecosystems at the land-sea margin: drainage basin to coastal sea. American Geophysical Union, Washington, District of Columbia, USA.

Takeda, S. and Y. Kurihara. 1988. The effects of reed, *Phragmites australis* (Trin.), on substratum grain-size distribution in a salt marsh. Journal of the Oceanographical Society of Japan **44**:103-112.

Tucker, G. C. 1990. The genera of Arundinoideae (Gramineae) in the Southeastern United States. Journal of the Arnold Arboretum **71**:145-177.

Wigley, T. M. L., R. Ricchels and J.A. Edmonds. 1996. Economic and environmental choices in the stabilization of atmospheric CO_2 concentrations. Nature **379**:240-243.

Windham, L. 1995. Effects of *Phragmites australis* invasion on aboveground biomass and soil properties in brackish tidal marsh of the Mullica River, New Jersey. Thesis, Rutgers - The State University of New Jersey, New Brunswick, New Jersey, USA.

SOIL ORGANIC MATTER (SOM) EFFECTS ON INFAUNAL COMMUNITY STRUCTURE IN RESTORED AND CREATED TIDAL MARSHES

STEPHEN W. BROOME
CHRISTOPHER B. CRAFT
WILLIAM A. TOOMEY, JR.
North Carolina State University
Box 7619
Raleigh, NC 27695-7619 USA

Abstract
Accumulation of significant quantities of organic matter is an important characteristic of anaerobic soils that influences the physical, chemical and biological processes of wetland ecosystems. Organic matter effects include soil water holding capacity, porosity, nutrient storage, nutrient cycling and species composition and abundance of sediment-dwelling invertebrates. These infauna are thought to be important links in transferring primary production from the marsh to the estuarine food web. Tidal marsh restoration and creation often occur on mineral soils that contain little or no organic carbon, and research results indicate that low SOM contents are associated with lower functional value of wetlands. The objectives of this paper are to review the literature and assess the relationship of SOM quantity and quality to functional value of created and restored tidal marshes relative to natural reference marshes. This assessment includes rate of accumulation of organic matter, comparison of carbon and nutrient pools in natural and created marshes, the relationship of organic matter to species composition and abundance of infauna, and the potential for accelerating functional development by adding organic amendments to the soil.

1. Introduction

Salt and brackish water marshes are transitional zones between the marine and terrestrial environments. Salt water marshes are among the most productive ecosystems in the world, and annually produce up to 80 metric tons per hectare per year of plant material (8,000 gm^{-2} y^{-1}) (Mitsch and Gosselink 1993). Salt marshes provide many valuable functions such as protecting shorelines from erosion, stabilizing deposits of dredged material, dampening flood effects, trapping water borne sediments, serving as nutrient reservoirs, acting as tertiary water treatment systems to reduce contaminants in coastal waters, capturing solar energy to provide detritus (carbon energy) to the estuarine food web, serving as nurseries for many juvenile fish and shellfish species, and as habitat for various wildlife species (Kusler and Kentula 1989).

Despite their ecological significance, wetland areas continue to be lost due to development and population pressures. The National Wetlands Inventory, which was conducted in the mid 1980s, estimated that since 1780 over 53% of the wetlands in the continental United States have been lost and approximately 44% of North Carolina wetlands were lost (Dahl 1990). Marsh creation, a technique initially developed to stabilize dredged material and eroding shorelines, has come into increasing use as a means to mitigate the loss of this habitat (Race and Christie 1982). The primary goal of wetland creation and restoration is to establish wetland ecosystems that are similar in structure and community composition and perform functions like the natural systems that they were designed to replace (Broome 1990, Zedler 1993).

Establishment of emergent salt marsh vegetation has been used to replace lost or damaged marsh habitat (Broome et al. 1983, 1986, Lindau and Hossner 1981, Woodhouse et al. 1974). Efforts have been made to explore different transplanting methods in order to decrease the time required for the establishment of salt marsh functions equivalent to natural marshes. The amount of time required to achieve comparable primary productivity rates in transplanted marshes depends on many factors including elevation and soil physical and chemical properties. Although it has been shown that re-establishment of emergent vegetation (primarily *Spartina alterniflora*) to a level comparable to adjacent natural reference marshes can be achieved within approximately 3 to 5 years (Broome et al. 1986, Craft et al. 1988b), the time required for a created salt marsh to exhibit other important features of natural marshes is not well documented. It may take 15 to 30 years to accumulate pools of organic matter similar to those found in natural marshes. (Craft et al. 1988b). Created salt marshes appear to differ from natural marshes in the following characteristics: lower sediment organic content, less below-ground biomass, lower densities of benthic infauna prey organisms (annelid worms, insect larvae, and small crustaceans) and lower densities of nekton of the marsh surface (Matthews and Minello, 1994).

Previous studies of infaunal communities in created marshes, which ranged in age from 1 to 17 years, indicated differences in infauna abundance and composition that apparently were related more to differences in elevation and substrate properties than marsh age (Moy and Levin 1991, Sacco et al. 1987, Sacco et al. 1994). Other factors that may influence infaunal abundance and composition in created marshes include soil texture (Lindau and Hossner 1981, Craft et al. 1991, Sacco et al. 1994), organic matter content and nutrients (Lindau and Hossner 1981, Craft et al. 1988a, Craft et al. 1991, Langis et al. 1991, Sacco et al. 1994, Simenstad and Thom 1996), and macro-organic matter (Craft et al. 1988a, Minello and Zimmerman 1992). Marshes of different ages and soil properties exhibit differences in density, species composition, faunal feeding modes (Cammen 1976, Minello and Zimmerman 1992, Moy and Levin 1991, Sacco et al. 1994, Scatolini and Zedler 1996), and diversity of benthic infauna (Levin and Thayer 1993, Moy and Levin 1991, Scatolini and Zedler 1996, Levin et al. 1996).

Some salt marsh functions may be related to the amount of organic matter that accumulates in the system. Detrital decomposition of plant material is a major pathway of energy utilization in the salt marsh (Mitsch and Gosselink 1993). The amount and quality of organic matter may play a role in development of the created marsh, and infaunal composition and abundance. Other studies that have examined infauna composition and abundance in created and natural marsh systems in North Carolina found that sediment organic content may affect macrofaunal species composition

(Cammen1976, Moy and Levin 1991, Sacco 1989, Sacco et al. 1987).

SOM refers to the organic fraction of the soil that includes plant and animal residues at various stages of decomposition, cells and tissues of soil organisms and substances synthesized by the soil population (Brady 1990). SOM consists of both non-humic (carbohydrates, proteins, amino acids, fats, waxes, and low molecular-weight acids) and humic substances (a series of high molecular weight, brown to black colored substances formed by secondary synthesis reactions) (Sparks 1995). Organic matter affects many soil physical and chemical reactions (Sparks 1995). Organic matter is typically dark in color, which enhances absorption of radiant energy. Organic matter increases soil water holding capacity and influences soil physical properties by serving as a cementing agent holding soil particles together and assisting in the formation of soil aggregates. SOM increases the cation exchange capacity (CEC) of the soil from 20 to 70% in many soils (Sparks 1995) and can be 2 to 30 times as great as the mineral colloids (Brady 1990). SOM increases buffer capacity in the soil and helps to maintain pH levels. It forms stable complexes with Cu^{2+}, Mn^{2+}, Zn^{2+}, and other polyvalent cations and can also combine with organic chemicals (Sparks 1995). SOM stores nutrients (N, P, S, and micronutrients) that become slowly available to plants as the organic matter decomposes.

The primary source of organic matter in tidal marshes is the above and below ground biomass of plants in the marsh. Additional sources can come from algae, sedimentation, bacteria and other detritovores living in the marsh, which assist in the decomposition process and can be responsible for the translocation of the organic matter within the marsh and its incorporation into the soil. The decomposition of the chemical constituents proceeds along a continuum from rapid to very slow decomposition (sugars, starches, and simple proteins, crude protein, hemicellulose, cellulose, fats, waxes and lignin) (Brady 1990). However, since saturated soils in wetland environments are generally reduced, decomposition rates of organic material are less than in aerobic soil systems. This can result in the accumulation of organic material with time because the rate of biomass inputs to the soil system exceeds the rates of decomposition of the more stable organic fractions.

In natural wetland systems, organic matter accumulation in sediment may occur over a period of centuries. Previous comparisons of created wetland systems with natural systems indicated that the SOM levels in created marshes were less than reference natural marshes (Craft et al. 1988a, Lindau and Hossner 1981). Studies indicated that low SOM contents were associated with low nutrient concentrations and slowed the rate of functional development in created marshes. While these studies found a trend of lower SOM in created marshes compared to natural marshes, they did not examine the qualitative composition of this organic matter.

2. Relationship of SOM to Infauna

Benthic infauna are important consumers of the detritus-based salt marsh food web. Infauna feed on vascular plant detritus and associated microflora, enhance soil porosity and aeration via bioturbation and serve as a link between salt marsh primary production and estuarine secondary productivity (Lopez and Levinton 1987, Levin et al. 1998). The abundance and distribution of infauna are governed by both biotic (predation, competition, dispersal, recruitment) and abiotic factors (oxygen availability, desiccation, sedimentation/food availability, disturbance, particle size, organic content, root density) (Daiber 1982, Lopez 1988, Kneib 1984, Marsh and Tenore 1990, Lana and Guiss 1992, Sarda et al. 1992, 1995). In many salt marshes, infauna densities are greatest near the marsh edge and decrease, along with silt and clay content, towards the terrestrial boundary (Kneib 1984). In a comparison of macrofauna communities (>300 um) of five southern California salt marshes, Levin et al. (1998) reported that infauna densities were positively associated with percent SOM and percent open area and negatively associated with percent sand. Likewise, Sarda et al. (1995) reported greater infauna biomass and productivity in organic rich sandy sediments as compared to sandy or muddy sediments. The abundance of fiddler crabs is also related to increased silt and clay content (Ringold 1979).

Infauna community structure is also influenced by stage of ecological succession. Early colonizers of salt marshes and mud flats are surface deposit or suspension feeders while, in the later stages of succession, subsurface deposit feeding infauna may dominate (Lopez 1988). In fact, the reduced density of benthic infauna observed in many created salt marshes has been attributed to the low soil organic content of these geologically young wetlands (Moy and Levin 1991, Sacco et al. 1994, Levin et al. 1996). Comparison of infauna densities in created and natural salt marshes reveals that, in most cases, created marshes contain fewer infauna as compared to natural marshes (Table 1). Young (<5 year-old) created marshes, in particular, have fewer numbers of epifauna and benthic infauna. However, infauna populations increase over time as documented by Levin et al. (1996) and Simenstad and Thom (1996), who followed infauna community development during the four years following marsh construction. Other studies comparing different age created marshes also have noted greater densities in older (8 to 16 year) versus younger (4 to 5 year) created marshes (LaSalle et al. 1991, Posey et al. 1997). By the time created marshes reach 10 to 20 years of age, infauna densities generally achieve or exceed those found in natural marshes (Table 1).

Species richness also tends to be lower in young constructed marshes than in natural marshes (Table 1). Minello and Zimmerman (1992) reported significantly lower infauna diversity in 2 to 5 year-old constructed marshes than in natural marshes. Levin et al. (1996) reported lower infauna species richness during the first 3 years following establishment of constructed marshes. However, by the end of the fourth year, species richness was comparable to natural salt marshes. Simenstad and Thom (1996) also observed an increase in species richness of infauna during the four years after establishment of a constructed marsh in Washington. Limited data from older constructed marshes suggests that, like density, species richness develops to levels comparable to natural marshes (Table 1). Craft and Reader (1997) reported that, on 20 to 25 year-old constructed marshes, species richness was similar to or exceeded levels found in nearby natural marshes.

Nearly all constructed salt marsh studies exhibit differences in the distribution of taxa between man-made and natural marshes. In many cases, the epifauna of young constructed marshes are dominated by fiddler and grapsid crabs, which are early successional colonists of marsh environments (LaSalle et al. 1991, Levin et al. 1996).

In contrast, older constructed marshes and natural marshes contain more grazing snails (Levin et al. 1996, Scatolini and Zedler 1996). Another common characteristic of constructed marshes is the predominance of early successional infauna such as the polychaetes, *Streblospio* and *Capitella*, which produce planktonic larvae (LaSalle et al. 1991, Minello et al. 1994, Levin et al. 1996, Craft and Reader 1997). Conversely, older marshes (natural and constructed) tend to be dominated by non-dispersing oligochaetes and *Manayunkia* (Moy and Levin 1991, Levin et al. 1996, Craft and Reader 1997, Posey et al. 1997). The trophic composition of constructed marshes also differs from natural marshes. Generally, the low organic matter constructed marsh soils contain greater numbers of surface deposit feeders (mostly polychaetes) while natural marshes contained more subsurface deposit feeders (oligochaetes) (Moy and Levin 1991, Levin et al. 1996). However, on older (>10 year) constructed marshes, the proportion of surface and subsurface deposit feeders converges to levels comparable to natural marshes (Sacco et al. 1994, Craft and Reader 1997).

TABLE 1. Mean density (/m^2) and species richness of epifauna and infauna of constructed and natural salt marshes.

	Constructed	Natural	Species Richness ≠
EPIFAUNA:			
North Carolina	0 to 128	72 to 248 ns	–
(1 to 2 yr old)[1]			
Texas	5 to 54	27 to 170*	ns
(4 yr)[2]			
California	11,570	31,960**	ns
(4 yr)[3]			
INFAUNA:			
North Carolina	15,100	17,640-21,610***	–
(1 to 3 yr)(>300 um)[4]	450,000	761,100 ns	–
(63 to 300 um)			
(1 to 2 yr)(>300 um)[1]	5,700 to 46,700	37,200 to 64,700*	N>C
(4 yr)	11,900 to 26,700	36,500 to 63,700*	ns
(1 yr)(>250 um)[5]	16,000	49,500*	–
(11 yr)	20,470	18,130 ns	–
(12 yr)	21,000	43,500*	–
(15 yr)	19,500 (15)	37,500*	–
(15 yr)	49,410 (15)	69,200 ns	–
(16 yr)	20,000 (16)	31,000 ns	–
(21 yr)(>250 um)[6]	30,980	18,670+	C>N
(25 yr)	102,200	30,520+	ns
Virginia (5 yr)(>500 um) [7]	1,200	1,030-1,470 ns	ns
Texas (94 yr)(>500 um)[2]	1,910 to 56,100	12,740 to 70,000 ns	N>C

[1] Levin et al. (1996)
[2] Minello and Zimmerman 1992)
[3] Scatolini and Zedler (1996); litter bags, >500 um
[4] Moy and Levin (1991)
[5] Sacco et al. (1994); >250 um
[6] Craft and Reader (1997)
[7] Havens et al. (1995)
*, significantly higher in the natural marsh.
+, significantly higher in the constructed marsh.
ns, nonsignificant.
**; significantly different at the lower elevation only.
***, significantly different from one of two marshes.
≠ Natural (N) vs. Constructed (C).

Nearly all studies suggest that low SOM contributes to the reduced density, diversity and taxonomic/trophic composition of constructed marshes (Moy and Levin 1991, Minello and Zimmerman 1992, Sacco et al. 1994, Scatolini and Zedler 1996, Levin et al. 1996). Other factors, including coarser texture and sparse plant cover, also are thought to be important. A long-term study following the succession of constructed marshes shows that both SOM and benthic infauna increase with time. Craft and Reader (1997) compared soil organic carbon pools and infauna communities in two 20 to 25 yr-old constructed salt marshes with nearby natural marshes and with historical data collected from these sites 11 years earlier (see Craft et al. 1988a, Sacco et al. 1994). Between 1984 and 1995, soil organic C (0 to 30 cm) nearly doubled in the constructed marshes. Likewise, benthic infauna densities also increased during this time. In 1984, the constructed marshes (11to 15 y-old) contained infauna densities that were similar to nearby natural marshes (Table 2). By 1995, infauna densities were significantly greater in the constructed marshes than the natural marshes.

TABLE 2. Mean infauna density (/m^2) in constructed and natural marshes sampled in 1984 and 1995 (Numbers in italics are percent soil organic C, 0 to 30 cm depth).

Site	Marsh Type	1984[1]	1995[2]
Pine Knoll Shores:	Constructed (1974)[3]	20,470 **a**	32,210 **a**
		(0.31%)	**(0.75%)**
	Natural	18,130 **a**	18,670 **b**
Snow's Cut:	Constructed (1970)	49,410 **a**	102,200 **a**
		(0.95%)	*(1.67%)*
	Natural	69,200 **a**	30,520 **b**

[1] From Sacco et al. (1994) [2] From Craft and Reader (1997)
[3] Year that the constructed marsh was established.
[4] 1984 soils data from Craft et al. (1988); 1995 data from Craft and Reader (1997).
Site means (within a given year) separated by the same letter are not significantly different (p=0.05) according to Student's T-test.

There was a significant relationship (p=0.001) between infauna density and soil organic C content in constructed salt marshes of North Carolina (Fig. 1). Clearly, as the constructed marsh ages, soil organic C increases (Craft et al. 1988a, Craft and Reader 1997) along with an increase in infauna density (see Fig. 1, Simenstad and Thom 1996, Levin et al. 1996). However, above a certain critical soil organic C level, infauna densities do not increase as the amount of organic C in the soil increases. In one 20 to 25 yr-old constructed marsh (Snow's Cut), infauna densities were nearly three times higher (102,200 organisms/m^2, organic C=4.01%) than in the natural marsh (30,520 organisms/m^2) even though soil organic C was twice as high in the natural marsh (9.55%). These findings suggest that, although constructed salt marsh soils require a minimum critical amount of SOM to support infauna densities comparable to natural marshes, other factors such as larval recruitment are important. Both of these factors rely on ecological succession (time) to develop although, depending on the proximity of other natural marshes (for recruitment), development of adequate SOM pools probably takes longer.

We also observed a significant relationship (P<0.01) between oligochaetes and soil organic C (Fig. 1b). As stated previously, the density of oligochaetes often is much lower in constructed than in natural marshes presumably because of the low soil organic C content (Moy and Levin 1991, Levin et al. 1996). Most oligochaetes are subsurface deposit feeders (Sacco et al. 1994) and, as such, depend on the development of an organic rich surface soil layer to provide suitable habitat. As with total infauna density, oligochaete populations in constructed marshes depend on a critical minimum amount of soil organic C, perhaps 1 to 3%, to support densities similar to those found in natural marshes. Above this ''threshold'' concentration of organic C, oligochaetes densities appear to be regulated by other environmental factors.

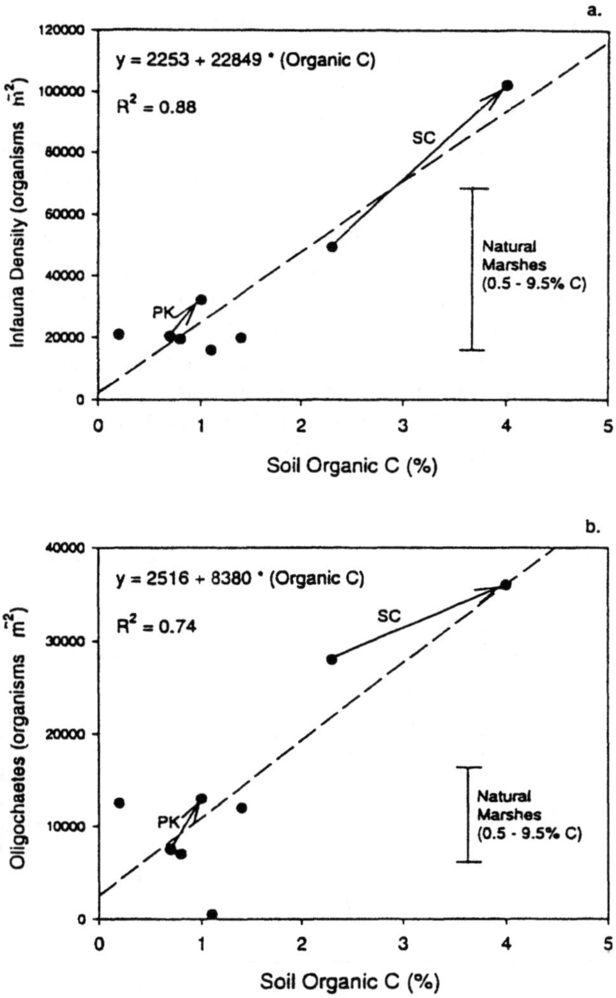

Figure 1. Regression of (a) total infauna and (b) oligochaete density (0 to 5 cm) versus soil organic C (0 to 10 cn) in constructed salt marshes of North Carolina. Arrows represent changes in soil organic C and infauna/oligochaete density over time in two constructed marshes. Data are from Craft and Reader (1997).

3. Effects of Organic Amendments on Functional Development

Since SOM is an important feature of tidal marshes, which influences many chemical, physical and biological properties, it is logical to hypothesize that adding organic matter as a soil amendment would accelerate functional development of created or restored marshes.

Gibson et al. (1994) tested the effects of adding organic matter and/or nitrogen to the soil of a newly excavated salt marsh mitigation site in Southern California. Alfalfa and straw were the organic amendments tilled into the soil before planting *Spartina foliosa*. Results of the experiment indicated that aboveground biomass of the transplanted *S. foliosa* was increased by the amendments in proportion to the amount of N added whether in the form of inorganic fertilizer or contained in the organic amendments.

A similar study in a North Carolina marsh, which was created on sandy dredged material, produced different results. Incorporating straw and alfalfa produced anoxic conditions in the substrate that resulted in complete mortality of the plants within two weeks. Decomposition of these organic amendments increased sulfides in pore water and lowered redox potential. Peat, which was also included as a treatment, did not significantly affect redox potential, sulfides and plant growth due to its refractory nature.

Soil redox potentials (10 cm depth) were measured weekly during June and July of the second growing season to assess the residual effects of the organic amendments. Redox potentials as low as -100 mv were recorded in the straw amended soil and as low as 50 mv in the alfalfa amended soil. The peat amended plots and the control plots were not significantly different with redox potentials around 0 to +40 mv (Fig. 2).

The California study (Gibson et al. 1994) found no significant differences in redox

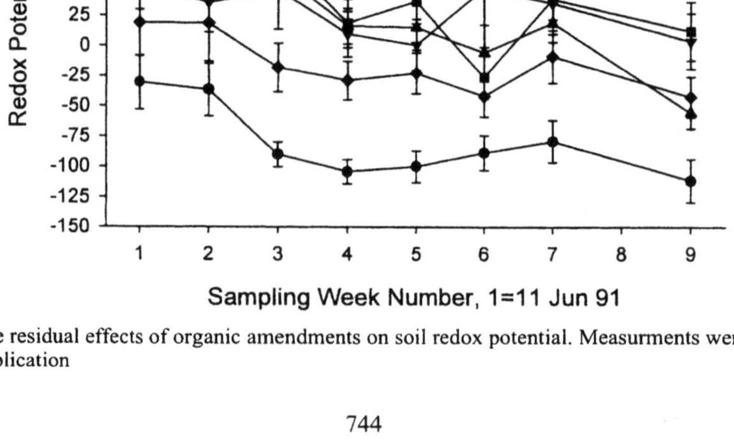

Figure 2. The residual effects of organic amendments on soil redox potential. Measurments were made one year after application

potentials between treatments. Means ranged between +13 and +150 mv, which is not low enough to produce sulfides. Apparently the less reduced soil in the California marsh prevented severe negative effects on growth of the vegetation.

These results indicate that care must be exercised in assessing soil and drainage conditions at a site before choosing the type of organic amendment that may be incorporated. Labile organic material is not suitable where strongly reduced soils are present but, well-composted refractory materials may be an alternative. Labile materials may also be applied to the surface without causing anoxia, and may serve as food for infauna. More research is needed to determine the effects of various types of organic amendments, and under what conditions their use might be desirable and cost effective.

4. Conclusions

When appropriate restoration and creation technology is applied, tidal marshes can be produced that have many of the attributes of natural marshes. The ideal goal is to produce systems that provide structure, values and functions that are equivalent to similar, relatively undistributed natural systems. Obviously, the created and restored systems require some period of time to achieve equivalence, and this time varies with the parameters considered. Success may be strictly defined as replacement of all aspects of a natural system, or replacement of certain key designated functions and values may be considered acceptable.

In many cases, the plant community develops quickly, achieving structural and functional equivalence within a few years. SOM pools and infaunal assemblages require more time to develop. More research and experience is needed to develop techniques to accelerate development of infauna numbers and species composition if this is considered to be an important goal and success criteria for tidal marsh restoration, creation and mitigation. Additional work is needed to determine the effects of incorporating various types and rates of organic matter. Other techniques to accelerate infaunal colonization of created Spartina marshes recommended by Levin et al. (1996) were marsh configuration, continuity with natural marshes, timing of marsh creation seeding with taxa that have poor dispersal ability, and attention to species habitat requirements.

5. Literature Cited

Brady, N.C. 1990. The nature and properties of soils. Second edition. Macmillan Publishing Company, New York, New York, USA.

Broome, S.W., E.D. Seneca and W.W. Woodhouse, Jr. 1983. Creating brackish-water marshes for possible mitigation of wetland disturbance. Pages 350-369 *in* J. Hernandez, editor. First environmental affairs. conference of the University of North Carolina Environmental Studies Council, UNC-Chapel Hill, Chapel Hill, North Carolina, USA.

Broome, S.W., E.D. Seneca and W.W. Woodhouse, Jr. 1986. Long-term growth and development of transplants of salt marsh grass *Spartina alterniflora*. Estuaries **9**: 63-74.

Broome, S.W. 1990. Creation and restoration of tidal wetlands of the southeastern United States. Pages 37-72 *in* J.E. Kusler and M.E. Kentula, editors. Wetland creation and restoration: the status of the science. Island Press, Washington, District of Columbia, USA.
Cammen, L.M. 1976. Abundance and production of macroinvertebrates from natural and artificially established salt marshes in North Carolina. America Midland Naturalist **96:** 487-493.
Craft, C.B. and J. Reader. 1997. Restored salt marshes: evolution of wetland structure and function over time. Final report to NOAA, National Estuarine Research Reserve Program.
Craft, C.B., S.W. Broome and E.D. Seneca. 1988a. Nitrogen, phosphorus and organic carbon pools in natural and transplanted marsh soils. Estuaries **11**: 272-280.
—— 1988b. Nitrogen, phosphorus and organic carbon between transplanted marshes and estuarine marshes using stable isotopes of carbon and nitrogen. Estuarine, coastal and shelf science **26**: 633-641.
Craft, C. B., E.D. Seneca and S.W. Broome. 1991. Porewater chemistry of natural and created marsh soils. Journal of Experimental Marine Biology and Ecology **152**: 187-200.
Dahl, T. 1990. Wetland losses in the United States, 1780-1980s. U.S. Department of Interior, Fish and Wildlife Service, Washington, District of Columbia, USA.
Daiber, F.C. 1982. Animals of the tidal marsh. Van Nostrand Reinhold Co., New York, New York, USA.
Gibson, K.D., J.B. Zedler and R. Langis. 1994. Limited response of cordgrass *Spartina foliosa* to soil amendments in a constructed marsh. Ecological Applications **4**: 757-767.
Kneib, R.T. 1984. Patterns of invertebrate distribution and abundance in the intertidal salt marsh: causes and questions. Estuaries **7**: 392-412.
Kusler, J. A. and M.E. Kentula. 1989. Wetland creation and restoration. the status of the science. Island Press, Washington, District of Columbia, USA.
Lana, P.C. and C. Guiss. 1992. Macrofauna-plant biomass interactions in an euhaline salt marsh in Paranagua Bay (SE Brazil). Marine Ecology Progress Series **80**:57-64.
Langis, R. M., M. Zalejko and J.B. Zedler. 1991. Nitrogen assessments in a constructed and a natural salt marsh of San Diego Bay. Ecological Applications **1**: 40-51.
LaSalle, M.W., M.C. Landin and J.G. Sims. 1991. Evaluation of the flora and fauna of *Spartina alterniflora* marsh established on dredged material in Winyah Bay, South Carolina. Wetlands **11**:191-208.
Levin, L.A., T.S. Talley and J. Hewitt. 1998. Macrobenthos of *Spartina foliosa* (Pacific cordgrass) salt marshes in southern California: community structure and comparison to a Pacific mudflat and a *Spartina alterniflora* (Atlantic smooth cordgrass) marsh. Estuaries **21**: 129-144.
Levin, L.A., D. Talley and G. Thayer. 1996. Succession of macrobenthos in a created salt marsh. Marine Ecology Progress Series **141**:67-82.
Lindau, C. W. and L.R. Hossner. 1981. Substrate characterization of an experimental marsh and three natural marshes. Soil Science Society of America Journal **45**: 1171-1176.
Lopez, G.R. 1988. Comparative ecology of the macrofauna of freshwater and marine muds. Limnology and Oceanography **33**: 946-962.
Lopez, G.R. and J.S. Levinton. 1987. Ecology of deposit feeding animals in marine sediments. Quarterly Review of Biology **62**: 235-260.
Marsh, A.G. and K.R. Tenore. 1990. The role of nutrition in regulating the dynamics of opportunistic, surface deposit feeders in a mesohaline community. Limnology and Oceanography **35**: 710-724.
Matthews, G. A. and T.J. Minello. 1994. Technology and success in restoration, creation and enhancement of *Spartina alterniflora* marshes in the United States. NOAA National Marine Fisheries Service, Galveston, Texas, USA.
Minello, T.J. and R.J. Zimmerman. 1992. Utilization of natural transplanted Texas salt marshes by fish and decapod crustaceans. Marine Ecology Progress Series **90**: 273-285.
Minello, T.J., R.J. Zimmerman and R. Medina. 1994. The importance of edge for natant macrofauna in a created salt marsh. Wetlands **14**: 184-198.
Mitsch, W. and J. Gosselink. 1993. Wetlands. Van Nostrand Reinhold, New York, New York, USA.
Moy, L.D. and L.A. Levin. 1991. Are *Spartina* marshes a replaceable resource? A functional approach to evaluation of marsh creation efforts. Estuaries **14**: 1-16.
Posey, M.H., T.D. Alphin and C.M. Powell. 1997. Plant and infaunal communities associated with a created marsh. Estuaries **20**: 42-47.
Race, M.S. and D.R. Christie. 1982. Coastal zone development: mitigation, marsh creation and decision-making. Environmental Management **6**: 317-328.

Ringold, P. 1979 Burrowing, root mat density and the distribution of fiddler crabs in the eastern United States. Journal of Experimental Marine Biology and Ecology **36**:11-21.
Sacco, J.N, E.D. Seneca and T. Wentworth. 1994. Infaunal community development of artificially established salt marshes in North Carolina. Estuaries **17**: 489-500.
Sacco, J. 1989. Infauna community development of artificially established salt marshes in North Carolina. Thesis. North Carolina State University, Raleigh, North Carolina, USA.
Sacco, J., F. Booker and E.D. Seneca. 1987. Comparison of the macrofaunal communities of a human-initiated salt marsh at two and fifteen years of age. Pages 282-285 *in* J. Zelanzny and S. Feierabend, editors. International Wetlands Symposium. National Wildlife Federation, Washington, District of Columbia, USA.
Sarda, R., K. Foreman and I. Valiela. 1992. Controls of benthic invertebrate populations and production of salt marsh tidal creeks: experimental enrichment and short- and long-term effects. Pages 85-91 *in* G. Colombo, I. Ferrari, V. Cecherelli and R. Rossi, editors. Marine eutrophication and population dynamics. Olsen and Olsen, Feredensborg, Denmark.
—— 1995. Macroinfauna of a southern New England salt marsh: seasonal dynamics and production. Marine Biology **121**: 431-445.
Scatolini, S.R. and J.B. Zedler. 1996. Epibenthic invertebrates of natural and constructed salt marshes of San Diego Bay. Wetlands **16**: 24-37.
Simenstad, C.A. and R.M. Thom. 1996. Functional equivalency trajectories of the restored Gog-Le-Hi-Te estuarine wetland. Ecological Applications **6**: 38-56.
Sparks, D. L. 1995. Environmental soil chemistry. Academic Press, San Diego, California., USA.
Woodhouse, W.W, E.D. Seneca and S.W. Broome. 1974. Propagation of *Spartina alterniflora* for substrate stabilization and salt marsh development. U.S. Army Corps of Engineers, Fort Belvoir, Virginia, USA.
Zedler, J.B. 1993. Canopy architecture of natural and planted cordgrass marshes: selecting habitat evaluation criteria. Ecological Applications **3**: 123-1.

INITIAL RESPONSE OF FISHES TO MARSH RESTORATION AT A FORMER SALT HAY FARM BORDERING DELAWARE BAY

K. W. ABLE
D. M. NEMERSON
P. R. LIGHT
R. O. BUSH
Marine Field Station, Institute of Marine and Coastal Sciences
Rutgers University
800 Great Bay Boulevard, c/o 132 Great Bay Boulevard
Tuckerton, N J 08087-2004 USA

Abstract

The success of salt marsh restoration, especially as it relates to the structural and functional role of fish populations, is poorly defined. In order to evaluate the effectiveness of the restoration of a former salt hay farm toward a functional marsh, we monitored the fish response to the restoration (resumed tidal flow, creation of creeks) from September 1996 to November 1997 and compared that to the prerestoration condition. During the post-restoration period we compared fish species richness, abundance, composition and size during the spring, summer and fall between the restored site and an adjacent reference marsh with similar physical characteristics (temperature, salinity, dissolved oxygen, depth, distance from the bay). Fish populations, primarily young-of-the-year, were characterized at both sites by monthly sampling with replicate (4 tows per site, 2 sites in each of two creeks) daytime otter trawls (4.9 m, 6 mm cod end mesh, n=375 two-minute tows) in large marsh creeks and with weirs (2.0 m x 1.5 m x 1.5 m, with 5.0 m x 1.5 m wings, 6.0 mm mesh, n=48) in smaller intertidal marsh creeks (2 sites in the restored marsh, 4 sites in reference marshes). Based on these observations, fish abundance was greater in the restored creeks while species richness, species composition, and average size of fishes were similar to the reference site. An analysis of fish assemblages at the same sites indicated that the reference and restored marshes were similar for large and small marsh creeks. Where differences occurred it was often the result of greater abundances of selected species at the restored marsh. Also, during this period the standing stock at the restored marsh may have exceeded that for the reference marsh. Thus, it appears that the fish responded quickly to the restoration.

1. Introduction

It is widely acknowledged that salt marshes serve as important nurseries for resident and transient fishes (Gunter 1956, Herke 1971, Nixon and Oviatt 1973, Daiber 1977, Weinstein 1979, Boesch and Turner 1984, Rozas et al. 1988, Rountree and Able 1992, Ayvazian et al. 1992, Minello and Zimmerman 1992, Baltz et al. 1993, Kneib 1997). In Delaware Bay our understanding of marsh fish populations (Able et al. 1997) has been complicated by the considerable influence man has had upon the flora and fauna of this area over the last 200 years (Sebold 1992), through activities including water level regulation, bulkheading, diking, and alteration of marsh plant communities. One of the common forms of alteration currently affecting these marshes is the construction and maintenance of diked salt hay farms. Restoration of these altered areas to *Spartina alterniflora* dominated salt marshes is underway (Strait 1997). One such project is currently being conducted by the Estuary Enhancement Program (EEP) of Public Service Electric and Gas Company (PSE&G) (Weinstein et al. 1997). In this program, marshes used as salt hay farms are being opened to tidal circulation by breaching the dikes and creating a network of constructed channels or creeks in an attempt to restore marsh structure and function.

Monitoring restored marshes to determine whether they have achieved functional equivalency with natural systems presents several challenges because too little background information is available on 1) the basic functions of marsh ecosystems (Thayer et al. 1978, Seneca and Broome 1992, Zedler 1992, Zedler 1996, Zedler this volume) or 2) the life histories of the fish that use the intertidal and subtidal portions of them (for reviews, see Rountree and Able 1992, Rountree and Able 1993, Kneib 1997, Kneib this volume). The objective of this research is, therefore, to evaluate the effectiveness of these restoration activities by determining the patterns of fish use for restored and reference marshes in lower Delaware Bay. More specifically, fish abundance, size, species richness, species composition and assemblage structure will be compared across habitat types (large and small marsh creeks) in restored and reference marshes.

2. Materials and Methods

2.1 STUDY SITES

The reference marsh at Moore's Beach and the restored marsh, a former salt hay farm at Dennis Township, are located in the mesohaline portion of lower Delaware Bay on the New Jersey side of the bay (Fig. 1, Table 1). Prior to restoration of the salt hay farm at Dennis Township, it, like other hay farms, consisted of a diked area that controlled access of tidal waters to the site. During the summer the perimeter dikes prevented normal tidal inundation and the site was maintained in a dry condition to allow for mowing of salt hay, *Spartina patens*. During the fall and winter, the site was allowed to flood through a series of linear ditches that are still present in portions of the site (Fig. 1). Some of these ditches were incorporated into the new flood/drain system when the dikes were opened during the restoration process. In addition, new created creeks were constructed as part of the restoration. These were typically 700 to 1,300 m in length, from the nearest natural creek to the upper limit of the created creek. The main channels were typically 500 to 700 m in length with 1 or 2 branches of 100 to 400 m in length. Each channel was 6 to 10 m in width from bank to bank, and 2 to 3 m deep at high tide. These are within the range of creek dimensions encountered for natural creeks in Delaware Bay and are similar to those in the reference creeks. The series of created creeks and existing

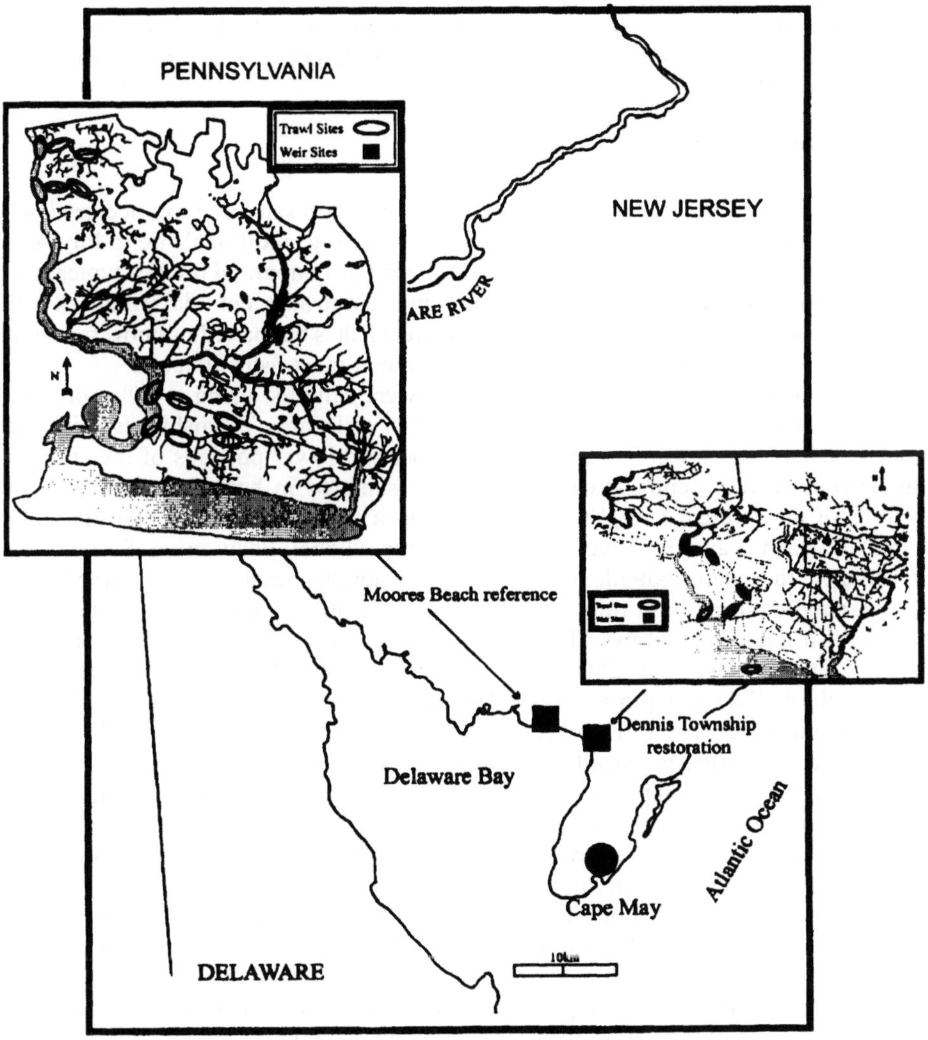

Figure 1. Location of restored (Dennis Township) and reference (Moore's Beach) marshes bordering Delaware Bay. Filled squares represent net weir sampling sites; open oblongs indicate trawl sampling sites.

ditches in the restored marsh were designed to provide the proper hydroperiod for revegetation by *Spartina alterniflora* (Weinstein et al. 1997). At the restored marsh in Dennis Township this consisted of six openings through the perimeter dike and 5,500 m of channels inside the dike (Strait 1997). Construction associated with the restoration began in January 1996 and was completed in August 1996.

The reference marsh at Moore's Beach was an operating salt hay farm until storms breached the perimeter dikes in 1979 and tidal flow was restored (Talbot et al. 1986). In the ensuing years regular marsh vegetation, dominated by *Spartina alterniflora*, returned to the site. This 521 ha site is composed of an upper area with natural creek drainage density (Upper Moore's Beach) and a lower area (Lower Moore's Beach) that was part of the original salt hay farm. Both the reference marshes at Moore's Beach and the restored marsh at Dennis Township are bordered by larger, natural creeks that connect the marshes to Delaware Bay.

2.2 SAMPLING TECHNIQUES

Prerestoration sampling for fishes at the Dennis Township salt hay farm occurred at five locations during October 1994 and the same sites plus one additional site in July/August 1995. The prerestoration habitats sampled were primarily nontidal ditches typical of salt hay farms and two ponds, one of which was connected to adjacent tidal waters. A variety of traps, hoop nets and dip nets were used to collect fishes but baited minnow traps with a conical opening and 0.6 mm mesh were the source of most fish collected.

Following restoration, large marsh creeks were sampled using a 4.9 m otter trawl with 6 mm cod end mesh. At the restored marsh at Dennis Township, two intertidal/subtidal marsh creeks were sampled in the upper and lower portions of the creeks and similar sites (width, depth, stream order) were sampled in both the upper and lower portions of the reference marsh at Moore's Beach (Fig. 1, Table 1). Starting and end points for each trawl tow were recorded using Global Positioning System (GPS) coordinates, to ensure that identical areas were sampled each month. Sampling took place around daytime high tides, and consisted of four replicate two-minute tows per station for a total of 375 tows (Table 2). All tows were made against the current, at a constant engine speed of 2500 rpm. Depth was measured at each site using a Hummingbird Wideye depth recorder. The ratio of tow line scope to water depth was usually maintained at 5:1. However, minor adjustments were occasionally made to compensate for current speed and tidal flow. Tows terminating early due to obstructions were eliminated from analyses if tow time was less than 1.5 min. All fish were identified, enumerated, and measured to the nearest millimeter for each replicate tow (up to 20 individuals of each species). Fork length (FL) was recorded for fish species with forked tails; total lengths (TL) were recorded for all other fish.

Small marsh creeks, in both the restored and reference marshes, were sampled using net weirs (2.0 m x 1.5 m x 1.5 m, with 5.0 m x 1.5 m wings, 6.0 mm mesh, n=48 sets, Table 2) at high tide and retrieved at low tide, approximately six hours later. At each intertidal creek sampled, the bag portion of the net was stretched across the channel with support poles embedded vertically in the sediment. The wings extended back onto the marsh surface from each end of the central part of the net, forming a funnel-shaped opening with the ground line buried in the bottom sediment to eliminate gaps. Local topography occasionally prevented the complete draining of creeks above the weir. Therefore, fish remaining in standing pools of water were seined into the block net although visual observations indicated that not all fish were captured with this technique, a problem typical of other salt marshes (Kneib 1997). Weir nets, at both reference and restored marshes, were deployed during the day at two intertidal creeks emptying into one of the marsh creeks adjacent to an otter trawling site whenever possible (Fig. 1).

Selected physical and chemical variables were measured at the end of each sample, for both otter trawl and weir samples. From April to November 1997, water temperature, dissolved oxygen concentration, and salinity were measured with a hand-held salinity, temperature, oxygen meter (YSI Model 85), by lowering the probe into the water and recording surface and bottom values. Water transparency was measured by lowering a Secchi disk in the water column until it was no longer visible, and recording the corresponding depth in 0.1 m increments.

2.3 ANALYSIS

Principal components analysis (PCA) was used to examine differences in the fish assemblage between the restored marsh at Dennis Township and reference marshes at Moore's Beach during April to November 1997. Separate analyses were performed on arcsin-square-root transformed monthly relative abundance data for the net weir (small marsh creeks) and the otter trawls (large marsh creeks). This resulted in eight observations per site entering each analysis, for a total of 24 observations per PCA. Any species that represented less than 5% of the abundance in all monthly samples was eliminated from the analyses to reduce the effect of rare individuals. Two species that are abundant both as age 0 and age 1 individuals, *Micropogonias undulatus* and *Anchoa mitchilli*, were first separated into age classes using monthly length-frequency plots to provide better definition of habitat use for these age classes. Multivariate analysis of variance (MANOVA) on the principal component scores was used to test the hypothesis that the fish assemblages did not differ among the restored and reference sites. For each analysis, the first five principal component axes were used in the MANOVA. This technique reduces the dimensionality of the fish assemblage data while capturing most of the variance present in the original variables. Analysis of variance (ANOVA) was used to test the hypotheses that total fish catch per unit effort and the values of selected environmental variables did not differ among the sites.

TABLE 1. Physical characteristics of Delaware Bay restored and reference marsh study sites during 1996 - 1997. See Fig. 1 for locations of individual sites.

Site	Type	Area (Ha)	Surface Salinity[1] (ppt)	Surface D.O.[1] (mg/l)	Surface Temp.[1] (°C)	Secchi Depth[1] (m)	Average Distance of Stations from Bay (km)	Depth at Trawling Stations[2] (m)	Percent Silt Clay Sediment at Sampling Stations (± SE)
Dennis Township	Salt Hay Farm Restoration	227	16.7	8.7	18.6	0.23	0.97	1.4 - 2.1	72.4 (8.6)
Upper Moore's Beach	Reference		16.3	6.8	17.6	0.32	3.6	1.4 - 1.9	61.4 (15.4)
Lower Moore's Beach	Reference	521	18.5	7.3	19.1	0.37	1.5	1.3 - 1.8	60.8 (13.4)

[1] The salinity, D.O., temperature and Secchi values were computed by finding a monthly average for each variable and then averaging across May to November.
[2] Values reported are the minimum to the maximum depth at each trawling station, averaged across the six trawling stations at each site.

TABLE 2. Sampling effort and selected results for the restored marsh at Dennis Township and the reference marshes at Moore's Beach during monthly postrestoration sampling from April to November 1997.

Study Sites	Number of Samples	Number of Fish Collected	Number of Species	Average Catch Per Unit Effort
Restored Marsh				
(Dennis Township)				
Otter Trawl	124	5,034	17	40.6
Weir	16	98,771	18	6,173
Reference Marshes				
(Upper Moore's Beach)				
Otter Trawl	123	1,992	20	16.2
Weir	16	21,345	11	1,334
(Lower Moore's Beach)				
Otter Trawl	128	1,784	16	13.9
Weir	16	6,255	13	390
Totals	423	135,181		

3. Results

3.1 PHYSICAL SETTING

The creation of creeks and the resumption of tidal flow to the former salt hay farm at Dennis Township drastically altered the marsh at this site. As a result, after restoration a tidal range of 1-2 m was typical and much of the marsh surface was consistently flooded on every high tide. The duration of inundation was relatively long because of the subsidence that resulted from previous dewatering, compaction, etc., while it was a salt hay farm. Water quality variables at the created creeks in the restored marsh and those at the reference sites were similar in most respects with a few exceptions (Table 1, Fig. 2). Average salinity and temperature overlapped and showed the same seasonal pattern at all the sites. Dissolved oxygen averaged significantly higher ($p<0.01$) at the restored marsh than at the reference marshes. The seasonal patterns of these variables were similar across all sites. However, the values were consistently lower in the Upper Moore's Beach creeks (Fig. 2). Turbidity was consistently and significantly higher ($p<0.001$) in the created creeks based on the lower average Secchi disk readings (Table 1).

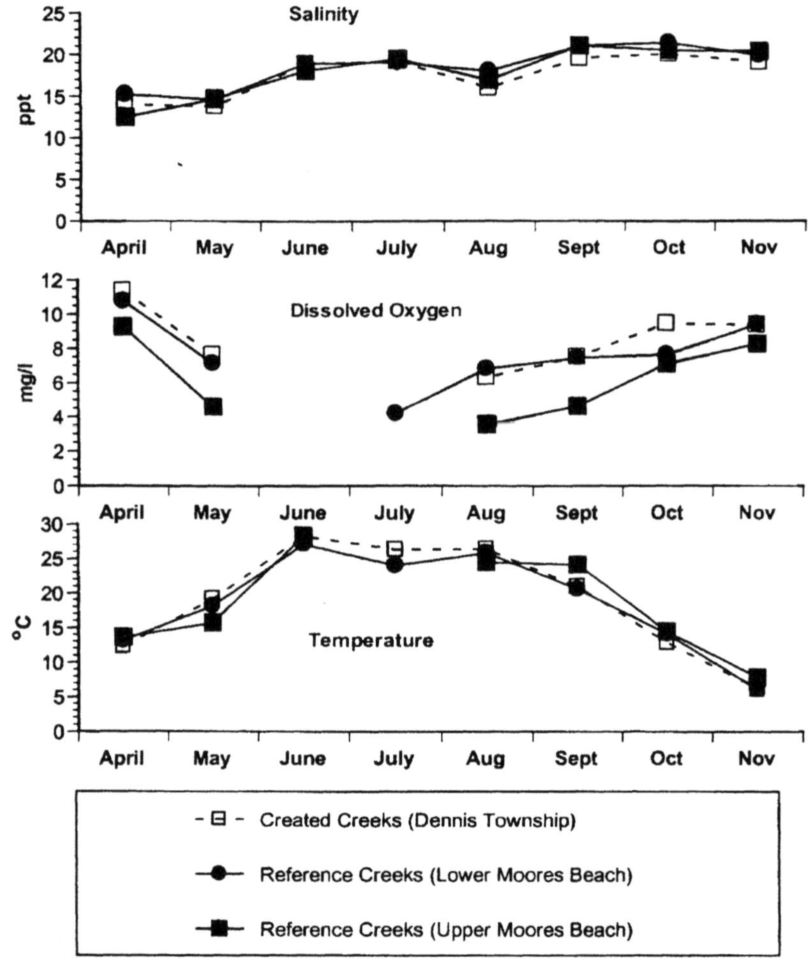

Figure 2. Selected physical variables at Delaware Bay marsh creeks in restored and reference marsh study sites during 1997. All values are measurements from the water surface.

3.2 PRERESTORATION VS POSTRESTORATION MARSH COMPARISONS

Fish species composition and abundance prior to restoration was typical of salt hay farms with only a few species present and in relatively low numbers. Samples collected in October 1994 and July/August 1995 were dominated by *Fundulus heteroclitus* (96.0% and 89.6% respectively) and *Cyprinodon variegatus* (2.2% and 10.3%). Several other species were rarely collected in 1994 including *Brevoortia tyrannus*, *Fundulus majalis*, *Anguilla rostrata*, *Menidia menidia* and *Leiostomus xanthurus*. *Menidia menidia* was the only other species collected in 1995.

Following restoration there were major changes in species richness, composition and abundance of fishes. Some of these postrestoration responses are difficult to compare to

TABLE 3. Composite species composition for small marsh creek net weir collections for created creeks at a restored marsh (Dennis Township) versus creeks in a reference marsh (upper and lower Moore's Beach), April to November 1997. Percent frequency of occurrence = percent of samples containing each species; relative abundance for fish = total number collected of the individual species/total number of fish collected; relative abundance for non-fish species = total number collected of the individual species/total number of fish and non-fish collected; catch per unit effort (CPUE) = mean number of individuals collected per sample. Dashes indicate zero catches.

	Created Creeks Dennis Twp.					Reference Creeks Lower Moore's Beach					Reference Creeks Upper Moore's Beach				
Species	% Freq. occ.	Rel. abun.	Catch per unit effort	Std. error	Total	% Freq. occ.	Rel. abun.	Catch per unit effort	Std. error	Total	% Freq. occ.	Rel. abun.	Catch per unit effort	Std. error	Total
Anchoa mitchilli	56	0.01	78.19	30.96	1251	25	<0.01	0.63	0.31	10	19	<0.01	0.25	0.14	4
Anguilla rostrata	6	<0.01	0.06	0.06	1	13	<0.01	0.13	0.09	2	6	<0.01	0.06	0.06	1
Bairdiella chrysoura	6	<0.01	0.06	0.06	1	13	<0.01	0.13	0.09	2	-	-	-	-	-
Brevoortia tyrannus	50	0.05	318.00	208.71	5088	25	0.05	28.13	16.79	450	19	<0.01	2.44	2.12	39
Cynoscion regalis	13	<0.01	0.25	0.19	4	25	<0.01	2.31	1.54	37	-	-	-	-	-
Cyprinodon variegatus	75	0.01	38.69	21.41	619	50	0.01	3.38	1.76	54	81	0.03	37.75	13.48	604
Fundulus heteroclitus	100	0.30	1868.63	1190.00	29898	81	0.58	334.25	170.04	5348	100	0.92	1222.56	642.54	19561
Fundulus majalis	25	<0.01	0.56	0.30	9	25	<0.01	1.56	1.13	25	6	<0.01	0.06	0.06	1
Gambusia holbrooki	6	<0.01	0.88	0.88	14	-	-	-	-	-	-	-	-	-	-
Gasterosteus aculeatus	6	<0.01	0.06	0.06	1	-	-	-	-	-	-	-	-	-	-
Gobiosoma bosc	25	<0.01	0.56	0.33	9	25	<0.01	0.38	0.20	6	-	-	-	-	-
Gobiidae spp.	6	<0.01	0.06	0.06	1	-	-	-	-	-	-	-	-	-	-

TABLE 3 *(cont'd)*

Species	Created Creeks Dennis Twp.					Reference Creeks Lower Moore's Beach					Reference Creeks Upper Moore's Beach				
	% Freq. occ.	Rel. abun.	Catch per unit effort	Std. error	Total	% Freq. occ.	Rel. abun.	Catch per unit effort	Std. error	Total	% Freq. occ.	Rel. abun.	Catch per unit effort	Std. error	Total
Leiostomus xanthurus	69	0.01	30.88	17.94	494	13	<0.01	2.75	2.62	44	44	<0.01	2.56	0.96	41
Lucania parva	6	<0.01	0.06	0.06	1	13	<0.01	0.25	0.14	4	-	-	-	-	-
Menidia beryllina	25	<0.01	2.56	1.31	41	<1	<0.01	1.13	1.13	18	13	<0.01	0.19	0.14	3
Menidia menidia	100	0.57	3531.56	2097.22	56505	69	0.31	179.25	84.97	2368	69	0.05	66.63	29.15	1066
Menidia spp.	6	<0.01	0.19	0.19	3	<1	<0.01	0.19	0.19	3	-	-	-	-	-
Micropogonias undulatus	63	0.05	300.25	200.42	4804	25	0.03	19.44	19.10	311	38	<0.01	1.44	0.71	23
Morone americana	-	-	-	-	-	-	-	-	-	-	13	<0.01	0.13	0.09	2
Pogonias cromis	6	<0.01	0.69	0.69	11	-	-	-	-	-	-	-	-	-	-
Syngnathus fuscus	13	<0.01	0.31	0.25	5	-	-	-	-	-	-	-	-	-	-
Unidentified	6	<0.01	0.94	0.94	15	-	-	-	-	-	-	-	-	-	-
Total - Fish			6173.44	2738.55	98775			573.88	192.76	9182			1334.06	674.92	21345
Limulus polyphemus	6	<0.01	0.19	0.19	3	-	-	-	-	-	-	-	-	-	-
Callinectes sapidus	88	0.01	38.31	11.89	613	75	0.04	24.88	7.29	398	100	0.05	75.88	25.00	1214
Total - All Species			6211.94	2740.94	99391			598.75	192.40	9580			1409.94	691.68	22559

TABLE 4. Composite species composition for large marsh creek otter trawl collections for created creeks at a restored marsh (Dennis Township) versus creeks at a restored marsh (upper and lower Moore's Beach), April to November 1997. Sampling locations indicated in Figure 1. Frequency of occurrence = percent of samples containing each species; relative abundance = total number collected of the individual species/total number of fish collected; catch per unit effort (CPUE) = mean number of individuals collected per sample. Dashes indicate zero catches.

	Created Creeks (Dennis Twp.)				Reference Creeks (Lower Moore's Beach)				Reference Creeks (Upper Moore's Beach)			
Species	Freq. Occ.	Rel. Abun.	Catch Per Unit Effort (Std. Error)	Total Coll.	Freq. Occ.	Rel. Abun.	Catch Per Unit Effort (Std. Error)	Total Coll.	Freq. Occ.	Rel. Abun.	Catch Per Unit Effort (Std. Error)	Total Coll.
Alosa aestivalis	-	-	-	-	-	-	-	-	2	0.00	0.02 (0.01)	2
Alosa sapidissima	-	-	-	-	-	-	-	-	1	<0.01	0.01 (0.01)	1
Alosa sp.	-	-	-	-	-	-	-	-	1	<0.01	0.01 (0.01)	1
Anchoa mitchilli	31	0.08	3.10 (0.84)	384	22	0.05	0.66 (0.18)	84	21	0.03	0.42 (0.12)	52
Anguilla rostrata	10	<0.01	0.15 (0.05)	18	4	<0.01	0.04 (0.02)	5	1	<0.01	0.01 (0.01)	1
Bairdiella chrysoura	5	<0.01	0.07 (0.03)	9	4	0.01	0.07 (0.04)	9	2	<0.01	0.02 (0.01)	3
Brevoortia tyrannus	15	0.13	5.38 (1.83)	667	29	0.85	11.86 (3.46)	1518	18	0.79	12.77 (5.22)	1571
Clupea harengus	-	-	-	-	2	<0.01	0.02 (0.01)	2	2	<0.01	0.02 (0.01)	2
Cynoscion regalis	19	0.02	0.80 (0.24)	99	2	<0.01	0.02 (0.01)	2	4	<0.01	0.04 (0.02)	5
Cyprinodon variegatus	1	<0.01	0.01 (0.01)	1	-	-	-	-	1	<0.01	0.01 (0.01)	1
Dorosoma cepedianum	1	<0.01	0.01 (0.01)	1	-	-	-	-	-	-	-	-

Table 4 *(cont'd)*

	Created Creeks (Dennis Twp.)				Reference Creeks (Lower Moore's Beach)				Reference Creeks (Upper Moore's Beach)			
Species	Freq. Occ.	Rel. Abun.	Catch Per Unit Effort (Std. Error)	Total Coll.	Freq. Occ.	Rel. Abun.	Catch Per Unit Effort (Std. Error)	Total Coll.	Freq. Occ.	Rel. Abun.	Catch Per Unit Effort (Std. Error)	Total Coll.
Fundulus heteroclitus	35	0.09	3.85 (1.76)	477	10	0.01	0.19 (0.07)	24	7	0.02	0.27 (0.15)	33
Fundulus majalis	-	-	-	-	1	<0.01	0.01 (0.01)	1	-	-	-	-
Gasterosteus aculeatus	-	-	-	-	-	-	-	-	1	<0.01	0.01 (0.01)	1
Gobiosoma bosc	4	<0.01	0.04 (0.02)	5	1	<0.01	0.01 (0.01)	1	2	<0.01	0.02 (0.01)	2
Leiostomus xanthurus	48	0.17	6.76 (1.40)	838	8	0.01	0.14 (0.06)	18	28	0.07	1.15 (0.30)	141
Menidia menidia	15	0.04	1.51 (1.07)	187	16	0.03	0.38 (0.12)	48	2	<0.01	0.02 (0.01)	2
Micropogonias undulatus	68	0.45	18.31 (4.32)	2270	18	0.02	0.33 (0.08)	42	38	0.07	1.07 (0.19)	132
Morone americana	7	<0.01	0.10 (0.05)	13	11	0.01	0.13 (0.03)	16	11	0.01	0.14 (0.04)	17
Morone saxatilis	6	<0.01	0.06 (0.02)	8	6	0.01	0.08 (0.03)	10	11	0.01	0.15 (0.05)	19
Pogonias cromis	15	0.01	0.39 (0.10)	48	2	<0.01	0.02 (0.02)	3	2	<0.01	0.02 (0.01)	2
Scophthalmus aquosus	-	-	-	-	-	-	-	-	1	<0.01	0.02 (0.02)	2
Syngnathus fuscus	6	<0.01	0.06 (0.02)	8	1	<0.01	0.01 (0.01)	1	2	<0.01	0.02 (0.01)	2
Trinectes maculatus	1	<0.01	0.01 (0.01)	1	-	-	-	-	-	-	-	-
Total			40.60 (7.13)	5034			13.94 (3.53)	1784			16.20 (5.35)	1992

prerestoration conditions because of the magnitude of the physical changes associated with restoration activities and the differences in the gears required to sample in the new habitats, as well as the increased sampling effort after restoration. Species richness increased dramatically, with two species making up almost all of the collections in the prerestoration marsh, while up to 12 species (of >10 individuals) were taken in trawl samples and 10 species (of >10 individuals) in weirs during postrestoration sampling (Table 3 and 4). If we limit the comparisons to the small marsh creeks, i.e. those sites most similar to the prerestoration habitats available to the fish, species richness was still much greater, with seven species making up 99.9% of the collections and a total of 20 species identified from these sites (Table 3).

Prior to the restoration, *F. heteroclitus* and *C. variegatus* made up most of the catch and these species were still well represented in postrestoration samples in the small creeks sampled with weirs (Table 3). Other abundant species in the postrestoration samples included *Anchoa mitchilli*, *B. tyrannus*, *L. xanthurus*, *M. menidia* and *Micropogonias undulatus*. Many of these same species were abundant in the deeper waters of the large marsh creeks but other taxa such as *Cynoscion regalis* and *Pogonias cromis* were common, while *F. heteroclitus* and *C. variegatus* were relatively much less abundant (Table 4).

3.3 POSTRESTORATION VS REFERENCE MARSH COMPARISONS

The average size and size distribution of fishes collected with weirs during the day in the small marsh creeks were similar between the reference marshes at Moore's Beach and the restored marsh at Dennis Township in 1997 with all sites being dominated by large numbers (Table 2) of small young-of-the-year fishes (Figs. 3 and 5). Similar patterns were evident for fishes collected with otter trawl during the day in large creeks at the restored and reference sites (Figs. 4 and 5); however the size distributions were more similar between the restored site and the Upper Moore's Beach reference site with a relatively large number of individuals >100 mm at these sites (Fig. 4). As with the small creeks, the catches were dominated by apparent young-of-the-year individuals but the otter trawl collections typically had average sizes 15 to 25 mm larger than those collected with the weir (Figs. 3, 4, and 5).

The overall abundance of fishes was greater in both large and small creeks at the restored marsh at Dennis Township than either of the reference marshes at Moore's Beach (Tables 2, 3, and 4; Fig. 5). In weir collections in the small creeks the overall abundance was an order of magnitude greater at the restored marsh than at the Lower Moore's Beach reference marsh and greater but at a similar order of magnitude as the Upper Moore's Beach marsh. The seasonal patterns of abundance for the small marsh creeks sampled with net weirs in the restored and reference marshes were somewhat similar in that the Lower Moore's Beach reference marsh and created creeks at the restored Dennis Township marsh had peaks in August (Fig. 6). However, the values for the creeks at the restored marsh were higher in most months and the summer peak extended longer, from July through September. The values were consistently low for the Upper Moore's Beach reference site. In otter trawl collections in the large creeks, the abundance averaged much higher at the restored site than at either of the reference sites (Fig. 5). The seasonal patterns of abundance for the reference and restored creeks were similar in that all sites had a pronounced peak in May but the values in the created creeks at the restored site were higher in almost every month with minor peaks in October 1996 and September 1997 (Fig. 7).

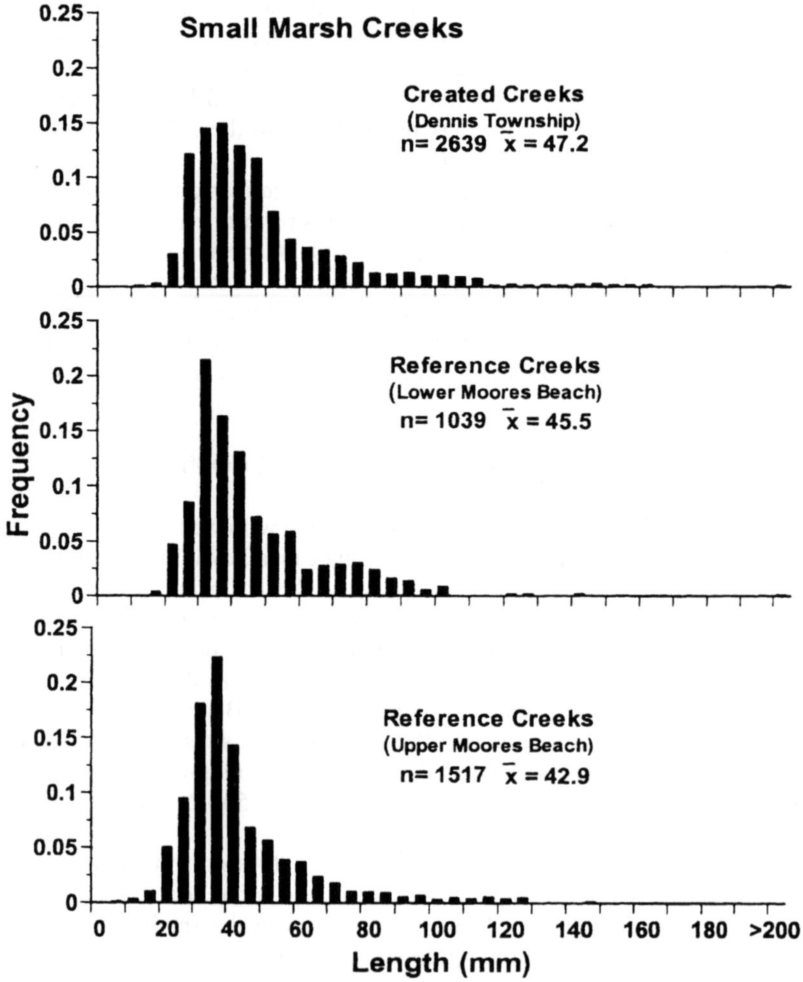

Figure 3. Length frequency diagram for all fishes collected in small marsh creeks by weir in restored (Dennis Township) and reference (upper and lower Moore's Beach) marsh sites bordering Delaware Bay during 1997.

Species richness in the small marsh creeks was highest in the restored creeks (18 species) and lower at the two reference sites (11 and 13 species) (Fig. 5). For the large marsh creeks, species richness was slightly higher (20 species) at the Upper Moore's Beach reference site and similar at the restored and Lower Moore's Beach reference site (17 and 16 species, respectively).

The overall species composition in both small and large creeks at both the restored and reference marshes was similar with some exceptions. In the newly created creeks sampled with weirs there were a number of species collected at the restored site that did not appear in similar collections in the reference sites (Table 3). These included several relatively rare forms such as *Gambusia holbrooki*, *Gasterosteus aculeatus*, *Pogonias cromis*, and *Syngnathus fuscus* and these partially account for greater species richness at the restored site. There were no species that occurred at both reference marshes that did not occur at the restored site. In the larger marsh creeks the species composition was more similar to the reference sites with single individuals of a few species (*Dorosoma cepedianum*, *Trinectes maculatus*) collected only in the restored site (Table 4). Other species occurred in one of the reference marshes and not in the restored marshes but these were typically rare.

The principal components analysis indicated a general similarity in the fish assemblage among the restored and reference sites. Plots of the monthly samples in principal component 1 (PC1) versus principal component 2 (PC2) space shows a strong seasonal pattern that was consistent across the reference and restoration sites for both large and small marsh creeks (Fig. 8b, 9b). The PCA did not appear to uncover any coordinated variance among the sites that would have resulted in a grouping of the sites in principal component space. For large marsh creeks, the first two components explained 21.5% and 16.3% of the variance in the fish assemblage data, respectively. Generally, early season samples had negative scores and later season samples had positive scores on PC1. The loadings of species on the axes confirms this interpretation, as spring dominants including age 1 *Micropogonias undulatus* and age 1 *Anchoa mitchilli* and *Brevoortia tyrannus* all loaded strongly negative on PC1. Similarly, summer/fall species including *Leiostomus xanthurus*, *Cynoscion regalis*, age 0 *A. mitchilli* and age 0 *M. undulatus* all loaded positively on PC1 (Fig. 8c, Table 5).

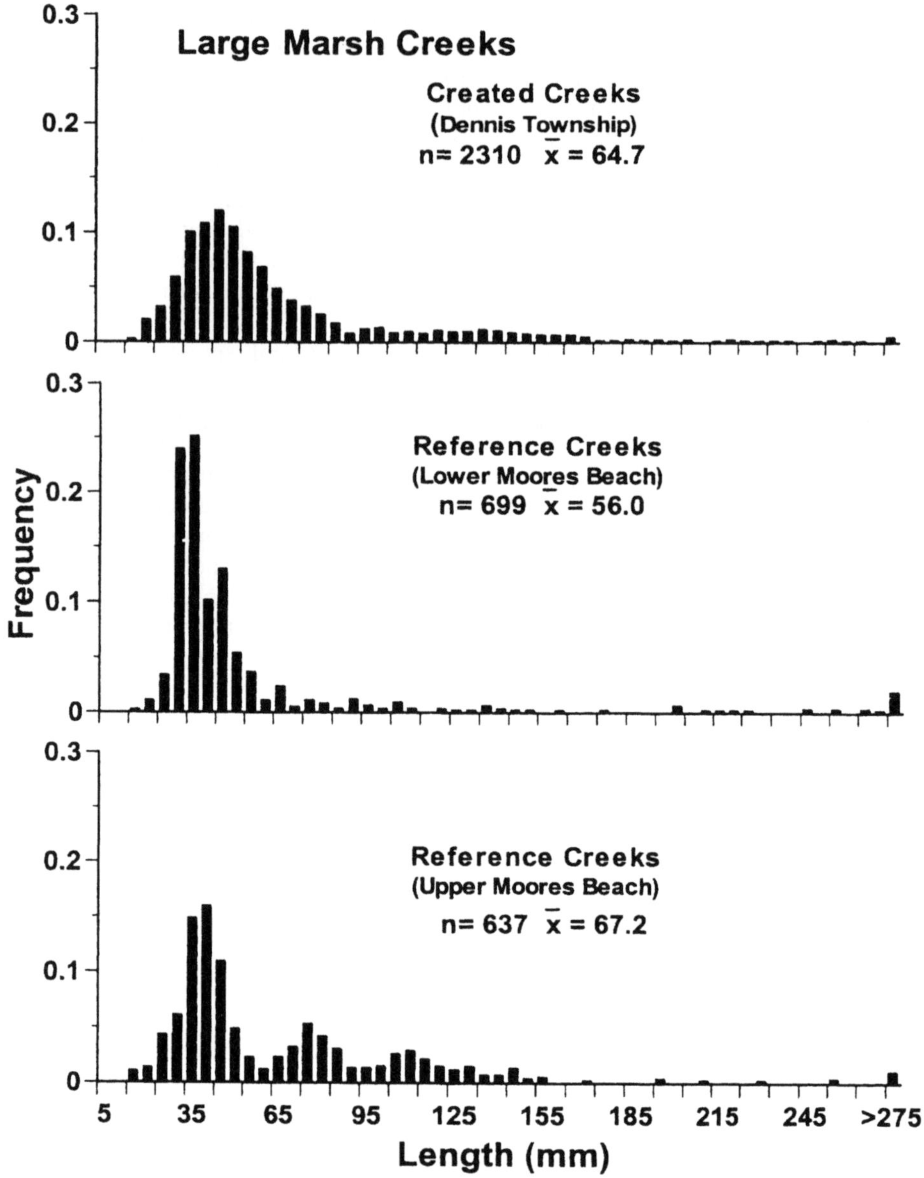

Figure 4. Length frequency diagram for all fishes collected by otter trawl in newly created marsh creeks at the Dennis Township restored marsh and the analogous creeks at Upper and Lower Moore's Beach reference marsh bordering Delaware Bay during 1997.

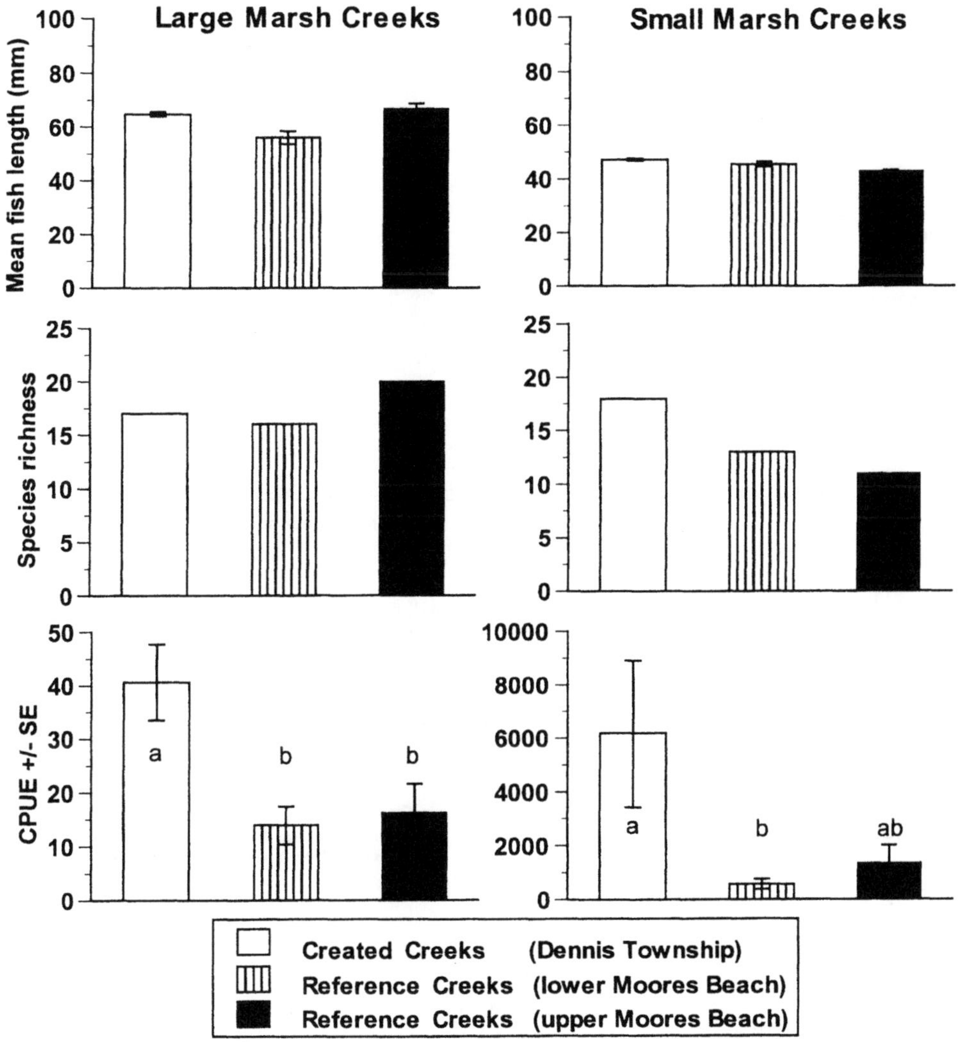

Figure 5. Comparison of mean fish length (± SE), species richness, and abundance (± SE) in large marsh creeks (otter trawl) in restored (Dennis Township) and reference (upper and lower Moore's Beach) marsh creeks and small marsh creeks (net weirs) during 1997. Sites with different letters in the abundance graph have significantly different CPUEs (ANOVO on log-transformed monthly station means, Tukey test, p.≤05).

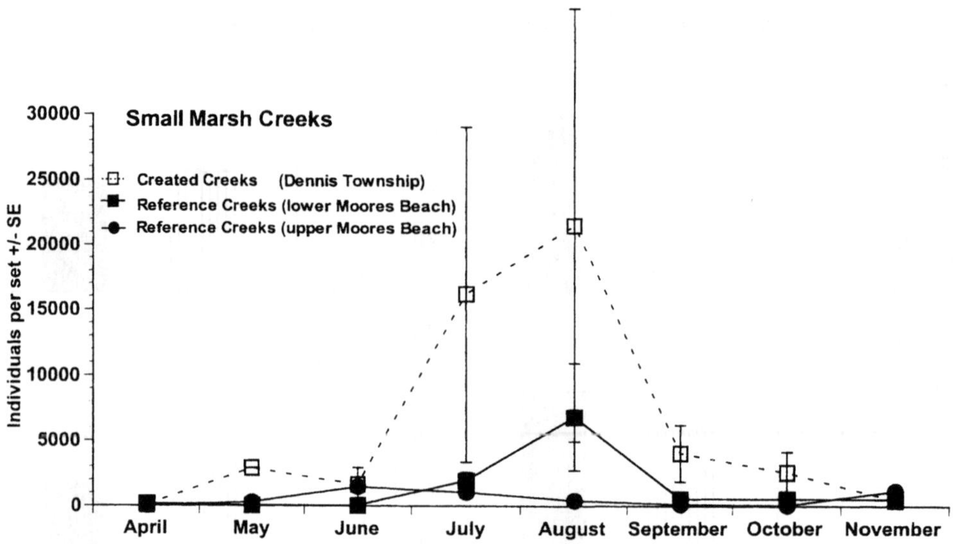

Figure 6. Abundance of fishes as collected by net weirs in small marsh creeks at the Moore's Beach reference marsh and the restored marsh at Dennis Township bordering Delaware Bay during 1997.

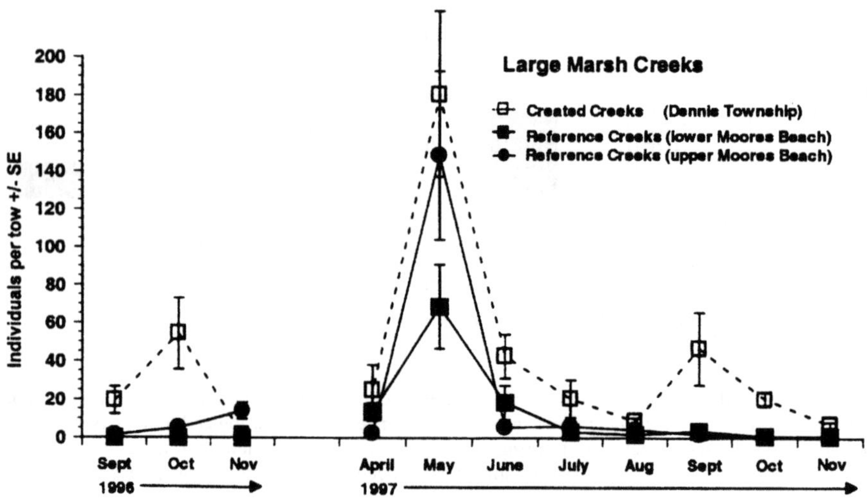

Figure 7. Abundance of fishes as collected by otter trawl in large marsh creeks at the Moore's Beach reference marsh creeks and the restored marsh at Dennis Township bordering Delaware Bay during 1997.

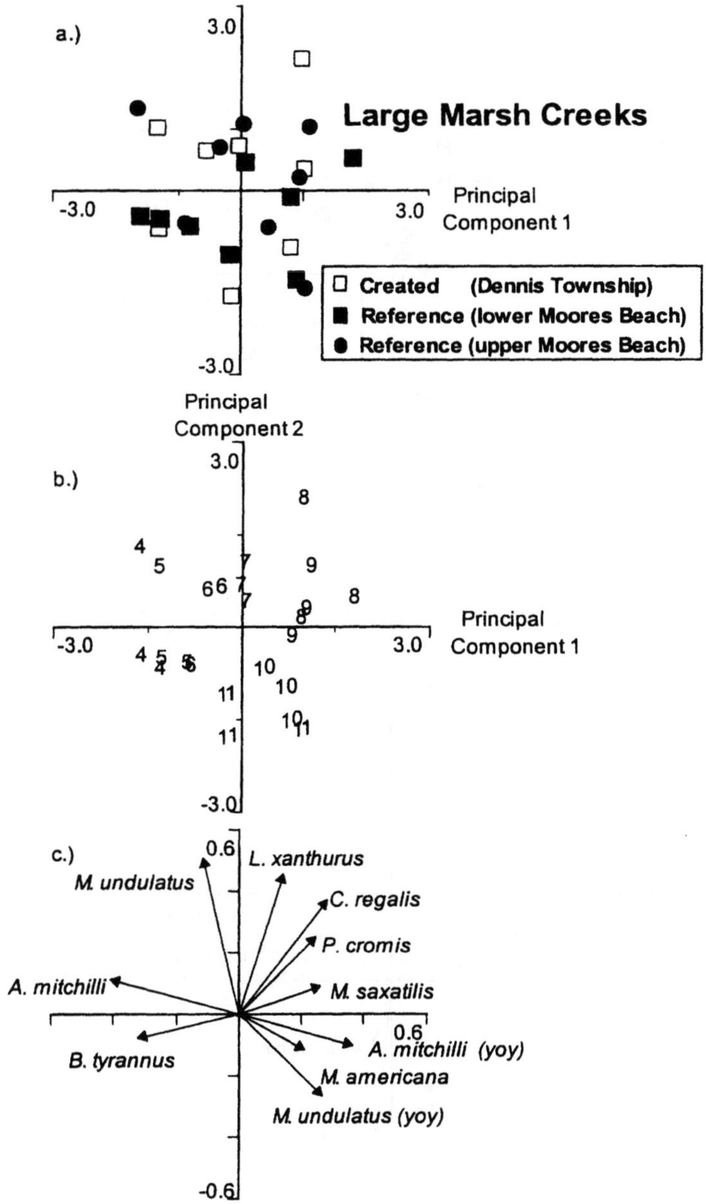

Figure 8. Comparison of fish assemblages at created large marsh creeks in the restored marsh at Dennis Township and reference creeks at Moore's Beach based on otter trawl samples during 1997 with principal components analysis a.) monthly large marsh creek (otter) collections plotted in principal component 2 versus principal component 1 space. Plot symbols correspond to the sampling sites. b.) as in a.), but plot symbols correspond to the month of the year. c.) species factor loadings responsible for location of sites and months in principal component space.

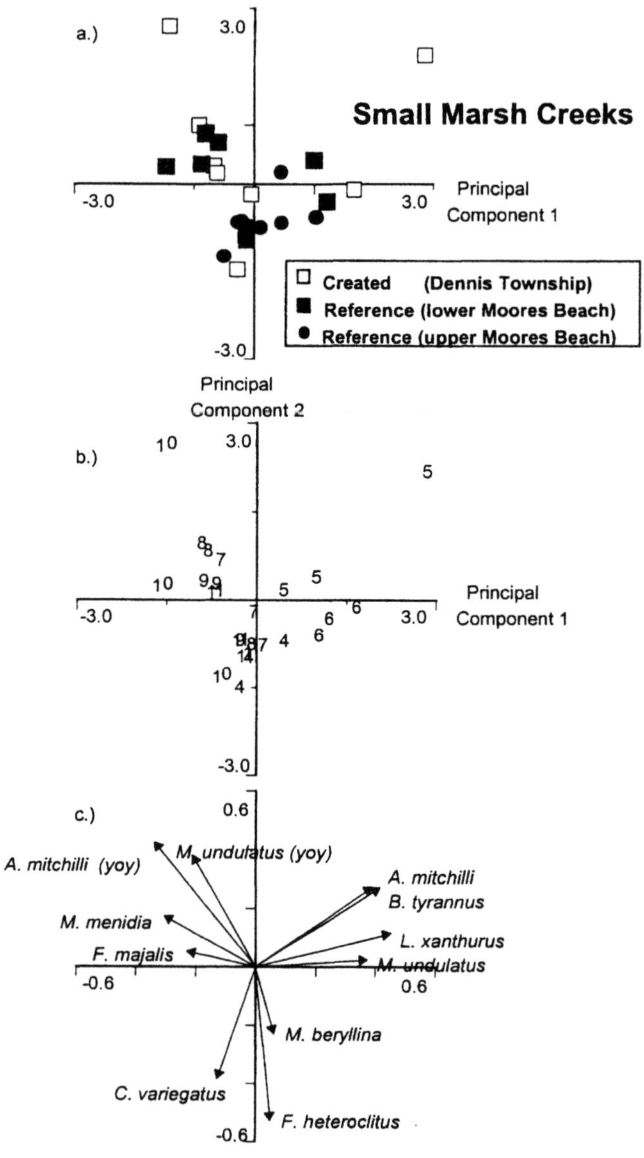

Figure 9. Comparison of fish assemblages at small marsh creeks at the restored marsh at Dennis Township and reference creeks at Moore's Beach based on weir samples during 1997 with principal components analysis. a.) monthly small marsh creek (weir) collections plotted in principal component 2 versus principal component 1 space. Plot symbols correspond to the sampling sites. b.) as in a.), but plot symbols correspond to the month of the year. c.) species factor loadings responsible for location of sites and months in principal component space.

TABLE 5. Principal Components Analysis for large marsh creeks (trawl samples) during 1997.

	Principal Component 1		Principal Component 2	
Species	Factor Loading	Pearson Correlation	Factor Loading	Pearson Correlation
Anchoa mitchilli (yoy)	0.368	0.683***	-0.104	-0.167
Anchoa mitchilli (1+)	-0.407	-0.755***	0.111	0.179
Anguilla rostrata	-0.099	-0.180	0.092	0.148
Bairdiella chrysoura	0.269	0.499*	0.177	0.285
Brevoortia tyrannus	-0.322	-0.597**	-0.079	-0.127
Cynoscion regalis	0.285	0.529**	0.370	0.597**
Fundulus heteroclitus	-0.096	-0.180	-0.292	-0.470*
Gobiosoma bosc	0.274	0.509*	-0.220	-0.354
Leiostomus xanthurus	0.142	0.260	0.455	0.734***
Menidia menidia	-0.027	-0.050	-0.037	-0.060
Micropogonias undulatus (yoy)	0.270	0.502*	-0.268	-0.433*
Micropogonias undulatus (1+)	-0.114	-0.210	0.506	0.816***
Morone americana	0.212	0.390	-0.117	-0.189
Morone saxatilis	0.265	0.492*	0.090	0.144
Pogonias cromis	0.249	0.462*	0.251	0.405*
Synodus foetens	-0.252	-0.469*	0.198	0.319
Eigenvalue	3.45		2.60	
Percent of Variance	21.5		16.3	

*<.0.05. **P<0.01, ***P<.001

The principal components analysis of the small marsh creek weir collections yielded fairly similar results, with a strong seasonal pattern in the plot of PC1 versus PC2. In this analysis, fall transient species (i.e., YOY *M. undulatus* and *A. mitchilli*) scored negatively; while spring/summer transients (i.e., *L. xanthurus*, *B. tyrannus*, and age-1+ *A. mitchilli* and *M. undulatus*) scored positively on PC1 (Table 6). The resident marsh species *Fundulus heteroclitus*, *Cyprinodon variegatus* and *Menidia beryllina* all scored near the origin on PC1 and negatively on PC2. The October and May collections from the restored marsh at Dennis Township scored higher on PC2 than any other observations (Fig. 9). This was due to very high catches of age-0 *M. undulatus* and *A. mitchilli* in October and age-1 *A. mitchilli* and *B. tyrannus* in May. For small marsh creeks, the first two components explained 25.9 % and 21.3 % of the variance in the fish assemblage data, respectively (Table 6).

For both gears, there was no significant site effect for the MANOVA on the first five principal components (weir multivariate F=1.65, p>0.1; trawl multivariate F=0.54, p>0.5). The first five principal components explained 83.1% and 70.5% of the variance for the weir and trawl collections, respectively.

TABLE 6. Principal Components Analysis for small marsh creeks (weir samples) during 1997.

Species	Principal Component 1 Factor Loading	Principal Component 1 Pearson Correlation	Principal Component 2 Factor Loading	Principal Component 2 Pearson Correlation
Anchoa mitchilli (yoy)	-0.337	-0.569**	0.424	0.648***
Anchoa mitchilli (1+)	0.391	0.660***	0.273	0.417*
Brevoortia tyrannus	0.418	0.706***	0.271	0.414*
Cyprinodon variegatus	-0.128	-0.216	-0.385	-0.590**
Fundulus heteroclitus	0.052	0.087	-0.530	-0.810***
Fundulus majalis	-0.229	-0.386	0.053	0.081
Leiostomus xanthurus	0.455	0.768***	0.114	0.174
Menidia beryllina	0.064	0.109	-0.232	-0.354
Menidia menidia	-0.304	-0.514**	0.174	0.266
Micropogonias undulatus (yoy)	-0.209	-0.353	0.378	0.579**
Micropogonias undulatus (1+)	0.375	0.634***	0.023	0.035
Eigenvalue	2.85		2.34	
Percent of Variance	25.9		21.3	

*<.0.05. **P<0.01, ***P<.001

4. Discussion

4.1 PRERESTORATION VS POSTRESTORATION MARSH COMPARISONS

The fish response to marsh restoration at the former salt hay farm at Dennis Township was immediate and dramatic. The prerestoration sampling was relatively limited but it was consistent with another earlier study of fish use of a salt hay farm at Moore's Beach (Talbot et al. 1986). In that study the two dominant species, *F. heteroclitus* and *C. variegatus*, were the same as observed in the prerestoration sampling at the Dennis Township salt hay farm. Additional species collected in the Talbot et al. (1986) study (*M. menidia* and *M. beryllina*) were not as abundant, or were absent, in the collections during the prerestoration sampling at the Dennis Township salt hay farm. Together these observations indicate that the fish fauna of salt hay farms, not surprisingly, is depauperate. This is in striking contrast to the diverse fauna that occurred in collections in small and large marsh creeks immediately after the Dennis Township site was restored to normal tidal fluctuations and the created creeks were flooded as part of extensive hydromodifications to the former salt hay farm (Tables 3 and 4).

It is interesting that a portion of the reference marsh (Lower Moore's Beach) sampled during this study in 1996 and 1997 was a former salt hay farm that was restored to tide by breaching the perimeter dikes in 1979 (Talbot et al. 1986). In the following year, the abundance of fish and the species composition were still similar to an adjacent, existing salt hay farm (Talbot et al. 1986) indicating the fish response did not occur over the seven months of that study. In contrast, an immediate response was evident in the marsh restoration at Dennis Township, undoubtedly because of the addition of fairly large subtidal creeks and the resulting hydromodification.

4.2 POSTRESTORATION VS REFERENCE MARSH COMPARISONS

Most structural parameters of the restored salt hay farm fish assemblages at Dennis Township were similar to those of the reference marsh creeks at Moore's Beach. These included the seasonal occurrence, average size, and size frequency distribution of fishes collected with otter trawl and weir at these sites. One interpretation of the similarity in size is that the fish fauna at the restored and reference marshes were derived from the same sources, i.e., local reproduction for resident species such as *F. heteroclitus* or immigration of young-of-the-year from spawning in the adjacent bay or the ocean. The latter included such species as *Brevoortia tyrannus*, *Cynoscion regalis*, *Leiostomus xanthurus*, and *Micropogonias undulatus* (see Able and Fahay 1998). Species richness and species composition were also similar, with the exceptions often being those species that occurred in the restored site but not in the reference sites. The principal components analysis showed considerable overlap in the composition and abundance of monthly samples from both the small and large creeks, thus further substantiating the similarity in species composition and abundance between the restored and reference sites.

One distinct difference between the restored and reference sites was the overall greater abundance of fishes at the restored marsh. Even though the seasonal pattern of abundance was similar between sites, the abundance of fishes was almost invariably greater at the restored marsh and this occurred in large and small creeks. It is probable that dewatering, compaction and oxidation of the sediments cause the relatively lower marsh surface elevation typical of this and other salt hay farms. This may cause a longer hydroperiod and deeper water on any high tide and thus allow greater access to intertidal sites, which may in turn result in larger catches in the net weirs. This is not likely to apply to the larger, deeper subtidal marsh creeks and thus does not help to explain the greater abundance in these habitats, which were sampled at high tide. Another possibility is that there was greater food available to fishes in the restored marsh at the time of sampling. This explanation is supported by preliminary observations of the diet of several representative species (Nemerson and Able, unpub. data).

The positive response of fishes to the marsh restoration activities at the Dennis Township salt hay farm indicates that a naturally appearing fish fauna was quickly established. Thus, the restored and reference marshes appear structurally similar. However, it is important that this response, as in other marsh restoration projects (Zedler 1992, Zedler this volume), be followed to ensure that these changes are persistent features. For example, in this restoration the greater abundance of fishes at the former salt hay farm could be due to greater food availability, which may be a short-term response by selected prey species. In addition, it is critical that these restored marshes be functionally equivalent to reference or natural marshes (Zedler 1992, Seneca and Broome 1992, Zedler this volume).

If we assume the major functions of marshes relative to estuarine nekton are for reproduction, as feeding areas, and as refuges from predation (Thayer et al. 1978, Boesch and Turner 1984) we will need further data on parameters such as fish growth and mortality in order to pronounce this restoration as functionally equivalent to natural marshes. Accurate measurements of these parameters require a detailed understanding of fish residence times and movements to be sure that the measures of these parameters are directly linked to the marsh in question.

5. Acknowledgments

Numerous individuals assisted in the field sampling efforts under somewhat difficult conditions or helped with data manipulations. Notable among these were Cathy McBride, Bertrand Lemasson, Steven Teo and James Chitty. John Balletto, Ken Strait, Jennifer Griffin and Joe Klein provided background information and logistical support in a variety of ways. Financial support for the project was made available by the Estuary Enhancement Program of Public Service Electric and Gas. This is Contribution No. 99-11 from the Institute of Marine and Coastal Sciences, Rutgers University.

6. Literature Cited

Able, K. W. and M. P. Fahay. 1998. The first year in the life of estuarine fishes in the Middle Atlantic Bight. Rutgers University Press, New Brunswick, New Jersey, USA.

Able, K. W., P. Light, D. Nemerson and R. Bush. 1997. Gradients in Delaware Bay (U.S.A.) marsh creek fish assemblages and habitats. ICES-97, Annual Science Conference, Baltimore, Md. CM 1997/S:01.

Ayvazian, S.G., L.A. Deegan and J.T. Finn. 1992. Comparison of habitat use by estuarine fish assemblages in the Acadian and Virginian zoogeographic provinces. Estuaries **15**: 368-383.

Baltz, D.M., C. Rakocinski and J.W. Fleeger. 1993. Microhabitat use by marsh-edge fishes in a Louisiana estuary. Environmental Biology of Fishes **36**: 109-126.

Boesch, D. F. and R. E. Turner. 1984. Dependence of fishery species on salt marshes: the role of food and refuge. Estuaries **7**: 460-468.

Daiber, F. C. 1977. Salt marsh animals: distributions related to tidal flooding, salinity and vegetation. Pages 79-108 *in* V. J. Chapman, editor. Wet Coastal Ecosystems. Elsevier Scientific Publishing Co., Amsterdam, The Netherlands.

Gunter, G. 1956. Some relations of faunal distribution to salinity in estuarine waters. Ecology **37**: 616-619.

Herke, W.H. 1971. Use of natural and semi-impounded Louisiana tidal marshes as nurseries for fishes and crustaceans. Dissertation, Louisiana State University, Baton Rouge, Louisiana, USA.

Kneib, R. T. 1997. The role of tidal marshes in the ecology of estuarine nekton. Oceanography and Marine Biology: An Annual Review **35**:163-220.

Minello,T.J. and R.J. Zimmerman. 1992. Utilization of natural and transplanted Texas salt marshes by fish and decapod crustaceans. Marine Ecology Progress Series **90**: 273-285.

Nixon, S. W. and C. Oviatt. 1973. Ecology of a New England salt marsh. Ecological Monographs **43**: 463-498.

——. 1993. Diel variation in decapod crustacean and fish assemblages in New Jersey polyhaline marsh creeks. Estuarine Coastal and Shelf Science **37**: 181-201.

Rountree, R. A. and K. W. Able. 1992. Fauna of polyhaline subtidal marsh creeks in southern New Jersey: Composition, abundance and biomass. Estuaries **15**: 171-185.

Rozas, L.P., C.C. McIvor and W.E. Odum. 1988. Intertidal rivulets and creek banks: corridors between tidal creeks and marshes. Marine Ecology Progress Series **47**: 303-307.

Sebold, K. R. 1992. From marsh to farm: the transformation of coastal New Jersey. New Jersey Coastal Heritage Trail. National Park Service, U. S. Department of the Interior, Washington D. C.

Seneca, E. C. and S. W. Broome. 1992. Restoring tidal marshes in North Carolina and France. Pages 53-78 *in* G.W. Thayer, editor. Restoring the Nation's Marine Environment. Maryland Sea Grant Program, College Park, Maryland, USA.

Strait, K. A. 1997. Diked salt hay farm wetland restoration. Proceedings: 84th Annual Meeting of the New Jersey Mosquito Control Association, Atlantic City, New Jersey, USA.

Talbot, C. W., K. W. Able and J. K. Shisler. 1986. Fish species composition in New Jersey salt marshes: effects of marsh alterations for mosquito control. Transactions of the American Fisheries Society **115**: 269-278.

Thayer, G. W., H. H. Stuart, W. J. Kenworthy, J. F. Ustach and A. B. Hall. 1978. Habitat values of salt marshes, mangroves and seagrasses for aquatic organisms. Pages 235-247 *in* P. E. Greeson, J. R. Clark and J. E. Clark, editors. Wetland Functions and Values: The State of Our Understanding. Proceedings of the National Symposium on Wetlands, American Water Research Association, Minneapolis, Minnesota, USA.

Weinstein, M. P. 1979. Shallow marsh habitats as primary nurseries for fish and shellfish, Cape Fear River, North Carolina. Fishery Bulletin 77: 339-357.

——, J. H. Balletto, J. M. Teal and D. F. Ludwig. 1997. Success criteria and adaptive management for a large-scale wetland restoration project. Wetlands Ecology and Management **4**:111-127.

Zedler, J. B. 1992. Restoring cordgrass marshes in southern California. Pages 7-52 *in* G.W. Thayer, editor. Restoring the nation's marine environment. Maryland Sea Grant Program, College Park, Maryland, USA.

——. 1996. Tidal Wetland Restoration: A Scientific Perspective and Southern California Focus. California Sea Grant College System, University of California, La Jolla, California. Report No. T-038.

SUCCESS CRITERIA FOR TIDAL MARSH RESTORATION

CATASTROPHES, NEAR-CATASTROPHES AND THE BOUNDS OF EXPECTATION: SUCCESS CRITERIA FOR MACROSCALE MARSH RESTORATION

MICHAEL P. WEINSTEIN
New Jersey Marine Sciences Consortium
Sandy Hook Field Station
Building 22
Fort Hancock, NJ 07732 USA

KURT R. PHILIPP
Wetlands Research Services
102 East Main Street, Suite 305
Newark, DE 19711 USA

PETER GOODWIN
College of Engineering
University of Idaho
800 Park Blvd., Suite 200
Boise, ID 83712 USA

Abstract

Most tidal marsh on Delaware Bay has a history of diking for purposes of salt hay *(Spartina patens)* production and wildlife management. Extensive ditching for drainage and mosquito control has also altered natural hydrological cycles, and combined with diking or other water control structures has provided suitable conditions for invasion by *Phragmites australis*. Where diking and other water control measures have been in place for extended periods, in some instances back to colonial times, marsh surfaces have subsided by oxidation and compaction, and high marsh species of plants are maintained artificially at low marsh elevations. These conditions lead to potential catastrophic ''blow-outs'' when dikes are rapidly breached, principally by storms. Although seawater may enter the breaches and fill the marsh, the absence of a typical fourth order drainage system (long filled by farming practices and sedimentation) prevents efficient return of tidal water to the adjacent bay. Massive circulation patterns and standing water combined with the low marsh plain elevation kill extant plants and prevent recolonization by low marsh species. The result may be destruction of the root mat and ''fluidization'' of the entire marsh surface—replaced by an open water lagoon environment. It may take many decades for the marsh to begin to reestablish itself, if ever. Without further intervention, the slowly recovering marsh is characterized by ''tree-like'' drainage configurations that appear to exhibit low drainage density downstream, low overall sinuosity and higher order intertidal streams. It is in this framework that the ecological engineering of macroscale marsh restoration and the criteria that determine its success, the ''bound of expectation,'' is undertaken.

1. Environmental Setting for Macroscale Marsh Restoration on Delaware Bay

Much of the Delaware Bay is fringed by emergent marshes dominated by smooth cordgrass, *Spartina alterniflora*, and at higher elevations by mixtures of *S. alterniflora, S. patens* and *Distichlis spicata*. These plant communities occur along tidal creeks or in broad open meadows. Tidal salt marshes comprise more than 72,845 ha in the brackish and lower estuary below the city of Philadelphia, Pennsylvania (Fig. 1). *Phragmites australis* has recently invaded many tidal marshes at higher elevations, or where tidal restrictions occur. Higher elevations in brackish marshes also contain stands of *Spartina cynosuroides*, *Spartina patens* and *Scirpus americanus*. Common freshwater and brackish species, such as *Peltandra virginica*, *Pontederia cordata, and Amaranthus cannabinus* are also abundant in upstream areas of brackish marshes. Although present for approximately 11,000 years (Daiber and Roman 1988, Philipp 1995), most tidal salt marshes of the Delaware Bay have experienced anthropogenic disturbance since colonial times.

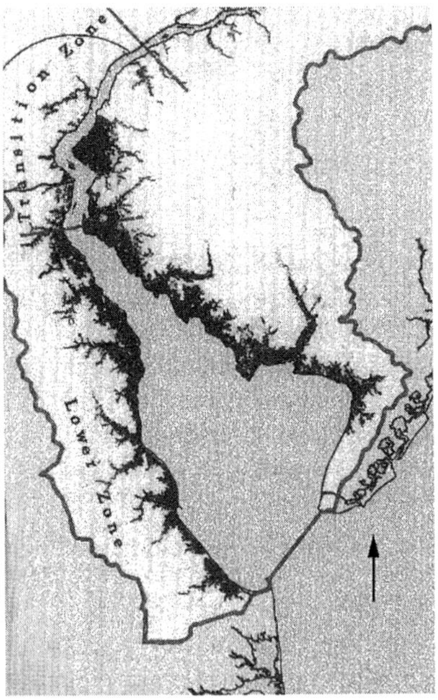

Figure 1. Tidal salt marshes comprise more than 72,845 ha in the brackish and lower Delaware Bay estuary, below the city of Philadelphia, Pennsylvania.

1.1 LAND RECLAMATION AND THE MEADOW BANKS

Any attempt at ecologically engineering large-scale wetland restorations, here generally defined as restoration of individual sites more than 500 ha, must consider the history of disturbance of the area. Although this is a common sense statement, we can not overemphasize the need for due diligence on the history of the site before undertaking any attempt at restoration.

The most common form of disturbance is impounding, a practice among the oldest and most extensive forms of marsh alteration on the Bay. Drainage ditches, dikes and tide gates were used to drain soils and exclude salt water for land reclamation, mosquito control, and waterfowl/muskrat (*Ondatra zibethicus*) management. Water management of the inundated shoreline has been practiced since the earliest recorded settlements (Daiber 1986).

Dutch and English colonists brought strong traditions in wetland reclamation to the New World (Daiber 1986, Orson et al. 1992, Seebold 1992). In 1675, Dutch magistrates under British rule began dike and sluice construction for a roadway near New Castle, Delaware. The St. Georges Marsh Company was formed in 1762 to manage an impoundment meadow, today known as the "1000 Acre Marsh" in northern Delaware. Early records from New Jersey and Pennsylvania describe the construction of impoundments near the populated headwaters of tidal creeks. Maps and records from the mid through late 1800s suggest that much of the Delaware River shoreline was reclaimed land, shown as meadow or cultivated land behind dikes (Warren 1911, Philipp 1995).

The construction of meadows was facilitated by cooperative ventures among "meadow companies." In 1788, for example, the New Jersey legislature promulgated laws to promote the formation and maintenance of meadow companies. By 1833, Salem County, New Jersey had registered 71 such companies. In 1866, a state geologist noted that the total area of impounded salt meadows in Cumberland and Salem Counties, New Jersey totaled approximately 31,567 ha (Seebold 1992). In contrast, reclaimed tidal salt marsh in Delaware (along the Bay) totaled about 7285 ha in 1885 (Nesbit 1885). In Pennsylvania, marsh reclamation was restricted to areas just south of Philadelphia, at the mouth of the Schuylkill River and along the river shoreline (Seebold 1992).

By the early twentieth century, the maintenance of impoundment meadows declined due to lack of cooperation among farmers, expense of maintenance, failure of individual farms, nationwide poverty introduced by the Great Depression followed by the upheaval of World War II, and conservationists' efforts to preserve marshland (Seebold 1992). Although many meadows were reclaimed by tides during the early twentieth century, some of the original impoundments have been continuously maintained for waterfowl, trapping, farming, flood protection, and salt hay production (Daiber 1986). The creation of impoundments for the control of salt marsh mosquitos began in the early 1940s and became widespread by the 1960s. Large impoundments with joint objectives for waterfowl management and mosquito control replaced smaller systems and by 1987, more than 4047 ha of salt marsh were enclosed by dikes and water control structures. Today, 41 major impoundments comprise more than 6,070 ha on Delaware Bay. Although the area of many impoundments exceed 405 ha, many private impoundments have been created under 8.1 ha.

1.2 IMPACTS OF SEA-LEVEL RISE

Confounding the challenge of macroscale marsh restoration is the process of sea-level rise. Tidal records for New Jersey indicate yearly increases in mean sea-level over the past century from between 2.74 and 3.85 mm y^{-1} (NOAA 1994). Many investigators believe that these rates will increase (National Research Council 1987, Houghton et al.1991, Titus and Narayanan 1995, Nuttle 1997). Whereas many wetlands adjusted to the slow rise of sea-level in the past 1000 years, the current higher rate and the predicted higher future rates appear to be inducing disequilibrium conditions in many estuaries (Psuty et al. 1993).

A rapid rise in sea-level over the next century may exceed the ability of some wetlands to keep pace. As sea-level rises, some tidal marshes will be increasingly inundated, while others may migrate landward or become filled with sediments. The net result would be a loss of wetlands due to in-place drowning, shoreline erosion, and other factors. Considering rising water levels alone, Kraft et al. (1992) predicted that up to 90% of the existing tidal marshes of Delaware (33,000 ha) would be lost by the year 2200. By including landward regression for some wetlands not constrained by developed shoreline, this number drops slightly to 80%. Existing time series for certain Delaware Bay shorelines reflect the combination of sea-level rise, land subsidence and anthropogenic effects on existing marsh areas (Fig. 2).

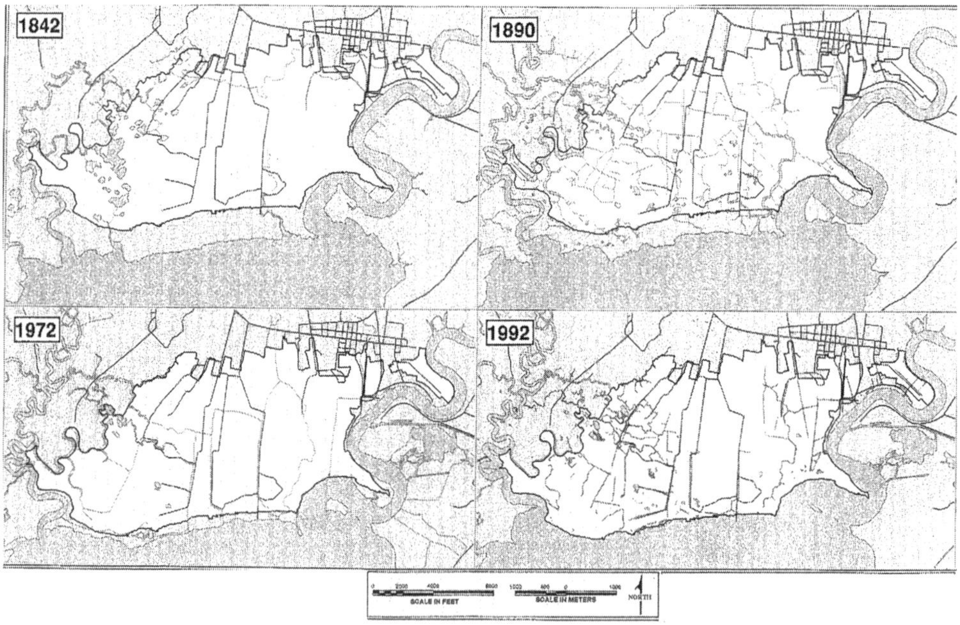

Figure 2. Times series depicting recession of historic shoreline at the Commercial Township wetland restoration site, Cumberland County, New Jersey. Shoreline erosion during the period 1842 to 1992 averaged 4.3 m y^{-1}.

Anthropogenic factors affecting coastal marshes include land use practices (conversion of forest to cropland), barrier island development, engineering of inlets, and dredging of tidal rivers. The area shown in Fig. 2 contains the 1620 ha Commercial Township wetland restoration site, one of the largest in the region. The character of the site has changed dramatically in response to shoreline erosion as a consequence of relative sea-level rise. While contributing to expansion of tidal marsh on the land margin, sea-level rise results in shoreline erosion on the bayfront. Shoreline erosion over an approximately 150 year period from 1842 to 1992 averaged 4.3 m y^{-1}, while current estimates range as high as 9+ m y^{-1}. Similarly, Chase (1979) calculated average shoreline erosion rates in the vicinity of the Commercial Township site to be 4.7 m y^{-1}, retreating nearly 150 m landward in a 32 y period.

In general, data collected by Phillips (1986) and others suggest that the Delaware Bay marshes are rapidly eroding and/or subsiding, resulting in a net loss of wetland area. To maintain a constant shoreline for an average erosion rate of 3.2 m y^{-1}, a mean vertical accretion rate of at least 10 cm y^{-1} would be required (Phillips 1986). This rate is 2 to 10 times that determined in Delaware Bay coastal wetlands (Chase 1979, Stumpf 1983), where values ranging from 0.1 to 0.5 cm y^{-1} were recorded.

Similarly, diked marshes in Cape May and Cumberland County have subsided by about 0.3 m and 0.4 to 0.8 m, respectively, over the past century. Rapid submergence is apparently associated with a combination of sea-level rise, and accelerated subsidence due to mosquito ditching, groundwater pumping, and biomass removal, oxidation, compaction, and reduced sedimentation associated with salt hay farming (Phillips 1987). Such rapid change must be incorporated into the engineering designs for wetland restoration in the long-term.

1.3 LAND ELEVATION EFFECTS OF IMPOUNDMENTS

Natural saltmarshes have generally kept pace with sea-level rise through sediment accretion, at least in relatively undisturbed systems. However, the creation of dikes for salt hay farming and other impoundments has virtually eliminated the inflow of sediments to the marsh, arresting this natural balance. Additionally, compaction by heavy mowing equipment and greater oxidation of organic sediments have combined to lower marsh plain elevation. The longer a site has been diked, the greater the difference in surface elevation between the diked marsh and nearby natural marshes. This relationship can be a precedent to potential catastrophes, or near-catastrophes in attempts to restore tidal wetlands on a large scale.

The planform morphology that evolves during controlled or uncontrolled dike breaching can be tied directly to the degree of prior land subsidence. At least four categories of marsh types can be described, but these are likely discrete points in a gradient of planform types.

1.3.1 Type A: "Catastrophes"

A substantial difference in elevation between the diked marsh surface and height of tide may cause the marsh to revert to open water and tidal flats (Fig. 3). This is largely because the marsh plain elevation is too low to support *Spartina alterniflora* or other marsh plants.

At the extreme, the marsh may revert to open water for extended periods, measured in decades or longer. In the case of Mill Creek/Goose Pond (Fig. 3), almost 70 years have passed between catastrophic breaching of the dikes and partial ''recovery'' of the system. Another example of this phenomenon occurred when storms in 1980 breached the dikes at Moore's Beach East, located near the Commercial Township site (Fig.4).

Figure 3. Mill Creek/Goose Pond marsh ''blowout'', Salem County, New Jersey. Catastrophic breaching of dikes due to storms led to an open water environment with loss of the entire marsh plain. Partial recovery of the site as shown, has taken at least 70 years.

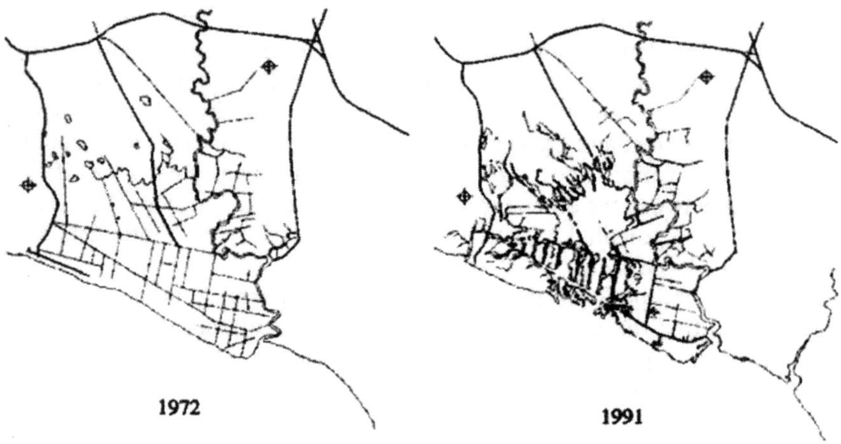

Figure 4. Catastrophic ''blowout'' at Moore's Beach East. Storms in the early 1970s breached dikes at the site. The open water environment has yet to show any signs of recovery.

A striking example of a catastrophic blowout of a natural dike system, and subsequent slow recovery of the site was described in the Delaware Conservationist:

> Following the storm [in 1878] the banks were breached just north of Woodland Beach [Delaware], and high tides from thence forward flowed into the marshlands creating a 'vast lake' . . .Old timers hereabouts remember when they could put on a pair of skates. . .and skate clear to the lighthouse [on the bayfront]. These were the halcyon years waterfowl-wise, when 'ducks were so thick they blotted out the sun', for the shallow flooded marshlands offered ideal wintering conditions to the vast flocks of migrating ducks. High tides would pour water into the marsh through the then shallow break, but the only way the water could get out was southward . . .to the bay . . . Since the distance to the bay was so far, by the time the water had dropped a foot or so, the tide had changed and was pouring back northward in the creek, thus precluding further ebbing. The result of all of this was a vast basin back of Woodland Beach with shallow water of more or less fixed level. . . As the years passed, however, the breaks enlarged in size, depth and number (there are now three). More and more waters flowed outward through these deepening channels, until in the early 1900s the tide ebbed and flooded full cycle in Broadway Meadows. The vast shallow lake became dry mud flats on low tide, and the cordgrasses began their inexorable march out onto these flats. . . now [many years later] a sea of grass.

The approximate recovery period was three to four decades.

1.3.2 Type B: "Near Catastrophes"

Intermediate degrees of subsidence with subsequent catastrophic dike loss, may result in formation of a "young marsh". The time-history for recovery of Type A and Type B systems is likely one of degree, and depends on depth of subsidence, local sediment accretion rates, sea-level rise and other factors. Some Type A marshes may never recover, Type B systems may evolve relatively rapidly, from less than a decade to more than 50 years (Fig. 3). Sediment accretion ultimately leads to development of a marsh drainage that is "tree-like" (dendritic) with a predominance of lower order channels (the "branches) near the upland edge, and large, rather linear third or fourth order streams (the "trunk") with little sinuosity and low bifurcation ratios extending toward the estuary. Sometimes these large channels are intertidal over most of their length. There are numerous examples of Type B marshes in Delaware Bay (Figs. 5 through 7).

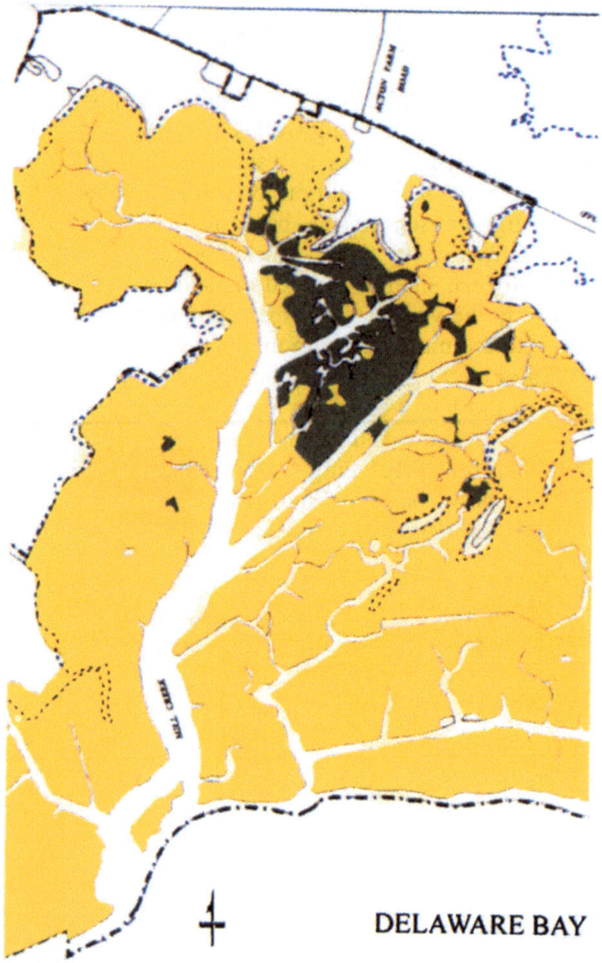

Figure 5. Mill Creek at Alloways marsh, Salem County, New Jersey. Note the "tree-like" configuration of the drainage system. Lower order creeks are most prevalent towards the upland, larger fourth order channels exhibiting little sinuousity are characteristic of the downstream reach.

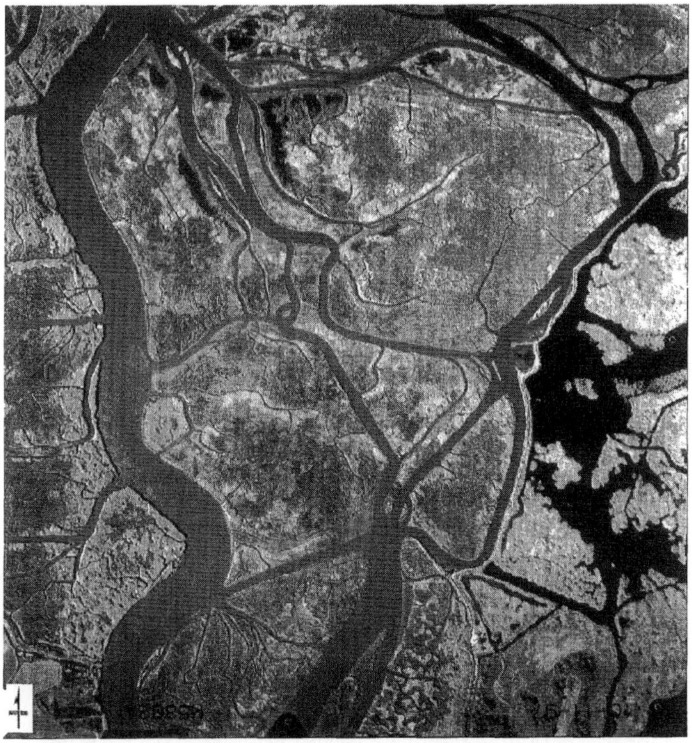

Figure 6. Abbotts Meadow at Alloways creek, Salem County, New Jersey. Note the "tree-like" configuration of the drainage system. Lower order creeks are most prevalent towards the upland, larger fourth order channels exhibiting little sinuousity are characteristic of the downstream reach.

1.3.3 Type C: Diked Marshes With Minimal Subsidence

Smaller degrees of subsidence can lead to two atypical drainage features that form during restoration. The first are "rectangular" creeks. Existing linear drainage features become higher order channels in the restored marsh, and only new lower order streams take on the typical sinuosity of natural tidal creeks. Rectangular creeks are distinctive of many formerly diked marshes that occur throughout the estuary (Fig. 8). The second configuration, characterized by extensive "lawnscapes" of *Spartina alterniflora* dominated marsh with few higher order channels and low drainage density appear to evolve where small breaches occur under controlled conditions (e.g., by activities of the local Mosquito Commission). Extensive ditching combined with lower initial rates of tidal flow allows these sites to drain effectively and *Spartina patens* is rapidly replaced at lower elevations by *S. alterniflora* (Figs. 9 and 10). Note that small areas of Type A and Type B creek patterns also occur in these examples.

Figure 7. Cedar Swamp in New Castle County, Delaware. Note the "tree-like" configuration of the drainage system. Lower order creeks are most prevalent towards the upland, larger fourth order channels exhibiting little sinuousity are characteristic of the downstream reach.

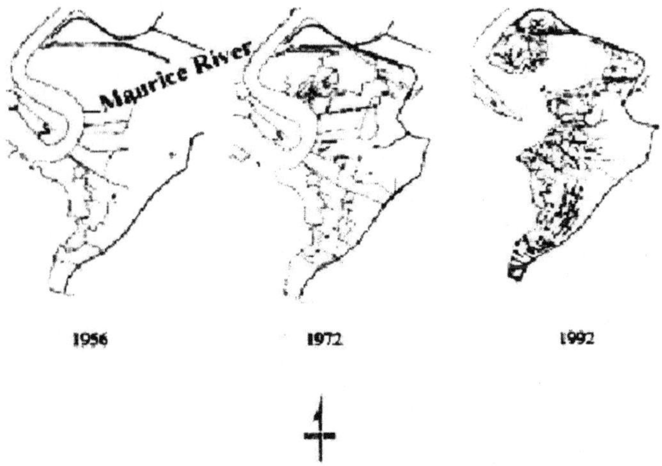

Figure 8. "Rectangular creeks" at Maurice River, Cumberland County, New Jersey that form as a result of existing linear drainage ditches becoming the higher order channels in the marsh. Only lower order streams exhibit the typical sinuousity of natural marshes.

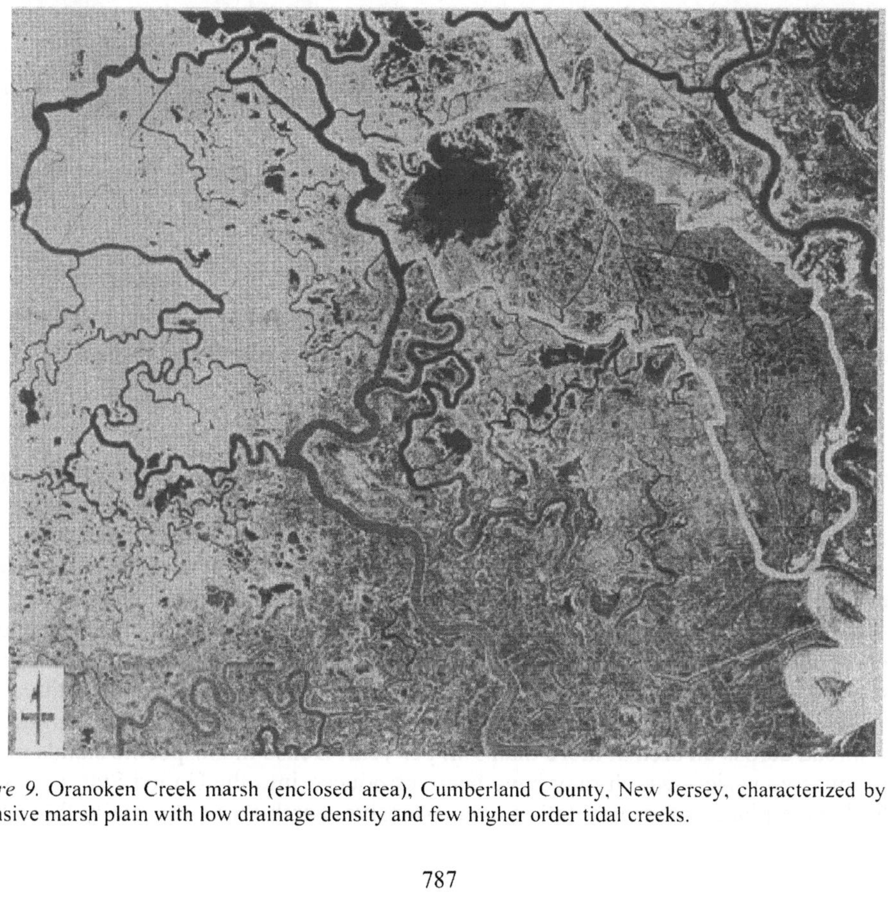

Figure 9. Oranoken Creek marsh (enclosed area), Cumberland County, New Jersey, characterized by an extensive marsh plain with low drainage density and few higher order tidal creeks.

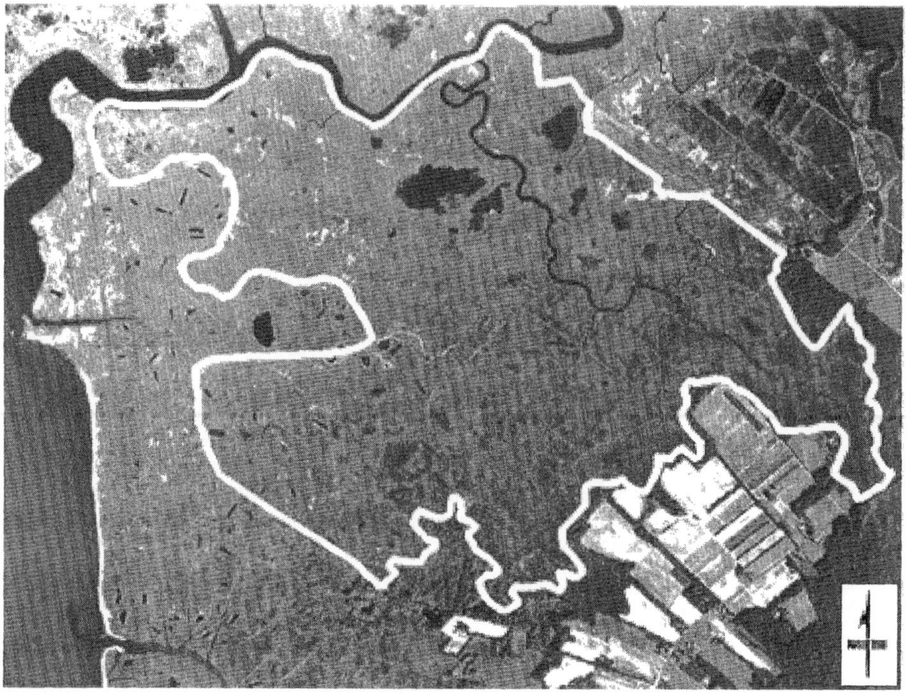

Figure 10. Wheeler Farm marsh, Cape May County, New Jersey, characterized by an extensive marsh plain with low drainage density and few higher order tidal creeks.

1.3.4 Type D: Relatively Undisturbed Marshes

Relatively undisturbed salt marshes on Delaware Bay are stable as a result of a balance between sediment accretion and submergence caused by sea-level rise (Figs. 11 and 12).

1.4 *PHRAGMITES AUSTRALIS* INVASION

Phragmites australis is an abundant macrophyte that favors brackish water, and is generally described as a plant of the upper level of the salt marsh. *Phragmites* is characterized by tall densely growing stems that reach heights of more than 4 m. The dense underground root (rhizome) mat may be more than 1 m thick in older stands. Sometimes *P. australis* is a stable, natural component of a wetland community, if the habitat is relatively undisturbed and the population does not appear to be expanding (Niering and Warren 1977, Marks et al. 1994). Indigenous populations that date back at least 3000 years can be relatively benign and do not appear to pose direct threats to wetlands. More often, however, an aggressive form of *Phragmites* dominates, one that may spread across an area at more than 9 m per year. Dense monospecific stands of this more invasive variety which may have been introduced into the United States (Besitka 1996) have replaced other vegetation over extensive marsh areas, especially those with

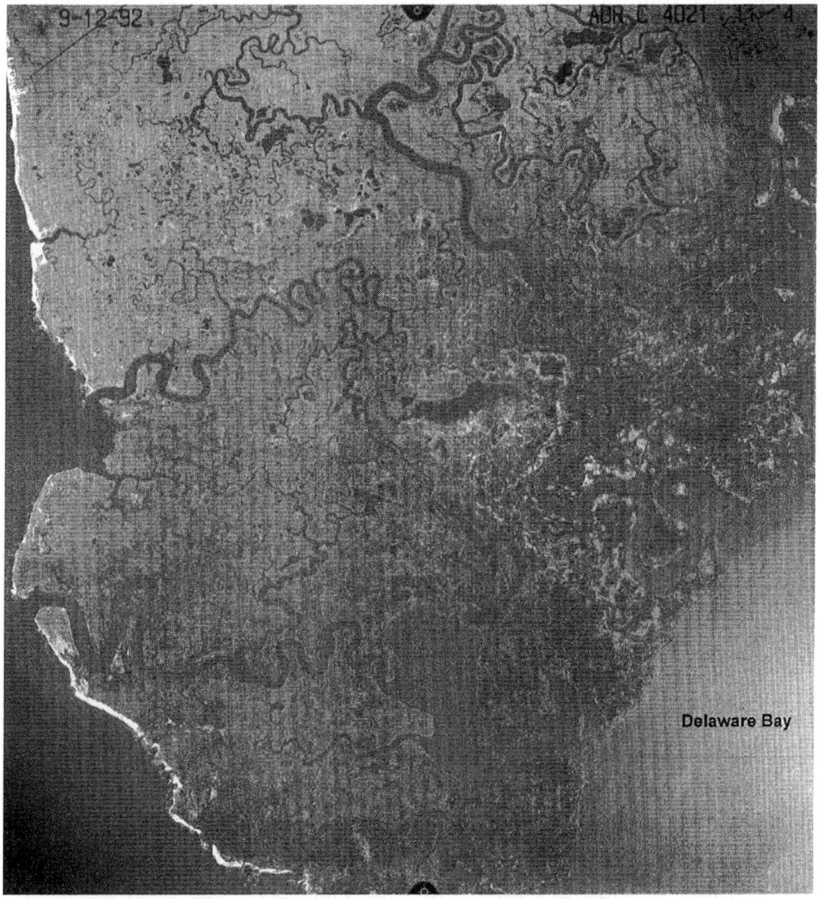

Figure 11. Fishing Creek, Cumberland County, New Jersey, a relatively undisturbed marsh with fourth order drainage system characterized by high sinuousity, drainage density and bifurcation ratios.

a previous disturbance history. Beginning about 50 years ago, more than 16,000 ha (about one third) of coastal marshes on the Delaware side of the Bay have become virtually monospecific stands of *Phragmites*, and this variety has also greatly expanded its range on the New Jersey side of the Bay.

A time series of aerial photographs dating back to 1951 for a large 1134 ha restoration area situated around Mill Creek and lower Alloways Creek (Fig.13) clearly depicts the increasing dominance by *P. australis* over a 41-year period. Based on these and other data for the region, *Phragmites* expansion on the marsh plain is occurring at the rate of 1 to 6 % per year (Windham 1995, R.S.Warren, pers. comm.). If this trend continues, most tidal wetlands in Delaware Bay that have average pore water densities of about $15^0/_{00}$ or less will be dominated by *Phragmites* within the next two or three decades.

Where *P. australis* covers extensive areas of the marsh, wildlife values appear to be reduced (Roman et al. 1984, Marks et al. 1994). In addition, *P. australis* appears to negatively influence hydrology and hydroperiod as well as drainage density, and other geomorphic features (e.g., stream bank slope and associated intertidal mudflats) of the marsh.

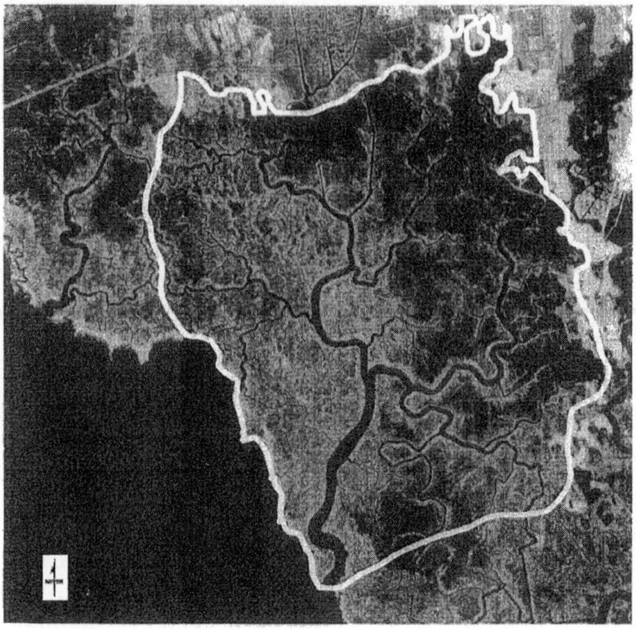

Figure 12. Mad Horse Creek, Cumberland County, New Jersey, a relatively undisturbed marsh with fourth order drainage system characterized by high sinuousity, drainage density and bifurcation ratios.

Two types of tidal restrictions in *Phragmites australis* degraded marshes are important to marsh function and the restoration process. First, some comparisons between the drainage of some degraded sites and relatively undisturbed nearby marshes have demonstrated generally greater tidal attenuation in the former (CH2MHill 1995a,b). Secondly, many first and second order tidal creeks appear to be filled, and drainage density is lower than in undisturbed marshes (Windham 1995, CH2MHill 1995a,b). Consequently, hydroperiod and sheet flow across the marsh plain may be restricted.

These restrictions are believed detrimental to natural marsh function for two reasons: a) because *P. australis* rapidly builds up the marsh plain (Windham 1995), creek channels are more often characterized by steep bank slopes rather than gentle ones, and b) combined with the reduced drainage density over the entire marsh, ready access to the marsh surface for many fishes may be denied (see below). These effects of *Phragmites australis* invasion are important because of the extent of this process in Delaware Bay and elsewhere.

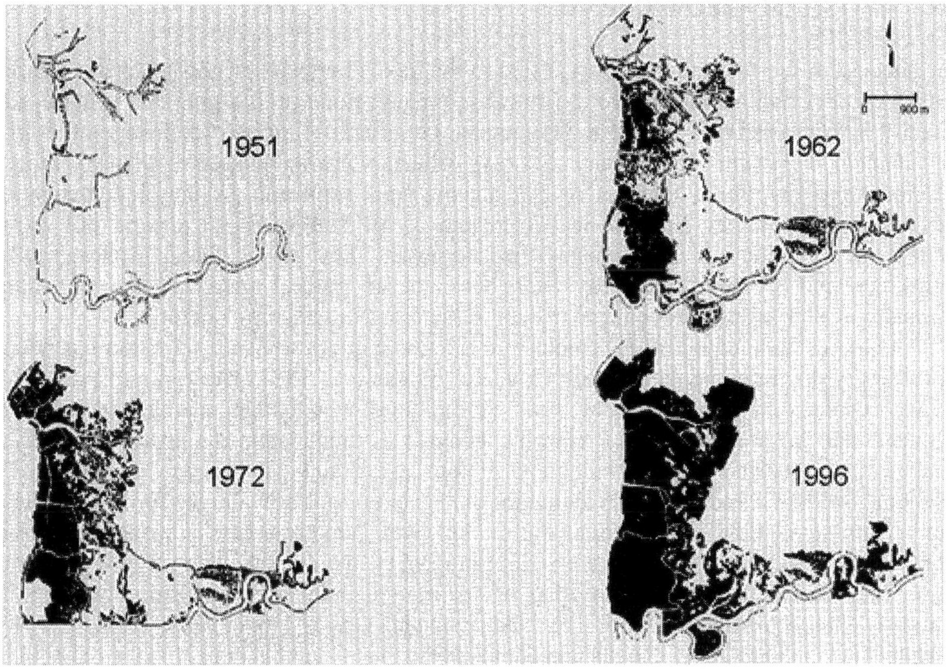

Figure 13. The time-course of *Phragmites australis* invasion at Mill Creek and lower Alloways Creek, Cumberland County, New Jersey between 1951 and 1992. *Phragmites* expansion is occurring at the rate of 1 to 6% per year on the marsh plain.

2. The "Bounds of Expectation"

Whether occurring by catastrophic events or by minimal intervention, many sites eventually come to resemble natural marshes. What constitutes an acceptable range of marsh types, the so-called "bound of expectation" (Weinstein et al. 1997) and the time-trajectory required to get there, depends on site-specific goals and compromises among all stakeholders—scientists, managers and the regulatory community.

On Delaware Bay, the largest salt marsh restoration project in the United States is underway—nearly 8300 ha of tidal salt marsh is being preserved, restored and/or enhanced. Both diked salt hay farms and previously diked, *Phragmites australis*-dominated marshes are being returned to a more natural state. The remainder of this chapter will focus on:

- Program specific goals that establish the bound of expectation; and
- Ecological engineering of the Commercial Township salt hay farm to minimize the potential for catastrophe, and shorten the restoration trajectory to a 10- to 15- year period.

2.1 PROGRAM SPECIFIC GOALS

The success of the restoration effort is tied to the nexus between primary and secondary (finfish) production. Re-establishing natural salt marsh function is a goal that is believed to result in improved fish habitat and the exchange/utilization of detrital production with the adjacent Delaware Bay estuary. Success criteria were developed to identify those structural and functional features of salt marshes that reinforce the coupling between primary and secondary production (Weinstein et al. 1997). These criteria not only include the production of emergent macrophytes, but also address hydroperiod, hydrology and geomorphological characteristics of natural marshes that enhance exchange of detrital production and access to the marsh surface by nekton.

The importance of intertidal creeks and the marsh plain to fish production has been increasingly documented in the past 15 years. As early as 1981, Weigert and Pomeroy (1981) commented, "our present view of the food web of the marsh and estuary suggests that the preservation of fisheries depends as much upon the protection of the smaller tidal creeks as upon protection of the marsh and its *Spartina* production." Predator avoidance and access to abundant prey are obvious advantages for those fishes that ascend small tidal rivulets onto the marsh plain. Although earlier work pioneered these efforts, e.g., Harrington and Harrington (1961), the advent of the flume weir (or flume net) made quantitative sampling of the marsh plain possible (Zimmerman and Minello 1984, Rozas and Odum 1987, McIvor and Odum 1986, 1988, Hettler 1989, Murphy 1991, Kneib 1991, Rozas and Reed 1993).

These studies demonstrated that fishes regularly follow the rising tide onto the vegetated marsh plain to feed or seek refuge from predators (see also Miller and Guillory 1980, Talbot and Able 1984, Targett 1985). Moreover, many fishes reach the marsh plain from sites within subtidal creeks that have gently sloping profiles, or preferentially through corridors created by small intertidal rivulets (first-order creeks) that drain the marsh surface. Rozas et al. (1988) observed in Virginia tidal marshes that:

- Fish catch was significantly greater in rivulet flumes, averaging more than three times that collected from natural creekbank sites; only bay anchovy *(Anchoa mitchilli*) was more abundant at creekbank sites;
- Intertidal marsh plain habitats were used extensively by several species; in addition, the two most dominant taxa, mummichogs (*Fundulus heteroclitus*) and banded killifish (*F. diaphanus*) moved into the marsh with an average of < 10% of their stomach volume filled—upon leaving the marsh, mummichogs averaged 80% gut fullness, banded killifish 60% (see also Weisberg and Lotrich 1982, Rountree and Able 1992, Kleypas and Dean 1983);
- Although rivulet entrances were preferred, rivulets occupied only about 3% of the distance along creeks studied; thus 59% of ingress to the marsh plain occurred over depositional creek banks.

Based on these data, most fishes would reach the marsh plain through rivulets when they comprised 19% of the distance along a tidal creek. As Frey and Basan (1978) postulated, the relative importance of rivulets as corridors would be greatest in low marsh environments.

2.1.1 Drainage Density and Marsh Edge

The importance of drainage density and the edge it creates was examined using drop samplers in the Barataria-Caminada Bay system in Louisiana (Baltz et al. 1993). Fishes were observed to concentrate near the interface between the *Spartina* marsh and open water, with habitat suitability steadily declining with distance from the marsh edge. Among the relevant taxa collected closer to the marsh edge were juvenile bay anchovy (*Anchoa mitchilli*) and young-of-year spot (*Leiostomus xanthurus*). Protection from predators, and the availability of food were thought to contribute to the use of marsh edge habitat.

Edge effects have also been evaluated recently in another Louisiana salt marsh by Browder et al. (1989). These authors noted that the production of fishery species may be more dependent on the land-water interface than on wetland acreage alone. They cited the studies of Faller (1979), Dow (1982) and Gosselink (1984) who found statistically significant relationships between fishery production and land-water interface in neighboring areas, and the work of Zimmerman et al. (1984), who found that brown shrimp (*Penaeus aztecus*) densities were highest in areas of higher shoreline ''reticulation''. Using a stochastic computer model, Browder et al. (1989) examined the relationship between land-water interface and shrimp catch during the period 1960 and 1967 (a period of rapid marsh disintegration). Browder et al. found a significant relationship, the regression coefficient (r^2) ''explaining'' 49% of the variation in catch during this period.'' Interface length (edge) alone accounted for 32% of the variance in the catch, a remarkable value given that fisheries effort was not included as an independent variable in their model, nor was the inherent variability in marsh systems and fisheries catch statistics.

Kneib (1994) studied the relationship between drainage density and marsh utilization by fishes. Effort was focused on the general issue of accessibility to forage sites by young nekton. On a scale of several kilometers, the prominent spatial feature in the marsh was a dense tidal creek system that channeled flows into the marsh interior and onto the marsh plain. The hypothesis was raised as to the relationship between creek drainage patterns and flood-tide use of intertidal marshes by fishes. Using flume weirs that were located at relative low and high intertidal elevations in two marshes with different drainage densities, Kneib demonstrated that:

- Fish densities at high tide were clearly greater in the high drainage density sites compared to low drainage density sites; and
- Fish were significantly more abundant in high than in low intertidal habitats at the high drainage density site, but no significant difference was observed in the use of low and high intertidal areas at the low drainage density site.

Kneib (1994) concluded that the spatial arrangement of creeks within marsh landscapes seemed to control the extent to which fish use potential foraging habitats in the intertidal marsh. By virtually eliminating available intertidal fish habitat, energy transfers (''trophic relays,'' Kneib 1994) are reduced and marsh function is severely degraded.

Moreover, marshes dissected by numerous tidal channels were used more by fishes than were the more mature marsh habitats characterized by low drainage densities.

Among the relevant taxa that Kneib (1994) and others (e.g., McIvor and Odum 1986, Hettler 1989, Kneib and Wagner 1994) observed in intertidal marshes were spot (*Leiostomus xanthurus*), anchovy (*Anchoa mitchilli*), spotted seatrout (*Cynoscion nebulosus*), and white perch (*Morone americana*). For spot, Hodson et al. (1981) demonstrated that young individuals entering the marsh had fuller stomachs than individuals captured in adjacent tidal creeks.

2.1.2 Site-Specific Considerations for Delaware Bay

Virtually all of the salt marshes on Delaware Bay have experienced varying degrees of disturbance including diking, extensive ditching and invasion by *Phragmites*. The planform characteristics that result when these marshes are restored, whether storm induced "self restoration" or by human intervention (or both), leads to a very wide range of variability in drainage configurations and other features (Figs. 5 through 10). Salt marshes in Delaware Bay (and elsewhere) differ in physiography, geomorphology, and relationships between vegetated areas, drainage, open water, and tidal flats and bars. All of this variability must be acknowledged early in the process of establishing restoration goals, and it is anticipated that the bound of expectation will encompass a relatively wide range of measured results. Some marsh types may produce more fish of certain kinds than others, the problem is that our current ability to predict which conditions are optimum is severely limited. At best, we can compare standing crops of resident and marine transient species in the restored sites with various types of reference marshes that represent the range of conditions described herein, and that is precisely what has been done for the Delaware Bay restoration project (Weinstein et al. 1997).

As part of a monitoring program to track restoration success, three reference marshes were selected to represent the range of conditions expected as end-points for the restoration process for the 14 sites comprising the project (Fig.14). Mad Horse Creek (Fig.12) is a relatively undisturbed marsh that is located in the oligo-mesohaline (0.5 to $10^{0}/_{00}$) portion of the estuary in New Jersey. Several portions of two previously diked areas where salt hay farming had been abandoned, Moore's Beach West and Wheeler's Farm, have undergone natural restorations in several phases (multiple breaches of the dikes over time) since 1975 and 1972 respectively. Both of these systems are generally meso-polyhaline (8.5 to $20^{0}/_{00}$). In the context of restoration goals, all of the reference sites are believed to have characteristics that represent desirable end-points for the restoration program, but they differ in the ratio of marsh plain to open water, and in the drainage configuration. Mad Horse Creek displays the typically sinuous drainage of relatively undisturbed marshes, while Moore's Beach West has more open water (as a result of a previously constructed pond on site), but otherwise, a general physiography that closely resembles the natural condition. Although transected by many drainage ditches, Wheeler's Farm is characterized by both naturally restored areas and relatively undisturbed marsh, and appears to have less open water than the other sites. By virtue of its more "natural" state, it is expected that Mad Horse Creek will fall near the upper end of the bound of expectation, while the latter two sites will represent the "average" condition. This premise remains to be tested.

Figure 14. Location of marsh restoration areas, and marsh reference sites.

Unfortunately, a large reference marsh characterized by the tree-like drainage configuration of Mill Creek (Fig.5), Abbott's Meadow (Fig.6) or Cedar Swamp (Fig.7) was not available. These sites, which are part of the current restoration program, were dominated by large monospecific stands of Phragmites australis so that fish production could not be measured as a baseline for a *Phragmatis australis*-dominated marsh. It will be extremely interesting to see the degree of production and how fish utilize this site when fully restored compared to, say, Mad Horse Creek, in the undisturbed condition.

2.1.3 Phragmites australis Degraded Marshes

Dense stands of *Phragmites australis* are thought to degrade habitat value and marsh functions (Roman et al. 1984). Whether this is true for fish habitat values is a subject of increasing debate. In our own research (Wainright et al., Weinstein et al., in preparation), we have early indications that *P. australis* carbon and nitrogen contributes directly to growth and production in *Fundulus heteroclitus* and marine transient taxa. In Connecticut salt marshes, macroinvertebrates on the marsh surface were shown to be equally abundant in *Phragmites*-dominated sites as at paired (along the salinity gradient) *Phragmites*-free marshes (Fell et al. in press). *Fundulus heteroclitus* were also equally abundant in Breder trap collections set 5 m from the creek edge surface on the surface of *Phragmites*-dominated and reference (*Phragmites*-free) marshes. Fish fed on benthic invertebrates on the marsh plain in both areas. Similarly, gut contents of fish collected in subtidal creeks suggested that *F. heteroclitus* fed extensively in both marsh types, and that diets were generally similar. The authors concluded that the abundance of marsh macroinvertebrates in the *Phragmites*-dominated marshes indicated that such marshes provide suitable physical habitat and usable food resources for these semi-aquatic detritus/algae feeders.

Litter bag studies at these sites also demonstrated that about one-half of the *P. australis* above-ground production in the form of leaves and leave sheaths broke down relatively rapidly. Between June and October, about 25% of leaf production entered the detrital food chain; the remainder did so before the next growing season (Fell et al., in press). Only the stems were noted to decompose slowly. Because *Phragmites* leaf production is substantial, the conversion of natural marsh vegetation to monotypic stands of *P. australis* may not dramatically change the amount of nutrients available to higher trophic levels (Warren and Fell 1996).

On the other hand, studies on the Mullica River estuary in New Jersey (K. Able, pers. commun.) appear to demonstrate that densities of *Fundulus heteroclitus* are much lower in monotypic stands of *Phragmites australis* than in adjacent stands of *Spartina alterniflora* marsh. These apparently contradictory results raise important questions concerning the relative secondary production that would be realized marsh-wide in these systems. Are the results from the Connecticut study applicable to Delaware Bay marshes, or vice-versa? How does the lower drainage density in *Phragmites*-dominated marshes (Windham 1995) affect total accessibility and feeding periodicity by fishes on flood tides? Are predation rates on young fishes greater in the steep banked *Phragmites* fringed tidal creeks than in *Spartina*-dominated creeks with their shallow side slopes (McIvor and Odum 1988). These are clearly questions for future research.

2.2 ECOLOGICAL ENGINEERING OF THE COMMERCIAL TOWNSHIP SITE

Topographic analysis of the Commercial Township site (Fig.15) suggests that more than 50% of the restoration area (shown in orange and yellow) is at elevations below optimum for *Spartina alterniflora* recolonization. The low elevations across much of the site also establish conditions of the Type A or Type B scenario, with a higher potential for a blow out, or other catastrophic event (perhaps like the anecdote described for Woodland Beach?). Thus, the conceptual design for the 1620 ha Commercial Township restoration site recognized the need for constructing high order channels to effectively drain the marsh surface, and avoid the potential for loss of much of the marsh plain. DIVAST (Depth Integrated Velocity and Solute Transport), a two-dimensional finite-difference model, was used to simulate flows and sediment transport at Commercial Township (Falconer 1986, Falconer and Chen 1991). Two advantages of this model are an ability to simulate an extensive intertidal area and an ability to use a database of calibration coefficients that have been developed for application in similar systems.

The use of DIVAST in developing a conceptual design for the Commercial Township site is discussed in Philip Williams & Associates (1995). The approach is summarized here to emphasize the need to prevent potential loss of the entire site, and to establish a wetting-drying cycle conducive to growth of *Spartina alterniflora*. The model was set up to simulate flows and sediment concentrations for all diked areas of the site. A 30.5 m grid with 34,000 nodes was used with a time step of 20 s to measure flow depth, velocity, and sediment concentrations at each node (Philip Williams & Associates 1995). Because the

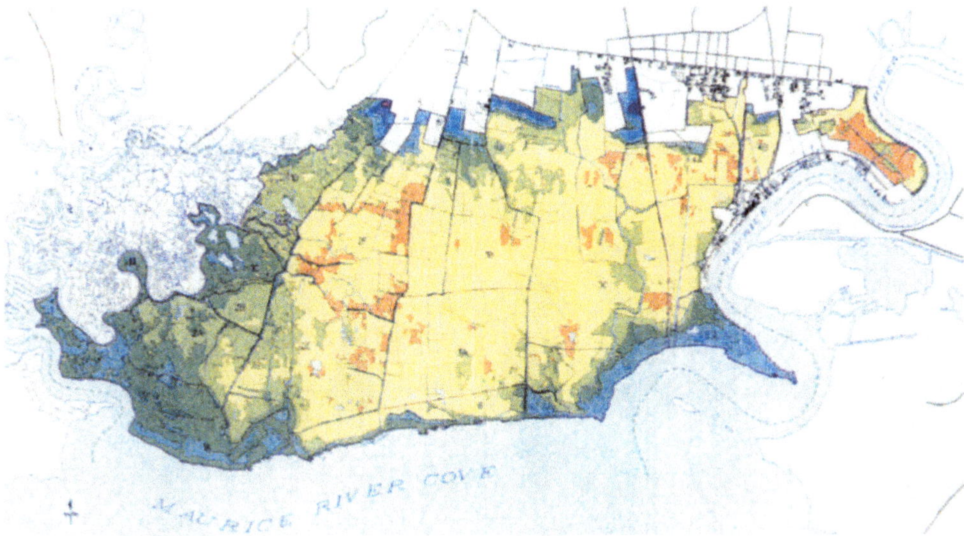

Figure 15. Topographic features of the Commercial Township restoration site, Cumberland County, New Jersey. From these data, it appears that more than 50 % of the area of the site (shown in orange and yellow) is below elevations that are suitable for colonization by *Spartina alterniflora*.

site was not subject to tidal flows, there were no data available to calibrate the model. Instead, DIVAST simulated flow resistance using the Colebrook-White equation (Falconer and Chen 1991) and a characteristic roughness length, ks. This formulation allowed roughness to vary with depth and simulate the very shallow, transitional and fully turbulent rough flows in a stable manner. A value of ks = 0.1 m was selected as a roughness representative of conditions in the early stages of transition from primarily salt hay (*Spartina patens*) to a marsh plain dominated by *Spartina alterniflora*. The value of ks was assumed to include the effects of surface and vegetation roughness, and was selected based on applications of the model to other similar sites where calibration data were available.

To simulate the wetting and drying of the marsh plain, a minimum depth of flow (hm) of 0.1 m was used. Hm is the depth of flow when a grid cell is assume dry; the model cannot simulate drainage below this minimum depth. The value of hm also has a physical interpretation because it should be greater than the typical elevation differences across a computational cell to ensure that the cell is not partially wet, and should not be less than ks. There is natural local ponding of water on the marsh plain due to natural undulations and deposition of detritus, so the choice of hm at 0.1 m is reasonable.

Other design criteria included six individual dike breaches (Fig.16)—four on the bayfront and two off of existing tidal creeks—minimum ebb velocities of 1.0 m sec^{-1}, selection of cross sectional areas of constructed higher order channels based on nearby unrestricted natural creeks, and shear stresses in the higher-order channels that exceed 5.0 N m^{-2} immediately following project implementation. Meeting these criteria would ensure stable inlet formation, and the continued erosion of the partially excavated higher order channels toward their historical alignment with geomorphological attributes mimicking those of natural systems (see Fig.12).

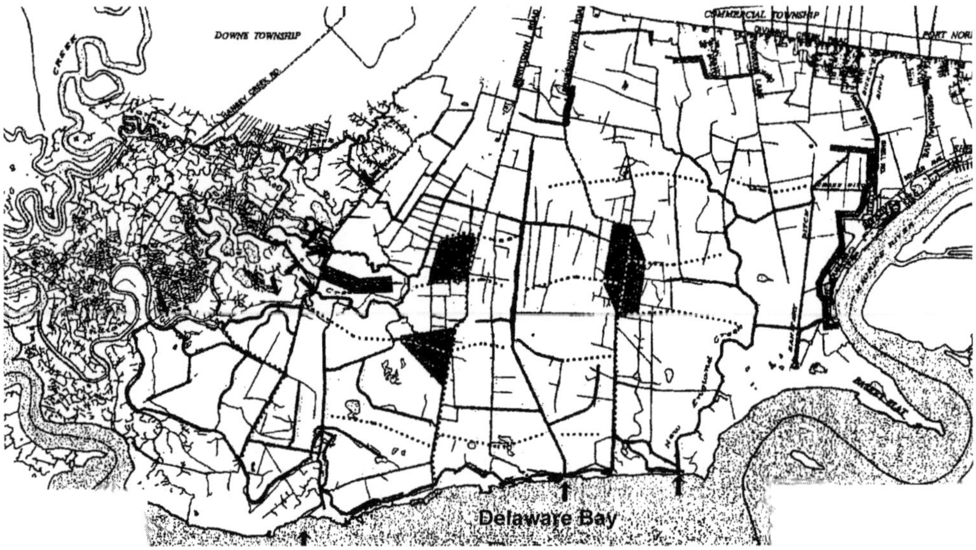

Figure 16. Engineering design criteria for the Commercial Township restoration site include six planned dike breaches—four on the bay front, and two off of existing tidal creeks—and partial excavation of higher order tidal channels.

The rate of temporal evolution of the site drainage system and accretion of sediments to elevations favoring *Spartina alterniflora* were also estimated with DIVAST. The model simulated the erosion, deposition and resuspension of cohesive sediments. It ran for a tidal month, and the deposition of material due to background suspended sediment concentrations and material scoured from the marsh channels was estimated, and used to predict the rate of evolution of the site. The choice of representative tidal heights was based on the peak growing period for *Spartina alterniflora* in Delaware Bay, usually occurring during the month of August.

2.2.1 Tidal Circulation Without Construction of Higher Order Channels and Berms

Under existing conditions which include measured time-lag and attenuation of tidal flows in adjacent creeks relative to the bayshore (Fig.17), and six planned dike breaches including two on existing tidal creeks, a massive circulation pattern with high tidal velocities would occur across the site. Pronounced jet flows are observed at all inlets, and suspended sediment concentrations are relatively high. The high velocities would inhibit sedimentation within the restoration areas, resulting in slower colonization by *Spartina alterniflora*. These conditions with the associated tidal lag, would lead to potential standing water within the site, that without rapid new equilibrium in tidal lags, might ultimately lead to permanent open water conditions. In addition, the modeling showed that the absence of regional berms would lead to substantial erosion in the tidal creeks to the west of the site.

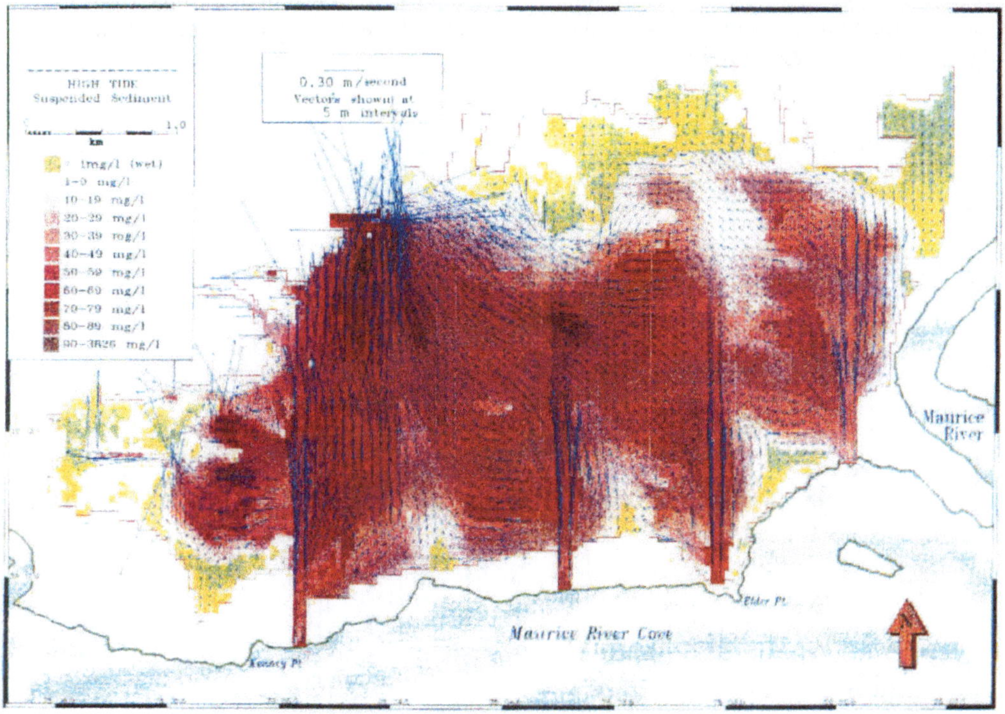

Figure 17. Without the partial excavation of tidal channels to drain the site, predicted massive circulation, and standing water threaten to cause a catastrophic blowout of the marsh at the Commercial Township site.

2.2.2 *Tidal Circulation With Partial Excavation of Higher Order Channels*

The selected design for Commercial Township included the partial excavation of third and fourth order streams linked to each of the six breach locations (Fig.18). This approach would allow for lateral erosion and headcutting of creeks to continue, and would:

- Facilitate efficient wetting and drying of the site without the development of large expanses of standing water;
- Guide the planform evolution of the site towards a natural form, not controlled by existing drainage ditches;
- Provide adequate bed shear stresses to initiate channel erosion and headcutting to create lower order tidal creeks; and
- Immediately provide subtidal fish habitat.

Flow simulations under spring tide conditions (Fig.18) show that nearly all of the site is inundated and that suspended sediment concentrations are distributed toward the headwaters. In this model run, background suspended sediment concentrations have been set to zero and the concentrations shown represent values arising only from the scour of the marsh channels. The analysis also showed that most eroded material would deposit on the marsh rather than in the Bay.

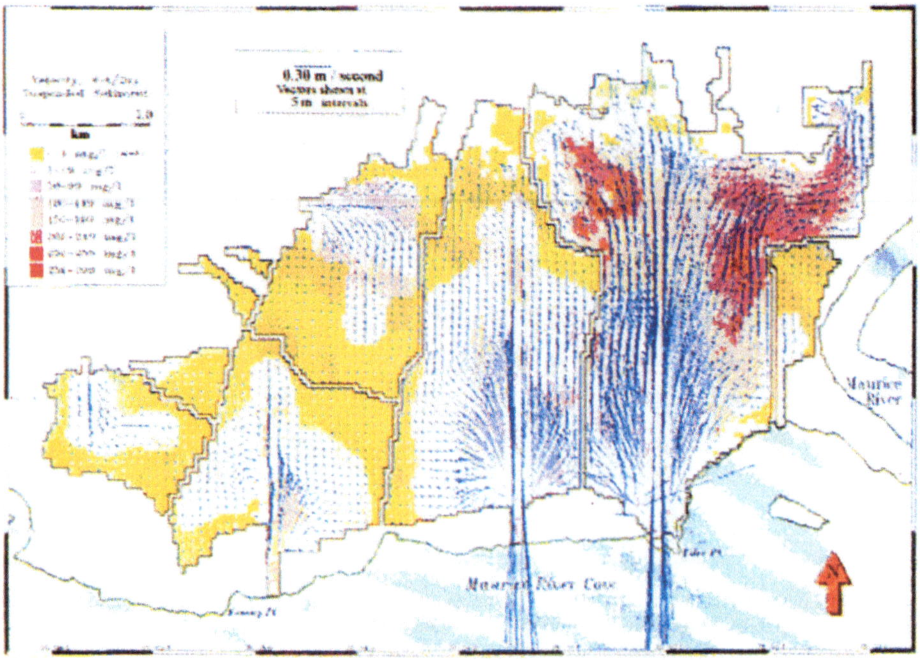

Figure 18. The partial excavation of third and fourth order tidal channels at the Commercial Township restoration site will promote efficient tidal exchange and drainage without the presence of standing water. It is expected that the marsh will progress towards a natural form, not controlled by existing drainages on site, and with minimal risk of a blowout.

The range of accretion rates for natural undiked marshes in Delaware Bay is about 3 to 5 mm y^{-1} (Church et al. 1987). Storm events may increase this rate many times, e.g., Church et al. 1987 noted that one extreme storm occurring in December 1986 deposited from 30% to 165% of annual demand (0.040 to 0.25 g cm^{-2}) with increasing amounts toward the bay-saltmarsh boundary at Kelly Island, Delaware. Combined with enhanced local sedimentation rates due to further scouring of constructed marsh channels, these background accretion rates suggest that the time-trajectory for raising low elevations (< 0.3 m NAVD88) at the Commercial Township site to a level conducive to *Spartina alterniflora* recolonization is on the order of 5 to 10 years (Philip Williams & Associates 1995, Weinstein et al.1997)

3. Summary and Conclusions

Much of the tidal salt marsh habitat of the Delaware Bay shore has been subject to water level management (gates, dikes, and ditching) and other anthropogenic disturbances. Subsidence of the marsh plain, combined with sea level rise, and invasion by *Phragmites australis* make restoration of these large sites particularly challenging. Without thoughtful engineering design, detailed topography and tidal elevation data and development of a wetting-drying cycle that prevents the evolution of standing water, it is possible to lose the entire marsh for extended periods. What constitutes an acceptable end-point, i.e., the bound of expectation, is equally challenging given the paucity of information on optimum fish habitats. The best that can be done is reach consensus among stakeholders on satisfactory end-points (Weinstein et al.1997), and develop success criteria and restoration trajectories to meet these expectations. Without far more detailed study of habitat utilization, the role of *Phragmites australis*, secondary production estimates and export of marsh products, there is no current way of knowing whether the end-points depicted by marsh types shown in Figs. 5 to 10 are comparable to relatively undisturbed systems (Figs. 11 and 12). Well-designed monitoring studies and targeted research must be included in permitting criteria to examine these premises.

If the goal of a large restoration program is the restoration of essential fish habitat, then certain criteria apply. Fish production and habitat utilization efficiency appear to be closely tied to hydrology (area, frequency and depth of inundation), drainage density, subtidal refugia (McIvor and Odum 1988, Rozas et al. 1988), ponding on the marsh surface, and edge (sinuosity, etc.). That is precisely why the restoration design for the diked salt hay farms seek to develop natural drainage configurations, minimizing rectangular creek formation, tree-like drainages, and too much open water. How much of the latter is acceptable can only be determined through additional monitoring and research.

4. Acknowledgments

The authors thank the engineers and scientists of Roy F. Weston, Inc., Woodward Clyde Consultants and the Public Service Electric and Gas Company for access to their engineering design reports and database. R. Hinkle (WCC) provided digital files for Figs. 5 and 6. Important suggestions for improving the engineering designs were offered by members of the Monitoring Advisory Committee, and the Management Plan Advisory Committee, particularly by M. Bruno, W. Mitsch and E. Turner. Without co-sponsorship of the Sea Grant College Program, the Port Authority of NY&NJ, the Delaware River Basin Commission, the USEPA and the NOAA Habitat Restoration Center, this symposium and book would not have been possible.

5. Literature Cited

Baltz, D.M., C. Rakocinski and J.W. Fleeger. 1993. Microhabitat use by marsh-edge fishes in a Louisiana estuary. Environmental Biology of Fishes **36:** 109-126.

Besitka, M.R. 1996. An ecological and historical study of *Phragmites australis* along the Atlantic coast. Thesis, Drexel University, Philadelphia, Pennsylvania, USA.

Browder, J.A., L.N. May, Jr., A. Rosenthal, J.G. Gosselink and R.H. Baumann. 1989. Modeling future trends in wetland loss and brown shrimp production in Louisiana using Thematic Mapper imagery. Remote Sensing of the Environment **28:**45-59.

Chase, C.M. 1979. The Holocene geologic history of the Maurice River Cove and its marshes, eastern Delaware Bay, New Jersey. Thesis, University of Delaware, Newark, Delaware, USA.

CH2MHill 1995a. Design alternatives analysis, Brown's Run site. Draft report to Public Service Electric and Gas Company, Nov. 7, 40 pp + App.

CH2MHill 1995b. Design alternatives analysis, Green Swamp site. Draft report to Public Service Electric and Gas Company, Nov. 7, 41 pp + App.

Church, T.M., R.B. Biggs and P. Sharma. 1987. The birth and death of salt marshes: geochemical evidence for sediment accumulation and erosion. EOS Abstracts **69**: 305.

Daiber, F.C. 1986. Conservation of Tidal Marshes. Van Nostrand Reinhold, New York, New York, USA.

Daiber, F.C. and C.T. Roman. 1988. Tidal marshes. Pages 95-109 *in* T.L. Bryant and J.R. Pennock, editors. The Delaware Estuary: rediscovering a forgotten resource. University of Delaware Sea Grant College Program, Newark, Delaware, USA.

Dow, D.D. 1982. Software programs to measure interface complexity with remote-sensing data, with an example of a marine ecosystem application. NASA Report No. **219**, NASA Earth Research Laboratory, National Space Technology Laboratory, NSTL, Mississippi, USA.

Falconer, R.A. 1986. A two dimensional mathematical model study of the nitrate levels in an inland natural basin. pp. 325-344. International Conference on Water Quality Model, Inland Natural Environment. BHRA, Bournemouth, England.

Falconer, R.A. and Y.P. Chen. 1991. An improved representation of flooding and drying and wind stress effects in a two-dimensional tidal numerical model. Proceedings of the Institute of Civil Engineering, Part 2, **91**: 659-678.

Faller, K.H. 1979. Shoreline as a controlling factor in commercial shrimp production. NASA Rep. No. 208, NASA Earth Research Laboratory, National Space Technology Laboratory, NSTL, Mississippi.

Fell P.F. et al., in press. Comparison of typical tidal marsh and reed grass (*Phragmites australis*) - dominated tidal marshes along the lower Connecticut River, selected macroinvertebrates and their predation by mummichogs. Journal of Experimental Marine Biolology and Ecology.

Frey, R.W. and P.B. Basan. 1978. Coastal salt marshes. Pages 101-169I *in* Davis, R.A. Jr., editor. Coastal sedimentary environments. Springer-Verlag, New York, New York, USA.

Gosselink, J.G. 1984. The ecology of delta marshes of coastal Louisiana: a community profile. FWS/OBS-**84/09**, Office Biological Services, U.S. Fish Wildlife Service, Washington, District of Columbia, USA.

Harrington, R.W. and E.S. Harrington. 1961. Food selection among fishes invading a high subtropical salt marsh: from onset of flooding through the progress of a mosquito brood. Ecology **42:** 646-666.

Hettler, W.F. 1989. Nekton use of regularly-flooded saltmarsh cordgrass habitat in North Carolina, USA. Marine Ecology Progress Series **56:**111-118.

Hodson, R.G., J.O. Hackman and C.R. Bennett. 1981. Food habits of young spots in nursery areas of the Cape Fear River, North Carolina. Transactions of the American Fisheries Society **110:** 495-501.

Houghton, J.T., G.J. Jenkins and J.J. Ephraums. 1991. Climate change: the IPCC scientific assessment. Cambridge University Press, Cambridge, England.

Kleypas, J. and J.M. Dean. 1983. Migration and feeding of the predatory fish, *Bairdiella chrysura* Lacepede, in an intertidal creek. Journal of Experimental Marine Biology and Ecology. **72:**199-209.

Kneib, R.T. 1991. Flume weir for quantitative collection of nekton from vegetated intertidal habitats. Marsh Ecology Progress Series **75:**29-38.

Kneib, R.T. 1994. Spatial pattern, spatial scale and feeding in fishes. Pages 171-185 *in* K. Fresh and D. Stouder, editors. Theory and application in fish feeding ecology. University of South Carolina Press, Columbia, South Carolina, USA.

Kneib, K.T. and S.L. Wagner. 1994. Nekton use of vegetated marsh habitats at different stages of tidal inundation. Marsh Ecology Progress Series **106:** 227-238.

Kraft, J.C., H.I. Yi and M. D. Khalequzzaman. 1992. Geologic and human factors in the decline of the tidal saltmarsh lithosome: the Delaware Estuary and Atlantic coastal zone. Sedimentary Geology **80:** 233-246.

Marks, M., B. Lapin and J. Randall. 1994. *Phragmites australis* (*P. communis*): Threats, management and monitoring. Natural Areas Journal **14**:285-294.

McIvor, C.C. and W.E. Odum. 1986. The flume net: a quantitative method for sampling fishes and macrocrustaceans on tidal marsh surfaces. Estuaries **9**:219-224.

McIvor, C.C. and W.E. Odum. 1988. Food, predation risk and microhabitat selection in a marsh fish assemblage. Ecology **69:**1341-1351.

Miller, C. and V. Guillory. 1980. A comparison of marsh fish communities using the Wegner ring. Proceedings of the Annual Conference of S. E. Associations of Fish and Wildlife Agencies **34:** 223-233.

Murphy, S.C. 1991. The ecology of estuarine fishes in southern Maine high salt marshes: access corridors and movement patterns. Thesis, University of Massachusetts, Amherst, Massachusetts, USA.

National Research Council. 1987. Responding to changes in sea-level. National Academy Press, Washington, District of Columbia, USA.

Nesbit, D.M. 1885. Tide marshes of the United States. U.S. Department of Agriculture Special Report No. **7**. Washington, District of Columbia, USA, GPO, 1885.

Nettle, W.K. 1997. Conserving coastal wetlands despite sea level rise. AGU Working Group, Sea Level Rise and Wetland Systems. EOS **78:** 257 pp.

Niering, W.A. and R.S. Warren. 1977. Our dynamic tidal marshes: vegetation changes as revealed by peat analysis. Connecticut College Arboretum Bulletin. **22**.

NOAA. 1994. Yearly mean sea-levels and monthly tidal summary reports for Atlantic City, NJ; Battery, NY; Lewes, DE; Philadelphia, PA; and Sandy Hook, NJ. U.S. Department of Commerce, National Ocean Service, Rockville, Maryland, USA.

Orson, R.A., R.L. Simpson and R.E. Good. 1992. The paleoecological development of a late Holocene tidal freshwater marsh of upper Delaware River estuary. Estuaries **15:** 130-146.

Philipp, K.R. 1995. Tidal wetlands characterization—then and now. Final Report, Delaware River Basin Commission.

Phillips, J.D. 1986. Coastal submergence and marsh fringe erosion. Journal of Coastal Research **2:** 427-436.

Phillips, J.D. 1987. Shoreline processes and establishment of *Phragmites australis* in a coastal plain estuary. Vegetation **71:** 139-144.

Psuty, N.P., M.P. DeLuca, R. Lathrop, K.W. Able, S. Whitney and J.F. Grassle. 1993. The Mullica River-Great Bay National Estuarine Reserve: a unique opportunity for research, preservation and management. Pages 1557-1568 *in* O.T. Magoon, W.S. Wilson, H. Converse and L.T. Tobin, editors. Coastal Zone '93, American Society of Civil Engineers, New York, New York, USA.

Roman, C.T., W.A. Niering and R.S. Warren. 1984. Salt marsh vegetation change in response to tidal restriction. Environmental Management **8:**141-150.

Rountree, R.A. and K.W. Able. 1992. Fauna of polyhaline subtidal marsh creeks in southern New Jersey: composition, abundance and biomass. Estuaries **15:**171-185.

Rozas, L.P. and W.E. Odum. 1987. Use of tidal freshwater marshes by fishes and macrofaunal crustaceans along a marsh stream-order gradient. Estuaries **10:**36-43.

Rozas, L.P. and D.J. Reed. 1993. Nekton use of marsh-surface habitats in Louisiana deltaic salt marshes undergoing submergence. Marsh Ecology Progress Series: **96**: 147-151.

Rozas, L.P., C.C. McIvor and W.E. Odum. 1988. Intertidal rivulets and creek banks: corridors between tidal creeks and marshes. Marsh Ecology Progress Series **47:** 303-307.

Seebold, K.R. 1992. From marsh to farm: the landscape transformation of coastal New Jersey. National Park Service, U.S. Department of Interior, Washington, District of Columbia, USA.

Stumpf, R.P. 1983. The process of sedimentation on the surface of a salt marsh. Estuarine and Coastal Shelf Science **17:** 495-508.

Stumpf, R.P. 1984. Analysis of suspended sediment distributions in the surface waters of Delaware Bay using remote sensing of optical properties. Dissertation, University of Delaware, Newark, Delaware, USA.

Targett, T. 1985. Utilization of salt marsh surface habitat by estuarine fishes (Abstract). Estuaries 8A.

Titus, J.G. and V.K. Narayan. 1995. The probability of sea-level rise. U.S. Environmental Protection Agency, Washington, District of Columbia. 186 pp.

Warren, G.M. 1911. Tidal marshes and their reclamation. U.S. Dept. Agriculture, Bulletin of the Experimental Station **240:** 1-99.
Warren, R.S. and P.E. Fell. 1996. *Phragmites australis* on the lower Connecticut River: impacts on emergent wetlands and estuarine waters. Final Report, Long Island Sound Research Fund, State of Connecticut, Department of Environmental Protection.
Weigert, R.G. and L.R. Pomeroy. 1981. The salt-marsh ecosystem: a synthesis. Pages 219-230 *in* L.R. Pomeroy and R.G. Weigert, editors. The ecology of a salt marsh. Springer-Verlag, New York, New York, USA.
Weinstein, M.P., J.H. Balletto, J.M. Teal and D.F. Ludwig. 1997. Success criteria and adaptive management for a large-scale wetland restoration project. Wetlands Ecology and Management **2:** 111-197.
Weisberg, S.B. and V.A. Lotrich. 1982. The importance of an infrequently flooded intertidal marsh surface as an energy source for the mummichog, *Fundulus heteroclitus*: an experimental approach. Marine Biology **66:** 307-310.
Phillip Williams & Associates, Ltd. 1995. Hydrologic evaluation of restoration alternatives, Commercial Township Salt Hay Farm, Cumberland County, New Jersey. Final Report. Public Service Electric and Gas Company, Newark, New Jersey, USA.
Windham, L. 1995. Effects of *Phragmites australis* invasion on abovcground biomass and soil properties in brackish tidal marsh of the Mullica River, New Jersey. Thesis, Rutgers University, New Brunswick, New Jersey, USA.
Zimmerman, R.J. and T.J. Minello. 1984. Densities of *Penaeus aztecus*, *Penaeus setiferus* and other natant macrofauna in a Texas salt marsh. Estuaries **7:** 421-433.
Zimmerman, R.J., T.J. Minello, G. Zamora, Jr. 1984. Selection of vegetated habitat by brown shrimp, *Penaeus aztecus*, in a Galveston Bay salt marsh. Fishery Bulletin, U.S. **82:**325-336.

REFERENCE IS A MOVING TARGET IN SEA-LEVEL CONTROLLED WETLANDS

ROBERT R. CHRISTIAN
LAURA E. STASAVICH
CASSONDRA R. THOMAS
MARK M. BRINSON
Department of Biology
East Carolina University
Greenville, NC 27858 USA

Abstract
Central to the restoration of wetlands is the concept of reference. Wetlands are grouped within a reference domain of the same hydrogeomorphic subclass and compared to reference standards of the least impacted members of the subclass. Restoration decisions can be based within this context. Sea-level controlled wetlands, including salt marshes, provide a challenge to establishing reference standards because they progressively change in response to rising sea-level and associated stressors. The natural progression of change is distinct from that induced by human activities. We review geomorphic classifications for sea-level controlled wetlands to identify a spatial scale appropriate for restoration. This scale encompasses an ecosystem state change model which accounts for the natural progression. We emphasize the importance of more proximal causes of change than sea-level rise itself (e.g., access to fresh and sea water, sediment, and space for transgression). Through examples from several marshes, we highlight the consequences of movement, note distinctions between low and high marshes, and describe the transition between them. These distinctions are made for hydrology in an intertidal and nontidal portion of a marsh and for nitrogen cycling in a northern and a southern marsh on the Atlantic coast of the USA. Further, we describe the nature of one change in state as the turf of a high marsh becomes unstable, producing a hollow and hummock pattern that is expected to transform into low marsh. Recognition of the consequences of movement to more explicit, hydrogeomorphology-based reference systems helps place restoration in a perspective which will improve both project design and probability of success.

1. Introduction

The term restoration, when used in the context of improving the functioning and condition of altered or impacted ecosystems, implies that the ecosystem in question will be directed toward a specific target or goal. To achieve such a goal, it is normally necessary to manipulate hydrology, sediments, and the biotic community to direct it toward one or several endpoints. Target, or ''reference standard,'' ecosystems should represent these endpoints and be self-sustaining (i.e., they will require little or no continuing maintenance other than protection from further human impacts). We should look to natural or relatively unaltered ecosystems for patterns and processes that contribute to sustainability. The use of such reference in restoration can provide stable endpoints toward which restoration projects can be designed and by which their success can be judged (Brinson and Rheinhardt 1996).

The concept of "reference" for ecosystems is implicit in the thinking of many field ecologists. Each ecologist has an understanding and preconception of how systems operate based on experience. This is very evident in coastal marsh ecology. Even though any coastal marsh may be relatively simple in community structure, and all are sea-level controlled, there is considerable diversity with respect to their hydrogeomorphic setting and associated functioning. A major purpose of using "reference" is to explicitly describe natural variation and to use such characterizations as a beginning point for assessing condition. Making reference explicit provides opportunity for acknowledging differences, making valid comparisons, controlling experiments, and incorporating the knowledge learned into both science and management.

Coastal ecosystems affected by rising sea level tend to move both vertically and horizontally (Hayden et al. 1995). Vertical accretion in response to sea-level rise allows the potential for horizontal movement across the landscape. One consequence of horizontal movement is seaward to landward shift in plant communities. For example, terrestrial forest converts to high marsh (i.e., inundated by estuarine water only by extreme spring and storm tides), and low marsh (i.e., intertidal) converts to mud flat. Here we help resolve the complexity and time-dependent scope of change in a way that will contribute to wetland restoration. As a starting point, we examine some of the premises and definitions concerning the nature of ecosystems in general and, more specifically, the extent to which coastal ecosystems vary over time. They include:

- Ecosystems are self-sustaining units of landscapes that perpetuate structure and functioning over time through the acquisition of energy and matter within a range of environmental conditions and disturbance regimes. When the acquisition of energy or matter changes, or environmental conditions or disturbance regimes shift, ecosystem structure and functioning often undergo fundamental changes to a new state. The model of "ecosystem state" and "state change" is more fully developed and explained in Hayden et al. (1995). An ecosystem state is a distinct landscape unit at a scale appropriate for analysis by the observer. For our purposes of analysis they are the zones defined by vegetation and soil type within a marsh or larger coastal landscape. A state change is the process of switching from one state to another. For example, a wetland of submersed aquatic plants may change to a forested wetland if the water depth decreases due to drainage or the accumulation of sediment. An ombrotrophic bog may change to a eutrophic cattail marsh with increases in nutrient loading.
- Coastal landscapes are arranged zonally by the proximity of terrestrial and marine endmembers such that ecosystem states somewhat predictably array themselves (e.g., mud flat, low marsh, transition marsh, high marsh, and terrestrial forest) between these extremes. Each state has characteristic functioning to maintain itself.

- Rising or falling sea level relative to the land surface is the fundamental cause of state change among zones of coastal ecosystems. Rates of change are a function of relative rate of sea-level change and the underlying slope of the land.
- The interaction of disturbance and stressors with the ecosystem state's abilities to function is the more proximal cause of state change.
- Minimally altered or natural conditions, including changes in ecosystem state, represent useful benchmarks from which restoration alternatives can be judged and evaluated relative to socioeconomic priorities and constraints.

Restoration projects that take into account the foregoing premises may improve the likelihood of meeting long-term success criteria. In this paper we address the ''movement of reference'' in sea-level controlled marshes and relate it to restoration. First, we provide a hydrogeomorphic classification of these systems and discuss the nature of the movement in the context of ecosystem state change. The concept of ecosystem state is further developed by assessing hydrogeomorphic relationships relative to marsh zonation and how these relationships affect ecosystem functioning within zones or states. Lastly, we relate these issues to the restoration process.

2. Classification

How many varieties of reference need to be identified as a starting point toward categorizing coastal wetlands for purposes of restoration? For example, sea-level controlled wetland types range from macrotidal salt marshes to freshwater wetlands without tides. They include boreal coastlines subjected to disturbance by ice and subtropical to tropical mangroves that cannot withstand the effects of frost. Restoration efforts in each of these conditions would be expected to have different limitations and opportunities. Consequently, ''success criteria'' would differ depending on these conditions.

One way of approaching the issue of wetland types is to develop a classification for coastal wetlands that takes into account the variety in a geographic region (e.g., Atlantic coast of the United States). The classification should meet two criteria: 1) that it classifies at a spatial scale that is relevant to restoration efforts and 2) that it recognizes the forces responsible for the self-sustaining properties of coastal wetland ecosystems. To our knowledge, there is no widely used and comprehensive classification for coastal wetlands that takes into account these two criteria. The widely used Cowardin et al. (1979) classification implicitly recognizes self-sustaining properties through water regime, water chemistry, and substrate, but does not provide a spatial context relative to other classes. Restoration plans must identify such details as the position of individual tracts of land slated for restoration (elevation and flood regime), the species composition if planting is to be used, and the long-term development of a site that is undergoing state change in response to rising sea level.

Salt marsh ecosystems are identified often by their species composition rather than by geomorphic setting explicitly. For example, the community profile on salt marshes of New England by Teal (1986), based on perhaps the most intensively studied group of tidal marshes in the world, indicates where marshes dominated by *Spartina alterniflora* are located. However, *S. alterniflora* marshes represent a biogeographic distribution of a species rather than a class that also incorporates abiotic factors. We will examine several classifications, beginning with the largest scales, to see if they meet the aforementioned criteria. Then, we will suggest a classification system that takes into account both the spatial scale and the dynamic response of coastal wetlands to rising sea level.

2.1 CLASSES BASED ON CLIMATE AND GEOGRAPHY

Odum et al. (1974) classified coastal ecological systems of the United States and placed salt marshes within the category of ''Natural temperate ecosystems with seasonal programming.'' In so doing, they separated the coastal marsh subset from tropical coastal ecosystems, most of which are dominated by mangroves. Thus temperate marshes and tropical mangroves were distinguished on the basis of seasonality of biotic responses to climate rather than on geomorphology. Salt marshes in the USA are further separated into east coast, south Atlantic and Gulf Coast, and west coast, with descriptions of irregularly flooded marshes limited to the east coast. However, the variety of species compositions, tidal regimes, and sedimentary environments within each of these regions is much too broad for most restoration projects.

2.2 CLASSES BASED ON REGIONAL GEOMORPHIC SETTINGS

Although Thom's (1982) classification was developed for mangroves, it also has relevance for coastal wetlands at higher latitudes. Major variables in the classification are sediment supply, tidal power, wave influence, and original coastline consistency. Thom identifies five mangrove types which we adapt more generally: 1) river-dominated allochthonous, 2) tide-dominated allochthonous, 3) wave-dominated barrier lagoon (autochthonous), 4) composite types: river- and wave-dominated, and 5) drowned bedrock valley. River-dominated allochthonous systems are abundant in deltaic settings, such as the Mississippi River. Because of dominance by fresh water sources, freshwater marshes and swamp forests are abundant in this geomorphic setting. Tide-dominated allochthonous wetlands occur in sediment rich areas and have higher salinity than the river-dominated allochthonous wetlands. They are represented by many of the marshes behind the barrier islands along the South Carolina and Georgia coast. Wave-dominated barrier lagoonal (autochthonous) wetlands occur in low energy environments, as found along North Carolina's outer banks and the eastern shore of Virginia where barrier islands are abundant and tidal amplitudes are moderate. Composite river- and wave-dominated types occur in the Mississippi River delta. Drowned bedrock valley types form near steep rocky coasts, similar to many of the coastal wetland sites on the Pacific coast of the United States.

Each of the five types represents very different sedimentary environments, but the scales of each type are so large as to encompass a broad group of local conditions ranging from rapidly prograding to rapidly eroding. From a practical perspective, classification based on these large geomorphic differences might be relevant for nationwide programs needing regionally calibrated success criteria. For example, restoration efforts in the Mississippi Delta region would be expected to yield large absolute increases in coastal wetland area, while efforts on the Pacific coast would be much smaller but have high local significance. For site-specific restoration projects, such regional classifications are of limited utility, yet they do provide a framework for assessing benefits of local restoration projects to the region or to the nation.

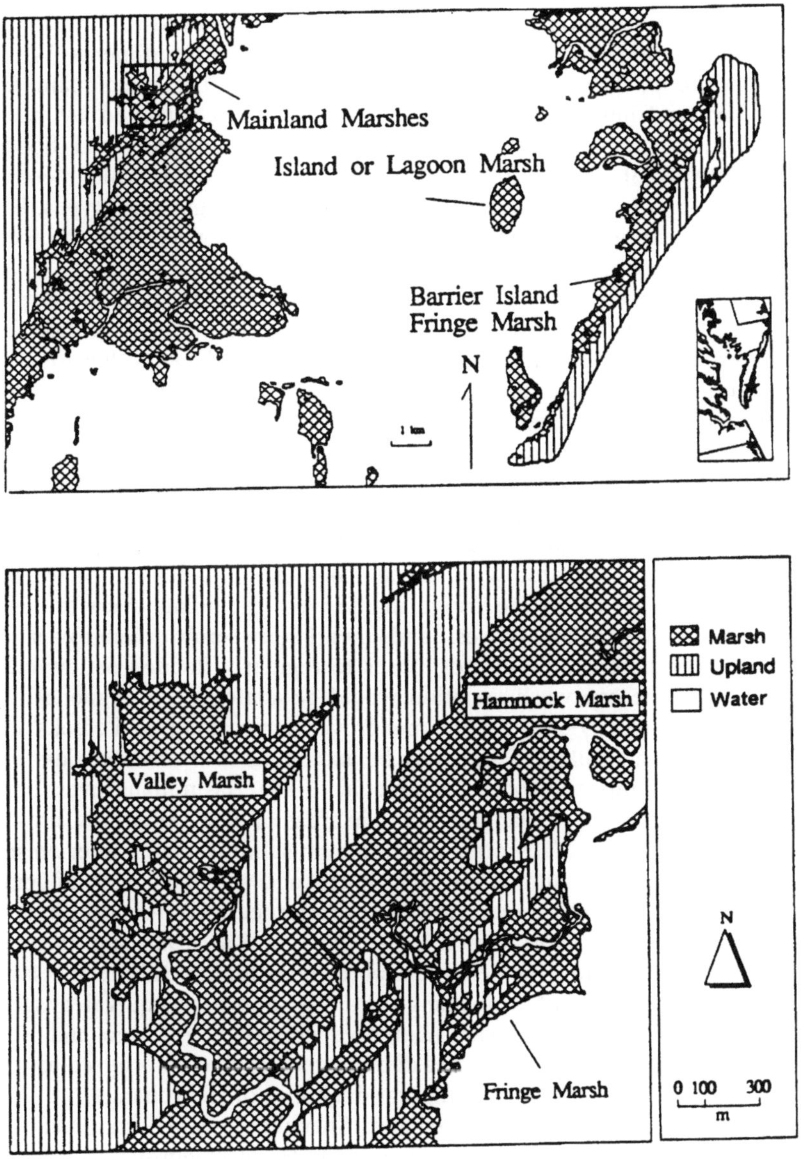

Figure 1. Saltmarsh geomorphic classes within the Virginia Coast Reserve, southeastern Delmarva Peninsula according to Oertel et al. (1992) and Oertel and Woo (1994). Adapted from Brinson et al. (1995b).

2.3 CLASSES BASED ON LOCAL GEOMORPHIC SETTINGS:

On the basis of wetlands found in the mid-Atlantic region, Oertel et al. (1992) and Oertel and Woo (1994) identified three geomorphic settings: backbarrier, lagoon, and mainland marshes (Fig. 1). The backbarrier marshes are influenced by storm overwash that transports sand to the marsh surface, and by flood tide deltas that provide intertidal flats for wetland vegetation to colonize. Lagoonal marshes are believed to form mainly on Pleistocene surfaces rather than by in-filling of the lagoon with sediments. Within the mainland marsh group, there are sloping interfluve settings (i.e., interstream divides) occupied by narrow fringe marshes, flat interfluve settings occupied by hammock marshes, and valley marshes that occupy former stream channels and floodplains (Brinson et al. 1995a). Stevenson et al. (1985), working at a similar scale, incorporated changing sea level by identifying emerging coast and submerging coast types. Because each class was not explicitly separated into zones useful to restoration projects, the criterion of relevant spatial scales is not met by these examples.

Bacon (1994) characterized coastal wetlands in an effort to predict their response to rise in sea level for Caribbean islands. He recognized relocation (migration inland), losses due to erosion or loss of protective sea barriers, and gains due to landward replacement and rejuvenation of salinas to mangroves. Although he used the classification of Lugo and Snedaker (1974) to characterize the mangrove vegetation, the classes were not recognized as parts of a continuum. Perhaps due to the small size of most wetlands and the steeper slopes associated with many of the islands, the idea of transition among states was not apparent or useful.

The classification of Lugo and Snedaker (1974) for mangroves provides some valuable parallels when compared with salt marshes. A simplification of the original six types (Cintron et al. 1985) are fringe mangroves (including overwash islands), riverine mangroves, basin mangroves, and dwarf (or scrub) mangroves. Fringe mangroves would be comparable in most cases to regularly flooded or low-elevation temperate marshes. The overwash variation would be analogous to lagoonal marshes described by Oertel et al. (1992). Riverine mangroves occur along tidal creeks and may extend several kilometers inland. These may be analogous to freshwater tidal marshes and swamps found in temperate zones (Odum et al. 1984). Basin mangroves, described as often occurring behind fringe mangroves and in areas of little water flow, are isolated from regular flooding events. They often develop high salinities and border on salt flats toward the upland boundary similar to high marsh and salt flat zones of temperate marshes. Scrub and dwarf mangroves, however, are respectively found in hypersaline and nutrient poor sites. The short form *S. alterniflora* or *Salicornia* spp. in temperate marshes develop in hypersaline pannes where porewaters high in sulfide and salt lower productivity similar to the case of scrub mangroves. We know of no equivalent salt marsh condition to the nutrient-limited dwarf mangrove described by Lugo and Snedaker (1974). While this classification recognizes distinct geomorphic units at scales useful to restoration, it does not incorporate the temporal dynamics of state change.

Twilley (1995) extended the Lugo and Snedaker (1974) classification, the geomorphic setting of Thom (1982), and other attributes of mangrove ecosystems into a composite collectively called the energy signature. By bringing other forcing functions to bear on mangroves (i.e., river flow, tidal amplitude, rainfall, solar radiation) he was able to explain such ecosystem properties as biomass production and organic matter export. In so doing, Twilley has developed a multivariate approach which serves as a functional classification, but not one that deals explicitly with state change.

Local geomorphic settings are linked together by a marsh-estuarine continuum concept proposed by Dame et al. (1992). They used some of the principles of the "river continuum concept" of Vannote et al. (1980) and the ecosystem development concept of Odum (1969) to examine the structure and function of estuaries typical of the Carolina-Georgia bight. In the marsh-estuarine continuum ephemeral creeks are considered the youngest endpoint and the ocean the mature endpoint, during periods of rising sea level. Gradients within the continuum include salinity, tidal action, material transport, habitat, and species composition. Dame et al. (1992) acknowledged that salt marshes migrate upslope and transform forest spodosols to sulfidic soils. Because they focused on tidally dominated systems (sensu Thom 1982), more emphasis was given to creek development, the role of oyster reefs, and especially the contrast of young ephemeral creeks that drain uplands and intertidal marsh zones with mature subtidal estuarine sites. The treatment of gradients allows one to place the continuum in the context of hydrogeomorphology and the dynamic nature of change within the coastal landscape and explains many aquatic ecosystem properties along this continuum. However, little attention is given to the nature of the marsh plain and composition of vegetation, both of which are germane to restoration.

2.4 PROPOSED CLASSIFICATION BASED ON STATE CHANGE CONTINUUM

For the purposes of formulating a classification for restoration of coastal marshes, the minimum set of conditions that should be considered in most restoration projects is the zones that comprise the continuum from terrestrial to marine end members. This is consistent with our premise that "Coastal landscapes are arranged zonally by the proximity of terrestrial and marine endmembers such that ecosystem states somewhat predictably array themselves (e.g., terrestrial forest, high marsh, transition marsh, low marsh, mud flat) between these extremes." This pattern provides a template upon which to recognize departures as being geomorphically controlled variations. Each variation would be considered a reference standard condition that would be the appropriate benchmark for comparison of restoration projects. We describe here 3 views of the state change continuum: (1) marsh zonation and factors promoting its change, (2) marsh edge movement and regulating factors, and (3) vegetation responses to flooding and salinity.

2.5 MARSH ZONATION AND FACTORS PROMOTING ITS CHANGE

The transgression template for zonation is an extension of one Brinson et al. (1995b) proposed for the mainland marshes of the Virginia Coast Reserve (VCR) along the eastern shore of Delmarva peninsula. In that paper we presented a conceptual model of how mainland marshes at the VCR would transgress across the landscape in response to sea-level rise. We identified 5 commonly recognized ecosystem states within these mainland marshes and assessed the mechanisms responsible for transitions from one state to another. The mainland marshes of the VCR have little terrigenous sediment supply and are generally protected from wave action and strong tidal currents.

Here we expand the previous model, and hence the template of transgression, to incorporate a broader range of marshes. We focus on salt marshes as indicated in the parallelogram in Fig. 2. Ecosystem states change sequentially during marsh transgression as the site initially occupied by forest progressively decreases in relative elevation as sea level rises. Beginning with terrestrial forest, marsh vegetation replaces woody plant species due to increases in soil salinity. Salinity levels are established by regional water sources that include precipitation, the amount of freshwater discharge in groundwater and through streams, and the conveyance of higher salinity tidal water. The interaction between these factors determines community structure, soil conditions and aspects of biogeochemistry (Gardner et al. 1992, Hmieleski 1994). The high marsh plant community becomes dominated by halophytes and biogeochemical processing becomes more dependent on sulfur cycling. Added to salinity effects are those of flooding from both tides and precipitation and waterlogging of soils (Eleuterius and Eleuterius 1979, Hmieleski 1994).

With more frequent inundation from rising sea level and with adequate allochthonous sediment supply, state change continues from high marsh through a transition to the low marsh state. Finally a mud flat develops when sea level rises beyond the sustainability of accretion by the marsh. (The term mid-marsh has been used in the literature. As we discuss later, we consider this zone to be transitory between high and intertidal, low marsh for the purposes of this paper.) Sediment availability arises from a combination of turbid waters, high tidal energies capable of moving sediments in suspension, and short distance from tidal water source to high marsh elevations.

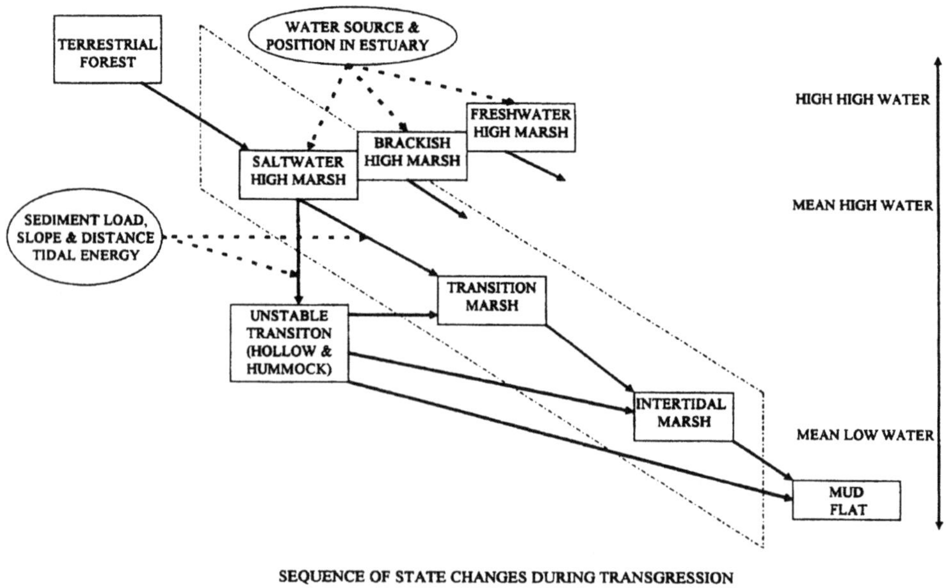

Figure 2. The state change continuum for coastal wetlands. Adapted from Brinson et al. (1995b).

Sediments can be deposited rapidly as water moves over the marsh surface and may not normally be taken deep within the marsh (Christiansen 1998). The contribution of sediment to the high marsh does not have to be in a continual fashion, however, but is likely to be episodic with large storms and tides (Stumpf 1983, Cahoon et al. 1995).

The state changes can progress via another route (Day et al. 1995). High marshes often develop organic rich soils (i.e., autochthonous inputs dominate) under the following conditions: 1) where water turnover is low on flat landforms due to poor drainage and where sites are distant from tidal flushing and sediment supplies; or 2) in microtidal regimes, where currents are not sufficient to cause regular tidal flushing, development of tidal creek channels, and sediment transport (Turner 1997). Such sites must undergo "unstable" transitions to other states and loss of relative elevation (Stevenson et al. 1985, Reed and Cahoon 1992, Nyman et al. 1993, Warren and Niering 1993, DeLaune et al. 1994). The loss of peat and relative elevation occurs through the formation of ponds or potholes. These areas become connected to headward eroding tidal creeks, mineral soils are retained, and transition or low marshes form.

Thus, movement of coastal marshes within the landscape and changes in state within a marsh may be perceived as being dependent on source and conveyance of relatively few hydrogeomorphic factors associated with the availability of fresh and saline water (including frequency and duration of flooding), sediment, and space (Fig. 2). The salinity regime depends on regional freshwater sources; position in the estuary, lagoon or coast; and conveyance of salt water by tidal forces. Sediment availability depends on magnitude of sediment source, the nature of the sediments (e.g., particle size) and the conveyance via tidal energy across slope and distance from sediment source. Space must be available at the terrestrial edge of the marsh. This margin must lack barriers to movement, have a low slope, and have land use that can be transformed into coastal wetland.

2.6 MARSH EDGE MOVEMENT AND REGULATING FACTORS

Superimposed upon this series of states are two variables having to do with local sediment supply and slope, also addressed by Brinson et al. (1995b). In that paper, we simplified terrestrial-estuarine interactions into four combinations from two possible conditions at the terrestrial-marsh margin (i.e., overland migrating versus stalling) and two at the marsh-estuarine margin (i.e., prograding versus eroding). These four combinations are a) migrating overland and prograding toward the estuary, b) migrating overland and eroding away from the estuary, c) stalling at the terrestrial margin and prograding toward the estuary, and d) stalling at the terrestrial margin and eroding away from the estuary. Two more directions may be added by including marsh regression from the terrestrial margin (Table 1). Regression occurs from human activity through fill operations and from natural phenomena through storm overwash of sand on barrier islands onto marshes (Hayden et al. 1995, Osgood et al. 1995) and glacial rebound. The area of marsh decreases at the landward edge as the "terrestrial state" rapidly spreads over it or the whole terrace rises.

The current and future status of a marsh depends not only on the opportunities to move horizontally, but also the associated ability to grow vertically in response to rising sea level. If a marsh does not have sufficient vertical accretion over the long

TABLE 1. Movement possibilities of coastal marshes with respect to sea-level change.

Relative to estuarine margin	Relative to terrestrial margin	Relative to sea level
Prograding	Transgressing	Accreting
X	X	
	Stalled	
Eroding	Regressing	Subsiding

term, it will be overwhelmed by rising sea level. However, not all locations in natural marshes are necessarily accreting. One can identify examples of whole marshes that are subsiding or subsiding areas within otherwise accreting marshes (Reidenbaugh et al. 1983, Reed and Cahoon 1992, Cahoon et al. 1995a,b). Subsidence may be a near surface process that is part of the changes in state (Fig. 2). In other cases, subsidence may occur from deeper processes such as ground water withdrawal and tectonic activity (White and Tremblay 1995).

All of these possibilities of states and state change may be held relative to a steady-state template. Such a template may be useful in establishing restoration criteria. A ''transgression template'' in this steady-state condition would have no change in surface area of marsh states over time. A gentle slope and eroding shoreline, a combination that accommodates the potential of landward migration at the same rate that the marsh shoreline is retreating, would characterize it. Departures from this neutral condition are ''losing'' marshes and ''gaining'' marshes, both defined by their tendency to lose or gain surface area.

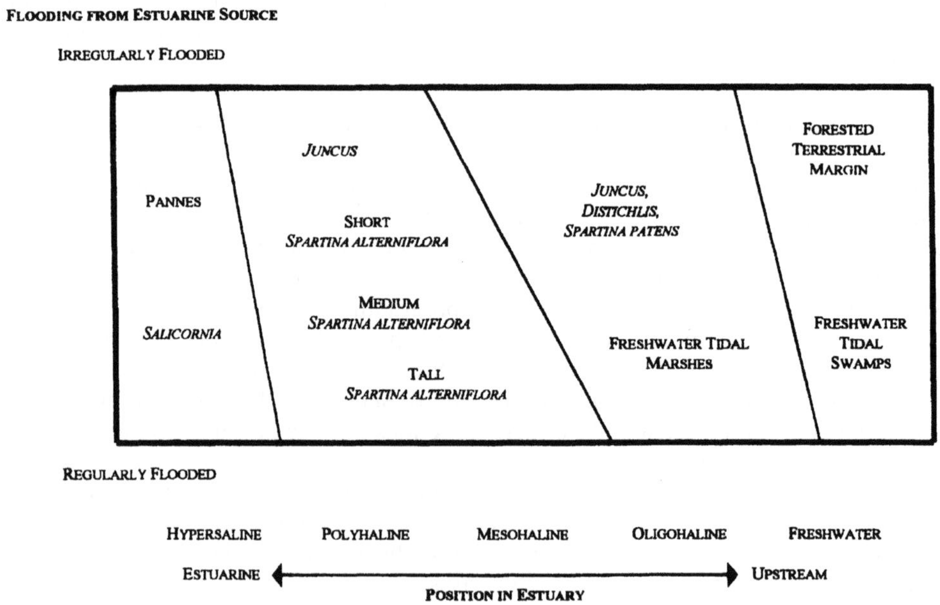

Figure 3. Anticipated dominance of vegetation for combinations of flooding frequency and salinity within the subtidal to terrestrial continuum for Atlantic coast marshes of the United States.

2.7 VEGETATION RESPONSES TO FLOODING AND SALINITY

The transgression template could also be adjusted upstream to tidal freshwater environments which become similar to the riverine subclass of Lugo and Snedaker (1974). These still can be described within the range of slope and local sediment supply conditions in Fig. 2. Furthermore, the influence of salinity and flooding pattern on vegetation is predictable at some scales. Fig. 3 illustrates for the eastern United States the anticipated dominance in vegetation for combinations of flooding frequency and salinity within the subtidal to terrestrial continuum. Other plant community distributions would be expected for other coastlines, and there would be differences in the details of plant response depending on latitude along the eastern coast.

Classification of coastal marshes at the scales necessary for restoration can then be done through the combined use of the information from Figs. 2 and 3 and Table 1. For example, a marsh may be recognized within the reference domain as being high and dominated by *Juncus*, *Distichlis* and *S. patens* (Figure 3). This high marsh will, through time, undergo a state change to low marsh (Fig. 2). Meanwhile, the accompanying low marsh is eroding (Table 1). Thus, the marsh is identified by its vegetation and factors affecting that vegetation, by its hydrogeomorphic position, and with respect to its probable future with respect to sea-level change. This designation can then be incorporated into establishing appropriate restoration criteria and actions, as discussed later.

3. Marsh Ecosystem States and their Functions

The reason for recognizing the continuum of coastal wetland state is to place restoration alternatives into a spatially broad and long-term context. While this broad-brush approach identifies the states, it does little to characterize differences in ecological functions among the states. For the purpose of marsh restoration, it is useful to know what processes and conditions are being restored so that success criteria can be based on more than just plant survival and cover. In the following sections, we compare low and high marsh zones, including the mechanisms involved in the transition between the two. We draw on examples from the Virginia Coast Reserve (VCR), the site from which the transgression template was developed. Comparisons are made of hydrology and the process of transition between high and low marsh. We also compare nitrogen cycling between the two states, but draw on data from two well-studied marshes: at Sapelo Island, Georgia and the Great Sippiwissett marsh, Massachusetts. The former are subject to tides of 2 to 3 m and much of the data comes from backbarrier marshes (Pomeroy and Wiegert 1981); whereas the latter is a mainland marsh with a maximum tidal amplitude of 1.6 m (Valiela and Teal 1979). Finally, we use the expansive high marsh of Cedar Island, within Pamlico Sound, North Carolina, to illustrate the range of variation that can occur within a single state.

3.1 TRANSITION OF VEGETATION FROM HIGH TO INTERTIDAL LOW MARSH

Extensive patchiness or fragmentation can characterize the unstable transitional zone between high and low marshes. We have observed this transition at the VCR. In 1990, Christian and Brinson (1999) established 8 permanent plots (3 x 8 m) strategically located to represent a range of marsh flooding regimes. These plots have been used to track the long-term interaction of *Juncus roemerianus* patches and surrounding plant communities dominated by either *S. alterniflora* in low marsh or *D. spicata* and *S. patens* in high marsh (Brinson and Christian 1999). (All communities produce root mats potentially contributing to peat formation and biogenic accretion.) Two plots were in each of 2 areas of high marsh. Two plots were in the zone of unstable transition that began with the same species as in the high marsh, and 2 were in a mineral low marsh. We compare these plots with the assumption that the transition zone represents the high marsh zone in an advanced state of transition to the low marsh state. Although the transition zone had a network of hollows at the beginning of the study separated by hummocks of *D. spicata* and *S. patens*, the plots were chosen to contain intact turf.

Over a 6-year period, microtopographic variation rose from the high marsh to the transition zone, but decreased in the mineral low marsh (Table 2). The intact turf of the *D. spicata* and *S. patens* community in the transition zone decreased to between 25% to 50% of its original surface area. The surface of the *J. roemerianus* community was more resistant to fragmentation, and its aboveground biomass in the transition was twice that in either of the other marsh zones. The *D. spicata* and *S. patens* community was replaced by depressional areas or hollows. Hollows coalesced with others such that the larger area collapsed into a shallow pothole covering tens of meters within an ever expanding network of hummocks and hollows. This network and accompanying coalescence appeared to be aided by the grazing activities of muskrat. Hollows were colonized by *Ruppia maritima*, algae, and an aquatic animal community of fish and invertebrates when flooded, as described for other marshes (Christian 1981, Kneib 1997). In late summer when precipitation was low and evapotranspiration was high, much of the hollows dried out. When flooded during warmer months, the water in hollows had significant diurnal fluctuations potentially from hypoxia to hyperoxia (Tarnowski 1997), as well as large fluctuations in nutrient concentrations. These resulted from the active benthic and shallow water communities and from *R. maritima* and its epiphytes. Thus, the transition area began to function differently than any of the other marsh states or the open tidal creek. In addition to changes in vegetation, functional differences included those of hydrology and water storage, biogeochemical cycling, organic matter processing, and habitat.

TABLE 2. Characteristics of the permanent plots within upper Phillips Creek marsh at the VCR.

	Organic high marsh	Unstable transiti-on	Near creek high marsh	Mineral low marsh
Relative elevation (msl)	1.02 m	0.84 m	0.93 m	0.86 m
Microtopographic variation as standard deviation (1990, 1995)	0.02 m, 0.04 m	0.10 m, 0.08 m	0.04 m, 0.02 m	0.04 m, 0.03 m
Percentage change in *J. roemerianus*[a]	+27%	+28%	+43%	-59%
Percentage change in *D. spicata* and *S. patens*[a]	-23%	-41%	-8%	-6%

[a] Based on ground cover in permanent 1 x 2 m plots at interface from 1990 through 1996. The difference in cover is between 1990 and 1996 was divided by the 1990 value and multiplied by 100

If our interpretations of the transition are correct, headward creek erosion will eventually connect this area to tidal flushing, providing conditions for development of a new low marsh. Formation of depressional areas may be viewed as intermediate and necessary in the change of state from high marsh to low marsh associated with rise in sea level. The distance from tidal source, tidal amplitude, and elevations of antecedent mineral surface would determine the rate of transition to low marsh.

3.2 COMPARISON OF HYDROLOGY

Information on the hydrology of low and high marsh zones can provide important information to a zone's functioning and to the transition process. We monitored water levels along a transect from low marsh to high marsh at the same VCR location (Stasavich 1998). A water level recorder was placed at each of the following: 1) a creekbank within the low marsh and dominated by *S. alterniflora*, 2) a transitional area at the edge of the high marsh, and 3) an interior region of high marsh. The latter two recorders were surrounded by *J. roemerianus* patches embedded in mixed communities of *S. patens* and *D. spicata*. Data from 1991 through 1996 were used for these analyses.

Considerable differences were found in hydroperiod among the 3 sites (Table 3). The low marsh site was flooded daily by tides with little variation across seasons or years. While water levels fell below ground surface for about 40% of the total time resulting from low tides, the soil was always saturated. Flooding or saturation occurred 93% of the time at the transitional site and was also fairly constant across season and year. Water levels dropped below ground surface during 20% of the summer and 5% of the fall season. Seasonal variation in water levels was greatest at the interior high marsh site. Flooding averaged 63%; however, when broken down by season, the site flooded 100% during winter and spring (duc to low evapotranspiration), 90% during fall, but only 31% during summer. Maximum flooding depth was 80 cm for the low marsh and 25 cm for the other sites. The number of flooding events decreased with distance inland, but the variation among years was greatest in the interior of the high marsh. Tidal inputs to the high marsh were associated with storm surges rather than astronomical tides.

We also analyzed hydrographs to distinguish the two water sources: estuarine water and precipitation. Precipitation events were recognized by 1) the nature of the shape of the peak in the hydrograph, 2) the common timing of peaks across the marsh zones, and 3) the correspondence to rainfall data at a meteorological station on the marsh. Data were expressed as percent cumulative rise above the ground surface. The low marsh received 99% of water sources from tides, the transition site was 86%, and the high marsh 30%.

TABLE 3. Hydrology of upper Phillips Creek marsh at the VCR, a regularly flooded system.

	Intertidal low marsh	Transitional marsh (edge with high)	High marsh
Distance from creek of water level recorder	0 m	25 m	250 m
Elevation from sea level	0.41 m	1.00 m	1.05 m
Percent time flooded or saturated to surface	99%	94%	63%
Maximum depth of flooding	0.80 m	0.25 m	0.25 m
Percentage contribution to water rise (tide, precipitation)	99%, 1%	86%, 14%	30%, 70%
Total number of tides per year	ca. 730	18	3
Average tidal amplitude (cm)	0.38 m	0.16 m	0.14 m

The high marsh site was extremely variable across seasons and years. For most of the time, inputs to this site were 100% from precipitation. However, when a tidal input did occur (usually in a fall month on the average of three per year), the magnitude was large enough to affect the relative percentage. Long periods of flooding and dry down varied seasonally. The elevation, relatively flat terrain, friction of vegetation, and distance from a tidal creek prevented water from both entering and leaving. Water from precipitation became ponded with little overland flow.

These differences in hydrology and water source promote differences in the characteristic functions of the marsh states. The low marsh's position and dependence on tidal waters supports sediment and organic matter exchange with the creek, whereas the high marsh has little exchange. In contrast surface water storage at the seasonal time scale would be greater in the high than in the low marsh. Furthermore, one might expect that nutrient cycling to correspond to the relatively open versus closed ecosystems. This change in function is addressed later.

3.3 HYDROLOGIC VARIATION WITHIN A SEA-LEVEL CONTROLLED, NON-TIDAL MARSH

As we have seen, frequent tidal flooding may only be a property of the edges of a coastal salt marsh. But some coastal marshes do not receive astronomic tides at all. One such marsh, at the Cedar Island (CI) National Wildlife Refuge, is located approximately 200 km south of the VCR in North Carolina. Information on this irregularly flooded marsh comes from the studies of Brinson and coworkers during the 1980s (Brinson et al. 1991). They focused on a 1700 m transect that passed through three distinct zones of vegetation (Table 4). However, all of these zones may be considered variations of the high marsh state. The dominant plant species is *J. roemerianus* in all zones, but subdominants and their distribution differ among zones (Knowles 1991). Zone 1 is largely monospecific with patches of *D. spicata* as a subdominant. In Zone 2 the patches within the *J. roemerianus* are larger on average and are dominated by *S. patens* and *Fimbristylis spadicea*. In Zone 3, plant associations become increasingly mixed with various combinations of *S. patens*, *Panicum virgatum*, and the shrub *Myrica cerifera*. The patches within the various zones appear to be maintained by wrack deposition (Knowles et al. 1991).

TABLE 4. Characteristics and hydrology of marsh at the Cedar Island National Wildlife Refuge, an irregularly flooded system dominated by Juncus roemerianus.

	Zone 1	Zone 2	Zone 3
Distance along transect (m)	Levee to 411	411 to 1108	1108 inward
Water level recorder position	200 m	800 m	1600 m
Elevation above sea level	687 mm	751 mm	832 mmm
Thickness of organic layer	1.2 m	0.6 m	0.3 m
Percent of soil organic carbon	18.0%	15.3%	17.0%
Percent of time flooded	99	85	30
Porewater salinity	29 ppt	22 ppt	10 ppt
Percent of weeks per year receiving flooding			
By tides	33	15	9
By precipitation	83	83	83
Average wrack patch size	590 m^2	871 m^2	1509 m^2
Nutrient limitation	stress limited	nitrogen	nitrogen

From the standpoint of the state-change continuum, Cedar Island marsh is a high marsh island in which both the intertidal low marsh and terrestrial forest are missing. As one moves inward from the seaward edge of this high marsh, hydrology becomes more similar to that of the high marsh studied at the VCR. At Cedar Island, Zone 1 is only flooded during storm events, occurring only 33% of weeks over the study (Table 4). When flooding does occur, Zone 1 remains inundated for an extended period because a natural shoreline levee and lack of tidal creeks restrict outward flow. In Zone 3, the major source of water is precipitation as with the high marsh at the VCR. Estuarine flooding only reaches either area during storms. Rain water brings fewer nutrients to the zone than tidal waters so that vegetation growth tends to be nitrogen limited. In both CI and VCR marshes, water levels are likely to be below the marsh surface during the growing season as a result of evapotranspiration. In winter the areas are flooded constantly. During this season, flooding by relatively frequent storm tides and precipitation continue to supply the marsh with surface water, and evapotranspiration is low and does not effectively remove the water. Surface water storage, retention of sediments and other materials, and nutrient cycling respond differently between zones as a result of water source and hydrology. The marsh continues to accrete vertically in response to rising sea level, but it is losing surface area through shoreline erosion that cannot be compensated by transgression.

3.4 COMPARISON OF NITROGEN CYCLING

Given the hydrologic differences between the low marsh zones (open with tidal exchange) and the high marsh zone (relatively closed), we postulated large differences in the way that nitrogen cycles. Network analysis is a group of algorithms designed to evaluate the qualitative and quantitative structure of a structured system (Wulff et al. 1989). Of particular utility is the ability to assess indirect relationships and systems-level attributes.

First, we constructed a standardized, 17-compartment network (box and arrow model) of the nitrogen cycle created for two areas of the low marsh, creekbank/tall *S. alterniflora* and low marsh/short *S. alterniflora*, and for the organic high marsh (Thomas 1998). For the various flows and standing stocks in the networks, we used data from the literature on two well-studied marshes, Great Sippewissett, MA, and Sapelo Island, GA. Great Sippewissett is a mainland marsh dominated by *S. alterniflora* in the creekbank and low marsh area and *D. spicata* in the high marsh. Sapelo Island is a barrier island with marshes 90 % covered by *S. alterniflora* and a small fringe of *J. roemerianus* in the high marsh. We used the software package, NETWRK4 for the execution of the network analysis (Ulanowicz 1987).

Various inputs of nitrogen to each area were followed through the system to export using input environs analysis (Ulanowicz 1987). Here we present the results from the import of tidal NH_4^+, tidal particulate nitrogen (PN), and precipitation. Tidal imports of other nitrogen species or N fixation were not considered. Tidal PN dominated import at each zone, but PN and tidal NH_4^+ decreased in magnitude with distance into the marsh from the 2 low marsh zones of tall and short *S. alterniflora* into the high marsh (Table 5). Precipitation imports were assumed to be the similar across the marsh. Percentage of imported nitrogen buried increased in importance moving from the tall *S. alterniflora* zone to the high marsh for each type of imported nitrogen species in each marsh. However, the burial rates (g N m^2y^{-1}) were generally similar across the marsh. There was also a trend for increased importance of transformations of imported dissolved nitrogen to occur in the high marsh compared to the low marsh. The percentage of tidal NH_4^+ that left a zone in the same form as entered was generally less for the high marsh than for the others. This may be due to increased duration of water column/marsh surface contact, increasing the probability of plant uptake and other transformations. Other trends across the marsh were less evident, but clear differences can be seen among zones (Thomas 1998).

Overall processing of nitrogen and internal cycling patterns reflects zonation (Table 5). The amount of total system throughput (sum of all flows within a system) decreased across the marsh. Finn Cycling Index represents the percentage of total system throughput that is involved in cycling (Finn, 1980). For both marshes, cycling was greater in the high marsh than in the low marsh areas. Primary productivity did not mirror the cycling patterns. Primary production often exceeded imports in each zone,especially in the high marsh; and some imports required transformations (e.g., tidal PN) prior to plant uptake. Therefore, the interdependency of flows between recycling and primary production depended on the position in the marsh. In the context of the state-change model, nitrogen cycling would be expected to be altered by the rise in

TABLE 5. Characteristics of nitrogen cycling for 3 Zones in 2 marshes. Tall and short refer to growth forms of *S. alterniflora* within the low marsh.

	Great Sippewissett			Sapelo Island		
	"Tall" low marsh	"Short" low marsh	High marsh	"Tall" low marsh	"Short" low marsh	High marsh
N imports ($g\ N\ m^{-2}y^{-1}$)						
Tidal NH4	5.9	1.6	0.5	1.6	0.8	0.2
Tidal PN	12.1	11.5	1.9	61.3	29.8	6.0
Precipitation	0.7	0.7	0.7	0.3	0.3	0.4
Total system throughput ($g\ N\ m^{-2}y^{-1}$)	413	323	221	903	628	404
Finn cycling index (%)	29.6	36.5	47.1	40.8	41.5	80.1
Primary production ($g\ N\ m^{-2}y^{-1}$)	30.2	27.0	29.0	51.6	38.3	53.5

sea level, but the alterations may not be a simple reflection of increased openness of the system with increased tidal exchange.

4. Movement of Reference

The use of reference in coastal wetland restoration is not new. It is implicit in compensatory mitigation of wetland regulatory programs through the choice of restoration site location (i.e., on-site versus off-site mitigation) and the choices of species composition, hydrological requirements, and local sediment regimes (i.e., in-kind versus out-of-kind restoration). However, the dynamics of coastal wetlands pose special challenges for restoration, especially if a goal is to achieve self-sustaining conditions over the long term. Not only does rising sea level force us to anticipate changing landscapes, but also to recognize that losses and gains of both area and functioning can occur without human intervention. This paper has identified a variety of conditions in coastal wetland dynamics that recognize the need for a long-term perspective. The process of state change in coastal ecosystems is especially relevant to policies that deal with land use.

The state-change continuum for coastal wetlands provides a suite of reference conditions for which self-sustaining restoration efforts can be evaluated. The transgression template highlighted in this paper, and the recognition of several departures from this template, are offered as a way to place restoration projects within the context of rising sea level. We suggest that restoration success can be enhanced if projects recognize state changes and respond to them through design. To illustrate this, we use two hypothetical conditions within the state-change continuum: 1) changes that occur toward the terrestrial endmember and 2) changes unique to the subtidal estuarine endmember.

4.1 APPLICATION AND LIMITATIONS OF REFERENCE TO RESTORATION

Highly altered sites are good candidates for restoration because they normally can achieve greater gains than minimally altered ones. Restoration of high marsh adjacent to the terrestrial endmember can be conducted in areas that have been 1) previously excluded from hydrologic connections with intertidal marsh, 2) connected through extensive ditching, or 3) filled to raise the elevation of the wetland surface. In the first case, reconnection to intertidal marsh may consist simply of breaching levees, resulting in state change (Burdick et al. 1997, Weinstein et al. 1997). The isolated wetland may change to one with lower or higher salinity relative to its state before isolation depending on water sources, tides and new hydroperiod. If relative sea level had risen significantly during the period of isolation, a state change from high marsh (the pre-levee condition) to low marsh may take place. This raises some questions that need consideration. What proportion of the high marsh would become intertidal low, and would additional high marsh displace terrestrial ecosystems through transgression?

The second case of an area available for restoration is typical of numerous mosquito control projects. Filling or plugging ditches, assuming pre-ditching conditions as a restoration goal, may restore extensively ditched marshes. This approach may raise questions about whether the original ditching remains a significant impact. For example, low marsh may have developed from high marsh anyway during the interval of rising sea level. In cases such as the Cedar Island National Wildlife Refuge, North Carolina, where no intertidal marsh exists, filling ditches may enhance the capacity of the high marsh to vertically accrete organic matter, a critical component of its self-sustaining condition.

For the last case of fill removal, the site may be converted from non-wetland to wetland by excavation down to the original marsh surface. But this surface may now have a different elevation relative to sea level than when originally filled. Again a new ecosystem state may replace the original one.

In all three cases, attention should be given to interactions at the terrestrial margin. For example, under potentially migrating conditions artificial barriers to migration (e.g., bulkheads) may have to be removed. In other cases land uses that prevent high marsh development (e.g., parking lots, agriculture) may have to be excluded near the margin. Furthermore, to be consistent with these migrating conditions, regulations that require buffers next to wetlands would have to accommodate migration.

Restoration at the subtidal estuarine endmember is perhaps more commonly encountered because of the emphasis on intertidal marshes normally dominated by *S. alterniflora*. First, one must consider whether the condition is prograding or eroding. If naturally prograding, it would be redundant of nature to use the location for restoration unless there was an interest in accelerating the rate of conversion of mud flat to low marsh. However, progradation cannot result in state change beyond the high marsh condition because high marsh cannot be transformed into upland under conditions of rising sea level. More commonly, however, eroding conditions are encountered at the low marsh margin. Consistent with the state-change continuum during transgression, a conversion from low marsh to mud flat would be anticipated. An example of restoration to reference conditions could involve the removal of barriers to erosion, such as seawalls and bulkheads. Because both sediment supply and the wave environment are often critical controlling forces at this end of the continuum (Turner 1997), restoration sites should be carefully chosen. Although sediment supplies are often controlled at great distance from a restoration site, supplies may be manipulated hydrologically through the dredging of tidal channels (Hackney and Cleary 1987) and the storage of sediments in reservoirs.

Manipulation of sediment supplies raises larger scale issues that vary geographically among coastal wetlands groups. Once restoration alternatives are raised to a landscape scale, the use of reference and the basic scientific principles that support its use become increasingly subjected to modification by policy concerns. Coastal wetland restoration is but one alternative within an array of socioeconomic choices to mitigating wetland loss or damage. One can argue that relatively unaltered reference standard conditions represent the default approach implicit in the regulatory program governed by Section 404 of the Clean Water Act. In other words, it is the alteration of wetlands that is regulated with the tacit assumption that active management beyond their protection is not required or necessarily desirable. Even though a "hands off" approach is the default condition in most regulatory programs, there are situations where socioeconomic goals for coastal resources eclipse the reference approach described here. Dredging tidal channels for navigation and building reservoirs for storage of sediments are examples previously mentioned. At the landward margin of transgressing marshes, urban and agricultural land uses may preclude restoration of high marsh or the conservation of land for state change to high marsh. Land use plans for coastal regions may benefit from acknowledging not only the consequences of rising sea level on coastal wetlands, but also the impact of other land uses. The state-change continuum can play a role in recognizing the variety of conditions under which coastal wetlands exist.

5. Acknowledgments

This research was supported in part by the National Science Foundation Long-Term Ecological Research grant DEB-9411974 and by Cooperative Agreement No. 14-16-0009-85-963 between the U. S. Fish and Wildlife Service and East Carolina University.

6. Literature Cited

Bacon, P. R. 1994. Template for evaluation of impacts of sea level rise on Caribbean coastal wetlands. Ecological Engineering **5**:171-186.

Brinson, M. M., P. B,. Hook and W. L. Bryant, Jr. 1991. Hydrologic environment of Cedar Island marsh. Pages 47-99 *in* M. M. Brinson, editor. Ecology of a nontidal brackish marsh in coastal North Carolina. U. S. Fish and Wildlife Service, National Wetlands Research Center, Open File Report 91-03.

Brinson, M. M., W. Kruczynski, L. C. Lee, W. L. Nutter, R. D. Smith, and. D. F. Whigham. 1994. Developing an approach for assessing the functions of wetlands. Pages 615-624 *in* W.J. Mitsch, editor. Global wetlands: old world and new. Elsevier Science B. V., Amsterdam, The Netherlands.

Brinson, M. M., F. R. Hauer, L. C. Lee, W. D. Nutter, R. D. Rheinhardt, R D. Smith and D. F. Whigham. 1995a. Guidebook for application of hydrogeomorphic assessments to riverine wetlands. Technical Report WRP-DE-11, an Operational Draft. U.S. Army Engineers Waterways Experiment Station, Vicksburg, M. S.

Brinson, M. M. and R. D. Rheinhardt. 1996. The role of reference wetlands in functional assessment and mitigation. Ecological Applications **6**: 69-76.
Brinson, M. M., and. R. R. Christian. 1999. Stability of *Juncus roemerianus* patches in a salt marsh. Wetlands **19**: 65-70.
Burdick, D.M., M. Dionne, R.M. Boumans and F.T Short. 1997. Ecological responses to tidal restorations of two northern New England marshes. Wetlands Ecology and Management **4**: 129-144.
Cahoon, D. R., D. J. Reed and J. W. Day, Jr. 1995a. Estimating shallow subsidence in microtidal salt marshes of the southeastern United States: Kaye and Barghoorn revisited. Marine Geology **128**: 1-9.
Cahoon, D. R., D. J. Reed, J. W. Day, Jr., G. D. Steyer, R. M. Boumans, J. C. Lynch , D. McNally and N. Latif. 1995b. The influence of hurricane Andrew on sediment distribution in Louisiana coastal marshes. Journal of Coastal Research **21**: 280-294.
Christian, R. R. 1981. Community metabolism of a salt-marsh pothole. Bulletin of the New Jersey Academy of Science **26**: 34-40.
Christiansen, T. 1998. Sediment deposition on a tidal salt marsh. Dissertation. University of Virginia, Charlottesville, Virginia, USA.
Cintron, G., A. E. Lugo and R. Martinez. 1985. Structural and functional properties of mangrove forests. Pages 53-66 *in* The botany and natural history of Panama. Missouri Botanical Garden, Saint Louis, Missouri, USA
Cowardin, L.M., V. Carter, F.C. Golet and E.T. LaRoe. 1979. Classification of wetlands and deepwater habitats of the United States. FWS/OBS-79/31, USDI Fish and Wildlife Service, U.S. Government Printing Office, Washington, D.C.
Dame, R., D. Childers and E. Koepfler. 1992. A geohydrologic continuum theory for the spatial and temporal evolution of marsh-estuarine ecosystems. Netherlands Journal of Sea Research **30**: 63-72.
DeLaune, R. D., J. A. Nyman and W. H. Patrick, Jr. 1994. Peat collapse, ponding and wetland loss in a rapidly submerging coastal marsh. Journal of Coastal Research **10**: 1021-1030.
Eleuterius, L. N. and C. K. Eleuterius. 1979. Tide levels and salt marsh zonation. Bulletin of Marine Science **29**: 394-400.
Finn, J.T. 1980. Flow analysis of models of the Hubbard Brook ecosystem. Ecology **61**: 562-571.
Gardner, L. R., B. R. Smith and W. K. Michener. 1992. Soil evolution along a forest-salt marsh transect under a regime of slowly rising sea level, southeastern United States. Geoderma **55**: 141-157.
Giese, G. L., H. B. Wilder and G. G. Parker, Jr. 1985. Hydrology of major estuaries and sounds of North Carolina. U. S. Geological Survey, Water Supply Paper **2221**.
Hackney, C. T. and W. J. Cleary. 1987. Saltmarsh loss in southeastern North Carolina lagoons: importance of sea level rise and inlet dredging. Journal of Coastal Research **3**: 93-97.
Hayden, B. P., C. F. V. Santos, G. Shao and R. C. Kochel. 1995. Geomorphological controls on coastal vegetation at the Virginia Coast Reserve. Geomorphology **13**: 283-300.
Hmieleski, J.I. 1994. High marsh-forest transitions in a brackish marsh: the effect of slope. Thesis. East Carolina University, Greenville, North Carolina, USA.
Kneib, R. T. 1997. The role of tidal marshes in the ecology of estuarine nekton. Oceanography and Marine Biology: An Annual Review **35**: 163-220.
Knowles, D. B. 1991. Vegetative analysis of Cedar Island marsh. Pages 125-146 *in* M. M. Brinson, editor. Ecology of a nontidal brackish marsh in coastal North Carolina. U. S. Fish and Wildlife Service, National Wetlands Research Center Open File Report **91-03**.
Knowles, D. B., W. L. Bryant, Jr. and E. C. Pendleton. 1991. Wrack as an agent of disturbance in an irregularly flooded brackish marsh. Pages 147-186 *in* M. M. Brinson, editor. Ecology of a nontidal brackish marsh in coastal North Carolina. U. S. Fish and Wildlife Service, National Wetlands Research Center Open File Report **91-03**.
Lugo, A. E. and S. C. Snedaker. 1974. The ecology of mangroves. Annual Review of Ecology and Systematics **5**: 39-64.
Nyman, J. A., R. D. DeLaûne, H. H. Roberts and W. H. Patrick, Jr. 1993. Relationship between vegetation and soil formation in a rapidly submerging coastal marsh. Marine Ecology Progress Series **96**: 269-279.
Odum, E. P. 1969. The strategy of ecosystem development. Science **164**: 262-270.
Odum, H. T., B. J. Copeland and E.A. McMahan. 1974. Coastal ecological systems of the United States. The Conservation Foundation, Washington, District of Columbia, USA.
Odum, W. E., T. J. Smith III, J. K. Hoover and C.C. McIvor. 1984. The ecology of tidal freshwater marshes of the United States East coast: community profile. U. S. Fish and Wildlife Service, FWS/OBS-**83/17**.

Oertel, G. F., J. C. Kraft, M. S. Kearney and H. J. Woo. 1992. A rational theory for barrier-lagoon development. Pages 77-87 *in* Quaternary coasts of the United States: marine and lacustrine systems. Society for Sedimentary Geology, SEPM Special Publication No. **48**.
Oertel, G. F. and H. J. Woo. 1994. Landscape classification and terminology for marsh in deficit coastal lagoons. Journal of Coastal Research **10**: 919-932.
Osgood, D. T., M. C. V. F. Santos and J. C. Zieman. 1995. Sediment physio-chemistry associated with natural marsh development on a storm-deposited sand flat. Marine Ecology Progress Series **120**: 271-283.
Patten, B. C. 1978. Systems approach to the concept of environment. Ohio Journal of Science **78**: 206-222.
Redfield, A. C. 1972. Development of a New England salt marsh. Ecological. Monographs **42**: 201-237.
Reed, D. J. and D. R. Cahoon. 1992. The relationship between marsh surface topography, hydrology and growth of *Spartina alterniflora* in a deteriorating Louisiana salt marsh. Journal of Coastal Research **8**: 77-87.
Reidenbaugh, T. G., W. C. Banta, M. Varricchio, R. P. Strieter and S. Mendoza. 1983. Short-term accretional and erosional patterns in a Virginia salt marsh. Gulf Research Reports **7**: 211-215.
Stasavich, L. E. 1998. Hydrodynamics of a coastal wetland ecosystem. Thesis. East Carolina University, Greenville, North Carolina, USA.
Stevenson, J. C., M. S. Kearney and E. C. Pendleton. 1985. Sedimentation and erosion in a Chesapeake Bay brackish marsh system. Marine Geology **67**: 213-235.
Stumpf, R. P. 1983. The process of sedimentation on the surface of a salt marsh. Estuarine, Coastal and Shelf Science **17**: 495-508.
Tarnowski, R. L. 1997. Effects of dissolved oxygen concentrations on nitrification in coastal waters. Thesis. East Carolina University, Greenville, North Carolina, USA.
Teal, J. M. 1986. The ecology of regularly flooded salt marshes of New England: a community profile. U.S. Fish Wildlife Service, Biological Report **85**(7.4).
Thom, B. G. 1982. Mangrove ecology: a geomorphical perspective. Pages 3-17 *in* B. F. Clough, editor. Mangrove ecosystems in Australia. Australian National University Press, Canberra, Australia.
Thomas, C. R. 1998. The use of network analysis to compare nitrogen cycles of three salt marsh zones experiencing relative sea-level rise. Thesis. East Carolina University, Greenville, North Carolina, USA.
Turner, R. E. 1997. Wetland loss in the northern Gulf of Mexico: a multiple working hypotheses. Estuaries **20**: 1-13.
Twilley, R.R. 1995. Properties of mangrove ecosystems related to the energy signature of coastal environments. Pages 43-62 *in* C.A.S. Hall, editor. Maximum power: the ideas and applications of H.T. Odum. University Press of Colorado, Niwot, Colorado, USA.
Ulanowicz, R.E. 1987. NETWRK4: a package of computer algorithms to analyze ecological flow networks. University of Maryland, Chesapeake Bay Laboratory, Solomons, Maryland, USA.
Valiela, I. and J. M. Teal. 1979. Inputs, outputs and interconversions of nitrogen in a salt marsh ecosystem. Pages 399-414 *in* R. L. Jefferies and A. J. Davy, editors. Ecological processes in coastal environments. Blackwell Scientific Publications, London, England.
Vannote R. L., G. W. Minshall, K. W. Cummins, J. R. Sedell and C. E Cushing. 1980. The river continuum concept. Canadian Journal of Fisheries and Aquatic Sciences **37**: 130-137.
Warren, R. S. and W. A. Niering. 1993. Vegetation change on a Northeast tidal marsh: interaction of sea-level rise and marsh accretion. Ecology **74**: 96-103.
White, W. A. and T. A. Tremblay. 1995. Submergence of wetlands as a result of human-induced subsidence and faulting along the upper Texas Gulf coast. Journal of Coastal Research **11**: 788-807.
Weinstein, M.P., J.H. Balletto, J. M. Teal and D.F. Ludwig. 1997. Success criteria and adaptive management for a large-scale wetland restoration project. Wetlands Ecology and Management **4**: 111-127.
Wulff, F., J.G. Field and K.H. Mann, editors. 1989. Network analysis in marine ecology: methods and applications. Coastal and Estuarine Studies **32**. Springer-Verlag, Heidelberg, Germany.

LINKING THE SUCCESS OF *PHRAGMITES* TO THE ALTERATION OF ECOSYSTEM NUTRIENT CYCLES

LAURA A. MEYERSON
Brown University, Center for Environmental Studies, Box 1943, 135 Angell Street, Providence, RI 02912-1943 USA

KRISTIINA A. VOGT
Yale University, School of Forestry and Environmental Studies
370 Prospect Street
New Haven, CT 06511 USA

RANDOLPH M. CHAMBERS
Fairfield University, Department of Biology
Fairfield, CT 06430 USA

Abstract
In the United States, *Phragmites australis* is often viewed as a pest species because it forms monocultures that dominate a site for longer time scales than other wetland plants and because it decreases biodiversity. While research has shown that *Phragmites* is successful in tidal marshes where it tolerates the effects of prolonged flooding and salinity, to date there is little evidence of its influence on ecosystem-level processes. We review the literature and suggest future areas of research by regarding the specific question of whether *Phragmites* can dominate a site through its ability to control the cycling of limiting nutrients. *Phragmites* sequesters nutrients in standing live and dead biomass that either accumulates in the soil or is removed from the system by tidal flushing. *Phragmites* reduces light and temperature under its dense growth that decreases decomposition rates and immobilizes nutrients in long-term storage pools. *Phragmites* outcompetes other species for increased nutrient inputs from various anthropogenic sources. We focus in particular on two possible mechanisms which may be related to *Phragmites* success: *Phragmites* immobilizes nitrogen into forms (e.g., DON) which cannot be utilized by other species; and/or *Phragmites* accumulates Si and/or Al which immobilize P making it unavailable to other plants. Identifying the relative importance of these mechanisms to *Phragmites* expansion could assist the development of successful wetland restoration and management plans.

1. Introduction

Ecologists have not satisfactorily explained the successful expansion of the reed grass

1 Hereinafter referred to as *Phragmites*

Phragmites australis during the last 100 years from its historical range in North America. To date, much of the research has focused on effects of *Phragmites* on floral and faunal diversity, its rate of expansion, and site characteristics that appear to favor its growth (Niering and Warren 1977, 1980, Sinicrope et al. 1990). Other studies have investigated methods to eradicate *Phragmites*, including tide gate removal to restore hydrologic function to pre-invasive conditions, and the application of herbicides in combination with cutting or mowing (Randall and Lapin 1995). While it is apparent that *Phragmites* is a formidable competitor for light and space, these mechanisms alone do not seem to fully explain its success. This paper reviews both the terrestrial and wetland literature regarding a variety of mechanisms that may contribute to the success of *Phragmites*, and concludes by suggesting future areas for research.

One possibility that has recently begun to receive attention by researchers is that *Phragmites* may be altering nutrient cycles in marsh ecosystems (Windham 1995, Ahearn 1996, Meijerson et al. 2000). Preliminary research results suggest that these changes can be detected. For example, Chambers (1997) and Windham (1995) measured lower nitrogen (N) concentrations in the pore-water of *Phragmites* compared to *Spartina* spp. dominated communities. This type of analysis needs to be duplicated in other *Phragmites* dominated systems to determine how generalizable these results are.

In addition to lowering porewater N concentrations, several questions should be answered to assess the impact of *Phragmites* on nutrient cycles: Does *Phragmites* sequester a larger pool of nutrients in its tissues through biomass accumulation on a percent of weight basis than other wetland species? Can the relatively recent expansion of *Phragmites* be explained by its ability to better exploit the increased nutrient inputs occurring in these systems than its competitors? Are elements such as Fe, Al, and Si important to the success of *Phragmites*, and are these minerals forming chemical complexes that are reducing the availability of limiting nutrients? Is *Phragmites* capable of using nutrient forms that its competitors cannot? How do these postulated mechanisms of *Phragmites* site dominance affect restoration efforts?

Understanding the mechanisms by which *Phragmites* is able to maintain dominance is crucial to restoring a site to pre-invasion conditions. This type of information is necessary to identify the minimum data needed to assess the potential success of restoration efforts. If the underlying mechanism is identified, it will be easier to focus resources to manipulate those variables most strongly controlling ecosystem resilience. This paper explores these questions through a review of the literature on nutrient cycling in both wetland and terrestrial systems to consider several alternative hypotheses that may account for the success of *Phragmites* as a species.

Phragmites australis has persisted as a minor component of tidal wetland communities for thousands of years in North America (Niering and Warren 1977). Over the last century on the Atlantic coast of the United States it has been perceived to be rapidly spreading beyond its historical range (Warren et al. 1995, Winogrond 1997, Chambers et al. 1999). It remains to be determined whether *Phragmites* expansion into new ecosystems simply escaped notice until relatively recently, whether its spread is a direct response to increased anthropogenic disturbance, whether a more aggressive genotype has been introduced, or whether it is a combination of these factors. Wetland scientists and managers are concerned because *Phragmites* expansion, typical of invasive plants, is leading to a substantial shift in the mosaic of wetland vegetation in the region (Niering and Warren 1980, Sinicrope et al. 1990, Meyerson et al. 2000). Components of biodiversity (habitat diversity and numbers of plant and animal species) are demonstrably decreased by ongoing expansion of plant monocultures (Stalter and Baden 1994, Windham, 1995, Meyerson et al., in press). Although the ''players'' in the wetland ecosystem may change, the impacts of invasive species on general ecosystem functions such as energy flows and nutrient cycling are less clear (Johnson et al. 1996, Tilman 1997, Tilman et al. 1997). In this context, we will examine how the success of *Phragmites* may be related to nutrient cycling in tidal wetlands.

In brackish tidal wetlands of the northeast (oligohaline to mesohaline), *Phragmites* expansion is via clonal growth and rhizome and seedling establishment. Typical of halophytes, *Phragmites* growth seems unaffected by salinity up to 15 ‰ or more (Hellings and Gallagher 1992, Lissner and Schierup 1997). Because of its rapid vertical and horizontal growth, *Phragmites* appears capable of overgrowing other wetland plant species by physical displacement. Once established at a site, the tall, dense shoots effectively shade out other species. In many brackish marshes, *Phragmites* replaces *Spartina,* spp. as the dominant vegetation type. Although competition for nutrients between *Spartina* and *Phragmites* has not been formally investigated, the dense, tall stands of *Phragmites* form a potentially large pool of nutrients in both living and dead tissues (Templer et al., 1998, Meyerson et al. 2000).

2. Limits to *Phragmites* Distribution

The ability of *Phragmites* to manifestly dominate many sites because its dense tall growth alters environmental conditions (e.g., light, space, and temperature) does not entirely explain its success. Recently, some researchers have begun investigating how *Phragmites* is affecting sites at the ecosystem level as well as the competitive interactions of *Phragmites* with other plants (Sinicrope et al. 1990, Meyerson et al. 2000). This type of study may demonstrate why *Phragmites* is currently expanding so rapidly in the northeastern and mid-Atlantic United States where its historical range was significantly more limited and why it has been difficult to eliminate *Phragmites* from the environment.

The limits of *Phragmites* distribution in tidal wetlands are set primarily by hydrological constraints. *Phragmites* cannot survive under regimes of extensive flooding by salt water (Hellings and Gallagher 1992). *Phragmites*, however, is capable of surviving in tidal marshes when the effects of prolonged flooding and salinity can be abated. Three general methods have been suggested for *Phragmites* survival: 1) avoidance: the plants grow where flooding is not extensive and salinity is low; 2) modification: the plants oxygenate the rhizosphere in flooded sites and accumulate sediments, eventually decreasing the flooding regime, and 3) accommodation: the plants extend deep tap roots to a freshwater lens to avoid salt; and use clonal integration from robust shoots to deliver necessary growth materials.

Expansion by *Phragmites* both horizontally and vertically (Warren et al. 1995) may be the key to its apparent, competitive superiority over other wetland species. Rapid growth (Hocking et al. 1983) requires rapid access to nutrients for growth, either from the soil and/or flooding tidal water. Although plausible mechanisms other than nutrient competition could account for the dominance of *Phragmites* in many wetlands, we consider here whether displacement of other wetland species occurs in part because *Phragmites* outcompetes them for limiting nutrients.

Nutrient availability in tidal wetland soils is difficult to assess since typically the concentrations of nutrients are extremely high, but edaphic conditions may make them inaccessible to plants (Chalmers et al., 1982, Shaver and Melillo 1984). For example, a suite of environmental factors including salt, sulfide and oxygen concentrations in the rhizosphere influences the ability of *Spartina alterniflora* to take up and assimilate N (Morris 1980, Bradley and Morris 1990, Chambers 1998). On the other hand, plant growth may influence soil environments of tidal wetlands: Howes et al. (1986) proposed a model of *Spartina* production that was positively linked to the extent to which *Spartina* was capable of modifying soil conditions to promote its growth.

How, then, is *Phragmites* able to exploit wetland environments occupied by species like *Spartina*? Whether or not a more aggressive genotype of *Phragmites* has been introduced from Europe (Besitka 1996, Chambers et al. 1999), *Phragmites* is now exploiting tidal wetland environments that for thousands of years were ''off-limits''. Some investigators have argued that a change in nutrient availability may have shifted the advantage to *Phragmites* over *Spartina* and other wetland species. Perhaps recent increases in nutrients delivered to estuarine environments have allowed *Phragmites* to expand its realized niche to include large sections of tidal wetlands once dominated by *Spartina* sp. Such a shift is not unusual. Valiela and Teal (1974), for example, found that plant community structure in a salt marsh was changed with increased fertilization. The invasion of cattail (*Typha*) into the Florida Everglades dominated by sawgrass (*Cladium jamaicense*) is occurring in part because of recent increases in phosphorus availability (Wu et al. 1997).

Undeniably, the recent and extensive eutrophication of coastal environments (Ryther and Dunstan 1971) has increased the exposure of wetland plants to dissolved and particulate forms of N and P from watershed runoff. Further, atmospheric deposition of N in coastal regions has increased dramatically with the burning of fossil fuels (Yang et al. 1996) so that approximately 15% of the total N load to Long Island Sound is from atmospheric sources (Stacy and Tedesco 1997). J.T. Morris (pers. commun.) has suggested that accelerated expansion of *Phragmites* in the northeastern United States may be directly related to eutrophication of coastal wetland environments. While incidences of wetland eutrophication occurred in earlier centuries in North America, *Phragmites* may not have been present to exploit the increase of resources. In Eastern, Central, and Northern Europe, *Phragmites* decline has been associated with the combined effects of eutrophication and water table management efforts (Van der Putten 1997). In other parts of Europe, namely the Mediterranean region, Denmark and Scandinavia, *Phragmites* die-back does not appear to be a problem, and in fact the species seems to be expanding in some systems (Van der Putten 1997).

Over the short term, *Phragmites* has responded to elevated nutrient concentrations by increasing its uptake of these nutrients and attaining higher biomass (Gries and Garbe 1989, Kuhl and Kohl 1993, Templer et al. 1998, Meyerson et al. 2000). On the other hand, no studies have demonstrated long-term retention of nutrients by *Phragmites* under high nutrient loads (Boar 1996), and therefore nutrient cycles do not appear to be tighter or more closed in *Phragmites* stands relative to other species. Chronic eutrophication may in fact be a stress on *Phragmites* growth and has been cited as a possible cause of *Phragmites* decline in some European locations (den Hartog et al. 1989, van der Putten 1997). Nutrient enrichment may provide short-term competitive advantages to *Phragmites*, but once established, continued nutrient enrichment may be detrimental.

The response of *Phragmites* to excess N has to be balanced by examining how it may control N availability to other plant species under N limiting conditions. One mechanism for *Phragmites* to decrease competition from other plants is to reduce the total amount of N available in the soil environment. For example, Chambers (1997) reported that porewater concentrations of ammonium are significantly lower in *Phragmites* stands relative to *Spartina alterniflora* stands, suggesting that *Phragmites* is capable of reducing the nitrogen concentrations in the soil. However, because *Phragmites* soils are so well ventilated by the plants (Armstrong et al. 1996a, Brix et al. 1996), the lower ammonium concentrations could be due to increased uptake by the plants and/or increased conversion to nitrate by nitrifying bacteria and their subsequent losses with leaching.

A study by Chambers et al. (1998) revealed that *Phragmites* and *S. alterniflora* have equal affinities for N under different salinity regimes and differ only in their response to sulfide in the rhizosphere. In these experiments, N uptake by *Phragmites* was effectively stopped by sulfide concentrations approaching 500 uM, whereas N uptake by *Spartina* was unaffected by sulfide concentrations of 1 mM. Expansion of *Phragmites* into lower tidal elevations of salt marshes is probably inhibited by the combined effects of prolonged submergence and high sulfide concentrations in the rhizosphere that cannot be overcome by root ventilation (Armstrong et al. 1996b).

3. Mechanistic Controls of Ecosystem Nutrient Cycles

The factors currently identified to limit *Phragmites* distribution do not account for its recently expanded realized niche. It is necessary, therefore, to investigate other potential mechanisms that may be giving *Phragmites* a competitive advantage over other marsh plant species and make it difficult to restore sites to pre-*Phragmites* conditions. Two types of nutrient transformations and immobilization mechanisms found in terrestrial systems may be especially interesting to study in wetlands because similar chemical processes may occur in both systems. Researchers in terrestrial systems have recently begun investigating how the dominance of the available N pools by dissolved organic nitrogen (DON) forms (Fig. 1) and the formation of chemical complexes (e.g., aluminum silicate phosphate complexes) (Fig. 2) can determine the outcome of competitive interactions between plants (Vogt et al. 1987a,b, Dahlgren et al. 1991a,b, Chapin et al. 1993, Northrup et al. 1995). This research may provide clues to the type of chemical complexes and transformations that should be further examined in wetland systems.

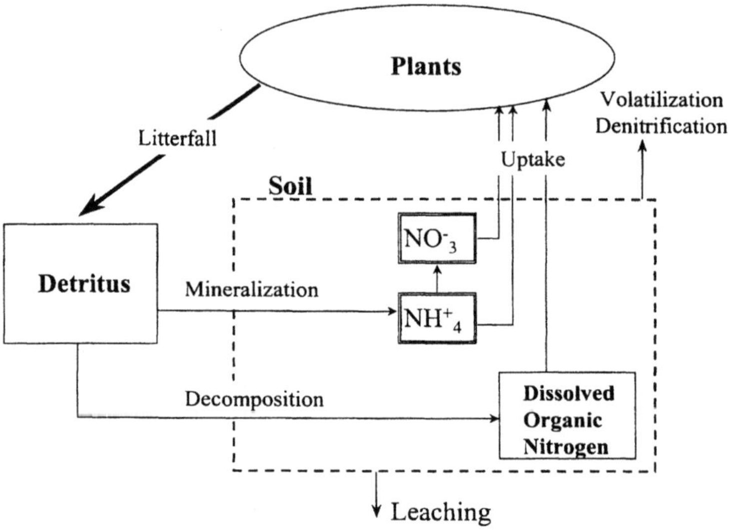

Figure 1. Conceptual model of the nitrogen cycle showing the potential different pathways and forms of N resulting from the decomposition of *Phragmites* detritus.

Many experiments have been conducted to determine the role of specific nutrients in controlling the function, health and resilience of salt marsh and other wetland ecosystems (see Mitsch and Gosselink 1993). Research on the role of nutrients in wetland ecosystem processes has mainly focused on nutrients identified to be either driving and/or limiting ecosystem function, particularly N. This has resulted in a comprehensive understanding of nutrient cycles, as well as nutrient limitations to plant growth. It has not explained some of the transformations measured in the field, e.g., what maintains species level plant productivity in mixed communities, or explored the possibility that some plants may maintain dominance by controlling the availability of limiting nutrients.

3.1 FORMS OF NITROGEN

Since nutrient availability and assimilation appear most important to plant growth in tidal wetland environments, one might expect that *Phragmites* is more efficient at obtaining nutrients than other wetland species. Most nutrient studies have focused on N as limiting the plant growth and in assessing the mineral forms (e.g., ammonium and nitrate) of available N. Other available N forms should be assessed for *Phragmites* communities given that terrestrial ecosystems having DON as a significant proportion of their available N pool have characteristics similar to *Phragmites* dominated sites (Fig. 1).

In certain types of terrestrial ecosystems, it has become apparent that DON may be an important N source for plants (Chapin et al. 1993, Northrup et al. 1995). The production of DON may be a mechanism by which some plants reduce the availability of N to other plants growing within the same ecological space, thus reducing the competition for abiotic resources (Northup et al. 1995). Since the 1980s it has been known that organic N is potentially present as part of the available N pool (Qualls et al. 1991, Yavitt and Fahey 1986). However, researchers had not realized that this pool could also be a major source of N to plants in some ecosystems. Prior to the publication of the Northup et al. (1995) article, it had been assumed that the most important part of the N cycle was the point where complex organic materials were converted to mineral forms (e.g., ammonium, nitrate) by microorganisms since this is where bottlenecks of N availability were recorded (Chapin et al. 1995). It was generally accepted that mineral N forms were the most important available N pools for plants and that their availability would determine the growth rates and resilience of plant communities. As a result, little research was conducted on the role of organic N forms in sustaining ecosystem processes.

A review of the literature indicates that there are similar generalizable characteristics for ecosystems where DON contributes significantly to the total plant available N (Swift et al. 1979, Chapin 1995, Northup et al. 1995, Vogt et al. 1995). These ecosystems have deep organic detrital accumulations on the soil surface or high quantities of organic matter within the soil profile (Vogt et al. 1995). These organic matter accumulations are found under conditions where certain abiotic site factors strongly control the quality of plant tissues produced, the decomposition rates of these tissues, and the end products produced as part of the decomposition process. The prevalent abiotic site factors that result in systems with deep organic matter accumulations are: 1) plant and microbial growth commonly limited by N (if other nutrients P, Ca, or K limit growth, they will have similar effects as N), 2) low temperatures, and 3) acid and infertile soils. These abiotic factors result in the production of plant tissues that decay slowly and have low rates of nutrient mineralization. Consequently, there is low nutrient availability from these decomposing tissues to other plants. The impoverished nutrient availability produces plant tissues with a chemical quality that is higher in lignin, phenolic compounds and tannins; as well as a tissue chemical composition which is more difficult for microorganisms to break down during decomposition (Swift et al. 1979, Kuiters 1990).

All three of these abiotic characteristics need not be present simultaneously for deep organic matter accumulations to occur. For example, low temperatures are not found in many tropical areas, but deep forest floor accumulations are present because soils are often acid and infertile (e.g., acid white sands) (Sanchez and Bandy 1982, Vogt et al. 1986, 1995). Slow decomposition rates of detrital materials occur in these tropical areas because (1) microorganisms are unable to produce the enzymes needed to decompose litter materials due to nutrient deficiencies (N, P, K or Ca) and/or (2) Al toxicity inhibits microbial activity and decreases decomposition rates (Swift et al. 1979, Cuevas and Medina 1986, Vogt et al. 1986, Dahlgren et al. 1991a,b, Vogt et al. 1995, Jingguo and Bakken 1997). Therefore, ecosystems potentially having the right conditions for DON to be present can be found in the following locations: coniferous and hardwood forests located in the cold temperate, boreal and arctic climatic zones,evergreen tropical or subtropical forests; pygmy forests; heathlands or wetlands (Vogt et al. 1986, Chapin et al. 1993, Northup et al. 1995, Vogt et al. 1996).

One of the early studies showing that DON contributed approximately one third of total plant available N in an ecosystem was conducted in a vegetative community dominated by conifers growing on an infertile and acid soil located in the cold temperate climatic zone (Yavitt and Fahey 1986). In this ecosystem, decomposition rates of litter were very low (< 10% decay of annual litterfall over a one year period; Cromack et al. 1982), the end-products of decomposition were dominated by small chain organic acids (Swift et al. 1979), plant available nutrients were present in low amounts in the rooting zone (Knight et al. 1985, Fahey and Knight 1986), and the plants were highly dependent on symbiotic associations on their root systems to acquire nutrients necessary for growth (Trappe 1962, Vogt et al. 1991).

A similar link between low decomposition rates and DON levels was recorded in acid and nutrient poor pygmy forests in California (Northup et al. 1995). In the pygmy forest, the amount of DON released from decaying litter material was directly related to the concentration of polyphenolics in plant tissues and therefore its decomposability (Northup et al. 1995). This study hypothesized that plants have adapted to strongly acid and infertile soils by producing a litter material high in phenolic compounds. Nitrogen becomes unavailable to other plants in the community because during decomposition, the phenolic compounds in the litter material form strong complexes with N (as DON) which then can only be acquired by plants with mycorrhizal associations able to take up organic N forms (Abuzinadah and Read 1986, Griffiths and Caldwell 1992, Vogt et al.1991, Northup et al. 1995).

This ability to convert a limiting nutrient to a form that cannot be utilized by other plants within the same community can considerably modify the competitive interactions between plants. This competitive advantage can be eliminated if plants have similar species of symbionts on their roots since the symbiont would equalize the nutrient uptake potentials for both plant species. Because it is unusual to find plants in nature without symbiotic associations on their roots (Harley and Smith 1983), the potential exists for equalizing nutrient competition between plants by forming symbiotic associations with the same species (Vogt et al. 1991). However, dissimilar species of mycorrhizal fungi are frequently found on different plants, and their efficiencies for nutrient acquisition are highly variable, with some species unable to utilize organic N forms. In this way the competitive advantage may be retained preferentially by one plant species if it has the only appropriate symbiotic association (Harley and Smith 1983, Vogt et al. 1993). While the importance of root symbionts for *Phragmites* has not yet been determined, Cooke and Lefor (1998) did find *Phragmites* and other salt marsh plant roots to be infected with VA mycorrhizae.

In those ecosystems characterized by slow decomposition rates and low nutrient availabilities, most plants are obligately dependent on symbionts for acquiring sufficient nutrients to maintain their growth (Harley and Smith 1983). Since many of the mycorrhizal fungi have been shown to be very effective at utilizing DON forms as the sole N source (Abuzinadah and Read 1986, Griffiths and Caldwell 1992), their importance in systems with deep organic matter accumulations has become even more apparent. Even if plants do not have symbionts on their root systems, the existence of other mechanisms for converting a limiting nutrient into a form not readily utilized by other plants can confer a significant advantage to a species when competing for resources in the same ecological space.

Phragmites can accumulate a substantial detrital layer even though its litter material has high N contents (Table 1). Normally tissues high in N decompose quite rapidly (Aber and Melillo 1982). However, when tissues are high in phenolic compounds, N becomes strongly complexed to these compounds and is converted to a form that is unavailable to microbes and to the plants themselves (Bloomfield et al. 1993, Northup et al. 1995b). Similar results have been recorded for some N-fixing plant species whose slowly decomposing litter tissues have high concentrations of both phenolic compounds and N. For those species the N remains unavailable from these decomposing tissues (Palm and Sanchez 1990, Bloomfield et al. 1993). This phenomenon has also been recorded for palm species that have high N concentrations in foliage but decompose slowly (Bloomfield et al. 1993).

TABLE 1: Aboveground biomass for *Phragmites* and non-*Phragmites* species in different communities.

	Aboveground Biomass (g/m^2)		
Marsh Type	***Phragmites***	**Non-*Phragmites***	**Reference**
Fresh, tidal			
Chapman Pond, CT	2050	900 (*Ta*)	Findlay (unpub data)
Chapman Pond, CT	1439	282 (*Ta*)	Meyerson (2000)
Chester Creek, CT	2642	152 (*Ta*)	Meyerson (2000)
Tivoli Bays, NY	1205	1513 (*Ls*)	Kiviat (unpub. data)
		644 (*Ta*)	
Brackish/Salt, Tidal			
Great Island, CT	1075	560*	Buck (1995)
Great Island, CT	2040	630*	Buck (1995)
Ben Marvin's Creek, CT	3663	1280*	Buck (1995)
Upper Island, CT	2875	1340*	Buck (1995)
Mullica River, NJ	1916	194 (*Sp, Ds*)	Windham (1995)
	1820	203 (Sp, Ds)	Windham (1995)
Blackbird Creek Marsh, DE	1883	727 (*Sp*)	Roman and Daiber (1984)
		652 (*Sa*)	

Ta = *Typha angustifolia*, Ls = *Lythrum salicaria*, Sp = *Spartina patens*, Ds = *Distichlis spicata*, Sa = *Spartina alterniflora*, * = *unspecified*

Many of the by-products of plant decomposition can affect the types and chemical characteristics of compounds formed in soils. For example, if the decomposition rates of tissues are slow, a greater potential exists for incomplete decomposition and organic acid end-product production (Swift et al. 1979). Phenolic acids have been shown to be effective competitors for sorption sites on clay surfaces and to cause the solubilization of fixed P (Davis 1982). This explains the increased phosphate availability that has been recorded in sites treated with organic acids (Hue et al. 1986). It is unknown whether organic acids are produced during the decomposition of slowly decaying *Phragmites* detrital tissues and if the residence time of these acids is long-term enough to result in P leaching. If organic acids are produced as by-products of the decomposition process, they can increase the solubilization of P and potentially its availability to plants. However, this has to be balanced by the other chemical reactions that are also potentially occurring at the same time in these ecosystems. For example, the high cycling of silica in vegetative detrital materials might result in the formation Si-Al-P complexes that would immediately tie up the solubilized P into unavailable forms.

While the mechanisms *Phragmites* utilizes to dominate a site over long time periods are not presently clear, it can be hypothesized that *Phragmites* controls the availability and form of N other plants can acquire within their growing space, allowing it to outcompete other species. It can also be hypothesized that *Phragmites* is controlling N availability by producing detrital materials with chemical compositions that decompose slowly. Several mechanisms are possible: 1) *Phragmites* may diminish the pool size of available N by immobilizing a greater proportion of ecosystem N in complex organic forms that decompose slowly and reduce the available pool of mineral N and/or 2) the slowly decomposing litter layer may result in the formation of DON forms that cannot be as readily used by the other plants growing in the same space (mechanism discussed earlier). Whether or not this DON can be utilized by *Phragmites* is not presently known.

If *Phragmites* immobilizes large amounts of N in its detrital material which also then decomposes slowly, this becomes by itself an effective manner for reducing total available ecosystem N. A decomposition system dominated by slowly decaying accumulated litter material that converted a site to a more N limited system by immobilizing N in the detrital pool could result in a greater proportion of the N being present as DON forms in wetland ecosystems. Both processes occurring in a system (slow decomposition, N sequestration) would together be an extremely effective mechanism by a plant species to change the competitive environment for limiting nutrients to its advantage.

Preliminary results support the idea that *Phragmites* litter tissues (i.e., stems) decompose more slowly than the tissues of other plants growing in the same space. This suggests that *Phragmites* may both remove limiting nutrients (e.g., N) from more readily available pools and that other plants will have difficulty acquiring sufficient nutrients for growth and maintenance of their tissues. If plants are not very efficient at acquiring these limiting nutrients, their ability to occupy the site will diminish. Since the mycorrhizal colonization of plants growing in wetlands is much lower, and the relationship is not as well developed when they do exist (Harley and Smith 1983), the ability of other plants to mitigate the nutrient limitations produced by *Phragmites* is limited.

The apparent inability of plants other than *Phragmites* to compete and reoccupy these sites after invasion suggests that *Phragmites* may be causing a bottleneck in the N cycle. *Phragmites* does not appear to affect its own efficiency for acquiring nutrients but may be changing resource availabilities for other plant species, and thereby improving its competitiveness.

3.2 NUTRIENT IMMOBILIZATION: HYDROXY-ALUMINUM SILICATE COMPLEXES

Phragmites tissues are high in silica, potentially increasing the cycling of this element in marsh ecosystems. Additionally, one pilot study in a freshwater marsh system found that *Phragmites* leaves and stems had substantially higher concentrations of Fe and Al than *Typha angustifolia* (Ahearn-Meyerson 1997). Soil formation is not only a process of differential movement, complexation, and precipitation of minerals released during the weathering of parent material but also includes the incorporation of organic material. It follows that these elements accumulated by *Phragmites* would also be found in high concentrations in the soil. Typically, the rate at which soils develop is strongly controlled by climate and the biologically produced chemical compounds that are highly reactive with minerals in the parent material. Factors controlling the rate of soil formation are typically expressed as five factors differentially contributing to soil formation: climate, parent material, organism, relief or topography, and time (Brady 1990). For example, during soil formation, the presence of a cool climate and silica-rich parent material will eventually result in the formation of an acid soil (Reuss and Johnson 1986). Under conditions where anthropogenic pollution is low, plants and microbes are the major sources of soil acidity and therefore producers of compounds which can cause the weathering of minerals and their movement out of a system (Brady 1990).

These soil-forming processes are particularly important for making P unavailable in terrestrial ecosystems. Several mechanisms have been identified in these ecosystems that can cause the availability of P to become limiting to plants and microorganisms (Ugolini et al. 1977ab, Vogt et al. 1987a,b, Dahlgren et al. 1991a,b): 1) P can become fixed to aluminum-silicate complexes because of the presence of plants which accumulate Si, or 2) P can become complexed to organic acid by-products produced during decomposition which are then leached from the site (Fig. 2). It would be prudent to determine whether these element movement and complexation processes recorded as part of soil formation for particular soil orders (e.g., spodosols and andisols) are also relevant for salt marsh and other wetland systems.

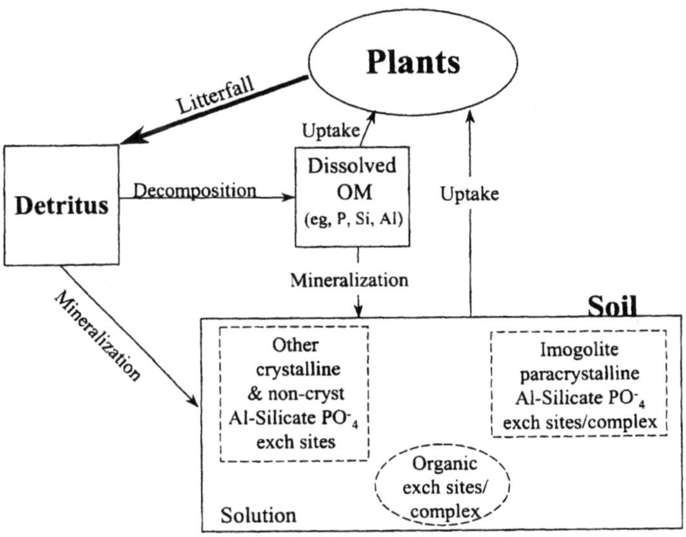

Figure 2. Conceptual model of the phosphorus cycle showing the potential different pathways and forms of P resulting from *Phragmites* detritus decomposition and soil chemistry.

For several soil orders (oxisols, spodosols, andisols), the movement and complexation of Al and Si with other minerals and with organic acids are the driving chemical reactions controlling the development of the soil (Ugolini et al. 1977a,b, Ugolini and Zasoski 1979, Dahlgren et al. 1991a,b, Dahlgren and Walker 1993). The mobility and translocation of Si in soils is well documented. For example, soils formed in the wet tropics are a result of the selective movement of Si out of the surface soil horizons with Al and Fe remaining in the soil matrix in which plants grow (Brady 1990). Other research has shown that Si is an important vehicle for translocation of minerals and in the development of andisols, but that its importance varies depending on what vegetation dominates the site (Dahlgren et al. 1991b). In Japanese pampas grass *(Miscanthus sinensis*), Dahlgren et al. (1991b) found aqueous concentrations of Si were sufficient for the formation of allophane/imogolite; however, the presence of oak (*Quercus serrata Thunb.*) on the same site resulted in organic acids becoming more important in translocating minerals. In these systems, Dahlgren et al. (1991b) found no detectable quantities of PO_4^{3-} in any of the solutions collected under the pampas grass or under the oak. Higher concentrations of Si in the upper horizons of a soil have been shown to be due to the presence of vegetation that accumulates Si (e.g., bamboo, pampas grass) and annually cycles significant quantities of Si (Shoji et al. 1990). This may also prove to be the case with *Phragmites*.

Andisols are characterized by their notable deficiencies of P and the extreme difficulty of increasing the availability of mineral P to plants by fertilizer additions of P; P is too quickly immobilized in chemical complexes in these sites (Ugolini and Zasoski 1979). Andisols have a strong affinity for P because of their amorphous Al-Si complexes, which reduce the translocation of mineral forms of P (Ugolini and Zasoski 1979). It has been suggested that strong fixation of P in the soil is probably occurring in association with imogolite, allophane and Fe and Al oxides (Ugolini and Zasoski 1979). It would be useful therefore to determine whether *Phragmites* or other silica accumulating plants have similar effects on nutrient cycles. Their ability to increase the cycling rate of Si and therefore the chemical complexes that can occur with Si, is significant since Si is an important in affecting soil chemical transformations.

Because *Phragmites* tissues have such high Si, it is important to determine whether the maintenance of high Si cycling in these ecosystems (as part of detrital transfers and the decomposition of these detrital materials) will contribute to the formation of Si-Al complexes (e.g., imogolite) in the rooting zone (Fig. 2). (Imogolite is a term that is used to define compounds comprised of hydrous aluminosilicate that can dominate the chemistry of some soils such as those derived from volcanic materials; Ugolini and Zasoski 1979). It will also be necessary to determine whether the presence of silica complexes will result in the strong retention of mineral P. Research conducted by soil scientists have shown that complexes comprised of organic acids and Al, Fe and Si are possible, and that these complexes accumulate in the soil with long mean residence times (Ugolini et al. 1977a,b, Dawson et al. 1978, Vogt et al. 1987a,b). Whether this is a potential mechanism for immobilizing limiting nutrients (e.g., P) in plant unavailable forms and changing the competitive interactions within the vegetative community in wetland ecosystems needs to be determined.

In addition to Si, Al has also been shown to have a dominant role in controlling P complexation and mobility in soils, but its contribution to P immobilization in wetlands remains unclear. In terrestrial soils, Al is known to react with phosphate anions and form complexes that are not very soluble below a pH of 6 (Lindsey and Vlek 1977). In the soil, P can exist in inorganic complexes with Ca, Al or Fe depending on the pH of the soil environment; with calcium phosphates being common at neutral pH (Khanna and Ulrich 1984). The availability of P to plants from the soil environment will depend on the form in which P is present (Khanna and Ulrich 1984). For example, at a neutral pH, Al and Fe are relatively insoluble, and therefore complexes associated with them should not be readily available to plants. However, Al displacement and high Al mobility can occur in soils if strong acid anions (i.e., Cl^- and SO_4^{2-}) are present (Dahlgren and Ugolini 1989).

4. Plants as Element Accumulators and Ecosystem Nutrient Cycles

Studies conducted in terrestrial ecosystems have shown the potential importance of certain plant species in modulating and controlling ecosystem nutrient cycles when they contribute disproportionately to the cycling of an element compared to other plants growing in the same environment (Vogt et al. 1987a,b, Dahlgren et al. 1991a, Vitousek 1990). When plants accumulate limiting nutrients in high concentrations in their tissues, they can have very positive effects on improving the nutrient availabilities at the whole ecosystem level. For example, there are palm species that accumulate K and Ca in their tissues, many early successional tropical tree species accumulate N, some cedars accumulate Ca, and many nitrogen fixing plants increase N accumulation in the system (Van Miegroet and Cole 1984, O'Hara, unpublished, Bloomfield et al. 1993, Vogt et al. 1993).

On the other hand, plants may accumulate and increase the cycling of a trace element (e.g., Al, Mn) which can be directly toxic to other plants or microbes or indirectly result in soil nutrients becoming unavailable because of the recalcitrant complexes they form with these nutrients (Vogt et al. 1987b, Dahlgren et al. 1991a). For instance, in a subalpine stand in Washington, *Tsuga mertensiana* was an Al accumulator and, even though it comprised only 20% of the basal area of this stand, this tree species controlled 80% of the Al annually cycled within this ecosystem. Since *T. mertensiana* accumulated higher concentrations of Al in its tissues and maintained a higher cycling rate of Al within the biological part of the ecosystem, higher amounts of P were complexed in less available plant pools and a greater potential occurred for Al to be toxic to other plants and microbes (Vogt et al. 1987a,b).

This accumulation of higher levels of Al in biological tissues indirectly and directly affected those parts of the nutrient cycles that were controlled or modified by Al, typically expressed as reduced nutrient availability to plants at the ecosystem level. By increasing the cycling rate of Al, *T. mertensiana* controlled 1) the rate at which soil forming processes occurred, 2) how much P was complexed in longer term residence pools and not readily available to other plants, 3) where plants acquired their nutrients because the soil environment was less favorable as a growth environment (e.g., Al toxicity, reduced P availability) than the surface organic horizons, 4) the decomposition rate of detrital materials because of nutrient deficiencies and Al toxicity to microbes, and 5) which plants could grow at this site since many plants cannot tolerate Al (Vogt et al. 1988a,b, Dahlgren et al. 1991a).

If it can be demonstrated that *Phragmites* accumulates trace elements such as Al or Mn, this would reveal a potential mechanism the species may use to control nutrients cycles in the marsh ecosystems it invades. Since it has been established that *Phragmites* sequesters limiting nutrients (e.g., N) in live and dead biomass (Meyerson et al. 2000) and also alters its environment by reducing light and temperature, an understanding of *Phragmites* control of limiting nutrients through the accumulation of trace metals could add a crucial piece in the puzzle of *Phragmites* expansion on the Atlantic coast of North America.

5. Conclusion

By expanding the focus of the current wetland research, investigators could identify whether other forms of nutrients (e.g., DON) or element complexes are mechanisms by which ecosystem functions and species dominance are maintained. In wetland systems invaded by *Phragmites,* two processes can be hypothesized to be important in changing (and in most cases reducing) nutrient availability: 1) the potential presence of particular nutrient forms (e.g., DON) because of the low available N levels and high immobilization of N in slowly decomposing detrital tissues of *Phragmites*, and 2) the potential formation of particular metal complexes (e.g., hydroxy-aluminum silicate complexes). This may be due to the high silica contents of *Phragmites* tissues that increase the cycling of Si and therefore the complexation of Ca and P in potentially recalcitrant forms.

These processes have received considerable study by soil scientists trying to understand soil development in terrestrial systems (e.g., Spodosols and Andisols) and ecosystem ecologists trying to understand plant control of ecosystem level nutrient cycles (Swift et al. 1979, Farmer et al. 1980, Yavitt and Fahey 1986, Vogt et al. 1986, Dahlgren et al. 1991a,b, Northup et al. 1995, Vogt et al. 1996). We can hypothesize that these same chemical processes controlling the cycling of limiting nutrients in terrestrial ecosystems may also occur in wetlands and that they may contribute to controlling their productivity, health and resilience.

When compared to other marsh plants, the *Phragmites* appears to sequester more nutrients in its tissues (both live and dead) on an area basis due to its high biomass. Other factors, such as the formation of chemical complexes like those found in some terrestrial soil orders (hydroxy-aluminum silicates), may also be causing the sequestration of nutrients, making these less available to other plants and therefore giving *Phragmites* a competitive advantage. If *Phragmites* increases the cycling of one or more elements (e.g., Fe, Al, Si), then *Phragmites* could potentially be controlling the sites that it invades at the ecosystem level. Since these types of nutrient cycling decoupling mechanisms have been documented in terrestrial systems with somewhat similar characteristics, they warrant investigation for wetland systems in general, and for sites invaded by *Phragmites* in particular.

Another possible reason for the success of *Phragmites* may be related to the effects of increased nutrient inputs to wetland systems. *Phragmites* may be better able to exploit high concentrations of nutrients than other wetland plants and therefore able to expand its realized niche because of this competitive advantage. In addition, unlike some other wetland plant species, *Phragmites* may be able to utilize organic nitrogen. If true, this ability would give *Phragmites* a great advantage over other species since N is usually limiting in wetland systems. Further, investigation of the effect of *Phragmites* on ecosystem nutrient cycles, particularly DON, Si, and chemical complexes can clarify which processes are driving and controlling *Phragmites* dominance of a site and reveal those processes that must be re-established in restoration efforts. Not only can this kind of research identify what factors need to be manipulated in restoring *Phragmites* dominated sites to pre-invasion plant communities, but it will also help to determine the minimum information needed to understand whether a site is sufficiently changed to be considered restored.

6. Acknowledgments

The ideas and the research presented here were partially supported by the Connecticut Department of Environmental Protection, the Connecticut Chapter of the Nature Conservancy, and a G.E. Hutchinson Fellowship from Yale University. The authors would also like to thank Frederick Meyerson for helpful comments on this manuscript, Daniel Vogt for discussions on the ideas presented in the paper and in preparing the figures, and Bruce Larson for facilitating the use of research sites.

7. References

Aber, J. D. and J. M. Melillo. 1982. Nitrogen immobilization in decaying hardwood leaf litter as a function of initial nitrogen and lignin content. Canadian Journal of Botany **60**: 2263-2269.

Abuzinadah, R. and D. Read. 1986. The role of proteins in the nitrogen nutrition of ectomycorrhizal plants. III. Protein utilization by *Betula*, *Picea* and *Pinus* in mycorrhizal association with *Hebeloma crustiliniforme*. New Phytologist **103**: 507-514.

Ahearn, L. L., F. A. B. Meyerson, K. A. Vogt. 1996. An ecosystem-level approach to biological invasions and ecological restoration. A *Phragmites* case study in eastern Connecticut. Bulletin of the Ecological Society of America 77 (3 suppl. Part 2).

Ahearn-Meyerson, L. and K.A. Vogt. 1997. Considerations for restoring structure, function, and diversity to an ecosystem colonized by invasive plants: A *Phragmites* case study. Pages 257-258 *in* K. MacDonald and F. Weinman, editors. Wetland and riparian restoration: taking a broader view. US EPA, Region 10, USA.

Armstrong, J. W., P. M. Armstrong, J. E. Beckett, S. Halder, H. R. Lythe and A. Sinclair. 1996a. Pathways of aeration and the mechanisms and beneficial effects of humidity- and Venturi convections in *Phragmites australis* (Cav.) Trin. ex Steud. Aquatic Botany **54**: 177-197.

Armstrong, J., F. Afreen-Zobayed and W. Armstrong. 1996b. *Phragmites* die-back: sulphide and acetic acid-induced bud and root death, lignifications and blockages within aeration and vascular systems. New Phytologist **134**: 601-614.

Bloomfield, J., K. A. Vogt and D. J. Vogt. 1993. Decay rate and substrate quality of fine roots and foliage of two tropical tree species in the Luquillo Experimental Forest, Puerto Rico. Plant and Soil **150**:233-245.

Boar, R. R. 1996. Temporal variations in the nitrogen content of *Phragmites australis* (Cav.) Trin. ex Steud. from a shallow fertile lake. Aquatic Botany **55**:171-181.

Bradley, P., M and J. T. Morris. 1990. Influence of oxygen and sulfide concentration on nitrogen uptake kinetics in *Spartina alterniflora*. Ecology **71**:282-287.

Brady, N. C. 1990. The Nature and Property of Soils. Tenth edition. Macmillan Publishing, New York, New York, USA.

Brix, H., B. K. Sorrell and H. H. Schierup. 1996. Gas fluxes achieved by *in situ* convective flow in *Phragmites australis*. Aquatic Botany **54**:151-163.

Broome, S. W., W. W. Woodhouse and E. D. Seneca. 1975. The relationship of mineral nutrients to growth of *Spartina alterniflora* in North Carolina II. The effects of N, P and Fe fertilizers. Soil Science Society of America Proceedings **39**:301-307.

Chalmers, A. G. 1982. Soil dynamics and the productivity of *Spartina alterniflora*. Pages 231-242 *in* V. S. Kennedy, editor. Estuarine comparisons. Academic Press, New York, New York, USA.

Chambers, R. M. 1997. Porewater chemistry associated with *Phragmites* and *Spartina* in a Connecticut tidal marsh. Wetlands **17**:360-367.

Chambers, R. M., T.J. Mozdzer and J.C. Ambrose. 1998. Effects of salinity and sulfide on the distribution of *Phragmites australis* and *Spartina alterniflora* in a tidal saltmarsh. Aquatic Botany **62**:161-169.

Chambers, R. M., L.A. Meyerson and K. Saltonstall. 2000. Expansion of *Phragmites australis* into tidal wetlands of North America. Aquatic Botany **64**:261-273.

Chapin, F. S. I., L. Moilanen and K. Kieland. 1993. Preferential use of organic nitrogen for growth by a non-mycorrhizal arctic sedge. Nature **361**:150-153.

Chapin, F. S. I. 1995. New cog in the nitrogen cycle. Nature **377**:199-200.

Cuevas, E. and E. Medina. 1986. Nutrient dynamics within Amazonian forest ecosystems I. Nutrient flux in fine litter fall and efficiency of nutrient utilization. Oecologia **68**:466-472.

Dahlgren, R. A. and F. C. Ugolini. 1989. Aluminum fractionation of soil solutions from unperturbed and tephra-treated Spodosols, Cascade Range, Washington, USA. Soil Society of America Journal **53**: 559-566.

——. 1989. Effects of tephra addition on soil processes in Spodosols in the Cascade Range, Washington, USA. Geoderma **45**: 331-355.

Dahlgren, R. A., K. A. Vogt and F. C. Ugolini. 1991. The influence of soil chemistry on fine root aluminum concentrations and root dynamics in a subalpine Spodosol, Washington State, USA. Plant and Soil **133**: 117-129.

Dahlgren, R. A., F. C. Ugolini, S. Shoji, T. Ito and R. S. Sletten. 1991. Soil-forming processes in Alic Melanudands under Japanese pampas grass and oak. Soil Society of America Journal. **55**:1049-1056.

Dahlgren, R. A. and W. J. Walker. 1993. Aluminum release rates from selected Spodosols Bs horizons: Effects of pH and solid-phase aluminum pools. Geochimica et Cosmochimica Acta **57**:57-66.

Davis, J. A. 1982. Absorption of natural dissolved organic matter at the oxide/water interface. Geochimica et Cosmochimica Acta **46**:2381-2393.

Dawson, H., F.C. Ugolini, B.F. Hrutfiord and J.M. Zachara. 1978. Role of soluble organics in the soil processes of a Podzol, central Cascades, Washington. Soil Science **126**:260-296.

DeLaune, R.D. and S.R. Pezeshki. 1988. Relationship of mineral nutrients to growth of *Spartina alterniflora* in Louisiana salt marshes. Northeast Gulf Science **10**:55-60.

Den Hartog, L.D., J. Kvet and H. Sukopp. 1989. Reed. A common species in decline. Aquatic Botany. **35**:1-4.

Fahey, T.J., J.B. Yavitt, J.A. Pearson and D. H. Knight. 1985. The nitrogen cycle in lodgepole pine forests, southeastern Wyoming. Biogeochemistry **3**:257-275.

Fahey, T.J. and D.H. Knight. 1986. The lodgepole pine forest ecosystem. BioScience **36**:610-617.

Farmer, V.C., J.D. Russell and M. L. Berrow. 1980. Imogolite and proto-imogolite allophane in spodic horizons: Evidence for a mobile aluminum silicate complex in podzol formation. Journal of Soil Science **31**: 673-684.

Gries, C. and D. Garbe. 1989. Biomass and nitrogen, phosphorous and heavy metal content of *Phragmites australis* during the third growing season in a root zone waste water treatment. Archiv Fuer Hydrobiologie **117**:97-106.

Griffiths, R.P. and B.A. Caldwell. 1992. Mycorrhizal mat communities in forest soils. Pages 98-105 *in* D.J. Read, D. H. Lewis, A. Fitter and I. Alexander, editors. Mycorrhizas in ecosystems. CAB International, Wallingford, England.

Harley, J.L. and S.E. Smith. 1983. *Mycorrhizal symbiosis*. Academic Press, London, England.

Hellings, S.E. and J.L. Gallagher. 1992. The effects of salinity and flooding on *Phragmites australis*. Journal of Applied Ecology **29**:41-49.

Hocking, P., C. Finlayson and A. Chick. 1983. The biology of Australian weeds. 12. *Phragmites australis* (Cav.) Trin. Ex Steud. The Australian Journal of Agricultural Science **49**:123-132.

Howarth, R.W. 1989. Nutrient limitation of net primary production in marine ecosystems. Annual Review of Ecology and Systematics **19**:89-110.

Howes, B.L., J.W. H. Dacey and D.D. Goeringer. 1986. Factors controlling the growth form of *Spartina alterniflora*: feedbacks between above-ground production, sediment oxidation, nitrogen and salinity. Journal of Ecology **74**:881-898.

Hue, N.V., G.R. Craddock and F. Adams. 1986. Effects of organic acids on aluminum toxicity in subsoils. Soil Society of America Journal **50**:28-34.
Jingguo, W. and L.R. Bakken. 1997. Competition for nitrogen during mineralization of plant residues in soil: Microbial response to C and N availability. Soil Biology and Biochemistry **29**:163-170.
Johnson, D.W. and D.E. Todd. 1983. Relationships among iron, aluminum, carbon and sulfate in a variety of forest soils. Soil Society of America Journal **47**:792-800.
Johnson K.H., K.A. Vogt, H.J. Clark, O.J. Schmitz and D.J. Vogt. 1996. Biodiversity and the productivity and stability of ecosystems. Trends in Ecology and Evolution **11**: 372-377.
Khanna, P.H. and B. Ulrich. 1984. Soil characteristics influencing nutrient supply in forest soils. Pages 79-117 *in* G.D. Bowen and E.K.S. Nambiar, editors. Nutrition of plantation forests. Academic Press, London, England.
Knight, D.H., T.J. Fahey and S.W. Running. 1985. Water and nutrient outflow from lodgepole pine forests in Wyoming. Ecological Monographs **55**: 29-48.
Kuhl, H. and J.G. Kohl. 1992. Seasonal nitrogen dynamics in reed beds (*Phragmites australis* (Cav.) Trin. ex Steudel) in relation to productivity. Hydrobiologia **251**:1-12.
Kuiters, A. T. 1990. Role of phenolic substances from decomposing forest litter in plant-soil interactions. Acta Botanica Neerlandica **39**:329-348.
Lindsey, W.L. and P.L.G. Vlek. 1977. Phosphate minerals. Pages 639-672 *in* J.B. Dixon and S.B. Weed, editors. Minerals in Soil Environments. Soil Science Society of America, Madison, Wisconsin, USA.
Lissner, J. and H.H. Schierup. 1997. Effects of salinity on the growth of *Phragmites australis* (Cav.) Trin. ex Steudel. Aquatic Botany **55**:247-260.
Meyerson, L.A., K. Saltonstall, L. Windham, E. Kiviat and S. Findlay (2000). A comparison of *Phragmites australis* in freshwater and brackish marsh environments in North America. Wetlands Ecology and Management.
Mitsch W.J. and J.G. Gosselink. 1993. Wetlands. Second edition. Van Nostrand Reinhold. New York, New York, USA.
Morris, J.T. 1980. The nitrogen uptake kinetics of *Spartina alterniflora* in culture. Ecology **61**:1114-1121.
Niering, W.A. and R.S. Warren. 1977. Our dynamic tidal marshes. Connecticut College Arboretum Bulletin **22**.
——. 1980. Vegetation patterns and processes in New England salt marshes. BioScience **30**: 301.
Northrup, R.R., Y. Zengshou, R.A. Dahlgren and K.A. Vogt. 1995. Polyphenol control of nitrogen release from pine litter. Nature **377**: 227-229.
Northrup, R. R., R. A. Dahlgren and Z. Yu. 1995. Interspecific variation of conifer phenolic concentration on a marine terrace soil acidity gradient: a new interpretation. Plant Soil **171**:255-262.
Palm, C. A. and P. A. Sanchez. 1990. Decomposition and nutrient release patterns of the leaves of three tropical legumes. Biotropica **22**:330-338.
Randall, J. and B. Lapin. 1995. The Nature Conservancy element stewardship abstract: *Phragmites australis*. The Nature Conservancy, Arlington, Virginia, USA.
Reuss, J.O. and D.W. Johnson. 1986. Acid deposition and the Acidification of Soils and Water. Springer-Verlag, New York, New York, USA.
Ryther, J.H. and W.M. Dunstan. 1971. Nitrogen, phosphorous and eutrophication in the coastal marine environment. Science **171**:1008-1013.
Sanchez, P.A. and D.D. Bandy. 1982. Amazon basin soils: management for continuous crop production. Science **216**: 821-827.
Shaver, G.R. and J.M. Melillo. 1984. Nutrient budgets of marsh plants: efficiency concepts and relation to availability. Ecology **65**:1491-1510.
Shoji, S., T. Kureybayashi and I. Yamada. 1990. Growth and chemical composition of Japanese pampas grass (*Micanthus sinesis*) with special reference to the formation of dark colored Andisols in northeastern Japan. Soil Science and Plant Nutrition **36:**105-120.
Sinicrope, T.L., P.G. Hine, R.S. Warren and W.A. Niering. 1990. Restoration of an impounded salt marsh in New England. Estuaries **13**: 20-25.
Stacey, P.E. and M.A. Tedesco. 1997. Development and use of atmospheric nitrogen loading estimates for management and planning to control Long Island Sound hypoxia. Estuarine Research Federation Meeting, Providence, Rhode Island, USA.
Stalter, R. and Baden, J. 1994. A twenty-year comparison of vegetation of three abandoned rice fields, Georgetown County, South Carolina. Castanea **59**: 69-77.
Sullivan, M. L. and F. C. Daiber. 1974. Response in production of cordgrass, *Spartina alterniflora*, to inorganic nitrogen and phosphorous fertilizer. Chesapeake Science **15**:121-123.

Swift, M.J., O.W. Heal and J.M. Anderson. 1979. Decomposition in Terrestrial Ecosystems. Blackwell Scientific, Oxford, England.

Templer, P., S. Findlay and C. Wigand. 1998. Sediment chemistry associated with native and non-native emergent macrophytes of a Hudson River marsh ecosystem. Wetlands **18**: 70-78.

Tilman, D. 1997. Biodiversity and ecosystem functioning. Pages 93-112 *in* G. Daily, editor. Nature's Services: Societal Dependence on Natural Ecosystems. Island Press, Washington, District of Columbia, USA.

Tilman, D., J. Knops, D. Wedin, P. Reich, M. Ritchie and E. Siemann. 1997. The influence of functional diversity and composition on ecosystem processes. Science **277**:1300-1302.

Ugolini, F.C., H. Dawson and J. Zachara. 1977. Direct evidence of particle migration in the soil solution of a Podzol. Science **188**:603-605.

Ugolini, F.C., R. Minden, H. Dawson and J. Zachara. 1977. An example of soil processes in the *Abies amabilis* zone of central Cascades, Washington. Soil Science **124**: 291-302.

Ugolini, F. C. and R. J. Zasoski. 1979. Soils derived from Tephra. Pages 83-124 in Payson D. Sheets and Donald K. Grayson, editors. Volcanic Activity and Human Ecology. Academic Press Inc., New York, New York, USA.

Valiela, I. and J.M. Teal. 1974. Nutrient limitation in salt marsh vegetation *in* R.J. Reimold and W.H. Queens, editors. Ecology of Halophytes. Academic Press, New York, New York, USA.

van der Putten, W.H. 1997. Die-back of *Phragmites australis* in European wetlands: an overview of the European Research Programme on Reed Die-Back and Progression (1993-1994). Aquatic Botany **59**:263-275.

Van Miegroet, H. and D.W. Cole. 1984. The impact of nitrification on soil acidification and cation leaching in a red alder ecosystem. Journal of Environmental Quality **13**:586-590.

Vitousek, P.M. 1990. Biological invasions and ecosystem processes: towards an integration of population biology and ecosystem studies. Oikos **57**: 7-13.

Vogt, K.A., C. Grier and D. Vogt. 1986. Production, turnover and nutrient dynamics of above- and below-ground detritus of world forests. Advances in Ecological Research **15**:303-377.

Vogt, K., R. Dahlgren, F. Ugolini, D. Zabowski, E.E. Moore and R. Zasoski 1987a. Aluminum, Fe, Ca, Mg, K, Mn, Cu, Zn and P in above- and belowground biomass I. Concentrations in subalpine *Abies amabilis* and *Tsuga mertensiana*. Biogeochemistry **4**:277-294.

——. 1987b. Aluminum, Fe, Ca, Mg, K, Mn, Cu, Zn and P in above- and belowground biomass: II. Pools and circulation in a subalpine *Abies amabilis* stand. Biogeochemistry **4**: 295-311.

Vogt, K. A., D. A. Publicover and D. J. Vogt. 1991. A critique of the role of ectomycorrhizas in forest ecology. Agriculture, Ecosystems and Environment **35**: 81-87.

Vogt, K., D. Vogt, S. Brown, J. Tilley, R. Edmonds, W. Silver and T. Siccama. 1995. Dynamics of forest floor and soil organic matter accumulation in boreal, temperate and tropical forests. Pages 159-178 *in* R. Lal, J. Kimble, E. Levine and B. Steward, editors. Soil Management and Greenhouse Effects. CRC Press, Boca Raton, Florida, USA.

Warren, R. S., E. L. Buck and P. E. Fell. 1995. *Phragmites australis* in the lower Connecticut River tidelands: rates of invasion, stand structure, productivity and decomposition in relation to salinity and hydroperiod. Society of Wetland Scientists, 16th Annual Meeting, Boston, Massachusetts, USA.

Windham, L. 1995. Effects of *Phragmites* australis invasion on aboveground biomass and soil properties in brackish tidal marsh of the Mullica River, New Jersey. Thesis, Rutgers University, New Brunswick, New Jersey, USA.

Winogrond, H. 1997. Invasion of *Phragmites australis* in the tidal marshes of the Hudson River. Thesis. Bard College, Annadale-on-Hudson, USA.

Wu, Y.F.H., Sklar and K. Rutchey. 1997. Analysis and simulations of fragmentation patterns in the Everglades. Ecological Applications **7**: 268-276.

Yang, X., D. R. Miller, X. Xu, L. H. Yang, H. Chen and N. P. Nikolaidis. 1996. Spatial and temporal variations of atmospheric deposition in interior and coastal Connecticut. Atmospheric Environment **30**: 3801-3810.

Yavitt, J. B. and T. J. Fahey. 1986. Litter decay and leaching from the forest floor in *Pinus contorta* (Lodgepole pine) ecosystems. Journal of Ecology **74**: 525-545.

RESTORATION OF SALT AND BRACKISH TIDELANDS IN SOUTHERN NEW ENGLAND

Angiosperms, Macroinvertebrates, Fish, and Birds

PAUL E. FELL
Department of Zoology
Connecticut College
New London, Connecticut 06320 USA

R. SCOTT WARREN
WILLIAM A. NIERING
Department of Botany
Connecticut College
New London, Connecticut 06320 USA

Abstract

Tidal restriction, dredge spoil deposition, and other fill activities have converted about 2000 ha of Connecticut's tidal salt marshes to non-tidal or microtidal systems vegetated by near monocultures of *Phragmites australis* or *Typha angustifolia*. In addition, *Phragmites* is also expanding in certain undisturbed brackish tidelands, replacing the typical tidal marsh angiosperms. Returning normal tidal hydrology to formerly restricted polyhaline (18 to 30 ‰) and euhaline (30 to 35 ‰) marshes results over time in re-establishment of typical *Spartina*-dominated marsh vegetation and associated macroinvertebrate populations. Vegetation and invertebrates, along with full use of these systems by estuarine fish and salt marsh dependent birds, are collectively considered high level integrators of multiple, complex, interacting tidal marsh functions. These various attributes return at different rates, and full functional equivalence relative to undisturbed marshes may require decades. Excavating dredged spoil filled sites to low marsh elevations and restoring tidal action allows natural repopulation by marsh angiosperms and invertebrates. Within the first year these open sites support seedling populations of *Spartina alterniflora* and annuals such as *Salicornia europaea*. Within five years *Spartina alterniflora* dominates and annuals are rare. Initial invertebrate colonizers are those with planktonic larvae, such as *Melampus bidentatus*, *Geukensia demissa* and *Uca* spp. Invasion of brackish tidelands at the mouth of the Connecticut River by *Phragmites* appears to have little effect on macroinvertebrate populations or fish use. The vegetation of such *Phragmites*-dominated tidal wetlands can be restored, at least temporarily, by a combination of herbicide and mowing treatments.

1. Introduction

During the past century, about 2000ha (30%) of Connecticut's tidal marshes have been degraded or lost through coastal development, with the greatest impacts occurring in the western, more urbanized part of the state. Tidal flow to many marshes was restricted by the construction of highways, railroads and impoundments, producing microtidal environments in which *Phragmites australis* (reedgrass) or less frequently *Typha angustifolia* (narrow-leaved cattail) became established at the expense of typical tidal marsh angiosperms (Roman et al. 1984, Rozsa 1995, Niering 1997). In other cases tidal marshes were filled with dredged or other material, creating microtidal or supratidal areas that were colonized by *Phragmites* and upland vegetation. Such conversion of coastal marshes was accompanied by the loss of characteristic tidal marsh animal communities (Rozsa 1995).

In addition to these human influences, *Phragmites* has also invaded apparently undisturbed brackish tidal marshes in the lower Connecticut River system where salinity levels are often reduced by fresh water inputs. This lower riverine spread of *Phragmites* began in the mid to late 1960s and has progressed at a rapid rate to the present (Warren 1993, Buck 1996). This aggressively invasive *Phragmites* appears to be a relatively new genetic strain (Bestika 1996).

Restoration of Connecticut tidal marshes began in earnest in 1978 with the re-introduction of tidal flow to two impounded marshes at Barn Island. Since then the Connecticut Department of Environmental Protection (DEP) has been implementing a systematic program for restoring tidal flow to degraded marshes all along the Connecticut coast. In addition, experimental removal of dredged material from a former salt marsh at Mumford Cove in Groton was started in 1989, and plans are being made for large-scale removal of fill at other sites (Rozsa 1995). The Wetlands Restoration Unit of the Connecticut DEP has also begun to restore brackish tidelands invaded by *Phragmites* through the use of herbicides and mowing.

In order to reasonably assess the success of such restoration efforts, physical and biological indicators of marsh functions (e.g. hydroperiod, salinity, productivity, and community structure) must be examined. Although vegetation integrates a number of factors, this indicator by itself provides an incomplete measure of marsh restoration. Habitat and food chain support functions as manifested by macroinvertebrate, fish and bird populations are also critical features in assessing restoration success. It has also become apparent that different marsh attributes or functions may return at different rates and to some degree independently of one another (Burdick et al. 1997, Niering 1997).

2. Restoration by Removal of Tidal Restriction

The marshes of the Barn Island Wildlife Management Area in Stonington, Connecticut provide a good example of restoration following the return of tidal flooding. This state-owned complex includes a series of five valley marshes situated on Little Narragansett

Bay at the eastern end of Long Island Sound. During the late 1940s, beginning in 1946, the four westernmost valley marshes were impounded by earthen dikes in an attempt to increase waterfowl habitat (Fig. 1). Proceeding from west to east, Impoundment 1 converted primarily to a *Typha angustifolia*-dominated brackish marsh, Impoundments 2 and 3 changed to largely unvegetated mud flats with standing water, and Impoundment 4 became dominated by *Phragmites australis*. In 1978 the Connecticut DEP restored tidal flow to Impoundments 1 and 2 by placement of a 1.5-m diameter culvert in each of the impoundment dikes; and in 1982 a 2.1-m diameter culvert was added to Impoundment 1. Similarly, culverts were installed in the remaining dikes in 1987. However, weir boards placed behind the culvert in dike 3 continued to substantially restrict tidal flow until 1991 when some of the boards were removed. This review will consider restoration of Impoundments 1, 3 and 4.

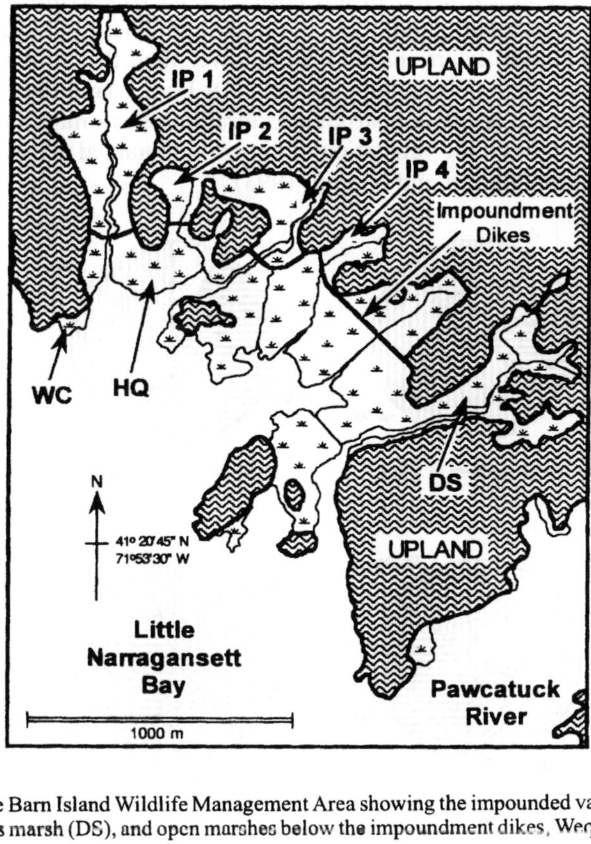

Figure 1. Map of the Barn Island Wildlife Management Area showing the impounded valley marshes (1-4), the unimpounded Davis marsh (DS), and open marshes below the impoundment dikes, Wequetequock Cove (WC) and Headquarters (HQ).

In 1976, 74% of Impoundment 1 was covered by *Typha* and 19% of it was unvegetated peat with standing water. By 1988, ten years after the restoration of tidal flow, *Typha* cover had declined to 16% and much of the remaining *Typha* was stunted

(Fig. 2). On the other hand, *Spartina alterniflora* (saltwater cordgrass) cover had dramatically increased from <1% to 45%. In addition, high marsh species including *Spartina patens* (saltmeadow cordgrass), *Distichlis spicata* (spikegrass), *Juncus gerardii* (blackgrass), and forbs had become re-established and covered another 20% of the marsh surface (fig. 3). Over the same period, live *Phragmites* cover nearly tripled (Sinicrope et al. 1990, Barrett and Niering 1993). Since 1988, salt marsh vegetation has increased to cover about 85% of the marsh area and *Phragmites* has been declining in abundance (R.S. Warren, unpublished data). In 1991, the primary productivity of Impoundment 1 was estimated to be about 73% of that of the pre-impounded marsh, as well as that of an existing reference marsh (Headquarters) situated immediately below the impoundment dike (Warren et al. 1993). This reference marsh has also undergone considerable vegetation change over the past several decades (Fig. 2). *S. patens* and *Juncus* have become much less abundant, whereas stunted *S. alterniflora* and forbs have increased. Large areas of marsh are now covered by mixtures of *S. patens*, *S. alterniflora*, other graminoids, and forbs (Warren and Niering 1993). The primary productivity of the Headquarters marsh is now lower than that of a stable *S. patens*-dominated reference marsh (Wequetequock Cove) at Barn Island (Warren et al. 1993). Although Impoundment 1 exhibits a striking restoration of salt marsh vegetation, point for point agreement between the vegetation of the pre-impounded marsh and the restored marsh is only moderate (Barrett and Niering 1993).

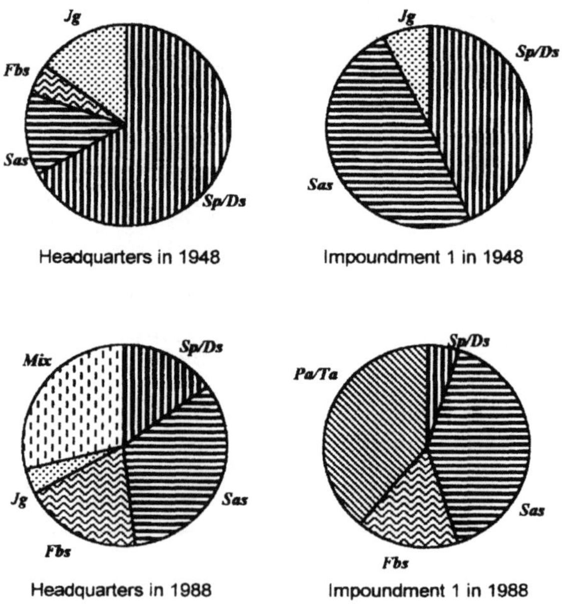

Figure 2. Plant community composition by percentage of total area on Impoundment 1 and on the adjacent open marsh below the dike (Headquarters) at Barn Island. (Sp/Ds - *Spartina patens* dominated community, often with significant *Distichlis spicata*; Sas-stunted *Spartina alterniflora* dominated community; Fbs - mixed forb community; Jg - *Juncus gerardii* community; Mix - mixed community of *S. patens*, stunted *S. alterniflora*, and forbs; Pa/Ta - unrestored *Phragmites australis* and *Typha angustifolia* dominated areas. (from R.S. Warren et al. 1993).

Figure 3. Aerial photograph of Impoundment 1 at Barn Island, Connecticut thirteen years after the re-establishment of tidal flow. Much of the marsh is covered by *Spartina alterniflora* and other saltmarsh angiosperms; *Phragmites australis* occurs along the upland border on each side. The impoundment dike is indicated by an arrow head.

There has been no detailed study of the vegetation in Impoundment 3. However, in 1996, after five years of substantial tidal flushing, large areas of the marsh (54% of the quadrats sampled) were dominated by *Spartina alterniflora* (E.T. Olson, P.E. Fell and R.L. Avery, unpublished data; also see Brawley et al. 1998)

Very little information concerning invertebrate populations in the impounded marshes of Barn Island is available, but it appears likely that typical salt marsh invertebrates were absent from most, if not all, regions of these marshes. By 1990, twelve years after the re-establishment of tidal exchange, a characteristic assemblage of macroinvertebrates including *Melampus bidentatus* (snail), *Geukensia demissa* (ribbed mussel), *Orchestia grillus* and *Uhlorchestia spartinophila* (amphipods), *Philoscia vittata* (isopod), and *Uca* spp. (fiddler crabs) had recolonized Impoundment 1. Quantitative sampling of *Melampus* indicated that its population density in Impoundment 1 was not significantly different from those in two reference marshes below the dike (Fig. 4) (Fell et al. 1991). However, *Melampus* was less abundant in the restored marsh than in the nearby unimpounded (Davis) marsh (Peck et al. 1994). This snail was generally larger in Impoundment 1 than in the reference marshes, perhaps because of a larger food resource that may have supported a more rapid rate of growth; and it produced more, larger egg masses in accordance with its greater size (Fell et al. 1991, Spelke et al. 1995, Helvenston et al. 1995).

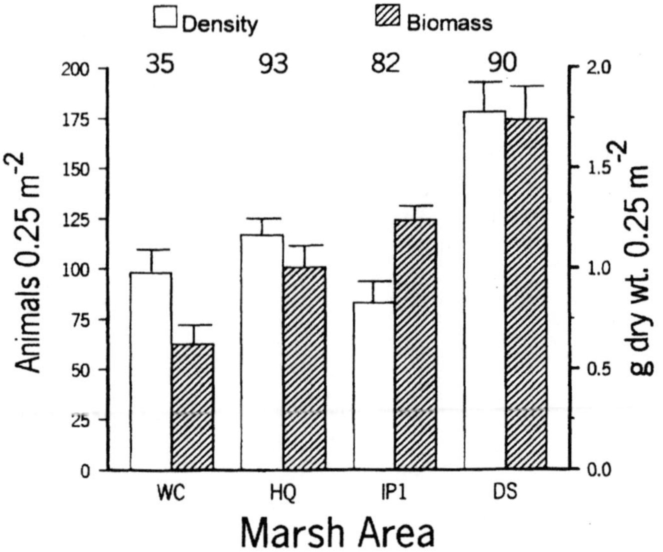

Figure 4. Density (open bars) and shell-free biomass (hatched bars) of *Melampus bidentatus* in different marsh areas at Barn Island (WC and HQ - open marshes below impoundment 1; IP1 - Impoundment 1; DS - unimpounded Davis marsh). The number of quadrats sampled in each marsh area is indicated above the bars and standard errors are represented by vertical lines above the bars (from Peck et al. 1994).

By 1996, five years after the tidal flow was restored to Impoundment 3, typical salt marsh invertebrates had repopulated the marsh in small numbers (Table 1). The densities of *Melampus*, *Geukensia, Orchestia*, *Uhlorchestia*, and *Philoscia* were significantly lower in this restored marsh than in the largely *S. alterniflora*-dominated reference marsh below the dike. On the other hand, the density of *Gammarus palustris*, a low marsh amphipod, was significantly greater in Impoundment 3. This marsh appears to be in an early stage of recovery and full restoration, if achieved, may require a decade or more.

By 1991 the fish assemblage using a mosquito-control ditch in Impoundment 1 was similar to those using ditches in the reference marshes (Table 2). The common mummichog, *Fundulus heteroclitus*, was the numerically dominant fish at all sites; however, the relative abundances of various fish species in the restored and reference

TABLE 1. Densities (No. $\pm$ SE/0.25m^2) of macroinvertebrates in Impoundment 3 and the reference marsh below the impoundment dike at Barn Island. (E.T. Olson, P.E. Fell and R.L. Avery, unpublished data).

Species	Impoundment 3 n = 26	Reference Marsh n = 18	Significance (t test)
Melampus bidentatus	32 ± 7.8	320 ± 42	t = 8.02, df = 42, p < 0.0001
Geukensia demissa	1.7 ± 0.7	6.3 ± 2.3	t = 2.24, df = 42, p = 0.0300
Orchestia grillus	0.7 ± 0.6	4.0 ± 1.1	t = 2.98, df = 42, p = 0.0048
Uhlorchestia spartinophila	19 ± 4.0	42 ± 5.0	t = 3.69, df = 42, p = 0.0006
Gammarus palustris	66 ± 17	0.7 ± 0.5	t = 3.23, df = 42, p = 0.0024
Philoscia vittata	0.5 ± 0.4	2.0 ± 0.6	t = 2.12, df = 42, p = 0.0398

marshes differed. Analyses of mummichog gut contents showed that in both the restored and reference marshes the major components of the diet were copepods, amphipods, tanaids, insects, algae and detritus. Of these food items, insects were more prominent and algae and detritus were less well represented in the diet of mummichogs caught in Impoundment 1 than in those of fish trapped in the reference marshes. In addition, it appeared that mummichogs consumed less food per unit body weight in Impoundment 1 than in the other marshes (Allen et al. 1994).

Fishes in the ditches and tidal creeks moved onto the flooded surface of Impoundment 1 and a reference marsh (Headquarters) during spring tides (Table 2). Gut content analyses indicated that mummichogs trapped in both marshes had fed extensively on amphipods, insects, algae and detritus and consumed *Orchestia*, *Uhlorchestia*, *Philoscia* and *Melampus* in small numbers (E.M. Tarlow, P.E. Fell and J.K. Shain, unpublished data).

Bird use of Impoundment 1 was extensive 16 years after tidal flushing was restored; in fact, it was greater than that of the open reference marshes below the dike. Among the birds observed in Impoundment 1 were shorebirds [killdeer (*Charadrius vociferus*), sandpipers (*Scolopacidae*) and yellowlegs (*Totanus* sp.)], waders [egrets (*Egretta* sp.) and ibises (*Plegadis* sp.)], a marsh generalist [red-winged blackbird (*Agelaius phoeniceus*)], and marsh specialists including the saltmarsh sharp-tailed sparrow (*Ammodramus caudacutus*) and seaside sparrow (*A. maritima*), which are two Connecticut species of Special Concern (Fig. 5). Three years following re-establishment of tidal flushing to Impoundment 3, use of this site by shorebirds and waders was approaching that of other marshes within the Barn Island system; however, saltmarsh sparrows were rare (Brawley et al. 1998).

TABLE 2. Comparison of fish caught in Impoundment 1 (IP1), the marsh immediately below Impoundment 1 (HQ), and the nearby unimpounded Davis marsh (DS) at the Barn Island Wildlife Management Area in Stonington, Connecticut. At two-week intervals from late May through late July 1991, fish were trapped in ditches using 6-mm mesh Fyke nets. In late September and late October 1995, fish were caught on the flooded high marsh using Plexiglas Breder traps. Data for all samples from each marsh are pooled and the percentage of the total catch represented by each species is given in parentheses. (P.E. Fell, unpublished data)

Species	Caught in Mosquito Ditches (1991)			Caught on Flooded High Marsh (1995)	
	IP1	HQ	DS	IP1	HQ
Fundulus heteroclitus (Common mummichog)	1550(75)	3054(90)	3261(88)	1301(56)	1219(76)
Fundulus majalis (Striped killifish)	30(1)	118(3)	99(3)	20(<1)	74(5)
Lucania parva (Rainwater killifish)	8(<1)	10(<1)	0	321(14)	160(10)
Cyprinodon variegatus (Sheepshead minnow)	458(22)	109(3)	274(7)	661(28)	80(5)
Menidia spp. (Silverside)	5(<1)	89(3)	80(2)	3(<1)	50(3)
Apeltes quadracus (Fourspine stickleback)	0	0	0	27(1)	13(<1)
Total number of fish	2054	3380	3717	2336	1596

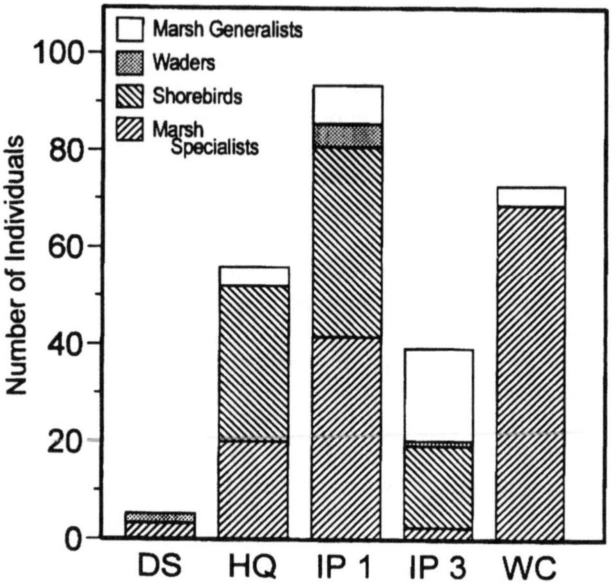

Figure 5. Bird use of different marsh areas at Barn Island (IP1Impoundment 1; IP3Impoundment 3; HQ and WC reference marshes below the Impoundment 1, DS unimpounded Davis marsh) (from Brawley et al. 1998).

Collectively, the data indicate that Impoundment 1 is now in a relatively advanced stage of restoration and equivalent in many respects to the adjacent unimpounded marsh. However, the mosaic of salt marsh vegetation differs from that of the pre-impounded valley marsh. This appears to be related to the slightly lower marsh elevations resulting from impoundment over several decades. In addition, it appears that Impoundment 3 is developing in a similar direction.

In 1996 recovery of five degraded, *Phragmites*-dominated marshes distributed along approximately 100 km of the Connecticut shoreline was examined. These formerly *Spartina*-dominated marshes had converted to *Phragmites* monocultures as a result of prolonged tidal restriction. Following 6 to 11 years of renewed tidal flooding, two of the marshes (Barn Island Impoundment 4 and Great Creek) were still dominated by *Phragmites*, whereas the others (Hammock River, Long Cove, and Great Meadows) exhibited substantial restoration of tidal marsh vegetation. The estimated rates of recovery ranged from 0.4 to 8.6% y^{-1}. At these sites *Phragmites* cover, height and stem density were negatively correlated with soil-water salinity (Fig 6). Furthermore, as salinity increased from 22 to 26 ‰, the frequency of occurrence of *Phragmites* plummeted from almost 100% to nearly 0%. Hydroperiod appeared to be another important factor. *Phragmites* tended to persist at higher elevations even when soil-water salinities were high. Marsh macroinvertebrates have become re-established on at least portions of every marsh; however, this has occurred to a large degree independently of changes in vegetation (A.C. Orsted, R.S. Warren, and W.A. Niering, unpublished data).

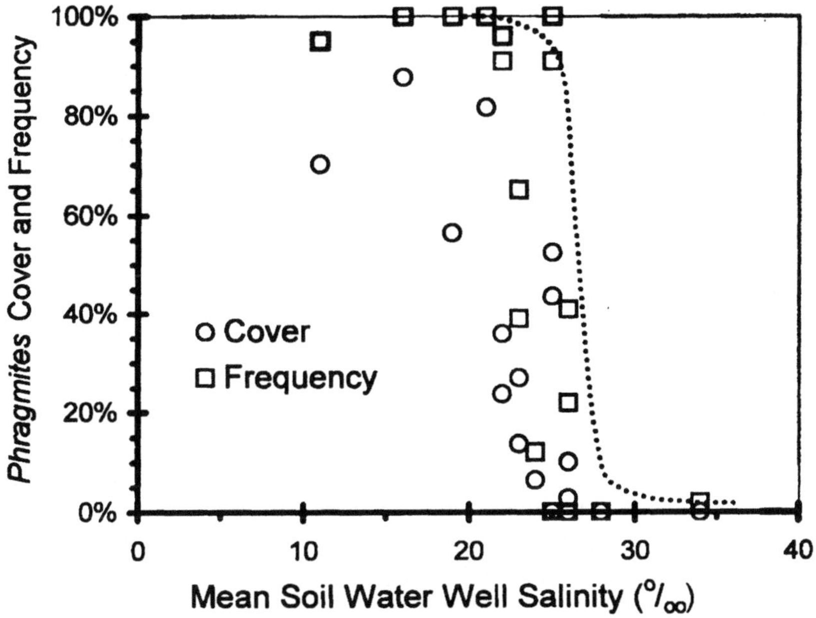

Figure 6. Mean percentage cover and frequency of *Phragmites australis* on restoring coastal marshes in relation to mean soil-water salinity. (A.C. Orsted, R.S. Warren and W.A. Niering, unpublished data).

Nine years after the re-establishment of tidal flow to Barn Island's Impoundment 4, about 97% of the system was still covered by *Phragmites*, although much of it was stunted. Even though typical salt marsh vegetation was absent from most of this area, invertebrates such as *Melampus, Uhlorchestia, Orchestia* and *Philoscia* had repopulated the site. The densities of first two species were significantly lower in the restored marsh than in the largely *S. patens*-dominated reference marsh below the dike ($t = 3.49$, $df = 36$, $p = 0.0013$ and $t = 3.23$, $df = 36$, $p = 0.0026$, respectively), but the latter two species exhibited comparable population densities ($\approx 12/0.25\ m^2$ in each case) on the two marshes (E.T. Olson, P.E. Fell and R.L. Avery, unpublished data).

3. Restoration Following Removal of Dredged Spoil

The Mumford Cove marsh in Groton, Connecticut has been restored by removal of dredged spoil material. In the 1950s, an earthen dike was constructed around the marsh, and sediments removed from the cove were spread over the marsh to depths of 0.6 to 1.2 m. The marsh area, which was no longer flooded by the tides, became essentially a *Phragmites* monoculture. Beginning in 1989, the fill was removed from much of the marsh surface, tidal creeks were relocated, and the area was graded to favor low marsh with scattered pools. Work on this project was carried out over a four-year period with much of it done during 1990. Except for the planting of about 12 plugs of *Spartina alterniflora* at one end of the marsh, recolonization of the site by salt marsh vegetation was natural (Rozsa 1995, Waters 1995, Niering 1997).

Seedlings of *Spartina alterniflora* became established during the first growing season, but large areas of marsh were dominated by the annual *Salicornia europaea* (saltwort). The mud snail *Ilyanassa obsoleta* occurred in the creeks and ditches. After three years, about 75% of the area at some low marsh elevations was covered by *Spartina alterniflora*; *Phragmites* or sparse salt marsh vegetation occurred at higher elevations. *Uca* sp. was present. After 8 years, *Spartina alterniflora* covered essentially all of the low marsh areas and had begun to invade open sites higher in the marsh (Waters 1995, Niering 1997, R.S. Warren and J. Roy, unpublished data). In addition, *Ruppia maritima* (widgeon grass) had become established in some of the pools (Rozsa 1995).

In 1998, *Melampus* was present in 82% of the 38 quadrats sampled in areas dominated by short *Spartina alterniflora* with a mean density of only $28 \pm 4.6/0.25\ m^2$. As in a number of other restored marshes, this snail tended to be large with 65% of the population being more than 1 cm in shell length. The amphipods *Orchestia* and *Uhlorchestia* each occurred in more than 85% of the quadrats at mean densities of $7.4 \pm 1.4/0.25\ m^2$, and $25 \pm 4.7/0.25\ m^2$, respectively, whereas *Gammarus palustris* was found in only 32% of the quadrats with a mean density of $7.8 \pm 4.2/0.25\ m^2$. The isopod *Philoscia* was rare as was another isopod *Trachelipus rathkei*. In drier areas covered by stunted *Phragmites*, *Orchestia* and *Philoscia* were common but few *Melampus* were present. A total of 9 species of fish were caught in the recreated tidal creeks. The more numerous species in order of decreasing abundance included Atlantic silverside (*Menidia menidia*), mummichog, striped killifish (*Fundulus majalis*), sheepshead minnow (*Cyprinodon variegatus*), and young-of-the-year winter flounder (*Pleuronectes americanus)* and white mullet (*Mugil curema*). Additionally, the grass shrimp (*Palaemonetes pugio*) was present in large numbers, and the green crab (*Carcinus maenas*) was common (P.F. Fell, J.J. Doe and S.A. Spiegal, unpublished data).

4. Restoration of Riverine Marshes Invaded by Phragmites

During the last 20 years, *Phragmites* has been displacing typical tidal marsh angiosperms in the brackish wetlands (<18 ppt early in growing season) of the lower Connecticut River at rates of about 1 to 3% y^{-1}, estimated from aerial photographs. Locally this plant now dominates large areas within the system, often forming dense monocultures (Warren 1993, Buck 1996). Although the domination of *Phragmites* reduces plant species diversity in the affected areas, its overall impact has not been extensively studied. In the Great Island - Upper Island - Lieutenant River marshes, typical tidal marsh invertebrates including *Melampus, Orchestia, Philoscia* and *Succinea* sp. (snail) often appeared to be at least as abundant in the *Phragmites*-dominated marshes as in adjacent marsh vegetation largely free of this invader (Fell et al. 1998). Furthermore, the use of *Phragmites*-dominated marshes as foraging areas by mummichogs appeared to be similar to that of reference marshes with little *Phragmites* (Fig. 7). In both types of marshes, mummichogs moved up onto the flooded marsh surface during spring tides and fed heavily upon *Orchestia, Philoscia, Succinea* and other invertebrates (Fell et al. 1998,

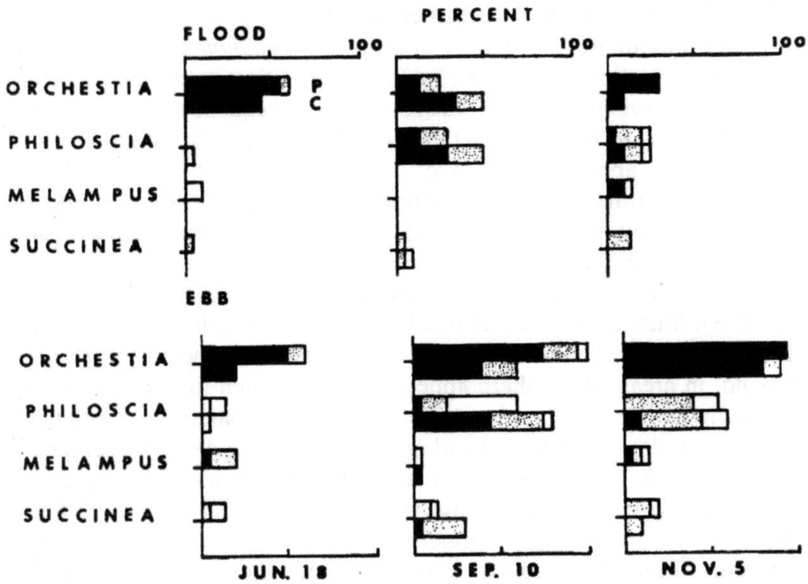

Figure 7. Frequency (%) with which various high marsh prey occurred in the guts of *Fundulus heteroclitus* trapped in a *Phragmites* marsh (P) and reference (*Juncus/Spartina patens*) marsh (C) on Great Island during flooding and ebbing spring tides. (Black-prey represented more than half of the total gut content volume; stippled-prey made up 10 to 50% of the total gut content volume; open-prey constituted less than 10% of the total gut content volume.) The guts of 20 fish were examined for each tidal phase sample from each marsh on each date. (from Fell et al. 1998).

Rilling et al. 1998). On the other hand, bird use of *Phragmites*-dominated marshes is often reduced compared to that of other types of marshes. For example, seaside sparrows, sharp-tailed sparrows and willets (*Catoptrophorus semipalmatus*), as well as waders, shorebirds and ducks, were less abundant (or absent) at *Phragmites*-dominated sites than in short-grass (*Spartina, Juncus, Distichlis)* marshes (Brawley 1994, Benoit and Askins 1999). Furthermore, although the avifauna of *Typha* and *Phragmites* marshes were similar, Virginia rails (*Rallus limicola*) appeared to prefer the former (Benoit and Askins 1999). Marsh wrens (*Cistothorus palustris*) and swamp sparrows (*Melospiza georgiana*), which are marsh specialists that prefer tall reedy vegetation, were common in both *Phragmites*-dominated and *Typha* marshes as were red-winged blackbirds and tree (*Tachycineta bicolor*) and barn swallows (*Hirundo rustica*) (Brawley 1994, Benoit and Askins 1999).

Concern over the loss of plant species diversity and diminished bird use, along with other potential effects resulting from *Phragmites* invasion, has provided impetus for restoration of former meadow marshes that are now dominated by *Phragmites*. The Connecticut Wetland Restoration Unit tested 3 methods for controlling *Phragmites*: mowing and mulching (fragmenting), spraying with the herbicide Rodeo® (≈5.6 liters ha^{-1}), and treating with Rodeo® followed by mowing and mulching. The last method appeared to yield the best results.

A *Phragmites*-dominated marsh near the mouth of the Lieutenant River was sprayed with Rodeo® in September 1995 and mowed/mulched in the spring of 1996. By the summer of 1997, the site had largely returned to a brackish meadow consisting of a mixture of *Agrostis stolonifera* (bentgrass), *Spartina patens*, *Juncus gerardii, Panicum virgatum* (switchgrass) and other species (Table 3). *Agrostis* was especially prominent following treatment. *Phragmites* cover was low but its frequency of occurrence was still very high. It remains to be seen whether the re-established brackish meadow will persist. This treatment appeared to have little effect on the macroinvertebrate community which was similar to that of the untreated *Phragmites*-dominated marshes (J.L. Grimsby and P.E. Fell, unpublished data); and foraging by mummichogs and eels (*Anguilla rostrata*) in the flooded restored marsh was similar to that in a nearby *Phragmites*-dominated marsh, although the relative importance of various prey differed (Rilling et al. 1998). Such restoration of brackish meadow marsh evidently took place where a sparse understory of meadow species had persisted beneath the *Phragmites*. On the other hand, in areas where there appeared to have been little or no understory, annual plants such as *Pluchea purpurascens* (marsh fleabane) and the perennial *Eleocharis parvula* (spike rush) invaded and only few invertebrates were present.

In late summer 1996, other *Phragmites*-dominated marsh areas were only treated with herbicide. After one growing season, these areas exhibited very little above ground vegetation (< 10% cover), but were covered with dense leaf and stem litter. The population densities of marsh macroinvertebrates at these sites were similar to those of untreated *Phragmites* stands (J.L. Grimsby and P.E. Fell, unpublished data).

TABLE 3. Abundance of plant species occurring in more than one third of the sampling quadrats in the restored meadow marsh near the mouth of the Lieutenant River in August 1997. (R.S. Warren, unpublished data)

		Mean Cover	
Species	Frequency	Where Found	All Points
Agrostis stolonifera	94%	34%	32%
Phragmites australis	62%	7%	5%
Juncus gerardii	52%	15%	13%
Solidago sempervirens	48%	1%	1%
Polygonun punctatum	47%	1%	1%
Spartina patens	47%	34%	30%
Scirpus spp.	40%	2%	1%
Panicum virgatum	34%	12%	6%
Total cover by these dominants		107%	90%

Mowing/mulching only of a *Phragmites*-dominated marsh area in August 1996 did not diminish *Phragmites* cover one year later but did dramatically reduce the amount of litter on the peat surface. In such areas, there tended to be lower densities of macroinvertebrates than in the reference and sprayed *Phragmites* sites. Increasing the amount of litter in plots of mowed/mulched *Phragmites* marsh led to an increased abundance of *Orchestia* and *Philoscia* compared to plots without added litter (ANOVA, $F = 7.1$, $p = 0.0182$ and $F = 16.4$, $p = 0.0012$, respectively. This result indicates the importance of ground cover for shelter and possibly food (J.L. Grimsby and P.E. Fell, unpublished data).

5. Conclusions

Restoring tidal flushing to degraded coastal marshes results in the return of a suite of organisms and associated functions, provided site elevations are favorable. These attributes may return at different rates and to some degree independently of one another. Attainment of full functional equivalence relative to reference marshes may require at least 1 or 2 decades, if it is ever achieved. While it is desirable to assess as many marsh attributes as possible in judging the success of restoration efforts, vegetation and macroinvertebrates together with fish and bird populations appear to be reasonable integrators of multiple wetland functions. Monitoring restoration over a decade or more can provide important basic information on individual species, as well as community and ecosystem processes, and should be part of any restoration project.

6. Acknowledgments

The authors are grateful to Audrey Goldstein for her help in preparing the manuscript and to numerous undergraduates who have participated in these studies over the years. Much of the work summarized here was supported by the Connecticut Department of Environmental Protection, the Nature Conservancy Connecticut Chapter, and Connecticut College.

7. Literature Cited

Allen, E.A., P.E. Fell, M.A Peck, J.A Gieg, C.R. Guthke and M.D. Newkirk. 1994. Gut contents of common mummichogs, *Fundulus heteroclitus* L., in a restored impounded marsh and in natural reference marshes. Estuaries **17**: 462-471.

Barrett, N.E. and W.A. Niering. 1993. Tidal marsh restoration: trends in vegetation change using a geographical information system (GIS). Restoration Ecology **1**: 18-28.

Benoit, L.K. and R.A. Askins. 1999. Impact of the spread of *Phragmites* on the distribution of birds in Connecticut tidal marshes. Wetlands **19**: 194-208.

Besitka, M.A.R. 1996. An ecological and historical study of *Phragmites australis* along the Atlantic Coast. Dissertation. Drexel University, Philadelphia, Pennsylvania, USA.

Brawley, M.A. 1994. Birds of the Connecticut River estuary: relating patterns of use to environmental conditions, Technical Report to the Nature Conservancy Connecticut Chapter Conservation Biology Research Program.

Brawley, M.A., R.S Warren and R.A. Askins. 1998. Bird use of restoration and reference marshes within the Barn Island Wildlife Management Area, Stonington, Connecticut, USA. Environmental Management **22**: 625-633.

Buck, E.L. 1996. Selected environmental factors and the spread of *Phragmites australis* (common reed) on the tidelands of the lower Connecticut River. Pages 1-67 *in* R.S. Warren and P.E. Fell, editors. *Phragmites australis* on the lower Connecticut River: impacts on emergent wetlands and estuarine waters. Technical Report to Connecticut DEP, Office of Long Island Sound Programs.

Burdick, D.M., M. Dionne, R.M. Boumans and F.T. Short. 1997. Ecological responses to tidal restorations of two northern New England salt marshes. Wetlands Ecology and Management **4**: 129-144.

Fell, P.E., K.A. Murphy, M.A. Peck and M.L. Recchia. 1991. Re-establishment of *Melampus bidentatus* (Say) and other macroinvertebrates on a restored impounded tidal marsh: comparison of populations above and below the impoundment dike. Journal of Experimental Marine Biology and Ecology **152**: 33-48.

Fell, P.E., S.P. Weissbach, D.A. Jones, M.A. Fallon, J.A. Zeppieri, E.K. Faison, K.A. Lennon, K.J. Newberry and L.K. Reddington. 1998. Does invasion of oligohaline tidal marshes by reed grass, *Phragmites australis* Trin. ex Steud., affect the availability of prey resources for the mummichog, *Fundulus heteroclitus* L.? Journal of Experimental Marine Biology and Ecology **222**: 59-77.

Helvenston, L.L., P.E. Fell and C.E Wood. 1995. Patterns of egg laying by the tidal salt marsh snail, *Melampus bidentatus* (Say), in relation to lunar phase. Invertebrate Reproduction and Development **27**: 159-166.

Niering, W.A. 1997. Tidal wetlands restoration and creation along the east coast of North America. Pages 260-285 *in* K.M. Urbanska, N.R. Webb and P.J. Edwards, editors. Restoration Ecology and Sustainable Development. Cambridge University Press, Cambridge, England.

Peck, M.A., P.E. Fell, E.A. Allen, J.A. Gieg, C.R. Guthke and M.D. Newkirk. 1994. Evaluation of tidal marsh restoration: comparison of selected macroinvertebrate populations on a restored impounded valley marsh and an unimpounded valley marsh within the same salt marsh system in Connecticut, USA. Environmental Management **18**: 283-293.

Rilling, G.C., P.E. Fell and R.S. Warren. 1998. Fish use of brackish high marsh areas in the lower Connecticut River: a comparison of *Phragmites australis*-dominated vs. a restored marsh. Technical Report to Connecticut DEP, Office of Long Island Sound Programs.

Roman, C.T., W.A. Niering and R.S. Warren. 1984. Salt marsh vegetation change in response to tidal restriction. Environmental Management **8**: 141-150.

Rozsa, R. 1995. Tidal wetland restoration in Connecticut. Pages 51-65 *in* G.D. Dreyer and W.A. Niering, editors. Tidal marshes of Long Island Sound — ecology, history and restoration. Connecticut College Arboretum Bulletin **34**.

Sinicrope, T.L., P.G. Hine, R.S Warren and W.A Niering. 1990. Restoration of an impounded salt marsh in New England. Estuaries **13**: 25-30.

Spelke, J.A., P.E. Fell and L.L. Helvenston. 1995. Population structure, growth and fecundity of *Melampus bidentatus* (Say) from two regions of a tidal marsh complex in Connecticut. The Nautilus **108**: 42-47.

Warren, R.S. 1993. *Phragmites australis* on the tidelands of the lower Connecticut River: patterns of invasion and spread. Technical Report to the Nature Conservancy, Connecticut Chapter Conservation Biology Research Program.

Warren, R.S. and W.A. Niering. 1993. Vegetation change on a northeast tidal marsh: interaction of sea-level rise and marsh accretion. Ecology **74**: 96-103.

Warren, R.S., P.E. Fell and W.A. Niering. 1993. Biotic changes at the Barn Island tidal marshes (Stonington, CT): sea-level rise and restoration. Pages 131-139 *in* M.S. Van Patten, editor. Long Island Sound Research Conference Proceedings. Connecticut Sea Grant Publication No. **CT-SG-93-03**.

Waters, R.F. 1995. Tidal wetland restoration on a dredge disposal site: patterns of initial plant species establishment, Mumford Cove, Groton, Connecticut. Dissertation, Connecticut College, New London, Connecticut, USA.

Subject Index

MIX
Papier aus verantwortungsvollen Quellen
Paper from responsible sources
FSC® C105338

If you have any concerns about our products,
you can contact us on
ProductSafety@springernature.com

In case Publisher is established outside the EU,
the EU authorized representative is:
Springer Nature Customer Service Center GmbH
Europaplatz 3, 69115 Heidelberg, Germany

Printed by Libri Plureos GmbH
in Hamburg, Germany